Space Sciences Series of ISSI

Volume 53

For further volumes:
www.springer.com/series/6592

André Balogh • Hugh Hudson • Kristóf Petrovay • Rudolf von Steiger
Editors

The Solar Activity Cycle

Previously published in *Space Science Reviews* Volume 186, Issues 1–4, 2014

Springer

Editors
André Balogh
The Blackett Laboratory
Imperial College London
Hertfordshire, UK

Hugh Hudson
Space Sciences Lab
University of California
Berkeley, California, USA

Kristóf Petrovay
Department of Astronomy
Eötvös University
Budapest, Hungary

Rudolf von Steiger
International Space Science Institute
Bern, Switzerland

ISSN 1385-7525 Space Sciences Series of ISSI
ISBN 978-1-4939-2583-4 ISBN 978-1-4939-2584-1 (eBook)
DOI 10.1007/978-1-4939-2584-1

Library of Congress Control Number: 2015937181

Springer New York Heidelberg Dordrecht London

Cover Image: The magnetic butterfly diagram illustrates the evolution and periodic reversal of the solar magnetic field through nearly four solar activity cycles. The strength and polarity of the magnetic field is illustrated by the intensity of yellow and blue colours. The horizontal axis is time, between 1975 and 2014, the vertical axis is heliolatitude from the South to the North pole of the Sun. (Image courtesy of Dr. David Hathaway, NASA's Ames Research Center; see http://solarscience.msfc.nasa.gov/images/magbfly.jpg)

Printed on acid-free paper

Springer is part of Springer Science+Business Media (www.springer.com)

Contents

DOI 10.1007/978-1-4939-2584-1_1
Reprinted from *Space Science Reviews* Journal, DOI 10.1007/s11214-014-0125-8

Introduction to the Solar Activity Cycle: Overview of Causes and Consequences

A. Balogh · H.S. Hudson · K. Petrovay · R. von Steiger

Published online: 5 December 2014

Abstract The 11-year activity cycle is a dominant characteristic of the Sun. It is the result of the evolution in time the solar dynamo that generates the solar magnetic field. The nearly periodic variation in the sunspot number has been known since the mid-1800s; as the observations of the Sun broadened to cover an increasing number of phenomena, the same 11-year periodicity was noted in most of them. The discovery of solar magnetic fields introduced a 22-year periodicity, as the magnetic polarities of the polar regions change sign every 11 years. Correlations have been identified and quantified among all the measured parameters, but in most cases such correlations remain empirical rather than grounded in physical processes. This introductory paper and the reviews in the volume describe and discuss the current state of understanding of the causal chains that lead from the variable nature of the solar magnetic fields to the variability of solar phenomena. The solar activity cycle is poorly understood: predictions made for the current Cycle 24 have proved to be generally wrong. However, the re-evaluation of the relationships in the light of unexpected shortcomings is likely to lead to a better physical understanding of solar physics. This will help in the systematic reassessment of solar activity indices and their usefulness in describing and predicting the solar activity cycle.

Keywords Sun · Solar activity · Solar cycle · Solar indices · Solar magnetic field

A. Balogh (✉)
Imperial College London, London SW7 2AZ, UK
e-mail: a.balogh@imperial.ac.uk

H.S. Hudson
Space Sciences Laboratory, University of California, 7 Gauss Way, Berkeley, CA 94720-7450, USA
e-mail: hhudson@ssl.berkeley.edu

K. Petrovay
Department of Astronomy, Eötvös Loránd University, Pf. 32., 1518 Budapest, Hungary
e-mail: K.Petrovay@astro.elte.hu

R. von Steiger
International Space Science Institute (ISSI), Hallerstrasse 6, 3012 Bern, Switzerland
e-mail: vsteiger@issibern.ch

1 Introduction: Correlations and Causal Relations

The solar activity cycle, a near-periodic variation generally close to a period of 11 years, appears prominently in most parameters that describe solar phenomena. Records of variations, originally based on sunspot counts, show that this has been the case for historic times, even if during certain intervals lasting decades and longer sunspots were rare. This demonstrates the existence of a timing engine in the Sun that appears to control or influence all aspects of solar phenomena and also beyond, including phenomena elsewhere in the solar system that are affected by the Sun. In recent decades, understanding the solar activity cycle has progressed from being of purely scientific interest to a practical need because of the recognition that solar phenomena affect the Earth's space environment in ways that can be potentially damaging. It is generally agreed that the timing engine in the Sun causes a nearly periodic generation and evolution of magnetic fields in the solar interior and on the solar surface, more explicitly, that a dynamo mechanism generates the Sun's magnetic field. In the past half century, great improvements in ground-based solar observations have been combined with observations from space—together, these advances have provided an almost unlimited data volume describing solar variability, on all time scales and at all spatial, spectral resolution.

The nature of the solar dynamo remains difficult to determine. But for solar activity, Leighton (1969) suggested that "*... a comprehensive test of any model of the solar cycle will ultimately require the knowledge of the strength, spatial distribution, and history of the solar magnetic field spanning a considerable time period. The existing indices of solar activity, while forming a valuable continuous record covering several complete cycles, will probably be of diminishing utility unless they can be reliably related to physically significant quantities*". The very broad range of indices now available, following the great increase in the quality and quantity of solar observations, remains subject to this assessment. Although progress in recent decades has been significant in delineating possible interdependences of solar activity indicators as well as the physical basis of such interdependences, in most instances the relationships remain on the phenomenological level and do not reach, generally speaking, a strict physical causal level of rigour.

The best test of causal understanding between related phenomena is the ability to make predictions that can then be verified by observations. The understanding of the solar activity cycle has largely failed this test by not being able to predict the dominant parameters (in particular sunspot numbers) for the current Cycle 24 that followed the already unusual Cycle 23, in particular the long-duration and deep minimum in activity levels between these cycles. See Pesnell (2012) for an overview of the methodology of the predictions used for forecasting the maximum sunspot number in Cycle 24. The summary of the numerous predictions for the maximum of the current Cycle 24 is shown in Fig. 1 (based on Pesnell 2012), together with the sunspot numbers actually observed in Cycle 24 to late 2014. It is generally agreed that Cycle 24 has a smaller maximum than previous recent cycles. Although maximum activity in Cycle 24 has probably, if not certainly, been reached by late 2014, the maximum sunspot number in the cycle remains to be agreed. In any case, Cycle 24 has already proved to be quite different from recent cycles. For a detailed review of the methodologies used in the predictions and their phenomenological and/or physical basis, see Hathaway (2010), Petrovay (2010) and Pesnell (2012).

The striking feature of many of the prediction techniques is their grounding in phenomenology, rather than in Leighton's "physically significant quantities". Given data sets that describe the solar cycle in different phenomena over as many cycles as possible, these methods seek correlations that can be seen to hold over a number of cycles. These correlations are then used as predictors, usually to describe the coming cycle in terms of the

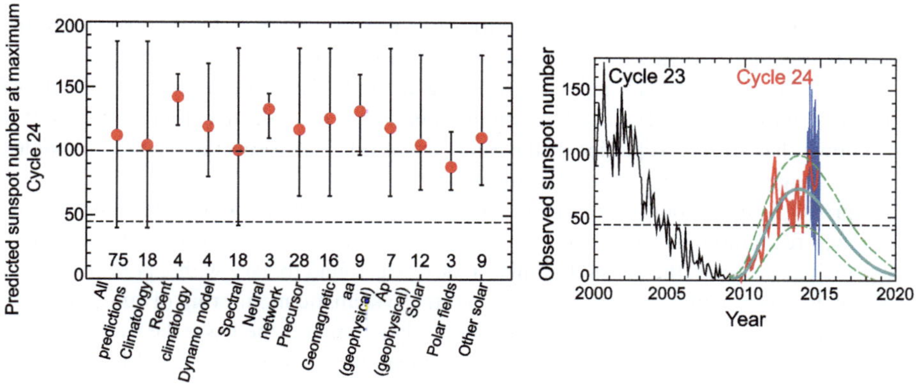

Fig. 1 *In the left panel*, the predictions made for the solar activity Cycle 24 are shown, using different methodologies (from Pesnell 2012). The methodologies are listed on the figure; the numbers of predictions that were made using the given methodologies are also given. *The red symbols* represent the mean value of the predictions, and *the vertical bars* represent the range of values that were predicted by the different methodologies. *In the right panel*, the actual values of the monthly sunspot numbers are plotted (in *red*), together the daily sunspot numbers for 2014 (in *blue*) indicating the very large scatter that will be smoothed out as the latest data are processed. *The green lines* represent the latest predicted sunspot numbers for cycle 24 (from http://solarscience.msfc.nasa.gov/images/ssn_predict_l.gif). *The dashed black lines* in the two plots correspond to the largest and smallest predicted maximum in Cycle 24

expected maximum value of the sunspot number and its timing. Predictions for Cycle 24 have uncovered the many potential limitations of the algorithms used and the correlations on which they have been based. It is expected that the deviations from the expected extrapolations will lead to a reassessment of the way in which the physically (at least directly) unrelated phenomena are used in empirical analyses to interpret and to predict solar activity variations.

It is important to review and assess the status of causal understanding of solar activity cycle indicators. This is the task that was undertaken in a Workshop organised by the International Space Science Institute (ISSI), Bern, in November 2013. The resulting reviews (this Issue of Space Science Reviews) provide a comprehensive overview of all aspects of the solar activity cycle and the indices that are used to describe it. A brief introduction to the issues, their historic context and key references are given in this summary paper.

Finally, we restate Leighton's remark in the modern context. We have many well-defined indices that reflect solar magnetic activity in various ways. They are all proxies of some property of the solar magnetic field, but what property is that? The papers in this volume all deal implicitly with this question. We are far from getting answers, but it seems clear that further progress will only come from a self-consistent and physically based combination of improved data and cleverer theory.

2 The Schwabe and Hale Solar Activity Cycles

The Sun may be only one star among the quasi-uncountable number of stars in the Universe, but it is of unique, immediate and vital interest to all life on Earth. Among stars, it has unexceptional properties, but its study in the scientific era has been a central concern in astronomy—the observational and descriptive aspects of the studies, and also in astrophysics—the research into the physical description of its observed properties. One of the many remarkable observational properties of the Sun is its variability, noted from the earliest telescopic observations four centuries ago. Variability was seen, almost exclusively, in the sunspots—their sizes, shapes, numbers, as the only departure from the "perfection" that had been hypothesised (indeed becoming often an act of faith) in earlier centuries. It is now known that all the properties of the Sun vary in time, with several chains of complex causality suggested in the past century and a half.

Surprisingly, it was only from the mid-19th century that the periodic nature of the variability was recognised as a key feature (for a historical review, see Cliver 2014, this issue). The principal measure, the sunspot number index, was found to vary with a period of about 11 years. Reconstructing past observations and the systematic monitoring of sunspots since that time has provided a time series that, even to date, remains the basic data set on which rests the science of solar variability and much of solar physics (Clette et al. 2014, this issue). Of course, the data set itself, given its importance, has been and remains still now the subject of scrutiny, reworking and even disagreements, but the underlying near-periodicity in this index conditions the way all aspects of solar variability are described and studied. It was the gradually increasing diversity of observational data sets that described other solar phenomena (and even their terrestrial effects), all in greater or smaller ways related to the variations in sunspot numbers, that has led to the concept of solar activity. This concept has become very complex, with large numbers of observables beyond sunspot numbers that can be used to describe the aspects of activity. Progress in solar physics and the science of the Sun's variability and activity have come, over the past century and a half, first from the range, volume and quality of solar observations and the general increase in scientific effort devoted to the Sun, to the interpretation of the observations as well as from the investigations of the underlying physical processes.

Measures of solar activity can be based on many observable and measurable phenomena. All of them have a common characteristic: they are all varying (at least nearly) in phase or anti-phase with the sunspot numbers. Generally speaking, all the observed and measured quantities can be taken to be proxies of some key aspect of the physical causes of variability in solar phenomena. Equally generally, the physical relationships between these measurable parameters are rarely easy to establish—in particular, it is found that the relationships are usually neither simple nor linear (Ermolli et al. 2014, this issue).

The 11-year solar activity cycle, based on the quasi-periodic variations of the sunspot numbers, has been named, after its discoverer, the Schwabe activity cycle (Schwabe 1844). Sunspot cycles are shown in Fig. 2, numbered according to convention, with Cycle 1 starting in 1755. Cycles are counted from one minimum in sunspot numbers to the next minimum. Early in the last century, the discovery and observation of the magnetic fields on the Sun, starting with the interpretation of spectral shifts as caused by the Zeeman effect due to intense magnetic fields in sunspots (Hale 1908), have led to the identification of a cycle with twice the duration of the Schwabe cycle, the solar magnetic, or Hale cycle (Babcock 1959). This 22-year cycle, manifested in magnetic phenomena including the reversal of the poles of the solar magnetic dipole, underlines the fundamental role played by solar magnetism in controlling the solar activity cycle (Petrie et al. 2014, this issue; Wang 2014, this issue; Cliver 2014, this issue). Figure 2 illustrates the two Hale cycles for the recent decades. For the Hale cycle, the magnetic field intensity observed in the polar regions is shown, indicating a reversal of the magnetic polarity of the Sun at about the time in the Schwabe cycle when the number of sunspots has a maximum.

While the global measures of solar activity, primarily the sunspot number, indicate the quasi-periodicity of the underlying mechanism (the solar dynamo, in effect), there are intriguing differences between the two solar hemispheres in detail. Activity is generally not

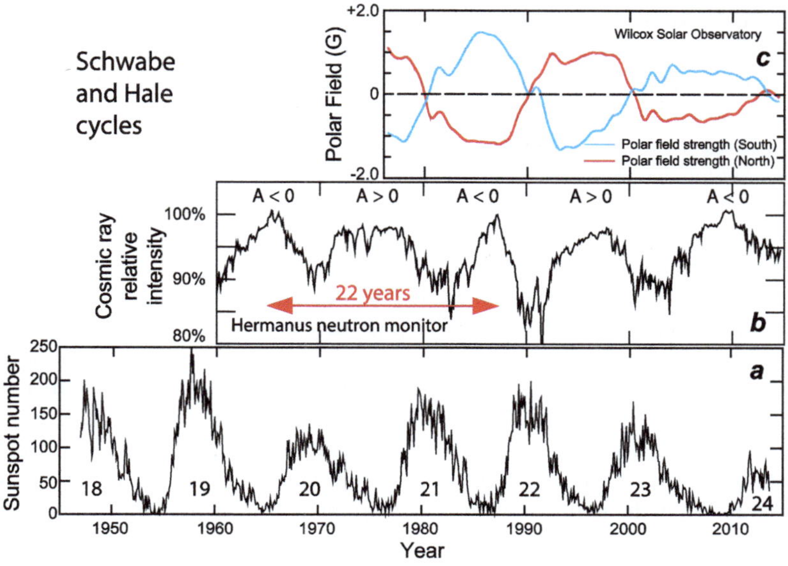

Fig. 2 In panel **a** the sunspot number is shown as the most commonly used indicator of solar activity. In panel **b**, the intensity of the galactic cosmic rays measured at Earth is shown, illustrating the dominant 11-year modulation with the Schwabe cycle, and indicating the different modulation patterns in alternate cycles, in phase with the 22-year Hale cycle. In panel **c**, the strength of the solar polar magnetic fields is shown. The polarities reverse every 11 years, producing the 22-year Hale cycle. Note the much reduced strength of the polar fields in the minimum activity phase between Cycles 23 and 24

equal between North and South: the sunspot numbers differ and the two hemispheres show slightly different evolution. Generally speaking, the reversal of the signs of the magnetic poles is not simultaneous, but can differ by many months. The asymmetry and asynchronicity between the hemispheres has been noted to extend to the solar atmosphere and into the heliosphere (Virtanen and Mursula 2014). The question of coupling between the solar hemispheres and its relationship with the evolution and solar activity, together with the wealth of observations that relate to the differences and similarities in activity indices is quite central to the understanding of the whole solar activity process (see Norton et al. 2014, this issue, and references therein).

The solar magnetic field extends into the heliosphere (see, e.g. Balogh and Erdős 2013) where it is entrained by the solar wind, resulting in complex evolutionary structures following the dynamic evolution of the solar wind. A particularly clear link between the solar origin of the magnetic field and the large-scale structures of the heliospheric magnetic field is the interplanetary sector structure. In interplanetary space, a global current sheet (the Heliospheric Current Sheet, or HCS, see Smith 2001), the sector boundary, separates the two large-scale magnetic polarities present in the heliosphere. This structure also has clear signatures at the level of the photosphere, and in the corona in between. The sector structure and the topology of the HCS are related to the evolution, over the Hale cycle, of the solar magnetic field and are important indicators of the state of the heliosphere as a function of solar activity (Hudson et al. 2014, this issue).

An interesting consequence of the extension of solar activity into the heliosphere is the periodic modulation of the flux intensities of high energy cosmic rays, as illustrated in Fig. 2. Galactic cosmic rays entering the heliosphere encounter propagation conditions in the interplanetary medium that vary with the solar activity cycle (see Potgieter 2013; Bazilevskaya et al. 2014a, this issue). Simply stated, the propagation from the boundaries to the inner heliosphere where the Earth is located involves a greater decrease in cosmic ray intensity at sunspot maximum than at sunspot minimum. Hence the cosmic ray intensity varies in antiphase with the sunspot number, in other words, with solar activity, as can be clearly seen in Fig. 2. However, as can be observed also in Fig. 2, the appearance of modulation cycles of cosmic ray intensities change in alternate 11-year cycles, with a peaked intensity profile followed by a flat-topped intensity profile. This, together with the satisfactory explanation of the observations linking them to the dominant dipole polarity of the sun, is evidence for a 22-year Hale cycle modulation effect.

Cosmic ray intensities are modulated by diffusion and convection effects in interplanetary medium; in effect irregularities in the interplanetary magnetic field and propagating structures lead, statistically, to the diffusion of the high energy particles. The properties of the interplanetary medium change due to solar activity, hence the 11-year variation in intensity. The explanation for the 22-year Hale cycle modulation is to be found in gradient and curvature drift effects on the cosmic ray particle trajectories of the large-scale heliospheric magnetic field. The effects depend on the direction of the magnetic field in the Parker spiral. When the northern heliospheric magnetic field is outward (the polar field is positive, denoted by $A > 0$), positively charged particles drift inward from the polar regions of the heliosphere to the equatorial plane in the inner heliosphere and the equatorial propagation conditions influence to a lesser extent the access of the particles. When the northern field is directed inward (the polar field is negative, represented by $A < 0$), positively charged particles enter the inner heliosphere by drifting towards the equatorial regions of the heliosphere, where propagation conditions are controlled by the properties of structures in the heliospheric magnetic field that are co-rotating with the sun. (The constant A is a multiplier in the heliospheric magnetic field model used for cosmic ray modulation studies, see Jokipii et al. 1977.) These effects have been largely confirmed by extensive modelling of the three-dimensional propagation of cosmic rays (Jokipii and Thomas 1981; Kota and Jokipii 1983; Webber et al. 1990).

Although this effect of the 22-year magnetic cycle of solar activity on cosmic rays is well understood, it is interesting to note that a now almost forgotten 22-year recurrent effect had been observed much earlier in the diurnal anisotropy of very high energy cosmic rays (Thambyahpillai and Elliot 1953; Forbush 1969). This effect was also found to be related to the gradients of cosmic rays in the inner heliosphere, and related in turn to the orientation of the sun's magnetic polarity at the activity cycle minima (McCracken et al. 2004).

Generally speaking, many of the intriguing problems in solar physics are related to the causal connections between the Schwabe and Hale cycles. Most of the numerical indicators of solar activity are associated with the 11-year Schwabe cycle, while the Hale cycle seems to reflect the quasi-periodic recurrence of generally complex magnetic phenomena. However, as can be seen in Fig. 3, even the most commonly used and most direct measures of solar activity present a great variability superimposed on the 11- and 22-year periodicity; this is the case for all the measured or derived parameters describing different aspects of solar activity (Cliver 2014, this issue; Ermolli et al. 2014, this issue). Over historical timescales, the Sun's activity has been variable over many timescales, with longer intervals of recurrence (see, e.g. Usoskin 2013), although the 11-year periodicity appears to have survived through all other modulation effects of solar activity. The duration of the cycles is itself quite variable. The average period is close to 11 years, but with a broad distribution around the mean. The cause of such variability in the period remains unclear, but is clearly related to the operation of the solar dynamo which arguably behaves chaotically (Arlt and Weiss 2014, this issue). Equally, the variability in the "strength" of the cycles, measured either by

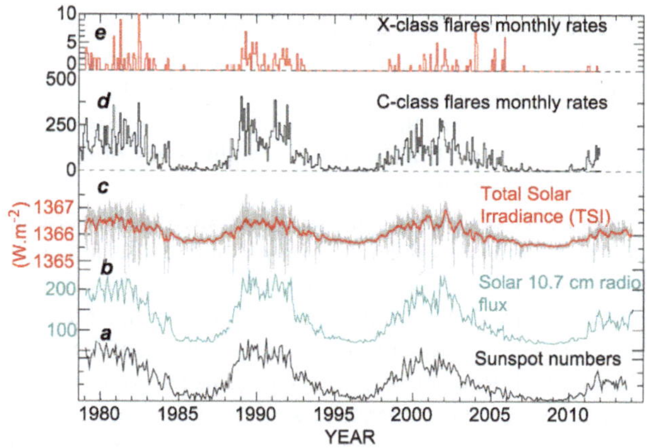

Fig. 3 Panel **a** shows the sunspot number, panel **b** the monthly mean intensity of the 10.7 cm radio flux, panel **c** the Total Solar Irradiance, panels **d** and **e** the monthly rates of C-class and X-class solar flares, respectively. The four quantities correspond to very different physical processes on the Sun, with no simple causal connections among them, but all four show the same 11-year periodicity, even though there are differences in the short term variabilities (The sunspot numbers are from the WDC-SILSO, Royal Observatory of Belgium, Brussels; the solar radio flux monthly averages are from Natural Resources Canada, http://www.spaceweather.ca/solarflux/sx-eng.php; the TSI figure shows data from Physikalisch-Metorologisches Observatorium Davos (PMOD); the solar flare rates are adapted from Fig. 3 in Aschwanden and Freeland 2012.)

the number of sunspots at maximum, or by any other activity measure, is unclear. These uncertainties and a lack of understanding of the causal chains of solar phenomena make the prediction of activity indices quite a challenge, as already discussed. Another approach to assess the meaning of the quasi-periodicity is to consider the low order modes of the solar dynamo and use observational evidence (both from sunspot numbers and from the insights about the solar interior provided by helioseismology, see Sect. 4) and explore the nature of the oscillator that matches the observations (Lopes et al. 2014, this issue). A better understanding of a first order dynamo model could lead even to an improvement in predictions, based on the accurate determination of the sunspot numbers, as this is becoming available (Clette et al. 2014, this issue).

A well-established shorter timescale variability in solar activity, the so-called quasi-biennial oscillations (QBOs) with the time scale of 0.6–4 years can be clearly seen in such basic indices as sunspot numbers and sunspot areas. Although basically unexplained, this phenomenon has allowed very detailed tests of the operation of the solar dynamo (see Bazilevskaya et al. 2014b, this issue, and references therein). A particularly attractive and potentially fruitful concept is that the QBOs are the effect of the interaction of the dipole and quadrupole terms of the solar dynamo.

3 A Selection of Solar Activity Indicators

The diversity of the observable parameters that can be linked to solar activity is very large, and they are all related at least in the sense that their variations follow the nearly 11-year periodicity, and therefore they are all cross-correlated to a greater or smaller extent. However, the correlations do not imply causal relationships, but are only indicative of the underlying driver of the variability that is the solar magnetic field. In Fig. 3, five measured parameters are shown, all five can be seen to vary in (near) synchronism at the rhythm of the activity cycle. The parameters have been chosen to illustrate that the variations in the parameters are correlated, despite the lack of known physical relationships between them. The parameters are (a) the sunspot number, (b) the integrated solar radio noise at 2800 MHz (the 10.7 cm radiation), (c) the Total Solar Irradiance, (d) monthly rates of C-class solar flares and (e) monthly rates of X-class solar flares.

The historical importance of sunspot counts as a measure of solar variability and the solar activity cycle has already been described. Even if calibrated, it can hardly be said that the Sunspot Number (SSN) is understood physically. In its modern state, it is relatively precise and reproducible. But the physical phenomenon of the sunspot itself has no simple physical theory that explains its nature, for example in the diminution of convective flux that results in its darkness (Archontis 2012; Rempel 2012). Smaller-scale but comparably intense magnetic fields appear in faculae, where the consequences on the emergent intensity differ radically. This juxtaposition of a careful set of observations with an ill-understood quantity, which we term solar magnetic activity, persists throughout the remaining list of solar activity indices. The topic of magnetic flux emergence through the solar cycle is reviewed by Schmieder et al. (2014, this issue). Given that magnetic fields on the Sun were first discovered in sunspots as already mentioned (Hale 1908), the magnetic fields of sunspots remains a topic of interest as a key observable that is quite intimately related to solar activity (Borrero and Ichimoto 2011). As discussed below, the recent decrease in the average magnetic field of sunspots is a significant indicator of more profound changes in solar activity (Livingston et al. 2012).

The solar radio flux, at a wavelength of 10.7 cm (2.8 GHz), has been measured continuously since 1947 and has been widely used as an indicator (often called F10.7) of solar activity (see, for a recent review, Tapping 2013, and in the context of other solar activity indices, Ermolli et al. 2014, this issue). The match between the SSN and the 10.7 cm radio flux had been very close, once the correct non-linear relationship between the two has been established (see, e.g., Tapping 2013). The 10.7 cm radiation comes from regions of strong magnetic fields on the solar surface and from strong field regions in the corona. These sources provide the variation in synchronism with the activity cycle. Additionally, there is a non-zero flux even at solar minimum, mostly from thermal free-free emission. The close relationship between sunspot numbers and 10.7 cm radio flux, however, has appeared to change over the last complete solar cycle, Cycle 23. The cause of this, briefly discussed below in this Section, may provide important clues to the causal connection between these two indices.

The Total Solar Irradiance (TSI) is a measure of the solar output; it is of immediate relevance as a fundamental parameter in the Earth's energy balance. TSI is the amount of solar electromagnetic radiation per unit area, integrated over all wavelengths, incident on the Earth (measured as $\mathrm{W\,m^{-2}}$, and normalised at 1 AU). It has been measured by space instruments for the past almost 40 years, but due to the difficulty of making calibrated absolute measurements, the construction of a generally accepted composite time series of the observations (using data from the different instruments not overlapping in time) has only been possible in the recent past (see, e.g. Frölich 2012; Ball et al. 2012). TSI varies on many timescales, from minutes to years, in particular in phase with the sunspot numbers. Figure 3, panel *c* shows the daily variations in TSI, together with a smoothed average; the amplitude of the daily variability is clearly much greater than the long-term variability. Most of its variability correlates well with sunspot and facular areas. However, the very complex variations in the spectrum of the solar irradiation as a function of the activity cycle, in particular as a function of the strength, distribution and short- and medium-term variability of the magnetic features remain to be explained (Yeo et al. 2014, this issue). Progress will be made by more accurate measurements, in particular spectral measurements, combined with more detailed models of the spectral irradiance.

Solar flares are short duration energetic outbursts; here short duration means minutes to tens of minutes, when magnetic energy stored in topologically complex magnetic fields above the photosphere is suddenly released. Solar flares produce energetic particles, plasma heating and a broad spectrum of electromagnetic radiation. Solar flare observations have provided considerable quantities of data that have increased in volume and complexity as new observing facilities have become operational (Fletcher et al. 2011). The occurrence rate of solar flares is linked to solar activity as measured by, for instance, the sunspot number, but in fact there are differences in terms of the phase of the sunspot cycle, as well as its intensity. Although, as shown in Fig. 3, panel d, the rates of C-class solar flares (relatively small flares, without noticeable terrestrial effects) varies with an approximate periodicity of 11 years, the shape of the frequency distribution differs significantly from other measures of solar activity. The relatively small number of the largest, X-class flares, shown in Fig. 3, panel e, have an occurrence frequency that only indicates that such flares occur away from solar minimum activity, but not matching the profile of other activity indicators, such as the sunspot numbers. (C- and X-class flare occurrence frequency in Fig. 3 is based on Fig. 3 in Aschwanden and Freeland 2012.)

This brief and illustrative discussion of apparently very different phenomena and processes that show the periodic solar activity variations indicates strongly that the underlying mechanism driving solar activity is an all-pervasive process that affects all the characteristic observable solar phenomena. Even though the physical causal relationship among the indicators is largely unknown, and in fact in many cases may not even exist, there are, undoubtedly, some dominant causal chains that may act in parallel. This means that the observed phenomena are driven by the yet to be established physical processes, some of which operate independently, but eventually all of them originate in a single, central driving mechanism which can only be the generation mechanism of the Sun's magnetic field.

An example of the way in which future progress can be made concerns the study of the deviation of the empirical function linking the sunspot number and the 10.7 cm solar radio noise, F10.7, during the last Cycle 23 (Svalgaard and Hudson 2010). The deviation is indicated in Fig. 4(left panel). As the F10.7 data and the sunspot number had been considered to be a good proxy for each other and for quantifying solar activity, the deviation observed shows that the physical link between "solar activity" and sunspot numbers on the one hand, and the F10.7 flux on the other is neither direct nor single-valued. Given the dependence of the of the radio noise on total magnetic field, the result has been linked to the decrease in the magnetic field of sunspots (Penn and Livingston 2006; Svalgaard and Hudson 2010; Pevtsov et al. 2011; Livingston et al. 2012). The sunspot formation fraction (SFF), a parameter introduced by Livingston et al. (2012) is the ratio of the sunspot number to that predicted using the previously established relationship to the F10.7 flux. As can be seen from Fig. 4, the SFF remained close to 1 (indicating a high degree of predictability) until about 1995. After that time, the sunspot number decreases relative to the prediction; this change occurs coincidentally with qualitative and quantitative changes in other solar activity indicators. The close relationship observed previously, from 1947 when the F10.7 data series starts, covered almost five Schwabe cycles and more than two Hale magnetic activity cycles and could be considered to be well established, even if not completely understood. The significant deviation through the last cycle from the previous behaviour can be used to investigate the underlying parametric causes of it and hence to clarify better the causal links to the extent that they can be identified. Livingston et al. (2012) found a systematic decrease in the magnetic field strength of sunspot umbrae since about 1998. A consequence of this is

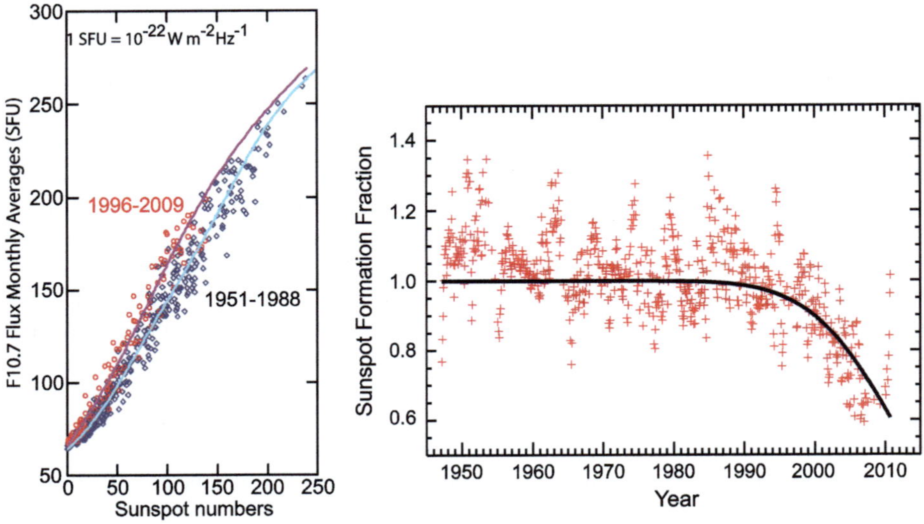

Fig. 4 *In the left panel*, the solar 10.7 cm radio flux is plotted vs. sunspot numbers. The radio flux has been used recently to indicate the level of solar activity, partly because of its close numerical match to the sunspot number variations (once normalized) and partly because it is closely linked to the solar effects on the Earth's upper atmosphere. However, the relationship between the two quantities is non-linear and appears to have changed significantly during the last activity cycle (based on Fig. 12 in Svalgaard 2009). *In the right panel*, the possible cause of the change is shown, based on a new parameter, the sunspot formation fraction introduced by Livingston et al. (2012)

the decrease in the detectability of weak-field sunspots that correlates well with the decrease in the SFF parameter over the same epoch.

4 The Source of Variability: The Solar Magnetic Field

The engine of solar variability is undoubtedly the solar magnetic field. However, the range of magnetic structures effectively covers all temporal and spatial scales, from still unresolved photospheric structures to the global field which, if not simple, nevertheless provides evidence for global organisation. The dominant quasi-periodic variations represented by both Schwabe and Hale activity cycles appear to be strong evidence for global structuring and evolution. A solar dynamo is essential for understanding these determining features of the phenomenology of solar magnetism and solar activity. However, despite steady progress observationally, theoretically and, most importantly, in realistic numerical simulations, there are fundamental areas of ignorance concerning the generation and evolution of magnetic fields. In the recent past, the paradigm of Flux Transport Dynamos (FTD) has dominated research efforts. Originally proposed by Wang and Sheeley (1991) to build on the generally well-regarded model by Babcock and Leighton (for a recent assessment, see Charbonneau 2007), it has provided a framework in which the kinematics and dynamics of magnetic field generation have been successfully investigated both observationally and in modelling terms (Karak et al. 2014, this issue). A link between the magnetic field dynamics in the solar interior and in the photosphere and corona is provided by the concept of magnetic helicity (Pevtsov et al. 2014, this issue). This concept has been fruitfully used to explain such activity related phenomena as Joy's Law on the directional orientation of sunspot pairs of opposite polarities and the orientation of solar active regions that evolve on timescales much shorter than the activity cycle.

Nevertheless, even if much of the theoretical basis and many of the numerical tools for modelling the solar dynamo have been developed, the problem remains very complex (see, e.g., Charbonneau 2010). The outer third or so of the Sun is convective, involving the transport of solar material towards the surface in an ill-understood manner. The relevant physical parameters of the solar material are not known in sufficient detail, and the kinetics cannot be as yet deduced or modelled to the extent that the magnetic field generation process can be unambiguously deduced. The key to progress is a better understanding of the Sun's convection zone, as well as the shear layers at the bottom of the convection zone and close to the solar surface. Sounding the material properties inside the solar interior, as well as the kinematics and dynamics of the solar convection zone have brought many new insights that help building up increasingly detailed numerical models of the solar dynamo. The study of the differences between the solar hemispheres and the way the development of activity is coupled between North and South, as already mentioned (Norton et al. 2014, this issue), highlights the complexity of the physical modelling the dynamo, while it also provides observational tools and even constraints on the models.

For the past 40 years, the use and continued improvement of a fundamentally important tool, helioseismology, has proved exceptionally powerful for investigating the interior of the Sun (see Broomhall et al. 2014, this issue). Helioseismology, both ground- and space-based, has provided the data on the internal structure, dynamics and physical properties of the Sun that have opened up the possibility to link structures and processes in its interior to its visible properties and its activity cycle (Thompson 2010; Komm 2010).

Given the observations of the changes in activity between the earlier cycles in the space age and in the declining phase of Cycle 23 leading to the current Cycle 24, it is possible to explore corresponding changes in a range of previously well-established relationships between solar activity indicators, as already shown in Sect. 3 with reference to the sunspot number and the solar 10.7 cm radio noise intensity. These studies will help in identifying actual physical processes behind the empirical relationships and thus better to distinguish between causes and consequences in solar activity phenomena.

The fundamental importance of the solar dynamo mechanism in the generation of all the activity phenomena and, in turn, of the role of the dynamics and physical characteristics of the solar interior, helioseimologic results concerning the changes through the solar cycle are primary indicators of the underlying causes of changes in activity. Shifts in the frequencies of solar oscillations track the 10.7 cm radio noise quite closely, see Fig. 5(a). The frequency shifts have been compared in fact to a range of solar activity indicators by Chaplin et al. (2007). Their conclusion, based on data sets that ended as the minimum activity phase between Cycles 23 and 24 just started, was that the frequency shifts were best correlated with activity indices that are sensitive to both weak and strong surface magnetic fields, e.g. the 10.7 cm radio flux, rather than those which are only sensitive to strong fields, e.g. the sunspot number. The good match is shown in Fig. 5(a) (after Broomhall et al. 2009). This implies that the frequency shifts correspond to magnetic field changes globally, rather than in the high-activity belt in the Sun's equatorial regions. Basu et al. (2012) have investigated the dependence of activity-related frequency shifts as a function of the low-, mid- and high-frequency modes of solar oscillations. Their results, shown in Fig. 5(b) indicate strongly that some changes in solar structure and the magnetic field were already taking place during Cycle 23, when other indicators had not yet clearly foreseen the long minimum between Cycles 23 and 24 and the very low activity level of Cycle 24. Their interpretation of these results is (also exposed in detail by Broomhall et al. 2014, this issue) that the Sun's subsurface

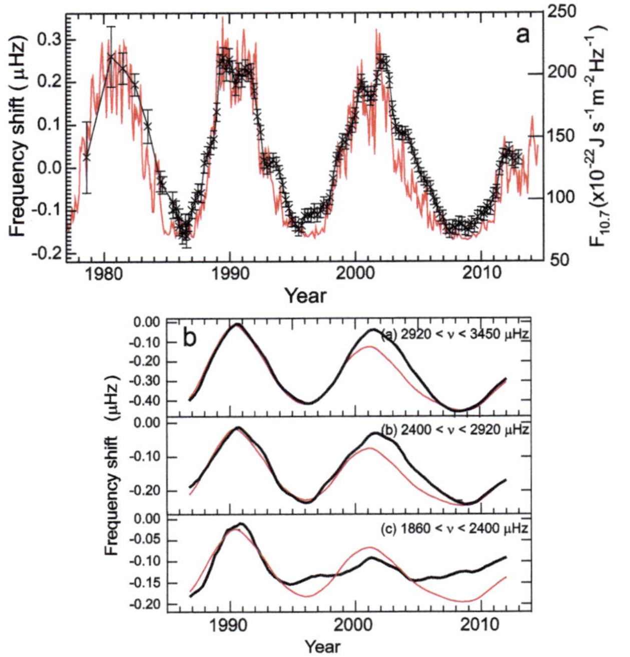

Fig. 5 (**a**) The frequency shift in solar oscillations over more than three solar cycles. The data are compared to the 10.7 cm radio flux (*red line*) that is a well matched indicator of solar activity (after Broomhall et al. 2009, extended data set from A.-M. Broomhall). (**b**) The behaviour of the frequency shifts is shown for high-, mid- and low frequency oscillations (after Basu et al. 2012), this time together with the sunspot number (*red line*). As can be seen, the changes that occurred at low frequencies started to deviate significantly from previous behaviour already from the start of Cycle 23 in 1996, signalling the forthcoming changes in solar activity indicators that became more generally obvious during the declining phase of Cycle 23

magnetic layer (which is thought to be where structural differences and magnetic fields lead to the manifestations of solar activity) thinned, starting from Cycle 22. The implications of this result are far-reaching, even if at this stage tentative. The solar dynamo appears to rely, for operation, on the subsurface flows to carry magnetic flux from the equatorial regions to the poles. Therefore a much better knowledge of solar structure near the surface which can be brought by more and more fully analysed and modelled helioseismic observations is likely to lead to better and more physical constraints on the causes of solar activity and its variations.

The data presented in Figs. 3–5 are intriguing, suggesting that important changes occurred in the Sun's behaviour early on during Cycle 23. However, important pieces of this puzzle are still missing. Among others, the alleged change in sunspot magnetic fields and temperatures claimed by Penn and Livingston (2006) should be revisited in the context of the statistical distribution of sunspot areas in various solar cycles; flare occurrence rates based on peak X-ray flux should be contrasted with changes in flare areas, H_alpha flare index and CME rates, free magnetic energy, current helicity, etc. In view of such missing information it is probably premature to put forward a physical interpretation of these changes. Our objective here has been just to call attention to these puzzling findings and to place the papers in this volume in context.

5 Activity in Other Stars

While the study of solar activity is the primary topic of this issue of Space Science Reviews, it is worth recognising that the study of stellar activity is also a rich field of research. Of course, the observational means in stellar research are mostly fundamentally different and considerably more limited in capabilities from those used to study solar phenomena. Although severely limited by the observational difficulties, comparative studies of stellar and solar activity phenomena have been attempted to better understand the dynamo operating in the Sun and in the active stars.

Photometric observations of stars (changes in light output) have been used to deduce the presence of "starspots" on the stellar surface, implying strong magnetic fields and the operation of stellar dynamos (for a review, see Strassmeier 2009). Starspots are used to determine the rotation rate of stars, the photometric variability over intervals longer than the deduced rotation period is used to imply magnetic variability. This objective is generally difficult to achieve, as the continuous observation of target stars over their presumed activity cycles may require many years or even decades. Shorter-term photometric observations of starspots have been achieved. Such observations, together with modelling of the modulation of the photometric light curves has provided insights, if not certainties into the range of magnetic activity of different categories of solar-like stars. The potential for gaining enough knowledge into stellar activity is high and will increase as observational capabilities are developed, together with data analytical techniques which optimise the interpretation of what remains an incomparably smaller data set when compared to what is available for the Sun (Kővári and Oláh 2014, this issue).

Similarly to using helioseismology to study solar activity by studying changes to the internal structure of the Sun, asteroseismology or the detection of stellar oscillations has been used to study stellar activity. New results have been, to some extent, the by-product of the search for extra-solar planets. Indeed, the manifestations of stellar activity can be seen as a noise source for the detection of planets. The Kepler space mission which has observed a fixed field of a very large number of stars for a long interval first demonstrated that oscillations of solar-type stars can indeed be observed (Chaplin et al. 2010). A larger study identified up to 500 solar like stars with oscillations (Chaplin et al. 2011), a data set that can be used for the statistical study of stellar properties using the powerful methods that have proved so useful for the study of the solar interior. The question whether such asteroseismic observations can be interpreted in terms of stellar magnetic activity, however, is more complex and will require extensive modelling and the search for observations that can be used to constrain the models (Chaplin and Basu 2014, this issue).

Acknowledgements The Editors are greatly indebted to all the participants of the Workshop held in ISSI Bern on 11–15 November 2013 who brought their broad range of expertise and interest in solar physics to deepen our understanding of the issues related to the causes and consequences of the complex phenomenology of solar activity. We thank the staff of ISSI for their dedicated support: Prof. Rafael Rodrigo, Executive Director, and his colleagues Jennifer Fankhauser, Andrea Fischer, Saliba Saliba and Sylvia Wenger. The resulting collection of review papers was the outcome of the exchanges and fruitful collaboration among the participants; we thank them for their successful efforts to integrate the lessons learned in the different topics, as the reviews in the volume testify. Thanks are also due to the reviewers of the papers; in all cases the reviews were thorough and constructive and the volume bears witness to their contribution. Finally the Editors thank the staff of Space Science Reviews, as well as the production staff for their patience on occasion and for an excellently produced volume.

References

V. Archontis, Magnetic flux emergence and associated dynamic phenomena in the Sun. Philos. Trans. R. Soc. A **370**, 3088–3113 (2012). doi:10.1098/rsta.2012.0001

R. Arlt, N. Weiss, Solar activity in the past and the chaotic behaviour of the dynamo. Space Sci. Rev. (2014, this issue). doi:10.1007/s11214-014-0063-5

M.J. Aschwanden, S.L. Freeland, Automated solar flare statistics in soft X-rays over 37 years of GOES observations: the invariance of self-organized criticality during three solar cycles. Astrophys. J. **754**, 112 (2012). doi:10.1088/0004-637X/754/2/112

H.D. Babcock, The Sun's polar magnetic field. Astrophys. J. **130**, 364–365 (1959)

W.T. Ball, Y.C. Unruh, N.A. Krivova, S. Solanki, T. Wenzler, D.J. Mortlock, A.H. Jaffe, Reconstruction of total solar irradiance 1974–2009. Astron. Astrophys. **541**, A27 (2012). doi:10.1051/0004-6361/201118702

A. Balogh, G. Erdős, The heliospheric magnetic field. Space Sci. Rev. **176**, 177–215 (2013). doi:10.1007/s11214-011-9835-3

S. Basu, A.-M. Broomhall, W.J. Chaplin, Y. Elsworth, Thinning of the Sun's magnetic layer: the peculiar solar minimum could have been predicted. Astrophys. J. **758**(1), 43 (2012), 6 pp. doi:10.1088/0004-637X/758/1/43

G. Bazilevskaya, A.-M. Broomhall, Y. Elsworth, V.M. Nakariakov, A combined analysis of the observational aspects of the quasi-biennial oscillation in solar magnetic activity. Space Sci. Rev. (2014a, this issue). doi:10.1007/s11214-014-0068-0

G.A. Bazilevskaya, E.W. Cliver, G.A. Kovaltsov, A.G. Ling, M.A. Shea, D.F. Smart, I.G. Usoskin, Solar cycle in the heliosphere and cosmic rays. Space Sci. Rev. (2014b, this issue). doi:10.1007/s11214-014-0084-0

J.M. Borrero, K. Ichimoto, Magnetic structure of sunspots. Living Rev. Sol. Phys. **8**, 4 (2011). http://www.livingreviews.org/lrsp-2011-4

A.-M. Broomhall, W.J. Chaplin, Y. Elsworth, S.T. Fletcher, R. New, Is the current lack of solar activity only skin deep? Astrophys. J. **700**, L162–L165 (2009). doi:10.1088/0004-637X/700/2/L162

A.-M. Broomhall, P. Chatterjee, R. Howe, A.A. Norton, M.J. Thompson, The Sun's interior structure and dynamics, and the solar cycle. Space Sci. Rev. (2014, this issue). doi:10.1007/s11214-014-0101-3

W.J. Chaplin, S. Basu, Inferences on stellar activity and stellar cycles from asteroseismology. Space Sci. Rev. (2014, this issue). doi:10.1007/s11214-014-0090-2

W.J. Chaplin, Y. Elsworth, B.A. Miller, G.A. Verner, R. New, Solar p-mode frequencies over three solar cycles. Astrophys. J. **659**, 1749–1760 (2007). doi:10.1086/512543

W.J. Chaplin, T. Appourchaux, Y. Elsworth et al., The asteroseismic potential of Kepler: first results for solar type stars. Astrophys. J. Lett. **713**, L169–L175 (2010). doi:10.1088/2041-8205/713/2/L169

W.J. Chaplin, H. Kjeldsen, J. Christensen-Dalsgaard et al., Ensemble asteroseismology of solar-type stars with the NASA Kepler mission. Science **332**, 213–216 (2011). doi:10.1126/science.1201827

P. Charbonneau, Babcock–Leighton models of the solar cycle: questions and issues. Adv. Space Res. **39**, 1661–1669 (2007)

P. Charbonneau, Dynamo models of the solar cycle. Living Rev. Sol. Phys. **7**, 3 (2010). doi:10.12942/lrsp-2010-3

F. Clette, L. Svalgaard, J.M. Vaquero, E.W. Cliver, Revisiting the sunspot number: a 400-year perspective on the solar cycle. Space Sci. Rev. (2014, this issue). doi:10.1007/s11214-014-0074-2

E.W. Cliver, The extended cycle of solar activity and the Sun's 22-yr magnetic cycle. Space Sci. Rev. (2014, this issue). doi:10.1007/s11214-014-0093-z

I. Ermolli, K. Shibasaki, A. Tlatov, L. van Driel-Gesztelyi, Solar cycle indices from the photosphere to the corona: measurements and underlying physics. Space Sci. Rev. (2014, this issue). doi:10.1007/s11214-014-0089-8

L. Fletcher, B.R. Dennis, H.S. Hudson et al., An observational overview of solar flares. Space Sci. Rev. **159**, 19–106 (2011). doi:10.1007/s11214-010-9701-8

S. Forbush, Variation with a period of two solar cycles in the cosmic-ray diurnal anisotropy and the superposed variations correlated with magnetic activity. J. Geophys. Res. **74**, 3451–3468 (1969)

C. Frölich, Total solar irradiance observations. Surv. Geophys. **33**, 453–473 (2012). doi:10.1007/s10712-011-9168-5

G.E. Hale, On the probable existence of a magnetic field in sunspots. Astrophys. J. **28**, 315–343 (1908)

D.H. Hathaway, The solar cycle. Living Rev. Sol. Phys. **7**, 1 (2010). doi:10.12942/lrsp-2010-1. http://www.livingreviews.org/lrsp-2010-1

H. Hudson, L. Svalgaard, I. Hannah, Solar sector structure. Space Sci. Rev. (2014, this issue). doi:10.1007/s11214-014-0121-z

J.R. Jokipii, B. Thomas, Effects of drift on the transport of cosmic rays. IV. Modulation by a wavy interplanetary current sheet. Astrophys. J. **243**, 1115–1122 (1981)

J.R. Jokipii, E.H. Levy, W.B. Hubbard, Effects of particle drift on cosmic-ray transport. I. General properties, application to solar modulation. Astrophys. J. **213**, 861–868 (1977)

B.B. Karak, J. Jiang, M.S. Miesch, P. Charbonneau, A.R. Choudhuri, Flux transport dynamos: from kinematics to dynamics. Space Sci. Rev. (2014, this issue). doi:10.1007/s11214-014-0099-6

R. Komm, The acoustic and magnetic solar cycle. Astron. Nachr. **331**(9/10), 873–878 (2010). doi:10.1002/asna.201011417

J. Kota, J.R. Jokipii, Effects of drift on the transport of cosmic rays. VI. A three-dimensional model including diffusion. Astrophys. J. **265**, 573–581 (1983)

Z. Kővári, K. Oláh, Observing dynamos in cool stars. Space Sci. Rev. (2014, this issue). doi:10.1007/s11214-014-0092-0

R.B. Leighton, A magneto-kinematic model of the solar cycle. Astrophys. J. **156**, 1–26 (1969)
W. Livingston, M.J. Penn, L. Svalgaard, Decreasing sunspot magnetic fields explain unique 10.7 cm radio flux. Astrophys. J. Lett. **757**, L8 (2012), 4 pp. doi:10.1088/2041-8205/757/1/L8
I. Lopes, D. Passos, M. Nagy, K. Petrovay, Oscillator models of the solar cycle. Space Sci. Rev. (2014, this issue). doi:10.1007/s11214-014-0066-2
K.G. McCracken, J. Beer, F.B. McDonald, Variations in the cosmic radiation, 1890–1986, and the solar and terrestrial implications. Adv. Space Res. **34**, 397–406 (2004)
A.A. Norton, P. Charbonneau, D. Passos, Hemispheric coupling: comparing dynamo simulations and observations. Space Sci. Rev. (2014, this issue). doi:10.1007/s11214-014-0100-4
M.J. Penn, W. Livingston, Temporal changes in sunspot umbral magnetic fields and temperatures. Astrophys. J. Lett. **649**, L45–L48 (2006). doi:10.1086/508345
W.D. Pesnell, Solar cycle predictions. Sol. Phys. **281**, 507–532 (2012). doi:10.1007/s11207-012-9997-5
G.J.D. Petrie, K. Petrovay, K. Schatten, Solar polar fields and the 22-year activity cycle: observations and models. Space Sci. Rev. (2014, this issue). doi:10.1007/s11214-014-0064-4
K. Petrovay, Solar cycle prediction. Living Rev. Sol. Phys. **7**, 6 (2010). doi:10.12942/lrsp-2010-6. http://www.livingreviews.org/lrsp-2010-6
A.A. Pevtsov, Y.A. Nagovitsyn, A.G. Tlatov, A.L. Rybak, Long-term trends in sunspot magnetic fields. Astrophys. J. Lett. **742**, L36 (2011), 4 pp. doi:10.1088/2041-8205/742/2/L36
A.A. Pevtsov, M.A. Berger, A. Nindos, A.A. Norton, L. van Driel-Gesztelyi, Magnetic helicity, tilt, and twist. Space Sci. Rev. (2014, this issue). doi:10.1007/s11214-014-0082-2
M.S. Potgieter, Solar modulation of cosmic rays. Living Rev. Sol. Phys. **10**, 3 (2013). doi:10.12942/lrsp-2013-3. http://solarphysics.livingreviews.org/Articles/lrsp-2013-3/
M. Rempel, Numerical sunspot models: robustness of photospheric velocity and magnetic field structure. Astrophys. J. **750**, 62 (2012). doi:10.1088/0004-637X/750/1/62
B. Schmieder, V. Archontis, E. Pariat, Magnetic flux emergence along the solar cycle. Space Sci. Rev. (2014, this issue). doi:10.1007/s11214-014-0088-9
H. Schwabe, Sonnen-Beobachtungen im Jahre 1843. Astron. Nachr. **21**(495), 233–236 (1844)
E.J. Smith, The heliospheric current sheet. J. Geophys. Res. **106**, 15819–15831 (2001)
K.G. Strassmeier, Starspots. Astron. Astrophys. Rev. **17**, 251–308 (2009)
L. Svalgaard, The solar radio microwave flux (2009). http://wattsupwiththat.files.wordpress.com/2009/05/svalgaard-solar-radio-flux.pdf
L. Svalgaard, H.S. Hudson, The solar microwave flux and the sunspot number, in *SOHO-23: Understanding a Peculiar Solar Minimum*, ed. by S.R. Cranmer, J.T. Hoeksema, J.L. Kohl. ASP Conf. Ser., vol. 428 (Astronomical Society of the Pacific, San Francisco, 2010), p. 325
K.F. Tapping, The 10.7 cm solar radio flux (F10.7). Space Weather **11**, 394–406 (2013). doi:10.1002/swe.20064
T. Thambyahpillai, H. Elliot, World-wide changes in the phase of the cosmic ray solar daily variation. Nature **171**, 918–920 (1953)
M.J. Thompson, Helioseismology over the Solar Cycle, in *SOHO-23: Understanding a Peculiar Solar Minimum*, ed. by S.R. Cranmer, J. Todd Hoeksema, J.L. Kohl. ASP Conference Series, vol. 428 (Astronomical Society of the Pacific, San Francisco, 2010), pp. 23–28
I.G. Usoskin, A history of solar activity over millennia. Living Rev. Sol. Phys. **10**, 1 (2013). doi:10.12942/lrsp-2013-1. http://www.livingreviews.org/lrsp-2013-1
I.I. Virtanen, K. Mursula, North-south asymmetric solar cycle evolution: signatures in the photosphere and consequences in the corona. Astrophys. J. **781**, 99 (2014), 8 pp. doi:10.1088/0004-637X/781/2/99
Y.-M. Wang, Solar cycle variation of the Sun's low-order magnetic multipoles: heliospheric consequences. Space Sci. Rev. (2014, this issue). doi:10.1007/s11214-014-0051-9
Y.-M. Wang, N.R. Sheeley Jr., Magnetic flux transport and the sun's dipole moment—new twists to the Babcock-Leighton model. Astrophys. J. **375**, 761–770 (1991)
W.R. Webber, M.S. Potgieter, R.A. Burger, A comparison of predictions of a wavy neutral sheet drift model with cosmic-ray data over a whole modulation cycle: 1976–1987. Astrophys. J. **349**, 634–640 (1990)
K.L. Yeo, N.A. Krivova, S.K. Solanki, Solar cycle variation in solar irradiance. Space Sci. Rev. (2014, this issue). doi:10.1007/s11214-014-0061-7

DOI 10.1007/978-1-4939-2584-1_2
Reprinted from *Space Science Reviews* Journal, DOI 10.1007/s11214-014-0121-z

Solar Sector Structure

Hugh S. Hudson · Leif Svalgaard · Iain G. Hannah

Received: 4 September 2014 / Accepted: 18 November 2014 / Published online: 10 December 2014

Abstract The interplanetary magnetic field near 1 AU has a characteristic "sector" structure that reflects its polarity relative to the solar direction. Typically we observe large-scale coherence in these directions, with two or four "away" or "towards" sectors per solar rotation, from any platform in deep space and near the ecliptic plane. In a simple picture, this morphology simply reflects the idea that the sources of the interplanetary field lie mainly in or near the Sun, and that the solar-wind flow enforces a radial component in this field. The sector boundaries are sharply defined in the interplanetary field near one AU, but have more complicated sources within the Sun itself. Recent evidence confirms that the origins of this pattern also appear statistically at the level of the photosphere, with signatures found in the highly concentrated fields of sunspots and even solar flares. This complements the associations already known between the interplanetary sectors and large-scale coronal structures (i.e., the streamers). This association with small-scale fields strengthens at the Hale sector boundary, defining the Hale boundary as the one for which the polarity switch matches that of the leading-to-following polarity alternation in the sunspots of a given hemisphere. Surface features that appear 4.5 days prior to the sector crossings observed at 1 AU correlate with this sense of polarity reversal.

Keywords First keyword · Second keyword · More

1 Introduction

In the absence of solar rotation (and other complexities), the solar wind would cause the heliospheric magnetic field to have a strictly radial orientation, with the two hemispheres

H.S. Hudson (✉)
Space Sciences Laboratory, University of California, Berkeley, CA 94720, USA
e-mail: hhudson@ssl.berkeley.edu

L. Svalgaard
HEPL, Via Ortega, Stanford University, Stanford, CA 94305, USA

I.G. Hannah
SUPA School of Physics & Astronomy, University of Glasgow, Glasgow, UK G12 8QQ

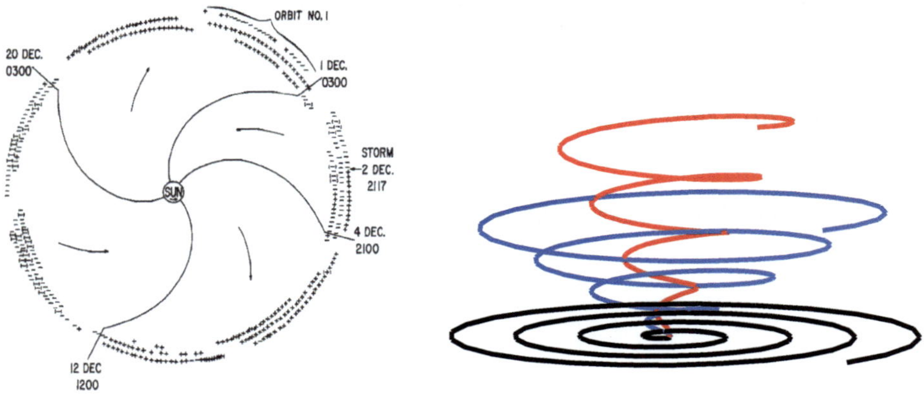

Fig. 1 *Left*, an early view of the interplanetary magnetic sector structure, from Wilcox and Ness (1965). The symbols reflect the predominant orientation of the interplanetary field in 3-hour data segments from the magnetometer on board the IMP-1 spacecraft, anchored at Earth but with an apogee (31.7 R_E, orbital period about 3.9 d) large enough to give clear views of the solar wind in the ecliptic plane. Plus and minus conventionally represent outward and inward field directions; this early sample shows a persistent four-sector pattern. *Right*, a sketch of how field lines at 0, 30, and 60° heliolatitude must look in an isotropic high-beta solar wind with constant radial speed (from Owens and Forsyth 2013)

having inward and outward directions. These complexities cause the current sheet separating the polarities to have a large-scale warp, and this appears to spacecraft exploring the heliosphere as the *sector structure*. In this introductory view for the present volume we provide a roughly chronological view of our development of ideas about the sector phenomenon. This structure originates below the photosphere, and so this review necessarily needs to touch on many of the topics dealt with in greater detail in the remaining chapters of this volume. We hope that it can serve as a good introduction to some of them, and try to establish the basic concepts here.

The early years of human exploration of space beyond the Earth system saw the development of instrumentation capable of detecting the then-hypothetical solar wind, and then the magnetic field embedded within it. Qualitative physical arguments had made it clear that the field stretches out in the radial direction, and that it eventually must adopt a spiral pattern in the ecliptic plane (Biermann 1957; Alfvén 1957; Parker 1958). The mean speed of the observed solar wind dictated that this spiral should have an angle of about 45° to the radial at 1 AU, and that the essentially radial flow should take 4–5 days in transit. Observations generally confirmed these rough ideas.

The first interplanetary space probes capable of sufficiently sensitive magnetic measurements revealed the existence of the sector structure (Wilcox and Ness 1965). Figure 1 (left) shows a data representation from that era, which illustrates the basic idea. This remarkable feature of the solar magnetic field, as evidenced deep in the heliosphere but at heliolatitudes near the ecliptic plane, immediately demanded an understanding of the relationship between the different domains that the "towards" and "away" sectors represented. The nature of the underlying solar magnetic field as revealed by magnetographic observations of the photosphere became a hot topic. The 3D nature of the heliosphere (Fig. 1, right) compounded the complexity of these questions, since the geometry of a sector could not be readily inferred from near-ecliptic observations. Note that systematic mapping of the photospheric line-of-sight magnetic field had not begun much earlier than the space age (Babcock 1953).

In a major development, Svalgaard (1972) established that one can use polar geomagnetic records to determine the times when Earth passed through a sector boundary. This opened the possibility of a proxy record of the global development of the heliospheric field that in principle can extend into the middle of the 19th century. We discuss this in more detail in Sect. 3 here. Shortly thereafter Gulbrandsen (1973) and Antonucci and Svalgaard (1974) established a correlation between the sector pattern at 1 AU with coronal green-line (Fe XIV) structures (Sect. 4) in the middle corona. This finding directly related the heliospheric sector structure to magnetic fields actually in the solar corona.

In this review we describe (Sect. 5) the recent further identification of the solar footprints of the sector patterns to the level of the photosphere (Svalgaard et al. 2011). This involves the concept of the "Hale sector boundary," defined by Svalgaard and Wilcox (1976) as that part of the sector boundary at which the polarity change matches that of the change of polarity from preceding to following sunspot (Sect. 5). Because the solar polar fields reverse polarity roughly every 22 years (the Hale cycle), the Hale sector boundaries also reverse to keep up. The extended proxy record of sector-boundary crossings greatly enhances our knowledge of this basic solar morphology; the space age has barely witnessed two full Hale cycles thus far.

In retrospect, it may seem reasonable that a heliospheric magnetic domain structure must exist, and that the domains must simplify into sector patterns at great distances from the solar surface (see e.g. Wang 2014, in this volume). The Sun possesses strong magnetic fields that originate in current systems largely contained below the photosphere; we know this because of the fairly good typical match with simple potential-field expansions in the lower corona (Sect. 4). This means that if one makes a multipole harmonic extrapolation of the very complicated pattern of photospheric magnetic fields—which evolves with large-scale organization, but also in seemingly random ways—one predicts that the lowest-order term, the dipole, will eventually dominate the field and become the only surviving component at large radial distances. This reasonable viewpoint ignores the fact that the field finds itself embedded in an active plasma, through which large-scale currents flow even far from the Sun. In fact, the heliospheric magnetic field—as opposed to the coronal field—has a maximally non-potential structure according to the Aly-Sturrock theorem (Aly 1991; Sturrock 1991). It contains the large-scale current sheet needed to reverse the field direction between hemispheres.

2 Solar Magnetography and the Sector Structure

Hale's discovery of Zeeman splitting in solar spectra has led to extensive, but tantalizingly incomplete, knowledge of the patterns of solar magnetism in the photosphere, and now to major efforts in the observational characterization of the magnetic field in the upper solar atmosphere; we return to this in Sect. 4. In the era of discovery of the interplanetary sector structure, our remote-sensing knowledge of the solar magnetic field mainly consisted of synoptic maps, at low resolution, of the line-of-sight component. Systematic measurements of this kind began in the 1950s via the production of whole-disk magnetograms (e.g. Babcock 1963). Measurement of the full vector field has also now become routine, and of course resolution and data quality have steadily increased, but problems of interpretation remain; Borrero and Ichimoto (2011) discuss many of them in a recent review. We return to this topic in Sect. 4, and continue roughly historically here.

One essential problem is that resolution has improved to the point where the atmospheric dynamics and the characterization of the magnetic field require a simultaneous solution via numerical simulations. The numerical approach will eventually help to solve the fundamental uncertainty in the physical heights of atmospheric structures. The standard semi-empirical modeling of the solar atmosphere (e.g., the VAL-C model of Vernazza et al. 1981)

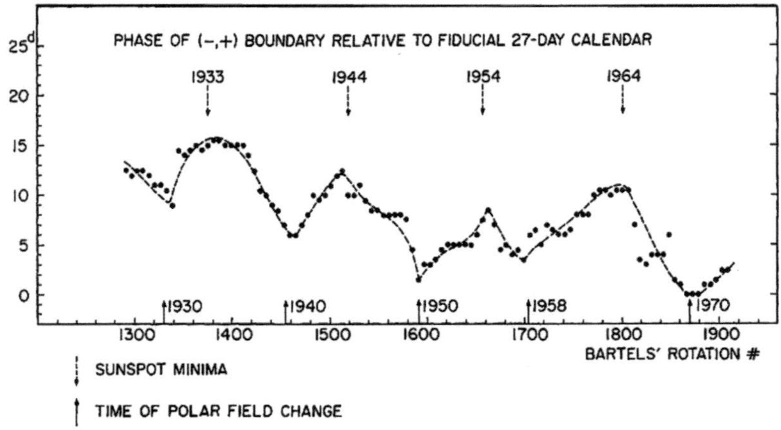

Fig. 2 Phase of sector-boundary crossing, relative to a fiducial 27-day rotation law (Svalgaard and Wilcox 1975). Note that these crossing times come from the geomagnetic proxy record of polarity reversals in the solar wind. The *arrows* mark times of polarity reversal, showing an extension of about 1.5 Hale cycles prior to the availability of *in situ* observations from space

scales with optical depth, rather than height directly, and has an upper boundary at 2543 km above the photosphere as normally defined in terms of optical depth unity at 5000 Å. Such models make many strong physical assumptions, for example in imagining a static structure that can be described as one-dimensional. The density structure of the corona is best known only from the ill-posed 3D reconstructions of its projected brightness from projected data in one or two planes (e.g., Aschwanden et al. 2008a, 2008b), or via rotational synthesis that has inevitable confusion with coronal time variations (Frazin et al. 2010). Any quantitative understanding of field gradients must of course have precise geometrical information. Nevertheless even the line-of-sight Zeeman measurements provide a wealth of morphological detail across the entire face of the Sun; we understand a great deal even at this level of observational technique.

Typical line-of-sight solar magnetograms turned out to give no clues at all to the physical origin of the domains that the sectors represent, and early representations searched in vain for simple rigidly-rotating zonal patterns in the photosphere—rigidly rotating, because one can derive a precise angular velocity from studying the phases of sector-boundary crossings. Figure 2 shows the stability of the sector pattern over a period of many years. Note that the epoch precedes the space age; these crossing times come from the geomagnetic proxy, which we discuss in more detail in Sect. 3. The figure shows the deviation of sector-boundary arrivals relative to a fiducial 27-day synodic rotation rate (Svalgaard and Wilcox 1975). The stability of these phase measurements suggests a source that has only a slowly-varying rotation law.

Some of the mystery regarding the solar origins had been resolved by the discovery of the heliolatitude dependence of the sector patterns (the Rosenberg-Coleman effect), clearly recognizable in the early data from outside the ecliptic plane (Rosenberg and Coleman 1969): the sector durations depend on the heliolatitude of the observation. This led to the concept of a single warped current sheet (e.g. Schulz 1973; Svalgaard et al. 1975) extending to large radial distances; its intersections with the ecliptic plane defines the sector boundaries seen there, and at sufficiently high heliolatitudes it must have a unipolar character; this concept extended the model proposed by Svalgaard et al. (1974) and attained great prominence as

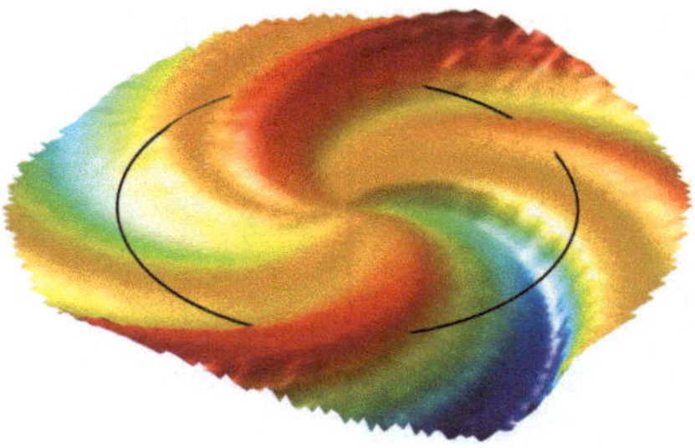

Fig. 3 Numerical simulation of the heliospheric structure at a time (Carrington rotation 1912) corresponding to solar minimum (Odstrčil et al. 2004). The *black circle* shows the intersection of Earth's orbit through the warped and fluted heliospheric current sheet, with the colors describing the out-of-the-ecliptic excursions of the solar wind over a range $\pm 10^\circ$. An observer at higher or lower heliolatitudes would see greater or lesser durations of the corresponding (*red* or *blue*) polarity

the heliospheric "ballerina skirt." Figure 3 illustrates the basic explanation of the Rosenberg-Coleman effect via a numerical simulation of the heliospheric structure at a specific epoch (Odstrčil et al. 2004). The Ulysses mission triumphantly confirmed this picture (e.g., Balogh and Erdős 2013), consistent with other data such as the radio (interplanetary scintillations) finding of high-speed solar wind streams at high heliographic latitudes (Coles and Rickett 1976) and the X-ray recognition of the polar coronal holes (Vaiana et al. 1973).

The presence of an initially mysterious sector structure in the interplanetary field (e.g., Wilcox and Howard 1968) led to the development of synoptic observing programs (see also Sect. 5). Wilcox and his collaborators at Stanford University created what is now called the Wilcox Solar Observatory, which has provided a long-running data set of low resolution but great stability. This facility began collecting data in 1975 and continues to the present time.

3 Geomagnetism and the Sector Structure

The interplanetary magnetic field in the vicinity of the Earth interacts with the Earth's own field in a complicated manner; the orientation and strength of the field in the incoming solar wind perturb the geomagnetic field as measured on the surface of the Earth. Given a terrestrial field of order 1 G and an interplanetary field of a only a few $\times 10^{-5}$ G, one might expect little effect, but in fact the signatures could readily be measured even with the techniques available in the 19th century (the Gauss-Weber variometer). The sector structure thus has a proxy record that extends quite far back in time. Svalgaard (1968) and Mansurov (1970) showed that the interplanetary magnetic field produced a characteristic pattern of diurnal variation at high-latitude sites such as Thule (12.5° geographic colatitude) or Vostok (168.5°). The Thule station has a geomagnetic colatitude of only about 5°, and Svalgaard (1973) notes that the major effects of this perturbing current system lie with 15° of the geomagnetic pole; he attributes the characteristic variation to a Hall current (Baker and Martyn 1953) typically extending about that distance from the pole, and circulating with opposite senses for away and towards sectors. The morphology of this diurnal effect distinguishes it sharply from others such as those of auroral electrojets.

This Svalgaard-Mansurov effect (Wilcox 1972) is strong enough to produce reliable determinations of the polarity of the interplanetary field (Friis-Christensen et al. 1971), usually even on a daily basis (Vokhmyanin and Ponyavin 2013), and so it serves to extend the record of Earth's sector-boundary crossing times back into the 19th century. This relies upon the understanding of the available geomagnetic observations at sufficiently high

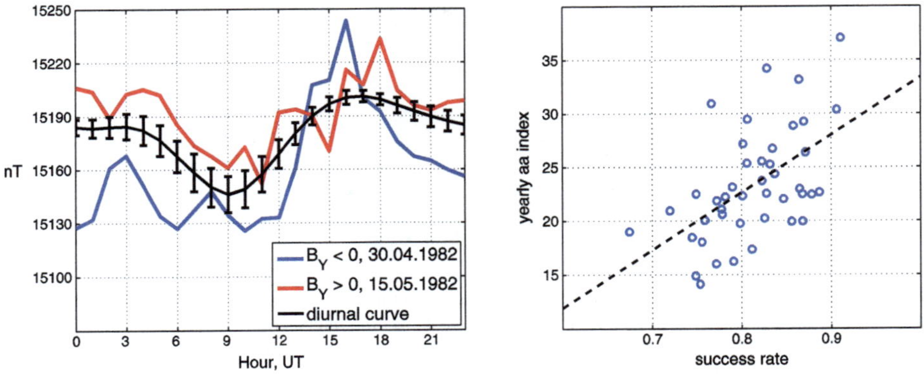

Fig. 4 *Left*, illustration of the Svalgaard-Mansurov effect using modern (1982) data from the Nurmijärvi station, which continued the 19th-century Helsinki observations. The *black line* shows the mean diurnal variation, and the *red* and *blue lines* represent data from days of known away (red) and towards (blue) sectors, showing the clear distinction between the two orientations. *Right*, estimates of the (diurnal) success rate for the inference of the field polarity, via comparison with satellite observations over the period 1966–2008 (Vokhmyanin and Ponyavin 2013)

latitudes. Vokhmyanin and Ponyavin (2013) published sector-boundary data for Carrington cycles 9–13 (1843–1902) in this way, for example, by making use of the records from the now-discontinued Helsinki (29.8° geographic colatitude) and St. Petersburg (30.0°) geomagnetic observatories. Figure 4 illustrates this heliospheric effect on the Earth's polar field, calibrating the residuals away from the mean diurnal variation of Nurmijärvi data by reference to known sector orientations. This observatory produces results closely matching those of the original Helsinki observatory (Nevanlinna 1997).

4 The Corona, the Solar Wind, and the Sector Structure

In this section we discuss three perspectives on the sector structure: from the heliosphere, from the corona, and from the photosphere. The information from these domains have different qualities (*in situ* or remote-sensing), and are never complete, but generally agree now upon most of the basic structure. We discuss one exception to this understanding in Sect. 5.

4.1 As Viewed in the Heliosphere

4.1.1 Observations In Situ

The identification of solar sources of the interplanetary sector structure began with Gulbrandsen (1973) and Antonucci and Svalgaard (1974), who found a remarkable long-term agreement between the sector boundaries, appropriately time-shifted, and coronal photometry in the "green line" of Fe XIV (5303 Å). We return to this important result in the next section, but first briefly describe the nature of the corona as seen in the green line or in various other ways.

One description of the solar corona envisions it as a concentric spherical shell of relatively low plasma beta (i.e., in force balance mainly via the Maxwell stress tensor); on intermediate time scales it is indeed quite stable but is prone to sudden and drastic disruptions, often on large scales, related to the domain structure of its magnetic connectivity. The

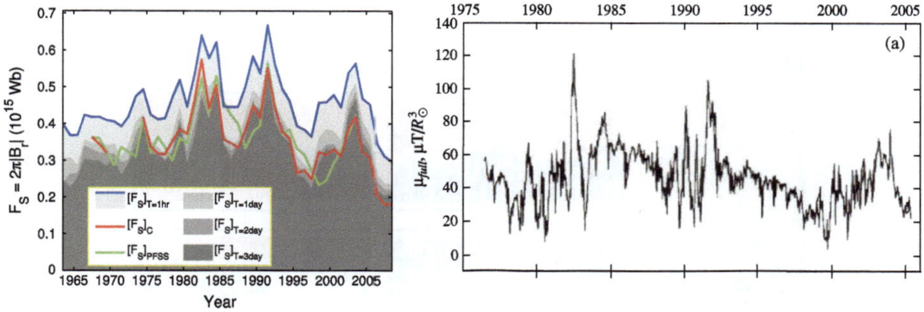

Fig. 5 *Left*, the solar open flux estimated via the PFSS method (*red line*) at the Sun, and by using different methods (*gray shading*) for heliospheric sampling (image reproduced by permission from Lockwood et al. 2009, copyright by AGU; see this paper for detailed definitions). *Right*, estimates of the amplitude of the best-fit dipole term for the photospheric field (Livshits and Obridko 2006), showing similar variability

lower boundary—subject to any of several possible definitions—has corrugations imposed by the dynamics of the lower solar atmosphere, and the upper boundary fades into the solar wind in a manner that probably defies observation; both the upper and lower boundaries thus presumably have distinctly non-spherical shapes.

This complicated-sounding region has a simple and attractive model that describes its geometry fairly well: the "potential-field source-surface" (PFSS) model of Schatten et al. (1969) and Altschuler and Newkirk (1969). Such models typically have a driver at the lower boundary from the line-of-sight Zeeman magnetic measurements, interpreted globally via a synoptic map of the measurements at central meridian. Given the foreshortening toward the polar regions, this approach clearly omits a great deal of spatial and temporal structure that the EUV or X-ray movies, for example, reveal on many scales. Based on this boundary, the PFSS model consists of a potential-field extrapolation out to an *ad hoc* "source surface," typically a spherical boundary at which a fictitious current system forces the exterior field into a strictly radial configuration. This radial field then constitutes the "open" magnetic flux that somewhat mysteriously fills the heliosphere via advection in the solar wind. See for example Lockwood (2013) for a discussion of subtle points in the estimation of heliospheric open field and its significance. A common choice for the radius of the source surface, 2.5 $R_\odot$ (Hoeksema et al. 1983), yields an approximate match to the heliospheric flux, which appears to vary by a small factor across the solar cycle.

We note that the heliospheric data make it possible to measure the open flux directly, in the sense of identifying it with the radial component B_r of the field. The Ulysses data showed, remarkably, that there was a minimal latitude dependence of this quantity (Smith and Balogh 1995), so with that fact and the inverse square law, one can derive a good estimate of the open flux from any point in the heliosphere. Again remarkably, given the weaknesses inherent in the photospheric data, the direct view in terms of B_r turns out to agree reasonably well with that derived from PFSS with a fixed source surface (Owens et al. 2008; Wang and Sheeley 2003). Figure 5 sketches the long-term variations of solar open flux as determined by these methods.

The Ulysses "fast latitude scans" also provided an excellent overview of the 3D sector structure as it related to the solar wind, as illustrated in Fig. 6. This view nicely suggests how the basic pictures in Fig. 1 work themselves out in three dimensions, at least with this crude sampling; each map corresponds to the entire year of the "fast" scan and thus presupposes great long-term stability (Owens et al. 2011). Note that the small regions of minority polarity

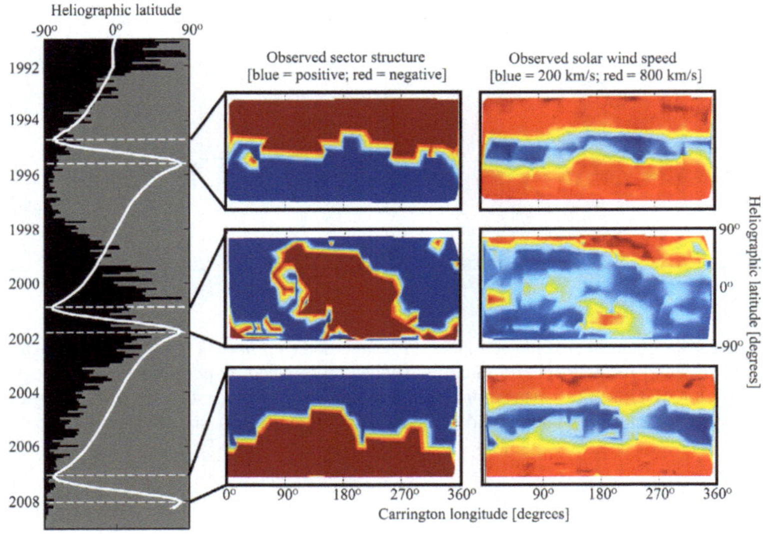

Fig. 6 The three rapid scans of heliolatitude executed by the Ulysses spacecraft: *left*, the heliolatitude and sunspot number; *middle* and *right*, source-surface projections of polarity and wind speed, respectively (from Owens and Forsyth 2013). The Ulysses orbital inclination of 80.2° and perihelion of 1.3 AU meant that each full scan required about one year

in the middle panels may just reveal artifacts resulting from this sampling; see Antiochos et al. (2012) for a theoretical discussion of the heliospheric magnetic connectivity.

4.1.2 Remote-Sensing Observations

The idea of using the cosmic-ray shadows of large structures (Clark 1957) has developed into a method for probing the coronal magnetic field (Amenomori et al. 1993, 2013). Cosmic rays of sufficiently high energy have Larmor radii large enough to avoid much deflection. The observation thus makes use of TeV-energy primary cosmic rays detected via their extensive air showers, and the first results immediately showed a significant modification of the solar cosmic-ray shadow depending on the presence of towards and away sectors. Since the original discovery (1993), the data have accumulated and improved to the point where more sophisticated analyses for coronal magnetic-field structure have become possible. Figure 7, from Amenomori et al. (2013), shows the development of the cosmic-ray shadow detection over a solar-cycle time span.

A quantitative understanding of the solar magnetic shadow involves complex forward-method simulation. Amenomori et al. estimate the FWHM angular resolution and modal energy of the Tibet air-shower data as 0.9° and 10 TeV, respectively. Because of the small event rate, and the shallowness of the shadow (a consequence of the low angular resolution), long integrations are required to obtain significant results. The data support an inference of the magnetic structure of the solar on this angular scale. The results depend upon Monte Carlo simulations of cosmic-ray transport and interaction within the inner heliosphere, for which a time-resolved magnetic model (e.g., PFSS) is necessary. The observations shown in Fig. 7 permitted the authors to distinguish a standard PFSS model from a more elaborate current-sheet model Zhao and Hoeksema (1995). More elaborate data analyses with improved data may allow us to follow the structure of the heliospheric field in the relatively

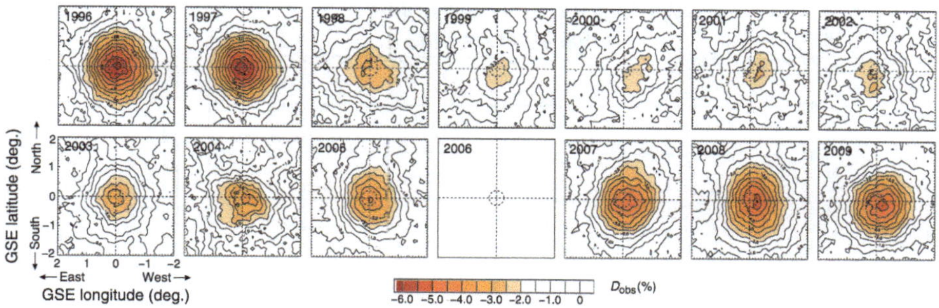

Fig. 7 The time development of the observed cosmic-ray shadow of the Sun, with its coronal magnetic field, from the Tibet air-shower array (Amenomori et al. 2013). Each panel shows a year's worth of data in the form of contours of the deficit in cosmic-ray intensity. Note that the maximum deficit at current resolution is only a few percent. Year 2006 had insufficient statistics and the figure omits its image

unknown domain at the distance of the standard source surface, 2.5 $R_{\odot}$, and of course it would be most interesting to be able to characterize the development of the warp structure of the heliospheric current sheet in this region.

4.2 As Viewed in the Corona

Understanding the magnetic field in the solar corona directly requires remote-sensing astronomical measurements (e.g., for Zeeman splitting), which can be exceedingly difficult. One can also interpret the images to obtain geometrical information from the orientations of the striations (e.g., Aschwanden 2013). The image-interpretation approach has improved continuously with the development of higher spatial resolution and the better sampling afforded by modern observations, because the images observable in the emission corona have striations at the finest observable scales. These striations arguably guide us to the orientation of the field threading the plasma. The gyroresonance condition also allows magnetography via microwave emission, especially in active regions; for example, Brosius and White (2006) found an example of kG fields at Mm heights via this technique. None of these approaches though provides detailed direct information about **B** everywhere in the corona.

Discerning the full geometry of the coronal field thus requires modeling. The corona presents a serious geometrical problem here; for the most part it is optically thin, and so the line-of-sight depth cannot be known observationally. The PFSS approach described in the previous section is the most frequently used method, but there are many other approaches with various degrees of sophistication (e.g., De Rosa et al. 2009; Mackay and Yeates 2012), mostly involving the assumption of a force-free field. Beyond the purely mathematical treatments, which might suffice for an idealized corona at low plasma beta and no chromosphere, there are various efforts to incorporate plasma physics in various approximations, as also reviewed by Mackay & Yeates.

The concept of the "web of separatrices" or S-web (Antiochos et al. 2007, 2011, 2012) may simplify our view of how the complicated magnetic field of the lower solar corona maps into the relatively simple one in the heliosphere; in the ideal case of no sunspot activity whatsoever, this heliospheric field tends to become radial while retaining its bipolar character. Figure 8 shows the S-web for a snapshot synoptic diagram at solar minimum. The diagram represents the "squashing factor" Q, derived from the Jacobian of the mapping of a given field line to its sources in the computational boundary. Within the S-web structure one envisions magnetic reconnection to occur in such a manner that closed-field

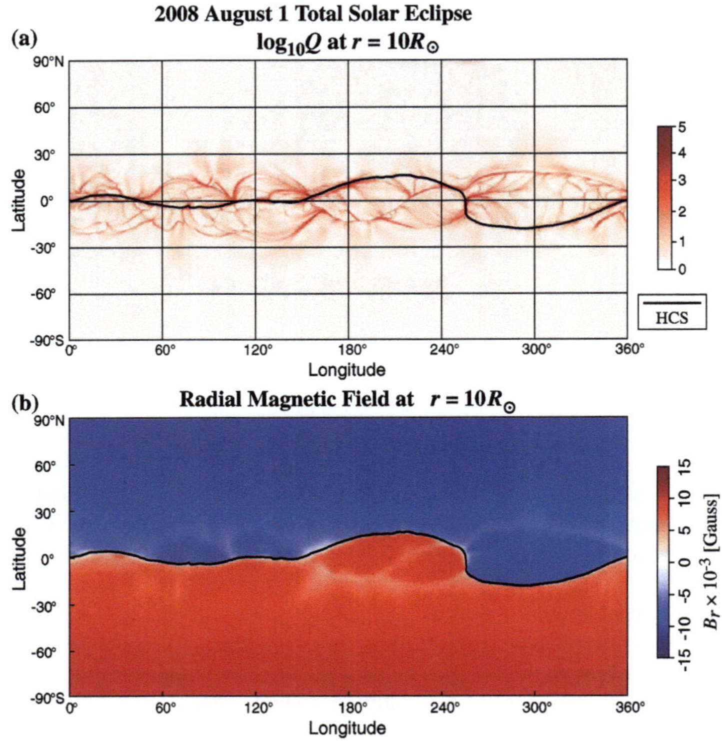

Fig. 8 Synoptic charts illustrating the "Web of separatrices" (Antiochos et al. 2011) for a snapshot at solar minimum (August 1, 2008). The *upper panel* shows the "squashing factor" Q, essentially the Jacobian of the mapping, with high values representing quasi-separatrix layers (Priest and Démoulin 1995). The *black line* shows the location of the heliospheric current sheet, which would be a true separatrix if the numerical model could resolve it. Otherwise, the quasi-separatrix layers represent the likeliest locations for the magnetic reconnection needed to mediate between the simple large-scale field in the solar wind, and the complicated small-scale field at the photosphere. The *lower panel* shows the radial field component B_r

plasma, enriched in high-FIP elements, can escape into the slow solar wind over the its full breadth in heliographic latitude (e.g., Zurbuchen 2007). A diffusive process with similar consequences underlies the "interchange reconnection" theories (cf. Fisk and Schwadron 2001; Fisk and Zhao 2009), but the details of both pictures remain somewhat unclear owing perhaps to the difficulty of modeling the local physics of magnetic reconnection within the MHD framework. Note that the MHD simulations represented in the figure cannot accurately describe the time variations of the structures involved, but should adequately capture the main physics of the steady-state Parker wind.

The synoptic chart in Fig. 8 shows an interesting feature of the S-web: sometimes it appears to form mainly on one side of the heliospheric current sheet. This asymmetric azimuthal dependence may be related to the large-scale organization of the field embodied in the Hale sector boundary (Sect. 5), but we are unaware of any work describing this possible association or its significance.

4.3 As Viewed at the Photosphere

A substantial development of solar magnetic-field models has proceeded entirely from their photospheric (surface) manifestations, the so-called flux-transport models following Leighton (1964, 1969). Leighton's "magnetokinematic model" contains a few free parameters, including descriptions of a diffusive motion of vertical magnetic flux at the photosphere, and a meridional circulation there, and successfully describes the main properties of the solar cycle on this basis. These models take their inspiration from the Babcock (1961) picture of how dynamo action may create the solar cycle deep in the interior of the Sun. Curiously, the Leighton flux-transport models succeed quite will in describing the surface manifestations without reference to the interior field structure at all. The coronal field can be modeled in this way (e.g. Wang and Sheeley 1991, 2003), and thence used to predict the properties of the heliospheric fields (Hoeksema et al. 1983), including the locations of the sector boundaries. These predictions do not require us to specify the interior volume threaded by the currents that give rise to the field outside the photosphere: deep-seated, as Babcock envisioned, or superficial, as modeled.

The interior-to-corona continuity of the large-scale solar field obviously has a descriptive representation in terms of spherical harmonics, and the behavior of the harmonic terms might give a guide to the physics involved. As noted for example by Wang et al. (2000), the lowest-order terms of the multipole expansion have the strongest effect at the source surface of a PFSS model because it is at the outer boundary of the domain. By construction, these terms determine the regions of open flux. The variations of the axisymmetric and equatorial dipole terms therefore must dominate the open flux, as noted by many, if the main source of the interplanetary field lies within the body of the Sun. Note the important caveat that we do not actually know how deep these sources may be (or what creates them), as opposed to the case of the Earth with its well-defined core.

5 Photospheric Identification of the Sector Boundaries

The early proxy inference of sector-boundary crossings showed them to have stable phases relative to fixed periods (Fig. 1). This suggested a rigid rotation pattern, rather than a differential one, and yet no solar surface feature has such a property. Subsequently the three-dimensional nature of the sector domains in the heliosphere became more apparent, as discussed above and illustrated in Fig. 6. At this point one could have inferred that the domain structure had a relatively simple interpretation: the waxing and waning of the polar coronal holes, and their pattern of polarity reversal, plus the simplification of the harmonic structure imposed by the expansion of the solar wind, could lead to a simplistic view of the structure formation. The theoretical ideas discussed in Sect. 4 confuse the issue by invoking differing views of the microphysics and its site.

Recent work has opened new issues regarding the solar origins of the magnetic domains of the interplanetary sectors, extending the association first noted by Gulbrandsen (1973) via the green-line corona and its correlation with the sector boundaries. The new work goes right to the level of the photosphere and identifies sector boundaries directly with small-scale features there, as we describe below. The key to these new associations is the concept of the Hale sector boundary (Svalgaard and Wilcox 1976), as illustrated in Fig. 9. The recognition that the Hale sector boundary systematically correlated with signatures of solar activity evidently was not anticipated, and still has no clear interpretation. Geometrically, the sector structure itself seems fixed with respect to the solar rotation (see Fig. 1); the various aspects

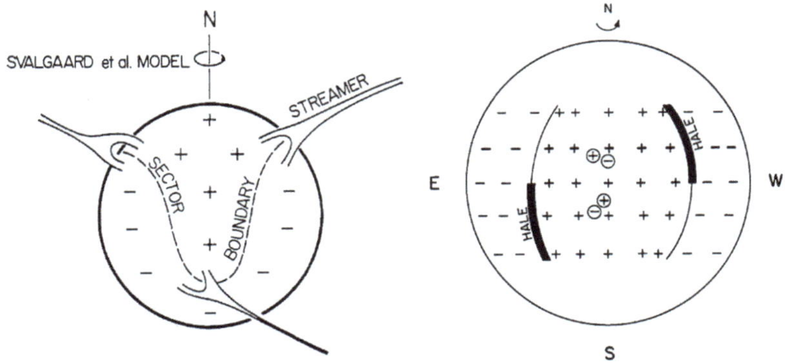

Fig. 9 *Left*, the now-accepted view of the solar origin of the interplanetary sectors, as proposed originally by Svalgaard et al. (1974). *Right*, the "Hale boundary" (Antonucci and Svalgaard 1974), defined as that part of the sector boundary at which the polarity switch matches that of the leading sunspot polarity in the corresponding hemisphere. In both panels the disk represents the photosphere, rather than a coronal or heliospheric surface

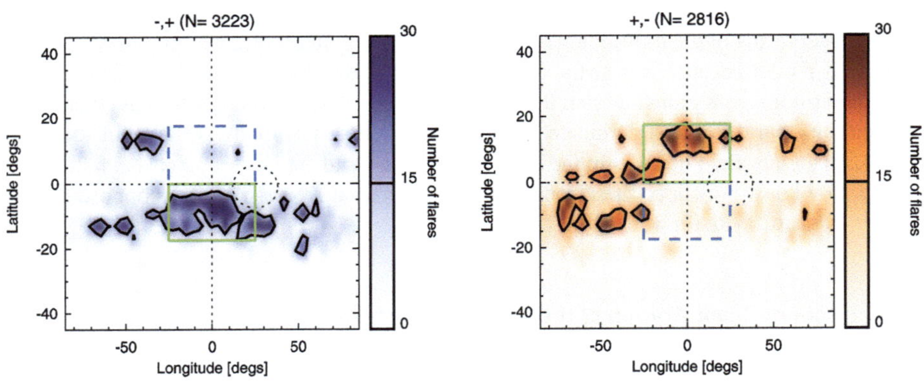

Fig. 10 RHESSI's view of the Hale boundary, for flare occurrence 2002–2010. The *colored contours* show the locations of flares relative to the times of sector boundary crossings at 1 AU; the *boxes with solid lines* show the locus expected for the Hale boundary segment, and the *dashed* the disfavored. The *left and right panels* show the (+, −) and (−, +) crossings, respectively (adapted from Svalgaard et al. 2011)

of solar activity also do, though with latitude dependence reflecting the photospheric differential rotation. So, what singles out the Hale boundary portion as a locus of solar activity? Or, what is the characteristic of solar activity that causes this association?

Further and more specific evidence for the linkage of the Hale sector boundary with solar activity came with the discovery that solar flares preferentially occur at the Hale boundary (Dittmer 1975; Svalgaard et al. 2011), the latter result making use of the RHESSI flare catalog of flare locations (J. McTiernan, private communication 2014). This analysis (shown in Fig. 10) reflects only the first few years of RHESSI observations, which began in 2002 and thus come from Cycle 23, but a further analysis (Iain Hannah, private communication 2013) extends this to Cycle 24 flares. As noted by Svalgaard et al. (2011), RHESSI is not central here, and the ordinary NOAA flare listings also show the same effect but with reduced spatial resolution; this had been the basis of Dittmer's original discovery of the effect.

The flare associations shown in Fig. 10 represent the superposition of RHESSI flare coordinates relative to the (crude) backwards extrapolation, by 4.5 days, from the times of the observed heliospheric sector boundary crossings.[1] The flares just come from the RHESSI catalog, with its particular biases. They include many lesser events not associated with CMEs (e.g. Yashiro et al. 2006). This is interesting from the point of view of the microphysics: do flares with no obvious opening of the field actually participate physically in maintaining the sector-boundary definition?

6 Long-Term Evolution of the Sector Structure

The long-term behavior of the solar magnetic field, extending many Hale cycles into the past, has theoretical as well as practical interest (the latter since some prediction techniques involve solar activity with patterns whose "memory" extends to the preceding cycle or even earlier). Regularities and irregularities exist on these longer scales, for example in the NS asymmetry discussed by Bell (1962). This paper also made the link between sunspot asymmetry and an even stronger asymmetry in the occurrence of great magnetic storms over the solar cycles between 1833 and 1960. A corresponding asymmetry appears in the Ulysses cosmic-ray data (Simpson et al. 1996), and Mursula and Hiltula (2003) showed this cone-like distortion to be a persistent pattern of behavior. Wang and Robbrecht (2011) then described this in terms of the additional magnetic pressure created in the low corona by the asymmetric eruption of field.

To the patterns of such long-term variations of sunspot, geomagnetic, and heliospheric variation we can now add the morphology of the Hale sector boundary, since it too can be traced well into the 19th century now. Figure 11 shows the result of extending the historical record back to Cycle 9, in the form of summed-epoch analyses similar to those shown in Fig. 10: maps of occurrence summed on the key times given by the sector-boundary crossings identified geomagnetically, each showing the distribution as a function of central meridian distance and heliographic latitude. The figure shows four panels for each of the Even cycles (16, 18, ..., 24) and Odd cycles (17, 19, ..., 23), distinguishing the odd/even crossings from the even/odd crossings and in each case identifying the hemisphere of the Hale boundary. This confirms the association of sector boundaries with flux concentrations and the coincidence of Hale boundaries with the correct hemisphere in each case, and shows the result to be stable over multiple Hale cycles.

7 What Are the Solar Sources of the Heliospheric Field?

An equivalent question might read "Why do sector boundaries correlate with solar activity?" The problem is that the small-scale closed fields of solar active regions, which lead through time to the structure of the streamer belt and the green-line observations (e.g., Antonucci and Svalgaard 1974), must appear as dome-shaped inclusions of more intense field in the low corona. Conceptually, the dome becomes larger as more flux emerges, but then flattens out as the active-region magnetism disperses. The streamer belt connects the active-region domes as illustrated in Fig. 9, left panel. The solar wind forms above these magnetic domes merged into the streamer belt; impelled by the large-scale sources of energy and momentum

[1] http://www.leif.org/research/sblist.txt.

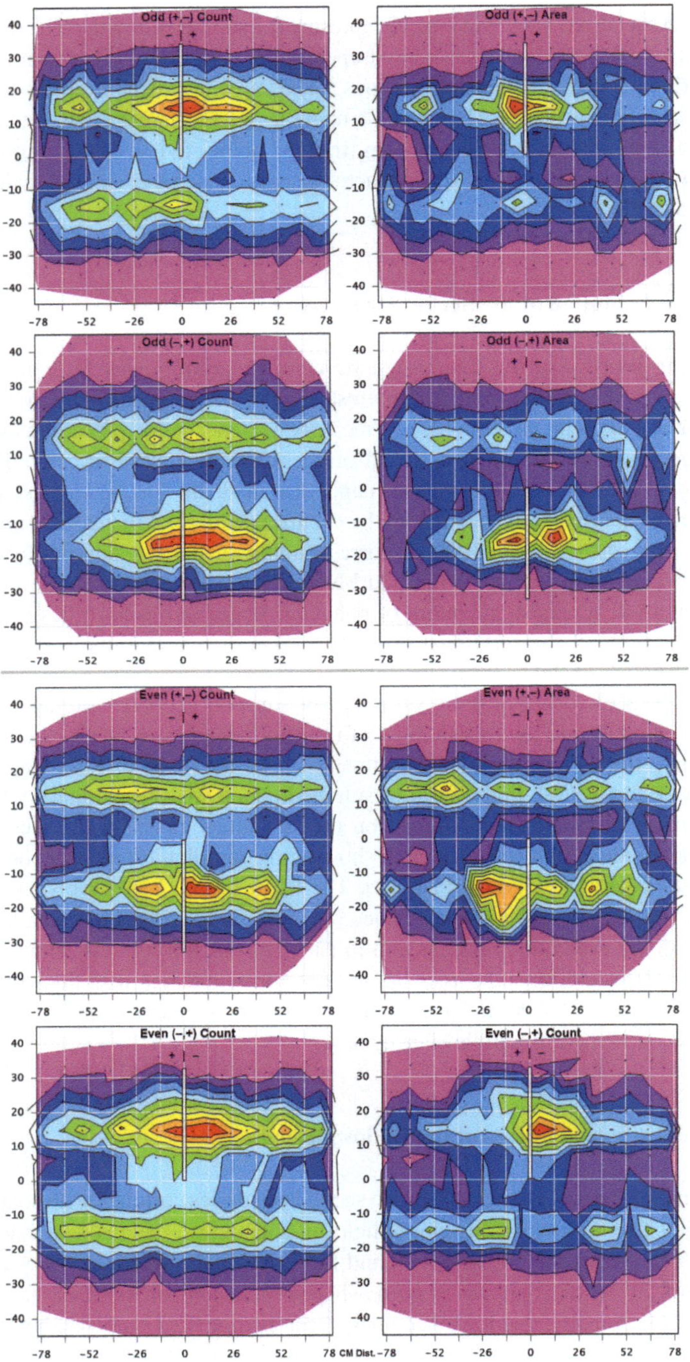

Fig. 11 Summed-epoch analysis for sunspot Group data, based on the Web archive maintained by David Hathaway, comprising some 600 sector boundaries identified for nine solar cycles (16 through 24). The eight panels each show synoptic maps of relative longitude and latitude; the *left column* is for Group sunspot counts, and the *right* for areas. The *top four* show cycles identified as "odd", and the *bottom* for the "even" ones, and in each case the Hale sector boundary is marked by a *vertical rectangle*

for the solar wind, the interplanetary field then develops its ballerina-skirt structure at higher altitudes.

The warp in the heliospheric current sheet has a natural explanation in terms of the intermittent dominance of activity in the N or S; roughly we could imagine that bigger magnetic domes or a more inflated streamer zone would simply push the heliospheric current sheet into the opposite hemisphere, considering only the structure resulting from magnetic pressure. This however would put the active regions, sunspots, and flares in between the sector boundaries, not at their very location. We do not have a ready explanation for this but suggest that it may have to do with the inherent latency of large-scale structure development in the corona. Yeates (2014) finds better fits to models with historical knowledge of the field development, rather than an instantaneous vacuum-field model. In this case, the match between activity and sector boundary would be a coincidence, and a manifestation of the time taken for global coronal relaxation.

8 Conclusions

The interplanetary magnetic sector structure offers opportunities and puzzles, as we have described. One of the great opportunities lies in its proxy record, which when unfolded will give us a record of the largest-scale solar magnetic fields extending to at least Carrington cycle 9 (from about 1845). The proxy extensions basically confirm that the Hale pattern of polar reversals persists over much longer time scales than the era of direct interplanetary observation, and these records may lead to further discovery. For example, the Hale-boundary segregation extends back in time at least to Cycle 16 already, as described above.

As regards puzzles, we have the basic issue of the connectivity of the solar magnetic field between the photosphere and the heliosphere. This broad subject involves several research communities, specifically those interested in the physical nature of the solar wind. We require intricate 3D patterns of the sort represented in Fig. 8, and these must be dynamical rather than static. Is there a link between the S-web (Antiochos) or interchange models (Fisk) and the Hale sector boundary? The recent discovery of a strong association between the Hale boundaries and the small-scale fields of sunspots and their dynamics (flares) suggests that there might be. The simplification of the photospheric fields as they map into the heliosphere has been found to be inherently dynamic, and so this connection seems like a reasonable one to suggest. Another open puzzle has to do with the structure of the four-sector pattern. The literature may not convincingly describe the physics behind this persistent corrugation of the heliospheric current sheet (the higher-order fluting of the ballerina's skirt).

Finally, the heliospheric sector structure and its clearly-identified sub-photospheric origin may bring clarity to the concept of "active longitudes." Large-scale coherent patterns at the limit of statistical significance had long been noted, and Howard (1996) remarked that "...there is a very long and somewhat confusing literature on this subject...". We see from the Hale boundary identification that the coherent patterns probably reflect large-scale internal structures within the Sun that require both latitude and longitude to define them.

Acknowledgements We thank the International Space Science Institute for support during the preparation of this chapter. Author Hudson thanks NASA for support under contract NAS 5-98033 for RHESSI, and acknowledges hospitality at the University of Glasgow.

References

H. Alfvén, On the theory of comet tails. Tellus **9**, 92 (1957)

M.D. Altschuler, G. Newkirk, Magnetic fields and the structure of the solar corona. I: Methods of calculating coronal fields. Sol. Phys. **9**, 131–149 (1969). doi:10.1007/BF00145734

J.J. Aly, How much energy can be stored in a three-dimensional force-free magnetic field? Astrophys. J. Lett. **375**, 61–64 (1991). doi:10.1086/186088

M. Amenomori, Z. Cao, L.K. Ding, Z.Y. Feng, K. Hibino, N. Hotta, Q. Huang, A.X. Huo, H.G. Jia, G.Z. Jiang, S.Q. Jiao, F. Kajino, K. Kasahara, Labaciren, D.M. Mei, L. Meng, X.R. Meng, Mimaciren, K. Mizutani, J. Mu, H. Nanjo, M. Nishizawa, Nusang, A. Oguro, M. Ohnishi, I. Ohta, T. Ouchi, J.R. Ren, T. Saito, M. Sakata, Z.Z. Shi, M. Shibata, T. Shirai, H. Sugimoto, X.X. Sun, K. Taira, Y.H. Tan, N. Tateyama, S. Torii, H. Wang, C.Z. Wen, Y. Yamamoto, X.Y. Yao, G.C. Yu, P. Yuan, T. Yuda, J.G. Zeng, C.S. Zhang, H.M. Zhang, L. Zhang, Zhasang, Zhaxiciren, W.D. Zhou, Direct evidence of the interplanetary magnetic field effect on the cosmic-ray shadow by the Sun. Astrophys. J. Lett. **415**, 147 (1993). doi:10.1086/187054

M. Amenomori, X.J. Bi, D. Chen, T.L. Chen, W.Y. Chen, S.W. Cui, Danzengluobu, L.K. Ding, C.F. Feng, Z.h. Feng, Z.Y. Feng, Q.B. Gou, Y.Q. Guo, K. Hakamada, H.H. He, Z.T. He, K. Hibino, N. Hotta, H. Hu, H.H. Hu, J. Huang, H.Y. Jia, L. Jiang, F. Kajino, K. Kasahara, Y. Katayose, C. Kato, K. Kawata, Labaciren, G.M. Le, A.F. Li, H.J. Li, W.J. Li, C. Liu, J.S. Liu, M.Y. Liu, H. Lu, X.R. Meng, K. Mizutani, K. Munakata, H. Nanjo, M. Nishizawa, M. Ohnishi, I. Ohta, H. Onuma, S. Ozawa, X.L. Qian, X.B. Qu, T. Saito, T.Y. Saito, M. Sakata, T.K. Sako, J. Shao, M. Shibata, A. Shiomi, T. Shirai, H. Sugimoto, M. Takita, Y.H. Tan, N. Tateyama, S. Torii, H. Tsuchiya, S. Udo, H. Wang, H.R. Wu, L. Xue, Y. Yamamoto, Z. Yang, S. Yasue, A.F. Yuan, T. Yuda, L.M. Zhai, H.M. Zhang, J.L. Zhang, X.Y. Zhang, Y. Zhang, Y. Zhang, Y. Zhang, Zhaxisangzhu, X.X. Zhou, Probe of the solar magnetic field using the "cosmic-ray shadow" of the Sun. Phys. Rev. Lett. **111**, 011101 (2013). doi:10.1103/PhysRevLett.111.011101. http://link.aps.org/doi/10.1103/PhysRevLett.111.011101

S.K. Antiochos, C.R. DeVore, J.T. Karpen, Z. Mikić, Structure and dynamics of the Sun's open magnetic field. Astrophys. J. **671**, 936–946 (2007). doi:10.1086/522489

S.K. Antiochos, Z. Mikić, V.S. Titov, R. Lionello, J.A. Linker, A model for the sources of the slow solar wind. Astrophys. J. **731**, 112 (2011). doi:10.1088/0004-637X/731/2/112

S.K. Antiochos, J.A. Linker, R. Lionello, Z. Mikić, V. Titov, T.H. Zurbuchen, The structure and dynamics of the corona–heliosphere connection. Space Sci. Rev. **172**, 169–185 (2012). doi:10.1007/s11214-011-9795-7

E. Antonucci, L. Svalgaard, Green corona and solar sector structure. Sol. Phys. **36**, 115–120 (1974). doi:10.1007/BF00151551

M.J. Aschwanden, Nonlinear force-free magnetic field fitting to coronal loops with and without stereoscopy. Astrophys. J. **763**, 115 (2013). doi:10.1088/0004-637X/763/2/115

M.J. Aschwanden, N.V. Nitta, J.-P. Wuelser, J.R. Lemen, First 3D reconstructions of coronal loops with the STEREO A+B spacecraft. II. Electron density and temperature measurements. Astrophys. J. **680**, 1477–1495 (2008a). doi:10.1086/588014

M.J. Aschwanden, J.-P. Wülser, N.V. Nitta, J.R. Lemen, First three-dimensional reconstructions of coronal loops with the STEREO A and B spacecraft. I. Geometry. Astrophys. J. **679**, 827–842 (2008b). doi:10.1086/529542

H.W. Babcock, The solar magnetograph. Astrophys. J. **118**, 387 (1953). doi:10.1086/145767

H.W. Babcock, The topology of the Sun's magnetic field and the 22-YEAR cycle. Astrophys. J. **133**, 572 (1961). doi:10.1086/147060

H.W. Babcock, The Sun's magnetic field. Annu. Rev. Astron. Astrophys. **1**, 41 (1963). doi:10.1146/annurev.aa.01.090163.000353

W.G. Baker, D.F. Martyn, Electric currents in the ionosphere. I. The conductivity. R. Soc. Lond. Philos. Trans. Ser. A **246**, 281–294 (1953). doi:10.1098/rsta.1953.0016

A. Balogh, G. Erdős, The heliospheric magnetic field. Space Sci. Rev. **176**, 177–215 (2013). doi:10.1007/s11214-011-9835-3

B. Bell, A long-term North-South asymmetry in the location of solar sources of great geomagnetic storms. Smithson. Contrib. Astrophys. **5**, 187 (1962)

L. Biermann, Solar corpuscular radiation and the interplanetary gas. Observatory **77**, 109–110 (1957)

J.M. Borrero, K. Ichimoto, Magnetic structure of sunspots. Living Rev. Sol. Phys. **8**, 4 (2011). doi:10.12942/lrsp-2011-4

J.W. Brosius, S.M. White, Radio measurements of the height of strong coronal magnetic fields above sunspots at the solar limb. Astrophys. J. Lett. **641**, 69–72 (2006). doi:10.1086/503774

G.W. Clark, Arrival directions of cosmic-ray air showers from the northern sky. Phys. Rev. **108**, 450–457 (1957). doi:10.1103/PhysRev.108.450. http://link.aps.org/doi/10.1103/PhysRev.108.450

W.A. Coles, B.J. Rickett, IPS observations of the solar wind speed out of the ecliptic. J. Geophys. Res. **81**, 4797–4799 (1976). doi:10.1029/JA081i025p04797

M.L. De Rosa, C.J. Schrijver, G. Barnes, K.D. Leka, B.W. Lites, M.J. Aschwanden, T. Amari, A. Canou, J.M. McTiernan, S. Régnier, J.K. Thalmann, G. Valori, M.S. Wheatland, T. Wiegelmann, M.C.M. Cheung, P.A. Conlon, M. Fuhrmann, B. Inhester, T. Tadesse, A critical assessment of nonlinear force-free field modeling of the solar corona for active region 10953. Astrophys. J. **696**, 1780–1791 (2009). doi:10.1088/0004-637X/696/2/1780

P.H. Dittmer, The relationship between solar flares and solar sector boundaries. Sol. Phys. **41**, 227–231 (1975). doi:10.1007/BF00152969

L.A. Fisk, N.A. Schwadron, The behavior of the open magnetic field of the Sun. Astrophys. J. **560**, 425–438 (2001). doi:10.1086/322503

L.A. Fisk, L. Zhao, The heliospheric magnetic field and the solar wind during the solar cycle, in *IAU Symposium*, ed. by N. Gopalswamy, D.F. Webb. IAU Symposium, vol. 257 (2009), pp. 109–120. doi:10.1017/S1743921309029160

R.A. Frazin, P. Lamy, A. Llebaria, A.M. Vásquez, Three-dimensional electron density from tomographic analysis of LASCO-C2 images of the K-corona total brightness. Sol. Phys. **265**, 19–30 (2010). doi: 10.1007/s11207-010-9557-9

E. Friis-Christensen, K. Lassen, J.M. Wilcox, W. Gonzalez, D.S. Colburn, Interplanetary magnetic sector polarity from polar geomagnetic field observations. Nat. Phys. Sci. **233**, 48–50 (1971). doi:10.1038/physci233048a0

A. Gulbrandsen, On the variation of the coronal λ5303 intensity relative to the interplanetary and solar magnetic sector structure, and to geomagnetic activity. Planet. Space Sci. **21**, 703–707 (1973). doi:10.1016/0032-0633(73)90086-X

J.T. Hoeksema, J.M. Wilcox, P.H. Scherrer, The structure of the heliospheric current sheet—1978–1982. J. Geophys. Res. **88**, 9910–9918 (1983). doi:10.1029/JA088iA12p09910

R.F. Howard, Solar active regions as diagnostics of subsurface conditions. Annu. Rev. Astron. Astrophys. **34**, 75–110 (1996). doi:10.1146/annurev.astro.34.1.75

R.B. Leighton, Transport of magnetic fields on the Sun. Astrophys. J. **140**, 1547 (1964). doi:10.1086/148058

R.B. Leighton, A magneto-kinematic model of the solar cycle. Astrophys. J. **156**, 1 (1969). doi:10.1086/149943

I.M. Livshits, V.N. Obridko, Variations of the dipole magnetic moment of the sun during the solar activity cycle. Astron. Rep. **50**, 926–935 (2006). doi:10.1134/S1063772906110060

M. Lockwood, Reconstruction and prediction of variations in the open solar magnetic flux and interplanetary conditions. Living Rev. Sol. Phys. **10**, 4 (2013). doi:10.12942/lrsp-2013-4

M. Lockwood, M. Owens, A.P. Rouillard, Excess open solar magnetic flux from satellite data: 2. A survey of kinematic effects. J. Geophys. Res. (Space Phys.) **114**, 11104 (2009). doi:10.1029/2009JA014450

D. Mackay, A. Yeates, The Sun's global photospheric and coronal magnetic fields: observations and models. Living Rev. Sol. Phys. **9**, 6 (2012). doi:10.12942/lrsp-2012-6

S.M. Mansurov, New evidence of a relationship between magnetic fields in space and on Earth. Geomagn. Aeron. **9**, 622 (1970)

K. Mursula, T. Hiltula, Bashful ballerina: southward shifted heliospheric current sheet. Geophys. Res. Lett. **30**, 2135 (2003). doi:10.1029/2003GL018201

H. Nevanlinna, Gauss's H-variometer at the Helsinki magnetic observatory 1844–1912. J. Geomagn. Geoelectr. **49**(10), 1209–1215 (1997). doi:10.5636/jgg.49.1209

D. Odstrčil, V.J. Pizzo, J.A. Linker, P. Riley, R. Lionello, Z. Mikić, Initial coupling of coronal and heliospheric numerical magnetohydrodynamic codes. J. Atmos. Sol.-Terr. Phys. **66**, 1311–1320 (2004). doi: 10.1016/j.jastp.2004.04.007

M.J. Owens, R.J. Forsyth, The heliospheric magnetic field. Living Rev. Sol. Phys. **10**, 5 (2013). doi:10.12942/lrsp-2013-5

M.J. Owens, N.U. Crooker, M. Lockwood, How is open solar magnetic flux lost over the solar cycle? J. Geophys. Res. (Space Phys.) **116**, 4111 (2011). doi:10.1029/2010JA016039

M.J. Owens, C.N. Arge, N.U. Crooker, N.A. Schwadron, T.S. Horbury, Estimating total heliospheric magnetic flux from single-point in situ measurements. J. Geophys. Res. (Space Phys.) **113**, 12103 (2008). doi:10.1029/2008JA013677

E.N. Parker, Dynamics of the interplanetary gas and magnetic fields. Astrophys. J. **128**, 664 (1958). doi:10.1086/146579

E.R. Priest, P. Démoulin, Three-dimensional magnetic reconnection without null points. 1. Basic theory of magnetic flipping. J. Geophys. Res. **100**, 23443–23464 (1995). doi:10.1029/95JA02740

R.L. Rosenberg, P.J. Coleman Jr., Heliographic latitude dependence of the dominant polarity of the interplanetary magnetic field. J. Geophys. Res. **74**, 5611 (1969). doi:10.1029/JA074i024p05611

K.H. Schatten, J.M. Wilcox, N.F. Ness, A model of interplanetary and coronal magnetic fields. Sol. Phys. **6**, 442–455 (1969). doi:10.1007/BF00146478

M. Schulz, Interplanetary sector structure and the heliomagnetic equator. Astrophys. Space Sci. **24**, 371–383 (1973). doi:10.1007/BF02637162

J.A. Simpson, M. Zhang, S. Bame, A solar polar North-South asymmetry for cosmic-ray propagation in the heliosphere: the ULYSSES pole-to-pole rapid transit. Astrophys. J. Lett. **465**, 69 (1996). doi:10.1086/310127

E.J. Smith, A. Balogh, Ulysses observations of the radial magnetic field. Geophys. Res. Lett. **22**, 3317–3320 (1995). doi:10.1029/95GL02826

P.A. Sturrock, Maximum energy of semi-infinite magnetic field configurations. Astrophys. J. **380**, 655–659 (1991). doi:10.1086/170620

L. Svalgaard, Sector structure of the interplanetary magnetic field and daily variation of the geomagnetic field at high latitudes. Danish Met. Inst. Geophys. Papers R-6 (1968)

L. Svalgaard, Interplanetary magnetic-sector structure, 1926–1971. J. Geophys. Res. **77**, 4027 (1972). doi:10.1029/JA077i022p04027

L. Svalgaard, Polar cap magnetic variations and their relationship with the interplanetary magnetic sector structure. J. Geophys. Res. **78**, 2064 (1973). doi:10.1029/JA078i013p02064

L. Svalgaard, J.M. Wilcox, Long-term evolution of solar sector structure. Sol. Phys. **41**, 461–475 (1975). doi:10.1007/BF00154083

L. Svalgaard, J.M. Wilcox, The Hale solar sector boundary. Sol. Phys. **49**, 177–185 (1976). doi:10.1007/BF00221492

L. Svalgaard, I.G. Hannah, H.S. Hudson, Flaring solar Hale sector boundaries. Astrophys. J. **733**, 49 (2011). doi:10.1088/0004-637X/733/1/49

L. Svalgaard, J.M. Wilcox, T.L. Duvall, A model combining the polar and the sector structured solar magnetic fields. Sol. Phys. **37**, 157–172 (1974). doi:10.1007/BF00157852

L. Svalgaard, J.M. Wilcox, P.H. Scherrer, R. Howard, The Sun's magnetic sector structure. Sol. Phys. **45**, 83–91 (1975). doi:10.1007/BF00152219

G.S. Vaiana, A.S. Krieger, A.F. Timothy, Identification and analysis of structures in the corona from X-ray photography. Sol. Phys. **32**, 81–116 (1973). doi:10.1007/BF00152731

J.E. Vernazza, E.H. Avrett, R. Loeser, Structure of the solar chromosphere. III—Models of the EUV brightness components of the quiet-Sun. Astrophys. J. Suppl. **45**, 635–725 (1981). doi:10.1086/190731

M.V. Vokhmyanin, D.I. Ponyavin, Sector structure of the interplanetary magnetic field in the nineteenth century. Geophys. Res. Lett. **40**, 3512–3516 (2013). doi:10.1002/grl.50749

Y.-M. Wang, Solar cycle variation of the Sun's low-order magnetic multipoles: heliospheric consequences. Space Sci. Rev. (2014). doi:10.1007/s11214-014-0051-9

Y.-M. Wang, E. Robbrecht, Asymmetric sunspot activity and the southward displacement of the heliospheric current sheet. Astrophys. J. **736**, 136 (2011). doi:10.1088/0004-637X/736/2/136

Y.-M. Wang, N.R. Sheeley Jr., Magnetic flux transport and the Sun's dipole moment—new twists to the Babcock-Leighton model. Astrophys. J. **375**, 761–770 (1991). doi:10.1086/170240

Y.-M. Wang, N.R. Sheeley Jr., On the fluctuating component of the Sun's large-scale magnetic field. Astrophys. J. **590**, 1111–1120 (2003). doi:10.1086/375026

Y.-M. Wang, J. Lean, N.R. Sheeley, The long-term variation of the Sun's open magnetic flux. Geophys. Res. Lett. **27**(4), 505–508 (2000). doi:10.1029/1999GL010744. http://dx.doi.org/10.1029/1999GL010744

J.M. Wilcox, Inferring the interplanetary magnetic field by observing the polar geomagnetic field. Rev. Geophys. Space Phys. **10**, 1003–1014 (1972). doi:10.1029/RG010i004p01003

J.M. Wilcox, R. Howard, A large-scale pattern in the solar magnetic field. Sol. Phys. **5**, 564–574 (1968). doi:10.1007/BF00147021

J.M. Wilcox, N.F. Ness, Quasi-stationary corotating structure in the interplanetary medium. J. Geophys. Res. **70**, 5793–5805 (1965). doi:10.1029/JZ070i023p05793

S. Yashiro, S. Akiyama, N. Gopalswamy, R.A. Howard, Different power-law indices in the frequency distributions of flares with and without coronal mass ejections. Astrophys. J. Lett. **650**, 143–146 (2006). doi:10.1086/508876

A.R. Yeates, Coronal magnetic field evolution from 1996 to 2012: continuous non-potential simulations. Sol. Phys. **289**, 631–648 (2014). doi:10.1007/s11207-013-0301-0

X. Zhao, J.T. Hoeksema, Predicting the heliospheric magnetic field using the current sheet-source surface model. Adv. Space Res. **16**, 181 (1995). doi:10.1016/0273-1177(95)00331-8

T.H. Zurbuchen, A new view of the coupling of the Sun and the heliosphere. Annu. Rev. Astron. Astrophys. **45**(1), 297–338 (2007). doi:10.1146/annurev.astro.45.010807.154030. http://www.annualreviews.org/doi/abs/10.1146/annurev.astro.45.010807.154030

DOI 10.1007/978-1-4939-2584-1_3
Reprinted from *Space Science Reviews* Journal, DOI 10.1007/s11214-014-0074-2

Revisiting the Sunspot Number

A 400-Year Perspective on the Solar Cycle

Frédéric Clette · Leif Svalgaard · José M. Vaquero · Edward W. Cliver

Received: 28 April 2014 / Accepted: 11 July 2014 / Published online: 5 August 2014

Abstract Our knowledge of the long-term evolution of solar activity and of its primary modulation, the 11-year cycle, largely depends on a single direct observational record: the visual sunspot counts that retrace the last 4 centuries, since the invention of the astronomical telescope. Currently, this activity index is available in two main forms: the International Sunspot Number initiated by R. Wolf in 1849 and the Group Number constructed more recently by Hoyt and Schatten (Sol. Phys. 179:189–219, 1998a, 181:491–512, 1998b). Unfortunately, those two series do not match by various aspects, inducing confusions and contradictions when used in crucial contemporary studies of the solar dynamo or of the solar forcing on the Earth climate. Recently, new efforts have been undertaken to diagnose and correct flaws and biases affecting both sunspot series, in the framework of a series of dedicated Sunspot Number Workshops. Here, we present a global overview of our current understanding of the sunspot number calibration.

After retracing the construction of those two composite series, we present the new concepts and methods used to self-consistently re-calibrate the original sunspot series. While the early part of the sunspot record before 1800 is still characterized by large uncertainties due to poorly observed periods, the more recent sunspot numbers are mainly affected by three main inhomogeneities: in 1880–1915 for the Group Number and in 1947 and 1980–2014 for the Sunspot Number.

After establishing those new corrections, we then consider the implications on our knowledge of solar activity over the last 400 years. The newly corrected series clearly indicates a

F. Clette (✉)
World Data Center SILSO, Observatoire Royal de Belgique, Brussels, Belgium
e-mail: frederic.clette@oma.be

L. Svalgaard
W.W. Hansen Experimental Physics Laboratory, Stanford University, Stanford, CA, USA

J.M. Vaquero
Departamento de Física, Universidad de Extremadura, Mérida, Spain

E.W. Cliver
National Solar Observatory, Sunspot, NM, USA

progressive decline of solar activity before the onset of the Maunder Minimum, while the slowly rising trend of the activity after the Maunder Minimum is strongly reduced, suggesting that by the mid 18th century, solar activity had already returned to levels equivalent to those observed in recent solar cycles in the 20th century. We finally conclude with future prospects opened by this epochal revision of the Sunspot Number, the first one since Wolf himself, and its reconciliation with the Group Number, a long-awaited modernization that will feed solar cycle research into the 21st century.

Keywords Sun · Sunspots · Sunspot Number · Solar cycle · Solar activity

1 Introduction: The Sunspot Number Needs to Be Recalibrated

The sunspot number (SN) time series is the only direct record at our disposal to retrace the long-term evolution of the solar cycle and of the probable long-term influence of the Sun on the Earth environment. Therefore, it was and is still used as a key information in many fields of research, quite obviously in solar physics, but also in climate studies or even stock-market modeling. The relative sunspot number, as defined by Wolf (1851, 1856), is based on the total number of sunspots Ns and the number of sunspot groups Ng according to the well-known formula:

$$R = k(10 \times Ng + Ns) \tag{1}$$

The k scaling coefficient, usually called the personal coefficient of the observer, allows compensating for the differences in the number of recorded sunspots by different observers. It depends mainly on the ability of the observer to detect the smallest sunspots (telescope aperture, local seeing, personal experience) and on how groups are split by the observer. As Wolf was the primary observer for the newly-created sunspot number, his k personal coefficient was set to 1, which defines the scale of the whole series. The sunspot number is thus a synthetic index defined on an absolute but arbitrary scale (no physical unit), which is why Wolf called it "relative".

In most analyses and publications, the sunspot number series is assumed to be carved in stone, i.e. it is considered largely as a homogeneous, well-understood and thus immutable data set. This feeling was probably reinforced by the stately process through which it was produced by a single expert center at the Zürich Observatory during 131 years, from 1849 to 1980 (Waldmeier 1961).

Still, since the mid-20th century, e.g., with the introduction of the American sunspot number by Alan Shapley, the accuracy and validity of such a visual index has been regularly questioned (Shapley 1949). During the 1970's, the SN series even went through a crisis of confidence, with the advent of new modern measurements of solar activity like the $F_{10.7\,\mathrm{cm}}$ background radio flux. The objectivity of such measurements was contrasted with the subjectivity of purely visual and manual sunspot counts, which at that time almost led to the termination of the SN production, considered as unreliable and old-fashioned. Fortunately, action took place in different Commissions of the URSI and IAU, ensuring the continuation of the series after the transfer of the World Data Center for sunspots from Zürich to Brussels in 1981. (For an historical account of this transition, see Berghmans et al. 2006; Clette et al. 2007.)

Since then, this wave of skepticism has receded thanks to a few key findings. First, studies of the AAVSO data series and of its statistical method (Hossfield 2002; Schaefer 1997a, 1997b) showed that the differences between the American SN and the Zürich-International

SN could be traced to flaws in the American SN and that both series could be brought to a close match after correction. Moreover, with the advent of new data series derived from advanced techniques like space-based solar spectral irradiances (Lyα, MgII core-to-wing ratio, etc.) or automated image-based feature recognition (e.g. sunspot areas, flare detection), statistical tests accumulated showing the high degree of correlation between the traditional SN and those modern impersonal indicators (e.g., Bachmann et al. 2004; Rybanský et al. 2005; Wilson and Hathaway 2006; Tapping et al. 2007; Bertello et al. 2010; Hempelmann and Weber 2012; Stenflo 2012). This high correlation indicates that sunspot and group counts give an accurate measurement of the emergence rate of the toroidal magnetic flux at the solar surface by the action of the subsurface dynamo process (Stenflo 2012; Petrovay 2010).

However, when considering the accuracy of the SN, the fact that this index is distributed as a single time series is misleading, as most users assume that the statistical properties of the series are constant with time. As we will show later in this paper, the SN series was actually built from successive blocks using different base data and processing techniques, often with rather abrupt transitions between them. Over the past decades, only a few authors delved into the series and its base input data. Most of those studies focused on specific time intervals or on single base observers. Those valuable efforts led to proposed revisions of specific segments of the SN record (Kopecky et al. 1980; Letfus 1993; Vaquero et al. 2011; Leussu et al. 2013; Lockwood et al. 2014) and to the still ongoing debate about a missing short cycle between cycles 4 and 5 (Usoskin et al. 2001, 2009; Arlt 2009a, 2009b; Zolotova and Ponyavin 2007, 2011). Overall, all those proposed corrections were mostly local in time and were dispersed over multiple publications. Moreover, they also often lacked an independent validation and are still a topic of scientific controversy. Therefore, until now, none of them were included in the master SN series.

The main effort undertaken in recent times was actually the production of an entirely new series: the group number (hereafter GN; Hoyt and Schatten 1998a, 1998b) that will be discussed in Sect. 3 of this chapter (Fig. 1). This work involved a revision of the original data used for the SN and the recovery of many additional observations, in particular in the early period, between the first telescopic observation in 1610 and the start of the systematic sunspot census initiated by Wolf in 1849. However, as we will show in more detail in Sect. 3, the new series showed immediately a strong discrepancy with the SN before 1880, a disagreement that remained unexplained since the GN publication (Fig. 1).

This left the users of the sunspot number series in certain confusion, having to choose between two apparently equivalent and interchangeable but disagreeing series. In many cases, authors have settled for the GN series when studying phenomena before the mid-19th century, mainly assuming that the GN is then more reliable, as it benefits from a wider observation base and was derived from a single well-documented compilation. As this early period includes the Maunder Minimum (1645–1715; Spörer 1887; Eddy 1976), the GN number played a key role in many recent reconstructions of solar outputs (irradiance, solar wind, total open magnetic flux): e.g. Hathaway et al. (2002), Solanki and Krivova (2004), Wang et al. (2005), Krivova et al. (2007), Vieira et al. (2011), Shapiro et al. (2011) and Owens and Lockwood (2012). Hence, it was a key element in the resulting conclusions about the past solar forcing on the Earth climate since the Maunder Minimum. Given the importance of the reconstructed time series, the coexistence of two conflicting series is a highly unsatisfactory situation that should now be actively addressed.

Starting from this enduring problem, various studies have been recently undertaken on the initiative of the Sunspot Number Workshops (Cliver et al. 2013) to revisit the calibration of both series and as far as possible, to reconcile them by diagnosing the biases affecting

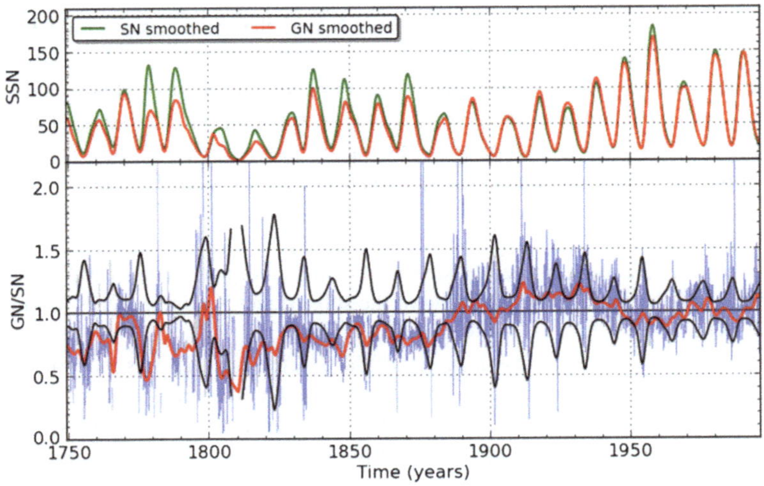

Fig. 1 *Top panel*: time series of the SN (*green*) and GN (*red*), processed with a 12-month Gaussian smoothing. *Lower panel*: the ratio GN/SN (*blue curve*) and the same with a 12-month Gaussian smoothing (*red*). Confidence intervals (*black*) are based on uncertainties given by Hoyt and Schatten (1998a). The ratio deviates significantly and systematically from unity before ~1880

them and identifying their underlying causes. By exploiting all the new sunspot data and the knowledge that were accumulated since Wolf and his successors established the official "heritage" SN series, those analyses have unraveled several major anomalies in both series.

For a large part, this chapter provides a review of the results harvested over the past three SN Workshops (Sacramento Peak, September 2011; Brussels, May 2012; Tucson, January 2013, with a 4th Workshop to be held in Locarno in May 2014). In Sect. 2, we first synthesize the history of the construction of the SN and GN series, emphasizing the key eras and dates that can leave an imprint in the resulting index values. Then, we will analyze in chronological order different biases and trends identified in those series, providing the corresponding diagnostics of their causes. In Sect. 3, we focus on the historical part of the series that was backward reconstructed, before the start of systematic observations in 1849. Section 4 is devoted to the Zürich era and finally Sect. 5 focuses on the most recent part of the series, derived by a new method by the SIDC-Brussels. In this section, we will also consider some peculiarities of the last two solar cycles and discuss how they can help us better interpret past trends in the historical series. We then come to an overall discussion (Sect. 6), bringing together the key corrections that have been established and assessing how the agreement of the SN with other data sets is improved after applying the corrections. As this is still work in progress, we finally conclude on the upcoming release of a fully revised SN series and on the possible implications of the corrected and reconciled SN and GN series on current solar issues.

2 The Sunspot Number in Time

2.1 Wolf's Historical Sunspot Number Reconstruction

Soon after Rudolph Wolf started the systematic census of sunspots in 1849, he turned to past observations in order to quickly extend his still very short series over several past solar cycles. First, he naturally turned to the very observations that triggered his interest in

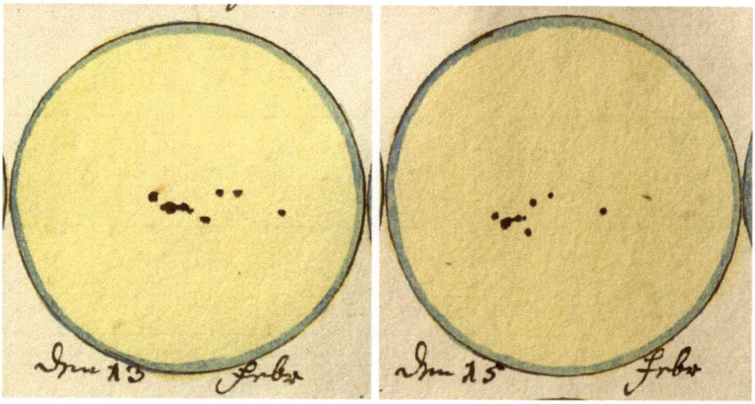

Fig. 2 Sunspot drawings by J.C. Staudach corresponding to the (*left*) 13th and (*right*) 15th of February 1760 (Image source: R. Arlt)

sunspots, namely the long record by Samuel Heinrich Schwabe (1789–1875), the discoverer of the (approximately) decadal cyclicity of the solar activity (Schwabe 1844). Schwabe was the most dedicated sunspot observer of his time (Johnson 1857; Hufbauer 1991; Cliver 2005), being active continuously from 1825 to 1868 (last preserved data in 1867). His original notes and sunspot drawings are preserved in the archives of the Royal Astronomical Society, London. Arlt (2011) provides a comprehensive inventory of the sunspot information from Schwabe's logbooks, which contain 8486 full-disk drawings with sunspots and 3699 additional verbal reports. It is interesting to note that, more than 140 years ago, de la Rue et al. (1869) used these drawings to estimate the time evolution of sunspot areas between 1832 and 1853 (Vaquero et al. 2002). As Schwabe was still observing in parallel with Wolf until 1867, his observations would later remain one of the main auxiliary set of observations used by Wolf to establish the daily Zürich sunspot number.

In 1857, Wolf extended further the historical reconstruction by using the longest continuous series of observations of the 18th century, produced by Johann Caspar Staudach (Wolf 1857; Svalgaard 2013a). Staudach, an amateur astronomer, made sunspot drawings from 15 February 1749 to 31 January 1796 (these drawings are currently stored in the library of the Leibniz Institute for Astrophysics Potsdam, Germany). There is a total of 1016 days giving sunspot positions, including dates with no spot observed. The average number of observations per year equals 21 but the distribution of observed dates is highly variable from year to year (Arlt 2008). Figure 2 shows two typical sunspot drawings by Staudach.

However, Wolf soon realized that Staudach's values were systematically lower than those for the recent cycles observed by Schwabe and he suspected that this was most probably due to the cruder instrument used by Staudach and the limited amount of details recorded in his small sketches (Fig. 2; Arlt 2008). Therefore, Wolf concluded that Staudach's observations had to be multiplied by a k personal coefficient of 2 in order to match his own 19th century observations (Wolf 1861a).

In 1874, Wolf added new observations mostly by Flaugergues, to bridge the gap between the two main series by Staudach and Schwabe (Wolf 1874). Indeed, the interval 1790 to 1826 is only sparsely covered by a few short sets of observations (Letfus 1999, 2000; Usoskin et al. 2003; Vaquero et al. 2012). Comparing observations by himself, Schwabe, Hornstein, and Carrington, Wolf had found already in 1861 that Schwabe's observations had to be increased by 25 %. This increase was effectively included only in the 1880 series (Wolf

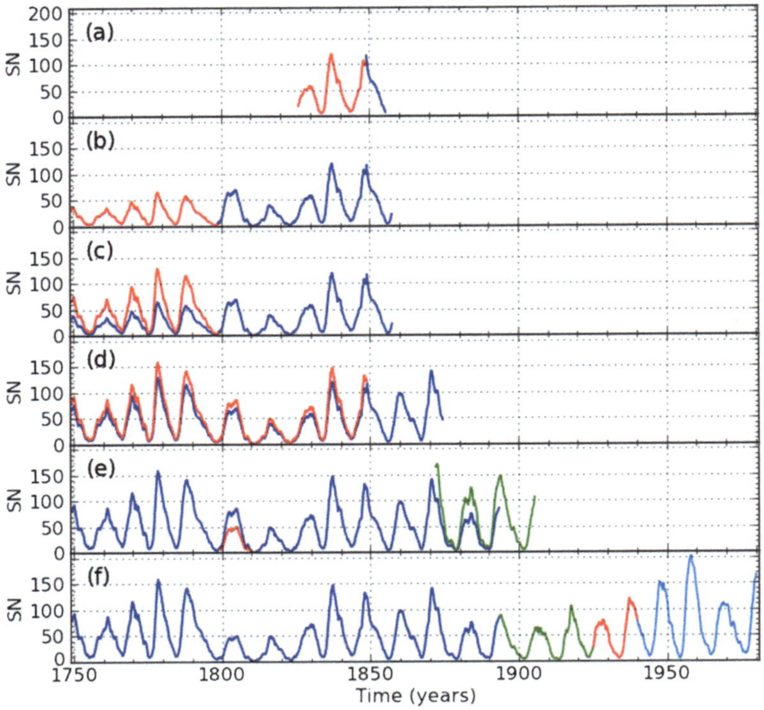

Fig. 3 The progressive construction of the Zürich sunspot number series by Rudolf Wolf. Panel **a**: pre-pending Schwabe's observation (1849); Panel **b**: pre-pending Staudach's observations (1857); Panel **c**: rescaling Staudach's data by a factor 2 (1861); Panel **d**: correction due to comparisons of Schwabe-Carrington-Hornstein-Wolf (1882); Panel **e**: revision of cycle 5 by Wolfer and new Wolfer counts from 1877 (1902); Panel **f**: extension of the original Wolf series (*blue*), successively by Wolfer (*green*), Brunner (*red*) and Waldmeier (*light blue*)

1880). As the scaling of the early historical sunspot numbers was based on those pre-1849 Schwabe numbers, Wolf thus applied a factor 1.25 to the entire reconstructed series before 1849. This was the last adjustment made by Wolf on the reconstructed part of the series. It is only much later in 1902 that his successor, Alfred Wolfer, applied a correction to cycle 5, based on new observations from Kremsmünster (1802–1830), lowering cycle 5 by a factor 0.58 (Wolfer 1902). Cycle 5 thus became, together with cycle 6, the weakest cycle of the Zürich series, forming what would later be named the Dalton minimum (see Fig. 3).

Wolf considered another way to estimate the strength of past solar cycles: magnetic needle readings. Far ultraviolet (FUV) radiation from the Sun, enhanced by solar activity, creates and maintains the E-layer of the ionosphere, where dynamo action from moving air causes an electric current to flow above the dayside of the Earth at about 100 km altitude (cf. Svalgaard 2014b, this issue). The magnetic effect of this current is readily measured by magnetometers on the ground and is best seen in the East Component of the geomagnetic field (Nevanlinna and Kataja 1993; Nevanlinna 1995; Svalgaard and Cliver 2007; Cliver and Svalgaard 2007). The current stays fixed with respect to the direction to the Sun and its magnetic effect, deflecting the "magnetic needle" at a right angle to the current, increases to a maximum at about 8 h local time, then disappears when the current is overhead, and finally increases again, but in the opposite direction, to a maximum at about 2 h. The range, rY, from the morning deflection to the afternoon deflection, depends essentially

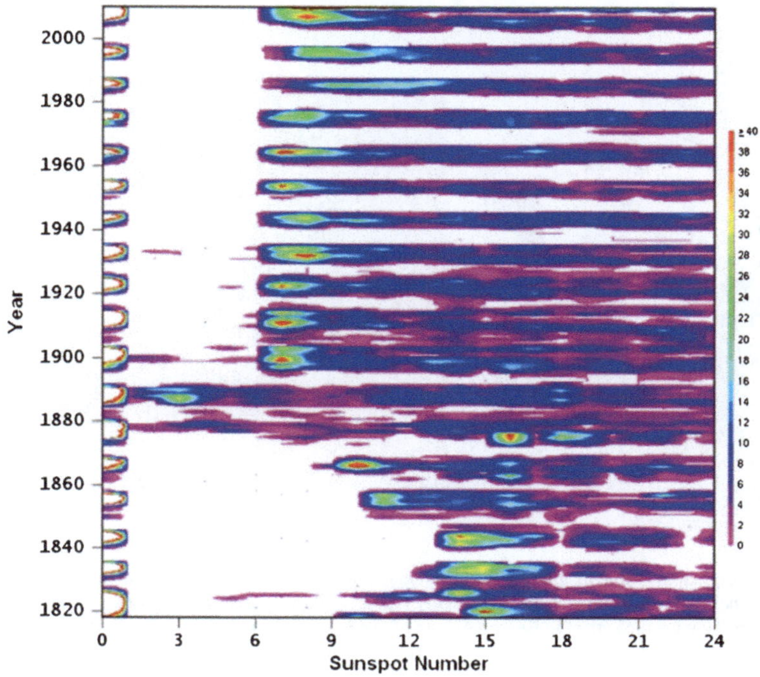

Fig. 4 Contour plots of the lower part of the SN histogram (horizontally) as a function of time (vertically by cycle) showing the variations of the lower values corresponding to a single spot (Wolf number = 11). This lower cut-off is constant at 7 (11 × 0.6) after 1893, while earlier values show step-wise changes that match the reported corrections brought by R. Wolf to the historical SN that he recovered

on the solar zenith angle and the FUV flux. In the yearly average, the zenith angle dependence averages out and the resulting index variations then essentially reflect the long-term variations of the solar FUV irradiance, i.e. the level of solar activity.

This magnetic effect was discovered as early as 1722 by George Graham and as it varied in step with the recently discovered 11-year solar cycle, Wolf considered that it provided a valid independent check on the amplitude of past solar cycles. However, published documents (Wolf 1861b, 1861c, 1862, 1875, 1882) indicate that the key modifications to the Wolf sunspot series were based only on sunspot counts from various observers progressively collected by Wolf over many years and that magnetic needle data were only used as a validation (NB: scanned versions of the original "Mitteilungen" can be found at: http://adsabs.harvard.edu/historical.html).

Those various modifications left clear traces in the standard sunspot number series. By exploiting the sharp lower boundary in the Wolf number for the first spot (i.e. $R = 10 + 1 = 11$, cf. Eq. 1), histograms of the lower range of SN values provide a direct confirmation of the time and magnitude of the corrections reported in the Zürich publications, i.e. of the k personal coefficients effectively adopted for the early observers (Fig. 4).

It should be noted that because of the scarcity of the data recovered by Wolf before the mid-18th century, he never extended his daily and monthly mean sunspot numbers before 1749. Only yearly means were derived back to 1700. The 17th century, including most of the Maunder minimum, was left out.

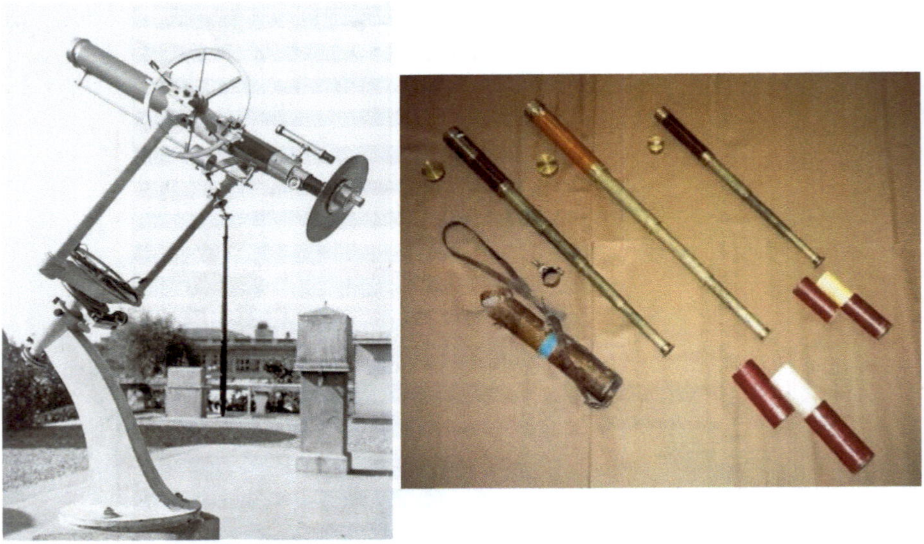

Fig. 5 The original instruments used by R. Wolf. On the *left*, the standard "4-foot" 80 mm Fraunhofer refractor, pictured here as it was set up at the Zürich Observatory (image source: Library of the ETH Zürich, Image Archive, http://www.e-pics.ethz.ch, Record N° Ans-05063-001). On the *right*, the three smaller portable refractors (apertures of 30 and 40 mm) used by Wolf while he was travelling

2.2 The Zürich Era

Until his death in 1893, Wolf established the base principles to calculate the Zürich sunspot number and guarantee its long-term stability. The base Zürich number was simply the raw Wolf number based on the sunspot and group counts made on the aerial image of the standard 80 mm "4-foot" Fraunhofer refractor installed on the grounds of the Zürich Observatory. As Wolf was often traveling across Switzerland for official duties, he also used smaller portable instruments, while an assistant was making counts with the standard 80 mm refractor (Fig. 5). A mean ratio between those simultaneous observations allowed deriving a k personal coefficient for the portable telescopes (Wolfer 1895; Svalgaard 2013a). Those k ratios were then applied to all observations made by Wolf with his travel telescopes to bring them to the standard scale. As those telescopes still exist and are still in regular use (Friedli and Keller 1993; Friedli 1997), a recent unpublished statistical analysis made by the WDC-SILSO over one year in 2012, shows that the scaling ratio for those instruments relative to the current sunspot number matches within 5 % the values derived by Wolf.

In addition, in order to fill in the daily gap due to bad weather in Switzerland, Wolf used a set of auxiliary observers in order to derive a sunspot number for the missing days. For this purpose, average k scaling coefficients were derived by yearly means of the ratios between the raw numbers from each station and the corresponding Zürich number. The missing sunspot number was then computed by an average of all auxiliary values, each multiplied by its k factor. Wolf's successors continued to follow those principles without much change except for a steady increase of the external collaborating stations, until the production of the Zürich sunspot number came to an end in 1980.

However, other very important aspects of the method for determining the sunspot number underwent changes following Wolf's death. Until then, as Wolf wanted to match his sunspot number with the early historical values and as he realized that his predecessors had smaller and cruder instruments, he deliberately applied restrictive rules to his counts, trying to mimic early observers, namely:

- small short-lived sunspots without penumbra ("pores") were not counted
- multiple umbrae within a common penumbra were counted as single sunspot

This lowered the resulting counts compared to what the standard 80 mm instrument could actually show. Although those reduced counts matched rather well what Wolf would also naturally count with his small portable telescopes, it imposed a sunspot selection process when using the 80 mm refractor and thus an additional interpretation prone to personal subjectivity. This "censored" count also dropped useful sunspot information that was plainly shown by the more advanced telescopes of the 19th century.

This is why a new counting series was started by Alfred Wolfer in 1877 in parallel with Wolf, where the above criteria were dropped and all spots (including pores) were included. Based on 16 years of parallel observations (1877–1893) between Wolfer at the 80 mm telescope and Wolf using his smaller portable telescopes and taking into account the average k ratio between the portable and standard telescopes (estimated by Wolf at 1.5), the average ratio between the two counts was established at 0.6 (Wolfer 1895). Once Wolfer took over the position of Director in 1893, he continued with the new counting method but applying the fixed factor 0.6 to the raw counts in order to match the original Wolf series. Ever since, all raw Zürich sunspot numbers were multiplied by this 0.6 factor. At that time, doing so was probably considered more convenient and less time-consuming than manually rescaling 150-years worth of past sunspot numbers. Although the 0.6 Zürich factor was initially equivalent to the Wolfer k personal coefficient relative to Wolf's numbers, it was later applied as a fixed factor allowing to join all recent SNs to the original Wolf series (thus assuming that all Zürich SNs produced since Wolfer have by construction a fixed k coefficient equal to 1 relative to Wolfer). Therefore, this fixed factor should not be confused with usual k coefficients in Eq. 1, which are used to rescale raw Wolf numbers from auxiliary stations to the Wolf numbers from the primary station, initially Zürich and later, Locarno. Those k coefficients can vary in time and are based on a continuous statistical recalculation.

The new Wolfer counting method was apparently applied uniformly during the following decades, at least by the next Director, William Brunner. Only when the last Director Max Waldmeier took over the SN compilation in 1945, a last major modification was introduced in the counting method, where individual sunspots are weighted according to their sizes (Waldmeier 1968, 1948). The consequences of this change are discussed in more detail in Sect. 4.2. While the timing is uncertain, the method itself can be properly reconstructed as it is still in use nowadays at the Specola Solare Ticinese station in Locarno. Indeed, this station was set up by Waldmeier in 1957 in order to complement the primary Zürich station. Sergio Cortesi, the main observer who was then recruited, was fully trained to count according to the Zürich method and has been continuously observing ever since (Fig. 6). He is thus a living witness of the Zürich "school". Part of the diagnostics presented in this paper is based on direct information from the Locarno station (personal communications, archives, drawings).

2.3 The Brussels Era

In 1980, when Max Waldmeier retired, the Zürich Observatory ceased to produce the sunspot number and this activity was taken over by the Royal Observatory of Belgium, leading to the foundation of the SIDC, "Sunspot Index Data Center". The context and circumstances

Fig. 6 *Left image*: view of the Specola Solare Ticinese Observatory overlooking the Lago Maggiore in Locarno. *Right image*: M. Waldmeier (*right*) and Sergio Cortesi (*left*), the primary observer at Locarno, in 1970. (Images courtesy Specola Solare Ticinese, Locarno)

of this transition are described by Berghmans et al. (2006) based on the original documents, including the correspondence between André Koecklenbergh, the founder of the SIDC, and A. Zelenka, one of Waldmeier's assistants. At that time, Zelenka was preparing a computerization of the Zürich processing and this codification was useful to ensure the continuity in the processing of the sunspot number, as it provided the base for the new SIDC method. The latter was developed in Brussels with two main objectives. First, the new computation should take advantage of a larger base of contributing stations. Second, it should include a new mechanism ensuring the stability of the index after the loss of the primary Zürich station.

The resulting method, which is summarized by Clette et al. (2007), features 3 key steps:

- Determination of the monthly average k coefficients relative to a pilot station, namely the Specola Locarno station, with statistical elimination of individual values when the daily k deviates abnormally relative to the monthly average. All values for each station are normalized by applying the corresponding monthly average k factor.
- Rejection of outlying values from the pilot station: daily values from the reference station are compared to the network average and values are rejected when they deviate significantly based on the standard deviation of all values for that day.
- Iterative calculation of the average daily sunspot index: the final sunspot number is computed by taking the average of k-normalized Wolf numbers from all available stations. It is done iteratively with a final elimination of outliers.

Part of the station statistics is a rather direct translation of what was done manually at Zürich, but the new method differs from the previous one in two main aspects:

- The network averaging: while most of Zürich sunspot numbers were simply the raw Wolf number of the Zürich station, the SIDC international sunspot number includes the information from all contributing stations.
- A validity control of the daily values from the pilot station: each daily reference value is compared against the values of all stations for the same day and can be eliminated in favor of the network average.

This new procedure leads to a reduction of the RMS dispersion of daily sunspot numbers, compared to daily Wolf numbers from a single station, which amounts to about 8 % rms (1.5 % rms for monthly averages). The number of observations available each day typically

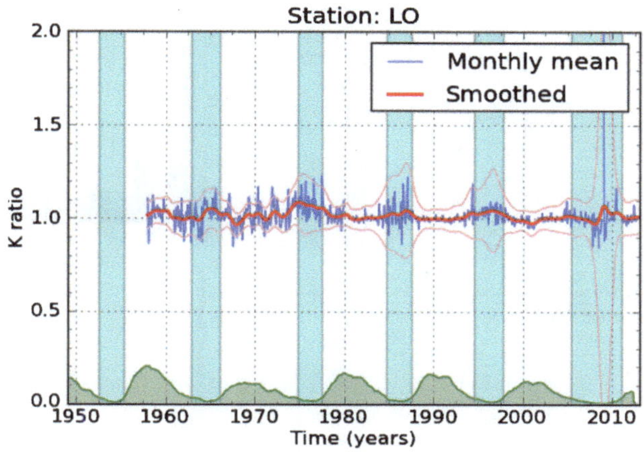

Fig. 7 Ratio between the raw Wolf number from the Locarno station (multiplied by the standard factor 0.6) and the SN. The *thin lines* indicate the 3 sigma confidence limits for the monthly average values (*blue*). The *red line* is the 13-month smoothed ratio. The sunspot number (*shaded green*) and the periods of minima (*shaded blue*), when the ratios are less accurate, are overlaid as time references. The ratio remains close to 1, showing that the scale of the SN followed closely the Locarno reference, in particular after 1981 when the Locarno station took the role of pilot station for the international sunspot index R_i (note the lower dispersion after 1981)

varies between 10 and 30. In order to mark this change of method, the resulting index was renamed to "International Sunspot Number", noted R_i, in order to make the distinction with the former Zürich Sunspot Number, noted R_Z.[1]

In spite of those differences and the higher complexity of the SIDC calculation, the adopted mechanism still ensures a close similarity with the earlier sunspot number. Indeed, as all values for each station are scaled relative to the pilot station though a monthly average of the k ratio, the final average of all normalized numbers leads to a value that is close to the Wolf number of a single station, namely the pilot station. The latter thus defines entirely the absolute scale of the index over timescales longer than one month, just like Zürich did previously. On the other hand, the statistical elimination process improves the precision of daily values, in particular by rejecting the outlying Locarno values and thus avoids biasing the network statistics due to a bad reference.

The outcome is illustrated in Fig. 7. Over the last 32 years, the ratio between the raw Locarno Wolf numbers and the international SN R_i remained perfectly flat around unity (or 0.6 taking into account the historical Wolfer scaling factor). The R_i scale is thus fully defined by the pilot station. On the other hand, small random deviations in the monthly mean k ($\sim$1.5 % rms) correspond to the statistical cleaning brought by the systematic use of the entire network compared to a single station. It gives also a measure of the gain in the RMS error of daily sunspot numbers associated with the global network statistics.

Finally, the choice of Locarno for the pilot station was quite natural. Its main observer, Sergio Cortesi, was trained into the Zürich observing method and in 1981, he had already carried out parallel observations with Zürich for more than 20 years. Locarno's equivalence to Zürich had thus been extensively checked before Zürich ceased observing. Moreover, as the Specola Observatory could continue observing for the SIDC-Brussels, it provided

[1]In this paper, we will follow this conventional notation for the SN, with R_Z and R_i corresponding to the periods before and after 1981, respectively.

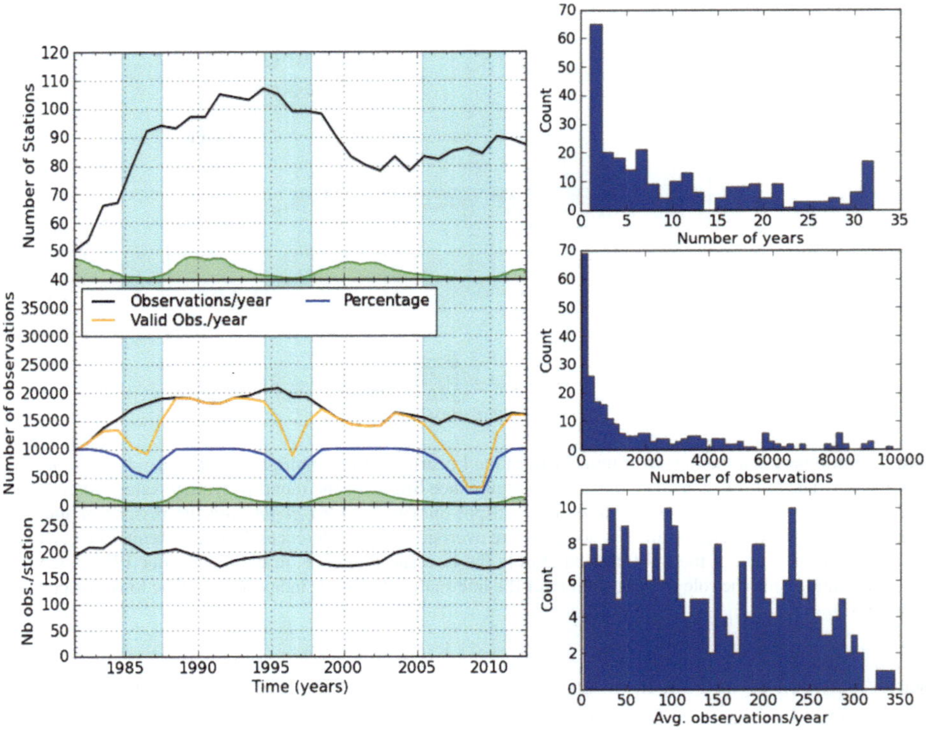

Fig. 8 (*Left*) Plots retracing the evolution of the worldwide sunspot network: number of contributing stations (*top panel*), annual number of collected observations (*middle panel*) and average number of observations per station per year (*lower panel*). The solar cycles (*shaded green*) and minima intervals (*shaded blue*) are overlaid as time reference. In the *middle panels*, valid observations are days when both a given station and the reference pilot station give a non-null Wolf number (i.e. when a personal k coefficient can be established). (*Right*) Overall statistical distributions of the contributions from all stations: duration over which stations have contributed (*top panel*), total number of contributed observations (*middle panel*) and average number of observations per year (*lower panel*)

an uninterrupted reference straddling the critical transition period between the Zürich and International SN around 1980. We will see in Sect. 6.2 that this choice indeed ensured a clean transition around 1980, although it brought other problems later on.

Figure 8 gives global statistics of the evolution of the world-wide sunspot network managed by WDC-SILSO in Brussels over the last 32 years. Since 1981, there have been a total of 270 contributing stations distributed over 30 countries. About 2/3 of stations are individual amateur astronomers and 1/3 professional observatories. All are subjected to the same quality control for stability and continuity. While many stations contributed only for a few years, there are 80 long-duration stations that provided data for more than 15 years (Fig. 8, right). After an initial rise of the number of stations, starting from the $\sim$40 auxiliary stations used formerly by the Zürich Observatory, the yearly number of contributing stations has remained between 80 and 100, providing from 15,000 to 20,000 observations each year (Fig. 8, left). Each station collects an average of 175 daily observations per year. The distribution of yearly averages is rather broad (Fig. 8, right), with a peak around 240 days/year (a typical value for dedicated stations limited only by weather conditions) and up to 360 days/year for a few stations, i.e. close to a 100 % time coverage. In total, since its creation in Brussels, the WDC has accumulated more than 450,000 sunspot counts that are now all preserved in a single database.

3 The Group Sunspot Number: A Better Number?

3.1 Group Counts

As can be seen from the previous section, the standard SN series is a composite time series assembled from very inhomogeneous material. Its accuracy decreases as we go back in time, in particular as the observer base shrinks dramatically before the 19th century. Moreover, Wolf's backwards reconstruction of the early sunspot numbers does not include at all the important Maunder minimum episode because of the dearth of sunspots prior to 1700.

Therefore, given the interest of extending the series and the knowledge of additional archived observations that were unknown to Wolf, a new long-term series was constructed by Hoyt and Schatten (1998a, 1998b). As early observations were rather sparse and crude, obtaining detailed counts of small spots proved to be difficult or even impossible. Moreover, a study by Schaefer (1993) suggested that the SN is largely proportional to the group count alone. This is why the new index included only the total count of sunspot groups. Hoyt and Schatten defined the group number as:

$$R_G = \frac{12.08}{N} \sum k_i N g_i \tag{2}$$

where Ng_i is the number of sunspot groups recorded by the ith observer, k_i is the ith observer's personal scaling factor, N is the number of observers used to compute the daily value and the 12.08 constant is a normalization factor chosen to bring the GN to the same scale as the Zürich sunspot number.

As a base for this normalization, they chose the Greenwich sunspot catalog based on the Greenwich photographic plate collection that spans the 1874–1976 interval (Willis et al. 2013) and the visual USAF/SOON group counts for 1976–1995. Starting from this reference, the scaling of earlier historical group counts is derived by working backward in time, using the average ratios between counts of parallel observers where the series are overlapping. Following this methodology, Hoyt and Schatten derived daily, monthly and yearly means from 1610 to 1995.

3.2 A New Extended and Improved Data Set

By approximately doubling the number of recovered observations and cleverly interpolating between sparse observations (Hoyt et al. 1994), it was possible to reduce the gaps in Wolf's original SN series and to extend it back to the very first telescopic observations (Fig. 9).

The first observations of sunspots were made by the Englishman Thomas Harriot on December 18, 1610 (Shirley 1983). Three other scientists made telescopic observations of sunspots independently about this time: Johannes Fabricius, Christoph Scheiner and Galileo Galilei. Thereafter, telescopic observations of sunspots were generalized using different techniques such as solar filters or projection of the solar disk on white screens. Preserved documents about those early observations allow us to retrace the evolution of sunspots since 1610, though with some gaps and indeterminacy (see the reviews by Hoyt and Schatten 1997 and by Vaquero and Vázquez 2009).

Hoyt and Schatten (1998a) made a huge effort to recover historical observations preserved in archives and libraries around the world. In general, they achieved a good coverage

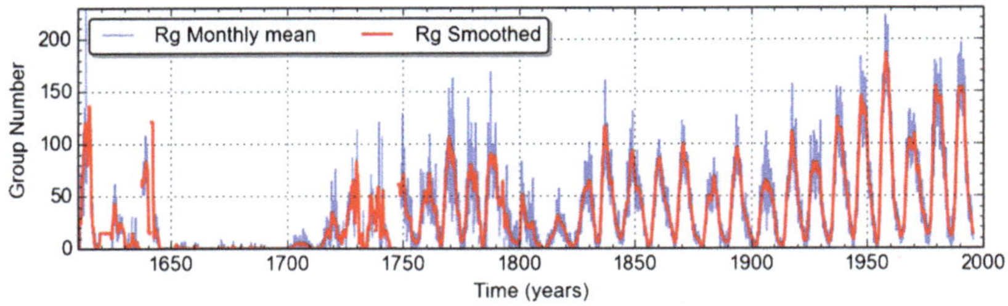

Fig. 9 The original group sunspot number series from 1610 to 1995 as published by Hoyt and Schatten (1998a, 1998b): monthly averages in *blue*, 12-month Gaussian smoothed values in *red*. It shows the Maunder Minimum (1645–1715) followed by a progressive rise up to the mid-20th century

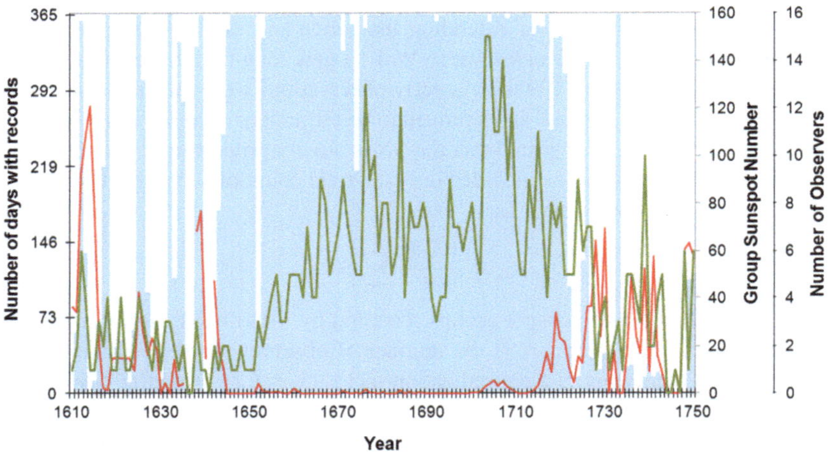

Fig. 10 Coverage of the original GN series in the time period 1610–1750: number of days with records per year (*blue bars*), number of sunspot observers per year (*green line*) and GN (*red line*)

of the series during the period 1610–1750: the mean number of sunspot observers per year is 5 and the mean number of days with records per year equals 237. However, these values have a large dispersion. Figure 10 shows the coverage of the original GN series in the time period 1610–1750, including the annual GN values (red line). The number of days with records per year is shown as blue bars and the number of sunspot observers per year as a green line.

In order to trace correctly the solar activity in the first decades of the 17th century, it is necessary to obtain more information from the most important observers. Based on the number of years with records, the most regular observers were Hevelius (39 years), La Hire (37), Flamsteed (32), Eimmart (25), Picard (23), Siverus (17), Kirch (17), Scheiner (16) and Derham (13). Moreover, we should note that some observers are contemporaries. For example, La Hire, Einmart and Flamsteed observed almost during the same period. Therefore, the main observers are Scheiner for the period before the Maunder Minimum (1610–1642), Hevelius for the first half of the Maunder minimum period (1643–1679) and La Hire for the second half of the Maunder minimum period (1680–1718). Note that there are very few records for the second quarter of the 18th century. However, we could highlight Plantade (423 records from 1704 to 1726), Kirch (394 records from 1716 to 1736), Alischer (524 records from 1719 to 1727) and Adelburner (367 records from 1730 to 1733).

We have additional information on some of the leading observers of this early period thanks to an encyclopedic work on sunspots, the book entitled *Rosa Ursina* by Christoph Scheiner published in 1630 (for a biographical account, see Kant 2007). In his book, Scheiner describes for instance the first telescope with equatorial mounting, called "Heliotelescopium", specially designed to observe sunspots. Unfortunately, this book was written in Latin and there is no translation to other modern language. However, Mitchell (1916) provides essential fragments translated into English. Recently, E. Reeves and A. van Elden have also translated into English some published documents with sunspot observations written by Scheiner and Galileo, with introduction and notes (Galilei and Scheiner 2010).

Sunspot observations by Johannes Hevelius (see MacPike 1937 for details about his life and astronomical background during this time period) are very important to obtain a picture of solar activity in the first decades of the Maunder minimum, because these are the only daily listings of solar observations during this time. Hevelius lists his daily solar observations, intended for the determination of the solar meridian altitude, in the last part of his book *Machina Coelestis*, published in 1679. Hoyt and Schatten (1995a) mention that Hevelius reports 19 sunspot groups during the period 1653–1679 and also provides the best record of the sunspot maximum of 1660 (when one sunspot group was active for seven solar rotations).

The most important observer after Hevelius was the French astronomer Philippe La Hire (1640–1718). The latter conducted solar observations for at least 35 years (from 1683 to 1718) at the Paris Observatory. These observations were studied by Ribes and Nesme-Ribes (1993) from the records and sunspot drawings preserved in the historical archive of this institution.

Another important observer supplementing the observations of La Hire is John Flamsteed. Hoyt and Schatten (1995b) studied the sunspot records that appear in his book *Historia Coelestis Britannica* which was published in three volumes in 1725, in the *Philosophical Transactions of the Royal Society* and in his letters to William Derham preserved in the Cambridge University Library. According to Hoyt and Schatten (1995b), in the period 1676–1700, Flamsteed only observed sunspots in the years 1766 (24 active days and 19 inactive days) and 1684 (17 active days and 32 inactive days).

In the second half of the 17th century, we can note that there is a large number of observations and observers documenting the Maunder Minimum (1645–1715; Eddy 1976). By contrast, the periods before and after the Maunder minimum are generally poorly covered (Fig. 9). In fact, there are even six years without any sunspot record (1636, 1637, 1641, 1744, 1745 and 1747) and there are many years with very few observations. The years 1610, 1614, 1640, 1723, 1731, 1732, 1734, 1737, 1738, 1746 and 1748 contain less than ten observations.

The high number of sunspot observations made during the Maunder Minimum needs a clarification. Hoyt and Schatten (1996) examined how well the Sun was observed during this Grand Minimum of solar activity. They compiled the specific dates of observations by Hevelius, Picard, La Hire, Flamsteed and others and they derived an estimate of the minimum fraction of the time over which the Sun was observed: namely, 52.7 % of all days have specific recorded observations. Moreover, they also compiled general comments mentioning the absence of sunspots during specific years or time intervals, obtaining an upper estimate of 98 %. Figure 11 shows the number of days with records per year, either for dates with explicit sunspot observations (blue line) or for all observations, including the general comments (red line), according to Hoyt and Schatten (1996, 1998a). We can see that the number of days with explicit observations is very low in the beginning of the Maunder Minimum and only becomes large in its late part. In any case, observers listed

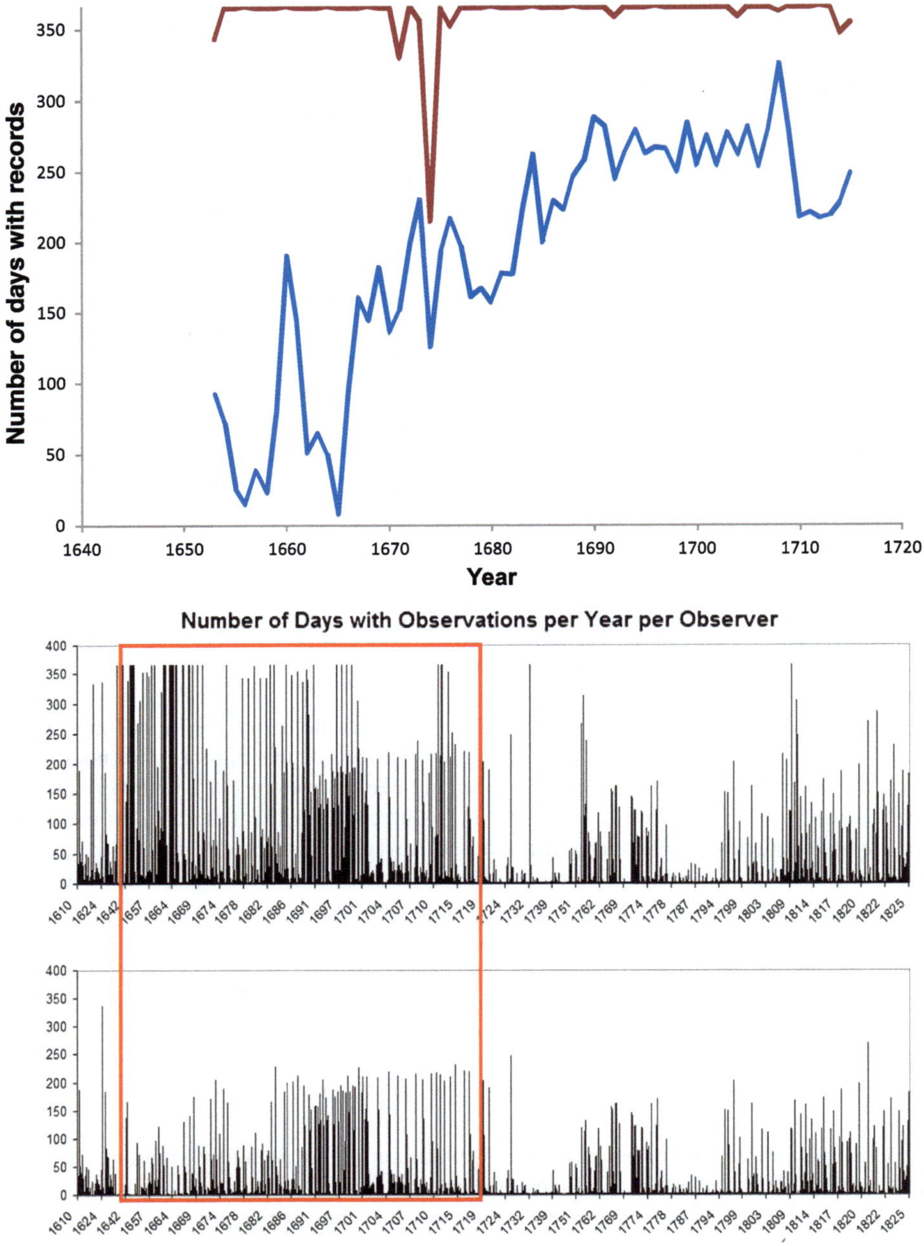

Fig. 11 (*Top*) Coverage of the GN series during the Maunder Minimum: number of days with records per year for specific dates of observation (*blue line*) and for all the observations including the general comments (*red line*). (*Bottom*) The number of days per observer for each year as assumed by Hoyt and Schatten (*upper panel*) and after removal of the observers who did not report specific observations. The *red box* indicates the Maunder Minimum corresponding to the *top part* of the figure

with ∼365 days of observations per year should be removed from any new reconstruction because these values (usually zero values) are based on general indirect comments and not on well-documented observations.

After the Maunder minimum, in the first half of the 18th century, the data coverage is particularly bad in the fifteen years from 1734 to 1748, during which the average annual number of days with records is only 12. This lack of interest in systematic observations of sunspots is common during the 18th century. Significantly, Lalande (1771) gave a very low priority to the regular surveying of sunspots in a list of astronomical observations (14th out of 18 duties). After the years 1744–1747 (without sunspot records), astronomers of the second half of the 18th and early 19th century also showed a limited interest for sunspot observations. Consequently, the yearly number of observations fluctuates around intermediate values, except for a new poorly observed interval from 1779 to 1794, before finally rising to almost 100 % coverage in the early 1810s. Thus, the annual average of days with sunspot records in the Hoyt and Schatten (1998a) database amounts to only 30.8 and 176.9 in the periods 1779–1794 and 1750–1810, respectively. Two observers stand out in the period 1750–1850: J.C. Staudach and S.H. Schwabe. Both observers were used by Wolf in his first reconstruction. Arlt (2008, 2011) recently localized, digitized and analyzed the original sunspot drawings made by those two observers. Other important observers of the second part of 18th and early 19th century were Herschel, Pastorff and Christian Horrebow.

Christian Horrebow and his colleagues of Copenhagen, Denmark, were active sunspot observers from 1761 to 1777. These observations were examined both by Thiele (1859) and by d'Arrest, who gave very different interpretations. Hoyt and Schatten (1995c) reexamined Horrebow's original notebooks and provided a more coherent interpretation of Horrebow's observations.

Another important set of sunspot observations is contained in the notebooks of the famous astronomer William Herschel (Crowe and Lafortune 2007). This documentation is preserved in the Churchill College in Cambridge, England. Hoyt and Schatten (1992a) edited and reproduced these solar observations made from 1779 to 1818 (although most of the observations were made in the period 1799–1806). Hoyt and Schatten (1992b) showed that these observations provide a better reconstruction of solar activity around solar cycle 5.

One of the main anomalies in the Wolf reconstruction was the exceptional length of solar cycle 4: 17 years, from the peak of solar cycle 4 to peak of solar cycle 5. Hoyt and Schatten (1992b) have shown that the date of the cycle 5 maximum falls in 1801–1803 instead of 1805 as originally proposed by Wolf. This reduced the cycle length from 17 to 14 years and partially solved the anomaly of solar cycle 4, although the new length was still the longest in the entire SN record.

The German astronomer J.W. Pastorff made about 1477 observations of sunspots between 1819 and 1833. Hoyt and Schatten (1995d) examined Pastorff's original observations providing more nearly correct values for the number of sunspot groups than the values published by Wolf (who got them indirectly in 1874 from A.C. Ranyard, who misinterpreted the original records).

3.3 Group Number Assets and Limitations

Compared to the sunspot number series, the new group number brought several key assets:

- The series rests on a larger and more refined observational base, as shown in the previous section.
- The group count is less dependent on the visibility of small spots in the early observations, reducing and hopefully eliminating the expected downward biases.
- The series was processed as a single batch by the same scientists, thus avoiding possible changes in practices over several generations of sunspot observers and compilers.

- The calibration is exclusively based on the backward propagation of cross-scaling factors between group counts from successive observers, without involving external indirect indicators like Wolf's magnetic needle corrections.

This is why the group number is often considered as the most reliable reference to retrace the past solar activity, in particular before the 19th century.

However, the group number has also some specific weaknesses:

- Methodological weaknesses:
  - The interpretation of some observing records is questionable. This will be shown in Sect. 4.
  - The base calibration of the series rests on a unique non-visual reference, the RGO photographic plates. Any flaw in the RGO group counts will thus affect the whole series before 1874.
- Truncated sunspot information:
  - The constant 12.08 scale factor assumes a constant average number of spots per group, namely: $Ns = 10.13Ng$, as $R_S = 0.6\,(10Ng + Ns)$. Yet, this ratio may actually vary over time, in the course of a solar cycle and over successive cycles, or according to the level of activity, as shown later in Sect. 6.6. The information on the actual group size, which is contained in the SN, is thus lost here.

Therefore, other global biases may affect the Group number in a different way and at other times than the SN, which suggests that both series can and should be used for a mutual cross-validation rather than assuming that one of them is the only true accurate reference. That is not to say that the two time series are equally well-constructed. As we shall show, the flaws in the Hoyt and Schatten GN series are more significant than those in the international SN.

Even more importantly, once the R_G series was published, it appeared that while the SN and GN series nicely agreed over the 20th century, there was a large discrepancy before ~1880, with R_G falling about 40 % below R_Z (Fig. 1). In the next sections, we will review in a chronological order the different flaws or ambiguities in both series that we have established or that have been identified by other studies.

4 Biases and Uncertainties in the Early Group and Sunspot Numbers

4.1 The Early Sunspot Number and Maunder Minimum (1610–1749)

After the publication of the GN series, a scenario for the occurrence of the Maunder Minimum emerged and was clearly based on the fact that solar activity dropped sharply at the onset of this Grand Minimum. However, the discovery and recovery of sunspot observations made by G. Marcgraf in 1637 and the revision of some earlier uncertain data for the period 1636–1642 now indicate a different picture of the onset of the Maunder Minimum (Vaquero et al. 2011). In particular, the additions and changes that Vaquero et al. (2011) made can be summarized as follows:

1. The Marcgraf sunspot records preserved in the Leiden Regional Archive were added. These observations were made in 1637, including three drawings of the solar disk. Note that there is no data for the year 1637 in the original GN series.
2. The estimated (not observed) values from Crabtree's comments (1638–1639) were eliminated. This problem has been discussed in Vaquero (2007).

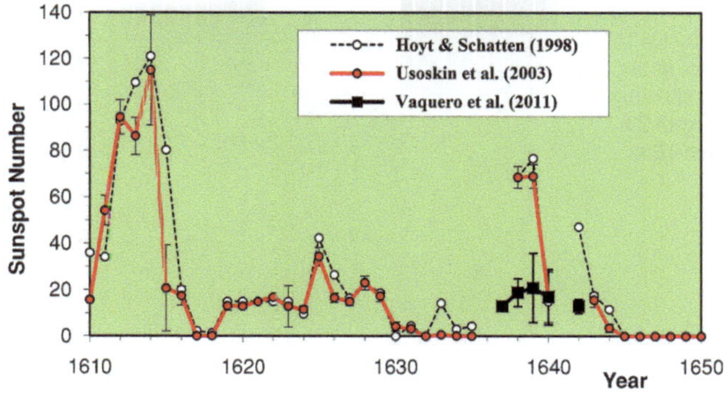

Fig. 12 The sunspot number before the Maunder minimum according to Hoyt and Schatten (1998a, 1998b) (*dashed line*, *white circles*), Usoskin et al. (2003) (*red line* and *circles*) and Vaquero et al. (2011) (*black line* and *squares*). Note that the peak value in 1614 was computed only with one sunspot observation made by F. Colonna in Naples (who reported 8 sunspot groups on the 3rd of October). Therefore, we have used a *broken line* around this date

3. The dates and numbers of sunspot groups from Horrox observations in Hoyt and Schatten (1998a, 1998b) (from Julian Calendar to Gregorian Calendar) were corrected.
4. One spurious observation by Gassendi on 1 Dec. 1638 was eliminated, because this record does not appear in his astronomical observations published in his *Opera Omnia* (1658, Tome IV).
5. The sunspot record by Rheita for 1642 were changed, after consulting the original source (pp. 242–243 of his book *Oculus Enoch et Eliae* published in 1645).
6. One sunspot record by Horrox in 4 December 1639 was added.

The final result of all these additions and corrections to the Hoyt and Schatten (1998a, 1998b) data series is shown in Fig. 12. The original GN series shows a high maximum for the solar cycle just preceding the Maunder Minimum in 1639. The new estimates by Vaquero et al. (2011) show a maximum in the same year but the amplitude of the cycle is much lower. Following those changes, the last two cycles preceding the Maunder minimum have only small amplitudes, in contradiction with the abrupt transition between states of normal solar activity and grand minimum that was assumed previously.

Despite the enormous effort by Hoyt and Schatten to locate all sunspot observations made during the Maunder minimum, it is still possible to find some additional records. As an example, Casas et al. (2006) analyzed hitherto ignored sunspot drawings by Nicholas Bion made in October and November 1672. Surprisingly, in October 1672, other astronomers also observed the Sun (Picard, Hevelius and Montanari) but did not detect any sunspots. A plausible explanation for this difference may be based on the fact that some of the observations compiled by Hoyt and Schatten (1998a, 1998b) were meridian observations of the Sun and, strictly speaking, these were not meant for establishing the presence or absence of sunspots.

For example, Montanari observed the Sun in 21 October 1672. However, he was using the great meridian line located at the Basilica of San Petronio in Bologna (Italy) (see Heilbron 1999 for more information about this gigantic camera obscura). His astrometric results were published by Manfredi (1736) and Hoyt and Schatten interpreted this observation as zero sunspot groups. However, the tables that we can consult in Manfredi (1736) never include information on sunspots. The negative records of Picard and Hevelius during October 1672 probably are also related to solar meridian observations. In fact, the information about sunspots reported by Hevelius on these dates are included in a table of solar meridian obser-

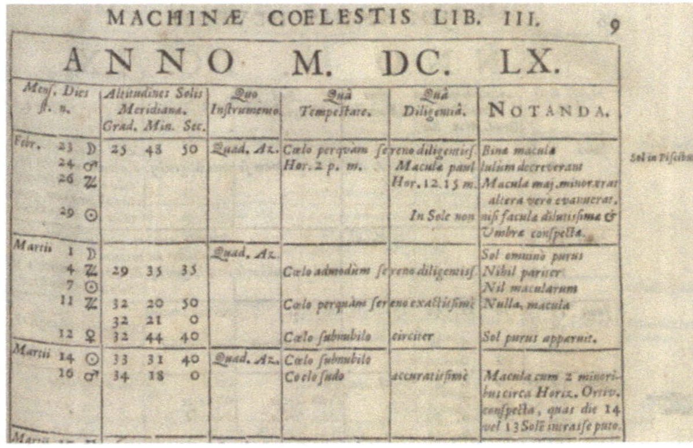

Fig. 13 A fragment of the table of solar meridian altitudes published by Hevelius in his *Machinae Coelestis*, showing records of sunspots from 23 February to 16 March 1660

vations published in the last part of his book *Machinae Coelestis* (1679). Figure 13 shows a fragment of this table, where the date, the solar meridian altitude, the instrument, the state of the sky, the quality of the observation and additional comments are listed. Note how, in this table, comments about sunspots are not exactly associated with the reported solar meridian altitude.

Accordingly, Vaquero and Gallego (2014) have reached the conclusion that solar meridian observations should be used with extreme caution to evaluate past solar activity. If no sunspots were mentioned in a meridian observation, it does not necessarily mean that they were absent. Vaquero and Gallego (2014) presented, as an example, the marginal notes about sunspot observations that were included in the manuscripts of the meridian solar observations made at the Royal Observatory of the Spanish Navy during the "modern" period 1833–1840. These meridian observations were made using the meridian telescope manufactured by Thomas Jones (meridian telescope: $D = 0.125$ m, $f = 3.05$ m). A simple analysis shows that there is not a clear relationship between the timing of recorded spots and solar activity indices. Additionally, no notable periodicities in these observations were detected. It is noteworthy that these records of sunspots only appear in the manuscripts of the observations and do not appear in the printed version of the meridian observations. Thus, the sunspots recorded in the manuscripts of the meridian solar observations made in the Royal Observatory of the Spanish Navy illustrate the difficulty to reconstruct the past solar activity from this kind of historical record.

The other important modification of the early part of the GN series is related to the anomalous shape of solar cycle -1 (approximately from 1733 to 1744). The original GN series shows three peaks in 1736, 1739 and 1741 (see Fig. 14, red line), i.e., a cycle shape completely different from all other observed solar cycles. The last of those peaks has the highest amplitude and should be considered as the maximum of the cycle. Although we still do not have a fully satisfactory solution, several recent publications find a more plausible shape for this solar cycle.

Vaquero et al. (2007a) improved the reliability of the GN series for the period 1736–1739 using the information about sunspot observations published in three of the most important scientific journals of that epoch: "Philosophical Transactions", "Histoire de l'Académie Royale des Sciences", and "Nova Acta Eruditorum". They identified 42 papers containing solar observations, including 30 papers with relevant information on sunspots, and from these, they provided corrected GN values for the years 1736–1739.

Moreover, Vaquero et al. (2007b) reviewed the sunspot observations included in the compilation by Maximiliano Hell (1768) of astronomical observations made by Jesuits in China

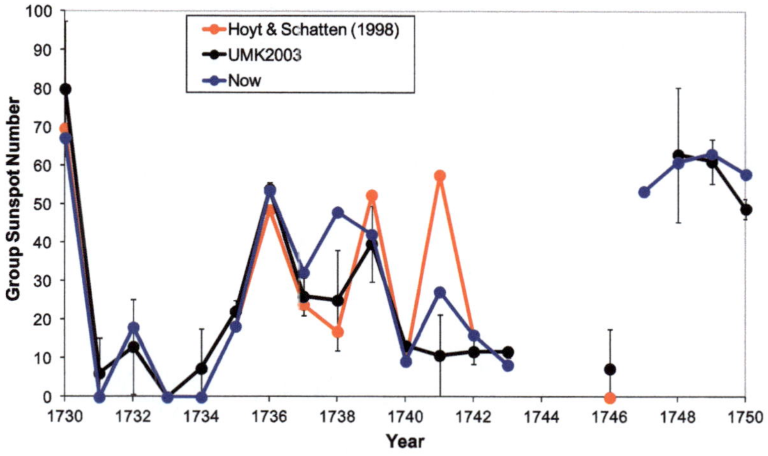

Fig. 14 Solar activity around the solar cycle −1: original GN series (*red*), weighed values by Usoskin et al. (2003) (*black line*) and the latest values from Vaquero et al. (2007a, 2007b), Vaquero and Trigo (2014) and Arlt (personal communication) in *blue*

during the period 1717–1752. Here, the sunspot information is based exclusively on solar eclipse observations. From Hell's compilation, Hoyt and Schatten (1998a, 1998b) deduced that there were no spots on the Sun on the day of an eclipse in 1731 and during another eclipse in 1746. Conversely, Vaquero et al. (2007b) pointed out that the correct interpretation is that no sunspot observations were made on these days. This is especially important in the case of 1746, because it could create a gap from 1744–1747 during which sunspot observations are not available. Fortunately, Rainer Arlt has found recently a "new" sunspot observer (Pehr Wargentin, from Uppsala) who left 17 drawings for 1747. The average group number is 3.53 (Arlt, personal communication). Therefore, we obtain a group sunspot number for this year equal to 53.5, assuming a calibration constant equal to 1.255 used by Hoyt and Schatten (1998a, 1998b) for similar cases. In any case, an important data gap remains in the period 1744–1746.

Usoskin et al. (2003) indicated the anomalous GN value for the year 1741 using a statistical analysis (see Fig. 14, black line). Recently, Vaquero and Trigo (2014) have shown that the original manuscript (preserved in the Harvard University Archives) of the report by John Winthrop dated 10 January 1741 is crucial to solve this problem. Interpreting this manuscript, they provide a new value significantly lower than the original GN value. With this revision, solar cycle −1 adopts a much more normal shape, more comparable to all the other observed cycles (see Fig. 14, blue line).

4.2 The Zürich Reconstruction and Anomalous Cycle 4 (1750–1849)

Corrections to the original GN series have also been proposed for the period 1775–1795, when there are very few sunspot observations. For instance, Vaquero (2004) and Arlt (2009b) revisited sunspot observations in the years 1784 and 1795–1797 respectively. However, the main issue is related to an anomaly in the late part of cycle 4, preceding the Dalton minimum, diagnosed by several independent studies (Loomis 1870; Gnevyshev and Ohl 1948; Sonett 1983). The most striking hypothesis invokes the presence of a "lost" cycle between solar cycles 4 and 5 (according to the numbering by Wolf), as proposed by Usoskin et al.

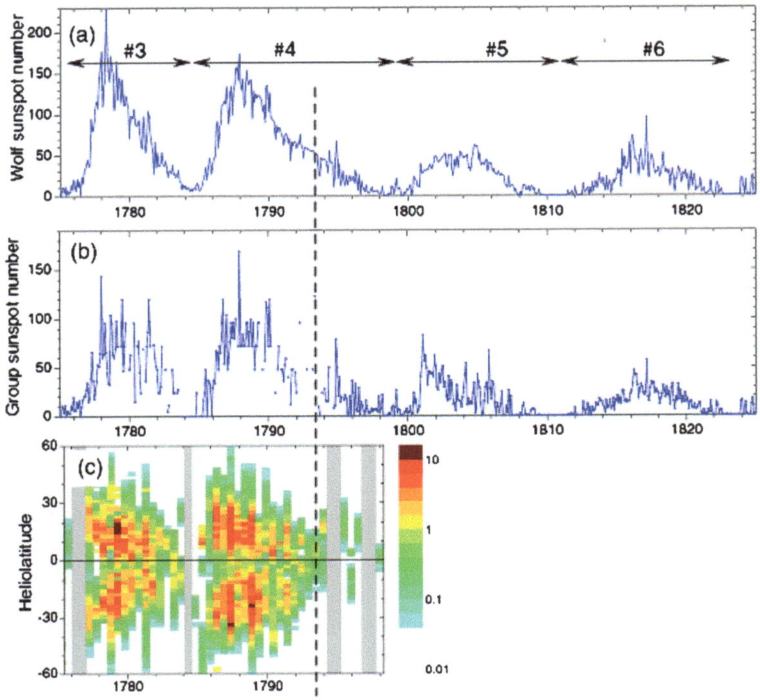

Fig. 15 Sunspot activity in the late 18th and early 19th century: (**a**) Monthly mean Wolf SN. (**b**) Monthly mean GN. (**c**) Reconstructed sunspot butterfly diagram. The color scale on the right gives the density (in $year^{-1}$ deg^{-1}) of sunspots in latitude-time bins and gray bars indicate that no latitudinal information is available. Each bin covers 2° in latitude and six months in time. The *vertical dashed line* marks the start of the "lost" cycle, late in 1793 (from Usoskin et al. 2009)

(2001). More recently, using recovered solar drawings by Staudach (Arlt 2008) and Hamilton (Arlt 2009b), Usoskin et al. (2009) managed to build the solar butterfly diagram for that period and found a sudden, systematic occurrence of sunspots at high solar latitudes in 1793–1796 (Fig. 15). This strengthened the case of a new cycle started in 1793, which would be missing in the original Wolf SN.

However, using another approach, Zolotova and Ponyavin (2011) conclude that the unusual length of cycle 4 can be the result of a late pulse of activity in the northern hemisphere during the declining phase, similar to what occurred in more recent cycles, like cycle 20. They conclude that splitting cycle 4 in two brief cycles leads to anomalously short durations, even less compatible with statistical properties of all known cycles than a long 14-year cycle 4. Therefore, this issue still remains open and new findings in historical archives and libraries are necessary to improve the database of historical solar observation and, thus, our knowledge on solar activity over this peculiar period.

A last issue was raised recently that questions the calibration of all early observations versus Wolf's own systematic sunspot numbers, which rests mostly on Schwabe's series. Using the Schwabe original sunspot data collected by Arlt (2011) and Arlt et al. (2013), Leussu et al. (2013) recently compared the original Wolf SN and the GN series with the long homogeneous sunspot record by Schwabe (1835–1867). Figure 16 gives a comparison of those yearly SNs and indicates the intervals when the main observers (Schwabe and Wolf) were active. They find that while the GN series is homogeneous and consistent with the

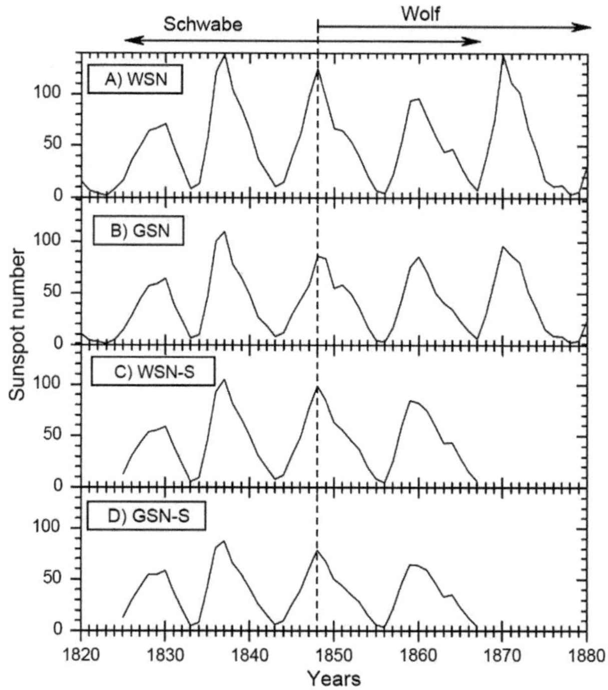

Fig. 16 Yearly mean SNs: (**a**—WSN) the original Wolf SN, (**b**—GSN) the GN, (**c**—WSN-S) the Wolf SN calculated from Schwabe data and (**d**—GSN-S) the GN calculated from Schwabe data. (Figure from Leussu et al. 2013)

Schwabe data throughout the studied period, the scale of the Wolf SN shows a jump around 1848. Based on their results, they conclude that all values in the original Wolf SN need to be lowered by 20 % before 1848. As we will see in the next section, the series resulting from other well established corrections does not support this conclusion. Further analysis will thus be required to clarify this apparent contradiction.

5 Biases in the Zürich Era: Causes and Diagnostics

5.1 The 1880 SN-GN Divergence (1850–1930)

A powerful way of comparing the two time series is to form the ratio between them, like in Fig. 1. In Fig. 17, we show again the ratio between the (Hoyt and Schatten) Group Number and the "Wolf" Sunspot Number, as preserved by the WDC-SILSO, but giving the mean ratio for each year in the interval 1700–1995 and only keeping the most accurate values by restricting the yearly ratio to years when the yearly values are above a suitable threshold, thus avoiding the smallest values, e.g. zeros. From this plot, it is clear that over the last two centuries, there are essentially two discontinuities in this ratio, as well as minor, short-lived drifts.

In this section we shall explore the first jump around 1885. We shall use yearly averages of the original Group numbers as reported by Hoyt and Schatten, calculated by averaging for each year all monthly values for which there is data.

The *backbone-method* developed by Svalgaard (2013b) for reconstructing the GN series starts by selecting a single "primary" observer for an interval of time. The selection should be based both on the length of the observational series (as long as possible) and on the perceived "quality" of the observations, taking into consideration such factors as regularity of

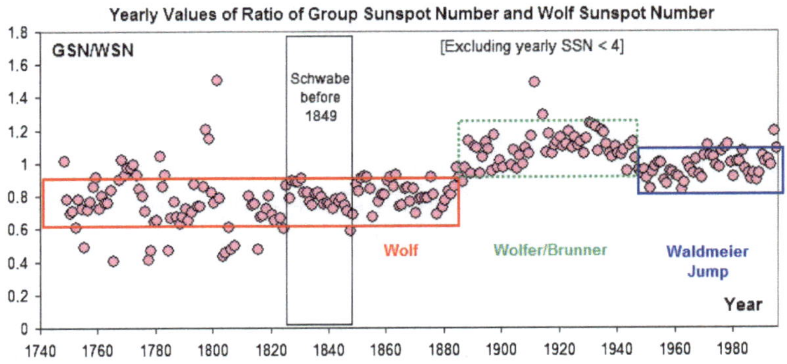

Fig. 17 Ratio between yearly averages of Group and "Wolf" Sunspot Numbers (when both are not less than 4)

observing, suitable telescope, and lack of obvious problems. Two backbones will first be discussed here, those of Schwabe (1794–1883) and Wolfer (1841–1944). The Schwabe backbone is centered on the observing interval for Schwabe and includes all "reliable" observers who overlap in time with Schwabe. The reliability is judged by how high the correlation is between simultaneous (on a yearly basis) observations by the observer and by Schwabe. Similarly, the Wolfer backbone includes all reliable observers who overlap with Wolfer. The two backbones overlap by 42 years so can be cross-calibrated with confidence. Figure 18 gives an overview of the time intervals observed by the observers listed.

For each Backbone, we regress the primary observer's group count against each observer's count for each year and we plot the result (for an example see Fig. 19). Experience shows that the regression line almost always very nearly goes through the origin, so we force it to do so and calculate the slope and various statistics, such as 1-σ uncertainty and the F-value.

The slope gives us what factor to multiply the observer's count by to match the primary's count. The right panel shows a result for the Wolfer Backbone: blue is Wolf's count (with his small telescope), pink is Wolfer's count (with the larger telescope), and the orange curve is the blue curve multiplied by the slope, bringing Wolf's observations on the same scale as Wolfer's. It is clear that the harmonization works well and that it shows that Wolfer with the larger telescope saw 65 % more groups than Wolf did with the small, handheld telescope (Fig. 5) as we would rightly expect. Applying this methodology yields the two backbones (Fig. 20). We stress that the backbones are independent and are based purely on solar observations with no empirical or *ad hoc* adjustments apart from the (necessary) harmonization just described.

It is of considerable interest to compare our "Schwabe" backbone with the Group Counts compiled by Hoyt and Schatten (Fig. 21). Apart from the very noisy period before 1815, the agreement is very good, as would be expected as the series are based on the same data. The minor disagreement around ~1838 for the maximum of solar cycle 8 needs to be resolved, and then, of course, we can see the beginning of the drift after 1882.

The next order of business is to harmonize the two backbones, i.e. bring them onto the same scale. We shall use the Wolfer scale as the base scale, because of its larger number of (better?) observers and choose the common interval 1860–1883 as the basis for the normalization (Fig. 22).

Assuming a normalization factor of 1.55 "explains" 98 % of the difference between the two backbones, with no clear systematic variation with time, we can thus produce a com-

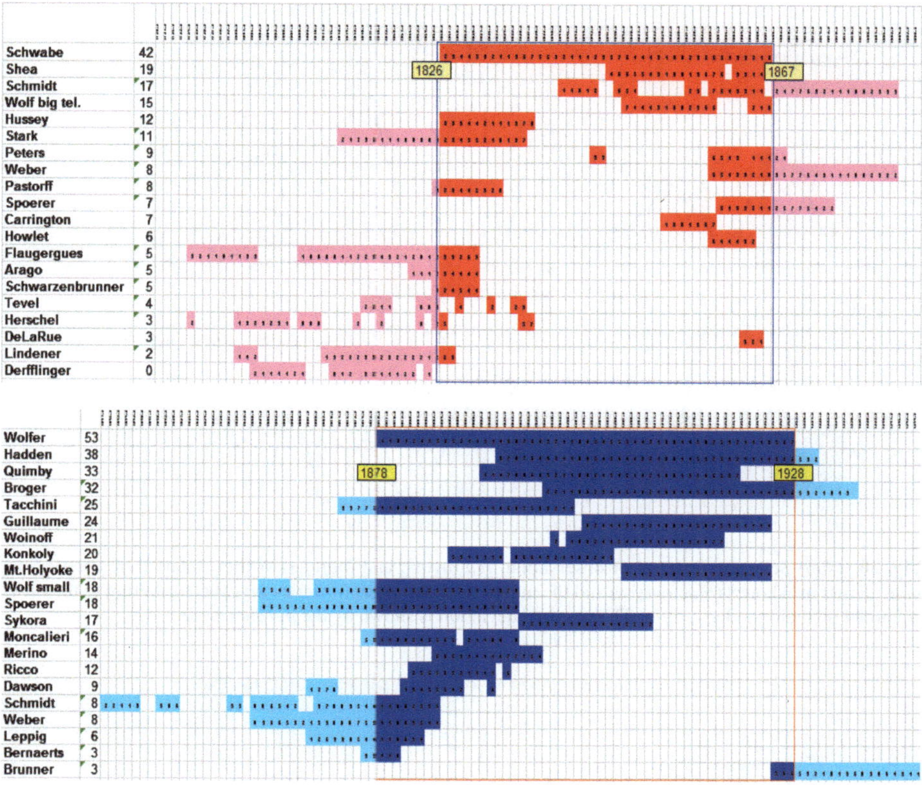

Fig. 18 (*Top*) Coverage and observers for the Schwabe Backbone (1794–1883). (*Bottom*) Coverage and observers for the Wolfer Backbone (1841–1944)

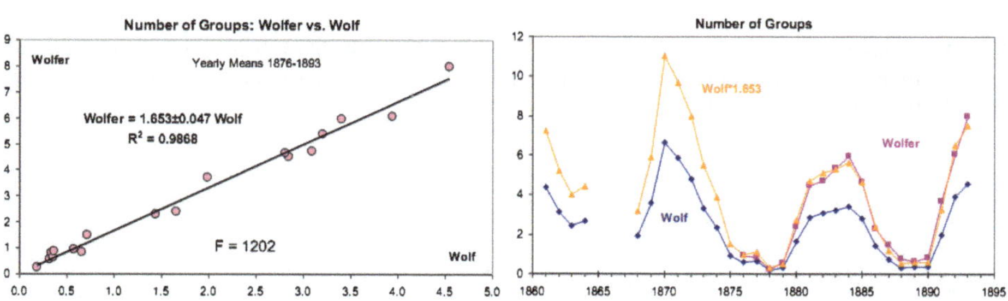

Fig. 19 Regression of number of groups observed by Wolfer (with standard telescope) against the number of groups observed by Wolf (with small telescope)

posite series by multiplying the Schwabe backbone values by 1.55 and then simply average the resulting, normalized Schwabe backbone and the Wolfer backbone (Fig. 23).

Hoyt and Schatten used the Group Count from RGO (Royal Greenwich Observatory) as their Normalization Standard. However, the ratio between the RGO group count and the Wolfer backbone count is not stable (Fig. 24).

This and the discrepancy between the Wolfer/Wolf ratio (Fig. 19) that we find (1.65) and that used by Hoyt and Schatten (1.02) seem to be the main reasons behind the large

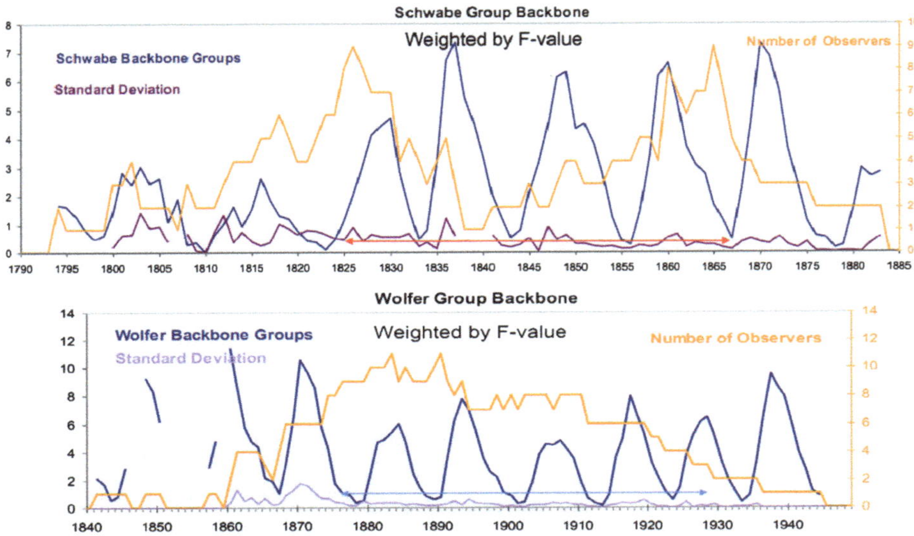

Fig. 20 (*Top*) Schwabe (interval of observation: *red arrow*) and (*bottom*) Wolfer (interval of observation: *blue arrow*) backbones (*blue curves*) and the numbers of observers (*orange curves*) contributing to the mean weighted by their goodness of fit. The standard deviation is shown by the *purple curves*

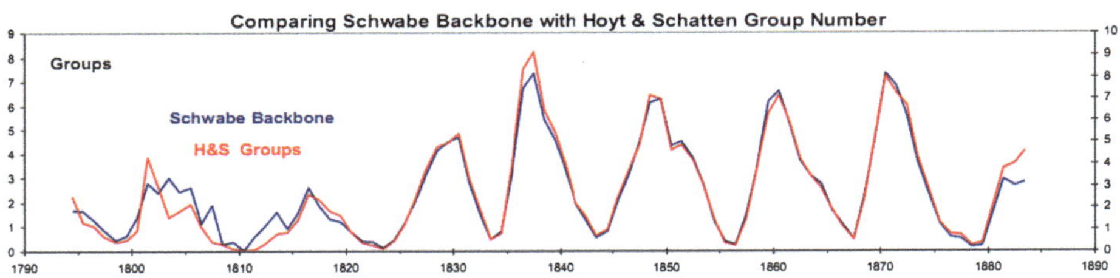

Fig. 21 Number of Groups per year reported by Hoyt and Schatten (1998b) (*red curve*; right-hand scale) and resulting from the Schwabe backbone (*blue curve*; left-hand scale). The scale of the backbone is at this point "free floating" and, in fact, 9/10 that of Hoyt and Schatten's

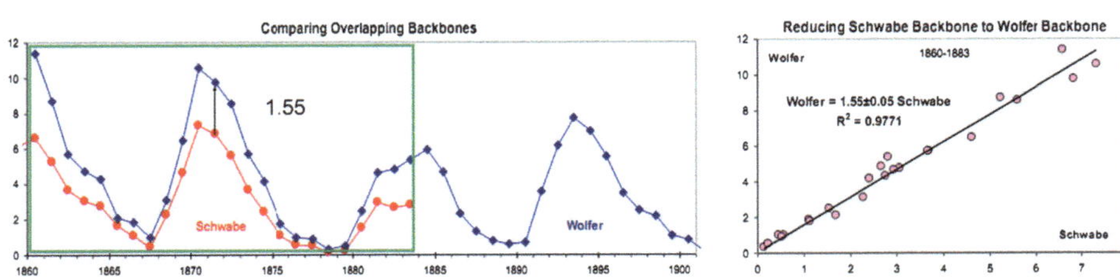

Fig. 22 (*Left*) overlapping backbones: Wolf (*blue*) and Schwabe (*red*). (*Right*) Linear correlation showing the best (least-square) fit for the interval 1860–1883

difference between the GN and the Zurich SN before and after ~1885. It is not clear why Hoyt and Schatten report an almost equal normalization factor for Wolf and for Wolfer with respect to the RGO group count, in spite of the fact that Wolf used the much weaker, handheld 37 mm telescope compared to the standard 80 mm telescope used by Wolfer.

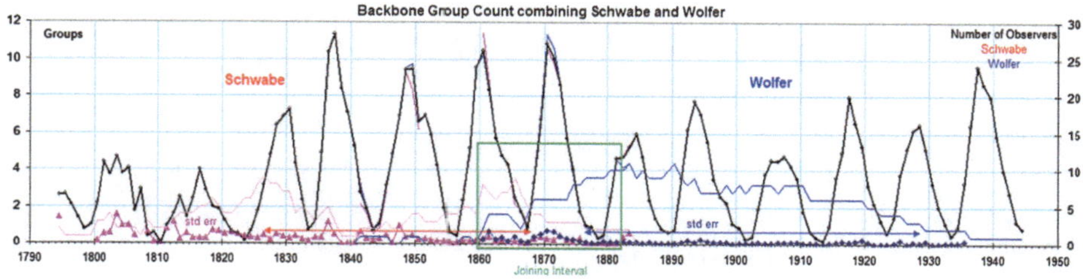

Fig. 23 Composite backbone 1794–1944, with the standard error of the mean

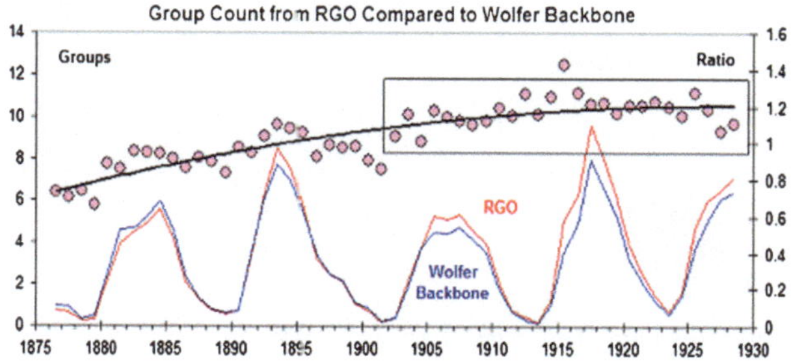

Fig. 24 Ratio (right-hand scale) between Group Count from RGO (*red curve*) and Wolfer Backbone (*blue curve*). After ~1900, the ratio is approximately constant, but before there is a clear progressive change. This change translates into the progressive drift of the SN/GN ratio over 1875–1915 seen in Fig. 17. Over this time interval, the number of observers per year is high (>5) and the standard deviation is comfortably low, as shown by Fig. 20

At this point, the composite backbone is still "free floating". We wish to connect it to "modern" observations so construct yet another backbone (from 1921–2000), based of the group counts by the National Astronomical Observatory of Japan. We name this backbone in honor of the principal observer, Ms. Hisako Koyama, 小山 ヒサ子 (1916–1997). See Fig. 25.

The resulting backbone is shown in Fig. 26 in the same format as used for Fig. 20. The correlation with the Wolfer backbone for the 24 years of overlap (1921–1944) is very high: Wolfer $= 1.0002$ Koyama ($R^2 = 0.9952$), i.e. indicating that the two backbones already match exactly and no re-normalization is needed to join them.

Finally, we also wish to extend the backbone reconstruction backwards to the 18th century. The original set of drawings constituting the long series of observations (1749–1796) by J.C. Staudach was examined by Wolf (1857), who determined group counts and sunspot counts for each drawing. Wolf's counts form the basis for the Staudach Backbone. The analysis of this is still ongoing but we shall here report a preliminary result (Fig. 27).

In his 1861 series, Wolf (1861a, 1861b) effectively doubled the counts that he had derived for Staudach, followed by a further factor of 1.25 increase in the 1882 series. Arlt (2008) suggests that Staudach missed all A- and B-groups (on the modern Waldmeier classification) on account of the relatively low quality of his telescope, perhaps justifying the doubling assumed by Wolf, as A&B-groups make up about 40 % of all groups. Comparisons with the geomagnetic and cosmic ray records are consistent with an overall (but still highly uncertain) factor of ~3 to match the combined Schwabe-Wolfer backbone. A task for the Fourth Sunspot Number Workshop is dedicated to improving the determination of that scaling factor, taking into account the recent digitization of Staudach's drawings by Arlt (2008).

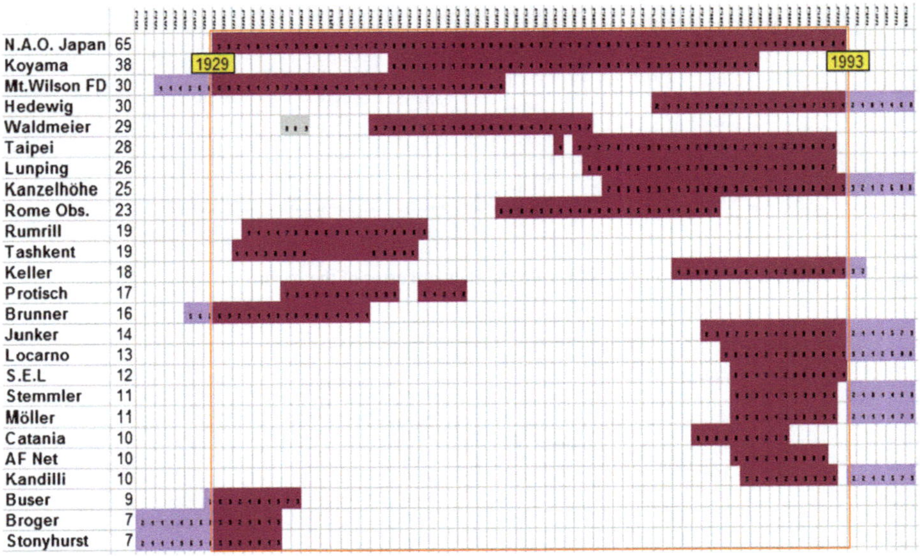

Fig. 25 Coverage and observers for the Koyama Backbone (1921–2000)

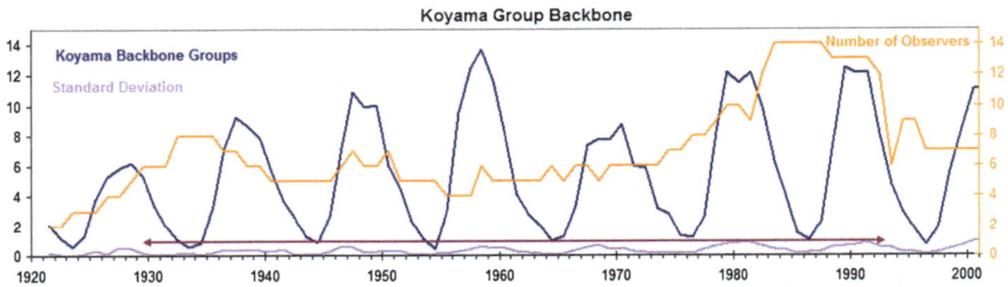

Fig. 26 The Koyama backbone (interval of observation: *purple arrow*) and its number of observers (*orange curve*). The standard deviation is shown by the *bottom purple curve*

Adding the raw group counts from SIDC's database to the Koyama backbone, allows us to present a preliminary synthesis of the evolution of a composite of the average number of groups per year back to 1749 derived from all four backbones, each backbone shown with a different color (Fig. 28).

In this section, we have striven to build a series based solely on solar observations. In another chapter of this volume (Svalgaard 2014b), we review what Geomagnetism can tell us about solar activity, but it is instructive already here to compare the number of groups with the range rY of the diurnal variation of the geomagnetic field, an indirect but fully independent tracer of solar activity introduced in Sect. 2 (Wolf's magnetic needle readings). Figure 29 shows the excellent agreement between our reconstructed number of groups and this diurnal range throughout the entire interval 1845–2013.

We can now compare the composite Group Number (GSN*) series with the official Zürich Sunspot Number (R_Z) and several geomagnetic indicators (Fig. 30).

Based on these comparisons, it does not seem reasonable to apply a wholesale decrease of the Wolf Numbers by 20 % before 1849, as advocated by Leussu et al. (2013) in Sect. 6.3.

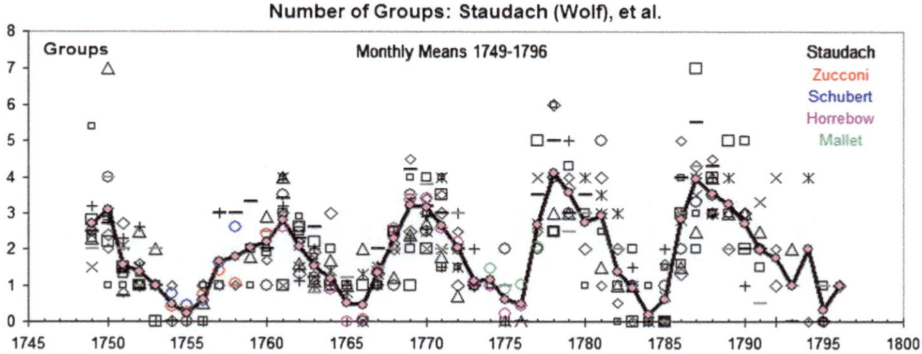

Fig. 27 *Open black symbols* show the counts of groups by month (by Staudach as determined by Wolf 1857). Yearly group counts by Zucconi (Venice), Schubert (Danzig), Horrebow (Copenhagen), and Mallet (Berlin) scaled to the yearly counts by Staudach are shown by *colored open circles*. The *full black curve with small pink circles* shows the yearly group counts, averaged over all observers, forming the Staudach Backbone, which is here "free-floating". A "group count" for an interval of time is the average of the number of groups observed by an observer on each day within the interval

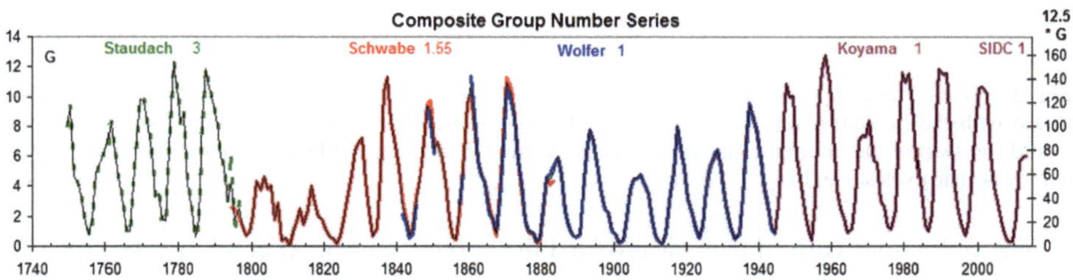

Fig. 28 A composite record of the yearly mean number of groups, back to 1749, derived from the four backbones constructed as described in this section. The numbers to the right of the backbone designations indicate the scale factor applied to each "raw" backbone to harmonize them to the Wolfer scale. The *green dashed line* marking the Staudach backbone reflects that it is uncertain. The right-hand scale is for the thin *black curve* showing the quantity 12.5 times the average group count, as an "equivalent" GN

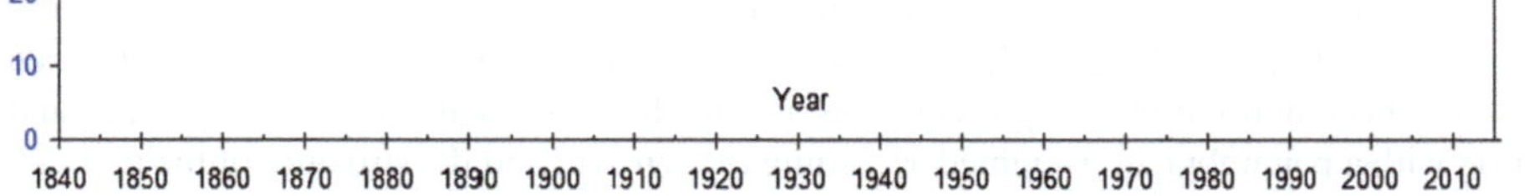

Fig. 29 Number of groups (*red curve*) compared with the range of the diurnal variation of the East Component (*blue curve*) of the geomagnetic field

If anything, R_Z is already too low (e.g. cycle 7). The size of cycle 10 (maximum in 1860.1) now becomes an important research problem, possibly for resolution at the Locarno 2014 Sunspot Number Workshop. This serves as a reminder that much work remains to be done.

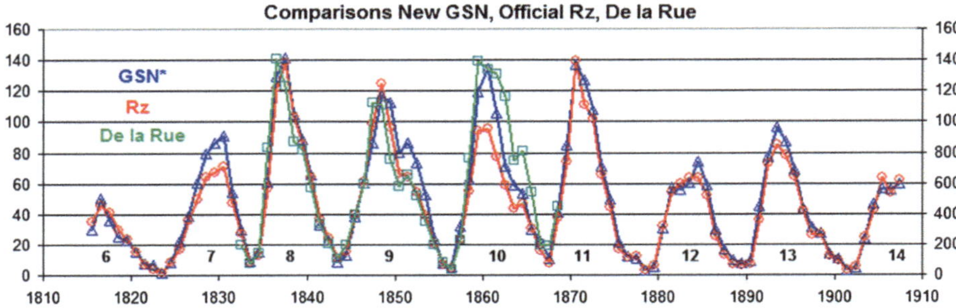

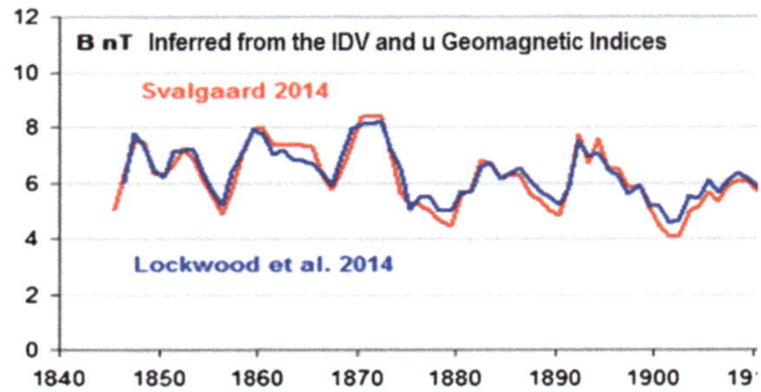

Fig. 30 (*Top*) The composite GSN* matches R_Z well, except for cycles 7 and 10. There is also a good match with the (scaled) sunspot areas determined by De la Rue (Vaquero et al. 2002). The geomagnetic record (*bottom*) of the inferred Heliospheric magnetic field (Svalgaard 2014a) also shows that cycle 10 was not lower than cycles 9 and 11

The corrections derived thus far involved the Group number. After 1876, when the GN relies entirely on photographic data and is scaled to the SN, the latter becomes the actual reference. It turns out that other inhomogeneities were found in that more recent part of the SN. Those inhomogeneities, described in the following sections, will thus influence the scale of both the SN and GN over the entire period preceding the 20th century, as the calibration is established by working backwards in time.

5.2 The 1950 Waldmeier Jump (1930–1980)

In this section, we address the discontinuity at ∼1947 in Fig. 17, which we have termed the "Waldmeier Jump" (Svalgaard 2010), because we attribute it to changes in the SN counting procedure implemented when Max Waldmeier became the Director of Zürich Observatory. In 1961, Max Waldmeier published the definitive Zürich sunspot numbers up until 1960 (Waldmeier 1961). He noted that "Wolf counted each spot—independent of its size—but single. Moreover, he did not consider very small spots, which are visible only if the seeing is good. In about 1882 Wolf's successors changed the counting method, which since then has been in use up to the present. This new method counts also the smallest spots, and those with a penumbra are weighted according to their size and the structure of the umbra". In 1968 Waldmeier (1968, 1948) codified the weighting scheme as follows "Später wurden den Flecken entsprechend ihrer Größe Gewichte erteilt: Ein punktförmiger Fleck wird einfach gezählt, ein größerer, jedoch nicht mit Penumbra versehener Fleck erhält das statistiche Gewicht 2, ein kleiner Hoffleck 3, ein größerer 5".[2] However, Wolfer in 1907 (Wolfer

[2] A spot like a fine point is counted as one spot; a larger spot, but still without penumbra, gets the statistical weight 2, a smallish spot within a penumbra gets 3, and a larger one gets 5.

Fig. 31 Drawing from Mount Wilson Observatory (MWO) of the single spot with penumbra on 21st Nov. 1920. The *insert at the left* shows a similar group observed at MWO on 5th Nov., 1922. For both groups, Wolfer should have recorded the observation as "1.3" if he had used the weighting scheme, but they were recorded as "1.1", thus counting the large spots only once (with no weighting)

1907) explicitly states: "Notiert ein Beobachter mit seinem Instrumente an irgend einem Tage g Fleckengruppen mit insgesamt f Einzelflecken, ohne Rücksicht auf deren Grösse, so ist die daraus abgeleitete Relativzahl jenes Tages $r = k(10g + f)$".[3] We can verify that Wolfer, contrary to Waldmeier's assertion that the Zürich observers began to use weighting "around 1882", did not weight the spots according to Waldmeier's scheme by comparing Wolfer's recorded count with sunspot drawings made elsewhere, e.g. Fig. 31.

L. Svalgaard uncovered many other such examples (e.g. 16th September, 1922 and 3rd March, 1924) in a systematic search of MWO drawings (from 1917–1925) for single sunspots with penumbra. He found no such spots during this period for which weighting was applied at Zürich. We thus consider it established that Wolfer did not apply the weighting scheme. This is consistent with the fact that nowhere in Wolf's and Wolfer's otherwise meticulous yearly reports in the *Mittheilungen über Sonnenflecken* series is there any mention of a weighting scheme.

We shall not here speculate about the motive or reason for Waldmeier ascribing the weighting scheme to Wolfer. Waldmeier himself was an assistant to Brunner since 1936 and performed routine daily observations with the rest of the team so would presumably have known what the rules were. Figure 32 shows that Brunner and Waldmeier were observing very close to the same scale in 1937.

All the Zürich SN observer records after 1937 appear to be lost, but in spite of the lack of original material, it is possible to perform a statistical analysis as follows. From the RGO series of sunspot group areas (Hathaway 2014), we select days where only one group was recorded on the disk. If that group had precisely one spot, the sunspot number for that day would be recorded as 11 by Wolf and as $7(0.6 * (10 * 1 + 1) = 6.6)$ by Wolfer and later observers, if there were no weighting by size and complexity. Figure 33 shows the distribution of solitary large spots over time. During the Wolf period, the largest single-spot groups had a sunspot number of 11 (there were scattered lower values in the 1880s due to averaging with Wolfer). Starting with Wolfer, there were many large groups with a single spot counted as just one spot (sunspot number 7), i.e. no weighting. With Brunner and later, the 7s effectively disappear. This seems to indicate that some weighting was done already by Brunner, explaining why Waldmeier matched Brunner's counts. On the other hand, there

[3]When an observer at his instrument on any given day records g groups of spots with a total of f single spots, without regard to their size, then the derived relative sunspot number for that day is $r = k(10g + f)$.

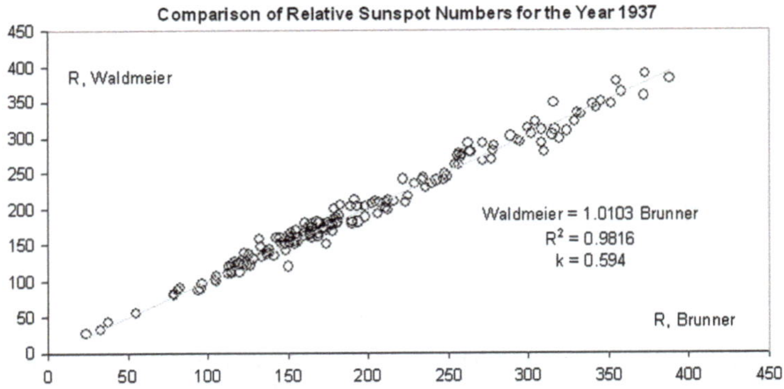

Fig. 32 Comparison of daily "raw" (i.e. with no k-factor applied) relative sunspot numbers derived by Waldmeier and Brunner for the year 1937. The k-factor for Waldmeier comes to $0.594 = 0.6/1.0103$ (Brunner reports 0.59)

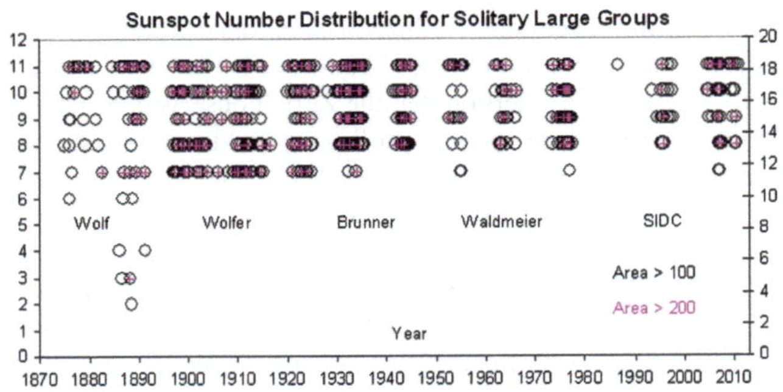

Fig. 33 For days where only one group was observed, the sunspot number (if less than 12) for that day (i.e. for that solitary group) is plotted if the projected area of the group is larger than 100 μHemisphere (*circles*) and larger than 200 μH (*pink "+" symbols*). The right-hand scale is for sunspot number divided by 0.6, i.e. on the original Wolf scale

are many 8s, so any weighting must have been slight and there simply were very few solitary spots during the active 1940s, so it is difficult to draw a definite conclusion from this analysis about the amount of weighting done by Brunner.

Brunner himself writes in 1936 (Brunner 1936) that "The subjective method of counting may also have an influence. In large centers of activity one is inclined—and this perhaps rightly—to give some single spots according to their sizes a different weight", but then continues "In the spot-statistics, introduced for our observatory by Rudolf Wolf 80 years ago, all these circumstances have been considered as far as possible by introducing a reduction-factor on Wolf's unit. The latter is determined by comparison of corresponding observations. In determining the Wolf relative-number a weight of ten is given for the groups of spots and a weight of one for the number of single spots or nuclei".[4] This seems to indicate that spots were not weighted, although Brunner at times might be inclined to do so. His assistant Max Broger (observed 1897–1936) appears to have weighted some of his counts, so it is conceivable that discussion was going on at Zürich about the preferred counting method.

[4]Presumably meaning umbrae (spots) within each penumbra.

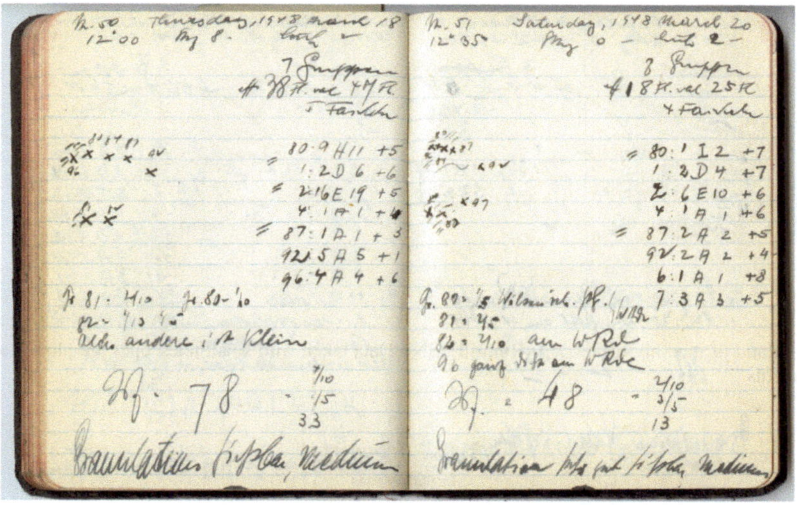

Fig. 34 Pages from Herbert Luft's notebook for March 18th and 20th, 1948. South is up and West is left. The letters in the columns at the right on each page show the Zürich Group Class for each numbered group, flanked left and right by the raw and the weighted count of spots in each group. The telescope was a superb 54 mm aperture Merz used at 96× magnification. In spite of the crude-looking drawings, the counts of groups and spots are of high quality, as can be seen from comparisons with MWO

The long-time observer Herbert Luft (1908, Breslau, Germany; 1988, New York, USA) was a corresponding contributor to the Zürich series. As a teenager, he joined various Amateur Associations and was mentored by the slightly older Wolfgang Gleissberg who suggested Luft concentrate on sunspots. Luft's notebooks are archived at AAVSO (http://www.aavso.org/herb-lufts-notebooks-new-science-aavso-archives) and L. Svalgaard recently digitized the observational material. The nearly 12,000 pages yielded 10,434 usable observations (when image quality was good enough) during 1924–1987. Interesting enough, Luft started using the weighting scheme on the 24th February 1947, but abandoned the scheme April 5th the next year. Figure 34 shows two pages from March 1948. Luft also started to use the Zürich Group Classification System at the same time (and did not later abandon the classification). The spot count for the A-groups were the same with (number to the right of the class letter) and without weighting (number to the left of the class), while the H-I, D, and E class groups had a weighted spot count on average 45 % higher.

We interpret this to indicate that Waldmeier was trying to get other observers to adopt the weighting scheme, but with little success. To our knowledge, the weighting was only adopted on a continuing basis by the Zürich observers and not by any others. Wolf and Wolfer published all raw observations from corresponding observers. Brunner reduced that to only the Zürich observers, and Waldmeier stopped the publication of all raw observations completely, noting that all data would be available in the archives of the observatory. As noted above, however, the post-1937 archives are apparently lost. We make a plea here that anybody who has archived correspondence with the Zürich observatory since 1925, to send copies of the material to us so that we can recover at least some of the raw data in order to investigate the effect of the weighting on the SN record.

One such case is that of Harry B. Rumrill (1867–1951) who was a friend of Rev. Quimby (American observer 1897–1921 whose data were utilized in the Group Backbone Construction). Rumrill continued Quimby's observations of sunspots through to 1951. His data and

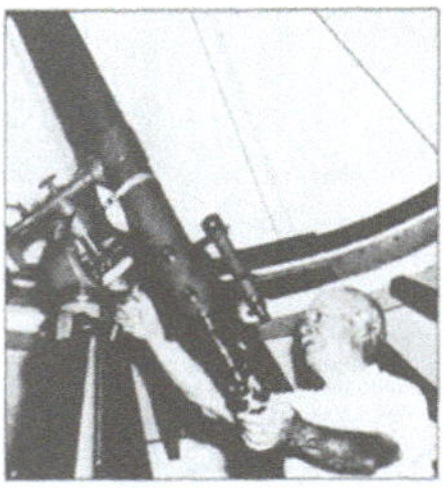

Harry Barlow Rumrill, president of the Rittenhouse Astronomical Society in 1932, with his 4-inch Brashear refractor. From *History of the Rittenhouse Astronomical Society*, courtesy Joy Crist.

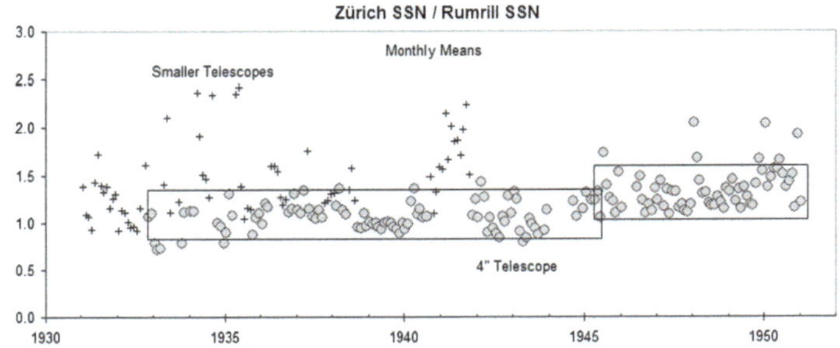

Fig. 35 Ratio of monthly means R_Z/(Rumrill SSN). Data taken with small telescopes are plotted as small "+" symbols

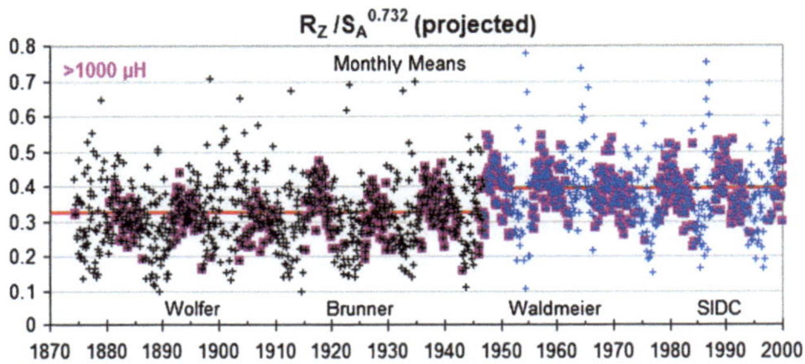

Fig. 36 The ratio $R_Z/S_A^{0.732}$ (see text) for monthly means. If the mean area, S_A, is in excess of 1000 μHemispheres, the data point (+) is marked by a *pink square*. Before 1947, a *horizontal red line* shows the average ratio at 0.3244. After 1947, the *red line* shows the average ratio at 0.3931. In addition, the individual monthly values are shown as *light blue plusses*

notebooks were considered lost until L. Svalgaard, with the help of "The Antique Telescope Society" (Bart Fried, Jack Koester), located most of them in early 2012. Rumrill used 2″ telescopes early on, but from 1942 employed exclusively a 4″ Brashear refractor (Fig. 35, left). The ratio between the Zürich Sunspot Number and Rumrill's indicates an increase of the Zürich values from ∼1945, by about 20 %.

The total sunspot area data available over the time of the Waldmeier jump can also provide a useful external reference. There is indeed a strong (slightly non-linear) relationship between the SN, R_Z, and the projected (i.e. observed) sunspot area, S_A. On average: $R_Z \approx K S_A^{0.732}$ with no linear offset, so we can meaningfully form the ratio between the quantities (Fig. 36); it seems clear that the ratio is lower than average before ∼1947 and higher thereafter. We ascribe the difference to introduction of the full weighting scheme, as there is no metadata indicating a change in derivation of the RGO sunspot areas at that time.

Using the value for the factor K derived from the pre-1947 data to calculate the monthly mean R_Z from S_A, we get excellent agreement before 1947 (Fig. 37) but a definite discrepancy thereafter, with the observed R_Z being larger by a factor of 1.22 than the calculated value, call it R_C, we would expect from S_A. This suggests that the weighting was fully implemented by 1947.

From ∼40,000 Ca-K spectroheliograms taken at the 60-foot tower at Mount Wilson between 1915 and 1985, a daily index of the fractional area of the visible solar disk occupied by plages and active network has been constructed (Bertello et al. 2010). Monthly averages

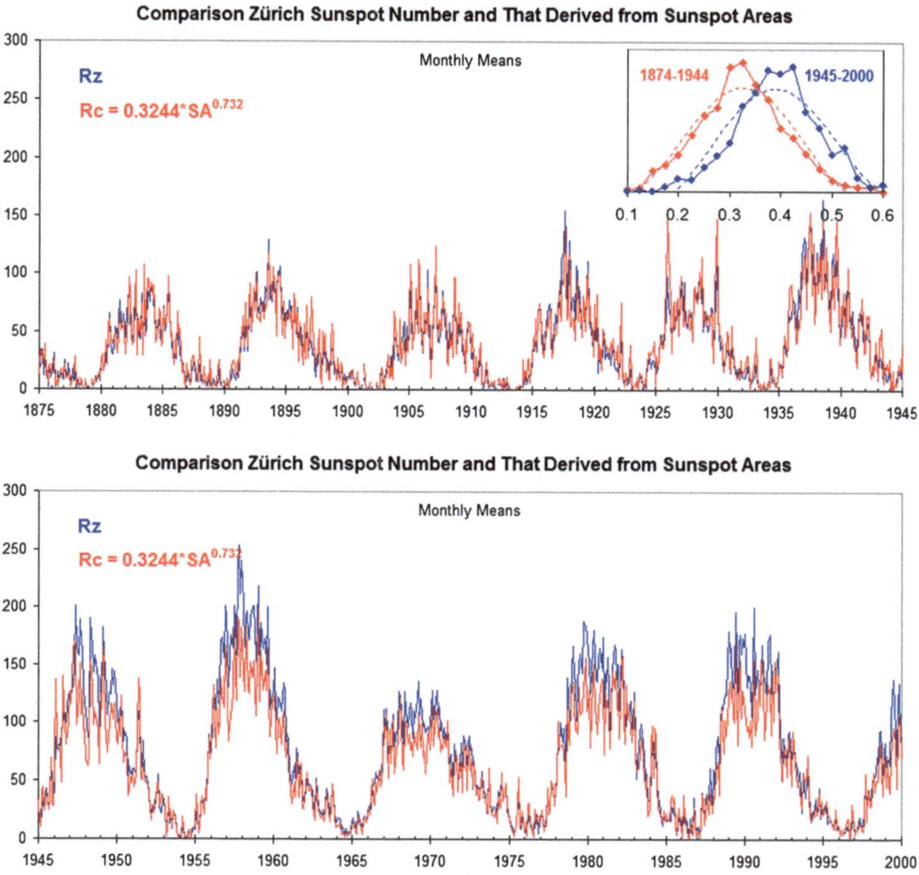

Fig. 37 Calculated R_C *red*, and observed, R_Z *blue*, Zürich Sunspot Number using the relationship derived before 1945. The constant $K = 0.3244$ is determined from the best fit before 1945. The *insert* shows the distribution of data points in Fig. 36 partitioned by the year 1945

of this index are strongly correlated ($R^2 = 0.8$) with the sunspot number. Using the correlation based on the Wolfer-Brunner era, we can calculate the expected sunspot number for the Waldmeier era from the Ca-K index (Fig. 38). On average, the observed sunspot number after 1945 is a factor 1.21 higher than the expected value, again showing the influence of the weighting of sunspots according to size, coinciding with the tenure of Waldmeier.

The effect of the weighting of individual spots can also be inferred by considering ionospheric measures of solar activity. The ionospheric F2-layer critical frequency, $f°$F2, is the maximum radio frequency that can be reflected by the F2-region at vertical incidence (that is, when the signal is transmitted straight up into the ionosphere). The critical frequency has been found to have strong solar cycle dependence (Sethi et al. 2002). Back in 1952, Ostrow and PoKempner (1952) compared the dependence of $f°$F2 on the Zürich sunspot number and concluded that there are "differences in the relationship between $f°$F2 and sunspot number for the current (18th) and preceding (17th) sunspot cycles" (Fig. 39). In Fig. 39, it is instructive to follow the dashed line (March 1944 to June 1947) from the low sunspot numbers of cycle 17 to the high values of cycle 18. Adapted after Qstrow and PoKempner (1952). The shift (red arrow) in sunspot number to bring the curves for cycles 17 and 18 to

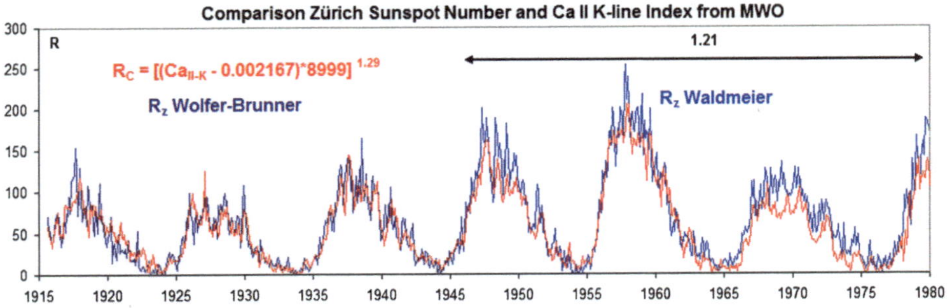

Fig. 38 Comparison of the MWO Ca II K-line index with the Zürich sunspot number

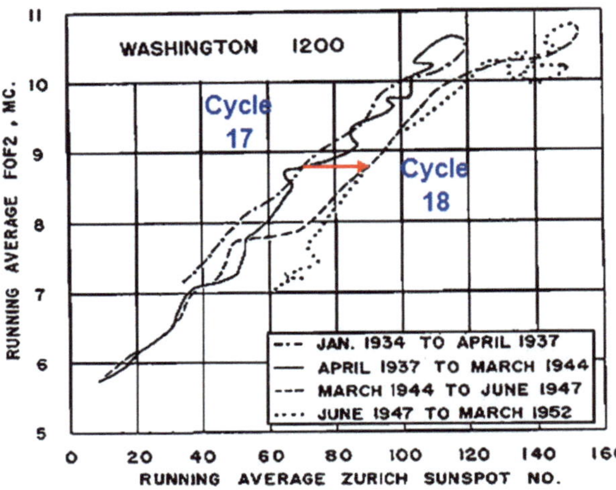

Fig. 39 12-month running average of the monthly median $f°$F2 at local noon against 12-month running average of the monthly Zürich sunspot number

overlap is the now familiar ~20 %. Today, we can ascribe their further conclusion that "the sunspot number is therefore not entirely satisfactory as an index for ionospheric variations" to the result of the introduction of Waldmeier's weighting scheme. The "fault" is not with the relationship, but with the sunspot number.

Another independent comparison can be done with the range rY of the East component of the geomagnetic field, described in Sect. 2. Considering the yearly averages of the rY index for geomagnetic stations in the latitude range 20°–60°, the residual differences from station to station are small and mainly due to local inhomogeneities in underground conductivity. Normalizing the range for a given station to a reference station (POT-SED-NGK) eliminates those small variations and allows us to make a composite of all stations. The result for observations since 1890 from a large number of observatories is shown in Fig. 40.

That Fig. 41 looks very much like a plot of the sunspot number is, of course, not a surprise as the linear correlation coefficient is in excess of 0.97 (see Fig. 29). With such high correlation, it should be possible to see the influence of weighting. Indeed, Fig. 41 shows that the slope of the regression line is different before and after 1947, and that in the ratio $6.217/5.005 = 1.24$, again suggesting that weighting increases the sunspot number by about that amount.

Calculating the slope of the correlation for each sunspot cycle (Fig. 41, right), we find, as before, that the increase in slope takes place between the 17th and 18th sunspot cycle (see

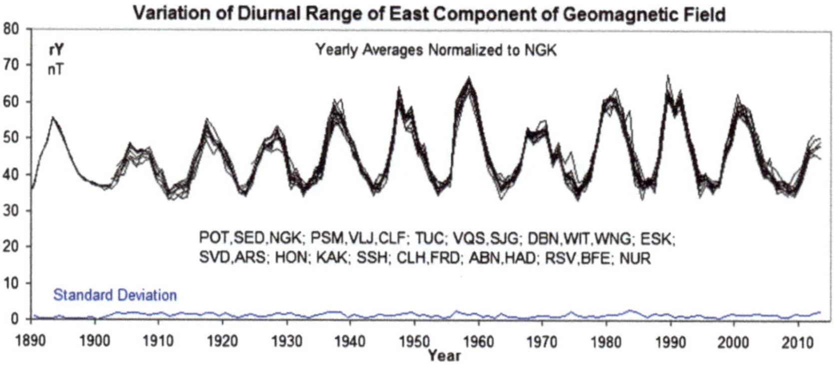

Fig. 40 Composite record of the ranges of the diurnal variation of the East Component of the geomagnetic field. The composite is the average of long series of observations from 14 "chains" of stations (identified by their standard station codes), each chain plotted with a *thin black line*. A chain is the combined record of a station and its replacement stations that were necessary over time to escape urban development, normalized to the POT-SED-NGK chain. The very small standard deviation is shown in *blue* at the bottom of the figure

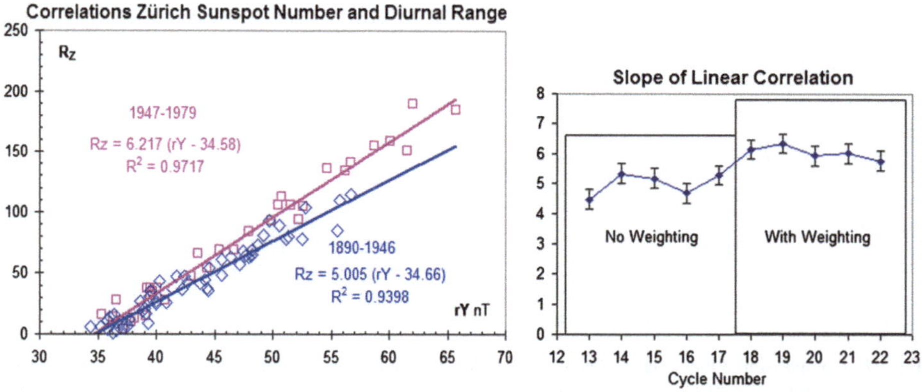

Fig. 41 (*Left*) Linear correlations R_Z as a function of rY for time before and after 1947. (*Right*) Variation of slope with time

also CAWSES Newsletter: http://www.bu.edu/cawses/calbrating_sunspot_number_using_mag_needle.pdf). There is therefore little doubt that Waldmeier introduced the weighting scheme in full force in or about 1947.

At the reference station Locarno, weighting has been used since the beginning in 1957, closely following Waldmeier's prescription (Sergio Cortesi, personal communication). To assess the magnitude of the increase due to weighting, Leif Svalgaard undertook to examine all the drawings and individual counts of groups and spots made at Locarno for the past decade and re-count the spots with and without weighting. There were 3229 observation days with 9532 groups containing 49,318 un-weighted spots at the time of writing. The weighted spot count was 72,548, for an excess of 47 %. The counts translate into an average sunspot number of $26.88[(10*9532+49318)/3229*0.6]$ without weighting and 31.19 with weighting, for an excess of 16 % for this rather low solar activity. It is, perhaps, noteworthy that the average number of (unweighted) spots per group for this period (2003–2014) is low, only 5.17.

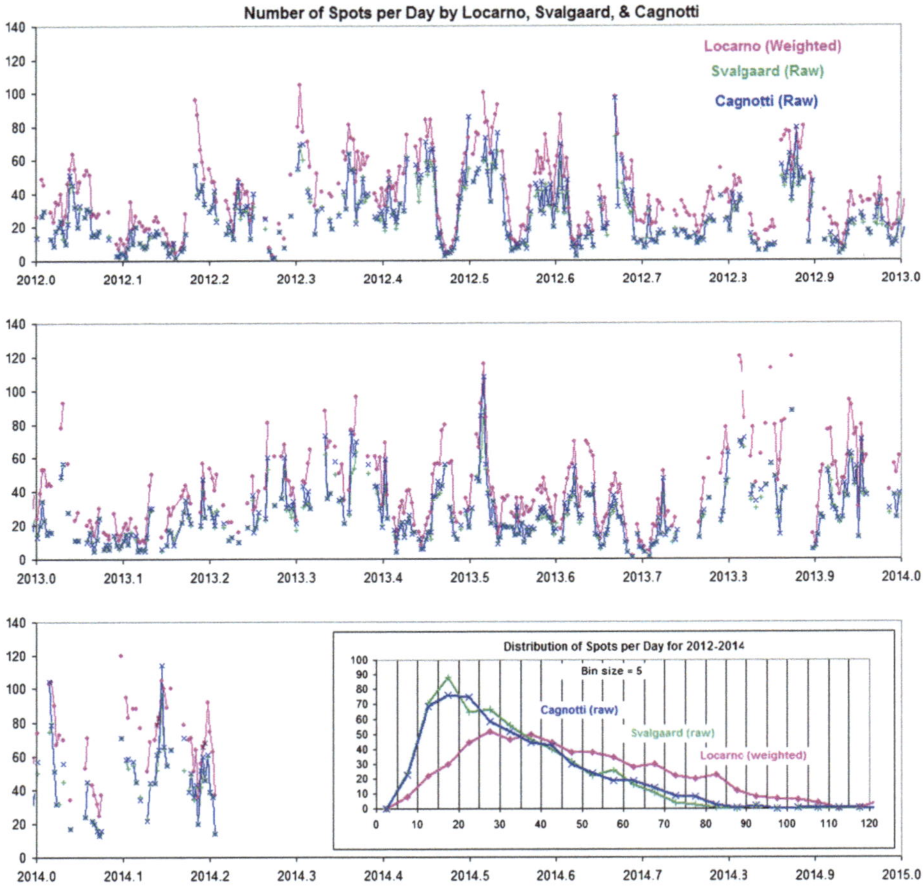

Fig. 42 Comparison of the number of sunspots per day determined by Cagnotti (*blue*) and Svalgaard (*green*) without weighting, i.e. by counting each spot singly as prescribed by Wolfer and Brunner, with the number reported by Locarno (*pink*) employing the Waldmeier weighting scheme. The *insert* shows the distribution of counts in bins of five

To verify that the re-count is valid, i.e. that Svalgaard has understood and applied correctly the Waldmeier weighting scheme, the observer Marco Cagnotti in Locarno agreed to maintain a parallel count of unweighted spots at a continuing basis since January 1st, 2012, following a brief trial in August 2011. We remind the reader that the sunspot count that Locarno is reporting is done visually at the telescope and not from the drawings. It is rare, though, that there is a difference.

Figure 42 shows that Svalgaard and Cagnotti very closely match each other in applying the weighting scheme, thus sufficiently validating the approach.

To determine the effect on the relative SN of the weighting, we evaluate the Relative SN = 10 Groups + Spots for Locarno, Cagnotti, and Svalgaard and compare that with the International SN R_i divided by the k-factor of 0.6 (Fig. 43). Ideally the ratio between the Locarno SN and $R_i/0.6$ should be unity, which it very closely is (0.99).

Determining the weight factor w, for years with different levels of solar activity allows us to quantify the relationship between w and the sunspot number as reported. Treating the values for the deep minimum 2008–2009 as outliers (as also the weight factor for very low

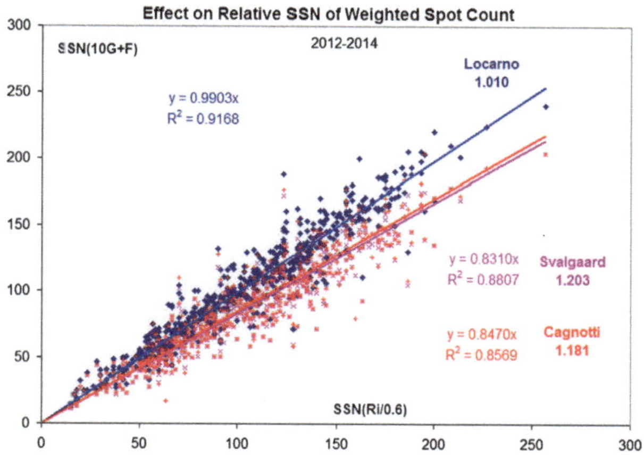

Fig. 43 The relative SN calculated from the weighted spot count reported by Locarno (*blue*) compared to the International SN without the standard 0.6 factor. The unweighted counts by Svalgaard (*pink*) and Cagnotti (*red*) agree very well with each other and translate into a SN that is only 0.839 of the International SN (which then is higher by a factor of 1.19 on average)

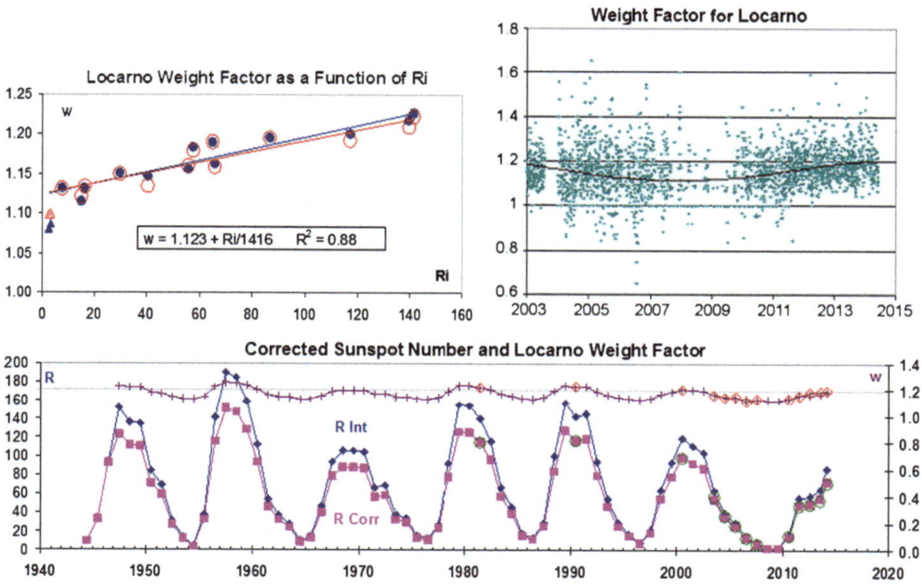

Fig. 44 (*Upper left*): The weight factor for the years 1981, 1990, 2000, 2003–2014 (other years still in preparation) as a function of the International Sunspot Number. The *blue filled circles* show the average weight factor for each year based on Locarno drawings. The *red open circles* show the ratio between the average weighted sunspot number and the average un-weighted sunspot number. Outliers are marked by *small triangles*. The best-fit regression lines (omitting the outliers) are shown with the equation giving the average coefficients. (*Upper right*): The weight factor at Locarno for each day since 2003. (*Bottom panel*): The weight factor calculated from the International Sunspot Number (*upper curve and scale at right*). *Red diamonds* show measured values. The *blue diamonds* and *blue curve* show yearly averages of the International Sunspot Number as reported by WDC-SILSO. The *pink squares* and *curve* show the resulting values when corrected for weighting. Values for years when the raw un-weighted averages were actually measured, rather than calculated, are marked by *open green circles*

sunspot numbers does not matter) yields $w = 1.123 \pm 0.006 + R_i/(1416 \pm 140)$, with a range of 1.126–1.264 for R_i in the range 4–200. The effective average weight factor for 1947–2014 is then readily determined to be 1.20 (Fig. 44). The bottom panel of Fig. 44 shows what the

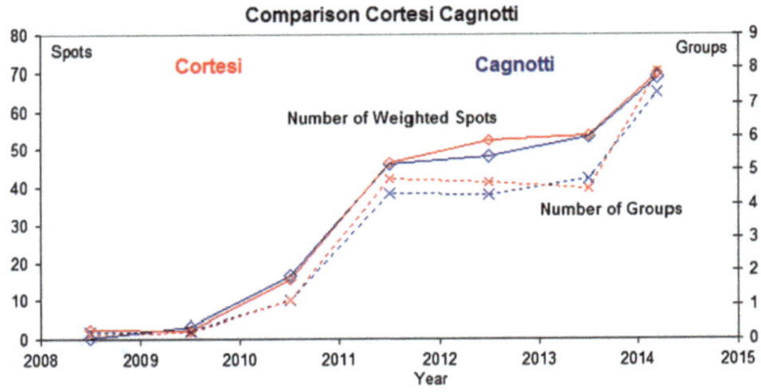

Fig. 45 The yearly average group count (*dashed lines* and *crosses*) and weighted sunspot count (*full lines* and *diamonds*) for observers Cortesi (*red*) and Cagnotti (*blue*)

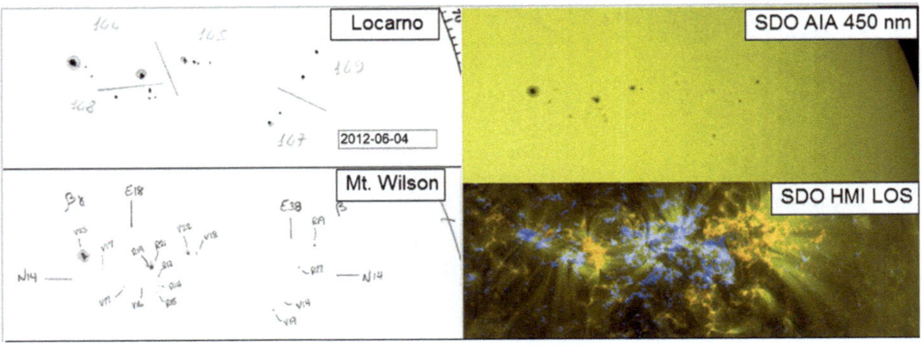

Fig. 46 Groups and spots observed 4th June 2012 at Locarno (five groups; *upper left*), MWO (two groups; *lower left* with polarities indicated), SDO AIA at 450 nm (*upper right*), and their magnetic fields (*blue* positive, *orange* negative; *lower right*). The two MWO groups are marked by the latitudes and longitudes of the centroids (N14, E18 and N14, E38)

sunspot number would be [pink squares] if we divide the International sunspot number by the weight factor just determined to correct the sunspot number for the over-counting caused by the weighting.

Is the weight factor observer dependent? With a novice one might be inclined to think so, but with training, observers tend to converge to agreement. We can compare the weighted counts made by the veteran Cortesi and the new observer Cagnotti from 2008 to the present (Fig. 45): there does not seem to be any systematic difference.

Waldmeier also introduced a new classification of groups (the Zürich classification) based on understanding of the evolution of the group rather than mere proximity of the spots. This tends to increase the number of groups over that mere proximity would dictate. We find that, on average, on a fifth of all days, an additional group is reported over what is observed at MWO, which means that the relative sunspot number increases by about 3 % due to this inflation of the group count brought about by the better understanding of what constitutes a group. Kopecky et al. (1980) quote the observer Zelenka suggesting a possible influence of the new Zürich classification of groups. This problem deserves a full, future investigation.

Figure 46 illustrates the problem. Today, we may use the magnetic field information to discriminate between groups, but such information was not available to observers in earlier times so proximity and longitudinal extent were the primary criteria for groups.

Recently, Lockwood et al. (2014) have suggested that the magnitude of the Waldmeier jump in R_Z in ~1947 is ~12 % rather than the ~20 % that we find in several indices and deduce from an analysis of the effect of the weighting. The difference is due to their

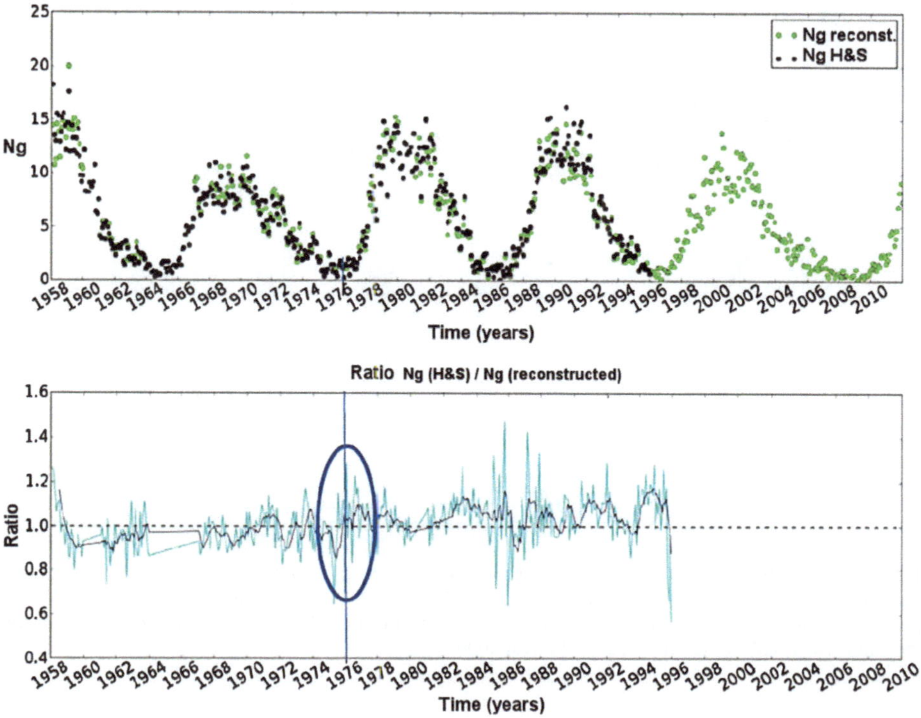

Fig. 47 *Top panel*: monthly mean group counts (*Ng*): in *black*, the original series from Hoyt and Schatten (GN divided by the 12.08 constant) and in *green*, new reconstructed average group counts based on the WDC-SILSO observations. *Lower panel*: ratio between the original and the new group counts: monthly mean ratio (*blue*) and 12-month smoothed ratio (*black*). The time when the original group number switches from the Greenwich to the SOON data set is marked by the *ellipse*

use of annual averages, rather than monthly means, of the SN and sunspot area in their determination of the non-linear relationship between these two parameters. This understates the size of the discontinuity and the corresponding correction factor.

5.3 The RGO-SOON Jump (1976)

As the WDC-SILSO data set contains the raw sunspot and group counts, it is possible to compute a GN by the same method as the International SN, using a statistics over many stations with one pilot station (Locarno). This GN offers the following advantages: it is built on a single continuous set of data and the base data are the same as the ones used for producing the SN, allowing a direct unambiguous comparison. Moreover, the group count is unaffected by the sunspot weighting adopted specifically at the Locarno station (cf. previous section).

We can then compare this reconstructed GN with the original GN from Hoyt and Schatten (1998a), which is based on RGO photographic data until 1975 and on sunspot drawings from the 4 stations of the USAF SOON network afterwards. As shown in Fig. 47 (straight group counts without the 12.08 scale factor), both series match rather well, with a global average ratio close to unity, although the monthly mean ratio is rather variable (~10 % rms). However, this average ratio shows a clear 10 % upward trend occurring over the interval

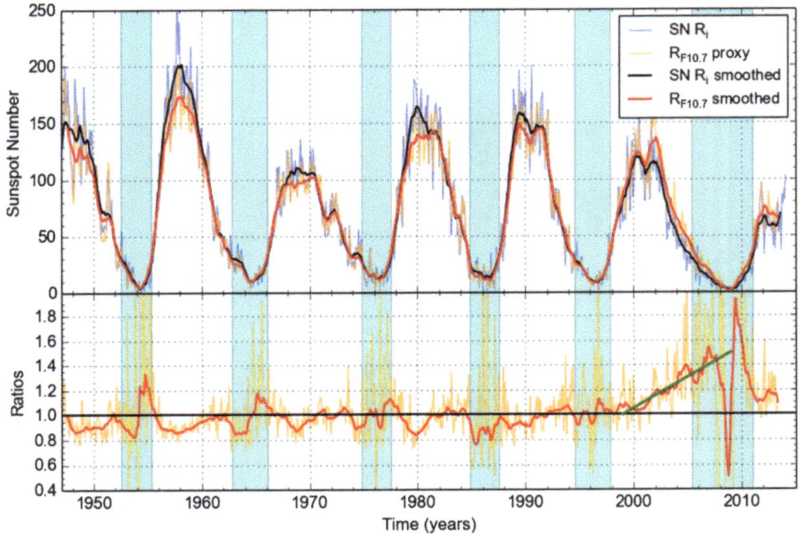

Fig. 48 Plot of the SN R_i (*blue line*) and of the SN proxy $R_{F10.7}$ (*red line*) based on the $F_{10.7}$ radio flux (Johnson 2011). Smoothed curves use the classical 13-month Zürich smoothing function. In the lower panel, the monthly mean ratio $R_{F10.7}/R_i$ (*yellow*) and its 13-month smoothed equivalent (*red line*), showing the unprecedented drift in cycle 23, after the year 2000

1974–1982, rising from 0.97 to 1.08. This change of scale occurs around the time of the transition from the RGO to the SOON data set and is probably due to the different underlying data and methods.

As the reconstructed GN is not affected by such a transition and its scale is expected to remain constant, the jump must therefore reside in the original GN series and be caused by a scaling mismatch between its two base data sets. Such a scaling shift parallels an equivalent mismatch in the RGO-SOON sunspot areas already identified by other analyses (Hathaway et al. 2002; Foukal 2013). Although the RGO scale is slightly lower than the reconstructed GN, it seems that the main bias is an overestimate of the group counts in the SOON catalog after 1975. This may be due to different group splitting rules applied for this data set (meant primarily for the real time numbering of active regions) versus the earlier RGO catalog, which also uses another group classification scheme than the McIntosh classification in SOON data (Willis et al. 2013).

Therefore, a 10 % reduction of the GN must be applied after 1975 to bring it to the same scale as the GN and SN in the first part of the 20th century.

6 The R_i-Locarno Drifts (1980–2014)

Following the unexpected low activity during the cycle 23–24 transition and the anomalous evolution of various solar, heliospheric or ionospheric indices, different comparisons were made recently between the SN and the $F_{10.7}$ radio flux (Svalgaard and Hudson 2010; Lukianova and Mursula 2011; Lefèvre and Clette 2011; Clette and Lefèvre 2012). They show an unprecedented divergence between those two indices starting around 2000, i.e. just after the maximum of cycle 23, with the SN falling 20 % below its standard $F_{10.7}$ proxy (Fig. 48). This motivated an investigation about a possible bias in the SN series.

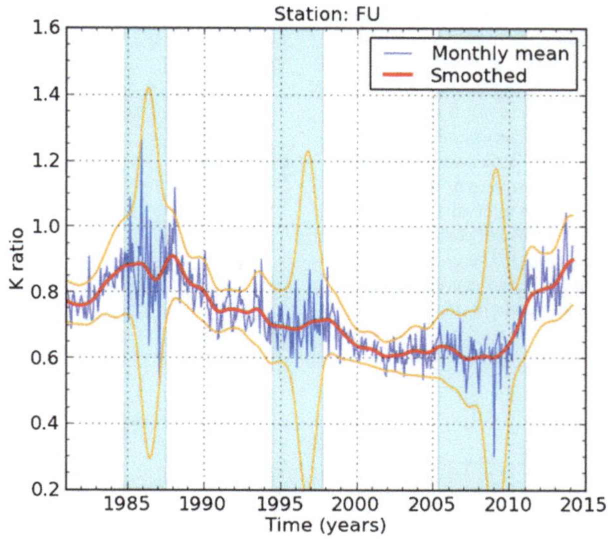

Fig. 49 Sample plot for one station (K. Fujimori, code FU) showing the variation of the k ratio relative to the Locarno-based SN: monthly means (*green line*), and 13-month smoothed values (*red line*). The standard rms error is indicated by the *orange lines* and reaches maxima at the times of solar cycle minima, when the SN becomes small (*blue-shaded bands*)

6.1 The Drift Diagnostics: The Global Network-Averaged k Coefficient

For this analysis, we used the entire 32 year archive of raw sunspot observations available at the WDC-SILSO. In order to ensure a good overlap between individual series, we restricted the data to the 80 stations that contributed continuously for more than 15 years. Among them, 16 stations provided data for the full 32 year period and sometimes even earlier, including e.g. Locarno, Uccle (Brussels) or individual observers like Kenichi Fujimori (Japan).

Our analysis consists in computing the monthly average k coefficient, i.e. the average of the daily ratios between the International SN and the raw Wolf number from each station for the same date. In order to filter out the monthly variability and extract the long-term trends, we smoothed this ratio with a Gaussian smoothing function with a FWHM of 13 months. As an example, Fig. 49 shows the resulting k series for the K. Fujimori station, one of the long-duration stations.

Knowing that the long-term scale of R_i is defined by the Locarno Wolf numbers (cf. Fig. 7), we wanted to check if a drift could be detected in the reference Locarno series. Therefore, we made global comparisons between network-wide averages of the raw k data and this reference series. We base our analysis on the assumption that scale variations of individual stations are uncorrelated. In this case, random individual errors cancel out in a global average and any remaining trend in the average k ratio can only reside in the common reference, i.e., the Locarno series. The absence of correlation among our sunspot observers is a reasonable assumption as they observe independently and have no way to communicate globally in real time and thus to influence each other, as they are dispersed all over the globe. This negligible mutual correlation is confirmed by simple cross-correlations between the actual individual k series.

In order to build the network average, in a first step, we scale all series so that their average k ratio is equal to unity over the time interval 1987–1995 (cycle 22). As the average k coefficient for different stations can differ by as much as a factor of 3, this first normalization ensures that all stations are included with the same weight in the subsequent global average k series. The time interval for this first normalization was chosen because this is

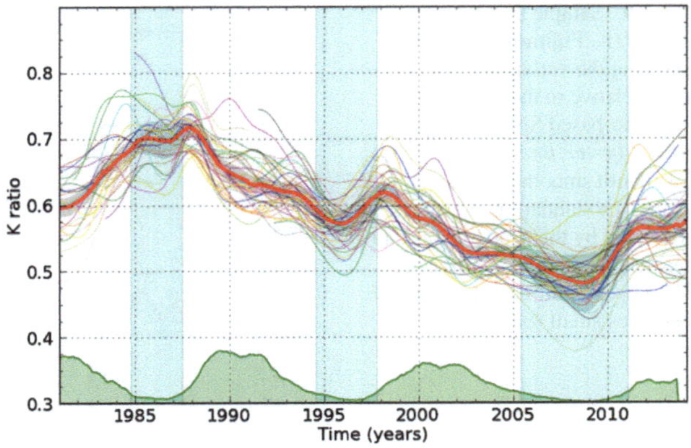

Fig. 50 Plot of the normalized 20-month Gaussian smoothed k ratios for all stations used for producing the average k profile (*thick red curve*). It shows the dispersion between individual series. All series were normalized by a least-square fit over the entire time interval 1981–2013 (single constant scaling factor for each series). The *gray shading around the red curve* gives the standard error on the average k value

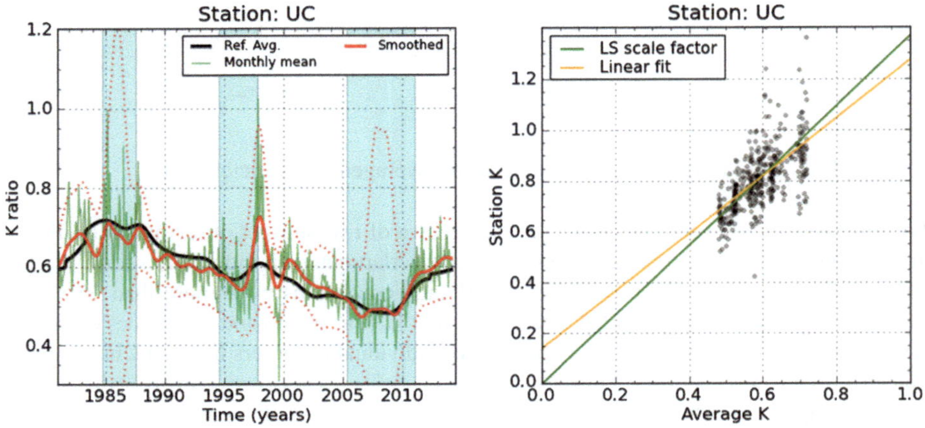

Fig. 51 Sample plot of the k coefficient series for the Uccle station (ROB, Brussels). *Left panel*: monthly means (*green line*), and 13-month smoothed values (*red line*) are compared to the average k curve based on multiple stations (*black line*; see text). The original UC k series has been rescaled only by using a constant factor to fit the average k curve. *Right panel*: scatterplot of the Uccle k coefficient versus the average multi--station k coefficient. Two linear fits are shown: linear fit with intercept at the origin (strict proportionality, *black line*) and normal linear fit (*orange line*). The close match between the two fits indicates that the Uccle k coefficient is closely proportional to the global multi-stations average k variations

when a maximum number of stations were simultaneously observing and it corresponds to a maximum of solar activity, thus providing more accurate average k ratios.

Then, in a second step, all normalized k profiles are averaged to obtain a first average k profile (Fig. 50). As this profile now spans the entire 32-year interval, all k series can be normalized more accurately by a least-square fit to the average profile for the entire observing period of each station. Some additional stations that were not observing in the 1987–1995 interval can also be added at this stage. Then, a final average k series can be computed by averaging all individual normalized k series.

In the process, we can check the quality of fit of each individual station relative to the average profile (linear correlation coefficient, intercept with origin). Figure 51 gives an example of the linear regression for the Uccle station. This fitting step also allowed us to rank the stations according to two criteria: length of the observations and stability of the k coefficient. Note that we used here a standard regression of the station k versus the average

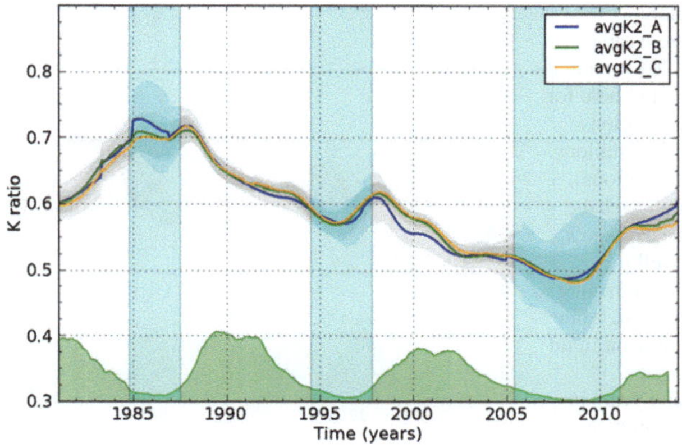

Fig. 52 Plot of three different multi-station average k ratios. Three subsets of stations have been chosen, starting with the most reliable and long-duration stations (set A; 9 stations, in *blue*) and then adding more stations of shorter duration and lower stability (set B, 37 stations, in *green*; set C: 79 stations in *orange*). The *gray shading* gives the standard error of the average k ratio, which increases at cycle minima (*blue shading*). Although all three sets are very different, all curves display the same drifts and reversals, including local features

network k, thus neglecting the error in the average network k, which is a fair approximation as the latter has a much smaller error than the k of an individual station.

In order to check the influence of the station quality, we repeated the above normalization and averaging process for various overlapping and non-overlapping subsets of stations. The resulting average k series are shown in Fig. 52, for successively larger sets of stations (but integrating progressively stations with shorter duration or lower stability). Except for local discrepancies of maximum 10 %, they all show the same reclining "Z" profile: a first rise from 1981 to 1987 is followed by a quasi monotonous decline up to 2008 and by a recent return to a value close to the initial 1981 value. The most reliable profiles (green and orange) give an amplitude of those drifts of ± 15 %. As this variation of the ratio is common to all stations, the source of the trends can only be attributed to the Locarno station. Given the close consistency of all average k profiles and assuming that the initial Locarno k coefficient in 1981 was matching the Zürich scale, we rescaled our average k profiles to start at 0.6 in 1981.

Finally, another feature of the average k profile are the dips coinciding with the minima of solar activity, which are superposed on the overall reclining "Z" shape. As we found a significant cycle dependence of the weight factor (Fig. 44) in our parallel-counts study over the 2003–2013 period and as we know that none of the SILSO stations except Locarno itself are using weighted counts, this ~ 15 % modulation can most probably be attributed to the Locarno weighting effect. Such a cycle modulation can be expected as during cycle minima, most of sunspot groups are small and contain small spots. There are almost no groups of Zürich type E, F, G and H, which contain spots with large penumbrae. The weighted counts should then hardly differ from unweighted counts, while over the rest of the solar cycle, they should lead to inflated k values, causing the observed dips. Further analysis will be required to fully confirm this connection.

6.2 Extended Study (1955–2013)

In order to verify if any similar scaling drift was present before 1981 and if the scaling experienced a jump during the 1981 Zürich–Brussels transition, we extended the analysis

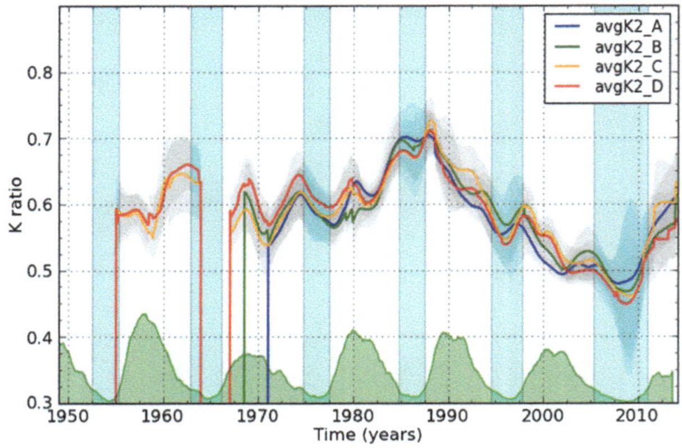

Fig. 53 Plot of different multi-station average *k* ratios spanning the 1955–2013 interval, equivalent to Fig. 52 but here for 4 different subsets of stations (A: 3 best stations; B: 6 stations; C: 16 stations; D: 22 stations). The *gray shading* indicates the standard error on the average *k*. All curves overlap well the interval 1981–2013, confirming the drifts over that period, while before 1981, no significant trend is present. The data gaps correspond to missing years in the recovered Zürich archives

back to 1955, by pre-pending raw data collected during the late Zürich period. For this purpose, we encoded original paper reports that were provided to the WDC-SILSO through the Specola Solare Observatory. Of course, the number of stations is much lower than after 1980 (22 long-duration stations). There are also gaps in the data series, in particular for years before 1965. The recovered data are only from auxiliary stations and from Locarno, as we were unable to locate original detailed counts from the Zürich station itself. Notably, a few stations provided very long series spanning almost the entire 60 year period, including Locarno, Catania, Uccle, Kanzelhöhe, etc. and a few other stations straddle the critical 1980 transition, providing a key reference for the continuity at that time.

By applying the same normalization and averaging method, we obtained an extended *k* series for four combinations of stations (Fig. 53). The series show larger random variations as can be expected from the lower sample of stations. However, the *k* ratio does not show any long-term trend before 1980, while the same reclining "Z" profile is again found after 1981. The drift thus starts once the Zürich reference is replaced by the Locarno station. As the average *k* ratio oscillates around 0.6 until 1981, without any significant discontinuity on the transition year, we have also a confirmation that the initial scale of the new Locarno-based International SN matched the scale of the preceding Zürich SN and that no significant scaling jump was associated with the Zürich–Brussels transition.

6.3 Possible Causes: Current Interpretation

As the above detected trends must be due to changes in the Locarno counts, we investigated directly the possible causes of such biases using the information in the Locarno observing logs. A first possible cause of long-term trends in the Locarno Wolf number is a change in the instrument. We had the confirmation that no modifications were made to the instrument since its installation and that there were no changes in the observing method, i.e., visual count on the aerial image at the eyepiece with the aperture stopped down to 80 mm (Sergio Cortesi, personal communication).

Another factor may be a systematic change in the image quality (local seeing conditions). As a quality index is attributed to each observation (Kiepenheuer scale, decreasing from 5 to 1 for improving sky conditions; Kiepenheuer 1962), the yearly average quality index can be plotted as a function of time (Fig. 54). This curve shows mainly a 0.25 jump around 1969. This sharp degradation was clearly associated with the construction of a large villa just south of the observatory. Thereafter, there is only a slight degradation by 0.15 mainly over the interval 1989–2000.

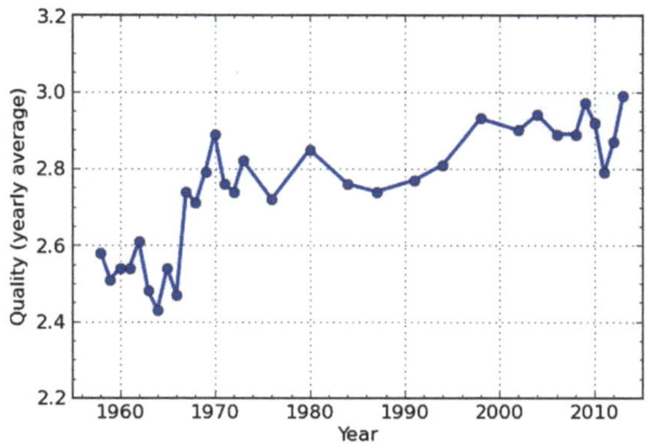

Fig. 54 The variation of the annual average quality index of the visual observations at the Specola Observatory in Locarno (Kiepenheuer scale: higher values correspond to lower image quality). A sharp jump occurs in 1968, followed by a smaller progressive degradation

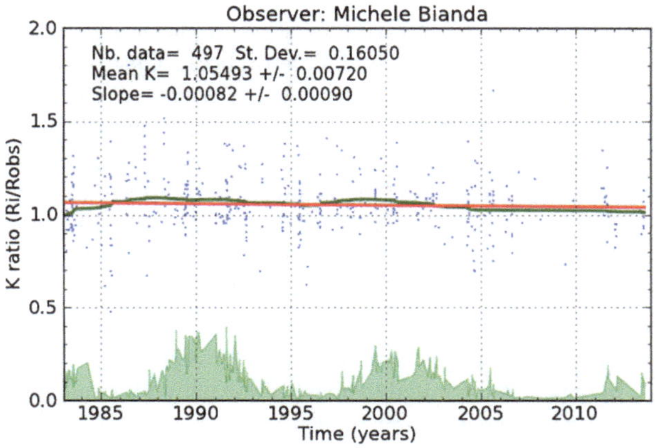

Fig. 55 Sample plot of the relative *k* coefficient of an auxiliary observer of the Specola station in Locarno (Michele Bianda) versus the SN (tied to the prime Locarno observer, Sergio Cortesi). The *red line* shows the linear fit to the data, while the *green curve* results from a 5-year running mean. No significant linear trend is found over the whole 1983–2009 interval. The parameters of the linear regressions are listed in the plot and the *green-shaded curve* gives the evolution of the solar cycle, as time reference

Overall, the changes in seeing conditions do not match the pattern of changes in the *k* ratio (times of inflexion and reversal of the trends). Moreover, the largest change takes place before 1980 and did not produce a significant jump in the Locarno *k* coefficient, making it unlikely that the smaller variations of the average seeing after 1980 could explain larger deviations of the Locarno scale.

A last factor could be a change in the eyesight of the lead observer, Sergio Cortesi, who observed during the remarkable duration of more than 55 years. The only material available to check this is a comparison between the Wolf numbers obtained by S. Cortesi and those obtained by auxiliary observers who occasionally replace him. Those auxiliary observations were often made for single days interleaved with those of the main observer. Figure 55 shows the *k* personal coefficient for the observer who accumulated the longest series (Michele Bianda). Just like similar *k* series for 3 other auxiliary observers, those *k* values only show very weak mutual trends. This comparison thus seems to contradict a possible change in the counts made by S. Cortesi.

However, when considering the high consistency between counts from all Locarno observers, we must take into account the fact that all Locarno observers are in close daily interaction, by contrast with the widely dispersed observers in the international sunspot network. Therefore, the observations made by auxiliary observers may be influenced by preceding observations made by other observers, primarily S. Cortesi, when trying to remain consistent with the counts of past days (M. Cagnotti, personal communication).

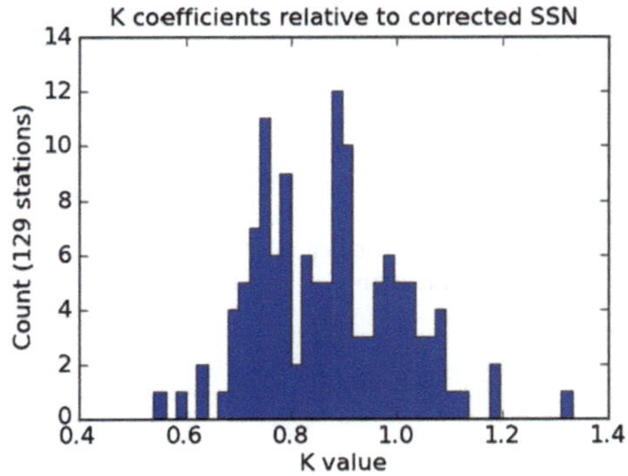

Fig. 56 Histogram of the k personal coefficients of all stations after correction by the average correction for the Locarno drift. It shows a multimodal distribution with several well-marked peaks

Moreover, a well-documented change in the observing routine took place progressively since 2005. As a careful preparation of the future replacement of S. Cortesi by Marco Cagnotti, the new Director of the Specola Observatory, the latter has been trained and is progressively taking over a larger fraction of the observations since 2005. Currently, M. Cagnotti has effectively become the prime observer for the Locarno station. This change matches well the observed recent reversal of the network-averaged k ratio, which increased systematically since 2005, indicating a progressive rise in the Locarno counts relative to the network average. This favors the following interpretation: over the last decades, the Wolf numbers reported by S. Cortesi decreased slowly due to subtle aging effect and eyesight degradation. The Wolf numbers started to increase again with the progressive contribution by a younger observer, who now essentially records the same sunspot counts as S. Cortesi when he was younger, more than 30 years ago.

In regard to the initial increase of the network k ratio between 1981 and 1984, we speculate, in the absence of direct quantitative evidence, that when Locarno took over as reference station, it started to overcount. Indeed, while in the previous 25 years, Locarno could take the Zürich Wolf numbers as a direct reference to maintain a common average scale on a daily basis, after 1980, it was suddenly in a stand-alone situation. Then maybe, a primary concern of missing small spots led to an opposite effect: more spots were included in the counts. The effect may have been amplified because it happened in the declining phase of cycle 21, when sunspots were actually decreasing in numbers.

6.4 The Need for a Full R_i Re-calculation

In order to assess the consequences of the extracted drift, we first applied a correction to the current standard R_i number by multiplying the series by the average network k ratio. Looking now at the network statistics after applying this drift correction, we can take advantage of the fact that the new k coefficients relative to the network average instead of Locarno will be much more constant over the full 32 year period. It then becomes meaningful to compute an average k over the full duration of the observations from each station. We can then obtain the overall distribution of k coefficients for all stations in the SILSO network (Fig. 56). Surprisingly, the distribution is highly non-uniform with three prominent peaks centered on 0.77, 0.88 and 1.00.

The first peak corresponds to a sub-population of stations that provide the highest sunspot counts. Their Wolf numbers are about 20 % lower than the Locarno values, probably reflecting the effect of the sunspot weighting used only by the Locarno station. The other peaks correspond to higher k values and thus to stations that give lower sunspot counts, with the upper end of the distribution near 1, i.e., stations that report SN numbers similar to the Wolf's original values in the 19th century. Further investigations at the level of individual stations will be required to fully understand what defines the well-marked sub-populations of observers appearing in the k distribution. At this stage, it just illustrates the prospects that are opened by the corrected R_i for refined analyses of the visual sunspot counting process based on the huge data set at our disposal at WDC-SILSO.

However, the corrected sunspot number used here is only an approximation of the true sunspot number that would be calculated using the standardized method described in Sect. 2.3 but based on a new unbiased reference. Fortunately, such an end-to-end re-calculation is possible, as the WDC-SILSO maintains a full archive of all raw observations that were used to derive the R_i number. As in our present analysis, our reference was the average over all available long-term stations, we probably get a close approximation of what would be this new standard R_i series. However, there are two main restrictions:

- Firstly, the use of a massive average prevents an in-depth understanding of the reference scaling, as the individual history of each station is diluted over a large sample. For instance, it would prevent the detection of a slow drift associated with the evolution of the network properties (e.g. evolution of telescopes) due to the steady renewal of its composition, with stations entering and leaving the network. In addition, the current analysis provides the relative variation of the k ratio but does not directly give the absolute scaling relative to the preceding Zürich series (NB: here, we attached the k scale to the Locarno numbers in 1981 but as the reference interval is very short, the statistical uncertainty is high).
- Secondly, by contrast with this retrospective bulk analysis, this approach is not really applicable for the progressive monthly computation of the future sunspot numbers. Diagnostics of systematic station flaws that can be easily established over a 30 year interval, cannot be done instantaneously and are lagging behind by months or even years, as information about the future evolution is lacking.

Our experience thus indicates that our new reference when fully re-calculating the sunspot number from all raw observations should be based on one or a few well-identified high-quality stations, for which we can obtain a good record of their daily practices.

We thus face the question: which reference can we use? Three main options can be envisioned:

- *Correcting the Locarno series*: the prospects for correcting the past observations are limited, because there is no material record for visual counts. We can consider instead using the Locarno sunspot drawings, but those drawings are made independently at full aperture (15 cm) and include all spots, even short lived ones that are neglected in the official Wolf number. Therefore, it will lead to a different Wolf number with a different k personal coefficient relative to the preceding Zürich–Locarno cross-calibration. We thus lose the direct scaling link that was intrinsic to the equivalence between the Zürich and Locarno SNs. Still, this alternate Locarno series may be more homogeneous and the counts could be done without the size-dependent weighting, while still retaining a strong affiliation with the preceding Zürich standards,
- *Choosing an alternate pilot station*: our analysis already hinted at rather stable long-duration stations that tracked well the network average, and thus do not suffer from large deviations. A more focused analysis of individual stations should be done to settle this critical selection. This alternate pilot station should however offer the same guarantees of long-term observations as Locarno for future years. This probably restricts the choice to professional observatories. The Uccle station seems then to be a prime candidate for three

reasons: absence of drifts, long overlap with the Zürich era (since 1940), and sunspot counts based on the drawings from which a detailed catalog is currently in construction and would allow fully documenting the counts.
- *Building a composite reference* based on an average over a core group of high-quality stations: this option would make it more difficult to understand the factors influencing the scaling, but it would avoid the risk of drifts going unnoticed as in the case of a single station. A continuous mutual comparison of the pilot stations would allow pinpointing and discarding any accidental flaw present in only one series. In order to be manageable, this core group should not include more than 4 to 10 stations. Here, as losing one of the stations does not cause a complete disruption, it relaxes the selection criteria for what concerns the longevity of the stations, opening the door to good and dedicated individual amateur observers.

The last two options will be tested in coming months, by comparing the output of different simulations (i.e., full recalculation of the SN over the last 32 years) based on different combinations of pilot stations. The first option first requires a recounting based on the Locarno sunspot drawings, before this new Locarno series can serve as reference and be tested for its stability.

6.5 Implication for the Solar Cycle: A Variable Sunspot Population?

Given the amplitude of the diagnosed Locarno drifts, they can lead to significant biases relative to other solar indices, in particular $F_{10.7}$. In order to assess if this drift can explain the cycle 23 divergence between the SN and $F_{10.7}$ mentioned at the start of this section, we tentatively used the corrected series R_i obtained as above by multiplying the series by the average k ratio.

Figure 57 shows a comparison between the original R_i number, the corrected number and the $F_{10.7}$ proxy $R_{F10.7}$ The corresponding ratios are shown on the lower panel. We find that the corrected series matches better the $F_{10.7}$ proxy up to 2002. This thus gives a confirmation that the correction derived self-consistently only from original sunspot data effectively eliminates a true bias in the SN series. However, while the corrected SN and $F_{10.7}$ come closer to each other after 2002, the difference is only reduced by half and is still significant over the rest of cycle 23. The persistence of a significant deviation suggests that the recent divergence between those indices cannot be simply explained by a flaw in the SN but must be associated with a true change on the Sun.

In order to diagnose the possible cause of this change, Lefèvre and Clette (2011, 2012) exploited the most extensive sunspot catalogs available for the last 28 years in order to derive the occurrence rate of groups or individual spots of different sizes. This study, together with parallel results by Kilcik et al. (2011), finds a significant decrease in the number of small sunspots, by about a factor 2, after about 2000, i.e. precisely when the above index deviation sets in (cf. de Toma et al. 2013a). The affected sunspot population consists of pores without penumbra and with a short lifetime of less than 2 days, both isolated (A and B-type groups) or inside groups containing larger sunspots.

Such a scale-dependent change in the sunspot population should be put in relation with two parallel findings derived by completely different measurements. Using spectroscopy, Penn and Livingston (2011) found a systematic decline of the average core magnetic field in sunspot umbrae over the 2002–2013 period. Combined with the determination of a sharp lower threshold in the magnetic field strength allowing the formation of a sunspot ($\sim$1500 Gauss), they conclude that, should this trend continue, the formation fraction of sunspot will be decreasing due to sunspots vanishing at the lower threshold, truncating the

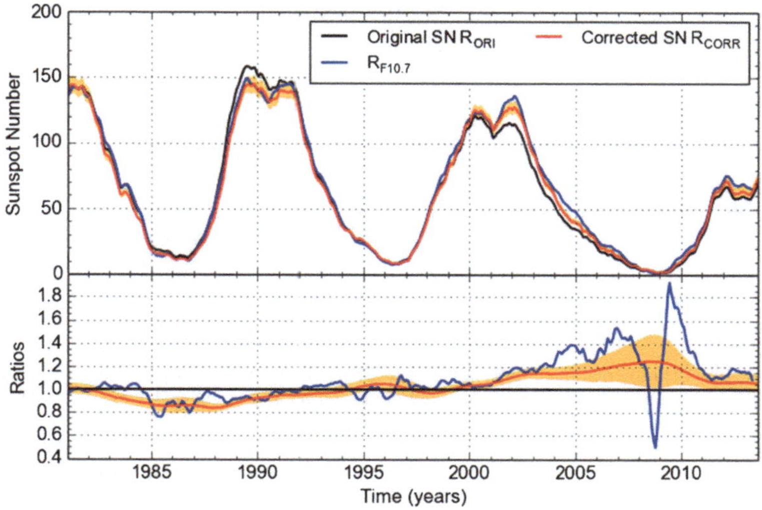

Fig. 57 *Upper panel*: comparison of the original SN series R_{ori} (*black*) with the new R_{corr} series after correction by the average network-wide k ratio (*red*) and the $F_{10.7}$-based proxy $R_{F10.7}$ (*blue*) over the last 3 solar cycles. *Lower panel*: the ratios $R_{\mathrm{corr}}/R_{\mathrm{ori}}$ (*red*) and $R_{F10.7}/R_{\mathrm{ori}}$ (*blue*), with the standard error in the corrected SN (*orange shading*). They show a better agreement between $F_{10.7}$ and the SN after the correction, but still with a significant though reduced divergence after 2002

lower part of the distribution of magnetic fields, i.e., the small sunspots. Although their speculation of a steady long-duration trend is debated (de Toma et al. 2013b), other studies still indicate that the recent trend is part of a particularly deep solar cycle modulation (e.g. Nagovitsyn et al. 2012; Pevtsov et al. 2014).

Using long-duration helioseismic data from the BISON network, Basu et al. (2012, 2013) detect the onset of a similar deviation between the frequency of high-frequency p modes and its SN-based proxy, after 2000. Such modes are confined near the surface and thus record the average magnetization at the surface due to the presence of active regions. They also find that modes having a deeper turning point start to deviate even earlier, already during cycle 22. They interpret this behavior as a thinning of the near-surface magnetized layer. The coincidence of those symptoms may indicate that this size-dependent variation of the sunspot population is real and reflects a physical change in the underlying dynamo processes or in the magnetic diffusion processes leading to the sunspot group decay. Finally, using image data from the RGO and SOON catalog, Javaraiah (2011) finds a significant change in the growth and decay rates of sunspot groups over past solar cycles. Although this analysis did not consider the change in sunspot size population, such global variations of the average rates can be influenced by the change in the relative fraction of small versus large sunspots, i.e. spots with very different lifetimes.

Other solar indicators also tend in the same direction. E.g., for each magnetogram taken at the 150-ft solar tower of Mount Wilson Observatory (MWO), a Magnetic Plage Strength Index (MPSI) value is calculated by summing the absolute values of the magnetic field strengths for all pixels where the absolute value of the magnetic field strength is between 10 and 100 gauss. This number is then divided by the total of number of pixels (regardless of magnetic field strength) in the magnetogram. The magnetic calibration after the instrument upgrade in 1982 is believed to be good, or at least stable (Parker et al. 1997). On average, there is a very nearly linear relationship between MPSI and the sunspot number:

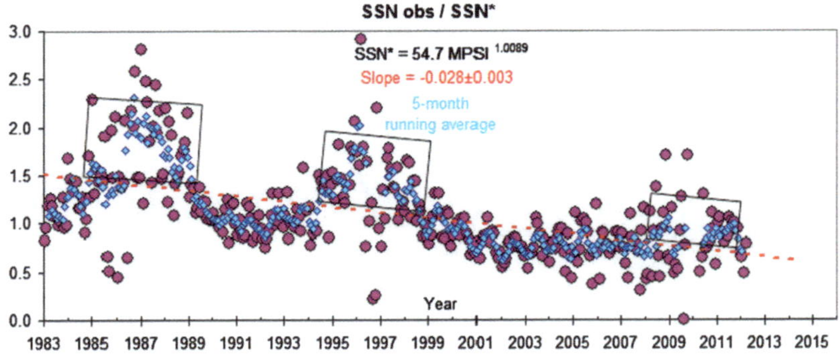

Fig. 58 The observed International Sunspot Number (SSN) divided by a synthetic sunspot number derived from the MWO Magnetic Plage Strength Index (*pink circles* for monthly values). A 5-month, centered running average is shown by *blue diamonds*. The ratio is high in the approach to solar minimum and in the very early part of the ascending phase of the cycle (*large boxes*) before settling down at solar maximum. For solar minimum years, the ratio is between two very small numbers and is thus very noisy and at times undefined. There is also a second-order annual variation of unknown origin, phased with solar distance

$SSN^* = 54.7\ MPSI^{1.0089}$. We can thus calculate a synthetic SSN^* for each (monthly) value of MPSI, and form the ratio between the observed sunspot number and the synthetic one derived from MPSI (Fig. 58). A 5-month, centered running average is shown by blue diamonds. The ratio is high in the approach to solar minimum and in the very early part of the ascending phase of the cycle (large boxes) before settling down at solar maximum where it is well-defined.

Again, we find that the ratio has been declining over the past two cycles, consistent with the similar findings mentioned above. Observations at or after the current maximum should settle the matter of the reality of a secular change. Unfortunately, MWO is currently not taking magnetograms, teaching us the detrimental effect of stopping a valuable synoptic observing program.

As noted by Lefèvre and Clette (2011), such a scale-dependent variation of the sunspot population could match the existence of two dynamo components, e.g. a deep and shallow dynamo. Various recent developments in dynamo theory provide schemes that could be tested against this peculiar evolution of sunspots, Babcock-Leighton near-surface flux diffusion mechanism (Muñoz-Jaramillo et al. 2010, 2011), role of a near-surface shear layer (Brandenburg 2005) or a near-surface magnetic flux aggregation mechanism (Schatten 2009). The selective disappearance of small spots can also have implications on the total and spectral irradiance reconstructions. Indeed, while vanishing sunspots will lead to lower sunspot blocking and thus should increase the TSI, the corresponding weakening magnetic fields should still be present but below the 1500G lower limit found by Livingston and Penn (2009). They may thus contribute to a corresponding excess in the near-UV and microwave emissions by the plage component. The result would be a reduction of the effective irradiance drop for weak solar cycles, which may be related to the existence of a base level in solar flux, as proposed by Schrijver et al. (2011). In any event, these scale and lifetime dependencies of sunspot magnetic fields should be a warning against any simple linear rescaling of the present irradiance proxies by a relation of proportionality.

6.6 Variations in the GN/SN Scaling Ratio

A scale-dependent deficit of sunspots such as the one found in the above studies, should logically lead to a corresponding change in the average number of sunspots per group. However, inherent in the definition of the Group Number is the assumption that the ratio between the number of spots and the number groups, i.e. the average number of spots per group is constant. Therefore, a new question arises: is the fixed 12.08 constant used in the GN definition a valid assumption?

We can investigate this assumption using data from the German SONNE network of sunspot observers (Sonne 2014; Andreas Bulling Personal Communication) and from the long-running Swiss station Locarno (Locarno 2014) supplemented by observations at Zürich by Waldmeier (1968) and Zelenka and Keller (Keller and Friedli 1995). From each data source, the number G, of groups, and the number S, of "spots" reported by the observers is extracted and tabulated. "Spots" is in quotation marks because Waldmeier, as we have shown, and to this day Locarno as well, weighted larger spots stronger than small spots. The SONNE observers do not employ weighting: each spot is counted only once. It is important that for both groups of observers, the counting methods (albeit different) have been unchanged over the period of interest.

If the Relative Number, R, and the Group count, G, are known, the spot count can be calculated as $S = R/k - 10G$, where k is the k-factor introduced by Rudolf Wolf to bring observers onto the same scale as Wolf himself, who by definition had $k = 1$. As explained in Sect. 2.2, for the later Swiss observers k was set by *adoption* to 0.60. The SONNE series is adjusted to match the Swiss k-factor, which, however, is also applied to the group numbers reported by SONNE so that a composite group count can be computed over many observers, effectively resulting in a spot/group ratio that is independent of the k-value. The published data for Waldmeier and SONNE gives us R and G, so S has to be calculated as detailed above. For Locarno, Zelenka, and Keller, both S and G are available directly. Given G and S, either determined directly or calculated from R and G, the average number of spots per group, S/G, can now be computed for each year.

Figure 59 shows that the average number of spots per group has been decreasing steadily for both SONNE and Locarno and is therefore not likely to be due to drifts of calibration or decreasing visual acuity of the primary Locarno observer (Sergio Cortesi since 1957). This is consistent with the conclusion in Sect. 4.2 (Fig. 45) where we compare the weighted counts made by the veteran Cortesi and the new observer Cagnotti from 2008 to the present, and find no systematic difference or variation with time.

Because the Locarno observers weight the spot count according to structure and size of spots, they report more spots that the SONNE observers. This is clearly seen in the bottom panel. Also, in Fig. 59, we plot (top panel) the variation of the ratio between the number of spots and the number of groups for the more than 431,000 SONNE individual daily observations without any correction for k-factors (green line with plus marks). The "raw" data show the same general variation and decline as the adjusted observations. Such a trend could be a solar phenomenon or due to an increasing group count brought about by sharper determination of what constitutes a "group". There seems to be a solar cycle variation as well.

When Wolf chose 10 as the weight for Groups in his definition of the Relative Sunspot Number, he remarked that he could as well have chosen 9 or 11, but that 10 was certainly "more convenient". For Wolfer the ratio spots/groups was on the average 9.0 (Fig. 60). In Fig. 60, the recent decrease of the ratio also seems to be seen at Kislovodsk, supporting a solar cause, rather than a drift due to an individual station.

On account of the weighting, one would expect a dramatic increase ($\sim$40 %) in the ratio when the weighting scheme was introduced. That this is not observed presents a puzzle which at the present time has not been resolved.

In order to confirm the decline over the recent cycles, we derived a reconstructed group number based on the WDC-SILSO archive over the last 6 cycles. The resulting values,

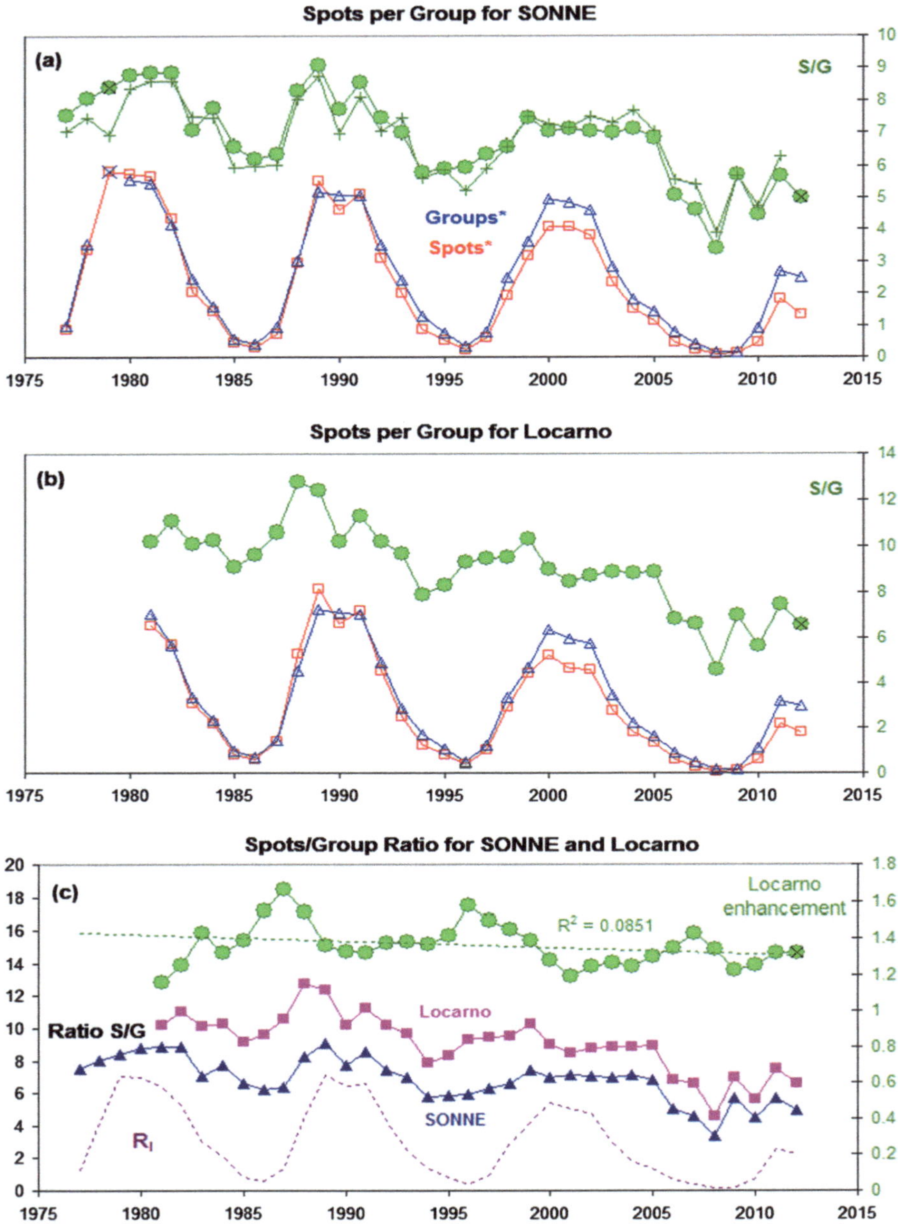

Fig. 59 (*Top*) The number of spots per group as a function of time (*green circles*) for SONNE. The *green curve* with pluses shows the ratio derived from the raw counts, not corrected with *k*-factors. The lower part of the panel shows the variation of number of groups (*blue triangles*) and the number spots (*red squares*) both scaled to match each other before 1992. Note the decreasing spot count, relative to the group count. (*Middle*) Same, but for Locarno. (*Bottom*) The decrease of the ratio Spots/Groups for Locarno (*pink squares*) and for SONNE (*blue triangles*) using the left-hand scale. The enhancement of the Locarno ratio over SONNE (see text) is shown by the *green circles* (right-hand scale). The trend indicated is not significant

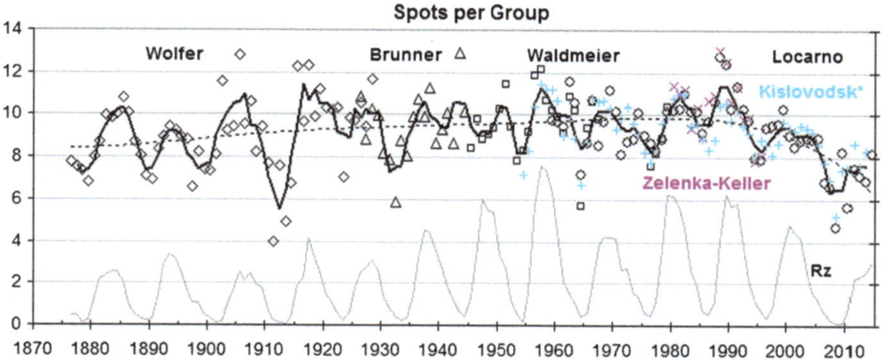

Fig. 60 The ratio between the number of single spots and the number of groups as recorded by the Zürich observers for each year of observation. There is a clear solar cycle dependence (sunspot number shown at *bottom*) with more spots per group at higher solar activity. The ratio observed at Kislovodsk (with no weighting) scaled to Locarno follows the same general trend

Table 1 Average ratio between sunspot counts and group counts and average ratio between the SN and GN over the last 6 solar cycles. Cycles 23 and 24 mark a sharp drop relative to earlier cycles

Cycle	Ns/Ng	North	South	SN/GN	North	South
19	11.3			12.8		
20	10.3			12.2		
21	10.4			12.2		
22	11.9	12.5	10.7	13.2	13.5	12.5
23	9.3	9.5	8.7	11.6	11.5	11.9
24	8.5	8.7	6.4	11.0	11.2	10.3

given in Table 1 and plotted in Fig. 61, the ratio GN/SN stays near 12.5 over cycles 19 to 22, corresponding to an average number of spots per group of 11, i.e. slightly higher than the standard value in the original group number definition. Again, those higher values can be attributed to the higher weighted counts at Locarno, the WDC pilot station. Then the ratio drops sharply between cycles 22 and 23. Thereafter, even during the present cycle, the GN/SN ratio remains at a lower value of 11, i.e. an average number of spots per group as low as 9, which is much below the standard Group Number scale.

A recent long-term reconstruction of the yearly average number of spots per group by Tlatov (2013) suggests that there were rather large variations of this quantity over the past century. In Fig. 61 (right plot), we included the last points from this study. It shows that the SONNE values nicely extend Tlatov's values and that all series share the same downward trend, except for the high value for cycle 22 and the globally higher values for the Locarno-based data set.

The above converging results give a strong indication that the diagnosed deficit of small spots is indeed directly reflected by a drop in the ratio between the SN and GN. Consequently, any interpretation of past disagreements and drifts of the SN relative to the GN, in which the information about the actual group size is absent, should take this significant variability into account. The above analyses indicate that the amplitude of the variations can reach up to 30 %. It may thus prove pointless to try matching the GN and SN series to better than a few percent, in particular for durations shorter than a solar cycle. Instead, the remaining differences may be considered as a useful indicator of true changes in the

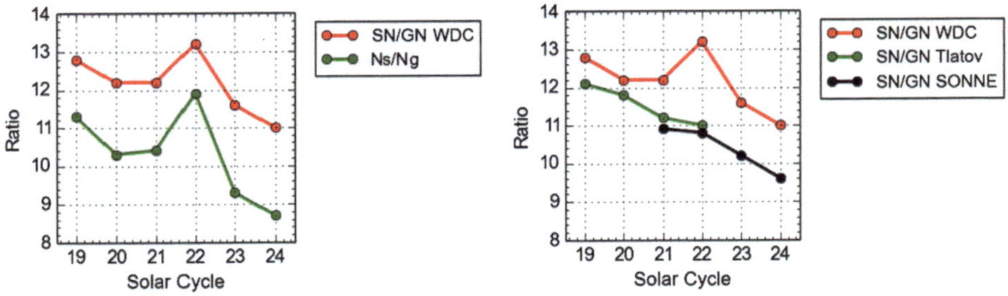

Fig. 61 (*Left*) Variations of the average number of sunspots per group (*green*) and SN/GN ratio (*red*) over cycles 19 to 24, showing a sharp decrease in cycles 23 and 24. (*Right*) Comparison of average number of sunspot per group for different data set: this study (*red*), data from the German SONNE network (*black*) and values from Tlatov (2013) in *green*, all showing a consistent decrease over recent solar cycles

activity regime of the Sun, like the ones that seem to occur right now during the cycle 23–24 transition, rather than irreducible flaws in either series.

Definitely, more work is needed to better track those changes in the past. It will require the use and probably the construction of improved sunspot catalogs, providing detailed properties of individual sunspot groups.

7 The Solar Cycle in a New 250-Year Perspective

7.1 Combining All Corrections: A Tentative Synthesis

In the previous sections, four primary corrections have been identified:

1. The RGO early drift affecting the group number
2. The Waldmeier sunspot weighting bias affecting the recent sunspot number
3. The RGO-SOON scaling bias occurring after 1975
4. The Locarno scaling drift affecting the recent sunspot number

Here, as a first step, we will apply the first three corrections to synthesize a corrected series from 1750 to the present. At this point, we will ignore the earlier part of the series, which is still affected by larger uncertainties. Based on the measured amplitudes and drift intervals, we applied the following modifications to the series:

1. A 40 % progressive ramp between 1880 and 1915, raising all GN values before 1915. The chosen ramp matches the progressive rise in the k ratios between reference visual observers and RGO-based group counts.
2. A sharp 20 % jump in 1947, lowering all SN values after that year. Note that this jump falls near the time of the solar minimum between cycles 17 and 18, i.e. a period when the effect of weighting should be limited.
3. The variable 15 % Locarno trend in the SN starting in 1981.
4. A sharp 10 % jump in 1976, lowering all SOON-based GN values after that year.

As can be seen in Fig. 62, with only those four corrections, the resulting series largely match over the entire 1820–2013 time interval, within the limits of statistical uncertainties. Only before 1820, when observations become sparse and the uncertainties grow and also become more difficult to assess, there are still significant disagreements, in particular for cycles 2, 3 and 4, which will require further analyses.

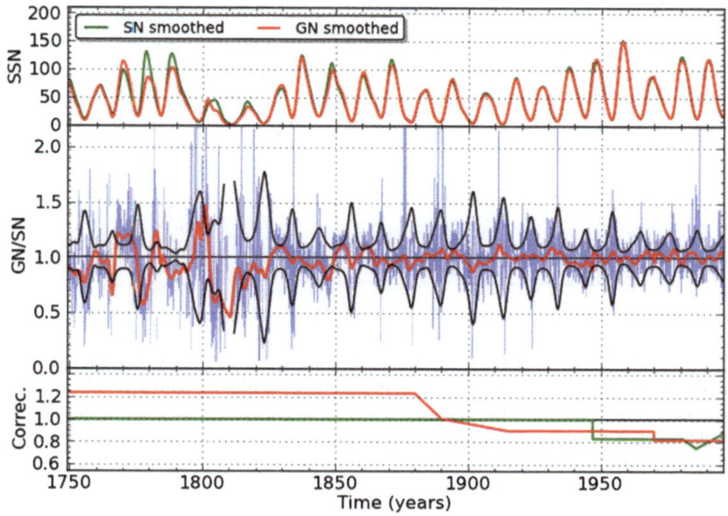

Fig. 62 Comparison of the SN and GN series after correction of the main biases identified by recent analyses (see main text). The plot corresponds to Fig. 1, with the *lower panel* showing an initial version of the corrections (*red* for correction to the GN, and *green* for the SN). The ratio in the *middle panel* is largely uniform from the early 19th century onwards, now remaining within the uncertainties. Local significant deviations still remain in the 18th century, when observations are sparse and uncertainties become large and difficult to estimate precisely

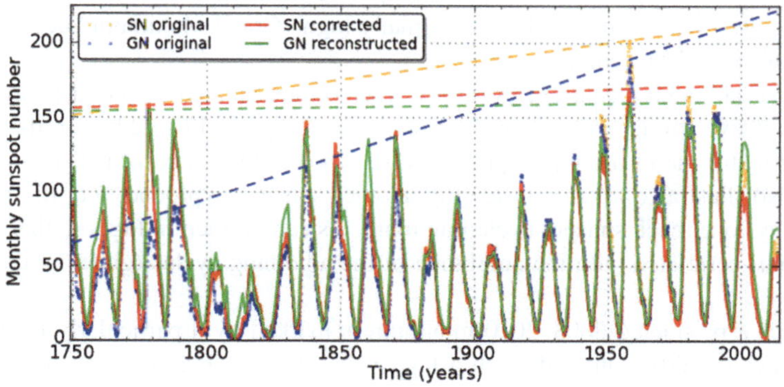

Fig. 63 Comparison of the original and corrected SN and GN series over the entire interval 1749–2013, showing the limited difference in maximum cycle amplitudes between the 20th century and previous centuries after the new corrections. In order to better visualize the trends, *dashed lines* connect the highest maxima of the 18th and 20th century, for each series of the corresponding color

7.2 Implications for Long-Term Solar Activity: No Trend in Solar Cycle Amplitude?

After corrections, the GN and SN series mostly agree over the last 250 years. Moreover, when compared with the base SN series and original GN series, both the backbone recalibrated GN, shown in Fig. 28 and the equivalent corrected SN series plotted in Fig. 63, indicate an important consequence: the secular trend in solar cycle amplitude, shown by the dashed in Fig. 63, is strongly reduced after applying the corrections.

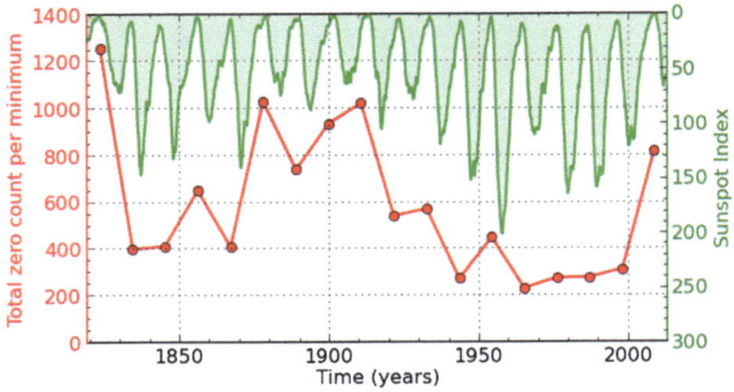

Fig. 64 Cycle-to-cycle variation of the total number of spotless days from cycle 6 to 24, for which daily sunspot numbers are available (*red curve*). The SN series is over-plotted with a reversed scale to highlight the strong anti-correlation between this indicator and the amplitude of the solar cycle. The count for the cycle 23–24 minimum is similar to the late 19th century. The first value for the cycle 5–6 minimum (Dalton minimum) is much larger

Now, except for the highest recorded cycle (19), the maxima of highest cycles of the past centuries are essentially the same as the recent maxima of the late 20th century. We note that recent independent reconstructions of the Sun's open magnetic field, based on the geomagnetic record, also show a very limited difference of the highest peak 11-year amplitudes occurring in the 19th and 20th centuries over the available 1840–2010 interval (see Fig. 30 in Lockwood 2013). Therefore, the upward trend in solar activity levels between the 18th and 20th that was adopted in many past interpretations and models is now questioned, as well as the associated concept of an abnormally high "Grand Maximum" occurring in the second half of the 20th century.

However, although recent cycles do not reach unprecedented amplitudes anymore, the repetition of five strong cycles over the last 60 years (cycles 17 to 22, with the exception of cycle 20) still marks a unique episode in the whole 400-year record. This unique character is also illustrated when considering another sunspot byproduct, i.e. the number of spotless days over each sunspot cycle minimum. As can be seen in Fig. 64, this number is strongly anti-correlated with the amplitude of the adjoining cycles (given by the reversed green curve).

The recent protracted 2008–2010 minimum is marked by a particularly high count of spotless days (~800). Although such a high value was not reached since the early 20th century, the plot shows that it was largely exceeded in the cycle 5–6 minimum belonging to the small Dalton minimum. Therefore, although it marked a contrast with recent minima, this last minimum was not particularly exceptional in a long-term perspective. On the other hand, the uninterrupted series of low spotless day counts during the last 6 cycles stands out as a unique episode over the last 250 years and most probably the last 400 years.

7.3 Cycle 24 in a 250 Year Perspective

As the cycle 23–24 minimum seems to mark a transition in the long-term solar activity, it is interesting to put the current evolution of cycle 24 in context, in order to identify the range of possible scenarios that may develop for the rest of this cycle. Unfortunately, current statistical or physics-based models of the solar cycle can only reproduce global properties of

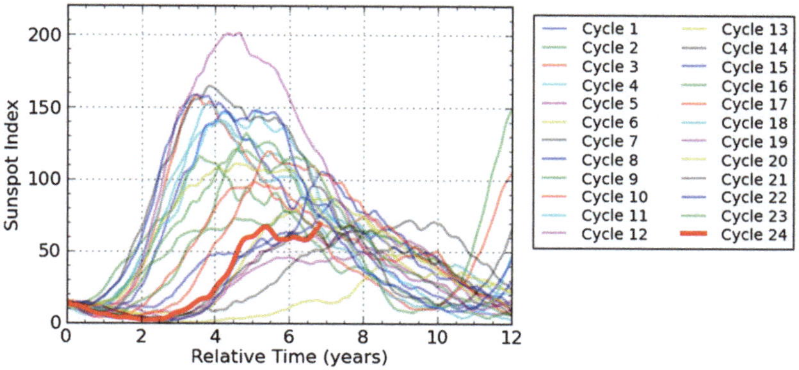

Fig. 65 Plot of the 12-month Gaussian smoothed SN for all cycles (1 to 24) after aligning each cycle on the SN = 13 crossing point in the final decline of the preceding cycles. Cycle 24 (*thick red line*) clearly matches cycles that have a late start and moderate amplitude

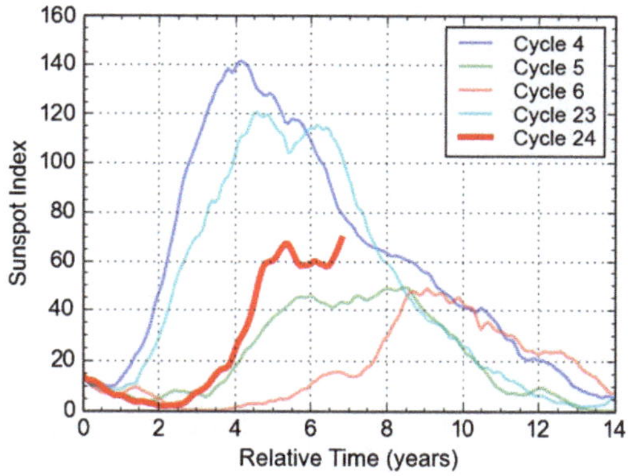

Fig. 66 Plot of the 12-month Gaussian smoothed SN for cycles 4, 5 and 6, compared to cycle 24 (*thick red line*). Like in Fig. 65, cycles have been aligned on the SN = 13 crossing point in the final decline of the preceding cycle. Cycles 5 and 6 belong to the Dalton minimum, while cycle 4 was the last cycle preceding the Dalton minimum

the cycle, with typically a smooth temporal evolution with a single maximum preceded and followed by a monotonous rise and declining phase. This is where the SN series provides unique information about the detailed evolution of the 24 past solar cycles.

In Fig. 65, using the original SN series, we aligned all cycles (from 1 to 24) at a common tie point (SN = 13) on the declining phase of the previous cycle. As expected, cycle 24 is among late cycles preceded by a protracted minimum. Although the cycles are spread over a continuous range of amplitudes, the plot also shows two concentrations of cycle profiles that share similar rise phases: fast rising cycles reaching high maxima (SN > 130) and slowly rising cycles reaching moderate amplitude (SN $\sim$ 60). Cycle 24 clearly belongs to this second group.

In Fig. 66, we specifically compare the current cycle with weak cycles 5 and 6 belonging to the Dalton minimum. The rise of cycle 24 is much steeper and even though it is still unsure whether the maximum has actually been reached, the recent SN values clearly exceed the maxima of cycles 5 and 6. Considering now cycle 4 that preceded the onset of the Dalton minimum, while cycle 23 had a similar amplitude, cycle 24 is again strikingly different. Therefore, the peculiar evolution of the current cycle does match the characteristics of the Dalton minimum and cannot be interpreted either as heralding a subsequent extended minimum.

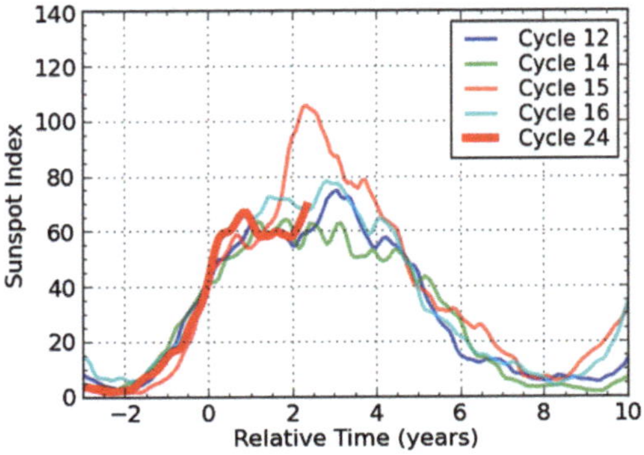

Fig. 67 Plot of the 12-month Gaussian smoothed SN for cycles 12, 14, 15, 16, which provide the best match with cycle 24 (*thick red line*). Here, cycles have been aligned on the middle of the rise phase (at SN = 40). Except for the sharp peak in cycle 15, the other cycles all show the same evolution with a rather long and flat maximum

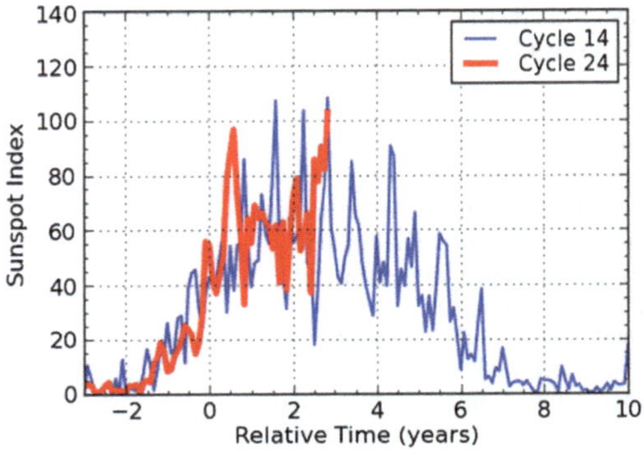

Fig. 68 Plot of the monthly mean SN for cycles 14 and 24, after aligning them according to the SN = 40 crossing point of the 12-month smoothed SN. Both cycles show a rather flat maximum formed by successive short peaks of activity, lasting each only about 2 to 4 months

Actually, the only cycles that closely match cycle 24 are shown in Fig. 67, where we aligned cycles in a tie point (SN = 40) in the rise phase of the cycles. Those cycles (12, 14, 15 and 16) all belong to the late 19th and early 20th century. Cycle 24 thus seems to be a return to an average level of activity that prevailed during a 60 year period. Those moderate cycles are characterized by extended and rather flat maxima, lasting up to 4 years. The maximum phase is typically marked by a plateau, on which several peaks are superimposed. Consequently, the absolute maximum of the cycle can then occur quite late, like in cycles 12 and 16, when it occurred more than two years after reaching the plateau phase and more than 5 years after the start of the cycle.

This is better shown if we avoid any smoothing. Although the overall trends then become more difficult to discern, the detailed progression of the activity is fully preserved. As illustrated for cycle 14 (Fig. 68), the culmination of moderate cycles is actually formed of a succession of multiple peaks or activity surges, up to 6, making the determination of an actual maximum quite elusive.

Cycle 15 is of particular interest, as it shares many characteristics with cycle 24 (Fig. 69): a steep rise to a first peak followed by an extended plateau near $R_i = 60$, lasting for almost 2 years. Cycle 15 then had a second rise towards a much higher peak, at twice the level of the initial plateau.

The broad sample of solar cycles recorded in the SN series thus brings unique information that should remind us of evolutionary scenarios that should not be overlooked when attempting solar cycle predictions.

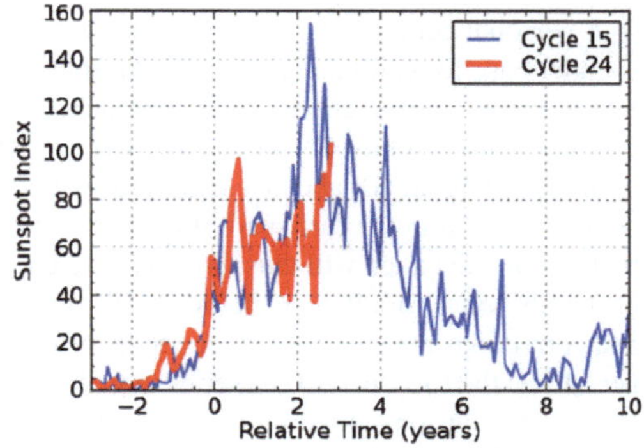

Fig. 69 Plot of the monthly mean SN for cycles 15 and 24, after aligning them according to the SN = 40 crossing point of the 12-month smoothed SN. Both cycles were characterized by an initial steep rise that is abruptly interrupted and is followed by a long constant plateau lasting for about 18 months. Cycle 15 then entered a second late rise phase to the actual maximum

8 Conclusions

8.1 New Series and New Insights

In the course of the preceding sections, we tried to show that the sunspot record provided by the SN and GN series is currently the focus of revived interest, far from the frozen image evoked in the introduction. This revival is not only due to a renewed interest in the sunspot record itself, to the availability of newly recovered historical data or to new data processing capabilities. The prime drivers are the recent realization of our very limited capacity to predict the long-term evolution of the solar cycles and the need to quantify the role of the main natural forcing on Earth's climate. Although the solar physics community now clearly identified those important motivations (development of "Space Climate" as a new discipline), the awareness at the level of science funding organizations is still lagging behind. This is why the first initiative aiming at a revision of the SN took the form of Sunspot Number Workshops, an informal and unfunded coordinated effort gathering a community of experts in the field. Let this be our plea for increased support to these fundamental long-term observations and studies of solar and Sun-Earth processes.

Here, we reported about the main advances that were brought by those SN Workshops in just three years, since September 2011. The main biases and drifts have been identified, affecting either the sunspot number or the group number. This first end-to-end recalibration of the SN series since its creation by R. Wolf thus marks an epochal step. It should be pointed out here that only intrinsic diagnostics were used, i.e. analyses based exclusively and directly on the sunspot observations themselves and on original information about the procedures used to create those indices. Comparisons with other parallel and more indirect indices were used when possible but only to provide support and confirmation to the primary diagnostics and not as calibration references. This allowed us to establish the robustness of the diagnosed corrections and to provide an essential validation, given the scarcity of original sunspot data. While the main corrections are now well established, various results discussed in this chapter are still "work in progress". Therefore, some of the discrepancies and puzzling findings identified here point to areas where more effort will still be needed in the future.

Using the new corrections, we show that the sunspot number and group number series can be largely reconciled after applying mutually-independent corrections to each series. Still, discrepancies remain for the early series, before the 19th century. More work is thus still needed to recover new information and improve the interpretation of this early part. Still, as the scaling of the sunspot series is largely established by a backward reconstruction of k personal coefficients, the new improvements to the last two centuries provide a new sounder base for calibrating the early sunspot number, back to the Maunder Minimum and first telescopic observations. New approaches like the "backbone" method presented here can now lean on robust SN values over the last 200 years.

Reconciling the SN and GN series is also essential because almost only group counts can be recovered from the early part of the telescopic sunspot record. The GN series assembled by Hoyt and Schatten will thus remain a reference, though now as the basis for an extension of the SN series for all years before 1750. Here however, we showed that various revisions and corrections are still needed, due to incorrect interpretations of original manuscripts, especially in the early part of the series (e.g., from 1610 to 1750 approximately). In particular, significant revisions based on the work of Vaquero and others are needed for observations within the Maunder Minimum (1645–1715 approximately). We also demonstrated that part of the early sunspot measurements were only meant for the astronomy of position (e.g. solar meridian altitudes), making accurate sunspot counts from such measurements (even spotless dates) extremely difficult. In fact, a "modern" example (using data recorded in the 19th century) has shown that the solar meridian altitudes should be used with extreme caution for the reconstruction of solar activity.

Regarding the impact of the new SN recalibration, the most prominent implication is the significant reduction of the upward trend in the average amplitude of solar cycles that was present in the original GN series between the 18th and the late 20th century. The recalibrated series indicates that 11-yr peak SNs during the 18th and 19th century were comparable to those observed during the recent interval of strong activity during the second half of the 20th century. The scenario of the initial post-Minimum recovery is still uncertain, as the exact amplitude of the first cycles of the 18th century remains difficult to establish given the scarcity of observations over that period. Still, the vanishing upward trend over the last 250 years questions the existence of a modern "Grand Maximum" in the 20th century (Solanki et al. 2004; Abreu et al. 2008; Lockwood et al. 2009; Usoskin et al. 2012, 2014), which resulted primarily from the erroneous transition between Wolf and Wolfer in the Hoyt and Schatten GN time series. As this "Grand Maximum" concept rests on the occurrence of out-of-range amplitudes of the solar cycle, it is definitely contradicted by the re-calibrated and reconciled SN and GN series.

Still, although the levels of activity were not exceptional except maybe for cycle 19, the particularly long sequence of strong cycles in the late 20th remains a noteworthy episode. Indeed, the 400-year sunspot record and one of its by products, the number of spotless days, show that such a tight sequence of 5 strong cycles over 6 successive cycles (from 17 to 22, except 20), which we can call the "Modern Maximum", is still unique over at least the last four centuries. Given the inertia of natural systems exposed to the solar influences, like the Earth atmosphere-ocean system, this cycle clustering could still induce a peak in the external responses to solar activity, like the Earth climate. However, we conclude that the imprint of this Modern Maximum (e.g. Earth climate forcing) would essentially result from time-integration effects (system inertia), since exceptionally high amplitudes of the solar magnetic cycle cannot be invoked anymore. In this suggested revision, the estimated or modeled amplitude of the effects, including the response of the Earth environment, can be quite different, necessarily smaller, and should thus be re-assessed.

The recalibrated series may thus indicate that a Grand Maximum needs to be redefined as a tight repetition/clustering of strong cycles over several decades, without requiring exceptionally high amplitudes for those cycles compared to other periods.

The residual differences between the SN and GN may be considered as random noise. However, based on the abundant sunspot data from recent years, recent statistics indicate that a substantial change in the sunspot population occurred during cycle 23 and continues into the current cycle 24. This change is reflected by a 30 % decrease of the average number of sunspots per group. This recent and well documented evolution is concomitant with parallel changes in other solar proxies and suggests that in different regimes of activities, the mutual relation between solar reference indices and proxies may vary. Therefore, the ongoing changes in sunspots invite us to consider the possibility that past variations in the ratio between the SN and GN may reflect true changes in the solar dynamo. This also warns us that many standard proxies used nowadays and mostly based on the last 3 to 5 cycles may not be valid for earlier times and should not be simply extrapolated from recent solar conditions.

Finally, the sunspot number series also provides us a direct picture of the actual evolution of a large sample of past cycles. A direct comparison of the current cycle 24 to equivalent mid-amplitude cycles, indicate that this evolution can strongly differ from the smooth profiles typically delivered by statistical or physical models of the solar cycle. Prominent features like multiple surges in a broad maximum or a two-step rise should not be overlooked when attempting forecasts and should ultimately be reproduced by theoretical models, before we can claim any reliability in future cycle forecasts. In that sense, the sunspot number series does not only provide a numerical reference for the global analysis of past cycles, it also feeds the current and future quest of new concepts to explain the complexity of the solar dynamo. Moreover, by providing a direct measure of sunspot activity spanning multiple centuries, the sunspot number provides a link to millennia of cosmogenic nuclide data. If we are able to understand the relationship between these time series for the last 400 years, then we have a potential key to unlock a much longer record of solar activity.

8.2 Future Work and Prospects

Now, in order to move forward and assemble a fully revised and extended SN series, three main steps are required:

- Combining all corrections obtained independently and verifying the global consistency of the resulting series.
- Determining more accurately the magnitude and time domain of each correction, and deriving estimates of the uncertainties.
- Extending the original SN series, so far limited to 1750, back to the first telescopic observations. As pointed out earlier, the scarcity of data will most probably impose the use of the group number for this early part. It is thus essential that the GN and SN series are brought first in full agreement for the whole interval following 1750.

As pointed out on several occasions in this chapter, several issues remain open and require deeper analyses that may still span many years. The revised series will thus be open to future improvements as new results are published and new historical documents are progressively recovered. Therefore, in order to properly document future occasional modifications, the WDC-SILSO will implement a versioning system, with an incremental description of changes added to each version.

As the recovery of past observations remains a critical element for future progress, we issue here a general call: readers and their hosting institutions who have raw sunspot observations in their archives are invited to send copies (or even originals if their local preservation is not guaranteed) either to the WDC-SILSO (F. Clette; http://www.sidc.be/silso) or to the newly created Historical Archive of Sunspot Observations (HASO) at the Universidad de Extremadura (J.M. Vaquero; http://haso.unex.es). In particular, past reports or corre-

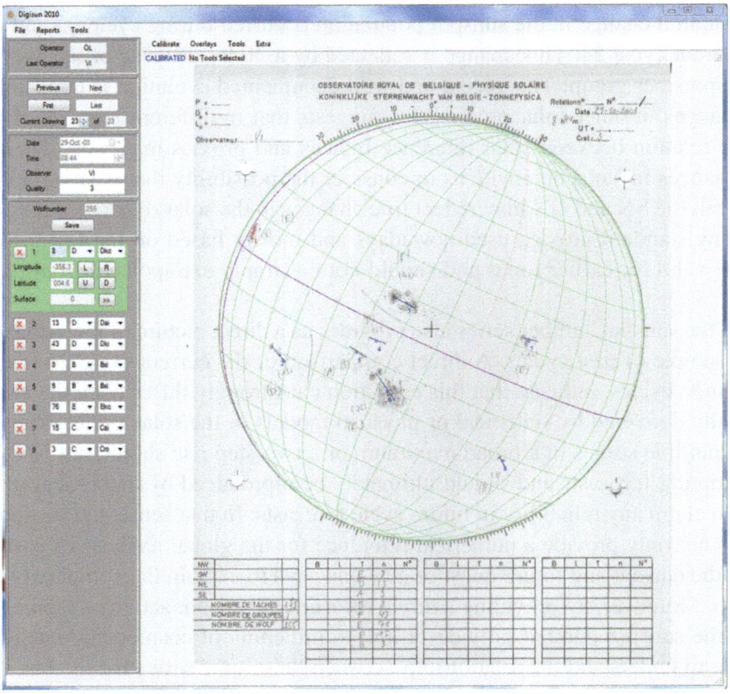

Fig. 70 Screenshot showing the main user interface of the DigiSun application developed at the WDC-SILSO (Brussels) for the measurements of individual sunspot groups in sunspot drawings. This tool will lead to a new detailed sunspot catalog spanning the last 70 years, which is currently under construction

spondence with original sunspot data that were sent to Wolfer or Waldmeier at the Zürich Observatory are of crucial importance.

Looking further into the future, the next step will be to complement the SN series, which only provides 1-dimensional temporal information, with spatial information though probably over a more restricted period starting only in the mid-18th or early 19th century. This could include the global distributions in latitude (hemispheric separation) and sizes, the topology of the magnetic dipoles (width, tilt), and properties of individual groups (growth, decay, morphology). Recent reconstructions of the Butterfly diagrams based on original Staudach and Schwabe drawings by Arlt (Arlt 2009a; Arlt and Abdolvand 2011; Arlt et al. 2013) are tracing the way for future such analyses. New recovered information on sunspot positions in 18th and 19th centuries by Arlt (2009b), Cristo et al. (2011) and Casas and Vaquero (2014) should soon allow further extensions.

The base data exist (drawings, photographic images) although they are dispersed and often not accessible in digital form. Recently, prototypes of sunspot drawing analysis tools have been developed (e.g. Cristo et al. 2011; Tlatov et al. 2014; DigiSun, Fig. 70). Therefore, a major endeavor for the coming years will be the bulk digitization of thus-far unexploited observations and the development of automated feature extraction software. The ultimate goal will be the construction of standardized long-term catalogs of individual sunspots and/or sunspot groups. Such catalogs would allow the creation of new advanced sunspot-based indices and proxies, extending back well before the 20th century.

Thanks to the diversity of the recovered information, various application-oriented indices could be envisioned, complementing the base SN standard reference, for e.g., improved solar irradiance reconstructions or the identification of precursors of extreme solar activity events. Therefore, historical sunspot observations will thus continue to play a central role for our understanding and our forecasting capability of the long-term solar activity and Sun-Earth relations.

Acknowledgements Part of the work contributed here by F. Clette was developed in the framework of the TOSCA project (ESSEM COST action ES1005 of the European Union; http://lpc2e.cnrs-orleans.fr/~ddwit/TOSCA/Home.html), of the SOLID project (EU 7th Framework Program, SPACE collaborative projects; http://projects.pmodwrc.ch/solid/) and of the Solar-Terrestrial Center of Excellence (http://www.stce.be). F. Clette would like to acknowledge the personal contributions from Laure Lefèvre and Laurence Wauters.

L. Svalgaard acknowledges support from Stanford University.

J.M. Vaquero acknowledges support from the Junta de Extremadura (Research Group Grant No. GR10131) and the Ministerio de Economía y Competitividad of the Spanish Government (AYA2011-25945).

The authors also wish to thank Sergio Cortesi, Marco Cagnotti and Michele Bianda for their hospitality and help for our various investigations on the Locarno and Zürich observations.

The Sunspot Number Workshop series has benefitted from the support of the Royal Observatory of Belgium, the National Solar Observatory, Stanford University, the Air Force Research Laboratory and the Specola Solare Ticinese.

Finally, the authors are also grateful to Rainer Arlt for his detailed reviewing work and constructive comments.

References

J.A. Abreu, J. Beer, F. Steinhilber, S.M. Tobias, N.O. Weiss, For how long will the current grand maximum of solar activity persist? Geophys. Res. Lett. **35**, 20109 (2008)

R. Arlt, Digitization of sunspot drawings by Staudacher, in 1749–1796. Sol. Phys. **247**, 399–410 (2008)

R. Arlt, The solar butterfly diagram in the 18th century. Sol. Phys. **255**, 143–153 (2009a)

R. Arlt, The solar observations at Armagh Observatory in 1795–1797. Astron. Nachr. **330**, 311–316 (2009b)

R. Arlt, The sunspot observations by Samuel Heinrich Schwabe. Astron. Nachr. **332**, 805–814 (2011)

R. Arlt, A. Abdolvand, First solar butterfly diagram from Schwabe's observations in 1825–1867, in *Physics of Sun and Star Spots*, ed. by D.P. Choudhary, K.G. Strassmeier. IAU Symposium 273, Ventura, California, USA, August 22–26, 2010. Proc. of IAU Symposium, vol. 273 (Cambridge University Press, Cambridge, 2011), pp. 286–289

R. Arlt, R. Leussu, N. Giese, K. Mursula, I.G. Usoskin, Sunspot positions and sizes for 1825–1867 from the observations by Samuel Heinrich Schwabe. Mon. Not. R. Astron. Soc. **433**, 3165–3172 (2013)

K.T. Bachmann, H. Maymani, K. Nautiyal, V. te Velde, An analysis of solar-cycle temporal relationships among activity indicators. Adv. Space Res. **34**, 274–281 (2004)

S. Basu, A.-M. Broomhall, W.J. Chaplin, Y. Elsworth, Thinning of the Sun's magnetic layer: the peculiar solar minimum could have been predicted. Astrophys. J. **758**, 43 (2012)

S. Basu, A.-M. Broomhall, W.J. Chaplin, Y. Elsworth, G.R. Davies, J. Schou, T.P. Larson, Comparing the internal structure of the Sun during the cycle 23 and cycle 24 minima. Astron. Soc. Pac. Conf. Ser. **478**, 161 (2013)

D. Berghmans, R.A.M. van der Linden, P. Vanlommel, F. Clette, E. Robbrecht, History of the Sunspot Index: 25 years SIDC. Beitrage zur Geschichte der Geophysik und Kosmischen Physik **VII**(1), 7, 288 (2006)

L. Bertello, R.K. Ulrich, J.E. Boyden, The Mount Wilson CaII K plage index time series. Sol. Phys. **264**, 31–44 (2010)

A. Brandenburg, The case for a distributed solar dynamo shaped by near-surface shear. Astrophys. J. **625**, 539–547 (2005)

W. Brunner, Letter to editor from Zurich Observatory. Terr. Magn. **41**(2), 210 (1936)

R. Casas, J.M. Vaquero, The sunspot catalogues of Carrington, Peters, and de la Rue: quality control and machine-readable versions. Sol. Phys. **289**, 79–90 (2014)

R. Casas, J.M. Vaquero, M. Vázquez, Solar rotation in the 17th century. Sol. Phys. **234**, 379–392 (2006)

F. Clette, L. Lefèvre, Are the sunspots really vanishing? Anomalies in solar cycle 23 and implications for long-term models and proxies. J. Space Weather Space Clim. **2**, A260000 (2012)

F. Clette, D. Berghmans, P. Vanlommel, R.A.M. Van der Linden, A. Koeckelenbergh, L. Wauters, From the Wolf number to the international sunspot index: 25 years of SIDC. Adv. Space Res. **40**, 919–928 (2007)

E.W. Cliver, Carrington, Schwabe, and the gold medal, Eos. Trans. Am. Geophys. Union **86**, 413–418 (2005)

E.W. Cliver, L. Svalgaard, in *Origins of the Wolf Sunspot Number Series: Geomagnetic Underpinning, AGU Fall Meeting, Abstract SH13A-1109* (2007)
E.W. Cliver, F. Clette, L. Svalgaard, Recalibrating the Sunspot Number (SSN): The SSN workshops. Cent. Eur. Astrophys. Bull. **37**(2), 401–416 (2013)
A. Cristo, J.M. Vaquero, F. Sánchez-Bajo, HSUNSPOTS: a tool for the analysis of historical sunspot drawings. J. Atmos. Sol.-Terr. Phys. **73**, 187–190 (2011)
M.J. Crowe, K.R. Lafortune, Herschel, (Friedrich) William [Wilhelm], in *The Biographical Encyclopedia of Astronomers*, ed. by T. Hockey (Springer, Berlin, 2007), pp. 494–496
W. de la Rue, B. Stewart, B. Loewy, Researches on solar physics. heliographical positions and areas of sunspots observed with the kew photoheliograph during the years 1862 and 1863. Philos. Trans. R. Soc. Lond. **159**, 1–110 (1869)
G. de Toma, G.A. Chapman, D.G. Preminger, A.M. Cookson, Analysis of sunspot area over two solar cycles. Astrophys. J. **770**, 89 (2013a) (13 pp.)
G. de Toma, G.A. Chapman, A.M. Cookson, D. Preminger, Temporal stability of sunspot umbral intensities: 1986–2012. Astrophys. J. Lett. **771**, L22 (2013b) (4 pp.)
J.A. Eddy, The Maunder minimum. Science **192**, 1189–1202 (1976)
P. Foukal, An explanation of the differences between the sunspot area scales of the Royal Greenwich and Mt. Wilson Observatories, and the SOON program. Sol. Phys. **289**, 1517–1529 (2013)
T.K. Friedli, Erste beobachtungen am Wolfschen Normal refraktor in Schliern b, Köniz. Mitt. Rudolf-Wolf-Ges. **3**(9), 3–46 (1997)
T.K. Friedli, H.U. Keller, Rudolf Wolf als Pionier der Sonnenfleckenforschung. Vierteljahrsschr. Nat.forsch. Ges. Zür. **138**(4), 267–281 (1993)
G. Galilei, J.C. Scheiner, *On Sunspot Translated and Introduced by, Reeves, E. and van Helden, A.* (The University of Chicago Press, Chicago and London, 2010)
M.N. Gnevyshev, A.I. Ohl, Astron. Zh. **25**, 18 (1948)
D.H. Hathaway, Royal Observatory, Greenwich—USAF/NOAA Sunspot Data (2014). http://solarscience.msfc.nasa.gov/greenwch.shtml
D.H. Hathaway, R.M. Wilson, E.J. Reichmann, Group sunspot numbers: sunspot cycle characteristics. Sol. Phys. **211**, 357–370 (2002)
J.L. Heilbron, *The Sun in the Church* (Harvard University Press, Cambridge, 1999), 366 p.
A. Hempelmann, W. Weber, Correlation between the sunspot number, the total solar irradiance, and the terrestrial insolation. Sol. Phys. **277**, 417–430 (2012)
C.H. Hossfield, A history of the Zurich and American relative sunspot number indices. J. Am. Assoc. Var. Star Obs. **31**, 48–53 (2002)
D.V. Hoyt, K.H. Schatten, A new look at Wolf sunspot numbers in the late 1700's. Sol. Phys. **138**, 387–397 (1992a)
D.V. Hoyt, K.H. Schatten, New information on solar activity, 1779–1818, from Sir William Herschel's unpublished notebooks. Astrophys. J. **384**, 361–384 (1992b)
D.V. Hoyt, K.H. Schatten, Overlooked sunspot observations by Hevelius in the early Maunder Minimum, 1653–1684. Sol. Phys. **160**, 371–378 (1995a)
D.V. Hoyt, K.H. Schatten, Observations of sunspots by Flamsteed during the Maunder Minimum. Sol. Phys. **160**, 379–385 (1995b)
D.V. Hoyt, K.H. Schatten, A new interpretation of Christian Horrebow's sunspot observations from 1761 to 1777. Sol. Phys. **160**, 387–392 (1995c)
D.V. Hoyt, K.H. Schatten, A revised listing of the number of sunspot groups made by Pastorff, 1819 to 1833. Sol. Phys. **160**, 393–399 (1995d)
D.V. Hoyt, K.H. Schatten, How well was the Sun observed during the Maunder minimum? Sol. Phys. **165**, 181–192 (1996)
D.V. Hoyt, K.H. Schatten, *The Role of the Sun in Climate Change* (Oxford University Press, New York, 1997)
D.V. Hoyt, K.H. Schatten, Group sunspot numbers: a new solar activity reconstruction. Sol. Phys. **179**, 189–219 (1998a)
D.V. Hoyt, K.H. Schatten, Group sunspot numbers: a new solar activity reconstruction. Sol. Phys. **181**, 491–512 (1998b)
D.V. Hoyt, K.H. Schatten, E. Nesme-Ribes, The one hundredth year of Rudolf Wolf's death: Do we have the correct reconstruction of solar activity? Geophys. Res. Lett. **21**, 2067–2070 (1994)
K. Hufbauer, *Exploring the Sun: Solar Science since Galileo* (Johns Hopkins University Press, Baltimore, 1991)
J. Javaraiah, Long-term variations in the growth and decay rates of sunspot groups. Sol. Phys. **270**, 463–483 (2011)
M.J. Johnson, Address delivered by the President, M.J. Johnson, Esq. on presenting the Medal of the Society to M. Schwabe. Mon. Not. R. Astron. Soc. **17**, 126–132 (1857)

R.W. Johnson, Power law relating 10.7 cm flux to sunspot number. Astrophys. Space Sci. **332**, 73–79 (2011)
H. Kant, Scheiner, Christoph, in *The Biographical Encyclopedia of Astronomers*, ed. by T. Hockey (Springer, Berlin, 2007), p. 1018
H.U. Keller, T.K. Friedli, The sunspot-activity in the years 1976–1995. Mitt. Rudolf-Wolf-Ges. **3**(7), 1–46 (1995)
K.O. Kiepenheuer, Solar site testing, in *Le choix des sites d'observatoires astronomiques (site testing)*, ed. by J. Rösch. IAU Symposium, vol. 19 (1962), pp. 193–219
A. Kilcik, V.B. Yurchyshyn, V. Abramenko, P.R. Goode, A. Ozguc, J.P. Rozelot, W. Cao, Time distributions of large and small sunspot groups over four solar cycles. Astrophys. J. **731**, 30 (2011)
M. Kopecky, B. Ruzickova-Topolova, G.V. Kuklin, On the relative inhomogeneity of long-term series of sunspot indices. Bull. Astron. Inst. Czechoslov. **31**, 267–283 (1980)
N.A. Krivova, L. Balmaceda, S.K. Solanki, Reconstruction of solar total irradiance since 1700 from the surface magnetic flux. Astron. Astrophys. **467**, 335–346 (2007)
J.F. Lalande, *Astronomie* (1771), Paris, 4 Vols.
L. Lefèvre, F. Clette, A global small sunspot deficit at the base of the index anomalies of solar cycle 23. Astron. Astrophys. **536**, L11 (2011)
V. Letfus, Solar activity in the sixteenth and seventeenth centuries—A revision. Sol. Phys. **145**, 377–388 (1993)
V. Letfus, Daily relative sunspot numbers 1749–1848: reconstruction of missing observations. Sol. Phys. **184**, 201–211 (1999)
V. Letfus, Relative sunspot numbers in the first half of eighteenth century. Sol. Phys. **194**, 175–184 (2000)
R. Leussu, I.G. Usoskin, R. Arlt, K. Mursula, Inconsistency of the Wolf sunspot number series around 1848. Astron. Astrophys. **559**, A28 (2013)
W. Livingston, M. Penn, Are sunspots different during this solar minimum? EOS Trans. **90**, 257–258 (2009)
Locarno, Specola Solare Ticinese Drawings, Archive (2014). http://www.specola.ch/e/drawings.html
M. Lockwood, Reconstruction and prediction of variations in the open solar magnetic flux and interplanetary conditions. Living Rev. Sol. Phys. **10**, 4 (2013)
M. Lockwood, A.P. Rouillard, I.D. Finch, The rise and fall of open solar flux during the current grand solar maximum. Astrophys. J. **700**, 937–944 (2009)
M. Lockwood, M.J. Owens, L. Barnard, Centennial variations in sunspot number, open solar flux, and streamer belt width: 1. Correction of the sunspot number record since 1874. J. Geophys. Res. Space Phys. **119** (2014, in press). doi:10.1002/2014JA019970, http://onlinelibrary.wiley.com/doi/10.1002/2014JA019970/pdf
E. Loomis, Am. J. Sci. Ser **50**, 153 (1870)
R. Lukianova, K. Mursula, Changed relation between sunspot numbers, solar UV/EUV radiation and TSI during the declining phase of solar cycle 23. J. Atmos. Sol.-Terr. Phys. **73**, 235–240 (2011)
E.F. MacPike, *Hevelius, Flamsteed and Halley: Three Contemporary Astronomers and Their Mutual Relations* (Taylor and Francis, London, 1937)
E. Manfredi, De Gnomone Meridiano Bononiensi ad Divi Petronii, Laeli a Vulpa, Bononiae (1736, 397 pp.)
W.M. Mitchell, The history of the discovery of the solar spots. Pop. Astron. **24**, 22–31, 82–96, 149–162, 206–218, 290–303, 341–354, 428–440, 488–499 (1916)
A. Muñoz-Jaramillo, D. Nandy, P.C.H. Martens, A.R. Yeates, A double-ring algorithm for modeling solar active regions: unifying kinematic dynamo models and surface flux-transport simulations. Astrophys. J. **720**, L20–L25 (2010)
A. Muñoz-Jaramillo, D. Nandy, P.C.H. Martens, Magnetic quenching of turbulent diffusivity: reconciling mixing-length theory estimates with kinematic dynamo models of the solar cycle. Astrophys. J. **727**, L23 (2011)
Y.A. Nagovitsyn, A.A. Pevtsov, W.C. Livingston, On a possible explanation of the long-term decrease in sunspot field strength. Astrophys. J. **758**, L20 (2012)
H. Nevanlinna, New geomagnetic activity index series published for 1844–1880. EOS Trans. **76**, 233 (1995)
H. Nevanlinna, E. Kataja, An extension of the geomagnetic activity index series aa for two solar cycles (1844–1868). Geophys. Res. Lett. **20**, 2703–2706 (1993)
S.M. Ostrow, M. PoKempner, The differences in the relationship between ionospheric critical frequencies and sunspot number for different sunspot cycles. J. Geophys. Res. **57**(4), 473–481 (1952). doi:10.1029/JZ057i004p00473
M.J. Owens, M. Lockwood, Cyclic loss of open solar flux since 1868: The link to heliospheric current sheet tilt and implications for the Maunder Minimum. J. Geophys. Res. **117**, 4102 (2012)
D.G. Parker, J.M. Pap, R.K. Ulrich, L.E. Floyd, D.K. Prinz, in *Developing New Mount Wilson Magnetic Indices to Model Solar UV Variations, SPD meeting #28, #02.54*. Bulletin of the American Astronomical Society, vol. 29 (American Astronomical Society, Washington, 1997), p. 902
M.J. Penn, W. Livingston, Long-term evolution of sunspot magnetic fields. IAU Symp. **273**, 126–133 (2011)

K. Petrovay, Solar cycle prediction. Living Rev. Sol. Phys. **7**, 6 (2010)
A.A. Pevtsov, L. Bertello, A.G. Tlatov, A. Kilcik, Y.A. Nagovitsyn, E.W. Cliver, Cyclic and long-term variation of sunspot magnetic fields. Sol. Phys. **289**, 593–602 (2014)
J.C. Ribes, E. Nesme-Ribes, The solar sunspot cycle in the Maunder minimum AD1645 to AD1715. Astron. Astrophys. **276**, 549–563 (1993)
M. Rybanský, V. Rušin, M. Minarovjech, L. Klocok, E.W. Cliver, Reexamination of the coronal index of solar activity. J. Geophys. Res. **110**, 8106 (2005)
B.E. Schaefer, Visibility of sunspots. Astrophys. J. **411**, 909–919 (1993)
B.E. Schaefer, Automatic inflation in the AAVSO sunspot number. J. Am. Assoc. Var. Star Obs. **26**, 40–46 (1997a)
B.E. Schaefer, Improvements for the AAVSO sunspot number. J. Am. Assoc. Var. Star Obs. **26**, 47–49 (1997b)
K.H. Schatten, Modeling a shallow solar dynamo. Sol. Phys. **255**, 3–38 (2009)
C.J. Schrijver, W.C. Livingston, T.N. Woods, R.A. Mewaldt, The minimal solar activity in 2008–2009 and its implications for long-term climate modeling. Geophys. Res. Lett. **38**, 6701 (2011)
M. Schwabe, Sonnen-Beobachtungen im Jahre 1843. Astron. Nachr. **21**, 233 (1844)
N.K. Sethi, M.K. Goel, K.K. Mahajan, Solar cycle variations of f°F2 from IGY to 1990. Ann. Geophys. **20**, 1677–1685 (2002). doi:10.5194/angeo-20-1677-2002
A.I. Shapiro, W. Schmutz, E. Rozanov, M. Schoell, M. Haberreiter, A.V. Shapiro, S. Nyeki, A new approach to the long-term reconstruction of the solar irradiance leads to large historical solar forcing. Astron. Astrophys. **529**, A67 (2011)
A.H. Shapley, Reduction of sunspot-number observations. Publ. Astron. Soc. Pac. **61**, 13–21 (1949)
J.W. Shirley, *Thomas Harriot: A Biography* (Clarendon Press, Oxford, 1983)
S.K. Solanki, N.A. Krivova, Solar irradiance variations: from current measurements to long-term estimates. Sol. Phys. **224**, 197–208 (2004)
S.K. Solanki, I.G. Usoskin, B. Kromer, M. Schüssler, J. Beer, Unusual activity of the Sun during recent decades compared to the previous 11,000 years. Nature **431**, 1084–1087 (2004)
C.P. Sonett, The great solar anomaly ca. 1780–1800—an error in compiling the record. J. Geophys. Res. **88**, 3225–3228 (1983)
Sonne, Fachgruppe Sonne (2014). http://www.vds-sonne.de/index.php
G. Spörer, Vierteljahrsschr. Astron. Ges. **22**, 323 (1887)
J.O. Stenflo, Basal magnetic flux and the local solar dynamo. Astron. Astrophys. **547**, A93 (2012)
L. Svalgaard, Updating the historical sunspot record, in *SOHO-23: Understanding a Peculiar Solar Minimum, Proceedings of a workshop held 21–25 September 2009 in Northeast Harbor, Maine, USA*, ed. by S.R. Cranmer, J.T. Hoeksema, J.L. Kohl. ASP Conference Series, vol. 428 (Astronomical Society of the Pacific, San Francisco, 2010), p. 297
L. Svalgaard, Solar activity—past, present, future. J. Space Weather Space Clim. **3**, A24 (2013a)
L. Svalgaard, Building a Sunspot Group Number Backbone Series SSN-Workshop 3 (Tucson) (2013b). http://www.leif.org/research/SSN/Svalgaard11.pdf
L. Svalgaard, Corrections of errors in scale values for magnetic elements for Helsinki. Ann. Geophys. **32**, 633–641 (2014a). doi:10.5194/angeo-32-633-2014
L. Svalgaard, What can geomagnetism can tell us about the solar cycle? Space Sci. Rev. (2014b, this volume)
L. Svalgaard, E.W. Cliver, in *Calibrating the Sunspot Number: Using "the Magnetic Needle", AGU Spring Meeting Abstracts 5454B-02* (2007)
L. Svalgaard, H.S. Hudson, The solar microwave flux and the sunspot number, in *SOHO-23: Understanding a Peculiar Solar Minimum, Proceedings of a Workshop Held 21–25 September 2009 in Northeast Harbor, Maine, USA*, ed. by S.R. Cranmer, J.T. Hoeksema, J.L. Kohl. ASP Conference Series, vol. 428 (Astronomical Society of the Pacific, San Francisco, 2010), p. 325
K.F. Tapping, D. Boteler, P. Charbonneau, A. Crouch, A. Manson, H. Paquette, Solar magnetic activity and total irradiance since the Maunder minimum. Sol. Phys. **246**, 309–326 (2007)
T.N. Thiele, De Macularum Solis antiquioribus quibusdam obstervationibus Hafniae institutis, scripsit Th.N. Thiele, astr. stud. Astron. Nachr. **50**, 257 (1859)
A.G. Tlatov, Long-term variations in sunspot characteristics. Geomagn. Aeron. **53**, 953–956 (2013)
A.G. Tlatov, V.V. Vasileva, V.V. Makarova, P.A. Otkidychev, Applying an automatic image-processing method to synoptic observations. Sol. Phys. **289**, 1403–1412 (2014)
I.G. Usoskin, K. Mursula, G.A. Kovaltsov, Was one sunspot cycle lost in late XVIII century? Astron. Astrophys. **370**, L31–L34 (2001)
I.G. Usoskin, K. Mursula, G.A. Kovaltsov, Reconstruction of monthly and yearly group sunspot numbers from sparse daily observations. Sol. Phys. **218**, 295–305 (2003)
I.G. Usoskin, K. Mursula, R. Arlt, G.A. Kovaltsov, A solar cycle lost in 1793–1800: Early sunspot observations resolve the old mystery. Astrophys. J. **700**, L154–L157 (2009)

I.G. Usoskin, S.K. Solanki, G.A. Kovaltsov, Grand minima of solar activity during the last millennia. IAU Symp. **286**, 372–382 (2012)
I.G. Usoskin, G. Hulot, Y. Gallet, R. Roth, A. Licht, F. Joos, G.A. Kovaltsov, E. Thébault, A. Khokhlov, Evidence for distinct modes of solar activity. Astron. Astrophys. **562**, L10 (2014)
J.M. Vaquero, On the solar activity during the year 1784. Sol. Phys. **219**, 379–384 (2004)
J.M. Vaquero, Historical sunspot observations: a review. Adv. Space Res. **40**, 929–941 (2007)
J.M. Vaquero, M.C. Gallego, Reconstructing past solar activity using meridian solar observations: the case of the Royal Observatory of the Spanish Navy (1833–1840). Adv. Space Res. **53**, 1162–1168 (2014)
J.M. Vaquero, R.M. Trigo, Revised group sunspot number values for 1640, 1652, and 1741. Sol. Phys. **289**, 803–808 (2014)
J.M. Vaquero, M. Vázquez, *The Sun Recorded Through History*. Astrophysics and Space Science Library, vol. 361 (Springer, Berlin, 2009), 382 p.
J.M. Vaquero, F. Sánchez-Bajo, M.C. Gallego, On the reliability of the de la Rue sunspots areas measurements. Sol. Phys. **209**, 311–319 (2002)
J.M. Vaquero, M.C. Gallego, R.M. Trigo, The sunspot numbers during 1736–1739 revisited. Adv. Space Res. **40**, 1895–1903 (2007a)
J.M. Vaquero, M.C. Gallego, F. Sánchez-Bajo, Improving sunspot records: observations by M. Hell revisited. Observatory **127**, 221–224 (2007b)
J.M. Vaquero, M.C. Gallego, I.G. Usoskin, G.A. Kovaltsov, Revisited sunspot data: a new scenario for the onset of the Maunder minimum. Astrophys. J. Lett. **731**, L24 (2011)
J.M. Vaquero, R.M. Trigo, M.C. Gallego, A simple method to check the reliability of annual sunspot number in the historical period 1610–1847. Sol. Phys. **277**, 389–395 (2012)
L.E.A. Vieira, S.K. Solanki, N.A. Krivova, I. Usoskin, Evolution of the solar irradiance during the Holocene. Astron. Astrophys. **531**, A6 (2011)
M. Waldmeier, Astron. Mitteil. Eidgn. Sternw. Zürich, No. 152 (1948)
M. Waldmeier, The sunspot-activity in the years 1610–1960 (Schulthess & Co., Swiss Federal Observatory, Zürich, 1961)
M. Waldmeier, Die Beziehung zwischen der Sonnenflecken-relativ-zahl und der Gruppenzahl. Astr. Mitteil. Eidgn. Sternw Zürich No. 285 (1968)
Y.-M. Wang, J.L. Lean, N.R. Sheeley Jr., Modeling the Sun's magnetic field and irradiance since 1713. Astrophys. J. **625**, 522–538 (2005)
D.M. Willis, H.E. Coffey, R. Henwood, E.H. Erwin, D.V. Hoyt, M.N. Wild, W.F. Denig, The Greenwich photo-heliographic results (1874–1976): summary of the observations, applications, datasets, definitions and errors. Sol. Phys. **288**, 117–139 (2013)
R.M. Wilson, D.H. Hathaway, On the Relation Between Sunspot Area and Sunspot Number. NASA STI/Recon Technical Report N, 6, 20186 (2006)
R. Wolf, Sonnenflecken Beobachtungen in der zweiten Hälfte des Jahres 1850. Mitt. Nat.forsch. Ges. Bern **207**, 89–95 (1851)
R. Wolf, Mitteilungen über die Sonnenflecken, I. Astron. Mitteil. Eidgn. Sternw. Zürich **1**, 3–13 (1856)
R. Wolf, Mitteilungen über die Sonnenflecken, IV. Astron. Mitteil. Eidgn. Sternw. Zürich **1**, 51–78 (1857)
R. Wolf, Abstract of his latest results. Mon. Not. R. Astron. Soc. **21**, 77–78 (1861a)
R. Wolf, Mittheilungen über die Sonnenflecken, XII. Astron. Mitteil. Eidgn. Sternw. Zürich **2**, 41–82 (1861b)
R. Wolf, Mittheilungen über die Sonnenflecken, XIII. Astron. Mitteil. Eidgn. Sternw. Zürich **2**, 83–118 (1861c)
R. Wolf, Mittheilungen über die Sonnenflecken, XIV. Astron. Mitteil. Eidgn. Sternw. Zürich **2**, 119–132 (1862)
R. Wolf, Astronomische Mittheilungen, XXXV. Astron. Mitteil. Eidgn. Sternw. Zürich **4**, 173–250 (1874)
R. Wolf, Astronomische Mittheilungen, XXXVIII. Astron. Mitteil. Eidgn. Sternw. Zürich **4**, 375–405 (1875)
R. Wolf, Astronomische Mittheilungen, LI. Astron. Mitteil. Eidgn. Sternw. Zürich **6**, 1–32 (1880)
R. Wolf, Astronomische Mittheilungen, LVI. Astron. Mitteil. Eidgn. Sternw. Zürich **6**, 177–220 (1882)
A. Wolfer, Astronomische Mittheilungen, LXXXVI. Astron. Mitteil. Eidgn. Sternw. Zürich **9**, 187–232 (1895)
A. Wolfer, Revision of Wolf's sunspot relative numbers. Pop. Astron. **10**, 449–458 (1902)
A. Wolfer, Astronomische Mittheilungen, XCVIII. Astron. Mitteil. Eidgn. Sternw. Zürich **10**, 251–281 (1907)
N.V. Zolotova, D.I. Ponyavin, Was the unusual solar cycle at the end of the XVIII century a result of phase asynchronization? Astron. Astrophys. **470**, L17–L20 (2007)
N.V. Zolotova, D.I. Ponyavin, Enigma of the solar cycle 4 still not resolved. Astrophys. J. **736**, 115 (2011)

DOI 10.1007/978-1-4939-2584-1_4
Reprinted from *Space Science Reviews* Journal, DOI 10.1007/s11214-014-0089-8

Solar Cycle Indices from the Photosphere to the Corona: Measurements and Underlying Physics

Ilaria Ermolli · Kiyoto Shibasaki · Andrey Tlatov · Lidia van Driel-Gesztelyi

Received: 20 May 2014 / Accepted: 3 August 2014 / Published online: 2 September 2014

Abstract A variety of indices have been proposed in order to represent the many different observables modulated by the solar cycle. Most of these indices are highly correlated with each other owing to their intrinsic link with the solar magnetism and the dominant eleven year cycle, but their variations may differ in fine details, as well as on short- and long-term trends. In this paper we present an overview of the indices that are often employed to describe the many features of the solar cycle, moving from the ones referring to direct observations of the inner solar atmosphere, the photosphere and chromosphere, to those deriving from measurements of the transition region and solar corona. For each index, we summarize existing measurements and typical use, and for those that quantify physical observables, we describe the underlying physics.

I. Ermolli (✉)
INAF Osservatorio Astronomico di Roma, via Frascati 33, 00040 Monte Porzio Catone, Italy
e-mail: ermolli@oaroma.inaf.it

K. Shibasaki
Nobeyama Solar Radio Observatory NAOJ, 462-2 Nobeyama, Minamimaki, Minamisaku, Nagano 384-1305, Japan
e-mail: shibasaki.kiyoto@nao.ac.jp

A. Tlatov
Kislovodsk Mountain Astronomical Station of the Pulkovo Observatory, Kislovodsk, Russia
e-mail: tlatov@mail.ru

L. van Driel-Gesztelyi
University College London, Mullard Space Science Laboratory, Holmbury St. Mary, Dorking, Surrey RH5 6NT, UK
e-mail: Lidia.vanDriel@obspm.fr

L. van Driel-Gesztelyi
LESIA, Observatoire de Paris, CNRS, UPMC, Université Paris Diderot, Paris, France

L. van Driel-Gesztelyi
Konkoly Observatory of the Hungarian Academy of Sciences, 1121 Budapest, Hungary

Keywords Solar cycle · Solar atmosphere · Solar magnetism

1 Introduction

The solar cycle, i.e. the cyclic regeneration of the appearance of the Sun that occurs with a dominant eleven year period, originates below the visible surface of the Sun. In a thin shear layer at the bottom of the convection zone, the turbulent convection operates, jointly with rotational shear, global circulations and boundary layers, to produce magnetic fields by dynamo processes. These fields thread their way to the solar surface, where they manifest themselves with the rich variety of features observed on the solar atmosphere. The evolution of these features constitutes the solar activity cycle. Most recent models and observations of the solar cycle are reviewed e.g. by Charbonneau (2010), Charbonneau et al. (2014) and Hathaway (2010).

Cyclic variability has been detected in cool stars like the Sun (Hall 2008; Oláh et al. 2009; Reiners 2012; Chaplin and Basu 2014). However, the Sun is the only star where its activity cycle has been studied in detail. Indeed, the Sun has been monitored systematically since the advent of the telescope in the early 17th century.

Early observations have shown that the Sun has sunspots that move on the solar disk due to the *solar differential rotation* (e.g. Galilei 1613; Scheiner 1626–1630). More that two centuries later the number of sunspots was recognized to have a cyclic variation (Schwabe 1843). Then, it was found that the latitudinal distribution of sunspots and its progression over the sunspot cycle follow a "butterfly diagram" (Maunder 1904). This diagram is now a representative image of the solar cycle.

Sunspots were give a foundation of physics in the early decades of 20th century, when they were demonstrated to be the seat of strong magnetic fields (Hale 1908). Later on, the magnetic laws of the solar cycle became apparent (Hale et al. 1919; Hale 1924; Hale and Nicholson 1925) once the measurements of magnetic fields in sunspots were extended over more than a single sunspot cycle. Sunspot and later polar magnetic field measurements also revealed the reversal of the global magnetic field of the Sun with the period of 22 years, the solar (or sunspot) cycle being half of the magnetic cycle (Babcock 1959).

Sunspots are the most commonly known, but only one, among the many manifestations of the solar magnetic fields, which structure the Sun's atmosphere and make it variable (e.g. Solanki et al. 2006; Stenflo 2008, 2013). This is shown in Fig. 1, which presents three simultaneous views of the Sun at different heights in the solar atmosphere, from the photosphere to the corona, and the contemporaneous measurements of the magnetic field in the photosphere.

The cyclic variability of sunspots was found to affect all the manifestations of the solar magnetism on the Sun's atmosphere, including striking eruptive events that are driven by the evolution of magnetic fields at small scales, such as flares, prominences, and coronal mass ejections. The number of these events follows the rise and decay of the solar cycle (e.g. Hudson et al. 2014a, and references therein). The solar cycle also characterizes the radiant behavior of the Sun (Willson et al. 1981; Hudson et al. 1982; Domingo et al. 2009) and there is growing evidence that changes in solar irradiance affect the Earth's middle and lower atmosphere (e.g. Ermolli et al. 2013; Solanki et al. 2013, and references therein). Besides, as solar magnetic fields are dispersed into the heliosphere with the solar wind, the solar cycle modulates the particulate and magnetic fluxes in the heliosphere, by determining the interplanetary conditions and imposing electromagnetic forces that affect the planetary atmospheres, including the Earth's atmosphere (Cranmer

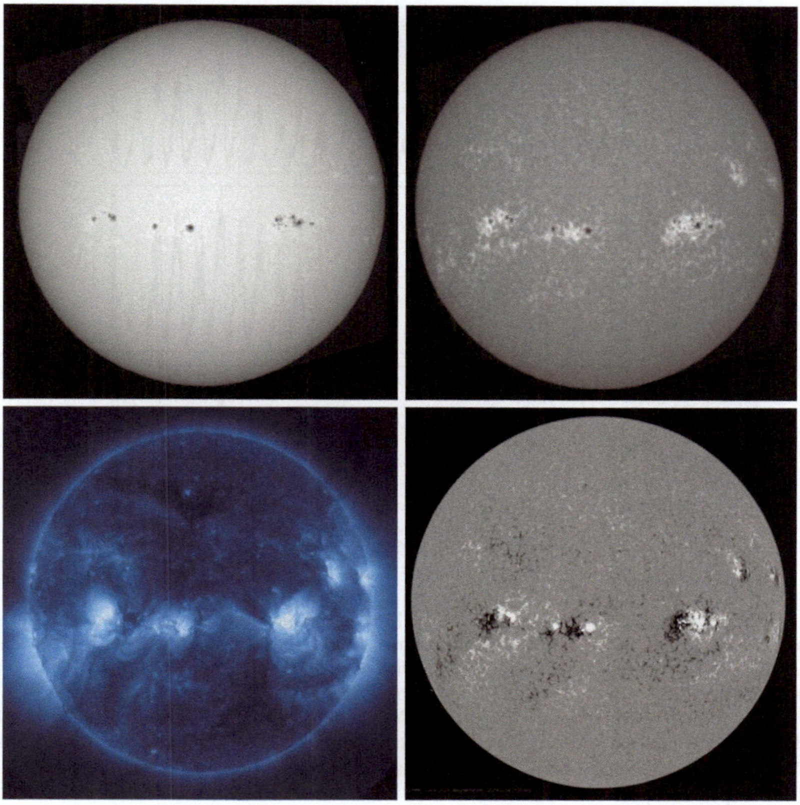

Fig. 1 Simultaneous views of the Sun at different heights in the solar atmosphere and measurements of magnetic field in the photosphere. Top panels: The photosphere (*left*) observed in the light of the wavelength 430 nm of the CH G-band on 14 February 2014, showing sunspots and faculae, the latter being apparent as bright regions mostly near the disk limb. The chromosphere (*right*) observed at the Ca II K line radiation at wavelength 393 nm showing the enhanced chromospheric emission from the photospheric magnetic regions. The chromospheric extensions of the faculae (the so-called "plages") are readily visible across the whole disk (images courtesy INAF Osservatorio Astronomico di Roma). *Bottom panels*: The contemporaneous views of active regions and full-disk solar corona (*left*) observed at wavelength 33.5 nm of the Fe XVI line and magnetogram (*right*) measurement of the line-of-sight (LOS) magnetic field (images courtesy SDO/AIA and SDO/HMI)

2009). Indeed, the large scale structure and dynamics of the magnetic field in the heliosphere is governed by the solar wind flow, which has its origin in the magnetic structure of the solar corona driven by the emergence of magnetic flux (e.g. Schmieder et al. 2014, and references therein), plasma motions in the photosphere and transient solar eruptions in the corona. Most of the flux emerged in the photosphere forms chromospheric or coronal loops that do not contribute to the heliospheric magnetic field carried by the solar wind, but a fraction of flux extends out to form the heliospheric magnetic field, whose variations are an important source of geomagnetic activity (e.g. Pulkkinen 2007; Pulkkinen et al. 2007). Interplanetary transients and geomagnetic disturbances are found to be related to the changing magnetic fields on the solar surface (Lockwood 2013; Owens and Forsyth 2013; Wang 2014), which indirectly modulate also the flux of high-energy galactic cosmic rays entering the solar system from elsewhere in the galaxy (e.g.

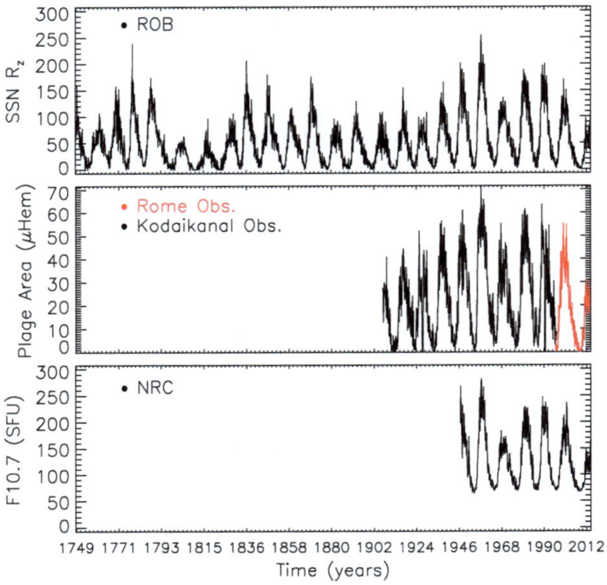

Fig. 2 Comparison of three indices of solar cycle. Monthly averaged values of the international sunspot number (*top panel*) from the *ROB-SILSO* archive (January 1749–March 2014), the Ca II K plage area (*middle panel*) from observations carried out at the Kodaikanal (*black symbols*) and Rome (*red symbols*) observatories (February 1907–August 1996 and May 1996–November 2013, respectively), and the F10.7 index (*bottom panel*) from the NRC series (February 1947–March 2014). The plage area is expressed in millionths of a solar hemisphere (μHem), whereas the F10.7 index values are given using the Solar Flux Units (SFU, 10^{-22} $\mathrm{Wm^{-2}Hz^{-1}}$). All indices show common features (eleven year cycle, long-term trends), but each index also reflects a specific aspect of solar activity

Usoskin 2013). Besides, as ^{14}C and ^{10}Be radioisotopes are produced in the Earth's stratosphere by the impact of galactic cosmic rays on ^{14}N and ^{16}O, the solar cycle modulation of the cosmic ray flux also leads to solar cycle related variations in the abundances of ^{14}C (Stuiver and Quay 1980) and ^{10}Be (Beer et al. 1990) in the Earth's atmosphere. Records of geomagnetic and cosmogenic isotope abundance variations can then be used to infer the near-Earth solar wind conditions and the magnetic field structure and intensity (Lockwood 2013; Svalgaard 2014; Hudson et al. 2014b). Lockwood (2013) gives detailed description of a large number of geomagnetic indices linked to the amount of heliospheric open flux, allowing its reconstruction of from historic geomagnetic activity observations.

A variety of indices have been proposed in order to represent the many different observables modulated by the solar cycle. Most of these indices are highly correlated with each other as they are intrinsically inter-linked through solar magnetism and its dominant eleven year cycle. However, they may differ in many features, as well as on short- and long-term trends. This is shown in Fig. 2, which presents the time series of monthly averaged values of three solar cycle indices discussed in the following. All measurements show the eleven year solar cycle, a long-term trend with increasing values in time from 1900 to 1960 and then a reduced solar activity. Apart from these common features, these indices also show clear differences, in particular during the solar minima and on the long-term trends. This is because the compared indices are related in various ways to different aspects of magnetic processes taking place on the Sun.

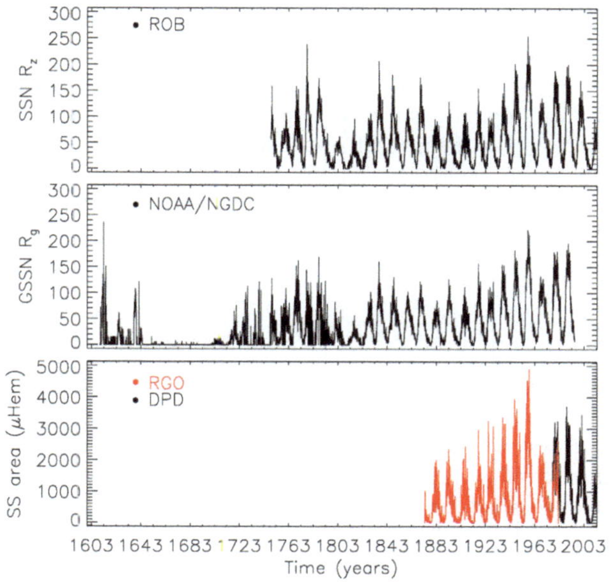

Fig. 3 Comparison of sunspot indices. Monthly averaged values of the international sunspot number (*top panel*) from the *ROB-SILSO* archive (January 1749–March 2014), group sunspot number (*middle panel*) from the NOAA/NGDC archive (February 1610–December 1995) and sunspot area (*bottom panel*) from the RGO (*red symbols*) and DPD (*black symbols*) series (January 1874–December 1982 and January 1974–April 2014, respectively). The sunspot area is expressed in millionths of a solar hemisphere (μHem)

We here give an overview of the indices that are often employed to represent solar cycle properties, moving from the ones resulting from direct observations of the inner solar atmosphere, the photosphere (Sect. 2) and chromosphere (Sect. 3), to those deriving from measurements of the transition region and solar corona (Sect. 4). For each index, we summarize existing measurements and typical use, and for those that quantify a directly-measurable value of a physical observable, we describe the underlying physics, before our conclusion are given (Sect. 5). The indices discussed in this paper derive only from observations of the solar atmosphere. Most of these observations have been obtained with ground-based instruments over many decades. A review of the indices that quantify effects caused by the solar magnetism on the heliospheric and terrestrial environments, mostly due to the varying properties of the solar-wind and interplanetary magnetic field, can be found in this volume, e.g., in Usoskin et al. (2014) and Svalgaard (2014), respectively. The solar cycle variation in the total and spectral solar irradiance is reviewed in this volume by Yeo et al. (2014).

2 Indices from Observations of the Photosphere

Since the advent of the telescope, *sunspots* have been monitored regularly by amateur astronomers and at many observatories. Several indices have been defined from the observed sunspots, e.g. the international (earlier, Zürich or Wolf) sunspot numbers, the group sunspot numbers, and the sunspot areas. The time series of these indices, which are shown in Fig. 3 and further discussed below, are used mainly as proxies of solar activity and as tracers of the dynamo processes responsible for the build-up of large scale magnetic fields into the solar atmosphere.

Apart from sunspots, faculae[1] are the next obvious feature on the solar disk. They have been known since telescopes have been pointed at the Sun. Observations show their magnetic nature. They represent magnetic field dispersed from strong magnetic field concentrations in sunspots. Their origin leads facular properties to vary with the sunspot cycle.

Regular measurements of the magnetic field strength in sunspot and faculae have been carried out in addition to photometric observations since the advent of magnetographs. These measurements have been employed to produce line-of-sight (LOS), or radial, or vectorial estimates of the photospheric magnetic field in physical units, and to obtain some direct and indirect measurements of the magnetic field properties.

2.1 Sunspot Number

The sunspot number indicates a weighted estimate of individual sunspots and sunspot groups derived from visual inspection of the solar photosphere in white-light[2] integrated radiation. The sunspot number was introduced in the middle of the 19th century by Rudolf Wolf, who also started the program of synoptic measurements carried out at the Zürich Observatory that now constitutes the so-called Zürich (or Wolf) sunspot number series. This measurement series covers the period 1849–1981, but it was extended for several solar cycles backwards by using data available from earlier observations.

In this series, the sunspot number R_z is defined as:

$$R_z = k(10G + N), \tag{1}$$

where G is the number of sunspot groups, N is the number of individual sunspots in all groups visible on the solar disk and k denotes a correction factor that compensates for differences in observational techniques and instruments used by the observers.

The Zürich sunspot number series is based on observations carried out by a single astronomer for each day, by using almost the same technique for the whole period, but an offset due to the changes of the weighting procedure introduced in 1945–1946 and other corrections applied for earlier data. However, it is known that data gaps were filled with interpolation between available measurements without note on the series, by leading to possible errors and inhomogeneities in the data. Besides, the single observers have changed in time, as well as also observatory instruments and practices. These changes raise questions about the accuracy and integrity of the sunspot number data series available to date. We omit discussion of this important aspect of the sunspot data and refer the reader to the review in this volume by Clette et al. (2014) for further information.

The Zürich sunspot number data set is available online at e.g. the archive of the National Oceanic and Atmospheric Administration's National Geophysical Data Center (NOAA/NGDC).[3]

Since the termination of the sunspot observations at the Zürich Observatory the sunspot number series has been routinely updated as the International Sunspot Number (ISN) by the World Data Center for the production, preservation and dissemination of the international sunspot number (SILSO, Sunpot Index and Long-term Solar Observation[4]) at the Royal

[1] Faculae is the name given to brightenings seen in photospheric radiation mainly near the solar limb and in the general vicinity of sunspots. Find more information in e.g. Solanki and Krivova (2009).

[2] The term white-light indicates the sum of all visible wavelengths of solar radiation from 400 to 700 nm, so that all colors are blended to appear white to the eye.

[3] http://www.ngdc.noaa.gov/stp/solar/ssndata.html.

[4] http://www.sidc.be/silso.

Observatory of Belgium (ROB, Clette et al. 2007). The ISN series is computed using the same definition as employed for the Zürich series, but it represents the weighted average of sunspot numbers determined by several approved observers from more than 30 countries instead of that by a single observer. It is worth noting that sunspot numbers provided by individual stations, e.g. the Zürich series, are often 20 to 50 % higher than the ISN value, due to a multiplicative factor introduced to reduce results of modern observations to the scale of the original series started by Rudolf Wolf. Find information on the survey and merging of sunspot catalogs in e.g. Lefevre and Clette (2014). The ISN data are available online at the *ROB-SILSO* web site.[5]

In addition to the data presented above there are series of hemispheric sunspot numbers that account for spots only in the Northern and Southern solar hemispheres. These series have been used to study e.g. the North–South asymmetry of solar activity (e.g. Temmer et al. 2002, 2006; Virtanen and Mursula 2014; Norton and Charbonneau 2014). The data are available at e.g. the *ROB-SILSO* web page.

2.2 Group Sunspot Number

Analysis of the Zürich series and of early sunspot data by Douglas Hoyt and Kenneth Schatten lead to the definition of the group sunspot number and creation of a time series of this index based on all the available archival sunspot records since 1610 (Hoyt and Schatten 1996, 1998a, 1998b).

In this series, the group sunspot number R_g is defined as:

$$R_g = \frac{12.08}{n} \sum_i k'_i G_i, \tag{2}$$

where G_i is the number of sunspot groups recorded by the i-th observer, k' is the observer's individual correction factor, n is the number of observers for the particular day, and 12.08 is a normalization number scaling the group sunspot number to sunspot number values for the period of 1874–1976. The group sunspot number is more robust than the sunspot number since it is based on the easier identification of sunspot groups than of individual sunspots. Besides, it does not include the number of individual sunspots that is strongly influenced by instrumental and observational conditions. Therefore, although the group sunspot number series suffers some uncertainties (see, e.g. Vaquero et al. 2012; Leussu et al. 2013), it is more reliable and homogeneous than the Zürich series. The two series are nearly identical after the 1880s (e.g. Hathaway and Wilson 2004).

The group sunspot number data set is available online at the NOAA/NGDC archive.[6]

In the recent years, the recovery, digitization, and analysis of archived sunspot drawings regularly produced by several observers from early decades of 17th to late 19th centuries (e.g. Arlt et al. 2013; Vaquero and Trigo 2014; Arlt and Weiss 2014, and references therein) has allowed improvement of earlier uncertain sunspot data and extension for several cycles back in time of the butterfly diagram.

2.3 Sunspot Area

Sunspot areas and positions were diligently recorded by the Royal Observatory in Greenwich (RGO) from 1874 to 1976, by using measurements from photographic plates obtained

[5] http://www.sidc.be/silso/datafiles.

[6] http://www.ngdc.noaa.gov/stp/solar/ssndata.html.

at the RGO and other observatories (Cape Town, South Africa, Kodaikanal, India, and Mauritius). Both umbral areas and whole spot areas were measured and corrected for apparent distortion due to the curvature of the solar surface (Willis et al. 2013b, 2013a). Sunspot areas are given in units of millionths of a solar hemisphere.

Since 1976 the RGO measurements have been continued in the Debrecen Photoheliographic Data (DPD) sunspot catalogue that is compiled by the Debrecen Heliophysical Observatory, as commissioned by the International Astronomical Union. The DPD catalogue includes the heliographic positions and the areas of the sunspots on the full-disk white-light images obtained at the Debrecen and Gyula observatories, as well as from other observatories. The DPD sunspot data are available online[7] together with images of sunspot groups, scans of full-disk white-light observations, and magnetic observations. Moreover, the US Air Force also started compiling sunspot data from the Solar Optical Observing Network of telescopes (USAF/SOON)[8] since RGO ceased its program. Measurements of sunspot area are also available from a number of solar observatories around the world e.g. Catania (1978–present), Kislovodsk (1954–present), Kodaikanal (1906–present), Mt. Wilson (1917–1985), Rome (1958–present), and Yunnan (1981–present).

The data recorded at the RGO and the other observatories listed above are available at e.g. the NOAA/NGDC archive.[9] This archive includes some fragmentary data of sunspot areas obtained for earlier periods from solar drawings (e.g. Lepshokov et al. 2012; Arlt et al. 2013; Vaquero and Trigo 2014).

Sunspot areas are considered to be a more physical characterisation of the solar cycle than sunspot or group numbers, due to the linear relation between sunspot area and total magnetic flux of the sunspot (see e.g. Kiess et al. 2014, and references therein, and see Preminger and Walton 2007 for the underlying cause of non-linearities). Besides, sunspot area data have the additional information on the disc position of the observed features with respect to sunspot number series. For example, Fig. 8 of Hathaway (2010) shows the butterfly diagram of the daily sunspot area values as a function of latitude and time from 1874 to 2010 from the RGO and USAF archives.

A number of studies have attempted to inter-calibrate the sunspot area measurements available from the various observatories. The data were found to be not uniform across the various sets and even within a given set (e.g. Baranyi et al. 2001, 2013; Balmaceda et al. 2009, and references therein). For example, the values reported by USAF/SOON series resulted to be significantly smaller than those from RGO measurements (Balmaceda et al. 2009). Li et al. (2009) found that the time series of area values corrected for foreshortening on the solar disk and the ISN values are highly correlated, with a high level of phase synchronization between the compared indices in their low-frequency components (7–12 years), and a noisy behavior with strong phase mixing in the high-frequency domain, mainly around the minimum and maximum times of a cycle.

Sunspot area measurements are main input data e.g. to models of total magnetic flux and solar irradiance variations (e.g. Preminger and Walton 2006, 2007; Pagaran et al. 2009a; Krivova et al. 2010, and references therein).

2.4 Magnetic Field Measurements

Magnetograms are measurements of the net magnetic field strength averaged over the resolution element of the observation and polarity at a given height in the solar atmosphere.

[7] http://fenyi.solarobs.unideb.hu/DPD/index.html.

[8] http://www.ngdc.noaa.gov/stp/solar/sunspotregionsdata.html.

[9] http://www.ngdc.noaa.gov/stp/solar/sunspotregionsdata.html.

The instruments employed to produce these data rely on a variety of techniques applied to measure the polarization of light at various wavelength positions within a solar spectral line. Circular polarization in the opposite sense on either side of a magnetically sensitive spectral line gives a measure of the longitudinal component of the magnetic field vector. Linear polarization provides information on the strength and direction of the magnetic field transverse to the LOS. These measurements have been employed to produce LOS, or radial, or vectorial estimates of the observed magnetic flux density in physical units (either Tesla or Gauss, by using SI or cgs units, respectively).

Full-disk solar magnetograms have been recorded on a daily basis, starting at the Mt. Wilson Observatory[10] in the late 1950s, then with higher spatial resolution at the National Solar Observatory (NSO) Kitt Peak since the early 1970s, with the Kitt Peak Vacuum Telescope (KPVT, 1974–2003)[11] and since 2003 with the Synoptic Optical Long-term Investigations of the Sun (SOLIS) vector-spectromagnetograph (VSM, 2003–present),[12] at the Wilcox Solar Observatory (WSO, 1975–present),[13] and more recently by instruments of the Global Oscillations Netowrk Group (GONG, 2006–present).[14] The measurements have been also performed in space with the SOHO Michelson Doppler Imager magnetograph (SOHO/MDI, 1996–2011),[15] and presently by SDO Helioseismic and Magnetic Imager (SDO/HMI, 2010–present).[16] The first KPVT observations are characterized by 4 arcsec resolution and daily cadence, while the latter data obtained with the SDO/HMI are characterized by 1 arcsec resolution and 45 second cadence. Most of the magnetogram time series listed above refer to photospheric observations of the either the LOS or radial photospheric magnetic field strength, but the recent SOLIS observations that are taken by sampling two spectral lines originating at photospheric and chromospheric heights. SOLIS and SDO/HMI observations also produce vector magnetic field measurements. For example, the synoptic magnetic field movie linked to Fig. 13 of Hathaway (2010) shows a magnificent evolution of the solar magnetism during the last two and half solar cycles using NSO magnetograms between 1980–2009.

The results of magnetic field measurements are available at the web page of the various institutes and space missions listed above. An online database at the Pulkovo Observatory[17] contains data of magnetic field measurements obtained since 1957 at a number of observatories of the former Soviet Union.

The field measurements have been stored at the various observatories as are and further processed to produce synoptic charts. These charts show the latitude-longitude distribution of the magnetic field during a complete solar rotation and the butterfly diagram patterns. For example, Fig. 1 of Petrie (2013) shows the butterfly diagram obtained from NSO Kitt Peak data, summing up the photospheric radial field distributions derived from the longitudinal photospheric field measurements from 1974 to 2013. These charts are the main input data to numerical models of e.g. the solar dynamo and of the solar outer atmosphere and inner heliosphere (e.g. Riley et al. 2011).

[10] http://obs.astro.ucla.edu/intro.html.

[11] http://diglib.nso.edu/ftp.html.

[12] http://solis.nso.edu/0/index.html.

[13] http://wso.stanford.edu.

[14] http://gong.nso.edu.

[15] http://soi.stanford.edu/data/.

[16] http://jsoc.stanford.edu.

[17] http://www.gao.spb.ru/database/mfbase/main_e.html.

The results of the synoptic measurements carried out at the WSO,[18] NSO, and other observatories have been also employed to derive the time series of the monopole, dipole, and higher order coefficients of the spherical harmonics that decompose the magnetic field observations as a function of the surface latitude and longitude. The multipole coefficients have been used to study the dominant length scales and symmetries of the observed field (e.g. Petrie 2013), as well as to predict the value of the sunspot numbers at the next solar maximum (see e.g. Pesnell 2012). The heliospheric consequences of the solar cycle variation of the Sun's low-order magnetic multipoles are discussed in this volume by e.g. Wang (2014).

Measurements carried out at the Mt. Wilson Observatory have been also analyzed to produce proxies of sunspot and plage properties, specifically the Mount Wilson Sunspot and Magnetic Plage Strength Indices, which represent the fractional solar surface covered by magnetic fields exceeding 100 Gauss and between 10 and 100 Gauss, respectively. These indices have been employed e.g. to model solar irradiance variations (Jain and Hasan 2004). Moreover, local helioseismology methods have provided a way to map medium-to-large magnetic regions on the far-side hemisphere of the Sun (Lindsey and Braun 1997; González Hernández et al. 2007). The sound waves that pass through areas of strong magnetic fields experience a phase shift (Braun et al. 1992) that can be estimated with respect to the propagation in a non-magnetized atmosphere. Maps of the difference between the model and the measured phase shift have been derived regularly during the last decade from 24 hours of solar surface velocity data obtained by the GONG[19] and SOHO/MDI observations. These maps show the locations of shorter travel times due to change of sound speed in strong magnetic-field regions, whose presence modify the local plasma temperature. SDO/HMI[20] is providing data continuation after the decommission of SOHO/MDI. Recent improvements have turned the far-side maps into a useful space weather forecasting tool (González Hernández et al. 2014).

A number of studies have attempted to inter-calibrate the LOS or radial magnetograph full-disk measurements from one or more observatories (e.g. Liu et al. 2012, and references therein). Riley et al. (2014) compared maps from several of the observatories listed above to identify consistencies and differences among them. They found that while there is a general qualitative agreement among the maps produced by the various programs, there are also some significant differences. They also computed conversion factors that relate measurements made by one observatory to another using both synoptic map pixel-by-pixel and histogram-equating techniques. Besides, they explored the relationship between the various data sets over more than a solar cycle, by finding that the conversion factors remain relatively constant for most of the studied series.

2.5 White-Light Facular Indices

White-light faculae were routinely measured at the RGO from 1874 to 1976, by producing a record of over 100 years of continuous daily measurements of the position and projected area of these features. These measurements are available at the NOAA/NGDC archive.[21] Besides, polar faculae, i.e. faculae seen at high latitudes (50–90°) in white-light observations, have

[18] http://wso.stanford.edu.

[19] http://farside.nso.edu.

[20] http://stereo-ssc.nascom.nasa.gov/beacon/beacon_farside.shtml.

[21] http://www.ngdc.noaa.gov/stp/solar/wfaculae.html.

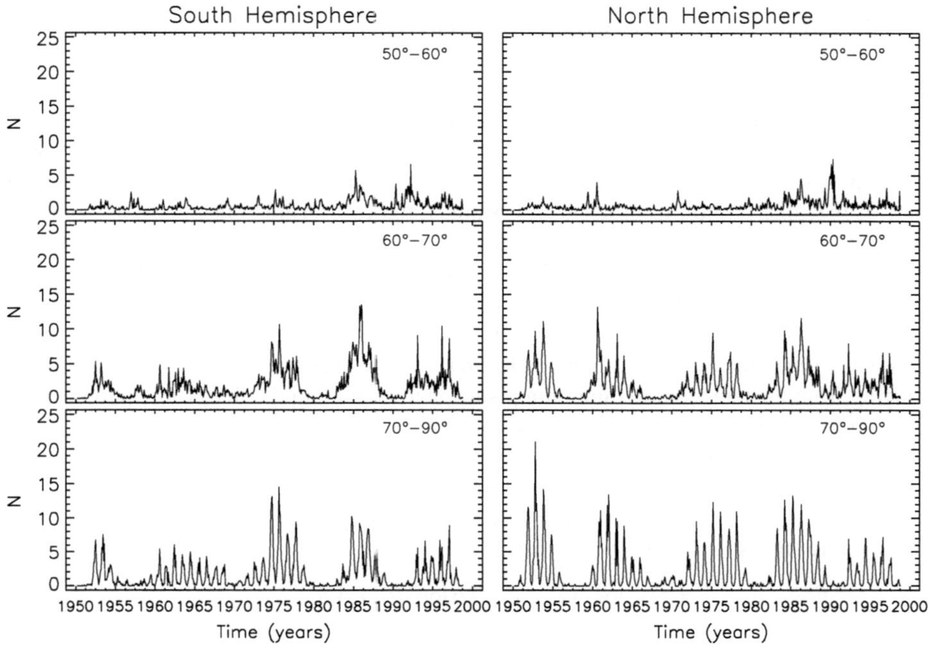

Fig. 4 Monthly averaged value of the number of white-light polar faculae measured from January 1951 to December 1998 at three high-latitude bands: 50–60° (*top panels*), 60–70° (*middle panels*), and 70–90°(*bottom panels*) on the South (*left panels*) and North (*right panels*) solar hemispheres. These measurements are from the archive of the National Astronomical Observatory of Japan

been also monitored since late 19th century. Regular observations of these features have been carried out at e.g. the RGO (1880–1954), Mt. Wilson (1906–1990), Zürich (1945–1965), National Astronomical Observatory of Japan (1951–1997),[22] and Kislovodsk (1960–present)[23] observatories. These observations have shown the counter phase variation of the number of polar faculae with respect to the sunspot cycle. Most recent studies indicate that the phase relationship between numbers of polar faculae and sunspots is both time- and frequency- dependent (Deng et al. 2013). However, the number of faculae at the poles of the Sun appeared to be well correlated with the LOS component of the polar magnetic field measured at the WSO (Sheeley 2008) and by the SOHO/MDI (Muñoz-Jaramillo et al. 2012), suggesting that the polar faculae number is a good proxy to study the evolution of the polar magnetic field. Observations and models of the solar polar fields are discussed in this volume by e.g. Petrie et al. (2014).

Figure 4 shows the monthly averaged values of the number of polar faculae at three high-latitude bands on both solar hemispheres from the archive of the National Astronomical Observatory of Japan.

[22] http://solarwww.mtk.nao.ac.jp/solar/faculae/.

[23] http://158.250.29.123:8000/web/P_fac/.

3 Indices from Observations of the Chromosphere

Monitoring programs of full-disk observations in the Ca II K and H_α resonance lines have been carried out at various observatories since the start of the 20th century.

The Ca II K line at 393.37 nm is among the strongest and broadest lines in the visible solar spectrum and thus easily accessible to ground-based observations, though in the violet part of the solar spectrum. In solar observations, the core of this line shows an emission reversal, with a central absorption minimum. In standard notation, K_3, $K_{2V,2R}$, and $K_{1V,1R}$ mark the core, the reversal (emission peaks), and the secondary minima of the doubly reversed profile of the line, in the violet (V) and the red (R) wings of the line, respectively. All these line features occur within a spectral range less than 1 Å wide. They result from a complex formation of the line (Uitenbroek 1989), which originates over atmospheric heights ranging from the temperature minimum photospheric region in the line wing up to the high chromosphere in the core (e.g. Leenaarts et al. 2013).

Observations at the Ca II K have long served as diagnostics of the solar chromosphere (e.g. Rutten 2007, and references therein). They show that the Ca II K line becomes brighter with non-spot magnetic flux. Quantitatively, the line core excess flux density, $\Delta F_{\mathrm{Ca\,II}}$ ($\mathrm{erg\,cm^{-2}s^{-1}}$), can be written as:

$$\log \Delta F_{\mathrm{Ca\,II}} = 0.6 \log\langle B\rangle + 4.8, \tag{3}$$

where $\langle B\rangle$ indicates the spatially averaged magnetic flux density in the resolution element of the observation (Schrijver et al. 1989). Therefore, the Ca II K line emission can be used as a good proxy of the LOS magnetic flux density over the whole solar disk (e.g. Ermolli et al. 2010, and references therein).

The H_α line at 656.28 nm has been also widely employed for studying the solar chromosphere, which was defined as what is seen in this line (Lockyer 1868). The formation of this deep-red line is complicated (Leenaarts et al. 2012) and originates over atmospheric heights ranging from the photosphere in the line wing up to the middle chromosphere in the core (e.g. Leenaarts et al. 2013). Observations at this line show a wealth of magnetically dominated solar features and processes, e.g. flares, filaments, prominences, varying pattern of large-scale magnetic polarity, and the fine structure of magnetic regions. The appearance of these solar features depend on the sensitivity of the H_α opacity to the mass density and temperature of the plasma traced by magnetic fields at different heights into the solar atmosphere.

Regular measurements of the solar Mg II h and k resonance lines at 280.27 and 279.55 nm have also been registered from space, as a measure of solar chromospheric variability and good proxies of the solar UV emission (e.g. Dudok de Wit et al. 2009). These lines sample higher layers of the solar chromosphere than the other diagnostics listed above (e.g. Leenaarts et al. 2013, and references therein). The Mg II core-to wing index has been derived from the Mg II measurements, by taking the ratio of the h and k lines of the solar Mg II emission at 280 nm to the background solar continuum near 280 nm (Heath and Schlesinger 1986). The solar Mg II index has been monitored on a daily basis by NOAA spacecraft since 1978. The data are available at the NOAA/NGDC archive.[24]

The Mg II index has been widely employed e.g. to parameterize the facular and network contribution to short- and long-term solar irradiance variations (Pagaran et al. 2009a; Lean et al. 2011; Fröhlich 2013).

[24] ftp://ftp.ngdc.noaa.gov/STP/SOLAR_DATA/SOLAR_UV/NOAAMgII.dat.

3.1 Ca II K Line and Plage Area Indices

Daily full-disk Ca II K observations have been obtained with either spectrographs, Lyot-type, or interference filters at various observatories, e.g. Kodaikanal (1904–present), Meudon (1909–present),[25] Mt. Wilson (1915–1985),[26] Arcetri (1934–1975),[27] Big Bear (1942–1987), Kislovodsk (1957–present),[28] Rome (1964–present),[29] San Fernando (1984–present),[30] Kanzelhöhe (2010–present).[31] The observations at the various sites have been reduced to measure e.g. the position and area of Ca II K plage[32] regions, or to derive measurements of the Ca II K line indices described in the following. The data have been also employed to produce synoptic charts of the Ca II K enhanced emission, which reflects the magnetic flux distribution on the solar disk during a complete solar rotation.

Measurements on modern and historical full-disk Ca II K observations show both cyclic and long-term variations of plage properties (e.g. Ermolli et al. 2009a, 2009b; Foukal et al. 2009; Chapman et al. 2011; Priyal et al. 2014, and references therein). However, similarly to sunspot number series, caution is needed when considering results derived from analysis of long time series of Ca II K full-disk observations, without careful analysis of their problems and intrinsic instrumental variations. In fact, simple visual inspection of the data available at the various archives reveals considerable differences between the images from the various time series, due to different observational and instrumental characteristics. For example, Ermolli et al. (2009a) show that the yearly median values of the plage area measurements derived from the Mt. Wilson, Kodaikanal, and Arcetri time series of historical Ca II K observations agree within 40 %, with a Pearson correlation coefficient ranging from 0.85 to 0.93. However, the values derived from the three series differ considerably for cycles 15, 17, 19, with a relative difference as high as 140 %. Besides, the time series of measured plage area values derived from the three archives show the eleven-year solar cycle variation of the sunspot numbers, but differences at short time scales are found when comparing sunspot and plage measurements, in particular, subtle but systematic differences around the minima of solar activity, from 1945 to the present. Middle panel of Fig. 2 shows the time series of monthly averaged values of the plage area derived from historical and modern Ca II K full-disk observations carried out at the Kodaikanal and Rome observatories, respectively.

Some Ca II K observations listed above have been also processed to produce time series of indices based on the intensity distribution measured over the solar disk (e.g. Caccin et al. 1998; Bertello et al. 2010). These indices resulted to be well correlated to the fractional area of the solar disk occupied by plages and network. Besides, the full-disk Ca II K observations have been employed to study e.g. the onset and end of solar cycles and cycle properties (e.g. Harvey 1992; Ermolli et al. 2009b). In a recent study Sheeley et al. (2011) computed

[25]http://bass2000.obspm.fr/.

[26]http://www.astro.ucla.edu/~ulrich/MW_SPADP.

[27]http://www.oa-roma.inaf.it/solare/index.html.

[28]http://old.solarstation.ru/.

[29]http://www.oaroma.inaf.it/solare/index.html.

[30]http://www.csun.edu/SanFernandoObservatory/.

[31]http://cesar.kso.ac.at.

[32]Plage is the name given to the brightening seen in chromospheric radiation corresponding to photospheric faculae. In contrast to faculae, plage are seen over the whole disk, in active regions and in the quiet sun, on the network pattern formed at the borders of supergranular cells. Find more information in e.g. Solanki and Krivova (2009).

butterfly diagrams of the longitudinally averaged Ca II K intensity from the Mt. Wilson observations for the years 1915–1985. From analysis of these charts they found that cycle 19 is remarkable for its broad latitudinal distribution of active regions, its giant poleward surges of flux, and for the emergence of a North–South asymmetry that lasted 10-years. It is wort noting that the CaII K observations exist also for prior sunspot cycles when magnetograms were not available.

Since late 1960s Ca II K line profiles of the Sun were obtained both integrated over the solar disk and at given latitudinal bands at e.g. the NSO in Sacramento Peak and Kitt Peak, and at the Kodaikanal Observatory, by using high resolution spectrographs. For the subsequent decades, these K-line monitoring programs have produced almost daily measurements of several parameters characterizing the Ca II K-line measurements. The daily observations of the disk integrated Ca II K emission at the NSO Kitt Peak have been obtained with the Integrated Sunlight Spectrometer (ISS) of the SOLIS telescope since 2006. The ISS monitors the solar emission at nine different wavelength bands regularly.[33]

The Ca II K line parameters measured to date include the Ca K emission index, which is defined as the equivalent width of a 1 Å band centered on the K line core, and various measures of line features (e.g. relative strength of the blue K_2 emission peak with respect to the K_3 intensity, separation of the two emission maxima, separation of the blue and red K_1 minima) and asymmetry (ratio of the blue and red K_2 emission maxima). The data derived from the NSO program are available online.[34]

Figure 5 shows the time series of monthly averaged values of the Ca II K line parameters measured at the NSO. A recent analysis of these data by Scargle et al. (2013) indicates that the temporal variation of the measured line parameters consists of five components, including the solar cycle eleven year period, a quasi periodic variation with 100 day period, a broad band stochastic process, a rotational modulation, and random observational errors. Pevtsov et al. (2013) found a weak dependency of intensity in the Ca II K line core measured in the quiet chromosphere with the phase of the solar cycle. This dependency has been attributed to the signature of changes in thermal properties of basal chromosphere with the solar cycle.

3.2 H_α Filament and Prominence Data

Daily full-disk H_α observations have been obtained with either filters or spectroheliographs at various observatories, e.g. Meudon (1909–present),[35] Arcetri (1927–1969), Catania (1906–present),[36] Kanzelhöhe (1973–present),[37] and Big Bear (1982–present),[38] to mention a few. These observations were taken regularly also to support solar-terrestrial prediction services.

The data obtained at various observatories have been stored as are and further processed to produce H_α synoptic charts, which incorporates information about evolving features observed in the solar chromosphere at the H_α radiation into a single map for each rotation of the Sun. These charts show e.g. the boundaries separating positive and negative magnetic

[33] http://solis.nso.edu/iss.

[34] http://nsosp.nso.edu/cak_mon/.

[35] http://bass2000.obspm.fr/home.php.

[36] http://www.oact.inaf.it.

[37] http://cesar.kso.ac.at/synoptic/ha4m_years.php.

[38] http://www.bbso.njit.edu/Research/FDHA/.

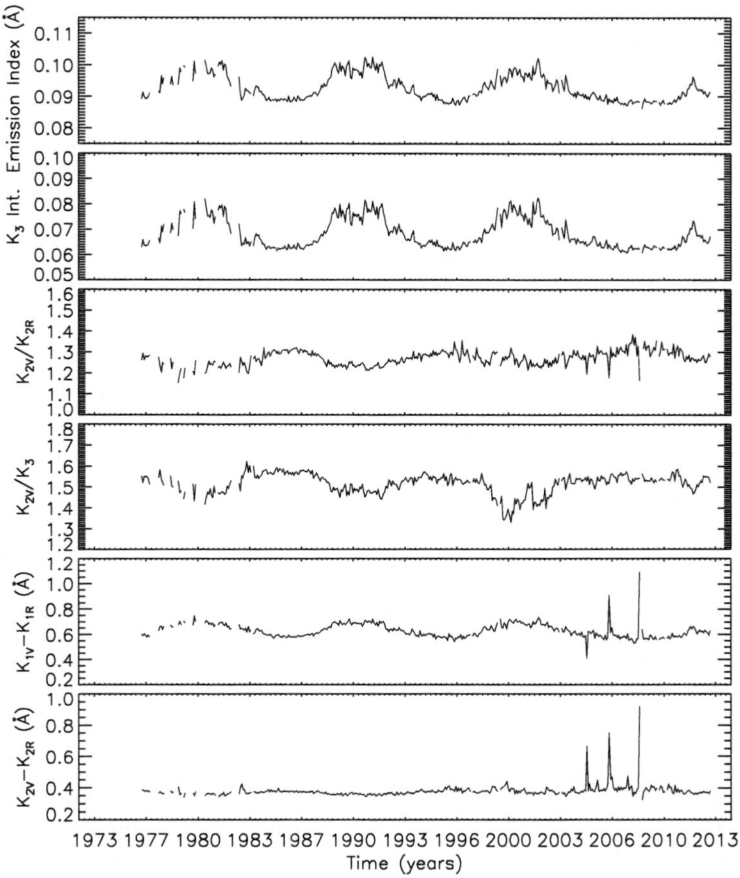

Fig. 5 Time series of monthly averaged values of the Ca II K line parameters measured from November 1976 to April 2014 at the NSO. Details are given in the text

polarities on the solar surface, and data of the filaments and filament channels. At present, the summary series of these charts cover the period from 1887 up to present (e.g. Vasil'Eva et al. 2002), though earlier data are rather uncertain.

Figure 6 shows the H_α map for a solar rotation at the cycle 23 maximum, from 26 September to 23 October 2003, obtained at the Meudon Observatory. Local H_α and Ca II K observations have been processed (Mouradian 1998) since late 1989 for drawing maps and for computing active region and filament tables. The produced maps show the properties of observed features during their transit across the solar disk. They display active regions (sunspots and plage) at their activity maximum, i.e. maximum area, and filaments at their maximum spread at the average position of daily baselines. Other synoptic maps derived from Meudon observations show daily information of filament traces and prominence positions observed on the limb. Besides, there are data of filaments available at the NOAA/NGDC archive,[39] dating back 1919 and consisting of the position, shape, and life time of the observed features and

[39]http://www.ngdc.noaa.gov/stp/space-weather/solar-data/solar-features/prominences-filaments/filaments/.

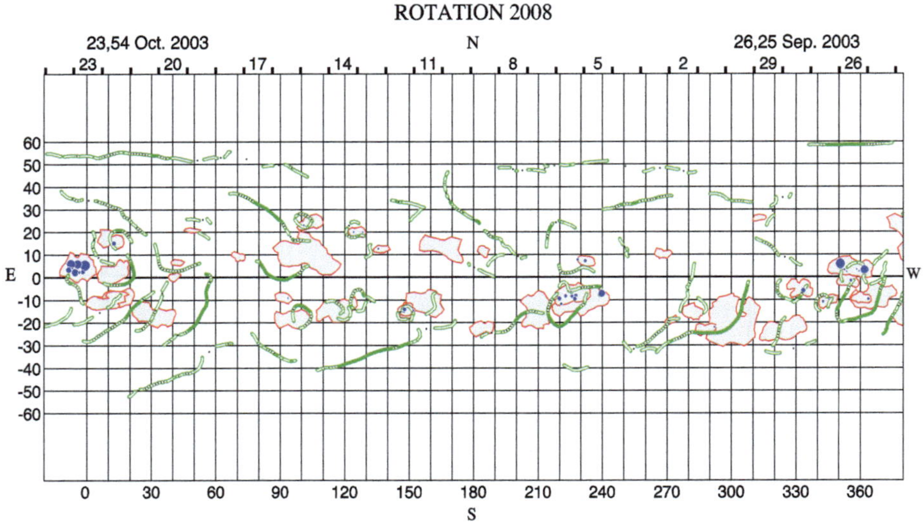

Fig. 6 Synoptic map of solar activity from H_α and Ca II K observations obtained at the Meudon Observatory. This map summarizes the properties of solar features observed during their transit across the solar disk from 26 September to 23 October 2003, i.e. for a solar rotation during the last activity maximum. The active regions (sunspots and plage marked with *blue* and *red areas*, respectively) and filaments (*green areas*) observed on the disk are shown at their activity maximum and maximum spread, respectively. The position on the map of the observed features represents the average position of daily baselines

of their different parts. Other available data include the area of prominences (1957–present) and filaments (1959–present) observed at the Kislovodsk station of the Pulkovo Observatory.

Figure 7 shows the coefficient of the spherical harmonic decomposition of the large scale solar magnetic field derived from the field polarity data deduced from the H_α synoptic charts produced at the Kodaikanal, Meudon, NSO Sacramento Peak, and Kislovodsk observatories (see e.g. Makarov and Tlatov 2000; Makarov et al. 2001, for a description of the methods applied). It is noteworthy that the period covered by these charts is comparable to the length of sunspot group series, exceeding the period covered by direct measurements of the large-scale solar magnetic field.

3.3 Flare Indices

Flares[40] are complex multi scale phenomena (Shibata and Magara 2011) that affect different spatial and temporal scales at various heights in the solar atmosphere, from the photosphere to the corona and beyond. Therefore, they are observed at all wavelengths from decameter radio waves to gamma-rays at 100 MeV. Occasionally, flares are also seen in white-light photospheric observations (Benz 2008). The X-ray and particle flux emitted during solar flares have been measured regularly since the advent of space missions. However, before the space age, flares were monitored for many years via H_α chromospheric observations carried out at many observatories, by visual, photographic, or digital inspection of the solar

[40] Flare is the name given to a sudden, rapid, and intense brightening observed over the solar disk or at the solar limb, due to a release of magnetic energy (up to 10^{32} erg on the timescale of hours), followed by ejection of solar plasma through the corona into the heliosphere. Find more information e.g. in Benz (2008).

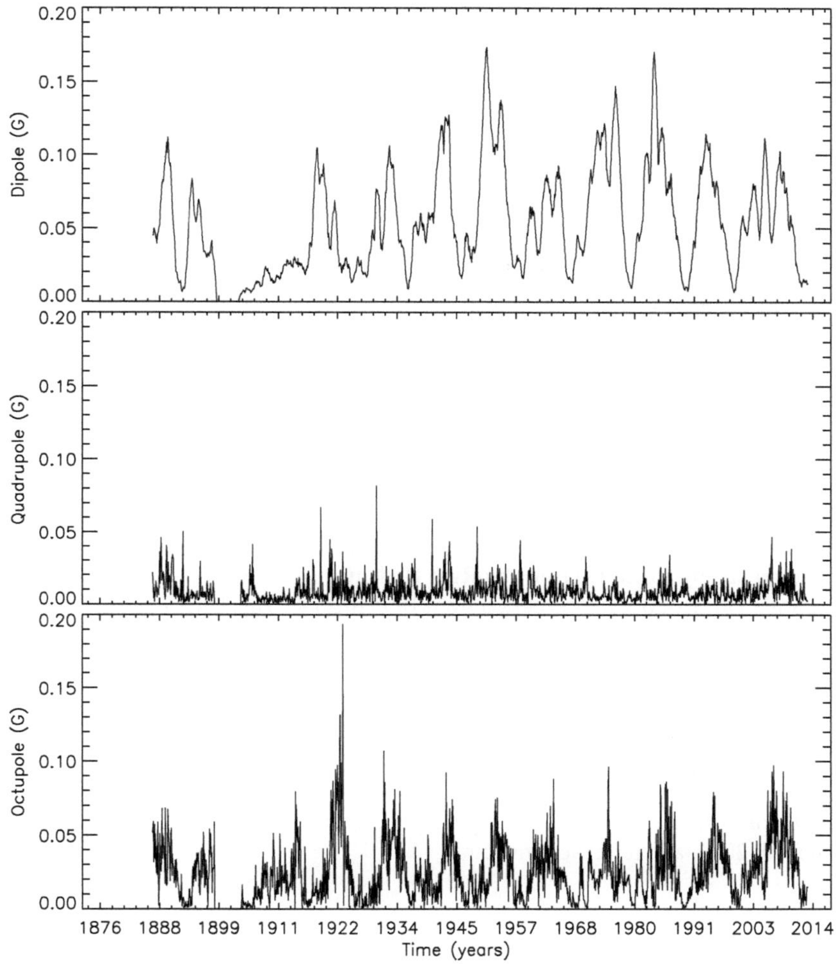

Fig. 7 Monthly averaged values of the dipole, quadrupole, and octupole coefficients of the spherical harmonics that describe the large scale solar magnetic field from the measurements of the filaments, filament channels, and prominences in the H_α synoptic charts (January 1887–December 2013) produced at the Kodaikanal, Meudon, NSO Sacramento Peak, and Kislovodsk observatories

disk. Although the full effect of the solar cycle on flares is depicted by the measurements of these events recorded at the various wavelength bands, we focus in the following on the flare indices from synoptic observations of the solar chromosphere, because of the remarkable archival data. Indeed, reports of the H_α solar flare patrol programs are available from 1938 to present at the NOAA/NGDC archive,[41] which holds data for about 80 observing stations. Currently five observatories send their data to the NOAA/NGDC archive on a routine monthly basis. The information stored in these reports include time of beginning-, maximum brightness-, secondary maxima-, and end- of the event, heliographic coordinates of center of gravity of flare at maximum brightness, optical brightness, area of the flaring region at

[41] http://www.ngdc.noaa.gov/stp/space-weather/solar-data/solar-features/solar-flares/h-alpha/reports/.

time of maximum brightness. These data are integrated by the X-ray flux of the events as measured from the NOAA satellites SOLRAD (1968–1974) and GOES (1975–present). For GOES X-ray events, the flare is assumed to start when four consecutive 1-minute X-ray flux measurements are above a threshold value, are strictly increasing, and the last value is at least 1.4 times larger than the value measured three minutes earlier. The maximum flux value measured during the flare defines the event size, which is classified by the C-, M-, X-class scale. The event ends when the flux reading returns to half the sum of the flux at maximum plus the flux value at the start of the event.

The flare activity shown by the H_α observations has been also described with the "flare index", whose concept was introduced by Josip Kleczek in early 1950s (Kleczek 1952). This quantity is defined as the product of the intensity scale of the flare observed in the H_α radiation and its duration in minutes. It is assumed to be roughly proportional to the total energy emitted by the flare (e.g. Özgüç et al. 2003, and references therein). Daily data of the flare index are available at e.g. the NOAA/NGDC archive[42] and, for solar cycles 20 to 23, also at the site of the Bogazici University.[43]

Alternative definitions of the flare index exist with respect to that presented above, e.g. the "X-ray flare index" as described e.g. by Criscuoli et al. (2009), based on X-ray flux measurements in the 1–8 Å range of the solar spectrum, and the "Comprehensive Flare Index" computed by Helen Dodson and Ruth Hedeman from observations of the McMath–Hulbert Solar Observatory for the period from 1955 to 2008. This latter index is based on H_α observations and measurements of the magnitude of the event measured at radio frequencies, as well as of the importance of sudden ionospheric disturbance. The data are available at the NOAA/NGDC archive.

The flare index has been widely employed to single out signatures of the flaring activity in magnetic regions. For example, it has been recently used to test the efficiency of measurements of the topological complexity of the magnetic field concentrations to discriminate between flaring and non-flaring regions (e.g. Georgoulis 2013; Ermolli et al. 2014, and references therein).

4 Indices from Observations of the Transition Region and Solar Corona

The solar corona has been observed regularly since the advent of the coronographs by Bernard Lyot in the early 1930s and the contemporary discovery of the He I 1083 nm infrared line in the solar spectrum by Harold and Horace Babcock (Babcock and Babcock 1934). Since then several coronal quantities have been measured daily. These quantities have been supplemented by the huge data set of imaging observations taken during the space era. These data allow to evince the magnetic structure of the corona that originates the solar wind governing the magnetic field in the heliosphere. Direct measurements of the coronal magnetic field are still scarce to date. However, the coronal field can be modeled by extrapolating the magnetic field observed in the inner solar atmosphere. To this purpose, sunspot data series have been employed as input data to e.g. solar surface transport models coupled with extrapolations of the heliospheric field (e.g. Jiang et al. 2011, and references therein).

The He I line is seen in absorption on the solar disk, but in emission under some special conditions, e.g. during solar flares. The helium absorption is enhanced with respect to the

[42] http://www.ngdc.noaa.gov/stp/solar/solarflares.html.

[43] http://www.koeri.boun.edu.tr/astronomy/fi_nedir.htm.

quiet Sun above active regions and H_α filaments (e.g. Brajša et al. 1996), and reduced in coronal holes (CH).[44] Due to a complex line-formation mechanism (Avrett et al. 1994), the patterns seen in He I 1083 nm line observations are affected by processes in the upper chromosphere, transition region, and low corona. In particular, model calculations show a strong dependence of the line absorption on coronal illumination. Lagg (2007) summarizes the results of magnetic field measurements using the He I line and discusses the potential of this line as a diagnostic of the solar outer atmosphere. Observations at this line have been taken at the NSO Kitt Peak since 1974 on a daily basis.

The solar upper atmosphere also emits radio flux. It consists of free-free emission from quiet sun coronal plasma and gyromagnetic emission from sunspots in active regions (e.g. Shibasaki et al. 2011, and references therein).

The integrated radio flux from the Sun is defined as:

$$F = \frac{2k_B}{\lambda^2} \int T_b d\Omega,$$

where, F, k_B, λ, Ω, are radio flux ($\mathrm{Wm^{-2}Hz^{-1}}$), the Boltzmann constant, the observing wavelength, and the solid angle of the source respectively. The solar radio flux is measured using the Solar Flux Unit (SFU, 10^{-22} $\mathrm{Wm^{-2}Hz^{-1}}$). T_b is called radio brightness temperature and is related to the plasma temperature T as follows:

$$T_b = \int T e^{-\tau} d\tau,$$

where, τ is the optical depth, which is related to the absorption coefficient κ by the relation $d\tau = \kappa d\ell$, where ℓ is the path length along the line of sight. In the case of thermal free-free emission (collisions of electrons with ions), the absorption coefficient is a function of the plasma density n, the temperature, and the observing wavelength: $\kappa = \xi c^{-2} \lambda^2 n^2 T^{-3/2}$, where ξ is a constant (0.1 in the chromosphere and 0.2 in the corona) and c is the light speed. In an optically thick case ($\tau \gg 1$), the brightness temperature is equal to the plasma temperature and the integrated flux is inversely proportional to square of the measured wavelength. Gyro-resonance is prominent around 3 GHz compared to free-free contribution.

Among the various radio measurements, those pertaining the flux in the wavelength range of 2.8 GHz or 10.7 cm, near the peak of the observed solar radio emission, constitute the longest, most stable and almost uninterrupted record of direct physical data of solar activity available to date. This is because the radio flux measurements are rather insensitive to weather and disturbances in the ionosphere. Historically, this index, which provides the so-called F10.7 index series, has been used as an input to ionospheric models as a surrogate for the solar UV output that produces photoionization in the Earth's ionosphere.

4.1 He 1083 nm Line Observations

Regular observations of the solar disk at the He I line started in late 1930s. These observations were carried out with spectroheliographs until the availability of spectropolarimetric measurements. The results were mostly employed to investigate the chromospheric topology and dynamics. After their first detection from space in the late 1960s and early 1970s, most of CH data available to date were later deduced from available He I 1083 nm observations.

[44] Coronal holes are areas where the Sun's corona is darker, and colder, and has lower-density plasma than average. They are associated with rapidly expanding open magnetic fields and the acceleration of the high-speed solar wind. Find more information in e.g. Potgieter (2013).

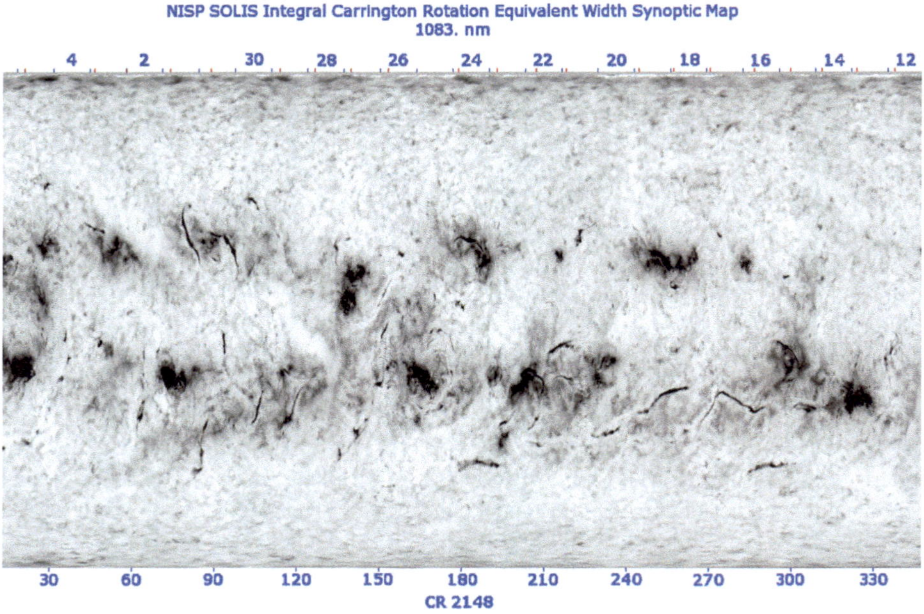

Fig. 8 Map derived from daily full-surface intensity values of the equivalent width as measured from profiles of the spectral line of He I 1083.0 nm. These measurements are from the NSO archive

Full-disk observations at the He I 1083 nm line have been taken at the NSO Kitt Peak since 1974 (Harvey and Sheeley 1977) on a daily basis, by using the KPVT magnetograph. The data were published by the NOAA/NGDC archive[45] as Helium synoptic charts for each solar rotation or as CH contours plotted on H_α synoptic charts. In particular, the latter method was used to determine the polarity of each observed CH, and to identify it over several solar rotations. Magnetograms from NSO Kitt Peak were also used for this purpose. The NSO He I 1083 nm observations have been continued by the SOLIS VSM since 2003. The data obtained consist of full-disk intensity maps representing the equivalent width as measured over the solar disk from profiles of the spectral line of He I 1083 nm.

Figure 8 shows a SOLIS He I synoptic chart for the period from 12 March to 5 April 2014 from the NSO archive. The data in this archive have been employed to derive information on CH positions. CHs were found at the Sun's polar regions at solar minimum, and located anywhere on the Sun during solar maximum. Besides, transient "dark points" in He I 1083 nm observations were found to be associated with small magnetic bipoles. The number of these dark points resulted to vary inversely with the sunspot number (Harvey 1985).

The NSO Kitt Peak He I 1083 nm intensity maps have been used e.g. by Harvey and Recely (2002) to identify and measure the evolution of polar CHs during cycles 22 and 23. They found that polar CHs, which are the largest at cycle minimum, evolve from high-latitude ($\approx 60°$) isolated holes. During the initial 1.2–1.4 years following the polar polarity reversal, the polar CHs develop asymmetric lobes extending to active latitudes and the area and magnetic flux of the CH increase rapidly.

[45] http://www.ngdc.noaa.gov.

4.2 Radio Flux Data

Synoptic radio measurements have been made at the various observatories since 1945 (Sullivan 1984), by using radio polarimeters, to measure the total coming flux and the circular-polarization degree at various frequencies, and by employing radio heliographs made of many antennas that allow to produce interferometric measurements at various frequencies. The radio emission data available to date include daily values of solar flux density at 30 different frequencies ranging from 0.1 to 15 GHz, and tables of distinctive events. Radio charts of active regions on the Sun have been also derived from the measurements carried out at e.g. the Nancay and Nobeyama observatories.

Radio flux from the Sun consists of three components that can be discriminated by the temporal scale of flux variations (e.g. Tapping 2013, and references therein). The burst component is the flux variation associated with flares on typical time scales from seconds to hours. It is employed to diagnose high energy particle acceleration and production of thermal plasma associated with flares. The slowly varying (S) component is linked with the evolution of active regions on typical time scales from a day to weeks. The base (B) component is the residual flux after subtraction of the burst and S components. Daily quiet time total flux (daily flux) data consists of S and B components.

As stated above, the time series of the radio flux measured at 2.8 GHz or 10.7 cm, the F10.7 index constitutes the longest record of a physical data of the solar cycle available to present. This index is a measurement of the integrated emission at 10.7 cm wavelength from all sources present on the disk. Daily measurements at local noon of this flux have been made by the National Research Council (NRC) of Canada from 1947 to 1991 in Ottawa and thereafter in Penticton. The flux values (Covington 1969) are expressed in the SFU units introduced above. The data, which are available at the site of the NRC[46] and NOAA/NGDC archive,[47] include daily F10.7 flux values, monthly averages, and rotational averages computed over a solar rotation. The NOAA/NGDC archive also contain radio measurements obtained at various frequencies e.g. since 1956 at the Astronomical Observatory of the Jagellonian University in Cracow, since 1966 with the USAF Radio Solar Telescope Network operated at various observatories, and from 1962 to 1973 with the Stanford Radio Astronomy Institute telescope.

The second longest record of solar radio measurements is from Japan (Tanaka et al. 1973). Measurement at 3.75 and 9.4 GHz started there in 1951 and 1956, respectively; they were supplemented with measurements at 1.0 and 2.0 GHz in 1957. The measurements were carried out at the Toyokawa observatory till 1994, and thereafter continued at the Nobeyama Solar Radio Observatory (NSRO). Results of these and of the earlier synoptic measurements carried out from 1951 to 1994 are available at the NSRO site.[48]

Figure 9 shows the time series of monthly averaged values of the radio flux measurements from the NSRO archive, superposed to scaled values of the ISN from the *ROB-SILSO* series. Pearson's correlation coefficients between the radio flux and ISN values are range between ≈ 0.94 and 0.98.

Additional radio data include those obtained with the Radio Solar Telescope Network started by the Sagamore Hill Solar Radio Observatory in 1966, with stations at Palehua/Kaena, Learmonth, and SanVito, whose observing frequencies are 245, 410, 610, 1415,

[46] http://www.spaceweather.gc.ca/solarflux/sx-5-eng.php.

[47] http://www.ngdc.noaa.gov/stp/space_weather/solar_data/solar_features/solar_radio/noontime_flux/.

[48] http://solar.nro.nao.ac.jp/norp/.

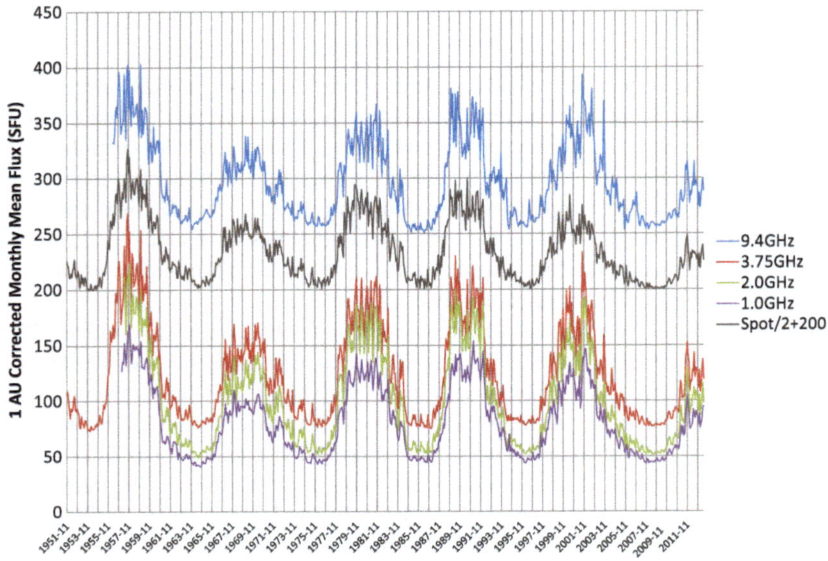

Fig. 9 Monthly averaged values of the radio flux measurements at the 9.4, 3.75, 2, 1 GHz from the NSRO archive (January 1951–December 2013), corrected for 1 AU distance, superposed to scaled values of the ISN from the *ROB-SILSO* series

2695, 4995, 8800, and 15400 MHz, and by the Kislovodsk station of the Pulkovo Observatory,[49] which started operation in 1960 at wavelengths of 5, 10, 15 GHz (or 2, 3, 5 cm). Besides, since 1996 the Nancay Radio Heliograph (Kerdraon and Delouis 1997) and decametric array telescopes [50] have been acquiring thousands of interferometric images each day of the solar corona and at very low frequency. The frequencies monitored regularly with these telescopes cover the range 5–10 Mhz, 10–100 MHz (from 20 to 75 MHz), higher than 100 MHz (at 164 MHz and 327 MHz). It is worth noting that 1D scans of the Sun were also taken from 1967 to 1996, but the data are not available online. Data from the decametric telescope goes back to 1991.[51]

The daily F10.7 index was found to be well correlated to sunspot number and area (Denisse 1949). It was then noticed that the radio emission in the range of 3–30 cm wavelength, or 10–1 GHz frequency range, correlates well with E-layer ionization (Kundu 1970). Since then the daily F10.7 index has been used as a proxy of solar cycle in models of the Earth's upper atmosphere. Recent studies by Tapping and Valdés (2011) and Svalgaard and Hudson (2010) suggested that the relation between sunspot numbers and radio flux has started to deviate from the previous relation since the 23 solar cycle. However, Henney et al. (2012) found that the relation between total magnetic flux on the solar surface and radio flux values does not show any noticeable change in the last cycles. The results suggest that the F10.7 index is a better proxy of the total magnetic flux on the solar surface than sunspot numbers.

[49] http://www.gao.spb.ru/english/database/sd/tables.htm.

[50] http://secchirh.obspm.fr/nrh_data.php.

[51] http://realtime.obs-nancay.fr/dam/data_dam_affiche/.

In a recent study, Dudok de Wit et al. (2014) merged daily observations from the Ottawa/Penticton and Toyokawa/Nobeyama observatories into a single homogeneous data set of the solar flux at wavelengths of 1, 2, 2.8, 3.75, 9.4 GHz (or 30, 15, 10.7, 8 and 3.2 cm), spanning from 1957 to present. They found that most solar proxies, in particular the MgII index, are remarkably well reconstructed by simple linear combination of radio fluxes at various wavelengths. Their results also indicate that the flux at 1 GHz (or 30 cm) stands out as an excellent proxy of solar cycle and is better suited than the F10.7 index for the modelling the Earth's thermosphere-ionosphere system.

4.3 Coronal Indices

The intensity of the coronal green radiation from the Fe XIV emission line at 530.3 nm has been observed since 1939 at the Arosa observatory and then recorded from 1947 onwards first at the Climax and Pic du Midi observatories and thereafter at other sites (e.g. at Norikura Observatory[52] between 1951–2009). The emission of this line can be observed above the entire solar limb during the whole solar cycle, contrary to the emission of other coronal lines, e.g. the Fe X red line at 637.4 nm and the Ca XV yellow line at 569.4 nm, which can be observed only occasionally. The intensity of the green corona depends on both the density and temperature of the plasma in the outer solar atmosphere, and both these quantities are modulated by the local magnetic fields. At present observations of the green corona are made at the Kislovodsk, Lomnicky Stit, and Sacramento Peak observatories. The measurements recorded at the various sites differ due to time difference, method of observation, height of the observation above the solar limb, and other parameters. However, the data obtained at these observatories have been processed since 1939 to obtain a physical index of the solar cycle signature on the outer solar atmosphere, the so-called coronal index (CI, e.g. Rybanský et al. 2005, and references therein), and a homogeneous data set (HDS, Minarovjech et al. 2011, see Fig. 10).

This data set, which has been obtained by scaling all the available measurements to the photometric scale of the Lomnicky Stit observations, is available for the period of 1939–2008 at the NOAA/NGDC web page.[53] Recently, Dorotovič et al. (2014) developed a method to substitute the ground-based observations by space-borne 28.4 nm (Fe XV) observations from the SOHO/EIT data. The dataset, named Modified Homogeneous Dataset (MHDS), extends the HDS beyond 2008. For the period up to 1996, the MHDS is identical with the former HDS. The MHDS is available online[54] for the 1996–2010 period.

The coronal index represents the averaged daily irradiance in the green coronal line emitted in 1 steradian towards the Earth, and it is expressed in ($\mathrm{W\,sr^{-1}}$) units.[55] Rušin and Rybansky (2002) analyzed these data to study the relationship between the intensity of the green corona and strength of photospheric magnetic flux over the period 1976–1999, by finding a relation between these indices that allowed them to extend solar surface magnetic fields estimates since 1976 back to 1939, when the green corona began to be observed. A discussion of the coronal index with respect to other solar cycle indices can be found in e.g. Rybanský et al. (2001).

In addition to the integrated flux measurements of the coronal green line emission, other measurements have been also recorded systematically, e.g. the intensity measured around

[52]http://solarwww.mtk.nao.ac.jp/en/db_gline1.html.

[53]http://www.ngdc.noaa.gov/stp.

[54]http://www.suh.sk.

[55]Conversion: $1 \times 10^{16}\ \mathrm{W\,sr^{-1}} = 4.5 \times 10^{-7}\ \mathrm{W\,m^{-2}} = 1.2 \times \mathrm{photons\ cm^{-2}s^{-1}}$.

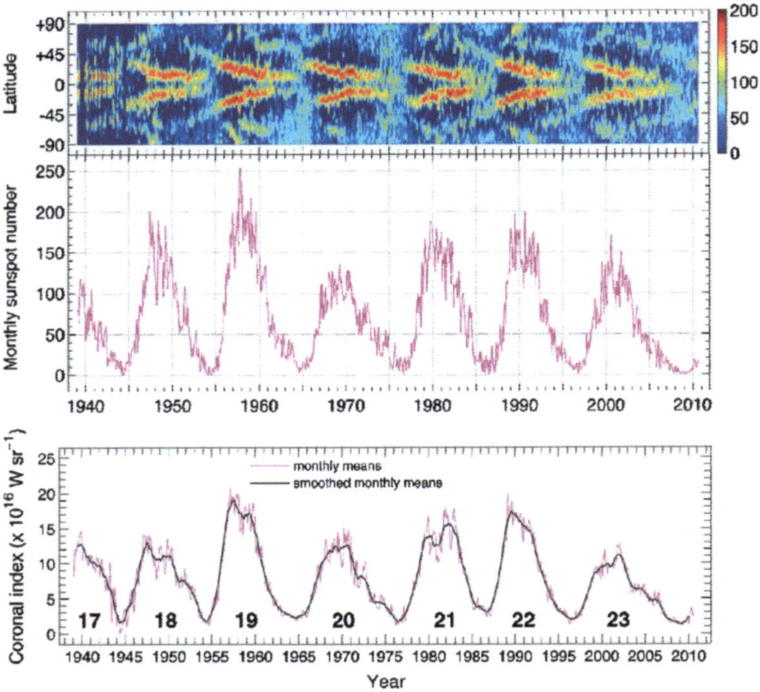

Fig. 10 Time series of the coronal index (CI, *bottom panel*), the sunspot number (*middle*), and the Fe XIV 530.3 nm coronal green emission line brightness maxima (*top*). The CI is computed from the homogeneous dataset (HDS). Panels adapted from Minarovjech et al. (2011)

the solar limb, within intervals of 5°, at the coronal lines at 637.4 nm (Fe X), 530.3 nm (Fe XIV) and 569.4 nm (Ca XV). These lines are formed at approximate temperatures of 1, 2 and 3 MK, respectively. These measurements have been also processed to produce full-disk maps from 14 days of measurements projected onto a sphere.

5 Conclusions

For hundreds of years the evolution of the solar features driven by the cyclic variation of the solar magnetic field has been monitored systematically. A variety of indices have been introduced in order to represent the many different results derived from observations of the solar features in time. These indices include values from measurements of physical quantities, e.g. the Ca II K line emission and radio flux, as well as of parameters derived from the observations, the most common being the sunspot numbers. The time series of the sunspot numbers, which constitute one of the longest continuous measurement programs in the history of science, continue to be used as the most common index to describe solar cycle properties. However, additional measurements of more physical quantities modulated by the solar cycle are also available online to enter the models of e.g. the solar dynamo and of the Earth's climate response to solar variability, though the time series of the physical quantities are shorter than that of sunspot numbers.

Continuation of the time series of the many indices of solar cycle available to date in the next decades is especially important since the long-term trends of solar activity have become a topic of great interest and research after the last activity minimum. Besides, continuous improvement of the accuracy of earlier data and of the whole series, precise long-term data calibration, and extension of the time series back in time, can also lead to a better knowledge of the solar cycle and its effects on the whole heliosphere. To this regard, the recovery, digitization, and analysis of historical observations started in recent years, consisting of both full-disk drawings dating back early 17th century and photographic observations taken since 1876 at several observatories, promise to extract from these unexploited data far more detailed information on solar magnetism than just the sunspot number and area records available till recent times.

The availability of long and accurate time series of solar cycle indices has a large potential impact not only on solar research but also on space weather and space climate studies. Indeed, sunspot number, UV and radio flux series are the key input data to studies of the solar activity impact on the Earth's upper atmosphere and ionosphere (Maruyama 2010), through e.g. multiscale models of traveling ionospheric disturbances (Fedorenko et al. 2013) and data-driven analysis (Kutiev et al. 2013; Scott et al. 2014). Besides, sunspot and plage area, Mg II and F10.7 measurement series enter models of total and spectral solar irradiance variations on long time scales (e.g. Krivova et al. 2010; Lean et al. 2011). These models have been recently employed to study e.g. the influence of spectral solar irradiance variations on stratospheric heating rates (Oberländer et al. 2012; Anet et al. 2013; Thuillier et al. 2014), to estimate the tendency of the Northern Hemisphere temperature for the next decades by using a thermodynamic climate model (Mendoza et al. 2010), to describe the long-term trends in the North Atlantic Oscillation Index (van Loon et al. 2012), and to carry out chemistry climate model experiments of the impact of a time-varying solar cycle and quasi-biennial oscillation forcings on the Earth's atmosphere and the ocean (Petrick et al. 2012; Matthes et al. 2013).

Finally, the availability of long and accurate time series of solar cycle indices may also improve our ability to predict the future evolution of the solar activity.

Acknowledgements The authors are grateful to the International Space Science Institute, Bern, for the organization of the workshop "The Solar Activity Cycle: Physical Causes and Consequences", the invitation to contribute to it, and the kind support received to the purpose. The authors thank Fabrizio Giorgi for preparing Figs. 1 to 8. This study received funding from the European Union's Seventh Programme for Research, Technological Development and Demonstration, under the Grant Agreements of the eHEROES (No. 284461, www.eheroes.eu), SOLARNET (No. 312495, www.solarnet-east.eu), and SOLID (No. 313188, projects.pmodwrc.ch/solid/) projects. It was also supported by COST Action ES1005 "TOSCA" (www.tosca-cost.eu). LvDG's work was supported by the Hungarian Research grants OTKA K-081421 and K-109276, and by the STFC Consolidated Grant ST/H00260/1.

Final acknowledgements go to the many observers and astronomers, both amateur and professional, for performing the regular observations of the solar atmosphere and creating the databases of solar indices described in this paper.

References

J.G. Anet, E.V. Rozanov, S. Muthers, T. Peter, S. BröNnimann, F. Arfeuille, J. Beer, A.I. Shapiro, C.C. Raible, F. Steinhilber, W.K. Schmutz, Impact of a potential 21st century "grand solar minimum" on surface temperatures and stratospheric ozone. Geophys. Res. Lett. **40**, 4420–4425 (2013). doi:10.1002/grl.50806

R. Arlt, R. Leussu, N. Giese, K. Mursula, I.G. Usoskin, Sunspot positions and sizes for 1825–1867 from the observations by Samuel Heinrich Schwabe. Mon. Not. R. Astron. Soc. **433**, 3165–3172 (2013). doi:10.1093/mnras/stt961

R. Arlt, N. Weiss, Solar activity in the past and the chaotic behaviour of the dynamo. Space Sci. Rev., 1–9 (2014). doi:10.1007/s11214-014-0063-5

E.H. Avrett, J.M. Fontenla, R. Loeser, Formation of the solar 10830 A line 1994, pp. 35–47

H.D. Babcock, The Sun's polar magnetic field. Astrophys. J. **130**, 364 (1959). doi:10.1086/146726
H.D. Babcock, H.W. Babcock, Some new features of the solar spectrum. Publ. Astron. Soc. Pac. **46**, 132 (1934). doi:10.1086/124428
L.A. Balmaceda, S.K. Solanki, N.A. Krivova, S. Foster, A homogeneous database of sunspot areas covering more than 130 years. J. Geophys. Res. **114**, 7104 (2009). doi:10.1029/2009JA014299
T. Baranyi, S. Király, H.E. Coffey, Indirect comparison of Debrecen and Greenwich daily sums of sunspot areas. Mon. Not. R. Astron. Soc. **434**, 1713–1720 (2013). doi:10.1093/mnras/stt1134
T. Baranyi, L. Gyori, A. Ludmány, H.E. Coffey, Comparison of sunspot area data bases. Mon. Not. R. Astron. Soc. **323**, 223–230 (2001). doi:10.1046/j.1365-8711.2001.04195.x
J. Beer, A. Blinov, G. Bonani, H.J. Hofmann, R.C. Finkel, Use of Be-10 in polar ice to trace the 11-year cycle of solar activity. Nature **347**, 164–166 (1990). doi:10.1038/347164a0
A.O. Benz, Flare observations. Living Rev. Sol. Phys. **5**, 1 (2008). doi:10.12942/lrsp-2008-1
L. Bertello, R.K. Ulrich, J.E. Boyden, The Mount Wilson Ca ii K plage index time series. Sol. Phys. **264**, 31–44 (2010). doi:10.1007/s11207-010-9570-z
R. Brajša, S. Pohjolainen, V. Ruždjak, T. Sakurai, S. Urpo, B. Vršnak, H. Wöhl, Helium 10830 Å measurements of the Sun. Sol. Phys. **163**, 79–91 (1996). doi:10.1007/BF00165457
D.C. Braun, C. Lindsey, Y. Fan, S.M. Jefferies, Local acoustic diagnostics of the solar interior. Astrophys. J. **392**, 739–745 (1992). doi:10.1086/171477
B. Caccin, I. Ermolli, M. Fofi, A.M. Sambuco, Variations of the chromospheric network with the solar cycle. Sol. Phys. **177**, 295–303 (1998). doi:10.1023/A:1004938412420
W.J. Chaplin, S. Basu, Sounding stellar cycles. Space Sci. Rev. (2014)
G.A. Chapman, J.J. Dobias, T. Arias, Facular and sunspot areas during solar cycles 22 and 23. Astrophys. J. **728**, 150 (2011). doi:10.1088/0004-637X/728/2/150
P. Charbonneau, Dynamo models of the solar cycle. Living Rev. Sol. Phys. **7**, 3 (2010). doi:10.12942/lrsp-2010-3
P. Charbonneau, A. Choudhury, J. Jang, B. Karak, M. Miesch, Challenges for the solar dynamo. Space Sci. Rev. (2014)
F. Clette, E. Cliver, L. Svalgaard, The sunspot number in time. Space Sci. Rev. (2014)
F. Clette, D. Berghmans, P. Vanlommel, R.A.M. Van der Linden, A. Koeckelenbergh, L. Wauters, From the Wolf number to the international sunspot index: 25 years of SIDC. Adv. Space Res. **40**, 919–928 (2007). doi:10.1016/j.asr.2006.12.045
A.E. Covington, Solar radio emission at 10.7 cm, 1947–1968. J. R. Astron. Soc. Can. **63**, 125 (1969)
S.R. Cranmer, Coronal holes. Living Rev. Sol. Phys. **6**(3) (2009). doi:10.12942/lrsp-2009-3
S. Criscuoli, P. Romano, F. Giorgi, F. Zuccarello, Magnetic evolution of superactive regions. Complexity and potentially unstable magnetic discontinuities. Astron. Astrophys. **506**, 1429–1436 (2009). doi:10.1051/0004-6361/200912044
L. Deng, Z. Qi, G. Dun, C. Xu, Phase relationship between polar faculae and sunspot numbers revisited: wavelet transform analyses. Publ. Astron. Soc. Jpn. **65**, 11 (2013). doi:10.1093/pasj/65.1.11
J.F. Denisse, Microwave solar noise and sunspot. Astron. J. **54**, 183 (1949). doi:10.1086/106280
V. Domingo, I. Ermolli, P. Fox, C. Fröhlich, M. Haberreiter, N. Krivova, G. Kopp, W. Schmutz, S.K. Solanki, H.C. Spruit, Y. Unruh, A. Vögler, Solar surface magnetism and irradiance on time scales from days to the 11-year cycle. Space Sci. Rev. **145**, 337–380 (2009). doi:10.1007/s11214-009-9562-1
I. Dorotovič, M. Minarovjech, M. Lorenc, M. Rybanský, Modified homogeneous data set of coronal intensities. Sol. Phys. **289**, 2697–2703 (2014). doi:10.1007/s11207-014-0501-2
T. Dudok de Wit, S. Bruinsma, K. Shibasaki, Synoptic radio observations as proxies for upper atmosphere modelling. J. Space Weather Space Clim. **4**(26), 260000 (2014). doi:10.1051/swsc/2014003
T. Dudok de Wit, M. Kretzschmar, J. Lilensten, T. Woods, Finding the best proxies for the solar UV irradiance. Geophys. Res. Lett. **36**, 10107 (2009). doi:10.1029/2009GL037825
I. Ermolli, S.K. Solanki, A.G. Tlatov, N.A. Krivova, R.K. Ulrich, J. Singh, Comparison among Ca II K spectroheliogram time series with an application to solar activity studies. Astrophys. J. **698**, 1000–1009 (2009a). doi:10.1088/0004-637X/698/2/1000
I. Ermolli, E. Marchei, M. Centrone, S. Criscuoli, F. Giorgi, C. Perna, The digitized archive of the Arcetri spectroheliograms. Preliminary results from the analysis of Ca II K images. Astron. Astrophys. **499**, 627–632 (2009b). doi:10.1051/0004-6361/200811406
I. Ermolli, S. Criscuoli, H. Uitenbroek, F. Giorgi, M.P. Rast, S.K. Solanki, Radiative emission of solar features in the Ca II K line: comparison of measurements and models. Astron. Astrophys. **523**, 55 (2010). doi:10.1051/0004-6361/201014762
I. Ermolli, K. Matthes, T. Dudok de Wit, N.A. Krivova, K. Tourpali, M. Weber, Y.C. Unruh, L. Gray, U. Langematz, P. Pilewskie, E. Rozanov, W. Schmutz, A. Shapiro, S.K. Solanki, T.N. Woods, Recent variability of the solar spectral irradiance and its impact on climate modelling. Atmos. Chem. Phys. **13**, 3945–3977 (2013). doi:10.5194/acp-13-3945-2013

I. Ermolli, F. Giorgi, P. Romano, F. Zuccarello, S. Criscuoli, M. Stangalini, Fractal and multifractal properties of active regions as flare precursors: a case study based on SOHO/MDI and SDO/HMI observations. Sol. Phys. **289**, 2525–2545 (2014). doi:10.1007/s11207-014-0500-3

Y.P. Fedorenko, O.F. Tyrnov, V.N. Fedorenko, V.L. Dorohov, Model of traveling ionospheric disturbances. J. Space Weather Space Clim. **3**(26), 30 (2013). doi:10.1051/swsc/2013052

P. Foukal, L. Bertello, W.C. Livingston, A.A. Pevtsov, J. Singh, A.G. Tlatov, R.K. Ulrich, A century of solar Ca ii measurements and their implication for solar UV driving of climate. Sol. Phys. **255**, 229–238 (2009). doi:10.1007/s11207-009-9330-0

C. Fröhlich, Total solar irradiance: what have we learned from the last three cycles and the recent minimum? Space Sci. Rev. **176**, 237–252 (2013). doi:10.1007/s11214-011-9780-1

G. Galilei, *Istoria e Dimostrazioni Intorno Alle Macchie Solari* (Accad. Naz. Lincei, Rome, 1613)

M.K. Georgoulis, Toward an efficient prediction of solar flares: which parameters, and how? Entropy **15**, 5022–5052 (2013). doi:10.3390/e15115022

I. González Hernández, F. Hill, C. Lindsey, Calibration of seismic signatures of active regions on the far side of the Sun. Astrophys. J. **669**, 1382–1389 (2007). doi:10.1086/521592

I. González Hernández, M. Díaz Alfaro, K. Jain, W.K. Tobiska, D.C. Braun, F. Hill, F. Pérez Hernández, A full-Sun magnetic index from helioseismology inferences. Sol. Phys. **289**, 503–514 (2014)

G.E. Hale, On the probable existence of a magnetic field in Sun-spots. Astrophys. J. **28**, 315 (1908). doi:10.1086/141602

G.E. Hale, Sun-spots as magnets and the periodic reversal of their polarity. Nature **113**, 105–112 (1924). doi:10.1038/113105a0

G.E. Hale, S.B. Nicholson, The law of Sun-spot polarity. Astrophys. J. **62**, 270 (1925). doi:10.1086/142933

G.E. Hale, F. Ellerman, S.B. Nicholson, A.H. Joy, The magnetic polarity of Sun-spots. Astrophys. J. **49**, 153 (1919). doi:10.1086/142452

J.C. Hall, Stellar chromospheric activity. Living Rev. Sol. Phys. **5**, 2 (2008). doi:10.12942/lrsp-2008-2

J.W. Harvey, N.R. Sheeley Jr., A comparison of He II 304 A and He I 10,830 A spectroheliograms. Sol. Phys. **54**, 343–351 (1977). doi:10.1007/BF00159924

K.L. Harvey, The relationship between coronal bright points as seen in He I Lambda 10830 and the evolution of the photospheric network magnetic fields. Aust. J. Phys. **38**, 875–883 (1985)

K.L. Harvey, The cyclic behavior of solar activity, in *The Solar Cycle*, ed. by K.L. Harvey Astronomical Society of the Pacific Conference Series, vol. 27, 1992, p. 335

K.L. Harvey, F. Recely, Polar coronal holes during cycles 22 and 23. Sol. Phys. **211**, 31–52 (2002). doi:10.1023/A:1022469023581

D.H. Hathaway, The solar cycle. Living Rev. Sol. Phys. **7**, 1 (2010). doi:10.12942/lrsp-2010-1

D.H. Hathaway, R.M. Wilson, What the sunspot record tells us about space climate. Sol. Phys. **224**, 5–19 (2004). doi:10.1007/s11207-005-3996-8

D.F. Heath, B.M. Schlesinger, The Mg 280-nm doublet as a monitor of changes in solar ultraviolet irradiance. J. Geophys. Res. **91**, 8672–8682 (1986). doi:10.1029/JD091iD08p08672

C.J. Henney, W.A. Toussaint, S.M. White, C.N. Arge, Forecasting $F_{10.7}$ with solar magnetic flux transport modeling. Space Weather **10**, 2011 (2012). doi:10.1029/2011SW000748

D.V. Hoyt, K.H. Schatten, How well was the Sun observed during the maunder minimum? Sol. Phys. **165**, 181–192 (1996). doi:10.1007/BF00149097

D.V. Hoyt, K.H. Schatten, Group sunspot numbers: a new solar activity reconstruction. Sol. Phys. **179**, 189–219 (1998a). doi:10.1023/A:1005007527816

D.V. Hoyt, K.H. Schatten, Group sunspot numbers: a new solar activity reconstruction. Sol. Phys. **181**, 491–512 (1998b). doi:10.1023/A:1005056326158

H. Hudson, L. Fletcher, J. McTiernan, Cycle 23 variation in solar flare productivity. Sol. Phys. **289**, 1341–1347 (2014a). doi:10.1007/s11207-013-0384-7

H. Hudson, L. Svalgaard, E. Cliver, Solar sector structure. Space Sci. Rev. (2014b)

H.S. Hudson, S. Silva, M. Woodard, R.C. Willson, The effects of sunspots on solar irradiance. Sol. Phys. **76**, 211–219 (1982). doi:10.1007/BF00170984

K. Jain, S.S. Hasan, Modulation in the solar irradiance due to surface magnetism during cycles 21, 22 and 23. Astron. Astrophys. **425**, 301–307 (2004). doi:10.1051/0004-6361:20047102

J. Jiang, R.H. Cameron, D. Schmitt, M. Schüssler, The solar magnetic field since 1700. II. Physical reconstruction of total, polar and open flux. Astron. Astrophys. **528**, 83 (2011). doi:10.1051/0004-6361/201016168

A. Kerdraon, J.-M. Delouis, The Nançay radioheliograph, in *Coronal Physics from Radio and Space Observations*, ed. by G. Trottet Lecture Notes in Physics, Berlin Springer Verlag, vol. 483, 1997, p. 192. doi:10.1007/BFb0106458

C. Kiess, R. Rezaei, W. Schmidt, Properties of sunspot umbrae observed in Cycle 24. ArXiv e-prints (2014)

J. Kleczek, Ionospheric disturbances and flares in the 11-years cycle. Bull. Astron. Inst. Czechoslov. **3**, 52 (1952)

N.A. Krivova, L.E.A. Vieira, S.K. Solanki, Reconstruction of solar spectral irradiance since the Maunder minimum. J. Geophys. Res. **115**, 12112 (2010). doi:10.1029/2010JA015431

M.R. Kundu, Solar active regions at millimeter wavelengths. Sol. Phys. **13**, 348–356 (1970). doi:10.1007/BF00153556

I. Kutiev, I. Tsagouri, L. Perrone, D. Pancheva, P. Mukhtarov, A. Mikhailov, J. Lastovicka, N. Jakowski, D. Buresova, E. Blanch, B. Andonov, D. Altadill, S. Magdaleno, M. Parisi, J. Miquel Torta, Solar activity impact on the Earth's upper atmosphere. J. Space Weather Space Clim. **3**(26), 6 (2013). doi:10.1051/swsc/2013028

A. Lagg, Recent advances in measuring chromospheric magnetic fields in the He I 10830 Å line. Adv. Space Res. **39**, 1734–1740 (2007). doi:10.1016/j.asr.2007.03.091

J.L. Lean, T.N. Woods, F.G. Eparvier, R.R. Meier, D.J. Strickland, J.T. Correira, J.S. Evans, Solar extreme ultraviolet irradiance: present, past, and future. J. Geophys. Res. **116**, 1102 (2011). doi:10.1029/2010JA015901

J. Leenaarts, M. Carlsson, L. Rouppe van der Voort, The formation of the Hα line in the Solar chromosphere. Astrophys. J. **749**, 136 (2012). doi:10.1088/0004-637X/749/2/136

J. Leenaarts, T.M.D. Pereira, M. Carlsson, H. Uitenbroek, B. De Pontieu, The formation of IRIS diagnostics. II. The formation of the Mg II h&k lines in the solar atmosphere. Astrophys. J. **772**, 90 (2013). doi:10.1088/0004-637X/772/2/90

L. Lefevre, F. Clette, Survey and merging of sunspot catalogs. Sol. Phys. **289**, 545–561 (2014). doi:10.1007/s11207-012-0184-5

D.K. Lepshokov, A.G. Tlatov, V.V. Vasil'eva, Reconstruction of sunspot characteristics for 1853–1879. Geomagn. Aeron. **52**, 843–848 (2012). doi:10.1134/S0016793212070109

R. Leussu, I.G. Usoskin, R. Arlt, K. Mursula, Inconsistency of the Wolf sunspot number series around 1848. Astron. Astrophys. **559**, 28 (2013). doi:10.1051/0004-6361/201322373

K.J. Li, P.X. Gao, L.S. Zhan, Synchronization of sunspot numbers and sunspot areas. Sol. Phys. **255**, 289–300 (2009). doi:10.1007/s11207-009-9328-7

C. Lindsey, D.C. Braun, Helioseismic holography. Astrophys. J. **485**, 895 (1997). doi:10.1086/304445

Y. Liu, J.T. Hoeksema, P.H. Scherrer, J. Schou, S. Couvidat, R.I. Bush, T.L. Duvall, K. Hayashi, X. Sun, X. Zhao, Comparison of line-of-sight magnetograms taken by the solar dynamics observatory/helioseismic and magnetic imager and solar and heliospheric observatory/Michelson Doppler imager. Sol. Phys. **279**, 295–316 (2012). doi:10.1007/s11207-012-9976-x

M. Lockwood, Reconstruction and prediction of variations in the open solar magnetic flux and interplanetary conditions. Living Rev. Sol. Phys. **10**(4) (2013). doi:10.12942/lrsp-2013-4. http://www.livingreviews.org/lrsp-2013-4

J.N. Lockyer, Supplementary note on a spectrum of a solar prominence. Proc. R. Soc. Lond., Ser. A **17**, 128 (1868)

V.I. Makarov, A.G. Tlatov, The large-scale solar magnetic field and 11-year activity cycles. Astron. Rep. **44**, 759–764 (2000). doi:10.1134/1.1320502

V.I. Makarov, A.G. Tlatov, D.K. Callebaut, V.N. Obridko, B.D. Shelting, Large-scale magnetic field and sunspot cycles. Sol. Phys. **198**, 409–421 (2001). doi:10.1023/A:1005249531228

T. Maruyama, Solar proxies pertaining to empirical ionospheric total electron content models. J. Geophys. Res. **115**, 4306 (2010). doi:10.1029/2009JA014890

K. Matthes, K. Kodera, R.R. Garcia, Y. Kuroda, D.R. Marsh, K. Labitzke, The importance of time-varying forcing for QBO modulation of the atmospheric 11 year solar cycle signal. J. Geophys. Res. **118**, 4435–4447 (2013). doi:10.1002/jgrd.50424

E.W. Maunder, Note on the distribution of Sun-spots in heliographic latitude, 1874–1902. Mon. Not. R. Astron. Soc. **64**, 747–761 (1904)

B. Mendoza, V.M. Mendoza, R. Garduño, J. Adem, Modelling the northern hemisphere temperature for solar cycles 24 and 25. J. Atmos. Sol.-Terr. Phys. **72**, 1122–1128 (2010). doi:10.1016/j.jastp.2010.05.018

M. Minarovjech, V. Rušin, M. Saniga, The green corona database and the coronal index of solar activity. Contrib. Astron. Obs. Skaln. Pleso **41**, 137–141 (2011)

Z. Mouradian, Synoptic data findings, in *Synoptic Solar Physics*, ed. by K.S. Balasubramaniam, J. Harvey, D. Rabin Astronomical Society of the Pacific Conference Series, vol. 140, 1998, p. 181

A. Muñoz-Jaramillo, N.R. Sheeley, J. Zhang, E.E. DeLuca, Calibrating 100 years of polar faculae measurements: implications for the evolution of the heliospheric magnetic field. Astrophys. J. **753**, 146 (2012). doi:10.1088/0004-637X/753/2/146

A. Norton, P. Charbonneau, Observed solar N–S asymmetry in relation to dynamo modeling. Space Sci. Rev. (2014)

S. Oberländer, U. Langematz, K. Matthes, M. Kunze, A. Kubin, J. Harder, N.A. Krivova, S.K. Solanki, J. Pagaran, M. Weber, The influence of spectral solar irradiance data on stratospheric heating rates during the 11 year solar cycle. Geophys. Res. Lett. **39**, 1801 (2012). doi:10.1029/2011GL049539

K. Oláh, Z. Kolláth, T. Granzer, K.G. Strassmeier, A.F. Lanza, S. Järvinen, H. Korhonen, S.L. Baliunas, W. Soon, S. Messina, G. Cutispoto, Multiple and changing cycles of active stars. II. Results. Astron. Astrophys. **501**, 703–713 (2009). doi:10.1051/0004-6361/200811304

M.J. Owens, R.J. Forsyth, The heliospheric magnetic field. Living Rev. Sol. Phys. **10**(5) (2013). doi:10.12942/lrsp-2013-5. http://www.livingreviews.org/lrsp-2013-5

A. Özgüç, T. Ataç, J. Rybák, Temporal variability of the flare index (1966–2001). Sol. Phys. **214**, 375–396 (2003). doi:10.1023/A:1024225802080

J. Pagaran, M. Weber, J. Burrows, Solar variability from 240 to 1750 nm in terms of faculae brightening and sunspot darkening from SCIAMACHY. Astrophys. J. **700**, 1884–1895 (2009a). doi:10.1088/0004-637X/700/2/1884

W.D. Pesnell, Solar cycle predictions (Invited review). Sol. Phys. **281**, 507–532 (2012). doi:10.1007/s11207-012-9997-5

C. Petrick, K. Matthes, H. Dobslaw, M. Thomas, Impact of the solar cycle and the QBO on the atmosphere and the ocean. J. Geophys. Res. **117**, 17111 (2012). doi:10.1029/2011JD017390

G.J.D. Petrie, Solar magnetic activity cycles, coronal potential field models and eruption rates. Astrophys. J. **768**, 162 (2013). doi:10.1088/0004-637X/768/2/162

G.J.D. Petrie, K. Petrovay, K. Schatten, Solar polar fields and the 22-year activity cycle: Observations and models. Space Sci. Rev., 1–33 (2014). doi:10.1007/s11214-014-0064-4

A.A. Pevtsov, L. Bertello, H. Uitenbroek, On possible variations of basal Ca II K chromospheric line profiles with the solar cycle. Astrophys. J. **767**, 56 (2013). doi:10.1088/0004-637X/767/1/56

M.S. Potgieter, Solar modulation of cosmic rays. Living Rev. Sol. Phys. **10**(3) (2013). doi:10.12942/lrsp-2013-3

D.G. Preminger, S.R. Walton, Modeling solar spectral irradiance and total magnetic flux using sunspot areas. Sol. Phys. **235**, 387–405 (2006). doi:10.1007/s11207-006-0044-2

D.G. Preminger, S.R. Walton, From sunspot area to solar variability: a linear transformation. Sol. Phys. **240**, 17–23 (2007). doi:10.1007/s11207-007-0335-2

M. Priyal, J. Singh, B. Ravindra, T.G. Priya, K. Amareswari, Long term variations in chromospheric features from Ca-K images at Kodaikanal. Sol. Phys. **289**, 137–152 (2014). doi:10.1007/s11207-013-0315-7

T. Pulkkinen, Space weather: terrestrial perspective. Living Rev. Sol. Phys. **4**, 1 (2007). doi:10.12942/lrsp-2007-1

T.I. Pulkkinen, M. Palmroth, E.I. Tanskanen, N.Y. Ganushkina, M.A. Shukhtina, N.P. Dmitrieva, Solar wind—magnetosphere coupling: a review of recent results. J. Atmos. Sol.-Terr. Phys. **69**, 256–264 (2007). doi:10.1016/j.jastp.2006.05.029

A. Reiners, Observations of cool-star magnetic fields. Living Rev. Sol. Phys. **9**, 1 (2012). doi:10.12942/lrsp-2012-1

P. Riley, R. Lionello, J.A. Linker, Z. Mikic, J. Luhmann, J. Wijaya, Global MHD modeling of the solar corona and inner heliosphere for the whole heliosphere interval. Sol. Phys. **274**, 361–377 (2011). doi:10.1007/s11207-010-9698-x

P. Riley, M. Ben-Nun, J.A. Linker, Z. Mikic, L. Svalgaard, J. Harvey, L. Bertello, T. Hoeksema, Y. Liu, R. Ulrich, A multi-observatory inter-comparison of line-of-sight synoptic solar magnetograms. Sol. Phys. **289**, 769–792 (2014). doi:10.1007/s11207-013-0353-1

R.J. Rutten, Observing the solar chromosphere, in *The Physics of Chromospheric Plasmas*, ed. by P. Heinzel, I. Dorotovič, R.J. Rutten Astronomical Society of the Pacific Conference Series, vol. 368, 2007, p. 27

V. Rušin, M. Rybansky, The green corona and magnetic fields. Sol. Phys. **207**, 47–61 (2002). doi:10.1023/A:1015587719072

M. Rybanský, V. Rušin, M. Minarovjech, Coronal index of solar activity—solar-terrestrial research. Space Sci. Rev. **95**, 227–234 (2001)

M. Rybanský, V. Rušin, M. Minarovjech, L. Klocok, E.W. Cliver, Reexamination of the coronal index of solar activity. J. Geophys. Res. **110**, 8106 (2005). doi:10.1029/2005JA011146

J.D. Scargle, S.L. Keil, S.P. Worden, Solar cycle variability and surface differential rotation from Ca II K-line time series data. Astrophys. J. **771**, 33 (2013). doi:10.1088/0004-637X/771/1/33

C. Scheiner, *Rosa Ursina Sive Sol* 1626–1630

B. Schmieder, V. Archontis, M. Schuessler, E. Pariat, Magnetic flux emergence. Space Sci. Rev. (2014)

C.J. Schrijver, J. Cote, C. Zwaan, S.H. Saar, Relations between the photospheric magnetic field and the emission from the outer atmospheres of cool stars. I—The solar CA II K line core emission. Astrophys. J. **337**, 964–976 (1989). doi:10.1086/167168

M. Schwabe, Die Sonne. Von Herrn Hofrath Schwabe. Astron. Nachr. **20**, 283 (1843). doi:10.1002/asna.18430201706

C.J. Scott, R.G. Harrison, M.J. Owens, M. Lockwood, L. Barnard, Evidence for solar wind modulation of lightning. Environ. Res. Lett. **9**(5), 055004 (2014). doi:10.1088/1748-9326/9/5/055004

N.R. Sheeley Jr., A century of polar faculae variations. Astrophys. J. **680**, 1553–1559 (2008). doi:10.1086/588251

N.R. Sheeley Jr., T.J. Cooper, J.R.L. Anderson, Carrington maps of Ca II K-line emission for the years 1915–1985. Astrophys. J. **730**, 51 (2011). doi:10.1088/0004-637X/730/1/51

K. Shibasaki, C.E. Alissandrakis, S. Pohjolainen, Radio emission of the quiet Sun and active regions (Invited review). Sol. Phys. **273**, 309–337 (2011). doi:10.1007/s11207-011-9788-4

K. Shibata, T. Magara, Solar flares: magnetohydrodynamic processes. Living Rev. Sol. Phys. **8**, 6 (2011). doi:10.12942/lrsp-2011-6

S. Solanki, N. Krivova, Faculae and Plague. Landolt Börnstein, 4124 (2009). doi:10.1007/978-3-540-88055-4_9

S.K. Solanki, B. Inhester, M. Schüssler, The solar magnetic field. Rep. Prog. Phys. **69**, 563–668 (2006). doi:10.1088/0034-4885/69/3/R02

S.K. Solanki, N.A. Krivova, J.D. Haigh, Solar irradiance variability and climate. Annu. Rev. Astron. Astrophys. **51**, 311–351 (2013). doi:10.1146/annurev-astro-082812-141007

J.O. Stenflo, Solar magnetic fields. J. Astrophys. Astron. **29**, 19–28 (2008). doi:10.1007/s12036-008-0003-4

J.O. Stenflo, Solar magnetic fields as revealed by Stokes polarimetry. Astron. Astrophys. Rev. **21**, 66 (2013). doi:10.1007/s00159-013-0066-3

M. Stuiver, P.D. Quay, Changes in atmospheric carbon-14 attributed to a variable Sun. Science **207**, 11–19 (1980). doi:10.1126/science.207.4426.11

W.T. Sullivan, *The Early Years of Radio Astronomy—Reflections Fifty Years After Jansky's Discovery* 1984

L. Svalgaard, What geomagnetism can tell us about the solar cycle? Space Sci. Rev. (2014)

L. Svalgaard, H.S. Hudson, The solar microwave flux and the sunspot number, in *SOHO-23: Understanding a Peculiar Solar Minimum*, ed. by S.R. Cranmer, J.T. Hoeksema, J.L. Kohl Astronomical Society of the Pacific Conference Series, vol. 428, 2010, p. 325

H. Tanaka, J.P. Castelli, A.E. Covington, A. Krüger, T.L. Landecker, A. Tlamicha, Absolute calibration of solar radio flux density in the microwave region. Sol. Phys. **29**, 243–262 (1973). doi:10.1007/BF00153452

K.F. Tapping, The 10.7 cm solar radio flux ($F_{10.7}$). Space Weather **11**, 394–406 (2013). doi:10.1002/swe.20064

K.F. Tapping, J.J. Valdés, Did the Sun change its behaviour during the decline of cycle 23 and into cycle 24? Sol. Phys. **272**, 337–350 (2011). doi:10.1007/s11207-011-9827-1

M. Temmer, A. Veronig, A. Hanslmeier, Hemispheric sunspot numbers R_n and R_s: catalogue and N–S asymmetry analysis. Astron. Astrophys. **390**, 707–715 (2002). doi:10.1051/0004-6361:20020758

M. Temmer, J. Rybák, P. Bendík, A. Veronig, F. Vogler, W. Otruba, W. Pötzi, A. Hanslmeier, Hemispheric sunspot numbers $\{R_n\}$ and $\{R_s\}$ from 1945–2004: catalogue and N–S asymmetry analysis for solar cycles 18–23. Astron. Astrophys. **447**, 735–743 (2006). doi:10.1051/0004-6361:20054060

G. Thuillier, S.M.L. Melo, J. Lean, N.A. Krivova, C. Bolduc, V.I. Fomichev, P. Charbonneau, A.I. Shapiro, W. Schmutz, D. Bolsée, Analysis of different solar spectral irradiance reconstructions and their impact on solar heating rates. Sol. Phys. **289**, 1115–1142 (2014). doi:10.1007/s11207-013-0381-x

H. Uitenbroek, Operator perturbation method for multi-level line transfer with partial redistribution. Astron. Astrophys. **213**, 360–370 (1989)

I.G. Usoskin, A history of solar activity over millennia. Living Rev. Sol. Phys. **10**, 1 (2013). doi:10.12942/lrsp-2013-1

I. Usoskin, G. Bazilevskaya, E. Cliver, G. Kovaltsov, Solar cycle in the heliosphere and cosmic rays. Space Sci. Rev. (2014)

H. van Loon, J. Brown, R.F. Milliff, Trends in sunspots and North Atlantic sea level pressure. J. Geophys. Res. **117**, 7106 (2012). doi:10.1029/2012JD017502

J.M. Vaquero, R.M. Trigo, Revised group sunspot number values for 1640, 1652, and 1741. Sol. Phys. **289**, 803–808 (2014). doi:10.1007/s11207-013-0360-2

J.M. Vaquero, R.M. Trigo, M.C. Gallego, A simple method to check the reliability of annual sunspot number in the historical period 1610–1847. Sol. Phys. **277**, 389–395 (2012). doi:10.1007/s11207-011-9901-8

V.V. Vasil'Eva, V.I. Makarov, A.G. Tlatov, Rotation cycles of the sector structure of the solar magnetic field and its activity. Astron. Lett. **28**, 199–205 (2002). doi:10.1134/1.1458351

I.I. Virtanen, K. Mursula, North-South asymmetric solar cycle evolution: signatures in the photosphere and consequences in the corona. Astrophys. J. **781**, 99 (2014). doi:10.1088/0004-637X/781/2/99

Y.-M. Wang, Solar cycle variation of the Sun's low-order magnetic multipoles: Heliospheric consequences. Space Sci. Rev., 1–21 (2014). doi:10.1007/s11214-014-0051-9

D.M. Willis, R. Henwood, M.N. Wild, H.E. Coffey, W.F. Denig, E.H. Erwin, D.V. Hoyt, The Greenwich photo-heliographic results (1874–1976): procedures for checking and correcting the sunspot digital datasets. Sol. Phys. **288**, 141–156 (2013a). doi:10.1007/s11207-013-0312-x

D.M. Willis, H.E. Coffey, R. Henwood, E.H. Erwin, D.V. Hoyt, M.N. Wild, W.F. Denig, The Greenwich photo-heliographic results (1874–1976): summary of the observations, applications, datasets, definitions and errors. Sol. Phys. **288**, 117–139 (2013b). doi:10.1007/s11207-013-0311-y
R.C. Willson, S. Gulkis, M. Janssen, H.S. Hudson, G.A. Chapman, Observations of solar irradiance variability. Science **211**, 700–702 (1981). doi:10.1126/science.211.4483.700
K.L. Yeo, N.A. Krivova, S.K. Solanki, Solar cycle variation in solar irradiance. Space Sci. Rev., 1–31 (2014). doi:10.1007/s11214-014-0061-7

DOI 10.1007/978-1-4939-2584-1_5
Reprinted from *Space Science Reviews* Journal, DOI 10.1007/s11214-014-0061-7

Solar Cycle Variation in Solar Irradiance

K.L. Yeo · N.A. Krivova · S.K. Solanki

Received: 14 March 2014 / Accepted: 17 June 2014 / Published online: 4 July 2014

Abstract The correlation between solar irradiance and the 11-year solar activity cycle is evident in the body of measurements made from space, which extend over the past four decades. Models relating variation in solar irradiance to photospheric magnetism have made significant progress in explaining most of the apparent trends in these observations. There are, however, persistent discrepancies between different measurements and models in terms of the absolute radiometry, secular variation and the spectral dependence of the solar cycle variability. We present an overview of solar irradiance measurements and models, and discuss the key challenges in reconciling the divergence between the two.

Keywords Solar activity · Solar atmosphere · Solar cycle · Solar irradiance · Solar magnetism · Solar physics · Solar variability

1 Introduction

The 11-year solar activity cycle, the observational manifest of the solar dynamo, is apparent in indices of solar surface magnetism such as the sunspot area and number, 10.7 cm radio flux and in the topic of this paper, solar irradiance. The observational and modelling aspects of the solar cycle are reviewed in Hathaway (2010) and Charbonneau (2010), respectively. Solar irradiance is described in terms of what is referred to as the total and spectral solar irradiance, TSI and SSI. They are defined, respectively as the aggregate and spectrally resolved solar radiative flux (i.e., power per unit area and power per unit area and wavelength) above the Earth's atmosphere and normalized to one AU from the Sun. By factoring out the Earth's atmosphere and the variation in the Earth-Sun distance, TSI and SSI characterize the radiant behaviour of the Earth-facing hemisphere of the Sun.

The variation of the radiative output of the Sun with solar activity has long been suspected (Abbot et al. 1923; Smith and Gottlieb 1975; Eddy 1976). However, it was not observed di-

K.L. Yeo (✉) · N.A. Krivova · S.K. Solanki
Max-Planck-Institut für Sonnensystemforschung, 37077 Göttingen, Germany
e-mail: yeo@mps.mpg.de

S.K. Solanki
School of Space Research, Kyung Hee University, Yongin, 446-701 Gyeonggi, Korea

rectly till satellite measurements, free from atmospheric fluctuations, became available. TSI and SSI, at least in the ultraviolet, have been monitored regularly from space through a succession of satellite missions, starting with Nimbus-7 in 1978 (Hickey et al. 1980; Willson and Hudson 1988; Fröhlich 2006; DeLand and Cebula 2008; Kopp et al. 2012). A connection between variations in TSI and the passage of active regions across the solar disc was soon apparent (Willson et al. 1981; Hudson et al. 1982; Oster et al. 1982; Foukal and Lean 1986), leading to the development of models relating the variation in solar irradiance to the occurrence of bright and dark magnetic structures on the solar surface. While not the only mechanism mooted, models that ascribe variation in solar irradiance at timescales greater than a day to solar surface magnetism have been particularly successful in reproducing observations (Domingo et al. 2009). At timescales shorter than a day, excluded from this discussion, intensity fluctuations from acoustic oscillations, convection and flares begin to dominate (Hudson 1988; Woods et al. 2006; Seleznyov et al. 2011).

The measurement and modelling of the variation in solar irradiance over solar cycle timescales, a minute proportion of the overall level (about 0.1 % in the case of TSI), is a substantial achievement. Though significant progress has been made over the past four decades, considerable discrepancies remain between different measurements and models in terms of the absolute radiometry, secular variation and the spectral dependence of the cyclical variability (see the recent reviews by Ermolli et al. 2013; Solanki et al. 2013). In the following, we present a brief overview of the current state of solar irradiance observations (Sect. 2) and models (Sect. 3). We then discuss the key issues in reconciling measurements and models (Sect. 4) before giving a summary (Sect. 5). Our focus will be on the far-ultraviolet to the infrared region of the solar spectrum. This is the spectral range where the bulk of the energy in solar radiation is confined and commonly covered by present-day models aimed at returning both TSI and SSI. We refer the reader to Lilensten et al. (2008), Woods (2008) for an overview of solar cycle variability in the extreme-ultraviolet and shortwards, and Tapping (2013), Dudok de Wit et al. (2014) for radio wavelengths.

2 Measurements

2.1 Total Solar Irradiance, TSI

The measurements from the succession of TSI radiometers sent into space since 1978, collectively representing a nearly uninterrupted record, exhibit clear solar cycle modulation. This is illustrated in Fig. 1 by the comparison between the various TSI records and the monthly sunspot number. All these instruments are based on active cavity electrical substitution radiometry (Butler et al. 2008; Fröhlich 2010). Succinctly, TSI is measured by allowing solar radiation into a heated absorptive cavity intermittently and adjusting the heating power as necessary to maintain thermal equilibrium. While these observations are sufficiently stable over time to trace solar cycle variability, only about 0.1 % of the overall level, the measurements from the various instruments are offset from one another by a greater margin, reflecting the uncertainty in the absolute radiometry.

With the early instruments, specifically Nimbus-7/ERB (Hickey et al. 1980; Hoyt et al. 1992), SMM/ACRIM1 (Willson 1979) and ERBS/ERBE[1] (Lee et al. 1987), the spread in absolute radiometry arose mainly from the uncertainty in the aperture area (Fröhlich 2012;

[1] Nimbus-7/ERB denotes the Earth Radiation Budget instrument onboard Nimbus-7, SMM/ACRIM1 the Active Cavity Radiometer Irradiance Monitor onboard the Solar Maximum Mission and ERBS/ERBE the Earth Radiation Budget Experiment onboard the similarly named satellite.

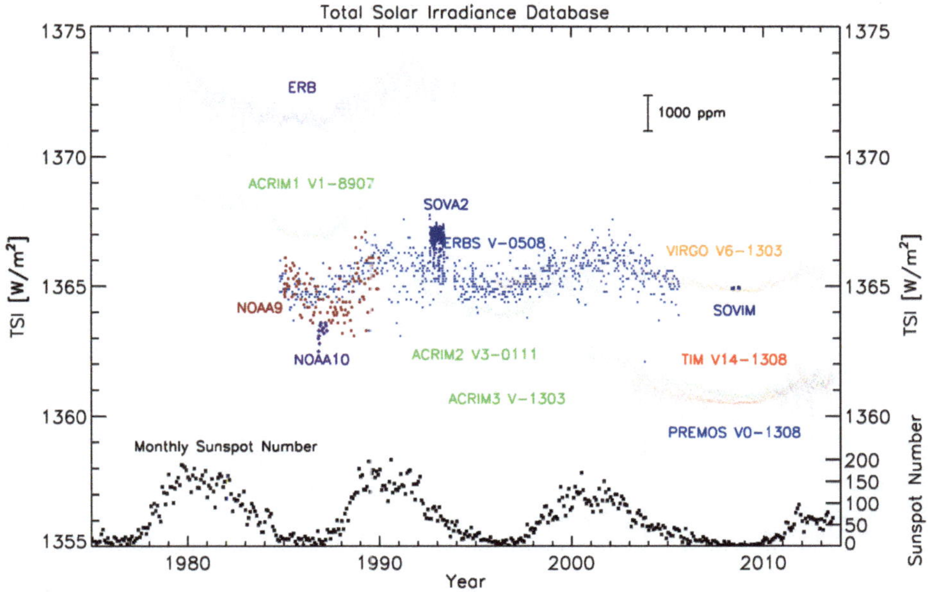

Fig. 1 The measurements from the succession of TSI radiometers sent into orbit since 1978 (*colour coded*) and the monthly mean of the sunspot number (*black, lower right axis*). Each TSI record is annotated by the name of the instrument and where applicable, the version number and/or the year and month of the version. Courtesy of G. Kopp (http://spot.colorado.edu/~koppg/TSI/)

Ermolli et al. 2013). As the determination of the aperture area improved, so the absolute radiometry from the succeeding missions converged. That is, up till the Total Irradiance Monitor, TIM (Kopp and Lawrence 2005; Kopp et al. 2005a,b) onboard the SOlar Radiation and Climate Experiment, SORCE, launched in 2003 (Rottman 2005).

The measurements from TIM were about 5 Wm^{-2} lower than the concurrent observations from ACRIMSAT/ACRIM3[2] (Willson and Mordvinov 2003) and SoHO/VIRGO[3] (Fröhlich et al. 1995, 1997). Tests conducted at the TSI Radiometer Facility, TRF (Kopp et al. 2007) with ground copies of ACRIM3, TIM and VIRGO revealed unaccounted stray-light effects in ACRIM3 and VIRGO (Kopp and Lean 2011; Kopp et al. 2012; Fehlmann et al. 2012). Correction for scattered light subsequently introduced to the ACRIM3 record based on the results of these tests brought it down to within 0.5 Wm^{-2} of the TIM record. The similar proximity between the measurements from the PREcision MOnitor Sensor, PREMOS onboard Picard, launched in 2010 (Schmutz et al. 2009, 2013; Fehlmann et al. 2012), with TIM radiometry provided further evidence that the lower absolute level first registered by TIM is likely the more accurate. In a first, PREMOS was calibrated in vacuum at full solar power levels prior to flight at two separate facilities, the National Physical Laboratory and the TRF. As such, it is considered to be more reliably calibrated than preceding TSI radiometers.

Due to the limited lifetime of TSI radiometers, there is no single mission that covered the entire period of observation. Apart from ERBS (1984 to 2003) and VIRGO (1996 to present), there are no records that encompass a complete solar cycle minimum-to-minimum.

[2]The ACRIM radiometer onboard the ACRIM SATellite.

[3]The Variability of IRradiance and Gravity Oscillations experiment onboard the Solar and Heliospheric Observatory.

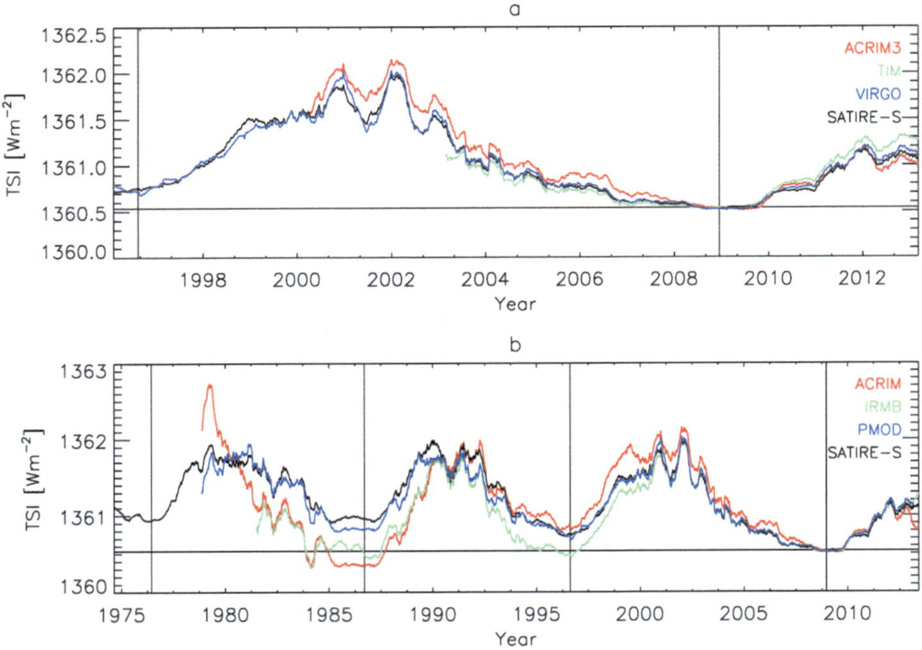

Fig. 2 (**a**) TSI measurements from ACRIMSAT/ACRIM3 (version 11/13), SORCE/TIM (level 3, version 14) and SoHO/VIRGO (level 2, version 6_002_1302). (**b**) The ACRIM (version 11/13), IRMB (version dated 19th December 2013, provided by S. Dewitte) and PMOD (version d41_62_1302) composite records of TSI. Also plotted is the SATIRE-S (Yeo et al. 2014b) reconstruction of TSI. *The vertical dashed lines* mark the position of solar cycle minima. All the time series were normalized to TIM at the 2008 solar cycle minimum (*horizontal dashed line*) and smoothed with a 181-day boxcar filter

Combining the measurements from the various missions into a single time series, obviously essential, is non-trivial due to ageing/exposure degradation, calibration uncertainty and other instrumental issues. The ACRIM, PREMOS, TIM and VIRGO instruments are designed with redundant cavities to allow in-flight degradation tracking. Even with this capability and the best efforts of the respective instrument teams, significant uncertainty persists over the long-term stability: conservatively, about 0.2 Wm^{-2} or 20 ppm per year (Solanki et al. 2013). This is visibly apparent in the discrepant amplitude of solar cycle variation in the ACRIM3, TIM and VIRGO records (Fig. 2(a)). Accounting for changes in instrument sensitivity, which are often particularly severe during early operation and can see discrete shifts such as that suffered by ERB and VIRGO, has proven to be particularly precarious (Hoyt et al. 1992; Lee et al. 1995; Dewitte et al. 2004a; Fröhlich 2006). It is worth mentioning here that the observations from the various radiometers do largely agree at solar rotational timescales, where apparent variability is much less affected by the instrumental influences discussed above.

There are, at present, three composite records of TSI, published by the ACRIM science team (Willson and Mordvinov 2003), IRMB[4] (Dewitte et al. 2004b; Mekaoui and Dewitte 2008) and PMOD/WRC[5] (Fröhlich 2000, 2006). These composites differ in terms of the

[4]IRMB is the francophone acronym of the Royal Meteorological Institute of Belgium. This composite is also variously referred to as the RMIB or SARR composite.

[5]The Physikalisch-Meteorologisches Observatorium Davos/World Radiation Center.

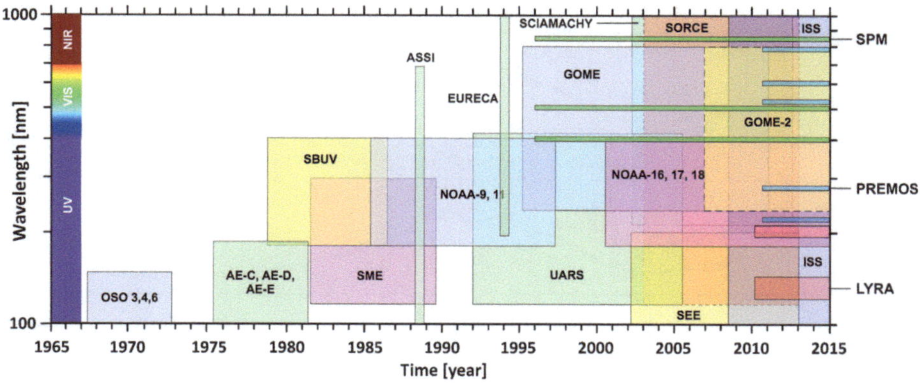

Fig. 3 Timeline and spectral range of the space missions making observations of the solar spectrum above 100 nm. Reproduced from Ermolli et al. (2013), Creative Commons Attribution 3.0 License

amplitude of solar cycle variation, most readily apparent in the conflicting secular trend of the solar cycle minima level (see Fröhlich 2006, 2012 and Fig. 2(c)). The TSI reconstructions presented by Wenzler et al. (2009), Ball et al. (2012), Yeo et al. (2014b) based on the SATIRE-S model (Fligge et al. 2000; Krivova et al. 2003, 2011) replicated, of these composites, the solar cycle and secular variation in the PMOD composite best (see Sect. 3.3.1).

2.2 Spectral Solar Irradiance, SSI

2.2.1 Ultraviolet Solar Irradiance

The solar spectrum has been probed through a miscellany of spaceborne instruments over the past five decades (Fig. 3) which differ in the regularity of measurements and the spectral range surveyed. As with TSI, SSI between around 120 and 400 nm has been monitored almost without interruption from space since 1978.

The key features of the body of ultraviolet solar irradiance measurements are similar to that of TSI, just discussed in Sect. 2.1. The observations from the various spectrometers display similar rotational variability but diverge in terms of the absolute radiometry and the amplitude of solar cycle variation, especially at wavelengths above 240 nm (see Fig. 4 and DeLand and Cebula 2012; Unruh et al. 2012; Ermolli et al. 2013; Yeo et al. 2014b). In particular, the solar cycle amplitude in the measurements from the SIM (Harder et al. 2005a,b) and SOLSTICE (McClintock et al. 2005; Snow et al. 2005a) experiments onboard SORCE is stronger than that registered by the SOLSTICE (Rottman et al. 2001) and SUSIM (Brueckner et al. 1993; Floyd et al. 2003) instruments onboard the predecessor mission, UARS[6] by a factor of three to ten, depending on wavelength (see Fig. 7 in DeLand and Cebula 2012). This disparity, while broadly within the long-term uncertainty of said instruments, is much greater than encountered between pre-SORCE instruments (DeLand and Cebula 2012; Ermolli et al. 2013). The long-term uncertainty of available ultraviolet solar irradiance observations is of similar magnitude as the variation over the solar cycle: on the order of 0.1 to 1 % per year, varying with wavelength and between instruments (Snow et al. 2005a; Merkel et al. 2011; DeLand and Cebula 2012). The long-term uncertainty is

[6]SIM denotes the Spectral Irradiance Monitor, SOLSTICE the SOLar STellar Irradiance Comparison Experiment, SUSIM the Solar Ultraviolet Spectral Irradiance Monitor and UARS the Upper Atmosphere Research Satellite.

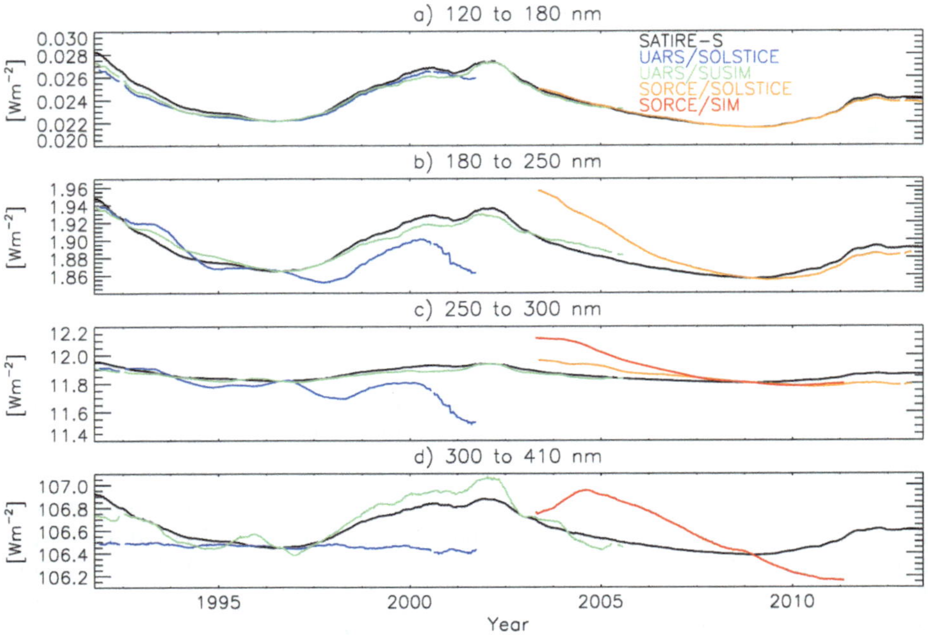

Fig. 4 Integrated solar irradiance in the annotated wavelength intervals in the SATIRE-S reconstruction of SSI (*black*). Also drawn are the measurements from the UARS and SORCE missions, rescaled to the SATIRE-S reconstruction at the 1996 and 2008 solar cycle minima, respectively. All the time series were smoothed by taking the 181-day moving average. Adapted from Yeo et al. (2014b)

also grossly greater than that afflicting TSI measurements. Again, due to the limited lifetime of spaceborne instrumentation, there is no record that extends beyond a complete solar cycle minimum-to-minimum with the exception of the observations from NOAA-9 SBUV/2[7] (DeLand and Cebula 1998).

The solar cycle modulation in and disparity between the various pre-SORCE records is illustrated in the time series plot by DeLand and Cebula 2008 (Fig. 2 in their paper), which is qualitatively analogous to Fig. 1. The authors presented the first published effort to compose the ultraviolet solar irradiance observations from these instruments into a single time series. The result, spanning 1978 to 2005, still contains overt instrumental trends for which the appropriate correction is not known. The challenge in the account of instrumental influences is substantially greater than with TSI, exacerbated by the wavelength dependence of instrumental effects and differences in the design, operation and calibration approach.

2.2.2 SORCE/SIM SSI

The series of GOME instruments, the first of which was launched onboard ERS-2 in 1996 (Weber et al. 1998; Munro et al. 2006) and ENVISAT/SCIAMACHY,[8] launched in 2002 (Skupin et al. 2005b) made regular measurements of the solar spectrum in the 240 to 790 nm

[7] The second generation Solar Backscatter UltraViolet spectrometer onboard the ninth National Oceanic and Atmospheric Administration satellite.

[8] GOME is short for the Global Ozone Monitoring Experiment, ERS-2 the second European Remote Sensing satellite and ENVISAT/SCIAMACHY the SCanning Imaging Absorption spectroMeter for Atmospheric CHartographY onboard the ENVIronmental SATellite.

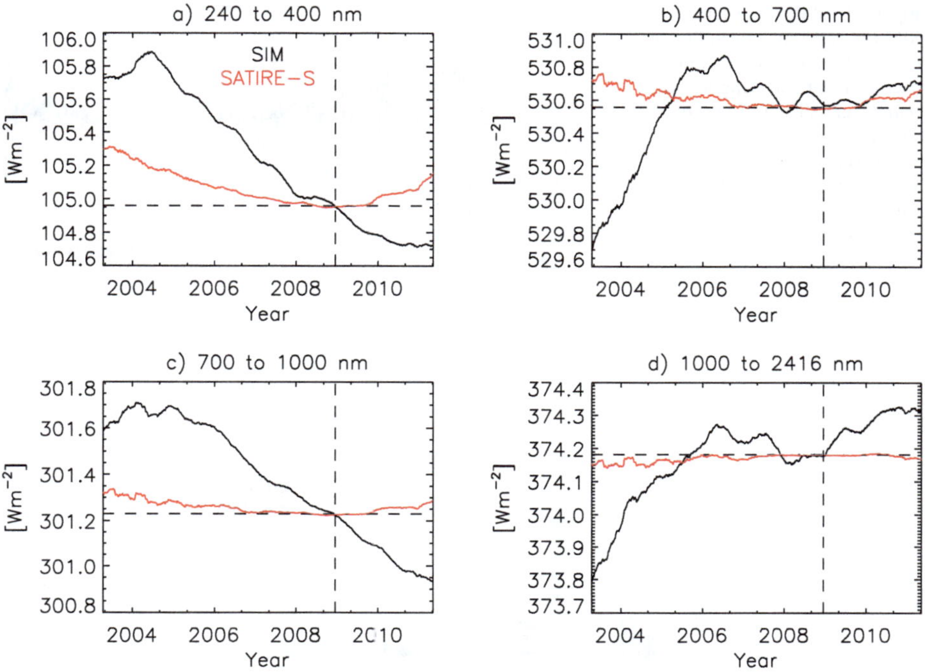

Fig. 5 181-day moving average of the integrated solar irradiance over the spectral intervals specified above each panel in the SORCE/SIM record (*red*) and in the SATIRE-S reconstruction (*black*). The SATIRE-S time series were rescaled to the corresponding SIM time series at the 2008 solar cycle minimum, the position and level at which is indicated by the dashed lines. Adapted from Yeo et al. (2014b)

and 240 to 2380 nm wavelength range, respectively. These instruments are designed primarily for atmospheric sounding measurements which do not require absolute radiometry rather than solar irradiance monitoring. As such, they lack the capability to track instrument degradation in-flight, rendering the observations unsuitable for tracing the solar cycle variation of the solar spectrum reliably. The long-term stability of the narrowband (FWHM of 5 nm) photometry at 402, 500 and 862 nm from the Sun PhotoMeter, SPM on SoHO/VIRGO (Fröhlich et al. 1995, 1997) is similarly problematic, though considerable progress has been made (Wehrli et al. 2013).

With these caveats in mind, one can say that regular monitoring of the solar spectrum longwards of the ultraviolet only started, in effect, with SIM. The instrument has been surveying the wavelength range of 200 to 2416 nm since 2003, providing the only continuous and extended record of the solar spectrum spanning the ultraviolet to the infrared presently available. The latest release (version 19, dated 23rd November 2013), covering the wavelength range of 240 to 2416 nm,[9] is depicted in Figs. 5 and 6.

SIM returned rather unexpected results, precipitating the ensuing debate on the overall trend registered by the instrument (Harder et al. 2009; Ball et al. 2011; DeLand and Cebula 2012; Unruh et al. 2012; Ermolli et al. 2013; Yeo et al. 2014b). Between 2004 and 2008,

[9]SIM measurement from between 200 and 240 nm are not made publicly available in consideration of the fact that below about 260 nm, SIM observations start to register a lower signal-to-noise ratio than the concurrent observations from the SOLSTICE experiment onboard the same mission (J. Harder, private communication).

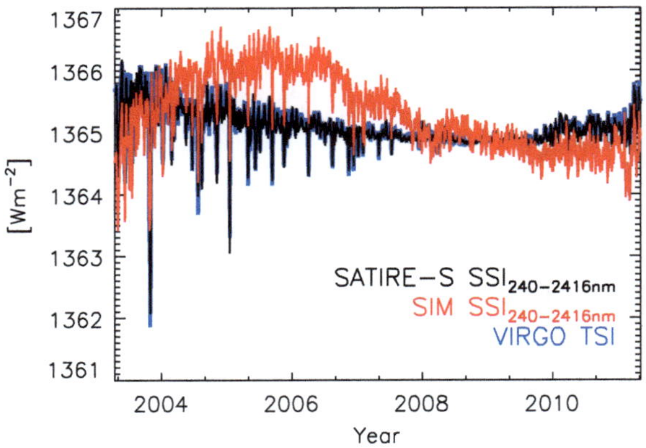

Fig. 6 Total flux registered by SIM (*red*), the integrated flux in the SATIRE-S reconstruction over a similar wavelength range (*black*) and VIRGO TSI (*blue*). The SIM and SATIRE-S time series were normalized to the VIRGO time series at the 2008 solar cycle minimum. Adapted from Yeo et al. (2014b)

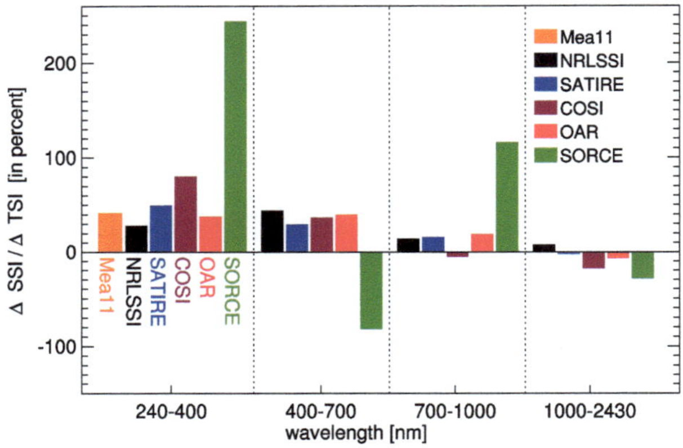

Fig. 7 Ratio of the variation in SSI over the same spectral intervals examined in Fig. 5, ΔSSI, and the corresponding variation in TSI, ΔTSI, between solar cycle maximum and minimum (2002 and 2008) in the models discussed in Sect. 3.3. For the model by Morrill et al. (2011), denoted Mea11, we took ΔTSI from the SATIRE-S model. Also depicted is the same for SORCE measurements between 2004 and 2008, with ΔSSI and ΔTSI from the SIM and TIM records, respectively. Adapted from Ermolli et al. (2013), Creative Commons Attribution 3.0 License

which is within the declining phase of solar cycle 23, SIM recorded a drop in ultraviolet flux (240 to 400 nm, Fig. 5(a)) that is almost double the decrease in TSI over the same period and multiple times greater than projections from pre-SORCE SSI measurements and models of solar irradiance (Fig. 7). Up to 2006, this pronounced downward trend in the ultraviolet is accompanied by a comparable increase in the visible (400 to 700 nm, Fig. 5(c)), in anti-phase with the solar cycle. This is in conflict with SPM photometry (see next paragraph) and present-day models, all of which (apart from Fontenla et al. 2011 see Sect. 3.3.2) point to visible solar irradiance varying in phase with the solar cycle (Fig. 7). The variation in the infrared (700 to 2416 nm, Figs. 5(c) and 5(d)) between 2004 and 2008 is also significantly stronger than in model reconstructions (Fig. 7). The investigations of Unruh et al. (2008, 2012), DeLand and Cebula (2012), Lean and DeLand (2012), Yeo et al. (2014b) did, however, note that the rotational variability in SIM SSI is similar to that in pre-SORCE measurements and models.

Looking now at the full length of the SIM record, there is no constancy in how the overall trend at a given wavelength relate to the solar cycle (Fig. 5). Measured solar irradiance is neither in phase nor anti-phase with the solar cycle. Apart from the solar cycle modulation evident in pre-SORCE ultraviolet solar irradiance measurements, this also runs counter to the positive correlation between SPM visible (500 nm) photometry and TSI over the similar period of 2002 to 2012 reported by Wehrli et al. (2013). There being no other extended record of SSI covering a similar spectral range, it cannot be ruled out completely that segments of the solar spectrum may vary in a non-cyclic manner as apparent in SIM SSI. It is, however, almost irrefutable that the integral of the solar spectrum over all wavelengths, TSI, does exhibit solar cycle variation. As the spectral range surveyed by SIM accounts for more than 97 % of the power in solar radiation, the total flux recorded by the instrument should already replicate most of the variability in TSI but that is evidently not the case (Fig. 6). In contrast, the reconstruction of the integrated solar irradiance over the spectral range of SIM from SSI models replicates most of the variability in TSI (see Fig. 6 and Ball et al. 2011; Lean and DeLand 2012; Yeo et al. 2014b).

The discrepancies between SIM SSI and other measurements and models reported in various studies, summarized above, were taken to indicate that there are unaccounted instrumental trends in the SIM record (Ball et al. 2011; DeLand and Cebula 2012; Lean and DeLand 2012; Unruh et al. 2012; Yeo et al. 2014b). This is favoured here over alternative interpretations such as the apparent trends between 2004 and 2008 implying a change in the physics of the Sun during this period compared to earlier times or that there are gaping insufficiencies in our understanding of the physical processes driving variations in solar irradiance. As we will argue in Sect. 3.1, the decline in visible flux between 2003 and 2006 (Fig. 5(c)) is not consistent with our current understanding of solar surface magnetism and its effect on solar irradiance.

Evidently, the direct observation of the variation in SSI over solar cycle timescales is afflicted by considerable uncertainty. The situation is set to improve with the continuing efforts to calibrate SCIAMACHY and SORCE spectrometry, and spectral measurements expected from ISS/SOLSPEC (Thuillier et al. 2009, 2014a) and the upcoming JPSS[10] mission (Richard et al. 2011). JPSS, set to be launched in 2017, will carry an improved version of the SIM instrument. For more extensive reviews of the measurement of solar irradiance, we refer the reader to DeLand and Cebula (2008, 2012), Domingo et al. (2009), Kopp et al. (2012), Fröhlich (2012), Ermolli et al. (2013), Solanki et al. (2013).

3 Models

The body of satellite measurements of solar irradiance, while evidently core to our understanding of the solar cycle variation in the radiative output of the Sun, cover a limited period in time and suffer significant uncertainty. Models of solar irradiance serve both to complement these observations and to advance our understanding of the physical processes driving the apparent variability.

3.1 Solar Surface Magnetism

A feature of the 11-year solar activity cycle is the cyclical emergence and evolution of kilogauss-strength magnetic concentrations in the photosphere (Solanki et al. 2006). The main properties of these magnetic concentrations can be explained by describing them as

[10]ISS/SOLSPEC denotes the SOLar SPECtrum experiment onboard the International Space Station and JPSS the Joint Polar Satellite System.

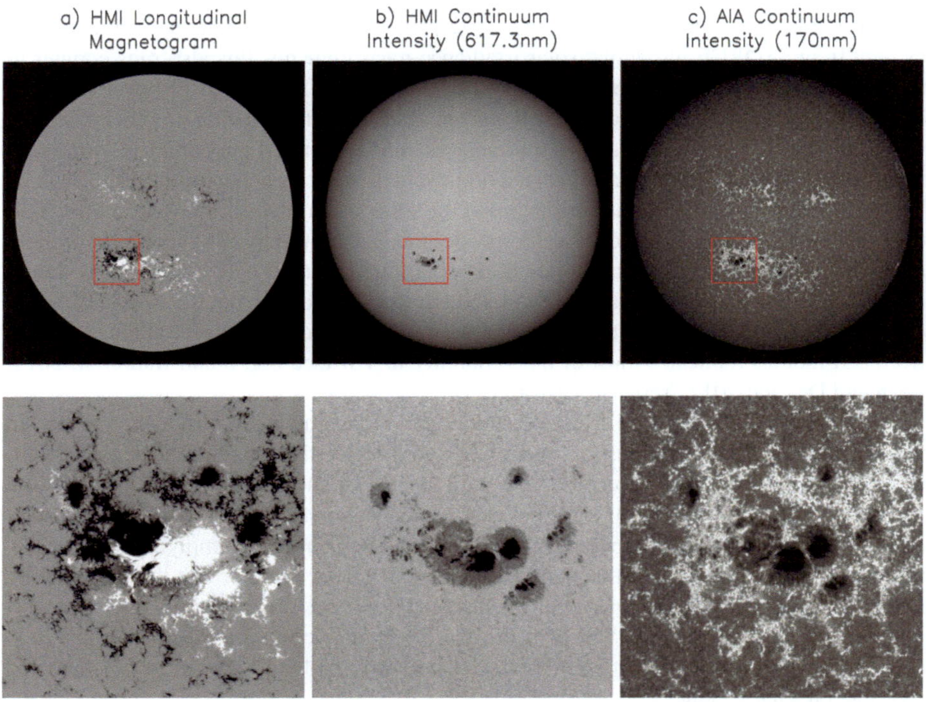

Fig. 8 *Top*: Concurrent observations, from 11th July 2012, of (**a**) the line-of-sight magnetic field and the continuum intensity at (**b**) 617.3 nm and (**c**) 170 nm from the Helioseismic and Magnetic Imager, HMI (Schou et al. 2012) and the Atmospheric Imaging Assembly, AIA (Lemen et al. 2012) onboard the Solar Dynamics Observatory, SDO. The AIA image has been resampled to register with the HMI images. *Bottom*: Blow up of the boxed region, featuring active region NOAA 11520. From (**a**) to (**c**), *the grey scale* is saturated at ±30 G, about 60 % and 120 %, and 20 % and 300 % of the mean quiet Sun level at disc centre

magnetic flux tubes (Spruit 1976; Spruit and Roberts 1983; Solanki 1993). They range in physical extent from on the order of 10^1 to 10^5 km in cross section. The lower end corresponds to the small-scale magnetic elements which make up active region faculae and quiet Sun network/internetwork (Lagg et al. 2010; Riethmüller et al. 2014), and the upper end to sunspots and pores.

Solar irradiance is modulated by photospheric magnetic activity from its effect on the thermal/radiant property of the solar surface and atmosphere. The influence of magnetic concentrations on the local temperature structure of the solar surface and atmosphere varies strongly with the size of the magnetic feature, as described below.

As a consequence of pressure balance, the interior of magnetic concentrations is evacuated. The lower density creates a depression in the optical depth unity surface and magnetic buoyancy, the result of which is flux tubes are largely vertical. The intensity contrast in the continuum is influenced by the competing effects of magnetic suppression of convection and radiative heating from surrounding granulation through the side walls of the depression. Sunspots and pores are dark from the magnetic suppression of convection within these features (see Fig. 8 and the reviews by Solanki 2003; Rempel and Schlichenmaier 2011). For small-scale magnetic concentrations, this is overcome by the lateral heating, rendering them bright (Spruit 1976; Spruit and Zwaan 1981; Grossmann-Doerth et al. 1994; Vögler et al. 2005), especially away from the disc centre as the side walls come into greater view

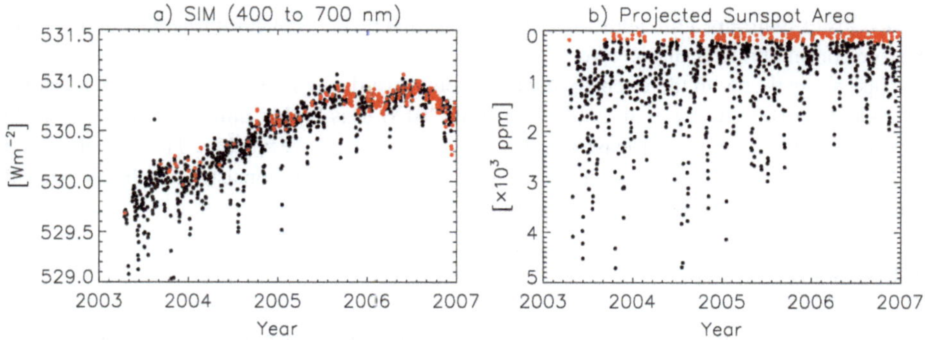

Fig. 9 (**a**) Integrated solar irradiance in the visible in SIM SSI and (**b**) the concurrent projected sunspot area from the composite record by Balmaceda et al. 2009 (version 0613). *Points* corresponding to days where the projected sunspot area is less than 200 ppm are highlighted in *red*. *The vertical axis* of the projected sunspot area plot is inverted to aid interpretation

(Spruit 1976; Keller et al. 2004; Carlsson et al. 2004; Steiner 2005). This feature gives rise to the bright faculae visible near the limb in white-light. The upper layers of the atmosphere enclosed within flux tubes are heated by mechanical and resistive dissipations (Musielak and Ulmschneider 2003; Moll et al. 2012), and radiation from deeper layers (Knölker et al. 1991). This enhances their intensity within spectral lines and in the ultraviolet, which are formed at greater heights than the visible continuum (see Fig. 8(c) and Frazier 1971; Mitchell and Livingston 1991; Morrill et al. 2001; Ermolli et al. 2007; Yeo et al. 2013; Riethmüller et al. 2010). A schematic of thin flux tubes and the mechanisms described above can be found in Fig. 5 of Solanki et al. (2013).

Models describing the variation in solar irradiance at timescales greater than a day by the intensity deficit and excess facilitated by photospheric magnetism have achieved substantial success in reproducing measured solar irradiance (see Sects. 3.2 and 3.3, and Domingo et al. 2009). While other plausible mechanisms have also been proposed (Wolff and Hickey 1987; Kuhn et al. 1988; Cossette et al. 2013), related to physical processes in the solar interior, supporting evidence is not straightforward to obtain and consequently still largely lacking.

The increase in visible solar irradiance registered by SIM between 2003 and 2006 (Fig. 5(c)) came at a time when solar activity was declining. For the solar cycle variation in photospheric magnetism to be compatible with this trend in visible solar irradiance, small-scale magnetic concentrations would have to be dark in the visible. The intensity deficit from sunspots and pores, while weakening over this period, is not driving the upward trend apparent in SIM visible solar irradiance. This is demonstrated in Fig. 9(a) by the variation in SIM visible flux over the days with minimal sunspot activity, highlighted in red. If the increase in SIM visible flux is indeed from sunspot darkening, then it should be absent or at least markedly weaker in the days with minimal sunspot activity. As depicted in the figure, the overall trend in the SIM visible flux time series is no less manifest in the days where the projected sunspot area is minimal, below 200 ppm.

Turning now to small-scale magnetic concentrations, various studies have noted negative contrasts in the visible continuum near disc centre at both low and high magnetogram signal levels (summarized in Table 2 of Yeo et al. 2013).

However, the results of recent investigations suggest that these apparent negative contrasts are associated with the fact that small-scale magnetic elements congregate mainly within dark intergranular lanes (Schnerr and Spruit 2011; Kobel et al. 2011) and observational effects related to the limited spatial resolutions and telescope diffraction (Röhrbein et al. 2011) than any indication that small-scale magnetic concentrations are dark in the visible continuum near disc centre.

Taken together with the intensity enhancement towards the limb and within spectral lines (Topka et al. 1997; Ortiz et al. 2002; Yeo et al. 2013), it is highly unlikely that small-scale magnetic concentrations, at least overall, might be dark in the visible. The upward trend in SIM visible solar irradiance between 2003 and 2006 is not consistent with our current understanding of solar surface magnetism and its effect on solar irradiance.

3.2 Model Architectures

There are two broad categories of solar irradiance models, distinguished by the modelling approach, commonly referred to as 'proxy' and 'semi-empirical'.

3.2.1 Proxy Models

As stated in the introduction, the measurement of solar irradiance from space quickly revealed an apparent connection between TSI and the passage of active regions across the solar disc. This was followed by the development of models aimed at reconstructing solar irradiance by the multivariate regression of indices of solar activity to measured solar irradiance. The index data serve as proxies of the effects of bright and dark magnetic structures on the radiative output of the Sun, therefore the term proxy models.

Sunspot darkening is usually represented by sunspot area or the photometric sunspot index, PSI (Hudson et al. 1982; Fröhlich et al. 1994) and facular brightening by chromospheric indices such as the Ca II K (Keil et al. 1998), Mg II (Heath and Schlesinger 1986) or F10.7 (10.7 cm radio flux, Tapping 1987, 2013) indices. In this context the term sunspot includes pores and the term faculae encompasses quiet Sun network. The Ca II K and Mg II indices are given by the ratio of the disc-integrated flux in the line core of the Ca II K line and the Mg II h and k doublet to that at nearby reference wavelengths. The line core to 'continuum' ratio is preferred over absolute fluxes as it is more robust to instrument degradation.

The reconstructions of TSI by the group at the San Fernando Observatory, SFO (Chapman et al. 1996, 2012, 2013; Preminger et al. 2002) employ sunspot and faculae indices derived from full-disc photometric images obtained at the observatory. These models proved to be particularly successful among proxy models. By employing full-disc imagery instead of Sun-as-a-star measures such as the indices listed above, they include the centre-to-limb variation of sunspot and faculae contrast, albeit only at the photometric bandpass. The latest iteration, based on visible red and Ca II K observations, reproduced most of the variation in TIM radiometry ($R^2 = 0.95$, Chapman et al. 2013).

Since the proxy model approach relies on reliable solar irradiance measurements, it is not straightforward to reconstruct SSI by this method due to the long-term uncertainty of available measurements and the relative paucity of observations outside the ultraviolet (Sect. 2.2). Certain proxy models make use of the fact that the effects of instrument degradation on apparent variability is relatively benign at shorter timescales to circumvent long-term stability issues. The idea is to fit index data to measured rotational variability, either by detrending the index and SSI data or confining the regression to rotational periods, and then assume the indices-to-irradiance relationships so derived to all timescales (Lean et al. 1997; Pagaran et al. 2009; Thuillier et al. 2012). As we will discussed in Sect. 4.1, the assumption that the underlying relationship between indices of solar activity and solar irradiance is similar at all timescales is not likely valid.

3.2.2 Semi-empirical Models

The next level of sophistication in the modelling of solar irradiance is realized by semi-empirical models. In these models, the solar disc is segmented by surface feature type, termed 'components'. The filling factor (proportion of the solar disc or a given area covered) and time evolution of each component is deduced from indices of solar activity or suitable full-disc observations. This information is converted to solar irradiance employing the intensity spectra of the various components. These are calculated applying spectral synthesis codes to semi-empirical model atmospheres of said feature types (Fontenla et al. 1999, 2009; Unruh et al. 1999; Shapiro et al. 2010). The reconstruction of the solar spectrum is given by the filling factor-weighted sum of the component intensity spectra. The semi-empirical model atmospheres describe the temperature and density stratification of the solar atmosphere within each component, constrained and validated by observations (therefore the term 'semi-empirical').

The semi-empirical approach has the advantage that it yields SSI independent of the availability of reliable measurements. Additionally, for the models that rely on full-disc observations for the filling factor of the solar surface components (i.e., when the exact disc position of magnetic features is known), the centre-to-limb variation of the radiant behaviour of each component can be taken into account by generating and applying the corresponding intensity spectra at varying heliocentric angles.

3.3 Present-Day Models

At present, there are five models aimed at reconstructing both TSI and SSI reported in the literature, reviewed in Ermolli et al. (2013). These models describe the solar spectrum over at least from the far-ultraviolet to the infrared, such that the bolometric value is a close approximation of TSI. They are,

- NRLSSI (Naval Research Laboratory Solar Spectral Irradiance, Lean et al. 1997; Lean 2000),
- SATIRE-S (Spectral And Total Irradiance REconstruction for the Satellite era, Unruh et al. 1999; Fligge et al. 2000; Krivova et al. 2003, 2006, 2011; Yeo et al. 2014b),
- SRPM (Solar Radiation Physical Modelling, Fontenla et al. 1999, 2004, 2006, 2009, 2011),
- OAR (Observatorio Astronomico di Roma, Ermolli et al. 2003, 2011, 2013; Penza et al. 2003) and
- COSI (COde for Solar Irradiance, Haberreiter et al. 2008; Shapiro et al. 2010, 2011, 2013).

Apart from the NRLSSI, these models adopt the semi-empirical approach. In the following, we discuss the recent results obtained with the SATIRE-S model by Yeo et al. 2014b (Sect. 3.3.1) before giving an overview of the other models listed (Sect. 3.3.2). For a broader review of models of solar irradiance, we refer the reader to Domingo et al. (2009), Ermolli et al. (2013), Solanki et al. (2013).

3.3.1 SATIRE-S

The SATIRE-S model (Fligge et al. 2000; Krivova et al. 2003, 2011) relies on spatially-resolved full-disc observations of magnetic field and intensity to segment the solar disc into quiet Sun, faculae, sunspot umbra and sunspot penumbra. It has been applied to longitudinal magnetograms and continuum intensity images collected at the KPVT (in operation from 1974 to 2003, Livingston et al. 1976; Jones et al. 1992), as well as from SoHO/MDI (1996 to 2011, Scherrer et al. 1995) and SDO/HMI[11] (2010 to the present, Schou et al. 2012) to reconstruct total and spectral solar irradiance over various periods between 1974 and 2013 (Krivova et al. 2003; Wenzler et al. 2006; Ball et al. 2012, 2014; Yeo et al. 2014b). In the latest iteration (Yeo et al. 2014b), KPVT, MDI and HMI magnetograms were cross-calibrated in such a way that the model input from all the data sets combine to yield a single consistent TSI/SSI time series covering the entire period of 1974 to 2013 as the output. Apart from the NRLSSI, which extends back to 1950, this is the only other daily reconstruction of the solar spectrum spanning the ultraviolet to the infrared from present-day models to extend over multiple solar cycles.

At present, the model employs the intensity spectra of quiet Sun, faculae, umbra and penumbra from Unruh et al. (1999), which were generated with the ATLAS9 spectral synthesis code (Kurucz 1993). As ATLAS9 assumes local thermodynamic equilibrium (LTE), it fails in the ultraviolet below approximately 300 nm. Ultraviolet solar radiation is formed in the upper photosphere and lower chromosphere, where the plasma is increasingly less collisional. The breakdown below 300 nm is accounted for by rescaling the 115 to 180 nm segment of the reconstructed spectra to SORCE/SOLSTICE SSI[12] and offsetting the 180 to 300 nm segment to the Whole Heliospheric Interval (WHI) reference solar spectra by Woods et al. (2009), as detailed in Yeo et al. (2014b). SATIRE-S is the only semi-empirical model to include the influence of departures from LTE through such a data-driven approach. The SRPM, OAR and COSI models take a more direct approach, making use of various non-LTE spectral synthesis codes to generate the intensity spectra of solar surface components. The various non-LTE codes differ from one another by the method non-LTE effects are approximated (see Fontenla et al. 1999; Uitenbroek 2002; Shapiro et al. 2010).

Recall, due to uncertainties in the amplitude of solar cycle variation, the three published TSI composites exhibit conflicting decadal trends (Sect. 2.1). The TSI reconstructed by SATIRE-S is a significantly closer match to the PMOD composite than to the ACRIM or the IRMB composite, replicating most of the variability ($R^2 = 0.92$) over the entire length of the composites (Fig. 2(c)). Reconstructed TSI also exhibits excellent agreement with the measurements from individual instruments such as ACRIM3, TIM and VIRGO. The record from the PMO6V[13] radiometer on VIRGO is particularly well-matched ($R^2 = 0.96$). The secular decline between the 1996 and 2008 solar cycle minima in VIRGO radiometry is reproduced to within 0.05 Wm^{-2} (Fig. 2(a)). This agreement between SATIRE-S TSI and VIRGO radiometry, which extends 1996 to the present, encompassing all of solar cycle 23, is significant. It implies that at least 96 % of the variability in solar irradiance over this period, including the secular variation between the 1996 and 2008 solar cycle minima can be explained by solar surface magnetism alone.

The bolometric facular brightening and sunspot darkening, $\Delta\mathrm{TSI}_{\mathrm{fac}}$ and $\Delta\mathrm{TSI}_{\mathrm{spt}}$ with respect to the TSI level of the magnetically quiet Sun, binned and averaged by month and

[11]In full, the Kitt Peak Vacuum Telescope (KPVT), the Michelson Doppler Imager onboard the Solar and Heliospheric Observatory (SoHO/MDI)), and the Helioseismic and Magnetic Imager onboard the Solar Dynamics Observatory (SDO/HMI).

[12]This is relatively unaffected by the long-term stability issues plaguing available SSI measurements discussed in Sect. 2.2.1 as they only emerge towards longer wavelengths (Fig. 4).

[13]VIRGO TSI is actually given by the combination of the measurements from two onboard radiometers, DIARAD and PMO6V. The solar cycle variation in the DIARAD and PMO6V records is nearly identical but they do differ at rotational timescales resulting in very different R^2 values on comparison with other measurements or models.

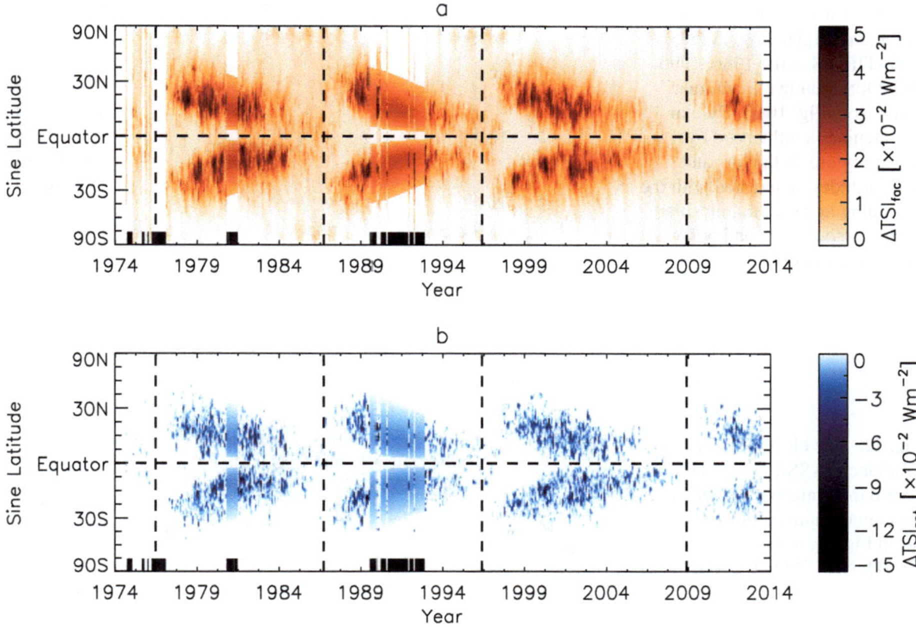

Fig. 10 Variation in TSI from (**a**) faculae brightening, ΔTSI_{fac} and (**b**) sunspot darkening, ΔTSI_{spt} in the SATIRE-S model, as a function of time and latitude (the monthly average in sine latitude intervals of 0.01). *The black bars* along the horizontal axes mark the months with no values from the lack of suitable magnetogram data. *The gaps* around the maxima of solar cycles 21 and 22 are filled by interpolation. *The horizontal and vertical dashed lines* denote the equator and epoch of solar cycle minima, respectively

sine latitude, is expressed in Fig. 10. This is an update of the similar figure by Wenzler (2005) based on an earlier SATIRE-S reconstruction that employed KPVT data alone. Since solar surface magnetism is concentrated in active regions, it follows then that the latitudinal distribution of the associated intensity excess/deficit demonstrate Spörer's law, resembling butterfly diagrams of sunspot area/position and magnetic flux (see e.g., Figs. 4 and 14 in Hathaway 2010). A diagram similar to Fig. 10(c), based on the PSI, was presented by Fröhlich 2013 (Fig. 5 in his paper).

Since sunspots are largely absent around solar cycle minima, the minimum-to-minimum variation in SATIRE-S TSI is dominantly from the change in facular brightening. In Fig. 11, we plot the latitudinal distribution of facular brightening at the last three solar cycle minima, taken from the butterfly diagram (Fig. 10(a)). Around the 1986 solar cycle minimum, facular brightening is elevated close to the equator, around mid-latitudes and towards high latitudes (red curve). The broad peaks near the equator and at mid-latitudes correspond to active regions associated with the preceding cycle and the succeeding cycle, respectively. The increase towards high latitudes relates to magnetic elements transported polewards by meridional circulation over the course of the previous cycle (i.e., polar flux). The differing latitudinal distribution at the three solar cycle minima illustrates how the minimum-to-minimum trend in faculae brightening and therefore TSI is modulated by the prevailing magnetic activity in the three latitude regions. The low flat profile of the blue curve, corresponding to the 2008 solar cycle minimum, reflects the near-complete absence of any form of activity during this period, which contributed significantly to the secular decline between the 1996 and 2008 solar cycle minima.

Fig. 11 Latitudinal distribution of facular brightening, ΔTSI_{fac} in SATIRE-S at the 1986, 1996 and 2008 solar cycle minima, taken from Fig. 10(a). Facular brightening is influenced by active regions in the low and mid-latitudes, associated with the preceding and succeeding solar cycles, respectively and polar flux at high latitudes

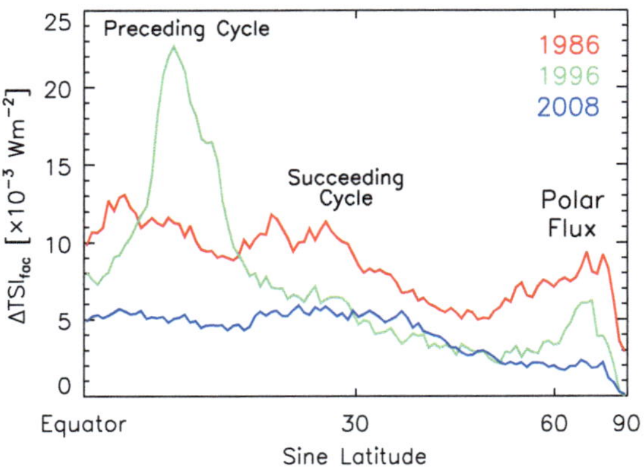

Fig. 12 The change in solar irradiance, ΔSSI, integrated over 10 nm intervals, between solar cycle maximum and minimum (as in Fig. 7), in the Morrill et al. (2011), NRLSSI and SATIRE-S models

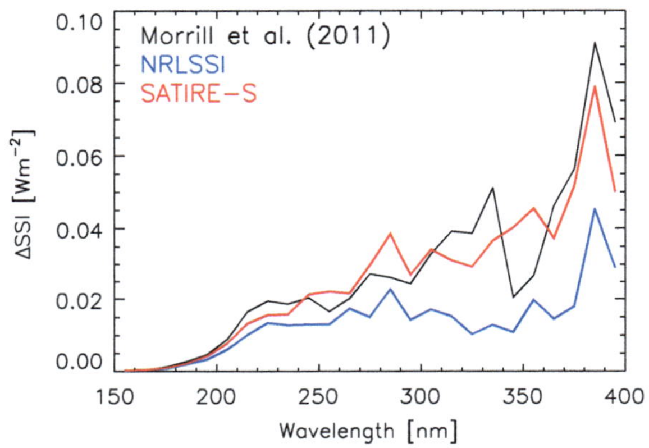

The ultraviolet solar irradiance measurements from the various instruments sent into space diverge in terms of the amplitude of solar cycle variation, especially above 240 nm and between SORCE and pre-SORCE missions (Sect. 2.2.1). Consequently, while the SATIRE-S reconstruction reproduces the cyclical variability in the ultraviolet solar irradiance observations from the UARS and SORCE missions closely below 180 nm, above this wavelength, it replicates certain records better than others (Fig. 4). The reconstruction is a close match to the empirical model by Morrill et al. (2011). This model, based on the matching of the Mg II index to SUSIM SSI, represents an estimation of SUSIM-like spectrometry with the stability corrected to that of the Mg II index. Notably, the amplitude of solar cycle variation is similar, even above 240 nm (Figs. 7 and 12). The reconstruction does not replicate the overall trend in SIM SSI (Fig. 5). It is worth noting here that there is no model reported thus far that is able to reproduce the solar cycle variation in SIM SSI and TSI simultaneously.

The Lyman-α irradiance of the reconstruction reproduces most of the variability ($R^2 = 0.94$), including the solar cycle trend (Fig. 13(a)), in the LASP[14] Lyman-α composite (Woods et al. 2000). The Mg II index derived from the reconstruction is highly correlated to the competing Mg II index composites by LASP (Viereck et al. 2004; Snow et al. 2005b) and by IUP[15] (Viereck and Puga 1999; Skupin et al. 2005a,c), in particular with the latter

[14]The Laboratory for Atmospheric and Space Physics.

[15]IUP is the germanophone acronym of the Institute of Environmental Physics at the University of Bremen.

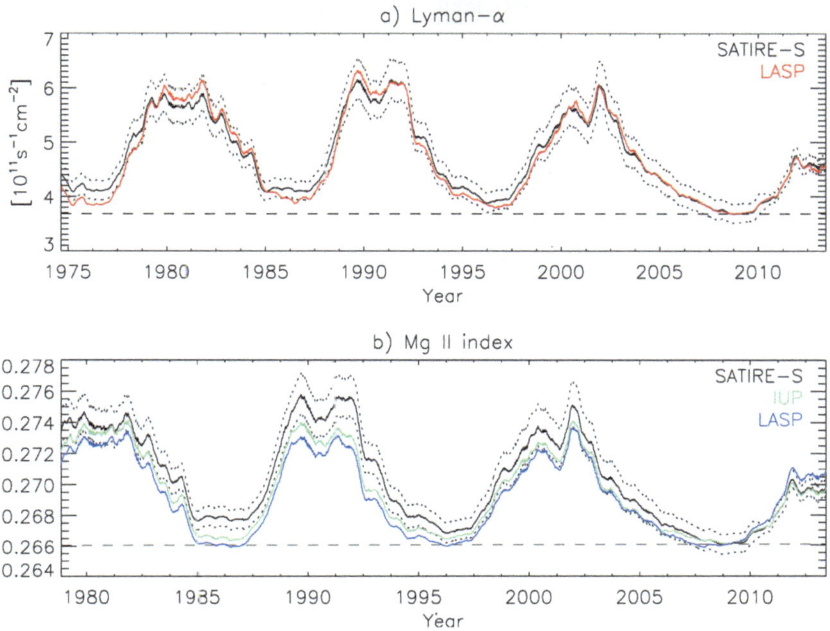

Fig. 13 (**a**) The Lyman-α irradiance and (**b**) Mg II index based on the SATIRE-S reconstruction of SSI (*black solid lines*). The reconstruction uncertainty is denoted by *the dotted lines*. Also illustrated, the LASP Lyman-α composite (*red*), and the IUP (*green*) and LASP (*blue*) Mg II index composites, regressed to the rotational variability and offset to the 2008 solar cycle minimum level (*dashed lines*) in the respective SATIRE-S time series. All the time series were smoothed with a 181-day boxcar filter. Adapted from Yeo et al. (2014b)

($R^2 = 0.96$). It was, however, less successful in replicating the decadal trend in these two composites (Fig. 13(c)). That said, it did reproduce, to within model uncertainty, the secular decline between the 1996 and 2008 solar cycle minima in the IUP Mg II index composite.

3.3.2 Other Present-Day Models

The present-day models capable of returning the solar spectrum over the ultraviolet to the infrared are, with the exception of the SRPM, broadly consistent with one another. They differ in certain aspects, most notably in terms of the spectral dependence of the cyclical variability (Fig. 7). In this section, we will give a brief description of the NRLSSI, SRPM, OAR and COSI models, and highlight the key discrepancies between these models and SATIRE-S.

The NRLSSI (Lean et al. 1997; Lean 2000) describes the effect of sunspot darkening and faculae brightening on a model spectrum of the quiet Sun. The time evolution of sunspot darkening is given by the PSI derived from sunspot area records and facular brightening by the Ca II K, Mg II and F10.7 indices. In the ultraviolet (120 to 400 nm), the variation in solar irradiance is inferred from the multivariate regression of the index data to the rotational variability in UARS/SOLSTICE SSI. This is achieved by detrending both index and SSI data prior to the regression. Above 400 nm, it is given by the sunspot and faculae contrast models from Solanki and Unruh (1998), modulated in time by the index data.

The models discussed here, apart from the SRPM, all see reconstructed SSI varying in phase with the solar cycle in the ultraviolet and visible (Fig. 7). In the infrared, facular contrast is weak and negative at certain wavelengths, depending on the model. This allows sunspot darkening to dominate such that the overall level at activity maximum can be lower than at minimum as illustrated for SATIRE-S in Fig. 5(d). Depending on the sunspot/facular contrast adopted by the various models, they differ in the wavelength range and strength of this effect (see Fig. 7 in Ermolli et al. 2013). This effect is relatively weak in the NRLSSI such that the integrated flux over the shortwave-infrared (1000 to 2430 nm) still varies in phase with the solar cycle, contrary to the other models (Fig. 7). The variation over the solar cycle between 240 and 400 nm is also weaker, almost half of that in SATIRE-S, attributed to confining the regression to rotational variability (Ermolli et al. 2013). The consistency between SATIRE-S and the SUSIM-based model of Morrill et al. (2011) gives further support to the amplitude of solar cycle variation in the ultraviolet exhibited by these two models (Fig. 12). While SATIRE-S replicates the secular decline between the 1996 and 2008 solar cycle minima in VIRGO TSI radiometry (Fig. 2(a)), the NRLSSI does not as this minimum-to-minimum variation is absent in the Mg II index composite employed, that released by LASP (Fig. 13(c)).

The SRPM denotes the set of data and tools for semi-empirical modelling of solar irradiance, including a non-LTE spectral synthesis code, developed by Fontenla et al. (1999, 2004, 2006, 2009, 2011). Currently, the package features semi-empirical model atmospheres for nine solar surface components: quiet Sun internetwork, quiet Sun network lane, enhanced network, plage, bright plage, sunspot umbra and sunspot penumbra, presented in Fontenla et al. (2009), and dark quiet Sun internetwork and hot facula, introduced later in Fontenla et al. (2011). Fontenla et al. (2011) made various adjustments to the Fontenla et al. 2009 model atmospheres guided by SORCE SSI. The solar irradiance reconstruction presented by the authors, extending 2000 to 2009, is based on these modified model atmospheres. The filling factors of the various components were derived from full-disc images in the red part of the visible spectrum and Ca II K acquired with the Precision Solar Photometric Telescope, PSPT at OAR (Coulter and Kuhn 1994; Ermolli et al. 1998).

The modifications to the Fontenla et al. (2009) model atmospheres resulted in the reconstructed visible solar irradiance registering lower levels during periods of higher activity (Figs. 6 and 9 in Fontenla et al. 2011), qualitatively consistent with the increase in SIM visible flux between 2003 and 2006 (Fig. 5(c)). However, the solar cycle variation in the ultraviolet is still weaker than what was recorded by SORCE/SOLSTICE. The model thus failed to reproduce the solar cycle variation in measured TSI, which all the other present-day models are able to do with reasonable success. These shortcomings were taken by the authors to imply that the number of solar surface components is insufficient, which runs counter to the OAR results.

The team at OAR also employed the full-disc images recorded with the PSPT at the observatory in the series of proxy and semi-empirical reconstructions of solar irradiance reported (Penza et al. 2003; Domingo et al. 2009; Ermolli et al. 2011). Work on a new semi-empirical model is in progress (Ermolli et al. 2013). This latest effort considers the seven solar surface components defined in Fontenla et al. (2009). The filling factors were obtained from PSPT observations covering the period of 1997 to 2012 following Ermolli et al. (2010). The intensity spectra corresponding to each component were calculated with the Fontenla et al. (2009) model atmospheres, without the more recent modifications introduced by Fontenla et al. (2011), using the non-LTE spectral synthesis code RH[16] (Uitenbroek 2002). The computation is therefore, apart from employing the Fontenla et al. (2009) model

[16] Based on the work of and abbreviated after Rybicki and Hummer (1991, 1992).

atmospheres as is, broadly analogous to the SRPM reconstruction presented by Fontenla et al. 2011.

In the ultraviolet and visible, the OAR reconstruction is roughly consistent with NRLSSI, SATIRE-S and COSI, exhibiting a similar disparity with SIM SSI (Fig. 7). Also in line with these other models, the TSI from the reconstruction replicates most of the variability in TIM radiometry and the PMOD composite. The agreement with measured TSI suggests that the seven solar surface components described in Fontenla et al. (2009) are sufficient for the semi-empirical modelling of at least TSI, in contradiction to the conclusion of Fontenla et al. (2011). The OAR results also support the conclusions of Ball et al. (2011), DeLand and Cebula (2012), Lean and DeLand (2012), Unruh et al. (2012), Yeo et al. (2014b), that the discrepancy between SIM SSI and models arise from unaccounted instrumental effects in the SIM record and does not warrant a significant rethink in how solar irradiance is modelled.

The non-LTE spectral synthesis code COSI (Haberreiter et al. 2008; Shapiro et al. 2010) has been utilised to generate intensity spectra of solar surface components for semi-empirical modelling of solar irradiance (Haberreiter et al. 2005). The current implementation utilizes intensity spectra generated with the model atmospheres by Fontenla et al. (1999). These were applied to sunspot number, ^{10}Be and neutron monitor data to reconstruct solar irradiance back to the Maunder minimum period and over the Holocene at 1-year and 22-year cadence, respectively (Shapiro et al. 2011). More recently, they were also applied to SRPM PSPT-based filling factors (Shapiro et al. 2013), and HMI full-disc longitudinal magnetograms and continuum intensity images (Thuillier et al. 2014b).

Contrary to the NRLSSI, SATIRE-S and OAR models, the Shapiro et al. (2011) reconstruction varied in anti-phase with the solar cycle in the near-infrared (700 to 1000 nm, Fig. 7), accompanied and compensated by an enhanced variability in the ultraviolet. This was attributed to the use of a single model atmosphere each for sunspots and for plage, so not distinguishing between sunspot umbra and penumbra and between plage and bright plage (Ermolli et al. 2013). To account for this simplification, Shapiro et al. (2013) reduced sunspot and plage contrast in such a manner that brought the reconstruction presented into alignment with SORCE/SOLSTICE measurements in the Herzberg continuum (190 to 222 nm). A more comprehensive approach is in development.

4 Reconciling Measurements and Models

4.1 Proxy Models

Solar irradiance observations, in particular SSI, suffer non-trivial long-term uncertainty (Sect. 2). As mentioned in Sect. 3.2.1, in the regression of indices of solar activity to measured SSI, certain proxy models confine the fitting to rotational variability to circumvent bias from instrumental trends. Such a step implicitly assumes that the relationship between the two is similar at all timescales.

While the rotational variability in solar irradiance is largely driven by the time evolution of active regions, the variability at longer timescales is dominated by the magnetic flux distributed in the quiet Sun (Foukal and Lean 1988; Fligge et al. 2000; Solanki et al. 2002). The response of chromospheric indices to magnetic flux in active regions and in the quiet Sun is evidently not the same (Tapping 1987; Solanki and Krivova 2004; Ermolli et al. 2010; Foukal et al. 2011). The weak solar cycle variation in the ultraviolet in the NRLSSI highlights the limitation of applying the relationship between chromospheric indices and solar irradiance at rotational timescales to longer timescales.

VIRGO registered a secular decline in TSI of over 0.2 Wm^{-2} between the 1996 and 2008 solar cycle minima, approximately 20 % of the solar cycle amplitude (Fröhlich 2009). The NRLSSI does not replicate this clear secular trend in VIRGO radiometry as it is absent in the LASP Mg II index composite. Fröhlich (2009, 2012, 2013) attributed the discrepant decadal trend in the LASP Mg II index composite and VIRGO TSI to a possible cooling/dimming of the photosphere between the two solar cycle minima. This disparity is more likely related to the non-linear relationship between chromospheric indices and solar irradiance, discussed in the previous paragraph (Foukal et al. 2011) and the long-term uncertainty of Mg II index data. While the LASP Mg II index composite is effectively level between the 1996 and 2008 solar cycle minima, the competing composite by IUP does exhibit a secular decline (Fig. 13(c)). This minimum-to-minimum drop in the IUP composite is qualitatively replicated in the Mg II index produced from the SATIRE-S reconstruction, the TSI from which reproduces the secular decline in VIRGO TSI (Sect. 3.3.1). The discrepancy between the IUP and LASP composites demonstrates how even for an activity proxy as robust to instrumental effects as the Mg II index, the long-term uncertainty can still be sufficient to obscure the underlying decadal variation.

The rigorous reconstruction of solar irradiance through proxy models would require a greater understanding of the relationship between indices of solar activity and solar irradiance, and of the long-term stability of index data, which is still largely unknown.

4.2 Semi-empirical Models

Present-day semi-empirical models, reviewed in Sect. 3.3, all employ one-dimensional or plane-parallel model atmospheres. Various studies have pointed out that the intensity spectra synthesized from one-dimensional representations of the spatially inhomogeneous solar atmosphere do not necessarily reflect the true average property (Uitenbroek and Criscuoli 2011; Holzreuter and Solanki 2013).

The intensity contrast of network and facular magnetic features varies with distance from disc centre and magnetic flux. In SATIRE-S, the magnetic flux dependence is linear with the magnetogram signal up to a saturation level, the free parameter in the model (Fligge et al. 2000). For the SRPM, OAR and COSI models, which employ full-disc Ca II K images, after identifying sunspots, the rest of the solar disc is segmented by the Ca II K intensity into multiple components. These measures are not only empirical but also do not properly account for the observation that the continuum and line core intensity contrast of small-scale magnetic concentrations scale differently with magnetogram signal (Yeo et al. 2013). As set out by Unruh et al. (2009), three-dimensional model atmospheres would allow the possibility to relate the appropriate calculated intensity spectra to the magnetogram signal or Ca II K intensity directly.

In the continuum, the intensity contrast of small-scale magnetic elements increases with distance from disc centre before declining again close to the limb while the converse is observed within spectral lines. This difference comes primarily from the differing interaction between the line-of-sight, and magnetic flux tubes and the intervening atmosphere at the continuum and spectral lines formation heights (Solanki et al. 1998; Yeo et al. 2013). In employing one-dimensional model atmospheres, present semi-empirical models do not capture these effects.

Three-dimensional model atmospheres based on observations (Socas-Navarro 2011) and magnetohydrodynamics (MHD) simulations (Vögler et al. 2005), while growing in sophistication and realism, cannot as yet reproduce observations at all heights (Afram et al. 2011). A limiting factor is our understanding of the effects of spatial resolution, that is, the point spread function and how it is sampled by the imaging array, on observations (Danilovic et al. 2008, 2013; Röhrbein et al. 2011). This is especially severe for the small-scale magnetic concentrations which make up network and faculae as they are largely unresolved in current observations. The increasing availability of atmospheric seeing-free observations from space and balloon-borne missions, in particular high spatial resolution imagery, such as those from SUNRISE (Solanki et al. 2010; Barthol et al. 2011), will provide stringent constraints on model atmospheres. Space and balloon-borne telescopes have the advantage that the point spread function can be well-constrained (Mathew et al. 2007, 2009; Wedemeyer-Böhm 2008; Yeo et al. 2014a), rendering them particularly useful for this purpose.

Another source of uncertainty is the treatment of non-LTE effects, which are highly complex and not fully understood. The SRPM, OAR and COSI models employ various non-LTE spectral synthesis codes which differ by the approach taken to approximate non-LTE effects. In SATIRE-S, which relies on an LTE spectral synthesis code, non-LTE effects are accounted for empirically instead. While inexact, the one-dimensional model atmospheres and the various measures taken to account for non-LTE effects, direct or empirical, in present-day semi-empirical models are a practical necessity. As these simplifications are tested against and so constrained by observations, current models are found to be reasonable for the purpose of solar irradiance reconstruction. This is demonstrated by the broad consistency between reconstructed solar irradiance and measurements (Sect. 3.3).

5 Summary

The TSI observations from the succession of radiometers sent into space since 1978 readily reveal solar cycle modulation. The clear detection of such modulations, only about 0.1 % of the overall level, is a remarkable achievement. The records from the various instruments do, however, differ in terms of the absolute level and the apparent amplitude of solar cycle variation. This is chiefly from the difficulty in accounting for instrument degradation. While the absolute radiometry of present-day instruments is converging due to the collaborative efforts of various teams, significant uncertainty persists over the long-term stability. This is evident in the conflicting decadal trends exhibited by the three published TSI composites.

Like TSI, ultraviolet (120 to 400 nm) solar irradiance has been monitored from space, almost without interruption, since 1978. Spectrometry is obviously a more complicated measurement. Not surprisingly, the uncertainty in the absolute radiometry and the amplitude of solar cycle variation is more severe than with TSI. Compounded by the wavelength dependence of instrumental influences, this translates into uncertainty in the spectral dependence of the cyclical variability. The problem is particularly acute above 240 nm and between measurements from the SORCE satellite and preceding missions.

The SIM instrument onboard SORCE provides what is still the only extended (2003 to 2011) record of SSI spanning the ultraviolet to the infrared (240 to 2416 nm) available. The measurements from the first few years of operation (2003 to 2008) saw ultraviolet solar irradiance declining almost twice as rapidly as TSI and visible solar irradiance ascending, in apparent anti-phase with the solar cycle. These trends conflict with projections from other measurements and models. Looking at the full period, the overall trend shows no obvious solar cycle modulation. The total flux recorded by the instrument, which surveys a wavelength range responsible for more than 97 % of the power in solar radiative flux, also fails to replicate the solar cycle variation evident in TSI.

Satellite monitoring of solar irradiance has been accompanied by the development of models aimed at recreating the observed variability. Solar irradiance is modulated by photospheric magnetism from its effect on the thermal structure and consequently the radiant behaviour of the solar surface and atmosphere. Models of solar irradiance based on the assumption that variations at timescales greater than a day are driven by solar surface magnetism have achieved considerable success in replicating observations.

There are two broad categories of solar irradiance models, termed proxy and semi-empirical. Proxy models are based on the regression of indices of solar activity to solar irradiance observations. Semi-empirical models employ the intensity spectra of solar surface features calculated from semi-empirical model atmospheres with spectral synthesis codes. These intensity spectra are combined with the apparent surface coverage of the various features, derived from index data or full-disc observations, to reconstruct the solar spectrum.

We discussed the present-day models capable of returning the spectrum over the ultraviolet to the infrared: NRLSSI, SATIRE-S, SRPM, OAR and COSI. Apart from the NRLSSI, these models adopt the semi-empirical approach.

In the regression of index data to solar irradiance observations, certain proxy models including the ultraviolet segment of NRLSSI restrict the fitting to the rotational variability to factor out any bias from the long-term uncertainty of the solar irradiance measurements employed. In doing so, these models implicitly assume that the relationship between chromospheric indices, utilized in these models as a proxy of facular brightening, and solar irradiance at rotational timescales is applicable to longer timescales. Likely a consequence of the fact that the relationship between chromospheric indices and solar irradiance is really non-linear, the amplitude of solar cycle variation in the ultraviolet in the NRLSSI is weaker than in other present-day models.

Another limitation of the proxy approach is the fact that the reconstructed solar irradiance adopts the variability of the index records used in the reconstruction, along with the associated uncertainty. We argued that the long-term uncertainty of Mg II index data might be the reason why the NRLSSI does not replicate the secular decline between the 1996 and 2008 solar cycle minima in VIRGO TSI radiometry.

The SRPM reconstruction of solar irradiance presented by Fontenla et al. (2011) is the only one where visible flux varied in anti-phase with the solar cycle, in qualitative agreement with early SIM observations. This was achieved with modifications to the Fontenla et al. (2009) model atmospheres. However, the TSI from the reconstruction failed to replicate the solar cycle variation in measured TSI. The other models reviewed see visible solar irradiance varying in-phase with the solar cycle and reproduced TSI variability, including the solar cycle modulation, with reasonable success. Significantly, the OAR computation is, apart from the use of the Fontenla et al. (2009) model atmospheres without any modifications, largely analogous to the Fontenla et al. (2011) study in terms of the approach.

Considering the role of photospheric magnetism in driving variations in solar irradiance, the increase in the visible registered by SIM during its early operation, coming at a time where solar activity is declining, requires that small-scale magnetic concentrations be darker than the quiet Sun in this spectral region. However, our current understanding of the radiant properties of these solar surface features point to the converse.

Apart from the NRLSSI, the semi-empirical model SATIRE-S, recently updated by Yeo et al. (2014b), gives the only other daily reconstruction of the solar spectrum spanning the ultraviolet to the infrared from present-day models to extend multiple solar cycles, covering 1974 to 2013. Of the three divergent TSI composites, the model found the greatest success in replicating the solar cycle variation in the PMOD composite. The TSI reconstruction is also a good match to present-day measurements, reproducing about 96 % of the variability in the PMO6V record and the secular decline between the 1996 and 2008 solar cycle minima in VIRGO radiometry. The SSI reconstruction replicates the solar cycle variation in the ultraviolet solar irradiance observations from the UARS and SORCE missions very closely below 180 nm. As the amplitude of solar cycle variation is poorly constrained in available SSI measurements at longer wavelengths, particularly above 240 nm, SATIRE-S, as with all other models, cannot exactly replicate SORCE solar cycle variation very well there. The amplitude of solar cycle variation in the reconstruction does however, match closely to the empirical model of Morrill et al. (2011), which represents an approximation of SUSIM-like SSI with the stability corrected to that of the Mg II index. The model also replicates the solar cycle variation in the LASP Lyman-α composite and the secular decline between the 1996 and 2008 solar cycle minima in the IUP Mg II index composite.

The intensity spectra of solar surface features employed in present-day semi-empirical models are derived from one-dimensional model atmospheres which do not capture all the complexities of the radiant behaviour of the solar surface and atmosphere. Three-dimensional model atmospheres, though increasingly realistic, still cannot reproduce observations at all heights. Their development is impeded by the limited availability of high spatial resolution observations and the challenge in understanding instrumental influences on apparent radiance. Current semi-empirical models account for non-LTE effects either empirically by offsetting/rescaling reconstructed spectra to measured SSI or directly by employing non-LTE spectral synthesis codes, neither of which is exact. Constrained by observations, the intensity spectra of solar surface features generated from one-dimensional model atmospheres and present non-LTE schemes, are still somewhat reliable for the intended purpose. This is demonstrated by the broad consistency between the various semi-empirical models and their success in replicating measurements.

The direct observation of solar irradiance is a challenging endeavour. At present, the body of spaceborne measurements is still afflicted by uncertainties in the absolute radiometry, secular variation and spectral dependence of the cyclical variability. However, one cannot discount the considerable progress made over the past four decades with the collective effort of the community. A good example is the collaborative efforts which led to the convergence of ACRIM3, TIM and PREMOS absolute TSI radiometry. Models of solar irradiance based on solar surface magnetism have proved to be an able complement, augmenting our understanding of the observations and the physical processes underlying solar cycle variation in solar irradiance. While open questions remain, continual observational and modelling efforts will undoubtedly see the emergence of a more cohesive picture of solar cycle variation in solar irradiance.

Acknowledgements We are thankful to Yvonne Unruh for her help in preparing Fig. 7, and Ilaria Ermolli and Karl-Heinz Glassmeier for the helpful discussions. The authors are also grateful to the International Space Science Institute for their invitation to and tremendous hospitality over the 'The Solar Activity Cycle: Physical Causes And Consequences' workshop. Acknowledgment goes to the various individuals and teams responsible for the observations and models featured in this review. K.L.Y. is a postgraduate fellow of the International Max Planck Research School for Solar System Science. This work is partly supported by the German Federal Ministry of Education and Research under project 01LG1209A and the Ministry of Education of Korea through the BK21 plus program of the National Research Foundation, NRF.

References

C.G. Abbot, F.E. Fowle, L.B. Aldrich, Chapter VI. Ann. Astrophys. Obs. Smithson. Inst. **4**, 177–215 (1923)

N. Afram, Y.C. Unruh, S.K. Solanki, M. Schüssler, A. Lagg, A. Vögler, Intensity contrast from MHD simulations and HINODE observations. Astron. Astrophys. **526**, A120 (2011). doi:10.1051/0004-6361/201015582

W.T. Ball, Y.C. Unruh, N.A. Krivova, S.K. Solanki, J.W. Harder, Solar irradiance variability: a six-year comparison between SORCE observations and the SATIRE model. Astron. Astrophys. **530**, A71 (2011). doi:10.1051/0004-6361/201016189

W.T. Ball, Y.C. Unruh, N.A. Krivova, S.K. Solanki, T. Wenzler, D.J. Mortlock, A.H. Jaffe, Reconstruction of total solar irradiance 1974–2009. Astron. Astrophys. **541**, A27 (2012). doi:10.1051/0004-6361/201118702

W.T. Ball, N.A. Krivova, Y.C. Unruh, J.D. Haigh, S.K. Solanki, A new SATIRE-S spectral solar irradiance dataset for solar cycles 21–23 and its implications for stratospheric ozone. J. Atmos. Sci. (2014, submitted)

L.A. Balmaceda, S.K. Solanki, N.A. Krivova, S. Foster, A homogeneous database of sunspot areas covering more than 130 years. J. Geophys. Res. **114**, 7104 (2009). doi:10.1029/2009JA014299

P. Barthol, A. Gandorfer, S.K. Solanki, M. Schüssler, B. Chares, W. Curdt, W. Deutsch, A. Feller, D. Germerott, B. Grauf, K. Heerlein, J. Hirzberger, M. Kolleck, R. Meller, R. Müller, T.L. Riethmüller, G. Tomasch, M. Knölker, B.W. Lites, G. Card, D. Elmore, J. Fox, A. Lecinski, P. Nelson, R. Summers, A. Watt, V. Martínez Pillet, J.A. Bonet, W. Schmidt, T. Berkefeld, A.M. Title, V. Domingo, J.L. Gasent Blesa, J.C. Del Toro Iniesta, A. López Jiménez, A. Álvarez-Herrero, L. Sabau-Graziati, C. Widani, P. Haberler, K. Härtel, D. Kampf, T. Levin, I. Pérez Grande, A. Sanz-Andrés, E. Schmidt, The SUNRISE mission. Sol. Phys. **268**, 1–34 (2011). doi:10.1007/s11207-010-9662-9

G.E. Brueckner, K.L. Edlow, L.E. Floyd IV, J.L. Lean, M.E. Vanhoosier, The Solar Ultraviolet Spectral Irradiance Monitor (SUSIM) experiment on board the Upper Atmosphere Research Satellite (UARS). J. Geophys. Res. **98**, 10695 (1993). doi:10.1029/93JD00410

J.J. Butler, B.C. Johnson, J.P. Rice, E.L. Shirley, R.A. Barnes, Sources of differences in on-orbit total solar irradiance measurements and description of a proposed laboratory intercomparison. J. Res. Natl. Inst. Stand. Technol. **113**, 187–203 (2008)

M. Carlsson, R.F. Stein, Å. Nordlund, G.B. Scharmer, Observational manifestations of solar magnetoconvection: center-to-limb variation. Astrophys. J. Lett. **610**, 137–140 (2004). doi:10.1086/423305

G.A. Chapman, A.M. Cookson, J.J. Dobias, Variations in total solar irradiance during solar cycle 22. J. Geophys. Res. **101**, 13541–13548 (1996). doi:10.1029/96JA00683

G.A. Chapman, A.M. Cookson, D.G. Preminger, Comparison of TSI from SORCE TIM with SFO ground-based photometry. Sol. Phys. **276**, 35–41 (2012). doi:10.1007/s11207-011-9867-6

G.A. Chapman, A.M. Cookson, D.G. Preminger, Modeling total solar irradiance with San Fernando Observatory ground-based photometry: comparison with ACRIM, PMOD, and RMIB composites. Sol. Phys. **283**, 295–305 (2013). doi:10.1007/s11207-013-0233-8

P. Charbonneau, Dynamo models of the solar cycle. Living Rev. Sol. Phys. **7**, 3 (2010). doi:10.12942/lrsp-2010-3

J.-F. Cossette, P. Charbonneau, P.K. Smolarkiewicz, Cyclic thermal signature in a global MHD simulation of solar convection. Astrophys. J. Lett. **777**, 29 (2013). doi:10.1088/2041-8205/777/2/L29

R.L. Coulter, J.R. Kuhn, RISE/PSPT as an experiment to study active region irradiance and luminosity evolution, in *Solar Active Region Evolution: Comparing Models with Observations*, ed. by K.S. Balasubramaniam, G.W. Simon. Astr. Soc. P., vol. 68 (1994), p. 37

S. Danilovic, A. Gandorfer, A. Lagg, M. Schüssler, S.K. Solanki, A. Vögler, Y. Katsukawa, S. Tsuneta, The intensity contrast of solar granulation: comparing Hinode SP results with MHD simulations. Astron. Astrophys. **484**, L17–L20 (2008). doi:10.1051/0004-6361:200809857

S. Danilovic, D. Röhrbein, R.H. Cameron, M. Schüssler, On the relation between continuum brightness and magnetic field in solar active regions. Astron. Astrophys. **550**, A118 (2013). doi:10.1051/0004-6361/201219726

M.T. DeLand, R.P. Cebula, Solar UV activity at solar cycle 21 and 22 minimum from NOAA-9 SBUV/2 Data. Sol. Phys. **177**, 105–116 (1998). doi:10.1023/A:1004931218139

M.T. DeLand, R.P. Cebula, Creation of a composite solar ultraviolet irradiance data set. J. Geophys. Res. **113**(A12), 11103 (2008). doi:10.1029/2008JA013401

M.T. DeLand, R.P. Cebula, Solar UV variations during the decline of cycle 23. J. Atmos. Sol.-Terr. Phys. **77**, 225–234 (2012). doi:10.1016/j.jastp.2012.01.007

S. Dewitte, D. Crommelynck, A. Joukoff, Total solar irradiance observations from DIARAD/VIRGO. J. Geophys. Res. **109**, 2102 (2004a). doi:10.1029/2002JA009694

S. Dewitte, D. Crommelynck, S. Mekaoui, A. Joukoff, Measurement and uncertainty of the long-term total solar irradiance trend. Sol. Phys. **224**, 209–216 (2004b). doi:10.1007/s11207-005-5698-7

V. Domingo, I. Ermolli, P. Fox, C. Fröhlich, M. Haberreiter, N. Krivova, G. Kopp, W. Schmutz, S.K. Solanki, H.C. Spruit, Y. Unruh, A. Vögler, Solar surface magnetism and irradiance on time scales from days to the 11-year cycle. Space Sci. Rev. **145**, 337–380 (2009). doi:10.1007/s11214-009-9562-1

T. Dudok de Wit, S. Bruinsma, K. Shibasaki, Synoptic radio observations as proxies for upper atmosphere modelling. J. Space Weather Space Clim. **4**(26), 260000 (2014). doi:10.1051/swsc/2014003

J.A. Eddy, The Maunder minimum. Science **192**, 1189–1202 (1976). doi:10.1126/science.192.4245.1189

I. Ermolli, S. Criscuoli, F. Giorgi, Recent results from optical synoptic observations of the solar atmosphere with ground-based instruments. Contrib. Astron. Obs. Skaln. Pleso **41**, 73–84 (2011)

I. Ermolli, M. Fofi, C. Bernacchia, F. Berrilli, B. Caccin, A. Egidi, A. Florio, The prototype RISE-PSPT instrument operating in Rome. Sol. Phys. **177**, 1–10 (1998). doi:10.1023/A:1004932431519

I. Ermolli, B. Caccin, M. Centrone, V. Penza, Modeling solar irradiance variations through full-disk images and semi-empirical atmospheric models. Mem. Soc. Astron. Ital. **74**, 603 (2003)

I. Ermolli, S. Criscuoli, M. Centrone, F. Giorgi, V. Penza, Photometric properties of facular features over the activity cycle. Astron. Astrophys. **465**, 305–314 (2007). doi:10.1051/0004-6361:20065995

I. Ermolli, S. Criscuoli, H. Uitenbroek, F. Giorgi, M.P. Rast, S.K. Solanki, Radiative emission of solar features in the Ca II K line: comparison of measurements and models. Astron. Astrophys. **523**, A55 (2010). doi:10.1051/0004-6361/201014762

I. Ermolli, K. Matthes, T. Dudok de Wit, N.A. Krivova, K. Tourpali, M. Weber, Y.C. Unruh, L. Gray, U. Langematz, P. Pilewskie, E. Rozanov, W. Schmutz, A. Shapiro, S.K. Solanki, T.N. Woods, Recent variability of the solar spectral irradiance and its impact on climate modelling. Atmos. Chem. Phys. **13**, 3945–3977 (2013). doi:10.5194/acp-13-3945-2013

A. Fehlmann, G. Kopp, W. Schmutz, R. Winkler, W. Finsterle, N. Fox, Fourth world radiometric reference to SI radiometric scale comparison and implications for on-orbit measurements of the total solar irradiance. Metrologia **49**, 34 (2012). doi:10.1088/0026-1394/49/2/S34

M. Fligge, S.K. Solanki, Y.C. Unruh, Modelling irradiance variations from the surface distribution of the solar magnetic field. Astron. Astrophys. **353**, 380–388 (2000)

L.E. Floyd, J.W. Cook, L.C. Herring, P.C. Crane, SUSIM'S 11-year observational record of the solar UV irradiance. Adv. Space Res. **31**, 2111–2120 (2003). doi:10.1016/S0273-1177(03)00148-0

J.M. Fontenla, J. Harder, G. Rottman, T.N. Woods, G.M. Lawrence, S. Davis, The signature of solar activity in the infrared spectral irradiance. Astrophys. J. Lett. **605**, 85–88 (2004). doi:10.1086/386335

J.M. Fontenla, E. Avrett, G. Thuillier, J. Harder, Semiempirical models of the solar atmosphere. I. The quiet- and active Sun photosphere at moderate resolution. Astrophys. J. **639**, 441–458 (2006). doi:10.1086/499345

J.M. Fontenla, W. Curdt, M. Haberreiter, J. Harder, H. Tian, Semiempirical models of the solar atmosphere. III. Set of non-LTE models for far-ultraviolet/extreme-ultraviolet irradiance computation. Astrophys. J. **707**, 482–502 (2009). doi:10.1088/0004-637X/707/1/482

J.M. Fontenla, J. Harder, W. Livingston, M. Snow, T. Woods, High-resolution solar spectral irradiance from extreme ultraviolet to far infrared. J. Geophys. Res. **116**(D15), 20108 (2011). doi:10.1029/2011JD016032

J. Fontenla, O.R. White, P.A. Fox, E.H. Avrett, R.L. Kurucz, Calculation of solar irradiances. I. Synthesis of the solar spectrum. Astrophys. J. **518**, 480–499 (1999). doi:10.1086/307258

P. Foukal, J. Lean, The influence of faculae on total solar irradiance and luminosity. Astrophys. J. **302**, 826–835 (1986). doi:10.1086/164043

P. Foukal, J. Lean, Magnetic modulation of solar luminosity by photospheric activity. Astrophys. J. **328**, 347–357 (1988). doi:10.1086/166297

P. Foukal, A. Ortiz, R. Schnerr, Dimming of the 17th century Sun. Astrophys. J. Lett. **733**, 38 (2011). doi:10.1088/2041-8205/733/2/L38

E.N. Frazier, Multi-channel magnetograph observations. III: Faculae. Sol. Phys. **21**, 42–53 (1971). doi:10.1007/BF00155772

C. Fröhlich, Observations of irradiance variations. Space Sci. Rev. **94**, 15–24 (2000). doi:10.1023/A:1026765712084

C. Fröhlich, Solar irradiance variability since 1978. Revision of the PMOD composite during Solar Cycle 21. Space Sci. Rev. **125**, 53–65 (2006). doi:10.1007/s11214-006-9046-5

C. Fröhlich, Evidence of a long-term trend in total solar irradiance. Astron. Astrophys. **501**, L27–L30 (2009). doi:10.1051/0004-6361/200912318

C. Fröhlich, Solar radiometry. ISSI Sci. Rep. Ser. **9**, 525–540 (2010)

C. Fröhlich, Total solar irradiance observations. Surv. Geophys. **33**, 453–473 (2012). doi:10.1007/s10712-011-9168-5

C. Fröhlich, Total solar irradiance: what have we learned from the last three cycles and the recent minimum? Space Sci. Rev. **176**, 237–252 (2013). doi:10.1007/s11214-011-9780-1

C. Fröhlich, J.M. Pap, H.S. Hudson, Improvement of the photometric sunspot index and changes of the disk-integrated sunspot contrast with time. Sol. Phys. **152**, 111–118 (1994). doi:10.1007/BF01473192

C. Fröhlich, J. Romero, H. Roth, C. Wehrli, B.N. Andersen, T. Appourchaux, V. Domingo, U. Telljohann, G. Berthomieu, P. Delache, J. Provost, T. Toutain, D.A. Crommelynck, A. Chevalier, A. Fichot, W. Däppen, D. Gough, T. Hoeksema, A. Jiménez, M.F. Gómez, J.M. Herreros, T.R. Cortés, A.R. Jones, J.M. Pap, R.C. Willson, VIRGO: experiment for helioseismology and solar irradiance monitoring. Sol. Phys. **162**, 101–128 (1995). doi:10.1007/BF00733428

C. Fröhlich, D.A. Crommelynck, C. Wehrli, M. Anklin, S. Dewitte, A. Fichot, W. Finsterle, A. Jiménez, A. Chevalier, H. Roth, In-flight performance of the VIRGO solar irradiance instruments on SOHO. Sol. Phys. **175**, 267–286 (1997). doi:10.1023/A:1004929108864

U. Grossmann-Doerth, M. Knölker, M. Schüssler, S.K. Solanki, The deep layers of solar magnetic elements. Astron. Astrophys. **285**, 648–654 (1994)

M. Haberreiter, W. Schmutz, I. Hubeny, NLTE model calculations for the solar atmosphere with an iterative treatment of opacity distribution functions. Astron. Astrophys. **492**, 833–840 (2008). doi:10.1051/0004-6361:200809503

M. Haberreiter, N.A. Krivova, W. Schmutz, T. Wenzler, Reconstruction of the solar UV irradiance back to 1974. Adv. Space Res. **35**, 365–369 (2005). doi:10.1016/j.asr.2005.04.039

J.W. Harder, J. Fontenla, G. Lawrence, T. Woods, G. Rottman, The spectral irradiance monitor: measurement equations and calibration. Sol. Phys. **230**, 169–204 (2005a). doi:10.1007/s11207-005-1528-1

J.W. Harder, G. Lawrence, J. Fontenla, G. Rottman, T. Woods, The spectral irradiance monitor: scientific requirements, instrument design, and operation modes. Sol. Phys. **230**, 141–167 (2005b). doi:10.1007/s11207-005-5007-5

J.W. Harder, J.M. Fontenla, P. Pilewskie, E.C. Richard, T.N. Woods, Trends in solar spectral irradiance variability in the visible and infrared. Geophys. Res. Lett. **36**, 7801 (2009). doi:10.1029/2008GL036797

D.H. Hathaway, The solar cycle. Living Rev. Sol. Phys. **7**, 1 (2010). doi:10.12942/lrsp-2010-1

D.F. Heath, B.M. Schlesinger, The Mg 280-nm doublet as a monitor of changes in solar ultraviolet irradiance. J. Geophys. Res. **91**, 8672–8682 (1986). doi:10.1029/JD091iD08p08672

J.R. Hickey, L.L. Stowe, H. Jacobowitz, P. Pellegrino, R.H. Maschhoff, F. House, T.H. Vonder Haar, Initial solar irradiance determinations from Nimbus 7 cavity radiometer measurements. Science **208**, 281–283 (1980). doi:10.1126/science.208.4441.281

R. Holzreuter, S.K. Solanki, Three-dimensional non-LTE radiative transfer effects in Fe I lines. II. Line formation in 3D radiation hydrodynamic simulations. Astron. Astrophys. **558**, A20 (2013). doi:10.1051/0004-6361/201322135

D.V. Hoyt, H.L. Kyle, J.R. Hickey, R.H. Maschhoff, The NIMBUS 7 solar total irradiance—a new algorithm for its derivation. J. Geophys. Res. **97**, 51–63 (1992). doi:10.1029/91JA02488

H.S. Hudson, Observed variability of the solar luminosity. Annu. Rev. Astron. Astrophys. **26**, 473–507 (1988). doi:10.1146/annurev.aa.26.090188.002353

H.S. Hudson, S. Silva, M. Woodard, R.C. Willson, The effects of sunspots on solar irradiance. Sol. Phys. **76**, 211–219 (1982). doi:10.1007/BF00170984

H.P. Jones, T.L. Duvall Jr., J.W. Harvey, C.T. Mahaffey, J.D. Schwitters, J.E. Simmons, The NASA/NSO spectromagnetograph. Sol. Phys. **139**, 211–232 (1992). doi:10.1007/BF00159149

S.L. Keil, T.W. Henry, B. Fleck, NSO/AFRL/Sac peak K-line monitoring program, in *Synoptic Solar Physics*, ed. by K.S. Balasubramaniam, J. Harvey, D. Rabin. ASP Conf. Ser., vol. 140 (1998), p. 301

C.U. Keller, M. Schüssler, A. Vögler, V. Zakharov, On the origin of solar faculae. Astrophys. J. Lett. **607**, 59–62 (2004). doi:10.1086/421553

M. Knölker, U. Grossmann-Doerth, M. Schüssler, E. Weisshaar, Some developments in the theory of magnetic flux concentrations in the solar atmosphere. Adv. Space Res. **11**, 285–295 (1991). doi:10.1016/0273-1177(91)90393-X

P. Kobel, S.K. Solanki, J.M. Borrero, The continuum intensity as a function of magnetic field. I. Active region and quiet Sun magnetic elements. Astron. Astrophys. **531**, A112 (2011). doi:10.1051/0004-6361/201016255

G. Kopp, G. Lawrence, The Total Irradiance Monitor (TIM): instrument design. Sol. Phys. **230**, 91–109 (2005). doi:10.1007/s11207-005-7446-4

G. Kopp, J.L. Lean, A new, lower value of total solar irradiance: evidence and climate significance. Geophys. Res. Lett. **38**, 1706 (2011). doi:10.1029/2010GL045777

G. Kopp, K. Heuerman, G. Lawrence, The Total Irradiance Monitor (TIM): instrument calibration. Sol. Phys. **230**, 111–127 (2005a). doi:10.1007/s11207-005-7447-3

G. Kopp, G. Lawrence, G. Rottman, The Total Irradiance Monitor (TIM): science results. Sol. Phys. **230**, 129–139 (2005b). doi:10.1007/s11207-005-7433-9

G. Kopp, K. Heuerman, D. Harber, G. Drake, The TSI radiometer facility: absolute calibrations for total solar irradiance instruments. Proc. SPIE **6677**, 667709 (2007). doi:10.1117/12.734553, 12 pp.

G. Kopp, A. Fehlmann, W. Finsterle, D. Harber, K. Heuerman, R. Willson, Total solar irradiance data record accuracy and consistency improvements. Metrologia **49**, 29 (2012). doi:10.1088/0026-1394/49/2/S29

N.A. Krivova, S.K. Solanki, L. Floyd, Reconstruction of solar UV irradiance in cycle 23. Astron. Astrophys. **452**, 631–639 (2006). doi:10.1051/0004-6361:20064809

N.A. Krivova, S.K. Solanki, Y.C. Unruh, Towards a long-term record of solar total and spectral irradiance. J. Atmos. Sol.-Terr. Phys. **73**, 223–234 (2011). doi:10.1016/j.jastp.2009.11.013

N.A. Krivova, S.K. Solanki, M. Fligge, Y.C. Unruh, Reconstruction of solar irradiance variations in cycle 23: is solar surface magnetism the cause? Astron. Astrophys. **399**, 1–4 (2003). doi:10.1051/0004-6361:20030029

J.R. Kuhn, K.G. Libbrecht, R.H. Dicke, The surface temperature of the sun and changes in the solar constant. Science **242**, 908–911 (1988). doi:10.1126/science.242.4880.908

R. Kurucz, ATLAS9 stellar atmosphere programs and 2 km/s grid. Kurucz CD-ROM no. 13. (Smithsonian Astrophysical Observatory, Cambridge, 1993)

A. Lagg, S.K. Solanki, T.L. Riethmüller, V. Martínez Pillet, M. Schüssler, J. Hirzberger, A. Feller, J.M. Borrero, W. Schmidt, J.C. del Toro Iniesta, J.A. Bonet, P. Barthol, T. Berkefeld, V. Domingo, A. Gandorfer, M. Knölker, A.M. Title, Fully resolved quiet-Sun magnetic flux tube observed with the SUNRISE/IMAX instrument. Astrophys. J. Lett. **723**, 164–168 (2010). doi:10.1088/2041-8205/723/2/L164

J. Lean, Evolution of the Sun's spectral irradiance since the Maunder minimum. Geophys. Res. Lett. **27**, 2425–2428 (2000). doi:10.1029/2000GL000043

J.L. Lean, M.T. DeLand, How does the Sun's spectrum vary? J. Climate **25**, 2555–2560 (2012). doi:10.1175/JCLI-D-11-00571.1

J.L. Lean, G.J. Rottman, H.L. Kyle, T.N. Woods, J.R. Hickey, L.C. Puga, Detection and parameterization of variations in solar mid- and near-ultraviolet radiation (200–400 nm). J. Geophys. Res. **102**, 29939–29956 (1997). doi:10.1029/97JD02092

R.B. Lee III, B.R. Barkstrom, R.D. Cess, Characteristics of the Earth radiation budget experiment solar monitors. Appl. Opt. **26**, 3090–3096 (1987). doi:10.1364/AO.26.003090

R.B. Lee III, M.A. Gibson, R.S. Wilson, S. Thomas, Long-term total solar irradiance variability during sunspot cycle 22. J. Geophys. Res. **100**, 1667–1675 (1995). doi:10.1029/94JA02897

J.R. Lemen, A.M. Title, D.J. Akin, P.F. Boerner, C. Chou, J.F. Drake, D.W. Duncan, C.G. Edwards, F.M. Friedlaender, G.F. Heyman, N.E. Hurlburt, N.L. Katz, G.D. Kushner, M. Levay, R.W. Lindgren, D.P. Mathur, E.L. McFeaters, S. Mitchell, R.A. Rehse, C.J. Schrijver, L.A. Springer, R.A. Stern, T.D. Tarbell, J.-P. Wuelser, C.J. Wolfson, C. Yanari, J.A. Bookbinder, P.N. Cheimets, D. Caldwell, E.E. Deluca, R. Gates, L. Golub, S. Park, W.A. Podgorski, R.I. Bush, P.H. Scherrer, M.A. Gummin, P. Smith, G. Auker, P. Jerram, P. Pool, R. Soufli, D.L. Windt, S. Beardsley, M. Clapp, J. Lang, N. Waltham, The Atmospheric Imaging Assembly (AIA) on the Solar Dynamics Observatory (SDO). Sol. Phys. **275**, 17–40 (2012). doi:10.1007/s11207-011-9776-8

J. Lilensten, T. Dudok de Wit, M. Kretzschmar, P.-O. Amblard, S. Moussaoui, J. Aboudarham, F. Auchère, Review on the solar spectral variability in the EUV for space weather purposes. Ann. Geophys. **26**, 269–279 (2008). doi:10.5194/angeo-26-269-2008

W.C. Livingston, J. Harvey, A.K. Pierce, D. Schrage, B. Gillespie, J. Simmons, C. Slaughter, Kitt peak 60-cm vacuum telescope. Appl. Opt. **15**, 33–39 (1976). doi:10.1364/AO.15.000033

S.K. Mathew, V. Zakharov, S.K. Solanki, Stray light correction and contrast analysis of Hinode broad-band images. Astron. Astrophys. **501**, L19–L22 (2009). doi:10.1051/0004-6361/200911975

S.K. Mathew, V. Martínez Pillet, S.K. Solanki, N.A. Krivova, Properties of sunspots in cycle 23. I. Dependence of brightness on sunspot size and cycle phase. Astron. Astrophys. **465**, 291–304 (2007). doi:10.1051/0004-6361:20066356

W.E. McClintock, G.J. Rottman, T.N. Woods, Solar-Stellar Irradiance Comparison Experiment II (SOLSTICE II): instrument concept and design. Sol. Phys. **230**, 225–258 (2005). doi:10.1007/s11207-005-7432-x

S. Mekaoui, S. Dewitte, Total solar irradiance measurement and modelling during cycle 23. Sol. Phys. **247**, 203–216 (2008). doi:10.1007/s11207-007-9070-y

A.W. Merkel, J.W. Harder, D.R. Marsh, A.K. Smith, J.M. Fontenla, T.N. Woods, The impact of solar spectral irradiance variability on middle atmospheric ozone. Geophys. Res. Lett. **38**, 13802 (2011). doi:10.1029/2011GL047561

W.E. Mitchell Jr., W.C. Livingston, Line-blanketing variations in the irradiance spectrum of the sun from maximum to minimum of the solar cycle. Astrophys. J. **372**, 336–348 (1991). doi:10.1086/169980

R. Moll, R.H. Cameron, M. Schüssler, Vortices, shocks, and heating in the solar photosphere: effect of a magnetic field. Astron. Astrophys. **541**, A68 (2012). doi:10.1051/0004-6361/201218866

J.S. Morrill, K.P. Dere, C.M. Korendyke, The sources of solar ultraviolet variability between 2765 and 2885 Å: Mg I, Mg II, Si I, and continuum. Astrophys. J. **557**, 854–863 (2001). doi:10.1086/321683

J.S. Morrill, L. Floyd, D. McMullin, The solar ultraviolet spectrum estimated using the Mg II index and Ca II K disk activity. Sol. Phys. **269**, 253–267 (2011). doi:10.1007/s11207-011-9708-7

R. Munro, C. Anderson, J. Callies, J. Callies, E. Corpaccioli, R. Lang, A. Lefebvre, Y. Livschitz, A. Pérez Albiñana, GOME-2 on MetOp, in *Atmospheric Science Conference*. ESA Sp. Pub., vol. 628 (2006)

Z.E. Musielak, P. Ulmschneider, Atmospheric oscillations in solar magnetic flux tubes. I. Excitation by longitudinal tube waves and random pulses. Astron. Astrophys. **400**, 1057–1064 (2003). doi:10.1051/0004-6361:20030023

A. Ortiz, S.K. Solanki, V. Domingo, M. Fligge, B. Sanahuja, On the intensity contrast of solar photospheric faculae and network elements. Astron. Astrophys. **388**, 1036–1047 (2002). doi:10.1051/0004-6361:20020500

L. Oster, K.H. Schatten, S. Sofia, Solar irradiance variations due to active regions. Astrophys. J. **256**, 768–773 (1982). doi:10.1086/159949

J. Pagaran, M. Weber, J. Burrows, Solar variability from 240 to 1750 nm in terms of faculae brightening and sunspot darkening from SCIAMACHY. Astrophys. J. **700**, 1884–1895 (2009). doi:10.1088/0004-637X/700/2/1884

V. Penza, B. Caccin, I. Ermolli, M. Centrone, M.T. Gomez, Modeling solar irradiance variations through PSPT images and semiempirical models, in *Solar Variability as an Input to the Earth's Environment*, ed. by A. Wilson. ESA Sp. Pub., vol. 535 (2003), pp. 299–302

D.G. Preminger, S.R. Walton, G.A. Chapman, Photometric quantities for solar irradiance modelling. J. Geophys. Res. **107**, 1354 (2002). doi:10.1029/2001JA009169

M. Rempel, R. Schlichenmaier, Sunspot modelling: from simplified models to radiative MHD simulations. Living Rev. Sol. Phys. **8**, 3 (2011). doi:10.12942/lrsp-2011-3

E.C. Richard, D. Harber, J.W. Harder, P. Pilewskie, S. Brown, A. Smith, K. Lykke, Future long-term measurements of solar spectral irradiance by JPSS TSIS. AGU Fall Meeting Abstracts 917 (2011)

T.L. Riethmüller, S.K. Solanki, V. Martínez Pillet, J. Hirzberger, A. Feller, J.A. Bonet, N. Bello González, M. Franz, M. Schüssler, P. Barthol, T. Berkefeld, J.C. del Toro Iniesta, V. Domingo, A. Gandorfer, M. Knölker, W. Schmidt, Bright points in the quiet Sun as observed in the visible and near-UV by the balloon-borne observatory SUNRISE. Astrophys. J. Lett. **723**, 169–174 (2010). doi:10.1088/2041-8205/723/2/L169

T.L. Riethmüller, S.K. Solanki, S.V. Berdyugina, M. Schüssler, V. Martínez Pillet, A. Feller, A. Gandorfer, J. Hirzberger, Comparison of solar photospheric bright points between SUNRISE observations and MHD simulations. Astron. Astrophys. (2014). doi:10.1051/0004-6361/201423892

D. Röhrbein, R. Cameron, M. Schüssler, Is there a non-monotonic relation between photospheric brightness and magnetic field strength in solar plage regions? Astron. Astrophys. **532**, A140 (2011). doi:10.1051/0004-6361/201117090

G. Rottman, The SORCE mission. Sol. Phys. **230**, 7–25 (2005). doi:10.1007/s11207-005-8112-6

G. Rottman, T. Woods, M. Snow, G. DeToma, The solar cycle variation in ultraviolet irradiance. Adv. Space Res. **27**, 1927–1932 (2001). doi:10.1016/S0273-1177(01)00272-1

G.B. Rybicki, D.G. Hummer, An accelerated lambda iteration method for multilevel radiative transfer. I—Non-overlapping lines with background continuum. Astron. Astrophys. **245**, 171–181 (1991)

G.B. Rybicki, D.G. Hummer, An accelerated lambda iteration method for multilevel radiative transfer. II—Overlapping transitions with full continuum. Astron. Astrophys. **262**, 209–215 (1992)

P.H. Scherrer, R.S. Bogart, R.I. Bush, J.T. Hoeksema, A.G. Kosovichev, J. Schou, W. Rosenberg, L. Springer, T.D. Tarbell, A. Title, C.J. Wolfson, I. Zayer (MDI Engineering Team), The solar oscillations investigation—Michelson Doppler imager. Sol. Phys. **162**, 129–188 (1995). doi:10.1007/BF00733429

W. Schmutz, A. Fehlmann, G. Hülsen, P. Meindl, R. Winkler, G. Thuillier, P. Blattner, F. Buisson, T. Egorova, W. Finsterle, N. Fox, J. Gröbner, J. Hochedez, S. Koller, M. Meftah, M. Meisonnier, S. Nyeki, D. Pfiffner, H. Roth, E. Rozanov, M. Spescha, C. Wehrli, L. Werner, J.U. Wyss, The PREMOS/PICARD instrument calibration. Metrologia **46**(4), 202 (2009)

W. Schmutz, A. Fehlmann, W. Finsterle, G. Kopp, G. Thuillier, Total solar irradiance measurements with PREMOS/PICARD. AIP Conf. Proc. **1531**(1), 624–627 (2013). doi:10.1063/1.4804847

R.S. Schnerr, H.C. Spruit, The brightness of magnetic field concentrations in the quiet Sun. Astron. Astrophys. **532**, A136 (2011). doi:10.1051/0004-6361/201015976

J. Schou, P.H. Scherrer, R.I. Bush, R. Wachter, S. Couvidat, M.C. Rabello-Soares, R.S. Bogart, J.T. Hoeksema, Y. Liu, T.L. Duvall, D.J. Akin, B.A. Allard, J.W. Miles, R. Rairden, R.A. Shine, T.D. Tarbell, A.M. Title, C.J. Wolfson, D.F. Elmore, A.A. Norton, S. Tomczyk, Design and ground calibration of the Helioseismic and Magnetic Imager (HMI) Instrument on the Solar Dynamics Observatory (SDO). Sol. Phys. **275**, 229–259 (2012). doi:10.1007/s11207-011-9842-2

A.D. Seleznyov, S.K. Solanki, N.A. Krivova, Modelling solar irradiance variability on time scales from minutes to months. Astron. Astrophys. **532**, A108 (2011). doi:10.1051/0004-6361/200811138

A.I. Shapiro, W. Schmutz, M. Schoell, M. Haberreiter, E. Rozanov, NLTE solar irradiance modelling with the COSI code. Astron. Astrophys. **517**, A48 (2010). doi:10.1051/0004-6361/200913987

A.I. Shapiro, W. Schmutz, E. Rozanov, M. Schoell, M. Haberreiter, A.V. Shapiro, S. Nyeki, A new approach to the long-term reconstruction of the solar irradiance leads to large historical solar forcing. Astron. Astrophys. **529**, A67 (2011). doi:10.1051/0004-6361/201016173

A.V. Shapiro, A.I. Shapiro, M. Dominique, I.E. Dammasch, C. Wehrli, E. Rozanov, W. Schmutz, Detection of solar rotational variability in the Large Yield RAdiometer (LYRA) 190–222 nm spectral band. Sol. Phys. **286**, 289–301 (2013). doi:10.1007/s11207-012-0029-2

J. Skupin, M. Weber, S. Noël, H. Bovensmann, J.P. Burrows, GOME and SCIAMACHY solar measurements: solar spectral irradiance and Mg II solar activity proxy indicator. Mem. Soc. Astron. Ital. **76**, 1038 (2005a)

J. Skupin, S. Noël, M.W. Wuttke, M. Gottwald, H. Bovensmann, M. Weber, J.P. Burrows, SCIAMACHY solar irradiance observation in the spectral range from 240 to 2380 nm. Adv. Space Res. **35**, 370–375 (2005b). doi:10.1016/j.asr.2005.03.036

J. Skupin, M. Weber, H. Bovensmann, J.P. Burrows, The Mg II solar activity proxy indicator derived from GOME and SCIAMACHY, in *Proceedings of the 2004 Envisat & ERS Symposium*. ESA Sp. Pub., vol. 572 (2005c)

E.V.P. Smith, D.M. Gottlieb, Solar flux and its variations. NASA Spec. Publ. **366**, 97–117 (1975)

M. Snow, W.E. McClintock, G. Rottman, T.N. Woods, Solar-Stellar Irradiance Comparison Experiment II (SOLSTICE II): examination of the solar stellar comparison technique. Sol. Phys. **230**, 295–324 (2005a). doi:10.1007/s11207-005-8763-3

M. Snow, W.E. McClintock, T.N. Woods, O.R. White, J.W. Harder, G. Rottman, The Mg II Index from SORCE. Sol. Phys. **230**, 325–344 (2005b). doi:10.1007/s11207-005-6879-0

H. Socas-Navarro, A high-resolution three-dimensional model of the solar photosphere derived from Hinode observations. Astron. Astrophys. **529**, A37 (2011). doi:10.1051/0004-6361/201015805

S.K. Solanki, Smallscale solar magnetic fields—an overview. Space Sci. Rev. **63**, 1–188 (1993). doi:10.1007/BF00749277

S.K. Solanki, Sunspots: an overview. Astron. Astrophys. Rev. **11**, 153–286 (2003). doi:10.1007/s00159-003-0018-4

S.K. Solanki, N.A. Krivova, Solar irradiance variations: from current measurements to long-term estimates. Sol. Phys. **224**, 197–208 (2004). doi:10.1007/s11207-005-6499-8

S.K. Solanki, Y.C. Unruh, A model of the wavelength dependence of solar irradiance variations. Astron. Astrophys. **329**, 747–753 (1998)

S.K. Solanki, B. Inhester, M. Schüssler, The solar magnetic field. Rep. Prog. Phys. **69**, 563–668 (2006). doi:10.1088/0034-4885/69/3/R02

S.K. Solanki, N.A. Krivova, J.D. Haigh, Solar irradiance variability and climate. Annu. Rev. Astron. Astrophys. **51**, 311–351 (2013). doi:10.1146/annurev-astro-082812-141007

S.K. Solanki, M. Schüssler, M. Fligge, Secular variation of the Sun's magnetic flux. Astron. Astrophys. **383**, 706–712 (2002). doi:10.1051/0004-6361:20011790

S.K. Solanki, O. Steiner, M. Bünte, G. Murphy, S.R.O. Ploner, On the reliability of Stokes diagnostics of magnetic elements away from solar disc centre. Astron. Astrophys. **333**, 721–731 (1998)

S.K. Solanki, P. Barthol, S. Danilovic, A. Feller, A. Gandorfer, J. Hirzberger, T.L. Riethmüller, M. Schüssler, J.A. Bonet, V. Martínez Pillet, J.C. del Toro Iniesta, V. Domingo, J. Palacios, M. Knölker, N. Bello González, T. Berkefeld, M. Franz, W. Schmidt, A.M. Title, SUNRISE: Instrument, mission, data, and first results. Astrophys. J. Lett. **723**, 127–133 (2010). doi:10.1088/2041-8205/723/2/L127

H.C. Spruit, Pressure equilibrium and energy balance of small photospheric fluxtubes. Sol. Phys. **50**, 269–295 (1976). doi:10.1007/BF00155292

H.C. Spruit, B. Roberts, Magnetic flux tubes on the sun. Nature **304**, 401–406 (1983). doi:10.1038/304401a0

H.C. Spruit, C. Zwaan, The size dependence of contrasts and numbers of small magnetic flux tubes in an active region. Sol. Phys. **70**, 207–228 (1981). doi:10.1007/BF00151329

O. Steiner, Radiative properties of magnetic elements. II. Center to limb variation of the appearance of photospheric faculae. Astron. Astrophys. **430**, 691–700 (2005). doi:10.1051/0004-6361:20041286

K.F. Tapping, Recent solar radio astronomy at centimeter wavelengths—the temporal variability of the 10.7-cm flux. J. Geophys. Res. **92**, 829–838 (1987). doi:10.1029/JD092iD01p00829

K.F. Tapping, The 10.7 cm solar radio flux (F10.7). Adv. Space Res. **11**(7), 394–406 (2013). doi:10.1002/swe.20064

G. Thuillier, T. Foujols, D. Bolsée, D. Gillotay, M. Hersé, W. Peetermans, W. Decuyper, H. Mandel, P. Sperfeld, S. Pape, D.R. Taubert, J. Hartmann, SOLAR/SOLSPEC: scientific objectives, instrument performance and its absolute calibration using a blackbody as primary standard source. Sol. Phys. **257**, 185–213 (2009). doi:10.1007/s11207-009-9361-6

G. Thuillier, M. Deland, A. Shapiro, W. Schmutz, D. Bolsée, S.M.L. Melo, The solar spectral irradiance as a function of the Mg II index for atmosphere and climate modelling. Sol. Phys. **277**, 245–266 (2012). doi:10.1007/s11207-011-9912-5

G. Thuillier, D. Bolsée, G. Schmidtke, T. Foujols, B. Nikutowski, A.I. Shapiro, R. Brunner, M. Weber, C. Erhardt, M. Hersé, D. Gillotay, W. Peetermans, W. Decuyper, N. Pereira, M. Haberreiter, H. Mandel, W. Schmutz, The solar irradiance spectrum at solar activity minimum between solar cycles 23 and 24. Sol. Phys. **289**, 1931–1958 (2014a). doi:10.1007/s11207-013-0461-y

G. Thuillier, G. Schmidtke, C. Erhardt, B. Nikutowski, A.I. Shapiro, C. Bolduc, J. Lean, N.A. Krivova, P. Charbonneau, G. Cessateur, M. Haberreiter, S. Melo, V. Delouille, B. Mampaey, K.L. Yeo,

W. Schmutz, Solar spectral irradiance variability in November/December 2012: comparison of observations by instruments on the International Space Station and models. Sol. Phys. (2014b, submitted)

K.P. Topka, T.D. Tarbell, A.M. Title, Properties of the smallest solar magnetic elements. II. Observations versus hot wall models of faculae. Astrophys. J. **484**, 479 (1997). doi:10.1086/304295

H. Uitenbroek, The effect of coherent scattering on radiative losses in the solar Ca II K line. Astrophys. J. **565**, 1312–1322 (2002). doi:10.1086/324698

H. Uitenbroek, S. Criscuoli, Why one-dimensional models fail in the diagnosis of average spectra from inhomogeneous stellar atmospheres. Astrophys. J. **736**, 69 (2011). doi:10.1088/0004-637X/736/1/69

Y.C. Unruh, W.T. Ball, N.A. Krivova, Solar irradiance models and measurements: a comparison in the 220–240 nm wavelength band. Surv. Geophys. **33**, 475–481 (2012). doi:10.1007/s10712-011-9166-7

Y.C. Unruh, S.K. Solanki, M. Fligge, The spectral dependence of facular contrast and solar irradiance variations. Astron. Astrophys. **345**, 635–642 (1999)

Y.C. Unruh, N.A. Krivova, S.K. Solanki, J.W. Harder, G. Kopp, Spectral irradiance variations: comparison between observations and the SATIRE model on solar rotation time scales. Astron. Astrophys. **486**, 311–323 (2008). doi:10.1051/0004-6361:20078421

Y.C. Unruh, S.K. Solanki, M. Schüssler, A. Vögler, D. Garcia-Alvarez, Towards long-term solar irradiance modelling: network contrasts from magneto-convection simulations, in *15th Cambridge Workshop on Cool Stars, Stellar Systems, and the Sun*, ed. by E. Stempels. AIP Conf. Proc., vol. 1094 (2009), pp. 768–771. doi:10.1063/1.3099228

R.A. Viereck, L.C. Puga, The NOAA Mg II core-to-wing solar index: construction of a 20-year time series of chromospheric variability from multiple satellites. J. Geophys. Res. **104**, 9995–10006 (1999). doi:10.1029/1998JA900163

R.A. Viereck, L.E. Floyd, P.C. Crane, T.N. Woods, B.G. Knapp, G. Rottman, M. Weber, L.C. Puga, M.T. Deland, A composite Mg II index spanning from 1978 to 2003. Adv. Space Res. **2**, 10005 (2004). doi:10.1029/2004SW000084

A. Vögler, S. Shelyag, M. Schüssler, F. Cattaneo, T. Emonet, T. Linde, Simulations of magneto-convection in the solar photosphere. Equations, methods, and results of the MURaM code. Astron. Astrophys. **429**, 335–351 (2005). doi:10.1051/0004-6361:20041507

M. Weber, J.P. Burrows, R.P. Cebula, GOME solar UV/VIS irradiance measurements between 1995 and 1997—first results on proxy solar activity studies. Sol. Phys. **177**, 63–77 (1998). doi:10.1023/A:1005030909779

S. Wedemeyer-Böhm, Point spread functions for the Solar optical telescope onboard Hinode. Astron. Astrophys. **487**, 399–412 (2008). doi:10.1051/0004-6361:200809819

C. Wehrli, W. Schmutz, A.I. Shapiro, Correlation of spectral solar irradiance with solar activity as measured by VIRGO. Astron. Astrophys. **556**, L3 (2013). doi:10.1051/0004-6361/201220864

T. Wenzler, Reconstruction of solar irradiance variations in cycles 21–23 based on surface magnetic fields. Ph.D. Thesis (University of Göttingen, 2005)

T. Wenzler, S.K. Solanki, N.A. Krivova, Reconstructed and measured total solar irradiance: is there a secular trend between 1978 and 2003? Geophys. Res. Lett. **36**, 11102 (2009). doi:10.1029/2009GL037519

T. Wenzler, S.K. Solanki, N.A. Krivova, C. Fröhlich, Reconstruction of solar irradiance variations in cycles 21–23 based on surface magnetic fields. Astron. Astrophys. **460**, 583–595 (2006). doi:10.1051/0004-6361:20065752

R.C. Willson, Active cavity radiometer type IV. Appl. Opt. **18**, 179–188 (1979). doi:10.1364/AO.18.000179

R.C. Willson, H.S. Hudson, Solar luminosity variations in solar cycle 21. Nature **332**, 810–812 (1988). doi:10.1038/332810a0

R.C. Willson, A.V. Mordvinov, Secular total solar irradiance trend during solar cycles 21–23. Geophys. Res. Lett. **30**, 1199 (2003). doi:10.1029/2002GL016038

R.C. Willson, S. Gulkis, M. Janssen, H.S. Hudson, G.A. Chapman, Observations of solar irradiance variability. Science **211**, 700–702 (1981). doi:10.1126/science.211.4483.700

C.L. Wolff, J.R. Hickey, Solar irradiance change and special longitudes due to r-modes. Science **235**, 1631–1633 (1987). doi:10.1126/science.235.4796.1631

T.N. Woods, Recent advances in observations and modelling of the solar ultraviolet and X-ray spectral irradiance. Adv. Space Res. **42**, 895–902 (2008). doi:10.1016/j.asr.2007.09.026

T.N. Woods, G. Kopp, P.C. Chamberlin, Contributions of the solar ultraviolet irradiance to the total solar irradiance during large flares. J. Geophys. Res. **111**(A10), 10 (2006). doi:10.1029/2005JA011507

T.N. Woods, W.K. Tobiska, G.J. Rottman, J.R. Worden, Improved solar Lyman α irradiance modelling from 1947 through 1999 based on UARS observations. J. Geophys. Res. **105**, 27195–27216 (2000). doi:10.1029/2000JA000051

T.N. Woods, P.C. Chamberlin, J.W. Harder, R.A. Hock, M. Snow, F.G. Eparvier, J. Fontenla, W.E. McClintock, E.C. Richard, Solar Irradiance Reference Spectra (SIRS) for the 2008 Whole Heliosphere Interval (WHI). Geophys. Res. Lett. **36**, 1101 (2009). doi:10.1029/2008GL036373

K.L. Yeo, S.K. Solanki, N.A. Krivova, Intensity contrast of solar network and faculae. Astron. Astrophys. **550**, A95 (2013). doi:10.1051/0004-6361/201220682

K.L. Yeo, A. Feller, S.K. Solanki, S. Couvidat, S. Danilovic, N.A. Krivova, Point spread function of SDO/HMI and the effects of stray light correction on the apparent properties of solar surface phenomena. Astron. Astrophys. **561**, A22 (2014a). doi:10.1051/0004-6361/201322502

K.L. Yeo, N.A. Krivova, S.K. Solanki, K.H. Glassmeier, Reconstruction of total and spectral solar irradiance since 1974 based on KPVT, SoHO/MDI and SDO/HMI observations. Astron. Astrophys. (2014b, submitted)

DOI 10.1007/978-1-4939-2584-1_6
Reprinted from *Space Science Reviews* Journal, DOI 10.1007/s11214-014-0093-z

The Extended Cycle of Solar Activity and the Sun's 22-Year Magnetic Cycle

E.W. Cliver

Received: 13 May 2014 / Accepted: 14 August 2014 / Published online: 18 September 2014

Abstract The Sun has two characteristic migrations of surface features—the equatorward movement of sunspots and the poleward movement of high-latitude prominences. The first of these migrations is a defining aspect of the 11-yr Schwabe cycle and the second is a tracer of the process that culminates in solar polarity reversal, signaling the onset of the 22-yr magnetic cycle on the Sun. Zonal flows (torsional oscillations of the Sun's differential rotation) have been identified for both of these migrations. Helioseismology observations of these zonal flows provide support for the extended (>11-yr cycle) of solar activity and offer promise of a long-term precursor for predicting the amplitude of the Schwabe cycle. We review the growth of observational evidence for the extended and 22-yr magnetic cycles and discuss: (1) the significance of latitude $\sim 50°$ on the Sun; (2) the "over-extended" cycle; and (3) the outlook for solar cycle 25.

Keywords Extended solar cycle · 22-Yr magnetic cycle · Schwabe cycle · Sunspots · Polar crown filaments

1 Introduction

"Solar periodicity is a most complex phenomenon. The more it is studied, the less it seems to be understood." Thus wrote Agnes Mary Clerke, the authoritative chronicler of 19th century astronomy, in 1903. In the intervening 100 plus years, our knowledge of the 11-yr solar cycle has increased dramatically (e.g., Hathaway 2010; Charbonneau 2010), but the understanding remains a challenge. For example, the 79 predictions for the peak smoothed sunspot number for cycle 24 reviewed by Pesnell (2012) range from 42 to 185, with the most probable value being nearer the lower end of that range (the current observed peak is ~ 75 at the time of this writing). "Prediction ...," as Clerke wrote, "remains at fault." Here we review the evolution of knowledge on the extended 18–22 year cycle of solar activity (Wilson et al. 1988) and the Sun's 22-yr magnetic cycle (Hale and Nicholson 1925; Babcock 1961). We

E.W. Cliver (✉)
National Solar Observatory, Sunspot, NM, USA
e-mail: ecliver@nso.edu

focus on the two characteristic migrations of surface features on the Sun: the equatorward movement of sunspot formation and the poleward movement of high-latitude prominence formation.

Sections 2 and 3 cover historical observations of sunspot and prominence migration and the discovery of the 22-yr magnetic cycle on the Sun, respectively. In Sect. 4, we review recent work on zonal flows of the Sun's differential rotation in relation to the extended and 22-yr cycles. Certain aspects of the extended activity cycle and the 22-yr magnetic cycle are discussed further in Sect. 5.

2 Cyclic Migrations of Sunspots and Prominences

2.1 Movement of Sunspot Formation Toward the Equator

Carrington (1858) discovered the latitude variation of sunspot formation over the 11-yr Schwabe (1844) cycle. He wrote,

> ...throughout the two years preceding the minimum of frequency in February 1856, the spots were confined to an equatorial belt, and in no instance passed the limits of 20° of latitude N. or S.; and that shortly after this epoch, whether connected with it or not, this equatorial series appears to have become extinct, and in seeming contradiction to the precept, *Natura non agit per saltum* [Nature makes no leaps], two new belts of disturbance abruptly commenced, the limits of which in both hemispheres, may be roughly set at between 20° and 40°, with exceptions in favor of the old equatorial region. The tendency at the present time appears to be contraction of the parallels.

In his book *Spots on the Sun*, Carrington (1863) wrote that the two fresh belts of sunspots that formed at high latitudes in the northern and southern hemisphere following the 1856 minimum "have in subsequent years shown a tendency to coalesce and ultimately contract as before to extinction. Whether this is what appears at each period of increase and decrease of frequency of the Spots must be left to observers who may follow me to show. At present it is only probable that such is the case, and another contribution made to facts on the broad scale which will ultimately elucidate the origin of this phenomenon and instruct us on the question, 'What is a Sun?' "

The pattern deduced by Carrington was codified by Spörer (1880) and illustrated by Maunder (1904) in the original butterfly diagram (Fig. 1) for cycles 12 and 13. Figure 2 gives portraits of Edward Maunder and his wife Annie Russell Maunder who assisted him in the production of this iconic figure. Many years later, Annie Maunder (1940) described the circumstances surrounding its creation, "We made this diagram in a week of evenings, one dictating and the other ruling these little lines. We had to do it in a hurry because we wanted to get it before the [Royal Astronomical] Society at the same meeting as the other sunspot observers, whose views we knew to be heretical. As it turned out, this paper, especially the diagram—wiped Father Cortis' and Prof. Lockyer's papers clean off the slate, and fully established the manner of procedure of the sunspot 11-year cycle."

2.2 Movement of Low-Latitude Prominences Toward the Equator and High-Latitude Prominences Toward the Poles

Two other husband and wife teams (Evershed and Evershed 1917; d'Azambuja and d'Azambuja 1948) helped to establish the cyclic migrations of prominences during the

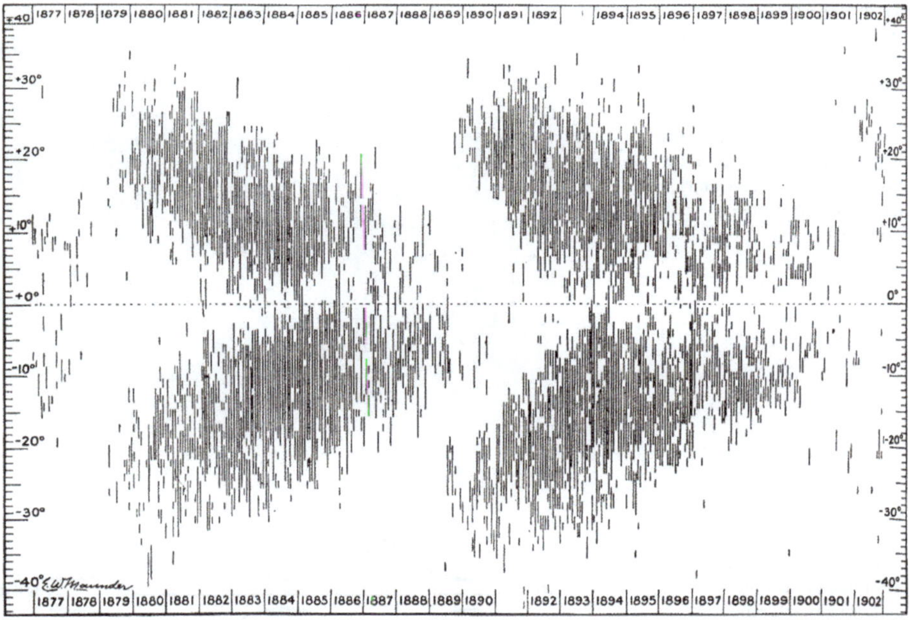

Fig. 1 The Maunder butterfly diagram showing the latitude distribution of sunspots for Schwabe cycles 12 and 13 (from Maunder 1904)

Fig. 2 Edward Walter Maunder (1851–1928) and Annie Russell Maunder (1868–1947). Photo from Kinder (2008), date unknown (With permission of the British Astronomical Association)

solar cycle—the equatorward movement of low-latitude prominence formation in concert with sunspots and the poleward movement of high-latitude prominence formation. An undated photograph circa 1915 of John and Mary Acworth Evershed (fifth and 6th from left in the second row) with the staff at Kodaikanal Observatory is given in Fig. 3. The Italian school of astronomers that included Respighi, Secchi, and Tacchini were the first to note that prominences appeared near the poles only at solar maximum (Lockyer 1931; Waldmeier 1973). The Eversheds credit Riccò with discovering the general laws of prominence distribution in latitude over the solar cycle and Lockyer and Lockyer (1902) with confirming them. A stylized diagram of the migrations of high- and low-latitude prominences from Riccò (1892) is shown in Fig. 4. Importantly, this figure shows oppositely directed migrations of high-latitude and low-latitude prominences starting from mid-latitudes near solar minimum. The depicted poleward migration of low-latitude prominences toward

Fig. 3 John Evershed (1864–1956) and Mary Acworth Evershed (1867–1949), fifth and sixth from right in the second row, with the staff of the Kodaikanal Observatory circa 1915 (With permission of the Observatory Science Centre, Herstmonceux/Terence Evershed)

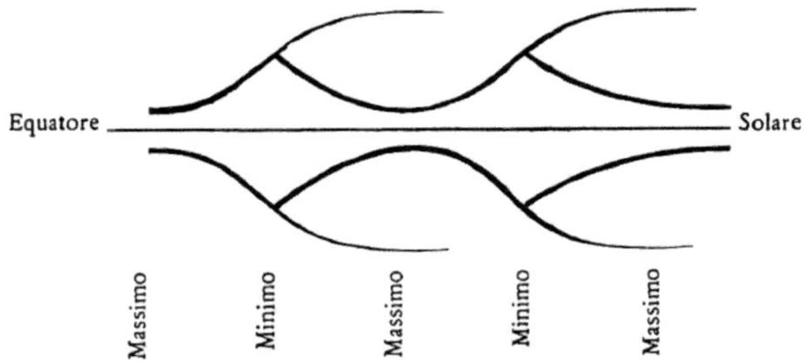

Fig. 4 Stylized diagram of the migrations of prominences over the solar cycle (from Riccò 1892). It captures the equatorward and poleward migrations of prominences that begin near solar minimum. The maximum to minimum phase depicted here is incorrect (see text)

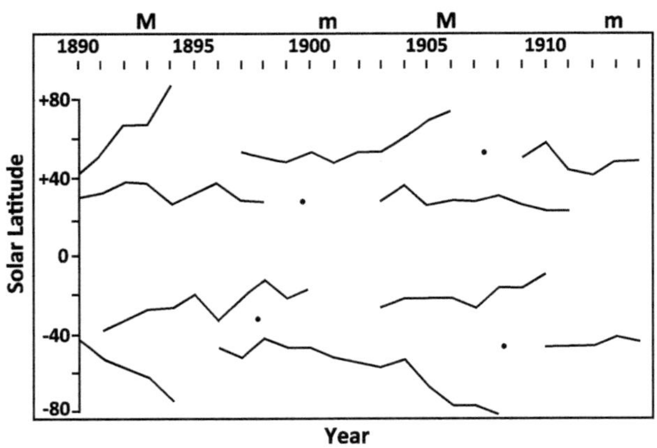

Fig. 5 Migration of prominences in latitude during cycles 13, 14, and the onset of cycle 15 (from Evershed and Evershed 1917), based on observations from Kenley and Kodaikanal. The *black dots* refer to "a few sporadic zones which do not seem to play a part in the history of [the] main zones". Times of solar minima (m) and maxima (M) are indicated

mid-latitudes following solar maximum, however, is erroneous. Figure 5, taken from Evershed and Evershed (1917) and covering the interval from 1890–1914 shows that, following the disappearance in each solar cycle of the polar prominences, a new high-latitude branch forms at about 50° latitude in both hemispheres where it resides during the decay of the cycle to begin migrating poleward near solar minimum. This poleward migration was referred to by the Eversheds as a "dash to the pole" and in current usage has come to be called "the

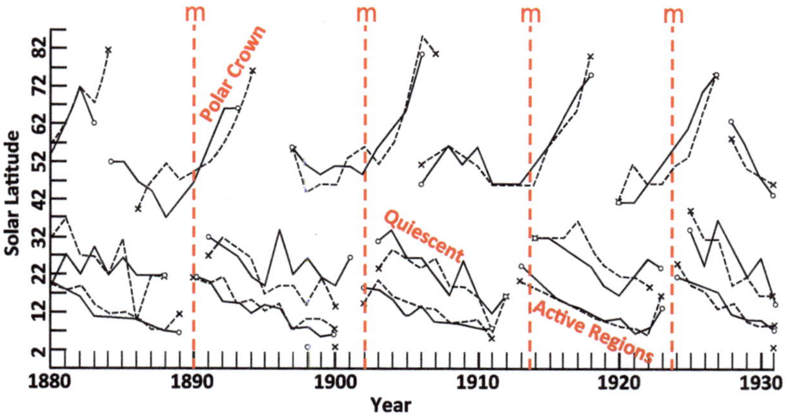

Fig. 6 Prominence migrations during cycles 12–15 (adapted from Bocchino 1933). Data for 1880–1911 are from Arcetri and for 1911–1930 are from various observatories including Catania, Madrid, and Zürich. Tracks are shown in each hemisphere for the polar crown filaments (*top*), the quiescent filaments that lay 10–15° poleward of the sunspots (*middle*), and active region filaments (*bottom*; based on sunspot observations). Solar minima (*m*) are marked by *dashed red lines*

rush to the poles" (e.g., Ananthakrishnan 1954; Altrock 1997, 2014). A separate low-latitude branch forms after solar minimum and moves equatorward.

An expanded plot of this behavior for the years 1880–1930 (adapted from Bocchino 1933) is given in Fig. 6. Here it can be seen that the time-averaged migratory behavior of the prominences (or filaments when viewed on the disk) is as systematic as that of the sunspots. For each solar cycle, three distinct branches of prominence activity can be seen or inferred in each hemisphere: (a) a high-latitude branch that moves poleward following solar minimum; (b) a lower-latitude branch which mimics the progression of sunspots toward the equator but is offset from the sunspot curve by ∼10–15° toward the poles; and (c) a branch that overlays the sunspot migration. The three tracks in Fig. 6 correspond, respectively, to the three classes of prominences defined by Waldmeier (1973): long-lived "polar crown" prominences/filaments that form at latitudes above 45° following sunspot maximum, long-lived quiescent prominences that have their maximum frequency of occurrence about 10–15° poleward of the sunspot zone, and active region prominences (which originally included such phenomena as sprays and post-eruption loop systems). The cyclic migrations of the high- and low-latitude quiescent filament branches in Fig. 6 have been supported/extended in subsequent studies by d'Azambuja and d'Azambuja (1948), Ananthakrishnan and Nayar (1954), Ananthakrishnan (1954), Li et al. (2008), and Li (2010). There are differences in details among the various works that can be attributed to different latitude/time averaging intervals and data sets. Recent developments in the automatic recognition of solar filaments (e.g., Bernasconi et al. 2005; Labrosse et al. 2010; Schuh et al. 2014) promise to standardize the results of such work.

The three tracks in Fig. 6 do not indicate motion of individual prominences but rather, just as is the case for sunspots, the migration of the locations of prominence formation (i.e., filament channels; Martin 1998; Mackay et al. 2014). Even the longest-lived prominences, those in the polar crown, have maximum individual lifetimes of only about eight months (Smith and Smith 1963)—short in comparison with their migration times to the pole.

3 Magnetic Polarity Reversals

3.1 Sunspots and the Hale-Nicholson Polarity Rule

In 1908 George Ellery Hale used the Zeeman (1897) effect to make his epochal discovery of strong magnetic fields in sunspots (Hale 1908). Hale et al. (1919) reported that before the minimum in 1912 preceding cycle 15, "the magnetic polarity of unipolar spots and of the preceding members of bipolar spots was positive [north] in the southern and negative [south] in the northern hemispheres of the Sun. Since the minimum these signs have reversed." At the sunspot number minimum in 1923, the sunspot polarities for new solar cycle 16 in each hemisphere reverted to the polarities of cycle 14. This enabled Hale and Nicholson (1925) to formulate the law of sunspot polarities: "The sun-spots of a new 11½-year cycle, which appear in high latitudes after a minimum of solar activity, are of opposite magnetic polarity in the northern and southern hemispheres. As the cycle progresses the mean latitude of the spots in each hemisphere steadily decreases, but their polarity remains unchanged. The high-latitude spots of the next 11½-year cycle, which begin to develop more than a year before the low-latitude spots of the preceding cycle have ceased to appear, are of opposite magnetic polarity." Anticipating the subsequent polarity reversal at the end of cycle 16, Hale and Nicholson noted that while the periodicity in the number of spots was 11½ years, the full sunspot period, which they called the "magnetic sun-spot period", was 23 years. Today, this magnetic cycle is generally referred to as the 22-yr, or Hale, cycle based on a nominal 11-yr Schwabe cycle.

3.2 Reversal of the Sun's Polar Field

The father and son team of Harold and Horace Babcock, shown in Fig. 7, used the newly-developed solar magnetograph (Babcock 1953) of the Hale Solar Laboratory to detect a general (now called polar) magnetic field of the Sun of $\sim$1 G based on magnetograms taken from the interval August 1952 to October 1954 that spanned the April 1954 solar minimum preceding cycle 19 (Babcock and Babcock 1955). Hale (1913) had been the first to attempt to look for such a field but, using the instrumentation available at the time, deduced a field of $\sim$50 G. The solar polar field (generally confined to latitudes $>55°$) that the Babcocks detected was positive in the north and negative in the south (opposite to that of Earth). Babcock and Babcock noted that the polar fields during this period showed no signs of reversal that might have been expected to accompany the change in the hemispheric polarity of sunspots over the 11-yr minimum. From the d'Azambuja's description of the poleward migration of the polar crown filaments, Babcock and Babcock presciently inferred that the general field of the Sun would reverse polarity at the maximum of cycle 19—out of phase with the polarity reversal for the sunspot curve. Babcock (1959) subsequently observed this reversal of the dipolar field of the Sun, with the field at the north heliographic pole changing in mid-1957 and the field in the south reversing in November 1958. The separation in time between the polar field reversal in the two hemispheres is now recognized to be a common feature of the reversal of the large-scale polar field (Svalgaard and Kamide 2013). As expected (Babcock 1961), the polar fields appeared to reverse, at least in the northern hemisphere, near the maximum of cycle 20 (Howard 1972) but they were weak relative to the magnetograph noise, making it difficult to determine the exact timing. Here was a different, more fundamental, 22-yr magnetic cycle, extending from (approximately) sunspot maximum to maximum over two $\sim$11 year intervals, rather than from minimum to minimum.

Fig. 7 Harold Delos Babcock (1882–1968) and Horace Welcome Babcock (1912–2003) (HDB image courtesy of the Archives, California Institute of Technology; HWB image courtesy of the Observatories of the Carnegie Institution for Science Collection at the Huntington Library, San Marino, California)

3.3 Reversal of the Flanking Magnetic Fields of the Polar Crown Filaments

There is evidence that the classical picture of the migration of the polar crown filaments in Fig. 6 is overly-simplistic. Topka et al. (1982) and Makarov and Sivaraman (1989a, 1989b) showed that in a given hemisphere for certain cycles there can be multiple rushes to the poles of bands of filaments rather than a single poleward rush leading to polarity reversal. Topka et al. start the migration of the secondary and tertiary polar crown branches, in such sequences, at latitudes above $\sim 20°$, similar to the poleward moving ripples in B_r reported by Ulrich and Tran (2013). Waldmeier (1973) had earlier noted the existence of a secondary zone of high-latitude northern-hemisphere prominences that reached the pole during cycle 20. No such occurrence had been observed for cycles 12 through 18, but Waldmeier allowed that a secondary rush of this type might have escaped notice in earlier cycles. Makarov and Sivaraman (1989a) reported even more complex behavior for cycle 20 in the north—a three-fold polarity reversal ($-$ to $+$, $+$ to $-$, and $-$ to $+$), consistent with Topka et al. (1982). Moreover, they deduced similar three-fold reversals in a single hemisphere for cycles 12, 14, 16, and 19. Subsequently, Dermendjiev et al. (1994) reported a secondary polar crown in the southern hemisphere in cycle 22. Presumably, such multiple poleward migrations are related to the quasi-periodic $\sim$1–3 year fluctuations seen in the sunspot record (Krivova and Solanki 2002) and in the stronger surges of magnetic flux to high latitudes seen in butterfly diagrams of photospheric magnetic fields (Fig. 8, taken from Ulrich and Tran 2013).

Instead of the usual method of obtaining a mean or average location for the various types of filaments (in which an arbitrary latitude—usually in the range from 40–60° (e.g., Bretz and Billings 1959; Waldmeier 1973; Li et al. 2008; Kong et al. 2014)—is used to separate the polar crown filaments from low-latitude filaments), Hansen and Hansen (1975) presented plots of all filaments observed on the Sun during several consecutive Carrington rotations. The composite plot they constructed by tracing outlines of filaments in the *Cartes Synoptiques* for the July–December 1967 interval is shown in Fig. 9 where it can be seen that the polar crown filament in each hemisphere at this time lies at a latitude of $\sim 60°$. Equatorward of the polar crown filaments lies a band of filaments that extend poleward to the east; these filaments are flanked on their north and south sides by opposite magnetic polarities to those in the polar crown and lie about $\sim 15°$ from the polar crown filaments with a relatively well-defined gap in between.

McIntosh (1992) constructed the plot in Fig. 10 by taking the latitudes of the two most poleward filaments for each solar rotation, in both hemispheres. This shows clearly, at least for cycles 21 and 22, the rush to the poles of the polar crown filament beginning near solar

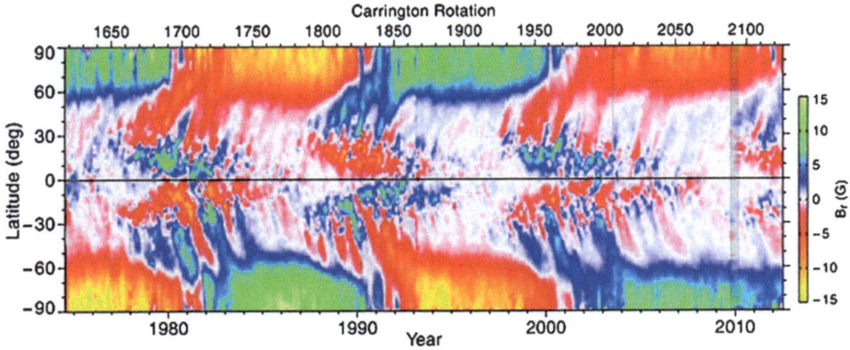

Fig. 8 Latitudinal variation of the radial component of the photospheric magnetic field from mid-1974 to mid-2012 (from Ulrich and Tran 2013) (©AAS. Reproduced with permission)

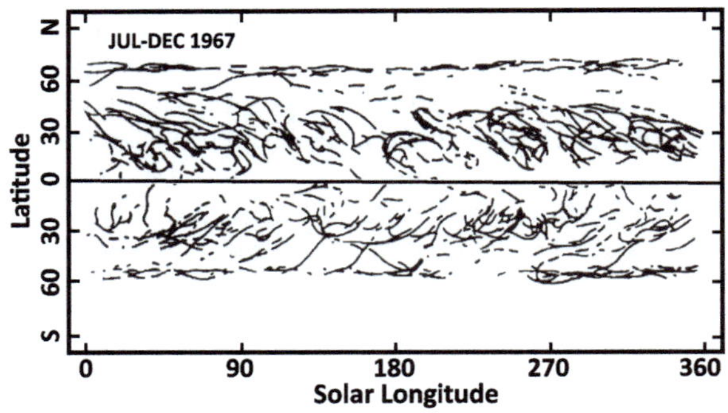

Fig. 9 Superposed tracings of all filaments in the *Cartes Synoptiques* for July–December 1967 (from Hansen and Hansen 1975) (With kind permission from Springer Science and Business Media)

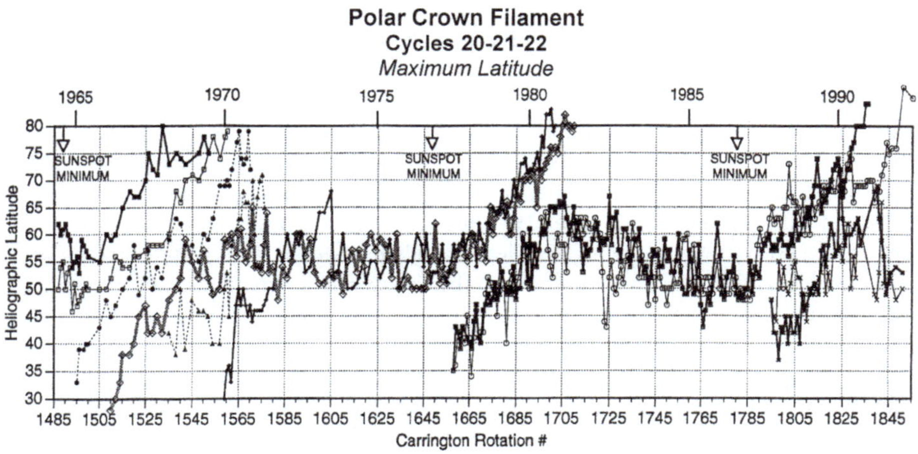

Fig. 10 Maximum latitudes of filaments in the true and replacement polar crowns for both hemispheres for cycles 20–22 (from Webb 1997; McIntosh 1992). For cycle 21, the true polar crown begins the rush to the poles in both hemispheres form ~55° in 1976. The replacement polar crown (true polar crown for cycle 22) begins moving poleward shortly thereafter form ~40° latitude. High-latitude filament behavior was more complex in cycle 20 (see text)

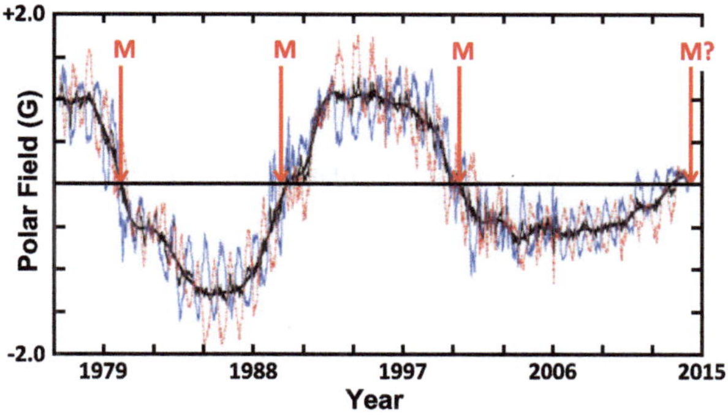

Fig. 11 Solar polar field strength from Wilcox Solar Observatory for 1976 to mid-2014 for the North (*solid blue*) and South (*dashed red*; negatived) hemispheres

minimum and the corresponding poleward progression of the replacement polar crown with opposite flanking polarities for the new magnetic cycle. As noted above, the march to the poles of the polar crown for cycle 20 and its replacement by the polar crown for cycle 21 was less stately. The onset of poleward movement of both types of filaments (true and replacement polar crown) in Fig. 10 occurs near sunspot minimum. The disappearance of the polar crown filaments in each hemisphere occurs near the time of the reversal of the polar fields (e.g., Waldmeier 1960) and sunspot maximum (Fig. 11). Here the reversal of the polar fields refers to the passage of the polar field strength through zero. As can be seen in Fig. 11, it takes several years for the field strength of the new field polarity to reach maximum. Ulrich and Tran (2013) note that the initial broad surges of poleward moving flux in the northern hemisphere in Fig. 8 are closely associated with the polarity reversals near 1980, 1990, and 2000. Reversal is more complicated in the south although in each case, it follows that in the north within ∼1 year. The replacement (new cycle) polar crown reaches its apex (at latitudes of ∼60–70°) near polarity reversal and then, as the new polar crown, drifts equatorward to ∼50–55° where it resides before beginning its own rush to the poles near solar minimum (Fig. 10). From a study of quiescent prominences at the end of cycle 20 (1973–1976) that employed the Hanle (1924) effect, Leroy (1978) showed that the magnetic polarities flanking the polar crown filaments were opposite to those deduced by Rust (1967) from magnetograph observations during 1964–1965 at the end of cycle 19.

4 Cyclic Zonal Flows

The notion of an extended solar cycle of 12–14 (rather than ∼11) years originated in the 19th century but it was the discovery of torsional oscillations of the Sun's surface rotation in Mt. Wilson Doppler data by Howard and LaBonte (1980) and LaBonte and Howard (1982) (see also Snodgrass 1985, 1992; Ulrich 2001; Howe et al. 2011) that brought the concept to wide attention. The torsional oscillations reported by Howard and LaBonte (1980) provided evidence for an even longer extended cycle. They moved from pole to equator, starting near solar minimum, over a ∼22-year interval. The torsional oscillations consist of alternating bands of faster and slower velocities, relative to the nominal differential rotation. The latitudinal derivative of the torsional oscillations, corresponding to maximal velocity shear, bisects the wings of the sunspot butterfly diagram in each hemisphere. As noted by LaBonte and Howard (1982), the torsional oscillations were the first large-scale velocity fields that were closely related to the solar cycle in space and time. The torsional oscillations are currently also referred to as migrating zonal (parallel to the equator) flow bands or simply zonal flows (Howe 2009). Similar, but slightly different, migrating patterns in the differential rotation have also been detected by tracking surface magnetic features (e.g., Komm et al. 1993). In this case the strong magnetic field regions coincide with bands of slower-than-average zonal flow.

Interest in the torsional oscillations or zonal flow bands was reinvigorated about twenty years after their 1980 discovery when they were detected in helioseismology studies (Schou 1999; Howe et al. 2000b; Toomre et al. 2000) based on inversion of rotational splittings of solar oscillation frequencies (e.g., Thompson et al. 1996; Schou et al. 1998; Howe 2009) observed by the *Michelson Doppler Imager* (MDI; Scherrer et al. 1995) on the *Solar and Heliospheric Observatory* (SOHO; Domingo et al. 1995) and by the *Global Oscillation Network Group* (GONG; Harvey et al. 1996). Subsequent studies revealed that the zonal flows persisted deep into the convection zone (Howe et al. 2000a; Antia and Basu 2000; Vorontsov et al. 2002). Moreover, the zonal flows detected by this method exhibited a high-latitude poleward-moving branch (Antia and Basu 2001; Vorontsov et al. 2002) in addition to the equatorward-moving branch discovered by Howard and LaBonte.

Generally, the torsional oscillations are viewed as a secondary or side effect of the magnetic activity cycle on the Sun's differential rotation (Gilman 1992), in part because the maximum amplitude of the flow is only $\sim 5\ \mathrm{m\,s^{-1}}$ against a solar mean rotation rate at the equator of $\sim 2000\ \mathrm{m\,s^{-1}}$ (and a more comparable mean velocity change across 1° of latitude of $\sim 15\ \mathrm{m\,s^{-1}}$ in the sunspot zone). Proposed mechanisms to account for the torsional oscillations (see Howe 2009, for a review) include: Lorentz force feedback from the magnetic cycle (Schüssler 1981; Yoshimura 1981), thermal feedback (Spruit 2003), magnetic quenching of (small-scale) turbulent angular momentum transport (Küker et al. 1999; Pipin 1999), or a combination of these effects (Rempel 2006). Recently, Beaudoin et al. (2013) attributed the zonal flows to modulation of angular-momentum transport by large-scale meridional flow.

4.1 The Extended Cycle of Solar Activity and the Equatorward-Propagating Branch of the Zonal Flows

There are several lines of evidence that the solar activity cycle is longer than the nominal 11-yr sunspot cycle. The first indication came from the sunspots themselves. Figure 12, taken from Clerke (1903) shows Spörer's observations of the mean latitude of sunspots during solar cycles 10 and 11. Quoting from Clerke, "spot production at minima [occurs in] two widely separated zones in each hemisphere ... the start of a new series [of sunspots] in high latitudes [anticipates] the termination of the old. ... [Figure 12] illustrates the nature of the progression. The overlapping of the curves at minimum brings before us the remarkable circumstance that, as a consequence of successive disturbances breaking out before those antecedent to them have expired, the full duration of each [cycle] is, not eleven but twelve to fourteen years." Harvey (1992) compiled and re-analyzed reports of the first and last magnetic bipoles, including ephemeral regions (Harvey and Martin 1973; Harvey et al. 1975; Martin and Harvey 1979) and Ca^+ plage regions, for solar cycles 14–22 (Table 1), assigning regions to cycles on the basis of latitude, magnetic polarity, and "an understanding of the characteristics of the magnetic configuration of active regions as a function of latitude and time." The entries for more recent cycles in the table, which are based on higher resolution data, indicate a latitude range of ~30–60° for the first bipoles in a cycle and extended cycle durations of ~14–16 years.

Observations of the 5303 Å coronal green line (e.g., Waldmeier 1960; Leroy and Noens 1983; Altrock 1997, 2014) reveal enhanced emission that appears at latitudes ~70° near the

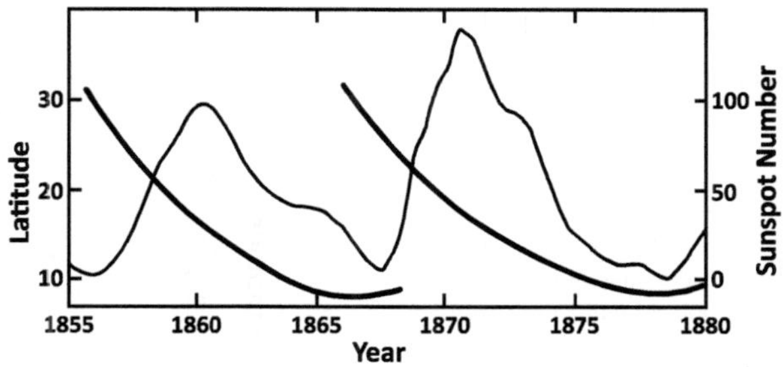

Fig. 12 Overlapping mean latitude curves of sunspots (*bold lines*) for cycles 10 and 11 as determined by Spörer (adapted from Young 1897; reproduced in Clerke 1903)

Table 1 First and last magnetic bipoles observed from cycles 14–22 and comparison of the resulting extended cycle with the corresponding minimum-to-minimum cycle (from Harvey 1992)

Cycle N	First region			Last region			Cycle duration (yrs)	
	Lat.	Year	Δt_N	Δt_{N+1}	Year	Lat.	Extended	Min-min
14	S21	1900.61	−1.1	+1.9	1915.45	N01	15.8	11.9
15	N20	1911.96	−1.6	+1.5	1925.13	N08	13.2	10.0
16	N34	1921.75	−1.8	+1.9	1935.51	N04	13.8	10.0
17	N48	1932.04	−1.6	+2.0	1946.19	N01	14.2	10.6
18	S30	1942.25	−2.0	+2.2	1956.53	N01	14.3	10.6
19	N42	1953.12	−1.2	+2.2	1967.12	S01	14.0	10.6
20	N46	1962.79	−2.1	+2.0	1978.45	S01	15.7	11.6
21	N61	1973.04	−3.5	+1.7	1988.48	S02	15.4	10.3
22	N32	1983.31	−3.5					

11-yr cycle maximum and migrates equatorward to apparently join and move in concert with the progression of sunspot formation to lower latitudes as can be seen in the composite in Fig. 13 for cycles 18–20 (Leroy and Noens 1983). The continuation of this behavior in cycles 21 through 23 is shown in Fig. 14 from Altrock et al. (2008). Extended cycle lengths deduced from green line measurements are ~17–18 years. As noted above, the surface torsional oscillations reported by Howard and LaBonte (1980) moved from pole to equator over a 22-year interval.

Wilson et al. (1988) compiled the lines of evidence available at that time to argue for an extended solar cycle of 18–22 years (Fig. 15). Since the overlap between 11-yr sunspot cycles had been known since the 19th century and the coronal green line has been routinely monitored since 1939 (Rybanský et al. 2005), the torsional oscillations along with the ephemeral regions of Harvey and Martin were the key new pieces of evidence that drew attention to an extended cycle, in this case up to twice as long as the Schwabe cycle.

Even before the Wilson et al. (1988) paper, however, the 22-yr duration of the torsional oscillations was being questioned. Snodgrass (1985) reanalyzed the Mt. Wilson Doppler data and concluded that the torsional oscillation began during ~1973.5–1974.5, roughly three years after the polar field reversal (Makarov and Sivaraman 1989a) and closer to latitude 45° than 90° in both hemispheres, although it appeared to end at ~10° near the maximum of the following cycle rather than at the equator near minimum.

The equatorward zonal flow discovered via helioseismology in cycle 23 began at ~45° in 2003 (Fig. 16; Howe et al. 2013), about three years after polarity reversal (Fig. 11). This delay from polarity reversal and mid-latitude onset were consistent with that inferred from

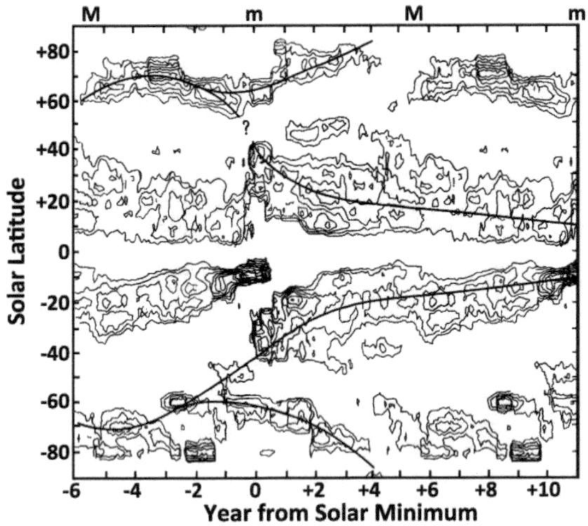

Fig. 13 Superposed epoch plot of the standard deviation of 5303 Å emission at each latitude recorded at Pic du Midi from 1944–1974 showing high-latitude (rush to the poles) and low-latitude (sunspot-related) branches of coronal emission (from Leroy and Noens 1983) (Reproduced with permission ©ESO)

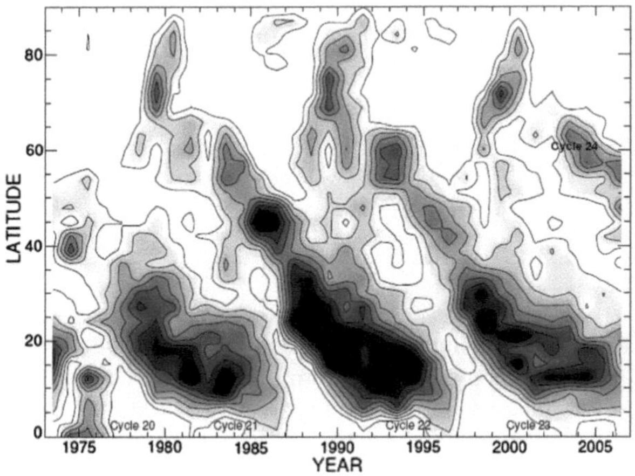

Fig. 14 Annual northern- plus southern-hemisphere averages of the number of coronal activity local maxima from 1973 through 2006 (from Altrock et al. 2008)

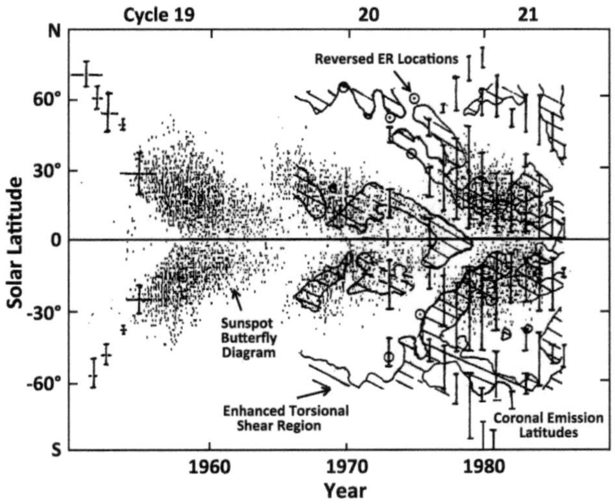

Fig. 15 The extended solar cycle for cycles 19–21 showing the time and latitudinal extension of the sunspot cycle inferred from reversed polarity ephemeral active regions (*circles with dots*), coronal green line emission (*vertical lines*), and the torsional oscillations (contoured hatching) (adapted from Wilson et al. 1988) (Adapted by permission from Macmillan Publishers Ltd: Nature)

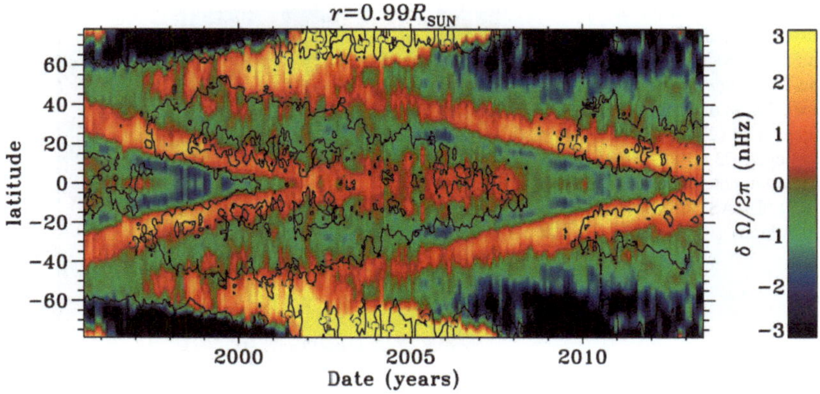

Fig. 16 Zonal flow pattern, symmetrized about the equator, at 0.99 $R_\odot$ for mid-1995 to mid-2013 based on GONG, MDI, and Heliospheric Magnetic Imager (HMI; Schou et al. 2012) observations showing the poleward- and equatorward-migrating branches (adapted and updated from Howe et al. 2013) (©AAS. Reproduced with permission)

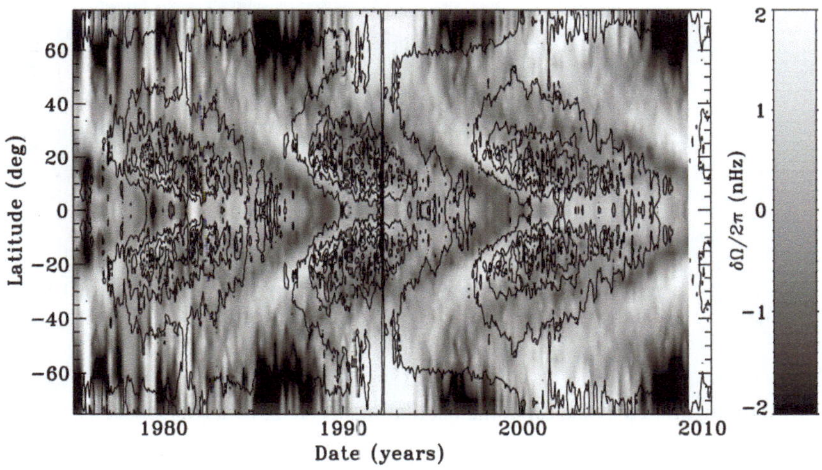

Fig. 17 Zonal flows at the Sun's surface deduced from Doppler measurements at Mt. Wilson 1975–2009; black contours give the unsigned magnetic field strength (adapted from Howe et al. 2011). Both data sets are symmetrized about the equator

spectroscopic observations by Snodgrass (1985) for the onset of the zonal flows in cycle 20. Figure 16 also shows that the low-latitude zonal flow for cycles 22 and 23 ended within about one year of the 11-yr minima in 1996 and 2008, in agreement with the finding of Howard and LaBonte (1980) for cycle 20. Thus the zonal flow observations yield a pattern for the equatorward flows that is more limited in latitude and time than that reported by Howard and LaBonte (1980). This can also be seen in Fig. 17 from Howe et al. (2011) that shows surface Doppler measurements from Mt. Wilson for cycles 21–23.

4.2 The Poleward-Propagating Zonal Flow and the 22-Yr Cycle of Solar Activity

Just as the rush to the poles of the high-latitude prominences was discovered later and is less well-known than the equatorward propagation of sunspots during the 11-yr cycle, so too the

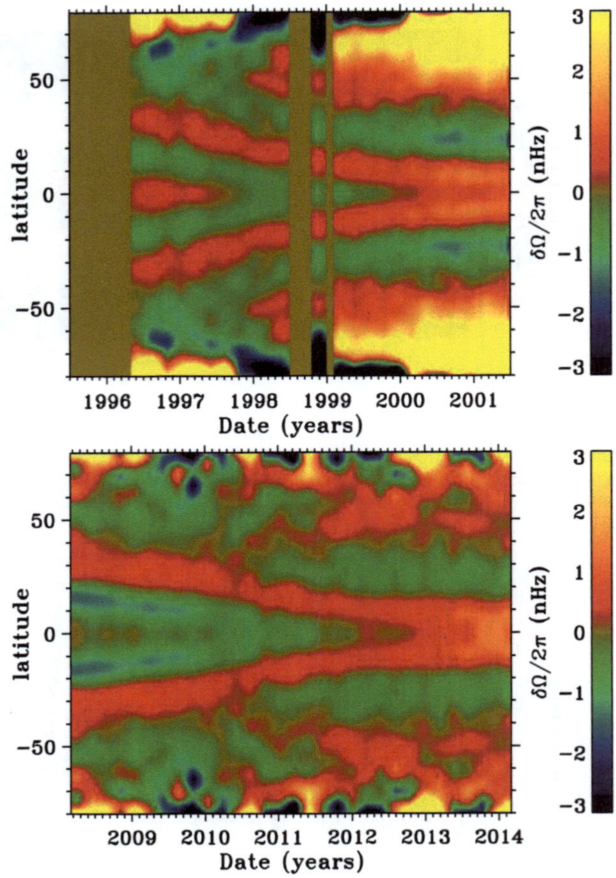

Fig. 18 Rotation rate residuals at 0.99 $R_\odot$ from MDI/HMI for comparable intervals in cycles 23 (*top*) and 24 (*bottom*). Note the relative weakness of the polar flows in cycle 24 (2009–present) relative to cycle 23 (1996–2008) (adapted and updated from Howe et al. 2013) (©AAS. Reproduced with permission)

poleward-moving branch of the torsional oscillations (Fig. 16) was discovered later than the equatorward moving branch and as a result has been less well-studied. Howard and LaBonte (1980) did not detect a poleward-propagating branch of the torsional oscillations, although Livingston and Duvall (1979) had reported a "spin-up" of the high-latitude rotation rate near the solar maximum of solar cycle 20. Subsequently, Snodgrass (1985) reported fast zonal flows at high latitudes at or near the maxima of both cycles 20 and 21. Discovery of the high-latitude poleward-moving branch by seismological methods (Antia and Basu 2001) was also delayed slightly relative to that of the low-latitude equatorward moving branch. The high-latitude branch was found to penetrate to the tachocline (Vorontsov et al. 2002; Basu and Antia 2003).

The polar zonal flow is intimately connected (either as a tracer or a driver) to the reversal of the Sun's polar fields, and thus the 22-yr magnetic cycle, in space and time. The rush to the poles of the polar crown filaments in cycles 20–22 lags the preceding 11-yr minima by intervals ranging from ~5–20 solar rotations (Fig. 10). For comparison the onset of the poleward moving branch of the zonal flows in cycles 23 and 24 lags the preceding minima by ~10 and ~20 rotations, respectively (Figs. 16 and 18). From approximately 2002 through 2004 the polar flow is a prominent feature above 60° (Fig. 16). The mean polar field reached its maximum field strength in 2004 (Fig. 11). The subsequent decay of the polar

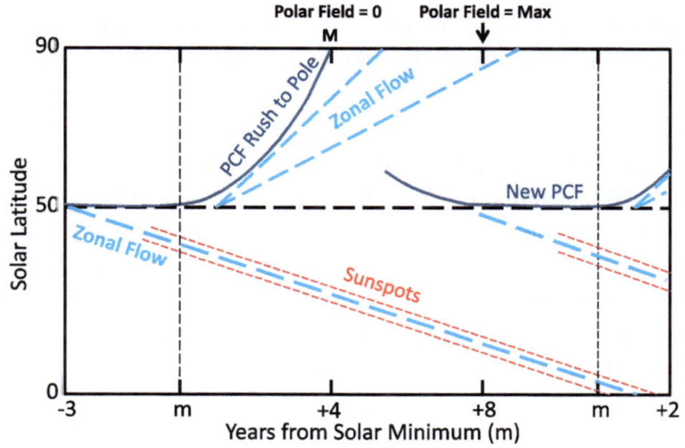

Fig. 19 Schematic diagram depicting zonal flows and solar surface migrations of sunspots and polar crown filaments (PCF) for a nominal 11-yr solar cycle. In this representation, the associated extended cycle has a duration of ∼15 years. $M =$ time of solar maximum

field corresponds roughly to the beginning of the disappearance of the zonal flow from the observable polar regions in 2005.

The apparent absence of the polar zonal flow early in cycle 24 (Howe et al. 2011; Hill et al. 2011) raised the possibility that cycle 25 would not appear and that we could be entering a grand minimum. However, further analysis by Howe et al. (2013) showed that the high-latitude zonal flow is present in cycle 24. Following the flux-transport model of Rempel (2006, 2012) which links weaker polar fields to a slower high-latitude rotation rate, Howe et al. (2013) subtracted a mean rotation rate based on the rise phase of cycle 24 (∼2009–2012.5) (rather than mean rate for all of cycle 23) from the helioseismic data to reveal the onset of the poleward propagating branch of the zonal flow in the residuals for cycle 24 (Fig. 18, bottom), although it is less well defined than was the case for cycle 23 (Fig. 18, top). Similarly, Altrock (2014) reported that the rush to the poles in the green line was unusual in cycle 24; in particular the branch in the south was either weak or absent in the southern hemisphere as late as early 2011.

5 Discussion

We have reviewed the evolution of thought/knowledge on the extended and 22-yr solar cycles in the framework of the two characteristic migrations of surface features on the Sun: (a) the equatorward motion of sunspots; and (b) the poleward motion of high-latitude prominences. In this section we comment on certain aspects of the extended and 22-year cycle that have the polar crown filament/rush to the poles phenomenon as a common thread. A schematic depicting the spatial/temporal relationship between the equatorward and poleward solar surface migrations and their corresponding zonal flows for a nominal 11-yr cycle is given in Fig. 19 as an aid to the discussion.

5.1 Latitude ∼50°

The latitude ∼50° (range from ∼40–60°) is a key fiducial on the Sun. Newton (1958) wrote, "… the prominences in the latitudes beyond the sunspot belt … reach their greatest frequency between sunspot minimum and maximum when their average latitude of about 50° north and south is beginning to move rapidly towards the Sun's polar regions, which they reach just after sunspot maximum. After a brief but distinct discontinuity they appear again around latitude 50°, to which they hold with lessened frequency throughout sunspot minimum, until with the rising activity of a new cycle, they rise to maximum frequency and 'rush to the pole' [Fig. 6]."

The diverging migrations of low- and high-latitude filaments beginning near solar minimum were qualitatively illustrated by Riccò in 1892, and documented in the first half of the 20th century by Evershed and Evershed (1917), Bocchino (1933), and d'Azambuja and d'Azambuja (1948). Nevertheless, the notion of a latitudinal divide in solar activity at ~50°, between the upper limit of the extended cycle and the solar minimum residence of the polar crown filaments has not taken hold. The transition at ~50° between the equatorward- and poleward migrations of zonal flows (Antia and Basu 2001; Vorontsov et al. 2002) brings renewed attention to this latitude (Fig. 19).

5.2 The Over-extended Cycle

The premise of the extended cycle is that the equatorward migration of sunspots during a given solar cycle can be extended to higher latitudes and earlier times by considering forms of solar activity that can be detected prior to sunspots. One such type of activity is the torsional oscillation. The Wilson et al. (1988) proposal of an 18–22 year cycle was based in large part on Howard and LaBonte's (1980) finding that the torsional oscillation extended from ~90° to ~0° over 22 years, beginning at solar minimum. However, helioseismology analysis of the zonal flows (Fig. 16) indicate a starting latitude of the equatorward-moving branch of ~45° and a corresponding reduction in length of the proposed extended cycle.

Wilson et al. also included coronal green line emission as evidence for a high-latitude (>70°) origin for the extended cycle. The green line emission exhibits both a high-latitude branch associated with the rush to the poles and an equatorward-moving branch associated with sunspot formation. Figure 13 shows how the two may be conflated. In this figure, at about the same time the high-latitude emission in the northern hemisphere is moving equatorward to ~60°, the green line emission associated with the extended cycle is making its appearance near 40°. One is inclined to connect the dots—thereby erroneously linking the two fundamentally different migrations of solar activity—at the time/latitude where Leroy and Noens (1983) cautiously inserted a question mark. The behavior of the high-latitude branch has been recently modeled by Robbrecht et al. (2010) using data from the Extreme-ultraviolet Imaging Telescope (Delaboudinière et al. 1995). They associate the high-latitude EUV emission with the boundary of the polar coronal hole (Wang et al. 1997; Benevolenskaya et al. 2001). In simple terms, the sequential equatorward, then poleward, motion of the band of high-latitude coronal emission (and the underlying polar crown filament, Fig. 10) following polarity reversal is viewed as a reflection of the waxing and waning of the polar coronal hole as the polar field first strengthens during the decline of the cycle and then begins to weaken near minimum (Fig. 11) in concert with the rush to the poles. In a latitude vs. time plot, the migration track of the polar crown filaments (and associated high-latitude green line emission) makes a U-turn (Fig. 19). Comparison of Figs. 10 and 11 shows that the stationary residence of the new polar crown filament at ~50–60° preceding the 1986 solar minimum corresponds to the broad maximum in polar field strength. Altrock (2014) has presented recent green-line evidence for the extended cycle beginning at 65° in ~2003 during cycle 24. However, like Robbrecht et al. (2010), Tappin and Altrock (2013) and Altrock (2014) link the high-latitude emission to streamers overlying polar crown filaments which do not migrate below ~50° to become part of the extended cycle.

Recently, Petrie et al. (2014) have re-visited the analysis of Robbrecht et al. (2010) and concluded that "The question of whether the extended cycle connects more to the main cycle or the march to the poles phenomena in green line variance remains open. The evidence more strongly indicates connection with the main cycle [as seen in their Fig. 7, our Fig. 16], and this would match the patterns of the interior torsional oscillations and photospheric ephemeral bipoles." Physically linking the contiguous high-latitude and low-latitude branches of the green line emission in Figs. 13 and 14 has the appeal of Occam's razor. The preponderance of evidence, however, supports the picture of Benevolenskaya et al. (2001) and Robbrecht et al. (2010). For example, in the contour diagram in Fig. 14 of 5303 Å peaks, consider the enhancements between ~67–77° centered on ~1979.5 and ~1989.5 that appear to be the highest latitude components of an extended cycle. Comparison with the McIntosh plot of polar crown filaments in Fig. 10 indicates that in 1979 and 1989 the polar crown filaments were located between ~65–75°, while undergoing their rush to the poles during cycles 21 and 22, respectively. Inspection of the magnetic butterfly diagram in Fig. 8 reveals strong poleward surges of trailing polarity flux with leading edges between ~60–65° in both hemispheres at these times. The relative locations of green line enhancements, polar crown filaments, and magnetic flux surges in these two cases are consistent with the picture that the green line emission overlies the polar crown filaments and results from the encroachment of trailing polarity magnetic flux on the opposite polarity flux of the polar coronal hole. Both Benevolenskaya et al. (2001) and Robbrecht et al. (2010) stress the role of trailing polarity flux in creating the enhanced high-latitude green line emission but it seems clear that poleward "counter-surges" of leading polarity flux (Ulrich and Tran 2013) will also play a role, particularly later in the cycle, for the new polar crown filament which has opposite flanking magnetic polarities to that of its predecessor (Fig. 19). Such counter-surges may contribute to the concentrations of 5303 Å peaks at ~55–65° centered at ~1983.5, ~1993, and ~2004 in Fig. 14. Finally, the interior torsional oscillations to which Petrie et al. (2014) refer appear to originate at ~45° latitude (Fig. 16) rather than the higher latitudes of these green line features while the new cycle ephemeral regions exhibit broad latitudinal scatter which straddles the polar crown magnetic inversion line (see Fig. 5 in Harvey 1994). Nonetheless, we agree with Petrie et al. (2014) that further study of the latitudinal limit of the green line manifestation of the extended cycle is in order.

To summarize, there is an extended cycle, as was known since the 19th century (Fig. 12) from sunspot data. An 18–22 year (~70°–0°) extended cycle is not supported by the data although a 13–16 year cycle (Table 1), assuming a nominal 11-yr Schwabe cycle, that extends poleward to the ~50° delineator between the equatorward motion of ephemeral regions/sunspots and the poleward movement of polar crown filaments is well-established, and is supported by observations of green line emission (Figs. 13 and 14) and zonal flows (Figs. 16 and 17).

5.3 The Poleward-Propagating Branch of Zonal Flow During Cycle 24 and the Amplitude of Cycle 25

The rush to the poles phenomenon first observed in solar prominences and subsequently in the coronal green-line emission and the zonal flows is intimately connected with polarity reversal and the ensuing development of the polar fields. The polar field strength at solar minimum is thought to be the most reliable precursor for predicting the peak sunspot number of the following cycle (e.g., Schatten et al. 1978; Svalgaard et al. 2005). The onset of the rush to the poles phenomenon precedes the peak of the cycle whose amplitude it is linked to by ~15 years (e.g., 1997 rush to poles onset in cycle 23 (Fig. 14) and ~2014 cycle 24 maximum), making it a potential long-term predictor of solar cycle amplitude. More realistically, if mid-term (3–5 years after onset) development of the high-latitude zonal flow has predictive value, the schematic in Fig. 19 indicates a lead time of ~10 years. Thus (echoing Komm et al. 2014), it will be of great interest to see if the stunted development of the poleward-propagating zonal flows (Fig. 18) and green line emission in cycle 24 (Altrock 2014), relative to cycle 23, translate into a further reduction in the cycle-to-cycle polar field strength at the next minimum and a prediction that cycle 25 will be weaker than cycle 24.

Acknowledgements I thank André Balogh, Hugh Hudson, Kristóf Petrovay, and Rudolf von Steiger for organizing a timely and stimulating workshop. I am grateful to: Dick Altrock, Rainer Arlt, Sara Martin, Alexei Pevtsov, and Leif Svalgaard for helpful comments/discussions, Tom Bogdan for providing the text of A. Maunder's letter, and Rachel Howe for providing updated/modified versions of Figs. 16, 17, and 18.

References

R.C. Altrock, An 'extended solar cycle' as observed in FeXIV. Sol. Phys. **170**, 411–423 (1997)
R.C. Altrock, Forecasting the maxima of solar cycle 24 with coronal Fe XIV emission. Sol. Phys. **289**, 623–629 (2014)
R. Altrock, R. Howe, R. Ulrich, Solar torsional oscillations and their relationship to coronal activity, in *Subsurface and Atmospheric Influences on Solar Activity*, ed. by R. Howe, R.W. Komm, K.S. Balasubramaniam, G.J.D. Petrie. ASP Conference Series, vol. 383 (2008), pp. 335–342
R. Ananthakrishnan, Prominence activity (1905–1952). Proc. Indian Acad. Sci. **40**, 72–90 (1954)
R. Ananthakrishnan, P.M. Nayar, Discussions of the results of observations of solar prominences made at Kodaikanal from 1905–1950. Bulletin No. 137 of Kodaikanal Observatory (1954)
H.M. Antia, S. Basu, Temporal variations of the rotation rate in the solar interior. Astrophys. J. **541**, 442–448 (2000)
H.M. Antia, S. Basu, Temporal variations of the solar rotation rate at high latitudes. Astrophys. J. Lett. **559**, L67–L70 (2001)
H.W. Babcock, The solar magnetograph. Astrophys. J. **118**, 387–396 (1953)
H.D. Babcock, The Sun's polar magnetic field. Astrophys. J. **130**, 364–365 (1959)
H.W. Babcock, The topology of the Sun's magnetic field and the 22-year cycle. Astrophys. J. **133**, 572–587 (1961)
H.W. Babcock, H.D. Babcock, The Sun's magnetic field, 1952–1954. Astrophys. J. **121**, 349–366 (1955)
S. Basu, H.M. Antia, Changes in solar dynamics from 1995 to 2002. Astrophys. J. **585**, 553–565 (2003)
P. Beaudoin, P. Charbonneau, E. Racine, P.K. Smolarkiewicz, Torsional oscillations in a global solar dynamo. Sol. Phys. **282**, 335–360 (2013)
E.E. Benevolenskaya, A.G. Kosovichev, J.R. Lemen, P.H. Scherrer, G.L. Slater, Detection of high-latitude waves of solar coronal activity in extreme-ultraviolet data from the Solar and Heliospheric Observatory EUV Imaging Telescope. Astrophys. J. Lett. **554**, L107–L110 (2001)
P.N. Bernasconi, D.M. Rust, D. Hakim, Advanced automated solar filament detection and characterization code: description, performance, and results. Sol. Phys. **228**, 97–117 (2005)
G. Bocchino, Migrazione delle protuberanze durante il ciclo undecennale dell'attivatà solare. Oss. Mem. Oss. Astrofis. Arcetri **51**, 5–47 (1933)
M.C. Bretz, D.E. Billings, Analysis of emission corona 1942–1955 from Climax spectrograms. Astrophys. J. **129**, 134–145 (1959)
R.C. Carrington, On the distribution of the solar spots in latitudes since the beginning of the year 1854, with a map. Mon. Not. R. Astron. Soc. **19**, 1–3 (1858)
R.C. Carrington, *Observations of the Spots on the Sun from November 9, 1853 to March 24, 1861* (Williams and Norgate, London, 1863), p. 17
P. Charbonneau, Dynamo models of the solar cycle. Living Rev. Sol. Phys. **7**, 3 (2010)
A.M. Clerke, *Problems in Astrophysics* (Black, London, 1903), pp. 150–152
L. d'Azambuja, M. d'Azambuja, A comprehensive study of solar prominences and their evolution from spectroheliograms obtained at the observatory and from synoptic maps of the chromosphere published at the observatory. Ann. Obs. Meudon **6**(Fasc. VII), 1–278 (1948)
J.-P. Delaboudinière, G.E. Artzner, J. Brunaud, A.H. Gabriel, J.F. Hochedez, F. Millier, X.Y. Song, B. Au et al., EIT: extreme-ultraviolet imaging telescope for the SOHO mission. Sol. Phys. **162**, 291–312 (1995)
V.N. Dermendjiev, K.Y. Stavrev, V. Rušin, M. Rybansky, Secondary polar zone of prominence activity revealed from Lomnicky Štít observations. Astron. Astrophys. **281**, 241–244 (1994)
V. Domingo, B. Fleck, A.I. Poland, The SOHO mission: an overview. Sol. Phys. **162**, 1–37 (1995)
J. Evershed, M.A. Evershed, Results of prominence observations. Mem. Kodaikanal Obs. **1**, 55–126 (1917)

P.A. Gilman, What can we learn about solar cycle mechanisms from observed velocity fields? in *The Solar Cycle*, ed. by K.L. Harvey. ASP Conference Series, vol. 27 (ASP, San Francisco, 1992), pp. 241–255
G.E. Hale, On the probable existence of a magnetic field in sun-spots. Astrophys. J. **28**, 315–343 (1908)
G.E. Hale, Preliminary results of an attempt to detect the general magnetic field of the Sun. Astrophys. J. **38**, 27–98 (1913)
G.E. Hale, S.B. Nicholson, The law of sun-spot polarity. Astrophys. J. **62**, 270–300 (1925)
G.E. Hale, F. Ellerman, S.B. Nicholson, A.H. Joy, The magnetic polarity of sun-spots. Astrophys. J. **49**, 153–178 (1919)
W. Hanle, Über magnetische Beeinflussung der Polarisation der Resonanzfluoreszenz. Z. Phys. **30**, 93–105 (1924)
R. Hansen, S. Hansen, Global distribution of filaments during solar cycle no. 20. Sol. Phys. **44**, 225–230 (1975)
K.L. Harvey, The cyclic behavior of solar activity, in *The Solar Cycle*, ed. by K.L. Harvey. ASP Conference Series, vol. 27 (ASP, San Francisco, 1992), pp. 335–367
K.L. Harvey, The solar magnetic cycle, in *Solar Surface Magnetism*, ed. by R.J. Rutten, C.J. Schrijver (Kluwer, Dordrecht, 1994), pp. 347–363
K.L. Harvey, S.F. Martin, Ephemeral active regions. Sol. Phys. **32**, 389–402 (1973)
K.L. Harvey, J.W. Harvey, S.F. Martin, Ephemeral active regions in 1970 and 1973. Sol. Phys. **40**, 87–102 (1975)
J.W. Harvey, F. Hill, R. Hubbard, J.R. Kennedy, J.W. Leibacher, J.A. Pintar, P.A. Gilman, R.W. Noyes et al., The global oscillation network group (GONG) project. Science **272**, 1284–1286 (1996)
D.H. Hathaway, The solar cycle. Living Rev. Sol. Phys. **7**, 1 (2010)
F. Hill, R. Howe, R. Komm, J. Christensen-Dalsgaard, T.P. Larson, J. Schou, M.J. Thompson, Large-scale zonal flows during the solar minimum – where is cycle 25? Bull. Am. Astron. Soc. **43**, 16.10 (2011)
R. Howard, Polar magnetic fields of the Sun: 1960–1971. Sol. Phys. **25**, 5–13 (1972)
R. Howard, B.J. LaBonte, The Sun is observed to be a torsional oscillator with a period of 11 years. Astrophys. J. Lett. **239**, L33–L36 (1980)
R. Howe, Solar interior rotation and its variation. Living Rev. Sol. Phys. **6**, 1 (2009)
R. Howe, J. Christensen-Dalsgaard, F. Hill, R.W. Komm, R.M. Larsen, J. Schou, M.J. Thompson, J. Toomre, Deeply penetrating banded zonal flows in the solar convection zone. Astrophys. J. Lett. **533**, L163–L166 (2000a)
R. Howe, R. Komm, F. Hill, Variations in solar sub-surface rotation from GONG data 1995–1998. Sol. Phys. **192**, 427–435 (2000b)
R. Howe, F. Hill, R. Komm, J. Christensen-Dalsgaard, T.P. Larson, J. Schou, M.J. Thompson, R. Ulrich, The torsional oscillation and the new solar cycle, in *GONG–SoHO 24: A New Era of Seismology of the Sun and Solar-Like Stars*. Journal of Physics: Conference Series, vol. 271 (IOP, Bristol, 2011), Issue 1, id. 012074
R. Howe, J. Christensen-Dalsgaard, F. Hill, R. Komm, T.P. Larson, M. Rempel, J. Schou, M.J. Thompson, The high-latitude branch of the solar torsional oscillation in the rising phase of cycle 24. Astrophys. J. Lett. **767**, L20 (2013), 4 pp.
A.J. Kinder, Edward Walter Maunder FRAS (1851–1928): his life and times. J. Br. Astron. Assoc. **118**, 21–42 (2008)
R.W. Komm, R.F. Howard, J.W. Harvey, Torsional oscillation patterns in photospheric magnetic features. Sol. Phys. **143**, 19–39 (1993)
R. Komm, R. Howe, I. González Hernández, F. Hill, Solar-cycle variation of subsurface zonal flow. Sol. Phys. **289**, 3435–3455 (2014)
D.F. Kong, Z.N. Qu, Q.L. Guo, Revisiting the question: does high-latitude solar activity lead low-latitude solar activity in time phase? Astron. J. **147**, 97 (2014), 7 pp.
N.A. Krivova, S.K. Solanki, The 1.3-year and 156-day periodicities in sunspot data: wavelet analysis suggests a common origin. Astron. Astrophys. **394**, 701–706 (2002)
M. Küker, R. Arlt, R. Rüdiger, The Maunder minimum as due to magnetic Λ-quenching. Astron. Astrophys. **343**, 977–982 (1999)
B.J. LaBonte, R. Howard, Torsional waves on the sun and the activity cycle. Sol. Phys. **75**, 161–178 (1982)
N. Labrosse, S. Dalla, S. Marshall, Automatic detection of limb prominences in 304 Å EUV images. Sol. Phys. **262**, 449–460 (2010)
J.L. Leroy, On the orientation of magnetic fields in quiescent prominences. Astron. Astrophys. **64**, 247–252 (1978)
J.-L. Leroy, J.-C. Noens, Does the solar activity cycle extend over more than an 11-year period? Astron. Astrophys. **120**, L1–L2 (1983)
K.J. Li, Latitude migration of solar filaments. Mon. Not. R. Astron. Soc. **405**, 1040–1046 (2010)

K.J. Li, Q.X. Li, P.X. Gao, X.J. Shi, Cyclic behavior of solar full-disk activity. J. Geophys. Res. **113**, A11108 (2008)

W. Livingston, T.L. Duvall Jr., Solar rotation, 1966–1978. Sol. Phys. **61**, 219–231 (1979)

W.J.S. Lockyer, On the relationship between solar prominences and the forms of the corona. Mon. Not. R. Astron. Soc. **91**, 797–809 (1931)

N. Lockyer, W.J.S. Lockyer, Solar prominence and spot circulation, 1872–1901. Proc. R. Soc. Lond. **71**, 446–452 (1902)

D.H. Mackay, C.R. DeVore, S.K. Antiochos, Global-scale consequences of magnetic-helicity injection and condensation on the Sun. Astrophys. J. **784**, 164 (2014), 15 pp.

V.I. Makarov, K.R. Sivaraman, Evolution of latitude zonal structure of the large-scale magnetic field in solar cycles. Sol. Phys. **119**, 35–44 (1989a)

V.I. Makarov, K.R. Sivaraman, New results concerning the global solar cycle. Sol. Phys. **123**, 367–380 (1989b)

S.F. Martin, Conditions for the formation and maintenance of filaments (Invited review). Sol. Phys. **182**, 107–137 (1998)

S.F. Martin, K.H. Harvey, Ephemeral active regions during solar minimum. Sol. Phys. **64**, 93–108 (1979)

E.W. Maunder, Note on the distribution of sun-spots in heliographic latitude, 1874–1902. Mon. Not. R. Astron. Soc. **64**, 747–761 (1904)

A.R. Maunder, Letter to Stephan Ionides 21 May 1940 (1940)

P.S. McIntosh, Solar interior processes suggested by large-scale surface patterns, in *The Solar Cycle*, ed. by K.L. Harvey. ASP Conference Series, vol. 27 (ASP, San Francisco, 1992), pp. 14–34

H.W. Newton, *The Face of the Sun* (Penguin, Baltimore, 1958), p. 116

W.D. Pesnell, Solar cycle predictions. Sol. Phys. **281**, 507–532 (2012)

G.J.D. Petrie, K. Petrovay, K. Schatten, Solar polar fields and the 22-year activity cycle: observations and models. Space Sci. Rev. (2014). doi:10.1007/s11214-014-0064-4

V.V. Pipin, The Gleissberg cycle by a nonlinear $\alpha\Lambda$ dynamo. Astron. Astrophys. **346**, 295–302 (1999)

M. Rempel, Flux-transport dynamos with Lorentz force feedback on differential rotation and meridional flow: saturation mechanism and torsional oscillations. Astrophys. J. **647**, 662–675 (2006)

M. Rempel, High-latitude solar torsional oscillations during phases of changing magnetic cycle amplitude. Astrophys. J. Lett. **750**, L8 (2012), 4 pp.

A. Riccò, Risultati delle osservazioni dell protuberanze solari nel periodo undecennale dell'attività solare dal 1880 al 1890. Mem. Soc. degli Spettro. Ital. **20**, 135–139 (1892)

E. Robbrecht, Y.-M. Wang, N.R. Sheeley Jr., N.B. Rich, On the "extended" solar cycle in coronal emission. Astrophys. J. **716**, 693–700 (2010)

D.M. Rust, Magnetic fields in quiescent solar prominences. I. Observations. Astrophys. J. **150**, 313–326 (1967)

M. Rybanský, V. Rušin, M. Minarovjech, L. Klocok, E.W. Cliver, Reexamination of the coronal index of solar activity. J. Geophys. Res. **110**, A08106 (2005), 9 pp.

K.H. Schatten, P.H. Scherrer, L. Svalgaard, J.M. Wilcox, Using dynamo theory to predict the sunspot number during solar cycle 21. Geophys. Res. Lett. **5**, 411–414 (1978)

P.H. Scherrer, R.S. Bogart, R.I. Bush, J.T. Hoeksema, A.G. Kosovichev, J. Schou, W. Rosenberg, L. Springer et al., The solar oscillations investigation – Michelson Doppler Imager. Sol. Phys. **162**, 129–188 (1995)

J. Schou, Migration of zonal flows detected using Michelson Doppler Imager f-mode frequency splittings. Astrophys. J. Lett. **523**, L181–L184 (1999)

J. Schou, H.M. Antia, S. Basu, R.S. Bogart, R.I. Bush, S.M. Chitre, J. Christensen-Dalsgaard, M.P. Di Mauro et al., Helioseismic studies of differential rotation in the solar envelope by the solar oscillations investigation using the Michelson Doppler Imager. Astrophys. J. **505**, 390–417 (1998)

J. Schou, P.H. Scherrer, R.I. Bush, R. Wachter, S. Couvidat, M.C. Rabello-Soares, R.S. Bogart, J.T. Hoeksema et al., Design and ground calibration of the Helioseismic and Magnetic Imager (HMI) instrument on the Solar Dynamics Observatory (SDO). Sol. Phys. **275**, 229–259 (2012)

M.A. Schuh, J.M. Banda, P.N. Bernasconi, R.A. Angryk, P.C.H. Martens, A comparative evaluation of automated solar filament detection. Sol. Phys. **289**, 2503–2524 (2014)

M. Schüssler, The solar torsional oscillation and dynamo models of the solar cycle. Astron. Astrophys. **94**, L17–L18 (1981)

S.H. Schwabe, Sonnen-Beobachtungen in Jahre 1843. Astron. Nachr. **21**, 233–236 (1844)

H.J. Smith, E.v.P. Smith, *Solar Flares* (Macmillan, New York, 1963), p. 30

H.B. Snodgrass, Solar torsional oscillations—a net pattern with wavenumber 2 as artifact. Astrophys. J. **291**, 339–343 (1985)

H.B. Snodgrass, Synoptic observations of large scale velocity patterns on the Sun, in *The Solar Cycle*, ed. by K.L. Harvey. ASP Conference Series, vol. 27 (ASP, San Francisco, 1992), pp. 205–240

G. Spörer, Beobachtungen der Sonnenflecken vom Januar 1874 bis December 1879. Publ. Astrophys. Obs. Potsdam **2**(5), 1–81 (1880)

H.C. Spruit, Origin of the torsional oscillation pattern of solar rotation. Sol. Phys. **213**, 1–21 (2003)

L. Svalgaard, Y. Kamide, Asymmetric solar polar field reversals. Astrophys. J. **763**, 23 (2013), 6 pp.

L. Svalgaard, E.W. Cliver, Y. Kamide, Sunspot cycle 24: smallest cycle in 100 years? Geophys. Res. Lett. **32**, L01104 (2005)

S.J. Tappin, R.C. Altrock, The extended solar cycle tracked high into the corona. Sol. Phys. **282**, 249–261 (2013)

M.J. Thompson, J. Toomre, E. Anderson, H.M. Antia, G. Berthomieu, D. Burtonclay, S.M. Chitre, J. Christensen-Dalsgaard et al., Differential rotation and dynamics of the solar interior. Science **272**, 1300–1305 (1996)

J. Toomre, J. Christensen-Dalsgaard, R. Howe, R.M. Larsen, J. Schou, M.J. Thompson, Time variability of rotation in solar convection zone from SOI-MDI. Sol. Phys. **192**, 437–448 (2000)

K. Topka, R. Moore, B.J. Labonte, R. Howard, Evidence for a poleward meridional flow on the Sun. Sol. Phys. **79**, 231–245 (1982)

R.K. Ulrich, Very long lived wave patterns detected in the solar surface velocity signal. Astrophys. J. **560**, 466–475 (2001)

R.K. Ulrich, T. Tran, The global solar magnetic field—identification of traveling, long-lived ripples. Astrophys. J. **768**, 189 (2013), 12 pp.

S.V. Vorontsov, J. Christensen-Dalsgaard, J. Schou, V.N. Strakhov, M.J. Thompson, Helioseismic measurement of solar torsional oscillations. Science **296**, 101–103 (2002)

M. Waldmeier, Zirkulation und Magnetfeld der solaren Polarzone. Z. Astrophys. **49**, 176–185 (1960)

M. Waldmeier, A secondary polar zone of solar prominences. Sol. Phys. **28**, 389–398 (1973)

Y.-M. Wang, N.R. Sheeley Jr., S.H. Hawley, J.R. Kraemer, G.E. Brueckner, R.A. Howard, C.M. Korendyke, D.J. Michels et al., The green line corona and its relation to the photospheric magnetic field. Astrophys. J. **485**, 419–429 (1997)

D.F. Webb, CMEs and prominences and their evolution over the solar cycle, in *New Perspectives on Solar Prominences, IAU Colloquium 167*, ed. by D. Webb, D. Rust, B. Schmieder. ASP Conference Series, vol. 150 (ASP, San Francisco, 1997), pp. 463–474

P.R. Wilson, R.C. Altrock, K.L. Harvey, S.F. Martin, H.B. Snodgrass, The extended solar activity cycle. Nature **333**, 748–750 (1988)

H. Yoshimura, Solar cycle Lorentz force waves and the torsional oscillations of the Sun. Astrophys. J. **247**, 1102–1112 (1981)

C.A. Young, in *The Sun*, 4th edn. (Appleton, New York, 1897), pp. 157

P. Zeeman, On the influence of magnetism on the nature of the light emitted by a substance. Astrophys. J. **5**, 332–347 (1897)

DOI 10.1007/978-1-4939-2584-1_7
Reprinted from *Space Science Reviews* Journal, DOI 10.1007/s11214-014-0101-3

The Sun's Interior Structure and Dynamics, and the Solar Cycle

A.-M. Broomhall · P. Chatterjee · R. Howe · A.A. Norton · M.J. Thompson

Received: 30 April 2014 / Accepted: 11 September 2014 / Published online: 1 October 2014

Abstract The Sun's internal structure and dynamics can be studied with helioseismology, which uses the Sun's natural acoustic oscillations to build up a profile of the solar interior. We discuss how solar acoustic oscillations are affected by the Sun's magnetic field. Careful observations of these effects can be inverted to determine the variations in the structure and dynamics of the Sun's interior as the solar cycle progresses. Observed variations in the structure and dynamics can then be used to inform models of the solar dynamo, which are crucial to our understanding of how the Sun's magnetic field is generated and maintained.

Keywords Sun: activity · Sun: helioseismology · Sun: interior · Sun: magnetic fields · Sun: oscillations

A.-M. Broomhall (✉)
Institute of Advanced Studies, University of Warwick, Coventry CV4 7HS, UK
e-mail: a-m.broomhall@warwick.ac.uk

A.-M. Broomhall
Centre for Fusion, Space, and Astrophysics, Department of Physics, University of Warwick, Coventry CV4 7AL, UK

P. Chatterjee · M.J. Thompson
High Altitude Observatory, National Center for Atmospheric Research, PO Box 3000, Boulder, CO 80307, USA

P. Chatterjee
e-mail: mppiyali@ucar.edu

M.J. Thompson
e-mail: mjt@ucar.edu

R. Howe
School of Physics and Astronomy, University of Birmingham, Edgbaston, Birmingham B15 2TT, UK
e-mail: rhowe@noao.edu

A.A. Norton
HEPL, Solar Physics, Stanford University, Stanford, CA 94305, USA
e-mail: aanorton@stanford.edu

1 Introduction to Helioseismology

Helioseismology is the study of the solar interior using observations of waves that propagate within the Sun. The Sun's natural acoustic resonant oscillations are known as solar p modes. The p stands for pressure as the main restoring force is a pressure differential. At any one time thousands of acoustic oscillations are traveling throughout the solar interior. Each individual solar p mode is trapped in a specific region of the solar interior, known as a cavity, and its frequency is sensitive to properties, such as temperature and mean molecular weight, of the solar material in the cavity. The sound waves can be considered as damped harmonic oscillators as they are stochastically excited and intrinsically damped by turbulent convection in the outer approximately 30 per cent by radius of the solar interior. The strongest p-mode oscillations have a periodicity of approximately 5 minutes. The solar oscillations can be observed in two ways: by line-of-sight Doppler velocity measurements over the visible disk; or by measuring the variations in the continuum intensity of radiation, which are caused by the compression of the radiating gas by the waves.

The horizontal structure of p modes can be modeled as spherical harmonics and so can be described by three main components: Harmonic degree, l, which indicates the number of node lines on the surface; azimuthal degree, m, which describes the number of planes slicing through the equator and takes values between $-l \leq m \leq l$; and radial degree, n, which gives the number of radial nodes in the solar interior. If the Sun was completely spherically symmetric the frequencies of the oscillations would be degenerate in m. However, asphericities in the solar interior, such as rotation, temperature variations and the presence of magnetic fields, lift the degeneracy and measurements of the differences in the frequencies of the m components allow inferences to be made concerning the effects responsible for the asphericities.

The frequency of a mode, $\nu_{n,l,m}$, can be expressed in terms of the $m = 0$ central frequency, $\nu_{n,l}$, and a polynomial expansion of splitting (or a) coefficients, $a_j(n,l)$:

$$\nu_{n,l,m} = \nu_{n,l} + \sum_{j=1}^{j_{\max}} a_j(n,l)\mathcal{P}_j^{(l)}(m), \tag{1}$$

where $\mathcal{P}_j^{(l)}(m)$ are the Ritzwoller-Lavely formulation of the Clebsch-Gordon expansion (Ritzwoller and Lavely 1991) and are given by

$$\mathcal{P}_j^{(l)}(m) = \frac{l\sqrt{(2l-j)!(2l+j+1)!}}{(2l)!\sqrt{2l+1}} C_{j0lm}^{lm}. \tag{2}$$

Here, C_{j0lm}^{lm} are the Clebsch-Gordon coefficients. The odd-order a coefficients express the difference in mode frequency caused by the modes' interaction with the rotation profile of the solar interior. The even-order coefficients are sensitive to all other departures from spherical symmetry. These could include local variations in the sound speed or cavity size (Balmforth et al. 1996), temperature variations (Kuhn 1988), second order rotation effects from the differential rotation (Dziembowski and Goode 1992), and the presence of magnetic fields (Gough and Thompson 1988b). Isolating the contributions of all these effects to the even-order coefficients is extremely difficult.

As sound waves travel from their excitation point near the surface into the solar interior they are refracted because of increasing pressure, and therefore sound speed (since $c_s^2 = \frac{\gamma_1 P}{\rho}$, where γ_1 is the first adiabatic exponent). Assuming the direction of travel is not radial the waves follow a curved trajectory which takes them back to the surface where they are reflected by the sharp drop in density. By the time the waves reach the surface they are approximately traveling in a radial direction and so the depth at which the near-surface reflection of the waves occurs depends only on the wave frequency. The p modes are set up by a superposition of such waves, and so the location of the upper edge of the acoustic cavity of the modes—the upper turning point—likewise depends only on frequency, with the upper turning point of low-frequency modes being deeper in the solar interior than the upper turning point of high-frequency modes. The radius of the lower turning point depends on the angle of trajectory and therefore l, with low-l modes turning deeper within the solar interior than high-l modes.

Helioseismology can be split into two categories: global and local. Global helioseismology studies the natural resonant acoustic oscillations of the solar interior that are able to form standing waves in the entire Sun, as opposed to local helioseismology, which studies propagating waves in part of the Sun. We now give a brief introduction to each category.

1.1 Global Helioseismology

Global helioseismology can itself be split into two sub-categories: Sun-as-a-star (unresolved) observations; and resolved observations. Sun-as-a-star observations are only sensitive to the lowest harmonic degrees ($0 \leq l \leq 3$, and occasionally $l = 4$ and 5). These modes are the truly global modes of the Sun as they travel right to the energy-generating core. However, by making resolved observations of the solar surface spatial filters can be applied that allow measurements of modes with degrees into the 100s or even 1000s.

The Birmingham Solar Oscillations Network (BiSON, Elsworth et al. 1995; Chaplin et al. 1996) has been making Sun-as-a-star line-of-sight velocity observations for over 30 years, covering cycles 21 (although with limited coverage), 22, 23, and 24. BiSON is an autonomous network of 6 ground-based observatories strategically positioned around the world so as to allow observations of the Sun to be made 24 hr a day. This is important as gaps in the time domain contaminate frequency spectra, making it more difficult to accurately obtain the parameters that describe the modes, such as their frequencies. One can also make continuous observations of the Sun from space and this is the approach taken by the SOlar and Heliospheric Observatory (SOHO) spacecraft, which was launched in 1995. Sun-as-a-star observing programs onboard SOHO include the Global Oscillations at Low Frequencies (GOLF; Gabriel et al. 1995) instrument, which also measures line-of-sight velocity, and the Variability of solar IRradiance and Gravity Oscillations (VIRGO; Fröhlich et al. 1995), which measures the changes in intensity of the Sun.

VIRGO is also capable of making resolved observations of the Sun, as was the Michelson Doppler Imager (MDI; Scherrer et al. 1995), also onboard SOHO. Global Oscillations Network Group (GONG) is a ground-based network of 6 sites (Harvey et al. 1996), which has been making resolved observations of the Sun since 1995. More recently, the Helioseismic and Magnetic Imager (HMI; Schou et al. 2012) and the Atmospheric Imaging Assembly (AIA; Lemen et al. 2012) onboard the Solar Dynamics Observatory (SDO) have been used for helioseismic studies Howe et al. (2011a, 2011b). SDO was only launched in 2010 and so these instruments cannot be used to study previous solar cycles. They are, however, likely to play an important role in helioseismic studies of cycle 24, and possibly beyond.

1.2 Local Helioseismology

Since the advent of helioseismic observations at high spatial resolution, a number of different data analysis techniques known collectively as "local helioseismology" have been developed. They do not rely on the global resonant-mode nature of the observed oscillations:

rather, the observations of wave motions are interpreted in terms of their local properties. Three methods in particular are widely used. Ring analysis (or ring-diagram analysis) works within a frequency-wavenumber framework for analysing the oscillations, albeit on localized patches of the solar surface, and may therefore be considered the local-helioseismic analogue of global-mode helioseismology. Time-distance helioseismology (or helioseismic tomography), and the closely related method of acoustic holography, work rather in the framework of interpreting the effect of heterogeneities and flows on propagating waves in terms of travel-time shifts or equivalently phase shifts. These two basically distinct approaches will be described briefly below.

In ring analysis (Hill 1989), measurements of wave properties are made in localized regions ("tiles") on the surface of the Sun. The observable—typically the Doppler velocity—over the pixels within the region are Fourier transformed in the two spatial horizontal directions and in time to give wave power as a function of spatial wavenumbers k_x, k_y and frequency. In cuts through the 3-D power spectrum at fixed frequency, the power is found to lie in rings corresponding to the different radial orders n of the modes: hence the name "ring analysis". In the absence of flows or any horizontal inhomogeneities, the dispersion relation of the waves is of the form $\omega = \omega_n(k_h)$, where $k_h \equiv \sqrt{k_x^2 + k_y^2}$ is the magnitude of the horizontal wavenumber vector. In other words, the frequencies do not depend on the direction of the horizontal wavenumber, only on its magnitude: thus, the rings of power are circular and centered on $(k_x, k_y) = (0, 0)$. Large-scale flows and magnetic fields, which can change the dispersion relation depending on the direction of propagation of the waves, shift and distort the rings. Further, local changes in the effective isotropic wave speed (such as could be caused by thermal anisotropies) change the frequency at fixed wavenumber or equivalently change the radius of the rings at fixed frequency. These various perturbations to the rings, measured at different frequencies and for different n values, permit the possibility of performing a one-dimensional inversion in depth to infer the large-scale flow, wave speed, etc., below each tile (Haber et al. 2002; Bogart et al. 2008; González Hernández et al. 2010). By combining inversion results under different tiles, a three-dimensional subsurface map can be built up. Recently, fully 3-D inversion methods have been developed and implemented that simultaneously invert the data from many different tiles to obtain 3-D maps of subsurface flows directly (Featherstone et al. 2011).

In time-distance helioseismology, travel times of waves that propagate beneath the surface between pairs of surface points are estimated by cross-correlating the observed oscillations between surface points. The cross-correlation function derived by cross-correlating the oscillations at two surface points A and B, say, exhibits a number of "wave packets". In a ray-theoretic interpretation, these packets correspond to waves that travel between A and B along different ray paths, either traveling directly without intermediate surface bounces, or bouncing at the surface one or more times before arriving at the point of observation. In practice, it can be necessary to average the cross-correlations between many similarly separated pairs of points before the wave packets are clearly visible. Partly for this reason, the observations are typically filtered in some way before the cross-correlations are made. The most common filters are a phase-speed filter or a filter that aims to isolate oscillations corresponding to a particular radial order n. Also, usually it is only the first-arrival wave packet—corresponding to propagation between A and B with no intermediate surface bounces—that is used in the subsequent analysis. Travel times are estimated by measuring the location of the wave packet, and a number of different methods are used to do that. Flows, magnetic fields and inhomogeneities experienced by the waves

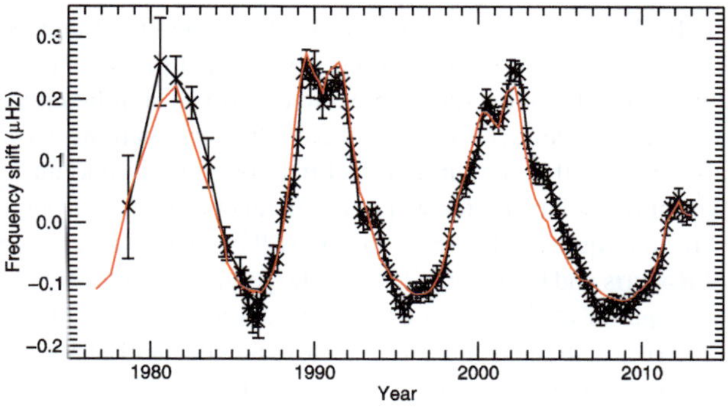

Fig. 1 Average frequency shifts of "Sun-as-a-star" modes with frequencies between 2.5 and 3.5 mHz. The results were obtained from 365 d BiSON time series that overlapped by 91.25 d. Also plotted is a scaled and shifted version of the 10.7 cm flux

during their propagation from A to B shift the location of the wave packet (and this in turn is measured as a travel-time shift) as well as potentially modifying the width and amplitude of the wave packet. Also, flows can cause a difference in travel time depending on whether the waves are traveling from A to B or from B to A. By measuring travel times between many different pairs of surface points, 3-D subsurface maps of flows and wave-speed inhomogeneities can be made using inverse techniques (Zhao and Kosovichev 2004).

For further discussion of the signatures of flows on the ring analysis and time-distance helioseismology measurements, see Sect. 5.1. For now, however, we return to concentrate on the global modes.

2 An Introduction to Global Helioseismology and the Solar Cycle

It has been known since the mid 1980s that p-mode frequencies vary throughout the solar cycle with the frequencies being at their largest when the solar activity is at its maximum (e.g. Woodard and Noyes 1985; Pallé et al. 1989; Elsworth et al. 1990; Libbrecht and Woodard 1990; Chaplin et al. 2007; Jiménez-Reyes et al. 2007).[1] For a low-l mode at about 3000 μHz the change in frequency between solar maximum and minimum is about 0.4 μHz. By examining the changes in the observed p-mode frequencies throughout the solar cycle we can learn about solar-cycle-related processes that occur beneath the Sun's surface. The 11 year cycle is seen clearly in Fig. 1, which shows the mean frequency shifts of the p modes observed by BiSON and the 10.7 cm flux[2] for comparison (also see Broomhall et al. 2009; Salabert et al. 2009; Fletcher et al. 2010).

What causes the observed frequency shifts? The magnetic fields can affect the modes in two ways: directly and indirectly. The direct effect occurs because the Lorentz force provides an additional restoring force resulting in an increase of frequency, and the appearance of new modes. Magnetic fields can influence the oscillations indirectly by affecting the physical properties of the cavities in which the modes are trapped and, as a result, the propagation of the acoustic waves within them. For example, changes in the thermal structure can affect both the propagation speed and the location of the upper turning point. Indirect effects can both increase or decrease the frequencies of p modes.

[1] Please note here that the references listed are not exhaustive, instead we have tried to include useful ones whose reference lists themselves are informative.

[2] http://www.ngdc.noaa.gov/stp/space-weather/solar-data/solar-features/solar-radio/noontime-flux/penticton/

To date the relative contributions from the direct and indirect effects remain uncertain. Dziembowski and Goode (2005) suggest that the magnetic fields are too weak in the near-surface layers for the direct effect to contribute significantly to the observed frequency shifts and that the indirect effects dominate the perturbations. However, Dziembowski and Goode also suggest that the direct effect may be more important for low-frequency modes deeper within the solar interior where the magnetic field is strong enough to produce a noticeable shift in frequency. These results are still controversial and appear to disagree with the results of Roberts and Campbell (1986), who found that field strengths of the order of 500 kG would be required at the base of the convection zone to explain the shift. Foullon and Roberts (2005) investigated the effect of a magnetic field at the base of the convection zone and a more shallow magnetic field, at a depth of 50 Mm, but found their direct influence on the frequencies of p modes was consistently smaller than the change in frequency that is observed. Changes in both the chromospheric magnetic field strength and temperature can explain the observed downturn in the magnitude of the frequency shifts at high frequencies (Jain and Roberts 1993). However, as we now describe, it is not only the mode frequencies that vary throughout the solar cycle.

2.1 Solar Cycle Variations in Mode Powers and Lifetimes

Solar p modes are excited and damped by turbulent convection beneath the solar surface. The process of excitation and damping varies throughout the solar cycle and these variations are observed as changes in the heights and widths of the mode profiles in a frequency-power spectrum. For typical low-l modes, as the surface activity increases the mode frequencies and widths increase but the mode heights decrease. An increase in the linewidths implies that the modes experience more damping at times of high activity and so the lifetimes decrease.

Numerous authors have observed that lifetimes decrease with solar activity and those observations have been made using data from different instrumental regimes and for a range of l (e.g. Jefferies et al. 1990; Pallé et al. 1990; Chaplin et al. 2000; Komm et al. 2000; Appourchaux 2001; Toutain and Kosovichev 2001; Jiménez et al. 2002; Howe et al. 2003; Jiménez-Reyes et al. 2004b; Salabert and Jiménez-Reyes 2006; Burtseva et al. 2009; Simoniello et al. 2010). The size of the variation appears to be dependent on l with lower-l modes showing a larger percentage change than high-l modes (e.g. Burtseva et al. 2009, and references therein). Burtseva et al. (2009) looked at the lifetimes of modes with $300 \leq l \leq 600$ and found that in active regions the lifetimes decrease as activity increases by about 13 % between minimum and maximum. In quiet regions the lifetimes still decrease with solar activity but to a lesser extent (the lifetimes at solar maximum are about 8 % of the lifetimes at solar minimum). This implies that the change to the damping is not just associated with the active regions that appear on the surface. In general p-mode linewidths increase with frequency, except for a plateau region at around 2800 μHz. Komm et al. (2000) found that the linewidths of modes in the plateau region were most sensitive to the level of solar activity.

Mode powers have been observed to decrease with increasing solar activity (e.g. Pallé et al. 1990; Anguera Gubau et al. 1992; Elsworth et al. 1993). The decrease in mode power is of the order of 20 % and has been now been observed by many authors (e.g. Chaplin et al. 2000; Komm et al. 2000; Jiménez-Reyes et al. 2001; Appourchaux 2001; Toutain and Kosovichev 2001; Jiménez et al. 2002; Jiménez-Reyes et al. 2004b; Simoniello et al. 2009). The total energy of the mode, E, can be determined by multiplying the total power of the mode by the mode mass (as defined by Christensen-Dalsgaard and Berthomieu 1991). The mode energy shows a similar decrease to the mode power, with a maximum to minimum variation of approximately 12 % (Jiménez et al. 2002).

The energy supply rate is determined by multiplying the mode energy by the mode width (e.g. Chaplin et al. 2000). Numerous studies have shown that while the mode energy decreases with activity the energy supply rate shows no solar cycle variation (e.g. Chaplin et al. 2000; Komm et al. 2000; Appourchaux 2001; Toutain and Kosovichev 2001; Jiménez et al. 2002; Howe et al. 2003; Jiménez-Reyes et al. 2004b; Salabert and Jiménez-Reyes 2006). Therefore, the observed variations in mode energy and mode power probably arise from an increase in damping only. As the energy of the modes decreases between solar minimum and solar maximum this implies that some energy has gone missing.

One potential source of damping is magnetic activity on the solar surface. Active regions are known to suppress the power of p modes, and effect lifetimes, and energy supply rates (e.g. Woods and Cram 1981; Lites et al. 1982; Brown et al. 1992; Rajaguru et al. 2001; Komm et al. 2002; Howe et al. 2004a). However, the associated mechanisms are not yet fully understood. One explanation is that strong-field magnetic regions, such as sunspots, are effective absorbers of p-mode power (e.g. Haber et al. 1999; Jain and Haber 2002). Komm et al. (2000) suggested that the energy could be in flux tubes, whose numbers increase with solar activity, and the energy could excite oscillations in magnetic elements. Another suggestion is that the efficiency of mode excitation is reduced in magnetic areas (e.g. Goldreich and Kumar 1988; Cally 1995; Jain et al. 1996). Other possible explanations as to why p-mode excitation is suppressed in sunspots include a different height of spectral line formation due to the Wilson depression or a modification of p-mode eigenfunctions by the magnetic field (Burtseva et al. 2009, and references therein).

Another possibility that could alter p-mode damping is variations in the convective properties near the solar surface, which are most likely to arise from the influence of magnetic structures (Houdek et al. 2001). Houdek et al. theorized that changes of parameters in the convection zone would affect linewidth shapes mainly in the plateau region of a frequency spectrum, as is observed. This, therefore, implies that during times of high-magnetic activity the convection zone is affected sufficiently to produce a measurable change in p-mode linewidths. Several authors have observed that the horizontal size of solar granules decreases from solar minimum to solar maximum (e.g. Macris et al. 1984; Muller 1988; Berrilli et al. 1999; Muller et al. 2007). Muller (1988) found that the horizontal granule size decreased by approximately 5 % from solar minimum to solar maximum. According to Houdek et al. (2001) this should result in an increase in damping rates of about 20 %, which is in reasonable agreement with the observed change in damping rates in BiSON data of approximately 24 ± 3 %.

We now return to look in more detail at the changes in frequencies of the p modes with solar cycle.

3 Dependence of Solar Cycle Frequency Shifts on l and Frequency

Solar cycle frequency shifts, $\delta\nu_{n,l}$, have well-known dependencies on both angular degree, l, and frequency, $\nu_{n,l}$, as demonstrated in Fig. 2 (also see e.g. Libbrecht and Woodard 1990; Elsworth et al. 1994; Chaplin et al. 1998, 2001; Howe et al. 1999; Jiménez-Reyes et al. 2001). We now look in more detail at these dependencies and discuss what they tell us about the origin of the perturbation. We note here that the frequency shifts plotted in Fig. 2 were determined for the central frequency of the mode i.e. the $m = 0$ frequency. In Sect. 3.3 we discuss the latitudinal dependence of the shifts and there we consider modes with different m.

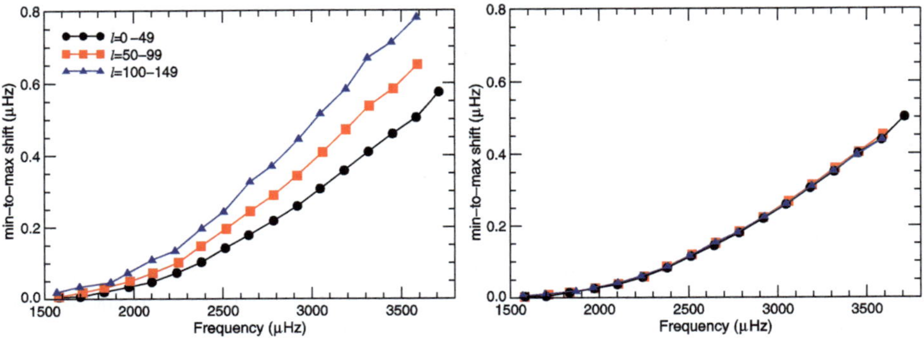

Fig. 2 *Left-hand panel*: Change in frequency between solar minimum and solar maximum, $\delta\nu$, as a function of mode frequency. The results have been obtained by averaging over different ranges in l (see *legend*), and by smoothing over 135 μHz. *Right-hand panel*: Frequency shifts that have been scaled by the inertia ratio, $Q_{n,l}$. The *error bars* on these figures are not visible because they are smaller than the symbol size. Results obtained using GONG frequencies

3.1 Dependence of Solar Cycle Frequency Shifts on Mode Inertia

The main l-dependence in $\delta\nu_{n,l}$ is associated with mode inertia. The normalized mode inertia is defined by Christensen-Dalsgaard and Berthomieu (1991) as

$$I_{n,l} = M_{\odot}^{-1} \int_{v} |\xi|^2 \rho \mathrm{d}V = 4\pi M_{\odot}^{-1} \int_{0}^{R_s} |\xi|^2 \rho r^2 \mathrm{d}r = \frac{M_{n,l}}{M_{\odot}}, \tag{3}$$

where ξ is the displacement associated with a mode, suitably normalized at the photosphere, V is the volume of the Sun, and $M_{\odot}$ is the mass of the Sun. $M_{n,l}$ is the 'mass' associated with a mode. The physical interpretation of the mode inertia is some measure of the interior mass affected by any given mode. At fixed frequency a decrease in l results in an increase in $M_{n,l}$ and therefore in $I_{n,l}$. In other words as l decreases a greater volume of the interior is associated with the motions generated by the mode. Therefore the high-l modes are more sensitive to a perturbation and so vary more throughout the solar cycle than low-l modes.

The inertia ratio, Q_{nl}, is defined by Christensen-Dalsgaard and Berthomieu (1991) as:

$$Q_{nl} = \frac{I_{nl}}{\bar{I}(\nu_{nl})}, \tag{4}$$

$\bar{I}(\nu_{nl})$ is the inertia an $l = 0$ modes would have at a frequency ν_{nl}. Multiplying the frequency shifts by Q_{nl} removes the l dependence of the frequency shifts at fixed frequency, as can be seen in Fig. 2. This makes the dependence a function of frequency alone. Collapsing the l dependence in this manner allows the frequency shifts of a wide range of l to be combined, thereby reducing any uncertainties associated with the shifts and allowing tighter constraints to be placed on the frequency dependence.

3.2 Dependence of Solar Cycle Frequency Shifts on Mode Frequency

The frequency dependence of the frequency shifts, which can be seen in the right-hand panel of Fig. 2, is a telltale indicator that the observed 11-year signal must be the result of changes in acoustic properties in the few hundred kilometres just beneath the visible surface of the Sun, a region to which the higher-frequency modes are much more sensitive than their lower-frequency counterparts because of differences in the upper boundaries of the cavities in which the modes are trapped (Libbrecht and Woodard 1990; Christensen-Dalsgaard and Berthomieu 1991). We now go into more detail.

As mentioned earlier, for a given l the upper turning point of low-frequency modes is deeper than the upper turning point of high-frequency modes. At a fixed frequency lower-l modes penetrate more deeply into the solar interior than higher-l modes. Therefore, higher-frequency modes, and to a lesser extent higher-l modes, are more sensitive to surface perturbations.

Libbrecht and Woodard (1990) discuss the origin of the perturbation. If the perturbations were to extend over a significant fraction of the solar interior, asymptotic theory implies that the fractional mode frequency shift would depend mainly on $\nu_{n,l}/l$, which is not what we observe. This implies that the relevant structural changes occur mainly in a thin layer. Thompson (1988) found that the effect of perturbing a thin layer in the propagating regions of modes, such as the layer where the second-ionization of helium occurs, is an oscillatory frequency dependence in $\delta\nu_{n,l}$. Such a frequency dependence is also not observed, implying that the dominant frequency dependence is not the direct result of, for example, changes in the magnetic field at the base of the convection zone. If the perturbation was confined to the centre of the Sun the size of the frequency shift would increase with decreasing l as low-l modes penetrate deeper into the solar interior than high-l modes. In fact, $\delta\nu_{n,l}$ increases with increasing l.

We can think of the frequency dependence as a power law where $\delta\nu_{n,l} \propto \nu_{n,l}^{\alpha}$. A perturbation from a layer strictly confined to the photosphere (but extending over less than one pressure scale height) is expected to result in an $\alpha = 3$ relationship (Cox 1980; Gough 1990; Libbrecht and Woodard 1990; Goldreich et al. 1991). If instead the perturbation extends beneath the surface, the frequency dependence will be weaker and α will get smaller (Gough 1990). Chaplin et al. (2001) determined that $\alpha < 3$, which implies that the significant contribution comes from a perturbation close to the surface, in the sub-photospheric layers. Rabello-Soares et al. (2008) observed that α is smaller for the lower-frequency modes, which suggests that the perturbation extends to greater depths the lower in frequency one goes. This is consistent with the results of Dziembowski and Goode (2005) (see Sect. 2).

In summary, the oscillations are responding to changes in the strength of the solar magnetic activity near the Sun's surface. As modes below approximately 1800 μHz experience almost no solar cycle frequency shift it is reasonable to conclude that the origin of the perturbation is concentrated in a region above the upper turning points of these modes. Figure 3 shows that the upper turning point of a mode with a frequency of $\sim$ 1800 μHz is about $0.996R_{\odot}$ (approximately 3 Mm) below the surface. Note that the upper turning point predicted by a model is strongly dependent on the properties of the model at the top of the convection zone.

Above 3700 μHz the relationship between frequency shift and activity changes, as the magnitude of the solar cycle frequency shift decreases and oscillates (e.g. Libbrecht and Woodard 1990; Goldreich et al. 1991; Jain and Roberts 1996). Furthermore, above $\sim$ 4100 μHz it appears that modes experience a decrease in frequency as activity increases (Ronan et al. 1994; Chaplin et al. 1998).

As we have seen the frequency dependence of the frequency shifts implies that the shifts can be associated with a near-surface perturbation. Therefore, we now move on to directly compare the change in mode frequency with the magnetic field that is observed at the solar surface (and beyond into the solar atmosphere).

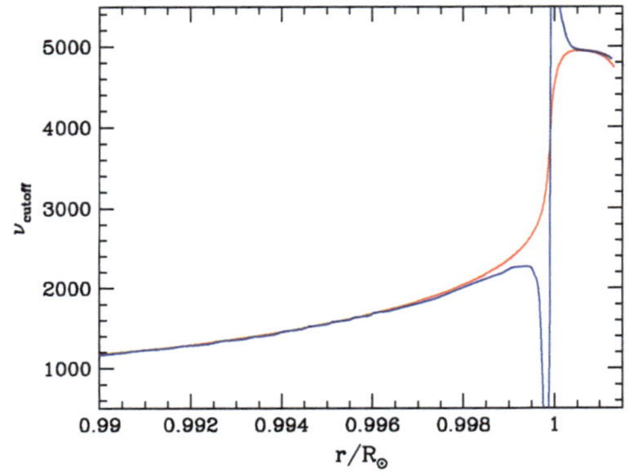

Fig. 3 Acoustic cut-off frequency as a function of solar radius. The upper turning point of the modes can be thought of in terms of the acoustic cut-off frequency because the upper turning point corresponds to the radius at which the acoustic cut-off frequency equals the mode frequency. The *blue curve* shows the cut-off frequencies determined from a standard solar model, while the *red curve* shows the acoustic cut-off frequency in an isothermal limit. Adapted from Basu et al. (2012)

3.3 Comparison Between Frequency Shifts and the Surface Magnetic Field

Numerous comparisons have been made between the p-mode frequency shifts and various proxies of the Sun's surface magnetic field (e.g. Jimenez-Reyes et al. 1998; Antia et al. 2001; Tripathy et al. 2001; Howe et al. 2002; Chaplin et al. 2004b, 2007; Jain et al. 2009, 2012), including the 10.7 cm and the International Sunspot Number[3] (ISN). Chaplin et al. (2007) compared six different proxies with the low-l mode frequency shifts and demonstrated that better correlations are observed with proxies that measure both the strong and the weak components of the Sun's magnetic field, such as the Mg II H and K core-to-wing data, the 10.7 cm radio flux, and the He I equivalent width data (as opposed to the ISN and the Kitt Peak Magnetic Index, which mainly sample the strong magnetic flux). The weak-component of the solar magnetic flux is distributed over a wider range of latitudes than the strong component and so a possible explanation of these results is in terms of the latitudinal distribution of the modes used in this study (i.e. low-l modes).

Figure 4 shows a comparison between two proxies of the Sun's magnetic field and low-l frequency shifts. Although the agreement is approximately linear the well-known hysteresis is clearly visible, particularly in the case of the sunspot number. This can be explained in terms of the variation in the latitudinal distribution of the surface magnetic field throughout the solar cycle (Moreno-Insertis and Solanki 2000): The different modes have different latitudinal dependencies, as described by the appropriate spherical harmonics (see Sect. 1), and so the relative influence of the surface magnetic flux on the modes changes throughout the solar cycle as the strong surface magnetic field migrates towards the equator.

As mentioned in Sect. 1 the even splitting coefficients can provide information about departures from spherical symmetry, such as those produced by the presence of a magnetic field. Although the correlation between the even a coefficients and global measures of the Sun's magnetic field is good, it is not linear. However, the even splitting coefficients are linearly correlated with the corresponding Legendre polynomial decomposition of the surface (e.g. Howe et al. 1999, 2002; Chaplin et al. 2003, 2004a, 2004b; Jiménez-Reyes et al.

[3]http://www.ngdc.noaa.gov/stp/space-weather/solar-data/solar-indices/sunspot-numbers/international/

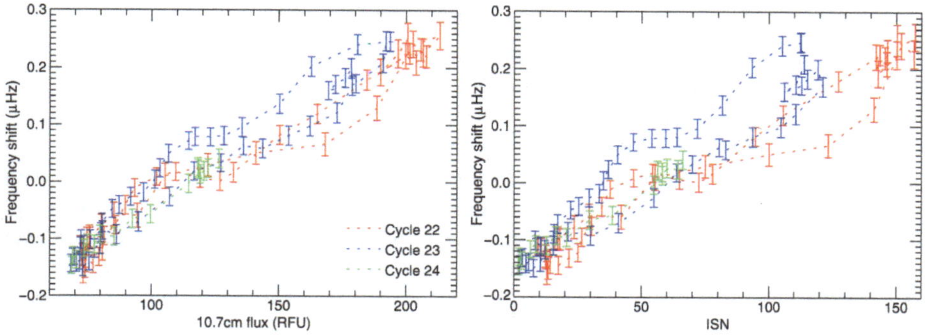

Fig. 4 Comparison of frequency shifts of low-l modes (observed by BiSON) with surface proxies of the Sun's magnetic field. The *left-hand panel* shows a comparison with the 10.7 cm flux, while the *right-hand panel* shows a comparison with the ISN. The different activity cycles are indicated by different *colours* (see *legend*)

Fig. 5 Frequency shift inversions using GONG data and modes with $40 \leq l < 80$, and $9 \leq n \leq 11$. The size of the shift is indicated by the *colour* and the *black contours* show the surface magnetic field at 5 G intervals. Updated from Howe et al. (2002)

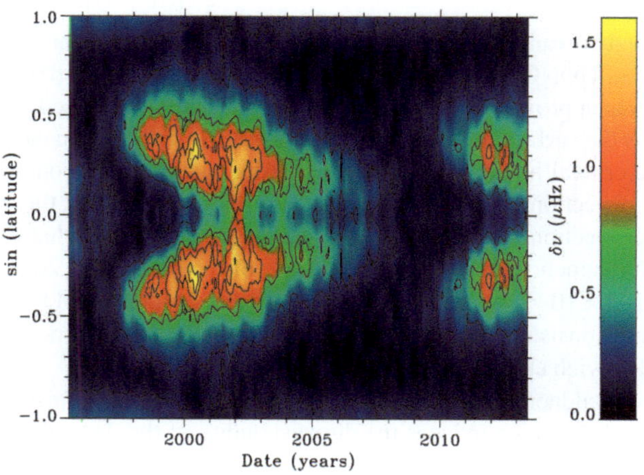

2004a) indicating that the size of the observed frequency shift experienced by a particular mode is dependent on the latitudinal distribution of the surface magnetic flux. Furthermore, Howe et al. (2002) showed that it is possible to use latitudinal inversion techniques to localize the frequency shifts in latitude and, in fact, reconstruct the evolution of the surface magnetic field. An up-to-date version of these inversions can be seen in Fig. 5, which clearly resembles the familiar butterfly diagrams usually associated with the surface magnetic field.

4 Evidence for Structural Changes due to Magnetic Fields Deeper in the Interior

All that has been discussed above concerns the near-surface magnetic field. However, one of the main advantages of helioseismology is that it allows the deeper interior to be studied. Therefore we now move on to discuss attempts to detect evidence of the solar magnetic field deeper in the solar interior. However, this is understandably hard. The plasma-β ($= P_{gas}/P_{mag}$) in the deep interior is significantly greater than unity. Furthermore, the sound speed in the interior increases substantially with depth, meaning the modes spend significantly longer in the near-surface regions than in the deep interior. All this means that the influence of a deep-seated magnetic field on the properties of the oscillations will be limited. However, the rewards for finding evidence of deep-seated magnetic fields are great. For example, many believe the solar dynamo is generated in the tachocline at the base of the convection zone (see Charbonneau 2010, for a recent review).

Numerous authors have looked but were unable to find helioseismic evidence for magnetic fields deeper in the solar interior (e.g. Gough and Thompson 1988a; Basu and Antia 2000, 2001, 2003; Basu and Schou 2000; Antia et al. 2001; Basu 2002; Eff-Darwich et al. 2002; Basu et al. 2003). They therefore resorted to putting limits on parameters at the base of the convection zone, such as a maximum change in field strength between solar minimum and maximum of 300 kG (Basu 1997; Antia et al. 2000) or a change in sound speed of $\delta c/c = 3 \times 10^{-5}$ (Eff-Darwich et al. 2002).

Chou and Serebryanskiy (2005) and Serebryanskiy and Chou (2005) considered the frequency shift, scaled by mode mass, as a function of the horizontal phase speed, which is given by

$$w = \frac{\nu}{2\pi [l(l+1)]^{1/2}}, \tag{5}$$

and so can be related to the lower turning point of the mode. The authors use both MDI and GONG data and use $190 \le w \le 1570$. They observed that the scaled frequency shift was approximately constant with horizontal phase speed around solar minimum. However, as the surface magnetic field increased, the scaled frequency shift decreased, but only above a critical horizontal phase speed value, which corresponds to a depth near the base of the convection zone. They interpret this as indicating that the wave speed near the base of the convection zone changes with activity and they find that this behaviour is consistent with a magnetic perturbation at the base of the convection zone. Further, they find that $\delta c/c =$ 1–3×10^{-5}, which implies a change in magnetic field of between 170–290 kG. These results are consistent with the upper limits set by earlier authors (Basu 1997; Antia et al. 2000; Eff-Darwich et al. 2002).

Baldner and Basu (2008) and Baldner et al. (2009) used a principal component analysis (PCA), which reduces the dimensionality of the data and consequently the noise, to find a small but statistically significant change in the frequencies of modes whose lower turning points are at or near the base of the convection zone. This change is tightly correlated with surface activity. If the change in frequency can be interpreted as a change in sound speed due to the presence of a magnetic field Baldner et al. (2009) find that their results imply a change in field strength between the maximum of cycle 23 and the preceding minimum of 390 kG, just above the limits set previously.

Moving slightly closer to the surface, Basu and Mandel (2004) showed evidence for solar structure changes around the zone associated with the second ionization of helium (approximately $0.98 R_\odot$). A spherically symmetric localized feature or discontinuity in the internal structure of the Sun causes a characteristic oscillatory component in mode frequencies (e.g. Vorontsov 1988; Gough 1990; Basu et al. 1994; Roxburgh and Vorontsov 1994). Basu and Mandel (2004) considered changes in the oscillatory signal caused by the zone in which the second ionization of helium occurs, using intermediate degree modes whose lower turning points were below the depth of the second ionization zone of helium but above the base of the convection zone, which is a discontinuity of its own right. Basu and Mandel found changes in the amplitude of the oscillatory signal and these variations scaled linearly with the surface magnetic field. They explained this in terms of changes in the equation of state, since magnetic fields contribute to both energy and pressure. These results were verified by Verner et al. (2006) using Sun-as-a-star data, and so using only low-l modes. However, we note that Christensen-Dalsgaard et al. (2011) found no evidence for a variation in the amplitude of the oscillatory signal.

Further evidence for changes in the solar interior came from Rabello-Soares (2012), who used the frequency differences of intermediate- and high-degree modes observed between the solar cycle 23 maximum and the preceding minimum to infer changes in the relative sound speed squared. Rabello-Soares found that the sound speed is larger at solar maximum than at solar minimum at radii greater than $0.8R_\odot$ and that the difference in sound speed increases with radial position from about $0.8R_\odot$ until about $0.985R_\odot$. Below $0.8R_\odot$ the uncertainties are too large and there is too much spurious variation introduced by the inversion process (Howe and Thompson 1996) to definitively state whether the sound speed is greater at solar minimum or maximum. However, at its peak ($\sim 0.985R_\odot$) the relative difference in sound speed squared is of the order of 10^{-4}, which is statistically significant. Above $0.985R_\odot$ the relative difference in the sound speed decreases with increasing radial position before passing through zero at $\sim 0.997R_\odot$ and then becoming negative. These results can be compared to those obtained using local helioseismology techniques to observe sound speed variations beneath a sunspot. For example, Bogart et al. (2008) also observed the change in sound speed beneath a sunspot goes from positive to negative with increasing radial position. Furthermore the locations at which the change in sign were observed to occur were in good agreement. Rabello-Soares therefore raises the question over how much of the change in sound speed observed in the global modes is due to local active regions, such as sunspots.

Although determining solar cycle variations in the internal structure of the Sun and uncovering evidence for a deep-seated magnetic field have proved to be very difficult far more success has been attained in measuring changes in the flow fields of the solar interior throughout the solar cycle.

5 Seismology of Flow Fields in the Convection Zone

5.1 Observational Signatures of Flows

Large-scale flows advect the acoustic-gravity waves that propagate in the interior of the Sun. This gives rise to a number of measurable effects in helioseismology that can then be used in turn to make inferences about the properties of the flows. Probably the best-known such effect is that of rotational splitting of the frequencies of global modes. The first-order effect on the frequencies is that the frequencies of modes of like n and l but different m are shifted by an amount that is given by m times a mode-weighted average of the internal solar rotation rate within the acoustic cavity of the mode. The effect is primarily due to advection: there is also a Coriolis contribution to the first-order frequency splitting, but for the observed p modes the Coriolis contribution is very small. In the particular case of a star that is rigidly rotating with uniform rotation rate Ω, the frequency shift experienced by a mode would be $m(1 - C_{nl})\Omega$, where the so-called Ledoux factor C_{nl} arises from the Coriolis contribution. The frequency shift due to a more general rotation profile can be written as $m\bar{\Omega}_{nlm}$ where $\bar{\Omega}_{nlm}$ denotes the mode-weighted average of the rotation rate. This is the principal effect of large-scale flows on the frequencies of the global modes. The mode dependence of the averages of the rotation rate is very useful: it enables inferences to be made about the spatial variation of the rotation rate, using inversion techniques.

As discussed in Sect. 1.2 above, in the local helioseismic ring analysis approach, in the absence of flows or any horizontal inhomogeneities, the dispersion relation of the waves is

$\omega = \omega_n(k_h)$, where $k_h \equiv \sqrt{k_x^2 + k_y^2}$ is the magnitude of the horizontal wavenumber vector. Thus the frequencies do not depend on the direction of the horizontal wavenumber, only on its magnitude, and so the rings of power are circular and centered on $(k_x, k_y) = (0, 0)$. In the presence of a flow, the rings are shifted. Suppose that there is a uniform flow of speed U in the x-direction. Then a Doppler shift of the frequencies changes the dispersion relation to become $\omega = \omega_n(k_h) + U K_x$. Provided the flow is weak (in the sense that $U \ll d\omega_n/dk_h$), the rings are still circular but their center is shifted in the $-k_x$ direction by an amount proportional to $U/(d\omega_n/dk_h)$. Thus the direction and magnitude of the ring shift can be used to infer the direction and magnitude of the flow. Analogously to the case of global-mode frequency shifts, for a non-uniform flow (e.g. one that varies with depth) the shifts will be a mode-dependent average and this allows inferences to be made about the spatial variation of subsurface flows.

In time-distance helioseismology, in the absence of flows the travel times in both directions between surface points A and B should be the same. Flows advect the waves and can cause the travel time in one direction (with the flow) to be shorter than the travel time in the opposite direction. In its simplest terms, the effect can be understood by thinking about travel times along ray paths: in one direction the waves travel at a speed $c + \mathbf{U}.\mathbf{s}$, where c is the sound speed, $\mathbf{U}$ is the flow speed and $\mathbf{s}$ is a unit vector along the ray in the direction of travel; whereas waves travelling in the opposite direction travel at a different speed in the presence of the flow because they have the sign of $\mathbf{s}$ reversed. The perturbations to the travel times due to the flow are measured in time-distance helioseismology and, by cross-correlating many different pairs of points, the magnitude and direction of subsurface flows can be inferred.

5.2 Rotation

The rotation rate in much of the solar interior has been inferred from global-mode frequency splittings. A typical result for the mean rotation profile, in this case using a Regularized Least Squares (RLS) inversion technique applied to data obtained with the HMI instrument on board SDO, is shown in Fig. 6. The latitudinal variation of the rotation rate that has long been observed at the surface of the Sun, with the rotation rate decreasing with increasing latitude, is seen to persist through the convection zone (the outer 30 per cent of the Sun). Near the base of the convection zone, there is a transition to what appears to be an essentially uniform rotation rate beneath, so that at the interface there is a region of strong rotational shear which has become known as the tachocline. There is also a region of rotational shear much closer to the Sun's surface, in about the outer five per cent by radius, so the maximum in the rotation rate occurs a few per cent beneath the surface. As remarked by Gilman and Howe (2003), in much of the convection zone the contours of isorotation make an angle with the rotation axis of about 25°, whereas there is a slight tendency for them to align parallel to the rotation axis ("rotation on cylinders") in the near-equatorial region (Howe et al. 2005).

5.3 Observations of Torsional Oscillations

There are temporal variations around the mean rotation profile. The most firmly established of these are the so-called torsional oscillations. At low latitudes, these manifest as weak but coherent bands of faster and slower rotation that start at mid-latitudes as, or even slightly before, sunspots appear at those latitudes during the solar cycle, and migrate equatorwards with the activity bands over a period of a few years. Helioseismology shows that these bands extend in depth at least a third of a way down into the convection zone (Antia and Basu

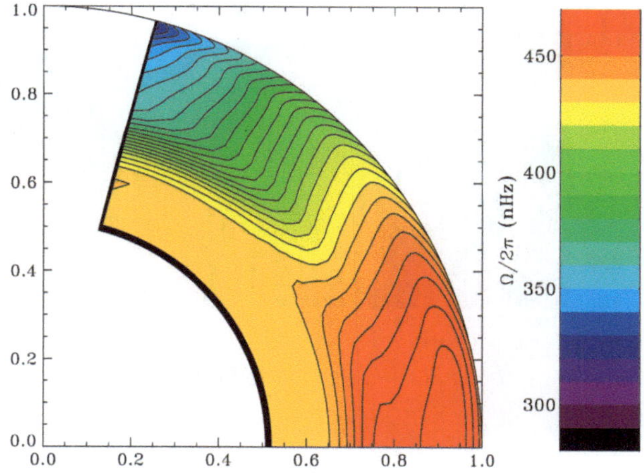

Fig. 6 Solar internal rotation as inferred from HMI observations using a Regularized Least Squares inversion. Contours of isorotation are shown

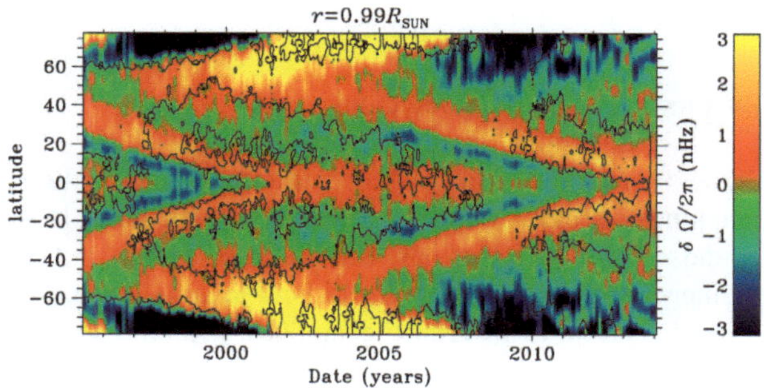

Fig. 7 Zonal flow residuals after a temporal mean at each radius/latitude has been subtracted. The residuals here are shown as functions of time and latitude, for a radial location of $0.99R_\odot$. The RLS inversion method was employed, and the results shown are a merge of inversions of HMI, MDI and GONG data. The black contours are of the photospheric magnetic field strength

2000; Howe et al. 2000, 2006b; Vorontsov et al. 2002). At high latitudes, helioseismology has revealed that there is a poleward-migrating branch of the torsional oscillation (Antia and Basu 2001; Vorontsov et al. 2002), which may extend over the whole depth of the convection zone. The signal is clearest in the near-surface layers: Fig. 7 shows the torsional oscillations at a depth of one per cent of the solar radius beneath the surface. As these results are based on global helioseismology they do not reflect any differences between the flows in northern and southern hemispheres.

The torsional oscillation was first observed in surface Doppler observations from Mount Wilson (Howard and Labonte 1980). These observations continued until very recently, and have been compared with the helioseismic observations by Howe et al. (2006a); when the Doppler observations are symmetrized across the equator the agreement makes it clear that the two techniques are detecting the same phenomenon. The better resolution of the helioseismic measurements at high latitudes makes the poleward-propagating nature of the high-latitude branch more obvious.

A complementary view on torsional oscillations is obtained by looking at the same results but as a function of time and depth at fixed latitude. Four such slices, at different latitudes, are shown in Fig. 8. At the equator and at 15° latitude, there is some hint in the first half of the time period that the torsional oscillation propagates upwards from the middle of the convection zone. At the higher latitudes, there seems to be no propagation in depth. In the second half of the period there is much weaker evidence for upward propagation, although there is still some hint of it at 15° latitude.

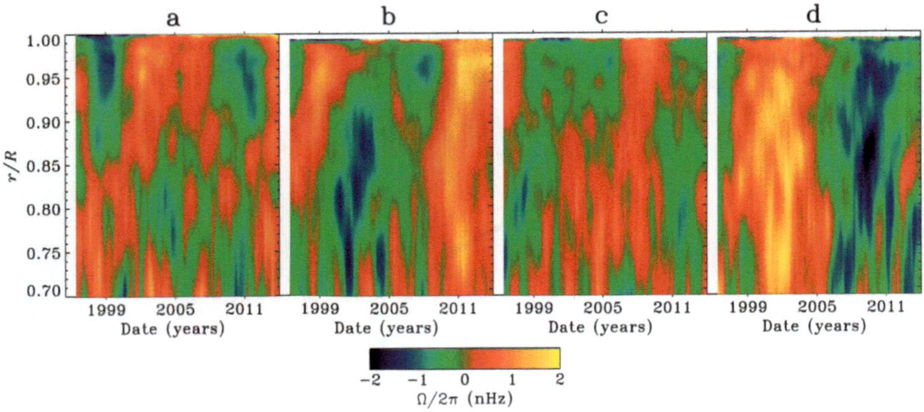

Fig. 8 Zonal flow residuals from Optimally Localized Averaging inversions of MDI and HMI data. A temporal mean at each radius/latitude has been subtracted. The residuals here are shown as functions of time and radial location, for latitudes (from *left* to *right*) 0°, 15°, 30° and 45°

5.4 Flows Around Active Regions

Convection motions manifest themselves at the Sun's surface and in the subsurface regions on a range of scales. Moreover, strong magnetic fields in sunspots and active regions modify those convective motions, and the maintenance and decay of sunspots likely involves an interplay between the magnetic field and the flows. These flows have been studied using various local helioseismic techniques: with time-distance helioseismology (Gizon et al. 2001), ring analysis (Hindman et al. 2009) and helioseismic holography (Braun and Wan 2011). Fig. 9 shows flows around and beneath an active region, obtained from ring analysis data derived from HMI observations. The inversion method employed is a 3-D RLS inversion developed by Featherstone et al. (2011). At the shallowest depths (0.2 Mm), the flows are dominated by the supergranular convection, and that scale is clearly visible. The flow speeds decrease with increasing depth, and the horizontal scale of the convective motions increases. Featherstone et al. also find that strong outflows are typical around sunspots at a depth of 5–6 Mm. Comparable studies with time-distance helioseismology (Gizon et al. 2009) agree roughly with the ring-analysis studies in terms of the magnitude of surface flows near sunspots. They are also in agreement that there are strong outflows beneath the surface, but time-distance helioseismology finds that the flow speeds are higher, by about a factor of two. Resolution differences may account for some of the apparent discrepancy.

5.5 Meridional Circulation: A Shallow Return Flow?

Meridional circulation is the large-scale flow in meridional planes—i.e., flows that are perpendicular to the rotational flow. Meridional circulation in the solar convection zone is an important ingredient in some models of the solar dynamo: in the near-surface regions it transports the remnant flux from old active regions polewards, where the flux is presumed to be subducted and carried down to the bottom of the convection zone, where again a suitably directed meridional circulation may aid the equatorward migration of toroidal magnetic field.

The meridional circulation near the surface of the Sun is only of order 10 m/s, much smaller than the rotation speed and the speeds of convective motions of granules, etc. Hence,

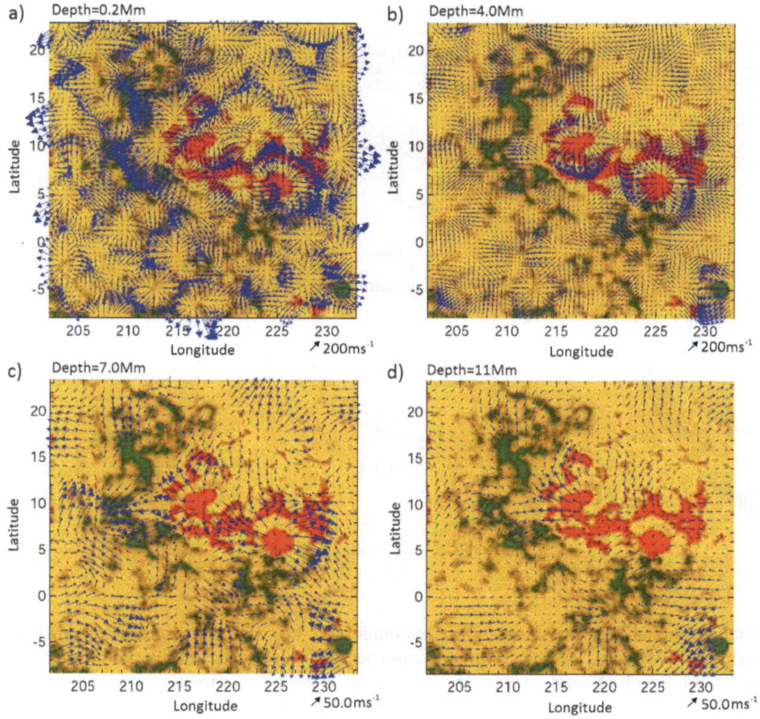

Fig. 9 Flows around an active region in January 2002, obtained by 3-D inversion of ring-analysis data using HMI observations. Flows are shown at depths (**a**) 0.2 Mm, (**b**) 4 Mm, (**c**) 7 Mm and (**d**) 11 Mm. Velocity vectors are shown in *blue*; *red* and *green* show the regions of strong positive and negative polarity photospheric magnetic field, respectively. (Adapted from Featherstone et al. 2011)

although there were pre-helioseismology surface measurements of poleward meridional flow at the surface, these measurements were difficult. In an early application of time-distance helioseismology, Giles et al. (1997) demonstrated that the near sub-surface meridional circulation in both hemispheres is polewards.

Following earlier pioneering attempts by Patron et al. (1995), Haber and colleagues have applied the ring analysis method extensively to obtain robust results regarding the meridional circulation in the outer few percent of the solar interior from the equator to nearly 60° latitude, e.g. Haber et al. (2002). The results show that the meridional circulation is generally poleward in this region. With a lesser degree of certainty, Haber and collaborators find that there are episodes when a submerged counter-flow develops at mid-latitudes.

As well as the possible development of a counter-cell, there are other temporal variations of the meridional circulation in the superficial subsurface layers that appear to be associated spatially and temporally with the torsional oscillations (Zhao and Kosovichev 2004; González Hernández et al. 2010)

Mass conservation demands that, corresponding to the poleward near-surface meridional flow, there must be an equatorward flow at some depth that closes the circulation. The location of this return flow is a key question in determining how and whether the flux-transport solar dynamo model operates. Ever since the work by Giles et al. (1997), there have been attempts that have hinted at a relatively shallow return flow. Mitra-Kraev and Thompson (2007) detected a transition from poleward to equatorward flow at a depth of

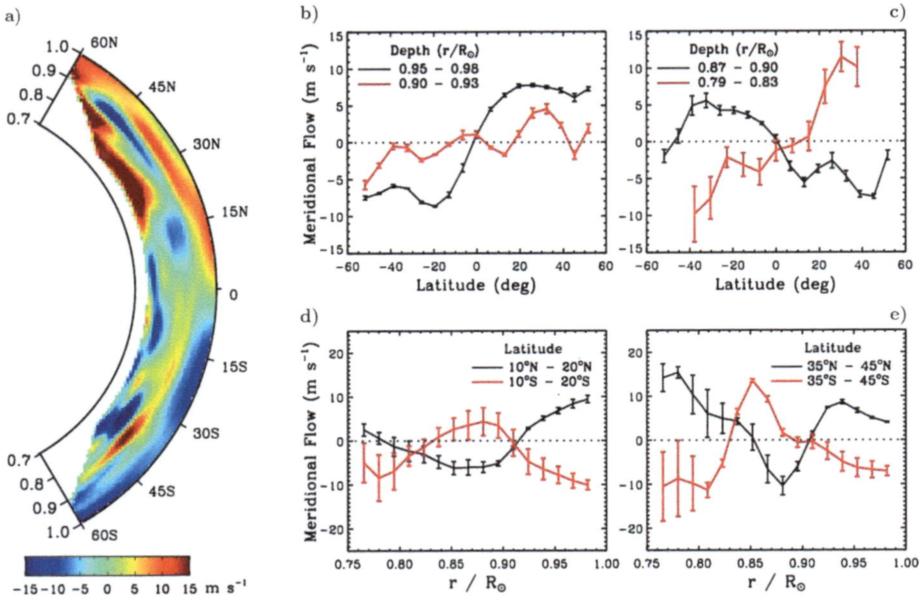

Fig. 10 The meridional flow profile is shown as obtained by inverting the measured acoustic travel times from two years of HMI data. Panel (**a**) shows a cross-section view of the profile with the positive velocity directed northward. Panels (**b**) and (**c**) show the velocity as functions of latitude averaged over several depths. Panels (**d**) and (**e**) show the velocity as functions of depth averaged over different latitudinal bands. From Zhao et al. (2013)

about 35 Mm, though with substantial error bars on the flow speeds. Hathaway (2012), using a cross-correlation method to track supergranules, found that the poleward meridional flow extends to about 50 Mm beneath the solar surface and reported a positive detection of equatorward flow at a depth of about 70 Mm. More recently still, arguably the most convincing helioseismic detection to date of a shallow return flow is the work by Zhao et al. (2013). These authors find that the poleward flow extends to about $0.91 R_\odot$ (about 63 Mm depth) with an equatorward flow between $0.82 R_\odot$ and $0.91 R_\odot$, with perhaps a poleward flow again below that, see panel a) in Fig. 10. However, even with the latest work, a note of caution is warranted. The result depends on the application of a correction for a systematic center-to-limb bias in the travel-time measurements: the correction seems justified from a data-analysis viewpoint, but as yet the origin of the systematic is not understood.

5.6 Hemispheric Asymmetry: Flows and Magnetic Field Distribution

The solar cycle appears to be strongly coupled across the equator as evident in the symmetry of the butterfly diagram. However, a snapshot of the Sun at any given time shows notable differences between the North (N) and South (S). Usually, the hemispheric asymmetry is most notable in the distribution of the surface magnetic field but it is also observed in flows recovered from local helioseismology. So the question arises—by what mechanism are the N and S hemispheres coupled?

Passive diffusion across the geometric equator is often considered the main mechanism of coupling (Charbonneau 2005). Results from dynamo simulations cause us to question magnetic diffusion (including turbulent diffusion) as the main coupling mechanism. Not only do the diffusion values incorporated into numerical models vary widely, the implementation as a function of depth and the effects of diffusion within the models are not well understood. Therefore, interest has been increasing regarding other, more active, hemispheric coupling mechanisms. N-S flows within latitudinally elongated convective cells (aka "banana cells") allow a mixing of electromagnetic flux from one hemisphere into the other (Passos and Charbonneau 2014) and can contribute to hemispheric coupling.

Local helioseismology techniques, such as ring diagram (Hill 1989) and time-distance (Duvall et al. 1993) analysis, are able to determine non-symmetric latitudinal structure in the solar interior. Results from local helioseismology highlight the differences in the hemispheric flows. These analyses have been used to measure distinct hemispheric differences in the meridional flows and zonal flows at a given time and depth in the interior (see Komm et al. 2011, and others). These measured asymmetries provide further quantitative constraints on the dynamo simulations in that the simulations must reproduce hemispheric asymmetries only within the range observed.

Specifically, the extent of hemispheric coupling as determined by surface magnetism is as follows. The N and S polar fields reverse their dominant polarity at distinctly different times, up to 14 months apart in some solar cycles (Norton and Gallagher 2010). The time of peak sunspot production of one hemisphere is usually lagged by the other, meaning the hemispheres experience solar maxima at different times. Recently, the first half of the 2014 calendar year shows the Southern hemisphere producing significantly more large active regions while the Northern hemisphere peaked in sunspot production earlier in 2013. The amplitudes of each hemispheric solar cycle as measured in sunspot number or area are within 20 % of each other and the timing (phase lags) between hemispheres are not more than 25 % of the total cycle length (Norton and Gallagher 2010). For more recent, in-depth analysis, Zolotova et al. (2010) and McIntosh et al. (2013) both analyze hemispheric asymmetry and interestingly enough find that a certain hemisphere appears to lead any given cycle and this lead persists for roughly 40 years on average. That is the surface magnetism characteristic of the hemispheric asymmetry.

The extent of the hemispheric coupling as determined by helioseismology is as follows. Zonal flows are seen as bands of faster and slower E-W flows that appear years prior to the appearance of activity on the solar surface. It is often thought that the flow patterns are caused by enhanced cooling by magnetic fields. Meridional flows are considered in many dynamo models as the crucial ingredient which sets the rate at which the toroidal magnetic band (and sunspots) move equatorward. Komm et al. (2011) determined the zonal and meridional flows as a function of latitude for cycle 23 (years 1996–2010), see Fig. 11. In the right panel of Fig. 11, note how the meridional flow at 10–15 Mm in the Northern hemisphere weakens in 2005 at 35 N latitude (seen as a break in green color) just before the Northern surface magnetic contour disappears in 2006. Similarly, the Southern hemisphere shows this behavior 2 years later in 2007 at 35 S latitude (again, a break in the green color) just before the Southern magnetic contour disappears in 2008. This ~ 2 year hemispheric phase lag observed in both the surface magnetism and the meridional flow is tantalizing.

More recently, Komm et al. (2014) investigated the behavior of the zonal flows as a function of latitude for the time period of 2001–2013 from the surface to a depth of 16 Mm using GONG and HMI. Many hemispheric differences are evident in the zonal flows. For example, see Fig. 12 showing the poleward branch of the zonal flow (at 50 degrees) is 6 $\mathrm{m\,s^{-1}}$ faster in the S than the N at a depth of 10–13Mm during cycle 23. In addition, Zhao et al.

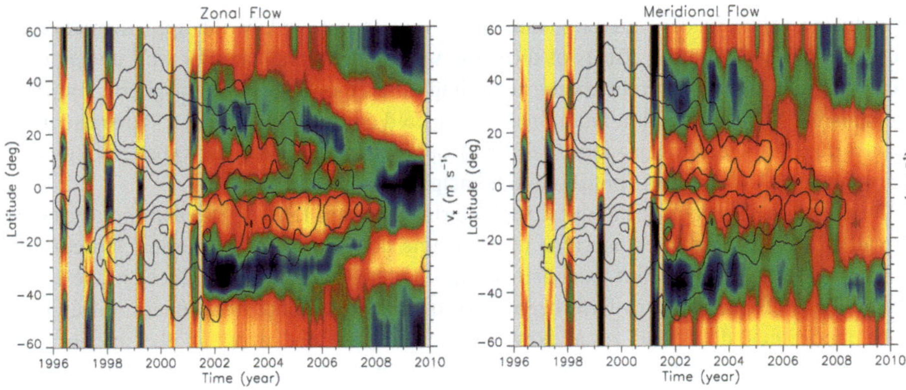

Fig. 11 Komm et al. (2011) recovered zonal (*left*) and meridional (*right*) flows as a function of time and latitude for a depth of 10.2–15.8 Mm

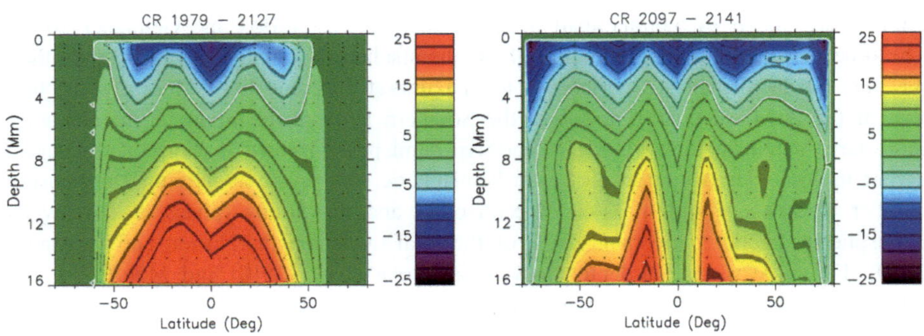

Fig. 12 Komm et al. (2014) recovered zonal flows as a function of latitude and depth. The zonal flow from GONG (*left*) is averaged over 149 Carrington rotations and from HMI (*right*) is averaged over 35 Carrington rotations. The white contour indicates a zero velocity. Errors are less than 0.1 $\mathrm{m\,s^{-1}}$ for GONG and less than 1.0 $\mathrm{m\,s^{-1}}$ for HMI. The poleward branch of the torsional oscillations is stronger in the S by 6 $\mathrm{m\,s^{-1}}$ at a depth of 10–13 Mm during Cycle 23

(2013) detected multiple cells in each hemisphere in the meridional circulation using acoustic travel-time differences. The double-celled profile shows a significant hemispheric asymmetry (see panels (d) and (e) in Fig. 10) in a range of latitudes. The profile asymmetry could be due to a phase lag in the hemispheres: does the meridional profile as a function of depth in the Southern hemisphere in 2013 look like the profile did in the Northern hemisphere two years earlier?

It is possible that a perturbation of meridional flow (presumably by convection since the meridional flow is a weak flow strongly driven by convection) in one hemisphere (but not in the other) sets a phase lag between the migration of activity belts, and hence, the sunspot production, that persists for years. Actively searching for correlated hemispheric asymmetric signatures in flows at depth and magnetic field distributions on the surface may provide insight as to which ingredients of the dynamo set the length and amplitude of the sunspot cycle. For an in-depth discussion of N-S hemispheric asymmetry from an observational perspective as compared to the results from dynamo simulations, see Norton et al. (2014) in this issue.

6 Mean Field Modelling of Solar Torsional Oscillations

Several authors (Durney 2000; Covas et al. 2000; Bushby 2006; Rempel 2006) have developed theoretical models of torsional oscillations, the observational aspects of which are described in Sect. 5.3, by assuming that the torsional oscillations are driven by the Lorentz force of the Sun's cyclically varying magnetic field, which is associated with sunspot cycle. If this is true, then one would expect the torsional oscillations to follow the sunspot cycles. The puzzling fact, however, is that the torsional oscillations of a cycle begin a couple of years before the sunspots of that cycle appear and at a latitude higher than where the first sunspots are subsequently seen. At first sight, this looks like a violation of causality. How does the effect precede the cause? The following section provides a possible explanation using the Nandy-Choudhuri hypothesis (Nandy and Choudhuri 2002) in the framework of flux transport dynamos.

6.1 Features of Solar Torsional Oscillations and Their Possible Explanations

We now summarize some of the other important characteristics of torsional oscillations, which a theoretical model should try to explain.

1. Apart from the weaker equatorward-propagating branch which moves with the sunspot belt after the sunspots start appearing, there is also a stronger poleward-propagating branch at high latitudes. The poleward branch lasts for about 9 years whereas the equatorward branch lasts for about 18 years (e.g. see Fig. 7).
2. The amplitude of the torsional oscillations near the surface is of the order of 4 nHz (20 $\mathrm{m\,s^{-1}}$) when averaged over all latitudes (e.g. see Fig. 7).
3. The torsional oscillations seem to be present throughout the convection zone, though they appear more intermittent and less coherent as we go deeper down into the convection zone (see Figs. 4, 5 and 6 in Howe et al. 2005 and Fig. 8 here).
4. In the equatorward-propagating branch at low latitudes, the torsional oscillations at the surface seem to have a phase lag of about 2 yr compared to the oscillations at the bottom of the convection zone (see Fig. 7 in Howe et al. 2005 and Fig. 8 here). The rate of upward movement appears to be about $0.05R_\odot\,\mathrm{yr}^{-1}$ or 1 $\mathrm{m\,s^{-1}}$. However, we recall from Sect. 5.3 that evidence for upward propagation is weaker in cycle 24 than observed previously.
5. Torsional oscillation contours of constant phase are inclined at 25° to the rotation axis. This is similar to the inclination of contours of constant rotation (see Fig. 6 and Howe et al. 2004b).

Let us now try to understand properties (3) and (4) listed above in some more detail. Property (4) of the torsional oscillations seems to suggest that the bottom of the convection zone is the source of the oscillations, which then propagate upwards. Property (3) then seems puzzling and contrary to common sense. One would expect the oscillations to be more coherent near the source, becoming more diffuse as they move upward further away from the source. However, we know that the magnetic field is highly intermittent within the convection zone and we need to take account of this fact when calculating the Lorentz force due to the magnetic field. Since the convection cells deeper down are expected to have larger sizes, Choudhuri (2003) suggested that the magnetic field within the convection zone would look as shown in Fig. 1 of that paper. In Fig. 1 of Choudhuri the convection granule size is a function of the pressure scale height in the solar convection zone, the intergranular lanes are more intermittent at the bottom of the convection zone and so is the magnetic field which is concentrated into tubes in these lanes. Since the velocity perturbations associated with the torsional oscillations are likely to be concentrated around the magnetic flux tubes, we expect the torsional oscillations to be spatially intermittent at the bottom of the convection zone, as seen in the observational data (property 3). Similarly, since the magnetic field near the surface is less intermittent, a torsional oscillation driven by the Lorentz stress would also appear more coherent there. This may explain the puzzling situation where the torsional oscillations seem to become more coherent as they move further away from the source which is at the footpoints of flux tubes at the bottom of the convection zone. This scenario therefore provides a natural explanation for property (3) of the torsional oscillations, as listed above. Accordingly, let us assume that the torsional oscillation gets initiated in the lower footpoints of the vertical flux tubes, where the Lorentz force builds up due to the production of the toroidal magnetic field. This perturbation then propagates upward along the vertical flux tubes at the Alfvén speed. If the magnetic field inside the flux tubes within the solar convection zone (not below it) is estimated as 500 G (see Sect. 3 of Choudhuri 2003), then the Alfvén speed at the bottom of the convection zone is of the order of 300 $\mathrm{cm\,s^{-1}}$ and the Alfvén travel time from the bottom to the top turns out to be exactly of the same order as the phase delay of torsional oscillations at the surface compared to the oscillations at the bottom of the convection zone. This may be an explanation for the property (4). In the next subsection, we shall try to incorporate these ideas into a mean field dynamo model of the solar cycle.

6.2 Numerical Modelling Efforts

We focus on the scenario where the Lorentz and Maxwell stresses on the flows in the solar interior give rise to the torsional oscillations. Poloidal flows, like the meridional circulation, are also affected. This is important because the meridional circulation is believed to set the clock for the 11 yr sunspot cycle. The magnetic feedback on the flows is one of the important non linearities of the solar dynamo. Some authors have proposed that such feedback on differential rotation alone could cause a modulation of solar cycle strengths and lead to grand minima like episodes (Kitchatinov et al. 1999).

Mean field models of the solar dynamo have been in existence for over two decades now and come in two main flavors: the $\alpha\Omega$ interface, and the flux transport dynamo. Depending on the values of parameters like turbulent diffusivity as compared to meridional circulation speed, the dynamo can also be characterized into advectively or diffusively dominated dynamo regimes (Jiang et al. 2007; Yeates et al. 2008). Such a mean field model of the solar dynamo can then be combined with the Reynolds-averaged Navier-Stokes (NS) equation, after including the effects of the Lorentz force and Maxwells stresses, to solve for the torsional oscillations. Several authors (Covas et al. 2000, 2004; Chakraborty et al. 2009) have solved for only the azimuthal component of the mean velocity, v_ϕ along with the dynamo equations while others (Rempel 2007) have solved for the evolution of entropy in addition to all three components of the mean velocity. Torsional oscillation models also differ in relation to the magnetic feedback term appearing in the NS equation. In some cases (Rempel 2007; Chakraborty et al. 2009; Covas et al. 2004) only a Lorentz feedback due to the mean magnetic field (macro-feedback), has been used whereas in other cases (Kitchatinov et al. 1999; Küker et al. 1999) only the Maxwell's stress due to the fluctuating magnetic field (micro-feedback) has been modeled. This is achieved using Λ-effect formulation of the non-diffusive component of the Reynolds stresses (Kitchatinov and Rüdiger 1993) and quenching the Λ coefficients using an algebraic dependence on the mean magnetic field. The saturation of the mean magnetic field in the above case has been achieved using two different mechanisms: In the first mechanism the Λ coefficients are quenched along with either the kinetic helicity or the α effect (achieved through the presence of super equipartition fields). The second mechanism was employed by Covas et al. (2004) who demonstrated that saturation of the solar dynamo can be obtained by means of the Lorentz feedback in the ϕ component of the NS equation without requiring any explicit α-quenching.

One of the popular models used for an explanation of the sunspot cycle is the flux transport dynamo, in which the meridional circulation carries the toroidal field produced from differential rotation in the tachocline equatorward and carries the poloidal field produced by the Babcock–Leighton mechanism at the surface poleward (Wang and Sheeley 1991; Choudhuri et al. 1995). Since the differential rotation is stronger at higher latitudes in the tachocline than at lower latitudes, the inclusion of solar-like rotation tends to produce a strong toroidal field at high latitudes rather than at the latitudes where sunspots are seen (Dikpati and Charbonneau 1999; Küker et al. 2001). Nandy and Choudhuri (2002) proposed a hypothesis to overcome this difficulty. According to their hypothesis, the meridional circulation penetrates slightly below the bottom of the convection zone and the strong toroidal field produced at the high-latitude tachocline is pushed by this meridional circulation into stable layers below the convection zone where magnetic buoyancy is suppressed and sunspots are not formed. Only when the toroidal field is brought into the convection zone by the meridional circulation rising at lower latitudes, does magnetic buoyancy take over and sunspots finally form. The torsional oscillation signals however cannot be buried below the convection zone by the meridional circulation since they can be transmitted out to the surface by Alfvén waves along vertical magnetic field lines, which intermittently thread the convection zone. It may be noted that there is a controversy at the present time as to whether the meridional circulation can penetrate below the convection zone since arguments having been advanced both against (Gilman and Miesch 2004) and for it (Garaud and Brummel 2008). The detailed dynamo model of Chatterjee et al. (2004) was based on this Nandy-Choudhuri hypothesis, which provided the correct sunspot emergence latitudes and phase relation between the low-latitude toroidal field and the polar fields at the surface.

Another conjecture proposed by Chakraborty et al. (2009) was that the torsional oscillations are initiated in the lower layers of the solar convection zone where toroidal flux tubes are formed due to differential rotation and are propagated upwards by Alfvén waves. The torsional oscillation signal therefore reaches the solar surface much ahead of the sunspot-forming toroidal magnetic field which is still buried by the downward meridional flow in the stable layers. These authors tried to model torsional oscillations using the Chatterjee et al. (2004) dynamo model after coupling it to the equation for the mean rotational velocity incorporating a very simple but insightful averaging of the Lorentz force feedback term. The ϕ component of the Navier–Stokes equation, is

$$\rho\left\{\frac{\partial v_\phi}{\partial t} + D_u[v_\phi]\right\} = D_\nu[v_\phi] + (\mathbf{F}_L)_\phi, \tag{6}$$

where

$$D_u[v_\phi] = \frac{v_r + v_{\text{alf}}}{r}\frac{\partial}{\partial r}(r v_\phi) + \frac{v_\theta}{r\sin\theta}\frac{\partial}{\partial\theta}(\sin\theta\, v_\phi) \tag{7}$$

is the term corresponding to advection by the meridional circulation (v_r, v_θ), and v_{alf} is a constant upward velocity of 300 cm s^{-1}, estimated in Sect. 6.1, to account for the upward transport by Alfvén waves when solving our basic equation Eq. (6). Note that this additional v_{alf} does not represent any actual mass motion and does not have to satisfy the continuity equation, unlike the meridional circulation.

$$D_\nu[v_\phi] = \frac{1}{r^3}\frac{\partial}{\partial r}\left[\nu\rho r^4 \frac{\partial}{\partial r}\left(\frac{v_\phi}{r}\right)\right] + \frac{1}{r^2\sin^2\theta}\frac{\partial}{\partial\theta}\left[\nu\rho\sin^3\theta\frac{\partial}{\partial\theta}\left(\frac{v_\phi}{\sin\theta}\right)\right] \tag{8}$$

is the diffusion term, and $(\mathbf{F}_L)_\phi$ is the ϕ component of the Lorentz force. The kinematic viscosity ν is primarily due to turbulence within the convection zone and is expected to be equal to the magnetic diffusivity. If the magnetic field in our model is assumed to have the standard form

$$\mathbf{B} = B(r,\theta,t)\mathbf{e}_\phi + \nabla \times \left[A(r,\theta,t)\mathbf{e}_\phi\right], \tag{9}$$

then the Lorentz force is given by the Jacobian

$$4\pi(\mathbf{F}_L)_\phi = \frac{1}{s^3} J\left(\frac{sB_\phi, sA}{r,\theta}\right), \tag{10}$$

where $s = r\sin\theta$. We, however, have to take some special care in averaging the term in Eq. (10), since this is the only term in our equations which is quadratic in the basic variables (A, B, v_ϕ) and has to be averaged differently from all the other terms due to intermittency of the magnetic field in the convection zone. The ϕ component of the Lorentz force primarily comes from the radial derivative of the magnetic stress $B_r B_\phi/4\pi$. This stress arises when B_r is stretched by differential rotation to produce B_ϕ and should be non-zero only inside the flux tubes. We assume that B_r, B_ϕ are the mean field values, whereas $(B_r)_{\text{ft}}$, $(B_\phi)_{\text{ft}}$ are the values of these quantities inside flux tubes. If f is the filling factor, which is essentially the fractional volume occupied by flux tubes, then we have $B_r = f(B_r)_{\text{ft}}$ and $B_\phi = f(B_\phi)_{\text{ft}}$, on assuming the same filling factor for both components for the sake of simplicity. It is easy to see that the mean Lorentz stress would be

$$f\frac{(B_r)_{\text{ft}}(B_\phi)_{\text{ft}}}{4\pi} = \frac{B_r B_\phi}{4\pi f}. \tag{11}$$

This suggests that the correct mean field expression for $(\mathbf{F}_L)_\phi$ should be given by the Eq. (10) divided by f. As pointed out by Chatterjee et al. (2004), the only non-linearity in our equations comes from the critical magnetic field B_c above which the toroidal field at the bottom of the convection zone is supposed to be unstable due to magnetic buoyancy. Jiang et al. (2007) found that we have to take $B_c = 108$ G (which is the critical value of the mean toroidal field and not the toroidal field inside flux tubes) to ensure that the poloidal field at the surface has correct values. Once the amplitude of the magnetic field gets fixed in this manner, we find that the amplitude of the torsional oscillations matches observational values only for a particular value of the filling factor f. The theoretical model of torsional oscillations proposed by Chakraborty et al. (2009) thus allows us to infer the filling factor of the magnetic field in the lower layers of the convection zone. Even though the conjecture of Chakraborty et al. looks very elegant, upon actual calculation they obtained a large phase lag of ~ 6 years instead of the observed 2 years between the onset of the low-latitude branch of the torsional oscillation and the first appearance of sunspots of the cycle. However, the radial dependence of the torsional signal was very close to the observed patterns and is shown in Fig. 13. It is clear in the plot of the torsional oscillation at a latitude of 20° (left panel of Fig. 13) that the Lorentz force is concentrated in the tachocline at $0.7R_\odot$, where the low-latitude torsional oscillations are launched to propagate upward. The plot for latitude 20° shows that the amplitude of the torsional oscillations becomes larger near the surface due to the perturbations propagating into regions of lower density, which is consistent with observational data. The physics of the high-latitude branch (right panel of Fig. 13) is, however, very different, with the Lorentz force contours indicating a downward propagation and not a particularly strong concentration at the tachocline. As the poloidal field sinks with the downward meridional circulation at the high latitudes, the latitudinal shear $d\Omega/d\theta$ in the convection zone acts on it to create the toroidal component and thereby the Lorentz stress. With the downward advection of the poloidal field, the region of Lorentz stress moves downward.

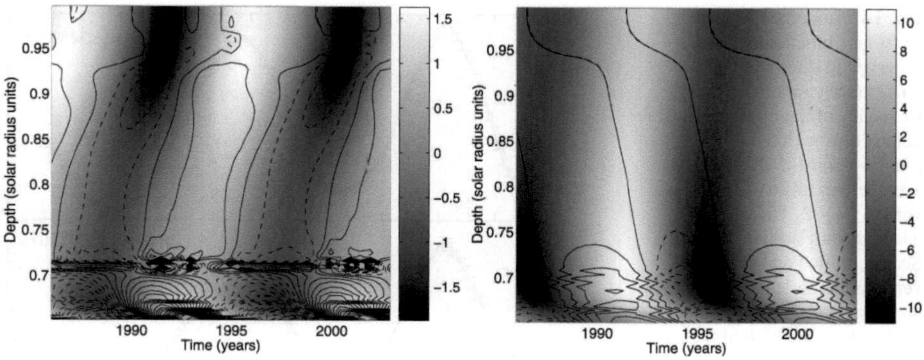

Fig. 13 Theoretical torsional oscillations (v_ϕ in $\mathrm{m\,s^{-1}}$) as functions of depth and time at latitudes 20° (*left*) and 70° (*right*). The plot for latitude 20° compares very well with Fig. 4(D) of Vorontsov et al. (2002), Fig. 7 (Howe et al. 2005), or Fig. 8 here. The *contours* indicate the Lorentz force $(\mathbf{F}_L)_\phi$, the *solid* and *dashed lines* indicating positive and negative values

Yet another baffling observation which has eluded modelers is the systematic deviation of the isolines of constant phase of the torsional oscillation from the Taylor-Proudman state, where the fluid velocity should be uniform along contours parallel to the rotation axis, as evident from Fig. 3B of Vorontsov et al. (2002). Howe et al. (2004b) first pointed out that lines of constant phase are inclined at 25° to the rotation axis, similar to the isorotation contours. Spruit (2003) suggested a thermal origin for the low-latitude branch of the torsional oscillation due to enhanced radiative losses in the active region belts. Rempel (2007) showed that only a localized thermal forcing can lead to such deviation from the Taylor-Proudman state. However since the thermal forcing is only concentrated near the surface, it does not explain the observational fact that torsional oscillations encompass almost the entire convection zone. Earlier, Rempel (2006, see Fig. 10 of that paper) showed that applying a surface cooling function confined to sunspot emerging latitudes can give rise to the low-latitude branch of the torsional oscillation where as the magnetic forcing is responsible for the high-latitude branch. According to our knowledge none of the authors have yet obtained the correct phase relation between the low-latitude band of the torsional oscillation and the sunspot migration. The precedence of the low-latitude branch over the first appearance of sunspots is still a mystery.

Note that, all the discussion above are for mean field models which give rise to regular solar cycle amplitudes for a relatively weak magnetic feedback. An exception is the model of Kitchatinov et al. (1999) which has a strong magnetic feedback and gives rise to significant modulation in the strengths of successive solar cycles. In reality, the solar cycle amplitudes are irregular or stochastic in nature. Naturally this would give rise to an irregular magnetic feedback on the solar flows and the resulting observed torsional oscillations. The irregular magnetic feedback can not only affect the amplitude of torsional oscillations but also the mean rotation rate of the Sun. It may be important to define the mean rotation rate as a solar cycle averaged quantity rather than a very long temporal average (see Howe et al. 2013). Thus, modelling the solar torsional oscillations still remains a very challenging problem.

7 Was Cycle 23 Unusual?

The minimum that preceded cycle 24 was considered to be unusually long and quiet, even for a solar minimum. What can the solar oscillations tell us about the structure of the Sun's

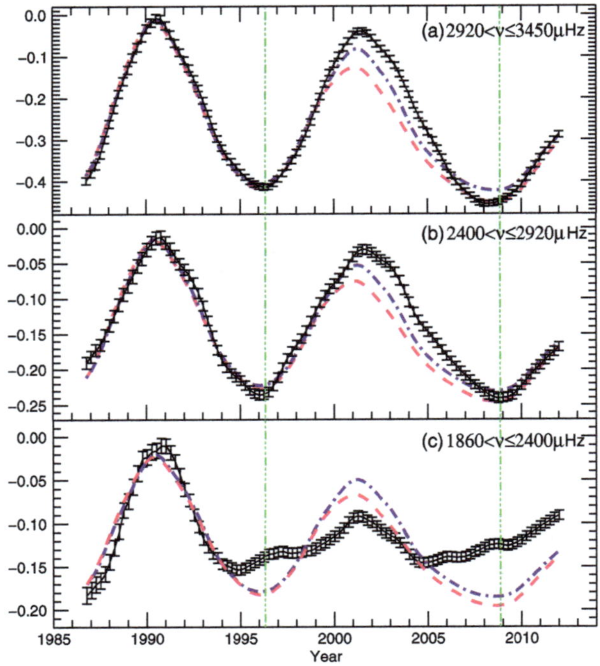

Fig. 14 Smoothed frequency shifts as a function of time observed in three frequency ranges (see captions). The frequency of the oscillations were obtained from 365 d Sun-as-a-star BiSON data. An average was taken over modes with $0 \leq l \leq 2$. This figure is adapted from Basu et al. (2012). Scaled, shifted and smoothed versions of the 10.7 cm flux (*blue dot-dashed line*) and the ISN (*red dashed line*) are plotted for comparison purposes

magnetic field during this time? Basu et al. (2012) examined the frequency dependence of the frequency shifts observed in Sun-as-a-star data during the last two solar cycles. Figure 14 shows the frequency shifts of low-l (≤ 2) modes in three frequency ranges. The shifts have been smoothed to remove the quasi-biennial variation (e.g. Fletcher et al. 2010). Figure 14 shows that the low-frequency modes behave unexpectedly, not just during the recent unusual solar minimum but for the entirety of cycle 23, experiencing little to no frequency shift during this time. Although it is expected that the low-frequency modes experience a smaller shift in frequency than the high-frequency modes (see Sect. 3.2), a comparison with the shifts observed in cycle 22 highlights the discrepancy. More precisely, while the behaviour of the high and intermediate ranges is consistent between cycles 22 and 23, the behaviour of the low-frequency modes changes. This can be explained in terms of the upper turning points of the modes, which as we have said previously, are dependent on the frequency of the mode. If we consider the perturbation responsible for the solar-cycle frequency shifts to be a near-surface magnetic layer. In cycle 22 the upper turning points of the low-frequency modes must lie within the magnetic layer, but in cycle 23 the upper turning points of the low frequency modes must lie beneath the layer, meaning their frequencies are not perturbed by it. These results, therefore, imply a thinning of the magnetic layer (or a change in the upper turning points with respect to the layer). Basu et al. (2012) therefore infer that the magnetic layer must be positioned above $0.9965R_\odot$ in cycle 23.

A change in behaviour has also been observed in the torsional oscillation. In Fig. 7 it appears that the high-latitude poleward-propagating spin-up, which was prominent in the 2000–2006 epoch, is absent in the present cycle. It has been speculated that the high-latitude branch is a precursor of the following solar cycle, and that its absence at the present time may indicate that cycle 25 may be delayed, weak, or non-existent. Another way to look at the data is to plot the inferred rotation at a single location, as a function of time (Fig. 15). It is clear from this representation that the rotation rate at mid- and high-latitudes has been increasing

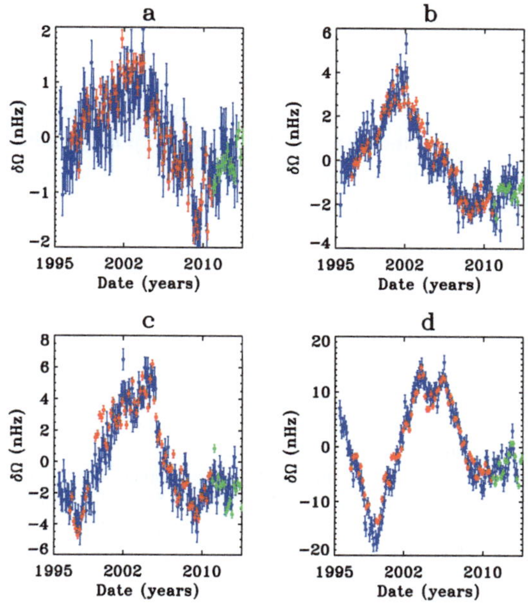

Fig. 15 Rotation rate from Regularized Least Squares inversions (after subtraction of a temporal mean) at latitudes (from *left* to *right*) 45°, 56°, 68° and 79°, at radius $0.99R$, inferred using an RLS inversion. *Colours* indicate the data used: GONG (*blue*), MDI (*red*) and HMI (*green*)

in the past 2–3 years, but more weakly than in the previous cycle. At mid-latitudes, it is also evident that the rotation rate dropped to a lower level than in the previous cycle, so even the weak increase is starting from a lower base (Howe et al. 2013).

8 Summarizing Remarks

The Sun's magnetic field, which is generated in the solar interior, varies on a time scale of 11 yrs from minimum to maximum and back again. Measures of the Sun's surface and atmospheric magnetic field are abundant. In order to really understand the interior of the Sun we use the Sun's natural oscillations. The frequencies, powers and lifetimes of these oscillations are dependent on the strength of the Sun's magnetic field, with the most significant influence on the oscillations arising from a near-surface perturbation. In fact, it has been shown that the change in frequency of the oscillations is tightly correlated with the surface magnetic field, once latitudinal distribution is taken into account. Evidence for a magnetic field deeper within the solar interior was hard to come by. However, there is now tentative evidence for solar cycle variations in the second-ionization zone of helium and at the base of the convection zone. Solar cycle variations in the dynamics of the solar interior have been far more forthcoming. For example, the link between flows and active regions, such as sunspots, have been well studied and reveal strong outflows around sunspots. The torsional oscillation has been shown to exist over a substantial fraction of the convection zone. The observational constraints of the torsional oscillations have proven to be key in recent developments in mean field dynamo models. This is particularly pertinent since one of the main aims in any solar cycle study is to improve our understanding of the mechanism by which the Sun's magnetic field is generated and maintained. Any model of the solar dynamo must explain observational features, including the North-South asymmetry, which is apparent, not only in surface measures of the Sun's magnetic field but in the flows seen in the solar interior. Finally it has been shown that, from a helioseismic standpoint the

Sun's activity appears to be changing from one cycle to the next. We must remember here that we only have two 11-yr solar cycles for comparison. Although the BiSON data do go back to cycle 21, the fill of the data is poor as the full 6-site network was not established until the early 1990s, meaning that it is hard to make helioseismic inferences about this cycle. It is, therefore, difficult to say, purely in helioseismic terms, which cycle is behaving unusually, or indeed if the behaviour is unusual at all: 11 yr only covers half a 22-yr Hale cycle. One can, therefore, only conclude that the helioseismic data imply that the behaviour of the Sun has changed. Hopefully these and other issues surrounding the Sun's magnetic field will become clearer as we continue to observe the Sun through cycle 24 and beyond.

Acknowledgements The paper was stimulated by the workshop "The solar activity cycle: physical causes and consequences". It is a pleasure to thank André Balogh, Hugh Hudson, Kristof Petrovay, Rudolf von Steiger and the International Space Science Institute for financial support, excellent organization and hospitality. This work utilizes Birmingham Solar Oscillations Network data which is run by School of Physics and Astronomy, University of Birmingham. This work utilizes GONG data obtained by the NSO Integrated Synoptic Program (NISP), managed by the National Solar Observatory, which is operated by AURA, Inc. under a cooperative agreement with the National Science Foundation. The data were acquired by instruments operated by the Big Bear Solar Observatory, High Altitude Observatory, Learmonth Solar Observatory, Udaipur Solar Observatory, Instituto de Astrofísica de Canarias, and Cerro Tololo Interamerican Observatory. *SOHO* is a mission of international cooperation between ESA and NASA. The Solar Oscillations Investigation (SOI) involving MDI was supported by NASA grant NNX09AI90G to Stanford University. The National Center for Atmospheric Research is sponsored by the National Science Foundation. HMI data courtesy of NASA and the HMI consortium; HMI is supported by NASA contract NAS5-02139 to Stanford University. NSO/Kitt Peak data used here were produced cooperatively by NSF/NOAO, NASA/GSFC, and NOAA/SEL; SOLIS data are produced cooperatively by NSF/NSO and NASA/LWS. A-MB thanks the Institute of Advanced Study, University of Warwick for their support. RH acknowledges computing support from the National Solar Observatory. A. Norton is supported by NASA Contract NAS5-02139 (HMI) to Stanford University. We thank N. Featherstone and R. Komm for their help.

References

M. Anguera Gubau, P.L. Palle, F. Perez Hernandez, C. Regulo, T. Roca Cortes, The low L solar p-mode spectrum at maximum and minimum solar activity. Astron. Astrophys. **255**, 363–372 (1992)

H.M. Antia, S. Basu, Temporal variations of the rotation rate in the solar interior. Astrophys. J. **541**, 442–448 (2000). doi:10.1086/309421

H.M. Antia, S. Basu, Temporal variations of the solar rotation rate at high latitudes. Astrophys. J. Lett. **559**, 67–70 (2001). doi:10.1086/323701

H.M. Antia, S.M. Chitre, M.J. Thompson, The Sun's acoustic asphericity and magnetic fields in the solar convection zone. Astron. Astrophys. **360**, 335–344 (2000)

H.M. Antia, S. Basu, F. Hill, R. Howe, R.W. Komm, J. Schou, Solar-cycle variation of the sound-speed asphericity from GONG and MDI data 1995–2000. Mon. Not. R. Astron. Soc. **327**, 1029–1040 (2001). doi:10.1046/j.1365-8711.2001.04819.x

T. Appourchaux, Results from the luminosity oscillations imager on board SOHO: low-degree p-mode parameters for a 4-year data set, in *SOHO 10/GONG 2000 Workshop: Helio- and Asteroseismology at the Dawn of the Millennium*, ed. by A. Wilson, P.L. Pallé. ESA Special Publication, vol. 464 (2001), pp. 71–74

C.S. Baldner, S. Basu, Solar cycle related changes at the base of the convection zone. Astrophys. J. **686**, 1349–1361 (2008). doi:10.1086/591514

C.S. Baldner, T.P. Larson, S. Basu, Solar-cycle related changes at the base of the convection zone, in *Solar-Stellar Dynamos as Revealed by Helio- and Asteroseismology: GONG 2008/SOHO 21*, ed. by M. Dikpati, T. Arentoft, I. González Hernández, C. Lindsey, F. Hill. Astronomical Society of the Pacific Conference Series, vol. 416 (2009), p. 477

N.J. Balmforth, D.O. Gough, W.J. Merryfield, Structural changes to the Sun through the solar cycle. Mon. Not. R. Astron. Soc. **278**, 437–448 (1996)

S. Basu, Seismology of the base of the solar convection zone. Mon. Not. R. Astron. Soc. **288**, 572–584 (1997)

S. Basu, What does helioseismology tell us about solar cycle related structural changes in the Sun? in *From Solar Min to Max: Half a Solar Cycle with SOHO*, ed. by A. Wilson. ESA Special Publication, vol. 508 (2002), pp. 7–14

S. Basu, H.M. Antia, Possible solar cycle variations in the convection zone. Sol. Phys. **192**, 449–458 (2000). doi:10.1023/A:1005232901528

S. Basu, H.M. Antia, A study of possible temporal and latitudinal variations in the properties of the solar tachocline. Mon. Not. R. Astron. Soc. **324**, 498–508 (2001). doi:10.1046/j.1365-8711.2001.04364.x

S. Basu, H.M. Antia, Temporal variations of solar structure, in *GONG+ 2002. Local and Global Helioseismology: The Present and Future*, ed. by H. Sawaya-Lacoste. ESA Special Publication, vol. 517 (2003), pp. 231–234

S. Basu, A. Mandel, Does solar structure vary with solar magnetic activity? Astrophys. J. Lett. **617**, 155–158 (2004). doi:10.1086/427435

S. Basu, J. Schou, Does the tachocline show solar cycle related changes? Sol. Phys. **192**, 481–486 (2000). doi:10.1023/A:1005231509224

S. Basu, H.M. Antia, D. Narasimha, Helioseismic measurement of the extent of overshoot below the solar convection zone. Mon. Not. R. Astron. Soc. **267**, 209 (1994)

S. Basu, J. Christensen-Dalsgaard, R. Howe, J. Schou, M.J. Thompson, F. Hill, R. Komm, A comparison of solar p-mode parameters from MDI and GONG: mode frequencies and structure inversions. Astrophys. J. **591**, 432–445 (2003). doi:10.1086/375331

S. Basu, A.-M. Broomhall, W.J. Chaplin, Y. Elsworth, Thinning of the Sun's magnetic layer: the peculiar solar minimum could have been predicted. Astrophys. J. **758**, 43 (2012). doi:10.1088/0004-637X/758/1/43

F. Berrilli, I. Ermolli, A. Florio, E. Pietropaolo, Average properties and temporal variations of the geometry of solar network cells. Astron. Astrophys. **344**, 965–972 (1999)

R.S. Bogart, S. Basu, M.C. Rabello-Soares, H.M. Antia, Probing the subsurface structures of active regions with ring-diagram analysis. Sol. Phys. **251**, 439–451 (2008). doi:10.1007/s11207-008-9213-9

D.C. Braun, K. Wan, Properties of near-surface flows around active regions from helioseismic holography. J. Phys. Conf. Ser. **271**(1), 012007 (2011). doi:10.1088/1742-6596/271/1/012007

A. Broomhall, W.J. Chaplin, Y. Elsworth, S.T. Fletcher, R. New, Is the current lack of solar activity only skin deep? Astrophys. J. Lett. **700**, 162–165 (2009). doi:10.1088/0004-637X/700/2/L162

T.M. Brown, T.J. Bogdan, B.W. Lites, J.H. Thomas, Localized sources of propagating acoustic waves in the solar photosphere. Astrophys. J. Lett. **394**, 65–68 (1992). doi:10.1086/186473

O. Burtseva, F. Hill, S. Kholikov, D. Chou, Lifetimes of high-degree p modes in the quiet and active sun. Sol. Phys. **258**, 1–11 (2009). doi:10.1007/s11207-009-9399-5

P.J. Bushby, Zonal flows and grand minima in a solar dynamo model. Mon. Not. R. Astron. Soc. **371**, 772 (2006)

P.S. Cally, Effects of weak-to-moderate vertical magnetic fields on solar f- and p-modes. Astrophys. J. **451**, 372 (1995). doi:10.1086/176226

S. Chakraborty, A.R. Choudhuri, P. Chatterjee, Why does the Sun's torsional oscillation begin before the sunspot cycle? Phys. Rev. Lett. **102**(4), 041102 (2009). doi:10.1103/PhysRevLett.102.041102

W.J. Chaplin, Y. Elsworth, R. Howe, G.R. Isaak, C.P. McLeod, B.A. Miller, H.B. van der Raay, S.J. Wheeler, R. New, BiSON performance. Sol. Phys. **168**, 1–18 (1996). doi:10.1007/BF00145821

W.J. Chaplin, Y. Elsworth, G.R. Isaak, R. Lines, C.P. McLeod, B.A. Miller, R. New, An analysis of solar p-mode frequencies extracted from BiSON data: 1991–1996. Mon. Not. R. Astron. Soc. **300**, 1077–1090 (1998)

W.J. Chaplin, Y. Elsworth, G.R. Isaak, B.A. Miller, R. New, Variations in the excitation and damping of low-l solar p modes over the solar activity cycle. Mon. Not. R. Astron. Soc. **313**, 32–42 (2000). doi:10.1046/j.1365-8711.2000.03176.x

W.J. Chaplin, T. Appourchaux, Y. Elsworth, G.R. Isaak, R. New, The phenomenology of solar-cycle-induced acoustic eigenfrequency variations: a comparative and complementary analysis of GONG, BiSON and VIRGO/LOI data. Mon. Not. R. Astron. Soc. **324**, 910–916 (2001). doi:10.1046/j.1365-8711.2001.04357.x

W.J. Chaplin, Y. Elsworth, G.R. Isaak, B.A. Miller, R. New, S. Thiery, P. Boumier, A.H. Gabriel, Observation of, and temporal variations in, solar p-mode multiplet frequency asymmetries at $l = 2$. Mon. Not. R. Astron. Soc. **343**, 343–352 (2003). doi:10.1046/j.1365-8711.2003.06686.x

W.J. Chaplin, T. Appourchaux, Y. Elsworth, G.R. Isaak, B.A. Miller, R. New, On comparing estimates of low-l solar p-mode frequencies from Sun-as-a-star and resolved observations. Astron. Astrophys. **424**, 713–717 (2004a). doi:10.1051/0004-6361:20040501

W.J. Chaplin, Y. Elsworth, G.R. Isaak, B.A. Miller, R. New, The solar cycle as seen by low-l p-mode frequencies: comparison with global and decomposed activity proxies. Mon. Not. R. Astron. Soc. **352**, 1102–1108 (2004b). doi:10.1111/j.1365-2966.2004.07998.x

W.J. Chaplin, Y. Elsworth, B.A. Miller, G.A. Verner, R. New, Solar p-mode frequencies over three solar cycles. Astrophys. J. **659**, 1749–1760 (2007). doi:10.1086/512543

P. Charbonneau, A Maunder minimum scenario based on cross-hemispheric coupling and intermittency. Sol. Phys. **229**, 345–358 (2005). doi:10.1007/s11207-005-8150-0

P. Charbonneau, Dynamo models of the solar cycle. Living Rev. Sol. Phys. **7**, 3 (2010). doi:10.12942/lrsp-2010-3

P. Chatterjee, D. Nandy, A.R. Choudhuri, Full-sphere simulations of a circulation-dominated solar dynamo: exploring the parity issue. Astron. Astrophys. **427**, 1019–1030 (2004)

D.-Y. Chou, A. Serebryanskiy, In search of the solar cycle variations of p-mode frequencies generated by perturbations in the solar interior. Astrophys. J. **624**, 420–427 (2005). doi:10.1086/428925

A.R. Choudhuri, On the connection between mean field dynamo theory and flux tubes. Sol. Phys. **215**, 31–55 (2003)

A.R. Choudhuri, M. Schussler, M. Dikpati, The solar dynamo with meridional circulation. Astron. Astrophys. **303**, 29 (1995)

J. Christensen-Dalsgaard, G. Berthomieu, Theory of solar oscillations, in *Solar Interior and Atmosphere* (1991), pp. 401–478

J. Christensen-Dalsgaard, M.J.P.F.G. Monteiro, M. Rempel, M.J. Thompson, A more realistic representation of overshoot at the base of the solar convective envelope as seen by helioseismology. Mon. Not. R. Astron. Soc. **414**, 1158–1174 (2011). doi:10.1111/j.1365-2966.2011.18460.x

E. Covas, R. Tavakol, D. Moss, A. Tworkowski, Torsional oscillations in the solar convection zone. Astron. Astrophys. **360**, 21–24 (2000)

E. Covas, D. Moss, R. Tavakol, The influence of density stratification and multiple nonlinearities on solar torsional oscillations. Astron. Astrophys. **416**, 775–782 (2004). doi:10.1051/0004-6361:20034046

J.P. Cox, *Theory of Stellar Pulsation* (Princeton University Press, Princeton, 1980)

M. Dikpati, P. Charbonneau, A Babcock-Leighton flux transport dynamo with solar-like differential rotation. Astrophys. J. **518**, 508–520 (1999)

B.R. Durney, On the torsional oscillations in Babcock-Leighton solar dynamo models. Sol. Phys. **196**, 1 (2000)

T.L. Duvall Jr., S.M. Jefferies, J.W. Harvey, M.A. Pomerantz, Time-distance helioseismology. Nature **362**, 430–432 (1993). doi:10.1038/362430a0

W.A. Dziembowski, P.R. Goode, Effects of differential rotation on stellar oscillations—a second-order theory. Astrophys. J. **394**, 670–687 (1992). doi:10.1086/171621

W.A. Dziembowski, P.R. Goode, Sources of oscillation frequency increase with rising solar activity. Astrophys. J. **625**, 548–555 (2005). doi:10.1086/429712

A. Eff-Darwich, S.G. Korzennik, S.J. Jiménez-Reyes, F. Pérez Hernández, An upper limit on the temporal variations of the solar interior stratification. Astrophys. J. **580**, 574–578 (2002). doi:10.1086/343063

Y. Elsworth, R. Howe, G.R. Isaak, C.P. McLeod, R. New, Variation of low-order acoustic solar oscillations over the solar cycle. Nature **345**, 322–324 (1990). doi:10.1038/345322a0

Y. Elsworth, R. Howe, G.R. Isaak, C.P. McLeod, B.A. Miller, C.C. Speake, S.J. Wheeler, R. New, The variation in the strength of low-l solar p-modes—1981–92. Mon. Not. R. Astron. Soc. **265**, 888 (1993)

Y. Elsworth, R. Howe, G.R. Isaak, C.P. McLeod, B.A. Miller, R. New, C.C. Speake, S.J. Wheeler, Solar p-mode frequencies and their dependence on solar activity recent results from the BISON network. Astrophys. J. **434**, 801–806 (1994). doi:10.1086/174783

Y. Elsworth, R. Howe, G.R. Isaak, C.P. McLeod, B.A. Miller, R. New, S.J. Wheeler, Techniques used in the analysis of solar oscillations data from the BiSON (University of Birmingham) network. I. Daily calibration. Astron. Astrophys. Suppl. Ser. **113**, 379 (1995)

N.A. Featherstone, B.W. Hindman, M.J. Thompson, Ring-analysis flow measurements of sunspot outflows. J. Phys. Conf. Ser. **271**(1), 012002 (2011). doi:10.1088/1742-6596/271/1/012002

S.T. Fletcher, A. Broomhall, D. Salabert, S. Basu, W.J. Chaplin, Y. Elsworth, R.A. García, R. New, A seismic signature of a second dynamo? Astrophys. J. Lett. **718**, 19–22 (2010). doi:10.1088/2041-8205/718/1/L19

C. Foullon, B. Roberts, The influence of internal magnetic layers on the frequencies of solar p-modes. Astron. Astrophys. **439**, 713–726 (2005). doi:10.1051/0004-6361:20041910

C. Fröhlich, J. Romero, H. Roth, C. Wehrli, B.N. Andersen, T. Appourchaux, V. Domingo, U. Telljohann, G. Berthomieu, P. Delache, J. Provost, T. Toutain, D.A. Crommelynck, A. Chevalier, A. Fichot, W. Däppen, D. Gough, T. Hoeksema, A. Jiménez, M.F. Gómez, J.M. Herreros, T.R. Cortés, A.R. Jones, J.M. Pap, R.C. Willson, VIRGO: experiment for helioseismology and solar irradiance monitoring. Sol. Phys. **162**, 101–128 (1995). doi:10.1007/BF00733428

A.H. Gabriel, G. Grec, J. Charra, J. Robillot, T. Roca Cortés, S. Turck-Chièze, R. Bocchia, P. Boumier, M. Cantin, E. Cespédes, B. Cougrand, J. Crétolle, L. Damé, M. Decaudin, P. Delache, N. Denis, R. Duc, H. Dzitko, E. Fossat, J. Fourmond, R.A. García, D. Gough, C. Grivel, J.M. Herreros, H. Lagardère, J.

Moalic, P.L. Pallé, N. Pétrou, M. Sanchez, R. Ulrich, H.B. van der Raay, Global oscillations at low frequency from the SOHO mission (GOLF). Sol. Phys. **162**, 61–99 (1995). doi:10.1007/BF00733427

P. Garaud, N.H. Brummel, On the penetration of meridional circulation below the solar convection zone. Astrophys. J. **674**, 498–510 (2008)

P.M. Giles, T.L. Duvall, P.H. Scherrer, R.S. Bogart, A subsurface flow of material from the Sun's equator to its poles. Nature **390**, 52–54 (1997). doi:10.1038/36294

P.A. Gilman, R. Howe, Meridional motion and the slope of isorotation contours, in *GONG+ 2002. Local and Global Helioseismology: the Present and Future*, ed. by H. Sawaya-Lacoste. ESA Special Publication, vol. 517 (2003), pp. 283–285

P.A. Gilman, M.S. Miesch, Limits to penetration of meridional circulation below the solar convection zone. Astrophys. J. **611**, 568–574 (2004)

L. Gizon, T.L. Duvall Jr., R.M. Larsen, Probing surface flows and magnetic activity with time-distance helioseismology, in *Recent Insights into the Physics of the Sun and Heliosphere: Highlights from SOHO and Other Space Missions*, ed. by P. Brekke, B. Fleck, J.B. Gurman. IAU Symposium, vol. 203 (2001), p. 189

L. Gizon, H. Schunker, C.S. Baldner, S. Basu, A.C. Birch, R.S. Bogart, D.C. Braun, R. Cameron, T.L. Duvall, S.M. Hanasoge, J. Jackiewicz, M. Roth, T. Stahn, M.J. Thompson, S. Zharkov, Helioseismology of sunspots: a case study of NOAA region 9787. Space Sci. Rev. **144**, 249–273 (2009). doi:10.1007/s11214-008-9466-5

P. Goldreich, P. Kumar, The interaction of acoustic radiation with turbulence. Astrophys. J. **326**, 462–478 (1988). doi:10.1086/166108

P. Goldreich, N. Murray, G. Willette, P. Kumar, Implications of solar p-mode frequency shifts. Astrophys. J. **370**, 752–762 (1991). doi:10.1086/169858

I. González Hernández, R. Howe, R. Komm, F. Hill, Meridional circulation during the extended solar minimum: another component of the torsional oscillation? Astrophys. J. Lett. **713**, 16–20 (2010). doi:10.1088/2041-8205/713/1/L16

D.O. Gough, Comments on helioseismic inference, in *Progress of Seismology of the Sun and Stars*, ed. by Y. Osaki, H. Shibahashi. Lecture Notes in Physics, vol. 367 (Springer, Berlin, 1990), p. 283. doi:10.1007/3-540-53091-6

D.O. Gough, M.J. Thompson, Magnetic perturbations to stellar oscillation eigenfrequencies, in *Advances in Helio- and Asteroseismology*, ed. by J. Christensen-Dalsgaard, S. Frandsen. IAU Symposium, vol. 123 (1988a), p. 155

D.O. Gough, M.J. Thompson, On the implications of the symmetric component of the frequency splitting, in *Advances in Helio- and Asteroseismology*, ed. by J. Christensen-Dalsgaard, S. Frandsen. IAU Symposium, vol. 123 (1988b), p. 175

D. Haber, R. Jain, E.G. Zweibel, Absorption/emission of solar p-modes: Michelson Doppler Interferometer/SOHO observations. Astrophys. J. **515**, 832–841 (1999). doi:10.1086/307038

D.A. Haber, B.W. Hindman, J. Toomre, R.S. Bogart, R.M. Larsen, F. Hill, Evolving submerged meridional circulation cells within the upper convection zone revealed by ring-diagram analysis. Astrophys. J. **570**, 855–864 (2002). doi:10.1086/339631

J.W. Harvey, F. Hill, R.P. Hubbard, J.R. Kennedy, J.W. Leibacher, J.A. Pintar, P.A. Gilman, R.W. Noyes, A.M. Title, J. Toomre, R.K. Ulrich, A. Bhatnagar, J.A. Kennewell, W. Marquette, J. Patron, O. Saa, E. Yasukawa, The global oscillation network group (GONG) project. Science **272**, 1284–1286 (1996). doi:10.1126/science.272.5266.1284

D.H. Hathaway, Supergranules as probes of the Sun's meridional circulation. Astrophys. J. **760**, 84 (2012). doi:10.1088/0004-637X/760/1/84

F. Hill, Solar oscillation ring diagrams and large-scale flows. Astrophys. J. Lett. **343**, 69–71 (1989). doi:10.1086/185513

B.W. Hindman, D.A. Haber, J. Toomre, Subsurface circulations within active regions. Astrophys. J. **698**, 1749–1760 (2009). doi:10.1088/0004-637X/698/2/1749

G. Houdek, W.J. Chaplin, T. Appourchaux, J. Christensen-Dalsgaard, W. Däppen, Y. Elsworth, D.O. Gough, G.R. Isaak, R. New, M.C. Rabello-Soares, Changes in convective properties over the solar cycle: effect on p-mode damping rates. Mon. Not. R. Astron. Soc. **327**, 483–487 (2001). doi:10.1046/j.1365-8711.2001.04749.x

R. Howard, B.J. Labonte, The sun is observed to be a torsional oscillator with a period of 11 years. Astrophys. J. Lett. **239**, 33–36 (1980). doi:10.1086/183286

R. Howe, M.J. Thompson, On the use of the error correlation function in helioseismic inversions. Mon. Not. R. Astron. Soc. **281**, 1385 (1996)

R. Howe, R. Komm, F. Hill, Solar cycle changes in GONG p-mode frequencies, 1995–1998. Astrophys. J. **524**, 1084–1095 (1999). doi:10.1086/307851

R. Howe, J. Christensen-Dalsgaard, F. Hill, R.W. Komm, R.M. Larsen, J. Schou, M.J. Thompson, J. Toomre, Deeply penetrating banded zonal flows in the solar convection zone. Astrophys. J. Lett. **533**, 163–166 (2000). doi:10.1086/312623

R. Howe, R.W. Komm, F. Hill, Localizing the solar cycle frequency shifts in global p-modes. Astrophys. J. **580**, 1172–1187 (2002). doi:10.1086/343892

R. Howe, W.J. Chaplin, Y.P. Elsworth, F. Hill, R. Komm, G.R. Isaak, R. New, A comparison of low-degree solar p-mode parameters from BiSON and GONG: underlying values and temporal variations. Astrophys. J. **588**, 1204–1212 (2003). doi:10.1086/374336

R. Howe, R.W. Komm, F. Hill, D.A. Haber, B.W. Hindman, Activity-related changes in local solar acoustic mode parameters from Michelson Doppler Imager and global oscillations network group. Astrophys. J. **608**, 562–579 (2004a). doi:10.1086/392525

R. Howe, M. Rempel, J. Christensen-Dalsgaard, F. Hill, R.W. Komm, J. Schou, M.J. Thompson, How sensitive are rotation inversions to subtle features of the dynamo?, in *ESA SP-559: SOHO 14 Helio- and Asteroseismology: Towards a Golden Future* (2004b), p. 468

R. Howe, J. Christensen-Dalsgaard, F. Hill, R. Komm, J. Schou, M.J. Thompson, Solar convection-zone dynamics, 1995–2004. Astrophys. J. **634**, 1405–1415 (2005). doi:10.1086/497107

R. Howe, R. Komm, F. Hill, R. Ulrich, D.A. Haber, B.W. Hindman, J. Schou, M.J. Thompson, Large-scale zonal flows near the solar surface. Sol. Phys. **235**, 1–15 (2006a). doi:10.1007/s11207-006-0117-2

R. Howe, M. Rempel, J. Christensen-Dalsgaard, F. Hill, R. Komm, R.M. Larsen, J. Schou, M.J. Thompson, Solar convection zone dynamics: how sensitive are inversions to subtle dynamo features? Astrophys. J. **649**, 1155–1168 (2006b). doi:10.1086/506931

R. Howe, F. Hill, R. Komm, A.-M. Broomhall, W.J. Chaplin, Y. Elsworth, Low-degree helioseismology with AIA. J. Phys. Conf. Ser. **271**(1), 012058 (2011b). doi:10.1088/1742-6596/271/1/012058

R. Howe, T.P. Larson, J. Schou, F. Hill, R. Komm, J. Christensen-Dalsgaard, M.J. Thompson, First global rotation inversions of HMI data. J. Phys. Conf. Ser. **271**(1), 012061 (2011a). doi:10.1088/1742-6596/271/1/012061

R. Howe, J. Christensen-Dalsgaard, F. Hill, R. Komm, T.P. Larson, M. Rempel, J. Schou, M.J. Thompson, The high-latitude branch of the solar torsional oscillation in the rising phase of cycle 24. Astrophys. J. Lett. **767**, 20 (2013). doi:10.1088/2041-8205/767/1/L20

R. Jain, D. Haber, Solar p-modes and surface magnetic fields: is there an acoustic emission? MDI/SOHO observations. Astron. Astrophys. **387**, 1092–1099 (2002). doi:10.1051/0004-6361:20020310

R. Jain, B. Roberts, Do p-mode frequency shifts suggest a hotter chromosphere at solar maximum? Astrophys. J. **414**, 898–907 (1993). doi:10.1086/173133

R. Jain, B. Roberts, Discrete high-frequency p-modes. Astrophys. J. **456**, 399 (1996). doi:10.1086/176662

R. Jain, B.W. Hindman, E.G. Zweibel, The influence of magnetism on p-mode surface amplitudes. Astrophys. J. **464**, 476 (1996). doi:10.1086/177337

K. Jain, S.C. Tripathy, F. Hill, Solar activity phases and intermediate-degree mode frequencies. Astrophys. J. **695**, 1567–1576 (2009). doi:10.1088/0004-637X/695/2/1567

R. Jain, S.C. Tripathy, F.T. Watson, L. Fletcher, K. Jain, F. Hill, Variation of solar oscillation frequencies in solar cycle 23 and their relation to sunspot area and number. Astron. Astrophys. **545**, 73 (2012). doi:10.1051/0004-6361/201219876

S.M. Jefferies, T.L. Duvall Jr., J.W. Harvey, M.A. Pomerantz, Helioseismology from the South pole: results from the 1987 campaign, in *Progress of Seismology of the Sun and Stars*, ed. by Y. Osaki, H. Shibahashi. Lecture Notes in Physics, vol. 367 (Springer, Berlin, 1990), p. 135

J. Jiang, P. Chatterjee, A.R. Choudhuri, Solar activity forecast with a dynamo model. Mon. Not. R. Astron. Soc. **381**, 1527 (2007)

A. Jiménez, T. Roca Cortés, S.J. Jiménez-Reyes, Variation of the low-degree solar acoustic mode parameters over the solar cycle. Sol. Phys. **209**, 247–263 (2002)

S.J. Jimenez-Reyes, C. Regulo, P.L. Palle, T. Roca Cortes, Solar activity cycle frequency shifts of low-degree p-modes. Astron. Astrophys. **329**, 1119–1124 (1998)

S.J. Jiménez-Reyes, T. Corbard, P.L. Pallé, T. Roca Cortés, S. Tomczyk, Analysis of the solar cycle and core rotation using 15 years of Mark-I observations: 1984–1999. I. The solar cycle. Astron. Astrophys. **379**, 622–633 (2001). doi:10.1051/0004-6361:20011374

S.J. Jiménez-Reyes, W.J. Chaplin, Y. Elsworth, R.A. García, Tracing the "acoustic" solar cycle: a direct comparison of BiSON and GOLF low-l p-mode variations. Astrophys. J. **604**, 969–976 (2004b). doi:10.1086/381936

S.J. Jiménez-Reyes, R.A. García, W.J. Chaplin, S.G. Korzennik, On the spatial dependence of low-degree solar p-mode frequency shifts from full-disk and resolved-Sun observations. Astrophys. J. Lett. **610**, 65–68 (2004a). doi:10.1086/423174

S.J. Jiménez-Reyes, W.J. Chaplin, Y. Elsworth, R.A. García, R. Howe, H. Socas-Navarro, T. Toutain, On the variation of the peak asymmetry of low-l solar p modes. Astrophys. J. **654**, 1135–1145 (2007). doi:10.1086/509700

L.L. Kitchatinov, G. Rüdiger, Λ-effect and differential rotation in stellar convection zones. Astron. Astrophys. **276**, 96 (1993)

L.L. Kitchatinov, V.V. Pipin, V.I. Makarov, A.G. Tlatov, Solar torsional oscillations and the grand activity cycle. Sol. Phys. **189**, 227–239 (1999)

R.W. Komm, R. Howe, F. Hill, Width and energy of solar p-modes observed by global oscillation network group. Astrophys. J. **543**, 472–485 (2000). doi:10.1086/317101

R. Komm, R. Howe, F. Hill, Localizing width and energy of solar global p-modes. Astrophys. J. **572**, 663–673 (2002). doi:10.1086/340196

R. Komm, R. Howe, F. Hill, I. González Hernández, D. Haber, Solar-cycle variation of zonal and meridional flow. J. Phys. Conf. Ser. **271**(1), 012077 (2011). doi:10.1088/1742-6596/271/1/012077

R. Komm, R. Howe, I. González Hernández, F. Hill, Solar-cycle variation of subsurface zonal flow. Sol. Phys. (2014). doi:10.1007/s11207-014-0490-1

J.R. Kuhn, Helioseismological splitting measurements and the nonspherical solar temperature structure. Astrophys. J. Lett. **331**, 131–134 (1988). doi:10.1086/185251

M. Küker, R. Arlt, G. Rüdiger, The Maunder minimum as due to magnetic Lambda-quenching. Astron. Astrophys. **343**, 977–982 (1999)

M. Küker, G. Rüdiger, M. Schultz, Circulation-dominated solar shell dynamo models with positive alpha-effect. Astron. Astrophys. **374**, 301–308 (2001). doi:10.1051/0004-6361:20010686

J.R. Lemen, A.M. Title, D.J. Akin, P.F. Boerner, C. Chou, J.F. Drake, D.W. Duncan, C.G. Edwards, F.M. Friedlaender, G.F. Heyman, N.E. Hurlburt, N.L. Katz, G.D. Kushner, M. Levay, R.W. Lindgren, D.P. Mathur, E.L. McFeaters, S. Mitchell, R.A. Rehse, C.J. Schrijver, L.A. Springer, R.A. Stern, T.D. Tarbell, J.-P. Wuelser, C.J. Wolfson, C. Yanari, J.A. Bookbinder, P.N. Cheimets, D. Caldwell, E.E. Deluca, R. Gates, L. Golub, S. Park, W.A. Podgorski, R.I. Bush, P.H. Scherrer, M.A. Gummin, P. Smith, G. Auker, P. Jerram, P. Pool, R. Soufli, D.L. Windt, S. Beardsley, M. Clapp, J. Lang, N. Waltham, The Atmospheric Imaging Assembly (AIA) on the Solar Dynamics Observatory (SDO). Sol. Phys. **275**, 17–40 (2012). doi:10.1007/s11207-011-9776-8

K.G. Libbrecht, M.F. Woodard, Solar-cycle effects on solar oscillation frequencies. Nature **345**, 779–782 (1990). doi:10.1038/345779a0

B.W. Lites, O.R. White, D. Packman, Photoelectric observations of propagating sunspot oscillations. Astrophys. J. **253**, 386–392 (1982). doi:10.1086/159642

C.J. Macris, R. Mueller, J. Rosch, T. Roudier, Variation of the mesh of the granular network along the solar cycle, in *Small-Scale Dynamical Processes in Quiet Stellar Atmospheres*, ed. by S.L. Keil (1984), p. 265

S.W. McIntosh, R.J. Leamon, J.B. Gurman, J.-P. Olive, J.W. Cirtain, D.H. Hathaway, J. Burkepile, M. Miesch, R.S. Markel, L. Sitongia, Hemispheric asymmetries of solar photospheric magnetism: radiative, particulate, and heliospheric impacts. Astrophys. J. **765**, 146 (2013). doi:10.1088/0004-637X/765/2/146

U. Mitra-Kraev, M.J. Thompson, Meridional flow profile measurements with SOHO/MDI. Astron. Nachr. **328**, 1009–1012 (2007). doi:10.1002/asna.200710873

F. Moreno-Insertis, S.K. Solanki, Distribution of magnetic flux on the solar surface and low-degree p-modes. Mon. Not. R. Astron. Soc. **313**, 411–422 (2000). doi:10.1046/j.1365-8711.2000.03246.x

R. Muller, Variability of solar granulation. Adv. Space Res. **8**, 159–167 (1988). doi:10.1016/0273-1177(88)90186-X

R. Muller, A. Hanslmeier, M. Saldaña-Muñoz, Variations of the granulation related to the solar cycle and with respect to its position on the solar disk. Astron. Astrophys. **475**, 717–722 (2007). doi:10.1051/0004-6361:20078387

D. Nandy, A.R. Choudhuri, Explaining the latitudinal distribution of sunspots with deep meridional flow. Science **296**, 1671–1673 (2002)

A.A. Norton, J.C. Gallagher, Solar-cycle characteristics examined in separate hemispheres: phase, Gnevyshev gap, and length of minimum. Sol. Phys. **261**, 193–207 (2010). doi:10.1007/s11207-009-9479-6

A.A. Norton, P. Charbonneau, D. Passos, Hemispheric coupling: comparing dynamo simulations and observations. Space Sci. Rev. (2014). doi:10.1007/s11214-014-0100-4

P.L. Pallé, C. Régulo, T. Roca Cortés, Solar cycle induced variations of the low L solar acoustic spectrum. Astron. Astrophys. **224**, 253–258 (1989)

P.L. Pallé, C. Régulo, T. Roca Cortés, The spectrum of solar p-modes and the solar activity cycle, in *Progress of Seismology of the Sun and Stars*, ed. by Y. Osaki, H. Shibahashi. Lecture Notes in Physics, vol. 367 (Springer, Berlin, 1990), p. 129

D. Passos, P. Charbonneau, Characteristics of magnetic solar-like cycles in a 3D MHD simulation of solar convection. Astron. Astrophys. **568**, A113 (2014). doi:10.1051/0004-6361/201423700

J. Patron, F. Hill, E.J. Rhodes Jr., S.G. Korzennik, A. Cacciani, Velocity fields within the solar convection zone: evidence from oscillation ring diagram analysis of Mount Wilson dopplergrams. Astrophys. J. **455**, 746 (1995). doi:10.1086/176620

M.C. Rabello-Soares, Solar-cycle variation of sound speed near the solar surface. Astrophys. J. **745**, 184 (2012). doi:10.1088/0004-637X/745/2/184

M.C. Rabello-Soares, S.G. Korzennik, J. Schou, Variations of the solar acoustic high-degree mode frequencies over solar cycle 23. Adv. Space Res. **41**, 861–867 (2008). doi:10.1016/j.asr.2007.03.014

S.P. Rajaguru, S. Basu, H.M. Antia, Ring diagram analysis of the characteristics of solar oscillation modes in active regions. Astrophys. J. **563**, 410–418 (2001). doi:10.1086/323780

M. Rempel, Flux-transport dynamos with Lorentz force feedback on differential rotation and meridional flow: saturation mechanism and torsional oscillations. Astrophys. J. **647**, 662–675 (2006). doi:10.1086/505170

M. Rempel, Origin of solar torsional oscillations. Astrophys. J. **655**, 651–659 (2007). doi:10.1086/509866

M.H. Ritzwoller, E.M. Lavely, A unified approach to the helioseismic forward and inverse problems of differential rotation. Astrophys. J. **369**, 557–566 (1991). doi:10.1086/169785

B. Roberts, W.R. Campbell, Magnetic field corrections to solar oscillation frequencies. Nature **323**, 603–605 (1986). doi:10.1038/323603a0

R.S. Ronan, K. Cadora, B.J. Labonte, Solar cycle changes in the high frequency spectrum. Sol. Phys. **150**, 389–392 (1994). doi:10.1007/BF00712900

I.W. Roxburgh, S.V. Vorontsov, Seismology of the solar envelope—the base of the convective zone as seen in the phase shift of acoustic waves. Mon. Not. R. Astron. Soc. **268**, 880 (1994)

D. Salabert, S.J. Jiménez-Reyes, Damping and excitation variations of the solar acoustic modes using LOWL observations. Astrophys. J. **650**, 451–460 (2006). doi:10.1086/507177

D. Salabert, R.A. García, P.L. Pallé, S.J. Jiménez-Reyes, The onset of solar cycle 24. What global acoustic modes are telling us. Astron. Astrophys. **504**, 1–4 (2009). doi:10.1051/0004-6361/200912736

P.H. Scherrer, R.S. Bogart, R.I. Bush, J.T. Hoeksema, A.G. Kosovichev, J. Schou, W. Rosenberg, L. Springer, T.D. Tarbell, A. Title, C.J. Wolfson, I. Zayer, MDI Engineering Team, The solar oscillations investigation—Michelson Doppler Imager. Sol. Phys. **162**, 129–188 (1995). doi:10.1007/BF00733429

J. Schou, P.H. Scherrer, R.I. Bush, R. Wachter, S. Couvidat, M.C. Rabello-Soares, R.S. Bogart, J.T. Hoeksema, Y. Liu, T.L. Duvall, D.J. Akin, B.A. Allard, J.W. Miles, R. Rairden, R.A. Shine, T.D. Tarbell, A.M. Title, C.J. Wolfson, D.F. Elmore, A.A. Norton, S. Tomczyk, Design and ground calibration of the Helioseismic and Magnetic Imager (HMI) instrument on the Solar Dynamics Observatory (SDO). Sol. Phys. **275**, 229–259 (2012). doi:10.1007/s11207-011-9842-2

A. Serebryanskiy, D.-Y. Chou, Comparison of solar cycle variations of solar p-mode frequencies from GONG and MDI. Astrophys. J. **633**, 1187–1190 (2005). doi:10.1086/491467

R. Simoniello, D. Salabert, R.A. García, Variation in p-mode power over solar cycle 23 as seen from BiSON and GOLF observations, in *Astronomical Society of the Pacific Conference Series*, ed. by M. Dikpati, T. Arentoft, I. González Hernández, C. Lindsey, F. Hill. Astronomical Society of the Pacific Conference Series, vol. 416 (2009), p. 281

R. Simoniello, W. Finsterle, R.A. García, D. Salabert, A. Jiménez, Y. Elsworth, H. Schunker, Acoustic power absorption and enhancement generated by slow and fast MHD waves. Evidence of solar cycle velocity/intensity amplitude changes consistent with the mode conversion theory. Astron. Astrophys. **516**, 30 (2010). doi:10.1051/0004-6361/200913091

H.C. Spruit, Origin of the torsional oscillation pattern of solar rotation. Sol. Phys. **213**, 1–21 (2003)

M.J. Thompson, Evidence for a thin perturbative layer near the base of the solar convection zone, in *Seismology of the Sun and Sun-Like Stars*, ed. by E.J. Rolfe. ESA Special Publication, vol. 286, 1988, pp. 321–324

T. Toutain, A.G. Kosovichev, Low-degree p-mode parameters: the solar-cycle dependence, in *SOHO 10/GONG 2000 Workshop: Helio- and Asteroseismology at the Dawn of the Millennium*, ed. by A. Wilson, P.L. Pallé. ESA Special Publication, vol. 464 (2001), pp. 123–126

S.C. Tripathy, B. Kumar, K. Jain, A. Bhatnagar, Analysis of hysteresis effect in p-mode frequency shifts and solar activity indices. Sol. Phys. **200**, 3–10 (2001). doi:10.1023/A:1010318428454

G.A. Verner, W.J. Chaplin, Y. Elsworth, BiSON data show change in solar structure with magnetic activity. Astrophys. J. Lett. **640**, 95–98 (2006). doi:10.1086/503101

S.V. Vorontsov, A search of the effects of magnetic field in the solar 5-minute oscillations, in *Advances in Helio- and Asteroseismology*, ed. by J. Christensen-Dalsgaard, S. Frandsen. IAU Symposium, vol. 123 (1988), p. 151

S.V. Vorontsov, J. Christensen-Dalsgaard, J. Schou, V.N. Strakhov, M.J. Thompson, Helioseismic measurement of solar torsional oscillations. Science **296**, 101–103 (2002). doi:10.1126/science.1069190

Y.-M. Wang, N.R. Sheeley, Magnetic flux transport and the sun's dipole moment—new twists to the Babcock-Leighton model. Astrophys. J. **375**, 761–770 (1991)

M.F. Woodard, R.W. Noyes, Change of solar oscillation eigenfrequencies with the solar cycle. Nature **318**, 449 (1985)

D.T. Woods, L.E. Cram, High resolution spectroscopy of the disk chromosphere. VII—Oscillations in plage and quiet Sun regions. Sol. Phys. **69**, 233–238 (1981). doi:10.1007/BF00149991

A.R. Yeates, D. Nandy, D.H. Mackay, Exploring the physical basis of solar cycle predictions: flux transport dynamics and persistence of memory in advection- versus diffusion-dominated solar convection zones. Astrophys. J. **673**, 544–556 (2008). doi:10.1086/524352

J. Zhao, A.G. Kosovichev, Torsional oscillation, meridional flows, and vorticity inferred in the upper convection zone of the sun by time-distance helioseismology. Astrophys. J. **603**, 776–784 (2004). doi:10.1086/381489

J. Zhao, R.S. Bogart, A.G. Kosovichev, T.L. Duvall Jr., T. Hartlep, Detection of equatorward meridional flow and evidence of double-cell meridional circulation inside the Sun. Astrophys. J. Lett. **774**, 29 (2013). doi:10.1088/2041-8205/774/2/L29

N.V. Zolotova, D.I. Ponyavin, R. Arlt, I. Tuominen, Secular variation of hemispheric phase differences in the solar cycle. Astron. Nachr. **331**, 765 (2010). doi:10.1002/asna.201011410

DOI 10.1007/978-1-4939-2584-1_8
Reprinted from *Space Science Reviews* Journal, DOI 10.1007/s11214-014-0088-9

Magnetic Flux Emergence Along the Solar Cycle

B. Schmieder · V. Archontis · E. Pariat

Received: 7 May 2014 / Accepted: 3 August 2014 / Published online: 11 September 2014

Abstract Flux emergence plays an important role along the solar cycle. Magnetic flux emergence builds sunspot groups and solar activity. The sunspot groups contribute to the large scale behaviour of the magnetic field over the 11 year cycle and the reversal of the North and South magnetic polarity every 22 years. The leading polarity of sunspot groups is opposite in the North and South hemispheres and reverses for each new solar cycle. However the hemispheric rule shows the conservation of sign of the magnetic helicity with positive and negative magnetic helicity in the South and North hemispheres, respectively. MHD models of emerging flux have been developed over the past twenty years but have not yet succeeded to reproduce solar observations. The emergence of flux occurs through plasma layers of very high gradients of pressure and changing of modes from a large β to a low β plasma (<1). With the new armada of high spatial and temporal resolution instruments on the ground and in space, emergence of magnetic flux is observed in tremendous detail and followed during their transit through the upper atmosphere. Signatures of flux emergence in the corona depend on the pre-existing magnetic configuration and on the strength of the emerging flux. We review in this paper new and established models as well as the recent observations.

Keywords Sun: sunspot · Sun: magnetic field

1 Introduction

The emergence of magnetic flux in the solar atmosphere is an important process in solar physics. Concentration of magnetic field occurs below the solar surface in the tachocline

B. Schmieder (✉) · E. Pariat
Observatoire de Paris, LESIA, 92190 Meudon, France
e-mail: Brigitte.Schmieder@obspm.fr

E. Pariat
e-mail: Etienne.Pariat@obspm.fr

V. Archontis
School of Mathematics and Statistics, University of St. Andrews, North Haugh, St. Andrews, Fife KY 169SS, UK
e-mail: vasilis@mcs.st-and.ac.uk

region (interface between the radiative region and the convection zone). The differential rotation of the sun transforms part of the global kinetic energy of the solar rotation into magnetic energy by means of solar dynamo processes. It is generally accepted that magnetic flux tubes are transported from the solar interior to the solar surface by buoyancy, forming Ω-shaped loops, and progressively the active regions (Zwaan 1987; Moreno-Insertis 2007; Fan 2009a). An active region is an extended bipolar configuration formed by the coalescence of small emerging flux tubes. Small scale emergence is strongly modulated by convection.

The emerging flux regions occur over a wide range of scales from the nano-emergence (0.01 Mm for a flux fiber) to the large sunspot (radius of the spot umbrae <11 Mm), and magnetic flux contents from 10^{18} Mx to 3×10^{22} Mx (Zwaan 1987). The contrast of the emerging flux compared to the quiet sun in white light is bright for tiny regions and dark as soon as a pore is formed. The excess brightness of small flux tubes (<300 km) is due to the lateral radiative heating from the side walls of the Wilson depression. Convection is always nearly suppressed in kG fields, irrespective of their size. Bigger flux concentrations become dark because the radiative heating only affects a boundary layer comparable to the mean free path of the photons. The interior of bigger structures therefore is not laterally heated and thus appears dark (Spruit 1976).

The emergence of magnetic flux occurs mainly during maximum of the solar activity cycle in the activity belts called in the past “royal zones”. Small ephemeral active regions emerge as plages at the beginning of each solar cycle at high latitudes, and often disappear before transformation to pores. Later in the solar cycle, coalescence of several pores leads to the formation of sunspots. The Coriolis force acts to orientate the bipole of active regions in the east-west direction, the inversion line between the two polarities being perpendicular to the equator (Hale law). At the beginning of the solar cycle bipoles appear at high latitudes and can be oriented in the north-south direction. Progressively as the solar cycle develops, they appear closer to the equator, as shown in the “butterfly diagram”. Close to the equator, the Coriolis force is larger, bipoles are changing in orientation, more rapidly during their emergence than when emerging at higher latitudes. A tilt angle between the axis of the bipolar active region and the east-west direction exists (Jow’s law), and is greater at high latitude than at the equator.

Flux tubes emerge through a complex region that couples the convection zone, where there are no observations, with the strongly stratified upper atmosphere. Helioseismology helps to detect the emergence before it appears in magnetograms (Hartlep et al. 2011; Toriumi et al. 2013; Barnes et al. 2014). This is relevant mainly for large active regions and not for ephemeral active regions. Understanding the crossing of the lowest layers of the solar atmosphere, photosphere and chromosphere is a challenge for theoreticians. In the subsurface layers (less than 3 Mm), the thermodynamics conditions (density, pressure, and temperature) change drastically by several orders of magnitude. Models of emergence of magnetic flux in the convection zone, involving dynamo action and convection, show that a minimum twist is a necessary condition for a flux tube to emerge (Jouve and Brun 2009; Pinto et al. 2011). The problem with such boundary conditions is that only low twists are commonly found in coronal field (1 to 2 turns) (Canou et al. 2009; Guo et al. 2010). Multiple MHD models to understand the emergence just below the solar surface (a few Mm) have been recently developed (Fan 2001; Moreno-Insertis 2007; Archontis and Hood 2012; MacTaggart and Hood 2009a; Leake et al. 2013; Rempel and Cheung 2014) showing that twist in the upper levels is not necessary. In these simulations, the upper part of the tube is emerging while the tube axis remains near the solar surface.

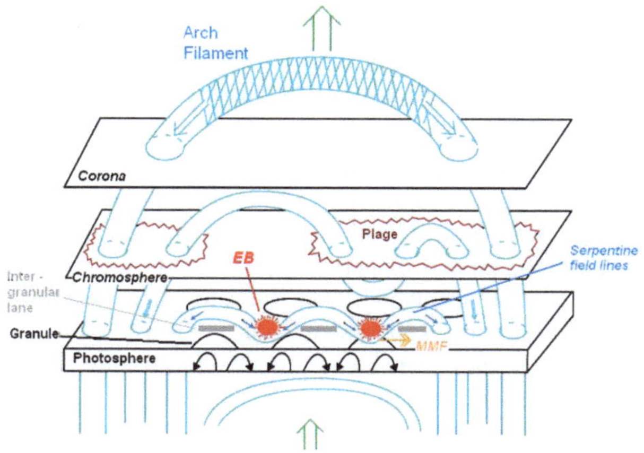

Fig. 1 Sketch of the emergence of magnetic flux through the lower layers of the solar atmosphere and related phenomenae: pores/sunspots (located at the intersection of the flux tubes with the photosphere), moving magnetic features (MMF), plages, Ellerman bombs (EBs), arch filaments systems (AFS), large coronal loops (Schmieder and Pariat 2007)

The plasma in the emerging flux tube transits through layers dominated by gas pressure to layers dominated by magnetic pressure. The ionization degree becomes very weak. Important radiative transfer mechanisms occur in such layers. Full 3D MHD simulations treating also the radiative transfer process have not yet been completely achieved for the passage of an emerging flux through such a complex atmosphere, although some attempts appear really successful (Martínez-Sykora et al. 2009).

The signatures of emerging flux tubes are various and can be summarized in the sketch in Fig. 1 presenting the evolution of an emerging tube from below the solar surface to the corona. The figure shows the relations between the different phenomena associated with flux emergence, such as plages, moving magnetic features (MMF), serpentine field lines, Ellerman bombs (EB) and arch filament system (AFS). The several dispersed pieces of the flux emergence puzzle fit together into a coherent picture that reconciles flux emergence, magnetic reconnection, and heating in the solar atmosphere (Pariat et al. 2004; Schmieder et al. 2004; Valori et al. 2012). The new solar missions (Hinode, SDO) show the dynamic nature of all these structures and particularly the persistent outflows at the edges of the active regions, as they are emerging (Bradshaw et al. 2011; Harra et al. 2012; Démoulin et al. 2013). These outflows could be reconnection jets or pressure driven outflows. The interaction of the emerging flux and the preexisting field lines leads to different solar activity events. Surges, anemones, eruptions, coronal mass ejections and jets are often observed (Shibata et al. 2007; Aulanier 2014). Flux emergence is very often associated with flux cancellation due to the magnetic environment of the emergence and therefore it is difficult to disentangle the role of each of them to produce the activity (Guo et al. 2013; Schmieder et al. 2013). Non-linear or linear magnetic extrapolations exhibit the topology of the region and show the existence of null point, and bald patch configuration where reconnection occurs (Mandrini et al. 2014; Schmieder et al. 2013). Observations have also revealed that emerging magnetic flux is inherent twisted and carries free energy which is released through flares and eruptions of large scales (e.g. coronal mass ejections) (van Driel-Gesztelyi and Culhane 2009).

Reviews on magnetic flux emergence have been presented by various authors in the recent past (e.g. Archontis 2008, 2012; van Driel-Gesztelyi and Culhane 2009). In the present review we focus on MHD simulations and observations of large scale structures (sunspots), and the consequent signatures in the upper atmosphere (Sect. 2), and on the small scale emergence with fragmentation of the magnetic field, and the topology of the region (Sect. 3).

2 Large Scale Emergence from the Solar Surface to the Corona

2.1 Numerical Experiments on Magnetic Flux Emergence

In this section, we present a short overview of the results of numerical experiments, which have been performed to study the process of magnetic flux emergence. We mainly focus upon the large-scale evolution of the emerging flux and the onset of dynamic events associated with this process. In particular, we describe the results of MHD simulations, which have investigated the emergence from the convection zone through the solar surface and across the highly stratified atmosphere in three dimensions.

A typical initial condition in these experiment consists of three major elements: (a) the sub-photospheric magnetic field, (b) the background atmosphere and (c) the magnetic field condition in the corona. For the sub-photospheric magnetic field, the majority of the numerical models have used a horizontal magnetic flux tube, which is usually twisted. In a three-dimensional Cartesian coordinated system, the axial magnetic field component (e.g. B_y) is given, as commonly used, by a simple Gaussian profile

$$B_y = B_0 \exp\left(-r^2/R^2\right), \tag{1}$$

where R is the radius of the tube, r is the radial distance from the tube axis ($r^2 = x^2 + (z + z_0)^2$), z_0 is the initial depth of the tube in the solar interior and B_0 is the initial field strength of the tube. For a uniformly twisted flux tube, the azimuthal component of the field is $B_\phi = \alpha r B_y$, with (constant) α the twist of the field lines about the tube's axis. To make the flux tube buoyant, one usually specifies a particular density deficit (or, equivalently, an entropy distribution) along its axis. In some models, the imposed specific density deficit varies along the axis of the tube, with the result that one part of the tube is buoyant. For instance, the density deficit could be in the form:

$$\Delta\rho = \left[p_t(r)/p(z)\right]\rho(z) \exp\left(-y^2/\lambda^2\right), \tag{2}$$

where p_t is the pressure within the flux tube, p and ρ are the background pressure and density and λ defines the length of the buoyant part of the tube. This differential magnetic buoyancy (Parker 1955) leads to the development of the magnetic buoyancy instability (Gilman 1970; Acheson 1979; Spruit and van Ballegooijen 1982) and of Ω-like loops of magnetic flux. When convective motions are present, it is not necessary to impose a variable entropy distribution along the tube in order to obtain Ω-like loops. Convective upflows (and downflows) naturally lead to the formation of loop-like structures, provided that the magnetic Reynolds number is large enough to ensure that the magnetic field lines are (at least partially) advected with the flow.

The second important element of the initial conditions is the background atmosphere. In the absence of a realistic convective envelope, the sub-photospheric layer is represented by an adiabatically stratified layer. The photosphere is simulated by an isothermal layer. Above it, the temperature firstly starts to increase slowly and then there is a steep gradient in the temperature distribution, which represents the chromosphere/transition region area. Finally, the uppermost layer is characterized by a constant temperature of the order of 1 million Kelvin, which is mimicking the solar corona.

The third element is the presence of a pre-existing magnetic field in the corona. The interaction of the emerging field with the ambient magnetic field results into the onset of dynamic events, such as jets and eruptions. This will be discussed in Sect. 2.3.

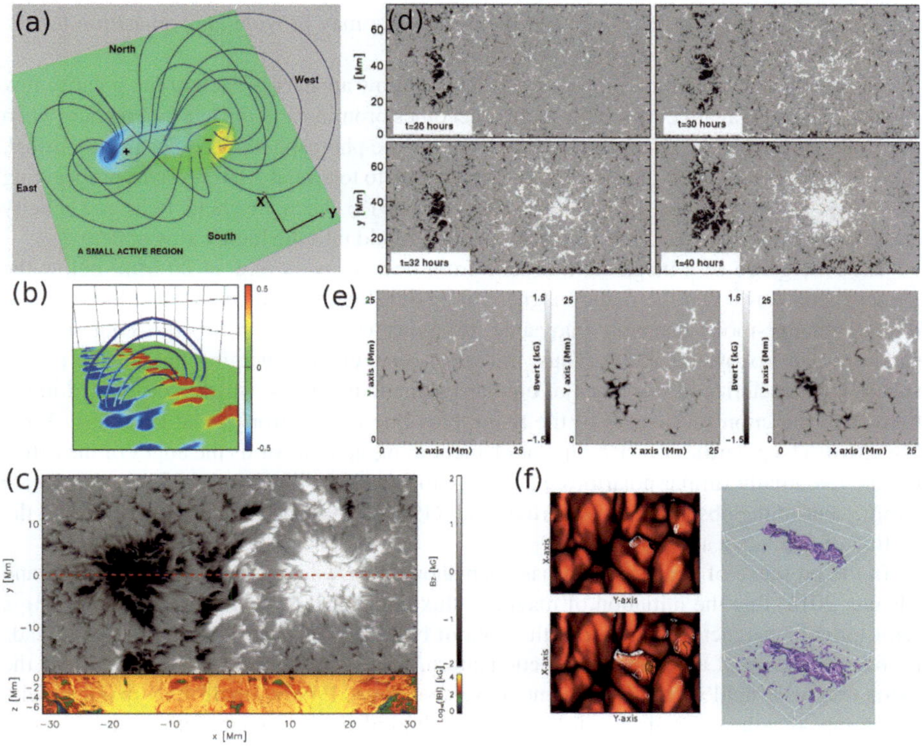

Fig. 2 Numerical simulations showing the emergence of magnetic flux at the solar surface. (**a**) The formation of sunspots and a small active region after the emergence of a large-scale twisted flux tube (Archontis and Hood 2012). (**b**) A series of small-scale bipolar regions form at the photosphere, due to the interchange-mode instability (Toriumi and Yokoyama 2012). (**c**) Sunspot formation due to kinematic emergence of a toroidal flux tube from the solar convection zone (Cheung et al. 2010). (**d**) The gradual build up of sunspots due to magnetic flux emergence (Rempel and Cheung 2014). (**e**) The combined action of turbulent convection and flux emergence of a magnetic flux sheet results in the formation of sunspots (Stein and Nordlund 2010). (**f**) The effect of granular convection on the emergence of a large scale magnetic flux tube. The flux tube undergoes convective shredding for large values of the magnetic Reynolds number. This is more highlighted for wider tubes (*bottom panel*) (Bushby and Archontis 2012)

2.2 Emergence at the Photosphere

The emergence of the Ω-shaped magnetic field at the photosphere is followed by the formation of a bipolar pair of sunspots and the development of an AR (Fig. 2(a)). Sunspots form where the axial magnetic field of the coherent, large-scale emerging flux tube intersects the photosphere (e.g. Fan 2001; Magara and Longcope 2001a; Archontis et al. 2004). If the twist of the sub-photospheric magnetic field is strong, the two sunspots appear almost perpendicular to the original axis of the tube (e.g. in a north-south orientation). Eventually, the opposite polarity sunspots move away from each other, towards an east-west direction (i.e. along the polarity inversion line of the active region, which is approximately parallel to the direction of the axis of the tube). The motion of the plasma in opposite directions, on the two sides of the polarity inversion line, results in a pronounced shear flow (e.g. Fan 2001; Manchester IV et al. 2004). The shearing plasma motions are magnetically driven by the tension of the emerging field lines and their existence may provide an explanation for the shear flows observed in emerging bipolar ARs.

As the sunspot separation proceeds, a rotational flow appears in each of the polarity flux concentrations (Magara 2006). This motion becomes prominent in the late emergence, when the flanks of the Ω-like emerging field intersecting the photosphere becomes more vertical. It has been suggested that the rotational flows are due to torsional Alfvén waves propagating along the twisted flux tube (Fan 2009b; Leake et al. 2013). Then, magnetic twist and helicity are steadily transported, from the twisted emerging field into the higher atmosphere.

The emergence of a magnetic flux tube from deeper layers (e.g. at 20 Mm) below the photosphere shows that the rising magnetic field develops several small-scale loops due to the interchange-mode magnetic buoyancy instability (while it is below the solar surface) (Toriumi and Yokoyama 2012). Thus, the appearance of the emerging field at the photosphere is characterized by a series of bipolar magnetic elements, which are oriented in the transverse direction to the axis of the rising tube due to the strong twist (see Sect. 3 and Fig. 2(b)). The polarities of each bipolar element move apart towards the edges of the active region. Eventually similar polarities may merge to form pore-like features. These results are reminiscent of the observations by Strous et al. (1996) and Strous and Zwaan (1999) on the motion of faculae in active regions.

In the presence of convection, it has been shown (e.g. Cheung et al. 2010; Rempel and Cheung 2014) that the initiation of magnetic flux emergence by kinematically advecting a semi-torus of magnetic field across the bottom boundary of the numerical domain, leads to the appearance of a complex magnetic pattern, which results from the interaction of the rising magnetic field with the turbulent convective flows. Firstly, small-scale magnetic elements appear at the solar surface, followed by their gradual coalescence into larger magnetic concentrations, which eventually results in the formation of a pair of opposite polarity spots. In these simulations, it has been found that spot formation is sensitive to the persistence of upflows at the bottom boundary footpoints and that the decay of the sunspots is due to subsurface flows that fragment the magnetic field under the action of turbulent diffusion (see Figs. 2(c), 2(d)).

It is widely accepted that bipolar magnetic regions at the solar surface are formed when loops of magnetic flux emerge from the solar interior and magnetic buoyancy plays a crucial role in this process. However it is not yet clear to what extent the local convective motions influence the evolution of the rising magnetic loops. Simulations of granular convection (Bushby and Archontis 2012) have determined the circumstances under which magnetic buoyancy enhances the flux emergence rate, which is otherwise driven solely by the convective up flows (Fig. 2(f)). One factor which has been taken into account in these simulations is that the magnetic Prandtl number (i.e. the ratio of the magnetic to the fluid Reynolds number) should be less than unity as would be expected for the solar convection zone. It has been shown that firstly, magnetic buoyancy contributes most effectively to the flux emergence process in the low plasma β, high magnetic Reynolds number regime. Secondly, for a given value of plasma β, the flux emergence rate is always lower in the higher magnetic Reynolds regime. This is due to the fact that convective disruption is much more efficient at higher magnetic Reynolds number (where field lines are more easily advected by the flow). At even higher Rm (e.g. larger than 300) this convective flow should be capable of sustaining a small-scale dynamo, although it is not clear how such a flow would interact with an emerging flux tube. This is a possible area for future numerical simulations of magnetic flux emergence.

Moreover, the reconnection between granular-scale emergence events and vertical magnetic concentrations in the lower atmosphere (e.g. chromosphere) may heat the local plasma, give rise to spicule-like jets and emit high-frequency waves that propagate into the corona (Isobe et al. 2008). These phenomena could be considered as likely candidates for the atmospheric heating and the acceleration of the solar wind.

2.3 Emergence Above the Photosphere and Associated Dynamic Events

The emergence of the magnetic flux at the surface is followed by the horizontal spreading of the field because photosphere does not support convective motions. Only the portion of the emerging field where the gradient of the magnetic field strength is larger than the subadiabatic excess in the photosphere can reach the layers above the photosphere (e.g. Archontis et al. 2004; Martínez-Sykora 2008), due to the magnetic buoyancy instability. The relation between the buoyancy instability and the superadiabatic excess has been studied by several authors (e.g. Acheson 1979; Archontis et al. 2004; Fan 2009b).

In addition, the majority of the simulations of an emerging horizontal flux tube have shown that only part of the original axis of the tube can emerge above the photosphere, while a considerable part of the axial flux remains at the photosphere. Only when the original shape of the emerging field is curved (e.g. a toroidal flux tube) and the draining of the dense plasma along the axis is efficient, a large amount of axial flux might emerge above the photosphere (Hood et al. 2009). On the other hand, there is a recent observation interpreted as being the full emergence of a helical flux rope from below the photosphere into the corona, along the polarity inversion line (PIL) of an active region (Okamoto et al. 2008). Therefore, the bodily emergence of a twisted flux tube above the photosphere requires further investigation.

On large scale, as the emergence of the magnetic flux proceeds through the highly stratified atmosphere, magnetic loops are formed that join the opposite polarities of the developed active region. The dynamic evolution of the magnetic field within the active region and the interaction of the emerging field lines with pre-existing magnetic fields are usually associated with the onset of various dynamic events. In the following, we describe results of numerical models of flux emergence followed by the onset of eruptions and the formation of jets.

2.3.1 Eruptions of Magnetic Flux Ropes in Emerging Flux Regions

As the emergence of a large scale twisted flux tube proceeds and the two opposite polarities move apart, shearing of the field lines occur along the PIL. The rapid expansion of the emerging field towards the upper solar atmosphere results in low pressure regions around the PIL. As a result, plasma motions towards the PIL develop self-consistency during the emergence (Archontis and Hood 2010). The combined action of horizontal motions (i.e. shearing and converging flows) leads to internal reconnection of the sheared field lines and the formation of a *new* flux rope (Fig. 3(a)) above the original axis of the emerging flux tube at photospheric/chromospheric heights (e.g. Magara and Longcope 2001a; Manchester IV et al. 2004; Archontis and Hood 2012).

The initial rise of the new flux rope is driven by the pressure gradient, which is build up due to the converging motions, and the Lorentz force. As the flux rope rises, reconnection underneath the rope leads to the acceleration of the rope by the tension of the reconnected field lines. Due to the rapid reconnection of the field, heating is produced at the site of internal reconnection forming flare loops under the erupting plasma. The erupting magnetic field comprises twisted field lines that carry dense plasma in their dips. The material emerged into the corona by the flux rope is generally 2–3 orders of magnitude heavier than the background plasma. The overall configuration consists of an envelope coronal magnetic field (i.e.

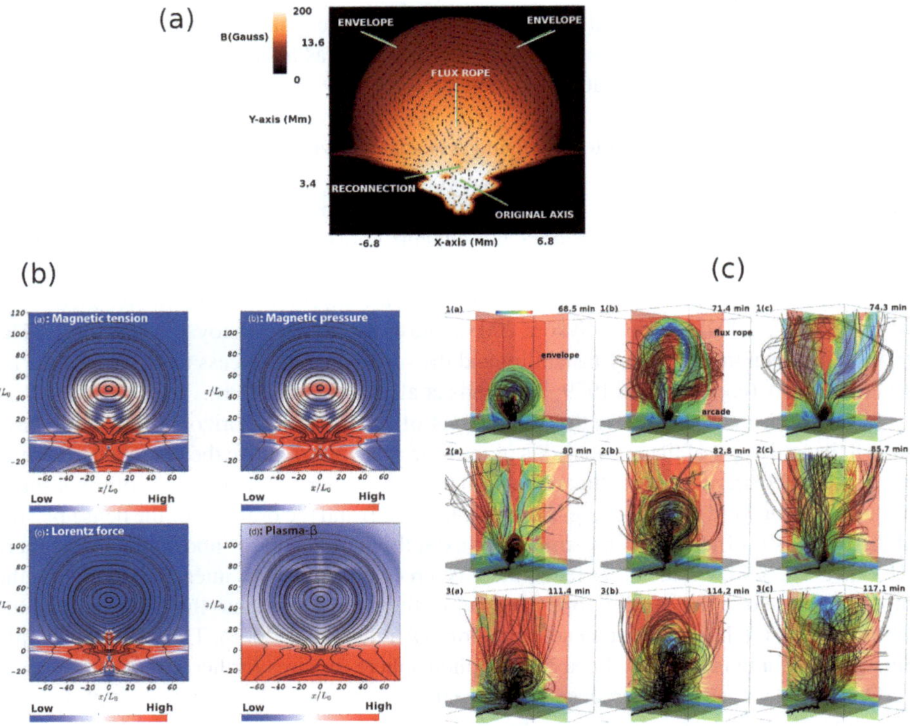

Fig. 3 A common feature in flux emergence simulations is the formation and rise of magnetic flux ropes. (**a**) Formation of a new flux rope above the axis of the emerging flux tube (Archontis and Hood 2012). (**b**) Formation of a stable coronal flux rope (Leake et al. 2013). (**c**) Recurrent CME-like eruptions in an emerging flux region (Archontis et al. 2014)

the uppermost emerging field lines), a fast-rising flux rope within the envelope (i.e. the core of the erupting plasma) and a hot cusp-like structure underneath it.

The dynamical evolution of the eruptions produced in emerging flux regions is subject to the parameters of the emerging field and its interaction with the pre-existing magnetic fields in the solar atmosphere. The analysis of the magnetic field configuration of the active region by using linear or non-linear force free extrapolation is important to detect the existence of null points, quasi separatrix layers and bald patches and understand the location of the electric currents which can be released during reconnection leading to different active phenomena: eruption, flare or jet (Fig. 4).

Previous simulations (e.g. Manchester IV et al. 2004; Archontis and Török 2008; Archontis and Hood 2012) have shown that the eruption of a flux rope into a field-free corona is likely to be confined by the envelope field. Moreover, there was no direct evidence that this expulsion of magnetic flux may evolve into a large-scale (e.g. CME-like) eruption. After the initial rising motion of the flux rope, the eruption slows down due to the downward tension force that the envelope field lines exert on the upcoming field. As a result, the rope rises into the corona and it approaches a quasi-static equilibrium, which is retained for a few Alfvén crossing times. However, more recent simulations (Archontis et al. 2014) have revealed that when the initial emerging field is strong enough (i.e. corresponding to plasma $\beta \approx 14$ at $z \approx 2.1$ Mm below the photosphere) the expanding field undergoes a severe vertical stretching and the envelope field lines reconnect via a tether-cutting mecha-

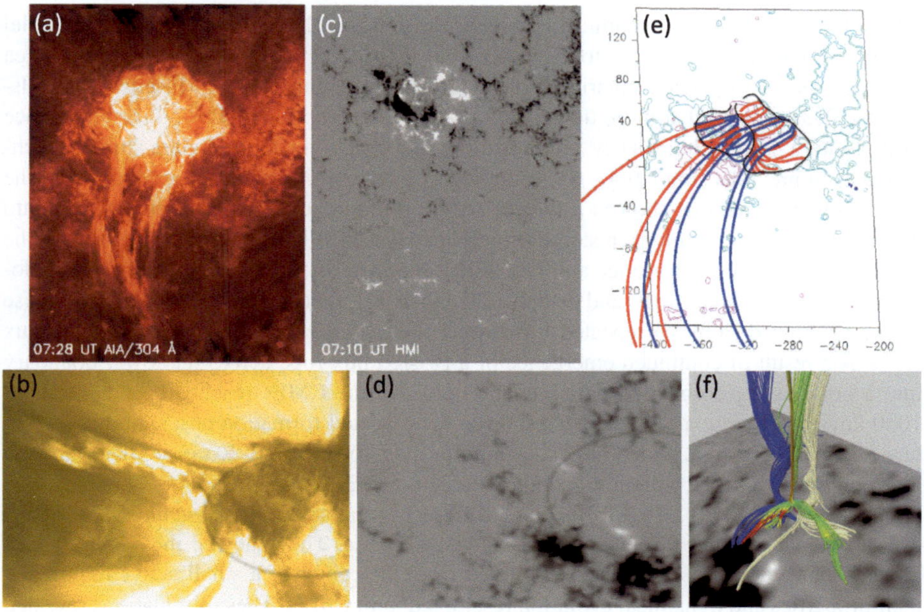

Fig. 4 Examples of signatures of flux emergence observed by SDO in the corona: (**a**) flare on November 10, 2010 in 304 Å, (**b**) jet on September 17, 2010, (**c**) and (**d**) are the HMI magnetograms corresponding respectively to (**a**) and (**b**), (**e**) linear force free extrapolations showing magnetic field lines and round-shaped separatrices as the observed ribbons in (**a**) (adapted from Mandrini et al. 2014). (**f**) Non-linear force free extrapolation showing the existence of a null point for the jet and two flux ropes (adapted from Schmieder et al. 2013)

nism. This process releases the tension of the envelope field lines and the flux rope erupts with high speed (up to 800 $\mathrm{km\,s^{-1}}$) in an ejective manner. The erupting plasma accelerates and expands considerably, which indicates that it might evolve into a CME-like eruption (Fig. 3(c)).

The interaction of the emerging field with a pre-existing magnetic field has a great impact on the evolution of the erupting flux rope. The relative orientation and the strength of the interacting fields plays a crucial role. If the pre-existing field is such that external reconnection is limited, the erupting flux rope will most likely follow a confined eruption (Fig. 3(b)). In this case, the erupting rope will be tethered by the remaining (i.e. which has not been reconnected) envelope or pre-existing field (e.g. Archontis and Török 2008; Leake et al. 2013). However, if the external reconnection is efficient and there is sufficient removal of flux from both the envelope and the pre-existing field, the flux rope will follow an ejective eruption in a CME-like manner. Also, there is the possibility that *all* the envelope field lines reconnect and, thus, the erupting flux rope starts to interact with the remaining pre-existing field during its rising motion. This interaction may lead to the deformation (or even annihilation) of the flux rope (e.g. Archontis and Hood 2012). An open problem in the afore-mentioned experiments is whether the final stage of the flux ropes may account for the stable configuration of a filament-like structure or for a large-scale eruption that is able to propagate towards the outer space.

Numerical experiments have also shown that as the emergence of the magnetic flux from the solar interior and the shearing/converging motions at the photosphere continue, there are more than one eruption triggered in the emerging flux region. MacTaggart and Hood

(2009b) showed that the dynamical emergence of a toroidal flux tube into an antiparallel horizontal magnetic field leads to the onset of at least two eruptions from the central area (between the two opposite polarities) of the emerging region. Moreno-Insertis and Galsgaard (2013) showed that more than 5 (mini CME-like) eruptions followed the emergence of flux into a coronal hole. They found that the first eruption occurred via a similar mechanism as in previous simulations (i.e. tether-cutting reconnection in a sheared arcade). The two following eruptions showed a twisted, Ω-loop-like rope with twist being turned into writhe, thus hinting at a kink instability (perhaps combined with a torus instability) as the cause of the eruption. Chatterjee and Fan (2013) reported on the development of a homologous sequence of three CMEs and demonstrated their so-called cannibalistic behavior. These CMEs originated from the repeated formations and partial eruptions of kink unstable flux ropes as a result of continued emergence of a twisted flux rope across the lower boundary into a pre-existing coronal potential arcade field. All the CMEs reached speeds of about 1000 $\mathrm{km\,s^{-1}}$. These experiments also showed the repeated formation of a sigmoidal structure after each eruption, which might account for the reformation of X-ray sigmoids and the "sigmoid-under-cusp" configuration in homologous CMEs. Similar results were obtained by Archontis et al. (2014) (Fig. 3(c)) using a different initial setup: the *dynamical* emergence of a horizontal twisted flux tube into a field-free corona. They showed the occurrence of at least four eruptions, which were triggered from the center of a small active region and their properties and dynamical evolution resembled mini CMEs. These simulations revealed that external reconnection is not a necessary condition for an ejective CME-like eruption in emerging flux regions. The twist inherent of the emerging flux brings enough free energy.

2.4 Standard and "Blowout" Jets in Emerging Flux Regions

Observations have shown that magnetic flux emergence associated with canceling flux lead to the ejection of hot X-ray jets (e.g. Shibata et al. 1992; Canfield et al. 1996; Schmieder et al. 2013). However their respective role is not clear and need to be discussed. This discussion is nevertheless out of the scope of this review. Jet-launching events in coronal holes have also been reported to follow the appearance of bipolar magnetic fields at the solar surface (Moreno-Insertis and Galsgaard 2013) and they have revealed some of the properties of the (e.g. coronal hole or polar X-ray) jets, such as the inverted Y shape or the apparent motions in the transverse direction of the main stream of the jets (e.g. Shibata et al. 2007; Cirtain et al. 2007; Nishizuka et al. 2008, see also Figs. 5(a), 5(b)).

A "standard" scenario, which has been extensively used to model such jets, was reported by Heyvaerts et al. (1977): a magnetic bipole emerges into a (unipolar) ambient magnetic field and reconnects to form hot and fast outflows (jets) that are emitted from the interface between the fields into contact. Numerical simulations of flux emergence in 2.5D (e.g. Yokoyama and Shibata 1995) and in 3D (e.g. Galsgaard et al. 2005, Fig. 5(c)) have used this scenario to investigate the onset and investigate some of the physical properties of these jets. In general, it was shown that reconnection between emerging and ambient magnetic fields results in a pair of high velocity (e.g. $O(100)$ $\mathrm{km\,s^{-1}}$) outflows that emanate from the rims of the current, which is build up at the interface. In the case of an oblique (open) ambient field (thus, mimicking a coronal hole environment), the emitted reconnection outflows (initially driven by the tension of the reconnected field lines) collide with the open field forming a fast shock. From there, the plasma flows are driven by the pressure gradient force to form a "standard" collimated jet as part of an overall slanted, inverse Y-shaped configuration. Due to reconnection, magnetic energy is partly released as heat via Joule dissipation and hot plasma is also ejected together with the field lines. The jets were found to be regions of

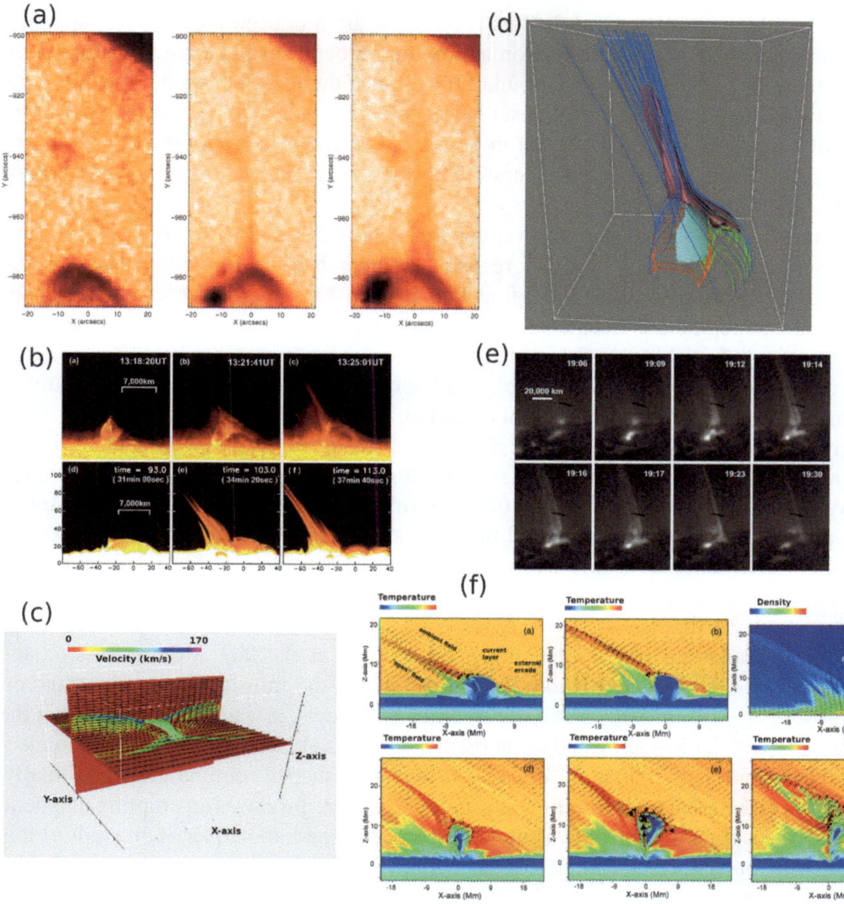

Fig. 5 Observations and simulations of solar jets. (**a**) Hinode XRT false-color images of three stages of a jet's evolution (Cirtain et al. 2007). (**b**) Comparison between a Ca jet (observed by Hinode SOT) and a simulated jet (shown is the density distribution) based on a reconnection model of flux emergence (Nishizuka et al. 2008). (**c**) Jets emitted from the rims of an interface current between emerging and pre-existing magnetic fields. (**d**) 3D visualization of a coronal hole jet. (**e**) Observation of a "blowout" jet (Moore et al. 2010). (**f**) A standard-to-blowout jet numerical model

high gas pressure, temperature, and velocity and they may account for the observed X-ray jets. In addition, Isobe et al. (2005) found that, as a result of the magnetic Rayleigh–Taylor instability, thin current sheets are formed in the emerging magnetic fields. Thus, magnetic reconnection between the emerging and the pre-existing coronal field occurs in a spatially intermittent way. This may explain the intermittent nature of coronal heating and the patchy brightenings in solar flares.

Similar 3D experiments (e.g. Moreno-Insertis et al. 2008, Fig. 5(d)) have also reported on the horizontal drift of the jet structure, with properties consistent with the statistics of this type of event. The produced jets reach speeds around 400 $\mathrm{km\,s^{-1}}$ and temperatures up to $\approx 3 \times 10^7$ K. Similar numerical models (e.g. Nishizuka et al. 2008) have succeeded in producing hot and cool jets, which can be accelerated simultaneously by magnetic reconnection in the transition region/upper chromosphere, driven by flux emergence. Another interesting result is that these simulations reveal that the produced jets undergo apparent motions perpendicular to the jet direction at amplitudes between 5–15 $\mathrm{km\,s^{-1}}$. The period of the oscillation is calculated to be around 200 s. Since the jet is believed to be along the magnetic field, the propagation of the oscillation could be interpreted as evidence of propagating Alfvén waves. It is possible that these waves are associated with reconnection, and may contribute to the heating and acceleration of the solar wind when the magnetic field is open (e.g. Parker 1988).

In the above simulations, the pre-existing magnetic field is simple: constant and uniform (horizontal or oblique). A more realistic ambient field is formed by the fan-like shape envelope field lines of an emerging flux region. New flux emergence at the surroundings of this region has been successfully incorporated into simulations investigating (a) the emission of jet-like up flows at the edges of ARs (e.g. Gontikakis et al. 2009; Murray et al. 2010) and (b) the effect of oscillatory reconnection in producing recurrent jets in three dimensions (e.g. Archontis et al. 2010).

Recent observations (e.g. Moore et al. 2010; Sterling et al. 2010; Liu et al. 2011; Shen et al. 2012, see also Fig. 5(e)) have revealed another type of jet that has, in addition to the X-ray emitting plasma, cooler (10^4–10^5 K) plasma that erupts along the jet. The eruptive chromospheric/transition-region material undergoes an ejective motion (100–300 $\mathrm{km\,s^{-1}}$), widens the surrounding field and becomes part of the jet. Due to the blast of plasma that opens the area in which the fast ejecta occur, these jets are known as "blowout" jets (e.g. Moore et al. 2010).

Numerical simulations (Pariat et al. 2009; Edmondson et al. 2009; Moreno-Insertis and Galsgaard 2013; Archontis and Hood 2013) have shown the transition from the standard to the "blowout" jets (Fig. 5(f)). Similarly to previous models, it was demonstrated that the interaction between an emerging and an ambient magnetic field leads to (external) reconnection and the formation of "standard" jets with an inverse Y-shaped configuration. Eventually, low-atmosphere (internal) reconnection of sheared field lines produces an erupting magnetic flux rope. The erupting plasma blows out the ambient field and, moreover, it unwinds as it is ejected into the outer solar atmosphere as in Török et al. (2009), Pariat et al. (2009, 2010). Thus, the simulations showed the twisted nature of the field lines along the jet, resulting from the reconnection between the eruptive twisted field and the pre-existing field, which consists of non-twisted oblique field lines. Therefore, an untwisting plasma motion was revealed during the emission of the "blowout" jet. The fast emission of the cool material that erupts together with the hot outflows due to external/internal reconnection form a wider "blowout" jet with a curtain-like configuration. Similar results were presented in the model of recurrent eruptions by Moreno-Insertis and Galsgaard (2013).

3 Small Scale Emergence Through the Solar Surface

3.1 Magnetic Properties

The development of higher spatial and temporal resolution instruments favored the observations of small scale magnetic flux during their emergence. Flare Genesis Telescope (FGE) with its spatial resolution of 0.5 arcsec, a balloon mission launched in Antarctica in 2000 was the first sub-arcsec instrument of the series to follow the evolution of magnetic vector during flux emergence (Bernasconi et al. 2002; Georgoulis et al. 2002). More recently Hinode/SOT, Sunrise a balloon mission (Solanki et al. 2010), the Swedish solar tower (SST) telescope, and the full disk continuous observing SDO/HMI provide fantastic data, showing the extremely intermittence and fragmentation of the magnetic field as it emerges (Fig. 6(a)).

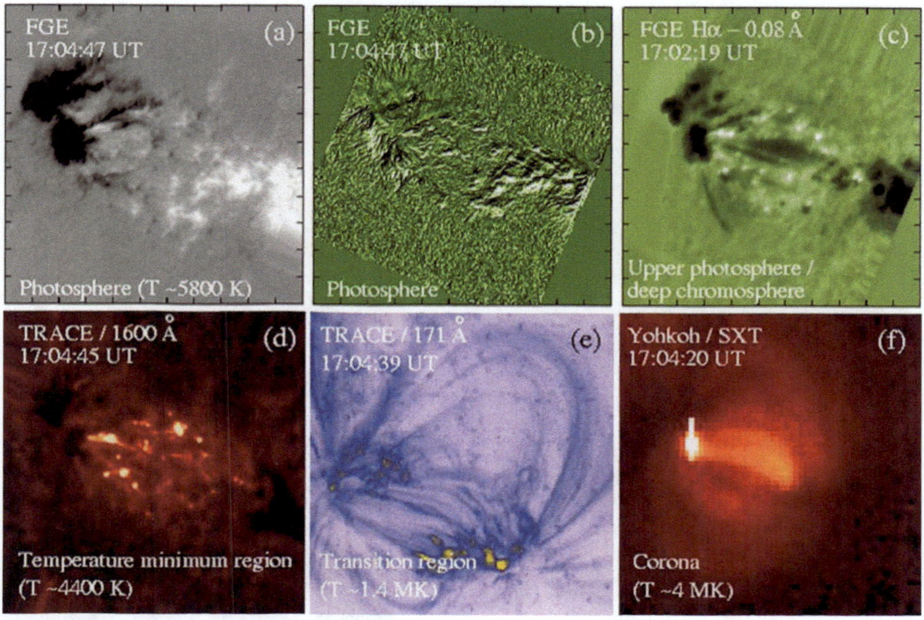

Fig. 6 Multi-wavelength observations of an emerging active region observed on 25 January 2000 by Flare Genesis Emerging telescope (FGE). (**a**) Vertical photospheric magnetic field (*white* is positive, *black* is negative magnetic polarity respectively). The magnetic field between the two main spots appears very fragmented. (**b**) Vertical component of the sheath current density calculated from the magnetic field vectors. (**c**) Off-band Hα filtergram showing EBs (Ellerman bombs as *bright dots*) and AFS (arch filament system as *dark elongated structures*), (**d**) deep chromosphere image with EBs. (e) EUV view of the transition region showing coronal loops. (**f**) The soft X-ray emission from the overlying corona. Tick mark interval = 7250 km. The 6 images represent the same field of view (adapted from Georgoulis et al. 2002; Schmieder et al. 2004)

Emergence of flux always starts as a circle of small polarities in longitudinal magnetic field maps. Granular-sized flux emergence has been detected using Sunrise with its Fabry Perrot (Palacios et al. 2012), and also the SST/CRISP spectropolarimeter due to their extreme fine spatial resolution (120 km) (Guglielmino et al. 2010, 2012; Ortiz et al. 2014). The emergence consisted in semi-spheres (cool bubbles of 3 arcsec diameter), a kind of 3D Ω loops with photospheric magnetic fields of the order of 500 Gauss (vertical in the half moon-shape legs and horizontal at the top of the bubble) (Fig. 7(a)). Ortiz et al. (2014) interpreted such observations as magnetized spherical structures rising due to low plasma pressure compared to the surrounding. They used the 3D numerical simulation—Bifrost code, and injected an untwisted magnetic flux sheet in the convection zone. These events were reported as isolated phenomena but frequent in regions of weak field. This kind of bubble is suspected to be of the same nature than the bubble observed below prominences (Dudík et al. 2012).

In larger events, rapidly a leading polarity is formed and an inversion polarity line, mainly north south oriented, is identifiable separating the strong leading polarity from the trailing ones. Continuous emergence occurs in the middle of these two main polarities until the formation of sunspots. Magnetograms show then a very patchy pattern with positive and negative polarities between the two main polarities. Magnetic vectors indicated that between these polarities exist a succession of U loops (magnetic field line tangent to the photosphere

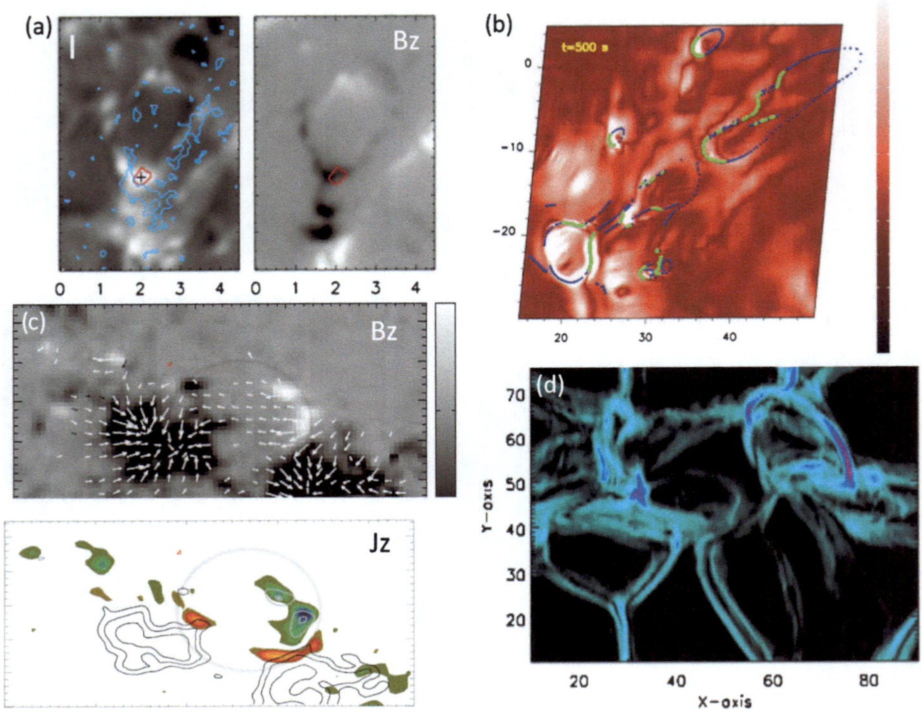

Fig. 7 (**a**) Tiny emergence of magnetic dome with a diameter below two arcsecs between granulation: black bubble surrounded by a bright lane in Ca II line wing (*left panel*), and longitudinal photosphere magnetic field in circle-shape (*right panel*). The axes are in arcsec (adapted from Ortiz et al. 2014). (**b**) Electric current sheet distribution (Jz) at the boundary of the simulation domain. Enhancement of currents is visible over bald-patches and separatrices (adapted from Pariat et al. 2009). The axes are in Mm. (**c**) Emerging flux at the edge of an active region (-800 G $< Bz <$ 800 G) (*red ellipse*) and associated vertical current density (Jz) computed from HMI vectors on September 17, 2010 (adapted from Guo et al. 2013). (**d**) Distribution of the total current density via emerging flux simulation in the low photosphere, consequence of the current layers created over emerging domes (adapted from Archontis and Hood 2009)

named "bald patch" regions), and Ω loops or semi-sphere structures (Bernasconi et al. 2002; Pariat et al. 2007). Bernasconi et al. (2002) showed that all these bipoles have a magnetic vector oriented from a negative polarity to a positive polarity indicating that they are sections of U-shaped (concave) magnetic loops stitched into the upper photosphere.

As soon as a sunspot is formed, it is frequent that the sunspot starts also to decrease in size and in net flux due to the emergence and cancellation of small bipoles surrounding the sunspot. These small bipoles are closely joined pairs of opposite polarities moving coherently, therefore they are called moving magnetic feature (MMF) or Moving Dipole Features (MDF) (Nelson et al. 2013; Vissers et al. 2013). An other kind of bipoles emerge in the center of large convection cells, and as the MDFs they quickly move as single units towards the cell edges (0.4 km/s) (Bernasconi et al. 2002). In that case they are part of the build-up of sunspots.

The observed U-shaped loops over magnetic bipoles related to emerging flux or MMF look like little wiggling disturbances embedded in a sea of mostly horizontal magnetic field lines. According to photospheric relative Doppler shift maps, persistent down-flows of material are found at the center of bipoles. This suggests that these flows could further distort the field lines, amplifying each concave dip in them. The bending of the field lines triggers further flow of material towards the center of each dip thus creating U-loops. It can be due to other instabilities (see Sect. 3.2). Each newly formed MMF is swept by the horizontal outward flows within the convection cell towards the edge of the cell where finally the entrained material can slide down into the sunspot or the network. The two polarities in a newly merging active region have elongated pattern which are traces of the horizontal component of the flux rope field (López Fuentes et al. 2000; Chandra et al. 2009; Schmieder et al. 2013). A zebra pattern is observed in Doppler velocities with a succession of upward and downward fields undulated pattern of horizontal magnetic field (Watanabe et al. 2008).

During the emergence, the evolution of the magnetic fields in the photosphere creates extremely complex patterns of strong vertical electric currents. Currents through reconnection can induce energy release. Such energy release can be detected in bright features surrounding emerging regions (Fig. 7(a)). These electric current patterns clearly reveal the filamentary nature of the emerged magnetic fields, which induce a highly structured three-dimensional configuration of magnetic loops or domes in the overlying solar corona. Vertical component of the sheath current density derived from the magnetic vectors show aligned structures between the two sunspots (Fig. 6(b)) or complex roundish structures (Fig. 7(c)). Electric vertical current density has been recently measured using SDO/HMI magnetic vectors in the photosphere during flux emergence, and cancellation at the edge of an active region leading to a jet (Guo et al. 2013).

3.2 Sea Serpentine, Topology and Simulation

The "sea-serpent" topology was introduced first by Harvey and Harvey (1973). At the photospheric level in the continuum and in the core of photospheric lines, during the emergence of flux the granulation pattern tends to be modified compared to the quiet-sun. The granulation looks fuzzy and transient dark threads, corresponding to elongated intergranular lanes, can be observed in the central part of the emerging flux region (Strous et al. 1996). Strous and Zwaan (1999) proposed a scheme of emerging undulatory structure. This kind of topology was confirmed by the works of Pariat et al. (2004, 2007, 2009), Otsuji et al. (2007), Watanabe et al. (2008), Valori et al. (2012) and summarized in a cartoon (Schmieder and Pariat 2010) (Fig. 1).

Pariat et al. (2004) applied a linear force-free extrapolation to the FGE magnetograms, and found that the emerging flux volume possesses a hierarchy of small undulating loops consisting in a series of concave and convex loops (Fig. 8(c)). This is confirmed by numerous simulations (Figs. 8(a), (b)) (Isobe et al. 2007; Cheung et al. 2007; Archontis and Hood 2009), and NLFF extrapolations (Valori et al. 2012).

In the quasi-static phase, the emergence of flux in the chromosphere is characterized by the formation of low-lying dark filaments, called Arch Filament System (AFS) (Fig. 6(c)). The solar plasma is compressed, and advected by the rising of new flux, and as it becomes denser, it flows down under gravity at both ends of the loops.

Downflows are commonly observed at the footpoints of the AFS and up-flows around the apex of the loops (Malherbe et al. 1998; Strous et al. 1996; Mandrini et al. 2002; Xu et al. 2010). Typical physical parameters of AFS are the rising speed about 10 km/s, the plasma downflows about 20–50 km/s, the electron density 5–10 $\times 10^{10}$ in cm^{-3}, the gas pressure 0.15–0.25 dyne cm^{-2}, the life time around 10–20 minutes and the magnetic field of the order of 50 Gauss. In the corona, large loops overlaid all the emerging active region (Figs. 6(e), (f)).

The prevailing, traditional viewpoint of a smooth emergence of convex active region magnetic loops (Ω-shaped loops) into the solar atmosphere is fundamentally incomplete. This Ω-shaped loop model is valid for large scale emergence and in the mid term of the emergence which has a typical life time of a day. At the solar surface, the observed Ω-shaped loops are formed by the resistive emergence of undulatory flux ropes (Pariat et al. 2004). The upper parts of the emerging flux ropes are significantly deformed and decelerated when reaching the photosphere thus creating a large number of MDFs and localized concave emerging magnetic structures. The serpentine line shave a typical scale of 2000 km. This distance is comparable to the convection scale and convection could be the main driver of undulation (Cheung et al. 2008). The convective motions produce a hierarchy of magnetic loops with at various spatial scales, adopting a sea-serpentine configuration. When a large loop approaches the surface, it produces a small active region with a compact leading spot and more diffuse following spots (Figs. 2(e) and (d)).

Magara and Longcope (2001b), Pariat et al. (2004) proposed an other scénario based on a MHD instability. Concave loops can be created by a buoyancy instability (Rayleigh–Taylor instability) when a heavier fluid overlies a lighter fluid due to the decrease of the magnetic field strength with height. This instability is developing if its time scale is shorter than the characteristic time scale of convective motions. Magnetic tension will prevent the instability for developing until a critical wavenumber.

The characteristic wavenumber is proportional to the inverse of two pressure scale heights. For wavelengths larger than 2 Mm, the magnetic tension in these curved field lines will be too low to prevent the instability (Parker instability). As the concave loops emerge, material is falling along the legs and fills dips which become deeper and deeper creating U-shaped loops. Topology analysis proves that the convex loops are tangent to the photosphere (bald patch regions) (Pariat et al. 2004, 2007). Bald patches are associated regions with field lines having dips. They are localized over magnetic inversion lines. For these U-loops to fully emerge into the solar atmosphere the action of magnetic reconnection (via resistive effects, i.e., energy dissipation in electric currents) is necessary. Magnetic reconnection between oppositely directed magnetic field lines of interacting magnetic loops allows the magnetic flux to be transported into the corona, leaving the dense plasma to dive from the lower atmosphere into the convection zone. Isobe et al. (2007) confirmed the role of resistive events in flux emergence. Simulations were necessary to show the distributions of the currents during emergence (Archontis and Hood 2009; Pariat et al. 2009). Pariat et al. (2009) used a data driven 3D MHD numerical simulation of an emerging active region and showed that currents were built up along the 3D serpentine magnetic field structure as a result of the shear between the two polarities of bipoles during emergence (Fig. 8(d)). The currents are found all over the separatrix domes structures and they are particularly intense at the photospheric level (Fig. 7(b), (d)). Reconnection can occur not only at the bald patches but at different locations of the separatrices (Fig. 7(b)).

3.3 Ellerman Bombs and Moving Magnetic Features

Ellerman Bombs are observed in many chromospheric lines (wings of Hα, Ca II 8542 Å, Ca H, and K lines, Mg II h and k lines, and UV continua 1600 Å and 1700 Å). They are intermittent brightenings roughly aligned with the axis of the active region. Their spectral signatures with brighter emission than the surroundings far away of the line center (H$\alpha \pm 1$ Å to 10 Å) indicate that hotter plasma is located in the photosphere and not in the chromosphere (see the extensive review of Rutten et al. 2013). Ellerman (1917) was the first to notice these intermittent brightenings in the wings of Hα line within young emerging active regions. It is

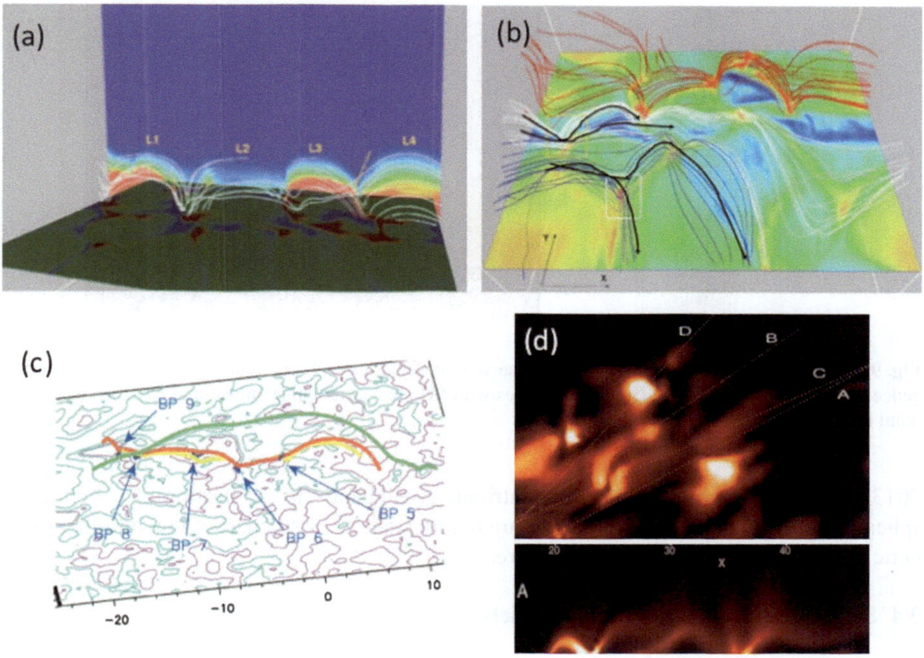

Fig. 8 (**a**) Snapshot of the magnetic field (Bz) during the emergence of flux tube with U and Ω between (L1, L2, ...) (adapted from Archontis and Hood 2009). (**b**) 3D topology of magnetic field lines as the undulatory magnetic field emerges through the atmosphere from simulation (adapted from Archontis and Hood 2009). (**c**) Magnetic field lines and Bald-patches from a lfff extrapolation (adapted from Pariat et al. 2004). (**d**) Distribution of the current jz viewed from above and side-view. Intense currents are at $z = 0$ and current layers all over the emergence domes (adapted from Pariat et al. 2009)

why they are called Ellerman bombs (EBs). Many works are studying their temporal, spatial behavior. EBs are observed in two different magnetic configurations: during the emergence of new active regions or during the decaying phase of sunspots. In the later case they are related to moving magnetic features (MMF). The local magnetic structure is similar. Sub-photospheric magnetic flux tubes are flat and emerge partly by instability. The EB spectral signatures of both cases are similar.

Nelson et al. (2013) reviewed the statistics concerning EBs due to MMF observing with IBIS (0.096 arcsec of resolution) and ROSA high spatial and temporal resolution instrument with images taken at Hα ± 0.7 Å and in the G band (proxy of the magnetic field). The previous statistics were obtained by Georgoulis et al. (2002). Comparing both statistics we notice that the number of EBs per arcsec^2 detected has increased due to higher spatial resolution and their characteristics number has decreased: the size decreases from 1 to 0.3 arcsec, the life time from 10 minutes to less than 2 min. However Rutten et al. (2013) claim that all the bright points seen around the decreasing sunspots are pseudo EBs, some of them might account for Ca bright points.

Detailed morphological and statistical analysis of the EBs show that they share the physics of small solar flares, obeying self-similar distribution functions in their total and peak released energies and their durations. Typical energies of EBs were estimated in the range 10^{27}–10^{28} ergs, which is comparable to that of microflares (Georgoulis et al. 2002). Smaller EBs could have an energy of nano flares as low as 10^{22}–10^{25} ergs (Nelson et al.

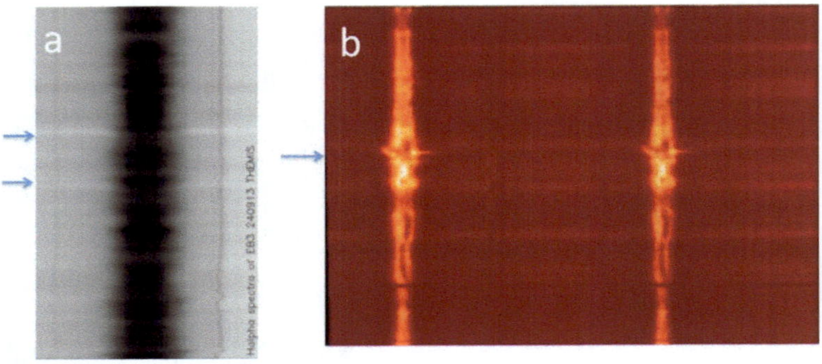

Fig. 9 Examples of Ellerman bomb spectra observed on September 24, 2013 during a magnetic flux emergence (**a**) in Hα with THEMIS, (**b**) in Mg II lines with IRIS spectrograph (Pontieu et al. 2014). The *arrows* point the EBs

2013). Large numbers of EBs could contribute significantly to the heating of the chromosphere in emerging active regions. EBs are manifestations of stochastic, low-altitude, magnetic reconnection in the solar atmosphere.

3.4 Spectral Signatures of EBs and Models

EBs are the signature of the resistive emergence of undulatory flux ropes in the solar atmosphere. EBs have a significant brief emission in chromospheric line wings (Hα, Ca II 8542 Å) (Fang et al. 2006; Pariat et al. 2007; Vissers et al. 2013; Berlicki and Heinzel 2014). The profiles are characterized by a deep absorption at the line center and strong emission in the wings (Fig. 9). Overlying arch filament system (AFS) may obscure the line centers and produce asymmetry of the core of the lines (Rutten et al. 2013). Realistic models of EB spectra are based on NLTE radiative-transfer modeling in a simple 1D plan parallel semi-infinite atmosphere (Kitai 1983; Fang et al. 2006; Socas-Navarro et al. 2006; Herlender and Berlicki 2011). Semi-empirical models assume the hydrostatic equilibrium. However due to readjustment of the kinetic temperature, heating is found not only in the photosphere but also in the upper chromosphere. There exist also MHD models of formation of EBs (Archontis and Hood 2009), but they consider full ionized plasma and heating is observed until the upper chromosphere. A new attempt of modeling considering non-fully ionized atmosphere was achieved by Xu et al. (2011), and it was shown that the temperature increase was due to the Joule dissipation caused by magnetic reconnection.

Recently Berlicki and Heinzel (2014) constructed a grid of models based on a model of quiet solar atmosphere and introduced a hot spot region at the photospheric level. By adjustments of four variable parameters for the hot spot (width, peak of temperature, density, position), they could fit the contrast of EBs observed in Ca II and Hα wings.

4 Conclusion

Magnetic flux emergence is an important part of the solar cycle activity. However it is still difficult to understand fully the physically mechanisms of the emergence of flux through the convection zone until the upper corona. We are very rich of observations with the new developments of solar instruments. With SDO the full disk of the Sun is continuously observed with a resolution of 0.5 arcsec and a time resolution of 45 s. It is always possible to understand how and when the first bipole has emerged to lead to an active phenomena. Nevertheless some pieces of the puzzle still are missing to have the full story from one end to the other. The new simulations help us to study the importance of the different mechanisms occurring during emergence in a large parameter space. However it is not yet possible to simulate the emergence through the convection until the upper layers due to the large gradient of the physical parameters in the atmosphere. It requires to solve radiative transfer equations as well as MHD in a 3D atmosphere.

In this review, we discussed the manifestation of magnetic flux emergence in the solar surface and the corresponding plasma response of the solar atmosphere. We have highlighted that understanding how buoyant magnetic fields emerge from the solar convection zone into the stably stratified, rarefied solar atmosphere and corona is vital for studying the formation of solar active regions and the development of dynamic events, over a large range of spatial and temporal scales, such as jets and precursor structures for solar eruptions such as CMEs. In recent years, a considerable progress has been achieved in understanding (a) the nature and dynamics of the flux emergence process and (b) the onset and evolution of associated (flux emergence driven) phenomena.

We expect that further advancement in computing power and numerical modeling in the coming years is likely to enable numerical models of solar active region formation, encompassing both realistic magneto-convection modeling of the formation of sunspots from emerging magnetic fields and the emergence of the active region flux into the solar corona. Such simulations will enable us to construct a comprehensive view of the atmospheric coupling as manifested via flux emergence.

Acknowledgements We would like to thank the ISSI institute in Bern which has allowed us to discuss deeply on emerging flux in the solar atmosphere in the framework of solar cycle activity.

References

D.J. Acheson, Instability by magnetic buoyancy. Sol. Phys. **62**, 23 (1979). doi:10.1007/BF00150129. 1979SoPh...62...23A

V. Archontis, Magnetic flux emergence in the Sun. J. Geophys. Res. **113**, 3 (2008). doi:10.1029/2007JA012422. 2008JGRA..11303S04A

V. Archontis, Magnetic flux emergence and associated dynamic phenomena in the Sun. Philos. Trans. R. Soc. Lond. A **370**, 3088 (2012). doi:10.1098/rsta.2012.0001. 2012RSPTA.370.3088A

V. Archontis, A.W. Hood, Formation of Ellerman bombs due to 3D flux emergence. Astron. Astrophys. **508**, 1469 (2009). doi:10.1051/0004-6361/200912455. 2009A%26A...508.1469A

V. Archontis, A.W. Hood, Flux emergence and coronal eruption. Astron. Astrophys. **514**, A56+ (2010). doi:10.1051/0004-6361/200913502. 2010A%26A...514A..56A

V. Archontis, A.W. Hood, Magnetic flux emergence: a precursor of solar plasma expulsion. Astron. Astrophys. **537**, A62 (2012). doi:10.1051/0004-6361/201116956. 2012A%26A...537A..62A

V. Archontis, A.W. Hood, A numerical model of standard to blowout jets. Astrophys. J. Lett. **769**, L21 (2013). doi:10.1088/2041-8205/769/2/L21. 2013ApJ...769L..21A

V. Archontis, T. Török, Eruption of magnetic flux ropes during flux emergence. Astron. Astrophys. **492**, L35 (2008). doi:10.1051/0004-6361:200811131. 2008A%26A...492L..35A

V. Archontis, F. Moreno-Insertis, K. Galsgaard, A. Hood, E. O'Shea, Emergence of magnetic flux from the convection zone into the corona. Astron. Astrophys. **426**, 1047 (2004). doi:10.1051/0004-6361:20035934. 2004A%26A...426.1047A

V. Archontis, K. Tsinganos, C. Gontikakis, Recurrent solar jets in active regions. Astron. Astrophys. **512**, L2+ (2010). doi:10.1051/0004-6361/200913752. 2010A%26A...512L...2A

V. Archontis, A.W. Hood, K. Tsinganos, Recurrent explosive eruptions and the "Sigmoid-to-arcade" transformation in the Sun driven by dynamical magnetic flux emergence. Astrophys. J. Lett. **786**, L21 (2014). doi:10.1088/2041-8205/786/2/L21. 2014ApJ...786L..21A

G. Aulanier, The physical mechanisms that initiate and drive solar eruptions, in *IAU Symposium*, vol. 300 (2014), p. 184. doi:10.1017/S1743921313010958. 2014IAUS..300..184A

G. Barnes, A.C. Birch, K.D. Leka, D.C. Braun, Helioseismology of pre-emerging active regions. III. Statistical analysis. Astrophys. J. **786**, 19 (2014). doi:10.1088/0004-637X/786/1/19. 2014ApJ...786...19B

A. Berlicki, P. Heinzel, Observations and NLTE modeling of Ellerman bombs. *ArXiv e-prints* (2014). 2014arXiv1406.5702B

P.N. Bernasconi, D.M. Rust, M.K. Georgoulis, B.J. Labonte, Moving dipolar features in an emerging flux region. Sol. Phys. **209**, 119 (2002). doi:10.1023/A:1020943816174. 2002SoPh..209..119B

S.J. Bradshaw, G. Aulanier, G. Del Zanna, A reconnection-driven rarefaction wave model for coronal outflows. Astrophys. J. **743**, 66 (2011). doi:10.1088/0004-637X/743/1/66. 2011ApJ...743...66B

P.J. Bushby, V. Archontis, Modelling magnetic flux emergence in the solar convection zone. Astron. Astrophys. **545**, A107 (2012). doi:10.1051/0004-6361/201015747. 2012A%26A...545A.107B

R.C. Canfield, K.P. Reardon, K.D. Leka, K. Shibata, T. Yokoyama, M. Shimojo, H alpha surges and X-ray jets in AR 7260. Astrophys. J. **464**, 1016 (1996). doi:10.1086/177389. 1996ApJ...464.1016C

A. Canou, T. Amari, V. Bommier, B. Schmieder, G. Aulanier, H. Li, Evidence for a pre-eruptive twisted flux rope using the themis vector magnetograph. Astrophys. J. Lett. **693**, L27 (2009). doi:10.1088/0004-637X/693/1/L27. 2009ApJ...693L..27C

R. Chandra, B. Schmieder, G. Aulanier, J.M. Malherbe, Evidence of magnetic helicity in emerging flux and associated flare. Sol. Phys. **258**, 53 (2009). doi:10.1007/s11207-009-9392-z. 2009SoPh..258...53C

P. Chatterjee, Y. Fan, Simulation of homologous and cannibalistic coronal mass ejections produced by the emergence of a twisted flux rope into the solar corona. Astrophys. J. Lett. **778**, L8 (2013). doi:10.1088/2041-8205/778/1/L8. 2013ApJ...778L...8C

M.C.M. Cheung, M. Schüssler, F. Moreno-Insertis, Magnetic flux emergence in granular convection: radiative MHD simulations and observational signatures. Astron. Astrophys. **467**, 703 (2007). doi:10.1051/0004-6361:20077048. 2007A%26A...467..703C

M.C.M. Cheung, M. Schüssler, T.D. Tarbell, A.M. Title, Solar surface emerging flux regions: a comparative study of radiative MHD modeling and hinode SOT observations. Astrophys. J. **687**, 1373 (2008). doi:10.1086/591245. 2008ApJ...687.1373C

M.C.M. Cheung, M. Rempel, A.M. Title, M. Schüssler, Simulation of the formation of a solar active region. Astrophys. J. **720**, 233 (2010). doi:10.1088/0004-637X/720/1/233. 2010ApJ...720..233C

J.W. Cirtain, L. Golub, L. Lundquist, A. van Ballegooijen, A. Savcheva, M. Shimojo, E. DeLuca, S. Tsuneta, T. Sakao, K. Reeves, M. Weber, R. Kano, N. Narukage, K. Shibasaki, Evidence for Alfvén waves in solar X-ray jets. Science **318**, 1580 (2007). doi:10.1126/science.1147050. 2007Sci...318.1580C

B. De Pontieu, A.M. Title, J.R. Lemen, G.D. Kushner, D.J. Akin, B. Allard, T. Berger, P. Boerner, M. Cheung, C. Chou, J.F. Drake, D.W. Duncan, S. Freeland, G.F. Heyman, C. Hoffman, N.E. Hurlburt, R.W. Lindgren, D. Mathur, R. Rehse, D. Sabolish, R. Seguin, C.J. Schrijver, T.D. Tarbell, J.-P. Wülser, C.J. Wolfson, C. Yanari, J. Mudge, N. Nguyen-Phuc, R. Timmons, R. van Bezooijen, I. Weingrod, R. Brookner, G. Butcher, B. Dougherty, J. Eder, V. Knagenhjelm, S. Larsen, D. Mansir, L. Phan, P. Boyle, P.N. Cheimets, E.E. DeLuca, L. Golub, R. Gates, E. Hertz, S. McKillop, S. Park, T. Perry, W.A. Podgorski, K. Reeves, S. Saar, P. Testa, H. Tian, M. Weber, C. Dunn, S. Eccles, S.A. Jaeggli, C.C. Kankelborg, K. Mashburn, N. Pust, L. Springer, R. Carvalho, L. Kleint, J. Marmie, E. Mazmanian, T.M.D. Pereira, S. Sawyer, J. Strong, S.P. Worden, M. Carlsson, V.H. Hansteen, J. Leenaarts, M. Wiesmann, J. Aloise, K.-C. Chu, R.I. Bush, P.H. Scherrer, P. Brekke, J. Martinez-Sykora, B.W. Lites, S.W. McIntosh, H. Uitenbroek, T.J. Okamoto, M.A. Gummin, G. Auker, P. Jerram, P. Pool, N. Waltham, The Interface Region Imaging Spectrograph (IRIS). Sol. Phys. **289**, 2733 (2014). doi:10.1007/s11207-014-0485-y. 2014SoPh..289.2733D

P. Démoulin, D. Baker, C.H. Mandrini, L. van Driel-Gesztelyi, The 3D geometry of active region upflows deduced from their limb-to-limb evolution. Sol. Phys. **283**, 341 (2013). doi:10.1007/s11207-013-0234-7. 2013SoPh..283..341D

J. Dudík, G. Aulanier, B. Schmieder, M. Zapiór, P. Heinzel, Magnetic topology of bubbles in quiescent prominences. Astrophys. J. **761**, 9 (2012). doi:10.1088/0004-637X/761/1/9. 2012ApJ...761....9D

J.K. Edmondson, B.J. Lynch, S.K. Antiochos, C.R. De Vore, T.H. Zurbuchen, Reconnection-driven dynamics of coronal-hole boundaries. Astrophys. J. **707**, 1427 (2009). doi:10.1088/0004-637X/707/2/1427. 2009ApJ...707.1427E

F. Ellerman, Solar hydrogen "bombs". Astrophys. J. **46**, 298 (1917). doi:10.1086/142366. 1917ApJ....46..298E

Y. Fan, The emergence of a twisted Ω-tube into the solar atmosphere. Astrophys. J. Lett. **554**, L111 (2001). doi:10.1086/320935. 2001ApJ...554L.111F

Y. Fan, Magnetic fields in the solar convection zone. Living Rev. Sol. Phys. **6**, 4 (2009a). doi:10.12942/lrsp-2009-4. 2009LRSP....6....4F

Y. Fan, The emergence of a twisted flux tube into the solar atmosphere: sunspot rotations and the formation of a coronal flux rope. Astrophys. J. **697**, 1529 (2009b). doi:10.1088/0004-637X/697/2/1529. 2009ApJ...697.1529F

C. Fang, Y.H. Tang, Z. Xu, M.D. Ding, P.F. Chen, Spectral analysis of Ellerman bombs. Astrophys. J. **643**, 1325 (2006). doi:10.1086/501342. 2006ApJ...643.1325F

K. Galsgaard, F. Moreno-Insertis, V. Archontis, A. Hood, A three-dimensional study of reconnection, current sheets, and jets resulting from magnetic flux emergence in the Sun. Astrophys. J. Lett. **618**, L153 (2005). doi:10.1086/427872. 2005ApJ...618L.153G

M.K. Georgoulis, D.M. Rust, P.N. Bernasconi, B. Schmieder, Statistics, morphology, and energetics of Ellerman bombs. Astrophys. J. **575**, 506 (2002). doi:10.1086/341195. 2002ApJ...575..506G

P.A. Gilman, Instability of magnetohydrostatic stellar interiors from magnetic buoyancy. I. Astrophys. J. **162**, 1019 (1970). doi:10.1086/150733. 1970ApJ...162.1019G

C. Gontikakis, V. Archontis, K. Tsinganos, Observations and 3D MHD simulations of a solar active region jet. Astron. Astrophys. **506**, L45 (2009). doi:10.1051/0004-6361/200913026. 2009A%26A...506L..45G

S.L. Guglielmino, L.R. Bellot Rubio, F. Zuccarello, G. Aulanier, S. Vargas Domínguez, S. Kamio, Multiwavelength observations of small-scale reconnection events triggered by magnetic flux emergence in the solar atmosphere. Astrophys. J. **724**, 1083 (2010). doi:10.1088/0004-637X/724/2/1083. 2010ApJ...724.1083G

S.L. Guglielmino, V. Martínez Pillet, J.A. Bonet, J.C. del Toro Iniesta, L.R. Bellot Rubio, S.K. Solanki, W. Schmidt, A. Gandorfer, P. Barthol, M. Knölker, The frontier between small-scale bipoles and ephemeral regions in the solar photosphere: emergence and decay of an intermediate-scale bipole observed with SUNRISE/IMaX. Astrophys. J. **745**, 160 (2012). doi:10.1088/0004-637X/745/2/160. 2012ApJ...745..160G

Y. Guo, M.D. Ding, B. Schmieder, H. Li, T. Török, T. Wiegelmann, Driving mechanism and onset condition of a confined eruption. Astrophys. J. Lett. **725**, L38 (2010). doi:10.1088/2041-8205/725/1/L38. 2010ApJ...725L..38G

Y. Guo, P. Démoulin, B. Schmieder, M.D. Ding, S. Vargas Domínguez, Y. Liu, Recurrent coronal jets induced by repetitively accumulated electric currents. Astron. Astrophys. **555**, A19 (2013). doi:10.1051/0004-6361/201321229. 2013A%26A...555A..19G

L.K. Harra, V. Archontis, E. Pedram, A.W. Hood, D.L. Shelton, L. van Driel-Gesztelyi, The creation of outflowing plasma in the corona at emerging flux regions: comparing observations and simulations. Sol. Phys. **278**, 47 (2012). doi:10.1007/s11207-011-9855-x. 2012SoPh..278...47H

T. Hartlep, A.G. Kosovichev, J. Zhao, N.N. Mansour, Signatures of emerging subsurface structures in acoustic power maps of the Sun. Sol. Phys. **268**, 321 (2011). doi:10.1007/s11207-010-9544-1. 2011SoPh..268..321H

K. Harvey, J. Harvey, Observations of moving magnetic features near sunspots. Sol. Phys. **28**, 61 (1973). doi:10.1007/BF00152912. 1973SoPh...28...61H

M. Herlender, A. Berlicki, Multi-wavelength analysis of Ellerman bomb light curves. Cent. Eur. Astrophys. Bull. **35**, 181 (2011). 2011CEAB...35..181H

J. Heyvaerts, E.R. Priest, D.M. Rust, An emerging flux model for the solar flare phenomenon. Astrophys. J. **216**, 123 (1977). doi:10.1086/155453. 1977ApJ...216..123H

A.W. Hood, V. Archontis, K. Galsgaard, F. Moreno-Insertis, The emergence of toroidal flux tubes from beneath the solar photosphere. Astron. Astrophys. **503**, 999 (2009). doi:10.1051/0004-6361/200912189. 2009A%26A...503..999H

H. Isobe, T. Miyagoshi, K. Shibata, T. Yokoyama, Filamentary structure on the Sun from the magnetic Rayleigh–Taylor instability. Nature **434**, 478 (2005). doi:10.1038/nature03399. 2005Natur.434..478I

H. Isobe, D. Tripathi, V. Archontis, Ellerman bombs and jets associated with resistive flux emergence. Astrophys. J. Lett. **657**, L53 (2007). doi:10.1086/512969. 2007ApJ...657L..53I

H. Isobe, M.R.E. Proctor, N.O. Weiss, Convection-driven emergence of small-scale magnetic fields and their role in coronal heating and solar wind acceleration. Astrophys. J. Lett. **679**, L57 (2008). doi:10.1086/589150. 2008ApJ...679L..57I

L. Jouve, A.S. Brun, Three-dimensional nonlinear evolution of a magnetic flux tube in a spherical shell: influence of turbulent convection and associated mean flows. Astrophys. J. **701**, 1300 (2009). doi:10.1088/0004-637X/701/2/1300. 2009ApJ...701.1300J

R. Kitai, On the mass motions and the atmospheric states of moustaches. Sol. Phys. **87**, 135 (1983). doi:10.1007/BF00151165. 1983SoPh...87..135K

J.E. Leake, M.G. Linton, T. Török, Simulations of emerging magnetic flux. I. The formation of stable coronal flux ropes. Astrophys. J. **778**, 99 (2013). doi:10.1088/0004-637X/778/2/99. 2013ApJ...778...99L

C. Liu, N. Deng, R. Liu, I. Ugarte-Urra, S. Wang, H. Wang, A standard-to-blowout jet. Astrophys. J. Lett. **735**, L18 (2011). doi:10.1088/2041-8205/735/1/L18. 2011ApJ...735L..18L

M.C. López Fuentes, P. Demoulin, C.H. Mandrini, L. van Driel-Gesztelyi, The counterkink rotation of a non-hale active region. Astrophys. J. **544**, 540 (2000). doi:10.1086/317180. 2000ApJ...544..540L

D. MacTaggart, A.W. Hood, Multiple eruptions from magnetic flux emergence. Astron. Astrophys. **508**, 445 (2009a). doi:10.1051/0004-6361/200913197. 2009A%26A...508..445M

D. MacTaggart, A.W. Hood, Multiple eruptions from magnetic flux emergence. Astron. Astrophys. **508**, 445 (2009b). doi:10.1051/0004-6361/200913197. 2009A%26A...508..445M

T. Magara, Dynamic and topological features of photospheric and coronal activities produced by flux emergence in the Sun. Astrophys. J. **653**, 1499 (2006). doi:10.1086/508926. 2006ApJ...653.1499M

T. Magara, D.W. Longcope, Sigmoid structure of an emerging flux tube. Astrophys. J. Lett. **559**, L55 (2001a). doi:10.1086/323635. 2001ApJ...559L..55M

T. Magara, D.W. Longcope, Sigmoid structure of an emerging flux tube. Astrophys. J. Lett. **559**, L55 (2001b). doi:10.1086/323635. 2001ApJ...559L..55M

J.M. Malherbe, B. Schmieder, P. Mein, N. Mein, L. van Driel-Gesztelyi, M. von Uexkull, Arch filament systems associated with X-ray loops. Sol. Phys. **180**, 265 (1998). doi:10.1023/A:1005092802593. 1998SoPh..180..265M

W. Manchester IV, T. Gombosi, D. DeZeeuw, Y. Fan, Eruption of a buoyantly emerging magnetic flux rope. Astrophys. J. **610**, 588 (2004). doi:10.1086/421516. 2004ApJ...610..588M

C.H. Mandrini, P. Démoulin, B. Schmieder, Y.Y. Deng, P. Rudawy, The role of magnetic bald patches in surges and arch filament systems. Astron. Astrophys. **391**, 317 (2002). doi:10.1051/0004-6361:20020745. 2002A%26A...391..317M

C.H. Mandrini, B. Schmieder, P. Démoulin, Y. Guo, G.D. Cristiani, Topological analysis of emerging bipole clusters producing violent solar events. Sol. Phys. **289**, 2041 (2014). doi:10.1007/s11207-013-0458-6. 2014SoPh..289.2041M

J. Martínez-Sykora, V. Hansteen, M. Carlsson, Twisted flux tube emergence from the convection zone to the corona. Astrophys. J. **679**, 871 (2008). doi:10.1086/587028. 2008ApJ...679..871M

J. Martínez-Sykora, V. Hansteen, M. Carlsson, Twisted flux tube emergence from the convection zone to the corona. II. Later states. Astrophys. J. **702**, 129 (2009). doi:10.1088/0004-637X/702/1/129. 2009ApJ...702..129M

R.L. Moore, J.W. Cirtain, A.C. Sterling, D.A. Falconer, Dichotomy of solar coronal jets: standard jets and blowout jets. Astrophys. J. **720**, 757 (2010). doi:10.1088/0004-637X/720/1/757. 2010ApJ...720..757M

F. Moreno-Insertis, Three-dimensional numerical experiments of flux emergence into the corona, in *New Solar Physics with Solar-B Mission*, ed. by K. Shibata, S. Nagata, T. Sakurai. Astronomical Society of the Pacific Conference Series, vol. 369 (2007) p. 335. 2007ASPC..369..335M

F. Moreno-Insertis, K. Galsgaard, Plasma jets and eruptions in solar coronal holes: a three-dimensional flux emergence experiment. Astrophys. J. **771**, 20 (2013). doi:10.1088/0004-637X/771/1/20. 2013ApJ...771...20M

F. Moreno-Insertis, K. Galsgaard, I. Ugarte-Urra, Jets in coronal holes: hinode observations and three-dimensional computer modeling. Astrophys. J. Lett. **673**, L211 (2008). doi:10.1086/527560. 2008ApJ...673L.211M

M.J. Murray, D. Baker, L. van Driel-Gesztelyi, J. Sun, Outflows at the edges of an active region in a coronal hole: a signature of active region expansion? Sol. Phys. **261**, 253 (2010). doi:10.1007/s11207-009-9484-9. 2010SoPh..261..253M

C.J. Nelson, J.G. Doyle, R. Erdélyi, Z. Huang, M.S. Madjarska, M. Mathioudakis, S.J. Mumford, K. Reardon, Statistical analysis of small Ellerman bomb events. Sol. Phys. **283**, 307 (2013). doi:10.1007/s11207-012-0222-3. 2013SoPh..283..307N

N. Nishizuka, M. Shimizu, T. Nakamura, K. Otsuji, T.J. Okamoto, Y. Katsukawa, K. Shibata, Giant chromospheric anemone jet observed with hinode and comparison with magnetohydrodynamic simulations: evidence of propagating Alfvén waves and magnetic reconnection. Astrophys. J. Lett. **683**, L83 (2008). doi:10.1086/591445. 2008ApJ...683L..83N

T.J. Okamoto, S. Tsuneta, B.W. Lites, M. Kubo, T. Yokoyama, T.E. Berger, K. Ichimoto, Y. Katsukawa, S. Nagata, K. Shibata, T. Shimizu, R.A. Shine, Y. Suematsu, T.D. Tarbell, A.M. Title, Emergence of a helical flux rope under an active region prominence. Astrophys. J. Lett. **673**, L215 (2008). doi:10.1086/528792. 2008ApJ...673L.215O

A. Ortiz, L.R. Bellot Rubio, V.H. Hansteen, J. de la Cruz Rodríguez, L. Rouppe van der Voort, Emergence of granular-sized magnetic bubbles through the solar atmosphere. I. Spectropolarimetric observations and simulations. Astrophys. J. **781**, 126 (2014). doi:10.1088/0004-637X/781/2/126. 2014ApJ...781..126O

K. Otsuji, K. Shibata, R. Kitai, S. Ueno, S. Nagata, T. Matsumoto, T. Nakamura, H. Watanabe, S. Tsuneta, Y. Suematsu, K. Ichimoto, T. Shimizu, Y. Katsukawa, T.D. Tarbell, B. Lites, R.A. Shine, A.M. Title, Small-scale magnetic-flux emergence observed with hinode solar optical telescope. Publ. Astron. Soc. Jpn. **59**, 649 (2007). doi:10.1093/pasj/59.sp3.S649. 2007PASJ...59S.649O

J. Palacios, J. Blanco Rodríguez, S. Vargas Domínguez, V. Domingo, V. Martínez Pillet, J.A. Bonet, L.R. Bellot Rubio, J.C. Del Toro Iniesta, S.K. Solanki, P. Barthol, A. Gandorfer, T. Berkefeld, W. Schmidt, M. Knölker, Magnetic field emergence in mesogranular-sized exploding granules observed with sunrise/IMaX data. Astron. Astrophys. **537**, A21 (2012). doi:10.1051/0004-6361/201117936. 2012A%26A...537A..21P

E. Pariat, G. Aulanier, B. Schmieder, M.K. Georgoulis, D.M. Rust, P.N. Bernasconi, Resistive emergence of undulatory flux tubes. Astrophys. J. **614**, 1099 (2004). doi:10.1086/423891. 2004ApJ...614.1099P

E. Pariat, B. Schmieder, A. Berlicki, Y. Deng, N. Mein, A. López Ariste, S. Wang, Spectrophotometric analysis of Ellerman bombs in the Ca II, Hα, and UV range. Astron. Astrophys. **473**, 279 (2007). doi:10.1051/0004-6361:20067011. 2007A%26A...473..279P

E. Pariat, S.K. Antiochos, C.R. DeVore, A model for solar polar jets. Astrophys. J. **691**, 61 (2009). doi:10.1088/0004-637X/691/1/61. 2009ApJ...691...61P

E. Pariat, S. Masson, G. Aulanier, Current buildup in emerging serpentine flux tubes. Astrophys. J. **701**, 1911 (2009). doi:10.1088/0004-637X/701/2/1911. 2009ApJ...701.1911P

E. Pariat, S.K. Antiochos, C.R. DeVore, Three-dimensional modeling of quasi-homologous solar jets. Astrophys. J. **714**, 1762 (2010). doi:10.1088/0004-637X/714/2/1762. 2010ApJ...714.1762P

E.N. Parker, The formation of sunspots from the solar toroidal field. Astrophys. J. **121**, 491 (1955). doi:10.1086/146010. 1955ApJ...121..491P

E.N. Parker, Nanoflares and the solar X-ray corona. Astrophys. J. **330**, 474 (1988). doi:10.1086/166485. 1988ApJ...330..474P

R.F. Pinto, A.S. Brun, L. Jouve, R. Grappin, Coupling the solar dynamo and the corona: wind properties, mass, and momentum losses during an activity cycle. Astrophys. J. **737**, 72 (2011). doi:10.1088/0004-637X/737/2/72. 2011ApJ...737...72P

M. Rempel, M.C.M. Cheung, Numerical simulations of active region scale flux emergence: from spot formation to decay. Astrophys. J. **785**, 90 (2014). doi:10.1088/0004-637X/785/2/90. 2014ApJ...785...90R

R.J. Rutten, G.J.M. Vissers, L.H.M. Rouppe van der Voort, P. Sütterlin, N. Vitas, Ellerman bombs: fallacies, fads, usage. J. Phys. Conf. Ser. **440**(1), 012007 (2013). doi:10.1088/1742-6596/440/1/012007. 2013JPhCS.440a2007R

B. Schmieder, E. Pariat, Magnetic flux emergence in scholarpedia. Scholarpedia **2**(12), 4335 (2007)

B. Schmieder, E. Pariat, Vector magnetic field in emerging flux regions, in *Magnetic Coupling Between the Interior and Atmosphere of the Sun*, ed. by S.S. Hasan, R.J. Rutten (2010) p. 505. doi:10.1007/978-3-642-02859-5_70. 2010mcia.conf..505S

B. Schmieder, D.M. Rust, M.K. Georgoulis, P. Démoulin, P.N. Bernasconi, Emerging flux and the heating of coronal loops. Astrophys. J. **601**, 530 (2004). doi:10.1086/380199. 2004ApJ...601..530S

B. Schmieder, P. Démoulin, G. Aulanier, Solar filament eruptions and their physical role in triggering coronal mass ejections. Adv. Space Res. **51**, 1967 (2013). doi:10.1016/j.asr.2012.12.026. 2013AdSpR..51.1967S

B. Schmieder, Y. Guo, F. Moreno-Insertis, G. Aulanier, L. Yelles Chaouche, N. Nishizuka, L.K. Harra, J.K. Thalmann, S. Vargas Dominguez, Y. Liu, Twisting solar coronal jet launched at the boundary of an active region. Astron. Astrophys. **559**, A1 (2013). doi:10.1051/0004-6361/201322181. 2013A%26A...559A...1S

Y. Shen, Y. Liu, J. Su, Y. Deng, On a coronal blowout jet: the first observation of a simultaneously produced bubble-like CME and a jet-like CME in a solar event. Astrophys. J. **745**, 164 (2012). doi:10.1088/0004-637X/745/2/164. 2012ApJ...745..164S

K. Shibata, Y. Ishido, L.W. Acton, K.T. Strong, T. Hirayama, Y. Uchida, A.H. McAllister, R. Matsumoto, S. Tsuneta, T. Shimizu, H. Hara, T. Sakurai, K. Ichimoto, Y. Nishino, Y. Ogawara, Observations of X-ray jets with the YOHKOH soft X-ray telescope. Publ. Astron. Soc. Jpn. **44**, L173 (1992). 1992PASJ...44L.173S

K. Shibata, T. Nakamura, T. Matsumoto, K. Otsuji, T.J. Okamoto, N. Nishizuka, T. Kawate, H. Watanabe, S. Nagata, S. UeNo, R. Kitai, S. Nozawa, S. Tsuneta, Y. Suematsu, K. Ichimoto, T. Shimizu, Y. Katsukawa, T.D. Tarbell, T.E. Berger, B.W. Lites, R.A. Shine, A.M. Title, Chromospheric anemone jets as evidence of ubiquitous reconnection. Science **318**, 1591 (2007). doi:10.1126/science.1146708. 2007Sci...318.1591S

H. Socas-Navarro, V. Martínez Pillet, D. Elmore, A. Pietarila, B.W. Lites, R. Manso Sainz, Spectro-polarimetric observations and non-lte modeling of Ellerman bombs. Sol. Phys. **235**, 75 (2006). doi:10.1007/s11207-006-0049-x. 2006SoPh..235...75S

S.K. Solanki, P. Barthol, S. Danilovic, A. Feller, A. Gandorfer, J. Hirzberger, T.L. Riethmüller, M. Schüssler, J.A. Bonet, V. Martínez Pillet, J.C. del Toro Iniesta, V. Domingo, J. Palacios, M. Knölker, N. Bello González, T. Berkefeld, M. Franz, W. Schmidt, A.M. Title, SUNRISE: instrument, mission, data, and first results. Astrophys. J. Lett. **723**, L127 (2010). doi:10.1088/2041-8205/723/2/L127. 2010ApJ...723L.127S

H.C. Spruit, Pressure equilibrium and energy balance of small photospheric fluxtubes. Sol. Phys. **50**, 269 (1976). doi:10.1007/BF00155292. 1976SoPh...50..269S

H.C. Spruit, A.A. van Ballegooijen, Stability of toroidal flux tubes in stars. Astron. Astrophys. **106**, 58 (1982). 1982A%26A...106...58S

R.F. Stein, Å. Nordlund, On the formation of active regions. Astrophys. J. Lett. **753**, L13 (2010). doi:10.1088/2041-8205/753/1/L13. 2012ApJ...753L..13S

A.C. Sterling, L.K. Harra, R.L. Moore, Fibrillar chromospheric spicule-like counterparts to an extreme-ultraviolet and soft X-ray blowout coronal jet. Astrophys. J. **722**, 1644 (2010). doi:10.1088/0004-637X/722/2/1644. 2010ApJ...722.1644S

L.H. Strous, C. Zwaan, Phenomena in an emerging active region. II. Properties of the dynamic small-scale structure. Astrophys. J. **527**, 435 (1999). doi:10.1086/308071. 1999ApJ...527..435S

L.H. Strous, G. Scharmer, T.D. Tarbell, A.M. Title, C. Zwaan, Phenomena in an emerging active region. I. Horizontal dynamics. Astron. Astrophys. **306**, 947 (1996). 1996A%26A...306..947S

S. Toriumi, T. Yokoyama, Large-scale 3D MHD simulation on the solar flux emergence and the small-scale dynamic features in an active region. Astron. Astrophys. **539**, A22 (2012). doi:10.1051/0004-6361/201118009. 2012A%26A...539A..22T

S. Toriumi, S. Ilonidis, T. Sekii, T. Yokoyama, Probing the shallow convection zone: rising motion of subsurface magnetic fields in the solar active region. Astrophys. J. Lett. **770**, L11 (2013). doi:10.1088/2041-8205/770/1/L11. 2013ApJ...770L..11T

T. Török, G. Aulanier, B. Schmieder, K.K. Reeves, L. Golub, Fan-spine topology formation through two-step reconnection driven by twisted flux emergence. Astrophys. J. **704**, 485 (2009). doi:10.1088/0004-637X/704/1/485. 2009ApJ...704..485T

G. Valori, L.M. Green, P. Démoulin, S. Vargas Domínguez, L. van Driel-Gesztelyi, A. Wallace, D. Baker, M. Fuhrmann, Nonlinear force-free extrapolation of emerging flux with a global twist and serpentine fine structures. Sol. Phys. **278**, 73 (2012). doi:10.1007/s11207-011-9865-8. 2012SoPh..278...73V

L. van Driel-Gesztelyi, J.L. Culhane, Magnetic flux emergence, activity, eruptions and magnetic clouds: following magnetic field from the Sun to the heliosphere. Space Sci. Rev. **144**, 351 (2009). doi:10.1007/s11214-008-9461-x. 2009SSRv..144..351V

G.J.M. Vissers, L.H.M. Rouppe van der Voort, R.J. Rutten, Ellerman bombs at high resolution. II. Triggering, visibility, and effect on upper atmosphere. Astrophys. J. **774**, 32 (2013). doi:10.1088/0004-637X/774/1/32. 2013ApJ...774...32V

H. Watanabe, R. Kitai, K. Okamoto, K. Nishida, J. Kiyohara, S. Ueno, M. Hagino, T.T. Ishii, K. Shibata, Spectropolarimetric observation of an emerging flux region: triggering mechanisms of Ellerman bombs. Astrophys. J. **684**, 736 (2008). doi:10.1086/590234. 2008ApJ...684..736W

Z. Xu, A. Lagg, S.K. Solanki, Magnetic structures of an emerging flux region in the solar photosphere and chromosphere. Astron. Astrophys. **520**, A77 (2010). doi:10.1051/0004-6361/200913227. 2010A%26A...520A..77X

X.-Y. Xu, C. Fang, M.-D. Ding, D.-H. Gao, Numerical simulations of magnetic reconnection in the lower solar atmosphere. Res. Astron. Astrophys. **11**, 225 (2011). doi:10.1088/1674-4527/11/2/010. 2011RAA....11..225X

T. Yokoyama, K. Shibata, Magnetic reconnection as the origin of X-ray jets and Hα surges on the Sun. Nature **375**, 42 (1995). doi:10.1038/375042a0. 1995Natur.375...42Y

C. Zwaan, Elements and patterns in the solar magnetic field. Annu. Rev. Astron. Astrophys. **25**, 83 (1987). doi:10.1146/annurev.aa.25.090187.000503. 1987ARA%26A..25...83Z

DOI 10.1007/978-1-4939-2584-1_9
Reprinted from *Space Science Reviews* Journal, DOI 10.1007/s11214-014-0100-4

Hemispheric Coupling: Comparing Dynamo Simulations and Observations

A.A. Norton · P. Charbonneau · D. Passos

Received: 15 April 2014 / Accepted: 7 September 2014 / Published online: 21 November 2014

Abstract Numerical simulations that reproduce solar-like magnetic cycles can be used to generate long-term statistics. The variations in north-south hemispheric solar cycle synchronicity and amplitude produced in simulations has not been widely compared to observations. The observed limits on solar cycle amplitude and phase asymmetry show that hemispheric sunspot area production is no more than 20 % asymmetric for cycles 17–23 and that phase lags do not exceed 20 % (or two years) of the total cycle period, as determined from Royal Greenwich Observatory sunspot data. Several independent studies have found a long-term trend in phase values as one hemisphere leads the other for, on average, four cycles. Such persistence in phase is not indicative of a stochastic phenomenon. We compare these observational findings to the magnetic cycle found in a numerical simulation of solar convection recently produced with the EULAG-MHD model. This long "millennium simulation" spans more than 1600 years and generated 40 regular, sunspot-like cycles. While the simulated cycle length is too long ($\sim$40 yrs) and the toroidal bands remain at too high of latitudes ($>30^\circ$), some solar-like aspects of hemispheric asymmetry are reproduced. The

A.A. Norton (✉)
W.W. Hansen Experimental Physics Laboratory, Stanford University, Stanford, CA 94305, USA
e-mail: aanorton@stanford.edu

P. Charbonneau · D. Passos
GRPS, Départment de Physique, Université of Montréal, C.P. 6128, Succ. Centre-ville, Montréal, QC H3C-3J7, Canada

P. Charbonneau
e-mail: paulchar@astro.umontreal.ca

D. Passos
e-mail: dariopassos@ist.utl.pt

D. Passos
CENTRA, Instituto Superior Técnico, Universidade de Lisboa, Av. Rovisco Pais 1, Lisboa 1049-001, Portugal

D. Passos
Departamento de Física, Universidade do Algarve, Campus de Gambelas, Faro 8005-139, Portugal

model is successful at reproducing the synchrony of polarity inversions and onset of cycle as the simulated phase lags do not exceed 20 % of the cycle period. The simulated amplitude variations between the north and south hemispheres are larger than those observed in the Sun, some up to 40 %. An interesting note is that the simulations also show that one hemisphere can persistently lead the other for several successive cycles, placing an upper bound on the efficiency of transequatorial magnetic coupling mechanisms. These include magnetic diffusion, cross-equatorial mixing within latitudinally-elongated convective rolls (a.k.a. "banana cells") and transequatorial meridional flow cells. One or more of these processes may lead to magnetic flux cancellation whereby the oppositely directed fields come in close proximity and cancel each other across the magnetic equator late in the solar cycle. We discuss the discrepancies between model and observations and the constraints they pose on possible mechanisms of hemispheric coupling.

Keywords Sun: magnetic fields · Hemispheric coupling · Sunspots · Solar activity

1 Introduction

The magnetic solar cycle is thought to be caused by a dynamo mechanism operating in the solar interior, by which inductive fluid flows generate a strong reservoir of magnetism at the bottom of the convection zone and/or coherent magnetic structures in the bulk of this unstable region. Many varieties of solar dynamo models have been designed based on varying levels of dynamical and geometrical simplifications, and consensus has yet to arise as to which best mimics solar behavior. All have in common an assumed solar structure, differential rotation and magnetic diffusivity profiles, although there are significant differences in the diffusivity profiles used in the various models. Some models invoke meridional circulation as highly important in the transport of magnetic fields, while others consider magnetic diffusivity and/or turbulent pumping to dominate over meridional circulation.

Alternately, dynamically self consistent numerical simulations of dynamo action in rotating, stratified convective turbulence can be run in order to compile statistics for dozens or more cycles, offering a unique laboratory to investigate how cycle outcomes depend on model assumptions. Although such simulations currently cannot be run in a solar-like parameter regimes, they can nonetheless reproduce some observed solar cycle behavior.

Hemispheric synchronicity and asymmetry is one aspect of cycle behavior that can inform modelers about the relative importance of meridional circulation and other physical mechanisms that participate in hemispheric coupling. In general, the Sun's hemispheres are considered moderately to strongly coupled since the wings of the butterfly diagram, see Fig. 1, show a high degree of mirroring across the geometric equator. Many models that reproduce the solar cycle assume a relatively high magnetic Reynolds number, which means that magnetic diffusion plays a small part in the transport of the magnetic field, which is mainly achieved through meridional circulation. In this case, the N and S hemispheres can decouple. Then the dynamo and the subsequent sunspot cycle can progress independently in each hemisphere (Dikpati and Gilman 2001; Chatterjee et al. 2004).

Hemispheric asymmetry was noted at least as early as the works of Spörer (1894) and Maunder (1904). Waldmeier (1955) wrote "the northern and southern hemispheres often show very unequal sunspot activity, and the inequality may persist for several years". Newton and Milsom (1955) also studied sunspot numbers and reported that the asymmetry in the activity of the two hemispheres from 1833–1954 did not appear to follow any definite laws. In addition to a moderate amplitude asymmetry measured in sunspot production, and

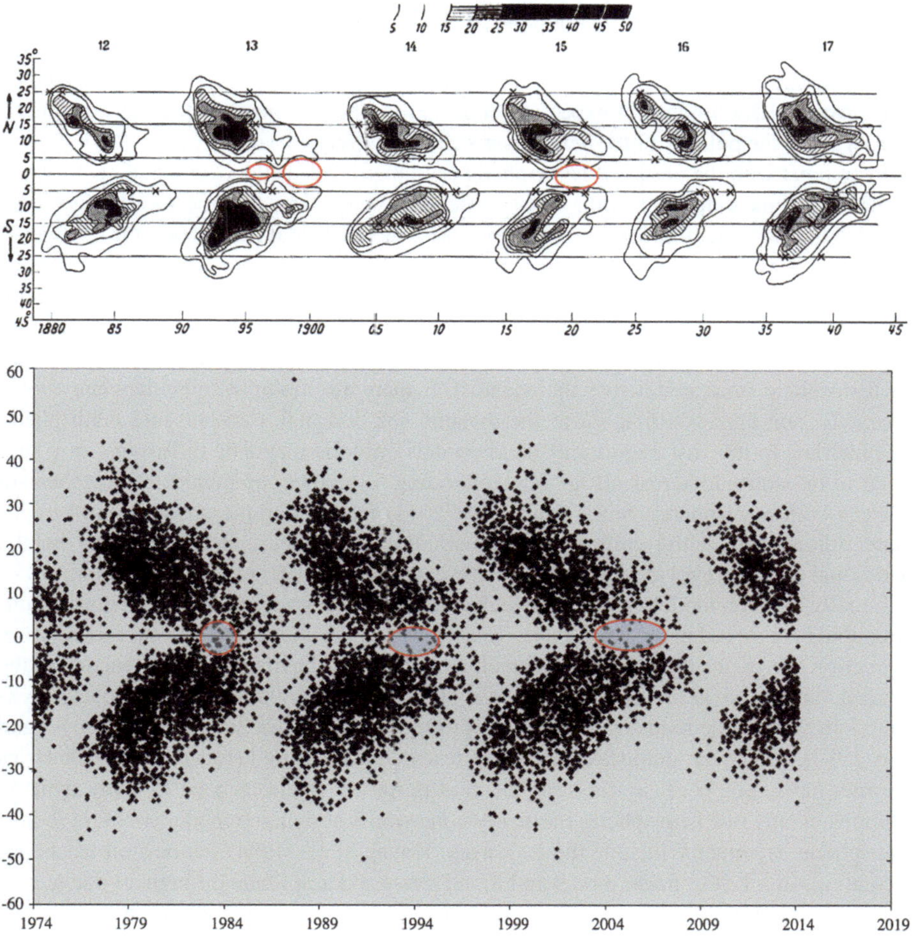

Fig. 1 Two versions of the butterfly diagram showing the latitudinal positions of sunspots as a function of time. *The top panel* (Becker 1955) plots iso-contours of equal spot frequency for cycles 12–17 from Greenwich Photoheliographic data while *the lower panel* plots each sunspot group position as an individual point from the end of cycle 20 until February 2014. Data for *the lower plot* is the USAF/NOAA sunspot data also from the Greenwich observatory prepared by Jan Janssen (STCE: Solar-Terrestrial Centre of Excellence) at http://users.telenet.be/j.janssens/SC24web/SC24.html. *Overplotted red ovals* are drawn on the plot to highlight periods when less sunspots appear near the magnetic equator than the period just preceding *the red ovals*. We propose that the time indicated by *the red ovals* may be when the leading edges of the N and S toroidal bands are undergoing a flux cancellation at depth

that sunspot maximum occurred at different times in the N and the S, Babcock (1959) noticed that the polar fields reversed at different times: the south in mid-1957 and the north in late-1958. Since then, significant hemispheric differences have been observed in many solar indices related to surface magnetism, and a more comprehensive list of these papers can be found in Sect. 2.1. In the past ten years, hemispheric differences have been observed in E–W zonal flows (Komm et al. 2013), N–S meridional flows (Komm et al. 2011) and Joy's law (McClintock and Norton 2013). The Sun's N–S asymmetry extends into the solar atmosphere, especially affecting the coronal magnetic equator and the deflection of the heliospheric current sheet, (see Virtanen and Mursula (2014) and included references). The southward shifted heliospheric current sheet has received attention as the "bashful ballerina" and has been persistently asymmetric with the northward component dominating the southward one during solar minimum for cycles 20–23 (Mursula and Hiltula 2003) which is noteworthy. One might not expect a hemisphere to be dominant instead of a magnetic polarity since the polarity in the hemispheres changed every 11 years.

In contrast with the observed hemispheric asymmetries detected over a half century ago in surface magnetism, only in the last two decades or so have dynamo modelling efforts included hemispheric differences. Under the classical mean-field framework, the α-effect—the action of cyclonic convection on magnetic field lines—depends only on the Coriolis force and stratification. By argument of symmetry, the α-effect should vanish at the equator allowing the possibility that the hemispheres become decoupled. The meridional circulation in flux transport dynamo models has, until recently, been thought to contain one cell per hemisphere (mirrored across the equator). If there are multiple cells, they can evolve separately, and their contribution to the dynamo solution may translate into hemispheric asymmetries. In the vast majority of these models only the magnetic diffusivity, η is supposed to be isotropic across all latitudes providing a reliable communication mechanism between both hemispheres. Since Zhao et al. (2013) reported on the observation of multiple meridional cells with significant hemispheric flow differences, we must assume that the meridional flows do evolve separately to some extent in the different hemispheres.

Usually dynamo modelers look for mechanisms that can produce the asymmetric modulation that is observed in the solar cycle. These mechanisms include stochastic forcing from convection (Hoyng et al. 1994), nonlinear effects caused by the magnetic backreaction of the induced magnetic field on the flows (Weiss et al. 1984) or even other more controversial explanations such as the remnant of a fossil or primordial field in the solar interior (Boyer and Levy 1984). Moreover, published modelling research addressing hemispheric asymmetries is somewhat scarce because, for simplification purposes, modelers tend to study dynamo solutions in only one hemisphere. Early, keystone works of dynamo modellers to reproduce hemispheric asymmetry include the following. Hoyng et al. (1994) focused on stochastic fluctuations in a 1.5D,[1] linear model and found phase and amplitude differences due to random fluctuations in the dynamo parameter α when applied with a correlation time on the order of the convective turnover time. Ossendrijver et al. (1996) furthered this type of work proposing that the stochastic fluctuations could be a consequence of giant convective cells and found some interesting evidence of transequatorial activity (not unlike that being found in current day models with "banana cells"). Other works that concentrated on backreactions of magnetic fields into flow components to cause hemispheric asymmetries were Sokoloff and Nesme-Ribes (1994) and Tobias (1997) who studied the effects on differential rotation. Another consideration is that an active coupling occurs deep in the interior when the leading edges of the N and S equatorward propagating toroidal bands "meet up" or come within a close enough range for flux cancellation and interaction to occur.

For the purpose of this paper, we review the literature dedicated to the observed hemispheric coupling (or lack thereof) in terms of amplitude and timing. We then discuss which models have been used to reproduce the observed hemispheric coupling but focus mainly on recent results from EULAG-MHD modeling efforts. We discuss which parameters or assumptions in the models most directly affect hemispheric coupling and address the following questions: are the observed hemispheric differences simply due to the possible stochastic na-

[1] In the technical jargon of this research field it is very common to use an integer number to represent spatial dimensions and 0.5 to make allusion to the time. Therefore 1.5D means one spatial dimension plus time.

ture of some dynamo ingredients? In the light of recent modeling results, which coupling mechanisms contribute significantly to maintaining the symmetry of the butterfly diagram?

2 Observations of N–S Hemispheric Asymmetry

2.1 Amplitude of Asymmetry Based on Sunspot Data

Temmer et al. (2006) created a sunspot data set separated by hemispheres from the Kanzelhöhe Solar Observatory (KSO) and Skalnaté Pleso Observatory for 1945–2004 (cycles 18–23). The Solar Influences Data Center (SIDC, now the World Data Center-SIDC) has hemispheric sunspot numbers from the Royal Observatory Belgium (ROB) from 1992 to present. These data have greatly assisted in the examinations of hemispheric asymmetry. Temmer et al. (2006) report that the solar-cycle minima from 1945–2004 were in phase while the maxima, declining and increasing phases were clearly shifted. They also found that the N–S asymmetry, based on an absolute N–S asymmetry index, was enhanced at cycle maximum, which is in contradiction with results obtained by Joshi and Joshi (2004) who found the N–S asymmetry was greatest as solar cycle minimum but they were using a normalized index. The asymmetry index for a cycle in terms of any index A (such as sunspot area or numbers) is considered normalized, Δ_{Norm}, if amplitudes A^N, A^S in the northern and southern hemispheres are normalized by the total in that cycle. As expected, they are considered absolute, Δ_{Abs}, when not normalized by the size of the cycle in that index.

$$\Delta_{Norm} = \frac{A^N - A^S}{A^N + A^S}, \qquad \Delta_{Abs} = A^N - A^S \tag{1}$$

Both the normalized and absolute asymmetry indices can be problematic. The absolute asymmetry index provides strong amplitudes around the times of sunspot maxima while the normalized asymmetry index provides strong signals around the times of minima.

Several studies of the statistical significance of hemispheric asymmetry have been carried out. Carbonell et al. (1993) report that in most cases, the N–S asymmetry is highly significant and cannot be obtained from a distribution of sunspot areas generated randomly from binomial or uniform distributions. Temmer et al. (2006) studied the monthly values of absolute asymmetry measures and found, for example, 63 % of the months during cycle 20 are asymmetric at the 99 % significance level. Carbonell et al. (2007) report that care should be taken when analyzing asymmetry due to different indices and binning techniques being employed. Ballester et al. (2005) remark that the time series generated from the normalized definition of asymmetry is misleading and that the correct asymmetry time series to be used is generated from an absolute difference between the northern and southern hemispheres. For these reasons, the plots in this paper do not depict either the absolute or normalized index, instead we show the absolute asymmetry index by plotting a hemispheric excess in relation to the sunspot cycle, see Figs. 2 and 3. No periodicities in the time series of the asymmetry indices are determined other than inferences on the timescale of the solar cycle. The asymmetry as measured by the hemispheric sunspot number is shown in Fig. 2 after smoothing with a 12-month window from 1945–2012 (McIntosh et al. 2013). The authors report it as a hemispheric "tug-of-war" for dominance. At any time, the differences in the hemispheric sunspot number is shown as red if the north has a higher sunspot number and blue if the south has a higher sunspot number. At any give time, the magnitude of the shaded portion in Fig. 2 is the absolute hemispheric asymmetry index in sunspot area; red if the Δ_{Abs} in Eq. (1) is positive or blue if negative. A slightly different measure, the absolute asymmetry index determined from sunspot area, is shown in Fig. 3 for a longer time span,

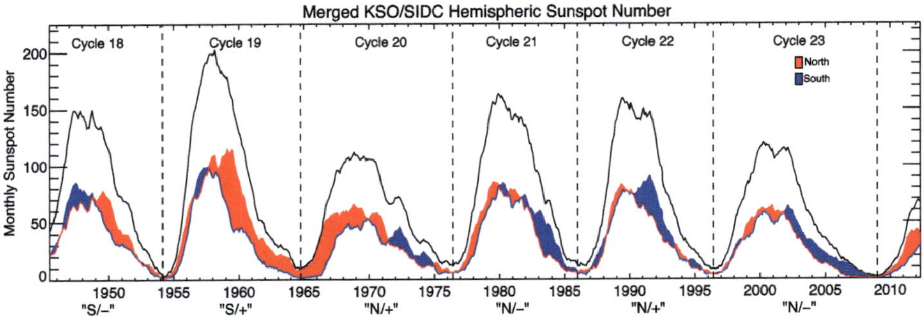

Fig. 2 International sunspot number by hemisphere combined from data provided by KSO and ROB/SIDC (e.g., Temmer et al. 2006) from 1945–2012. The N (*red*) and S (*blue*) hemispheric sunspot numbers are shown against the total (*black*). The shading of the difference between the N and S sunspot numbers indicates the excess between them. *Dashed vertical lines* are drawn at sunspot minima to delineate the cycles. The tags at *the bottom* indicate the hemisphere with most spots in the ascending phase of the cycle and the polarity of the magnetic field being advected into that hemisphere in the declining phase of the cycle. This figure reproduced from McIntosh et al. (2013)

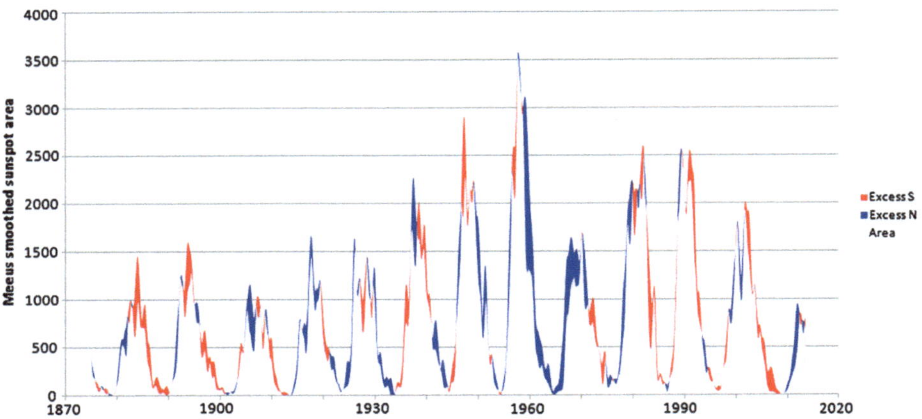

Fig. 3 Sunspot area by hemisphere combined from data provided by KSO and ROB/SIDC (e.g., Temmer et al. 2006) from 1876–2014. The excess N (*blue*) and S (*red*) (note: opposite colors compared to colors in Fig. 1) hemispheric sunspot areas are shown. This figure produced by J. Janssens (STCE: Solar-Terrestrial Centre of Excellence)

1870–2014. Unfortunately, the colors of N and S excess are reversed between Fig. 2 and 3 but the same behavior can be seen for cycles 18–23. Due to historical sunspot counts being somewhat contentious, the absolute sunspot area is preferred over the sunspot number as a diagnostic for asymmetry.

It may be useful to report on the amplitude of hemispheric asymmetry exhibited over an entire cycle. To do so, we report an average normalized asymmetry, Δ_{Norm}, which is determined from the difference in sunspot areas produced by each hemisphere over an entire cycle, then normalized by the total sunspot area produced in that cycle. For cycles 12–23, the average normalized asymmetry is 16 %, see Fig. 4 (Norton and Gallagher 2010). Using sunspot number instead of sunspot area, the average normalized asymmetry for cycles 18–23 is slightly different, $\sim$11 % (Norton and Gallagher 2010). Cycles 19 and 20 are the outliers showing much more hemispheric asymmetry than any other cycles. The N–S difference in sunspot area, $\sim$3000 μHem, divided by the total sunspot area produced in cycle 19 by both

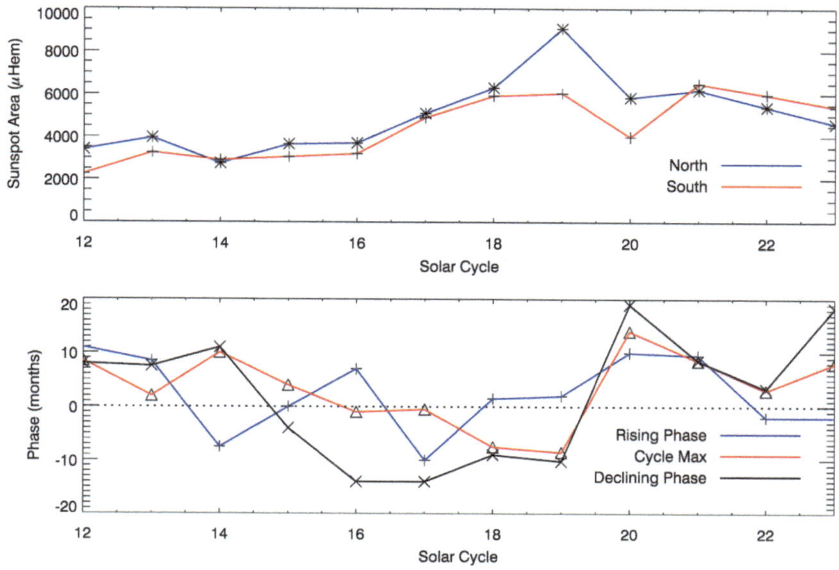

Fig. 4 *Top*: The strength, or size, of sunspot cycles 12–23 shown as the summed sunspot area for the N (*blue, asterisks*) and S (*red, crosses*) hemispheres. These data are from RGO sunspot area. *Bottom*: Hemispheric phase difference determined from RGO sunspot area for the rising, maximum and declining phases of each cycle. Positive values indicate the N is leading the S. Rising and declining phases are determined as the difference in time at which each hemisphere has produces 25 % of the maximum value for the rising and declining periods of the cycle. Phases associated with cycle maxima are determined as differences in time each hemisphere takes to produce half the total sunspot area

hemispheres, $\sim$15000 μHem, is 20 %. Cycle 20 is similarly very asymmetric while half of the cycles show a much smaller, roughly 10 %, difference. We provide an observed limit on the hemispheres not being more than 20 % asymmetric in sunspot area.

2.2 Hemispheric Phase Lag Based on Sunspot Data

Researchers calculate the hemispheric phase lag in a few different manners. McIntosh et al. (2013) finds the phase lag (reported in units of months) by a cross-correlation of a 120-month window of the N and S sunspot area time series, see Fig. 5. Norton and Gallagher (2010) found that the cross correlation method of determining phase lag was not optimal because it took the entire solar cycle into consideration, whereas it is evident that the phasing of the hemispheres can change midway through the cycle. They smoothed the sunspot area data over 24 months and normalized the resultant curves, then determined a hemispheric phase lag for the rising, maximum and declining epochs of each cycle. The rising phase is defined as the difference in time at which each hemisphere reaches 25 % of the maximum hemispheric sunspot area for that cycle (see Fig. 1 in Norton and Gallagher 2010). Similarly, the declining phase is defined as the difference in time when the sunspot area has fallen to 25 % of the maximum for that cycle and hemisphere. For cycle maxima, Norton and Gallagher (2010) summed the total sunspot area produced in each hemisphere and identified when half the total was produced. Hemispheric differences between these times are reported as the phases in units of months and plotted in Fig. 4, lower panel. Zolotova et al. (2010) measure the phase differences of sunspot cycles in the two hemispheres using the Cross-Recurrence Plot (CRP) method (Marwan et al. 2007), where two time series are embedded

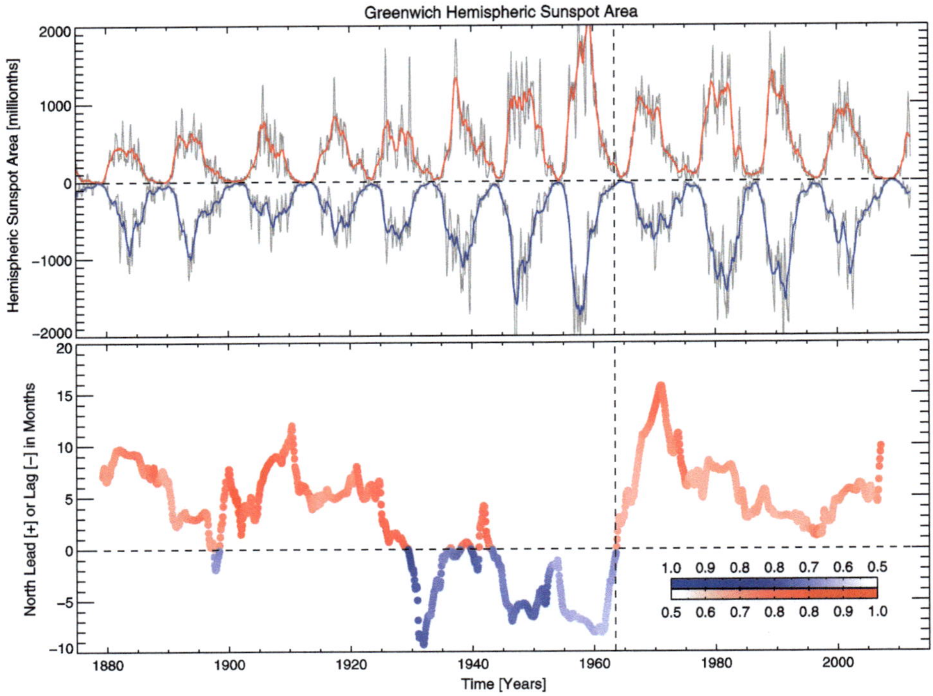

Fig. 5 *Top and bottom*: the Greenwich sunspot areas differentiated by hemisphere and the 120-month-averaged cross-correlation lag. Increasing color depth indicates the magnitude of the cross-correlation coefficient while the color itself dictates whether the north (*red*) or south (*blue*) is leading. This figure is reproduced from McIntosh et al. (2013), showing a proclivity towards the hemispheric lead to last for four cycles

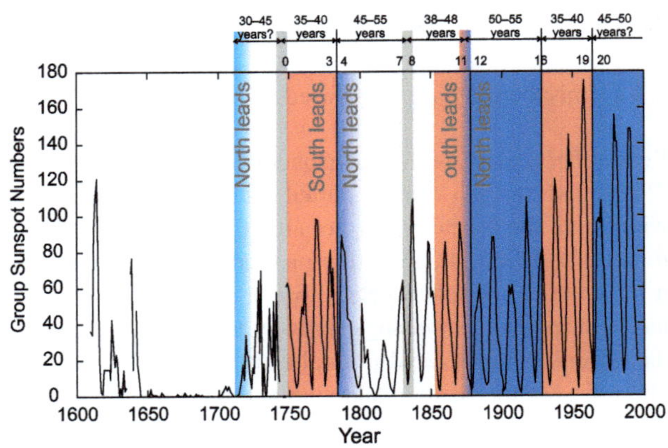

Fig. 6 Zolotova et al. (2010) plotted the yearly group sunspot numbers versus time. *Blue* indicated periods in which the northern hemisphere precedes the southern one. *Red* indicates that the southern hemisphere is leading in time. *Vertical lines* indicate a sign change of phase difference. When it is uncertain, *gray bars* indicate the probable time of a sign change

in the same phase space. They extracted phase information from sunspot data from the past four hundred years and show the persistence of phase over several cycles, see Fig. 6.

Zolotova et al. (2010) and McIntosh et al. (2013) independently came to similar conclusions about the hemispheric phase differences of the sunspot cycle. Namely: If the N–S phase difference exhibits a long-term tendency, a memory of sorts, it should not be regarded as a stochastic phenomenon. McIntosh et al. (2013) found a persistent phase lag that lasted roughly four cycles. Specifically, the N led for cycles 12–15, the S led for cycles 16–19 and then the N again led for cycles 20–23, see Fig. 5. Zolotova et al. (2010) investigated the phase difference of the sunspot cycles in the two hemispheres for a much longer duration as shown in Fig. 6, using historical records from Staudacher, Hamilton, Gimingham, Carrington, Spörer, and Greenwich observers, as well as the sunspot activity during the Maunder minimum reconstructed by Ribes and Nesme-Ribes. Zolotova et al. (2010) show that during the last 300 years, the persistence of phase-leading in one of the hemispheres exhibits a secular variation. Changes from one hemisphere leading to the other were registered near 1928 and 1968 as well as two historical ones near 1783 and 1875, see Fig. 6.

This is not the whole story, though. Other researchers (Li et al. 2002; Norton and Gallagher 2010) parsed the solar cycle into distinct periods of minimum, rising, maximum, and declining times, then found the phase relationship between the hemispheres during these times. (Of course, the smaller the time period analyzed, the smaller the statistical sample of data becomes, especially during cycle minimum). Li et al. (2002) found that less than half of the minimum time periods during cycles 12–23 (partial data for cycle 23) were significantly out of phase. In contrast, 75 % of the cycles were significantly out of phase at cycle maximum. Norton and Gallagher (2010) report the hemispheric phase difference for cycle 12–23 for the rising, maximum and declining cycle times, see Fig. 4. The phases shown for the declining periods are consistent with the picture painted by McIntosh et al. (2013) and Zolotova et al. (2010), with the last 4 cycles showing the N leading and hemispheric switches occurring at the end of cycle 19, and the cycles 16–19 being led by the S, etc. However, there is not a persistent 4 cycle phase found in the rising periods of the cycles. It is not clear from these reports whether the hemispheres truly are in-phase more often at cycle minimum and become out-of-phase as the cycle progresses, or whether the lack of data and the shorter duration of rise time simply make the phase difficult to ascertain at minimum and during the beginning of a solar cycle. It is worth keeping in mind, though, that the phase lag between the hemispheres may be set in the duration of each cycle, as only the sign of the phase is persistent for 4 cycles, not the phase value itself. The phase lags found by Norton and Gallagher (2010) for cycles 12–23 vary from 0–19 months translating into an upper limit of $\sim$15 % of the cycle length. Zolotova et al. (2010), who studied historical sunspot cycles back to 1750, found that the phase lag did not exceed 2 years, so we place an upper limit on the phase lag to be 20 %.

Hemispheres out of phase with one another may manifest in different latitudinal averages of sunspot locations. Phase-leading was found to be anticorrelated with the latitudinal distribution of the sunspots by Zolotova et al. (2009, 2010). They describe that the sunspots in the leading hemisphere show a butterfly wing emerging at slightly lower latitudes as compared to the butterfly wing of the delayed hemisphere, where sunspots emerged at slightly higher latitudes. The asymmetric placement of the butterfly wings suggests a magnetic equator offset from the geometric equator (Pulkkinen et al. 1999) which may result from one hemisphere leading the other in sunspot activity.

2.3 Polar Magnetic Fields

Polar fields are equally important since they present hemispherically asymmetric lag times of the field reversal and the different polar flux amplitudes. However, polar field reversals out of phase are not necessarily the best indicators of the hemispheres being out of phase because the high scatter in the tilt angles of bipolar sunspot pairs means that an emergence at high latitudes of an active region with a significant poloidal component (high tilt angle) and large amount of flux can influence the polar field reversal significantly. Basically, the low amount

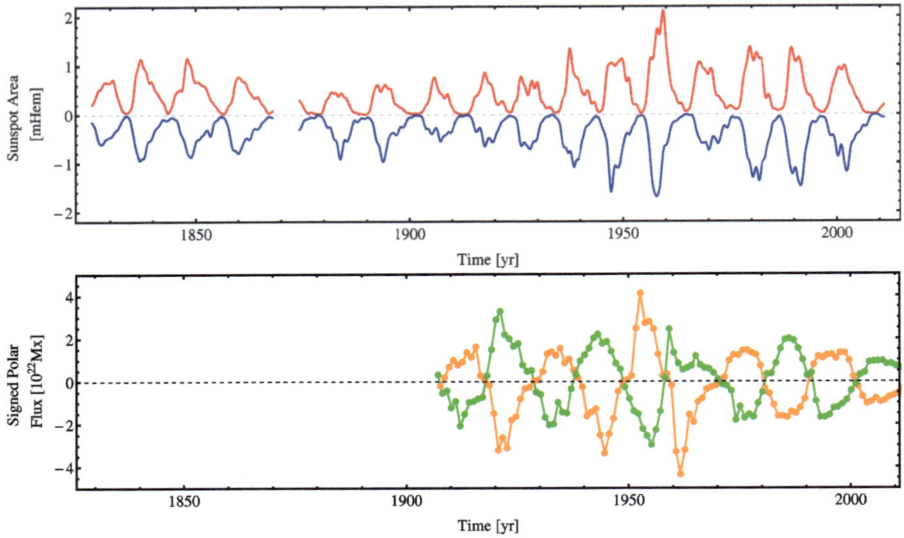

Fig. 7 *Top*: Smoothed daily sunspot area for the northern (*red line*) and southern (*blue line*) hemispheres. The gap corresponds to missing data on cycle 11. *Bottom*: Polar flux (based on magnetic and polar faculae observations) for the northern (*orange*) and southern (*green*) hemispheres. Adapted from Muñoz Jaramillo et al. (2013b)

of flux needed to reverse the poles means that the stochastic nature of flux emergence and tilt angles can affect it greatly (see Jiang et al. 2014, this issue). However, the surface dipole is used for predicting solar cycles and is a window into the possible magnetic behavior deep inside the convection zone.

Individual hemispheric data also show advantages over whole hemispheric data for prediction purposes. For example, Muñoz Jaramillo et al. (2013a, 2013b), showed that using the hemispheric polar flux as a precursor for the amplitude of the following cycle yields better results than when using whole sun measurements. Muñoz Jaramillo et al. (2013a) combine four campaigns observing the polar faculae from Mt. Wilson Observatory. The data are combined and calibrated with MDI /SOHO to obtain a consistent polar flux database. Using the polar faculae as a proxy for magnetic flux, Muñoz Jaramillo et al. (2013a) are able to provide a reliable time series of polar field dynamics from 1906 till 2013. In Fig. 7, we show the polar data used in these studies with an extended timeline of sunspot data also provided for reference. The top panel of this figure shows smoothed daily sunspot areas for each hemisphere. Cycles 7–10 were calibrated to RGO total area and SSN by Andrés Muñoz-Jaramillo using a compilation of Schwabe's data by Arlt et al. (2013). The details of the calibration and being prepared by Rainer Arlt and will be forthcoming later this year. For cycles 12 through 23, the data was reduced by Andrés Muñoz-Jaramillo using the data of Balmaceda et al. (2009). The polar flux curves are generated from the database of polar faculae observation observed at Mt. Wilson Observatory and calibrated from Muñoz Jaramillo et al. (2012). The details of this calibration are out of the scope of this review and we show the Fig. 7 for illustration purposes only.

How asymmetric and out of phase can the poles become? Babcock (1959) reported that the south polar field reversed its sign between March and July, 1957 and the north polar field, however, remained positive until November, 1958. He states that for "more than a year, the unexpected peculiarity was presented of two poles with the same sign". Cycle 19,

Table 1 Polar reversal lag times in months. Cycle 19 from Babcock (1959) while cycles 21–23 are from Wilcox Solar Observatory

Cycle	Leading hemisphere	Lag (months)
19	South	18
21	North	5
22	North	14
23	South (North)[a]	5 (5)
24	North	13

[a]Durrant and Wilson (2003) and Benevolenskaya (2007) report that the north reversed first using Kitt Peak and MDI data. There is a discrepancy since in the smoothed WSO data, it would appear the south reverses first.

then, had polar reversals that occurred eighteen months apart. Durrant and Wilson (2003) report that the northern polar field reversal occurred five months before the south in cycle 23 from analyzing Kitt Peak and Mt. Wilson Observatory data. Benevolenskaya (2007) confirmed this. Using Wilcox Solar Observatory (WSO) data, collected since 1975, one can estimate the polar reversal times for cycles 21–24 using field strength data smoothed over eighteen months (see http://www.solen.info/solar/polarfields/polar.html and Table 1). However, the times of polar reversal determined from the smoothed, low-resolution WSO data do not match the times of polar reversal determined from the higher-resolution Kitt Peak and Michelson Doppler Imager data. One might expect the lags (differences in times of reversal) to vary using different data, but in the case of cycle 23, the hemisphere that reversed first is not in agreement. Determining polar field reversal times can be tricky since it depends upon the defined latitude above which the summed flux would be either negative or positive as well as any polar field interpolation methods employed (Sun et al. 2011).

2.4 Miscellaneous: Phase Correlations with Cycle Characteristics, Dynamics

Other research turned up null results: Norton and Gallagher (2010) report no correlation between hemispheric phase lag and duration of cycle while Muraközy and Ludmány (2012) find no relationship between phase lag and strength of solar cycle. Norton and Gallagher (2010) also sought, but did not find, a correlation between the estimation of flux transport across the equator late in one cycle (based on Kitt Peak magnetogram data and an assumed surface diffusivity) and the phase of the hemispheres in the rising phase of next solar cycle. In terms of dynamics, Chowdhury et al. (2013) identified mid-duration and shorter periods in the asymmetry time-series data including a Rieger-type period ($\sim$155 days). Antonucci et al. (1990) used the low-resolution Wilcox Solar Observatory data to report that the central latitudes of sunspots were lower, the width of the latitudinal zone smaller and the rotation rate higher in the northern hemisphere (15°, 24° wide, 26.9 days synodic) compared to the southern hemisphere (26°, 32° wide, 28.1 days) for cycle 21. A similar finding was reported using Mt. Wilson cycle 20 data. The sunspots appearing at lower latitudes in the N for cycles 20 and 21 corroborates Zolotova et al. (2010) findings that the leading hemisphere produces sunspots at a slightly lower latitude than the delayed hemisphere.

2.5 Meridional and Zonal Flows

Local helioseismology techniques, such as ring diagram (Hill 1989) and time-distance (Duvall et al. 1993) analysis, are able to determine non-symmetric latitudinal structure in the solar interior. These analyses have been used to measure distinct hemispheric differences in the meridional flows and zonal flows at a given time and depth in the interior, see Komm et al. (2011) and Komm et al. (2013) and references included. These measured flow asymmetries provide further quantitative constraints on the dynamo simulations in that the simulations must reproduce asymmetries within the range observed.

The extent of the hemispheric coupling as determined by helioseismology is as follows. Zonal flows are seen as bands of faster and slower E–W flows that appear years prior to the appearance of activity on the solar surface. Zonal flow patterns may be caused by enhanced cooling by magnetic fields. Meridional flows are oft-considered the crucial ingredient which sets the rate at which the toroidal magnetic band (and sunspots) move equatorward. Komm et al. (2011) determined both the zonal and meridional flows as a function of latitude for cycle 23 (years 1996–2010), see Broomhall et al. (2014, this issue). They showed that meridional flow at 10–15 Mm in the northern hemisphere weakens in 2005 at 35°N latitude just before the northern surface magnetic contour disappears in 2006 (see Fig. 11 in Broomhall et al. 2014, this issue). Similarly, the southern hemisphere shows this behavior 2 years later in 2007 at 35°S latitude just before the southern magnetic contour disappears in 2008. This $\sim$2 year hemispheric phase lag observed in both the surface magnetism and the meridional flow is certainly suggestive.

Komm et al. (2013) also investigated the behavior of the zonal flows as a function of latitude for the time period of 2001–2013 from the surface to a depth of 16 Mm using GONG and HMI (as opposed to GONG and MDI data in the Komm et al. 2011 paper). Many hemispheric differences are evident in the zonal flows. For example, they find the poleward branch of the zonal flow (at $\sim$50° latitude) is 6 m s^{-1} stronger in the S at a depth of 10–13 Mm during cycle 23. In addition, Zhao et al. (2013) detected multiple cells in each hemisphere in the meridional circulation using acoustic travel-time differences. The double-celled profile shows an equatorward flow extending from approximately 0.92–0.83 $R_\odot$ with a speed of about 10 m s^{-1}. The poleward flow covers depths of 0.92–1.0 $R_\odot$ as well as 0.75–0.83 $R_\odot$. These flows show significant hemispheric asymmetry in a range of latitudes. The profile asymmetry could be due to a phase lag in the hemispheres. It begs the question: does the meridional profile as a function of depth in the southern hemisphere in 2013 look like the profile did in the northern hemisphere two years earlier?

A perturbation in the meridional flow (presumably by turbulence since the meridional flow is a weak flow strongly driven by convection) in one hemisphere (but not in the other) could set a phase lag between the migration of activity belts, and hence, the sunspot production, that persists for years. Although it is unlikely that the hemisphere that leads would be the same one for 4 cycles in this case. Actively searching for correlated hemispheric asymmetric signatures in flows at depth and magnetic field distributions on the surface may provide insight as to which ingredients of the dynamo set the length and amplitude of the sunspot cycle.

3 Dynamo Models: N–S Asymmetries and Hemispheric Coupling

Mean-field electrodynamics is well-covered in a number of textbooks and review articles (e.g., Moffatt 1978; Krause and Rädler 1980; Ossendrijver 2003; Brandenburg and Subramanian 2005; Charbonneau 2014). What follows is consequently brief, and focuses primarily on aspects of the theory concerned with turbulent transport of the mean magnetic field.

In mean-field and mean-field-like models of the solar cycle, the diffusive decay and transport of magnetic fields is almost always modeled as an isotropic Fickian (linear) diffusion, with the diffusion coefficient assuming large values, meant to reflect the enhanced dissipation by the small-scale turbulent flow pervading the convecting layers. Even within the framework of mean-field electrodynamics, the latter's impact on large-scale magnetic fields is in fact more complex. The mean-field dynamo equations derive from the assumption of scale separation, whereby the magnetic field and flow are decomposed as sums or large-scale (mean) and small-scale (turbulent) components, $\mathbf{B} = \langle\mathbf{B}\rangle + \mathbf{b}'$, $\mathbf{U} = \langle\mathbf{U}\rangle + \mathbf{u}'$, and the definition of an averaging operator such that $\langle\mathbf{u}'\rangle = 0$ and $\langle\mathbf{b}'\rangle = 0$. Under these conditions, introduction of scale separation into the MHD induction equation and averaging leads to the mean-field induction equation:

$$\frac{\partial\langle\mathbf{B}\rangle}{\partial t} = \nabla \times \big(\langle\mathbf{U}\rangle \times \langle\mathbf{B}\rangle + \boldsymbol{\xi} - \eta\nabla \times \langle\mathbf{B}\rangle\big), \tag{2}$$

where η is the magnetic diffusivity (inversely proportional to the electrical conductivity of the plasma) and $\boldsymbol{\xi} = \langle\mathbf{u}' \times \mathbf{b}'\rangle$ is the mean turbulent electromotive force (emf), appearing now in addition to the usual motional emf $\langle\mathbf{U}\rangle \times \langle\mathbf{B}\rangle$ associated with the large-scale flow and magnetic field. This turbulent emf can be developed as a Taylor series in terms of the mean magnetic field:

$$\mathcal{E}_i = \alpha_{ij}\langle B\rangle_j + \beta_{ijk}\frac{\partial\langle B\rangle_j}{\partial x_k} + \cdots, \tag{3}$$

where the tensors $\boldsymbol{\alpha}$, $\boldsymbol{\beta}$, etc, may depend on the properties of the flow, but not on $\langle\mathbf{B}\rangle$. Retaining only the first two terms in this expansion and separating the symmetric and anti-symmetric parts of the α-tensor as $\alpha_{ij} = \alpha^S_{ij} - \varepsilon_{ijk}\gamma_k$ then yields:

$$\mathcal{E}_i = \alpha^S_{ij}\langle\mathbf{B}\rangle_j + \big[\boldsymbol{\gamma} \times \langle\mathbf{B}\rangle\big]_i + \beta_{ijk}\frac{\partial\langle\mathbf{B}\rangle_j}{\partial x_k}, \tag{4}$$

with $\gamma_i = -(1/2)\varepsilon_{ijk}\alpha_{jk}$. Upon substitution into Eq. (2), $\boldsymbol{\gamma}$ emerges as a vectorially additive contribution to the large-scale flow $\langle\mathbf{U}\rangle$. This pseudo-flow is known as turbulent pumping, and acts as a transport agent for $\langle\mathbf{B}\rangle$. In the case of near-homogeneous, near-isotropic turbulence, the isotropic (diagonal) component of α^S_{ij} and β_{ijk} reduce to

$$\alpha = -\frac{\tau_c}{3}\langle\mathbf{u}' \cdot \nabla \times \mathbf{u}'\rangle, \tag{5}$$

$$\beta = \frac{\tau_c}{3}\langle(\mathbf{u}')^2\rangle, \tag{6}$$

and turbulent pumping becomes proportional to the gradient of turbulent intensity:

$$\boldsymbol{\gamma} = -\frac{\tau_c}{3}\nabla\langle(\mathbf{u}')^2\rangle, \tag{7}$$

where τ_c is the correlation time of the turbulent flow, usually assumed to be commensurate with the convective turnover time. The mean-field induction equation (Eq. (2)) then becomes:

$$\frac{\partial\langle\mathbf{B}\rangle}{\partial t} = \nabla \times \big((\langle\mathbf{U}\rangle + \boldsymbol{\gamma}) \times \langle\mathbf{B}\rangle + \alpha\langle\mathbf{B}\rangle - (\eta + \beta)\nabla \times \langle\mathbf{B}\rangle\big)\,. \tag{8}$$

Note that the isotropic part of the $\boldsymbol{\alpha}$ tensor describes an electromotive force aligned with the mean field $\langle\mathbf{B}\rangle$, and the $\boldsymbol{\beta}$-tensor appears as an additive contribution to the magnetic diffusivity η, with order-of-magnitude estimates based on Eq. (6) indicating that $\beta \gg \eta$. This is the turbulent diffusion introduced in virtually all mean-field and mean-field-like models of the solar cycle. Note already that β captures only one specific contribution of the small-scale flow to the transport and dissipation of $\langle \mathbf{B} \rangle$.

In the modelling of the solar cycle, the large-scale magnetic field $\langle \mathbf{B} \rangle$ is considered axisymmetric, and the natural averaging operator becomes a zonal average (i.e. in the ϕ direction). The large-scale flow $\langle \mathbf{U} \rangle$ includes (axisymmetric) contributions from differential rotation and meridional flows. Both are ultimately driven by turbulent Reynolds stresses and latitudinal temperature differences (see Sect. 5.3 in review by Karak et al. 2014, this issue) and so are symmetric about the equatorial plane on timescales longer than the convective turnover time. Turbulent pumping is directed primarily downwards throughout the bulk of the convection zone (Tobias et al. 2001), but in the physical regime where convection is significantly influenced by rotation, a latitudinal, equatorward component of turbulent pumping also materializes, reaching speeds comparable to the meridional flow (see, e.g., Ossendrijver et al. 2002; Racine et al. 2011; also Fig. 12 in Karak et al. 2014, this issue). This latitudinal component of $\boldsymbol{\gamma}$ may contribute significantly to the observed equatorward drift of the sunspot butterfly diagram.

3.1 Mean-Field Dynamo Models

In classical mean-field dynamo models of the solar cycle, differential rotation is producing the toroidal large-scale magnetic field, while the turbulent emf is responsible for the regeneration of the poloidal component. However, in the solar dynamo context, additional magnetic source terms may appear on the RHS of Eq. (8). In particular, the surface decay of bipolar active regions and subsequent poleward transport of decay products by surface diffusion and meridional flow can contribute to the dipole moment, thus acting as a source term for the poloidal field, a process known as the Babcock-Leighton mechanism. Various MHD instabilities of the strongly toroidal flux system accumulating in the tachocline, immediately beneath the base of the convection zone, can also act as sources of poloidal field. For further discussion of these dynamo models see Sect. 4 in Charbonneau (2010).

With the large-scale flows and tensors $\boldsymbol{\alpha}$ and $\boldsymbol{\beta}$ considered given, Eq. (8) becomes linear in $\langle \mathbf{B} \rangle$, and accepts eigensolutions with temporal dependence $\propto \exp((\sigma + i\omega)t)$. The growth rates σ for the lowest odd (dipolar) and even (quadrupolar) eigenmodes are often similar, with high diffusivity typically favoring the very lowest eigenmode (usually dipolar), while dominant transport by a quadrupolar meridional flow will often favor the quadrupolar eigenmode. The tendency for coexistence of dipolar and quadrupolar solutions can persist in the nonlinear regime, and in itself lead to hemispheric asymmetries and attendant modulation of the magnetic cycle's amplitude. The key is then to ensure that the non-dominant eigenmodes are continuously excited to some significant amplitude. It turns out that this can be achieved though a variety of mechanisms, reviewed in what follows.

3.2 Stochastic Fluctuations

Since the dynamo resides in the solar convection zone, a very turbulent place, it is only natural to assume that the physical mechanisms that operate in this region can be perturbed by convective turbulence (Hoyng 1988). The most common recipe applied to simulate this perturbative effect is to add stochastic fluctuations to the chosen physical mechanism, at a correlation time associated with the phenomenon's characteristic time-scale (Choudhuri 1992; Moss et al. 1992). In mean-field dynamo models the usual targets are the α-effect and the meridional circulation. By definition, the large scale meridional circulation arises from longitudinal averaging of the global velocity field. The flow speed associated with this weak flow (of the order of a few m s^{-1}) is much lower than the velocities of turbulent eddies, which makes it susceptible to fluctuations. The origin of fluctuations in the mean-field α-effect is based on the same principle and are supported by some highly turbulent global 3D MHD simulations of solar convection (Ossendrijver et al. 2001; Racine et al. 2011). In addition, significant fluctuations in the Babcock-Leighton mechanism are also expected because of the large variation in the tilt angle distribution of bipolar solar active regions (Longcope and Fisher 1996; Dasi-Espuig et al. 2010). This type of modelling methodology can produce hemispheric asymmetric solutions because the fluctuations applied are usually spatially uncorrelated (different fluctuations at different latitudes and depths or different in both hemispheres).

In Hoyng et al. (1994) the authors use a simplified 1.5D (θ, t) mean-field dynamo and study the effects of random fluctuations in their α source term (mean helicity). These fluctuations are a function of latitude and are applied at a correlation time of the order of the turnover time of convection (much shorter then the solar cycle period). They show that by increasing the level of fluctuations they produce larger hemispheric asymmetries between the wings of their simulated butterfly diagram. In this model the interaction between fundamental and higher order excited eigenmodes is the source of the hemispheric asymmetries observed. Ossendrijver et al. (1996) used an $\alpha\omega$ mean field dynamo in (r, θ, t) to further test this idea. They propose that the origin of the stochastic fluctuations in the α-effect could be a consequence of giant convective cells. Under certain parameter regime their model also produces transequatorial magnetic activity.

A different approach to the implementation of stochastic fluctuations was taken by Goel and Choudhuri (2009). These authors suggest that the random nature of the Babcock–Leighton mechanism can make the poloidal field in one hemisphere stronger than the other inducing therefore an hemispheric asymmetry. They test this idea by feeding into a BL flux transport dynamo model observational polar flux data and find a good correlation between the simulated cycle and observation. Their methodology is analog to that presented in Dikpati et al. (2006) and sets a preliminary framework for using mean field models for prediction purposes.

More recently (Passos et al. 2014) developed a mean-field flux transport dynamo model that incorporates a dual poloidal source formalism, namely a Babcock-Leighton mechanism, α_{BL} that acts on strong toroidal fields that buoyantly rise to the near surface layers and a mean-field classical poloidal source, α_{MF} that acts only in weak toroidal fields that diffuse through the convection zone. The Babcock-Leighton mechanism was held as the leading source, i.e. $\alpha_{BL} > \alpha_{MF}$. Nevertheless the relative contribution from both terms is a bit more difficult to evaluate because while α_{BL} is confined to a thin region near the surface, the α_{MF} although lower in amplitude spreads across a larger area, the bulk of the convection zone. The authors study several scenarios where stochastic fluctuations are added to these two source terms individually or simultaneously. According to flux tube simulations, e.g. Caligari et al. (1995), the rise time of these objects through the CZ is of the order of months. Since that is the time that flux tubes are exposed to turbulent buffeting, the authors chose a correlation time of 6 months for the fluctuations in the BL source term and 1 yr for fluctuations in the mean-field source term. As previously mentioned in the beginning of this section, the amplitudes of the fluctuations are motivated by the large variation of the tilt angle distribution in active regions and by the eddy velocities distributions and large scale flows amplitude variations measured in several 3D MHD global simulations of solar convection. This model returns a wide range of solutions, from high to low hemispheric coupling all the way to grand minima. For illustrative purposes we present in Fig. 8 an example of a solution computed with this model using the following level of fluctuations: 100 % for $\alpha_{MF} = 0.4$ and 25 % for $\alpha_{BL} = 21.0$ (see Passos et al. 2014 for details).

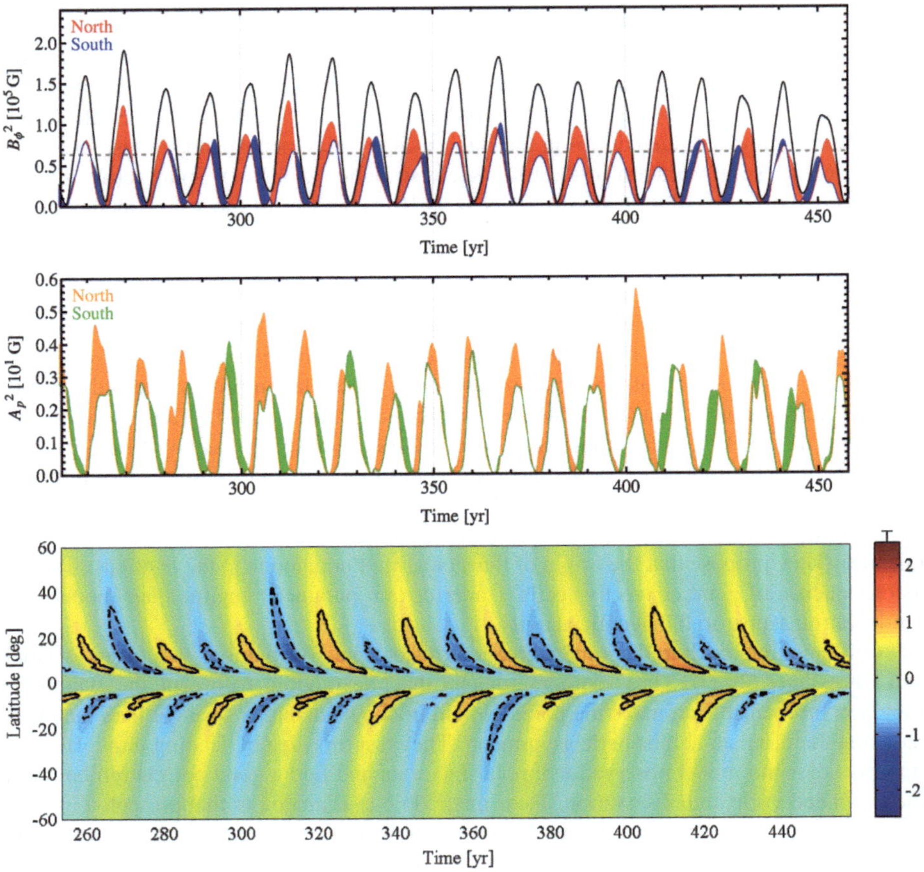

Fig. 8 Sample of a simulation using the model of Passos et al. (2014). Represented here are in *the top panel*, the squared toroidal field measured just above the tachocline at $r = 0.706\ R_\odot$ at active latitudes (14°N (*red*) and 14°S (*blue*)). *The black thick line* represents the sum of the two hemispheric contributions. *The thin dashed line* indicates the buoyancy threshold at $B_\phi = 0.8 \times 10^5$ G. *The middle panel* shows the squared poloidal field measured at near surface layers at $r = 0.96\ R_\odot$ and near the north (*orange*) and south (*green*) poles. *The bottom panels* displays the toroidal field measured at $r = 0.706\ R_\odot$ as a function of latitude and time. *The black contour lines* enclose the regions where the buoyant emergence of toroidal field takes place (analog to sunspot emergence locations). *Solid* (*dashed*) *contours* represent positive (negative) polarity. In this sample solution, dipolar parity is maintained

This dynamo model also incorporates a buoyancy algorithm that takes toroidal field exceeding a buoyancy threshold at the base of the convection zone and places it in the near surface layers, at the same latitude. This is done to emulate the fast rise of flux tubes (Chatterjee et al. 2004). It is then possible to plot the analog of a butterfly diagram with the emergence latitudes highlighted (see Fig. 8C). In this example we can observe that the two hemispheres are moderately coupled but we also see phase lags between both hemispheres (top panel) and different hemispheric emergence rates as seen in the different shapes of the butterfly wings of each cycle in Fig. 8.

In this work the decoupling between hemispheres increases with the fluctuation level. The authors also find evidence that hemispheric decoupling is naturally associated with, and symptomatic of grand minima episodes, a result that also was proposed by Olemskoy and

Kitchatinov (2013). The next step for this model its a complete mapping of the parameter space coupled with statistical constraints from long reconstructions of solar activity.

A more general scenario that considers a superposition of nonlinear quenching and stochastic forcing, was presented by Mininni and Gómez (2002). They subjected a 1.5D $\alpha\omega$ model, containing a nonlinearity in the form of α-quenching to stochastic fluctuations in this parameter. The solutions obtained with this model are in good agreement with observations and, according to the authors can be explained as the interaction of the deterministic part of the solution with overtones excited by the forced stochastic fluctuations. The asymmetries in their model arise as the combination of stochastic effects and deterministic coupling, and evolution of mixed-parity excited modes.

3.3 Non-kinematic Mean-Field Models

Non-kinematic dynamo models that include some aspects of the backreaction of the magnetic field on the large-scale flow components, can also show hemispheric asymmetries in their solutions. That is the case of the work by Sokoloff and Nesme-Ribes (1994), where the authors propose that the observed north-south asymmetry in the solar cycle amplitude and phasing could be a consequence of mix-parity solutions caused by nonuniform rotation. In these mix-parity solutions, the dipole and quadrupole components can interact leading to hemispheric asymmetries. Also using a nonlinear $\alpha\Omega$ mean-field model (although more complex) that included the Malkus-Proctor effect (the feedback of the magnetic field into the differential rotation), Tobias (1997) arrived at similar conclusions. Moreover, this author argues that relevant asymmetries only arise from mixed-parity solutions when the dynamo exhibits weak-field solutions (like in a grand minimum) and that the Malkus-Proctor effect has a more profound impact in the asymmetry for dipolar strong-field solutions. Building on the results and model of Tobias (1997), Bushby (2003) studied solutions with higher degree of asymmetry, and discussed these in the more general context of solar-like stellar dynamos.

Pipin (1999) considered a weakly nonlinear $\alpha\Lambda$ mean field in a spherical shell. In this type of model the differential rotation is maintained by non-dissipative sources in the angular momentum transport, the so-called Λ terms, parameterizing Reynolds stresses. Solutions produced by this model indicate that magnetic feedback on angular momentum fluxes produces a long term cyclic modulation that resembles the sunspot's Gleissberg cycle. The asymmetries between hemispheric activity comes from parity breaking, which in this model is always connected with breaking of the symmetry of differential rotation.

3.4 Global MHD Simulations of Convection

The production of large-scale axisymmetric magnetic fields undergoing polarity reversals in strongly turbulent global (full-sphere) MHD simulations of solar convection is a recent development. Computationally, the problem is quite challenging, as this large-scale magnetic field evolves on spatial and temporal scales much larger/longer than convection itself. When such large-scale fields build up and undergo polarity reversals, most often the latter occur at a irregular cadence, and often strong hemispheric asynchrony and asymmetries are present. Working with the PENCIL code in spherical wedge geometry Käpylä et al. 2012 find in some parameter regimes a fairly regular cycle with a well-defined period with sustained phase lags between hemisphere (see their Fig. 3); but more typically the large-scale magnetic cycles materializing in their simulations are extremely irregular (see Figs. 4 and 5 in Käpylä et al. 2013). The ASH simulations of Brown et al. (2011) have produced short-period (a few yr) cycles showing large hemispheric asymmetry and asynchrony (see their

Fig. 1B), a situation improved upon to a significant extent in later ASH simulations operating in a more strongly turbulent regimes (see Figs. 3, 6 and 15 in Nelson et al. 2013). The recent global MHD simulations of Fan and Fang (2014), on the other hand, produce polarity reversals on decadal timescales and well-synchronized across hemisphere, however with large cycle-to-cycle variations in the half-period (see their Fig. 5).

To the best of our knowledge, the global simulations of Ghizaru et al. (2010) and Racine et al. (2011) are currently producing the most regular cycles, with polarity reversals well-synchronized across hemispheres. Some tantalizingly solar-like secondary dynamical features are also reproduced, including rotational torsional oscillations and modulation of convective energy transport, both with the observed phasing and amplitude inferred on the sun (Beaudoin et al. 2013; Cossette et al. 2013). As a computational avatar of the real solar cycle, significant discrepancies remain: the cycle period is four times too long as compared to the sun, the deep-seated magnetic field is concentrated at mid-latitudes and exhibits very little equatorward propagation, and the dipole moment is over ten times stronger than the solar one and oscillates in phase with the deep toroidal component, in contrast to the $\pi/2$ phase lag observed in the sun.

In what follows we focus on one specific EULAG-MHD simulation spanning 1600 yr, in the course of which 39 polarity reversals have taken place. This is a low-resolution simulation ($128 \times 64 \times 48$ in longitude$\times$ latitude$\times$radius) spanning $0.604 \leq r/R \leq 0.96$, mildly superadiabatic above $r/R = 0.711$ and strongly subadiabatic below, with stress-free impenetrable upper and lower boundaries on which the horizontal magnetic field components are forced to vanish. The presence of a stably stratified fluid layer underlying the convectively unstable layers appears to be an important agent favoring self-organization of the magnetic field on large spatial scales in these simulations (on this point see also Browning et al. 2006). The magnetic cycle developing in this "millennium simulation" is analyzed in detail in Passos and Charbonneau (2014), to whom we refer the interested reader for a full description of cycle characteristics, period-amplitude relationships, etc.

Figure 9A and B shows two meridional cuts of the zonally-averaged toroidal magnetic field, at epochs of (A) cycle maximum and (B) minimum. The large-scale, axisymmetric toroidal magnetic component is antisymmetric about the equatorial plane, and accumulates immediately beneath the base of the convecting layers in response to downwards turbulent pumping, reaching there strengths of a few tenths of Tesla. Figure 9C shows a time-latitude diagram of the zonally-averaged toroidal magnetic field, extracted at the base of the simulated convection zone (dashed circular arc on Fig. 9A and B). This is the simulation's equivalent of the sunspot butterfly diagram, under the assumption that the sunspot-forming toroidal magnetic flux ropes form in regions of highest magnetic intensity and rise radially to the surface. Figure 9D is a similar time-latitude diagram, this time for the evolution of the zonally-averaged radial magnetic field in the subsurface layers of the simulation ($r/R = 0.94$, with the upper boundary of the simulation domain at $r/R = 0.96$). The long term spatiotemporal stability of the magnetic cycle is well evidenced on these time-latitude diagrams.

The sunspot number proxy constructed from the simulation output is the squared magnetic flux in an integration domain located at mid-latitudes and straddling the base of the convection zone, as indicated on Fig. 9 by the black box. Similarly, a polar field proxy is constructed by integrating the radial magnetic field at high latitudes in the subsurface layers, indicated here by the dashed box (see also Passos and Charbonneau 2014). Figure 10 shows time series of these two proxies, calculated independently for the northern (red and orange) and southern (blue and green) hemispheres. Only a 500 yr segment of the 1600 yr simulation is plotted, covering twelve cycles. We follow the usual convention of numbering

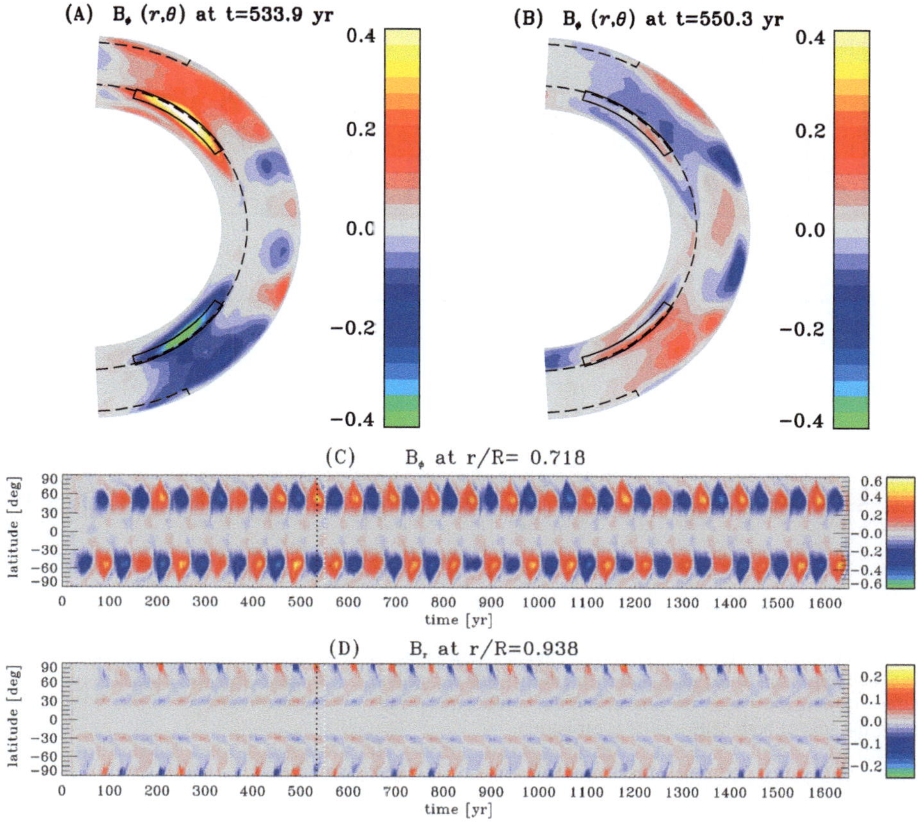

Fig. 9 Meridional snapshots of the zonally-averaged toroidal magnetic component taken at the time of solar maximum (**A**) and minimum (**B**). Panels (**C**) and (**D**) show time latitude diagrams for the toroidal component sampled at tachocline depth and the radial magnetic component in the near surface layers, respectively. The color scales codes the magnetic field strength, in Tesla, and the *dashed circular arc* in (**A**) and (**B**) indicates the boundary between the convectively unstable region and the underlying stable fluid layer, located at $r/R = 0.718$; the time-latitude diagram in (**C**) is extracted at that same depth, and that in (**D**) at $r/R = 0.94$. The upper boundary of the simulation domain is at $r/R = 0.96$. The meridional snapshots in (**A**) and (**B**) are extracted at the times marked in the time latitude diagrams with *the dotted vertical lines*, *black* for maximum and *white* for minimum. *The boxes* in these panels represent the areas of integration for the toroidal (*solid box*) and poloidal (*dashed box*) proxies. These results are taken from the global MHD simulation of solar convection analyzed in Passos and Charbonneau (2014)

cycles from one minimum to the next, and refer to the corresponding time span as the "cycle period", even though the period of the underlying magnetic cycle covers two such cycles.

Each pair of hemispheric proxy time series correlate rather well as a whole, reflecting primarily the good hemispheric synchrony of polarity reversals. However, the peak amplitudes for individual cycles in the north and south do not ($r = -0.24$ and $+0.23$ for the toroidal and dipolar proxies, respectively). Within an hemisphere, the peak amplitudes of the toroidal and dipolar proxies do correlate well with one another ($r = +0.58$ and $+0.66$ in the north and south, respectively). This is suggesting that once underway, cycles in each hemisphere rise and saturate independently of one another. We define as follows an asymmetry parameter Δ_n for cycle n in terms of the peak amplitudes A_n^N, A_n^S of the proxies in the northern and southern hemispheres. This parameter is similar to the normalized asymmetry

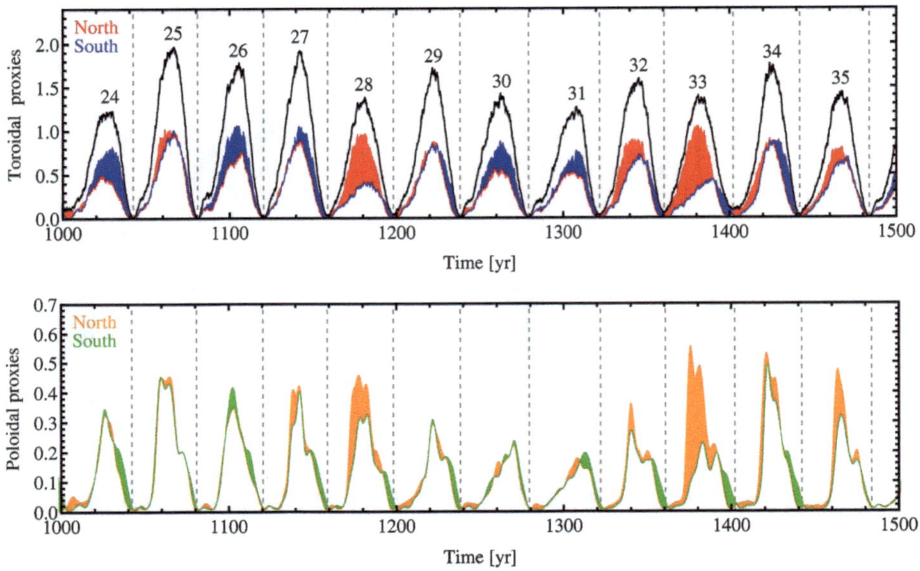

Fig. 10 A 500 yr time series segments of the hemispheric toroidal (**top**) and poloidal (**bottom**) proxies in the northern and southern hemispheres, in the EULAG-MHD "millennium simulation" analyzed by Passos and Charbonneau (2014). The simulated cycles numbers are shown for reference. Cycles are defined as spanning from one minimum to the next, and each corresponds to one half of the underlying full magnetic cycle. *The dashed vertical lines* mark the time of minimum measured for the whole sun toroidal proxy (*black thick line*) that is the sum of both hemispheric contributions

Fig. 11 Time sequence of the hemispheric asymmetry parameter Δ_n (in *red*), as defined through Eq. (9) for the toroidal proxy, calculated for the 39 cycles in the simulation. The corresponding time sequence of hemispheric lag, defined as the difference between the time of cycle minimum in the north minus the time of cycle minimum in the south, is plotted in *black*

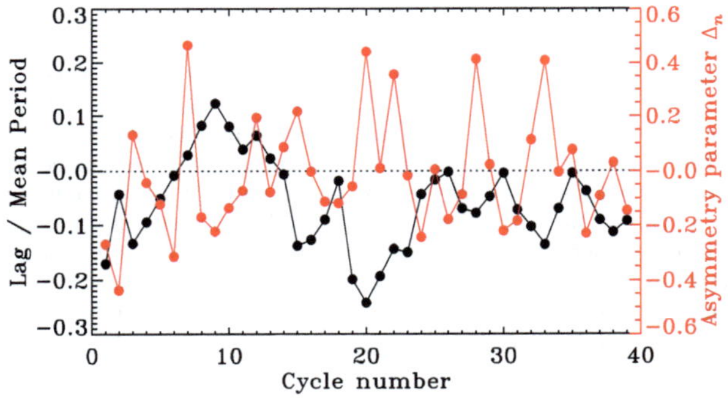

parameter presented in Sect. 2.1:

$$\Delta_n = \frac{A_n^N - A_n^S}{A_n^N + A_n^S}, \quad n = 1, \ldots, 39 \tag{9}$$

so that $\Delta_n \to +1$ (-1) if the northern (southern) hemisphere dominates, and $\Delta_n = 0$ if both hemispheres exhibit the same cycle amplitude. The time sequence of Δ_n values for their toroidal proxies is plotted in red on Fig. 11; the corresponding sequence for the polar field proxy is very much similar. The asymmetry parameter reaches values as large as ± 0.5 for some cycles, which is much larger than for the sunspot cycles on Fig. 5.

Over the 20 full magnetic cycles covered in the 1600 yr of this specific simulation, The average hemispheric (half-)cycle periods are virtually identical, 40.65 ± 1.81 yr in the north, and 40.57 ± 1.65 yr in the south. Moreover, as already noted, polarity reversals maintain a very good synchrony in both hemispheres, hinting at some significant level of cross-hemispheric coupling, even though the hemispheric cycle amplitudes are not well-correlated. The time sequence in black on Fig. 11 shows the time sequence of hemispheric lag values for successive cycles defined here as the time of cycle minimum in the north minus that in the south, as defined by the toroidal proxies (so that a negative lag indicate that the north leads the south). The average lag is 2.3 yr over the whole simulation, or about 6 % of the cycle period. The final cumulative lag after 39 cycles amounts to a mere 10 % of the cycle period for the toroidal proxy, and only 2 % for the polar field proxy (not shown). At first glance the lag seems to execute a form of random walk, with the lag increasing or decreasing by the same amount (within a factor of two) from one cycle to the next. The asymmetry parameter, in contrast, shows a much more abrupt cycle-to-cycle variation. The plot is also suggestive of an anticorrelation of the amplitude asymmetry parameter, Δ_n, with the lag values, but the correlation coefficient turns out to be rather low, $r = -0.31$.

4 Coupling Mechanisms

The various models and simulations reviewed in the preceding section exhibit varying levels of cross-hemispheric coupling, in part because different physical coupling mechanisms are at play, and/or operate in distinct physical regimes. In this section we focus specifically on these mechanisms, and discuss under which circumstances they can (or cannot) achieve cross-hemispheric coupling.

4.1 Magnetic Diffusivity

The last term on the RHS of Eq. (2) embodies resistive dissipation of the large-scale magnetic field. In a situation where $\beta \gg \eta$ and is dependent on position, this resistive term can be developed as

$$-\nabla \times \left(\beta \nabla \times \langle \mathbf{B} \rangle\right) = -(\nabla \beta) \times \left(\nabla \times \langle \mathbf{B} \rangle\right) + \beta \nabla^2 \langle \mathbf{B} \rangle, \tag{10}$$

The first term on the RHS is known as diamagnetic transport; it is fundamentally distinct from the turbulent pumping introduced in Sect. 3, as it can arise in a situation were turbulence is isotropic (the $\boldsymbol{\alpha}$ and $\boldsymbol{\beta}$ tensors are diagonal), provided the β coefficient varies with position. True turbulent pumping, in contrast, materializes only when turbulence is anisotropic, as captured in the off-diagonal component of the $\boldsymbol{\alpha}$-tensor. The second term on the RHS has the form of a classical Fickian (linear) diffusion, with β acting as a diffusion coefficient. This offers a prime mechanism to achieve cross-hemispheric coupling. However, of all dynamo ingredients required in mean-field and mean-field-like dynamo models, the turbulent magnetic diffusivity β is perhaps the most difficult quantity to estimate reliably from first principles, and currently the one for which the least direct observational constraints are available.

While most can agree that it is turbulent and anisotropic, i.e. there is greater meridional diffusion than radial diffusion due to the rotational influence on convection, we can agree on little else. For example, measurements of β at the surface determined from observations vary depending on the feature and size-scale being studied. The values for magnetic diffusion reported from observational studies are: 600 km^2 s^{-1} for supergranules (Simon et al.

1995), to 200 $km^2\,s^{-1}$ for small-scale magnetic elements (Komm et al. 1995), 60 $km^2\,s^{-1}$ for granular flows (Berger et al. 1998), and 1–5 $km^2\,s^{-1}$ for high-resolution plage flows (Chae et al. 2008). The disagreement of the correct range of magnetic diffusion as a function of radius is equally apparent in modeling and simulation efforts. Two examples of assumed diffusion values (in units of $cm^2\,s^{-1}$) as a function of solar radius are 6×10^{11} near the surface, peaking at 0.93 $R_\odot$ with a value of 14×10^{11}, declining gradually to 1×10^{11} at 0.72 $R_\odot$ (Pipin and Kosovichev 2013). In contrast, Dikpati et al. (2004) assume a diffusivity profile that peaks at the surface at 2×10^{12} (a supergranular value) declining to a constant value of 5×10^{10} from 0.73–0.90 $R_\odot$. A β of 2×10^{11} $cm^2\,s^{-1}$ means the magnetic field diffuses the depth of the convection zone in 50 years, but diffuses to mid-depth in 10 years.

Consider, in the descending phase of the solar cycle, the two toroidal flux systems located on either side of the equatorial plane, peaking at $\pm 8°$ latitude (say). This corresponds to a linear distance $L \simeq 0.07$ $R_\odot$. The timescale for these two structures to diffusively annihilate is $\tau = L^2/\beta$; this is equal to the solar cycle period $\simeq 10$ yr for $\beta \simeq 10^{13}$ $cm^2\,s^{-1}$. For a turbulent diffusivity in excess of this value, one can thus expect diffusive cross-hemispheric coupling to act on timescales shorter than the cycle period, i.e., strong coupling. Conversely, for values of β much smaller, the diffusive cross-hemispheric coupling is correspondingly weaker. In this latter situation, any hemispheric lag or amplitude asymmetry, of whatever origin, can endure for many magnetic cycles (see, e.g., Dikpati et al. 2004; Charbonneau 2007; Chatterjee et al. 2004). These two physical regimes have been dubbed "diffusion-dominated" and "advection-dominated" (see, e.g., Yeates et al. 2008). Transequatorial diffusive transport also has a strong impact on the parity of the dynamo modes, with high magnetic diffusivity favoring the odd (dipolar) mode, characterized by the longest spatial scale, while in the advection-dominated regime the even (quadrupolar) parity is imprinted on the dynamo (Hotta and Yokoyama 2010).

The distinction between these two physical regimes becomes particularly important in the so-called flux-transport dynamos (see Karak et al. 2014, this issue), in which the meridional flow drives the equatorward drift of the deep-seated toroidal magnetic field ultimately giving rise to sunspots (at least according to the prevalent view on the matter; see Schmieder et al. 2014, this issue). Even then high diffusivity is no universal panacea. For example, Chatterjee and Choudhuri (2006) have studied dynamo solutions using a Babcock-Leighton-type flux transport model in which small, temporally steady differences are introduced between hemispheres in the amplitude of the meridional flow (1 % hemispheric difference) and/or in the magnitude of the surface source term (2 % hemispheric difference). Even for such small imposed hemispheric differences, persistent phase lag and even distinct hemispheric periodicities can develop. Typically, in their model higher diffusivity is found to help couple the solutions in both hemispheres and establish a single period. These authors also point out that the presence of a weak turbulent α-effect operating throughout though the convection zone also helps to keep the solution locked in a dipolar parity, in agreement with earlier findings (Dikpati and Gilman 2001).

Intermittency in the advection-dominated regime also tends to show strong hemispheric asymmetries, with one or the other hemisphere entering Grand Minima state while the other remains in "normal" cyclic mode (see, Charbonneau 2007 and Passos et al. 2014 for specific examples). Once again diffusive cross-hemispheric coupling can alleviate in part this difficulty, and synchronize grand minima episodes in both hemispheres. This behavior was explored in Charbonneau (2005), using a simple but well-validated one-dimensional iterative map describing the variation of successive cycle amplitudes. More elaborate mean-field and mean-field-like models using higher magnetic diffusivity values, i.e. operating closer to or in the diffusion-dominated regime, tend to produce better hemispheric synchrony in

Grand Minima. They also show a tendency for recovery to start preferentially in one hemisphere (see Fig. 8 in Passos et al. 2014), as was observed at the end of the Maunder Minimum (Ribes and Nesme-Ribes 1993). This latter behavior had been invoked in support of the ideas that deterministically-driven parity modulation lead to Grand Minima (Sokoloff and Nesme-Ribes 1994; Beer et al. 1998). Here it materializes simply because as one hemisphere shuts down, diffusive leakage from the other can shut it down as well after some temporal delay depending on the magnitude of turbulent diffusivity, and likewise following recovery in one hemisphere.

4.2 Transequatorial Convective Flows and Poynting Flux

In mean-field and mean-field-like dynamo models, transequatorial transport of magnetic field by turbulent convection operates exclusively via the (enhanced) turbulent dissipation term $\propto \beta$ in Eq. (8). Sustained cross-equatorial meridional flows are ruled out by dynamical and symmetry considerations. Latitudinal turbulent pumping vanishes at the equator by symmetry, but can lead to the concentration of magnetic fields at low latitudes, with diffusion then taking over for cross-equatorial exchange.

In global MHD simulations of rotating stratified convection, the situation turns out to be more complex, particularly in physical parameter regimes where rotation alters convection significantly. Outside of a cylinder tangent to the base of the convection zone, where no inner boundaries can break the Taylor-Proudman constraint, convection organizes itself as a longitudinally-stacked series of latitudinally-elongated convective rolls (see Fig. 12A), with the roll axes parallel to the rotation axis and sense of roll alternating from one to the next. These "banana cells" extend across the equator, and drive an internal flow parallel to their axis alternating in their north-south direction from one roll to the next in longitude (viz. Fig. 12B; see also Busse 2002; Miesch and Toomre 2009).

Although this trans-equatorial flow does not lead to any net mass exchange between hemispheres, it can generate a net trans-equatorial Poynting flux ($\mathbf{S} = \mu_0^{-1}\mathbf{E} \times \mathbf{B}$) contributing to cross-hemispheric coupling, as illustrated on Fig. 13. The top panel shows the zonally-averaged latitudinal component of the Poynting flux in the equatorial plane, in the form of a radius-time diagram spanning here 5 activity cycles. The bottom panel shows a time series of the net latitudinal Poynting flux across the equatorial plane (i.e., the top diagram integrated over depth), together with a time series of magnetic energy associated with the large-scale field $\langle \mathbf{B} \rangle$ (orange line). The transequatorial Poynting flux is spatiotemporally very intermittent, and shows some clear short term quasi periodicities. Careful intercomparison of the top and bottom panels reveals that transequatorial activity is more pronounced in the descending phase of the large-scale magnetic cycle, and less so in the rising phase. Integrating the Poynting flux over all depths (solid line on bottom panel) yield a very intermittent signal with zero mean, without much obvious imprint of the 40 yr cycle period.

The overall fluctuation level of the transequatorial Poynting flux on Fig. 13 does show some cycle-to-cycle variations. One might expect that these variations can be traced to the level of hemispheric asymmetry characterizing the amplitude of the large-scale toroidal magnetic field building up in each hemisphere for this simulation (viz. Fig. 10 herein). More specifically, if the magnetic field is passively "mixed" across hemispheres by transequatorial flows, one would expect a net Poynting flux from the dominant hemisphere toward the weaker hemisphere. Figure 14 shows the transequatorial Poynting flux integrated over a 10 yr-wide temporal window covering the late descending phase of each cycle, plotted against the corresponding hemispheric asymmetry parameter Δ_n defined earlier (viz. Eq. 9). As in Fig. 13, a positive Poynting flux indicates electromagnetic energy flux from the southern into the northern hemisphere, and a positive asymmetry parameter indicates dominance

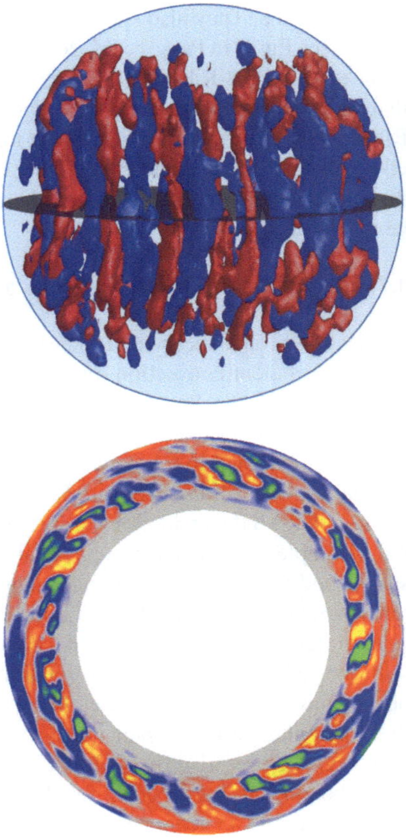

Fig. 12 *Top*: Snapshot of isosurfaces for the components of the vorticity parallel to the rotation axis (*red* is positive, *blue* negative) in the EULAG-MHD simulation described previously in Sect. 3.4. The viewpoint is from slightly above the equatorial plane, shown as *a dark disk*. *Bottom*: latitudinal flow in the equatorial plane at one specific time in the same simulation (*red-yellow*: northward, *blue-green* southward, going from -20 to $+20\ \mathrm{m\,s^{-1}}$). The view is from the N-pole, down along the rotation axis. This latitudinal flow crossing the equatorial plane ($\sim 1\ \mathrm{m\,s^{-1}}$) is of a magnitude similar to the roll speed of the cells in the equatorial plane, $\sim 10\ \mathrm{m\,s^{-1}}$

of the northern hemisphere toroidal magnetic field. The positive correlation ($r = +0.74$) runs contrary to the expectation that the stronger magnetic field of the dominant hemisphere leaks into the weaker hemisphere, which should translate here as an *anti*correlation. The observed correlation must instead materialize because the stronger magnetic field in the dominant hemisphere affects the turbulent transport coefficients in a manner such as to reduce the Poynting flux out of that hemisphere. Under this view, hemispheric coupling becomes a fully nonlinear magnetohydrodynamical phenomenon, with magnetically-mediated alterations of convective patterns dominating over passive, linear diffusive-like coupling.

Taken at face value, the positive correlation observed in Fig. 14 means that hemispheric asymmetries should be amplified by the transequatorial Poynting flux, which does not seem to be happening in this simulation, as strongly asymmetric cycles (e.g. cycle 28 in the top panel of Fig. 10) are seldom followed by similarly asymmetrical cycles. Knowing the total magnetic energy content E^N, E^S associated with the large-scale magnetic field in each hemisphere, and the total transequatorial Poynting flux $\bar{S}$, one can estimate the timescale τ_Δ required to equilibrate a 50 % difference (say) in hemispheric energy content:

$$\tau_\Delta = \frac{E^N + E^S}{4\bar{S}} . \tag{11}$$

For the EULAG-MHD simulation analyzed here, typical values are $E^N \simeq E^S \simeq 6 \times 10^{31}$ J and $\bar{S} \simeq 2 \times 10^{21}\ \mathrm{J\,s^{-1}}$ for the most asymmetric cycles on Fig. 14, leading to $\tau_\Delta \simeq 50$ yr,

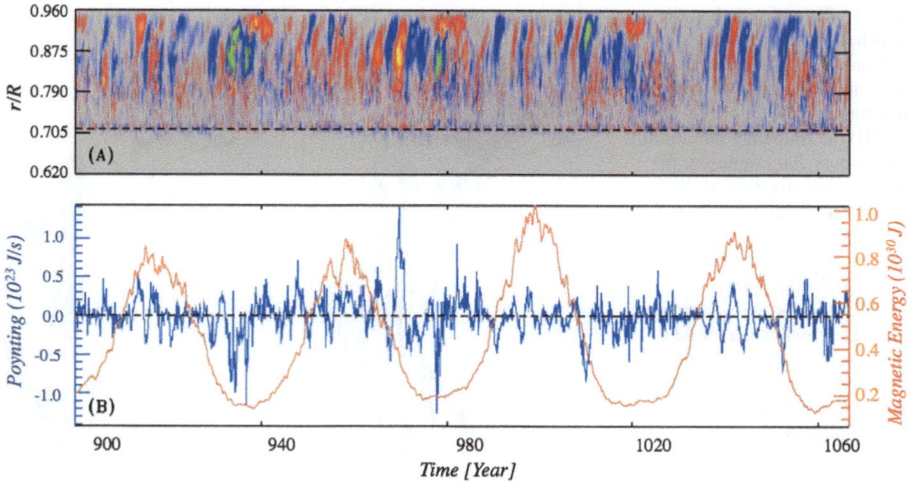

Fig. 13 *Top*: radius-time diagram of the zonally-averaged latitudinal component of the Poynting flux in the equatorial plane, for the same EULAG-MHD simulation as in previous figures. Positive values (in *red-yellow*) indicates a flux of electromagnetic energy from the southern into the northern hemisphere, and conversely for negative (in *blue-green*). *Bottom*: time series of the latitudinal Poynting flux integrated over the full equatorial plane (*blue line*), and of magnetic energy of the zonally-averaged toroidal magnetic field (*orange line*), the latter indicative of the phase of the large-scale magnetic cycle

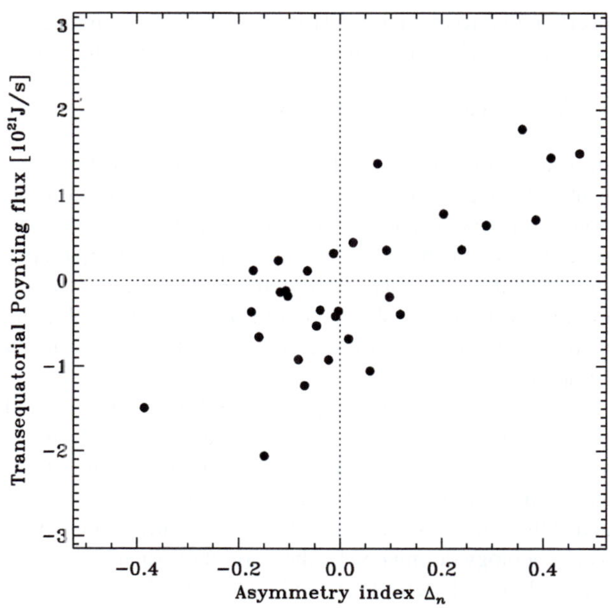

Fig. 14 Correlation plot of the trans-equatorial Poynting plot integrated over the equatorial plane in the descending phase of cycles, versus the asymmetry of hemispheric cycle amplitude, as defined in Eq. (9). On average (linear correlation coefficient $r = +0.74$), the Poynting flux carries electromagnetic energy from the weaker hemisphere into the dominant hemisphere, contrary to the "diffusive" expectation (see text). Numerical data from the EULAG-MHD simulation of Fig. 13 (see also Passos and Charbonneau 2014)

i.e., a little larger than the average cycle period in this simulation. One can but conclude that the transequatorial Poynting flux remains significant over cycle-timescales, at least in this specific simulation.

Interestingly, if the transequatorial Poynting flux integration is carried out only in the stable fluid layers underlying the convection zone, the expected "diffusive" anticorrelation

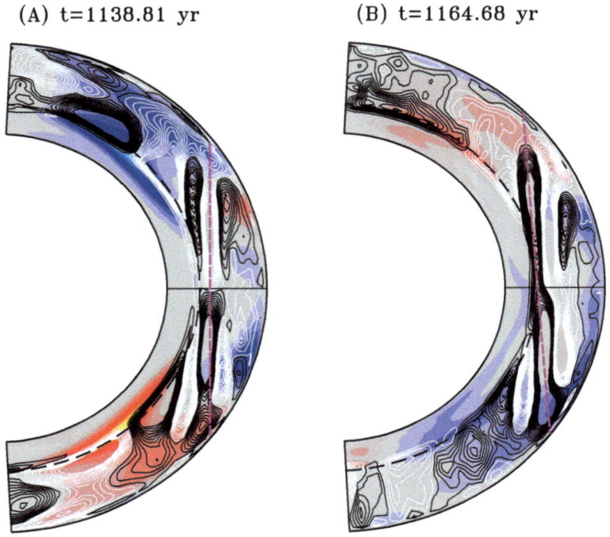

Fig. 15 *The two panels* show a snapshot of the meridional circulation stream function, ψ, taken at (**A**) the time of solar maximum (cycle 27 of Fig. 10) and (**B**) the following minimum. *The contour lines* represent the complex cell structure of the meridional circulation where *black* indicates counter-clockwise rotation and *white* clockwise. These meridional cells are overlaid on top of the zonal mean toroidal field (as on in panels (**A**) and (**B**) of Fig. 9). *The dashed pink line* is plotted to show deviations from vertical alignment

is recovered ($r = -0.57$), but the net Poynting flux is nearly two orders of magnitude lower than that crossing the equatorial plane within the convection zone. This is now more in line with purely diffusive behavior, reflecting the absence of rotationally-aligned banana cells in the stable layer, and small-scale motions being much weaker therein than they are in the overlying convection zone. However, the corresponding hemispheric coupling timescale is correspondingly longer, and likely irrelevant over the time span of this simulation.

4.3 Transequatorial Meridional Flows

In mean-field models, the large-scale meridional flow, i.e. the axisymmetric flow component contained in meridional $[r, \theta]$ planes, is usually considered steady and made of a single flow cell per hemisphere. This flow component can also be extracted from MHD numerical simulations, which usually reveals a more complex pattern of multiple flow cells per hemispheres and large deviations from the zonal average as a function of longitude and time. The meridional flow also shows significant temporal evolution, although on timescales longer than convection, so that it can be legitimately be considered a physically meaningful flow component.

Following a procedure typically used in mean-field models, the zonally averaged components of the velocity u_r and u_θ can be used to build a stream function ψ so that $\mathbf{v}_p = \frac{1}{\rho}(\nabla \times \psi \hat{e}_\phi)$, representing the cellular structure of the meridional flow, $\mathbf{v}_p$. Figure 15 shows this stream function taken at the maximum of simulated solar cycle 27 (see Fig. 10A) and at the following minimum. In this simulation, at low latitudes the meridional flow shows a cell topology dominated by the imprint of banana cells previously described (viz. Fig. 12). These flow cells tend to extend symmetrically across the equator, as highlighted by the dashed purple vertical line in the figure. This vertical alignment tends to break around cycle minima with the cells merging near the equator and forming *skewed* (not parallel to the rotation axis) trans-equatorial cells. The duration of the anti-symmetric coupling between cells is usually short, of the order of 3 yr in the simulation, which is much less than the cycle period but much more than the convective turnover time in the bulk of the convecting layers. Whether this is related to some of the quasi-periodicities apparent on Fig. 13B (viz. around $t = 1040$ yr), and the impact it may have on cycle amplitude, is still under investigation.

Looking at the latest measurements of Zhao et al. (2013), the situation on the Sun seems to be different, mainly due to the presence of the large vertical rolls (see Fig. 12, top panel) in the simulation affecting the development of solar-like meridional circulation. Nevertheless, it is completely plausible that a similar mechanism might be operating in the Sun on a much weaker/smaller scale.

4.4 Impulsive Versus Steady-State Coupling

Diffusive cross-hemispheric coupling, as embodied in the last term on the RHS of Eq. (2), is a continuous process operating gradually at a rate proportional to the local gradient of the large-scale magnetic field, with the turbulent diffusivity β providing the proportionality constant. Acting across the equatorial plane on the dipolar large-scale magnetic component generated in each hemisphere, diffusive coupling operates throughout the whole cycle, adjusting to whatever amplitude asymmetry or temporal lag may be developing across hemispheres.

The hemispheric coupling produced by convectively-driven transequatorial flows, in contrast, is anything but a slowly-varying, quasi-steady process. This quite evident upon cross-examination of Figs. 13 and 14 herein. The transequatorial Poynting flux averaged over the descending phase of the cycles is of order 10^{21} J s^{-1}, yet the corresponding time series on Fig. 13B reveals a high level of temporal intermittency, with the transequatorial flux often changing sign in the course of the same cycle and often peaking at values approaching 10^{22} J s^{-1} for time intervals of a few years. Moreover, Fig. 13A shows that these surges are well-localized spatially, and typically take place in the upper half of the convecting layers in this simulation.

A recent paper by Cameron et al. (2014) sheds light on how effective the occasional active region that emerges close to the equator can be. This is an impulsive, diffusive event which is contrary to the conventional picture retained of diffusion occurring slowly and steadily during the cycle. Cameron et al. (2014) explores surface flux transport in the Babcock-Leighton model with an emphasis on inflows into the active region belts and active region emergence near the equator. They show that the flux transported by a few cross-equatorial flux plumes by tilted sunspot groups near the equator is important for the polar field amplitude and open flux at minima. They argue that inflows to active regions (a nonlinear effect in the BL dynamo) and cross-equatorial flux plumes provides an explanation for the weakness of the polar fields at the end of solar cycle 23, which led to a weak cycle 24 (or more on these matters see review by Jiang et al. 2014, this issue).

4.5 Toroidal Flux Cancellation in the Interior Across the Equator

We propose that late in the solar cycle, the strong magnetic flux toroids at the bottom of the convection zone begin to interact across the equator. At this time, magnetic flux cancellation may actively couple the hemispheres in a manner distinctly different than the mechanisms described in Sects. 4.1–4.4. This proposed coupling process, wherein the flux toroids in the interior actively cancel across the equator, has not been explored in detail in the literature. It is our intention to follow-up this paper with future research including observations and modeling as to how this might proceed.

We limit ourselves to a brief depiction (with broad brush stokes) of the proposed process in the Sun. Exactly when in the cycle this leading edge flux cancellation across the equator would occur depends on how broad the toroidal bands (or magnetic wreaths) are in latitude. Using sunspot location patterns and a forward model to match sunspot data from cycles

21–23, Norton and Gilman (2005) report that the toroidal bands are roughly 10° wide in latitude at the beginning of the solar cycle but become broader, possibly due to magnetic drag, to ~20° wide late in the cycle. If the bands in each hemisphere are 20° in latitude, then when the butterfly wing centroids are at 10° latitude, the leading edge of the butterfly wings begins to reconnect across the equator. The observation that sunspots avoid the equator has been attributed to the meridional circulation decreasing its equatorward flow and turning upwards toward the surface (Hathaway et al. 2003). While the meridional flow is certainly seen at the surface to decrease to near-zero values (errors due to p-angle changes preclude absolute value determination) and increase dramatically as one moves away from the equator, we argue that sunspots avoid the equator more persistently after the toroidal bands meet at the equator (deep under the surface). Meaning, there is a V-shaped hollow in the butterfly diagram early in the cycle before maxima, but there is also a smaller, less notable hollow that appears after cycle maxima, see the location of the red ovals in Fig. 1. Simulations show that weak sunspots have a non-radial emergence flux trajectory through the convection zone i.e. they rise slightly pole-ward of the latitude in the interior that they were formed (Choudhuri and Gilman 1987). This may keep sunspots away from the equator. However, some spots do emerge on or very near the equator so it is worth considering other mechanisms that might explain it.

When the late–cycle void at the equator appears in Fig. 1, the butterfly wings are eroding from the equator upwards in latitude. We propose that an active flux cancellation at depth may prevent sunspots from emerging near the equator late in the cycle as well as assisting the coupling of the hemispheres. This erosion at the equator may explain why the latitudinal centroid of hemispheric sunspot locations is seen to move poleward (or retrograde) at times late in the cycle (see Figs. 2, 3 in Ternullo 2007, Fig. 4 in Norton and Gilman 2005). The centroid eventually moves equatorward again after a period of months or years. A "ghost thorax" of the butterfly diagram may indicate a time that the leading (or equatorward) edges of the toroidal bands cancel each other as opposing currents meet. This flux cancellation would not look exactly like ohmic diffusion (too slow) or X-point magnetic reconnection (too fast) but could be an effective mechanism of hemispheric coupling. When the hemispheres are significantly out of phase, the magnetic equator becomes offset in relation to the geometric equator (see Zolotova et al. 2010) in which case the bands meet up and cancel slightly above or below the geometric Equator. More research is necessary to determine under what conditions flux cancellation at depth may occur and what the observational signatures would be, including whether the absence of sunspots near the equator late in the cycle is significant.

5 Discussion: Observations Versus Models

To briefly point out the similarities and dissimilarities in hemispheric asymmetry in observations and the EULAG-MHD simulations, we provide the following list.

1. There is very good agreement in synchrony for polarity reversals as observed in the sunspot data and the simulations.
2. Low hemispheric phase lags are seen in both data and simulations. Phases are less than two years or 20 % of cycle time in observations and less than 25 % of cycle period in simulations.
3. Both observations and models show a persistent phase lag for four cycles (in observations) and more (for simulations).

4. Both observations and modelling agree that magnetic flux crosses the equator during the mid to late declining phase of the cycle.
5. Simulations do not produce correct cycle length nor do they produce toroidal fields close enough to the equator to create a solar-like butterfly diagram. It is conceivable that in this simulation, the banana cells and associated meridional flow structures play a significant role in preventing the toroidal field from migrating toward the equator, leading to deep-seated toroidal bands that effectively remain fixed in latitude, in stark contrast to the solar cycle. This may, in turn, lead to diminished hemispheric coupling and stronger amplitude asymmetries (next item).
6. Stronger amplitude asymmetries are found in the toroidal field in simulations, up to 40 % difference in the hemispheres in a single cycle. This is twice as asymmetric as the observed values which are maximally 20 % asymmetric.
7. In global MHD simulations, the dipole strength is usually not a successful precursor for next cycle toroidal proxy, whereas it is known observationally that the solar polar fields do have forecasting value (Charbonneau and Barlet 2011, (Muñoz Jaramillo et al. 2013b)). Because there is no surface flux transport in the simulation, the dipole source term as generated by active region decay with bipolar tilt angles is not incorporated. Instead the dipolar fields that are generated are a direct consequence of the inner dynamo mechanism. This could have an impact on the degree of hemispheric coupling characterizing the simulation.
8. Muñoz Jaramillo et al. (2013b) showed that polar flux becomes a better cycle predictor by taking advantage of the hemispheric polar field strength proxy. Since the hemispheric polar field strengths are a significantly better precursor than the whole-Sun measure, this is one more indication that the hemispheric coupling is imperfect.

6 Conclusions

A slow ($\sim$10 year timescale), whole-Sun magnetic diffusion rate alone may be too small to account for the strong coupling of the hemispheres as observed in the butterfly diagram. Examination of the butterfly diagram also shows an absence of sunspots that emerge near the equator, but only after the toroidal bands have met up at the magnetic equator, see the butterfly diagram in Fig. 1. The circles indicate the void, or the "ghost thorax" of the butterfly, where the wings have met at the equator, but then a sharp decrease in the number of low latitude spots being produced is evident. We propose that the suppression of low latitude sunspots late in the solar cycle (the lack of sunspots near the equator) is an indication that the bands are undergoing an active flux cancellation in the interior that decreases the amount of flux available to produce sunspots. It is a type of forced or confined ohmic diffusion that is not part of our current conventional picture of hemispheric coupling. The toroidal bands, at some point, meet across the equator and actively couple the hemispheres late in the solar cycle by cancelling out the flux in the opposite toroidal band. When one uses the distribution of surface flux and a surface diffusion rate to calculate the cross-equatorial flux, the estimate is found to peak slightly after cycle maximum and then decrease (Cameron and Schüssler 2007; Norton and Gallagher 2010) due to the avoidance of flux at the equator late in the cycle. Again, we argue that cross-equatorial flux cancellation at depth late in the solar cycle is causing a dearth of magnetic flux at the surface at this time right around the equator.

In the vast majority of extant mean-field dynamo models of the solar cycle, cross-hemispheric coupling is mediated by (linear) diffusive transport, i.e., the last term on the

RHS of Eq. (2). In models where this diffusive transport is very efficient, stochastically-driven amplitude asymmetries and hemispheric phase lags are rapidly erased unless the dynamo is operating in the mildly supercritical regime. Consequently, significant hemispheric asymmetries can only be sustained through parity modulation of odd/even modes, excited either stochastically by turbulent convection or deterministically through nonlinear magnetic backreaction on large-scale flows.

In contrast, in mean-field dynamo models operating in the weakly diffusive, advection-dominated regime, persistent hemispheric asymmetries and phase lags can be triggered and sustained by (relatively) weak, spatiotemporally uncorrelated stochastic forcing, even with a coherence time for these fluctuations much smaller than the cycle period. In either cases, the paucity of observational constraints means that many critical parameters and/or functionals introduced in such models can be freely adjusted so as to yield the desired dynamo behavior. This does not diminish the value of such models as exploratory thinking tools, but implies that their role is primarily descriptive, rather than predictive.

Global MHD simulations of solar convection producing large-scale magnetic cycles, such as the EULAG-MHD "millennium simulation" used herein for illustrative purposes, are offering a new perspective on the problem. Not only can they potentially replace observations to some extent in constraining the parameter space of simpler forms of dynamo models, but they also open a window into the fully dynamical, multi-scale 3D regime which, by definition, is inaccessible to the mean-field formulation. For example, it is through the careful analysis of MHD numerical simulations starting a little over a decade ago that turbulent pumping is resurfacing as a potentially key ingredient for the spatiotemporal evolution of the solar large-scale magnetic field. Another, discussed in this paper, is the importance for cross-hemispheric coupling of transequatorial flows associated with persistent convective structures, with it associated fully magnetohydrodynamical, non-diffusive transport of magnetic fields.

Arguably the most striking discrepancy between the hemispheric asymmetries characterizing the EULAG-MHD simulation used in this paper or comparison to the solar cycle lies with hemispheric cycle amplitudes, which show a N–S correlation much smaller than observed. On the one hand, this might be understood upon noting that the toroidal field bands in the simulation peak at mid-latitudes and show very little equatorward propagation (see Fig. 9); consequently, these "activity bands" never meet at the equator, in contrast to what is seen in the sunspot butterfly diagram (cf. Fig. 1). It is then perhaps natural to expect the real solar cycle to exhibit higher levels of cross-hemispheric coupling. On the other hand, the sun's ability to sustain over many cycles a cross-hemispheric lag in the sunspot counts (viz. Fig. 6) puts an upper bound on the efficiency of whichever physical mechanism is responsible for magnetic cross-hemispheric coupling. Ongoing efforts to extend the hemispheric sunspot number time series (Clette et al. 2014, this issue), as well as the sunspot butterfly diagram (Arlt and Weiss 2014, this issue) all the way back to the beginning of the SSN record, would be most helpful in providing tighter constraints to modelling and simulation efforts.

It is perhaps appropriate to close this review with a reality check. Despite staggering advances in computing power and algorithmic design, current global MHD simulations are still a long way from the solar parameter regime. Even the most solar-like large-scale magnetic cycles they produce still show important discrepancies with respect to the observed solar cycle. Moreover, many important processes such as sunspot emergence, with subsequent decay and surface flux dispersal, still cannot be captured within such simulations due to the extreme disparity of the spatial and temporal scales involved. Nonetheless, these simulations are dynamically consistent on spatial and temporal scales that they do resolve, and as such their output can be treated as "experimental data" to explore the intricacies of dynamo action in thick, stratified, rotating shells of convectively turbulent electrically conducting fluid. This is the approach we have adopted here in the context of cross-hemispheric coupling; but more generally, it represents a unique springboard towards broader investigations of dynamo action in the sun and stars.

Acknowledgements We thank Nicolas Lawson for producing some Figures and the analysis leading to Fig. 14. We thank S. McIntosh and N. Zolotova for allowing the reproduction of several figures regarding sunspot measures of hemispheric asymmetry. We thank J. Janssens for maintaining and updating the cycle 24 website. P. Charbonneau is supported by the Natural Sciences and Engineering Research Council of Canada. D. Passos acknowledges support from the Fundação para a Ciência e Tecnologia (FCT) grant SFRH/BPD/68409/2010, CENTRA-IST and the University of the Algarve for providing office space. A. Norton is supported by NASA Contract NAS5-02139 (HMI) to Stanford University. We thank the ISSI organizers for a most productive and insightful conference held in Bern in November 2013.

References

E. Antonucci, J.T. Hoeksema, P.H. Scherrer, Astrophys. J. **360**, 296 (1990)
R. Arlt, R. Leussu, N. Giese, K. Mursula, I.G. Usoskin, Mon. Not. R. Astron. Soc. **433**, 3165 (2013)
R. Arlt, N. Weiss, Space Sci. Rev. (2014, this issue). doi:10.1007/s11214-014-0063-5
H.D. Babcock, Astrophys. J. **130**, 364 (1959)
J.L. Ballester, R. Oliver, M. Carbonell, Astron. Astrophys. **431**, 5L (2005)
L.A. Balmaceda, S.K. Solanki, N.A. Krivova, S. Foster, J. Geophys. Res. **114**, A07104 (2009)
P. Beaudoin, P. Charbonneau, É. Racine, P.K. Smolarkiewicz, Sol. Phys. **282**, 335–360 (2013)
U. Becker, Z. Astrophys. **37**, 47 (1955)
J. Beer, S.M. Tobias, N.O. Weiss, Sol. Phys. **181**, 237 (1998)
E.E. Benevolenskaya, Highlights Astron. **14**, 273 (2007)
T.E. Berger, M.G. Löfdahl, R.A. Shine, A.M. Title, Astrophys. J. **506**, 439 (1998)
D.W. Boyer, E.H. Levy, Astrophys. J. **277**, 848 (1984)
A. Brandenburg, K. Subramanian, Phys. Rep. **417**, 1–209 (2005)
A.-M. Broomhall, P. Chatterjee, R. Howe, A A. Norton, M.J. Thompson, Space Sci. Rev. (2014, this issue). doi:10.1007/s11214-014-0101-3
B.P. Brown, M.K. Browning, A.S. Brun, M.S. Miesch, J. Toomre, Astrophys. J. **711**, 424 (2010)
B.P. Brown, M.S. Miesch, M.K. Browning, A.S. Brun, J. Toomre, Astrophys. J. **731**, 69 (2011)
M.K. Browning, M.S. Miesch, A.S. Brun, J. Toomre, Astrophys. J. Lett. **648**, L157–160 (2006)
P. Bushby, Mon. Not. R. Astron. Soc. **328**, 655 (2003)
F.H. Busse, Phys. Fluids **14**, 1301–1314 (2002)
P. Caligari, F. Moreno-Insertis, M. Schüssler, Astrophys. J. **441**, 886 (1995)
R.H. Cameron, M. Schüssler, Astrophys. J. **659**, 801 (2007)
R.H. Cameron, J. Jian, M. Schüssler, L. Gizon, J. Geophys. Res. **119**, 680 (2014)
M. Carbonell, R. Oliver, J.L. Ballester, Astron. Astrophys. **274**, 497 (1993)
M. Carbonell, J. Terradas, R. Oliver, J.L. Ballester, Astron. Astrophys. **476**, 951 (2007)
J. Chae, Y.E. Litvinenko, T. Sakurai, Astrophys. J. **683**, 1153 (2008)
P. Charbonneau, Sol. Phys. **229**, 345 (2005)
P. Charbonneau, Adv. Space Res. **40**, 899 (2007)
P. Charbonneau, Living Rev. Sol. Phys. **7**, 3 (2010)
P. Charbonneau, Annu. Rev. Astron. Astrophys. **52**(1), 251 (2014)
P. Charbonneau, G. Barlet, J. Atmos. Sol.-Terr. Phys. **7**(73), 198 (2011)
P. Chatterjee, A.R. Choudhuri, Sol. Phys. **239**, 29 (2006)
P. Chatterjee, D. Nandy, A.R. Choudhuri, Astron. Astrophys. **427**, 1019 (2004)
A.R. Choudhuri, Astron. Astrophys. **253**, 277 (1992)
A.R. Choudhuri, P.A. Gilman, Astrophys. J. **316**, 788 (1987)
P. Chowdhury, D.P. Choudhary, S. Gosain, Astrophys. J. **768**, 188 (2013)
F. Clette, L. Svalgaard, J.M. Vaquero, E.W. Cliver, Space Sci. Rev. (2014, this issue). doi:10.1007/s11214-014-0074-2
J.-F. Cossette, P. Charbonneau, P.K. Smolarkiewicz, Astrophys. J. Lett. **777**, L29 (2013)
M. Dasi-Espuig, S.K. Solanki, N. Krivova, R. Cameron, T. Peñuela, Astron. Astrophys. **518**, 10 (2010)
M. Dikpati, P.A. Gilman, Astrophys. J. **559**, 428–442 (2001)

M. Dikpati, G. de Toma, P.A. Gilman, C.N. Arge, O.R. White, Astrophys. J. **601**, 1136 (2004)
M. Dikpati, G. de Toma, P.A. Gilman, Geophys. Res. Lett. **33**, L05102 (2006)
C.J. Durrant, P.R. Wilson, Sol. Phys. **214**, 23 (2003)
T.L. Duvall Jr., S.M. Jefferies, J.W. Harvey, M.A. Pomerantz, Nature **362**, 430 (1993)
Y. Fan, F. Fang, Astrophys. J. **789**, 35 (2014)
M. Ghizaru, P. Charbonneau, P.K. Smolarkiewicz, Astrophys. J. Lett. **715**, L133–137 (2010)
A. Goel, A.R. Choudhuri, Res. Astron. Astrophys. **9**, 115 (2009)
D.H. Hathaway, D. Nandy, R.M. Wilson, E.J. Reichmann, Astrophys. J. **589**, 665 (2003)
F. Hill, Astrophys. J. **343**, L69 (1989)
H. Hotta, T. Yokoyama, Astrophys. J. Lett. **714**, L308 (2010)
R. Howe, D. Baker, L. Harra, L. van Driel-Gesztelyi, R. Komm, F. Hill, I. González Hernández, ASP Conf. Ser. **478**, 291 (2013)
P. Hoyng, Astrophys. J. **332**, 857 (1988)
P. Hoyng, D. Schmitt, L.J.W. Teuben, Astron. Astrophys. **289**, 265 (1994)
J. Jiang, D.H. Hathaway, R.H. Cameron, S.K. Solanki, L. Gizon, L. Upton, Space Sci. Rev. (2014, this issue). doi:10.1007/s11214-014-0083-1
B. Joshi, A. Joshi, Sol. Phys. **219**, 343 (2004)
P.J. Käpylä, M.J. Mantere, A. Brandenburg, Astrophys. J. Lett. **755**, L22 (2012)
P.J. Käpylä, M.J. Mantere, E. Cole, J. Warnecke, A. Brandenburg, Astrophys. J. **778**, 41 (2013)
B. Karak et al., Space Sci. Rev. (2014, this issue)
R.W. Komm, R.F. Howard, J.W. Harvey, Sol. Phys. **158**, 213 (1995)
R. Komm, R. Howe, I. González Hernández, F. Hill, D. Haber, J. Phys. Conf. Ser. **271**, 012077 (2011)
R. Komm, R. Howe, I. González Hernández, F. Hill, R.S. Bogart, D. Haber, ASP Conf. Ser. **478**, 217 (2013)
F. Krause, K.-H. Rädler, *Mean-Field Magnetohydrodynamics and Dynamo Theory* (Pergamon Press, Oxford, 1980), 271 pp.
K.J. Li, X.H. Liu, H.S. Yun, S.Y. Xiong, H.F. Liang, H.Z. Zhao, L.S. Zhan, X.M. Gu, Publ. Astron. Soc. Pac. **54**, 629 (2002)
D.W. Longcope, G.H. Fisher, Astrophys. J. **458**, 380 (1996)
N. Marwan, M.C. Romano, M. Thiel, J. Kurths, Phys. Rep. **438**, 237 (2007)
E.W. Maunder, Mon. Not. R. Astron. Soc. **64**, 747 (1904)
B.H. McClintock, A.A. Norton, Sol. Phys. **287**, 215 (2013)
S.W. McIntosh et al., Astrophys. J. **765**, 146 (2013)
M.S. Miesch, J. Toomre, Annu. Rev. Fluid Mech. **41**, 317–340 (2009)
M. Miesch, J.R. Elliott, J. Toomre, T.L. Clune, G.A. Glatzmaier, P.A. Gilman, Astrophys. J. **532**, 593 (2000)
P. Mininni, D.O. Gómez, Astrophys. J. **573**, 454 (2002)
H.K. Moffatt, *Magnetic Field Generation in Electrically Conducting Fluids* (Cambridge University Press, Cambridge, 1978), 343 pp.
D. Moss, A. Brandenburg, R. Tavakol, I. Tuominen, Astron. Astrophys. **265**, 843 (1992)
A. Muñoz-Jaramillo, N.R. Sheeley, J. Zhang Jr., E.E. Deluca, Astrophys. J. **753**, 146 (2012)
A. Muñoz-Jaramillo, M. Dasi-Espuig, L. Balmaceda, E. DeLuca, Astrophys. J. Lett. **767**, L25 (2013a)
A. Muñoz-Jaramillo, L. Balmaceda, E. DeLuca, Phys. Rev. Lett. **111**, 041106 (2013b)
J. Muraközy, A. Ludmány, Mon. Not. R. Astron. Soc. **419**, 3624 (2012)
K. Mursula, T. Hiltula, Geophys. Res. Lett. **30**, 2135 (2003)
N.J. Nelson, B.P. Brown, A.S. Brun, M.S. Miesch, J. Toomre, Astrophys. J. **762**, 73 (2013)
H.W. Newton, A.S. Milsom, Mon. Not. R. Astron. Soc. **115**, 398 (1955)
A.A. Norton, P.A. Gallagher, Sol. Phys. **261**, 193 (2010)
A.A. Norton, P.A. Gilman, Astrophys. J. **630**, 1194 (2005)
S.V. Olemskoy, L.L. Kitchatinov, Astrophys. J. **777**, 71 (2013)
M.A.J.H. Ossendrijver, Astron. Astrophys. Rev. **11**, 287–367 (2003)
A.J.H. Ossendrijver, P. Hoyng, D. Schmitt, Astron. Astrophys. **313**, 938 (1996)
A.J.H. Ossendrijver, M. Stix, A. Brandenburg, Astron. Astrophys. **376**, 713 (2001)
M.A.J.H. Ossendrijver, M. Stix, A. Brandenburg, G. Rüdiger, Astron. Astrophys. **394**, 735–745 (2002)
D. Passos, P. Charbonneau, Astron. Astrophys. **568**, A113 (2014). doi:10.1051/0004-6361/201423700
D. Passos, D. Nandy, S. Hazra, I. Lopes, Astron. Astrophys. **562**, A18 (2014)
V.V. Pipin, Astron. Astrophys. **346**, 295 (1999)
V.V. Pipin, A.G. Kosovichev, Astrophys. J. **776**, 36 (2013)
P.J. Pulkkinen, J. Brooke, J. Pelt, I. Tuominen, Astron. Astrophys. **341**, L43 (1999)
E. Racine, P. Charbonneau, M. Ghizaru, A. Bouchat, P.K. Smolarkiewicz, Astrophys. J. **735**, 46 (2011)
M. Rempel, in *Heliophysics*, ed. by C.J. Schrijver, G.L. Siscoe (2006), pp. 42–74
J.C. Ribes, E. Nesme-Ribes, Astron. Astrophys. **276**, 549 (1993)
B. Schmieder et al., Space Sci. Rev. (2014, this issue). doi:10.1007/s11214-014-0088-9

G.W. Simon, A.M. Title, N.O. Weiss, Astrophys. J. **442**, 886 (1995)
P.K. Smolarkiewicz, P. Charbonneau, J. Comput. Phys. **236**, 608–623 (2013)
D. Sokoloff, E. Nesme-Ribes, Astron. Astrophys. **288**, 293 (1994)
G. Spörer, *Publicationen des Astrophysikalischen Observatoriums zu Potsdam*. Nr., vol. 32, Vol. 10, part 1 (Wilhelm Engelmann, Leipzig, 1894)
X. Sun, Y. Liu, J.T. Hoeksema, K. Kayashi, X. Zhao, Sol. Phys. **270**, 9 (2011)
M. Temmer, J. Rybák, P. Bendík, A. Veronig, F. Vogler, W. Otruba, W. Pötzi, A. Hanslmeier, Astron. Astrophys. **447**, 735 (2006)
M. Ternullo, Mem. Soc. Astron. Ital. **78**, 596 (2007)
S. Tobias, Astron. Astrophys. **322**, 1007 (1997)
S.M. Tobias, N.H. Brummell, T.L. Clune, J. Toomre, Astrophys. J. **549**, 1183–1203 (2001)
I.I. Virtanen, K. Mursula, Astrophys. J. **781**, 99 (2014)
M. Waldmeier, *Ergebnisse und Probleme der Sonnenforschung*, 2nd edn. (Geest & Portig, Leipzig, 1955)
N.O. Weiss, F. Cattaneo, C.A. Jones, Geophys. Astrophys. Fluid Dyn. **30**, 305 (1984)
A. Yeates, D. Nandy, D.H. Mackay, Astrophys. J. **673**, 544 (2008)
J. Zhao, R.S. Bogart, A.G. Kosovichev, T.L. Duvall Jr., T. Hartlep, Astrophys. J. **774**, 29 (2013)
N.V. Zolotova, D.I. Ponyavin, N. Marwan, J. Kurths, Astron. Astrophys. **503**, 197 (2009)
N.V. Zolotova, D.I. Ponyavin, R. Arlt, I. Tuominen, Astron. Nachr. **331**, 765 (2010)

DOI 10.1007/978-1-4939-2584-1_10
Reprinted from *Space Science Reviews* Journal, DOI 10.1007/s11214-014-0082-2

Magnetic Helicity, Tilt, and Twist

Alexei A. Pevtsov · Mitchell A. Berger · Alexander Nindos · Aimee A. Norton · Lidia van Driel-Gesztelyi

Received: 2 May 2014 / Accepted: 4 August 2014 / Published online: 2 September 2014

Abstract Since its introduction to astro- and solar physics, the concept of helicity has proven to be useful in providing critical insights into physics of various processes from astrophysical dynamos, to magnetic reconnection and eruptive phenomena. Signature of helicity was also detected in many solar features, including orientation of solar active regions, or Joy's law. Here we provide a summary of both solar phenomena and consider mutual relationship and its importance for the evolution of solar magnetic fields.

Keywords Helicity · Joy's law · Magnetic field

A.A. Pevtsov (✉)
National Solar Observatory, Sunspot, USA
e-mail: apevtsov@nso.edu

M.A. Berger
University of Exeter, Exeter, UK

A. Nindos
University of Ioannina, Ipiros, Greece

A.A. Norton
Stanford University, Stanford, USA

L. van Driel-Gesztelyi
Mullard Space Science Laboratory, University College London, Holmbury St. Mary, Dorking, Surrey, RH5 6NT, UK

L. van Driel-Gesztelyi
Observatoire de Paris, LESIA, FRE 2461 (CNRS), 92195 Meudon Principal Cedex, France

L. van Driel-Gesztelyi
Konkoly Observatory of the Hungarian Academy of Sciences, Budapest, Hungary

1 Introduction

This article deals with two seemingly unlike phenomena: (magnetic, current) helicity and the orientation of solar active regions (Joy's Law). Joy's law represents one of the earliest observational tendencies discovered at the beginning of modern era of solar observations (Hale et al. 1919), while helicity is a more recent subject (Berger and Field 1984; Seehafer 1990; Martin et al. 1994; Pevtsov et al. 1994; Rust 1994). But are these two phenomena completely unrelated? Here we present overview of recent studies of helicity and Joy's law, investigate their possible relation, and discuss what the relation between helicity and active region tilt may tell us about the origin and the evolution of magnetic fields on the Sun.

Early studies of helicity on the Sun are well-represented by individual articles in Brown et al. (1999). Later developments were described in Büchner and Pevtsov (2003). A graphic summary of the hemispheric helicity rule encompassing magnetic fields on spatial scales from quiet Sun network to the interplanetary magnetic field was presented in Pevtsov (2002). A more recent update on the hemispheric helicity rule can be found in Pevtsov (2008). Some details omitted from the present article can be found in these earlier reviews.

The rest of the article is organized as follows. Section 2 presents a summary of the theoretical basis of the helicity concept. Sections 3–5 present an overview of the current observations of helicity on the Sun and the heliosphere. Section 6 considers different aspects relevant to orientation of solar active region (Joy's Law), and Sect. 7 summarizes the article.

2 Basic Theory

Magnetic helicity is a measure of important structural properties of a magnetic field such as twist, shear, linking, and kinking. It is conserved to an excellent approximation in the highly conducting plasmas present in the Sun and heliosphere. The large magnetic Reynolds numbers (10^8 to 10^{15}) in the solar corona produce ideal magnetohydrodynamic (MHD) behaviour on large scales. Resistive MHD processes on smaller scales (current sheet formations and magnetic reconnection) barely affect the conservation of total magnetic helicity (Berger 1984). Any detailed model of the Sun and heliosphere, in order to be consistent, must obey the helicity balance equations. In this sense, magnetic helicity can be considered a more robust invariant than total energy, as ideal motions convert energy back and forth between kinetic and magnetic forms, dissipative processes convert both forms into heat, and various plasma processes convert energy into particle acceleration.

Although magnetic helicity is conserved, it can be transported between regions; in addition helicity can be transferred between different length-scales. Helicity transport between regions obeys a Poynting-like equation; it involves either the bulk transfer of a helicity-carrying field from one region to the other, or propagation of twist and braiding along field lines crossing the boundary between the regions (Berger and Field 1984; Pariat et al. 2006).

We can distinguish two types of scale transfers. First, when there is a net helicity of one sign, there is a tendency for turbulent plasma to generate or maintain a large-scale magnetic field structure via the inverse cascade of magnetic helicity to longer wavelengths. Secondly, ideal processes can generate large-scale helicity of one sign balanced by small-scale helicity of the opposite sign. Thus a magnetic flux rope rising through the convection zone can kink due to Coriolis forces. The kinking (measured as *writhe*) represents large scale helicity of one sign, balanced by a small scale helicity (internal twisting of field lines) of the opposite sign.

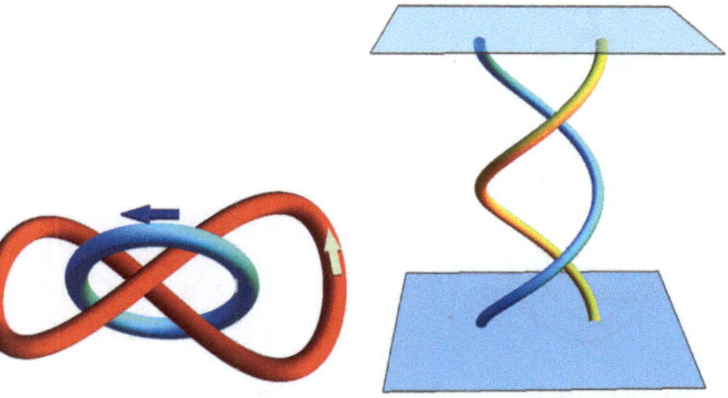

Fig. 1 *Left*: If these were magnetic flux tubes with net axial fluxes Φ_1 and Φ_2, then the helicity would be $H = 2L\Phi_1\Phi_2 + T_1\Phi_1^2 + T_2\Phi_2^2$ where T_1 and T_2 measure the net twisting of field lines within the tubes about their axes. For these tubes, the linking number is $L = 2$. *Right*: two tubes stretched between parallel planes. Here $H = 2w\Phi_1\Phi_2 + T_1\Phi_1^2 + T_2\Phi_2^2$, where w is their winding number. For the tubes shown, the winding number $w = 1$

Helicity affects the solar cycle and activity in several ways. First, in the interior of the Sun, the solar dynamo is responsible for regeneration of the magnetic field and the 11-year solar cycle. Differential rotation provides an essential contribution to the dynamo; the effect of differential rotation on the magnetic field can be directly quantified using magnetic helicity techniques (Berger and Ruzmaikin 2000). On the other hand, too much helicity can suppress turbulent dynamo (alpha effect) action.

Secondly, magnetic helicity build-up in a solar active region can lead to enhanced activity. Of particular importance for the space weather effects and forecasting potential is the realisation that increasing the pinch or helicity content in a solar filament or an active region can also increase the probability of eruption (e.g., Romano and Zuccarello 2011).

Third, if we go further into the heliosphere, helicity can also play an important role. The helicity of filaments and active regions can be examined before an eruption; after the eruption, this helicity can be compared to that of the remaining active region and of the corresponding magnetic cloud (e.g. Foullon et al. 2007).

Here we detail how helicity and helicity transport can be calculated, and how it relates to structural features of a magnetic field such as twist, kinking, and linking.

2.1 Linking, Twisting, and Helicity Integrals

Magnetic helicity integrals measure the net interlinking and twisting of magnetic field lines. Before thinking about magnetic fields in general, consider two simple cases: first, where the field-lines are all closed curves; and second, where the field-lines extend between parallel planes. In the first case, the helicity measures the net linking between all pairs of field-lines. In the second, the helicity measures the net twist between all pairs of field-lines. Of course, the number of field-lines in a magnetised volume is infinite, so the sums over linking and winding are weighted by magnetic flux.

The Gauss linking number (Ricca and Nipoti 2011) is a double line integral along two closed curves (see Fig. 1). This integral gives an integer, as long as the two curves are distinct (when they are the same curve, it gives the *writhe* as discussed below). The magnetic field counterpart of the linking number is the closed magnetic helicity.

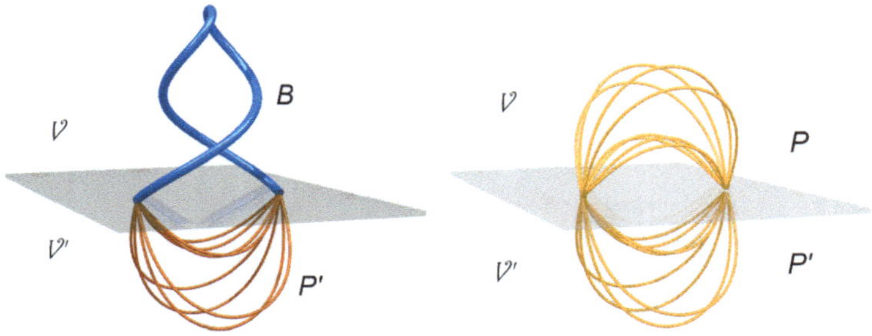

Fig. 2 *Left*: We wish to define the helicity of the open field in the *upper* part ($\mathcal{V}$) of the *left* figure. To accomplish this, we extend the field to the volume $\mathcal{V}'$ below the boundary. Let the total helicity, as measured by (4) be H_{tot}. Next, replace **B** in $\mathcal{V}$ by its corresponding potential field $\mathbf{B}_{pot}$; the helicity of this reference field (with the field inside $\mathcal{V}'$ unchanged) is H_{ref}. The relative helicity is then $H_{tot} - H_{ref}$

Let σ and τ parametrize curves 1 and 2, and label points on these curves as $\mathbf{x}(\sigma)$ and $\mathbf{y}(\tau)$. Also let $\mathbf{r} = \mathbf{y} - \mathbf{x}$. The Gauss linking number is

$$L_{12} = -\frac{1}{4\pi} \oint_1 \oint_2 \frac{d\mathbf{x}}{d\sigma} \cdot \frac{\mathbf{r}}{r^3} \times \frac{d\mathbf{y}}{d\tau}\, d\tau\, d\sigma. \tag{1}$$

The magnetic field counterpart is a double volume integral in a volume $\mathcal{V}$ with boundary S. Assuming the field is closed within the volume (i.e. $\mathbf{B} \cdot \hat{n} = 0$ at the boundary S) we have the magnetic helicity (Moffatt 1969)

$$H = -\frac{1}{4\pi} \int \int \mathbf{B}(\mathbf{x}) \cdot \frac{\mathbf{r}}{r^3} \times \mathbf{B}(\mathbf{y})\, \mathrm{d}^3x\, d^3y. \tag{2}$$

This double integral form of the helicity provides an important caution: helicity is a non-local quantity; it does not have a simple density like mass. However, this form is unwieldy and difficult to use. To simplify calculations, employ the Coulomb gauge vector potential

$$\mathbf{A}(\mathbf{x}) = -\frac{1}{4\pi} \int \frac{\mathbf{r}}{r^3} \times \mathbf{B}(\mathbf{y})\, d^3y, \tag{3}$$

which gives magnetic helicity the more widely known form (for a closed volume)

$$H = \int \mathbf{A} \cdot \mathbf{B}\, \mathrm{d}^3x. \tag{4}$$

Note, however, that gauge transformations of **A** may change the value of this integral when magnetic flux crosses the boundary (and also if the volume is multiply-connected, Berger 1999).

For magnetically open or multiply connected volumes (where $\mathbf{B} \cdot \hat{n} \neq 0$ at the boundary S), we measure helicity relative to some field with minimum structure. Let $\mathbf{B}_{pot}$ be the potential field in $\mathcal{V}$. Suppose the total magnetic field $\mathbf{B}_{tot}$ in all space (including $\mathcal{V}$ and its complement outside of $\mathcal{V}$) vanishes at infinity, so that its helicity $H(\mathbf{B}_{tot})$ is finite. Then (Berger and Field 1984) we can define a reference field $\mathbf{B}_{ref}$ which is the same as $\mathbf{B}_{tot}$ outside of $\mathcal{V}$ but equals the potential field **P** inside (Fig. 2). We then define the helicity inside the volume $\mathcal{V}$ to be

$$H(\mathcal{V}) = H(\mathbf{B}_{tot}) - H(\mathbf{B}_{ref}), \tag{5}$$

with helicities calculated using (4). This helicity only depends on what is happening inside $\mathcal{V}$; one can show (Berger and Field 1984; Finn and Antonsen 1983)

$$H(\mathcal{V}) = \int (\mathbf{A} + \mathbf{A}_{pot}) \cdot (\mathbf{B} - \mathbf{B}_{pot})\, \mathrm{d}^3x. \tag{6}$$

This generalized helicity can be shown to be gauge invariant for any boundary conditions or topology of the boundary S.

If the volume $\mathcal{V}$ lies between two planes, and all field lines stretch between the boundaries, then the helicity is a sum of winding numbers between the planes, as in Fig. 1, right. For other configurations, the interpretation can become more complicated. For most solar atmospheric problems, a more appropriate geometry would be loops with footpoints on a sphere (or at least a plane). Again the helicity measures the net amount the loops wind about each other. In addition, for a sheared arcade, the helicity will be related to the amount of shear. A detailed account of calculating the helicity of loops is given in Pariat et al. (2006).

2.2 Poloidal and Toroidal Fields

In Cartesian or spherical geometries it is often useful to decompose a magnetic field into toroidal and poloidal components. Solar dynamo action is often described in terms of differential rotation acting on a poloidal field to regenerate the toroidal field (the *omega* effect), along with helical turbulence acting on the toroidal field to regenerate the poloidal field (the *alpha* effect, Charbonneau 2005).

Let $\mathcal{L}$ be the operator

$$\mathcal{L} \equiv \begin{cases} -\hat{z} \times \nabla, & \text{Cartesian geometry;} \\ -\mathbf{r} \times \nabla, & \text{Spherical geometry.} \end{cases} \tag{7}$$

Then we can write

$$\mathbf{B} = \mathcal{L}T + \nabla \times \mathcal{L}P, \tag{8}$$

where T is the toroidal function and P is the poloidal function.

The operator $\mathcal{L}$ has two important properties: first, it's divergence vanishes,

$$\nabla \cdot \mathcal{L} = 0. \tag{9}$$

Secondly, it is transverse:

$$\begin{array}{ll} \hat{z} \cdot \mathcal{L} = 0, & \text{Cartesian geometry;} \\ \mathbf{r} \cdot \mathcal{L} = 0, & \text{Spherical geometry.} \end{array} \tag{10}$$

The vector potentials of the toroidal field $\mathbf{B}_T = \mathcal{L}T$ and the poloidal field $\mathbf{B}_P = \nabla \times \mathcal{L}P$ are

$$\mathbf{A}_T = T\hat{z} + \nabla \Psi_T; \tag{11}$$

$$\mathbf{A}_P = \mathcal{L}P + \nabla \Psi_P. \tag{12}$$

where $\Psi_T(x, y, z)$ and $\Psi_P(x, y, z)$ are gauge functions.

The functions P and T can be obtained from $\mathbf{B}$ by solving the equations

$$\frac{\partial^2 P}{\partial x^2} + \frac{\partial^2 P}{\partial y^2} = B_z; \tag{13}$$

$$\frac{\partial^2 T}{\partial x^2} + \frac{\partial^2 T}{\partial y^2} = (\nabla \times \mathbf{B})_z. \tag{14}$$

Theorem *Consider a magnetic field* $\mathbf{B} = \mathbf{B}_T + \mathbf{B}_P$ *in a region* $\mathcal{V}$. *Assume that* $\mathcal{V}$ *is either* 1) *all space,* 2) *a half space bounded by a plane,* 3) *a layer bounded by two planes,* 4) *the interior or exterior of a sphere, or* 5) *a spherical shell bounded by two concentric spheres.*

Then

1. *A purely poloidal field* ($T = 0$) *has helicity* $H = 0$.
2. *A purely toroidal field* ($P = 0$) *has helicity* $H = 0$.
3. *In general,*

$$H(\mathbf{B}) = 2 \int_{\mathcal{V}} \mathcal{L}T \cdot \mathcal{L}P \, \mathrm{d}^3x. \tag{15}$$

These results follow from the definitions of helicity and relative helicity, and (9) *and* (10). *Here we give an outline of the proof in a Cartesian geometry. As the helicity integrals are gauge invariant, we can ignore gauge functions* Ψ_P *and* Ψ_T. *For a purely toroidal field*

$$\int \mathbf{A}_T \cdot \mathbf{B}_T \, \mathrm{d}^3x = \int T\hat{z} \cdot \mathcal{L}T \, \mathrm{d}^3x = 0. \tag{16}$$

For purely poloidal fields, there is somewhat more work:

$$\begin{aligned}\int \mathcal{L}P \cdot \mathbf{B}_P \, \mathrm{d}^3x &= \int (\nabla \times P\hat{z}) \cdot \Big(\nabla \frac{\partial P}{\partial z} - \hat{z}\nabla^2 P\Big) \, \mathrm{d}^3x \\ &= \int \nabla \cdot P\hat{z} \times \nabla \frac{\partial P}{\partial z} \, \mathrm{d}^3x = \oint \hat{z} \cdot P\hat{z} \times \nabla \frac{\partial P}{\partial z} \, \mathrm{d}^2x \\ &= 0.\end{aligned} \tag{17}$$

For a field containing both toroidal and poloidal components, only the cross-terms give non-zero results. There will be two of these, corresponding to $\mathbf{A}_T \cdot \mathbf{B}_P$ *and* $\mathbf{A}_P \cdot \mathbf{B}_T$, *but both reduce to the integral of* $\mathcal{L}P \cdot \mathcal{L}T$. *When calculating relative helicities, the same results apply* (*note that the potential field is purely poloidal*).

2.3 Self and Mutual Helicity

Suppose we divide the magnetic field in a volume $\mathcal{V}$ into two or more distinct components, with the proviso that any individual field line stays entirely within one component. Recall that helicity can be described as a sum or integral over the linking or twist numbers of all pairs of field lines. Then we can ask what happens when we restrict the sum to pairs where both lines stay belong to one component. In this case we obtain the *self-helicity* of the component. If we look at all pairs where one line is in one component and a second line is in another component, then we obtain the *mutual helicity* between the two components. For example, Fig. 3 shows a cartoon of a prominence field (Martin et al. 2012), divided into components: an overlying arcade, the field along the filament axis, and low-lying barbs connecting the filament with the field below. Each component can have its own self-helicity due to its internal twist and shear. In addition, two different components will share a mutual helicity due to their interlinking or shear.

Reconnection between two components of a field can exchange self and mutual helicities, while keeping the total helicity approximately conserved. For example, consider a simple reconnection of two equal flux tubes of flux Φ. The mutual helicity of the new tubes will differ from that of the old tubes by $\Delta H_{mutual} = \pm\Phi^2$. In a simple reconnection (with one reconnection site, the final tubes will share this helicity equally, both obtaining a half twist $\Delta H_{self} = \pm\frac{1}{2}\Phi^2$ as a result (Wright and Berger 1989). Such sudden acquisitions of half twists may be observable. For example, a recent force-free modelling of an active region corona using vector SDO/HMI magnetograms (Thalmann et al. 2014) displays an apparent increase of a half turn twist in a flux rope.

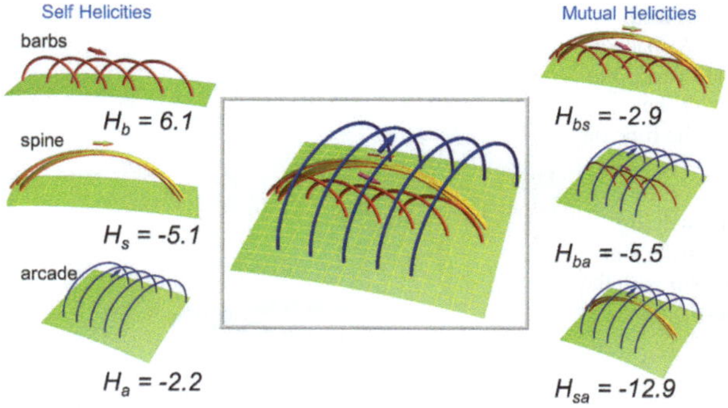

Fig. 3 A cartoon model of a prominence magnetic field, based on Martin et al. (2012). The magnetic flux has been divided into distinct regions corresponding to arcade fields on top, the axis field going all the way along the prominence, and barb fields below. Each single flux element shown has one unit of flux. The self helicities of each region are shown on the *left*, while their mutual helicities are shown on the *right*

Often in solar physics and MHD theory one considers individual flux ropes, for example an X-ray loop with footpoints on the photosphere. Let the coronal volume (here anything exterior to the photosphere) be denoted by $\mathcal{V}_c$. Suppose a particular flux rope occupies a volume $\mathcal{V}_r$ inside $\mathcal{V}_c$. There are a few ways of calculating the self-helicity of the rope. The self helicity should give the sum of twist inside the rope, plus the writhe of the rope's axis (see below). If we calculate the relative helicity of the rope volume $\mathcal{V}_c$ alone, we will obtain only the twist due to axial currents inside the rope, losing the writhe altogether (for example, if there are no currents inside the volume (potential field of $\mathcal{V}_c$) then the relative helicity $H(\mathcal{V}_c) = 0$). Alternatively, one could calculate the relative helicity of the entire corona, i.e. $H(\mathcal{V}_c)$, but with $\mathbf{B}$ set to 0 outside the rope. This will give the sum of twist and writhe. Detailed methods for calculating helicities of individual ropes or subvolumes are given in Low (2006), Longcope and Malanushenko (2008). These methods are especially useful when considering partial relaxation of a field to a piece-wise linear force free field.

2.4 Twist and Writhe

The self helicity of a flux tube can take two forms, *twist* and *writhe*: $H = Tw + Wr$ (Călugăreanu 1961; Moffatt and Ricca 1992; Berger and Prior 2006; Török et al. 2010). The writhe measures the winding and kinking of the axis, but is not itself a topological or ideal invariant. The concept of writhe is commonly used in biochemistry: DNA molecules must be highly coiled in order to fit inside the nucleus of a cell, and hence exhibit a very large value of Wr.

Writhe measures helical structure of the axis of a loop, but cannot be computed from the projection of the loop on the photosphere alone (see Fig. 4); the height of the loop is also important. During a kink instability, a twisted flux rope may convert some of its twist to writhe. Unfortunately, the amount of twist contained in the rope before the eruption cannot be accurately inferred from observing the amount of writhe produced by this conversion (Török et al. 2014). The amount of writhe produced depends mainly on the details of the instability, rather than the initial twist.

2.5 Magnetic Helicity Transport

We wish to give a simple expression for dH/dt, which may include both dissipation and transport across boundaries. Although we may use any gauges we please for $\mathbf{A}$ and $\mathbf{A}_{pot}$,

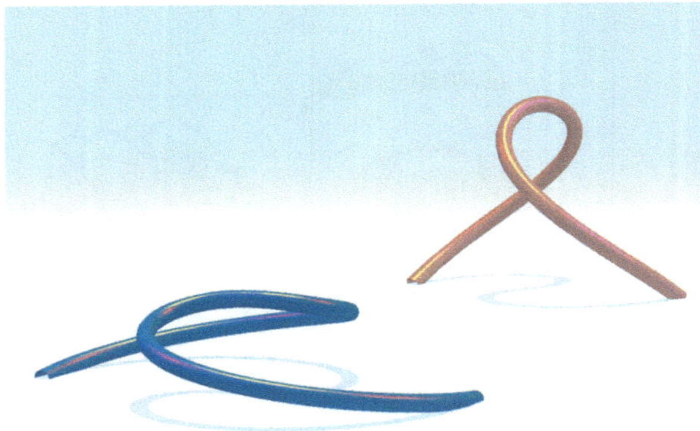

Fig. 4 Two flux ropes. Both ropes have an inverse S shape as seen in projection. However, the one on the *left* has a writhe $W_r = -0.2\Phi^2$ while the rope on the *right* has writhe $W_r = +0.2\Phi^2$. This demonstrates that a low lying loop can have the opposite sign of writhe to a taller loop with the same projection curve

one particular choice considerably simplifies the algebra (Berger 1984). First, we require that $\mathbf{A}_{pot}$ satisfies

$$\nabla \cdot \mathbf{A}_{pot} = 0; \qquad \mathbf{A}_{pot} \cdot \hat{\mathbf{n}}|_S = 0. \tag{18}$$

One then finds

$$\frac{\mathrm{d}H}{\mathrm{d}t} = -2 \int \mathbf{E} \cdot \mathbf{B}\, \mathrm{d}^3x + 2 \oint_S \mathbf{A}_{pot} \times \mathbf{E} \cdot \hat{\mathbf{n}} \mathrm{d}S, \tag{19}$$

where $\mathbf{E}$ is the electric field. This equation is analogous to Poynting's theorem for the dissipation and transport of electromagnetic energy. The first term corresponds to dissipation; magnetic helicity dissipation can be shown to be negligible in high magnetic Reynolds number plasmas (Berger 1984; Ortolani and Schnack 1993).

The second term governs transport across the boundary S. For the photosphere, the dominant contribution to $\mathbf{E}$ comes from ideal MHD flow. In this case $\mathbf{E} = \mathbf{B} \times \mathbf{V}$ (also see, Berger 1984)

$$\frac{\mathrm{d}H}{\mathrm{d}t} = 2 \oint_S \left((\mathbf{A}_{pot} \cdot \mathbf{V})\mathbf{B} - (\mathbf{A}_{pot} \cdot \mathbf{B})\mathbf{V} \right) \cdot \hat{\mathbf{n}} \mathrm{d}S. \tag{20}$$

The first term represents the effect of twisting motions on the boundary, while the second represents the bulk transport of helical field across the boundary.

The helicity transport equation takes an intuitive form in the simple case where N flux tubes pass through a planar boundary. Let flux tube i have flux Φ_i. Suppose the motion at boundary rotates flux tube i at an angular velocity ω_{ii}, while tube i and tube j rotate about each other at an angular velocity ω_{ij}. Then

$$\frac{\mathrm{d}H}{\mathrm{d}t} = \frac{1}{2\pi} \sum_{i=1}^{N} \sum_{j=1}^{N} \omega_{ij} \Phi_i \Phi_j. \tag{21}$$

3 Computations of Helicity in Finite Volumes

3.1 Computations of Instantaneous Helicity in the Corona

The application of the concept of relative magnetic helicity defined in Sect. 2 is problematic in active region (AR) magnetic field extrapolations because their volumes are finite. For linear force-free (LFF) fields, this problem can be bypassed by formulating the helicity content of an AR in terms of surface integrals applied to the lower boundary (Berger 1985; Georgoulis and LaBonte 2007). These calculations include summations over the Fourier modes of the magnetogram. When the force-free parameter α exceeds a certain critical value that depends on the horizontal size of the calculation box, helicity attains unphysical large values. Démoulin et al. (2002a) proposed to use the linearized Berger's (1985) expression in which the helicity is proportional to α. In several publications, (e.g. Démoulin et al. 2002a; Green et al. 2002; Nindos and Andrews 2004) the linearized formula has been used in conjunction with the single best value of α, α_{best}, for the AR (see Sect. 3.3). In framework of LFF field, magnetic helicity can also be computed via α and magnetic energy, E_m (Pevtsov and Canfield 1999; Pevtsov 2008), with E_m determined from the Virial theorem.

Georgoulis et al. (2012) extended the work by Georgoulis and LaBonte (2007) on LFF fields to non-linear force-free (NLFF) fields. They developed a method that depends on a lower-boundary connectivity matrix that can be inferred either by a NLFF field extrapolation or in some other way. Instead of using extrapolation results, these authors used a unique connectivity-matrix solution for a given flux-partition map. The solution relies on a simulated annealing algorithm designed to minimize the distances of connected opposite polarity partitions.

In several publications that treat the general problem of the calculation of the instantaneous helicity in the corona, NLFF field extrapolations are used, and the helicity in the computation box is estimated using methods that are based on the Coulomb gauge (e.g. Rudenko and Myshyakov 2011; Thalmann et al. 2011). The choice of gauge is irrelevant for the relative magnetic helicity value, but it may influence how computationally expensive the algorithm becomes. Valori et al. (2012) exploited the gauge freedom by choosing one that requires that one component of the vector potential vanishes; their method is a direct extension of an earlier work by DeVore (2000) to finite volumes.

3.2 Computations of the Injection Rate of Helicity

When high-cadence photospheric magnetograms are available, the horizontal velocity related to (20) can be computed. In the early studies, several authors (e.g. Chae 2001; Nindos and Zhang 2002; Moon et al. 2002a; Moon et al. 2002b; Nindos et al. 2003; Chae et al. 2004) utilized the local correlation tracking (LCT) technique (November and Simon 1988) for the computation of horizontal velocities. Démoulin and Berger (2003) have pointed out that with magnetograms one follows the photospheric intersection of the magnetic flux tubes but not the evolution of the plasma. Consequently, from the observed magnetic evolution we obtain the flux tube motion and not the plasma motion parallel to the photosphere. If $\mathbf{v}_t$ is the tangential component of the photospheric plasma velocity and v_n the velocity perpendicular to the photosphere, the tracking algorithm detects the velocity of the footpoints of the flux tube which is

$$\mathbf{u} = \mathbf{v}_t - \frac{v_n}{B_n}\mathbf{B}_t \qquad (22)$$

This formula should be correct if v_n is the vertical component of the pattern speed of the emerging flux system. The plasma velocity may differ from this as flows along the field lines are possible. Indeed, MHD simulations (Welsch et al. 2007) have shown that this formula is not always valid, and its use should be treated with caution especially during flux emergence episodes (see also Schuck 2008; Ravindra et al. 2008). This conclusion has been confirmed by Liu and Schuck (2012) who analyzed time series of HMI vector magnetograms from two emerging active regions (ARs).

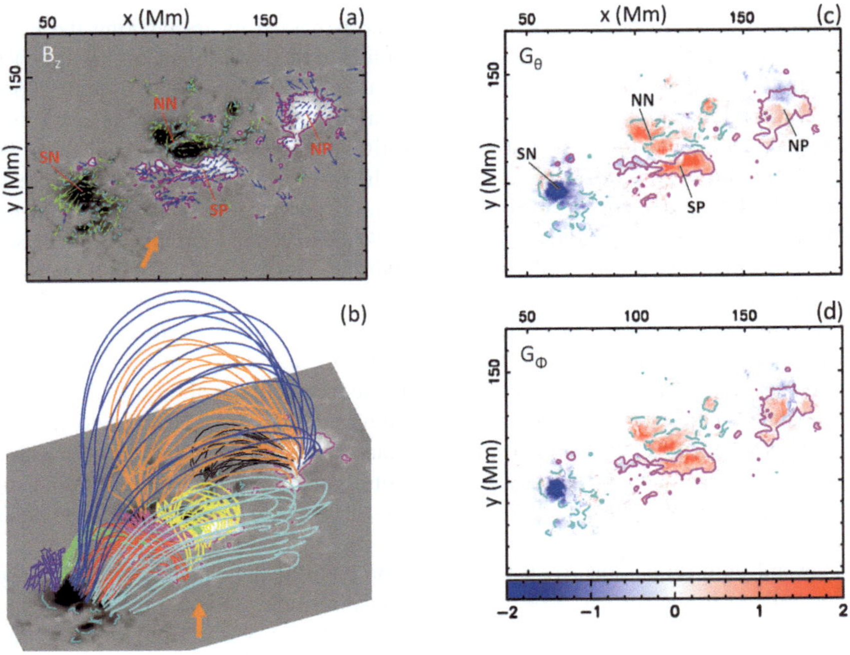

Fig. 5 AR11158 at 06:28 UT on 2011 February 14. (**a**) Photospheric vertical magnetic field, B_z, in *grayscale* overplotted with the flux transport velocity field (*blue/green arrows*) and the polarity labels. (**b**) 3D view of the NLFFF extrapolation with selected magnetic field lines. (**c**) G_θ map. (**d**) G_Φ map. In (**c**) and (**d**) the helicity flux density distributions are in units of 10^7 Wb^2 m^{-2} s^{-1} with the same color scale, and the ±500 G isocontours of B_z are overplotted (modified from Dalmasse et al. 2013)

The combination of (20) and (22) shows that the whole helicity flux density can be retrieved within the accuracy of the calculation. Consequently, one may use the quantity $G_A = -2\mathbf{u} \cdot \mathbf{A}_{pot} B_n$ as a proxy to the whole helicity flux density (e.g. Chae 2001; Nindos and Zhang 2002; Moon et al. 2002a, 2002b; Nindos et al. 2003; Chae et al. 2004). The G_A maps always appear extremely complex both in space and time, with polarities of both signs present at any time. Pariat et al. (2005) showed that G_A is not a real helicity flux density and that its properties introduce artificial polarities of both signs. The spurious signals appear due to the fact that helicity flux density per unit surface is not physical quantity. Due to the properties of helicity, only helicity flux density per unit of elementary magnetic flux has a physical meaning. But to estimate such quantity using observations would require to isolate flux tubes and determine their connectivity, which is very difficult. Thus any definition of a helicity flux density will only be a proxy of the helicity flux density per unit magnetic flux. Pariat et al. (2005) introduced a new proxy for helicity flux density, G_θ, which does not suffer from G_A's problems. G_θ implies that the helicity injection rate is the summation of the rotation rate of all pairs of elementary fluxes weighted by their magnetic flux. An example of a G_θ map is given in Fig. 5(c).

In order to define the real helicity flux density, the coronal linkage needs to be provided. With it one can represent how all elementary flux tubes move relatively to a given elementary flux tube, and the helicity flux density is defined per elementary flux tube. Using photospheric maps this can be achieved by distributing equally the helicity input between the two footpoints for each elementary flux tube. Then the helicity flux can be rewritten as

Fig. 6 (**a**) Time profile of the net flux of injected helicity, dH/dt, over the field of view of AR11158 that is presented in Fig. 5(b). (**b**) Time profile of the net accumulated change of helicity, ΔH, calculated from the measured dH/dt. The GOES soft X-ray flux time profile is also presented. (**c**) Temporal evolution of the magnetic free energy for AR11158 (modified from Nindos et al. 2012)

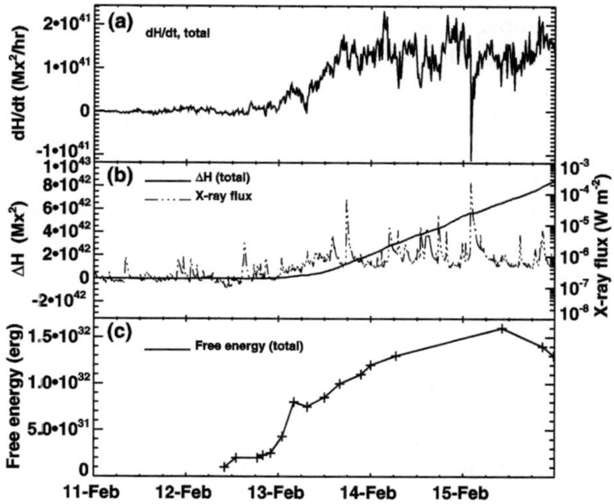

a flux of magnetic helicity per unit of surface, G_Φ, where G_Φ is a field-weighted average of G_θ at both photospheric footpoints, $\mathbf{x}_\pm$, of the photospheric connection.

While G_Φ provides the true helicity flux density, its practical use is presently limited by our ability to define the coronal linkage for all magnetic polarities. However, progress can be made if we obtain the coronal linkage from NLFFF extrapolations (Dalmasse et al. 2013, 2014). A G_Φ map resulted from such treatment appears in Fig. 5(d). This map corresponds to the same area whose G_θ map appears in Fig. 5(c).

It has been reported (e.g. Pariat et al. 2006, 2007) that unlike the usual G_A maps, most G_θ maps show almost unipolar spatial structures because the non-dominant helicity flux densities are significantly suppressed. However, as can be seen from Fig. 5(c), this is not always the case. In Fig. 5(d), the G_Φ map also displays mixed signals which implies that there are real mixed signs of the helicity flux in the AR. However, a comparison between Fig. 5(c) and Fig. 5(d) shows that the G_Φ distribution is different from the G_θ distribution, except in the regions of open magnetic fields (in these regions, Dalmasse et al. 2013, set $G_\Phi = G_\theta$). Overall, the comparison shows that the intensity of the G_θ signal tends to be overestimated.

When a cube of helicity flux density maps is available, one can calculate the time evolution of the total helicity flux, dH/dt. An example using a time series of G_θ maps which includes the G_θ map of Fig. 5(c), appears in Fig. 6 (top panel, see Nindos et al. 2012). The resulting time profile of the net accumulated change of helicity is shown in the middle panel of Fig. 6. Note that we expect that the helicity flux integrated using G_A to be identical to the one integrated using G_θ or G_Φ because all definitions are derived from (20) (see Liu and Schuck 2013).

3.3 Computations of Current Helicity

Due to the difficulties involved in the computation of magnetic helicity, often proxies of the current helicity, H_c, are used instead. The current helicity is defined as $H_c = \int_V \mathbf{B} \cdot \mathbf{j} dV$, with $\mu_0 \mathbf{j} = \nabla \times \mathbf{B}$. While the magnetic helicity is gauge-dependent through $\mathbf{A}$, there is no gauge freedom with H_c ($\nabla \cdot \mathbf{B} = 0$). Furthermore, magnetic helicity is a conserved MHD quantity, but the current helicity is not. However, the magnetic and current helicity are usually considered to have the same sign (although this has not been proven rigorously).

Observationally, we can derive values of the current helicity that represent only a fraction of its full value. This is because the volume integral of H_c can be written as $\int_V (B_x j_x + B_y j_y + B_z j_z) dV$, of which only the last component can be derived from observations. Though using photospheric vector magnetograms all three components of $\mathbf{B}$ are available, only the vertical component of $\mathbf{j}$ can be computed via the horizontal derivatives of $\mathbf{B}$. Therefore we can only determine a fraction of the whole current helicity density by calculating $h_c = B_z j_z$. Two different proxies based on current helicity have been used in statistical studies: the fractional imbalance of h_c (percentage of pixels of one sign of h_c in a given magnetogram) and area-averaged h_c $\langle h_c \rangle$ (e.g. Abramenko et al. 1996; Bao and Zhang 1998). In sunspots, the horizontal components of $\mathbf{j}$ were determined either using the assumption of axisymmetric model (Pevtsov and Peregud 1990) or from the full Stokes polarimetric inversions on the basis of known difference in height formation of spectral line winds and core (Puschmann et al. 2010; Ruiz Cobo and Puschmann 2012). These studies indicated that the horizontal components of $\mathbf{j}$ may exceed significantly its vertical component.

One can also employ the value of the force-free field parameter α under the LFF approximation, as helicity proxy. Using photospheric vector magnetograms, one can determine α either by fitting LFF field to the observed transverse field (so called, α_{best}), or by computing the vertical component of α for each pixel and then averaging it ($\langle \alpha_z \rangle$). Burnette et al. (2004) found a reasonably good correlation between α_{best} and $\langle \alpha_z \rangle$. When longitudinal magnetograms are available instead of vector magnetograms, the α_{best} can be determined by comparing the computed LFF field lines with the observed soft X-ray or EUV coronal structures (e.g. Nindos and Andrews 2004). It follows from the definitions (e.g., Hagyard and Pevtsov 1999) that helicity proxies h_c and α have the same sign.

4 Observations of Magnetic Helicity

4.1 Sources of Magnetic Helicity

The first term of the right-hand side of (20) corresponds to the injection of helicity by advection (i.e. emergence of field lines that cross the photosphere) while the second term (also known as shearing term) is the flux of helicity due to motions parallel to S (see Sect. 3.2). Such motions may come either from differential rotation and/or transient photospheric shearing flows.

Differential rotation was the first mechanism considered that injects helicity into ARs (DeVore 2000). Even when a single bipole is considered, differential rotation does not provide a monotonous input of magnetic helicity (DeVore 2000). This is because differential rotation rotates both magnetic polarities on themselves and also changes their relative positions, introducing twist and writhe helicity fluxes, respectively. These fluxes always have opposite signs and similar amplitudes, and therefore partially cancel (Démoulin et al. 2002b). Démoulin et al. (2002a) and Green et al. (2002) studied the long-term evolution of the helicity injected by differential rotation into the coronal part of two active regions which were followed from their birth for several months during their decay. These studies showed that the contribution of differential rotation to the helicity budget of active regions remains small even on the long term.

However, the conclusion is different in the convection zone for a solar cycle and a global spatial scale. Berger and Ruzmaikin (2000) calculated the magnetic helicity production by differential rotation using 22 years of magnetogram data and differential rotation curves. They found that the helicity production in the solar interior by differential rotation had the correct sign compared to observations of coronal structures (the magnetic helicity conservation is satisfied by the natural generation of the same amount, but of opposite sign, in both hemispheres). The net helicity flow into each hemisphere over the whole 22-year magnetic cycle was about 4×10^{46} Mx^2.

ARs cover only a tiny fraction of the solar surface. The quiet Sun is characterized by small magnetic flux density but its total magnetic flux is huge. Welsch and Longcope (2003) estimated the helicity flux in five time series of quiet Sun magnetograms located close to disk center. They extrapolated their results to the whole Sun and found that the helicity injection in the quiet Sun is negligible compared to the helicity injected even by a single AR. We note, however, that their measurements included the braiding component only and therefore they should be considered as lower limits (the twist component of individual flux tubes was not measured because it was not possible to resolve tiny flux tubes).

The study of the helicity budget of active regions requires knowledge of the helicity injected into them and of the helicity carried away from them. The former is computed using the methods described in Sect. 3 while CMEs are considered responsible for the latter. The helicity content of a CME can be estimated by the change of coronal helicity of the source region during the event (e.g. Mandrini et al. 2005; Tziotziou et al. 2013). Inside magnetic clouds, helicity is estimated from modeling of the *in situ* measurements of the magnetic field vector. In practice, in studies of the long-term evolution of helicity of ARs that are linked to at least one magnetic cloud at 1 AU one assumes that the helicity carried away by each CME is equal to the helicity content in the magnetic cloud. Nindos et al. (2003) and Lim et al. (2007) were able to partially reconcile the amount of helicity injected into the corona with the helicity carried away by the CMEs in the ARs they studied. However, the uncertainties of these studies are significant primarily due to the large uncertainties in the calculation of the helicity transported away by CMEs.

Using line-of-sight magnetograms together with the Démoulin and Berger (2003) formula (22), one cannot calculate separately the advection and the shearing terms of (20). Furthermore, the validity of the Démoulin and Berger hypothesis has been questioned (see Sect. 3.2). Theoretical studies (e.g. Démoulin et al. 2002a) indicate that on the AR scale transient shearing motions are a rather inefficient way to bring helicity into the corona, compared to the helicity carried by the emergence of a significantly twisted flux tube. Indeed, Pevtsov et al. (2003a) have shown that helicity in ARs typically increases during flux emergence while upon the termination of emergence the helicity increase usually stops. The separate computation of both the shearing and advection terms requires the use of photospheric vector magnetograms. Several methods have been developed towards this goal and contradictory results have been reported (e.g., Kusano et al. 2002; Welsch et al. 2004; Longcope 2004; Georgoulis and LaBonte 2006; Schuck 2008; Zhang et al. 2012; Liu and Schuck 2012). Furthermore, the methods which were developed before 2007 were checked against the same anelastic MHD simulation (Welsch et al. 2007) and produced different results.

On solar cycle time scales, the helicity injected into ARs has an upper limit of 6.6×10^{45} Mx^2 and is an inherently disorganized, impulsive and aperiodic process (Georgoulis et al. 2009). The whole solar-cycle value for the quiet Sun is about 10^{43} Mx^2 (Welsch and Longcope 2003).

4.2 Hemispheric Helicity Rule

4.2.1 Observations

All helicity proxies discussed in Sect. 3 reveal that solar magnetic fields, at different spatial scales, show primarily negative chirality (i.e. negative helicity sign) in the northern hemi-

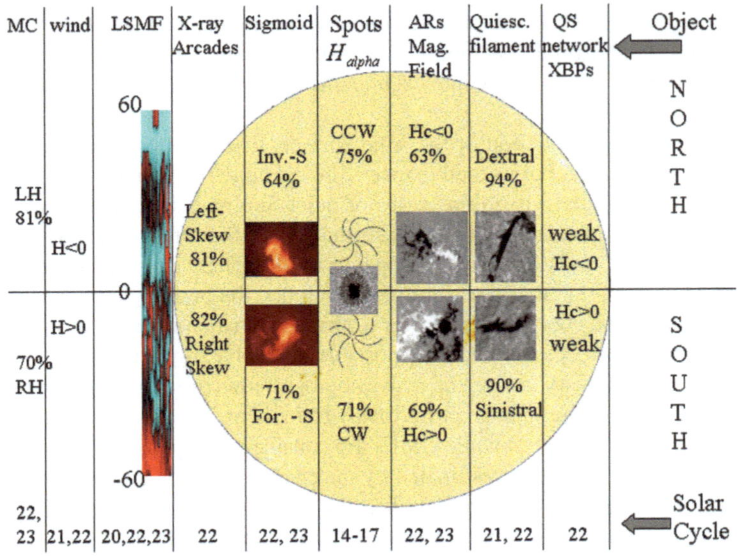

Fig. 7 Pictorial representation of the hemispheric helicity rule. Adopted from Pevtsov (2002)

sphere while those in the southern hemisphere show primarily positive chirality. This trend is called hemispheric helicity rule and a pictorial summary of it appears in Fig. 7.

Using vector magnetograph observations of ARs and the α_{best}, α_{av} or current helicity imbalance methods (see Sect. 3.3), it has been established (e.g. Seehafer 1990; Pevtsov et al. 1995, 2001; Abramenko et al. 1997; Bao and Zhang 1998; Bao et al. 2000; Hagino and Sakurai 2004, 2005; Zhang 2006; Hao and Zhang 2011; Liu et al. 2014) that the hemispheric helicity rule is a relatively weak statistical trend satisfied by 60–75 % of active regions.

Soft X-ray sigmoids have also been used as proxies to the chirality of ARs with S (reverse-S) shapes corresponding to positive (negative) chirality. Rust and Kumar (1996); Lim and Chae (2009) determined that the hemispheric rule was obeyed by 80–87 % of sigmoids whereas in a larger sample analyzed by Pevtsov and Canfield (1999) the fraction was 64 %. We note, however, that potential fields may occasionally show sigmoidal structures (Pevtsov et al. 1997); furthermore the apparent direction of curvature of a three-dimensional sigmoidal structure may depend on projection effects.

The hemispheric helicity rule is also observed in filaments, filament channels, and their overlying coronal arcades (Martin 1998). By visual inspection of the orientation of the "barbs" of both quiescent and active-region filaments Martin (1994); Rust and Martin (1994); Pevtsov et al. (2003); Yeates et al. (2007); Lim and Chae (2009) found that 76–82 % of them satisfy the hemispheric helicity rule. However, the corresponding percentage found by Bernasconi et al. (2005) who used an automated detection algorithm was only 68 %. Wang (2013) argued that the higher strength of the hemispheric rule in filaments reflects the difficulty in determining the twist in newly emerged ARs, and that the strength of the hemispheric rule in ARs should indeed be similar to that in quiescent filaments.

For the study of the chirality of large-scale solar magnetic fields, proxies of the current helicity were computed from the large-scale vector magnetic fields which, in turn, were derived from time sequences of longitudinal magnetograms (by tracking the flux as it rotated across the disk and using the changing projection angle to reconstruct its vector components under the assumption that the field did not change over several days, e.g. Pevtsov and Latushko 2000; Wang and Zhang 2010) or even from full-disk vector magnetograms directly (Gosain et al. 2013). Calculations of the helicity injection rate applied on series of synoptic maps derived from longitudinal magnetograms have also been used (Yang and Zhang 2012). Pevtsov and Latushko (2000) did not detect any significant hemispheric chirality asymmetry within $\pm 40^\circ$ of the equator. In their data, however, asymmetry was present at high latitudes where the current helicity was negative (positive) in the northern (southern) hemisphere. The studies by Wang and Zhang (2010); Yang and Zhang (2012); Gosain et al. (2013) found that the hemispheric helicity rule still holds in general, when large-scale fields are considered. Note also that from a large sample of AR vector magnetograms, Zhang (2006) found that weaker fields (< 500 G) followed the hemispheric rule, but stronger fields had the reverse helicity sign. However, using vector synoptic maps, Gosain et al. (2013) reached exactly the opposite conclusion.

The structures with the smallest spatial scales where the hemispheric helicity rule still holds are quiet Sun network elements (Pevtsov and Longcope 2001, 2007) while the structures with the largest spatial scales include magnetic clouds (e.g. Leamon et al. 2002).

The hemispheric helicity rule does not change from one solar cycle to the other. The study that covered the most time-extended data set was done by Pevtsov et al. (2008) who analyzed data from four different vector magnetograms for 19 years from solar cycles 21, 22, and 23 and found that the hemispheric rule is a weak tendency with significant scatter. However, there are published reports that the rule might change its sign in some periods of a solar cycle. For example, using current helicities Bao et al. (2000) argued that the hemispheric helicity rule was not present at the ascending phase of solar cycle 23. On the other hand, the α_{best} helicity proxy did not show any change in the hemispheric helicity rule for the same period. Hagino and Sakurai (2005) found that the rule is satisfied in the solar maximum phase but may not be satisfied in the solar minimum phase. Hao and Zhang (2011) studied 64 ARs that appeared in the descending phase of solar cycle 23 and the ascending phase of solar cycle 24. The 34 ARs of cycle 24 followed the hemispheric helicity rule, whereas the 30 ARs of cycle 23 did not. However, when combining all ARs as one sample, they followed the hemispheric helicity rule. Zhang et al. (2010) studied AR vector magnetograms from more than 20 years of observations and found that the "wrong" signs disappear at the ends of the butterfly wings as well as at their very beginnings. Furthermore, the large-scale magnetic helicity fluxes computed by Yang and Zhang (2012) showed the same sign in both hemispheres around 2001 and 2005. We note that when we investigate whether or not there is a change of the hemispheric rule with solar cycle we need to take into account how this might be affected by a possible lack of consistency between different magnetograms obtained in the years when the hemispheric rule is reported to change sign. For example, Pevtsov et al. (2008) argued that due to significant scatter in the data, years with low sunspot activity may show deviations from the rule because the statistical sample is insufficient.

4.2.2 *Interpretation*

Potential candidates for the origin of the hemispheric helicity rule include solar differential rotation, direct action of the Coriolis force, solar dynamo, and turbulent convection in the upper part of the convection zone.

Wang (2013) showed that the net twist introduced by the Coriolis force on the cross-sectional expansion of each leg of a rising Ω-loop may be substantial if the loop's rise time scale is sufficiently long; he also found that the sense of the twist satisfies the hemispheric rule. But the rule of helicity generation in the hydrodynamic dynamo that is driven by the Coriolis force dictates that the variation of α should be proportional to the sine of solar latitude; this trend is not supported by the observations (Pevtsov et al. 1995).

Differential rotation of the Sun acts over extremely long time scales during which it is unlikely that a single flux tube could survive. But even if we assume so, the maximum possible twist introduced to the tube will be well below observed values (e.g. Longcope et al. 1999; Démoulin et al. 2002b). Furthermore, a fundamental problem with both differential rotation and the Coriolis force is that they are steady mechanisms which lack inherent fluctuations while an intrinsic significant scatter is found in the hemispheric rule.

Choudhuri (2003) has investigated the relation between dynamo theory and the generation of magnetic helicity in flux tubes in the convection zone. He postulated that the toroidal and poloidal fields are generated in two different regions; at the bottom and at the top of the convection zone, respectively. When toroidal flux tubes move upward into the region near the surface where the poloidal field is present, the poloidal field gets wrapped around the flux tube giving rise to helicity. Choudhuri (2003) showed that this process generates helicities with signs that obey the hemispheric rule. Based on this model, computations by Choudhuri et al. (2004) show that at the beginning of a solar cycle the hemispheric rule reverses sign. This is in agreement with some observations (see Sect. 4.2.1). Choudhuri's idea of acquiring twist by the accretion of ambient poloidal field was supported by Chatterjee et al. (2006); their models predicted a change in the sign of helicity from the central, stronger parts of the flux tube to its periphery in apparent agreement with the findings of Gosain et al. (2013).

The significant scatter exhibited by the hemispheric rule implies that turbulence in the convection zone may play an important role in the generation of the observed chirality trends. Indeed, Longcope et al. (1998) showed that turbulence of the convection zone can introduce to a flux tube helical deformation on small scales (at the mixing length) if the turbulence contains a nonvanishing kinetic helicity. Longcope et al. (1998) called this effect the Σ-effect and showed that these helical deformations will be of the handedness appropriate to explain the scatter observed in the hemispheric helicity rule.

Nandy (2006) provided indirect support in favor of the Σ-effect mechanism. He found that the dispersion in the AR twist distribution is latitude-independent, implying that the amplitude of turbulent fluctuations does not vary with latitude in the convection zone. His data set also showed that the amplitude and dispersion of twist decreased with increasing magnetic size of active regions, supporting the conclusion that larger flux tubes are less affected by turbulence.

5 Magnetic Helicity and the Initiation of CMEs

5.1 Helicity as an Important Agent for CME Initiation

Whatever the mechanism that generates helicity on the Sun, it operates continuously. Thus helicity should be removed from the Sun at approximately the same rate as it is created. Accumulation of helicity would make it difficult for solar dynamo to operate—a problem referred to as dynamo quenching (e.g. Brandenburg and Sandin 2004).

CMEs are thought to be the primary agent through which the Sun gets rid of its excess helicity (e.g. Rust 1994; Low 1996; Zhang and Low 2005). The main arguments supporting this idea are: (1) solar magnetic fields obey the hemispheric helicity rule (see Sect. 4.2) which appears not to change from solar cycle to solar cycle. (2) A fraction of AR's helicity is created by the dynamo and then transported into the corona through the photosphere with the emerging magnetic flux. This process would constantly accumulate helicity into the corona because of helicity's property not to be destroyed under reconnection. Furthermore, cancellations of opposite helicity fluxes involves a small fraction of the magnetic flux (e.g. between ARs of opposite helicity sign, either within the same hemisphere or across the equator, Pevtsov 2000). Finally, it is speculated that a small fraction of coronal helicity is cancelled via magnetic reconnection between magnetic fields (ARs and coronal holes) of opposite helicity sign. However, this represents only a small fraction of the accumulated helicity, as the relevant magnetic fluxes are small.

Along the above lines, Low and Zhang (2002) and Zhang and Low (2001, 2003) have developed a unified view of CMEs as the last chain of processes that transfer helicity from the convection zone into the interplanetary medium. Their theory exploits Taylor's conjecture that the magnetic field will relax towards a LFF field state. A summary of their results is as follows. When new field enters the corona, repeated reconnections between the new and pre-existing field take place. This process simplifies the magnetic topology and the dissipated magnetic energy produces flares. The relaxation proceeds according to Taylor's conjecture and results in the formation of a flux rope which contains a significant fraction of the total helicity of the system. The fate of the flux rope is determined by the efficiency of its confinement by its surrounding anchored field. Flux rope ejection occurs when the magnetic energy it contains is sufficient to drive an outward expansion against the confining field.

In a series of articles, Zhang et al. (2006, 2012); Zhang and Flyer (2008) studied the theoretical upper limit of helicity that can be stored in various field configurations and its implications for coronal evolution. In the 2006 article they showed that in an open spherical volume like the corona, for a given boundary flux distribution, there is an upper bound on the magnitude of the total helicity of all axisymmetric power-law force-free fields. When the accumulated helicity exceeds this limit, a non-equilibrium situation is reached which mimics the initiation of CMEs. In the 2008 article they found that the helicity upper bound of force-free fields depends on the boundary flux distribution: multipolar photospheric configurations can have a helicity upper bound 10 times smaller than dipolar ones. In the 2012 article they studied the helicity of self-similar axisymmetric force-free fields and found that there may be an upper bound on the total helicity of all bipolar axisymmetric force-free fields. As the helicity increases, the fields open up forming a current sheet surrounded by Parker-spiral-like structures.

Amari et al. (2003a, 2003b) constructed a set of force-free fields having different magnetic flux and helicity contents and used them as initial conditions by applying converging motions or a turbulent diffusion-driven evolution. These processes can trigger eruptive events that may be either confined or global, depending on the value of the initial helicity. Amari et al. (2003b) concluded that helicity cannot be the only parameter controlling the triggering of an ejection, as its value is constant during the diffusion-driven evolution: having a large enough helicity seems a necessary condition for an ejection to occur, but not a sufficient one. Jacobs et al. (2006) performed MHD simulations by shearing an axisymmetric arcade outside a sphere, and found that a twisted flux tube forms and ejects when H/Φ^2 is typically above 0.2–0.3.

5.2 Observational Evidence

The physical view presented in Sect. 5.1 has been supported by several observations. Nindos and Andrews (2004) used LFF field extrapolations and the α_{best} method to model the pre-flare coronal field of 78 ARs that produced big flares. Only some 60 % of these flares were associated with CMEs. Then from the derived values of α_{best} they computed the corresponding coronal helicities. Their results indicated that in a statistical sense both the pre-flare

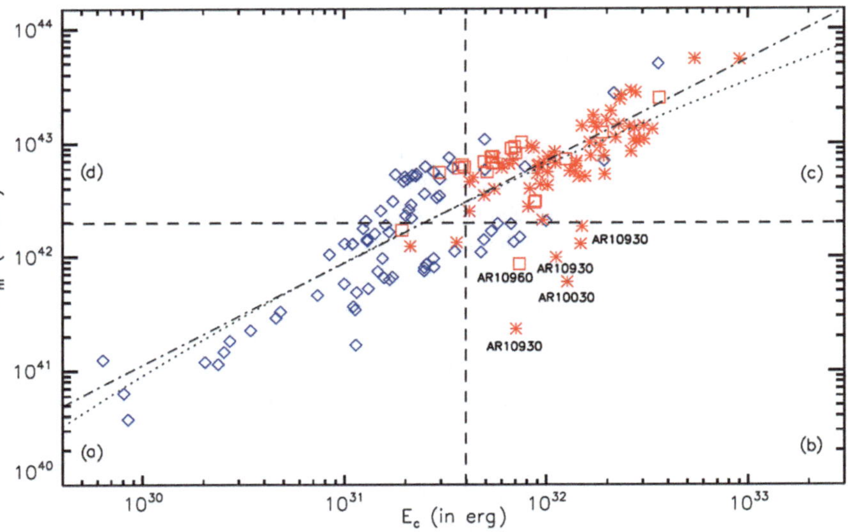

Fig. 8 Free energy-helicity diagram of ARs. *Diamonds*, *squares* and *asterisks* correspond to non-flaring, M- and X-class flaring ARs, respectively. *Dashed lines* indicate the estimated thresholds for helicity and magnetic free energy above which ARs give major flares. These thresholds divide the diagram into four regions, labeled a, b, c, and d. The *dotted* and *dashed-dotted lines* denote the least-squares best fit and the least-squares best logarithmic fit, respectively, between helicity and magnetic free energy (from Tziotziou et al. 2012)

absolute value of α and the corresponding coronal helicity of the ARs producing CME-associated big flares were larger than the absolute value of α and helicity of those that did not have associated CMEs.

The above results are consistent with the ones reported by LaBonte et al. (2007) who calculated the helicity flux in 48 ARs that produced X-class flares and in 345 non-X-flaring ARs. They found that a necessary condition for the occurrence of an X-class flare is that the peak helicity flux has a magnitude $> 6 \times 10^{36}$ Mx^2 s^{-1}.

Tziotziou et al. (2012) used the method developed by Georgoulis et al. (2012, see Sect. 3.1) to calculate the instantaneous magnetic free energy and helicity from 162 vector magnetograms in total of 42 different ARs. They found a statistically significant, monotonic correlation between the free energy and helicity. This correlation implies that, in addition to helicity, free energy may play a significant role in eruptive phenomena. In their study, the eruptive ARs appeared well segregated from the non-eruptive ones in both free energy and helicity (see Fig. 8).

In ARs, a primary constraining force that inhibits global eruptions is provided by the overlying background field. Using both line-of-sight and vector magnetograms, Nindos et al. (2012) studied the long-term evolution of the background field in AR11158 that produced three major CMEs. In their calculations they used the decay index of the magnetic field which is a parameter that quantifies how fast the field decreases with height. Their results indicated that the initiation of eruptions did not depend critically on the temporal evolution of the variation of the background field with height. On the other hand, they showed that both the magnetic free energy (computed from NLFF field extrapolations) and the accumulated helicity into the corona (computed from the helicity injection rate) contributed the most to the eruptions by their increase throughout the observations (by factors of more than 5 and more than two orders of magnitude, respectively; see Fig. 6).

5.3 Other Approaches

There are several other approaches to the initiation of CMEs and the role of helicity. Some models suggest that eruptive events can occur without any significant helicity accumulations. MacNeice et al. (2004) studied the evolution of helicity under the breakout model (Antiochos et al. 1999). In their simulation, the model was driven by a shear flow that injected both free energy and net helicity into the corona. Their results showed that the helicity shed by the plasmoid ejection was at least 80 % of the total originally injected into the system. They interpreted this result as an indication that although CMEs remove the bulk of the coronal helicity, some fraction remains behind. They suggested that some other mechanism (possibly small-scale diffusion) might be responsible for dissipating the rest of the helicity. Furthermore, Kliem et al. (2011) found that simulated flux rope CMEs carried away only a minor part of the initial helicity that was present in the simulation box; most of the helicity remained in the simulation box even after the departure of the CME from there. This result was interpreted as a consequence of the requirement that the current through an expanding loop must decrease if the magnetic energy of the configuration is to decrease as the loop rises, to provide the kinetic energy of the CME.

Phillips et al. (2005) presented simulations of the breakout model where eruption occurs even when no net helicity is injected into the corona. In their simulations the eruption occurs at a fixed magnitude of free energy in the corona, independent of the value of helicity. It would be desirable to check these results against computations of the helicity evolution in observed eruptions that appear to be due to breakout. The MHD simulations by Zuccarello et al. (2009) also showed that the injection of helicity is not a necessary constraint in the initiation of CMEs. However, the absence of significant net helicity accumulation prior to an eruption might result from the accumulation of similar amounts of both positive and negative helicity; in such case the "helicity annihilation" might be at work (Kusano et al. 2003). Indeed, Kusano et al. (2004) presented simulations where the introduction of a reverse helicity was essential for the eruption of a sheared arcade.

The helicity content of a twisted flux rope with a uniform twist across its section and having N number of turns is simply N, when it is measured in units of its flux to the second power. It is convenient to express helicity values in number of end-to-end twist and turns, i.e. in units of the square of the magnetic flux of the active region. For example in an active region with a total magnetic flux and helicity of $\Phi = 5 \times 10^{21}$ Mx and $H = 1.25 \times 10^{43}$ Mx^2, respectively, helicity can be expressed as $N = 0.5\Phi^2$, i.e. the global helicity of the flux rope forming the active region can be characterised as having 0.5 end-to-end turn, equivalent to 1π. In the solar corona twist values are usually low, below $N = 1$, or 2π. However, in the interplanetary medium magnetic clouds are significantly more twisted: values can reach $N = 10$ (Gulisano et al. 2005; Démoulin 2008). The increase in end-to-end twist likely takes place during CME eruption via magnetic reconnection between the erupting flux rope and the sheared arcade field below (Mandrini et al. 2005; Attrill et al. 2006).

Magnetic helicity of a thin flux rope can also be expressed via its twist (Tw) and writhe (Wr, see Sect. 2.4). Since magnetic helicity is conserved in ideal MHD, twist can be converted into an equal amount of writhe. Such a conversion takes place during kink instability (e.g. Török and Kliem 2005). The threshold of kink instability is reached at around a twist of 2π. As presently twist cannot be directly measured in the corona, Török et al. (2014) asked the question whether or not the amount of writhing of erupting flux ropes, proxyed by erupting filaments, which is a more directly measurable quantity, could provide quantitative information on the amount of twist in the pre-eruptive magnetic configuration. They carried out MHD simulations of kink instability with a range of internal twists (3.0–10.6π)

and found that up to a high initial twist of 7.5π the rope axis invariably develops a one-turn helix, i.e. the amount of twist converted to writhe being about 2π (for an observational example see Williams et al. 2005). Writhing only increased for the highest initial twist values, which are well above twist values usually observed on the Sun. (Although several examples were found by Vrsnak et al. 1991 of end-to-end twist of pre-eruption prominences being in the range $3\text{–}15\pi$, and another example by Romano et al. 2003 of about 10π.) Therefore from writhing of erupting filaments one can determine the sign of twist/helicity in the active region (Green et al. 2007), but can only deduce whether or not they are twisted less or more than about 6π.

Moreover, several proposed eruption mechanisms do not explicitly rely on helicity. These models include the tether cutting (e.g. Moore et al. 2001), breakout (Antiochos et al. 1999), magnetic flux cancellation (e.g. van Ballegooijen and Martens 1989). On the other hand, helicity is at the heart of the helical kink instability (e.g. Rust and Kumar 1996; Török and Kliem 2005; Kliem et al. 2012), which has been invoked to explain several eruptive phenomena.

6 Active Region Tilt

6.1 Introduction of Joy's Law

Sunspots do not form randomly over the solar photosphere, but develop in groups, or active regions. Typically, active regions are elongated in East-West direction. Early observations of magnetic fields indicated that active regions are represented by bipolar patterns, with leading and following sunspots having opposite magnetic polarities. Leading sunspots in opposite hemispheres have opposite polarities: for example, in current cycle 24, leading sunspots of active regions in Northern hemisphere have negative polarity field, while leading spots in southern hemisphere have positive polarity field. In subsequent cycle, these polarities will reverse (leading positive polarity in the Northern hemisphere and negative in the southern hemisphere). Collectively, these patterns, are now known as Hale (or Hale-Nicholson) polarity rule. In both hemispheres, leading sunspots of all groups are situated closer to solar equator then the following sunspots. The average angle between the main axis of the active region and the line of parallel latitude (or equator) is a function of latitude. This dependence of active region tilt on latitude is commonly referred to as Joy's law (term Joy's law was introduced by H. Zirin in 1988; prior to that, the dependency was referred to simply as active regions tilt). Figure 9 shows examples of patterns associated with the Hale polarity rule and Joy's law.

Although active region tilt can be determined from white light images of active regions (Fig. 9, left), in some cases, the determination of tilt may have a high degree of uncertainty or even errors (e.g., AR2 and AR3, Fig. 9). Using longitudinal magnetograms alone may also cause a problem for active regions situated far from the central meridian because of apparent polarity reversal due to projection effects. Therefore, the tilt angle is best determined from the analysis of both white light and magnetic field data.

Both Hale's polarity rule and Joy's tilt law were first noted by Hale et al. (1919). However, the significance of trend in tilt angles was not recognized until much later.

6.2 Latitudinal Dependence of Tilt

Using sunspot drawings by Carrington (1856–1861) and Spörer (1861–1893), Joy (see, Hale et al. 1919) measured tilt of 2633 active regions over three and a half sunspot cycles. He

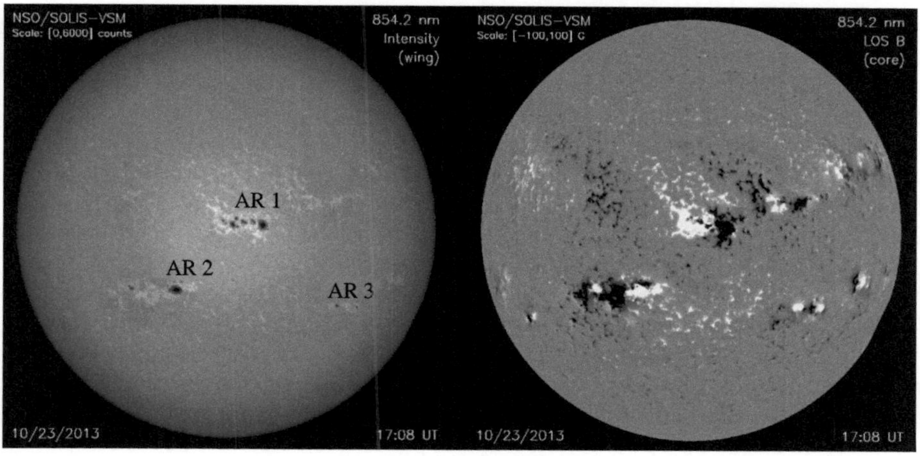

Fig. 9 White light image (**a**) and corresponding magnetogram (**b**) showing dark sunspots surrounded by bright plages (*left*) and the associated magnetic fields (*right*, *white color* corresponds to positive and *black* is negative polarity field). One of the active regions, AR1, provides example, when using white light image alone would allow to correctly identify active region and its tilt. AR2 does not show a well-developed sunspot in its trailing part, and thus, the determination of tilt is less certain. Furthermore, the small spot on the East of AR2 belongs to another bipole, representing its leading part, so AR2 could also be falsely identified as a bipole, giving incorrect tilt angle. In case of AR3, the observer may erroneously classify two close sunspots as an active region, which will result in incorrect tilt. In this case, two sunspots in area marked as AR3 have the same polarity field, and thus, belong to different groups. Data are from the Solar Optical Long-term Investigations of the Sun (SOLIS) system

Fig. 10 Joy's measurements of tilt (γ) vs. latitude (φ) averaged over 5 degree latitudinal intervals (0–4°, 5–9°, etc.) for rising and declining phases of cycles 10–13 from Hale et al. (1919). *Dashed* and *solid lines* show least-square fits by linear functions ($\gamma = (0.29 \pm 0.04) \cdot \varphi +$ (0.03 ± 0.01)) and non-linear ($\gamma = (0.31 \pm 0.04) \sin \varphi +$ (0.03 ± 0.01)) functions, where γ and φ are expressed in units of radians

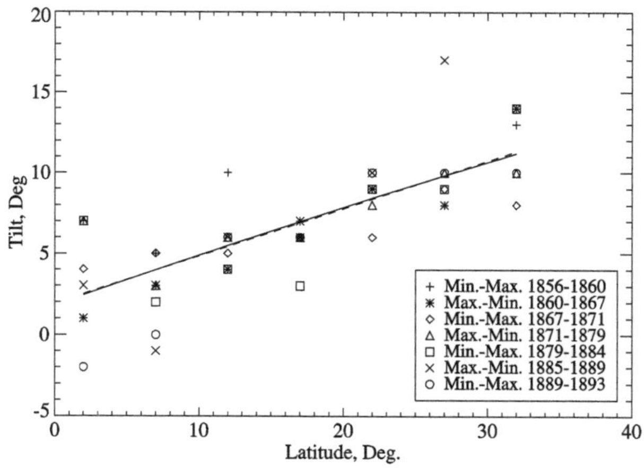

found that the tilt varies with the latitude, and that the tendency is independent of solar cycle. Figure 10 shows Joy's data (see, Table 2 in Hale et al. 1919) and the least-square fit by two functions (done by us). The two fits are nearly indistinguishable from each other. All tilt angles in Fig. 10 were determined from white light images only (since magnetogram data were not available prior to 1908) including erroneous tilts determined for anti-Hale and complex polarity spots.

Wang and Sheeley (1989) used line-of-sight magnetograms from the National Solar Observatory (NSO) at Kitt Peak full disk magnetograph to measure tilt of about 2700 bipolar magnetic regions (BMRs) from 1976–1986 (cycle 21). Although, the article provides no coefficients for least-square fit, judging from their Fig. 6a, the latitudinal dependence is much steeper than in Joy's data. Table 1 provides a summary of fitted functional dependencies to latitudinal variation of tilt angle, $f(\gamma) = A \cdot f(\varphi) + B$, where $f(\gamma)$ and $f(\varphi)$ are functions of tilt and latitude, and A and B are coefficients. For consistency, we converted all fitted coefficients taken from their respective articles to units of radians. We used data from Hale et al. (1919, Carrington and Spörer, C&S sunspot drawings) and Brunner (1930, Zurich sunspot measurements) to compute the functional approximates, which are shown in Table 1.

Three different functional dependencies used in these previous studies return very similar results. The first group (γ vs. φ) shows very consistent fits, with the exception of Zurich and CSA (Russian) datasets, for which the tilt is notably higher. We do not have an explanation for this inconsistency, but we speculate that this could be related to the sampling of active regions for two data sets. In both cases, the selection was done by the observer, and thus, it could be that the size of selected regions played a role. We note that in comparison with Mount Wilson (MWO) and Kodaikanal (KK) observatories, CSA data set show significantly smaller scatter in tilt angles (Ivanov 2012), which indirectly supports the idea that the determination of active regions in this data set was emphasizing larger regions.

In the second group (γ vs. $\sin\varphi$) fits to the Carrington and Spörer (C&S) and to MWO data are in agreement with the results for the first group. Latitudinal dependence of tilt derived from longitudinal magnetograms from SOHO/MDI (Stenflo and Kosovichev 2012) is steeper and it is in a better agreement with fit to Brunner (1930) data.

In the third group ($\sin\gamma$ vs. $\sin\varphi$), fit to C&S data is in agreement with the first and second groups, but fits to NSO/KP and Zurich data are much steeper. Overall, it appears that fitting the orientation of bipolar magnetic regions (BMRs) returns higher tilt as compared with the data based on white light images and sunspot drawings. The latter may be the effect of the difference in evolutionary state of active regions covered by the white-light and magnetic datasets of active region tilt angles. White-light signatures active regions (sunspots) have shorter life span than that of the magnetic signatures of BMRs distinguishable from background field. Therefore the difference between tilt angles obtained from white-light and magnetic datasets may rather result from the dominance of active regions in their decay phase in the magnetic data sets, while the white-light (sunspot) data is dominated by relatively young active regions.

We also note that more than half of the linear fits shown in Table 1 return a non-zero intersect point (corresponding to about 1–2 degrees) which suggests that even at the equator the active regions have a slight non-zero tilt. Many fits are forced through the origin on the assumption that active regions have a zero tilt angle at the equator. This is done because it is assumed that the Coriolis force is the only contribution to the tilt angle value. Since all the physical processes that contribute to the tilt angle are not yet understood, it is best to fit the data and not assume that tilt angles are zero at the equator. As an example, NOAA active region 11987 was a Northern hemispheric polarity group with the leader at -2 degrees southern latitude and the follower spot on or slightly above the equator. This group basically formed a bridge across the equator and maintained a healthy tilt angle of $\approx$ 15–18 degrees from when it was visible on 22 Feb 2014 until the follower spot broke apart on 27 Feb 2014. This is just one example of active regions near the equator not having a zero tilt angle. More studies should be done on the tilt angles of active regions that emerge within several degrees of the equator as their existence contradicts the conventional wisdom of Coriolis force imparting the largest portion of the tilt.

While in most cases tilt vs. latitude dependence is assumed to be monotonic with latitude, some data indicate a possible presence of maximum in tilt angles in mid-latitudes. The

Table 1 Summary of latitudinal functional dependence

Tilt function	A	Latitude function	B, radian	Type	Dataset	Years	Reference
γ	0.29±0.04	φ	0.03 ± 0.01	WL[a]	C&S[c]	1856–1893	Pevtsov et al. (2014)
γ	0.26±0.05	φ		WL	MWO[g]	1917–1985	Dasi-Espuig et al. (2010)
γ	0.28±0.06	φ		WL	KK[e]	1906–1987	Dasi-Espuig et al. (2010)
γ	0.26±0.02	φ		WL	MWO	1917–1985	Ivanov (2012)
γ	0.28±0.03	φ		WL	KK	1906–1987	Ivanov (2012)
γ	0.38±0.03	φ		WL	CSA[d]	1948–1991	Ivanov (2012)
γ	0.55±0.03	φ	-0.02 ± 0.01	WL	Zurich[i]	1894–1928	Pevtsov et al. (2014)
γ	0.50±0.20	φ	-0.02 ± 0.01	MG	MWO & MDI[f]	1976–2010	Li and Ulrich (2012)
γ	0.31±0.04	$\sin\varphi$	0.03 ± 0.01	WL	C& S	1856–1893	Pevtsov et al. (2014)
γ	0.56±0.01	$\sin\varphi$		MG[b]	MDI	1995–2011	Stenflo and Kosovichev (2012)
γ	0.27±0.01	$\sin\varphi$		WL	MWO	1917–1985	Fisher et al. (1995)
γ	0.58±0.04	$\sin\varphi$	-0.02 ± 0.01	WL	Zurich	1894–1928	Pevtsov et al. (2014)
$\sin\gamma$	0.50	$\sin\varphi$		WL	Zurich	1894–1928	Leighton (1969)
$\sin\gamma$	0.57±0.04	$\sin\varphi$	-0.02 ± 0.01	WL	Zurich	1894–1928	Pevtsov et al. (2014)
$\sin\gamma$	0.31±0.04	$\sin\varphi$	0.03 ± 0.01	WL	C&S	1856–1893	Pevtsov et al. (2014)
$\sin\gamma$	0.48	$\sin\varphi$	0.03	MG	NSO/KP[h]	1976–1986	Wang and Sheeley (1991)

[a]WL—white light filtergrams and/or drawings. [b]MG—line-of-sight magnetograms. [c]C&S—Carrington and Spörer datasets. [d]CSA—the Pulkovo observatory's "Catalogue of Solar Activity". [e]KK—Kodaikanal observatory. [f]MDI—Michelson Doppler Imager. [g]MWO—Mount Wilson Observatory. [h]NSO/KP—National Solar Observatory at Kitt Peak. [i]Zurich—sunspot drawings.

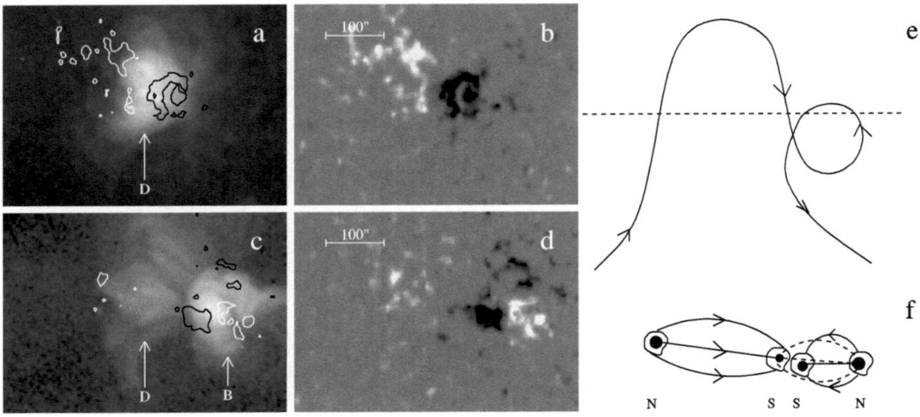

Fig. 11 *Left*: NSO/Kitt Peak line-of-sight magnetogram (**b** and **d**) and soft X-ray Yohkoh image (**a** and **c**) showing the overall structure of NOAA AR 7918 and AR 7926. *White* indicates positive (N) flux and *black* negative (S). The contours overlaying the coronal image correspond to ±100 G of the line-of-sight magnetic field. *Arrows* indicate two different sets of the loops. (**a**) NOAA 7918, 1995 October 28, 18:30:39 UT; (**b**) NOAA 7918, 1995 October 28, 17:54:37 UT; (**c**) NOAA 7926, 1995 November 21, 16:34:35 UT; (**d**) NOAA 7926, 1995 November 21, 16:47:50 UT. *Right*: Model of Ω-loop with stitch. *Arrows* show direction of the magnetic field in the loop. (**e**) Side view of the loop. The *dashed line* indicates the photosphere. (**f**) Top view of the loop schematically showing sunspots at the places where the Ω-loop crosses the photosphere. *Full lines* with *arrows* show magnetic field lines above the photosphere projected onto the horizontal plane. *Dashed lines* indicate subsurface parts of field lines. Based on Figs. 2 and 8 from Pevtsov and Longcope (1998)

maximum can be seen in MWO, KK and CSA data (Sivaraman et al. 1999; Ivanov 2012). Due to reduced size of statistical sample for high latitudes, standard deviations of the means are also large, and thus, this apparent decrease may be statistically insignificant. Still, this may need additional studies to confirm (or disprove) the presence of a peak in tilt angles in mid-latitudes.

6.3 Non-Hale Polarity Regions and δ-Spots as Highly Tilted Bipoles

Active regions' tilt is a weak tendency with significant scatter (for example, see Fisher et al. 1995). In contrast, Hale polarity law is a strong dependence. For example, in MWO data set from 1913–1917, annual fraction of non-hale polarity regions vary between 1.4–6.3 % with average of about 3.7 % (Hale et al. 1919). More recent data from SOHO/MDI reveal similar fraction of about 4 % (Stenflo and Kosovichev 2012). Li and Ulrich (2012) found a larger fraction between 6.5 % and 9.1 % for cycles 21–23. Non-Hale polarity regions could be interpreted as highly tilted regions, whose main axis is rotated by about 180°. In support of that interpretation, Pevtsov and Longcope (1998) described a linked appearance of Hale and non-Hale polarity regions (see Fig. 11). Two regions shared a common polarity, which first appeared as the leading polarity for Hale polarity BMR and on the sequential solar rotation, it was the following polarity of a non-Hale active region. López Fuentes et al. (2000) described evolution of anti-Hale bipolar active region, in which over four solar rotations one magnetic polarity was gradually moving around the other opposite polarity flux; with the corresponding BMR changing its orientation from Hale to non-Hale polarity. Based on the observation that the non-Hale polarity regions develop at any phase of sunspot cycle, Stenflo and Kosovichev (2012) speculated about the presence of two toroidal fluxes with opposite orientation in the convection zone.

δ-spots are another type of irregular sunspots, which show a significant deviation from Joy's tilt and Hale polarity rules. A δ-spot consists of two (or multiple) umbrae with the opposite polarity magnetic field within a single penumbra. Line connecting opposite polarity umbrae is usually highly tilted relative to solar equator. Based on change in orientation of δ-spots during their evolution, this type of sunspot groups has been interpreted in the framework of the emergence of highly kinked flux tubes. δ-spots show a well-defined pattern of electric currents with current flowing upward in one umbra and the downward in the other umbra (Pevtsov 2003). This pattern supports the notion that these sunspots are associated with single (but highly twisted) magnetic flux tubes.

6.4 Changes in the Tilt Angle During the Lifetime and Evolution of Active Regions

Observations suggest that active regions do not emerge with the "right" tilt, but the tilt develops rapidly during the first few days as the region emerges through the photosphere (for example, see Pevtsov et al. 2003a). As active regions decay, their tilt relaxes to a mean tilt for a given latitude (Howard 1996; Kosovichev and Stenflo 2008). The latter was interpreted as an indication that the tilt is not the result of the Coriolis force action. According to this argument, since the Coriolis force vanishes once the emergence stopped, the tilt should relax to East-West direction. Kosovichev and Stenflo (2008) argued that the relaxation of tilt to a mean (non-zero) value indicates that the tilt represents the orientation of toroidal field in the convection zone. However, Longcope and Welsch (2000) suggested that the evolution of tilt and twist may depend on how rapidly the active region emerges through the photosphere. This theoretical prediction was later confirmed by the observations (Pevtsov et al. 2003a). Subsequently, one can argue that tilt may represent writhe component of helicity in the magnetic flux tubes forming active regions. Then, the relaxation of active regions to a mean (non-zero) tilt may be related to helicity content of active regions for a given latitude (e.g. Canfield and Pevtsov 1998).

The presence of a global twist in emerging flux tube can modify the tilt angle of the emerging bipole when it is measured in magnetic data. Luoni et al. (2011) show that while the flux rope is breaking through the photosphere, a characteristic elongated asymmetric polarity pattern, dubbed magnetic tongues or tails, develops in the line-of-sight magnetograms owing to the presence of azimuthal field components. The pattern is dependent on the sign of the global twist and thus provides a simple proxy for it (López Fuentes et al. 2000; Luoni et al. 2011). In magnetograms magnetic tongues or tails may mask the true tilt angle of the bipole, since the tongue pattern resulting from e.g. negative/positive global twist on the northern/southern hemisphere shifts the centre of gravity of magnetic polarities introducing an angle which is opposite to that of Joy's law (cf. Fig. 12). Therefore the tilt angle of those bipoles, which obey the hemispheric helicity rule, may be cancelled or reduced (while the tilt angle of those bipoles, which disobey the hemispheric helicity rule are enhanced) by this effect during the emergence process and the bipole will only assume the Coriolis-induced tilt when the entire flux rope has crossed the photosphere and the tongues/tails have retracted.

6.5 Variations with Cycle, Size of Active Region, Magnetic Flux

Several studies searched for a correlation between the active region tilt and the amplitude of the next solar cycle. While the studies showed some indication of such dependency, overall the results are still not well-understood. Dasi-Espuig et al. (2010) found a strong negative correlation between the strength of solar cycle and the area-weighted mean tilt value normalized by latitude. Both MWO and KK data exhibited equally strong correlation. This is a

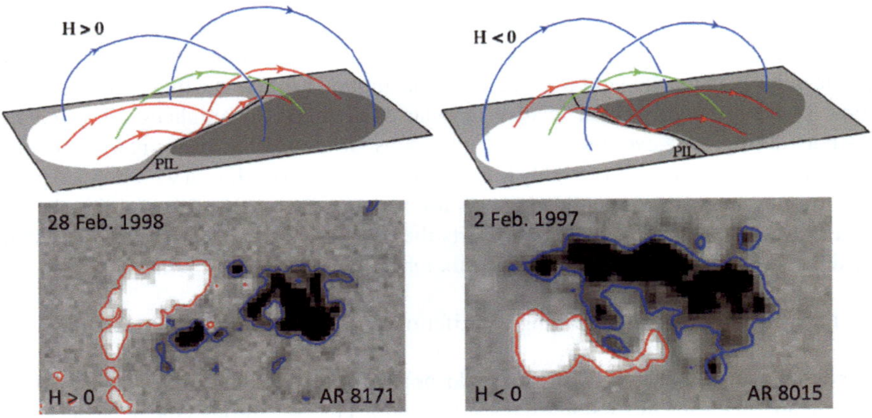

Fig. 12 *Top*: Sketches of magnetic "tongue" patterns in longitudinal magnetograms with (*left*) positive and (*right*) negative global twist in the emerging flux rope, which develop owing to the presence of the transverse magnetic field component. Representative field lines of the enveloping arcade and internal (dipped) field lines are also shown, so is the polarity inversion line (PIL). *Bottom*: Two examples of magnetic tongue pattern during the emergence of AR 8171 (*left*) and AR 8015 (*right*) corresponding to the helicity sign as the sketches above them. Both the sketches and the SOHO/MDI observations were adapted from Luoni et al. (2011)

very interesting idea that a stronger cycle would produce bipolar active regions with lower average tilt-angles and a weaker cycle would produce higher average tilt-angles. Later, however, Ivanov (2012) and McClintock and Norton (2013) were not able to reproduce the exact dependency found in Dasi-Espuig et al. (2010) for MWO data. Dasi-Espuig et al. (2013) revised their results in agreement with Ivanov (2012) and McClintock and Norton (2013) findings. Dependency between the amplitude of solar cycle and the tilt angle found by McClintock and Norton (2013) is in a general agreement with Ivanov (2012) and the revised findings by Dasi-Espuig et al. (2010). However, the McClintock and Norton (2013) analysis separated the data by hemisphere and found that the correlation of cycle strength and area-weighted mean tilt angle normalized by latitude only held for the Southern hemisphere for cycles 15–21. The dependence was insignificant for the Northern hemisphere data. One can question the validity of the proposed dependency if both hemispheres do not exhibit it. Also, the negative correlation between the amplitude of solar cycle and the normalized tilt angle for MWO data appears primarily due to the data point corresponding to solar cycle 19. If this data point is excluded, MWO data show no correlation with solar cycle. Cameron and Schüssler (2012) and Cameron et al. (2010) present a convincing, and important, argument that tilt angle variations are the main nonlinear feedback mechanism operating in the dynamo.

Action of the Coriolis force on emerging flux tube depends on several parameters including total magnetic flux and the field strength at the axis of flux tube. Fan et al. (1994) predicted that the tilt angle γ will depend on total magnetic flux Φ, maximum field strength B_0 at the bottom of the convection zone, and latitude φ as following: $\gamma \propto \Phi^{0.25} \cdot B_0^{-1.25} \cdot \sin\varphi$. Since both the maximum field strength and the magnetic flux in sunspots correlate with their size (e.g., Tlatov and Pevtsov 2014), one would also expect to see a correlation between tilt angle and active region area. Study of tilt properties of 27,701 active regions from MWO data set (Fisher et al. 1995) concluded that the functional dependency of tilt on foot-point separation and sine of latitude appears to be in agreement with the predictions of Fan et al. (1994) model. They also estimated the field strength in flux tubes at the bottom of the convection zone $B_0 = 20$–30 kG.

It has been noted (Howard 1991; Sivaraman et al. 1999) that it is large active regions that follow the Joy's law most closely. On the other hand, Ivanov (2012) found that CSA data set deviates from that dependency. Tlatov et al. (2010) analyzed the orientation of magnetic bipoles over three solar cycles 21–23 using NSO/KP and SOHO/MDI longitudinal magnetograms. The bipoles were classified by their area as quiet Sun bipoles (QSBs, area, A, smaller than 50 millionth of solar hemisphere, MSH), ephemeral regions (ERs, 100 MSH $\leq$ A $<$ 500 MSH), and regular active regions (ARs, A $\geq$ 500 MSH). QSBs were found to exhibit a random orientation independent of latitude, which suggests their formation via a random encounter of their foot-points. ERs and ARs followed the Joy's law in their orientation, with ERs showing larger scatter in their tilt angles relative to the mean. Scatter in tilt angles is usually contributed to convective buffeting, or interaction of magnetic flux tube with turbulent convection as it rises through the convection zone. Longcope and Fisher (1996) found that the amplitude of fluctuations in tilt angles should be inversely proportional to foot-point separation, and thus, smaller bipoles should exhibit much larger fluctuations in their tilts. The latter supports Tlatov et al. (2010) findings for ERs scatter in their orientation. Tlatov et al. (2010) also found that around the maximum of previous cycle, high latitude ERs had their polarity orientation (Hale polarity rule) corresponding to a subsequent solar cycle. The latter was interpreted as the indication of extended solar cycle (Wilson et al. 1988).

Time variation of active region tilt within solar cycle follows the properties of active regions. As new cycle starts, the active regions appear at high latitudes with their tilt and polarity orientation corresponding to this cycle. As cycle progresses, the latitude of active region emergence moves towards the equator. In that respect, the time variation of tilt angles is more reminiscent to the butterfly diagram of sunspots, and is dissimilar to a sunspot number plot.

A possible difference in Joy's law between northern and southern hemispheres was noted by several researchers (Dasi-Espuig et al. 2010; McClintock and Norton 2013; Li and Ulrich 2012). Based on the analysis of the tilt angle data separated by hemisphere, McClintock and Norton (2013) recommend a revision of Joy's law toward a weaker dependence on latitude (slope of 0.13–0.26) and without forcing the tilt to zero at the equator. They determined that the hemispheric mean tilt value of active regions varies with each solar cycle, although the noise from a stochastic process dominates and does not allow for a determination of the slope of Joy's law on an 11-year time-scale. The hemispheric difference in mean tilt angles, $1.1^\circ \pm 0.27$, over Cycles 16 to 21 was significant to a three-σ level, with average tilt angles in the northern and southern hemispheres of $4.7^\circ \pm 0.26$ and $3.6^\circ \pm 0.27$ respectively.

When analyzing the active regions of a single hemisphere in a single solar cycle, Joy's law only appears weakly (see Fig. 13, McClintock and Norton 2013). A linear function is a poor fit to the data in most cases. The linear correlation coefficients range from $r = 0.18$ (cycle 17 North, cycle 19 South) to $r = 0.86$ (cycle 20 North). The large amount of scatter and high noise apparent in Joy's law is interesting, because it indicates that a stochastic process is competing with the mechanism that determines the tilt angles. The stochastic process dominating Joy's law on the short time scale is considered to be turbulent convection imparting random tilt angles to the rising flux tubes (Fisher et al. 1995; Weber et al. 2011). Dasi-Espuig et al. (2010) also state that "no clear difference could be determined between the slopes of Joy's law from cycle to cycle," as can be seen in Fig. 13; therefore, we use the mean tilt value from each hemisphere for each cycle to analyze the hemispheric differences.

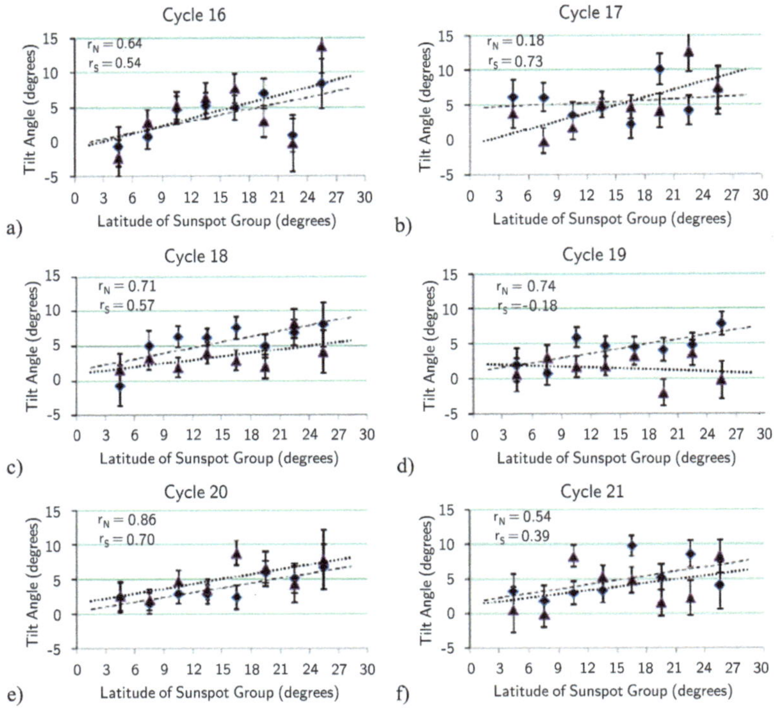

Fig. 13 Tilt angle of bipolar active regions as a function of latitude for the North (*diamond*) and South (*triangle*) hemispheres for cycles 16–21 are shown in panels (**a**)–(**f**), respectively. Data were binned in 3° of latitude. Standard errors of the mean are overplotted as *error bars*. Linear fits to Northern (*dash*) and Southern (*dot*) hemisphere data are shown with linear correlation coefficients [r_N, r_S] included in the *legends*. The fits were not forced through the origin. Courtesy McClintock and Norton (2013)

McClintock and Norton (2013) argue it is possible that the recovery of a mean bipolar region tilt angle and scatter for a given solar cycle can be used as a diagnostic for that cycle, i.e. the strength of the cycle as indicated by Dasi-Espuig et al. (2010) or the geometry/orientation of the toroidal fields from which the flux ropes begin their initial rise (Babcock 1961; Norton and Gilman 2005).

They also point out why specific bins in the Southern hemisphere in Fig. 13 showed such different behavior from the other bins. For example, late in all solar cycles (except cycle 20) aberrant activity occurred at the 18–21° latitudes. In particular, the Southern hemisphere during cycle 19 is very disorganized, with the high-latitude bins of 18–21° (304 regions, 24 %) and 24–27° (150 regions, 12 %) having negative mean tilt values, meaning that these bipolar regions have a following spot closer to the Equator than the leading spot. It would be interesting to study this in more detail and better understand the conditions favorable for aberrant configurations, i.e. anti-Hale and negative tilt angles, to occur.

McClintock and Norton (2013) found that cycles 18 and 19, as well as the data averaged over all cycles 16–21, show that the mean tilt angles differ significantly between the hemispheres. For example, they found an average value of mean tilt for cycle 18 to be 5.7 ± 0.61 in the Northern hemisphere while the mean tilt value was 2.9 ± 0.6 in the Southern hemisphere. The different hemispheric and cycle mean tilts, and different slopes of Joy's law, warrant closer scrutiny.

6.6 Tilt and Helicity

The action of an external force (e.g., Coriolis force) that distorts the shape of a flux tube will introduce writhe, Wr and twist components of helicity, Tw into the flux tube. If initially the flux tube had zero helicity $H = Tw + Wr = 0$, the twist and writhe will have an opposite sign ($\text{sign}(Tw) = -\text{sign}(Wr)$). If, on the other hand, the flux tube had a non-zero (say, positive) helicity, and the writhe developed as the result of kink-instability, then part of the internal twist was traded for writhe, and the two should have the same sign ($0 < H = Tw + Wr$; $\text{sign}(Tw) = \text{sign}(Wr)$). These arguments were first put forward by Canfield and Pevtsov (1998) and Pevtsov and Canfield (1999), who studied the correlation between tilt and helicity density proxy of 99 ARs observed by the Haleakala Stokes Polarimeter (HSP). (One should note that in these studies, the writhe was defined as being positive in counter-clockwise direction from E-W direction. This definition is similar to a definition of sign of an angle in regular Cartesian coordinate system. On the other hand, traditionally, active region's tilt is defined as positive angle in clockwise direction from W-E direction. Thus, positive tilt will correspond to negative writhe.) The helicity proxy in Canfield and Pevtsov studies was represented by average α coefficient in force-free field fit to the observed magnetic field. A weak negative correlation between sign of tilt and sign of α was found for active regions that significantly deviate from Joy's law. This negative correlation is contrary to the expected dependence introduced by the action of the Coriolis force on a rising flux tube. Hence, Pevtsov and Canfield (1999) concluded that at least in some cases, the tilt of active regions could be the result of kink-instability of active regions with non-zero helicity.

On the other hand, Tian et al. (2001) compared the sign of tilt and twist and found a positive correlation. They concluded that twist in active region magnetic fields is the result of the Coriolis force acting on the apex of emerging-loops, which introduces both twist and writhe in originally untwisted magnetic fields. In a later study, Tian and Liu (2003) found evidence of kink-instability in their study of tilt–twist relation for 86 flare-productive active regions. Study of tilt–twist relation for 22 active regions, López Fuentes et al. (2003) found both twist and writhe with the same sign (supporting kinking) and with opposite signs (Coriolis force action) in about 35 % and 41 % of cases, respectively. Holder et al. (2004) conducted an extensive study of tilt–twist (helicity) relation for 356 active regions. Helicity proxy α coefficient was computed using vector magnetograms from the Haleakala Stokes Polarimeter (HSP) and tilt angles were taken from MWO data set. They found weak (but statistically significant) negative correlation (opposite sign of tilt and twist) supporting the kink-instability scenario. The correlation was even stronger for active regions deviating from the general tendency for the hemispheric helicity rule, which further supports the kink-instability as the origin of tilt for these regions. Non-Hale polarity regions were found having a positive correlation (same sign of tilt and twist). Again, the latter supports the kink-instability scenario if non-Hale regions are interpreted as bipolar structures tilted in excess of 90 degrees relative to direction parallel to equator. For active regions that follow Joy's tilt law, Holder et al. (2004) found no twist-tilt dependence. They also concluded that the scatter in tilt angles and the dispersion in twist are uncorrelated with each other. Both of these results are in agreement with the effects of convective buffeting of initially untwisted and unwrithed flux tubes, or Σ-effect (Longcope et al. 1998).

Liu et al. (2014) used the relation between tilt and writhe in their study of the strength of hemispheric helicity rule. Observations of 151 active regions from Helioseismic and Magnetic Imager (HMI) were used to derive both twist and writhe. The data showed both the hemispheric helicity rule and the Joy's law. However, the hemispheric helicity rule was found to be significantly stronger for a sub-group of active regions with the same sign of twist and writhe. Liu et al. (2014) concluded that either the flux tubes do not have a hemispheric preference for twist prior to their emergence from the base of the convection zone, or that the initial twist is weak.

Finally, from the modeling Fan and Gong (2000) found that the twist arising as a result of helicity conservation from Coriolis force-induced writhe is insufficient to explain the observed values of the twist. As the magnetic flux tubes emerging from the bottom of the convection zone will inevitably acquire some tilt (writhe) due to the Coriolis force action (e.g., Petrovay and Christensen 2010), the findings by Fan and Gong (2000) suggest that the twist and the active region tilt may have independent origins.

6.7 Importance of Tilt for Solar Dynamo

In flux-transport models of solar dynamo (Wang and Sheeley 1991), the magnetic field of following polarity active regions is transported to solar poles, where it interacts with the existing polar field leading to polar field reversal. The polar field may also establish the strength of the next solar cycle (Upton and Hathaway 2014). In this process, tilt of active regions may play important role by either facilitating the creation of polar field (by placing following polarities higher in latitudes via larger tilt) or eroding it. In periods of high activity active regions are more closely packed, which increases the cancellation rate, including that of across the equator. With increasing tilt angles of active regions opposite polarities of neighboring active regions are more separated in latitude, which decreases inter-active region cancellations. However, cross-equatorial cancellation of the preceding polarities are enhanced.

This will allow more following-polarity flux to be transported to the poles. This was investigated in a parameter study by Baumann et al. (2004) whose model predicted a strong dependency between tilt angle (normalized to latitude of emerging bipole) and the strength of polar magnetic field. This model simulations catalysed search for correlation between mean tilt angle and the strength of sunspot cycle. So far, no conclusive correlation was found (see, Sect. 6.5). Even if present, the correlation between tilt angle and the strength of solar cycle can be masked by several factors. For example, the strength of polar field from previous cycle may negate the effect of higher tilt angle of active regions from the present cycle. Even though with higher tilts, the magnetic flux of trailing polarity will be transported to the poles in a more efficient manner, it would have to cancel out stronger polar flux. Such effects would need to be taken into consideration in future numerical modelling. Petrie (2012) found that in cycle 23 the net poloidal contribution of active regions effectively disappeared in 2004, thereby lowering the polar field strengths significantly as the Sun headed towards the minimum and the beginning of cycle 24. From his analysis he could not determine whether the reduction in net poloidal field was due to a lowering of average tilt angles of active regions or another cause. He notes, however, that there was a notable N-S hemispheric asymmetry present from 2004 and that faster-than-average meridional flows at active latitudes (another mechanism that could produce weak polar fields) were not detected.

6.8 Interpretation of Joy's Law

Active region tilt has been interpreted in the framework of three different main underlying causes: toroidal field orientation, action of Coriolis force, and more recently, as signature of kink-instability.

Tilt as a Signature of Toroidal Flux Orientation In the Babcock-Leighton model, at the beginning of cycle, the magnetic field is mostly poloidal (oriented in North-South direction parallel to solar meridian). As the solar equator rotates faster than higher latitudes, the differential rotation action gradually changes the orientation of magnetic field to be more aligned with latitudinal circles. When portion of subphotospheric toroidal flux tube erupts through the photosphere, it forms an active region, whose tilt may reflect a general orientation of the toroidal flux tube (Babcock 1961). Some researchers (e.g., Kosovichev and Stenflo 2008) argued that since the tilt of active regions after the complete emergence does not relax to East-West direction, the tilt must represent the orientation of toroidal field in the convection zone.

To evaluate the validity of this mechanism, we consider a very simplified model, in which two footprints of a flux tube are separated by 2 degrees in latitude and the orientation of the footpoints is that one is northward of the other (in other words, a N-S orientation). Note that we are not discussing the wreath of flux in the interior, but more a cartoon of a thin flux tube. If we assume that the tilt angle (or geometry of the footpoints) is determined only by the difference in rotation rates (differential rotation) for these two latitudes, then we can easily calculate how long it would take to turn a N-S flux tube into an E-W flux tube. This is a simplified example but is worth considering. Figure 14 shows the resulting tilt angles for different latitudes after 1, 5, and 10 solar rotations. After just five solar rations, the toroidal flux tube will be oriented at about 10 degrees relative to the equator, and after ten rotations, the tilt will be less than 5 degrees. Thus, in just a few years after the beginning of solar cycle, the toroidal flux tube should become oriented nearly parallel to solar equator over all latitudes of active region formation. In low latitudes, however, the tilt will remain higher in amplitude as compared with mid-latitudes, opposite to the trend observed. This is one argument against a tilted toroidal flux playing a significant role in orientation (tilt) of active regions at the photosphere. However, this does not take into account that the flux wreath may have an E-W orientation in general, but that individual tubes within the wreath may have a tilt or be more complexly oriented than the larger structure.

The idea that active region tilt may reflect the orientation of subphotospheric toroidal field was further explored by Norton and Gilman (2005) in respect to possible tipping and warping instabilities that may develop in a rotating ring-like magnetic flux tube due to an instability of differential rotation and concentrated magnetic fields. Development of these modes was carried out by Cally et al. (2003), Gilman and Fox (1997) and Gilman and Dikpati (2000) who found that narrow toroidal bands tip about their axis ($m = 1$) if they are ≈ 100 kG while toroidal bands of lesser field strengths may deform or warp ($m > 1$). A toroidal field tipped with respect to the equator would increase the scatter of the mean tilt angle, but not the average value of tilt angle. Norton and Gilman (2005) analyzed extensive data from longitudinal magnetograms from NSO/KP and SOHO/MDI and found the presence of tipping and warping instabilities from the location of active regions, but they did not examine the tilt angles. Note that this model and corresponding analysis is examining the non-axisymmetric nature of the solar dynamo. This is a further complexity in the model and analysis—not only are the hemispheres thought to progress slightly differently during the solar cycle, but the signatures are not averaged over all longitudes. McClintock and Norton (2013) searched for evidence of the tilt and warp (deformation) of the toroidal band from the active region tilt angle data. They found no significant signal in average tilt angle values or tilt angle scatter as a function of longitude. However, this does not prove that the geometry of the toroidal band does not influence the tilt angle, or that the toroidal band in the interior is axisymmetric, only that the variation of tilt angles as a function of longitude can not be distinguished from the noise in this data.

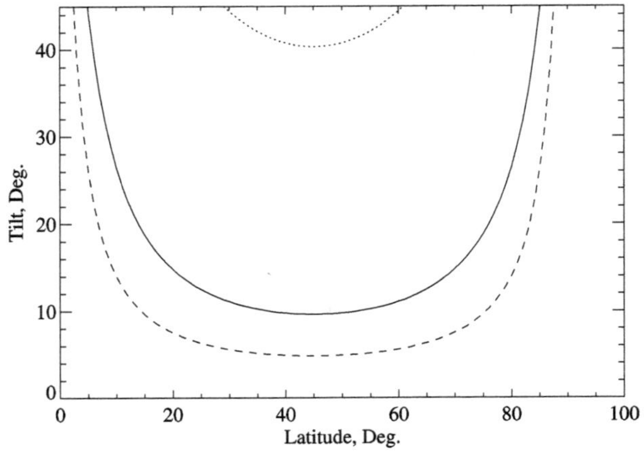

Fig. 14 We assume a simple flux tube were oriented in the N-S direction, horizontal to the solar surface, with the footpoints separated by two degrees of latitude, this plot shows the orientation of the flux tube from the effect of differential rotation after one (*dotted*), five (*solid*) and ten (*dashed line*) solar rotations. This plot is a simplification but is intended to show how quickly flux in the interior becomes purely toroidal and would not maintain a significant tilted geometry unless other factors were considered

Tilt as the Result of Coriolis Force Action The role of the Coriolis force in formation of active region tilt was mentioned by Babcock (1961) and Schmidt (1968). Wang and Sheeley (1991) proposed that as the magnetic flux tube rises through the convection zone and expands, the Coriolis force distorts the apex of the tube tilting it in the way it can explain Joy's law. The scenario was further developed in the framework of 3D thin flux tube model (D'Silva and Choudhuri 1993; Schüssler et al. 1994; Fan et al. 1994). The high scatter observed in tilt angles is generally thought to be caused by the turbulent nature of flow in the convection zone imparting a large variation into the tilt angle values. Later model developments were very successful in reproducing many observational properties of active regions including their tilt, asymmetry between leading and following polarity sunspots in compactness and inclination of magnetic field relative to vertical direction, and large scatter in tilt angles (Weber et al. 2011). By fitting the model predictions with the observations, field strength of toroidal flux is estimated at 40–50 kG at the bottom of the convection zone. The effect of the Coriolis force on active region tilt depends on total rise time of flux tube through the convection zone, which in turn depends on magnetic flux. By varying the magnetic flux, Weber et al. (2013) were able to explain some differences in tilt-latitude tendencies found by different studies. They argued, for example, that since magnetograms enable selecting structures with weaker magnetic fields, they may show steeper tilt vs. latitude trends as compared with white light images, which are more likely to select solar features with stronger magnetic flux. On the other hand, taking into account the scatter in tilts derived from model calculations, Weber et al. (2013) did not find a strong dependence of tilt on field strength. These later simulations also found about 6.9 % of emerging bipoles having non-Hale polarity orientation, which is in a very good agreement with the observations. One of the most interesting aspects of the work done by Weber et al. (2013) is that the simulations of the rise of a thin flux tube in a turbulent, convecting, rotating shell enables an estimation of the expected standard deviations of the tilt angles for a given combination of toroidal field strength in the tachocline and flux in the rising tube.

Tilt as the Result of Kink-Instability Leighton (1969) was the first to note that kinking of a twisted flux tube may form a tilted pattern that can explain the axial tilt of active regions. This idea was observationally explored by several researchers as described in Sect. 6.6. The presence of non-zero twist (helicity) in flux tubes forming active regions may explain why active region tilt does not relax to a purely East-West direction. Longcope et al. (1999) compared the amplitude of twist (helicity proxy α coefficient) derived from the observations of vector magnetic field in the photosphere with the estimated contribution from different mechanisms, and concluded that majority of helicity originates from the interaction of magnetic flux tubes with turbulent convection (Σ-effect). The contribution of subphotospheric (interface) dynamo is minor. While the Σ-effect only "creates" helicity in the upper portion of a flux tube, the amount of this injected helicity is sufficient to explain statistical properties of Joy's law. Other mechanism of helicity injection into rising flux tube was recently proposed by Wang (2013), who considered the effects of the Coriolis force separately on each foot-point (sunspot) of rising flux tube. He concluded that as the flux tube foot-points expand, the action of the Coriolis force will inject helicity in each foot-point independently. This mechanism could produce a strong hemispheric helicity rule.

7 Summary

As magnetic field propagates from the dynamo region, evolves and dissipates, its changes also reflect the transport of helicity through solar interior and upper atmosphere. While helicity does not play a major role in all solar processes, it may affect many important processes from dynamo, magnetic reconnection, stability of magnetic systems and their eruption. In respect to tilt of active regions (Joy's law), the presence of non-zero helicity may explain why in their evolution, active regions evolve to a mean tilt (but not to a zero tilt), or why active regions at equatorial areas may still exhibit non-zero tilt. It seems that helicity concept adds an additional level of "fine detail" to many solar processes, and this level of details requires considering the solar magnetic fields as a unified system, not as a collection of unrelated features. As such, potentially the concept of helicity may bring a much better understanding of "how our nearest star works."

Acknowledgements The National Solar Observatory (NSO) is operated by the Association of Universities for Research in Astronomy, AURA Inc under cooperative agreement with the National Science Foundation (NSF). AN's research has been partly co-financed by the European Union (European Social Fund—ESF) and Greek national funds through the Operational Program "Education and Lifelong Learning" of the National Strategic Reference Framework (NSRF)—Research Funding Program: "Thales. Investing in knowledge society through the European Social Fund." LvDG acknowledges support by STFC Consolidated Grant ST/H00260/1 and the Hungarian Research grants OTKA K-081421 and K-109276. AAP acknowledges support by NASA's NNH09AL04I inter agency transfer. Data used in Fig. 9 were acquired by SOLIS instruments operated by NISP/NSO/AURA/NSF.

References

V.I. Abramenko, T. Wang, V.B. Yurchishin, Analysis of electric current helicity in active regions on the basis of vector magnetograms. Sol. Phys. **168**, 75–89 (1996). doi:10.1007/BF00145826

V.I. Abramenko, T. Wang, V.B. Yurchishin, Electric current helicity in 40 active regions in the maximum of solar cycle 22. Sol. Phys. **174**, 291–296 (1997). doi:10.1023/A:1004957515498

T. Amari, J.F. Luciani, J.J. Aly, Z. Mikic, J. Linker, Coronal mass ejection: initiation, magnetic helicity, and flux ropes. I. Boundary motion-driven evolution. Astrophys. J. **585**, 1073–1086 (2003a). doi:10.1086/345501

T. Amari, J.F. Luciani, J.J. Aly, Z. Mikic, J. Linker, Coronal mass ejection: initiation, magnetic helicity, and flux ropes. II. Turbulent diffusion-driven evolution. Astrophys. J. **595**, 1231–1250 (2003b). doi:10.1086/377444

S.K. Antiochos, C.R. DeVore, J.A. Klimchuk, A model for solar coronal mass ejections. Astrophys. J. **510**, 485–493 (1999). doi:10.1086/306563

G. Attrill, M.S. Nakwacki, L.K. Harra, L. van Driel-Gesztelyi, C.H. Mandrini, S. Dasso, J. Wang, Using the evolution of coronal dimming regions to probe the global magnetic field topology. Sol. Phys. **238**, 117–139 (2006). doi:10.1007/s11207-006-0167-5

H.W. Babcock, The topology of the sun's magnetic field and the 22-year cycle. Astrophys. J. **133**, 572 (1961). doi:10.1086/147060

S. Bao, H. Zhang, Patterns of current helicity for the twenty-second solar cycle. Astrophys. J. Lett. **496**, 43 (1998). doi:10.1086/311232

S.D. Bao, A.A. Pevtsov, T.J. Wang, H.Q. Zhang, Helicity computation using observations from two different polarimetric instruments. Sol. Phys. **195**, 75–87 (2000). doi:10.1023/A:1005244700895

I. Baumann, D. Schmitt, M. Schüssler, S.K. Solanki, Evolution of the large-scale magnetic field on the solar surface: a parameter study. Astron. Astrophys. **426**, 1075–1091 (2004). doi:10.1051/0004-6361:20048024

M.A. Berger, Rigorous new limits on magnetic helicity dissipation in the solar corona. Geophys. Astrophys. Fluid Dyn. **30**(1–2), 79–104 (1984). doi:10.1080/03091928408210078

M.A. Berger, Structure and stability of constant-alpha force-free fields. Astrophys. J. Suppl. Ser. **59**, 433–444 (1985). doi:10.1086/191079

M.A. Berger, Magnetic helicity in space physics, in *Magnetic Helicity in Space and Laboratory Plasmas*. Geophysical Monograph Series, vol. 111 (American Geophysical Union, Washington DC, 1999), pp. 1–11.

M.A. Berger, G.B. Field, The topological properties of magnetic helicity. J. Fluid Mech. **147**, 133–148 (1984)

M.A. Berger, C. Prior, The writhe of open and closed curves. J. Phys. A, Math. Gen. **39**, 8321–8348 (2006). doi:10.1088/0305-4470/39/26/005

M.A. Berger, A. Ruzmaikin, Rate of helicity production by solar rotation. J. Geophys. Res. **105**, 10481–10490 (2000). doi:10.1029/1999JA900392

P.N. Bernasconi, D.M. Rust, D. Hakim, Advanced automated solar filament detection and characterization code: description, performance, and results. Solar Phys. **228**, 97–117 (2005). doi:10.1007/s11207-005-2766-y

A. Brandenburg, C. Sandin, Catastrophic alpha quenching alleviated by helicity flux and shear. Astron. Astrophys. **427**, 13–21 (2004). doi:10.1051/0004-6361:20047086

M.T. Brown, R.C. Canfield, A.A. Pevtsov (eds.), *Magnetic Helicity in Space and Laboratory Plasmas*. Geophysical Monograph Series (AGU, Washington, D.C., 1999)

W. Brunner, Gesetzmäßigkeiten in der Anordnung der Sonnenflecken zu Gruppen. Astron. Mitt. Eidgenöss. Sternwarte Zür. **13**, 67–78 (1930)

J. Büchner, A.A. Pevtsov (eds.), *Magnetic Helicity at the Sun, in Solar Wind and Magnetospheres*. Advances in Space Research, vol. 32 (Elsevier, Amsterdam, 2003)

A.B. Burnette, R.C. Canfield, A.A. Pevtsov, Photospheric and coronal currents in solar active regions. Astrophys. J. **606**, 565–570 (2004). doi:10.1086/382775

P.S. Cally, M. Dikpati, P.A. Gilman, Clamshell and tipping instabilities in a two-dimensional magnetohydrodynamic tachocline. Astrophys. J. **582**, 1190–1205 (2003). doi:10.1086/344746

G. Călugăreanu, On isotopy classes of three dimensional knots and their invariants. Czechoslov. Math. J. **11**, 588–625 (1961)

R.H. Cameron, M. Schüssler, Are the strengths of solar cycles determined by converging flows towards the activity belts? Astron. Astrophys. **548**, 57 (2012). doi:10.1051/0004-6361/201219914

R.H. Cameron, J. Jiang, D. Schmitt, M. Schüssler, Surface flux transport modeling for solar cycles 15–21: effects of cycle-dependent tilt angles of suns pot groups. Astrophys. J. **719**, 264–270 (2010). doi:10.1088/0004-637X/719/1/264

R.C. Canfield, A.A. Pevtsov, Helicity of solar active-region magnetic fields, in *Synoptic Solar Physics*, ed. by K.S. Balasubramaniam, J. Harvey, D. Rabin. Astronomical Society of the Pacific Conference Series, vol. 140 (1998), p. 131

J. Chae, Observational determination of the rate of magnetic helicity transport through the solar surface via the horizontal motion of field line footpoints. Astrophys. J. Lett. **560**, 95–98 (2001). doi:10.1086/324173

J. Chae, Y.-J. Moon, Y.-D. Park, Determination of magnetic helicity content of solar active regions from SOHO/MDI magnetograms. Sol. Phys. **223**, 39–55 (2004). doi:10.1007/s11207-004-0938-9

P. Charbonneau, Dynamo models of the solar cycle. Living Rev. Sol. Phys. **2**, 2 (2005)

P. Chatterjee, A.R. Choudhuri, K. Petrovay, Development of twist in an emerging magnetic flux tube by poloidal field accretion. Astron. Astrophys. **449**, 781–789 (2006). doi:10.1051/0004-6361:20054401

A.R. Choudhuri, On the connection between mean field dynamo theory and flux tubes. Sol. Phys. **215**, 31–55 (2003). doi:10.1023/A:1024874816178

A.R. Choudhuri, P. Chatterjee, D. Nandy, Helicity of solar active regions from a dynamo model. Astrophys. J. Lett. **615**, 57–60 (2004). doi:10.1086/426054

K. Dalmasse, E. Pariat, G. Valori, P. Démoulin, L.M. Green, First observational application of a connectivity-based helicity flux density. Astron. Astrophys. **555**, 6 (2013). doi:10.1051/0004-6361/201321999

K. Dalmasse, E. Pariat, P. Démoulin, G. Aulanier, Photospheric injection of magnetic helicity: connectivity-based flux density method. Sol. Phys. **289**, 107–136 (2014). doi:10.1007/s11207-013-0326-4

M. Dasi-Espuig, S.K. Solanki, N.A. Krivova, R. Cameron, T. Peñuela, Sunspot group tilt angles and the strength of the solar cycle. Astron. Astrophys. **518**, 7 (2010). doi:10.1051/0004-6361/201014301

M. Dasi-Espuig, S.K. Solanki, N.A. Krivova, R. Cameron, T. Peñuela, Sunspot group tilt angles and the strength of the solar cycle (corrigendum). Astron. Astrophys. **556**, 3 (2013). doi:10.1051/0004-6361/201014301e

P. Démoulin, A review of the quantitative links between CMEs and magnetic clouds. Ann. Geophys. **26**, 3113–3125 (2008). doi:10.5194/angeo-26-3113-2008

P. Démoulin, M.A. Berger, Magnetic energy and helicity fluxes at the photospheric level. Sol. Phys. **215**, 203–215 (2003). doi:10.1023/A:1025679813955

P. Démoulin, C.H. Mandrini, L. Van Driel-Gesztelyi, M.C. Lopez Fuentes, G. Aulanier, The magnetic helicity injected by shearing motions. Sol. Phys. **207**, 87–110 (2002a). doi:10.1023/A:1015531804337

P. Démoulin, C.H. Mandrini, L. van Driel-Gesztelyi, B.J. Thompson, S. Plunkett, Z. Kovári, G. Aulanier, A. Young, What is the source of the magnetic helicity shed by CMEs? The long-term helicity budget of AR 7978. Astron. Astrophys. **382**, 650–665 (2002b). doi:10.1051/0004-6361:20011634

C.R. DeVore, Magnetic helicity generation by solar differential rotation. Astrophys. J. **539**, 944–953 (2000). doi:10.1086/309274

S. D'Silva, A.R. Choudhuri, A theoretical model for tilts of bipolar magnetic regions. Astron. Astrophys. **272**, 621 (1993)

Y. Fan, D. Gong, On the twist of emerging flux loops in the solar convection zone. Sol. Phys. **192**, 141–157 (2000). doi:10.1023/A:1005260207672

Y. Fan, G.H. Fisher, A.N. McClymont, Dynamics of emerging active region flux loops. Astrophys. J. **436**, 907–928 (1994). doi:10.1086/174967

J.M. Finn, T.M. Antonsen Jr., Turbulent relaxation of compressible plasmas with flow. Phys. Fluids **26**, 3540–3552 (1983). doi:10.1063/1.864115

G.H. Fisher, Y. Fan, R.F. Howard, Comparisons between theory and observation of active region tilts. Astrophys. J. **438**, 463–471 (1995). doi:10.1086/175090

C. Foullon, C.J. Owen, S. Dasso, L.M. Green, I. Dandouras, H.A. Elliott, A.N. Fazakerley, Y.V. Bogdanova, N.U. Crooker, Multi-spacecraft study of the 21 January 2005 ICME. Sol. Phys. **244**(1–2), 139–165 (2007). doi:10.1007/s11207-007-0355-y

M.K. Georgoulis, B.J. LaBonte, Reconstruction of an inductive velocity field vector from Doppler motions and a pair of solar vector magnetograms. Astrophys. J. **636**, 475–495 (2006). doi:10.1086/497978

M.K. Georgoulis, B.J. LaBonte, Magnetic energy and helicity budgets in the active region solar corona. I. Linear force-free approximation. Astrophys. J. **671**, 1034–1050 (2007). doi:10.1086/521417

M.K. Georgoulis, D.M. Rust, A.A. Pevtsov, P.N. Bernasconi, K.M. Kuzanyan, Solar magnetic helicity injected into the heliosphere: magnitude, balance, and periodicities over solar cycle 23. Astrophys. J. Lett. **705**, 48–52 (2009). doi:10.1088/0004-637X/705/1/L48

M.K. Georgoulis, K. Tziotziou, N.-E. Raouafi, Magnetic energy and helicity budgets in the active-region solar corona. II. Nonlinear force-free approximation. Astrophys. J. **759**, 1 (2012). doi:10.1088/0004-637X/759/1/1

P.A. Gilman, M. Dikpati, Joint instability of latitudinal differential rotation and concentrated toroidal fields below the solar convection zone. II. Instability of narrow bands at all latitudes. Astrophys. J. **528**, 552–572 (2000). doi:10.1086/308146

P.A. Gilman, P.A. Fox, Joint instability of latitudinal differential rotation and toroidal magnetic fields below the solar convection zone. Astrophys. J. **484**, 439 (1997). doi:10.1086/304330

S. Gosain, A.A. Pevtsov, G.V. Rudenko, S.A. Anfinogentov, First synoptic maps of photospheric vector magnetic field from SOLIS/VSM: non-radial magnetic fields and hemispheric pattern of helicity. Astrophys. J. **772**, 52 (2013). doi:10.1088/0004-637X/772/1/52

L.M. Green, M.C. López fuentes, C.H. Mandrini, P. Démoulin, L. Van Driel-Gesztelyi, J.L. Culhane, The magnetic helicity budget of a CME-prolific active region. Sol. Phys. **208**, 43–68 (2002). doi:10.1023/A:1019658520033

L.M. Green, B. Kliem, T. Török, L. van Driel-Gesztelyi, G.D.R. Attrill, Transient coronal sigmoids and rotating erupting flux ropes. Sol. Phys. **246**, 365–391 (2007). doi:10.1007/s11207-007-9061-z

A.M. Gulisano, S. Dasso, C.H. Mandrini, P. Démoulin, Magnetic clouds: a statistical study of magnetic helicity. J. Atmos. Sol.-Terr. Phys. **67**, 1761–1766 (2005). doi:10.1016/j.jastp.2005.02.026

M. Hagino, T. Sakurai, Latitude variation of helicity in solar active regions. Publ. Astron. Soc. Jpn. **56**, 831–843 (2004). doi:10.1093/pasj/56.5.831

M. Hagino, T. Sakurai, Solar-cycle variation of magnetic helicity in active regions. Publ. Astron. Soc. Jpn. **57**, 481–485 (2005). doi:10.1093/pasj/57.3.481

M.J. Hagyard, A.A. Pevtsov, Studies of solar helicity using vector magnetograms. Sol. Phys. **189**, 25–43 (1999). doi:10.1023/A:1005215001514

G.E. Hale, F. Ellerman, S.B. Nicholson, A.H. Joy, The magnetic polarity of sun-spots. Astrophys. J. **49**, 153 (1919). doi:10.1086/142452

J. Hao, M. Zhang, Hemispheric helicity trend for solar cycle 24. Astrophys. J. Lett. **733**, 27 (2011). doi:10.1088/2041-8205/733/2/L27

Z.A. Holder, R.C. Canfield, R.A. McMullen, D. Nandy, R.F. Howard, A.A. Pevtsov, On the tilt and twist of solar active regions. Astrophys. J. **611**, 1149–1155 (2004). doi:10.1086/422247

R.F. Howard, Axial tilt angles of sunspot groups. Sol. Phys. **136**, 251–262 (1991). doi:10.1007/BF00146534

R.F. Howard, Axial tilt angles of active regions. Sol. Phys. **169**, 293–301 (1996). doi:10.1007/BF00190606

V.G. Ivanov, Joy's law and its features according to the data of three sunspot catalogs. Geomagn. Aeron. **52**, 999–1004 (2012). doi:10.1134/S0016793212080130

C. Jacobs, S. Poedts, B. van der Holst, The effect of the solar wind on CME triggering by magnetic foot point shearing. Astron. Astrophys. **450**, 793–803 (2006). doi:10.1051/0004-6361:20054670

B. Kliem, S. Rust, N. Seehafer, Helicity transport in a simulated coronal mass ejection, in *IAU Symposium*, ed. by A. Bonanno, E. de Gouveia Dal Pino, A.G. Kosovichev IAU Symposium, vol. 274 (2011), pp. 125–128. doi:10.1017/S1743921311006715

B. Kliem, T. Török, W.T. Thompson, A parametric study of erupting flux rope rotation. Modeling the "Cartwheel CME" on 9 April 2008. Sol. Phys. **281**, 137–166 (2012). doi:10.1007/s11207-012-9990-z

A.G. Kosovichev, J.O. Stenflo, Tilt of emerging bipolar magnetic regions on the sun. Astrophys. J. Lett. **688**, 115–118 (2008). doi:10.1086/595619

K. Kusano, T. Maeshiro, T. Yokoyama, T. Sakurai, Measurement of magnetic helicity injection and free energy loading into the solar corona. Astrophys. J. **577**, 501–512 (2002). doi:10.1086/342171

K. Kusano, T. Yokoyama, T. Maeshiro, T. Sakurai, Annihilation of magnetic helicity: a new model for solar flare onset. Adv. Space Res. **32**, 1931–1936 (2003). doi:10.1016/S0273-1177(03)90628-4

K. Kusano, T. Maeshiro, T. Yokoyama, T. Sakurai, The trigger mechanism of solar flares in a coronal arcade with reversed magnetic shear. Astrophys. J. **610**, 537–549 (2004). doi:10.1086/421547

B.J. LaBonte, M.K. Georgoulis, D.M. Rust, Survey of magnetic helicity injection in regions producing X-class flares. Astrophys. J. **671**, 955–963 (2007). doi:10.1086/522682

R.J. Leamon, R.C. Canfield, A.A. Pevtsov, Properties of magnetic clouds and geomagnetic storms associated with eruption of coronal sigmoids. J. Geophys. Res. **107**, 1234 (2002). doi:10.1029/2001JA000313

R.B. Leighton, A magneto-kinematic model of the solar cycle. Astrophys. J. **156**, 1 (1969). doi:10.1086/149943

J. Li, R.K. Ulrich, Long-term measurements of sunspot magnetic tilt angles. Astrophys. J. **758**, 115 (2012). doi:10.1088/0004-637X/758/2/115

E.-K. Lim, J. Chae, Chirality of intermediate filaments and magnetic helicity of active regions. Astrophys. J. **692**, 104–108 (2009). doi:10.1088/0004-637X/692/1/104

E.-K. Lim, H. Jeong, J. Chae, Y.-J. Moon, A check for consistency between different magnetic helicity measurements based on the helicity conservation principle. Astrophys. J. **656**, 1167–1172 (2007). doi:10.1086/510575

Y. Liu, P.W. Schuck, Magnetic energy and helicity in two emerging active regions in the sun. Astrophys. J. **761**, 105 (2012). doi:10.1088/0004-637X/761/2/105

Y. Liu, P.W. Schuck, A note on computation of relative magnetic-helicity flux across the photosphere. Sol. Phys. **283**, 283–294 (2013). doi:10.1007/s11207-012-0219-y

Y. Liu, J.T. Hoeksema, X. Sun, Test of the hemispheric rule of magnetic helicity in the sun using the helioseismic and magnetic imager (HMI) data. Astrophys. J. Lett. **783**, 1 (2014). doi:10.1088/2041-8205/783/1/L1

D.W. Longcope, Inferring a photospheric velocity field from a sequence of vector magnetograms: the minimum energy fit. Astrophys. J. **612**, 1181–1192 (2004). doi:10.1086/422579

D.W. Longcope, G.H. Fisher, The effects of convection zone turbulence on the tilt angles of magnetic bipoles. Astrophys. J. **458**, 380 (1996). doi:10.1086/176821

D.W. Longcope, A. Malanushenko, Defining and calculating self-helicity in coronal magnetic fields. Astrophys. J. **674**(2), 1130 (2008). http://stacks.iop.org/0004-637X/674/i=2/a=1130

D.W. Longcope, B.T. Welsch, A model for the emergence of a twisted magnetic flux tube. Astrophys. J. **545**, 1089–1100 (2000). doi:10.1086/317846

D.W. Longcope, G.H. Fisher, A.A. Pevtsov, Flux-tube twist resulting from helical turbulence: the sigma-effect. Astrophys. J. **507**, 417–432 (1998). doi:10.1086/306312

D.W. Longcope, M.G. Linton, A.A. Pevtsov, G.H. Fisher, I. Klapper, Twisted flux tubes and how they get that way, in *Magnetic Helicity in Space and Laboratory Plasmas*, ed. by M.R. Brown, R.C. Canfield, A.A. Pevtsov Geophysical Monographs Series, vol. 111 (AGU, Washington, D.C., 1999), pp. 93–101

M.C. López Fuentes, P. Démoulin, C.H. Mandrini, L. van Driel-Gesztelyi, The counterkink rotation of a non-Hale active region. Astrophys. J. **544**, 540–549 (2000). doi:10.1086/317180

M.C. López Fuentes, P. Démoulin, C.H. Mandrini, A.A. Pevtsov, L. van Driel-Gesztelyi, Magnetic twist and writhe of active regions. On the origin of deformed flux tubes. Astron. Astrophys. **397**, 305–318 (2003). doi:10.1051/0004-6361:20021487

B.C. Low, Solar activity and the corona. Sol. Phys. **167**, 217–265 (1996). doi:10.1007/BF00146338

B.C. Low, Magnetic helicity in a two-flux partitioning of an ideal hydromagnetic fluid. Astrophys. J. **646**(2), 1288 (2006). http://stacks.iop.org/0004-637X/646/i=2/a=1288

B.C. Low, M. Zhang, The hydromagnetic origin of the two dynamical types of solar coronal mass ejections. Astrophys. J. Lett. **564**, 53–56 (2002). doi:10.1086/338798

M.L. Luoni, P. Démoulin, C.H. Mandrini, L. van Driel-Gesztelyi, Twisted flux tube emergence evidenced in longitudinal magnetograms: magnetic tongues. Sol. Phys. **270**, 45–74 (2011). doi:10.1007/s11207-011-9731-8

P. MacNeice, S.K. Antiochos, A. Phillips, D.S. Spicer, C.R. DeVore, K. Olson, A numerical study of the breakout model for coronal mass ejection initiation. Astrophys. J. **614**, 1028–1041 (2004). doi:10.1086/423887

C.H. Mandrini, S. Pohjolainen, S. Dasso, L.M. Green, P. Démoulin, L. van Driel-Gesztelyi, C. Copperwheat, C. Foley, Interplanetary flux rope ejected from an X-ray bright point. The smallest magnetic cloud source-region ever observed. Astron. Astrophys. **434**, 725–740 (2005). doi:10.1051/0004-6361:20041079

S.F. Martin, Observational criteria for filament models, in *Solar Active Region Evolution: Comparing Models with Observations*, ed. by K.S. Balasubramaniam, G.W. Simon. Astronomical Society of the Pacific Conference Series, vol. 68 (1994), p. 264

S.F. Martin, Conditions for the formation and maintenance of filaments (invited review). Sol. Phys. **182**, 107–137 (1998). doi:10.1023/A:1005026814076

S.F. Martin, R. Bilimoria, P.W. Tracadas, Magnetic field configurations basic to filament channels and filaments, in *Solar Surface Magnetism*, ed. by R.J. Rutten, C.J. Schrijver (Kluwer, Dordrecht, 1994), p. 303

S.F. Martin, O. Panasenco, M.A. Berger, O. Engvold, Y. Lin, A.A. Pevtsov, N. Srivastava, The build-up to eruptive solar events viewed as the development of chiral systems, in *Second ATST-EAST Meeting: Magnetic Fields from the Photosphere to the Corona*, ed. by T.R. Rimmele, A. Tritschler, F. Wöger, M. Collados Vera, H. Socas-Navarro, R. Schlichenmaier, M. Carlsson, T. Berger, A. Cadavid, P.R. Gilbert, P.R. Goode, M. Knölker. Astronomical Society of the Pacific Conference Series, vol. 463 (2012), p. 157

B.H. McClintock, A.A. Norton, Recovering Joy's law as a function of solar cycle, hemisphere, and longitude. Sol. Phys. **287**, 215–227 (2013). doi:10.1007/s11207-013-0338-0

H.K. Moffatt, The degree of knottedness of tangled vortex lines. J. Fluid Mech. **35**, 117 (1969)

H.K. Moffatt, R.L. Ricca, Helicity and the Calugareanu invariant. Proc. R. Soc. Lond. Ser. A, Math. Phys. Sci. **439**, 411–429 (1992). doi:10.1098/rspa.1992.0159

Y.-J. Moon, J. Chae, G.S. Choe, H. Wang, Y.D. Park, H.S. Yun, V. Yurchyshyn, P.R. Goode, Flare activity and magnetic helicity injection by photospheric horizontal motions. Astrophys. J. **574**, 1066–1073 (2002a). doi:10.1086/340975

Y.-J. Moon, J. Chae, H. Wang, G.S. Choe, Y.D. Park, Impulsive variations of the magnetic helicity change rate associated with eruptive flares. Astrophys. J. **580**, 528–537 (2002b). doi:10.1086/343130

R.L. Moore, A.C. Sterling, H.S. Hudson, J.R. Lemen, Onset of the magnetic explosion in solar flares and coronal mass ejections. Astrophys. J. **552**, 833–848 (2001). doi:10.1086/320559

D. Nandy, Magnetic helicity and flux tube dynamics in the solar convection zone: comparisons between observation and theory. J. Geophys. Res. **111**(A10), 12 (2006). doi:10.1029/2006JA011882

A. Nindos, M.D. Andrews, The association of big flares and coronal mass ejections: what is the role of magnetic helicity? Astrophys. J. Lett. **616**, 175–178 (2004). doi:10.1086/426861

A. Nindos, H. Zhang, Photospheric motions and coronal mass ejection productivity. Astrophys. J. Lett. **573**, 133–136 (2002). doi:10.1086/341937

A. Nindos, J. Zhang, H. Zhang, The magnetic helicity budget of solar active regions and coronal mass ejections. Astrophys. J. **594**, 1033–1048 (2003). doi:10.1086/377126

A. Nindos, S. Patsourakos, T. Wiegelmann, On the role of the background overlying magnetic field in solar eruptions. Astrophys. J. Lett. **748**, 6 (2012). doi:10.1088/2041-8205/748/1/L6

A.A. Norton, P.A. Gilman, Recovering solar toroidal field dynamics from sunspot location patterns. Astrophys. J. **630**, 1194–1205 (2005). doi:10.1086/431961

L.J. November, G.W. Simon, Precise proper-motion measurement of solar granulation. Astrophys. J. **333**, 427–442 (1988). doi:10.1086/166758

S. Ortolani, D.D. Schnack, *Magnetohydrodynamics of Plasma Relaxation* (World Scientific, Singapore, 1993)

E. Pariat, P. Démoulin, M.A. Berger, Photospheric flux density of magnetic helicity. Astron. Astrophys. **439**, 1191–1203 (2005). doi:10.1051/0004-6361:20052663

E. Pariat, A. Nindos, P. Démoulin, M.A. Berger, What is the spatial distribution of magnetic helicity injected in a solar active region? Astron. Astrophys. **452**, 623–630 (2006). doi:10.1051/0004-6361:20054643

E. Pariat, P. Démoulin, A. Nindos, How to improve the maps of magnetic helicity injection in active regions? Adv. Space Res. **39**, 1706–1714 (2007). doi:10.1016/j.asr.2007.02.047

G.J.D. Petrie, Evolution of active and polar photospheric magnetic fields during the rise of cycle 24 compared to previous cycles. Sol. Phys. **281**, 577–598 (2012). doi:10.1007/s11207-012-0117-3

K. Petrovay, U.R. Christensen, The magnetic sun: reversals and long-term variations. Space Sci. Rev. **155**, 371–385 (2010). doi:10.1007/s11214-010-9657-8

A.A. Pevtsov, Transequatorial loops in the solar corona. Astrophys. J. **531**, 553–560 (2000). doi:10.1086/308467

A.A. Pevtsov, Sinuous coronal loops at the sun, in *Multi-Wavelength Observations of Coronal Structure and Dynamics*, ed. by P.C.H. Martens, D.P. Cauffman. COSPAR Colloquia Series, vol. 13 (Pergamon, Dordrecht, 2002), pp. 125–134

A.A. Pevtsov, Helicity generation and signature in the solar atmosphere, in *IAU Joint Discussion*. IAU Joint Discussion, vol. 3 (2003)

A.A. Pevtsov, What helicity can tell us about solar magnetic fields. J. Astrophys. Astron. **29**, 49–56 (2008). doi:10.1007/s12036-008-0006-1

A.A. Pevtsov, R.C. Canfield, Helicity of the photospheric magnetic field, in *Magnetic Helicity in Space and Laboratory Plasmas*, ed. by M.R. Brown, R.C. Canfield, A.A. Pevtsov. Geophysical Monographs Series, vol. 111 (AGU, Washington, D.C., 1999), pp. 103–110

A.A. Pevtsov, S.M. Latushko, Current helicity of the large-scale photospheric magnetic field. Astrophys. J. **528**, 999–1003 (2000). doi:10.1086/308227

A.A. Pevtsov, D.W. Longcope, NOAA 7926: a kinked omega-loop? Astrophys. J. **508**, 908–915 (1998). doi:10.1086/306414

A.A. Pevtsov, D.W. Longcope, Origin of helicity in the quiet sun, in *Advanced Solar Polarimetry—Theory, Observation, and Instrumentation*, ed. by M. Sigwarth. Astronomical Society of the Pacific Conference Series, vol. 236 (2001), p. 423

A.A. Pevtsov, D.W. Longcope, Helicity as the ultimate test to the surface dynamo problem, in *New Solar Physics with Solar-B Mission*, ed. by K. Shibata, S. Nagata, T. Sakurai. Astronomical Society of the Pacific Conference Series, vol. 369 (2007), p. 99

A.A. Pevtsov, N.L. Peregud, Electric currents in a unipolar sunspot, in *Physics of Magnetic Flux Ropes*. Geophysical Monograph Series, vol. 58 (American Geophysical Union, Washington DC, 1990), pp. 161–165

A.A. Pevtsov, R.C. Canfield, T.R. Metcalf, Patterns of helicity in solar active regions. Astrophys. J. **425**, 117–119 (1994). doi:10.1086/187324

A.A. Pevtsov, R.C. Canfield, T.R. Metcalf, Latitudinal variation of helicity of photospheric magnetic fields. Astrophys. J. Lett. **440**, 109–112 (1995). doi:10.1086/187773

A.A. Pevtsov, R.C. Canfield, A.N. McClymont, On the subphotospheric origin of coronal electric currents. Astrophys. J. **481**, 973 (1997). doi:10.1086/304065

A.A. Pevtsov, R.C. Canfield, S.M. Latushko, Hemispheric helicity trend for solar cycle 23. Astrophys. J. Lett. **549**, 261–263 (2001). doi:10.1086/319179

A.A. Pevtsov, K.S. Balasubramaniam, J.W. Rogers, Chirality of chromospheric filaments. Astrophys. J. **595**, 500–505 (2003). doi:10.1086/377339

A.A. Pevtsov, V.M. Maleev, D.W. Longcope, Helicity evolution in emerging active regions. Astrophys. J. **593**, 1217–1225 (2003a). doi:10.1086/376733

A.A. Pevtsov, R.C. Canfield, T. Sakurai, M. Hagino, On the solar cycle variation of the hemispheric helicity rule. Astrophys. J. **677**, 719–722 (2008). doi:10.1086/533435

A.A. Pevtsov, M. Berger, A. Nindos, A. Norton, L. van Driel-Gesztelyi, Magnetic helicity, tilt, and twist, in *The Solar Activity Cycle: Physical Causes and Consequences*, ed. by A. Balogh, H. Hudson, K. Petrovay, R. von Steiger. Space Science Series of ISSI, vol. 53 (Springer, Heidelberg, 2014)

A.D. Phillips, P.J. MacNeice, S.K. Antiochos, The role of magnetic helicity in coronal mass ejections. Astrophys. J. Lett. **624**, 129–132 (2005). doi:10.1086/430516

K.G. Puschmann, B. Ruiz Cobo, V. Martínez Pillet, The electrical current density vector in the inner penumbra of a sunspot. Astrophys. J. Lett. **721**, 58–61 (2010). doi:10.1088/2041-8205/721/1/L58

B. Ravindra, D.W. Longcope, W.P. Abbett, Inferring photospheric velocity fields using a combination of minimum energy fit, local correlation tracking, and Doppler velocity. Astrophys. J. **677**, 751–768 (2008). doi:10.1086/528363

R.L. Ricca, B. Nipoti, Gauss linking number revisited. J. Knot Theory Ramif. **20**, 1325–1343 (2011)

P. Romano, F. Zuccarello, Flare occurrence and the spatial distribution of the magnetic helicity flux. Astron. Astrophys. **535**, 1–5 (2011). doi:10.1051/0004-6361/201117594

P. Romano, L. Contarino, F. Zuccarello, Eruption of a helically twisted prominence. Sol. Phys. **214**, 313–323 (2003). doi:10.1023/A:1024257603143
G.V. Rudenko, I.I. Myshyakov, Gauge-invariant helicity for force-free magnetic fields in a rectangular box. Sol. Phys. **270**, 165–173 (2011). doi:10.1007/s11207-011-9743-4
B. Ruiz Cobo, K.G. Puschmann, Twist, writhe, and helicity in the inner penumbra of a sunspot. Astrophys. J. **745**, 141 (2012). doi:10.1088/0004-637X/745/2/141
D.M. Rust, Spawning and shedding helical magnetic fields in the solar atmosphere. Geophys. Res. Lett. **21**, 241–244 (1994). doi:10.1029/94GL00003
D.M. Rust, A. Kumar, Evidence for helically kinked magnetic flux ropes in solar eruptions. Astrophys. J. Lett. **464**, 199 (1996). doi:10.1086/310118
D.M. Rust, S.F. Martin, A correlation between sunspot whirls and filament type, in *Solar Active Region Evolution: Comparing Models with Observations*, ed. by K.S. Balasubramaniam, G.W. Simon. Astronomical Society of the Pacific Conference Series, vol. 68 (1994), p. 337
H.U. Schmidt, Magnetohydrodynamics of an active region, in *Structure and Development of Solar Active Regions*, ed. by K.O. Kiepenheuer. IAU Symposium, vol. 35 (1968), p. 95
P.W. Schuck, Tracking vector magnetograms with the magnetic induction equation. Astrophys. J. **683**, 1134–1152 (2008). doi:10.1086/589434
M. Schüssler, P. Caligari, A. Ferriz-Mas, F. Moreno-Insertis, Instability and eruption of magnetic flux tubes in the solar convection zone. Astron. Astrophys. **281**, 69–72 (1994)
N. Seehafer, Electric current helicity in the solar atmosphere. Sol. Phys. **125**, 219–232 (1990). doi:10.1007/BF00158402
K.R. Sivaraman, S.S. Gupta, R.F. Howard, Measurement of Kodaikanal white-light images—IV. Axial tilt angles of sunspot groups. Sol. Phys. **189**, 69–83 (1999). doi:10.1023/A:1005277515551
J.O. Stenflo, A.G. Kosovichev, Bipolar magnetic regions on the sun: global analysis of the SOHO/MDI data set. Astrophys. J. **745**, 129 (2012). doi:10.1088/0004-637X/745/2/129
J.K. Thalmann, B. Inhester, T. Wiegelmann, Estimating the relative helicity of coronal magnetic fields. Sol. Phys. **272**, 243–255 (2011). doi:10.1007/s11207-011-9826-2
J.K. Thalmann, S.K. Tiwari, T. Wiegelmann, Force-free field modeling of twist and braiding-induced magnetic energy in an active-region corona. Astrophys. J. **780**(1), 102 (2014). doi:10.1088/0004-637X/780/1/102
L. Tian, Y. Liu, Tilt and α_{best} of major flare-producing active regions. Astron. Astrophys. **407**, 13–16 (2003). doi:10.1051/0004-6361:20030977
L. Tian, S. Bao, H. Zhang, H. Wang, Relationship in sign between tilt and twist in active region magnetic fields. Astron. Astrophys. **374**, 294–300 (2001). doi:10.1051/0004-6361:20010701
A.G. Tlatov, A.A. Pevtsov, Bimodal distribution of magnetic fields and areas of sunspots. Sol. Phys. **289**, 1143–1152 (2014). doi:10.1007/s11207-013-0382-9
A.G. Tlatov, V.V. Vasil'eva, A.A. Pevtsov, Distribution of magnetic bipoles on the sun over three solar cycles. Astrophys. J. **717**, 357–362 (2010). doi:10.1088/0004-637X/717/1/357
T. Török, B. Kliem, Confined and ejective eruptions of kink-unstable flux ropes. Astrophys. J. Lett. **630**, 97–100 (2005). doi:10.1086/462412
T. Török, M.A. Berger, B. Kliem, The writhe of helical structures in the solar corona. Astron. Astrophys. **516**, 49 (2010). doi:10.1051/0004-6361/200913578
T. Török, B. Kliem, M.A. Berger, M.G. Linton, P. Demoulin, L. van Driel-Gesztelyi, The evolution of writhe in kink-unstable flux ropes and erupting filaments. Plasma Phys. Control. Fusion, **56**(6), 064012 (2014)
K. Tziotziou, M.K. Georgoulis, N.-E. Raouafi, The magnetic energy-helicity diagram of solar active regions. Astrophys. J. Lett. **759**, 4 (2012). doi:10.1088/2041-8205/759/1/L4
K. Tziotziou, M.K. Georgoulis, Y. Liu, Interpreting eruptive behavior in NOAA AR 11158 via the region's magnetic energy and relative-helicity budgets. Astrophys. J. **772**, 115 (2013). doi:10.1088/0004-637X/772/2/115
L. Upton, D.H. Hathaway, Predicting the sun's polar magnetic fields with a surface flux transport model. Astrophys. J. **780**, 5 (2014). doi:10.1088/0004-637X/780/1/5
G. Valori, P. Démoulin, E. Pariat, Comparing values of the relative magnetic helicity in finite volumes. Sol. Phys. **278**, 347–366 (2012). doi:10.1007/s11207-012-0044-3
A.A. van Ballegooijen, P.C.H. Martens, Formation and eruption of solar prominences. Astrophys. J. **343**, 971–984 (1989). doi:10.1086/167766
B. Vršnak, V. Ruždjak, B. Rompolt, Stability of prominences exposing helical-like patterns. Sol. Phys. **136**, 151–167 (1991). doi:10.1007/BF00151701
Y.-M. Wang, On the strength of the hemispheric rule and the origin of active-region helicity. Astrophys. J. Lett. **775**, 46 (2013). doi:10.1088/2041-8205/775/2/L46
Y.-M. Wang, N.R. Sheeley Jr., Average properties of bipolar magnetic regions during sunspot cycle 21. Sol. Phys. **124**, 81–100 (1989). doi:10.1007/BF00146521

Y.-M. Wang, N.R. Sheeley Jr., Magnetic flux transport and the sun's dipole moment—new twists to the Babcock-Leighton model. Astrophys. J. **375**, 761–770 (1991). doi:10.1086/170240

C. Wang, M. Zhang, A hemispheric helicity sign rule indicated by large-scale photospheric magnetic fields at three phases of solar cycle 23. Astrophys. J. **720**, 632–638 (2010). doi:10.1088/0004-637X/720/1/632

M.A. Weber, Y. Fan, M.S. Miesch, The rise of active region flux tubes in the turbulent solar convective envelope. Astrophys. J. **741**, 11 (2011). doi:10.1088/0004-637X/741/1/11

M.A. Weber, Y. Fan, M.S. Miesch, Comparing simulations of rising flux tubes through the solar convection zone with observations of solar active regions: constraining the dynamo field strength. Sol. Phys. **287**, 239–263 (2013). doi:10.1007/s11207-012-0093-7

B.T. Welsch, D.W. Longcope, Magnetic helicity injection by horizontal flows in the quiet sun. I. Mutual-helicity flux. Astrophys. J. **588**, 620–629 (2003). doi:10.1086/368408

B.T. Welsch, G.H. Fisher, W.P. Abbett, S. Regnier, ILCT: recovering photospheric velocities from magnetograms by combining the induction equation with local correlation tracking. Astrophys. J. **610**, 1148–1156 (2004). doi:10.1086/421767

B.T. Welsch, W.P. Abbett, M.L. De Rosa, G.H. Fisher, M.K. Georgoulis, K. Kusano, D.W. Longcope, B. Ravindra, P.W. Schuck, Tests and comparisons of velocity-inversion techniques. Astrophys. J. **670**, 1434–1452 (2007). doi:10.1086/522422

D.R. Williams, T. Török, P. Démoulin, L. van Driel-Gesztelyi, B. Kliem, Eruption of a kink-unstable filament in NOAA active region 10696. Astrophys. J. Lett. **628**, 163–166 (2005). doi:10.1086/432910

P.R. Wilson, R.C. Altrocki, K.L. Harvey, S.F. Martin, H.B. Snodgrass, The extended solar activity cycle. Nature **333**, 748–750 (1988). doi:10.1038/333748a0

A. Wright, M.A. Berger, The effect of reconnection upon the linkage and interior structure of magnetic flux tubes. J. Geophys. Res. **94**, 1295–1302 (1989)

S. Yang, H. Zhang, Large-scale magnetic helicity fluxes estimated from MDI magnetic synoptic charts over the solar cycle 23. Astrophys. J. **758**, 61 (2012). doi:10.1088/0004-637X/758/1/61

A.R. Yeates, D.H. Mackay, A.A. van Ballegooijen, Modelling the global solar corona: filament chirality observations and surface simulations. Sol. Phys. **245**, 87–107 (2007). doi:10.1007/s11207-007-9013-7

M. Zhang, Helicity observations of weak and strong fields. Astrophys. J. Lett. **646**, 85–88 (2006). doi:10.1086/506560

M. Zhang, N. Flyer, The dependence of the helicity bound of force-free magnetic fields on boundary conditions. Astrophys. J. **683**, 1160–1167 (2008). doi:10.1086/589993

M. Zhang, B.C. Low, Magnetic flux emergence into the solar corona. I. Its role for the reversal of global coronal magnetic fields. Astrophys. J. **561**, 406–419 (2001). doi:10.1086/323238

M. Zhang, B.C. Low, Magnetic flux emergence into the solar corona. III. The role of magnetic helicity conservation. Astrophys. J. **584**, 479–496 (2003). doi:10.1086/345615

M. Zhang, B.C. Low, The hydromagnetic nature of solar coronal mass ejections. Annu. Rev. Astron. Astrophys. **43**, 103–137 (2005). doi:10.1146/annurev.astro.43.072103.150602

M. Zhang, N. Flyer, B.C. Low, Magnetic field confinement in the corona: the role of magnetic helicity accumulation. Astrophys. J. **644**, 575–586 (2006). doi:10.1086/503353

H. Zhang, T. Sakurai, A. Pevtsov, Y. Gao, H. Xu, D.D. Sokoloff, K. Kuzanyan, A new dynamo pattern revealed by solar helical magnetic fields. Mon. Not. R. Astron. Soc. **402**, 30–33 (2010). doi:10.1111/j.1745-3933.2009.00793.x

M. Zhang, N. Flyer, B. Chye Low, Magnetic helicity of self-similar axisymmetric force-free fields. Astrophys. J. **755**, 78 (2012). doi:10.1088/0004-637X/755/1/78

Y. Zhang, R. Kitai, K. Takizawa, Magnetic helicity transported by flux emergence and shuffling motions in solar active region NOAA 10930. Astrophys. J. **751**, 85 (2012). doi:10.1088/0004-637X/751/2/85

F.P. Zuccarello, C. Jacobs, A. Soenen, S. Poedts, B. van der Holst, F. Zuccarello, Modelling the initiation of coronal mass ejections: magnetic flux emergence versus shearing motions. Astron. Astrophys. **507**, 441–452 (2009). doi:10.1051/0004-6361/200912541

DOI 10.1007/978-1-4939-2584-1_11
Reprinted from *Space Science Reviews* Journal, DOI 10.1007/s11214-014-0064-4

Solar Polar Fields and the 22-Year Activity Cycle: Observations and Models

G.J.D. Petrie · K. Petrovay · K. Schatten

Received: 31 March 2014 / Accepted: 20 June 2014 / Published online: 1 July 2014

Abstract We explore observations and models of the interacting, cyclical behavior of the active regions and the polar magnetic fields of the Sun. We focus on observational evidence of these fields interacting across the corridor between active and polar latitudes. We present observations of diverse magnetic signatures on, above and beneath the solar surface, and find much evidence of phenomena migrating in both directions across this corridor in each hemisphere, including photospheric fields, ephemeral bipoles, interior torsional oscillations, high-latitude filaments, and coronal green line intensity. Together these observations produce a complex physical picture of high-latitude solar magnetic field evolution in the photosphere, atmosphere and interior, and demonstrate their essential role in the solar cycle. The picture presented by these collected observations is consistent with the Babcock-Leighton phenomenological model for the cycle, and we discuss related efforts to predict cycle amplitudes based on polar field strengths and on combining activity and polar-field information in a single phase-independent, slowly-evolving index. We also briefly review related work on magnetic flux transport models for the solar cycle, with particular reference to the interaction between flux emergence patterns and meridional flows.

Keywords Solar magnetic fields · Activity cycle · Solar poles · Photosphere · Chromosphere · Corona · Solar interior · Magnetic flux-transport models

G.J.D. Petrie (✉)
National Solar Observatory, 950 N. Cherry Ave., Tucson, AZ 85719, USA
e-mail: gpetrie@nso.edu

K. Petrovay
Department of Astronomy, Eotvos University, Pf. 32, Budapest, 1518 Hungary
e-mail: K.Petrovay@astro.elte.hu

K. Schatten
a.i. solutions, Suite 215, 10001 Derekwood Lane, Lanham, MD 20706, USA
e-mail: ken.schatten@ai-solutions.com

1 Introduction

We review recent efforts to characterize the 22-year solar magnetic cycle, focusing in particular on observations of the polar fields and related high-latitude phenomena and on the role of the polar fields in modeling and predicting cycles. The solar cycle can be studied via observations of many different parts of the Sun besides the photosphere, from the interior to the corona. The basic goal of the review is to collect diverse observations from the solar surface, from several atmospheric layers and from the interior, and discuss them alongside related theoretical models. These observations and models together give us an opportunity to elucidate the interaction between active regions and polar fields.

We will begin our review by presenting full-surface observations of the Sun's photospheric magnetic field, and note the salient features of the cyclic relationship between the fields at high and low latitudes. The relationship between the active regions and the polar fields has been central to the study of the solar cycle since Babcock (1961) introduced his phenomenological model for the cycle based on full-disk photospheric magnetogram observations (Babcock 1959). A butterfly diagram of the photospheric magnetic field provides a global view of photospheric field evolution over multiple solar cycles, including both strong and weak fields at all latitudes. This kind of diagram is particularly useful in allowing us to study the motions and interactions of the two classes of field that dominate the cycle: the active regions and the polar fields.

Because of the intrinsic weakness of the polar fields and the large projection angle when observing polar latitudes from Earth, the polar fields are difficult to measure. Furthermore, full-disk magnetogram observations have been taken continuously only since the 1970s. For these reasons it is useful to study quantities that do not suffer from these problems, and that are proportional to the polar field strength and can therefore serve as proxies for the polar field strength. Microwave brightness temperature measurements above 15 GHz have been found to be sensitive to field strength and their measurement is not compromised by projection effects. Faculae are most easily identified near the limb, and polar faculae counts have also been found to correlate well with polar field strength. Observations of polar faculae have also been produced for a much longer time than direct measurements of polar field strengths. We will discuss how measurements of these quantities supplement our knowledge of the polar fields.

The high-latitude corridor between the activity belt and the polar cap is an especially important focus of study regarding the solar cycle since it is across this corridor that the active and polar fields can be observed to interact. As well as direct observations of the magnetic field, many proxies for the magnetic field have been studied at these latitudes. Some of these phenomena are observed to migrate poleward during the cycle in a "march to the poles". These features can give us more information on the magnetic connection between the active and polar fields that is so central to the solar cycle. On the other hand, some phenomena migrate equatorward towards active latitudes before the cycle proper begins, blending with it at the beginning of the cycle, and appearing as an "extended cycle" in butterfly diagrams, extensions of the activity cycle in both space (latitude) and time. The march to the poles is observed in butterfly diagrams of the photospheric magnetic field and in distributions of solar filaments. Lying along boundaries between fields of opposite sign, filaments are very useful markers of the solar field. The extended cycle is also observed in the photospheric field in distributions of ephemeral bipoles. Both forms of high-latitude migration are seen in interior torsional oscillations and in coronal green line emission. We will discuss the significance of these poleward- and equatorward-migrating phenomena to polar field reversals and the solar cycle.

Statistics for both polar and active fields and their proxies of past cycles are often used to predict the size of future cycles. An attempt to combine these statistics in a single, simple, well-behaved index is the SoDA (Solar Dynamo Amplitude) index (see Schatten and Pesnell 1993). The SoDA index is designed to represent the strength of the cycle at any given time, with the 11-year modulation suppressed by combining the squares of the polar field strength and F10.7 flux. In practice the behavior is slightly complicated by the non-sinusoidal variations of the polar fields and activity, with peaks in the SoDA index at activity maxima. We will discuss the behavior and performance of the SoDA index.

Besides direct observation of polar and active fields and the study of proxies, a third major effort has attempted to model the solar magnetic cycle in a simple way. Beginning with the theory of Parker (1955) relating the buoyant emergence of intense toroidal flux tubes from the interior, twisted by helical upflows subject to the Coriolis force, to the creation of poloidal field components and the polar fields, and the phenomenological solar cycle model of Babcock (1961) linking polar field creation to the emergence and dispersal of bipolar magnetic regions, quantitative modeling of the cyclical relationship between active regions and polar fields continues to this day. We review recent efforts to understand the relationship between active region flux emergence, large-scale flows and solar cycle behavior.

This review follows a traditional pattern in beginning with direct observations and ending with theory/models. In Sect. 2 we begin with a butterfly diagram (latitude-time plot) of multi-cycle full-disk solar magnetic measurements, and review its features. We then discuss in Sect. 3 indirect proxies of the magnetic field, polar faculae and microwave brightness temperature, that supplement our knowledge of the polar fields from direct measurement, and we discuss observed extended cycle phenomena from the corona to the interior, including coronal green-line patterns at high latitudes, ephemeral bipoles, and torsional oscillations. We also discuss the results of research into polar precursors of activity cycles, focusing on the SoDA index. Finally, in Sect. 5 we discuss kinematic models for the transport of solar magnetic flux and the solar cycle, with particular reference to the interaction between flux emergence patterns and meridional flows. We conclude in Sect. 6.

2 Observations of the Polar Magnetic Fields

2.1 Observed Relationship Between Polar and Active Fields: The Magnetic Butterfly Diagram

Measurement of the polar field is very difficult because of the foreshortening of the solar surface near the limb and the intrinsic weakness of the field near the poles. While evidence of strong (kG) flux has been found at polar latitudes in high-resolution vector data, shown in Fig. 1, from the Solar Optical Telescope (SOT) instrument on the Hinode satellite, these fields appear to be confined to compact, sparsely distributed structures so that, observed with the lower spatial resolutions characteristic of synoptic full-disk magnetographs, the polar fields appear weak (< 10 G). Yet these measurements are critical because the polar fields play a critical role in determining the basic global magnetic structure of the solar atmosphere, and they are also believed to play a central role in the global dynamo. Global models of the corona (potential-field source-surface and magnetohydrodynamics) are particularly sensitive to the polar fields because the dominant multipole components, the axial dipole and octupole, are highly correlated with the polar fields (see Sect. 2.2).

Using full-disk longitudinal photospheric magnetograms from the NSO's three Kitt Peak magnetographs and from the Mt. Wilson Observatory (MWO) 150-foot tower, Petrie (2012)

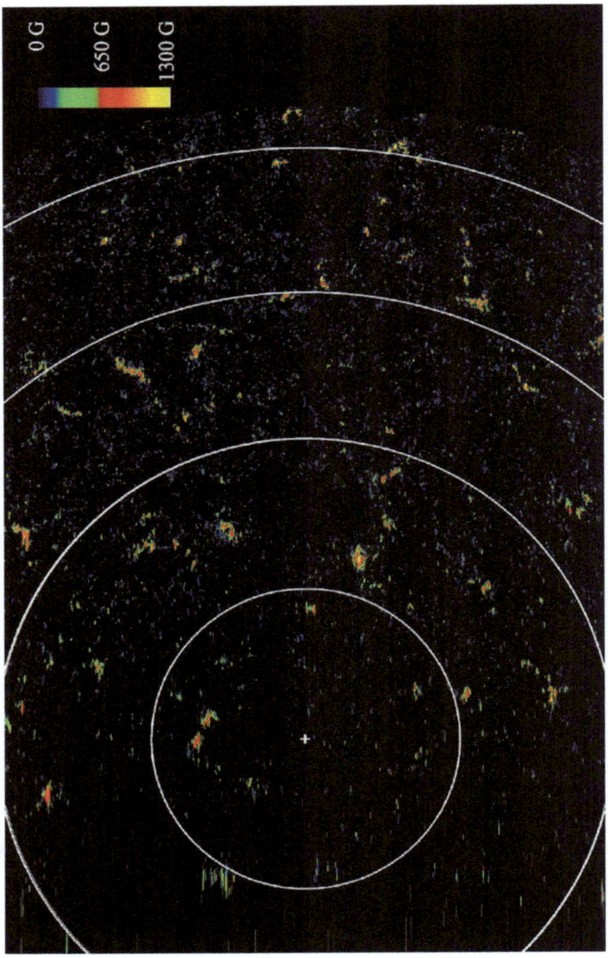

Fig. 1 South polar view of the magnetic field strength taken from 12:02:19–14:55:48 UT on 2007 March 16. The original observing field of view is 327.52″ (east-west) by 163.84″ (north-south) and was converted to a map seen from above the south pole. East is to the left, west is to the right, and the observation was carried out from the top down. Spatial resolution is lost near the extreme limb (i.e., near the *bottom* of the figure). The field of view is 327.52″ (east-west) by 472.96″ (north-south along the line of sight). The field of view for the line-of-sight direction (163.84″) expands to 472.96″ as a result of correction for foreshortening. The pixel size is 0.16″. Latitudinal lines for 85°, 80°, 75°, and 70° are shown as *large circles*, and *the plus sign marks* the south pole. From Tsuneta et al. (2008)

studied the evolution of the photospheric magnetic field over the 3 1/2 solar cycles from the beginning of cycle 21 to the maximum of cycle 24.

In Fig. 2, updated from Petrie (2012, 2013), a butterfly diagram, a sine(latitude)-time plot, of the radial field component is shown. This was derived from the longitudinal measurements by assuming that the photospheric field is approximately radial (based on evidence from, e.g., Svalgaard et al. 1978; Wang and Sheeley 1992; Petrie and Patrikeeva 2009) and dividing by the cosine of the heliocentric angle (the angle between the line of sight and the local solar radial vector).

Gosain et al. (2013) reported that photospheric fields have systematic patterns of tilt based on a comparison of a SOLIS/VSM vector synoptic map radial-field component compared to the corresponding scalar synoptic map for the radial field. This result includes an unphysical north-south asymmetry that seems to be caused by a heliospheric angle-dependence of the vector magnetogram measurements and work on this project continues. Jin et al. (2013) constructed synoptic maps for the radial chromospheric field from longitudinal measurements via a deprojection method for approximately static fields, and obtained results consistent with radial photospheric maps constructed using the radial field assumption in the usual way.

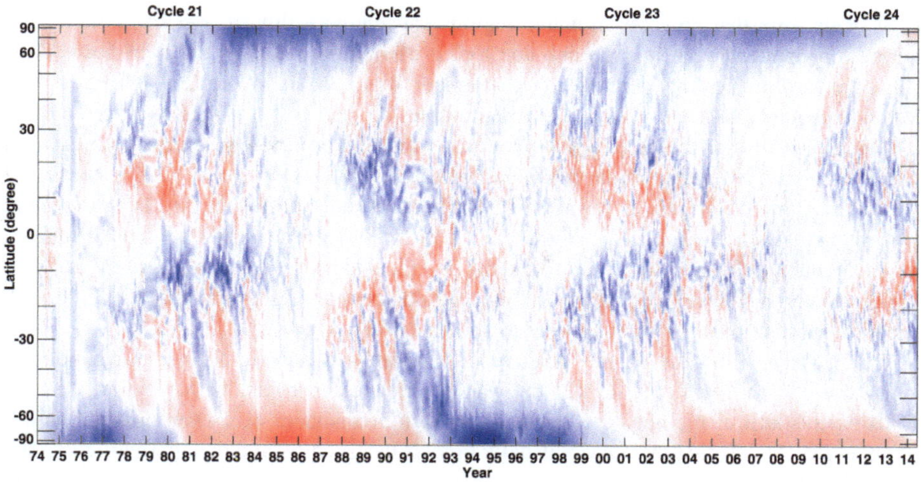

Fig. 2 Butterfly diagram based on Kitt Peak magnetogram data, summarizing the photospheric radial field distributions derived from the longitudinal photospheric field measurements. Each pixel is colored to represent the average field strength at each time and latitude. *Red/blue* represents positive/negative flux, with the color scale saturated at 20 G. Updated from Petrie (2012, 2013)

An important advantage of studying approximately radial fields is that the full magnetic flux through the photosphere can be estimated reasonably accurately over most of the solar disk. However, because the solar rotation axis is tilted at an angle of 7.25° with respect to the ecliptic plane, the fields near the solar poles are either observed with very large viewing angles or, for six months at a time, not observed at all. Also the noise level is inflated near the poles by the radial field correction discussed above. For these reasons we fill the locations in the butterfly diagram nearest the poles using estimated values for these fields. This is a well-known problem in the construction of synoptic magnetograms (e.g., Sun et al. 2011). Raouafi et al. (2007) reported that the latitudinal distribution of chromospheric flux elements, defined as regions of field with strength greater than 5 G and of size at least a few arcseconds in each direction, tend to be more concentrated at lower rather than higher latitudes within the polar cap. On the other hand, the polar flux is generally found to become stronger as one observes further poleward (Svalgaard et al. 1978; Wang and Sheeley 1988; Petrie and Patrikeeva 2009), and Fig. 2 reflects this. The fields at the highest latitudes in Fig. 2 were calculated from measurements taken with advantageous solar axis tilt (B_0) angles ($B_0 > 5°$ for northern and $B_0 < -5°$ for southern high-latitude fields). The resulting fields are well defined and regular functions of time for all but the two most poleward sin(latitude) bins at each pole. For these bins, simulated data were used based on a polynomial fit for each image. The simulated data were then blended with the measurements in a B_0-dependent fashion.

Figure 2 covers cycles 21–23 in their entirety, as well as the end of cycle 20 and the ascent of cycle 24. The diagram shows several distinctive patterns in the long-term behavior of the fields at active and polar latitudes, and at latitudes in between. The active fields begin each cycle emerging at latitudes around ±30° and subsequently emerge at progressively lower latitudes on average, creating the distinctive wings of the butterfly patterns first reported by (Maunder 1913). The diagram also shows the change of polarity of the polar fields once each cycle, coinciding with activity maximum. Between the active and polar latitudes, around ±50°, there is clear evidence of poleward flux transport of both polarities in each hemisphere during each cycle, which appears most intense during the most active phases of the cycle.

Most of the flux that emerges in active regions cancels with flux of opposite polarity but a critical proportion survives as weak flux that is carried poleward by the meridional flow. This poleward drift of the weak, decayed magnetic flux appears in Fig. 2 as plumes of one dominant polarity at high latitudes, between about 40° and 65°, sometimes referred to as a "march" or "rush to the poles". Such patterns were first reported by Bumba and Howard (1969) who named them unipolar magnetic regions (UMRs). Howe et al. (2013) compared the poleward migration rate estimated from the high-latitude surges in a Kitt Peak magnetic butterfly diagram to the subsurface meridional flow rates measured in helioseismic data from the GONG network since 2001, and found the two rates to be in reasonable agreement. Note that there is no evidence of equatorward high-latitude counter-cell meridional flow in Fig. 2. The high-latitude flux transport appears to be exclusively poleward.

The latitude centroids of the positive and negative magnetic fluxes in each hemisphere were found by Petrie (2012) to swap positions each cycle, with the positive centroids north/south of the negative centroids in each hemisphere over even-numbered/odd-numbered cycles. Active regions are usually bipolar in structure, and the alternating latitude centroid patterns reflect the well-known facts that bipolar regions are typically asymmetric and tilted so that the leading polarity is stronger, more compact and slightly closer to the equator than the following polarity (Hale et al. 1919, referred to as "Joy's law tilt"), and that the vast majority of bipoles in each hemisphere have the same leading polarity, the leading polarities are opposite in the two hemispheres during any given cycle, and all polarity patterns reverse each cycle (Hale and Nicholson 1925, referred to as "Hale's polarity rule"). The Joy's law tilt plays a central role in the phenomenological model for the solar activity cycle of Babcock (1961), where decaying flux from the following, generally more dispersed and more poleward polarity of the bipole contributes more to the polar fields than the leading, generally more compact and more equatorward polarity. The resulting cycle-dependent polarity bias and Joy's law tilt preference are generally believed to be responsible for creating the polar field cycle from the activity cycle. These ideas were used to develop a kinematic model for photospheric flux transport and the solar cycle by Leighton (1964, 1969), initiating a class of models known as "Babcock-Leighton" models (see Sect. 5). The Joy's law tilt and polarity biases are not strict. Even in averaged form, the poleward flux surges appear to be approximately periodic in time (Ulrich and Tran 2013) with frequent changes in sign. The streams originate from sizable densely packed groups of sunspots that survive for many months and are almost continuously refreshed during their lifetimes by the emergence of new bipoles (Gaizauskas et al. 1983; Schrijver and Zwaan 2000). The interaction of multiple tilted bipoles may produce large poleward surges of flux and large polar field changes in a relatively short time.

Figure 2 features significant north-south asymmetry in the strength of the activity. In particular, the southern hemisphere was more active than the north during the decline of cycle 23, and the north was the more active hemisphere during the rise of cycle 24. Also the south polar field developed more slowly than the north polar field during the decline of cycle 23 and began to reverse more slowly at the beginning of cycle 24. We will pick up the topic of north-south asymmetry again in Sect. 5.6 in the context of a flux transport dynamo model.

Babcock's phenomenological model suggests that the contribution to the polar field in a given hemisphere from decayed active region flux may be proportional to the total active region flux in that hemisphere and to the latitude displacement between the positive and negative active-region flux centroids in that hemisphere. Indeed, Petrie (2012) found a correlation between the annual-averaged high-latitude (approx. ±50°) poleward stream fluxes and the product of the annual-averaged latitude flux centroid displacements and total

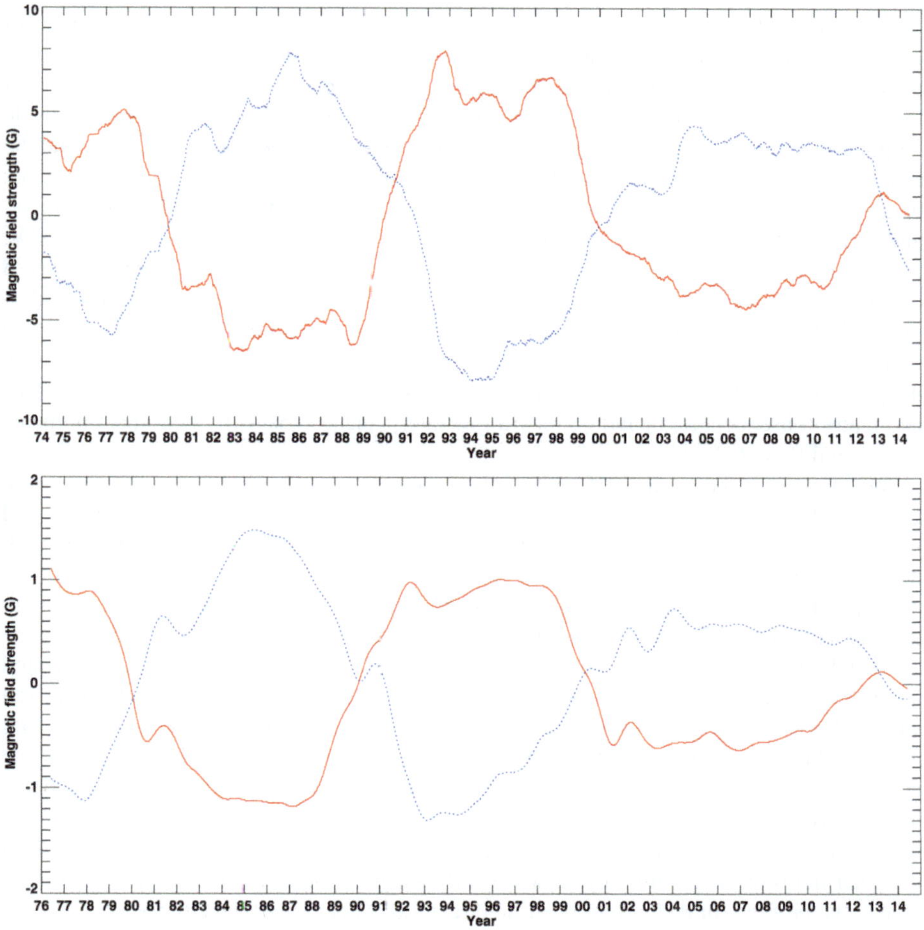

Fig. 3 10-day averages of the north (*red solid lines*) and south (*blue dotted lines*) polar fields measured at Kitt Peak (*top*) and Wilcox (*bottom*) are plotted against time. The Kitt Peak data are for the radial field component and derive from heliographic latitudes between about latitudes ranging from ±63° to ±70°. The Wilcox measurements are for the line-of-sight field component and come from the 3′ apertures nearest the poles, which cover between about ±55° and the poles. Updated from Petrie (2013)

active-region fluxes. A correlation was also found between the annual-averaged high-latitude poleward stream fluxes and the annual-average polar (±63–70°) field changes. These correlations were found in both hemispheres with both NSO Kitt Peak data and MWO data. Muñoz Jaramillo et al. (2013) has found an analogous correlation in proxy data covering 100 years—see Sect. 3.

Figure 3 shows estimates of the average field strengths at polar latitudes based on measurements from NSO Kitt Peak and Wilcox Solar Observatory (WSO). At WSO the north and south polar line-of-sight field strengths are measured daily in the 3′ apertures nearest the poles, north and south. The NSO/KP estimates are derived from the butterfly map shown in Fig. 2 and represent the radial field component. This is part of the reason why the NSO polar data are so much stronger than the WSO measurements. There are also well documented calibration issues with the WSO data (Svalgaard et al. 1978) causing the fields to be underestimated by a factor of 1.8. Nevertheless, the time series shown in the two panels of Fig. 3 are clearly well correlated with each other, boosting confidence that they represent real patterns in the polar field evolution. They agree, for example, the widely-reported fact that the polar fields, both north and south, have been only about 60 % as strong since the cycle 23 polarity reversal compared to before (e.g., Hoeksema 2010; de Toma 2011). Around the time of this reversal of polarity, 2002–2003, the positive and negative flux latitude centroids began to converge in each hemisphere (Petrie 2012). This forced the latitude displacement, and therefore the net flux contribution from the decaying active regions to the polar fields, to become significantly smaller than previously. It appears that the polar fields were starved of unipolar decayed active-region flux between the cycle 23 polar reversal and the end of cycle 23. The high-latitude surges carried weaker flux of more mixed polarity during cycle 23 maximum than during the maxima of cycles 21 and 22, as we saw in Fig. 2. Therefore the polar fields did not strengthen as much as during these previous polarity reversals. This had many consequences, including record-low interplanetary magnetic field measurements (Smith and Balogh 2008).

When cycle 24 began in 2010, the northern hemisphere developed ahead of the south. Figure 2 shows that positive field-carrying surges traveled poleward as shown by the red streaks, causing the north polar field to reverse first as shown in Figs. 2 and 3. During 2013, however, a surge of negative flux moved poleward, threatening to cancel the progress of the north polar field. In the south the opposite happened: the activity started late and initially sent mixed polarity-flux poleward, and the south polar field was slow to reverse. But the south polar field has been fed a steady diet of negative flux in the past few years, represented by the blue streaks in the bottom right of Fig. 2. In response the south polar field has reversed and has been strengthening ever since. The most recent active regions appear to be sending negative flux towards the south pole but a mixture towards the north pole. The fate of the north polar field reversal is therefore in doubt, and we may have asymmetric polar fields later this year.

The most successful predictions of solar activity cycle amplitudes have been based on the polar fields of the previous cycle, as we will discuss in Sect. 4.4. For this reason among others, the progress of the current polar field reversal is being watched closely.

2.2 The Effects of the Polar Fields on Global Coronal Magnetic Structure

The Sun's polar fields have long been known to have a profound influence on the global structure of the solar atmosphere. There is a close relationship between the polar fields and the axial dipole component such that the polar fields dominate the coronal structure over much of the cycle. Hoeksema (1984) used WSO synoptic magnetograms to demonstrate that during most of the solar cycle the quadrupole and occasionally octupole moments of the field are important for the large-scale structure of the coronal field, and especially for the structure in the ecliptic, producing tilts and warps in the streamer belt and equatorial current sheet and creating low-latitude coronal holes. He concluded from a study of the total strengths of the various multipolar orders that the complex field configuration near maximum does not correspond to a dipole rotating from north to south as the polar fields change as had been previously suggested. He also found that the large latitudinal extent of the current sheet over much of the cycle affects the propagation of cosmic rays.

DeRosa et al. (2012) studied the axisymmetric multipoles of different orders using WSO and Michelson Doppler Imager (MDI) magnetogram data and found a relationship between those symmetric and antisymmetric about the equator. Using kinematic Babcock-Leighton models they showed that both north-south symmetric and antisymmetric components need to be present for cyclical behavior to be maintained (See Sect. 5.6). This implies that imbalances between the polar fields play an important role in the cycle. Such imbalances result in net displacements of the equatorial current sheet from the equator.

Using Kitt Peak and WSO synoptic magnetograms, Petrie (2013) investigated the evolution of the observed photospheric magnetic field and the modeled global coronal magnetic field during the past 3 1/2 solar activity cycles observed since the mid-1970s. Polar field changes were found to be well correlated with active fields over most of the period studied, except between 2003 and 2006 when the active fields did not produce significant polar field changes. It was also found that the combined strength of the non-axisymmetric multipoles correlates very well with the sunspot number ($cc = 0.97$) and the axisymmetric fields correlate almost equally well with the polar field strength, the combined strengths of the north and south polar fields ($cc = 0.97$). Of the axisymmetric multipoles, only the dipole and octupole followed the poles whereas the higher orders followed the activity cycle, possibly perturbed by the Joy's law tilt of active regions. The non-axisymmetric orders all follow the activity cycle. The excellent overall correlation between the total axisymmetric field and the polar field is due to the axial dipole and octupole fields. And, because of the large strength and spatial scale of these components, they have a leading influence on the structure of the corona over the cycle, except when the polar fields reverse. During the long cycle 23/24 minimum the axisymmetric dipole and octupole components were about 40 % weaker than during the previous cycle minimum, whereas the non-axisymmetric multipoles did not become so comparatively weak until 2009. This resulted in a more complex coronal structure with more low-latitude coronal holes during the cycle 23/24 minimum than during the previous minimum (de Toma 2011, see Sect. 3.3). Besides playing a dominant role in structuring the solar atmosphere, the polar fields are also believed to be fundamental to the solar cycle, as discussed by DeRosa et al. (2012) and other modelers—see Sect. 5.

3 Proxies of the Polar Magnetic Field

Besides direct measurement of the photospheric magnetic field, high-latitude magnetic field-related phenomena in other parts of the Sun (interior or atmosphere) can be quantified. This can help to overcome some limitations of magnetic measurements (projection effects, temporal coverage, restriction to the photosphere) and also give new perspectives on these measurements. We review selected studies of proxy data sets in this section.

3.1 Polar Faculae

We have continuous high-quality full-disk observations of the photospheric line-of-sight magnetic field from the mid-1970s to the present. These 40 years or so represent only a few solar cycles, not enough data points to study the polar fields' role in the solar cycle for a statistical sample of cycles. One possible substitute for real polar field measurements is polar faculae data. Faculae are bright features on the surface of the Sun associated with intergranular accumulations of magnetic flux (Hale 1920). It is believed that the strong magnetic field causes depressions to form in the visible surface of the Sun so that relatively hot granular plasmas become visible as bright features at large projection angles. Faculae can therefore be used to study surface magnetic flux at the poles. A continuous data set from 1906 to the present is available from MWO. Strong correlations between polar faculae and WSO polar field strength measurements were found by Sheeley (1991, 2008). Muñoz Jaramillo et al. (2012) validated these data using automatically counted MDI facular statistics from

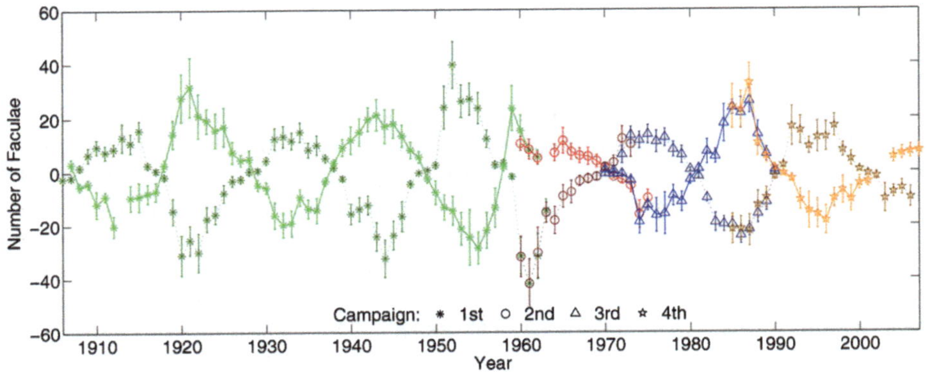

Fig. 4 Calibrated MWO facular measurements from four separate campaigns described by Sheeley (1991, 2008). Each campaign is marked with a different color and marker: *green asterisk* (1st), *red circle* (2nd), *blue triangle* (3rd), and *orange star* (4th). Measurements for the north (south) pole are shown using a *dark dashed* (*light solid*) *line*. All campaigns are calibrated to the 3rd campaign. From Muñoz Jaramillo et al. (2012)

MDI intensity images and, and calibrated them using polar field measurements from WSO and MDI. The calibrated data are plotted in Fig. 4. The MWO data can now provide proxy polar field data for over a century of time. Muñoz Jaramillo et al. (2013) have used these facular-based polar field proxy data to study the cyclical interrelationship between the polar fields and the active fields (sunspot area) between 1906 and the present time. They found sunspot area to be uncorrelated with cycle amplitude unless multiplied by area-weighted average Joy's law bipole tilt (see Sect. 2). This suggests that the assimilation of both bipole tilt and sunspot area is a better approach (with regard to cycle prediction) than assimilating sunspot area data alone, supporting the Babcock-Leighton mechanism for polar field generation. (Petrie (2012) drew a similar conclusion in terms of products of active region flux and latitude displacements of positive and negative active region flux centroids—see Sect. 2.) Using a similar approach, they also found solar cycle memory to be limited to only one cycle.

3.2 17 GHz Radio Intensity

While magnetogram measurements of the polar fields suffer from large projection effects, microwave imaging observations at frequencies above 15 GHz provide an alternative means of studying polar regions because coronal holes appear bright at these frequencies, including polar holes. The Nobeyama radioheliograph has been producing daily full-disk brightness temperature maps at 17 GHz since 1992, covering cycles 22–24, with spatial resolution around 10″. Gopalswamy et al. (2012) used butterfly diagrams of microwave brightness temperature and magnetic field strength to show that the polar coronal hole field strength and polar microwave brightness temperature are well correlated, sharing the same north-south hemispheric asymmetry and the same diminishment during the cycle 23/24 minimum relative to the cycle 22/23 minimum. This extended an earlier result, where Gopalswamy et al. (2000) found good correlation between the peak microwave brightness temperature and peak field strength of 71 equatorial coronal holes. Figure 5 shows low-latitude brightness features corresponding to active regions, and polar enhancements resembling polar field strengths. The microwave measurements appear more sensitive to the polar coronal holes than to the active regions. The fact that these measurements show good qualitative agreement

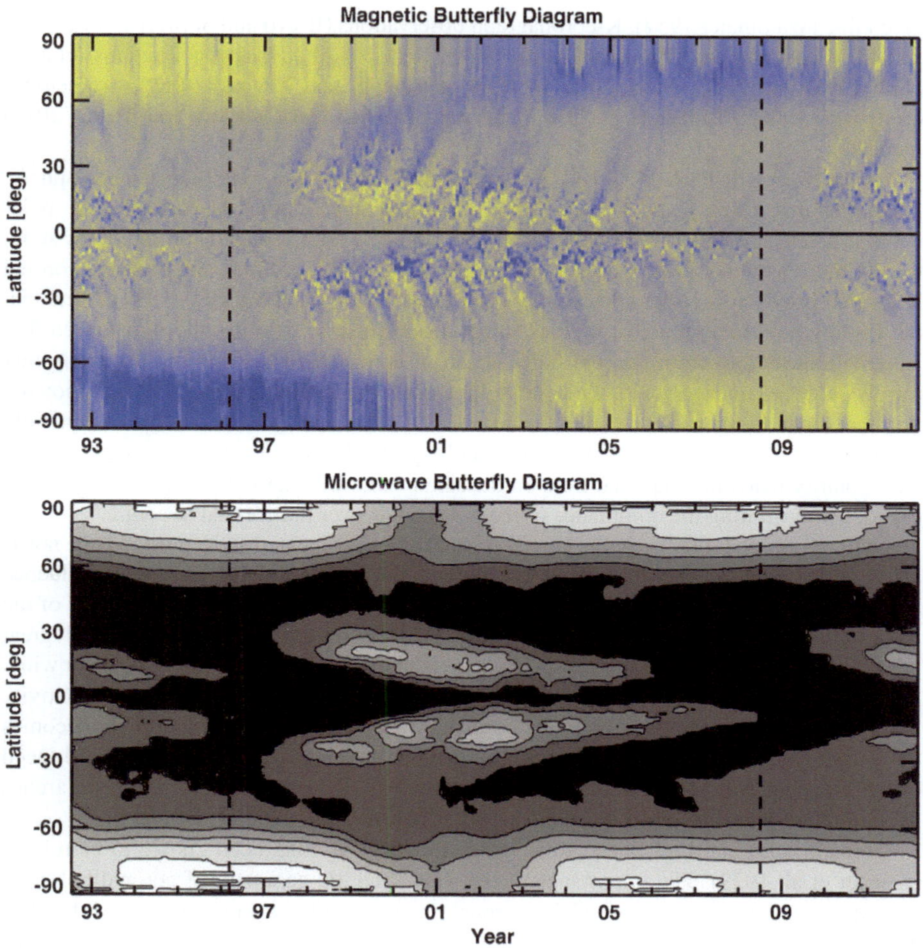

Fig. 5 *Top*: the magnetic butterfly diagram from 1992 to the present constructed from Kitt Peak National Observatory with SOHO/MDI data filling several data gaps. Blue and yellow shading denote positive and negative polarities, respectively. The magnetic field strength ranges from −10 G to +10 G. *Bottom*: a microwave butterfly diagram constructed from the Nobeyama radioheliograph images at 17 GHz. A 13-rotation smoothing was applied along the time axis to eliminate the periodic variation due to solar B_0-angle variation. The contour levels are at 10000, 10300, 10609, 10927, 11255, 11592, and 11940 K. *The vertical dashed lines* mark the ends of cycles 22 and 23 around 1996 March and 2008 July, respectively, determined from the low-latitude activity in microwaves. From Gopalswamy et al. (2012)

with the polar field measurements helps to validate these challenging measurements. We will discuss coronal hole observations in more detail next, in Sect. 3.3.

3.3 Coronal Holes

The existence of low-density regions, particularly prominent at the Sun's poles over most of the cycle, has been known for many decades (Waldmeier 1957) and it has become increasingly clear that these "coronal holes" are associated with open magnetic flux (Harvey 2013). Polar coronal holes have been studied using limb observations of the Fe XIV 5303 Å coronal green line (Waldmeier 1957), K-coronagraph observations (Bravo and Stewart 1994), X-ray and EUV observations (Broussard et al. 1978), He I 10830 Å spectroheliograms (Harvey et al. 1975) and PFSS (Schatten et al. 1969; Altschuler and Newkirk 1969), magnetohydrodynamics (e.g., Linker et al. 2011) and magnetofrictional (Mackay and Yeates 2012) models.

Using NSO Kitt Peak He I 10830 Å spectroheliograms and longitudinal photospheric magnetograms, Harvey and Recely (2002) identified three classes of coronal hole: polar coronal holes confined to high latitudes ($> 60°$ and $< -60°$), isolated coronal holes at active latitudes associated with the remnants of decaying active regions, and transient coronal holes that briefly form after coronal mass ejections. Harvey and Sheeley (1979) found that the leading polarities of active regions sometimes diffuse to form locally unbalanced flux patterns that tend to form coronal holes ahead of active regions while the following flux travels poleward to cancel with the polar field. During the declining phase it is the following flux that tends to open. Petrie and Haislmaier (2013) showed that coronal holes generally open without changing the global coronal topology: in 14 examples the hole always formed with polarity matching the polar hole on the side of the streamer belt where the region decayed, demonstrating the dominance of the polar fields over the coronal structure.

Over most of the solar cycle the largest coronal holes are located at the poles. While not as active as and much weaker than the active region fields, the polar fields have great influence in the heliosphere via the polar coronal holes. The large spatial scale and unipolarity of the polar fields prevent these fields from connecting back to the surface before they reach great heights, where the field is too weak to close against the pressure of the expanding solar wind. The existence of a super-Alfvénic solar wind, coupled with the high plasma conductivity leads to magnetic field openings in the corona. At these sites, the magnetic field becomes too weak to pull the magnetic field lines closed above the highest arches. The outward speed of the escaping wind exceeds the inward-directed Alfvén speed, and the magnetic arches open into interplanetary space at about one solar radius (Parker 1958).

Coronal holes and high-speed solar wind streams have long been known to be related (Krieger et al. 1973). The importance of the shape and 3D geometry of coronal holes for the expansion and acceleration of the coronal gas was explored in calculations related to de Laval nozzle theory in hydrodynamics by Wang and Sheeley (1990), whose work still forms the basis of most solar wind forecasting techniques today. The polar holes also supply the vast majority of heliospheric flux. According to Ulysses observations, the heliospheric field strength reached a record low during the cycle 23/24 minimum (Smith and Balogh 2008), reflecting the weakness of the polar fields since their cycle 23 reversal, as discussed in Sect. 2.

During the cycle 22/23 minimum the two polar coronal holes were large and dominant. Using SoHO/EIT Fe XII 195 Å EUV synoptic maps, de Toma (2011) estimated that the polar holes covered about 7.9–8.1 % of the solar surface in the north and about 6.9–7.1 % in the south, in good agreement with estimates by Harvey and Recely (2002) from He I 10830 Å data from Kitt Peak. Isolated low-latitude coronal holes were mostly absent, covering less than 0.5 % of the solar surface except during mid-1996 when a large extension from the northern polar hole developed, shaped like an elephant's trunk, and eventually detached itself from the polar hole (Zhao et al. 1999). Even over this period the low-latitude coronal area never exceeded 2 % of the solar surface. The cycle 23/24 minimum was quite different (Wang et al. 2009), characterized by smaller polar coronal holes and more low-latitude structure. From STEREO/SECCHI/EUVI Fe XII 195 Å EUV synoptic maps, de Toma (2011) estimated that the maximum areas of the polar coronal holes between 2006 and 2009 were 4.4–4.7 % of the solar surface in the north and 5.3–6.0 % in the south. Compared to the cycle

22/23 minimum this represents a 40–45 % decrease in the north and a 15–25 % decrease in the south. Meanwhile, the low-latitude holes covered 2–3 % of the Sun's surface during 2007 and 2008 and dropped to less than 1.5 % in 2009.

3.4 Polar Crown Filaments

Solar filaments (called prominences when observed at the limb) are long-lived, relatively cool and dense structures suspended by the magnetic field at heights of order 50 M above the solar surface (e.g., Mackay et al. 2010). Observations of filaments at high latitudes are, along with polar faculae, the longest available data set providing information on solar magnetism at high latitudes. Filaments and filament channels delineate neutral lines, so the overall structure of the magnetic field can be reconstructed from such data (McIntosh 1972). In particular, polar magnetic reversal times were determined by Makarov and Makarova (1996) from H_α synoptic maps back to 1870 (cycle 11).

The filaments provide no direct information on the distribution on the field strength in the unipolar areas between them. Yet, as higher order harmonic components of the solar magnetic field rapidly become unimportant with increasing distance, for an approximate determination of the interplanetary magnetic field structure it may suffice to assume a constant field strength of $|B| = 1$ G, with a sign depending on the polarity reconstructed from H_α maps. This approach was used by Obridko and Shelting (2008) to reconstruct the polar field strength at the source surface back to 1915.

Filaments are characterized as having dextral or sinistral chirality (handedness) according to the direction of the filament's axial field relative to the underlying photospheric polarity distribution. A significant preference for dextral prominences to form in the northern hemisphere and sinistral in the south has been observed at active and high latitudes (Leroy et al. 1983; Martin et al. 1994). In active regions the main polarity inversion lines (PILs) have chirality consistent with the overall helicity of the active region (Rust and Martin 1994), where dextral/sinistral chirality corresponds to negative/positive helicity, but the chirality outside active regions is determined by the complex interactions between active regions. Of particular interest to us are the "polar crown" filaments found at the high latitudes between active and polar latitudes. Differential rotation adds chirality of the "wrong" sign to neutral lines at high latitudes (Leroy 1978) and small-scale flux emergence and motions are not expected to add significant net chirality of either sign. Therefore it appears that chirality of the correct sign must be transported from active to high latitudes.

Using kinematic flux transport models for the photospheric field (such models are discussed in Sect. 5) based on NSO Kitt Peak photospheric longitudinal magnetograms, and magnetofrictionally relaxed non-potential equilibrium models for the coronal field, Yeates and Mackay (2012) modeled 15 years (1996–2011) of coronal field evolution. This simulation enabled the authors to study the formation of high-latitude filaments over an entire solar cycle. The results are shown in Fig. 6. The figure shows that the expected dominant chirality in each hemisphere (dextral in the north and sinistral in the south) was indeed dominant at active latitudes but that the chirality profile at high latitudes changed significantly with time. The expected dominant polarity was dominant at high latitudes during the ascent and the maximum of the activity cycle, including the "rush to the poles" and polar field reversal. Then during the declining phase the opposite chirality became dominant at high latitudes. Finally, late in the declining phase and throughout solar minimum the expected dominant polarity again became dominant at high latitudes. Yeates and Mackay (2012) explain this pattern in terms of a competition between flux transport from active latitudes, which tends to add chirality of the expected dominant chirality, and differential rotation, which tends to

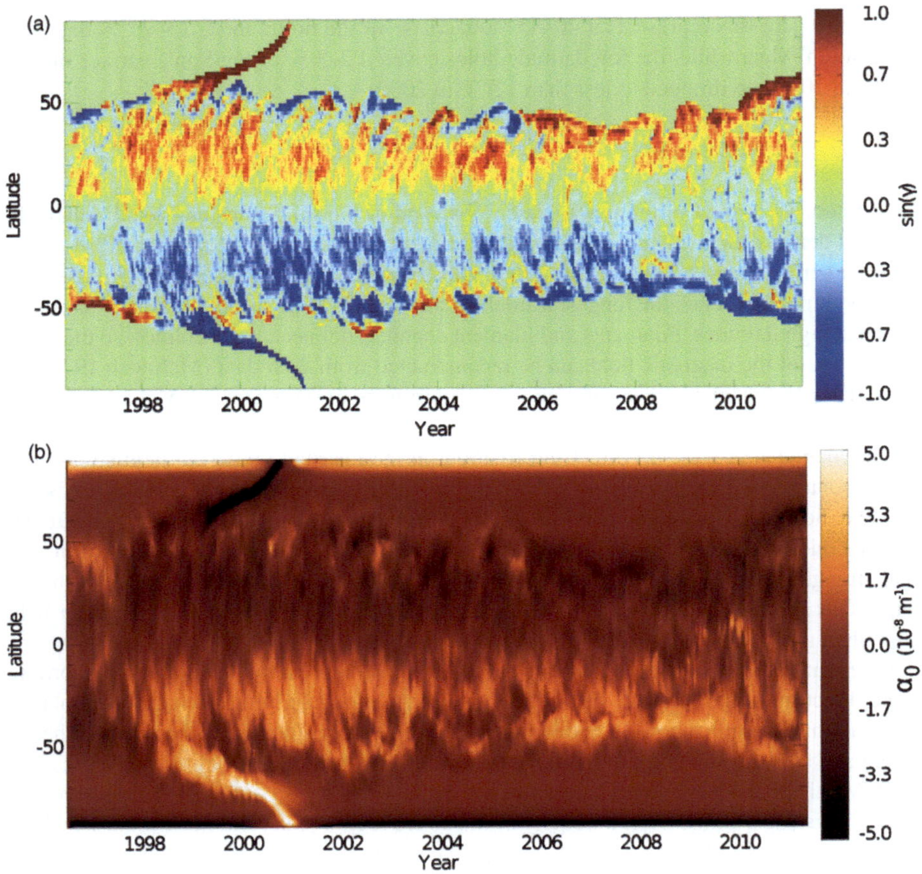

Fig. 6 Butterfly diagrams showing longitude-averages of (**a**) skew $\sin\gamma$, defined so that negative skew corresponds to sinistral chirality and positive skew to dextral chirality, and (**b**) current helicity density α_0, during the cycle 23 non-potential simulation. Both are measured at height 1.033 solar radii. From Yeates and Mackay (2012)

add chirality of opposite chirality. Flux transport is very influential and dominates the high-latitude chirality during the ascent and the maximum of the activity cycle. The poleward flux transport is too weak to overcome differential rotation as the cycle declines, and the average high-latitude chirality changes sign. Late in the cycle the highest-latitude PILs are closer to the equator, so that dominant chirality transported there from active latitudes can more easily dominate.

Gopalswamy et al. (2012) plotted the locations of prominence eruptions, as observed in Nobeyama radioheliograph full-disk microwave brightness observations at 17 GHz, over the microwave brightness temperature butterfly diagram plotted in Fig. 5. Over most of the solar cycle, eruptions are confined to latitudes between $\pm 60^\circ$. They occur at higher latitudes only during activity maximum, beginning when the polar fields begin to reverse and ending when the polar reversal is complete. Gopalswamy et al. (2012) also found a cycle-dependent relationship between the latitude offset of prominence eruptions compared to coronal mass ejections. There was a significant average positive offset (prominence eruptions were more poleward than coronal mass ejections) when the polar coronal holes were most prominent, i.e., during solar minimum. This average positive offset disappeared whenever significant activity appeared, apparently reducing the influence of the polar coronal holes on the eruptions.

4 The Extended Solar Cycle

4.1 Torsional Oscillations

Theories of magnetic cycles based on the phenomenological model of Babcock (1961) must balance the conversion of toroidal sunspot fields to poloidal polar fields. For a new activity cycle to be created, magnetic field must be generated by a dynamo process in the interior. Parker (1955), Babcock (1961) and others have put forward ideas based on spatial changes in the interior differential rotation process creating strong toroidal fields from weak poloidal fields. Consistent with this type of model, the torsional oscillation is a pattern of zonal flow bands, shown in Fig. 7, that migrate towards the equator and poles over a solar cycle, the

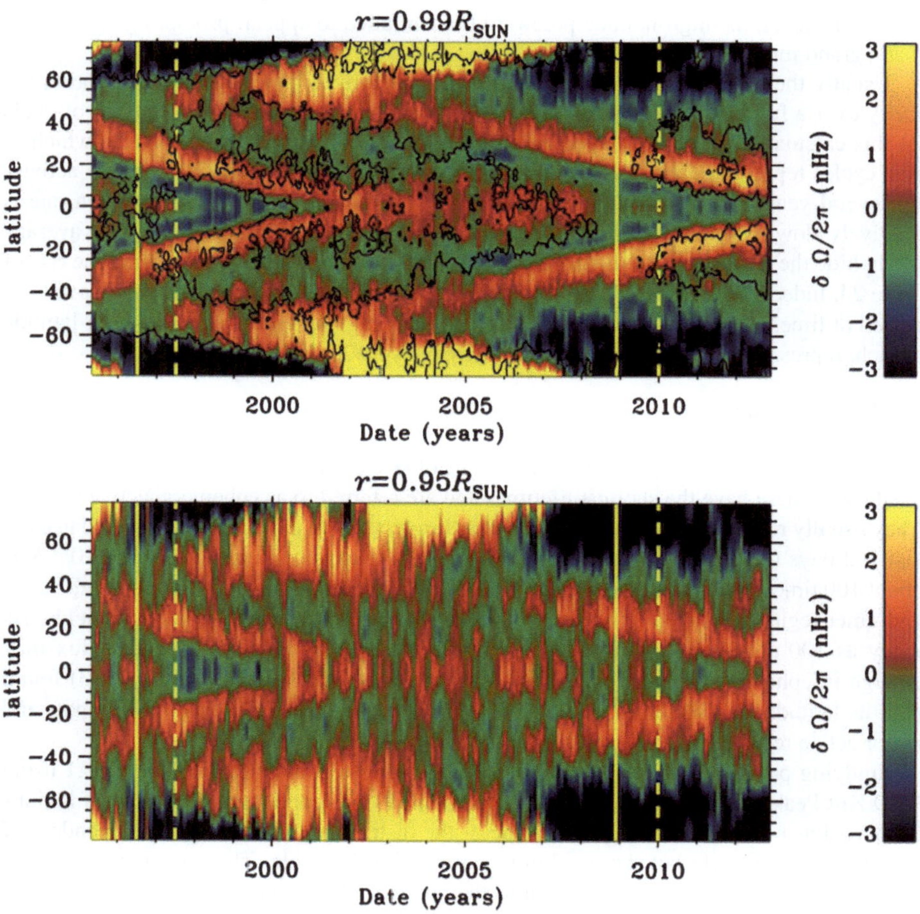

Fig. 7 Combined rotation-rate residuals from helioseismic inversions of GONG, MDI, and HMI Doppler data at selected depths. *The solid yellow vertical lines* mark the solar minima and *the dashed yellow lines* the onset of widespread activity at the start of the new cycle. *The top panel* also shows the 5, 25, and 45 G contours of unsigned magnetic field strength from Kitt Peak magnetograms. From Howe et al. (2013)

lower-latitude, equatorward-migrating band corresponding to the activity cycle observed on the surface. As the figure shows, the lower band begins years before the main magnetic cycle and, when it reaches latitudes around $\pm 25^\circ$, the magnetic activity appears at this latitude, marking the beginning of the new cycle. The lower torsional oscillation band and the activity then migrate equatorward together. The accompanying high-latitude, poleward-migrating branch corresponds to the "rush to the poles" discussed in previous sections. The torsional oscillation pattern was first observed in MWO surface Doppler flow data by Howard and LaBonte (1980) and was later found in continuous helioseismic interior flow observations from GONG and MDI to extend down through the convection zone. The polar part of the flow pattern is more obvious in the helioseismic observations because of the large projection angle in the surface measurements.

Using GONG, MDI and HMI data, Howe et al. (2011, 2013) showed that the torsional oscillation behaved strangely during the rising phase of cycle 24. The magnetic activity of cycle 24 was weaker and appeared later than expected, and the equatorward torsional oscillation branch migrated significantly more slowly than usual. But, as in past cycles, when this branch reached around 25° latitude the magnetic activity began to appear (Howe et al. 2011). However, the high-latitude poleward branch failed to appear, prompting speculation that a grand minimum was in progress.

Usually the torsional oscillation pattern is revealed by subtracting a mean rotation velocity over a full cycle at each latitude. However, as Fig. 7 shows, during the rise of cycle 24 this calculation did not produce the usual high-latitude branch of the pattern which in past cycles represented an early extension to the activity cycle that anticipated the activity by several years. Then Rempel (2012) demonstrated that the high-latitude rotation rate is relatively slow with relatively weak polar fields, so that subtraction of a long-term average might hide the appearance of the high-latitude torsional oscillation branch during the rise of cycle 24. Indeed, Howe et al. (2013) found that subtracting the mean velocity over a shorter period of time, a significantly different mean value at high latitudes, the usual high-latitude branch, representing an early extension of activity cycle 25, was revealed.

4.2 Ephemeral Active Regions

Bipolar active regions appear with a large range of sizes, magnetic fluxes and lifetimes. The smallest of these have the shortest lifetimes and are referred to as ephemeral active regions. They usually reach maximum development within less than a day of emergence and usually last 1–2 days (Harvey and Martin 1973). They have total maximum flux around 10^{20} Mx, about 100 times less than a major active region. They appear in large numbers in both active and quiet regions, distributed apparently independently of the existing network fields. As many as 100 ephemeral regions may emerge in a day, and as much magnetic flux may emerge in ephemeral regions as in large active regions. Harvey and Martin (1973) found that the latitude distribution of ephemeral regions was much broader than for sunspots and major active regions.

Studying patterns in the properties of bipolar active regions throughout cycle 21 using NSO Kitt Peak mangetograms, Harvey and Zwaan (1993) found that the number of regions decreased as a function of size. The shape of the distribution function was independent of the phase of the cycle but the amplitude varied with the cycle. The emergence rate was much higher within existing active regions than outside. The emergence rate of all size bins appeared to vary over the cycle in phase but the cycle variation was larger for larger regions than for smaller ones.

Hagenaar Hermance et al. (2003) used MDI magnetograms to study bipolar regions with a larger range of magnetic flux, from 5×10^{18} Mx to 10^{22} Mx, and found that the emergence

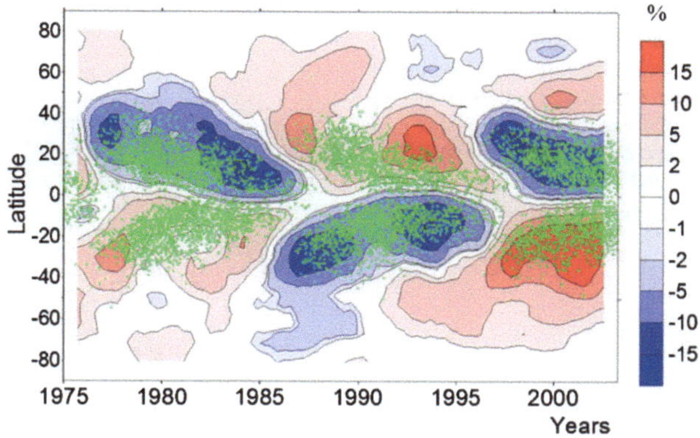

Fig. 8 Tilt preference in the distribution of bipoles with area 50–300 MSH (ephemeral regions) as a function of time using NSO Kitt Peak magnetograms. *Green dots* show the distribution of ARs. From Tlatov et al. (2010)

statistics of smaller regions evolved with much smaller cycle variations than the large regions, and in antiphase. It is therefore clear that the activity cycle of the bipole population as a whole must extend well beyond the usual activity cycle of active regions (Fig. 2) and sunspots.

Using Kitt Peak and MDI longitudinal magnetograms, Tlatov et al. (2010) studied the spatial tilt distributions of three different classes of bipole: in ascending size and lifetime, quiet-Sun bipoles (QSBs), ephemeral regions (ERs), and active regions (ARs). Both ERs and ARs followed the Hale-Nicholson polarity rule. While ARs tended to obey Joy's law, ERs showed the opposite preference on average, because of significantly larger tilts of opposite sense in each hemisphere. Figure 8 shows the ERs' tilt distribution in the form of a butterfly diagram. The ERs' orientation preference appeared to be related to the weaker field strength. Weaker flux tubes take more time to rise through the convection zone to the surface than stronger flux tubes (Fan and Fisher 1996), and the Coriolis force will therefore have time to have a greater effect on them, resulting in average tilt of opposite sense. This preference therefore appears to be a global phenomenon.

Figure 8 compares the latitude distributions of ERs and ARs. Both classes of bipole exhibit equatorward drift over the solar cycle. The ER distribution extends the AR butterfly wings to higher latitudes, representing an extended solar cycle such that the first ERs at high latitudes appear long before the first ARs appear at around $\pm 30°$ latitude, and only 2–3 years after the maximum and polar reversal of the previous cycle.

4.3 The Solar Cycle in the High Latitude Corona

Historically, the first claims for an extended solar cycle are due to Müller (1955) and Trellis (1957). These authors interpreted their observations of polar plumes and the coronal green line, respectively, as showing the early appearance of the main activity belts at high latitudes (up to 60°) up to 4 years before the start of the new sunspot cycle. This interpretation, however, ran against the views of the first observer of coronal activity at high latitudes, M. Waldmeier, who maintained the opinion that the high latitude "polar coronal zone" remains distinct from the low latitude "main zone" (Waldmeier 1964). In Waldmeier's interpretation this polar zone remains nearly stationary at a latitude of $\sim 60°$ during the decaying phase of a solar cycle, then, about a year before the start of the new cycle it starts to migrate towards the poles, arriving there shortly after the maximum (Fig. 9). Waldmeier's main argument against a connection between the early high latitude coronal activity and the main belt was that the polar filament crown seems to clearly separate both zones at all times.

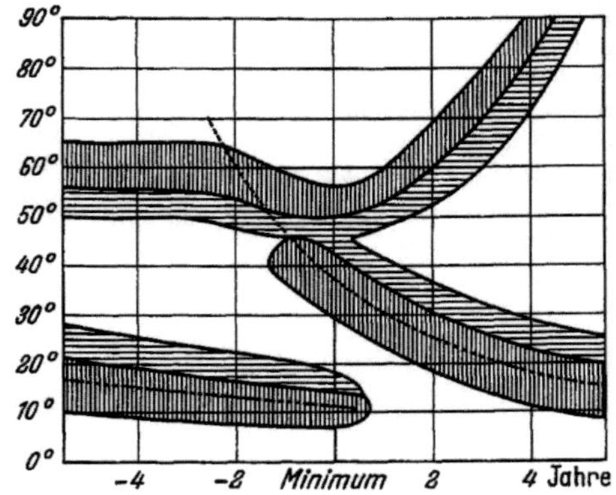

Fig. 9 Waldmeier's (1964) original sketch of the coronal butterfly diagram. *Vertical shading*: coronal activity belts. *Horizontal shading*: filament zones. *Dotted curve*: the main coronal activity zone according to Trellis (1957)

While Waldmeier's views apparently found more general acceptance at the time, the alternative view gained popularity in the 1980s, following the clear demonstration of a backwards extension of the main activity belts in the torsional oscillation pattern. The reintroduction of this idea for the coronal activity zone is due to Leroy and Noens (1983) who were also the first to plot the small scale variance of the coronal intensity rather than the intensity value itself—a method that significantly improves the visibility of the weak high latitude signal. Following this, it has become widely accepted that, along with torsional oscillation and ephemeral active regions (Sects. 4.1 and 4.2), the coronal activity at high latitudes also follows the extended cycle pattern. This gave rise to the widely held, though unproven assumption that the coronal emission is mainly due to loop structures above the small EAR bipoles. The poleward branch of the green line butterfly diagram, or as it came to be known, the "rush to the poles", in turn, remained unexplained.

With the recent work of Robbrecht et al. (2010) the pendulum is now swinging back again from the extended cycle paradigm to the polar coronal zone interpretation. In this work, the coronal emission is simulated on the basis of potential field models of the coronal magnetic field using the observed photospheric magnetograms as a boundary condition. The simulated green line synoptic maps are in good agreement with the observed ones demonstrating the viability of this approach. The polar activity zones in this model are shown to be situated above the foot points of the first closed loop structure outside the polar coronal holes, so their motion follows the expansion–contraction cycle of the polar coronal holes and they remain distinct from the main activity zone, as in Waldmeier's original suggestion.

While Robbrecht et al.'s modeling of the green line appears to be convincing at first sight, it still remains to be clarified why in some representations of the coronal butterfly diagram the equatorward part of the polar activity zone appears to be neatly connected to the main butterfly wings (e.g., Tappin and Altrock 2013; Altrock 2014). To clarify this, we subjected the observed green line intensity data taken from the homogenized data base compiled by Sykora and Parisi (1998) to the same unsharp masking procedure as used by Robbrecht et al. (2010), marking with colors the lowest contiguous contour lines enveloping 2 of the 3 branches (equatorial, polar equatorward and polar poleward) of the butterfly diagram. A comparison of this plot with Robbrecht et al.'s simulated plot for the minimum before cycle 22 is shown in Fig. 10. There is a clear, marked topological difference between the two plots: while the simulated emission is strongly suggestive of a polar "garland" of coronal emission neatly separated from the low latitude emission, observations indicate more organic connection between the high and low latitude equatorward branches, both being separated from the poleward branch by a deeper *graben*. One should note, however, that the

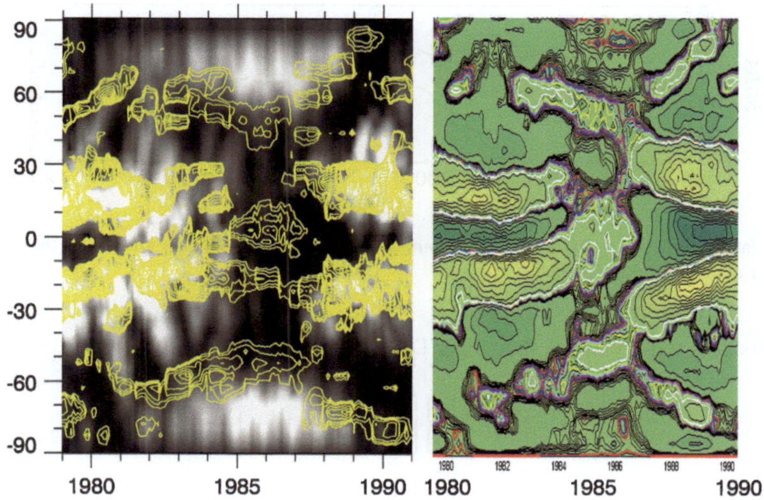

Fig. 10 Intensity of the coronal *green line* on the time–latitude plane after unsharp masking with a window of 25 degrees in latitude and smoothing with a window of 11 months in time for the period 1980–1990. *Left*: simulated data from Robbrecht et al. (2010) plotted over magnetic data. *Right*: observed intensities (figure courtesy of Melinda Nagy)

topology of the green line butterfly branches varies significantly from cycle to cycle. This problem is currently under study (Petrovay et al., in preparation).

4.4 High Latitude Activity as a Precursor of Sunspot Activity

Schatten et al. (1978) suggested the use of polar magnetic field strength around sunspot minimum as a precursor of the amplitude of the next activity cycle. Using various proxies including polar faculae and corona shape at eclipses, they predicted a value between ~ 120 and 180 for the maximal smoothed monthly SSN value in cycle 21. This forecast proved to be correct (the actual value being 165). Later, forecasts for subsequent cycles using this polar precursor method were also consistently in the right ballpark (Schatten and Sofia 1987; Schatten et al. 1996).

A shortcoming of the polar precursor method is that the prediction only becomes available at the time of solar minimum 3–4 years ahead of the maximum. In order to remedy this situation, Schatten and Pesnell (1993) suggested the use of a combined predictor along the following lines. Solar dynamo theories suggest that solar magnetism oscillates between poloidal (polar fields) and toroidal (active region) components—see Sects. 2 and 5. The poloidal and toroidal fields oscillate between maxima and minima, behaving like sine and cosine functions. These are roughly 90° out of phase, as the magnetic fields change in time, counting the 11 year cycle to be 180°. The identity $\sin^2 x + \cos^2 x = 1$ implies that for a pair of linear oscillators with a 90° phase lag, the sum of their squared amplitudes may be introduced as a phase independent variable subject to long term amplitude variations only. Adding the squares of the poloidal and toroidal fields together in this fashion, Schatten and Pesnell (1993) hoped to obtain a new index that varied minimally in time, and could capture the state of the Sun's dynamo at any cycle phase. Schatten and Pesnell (1993) suggested the

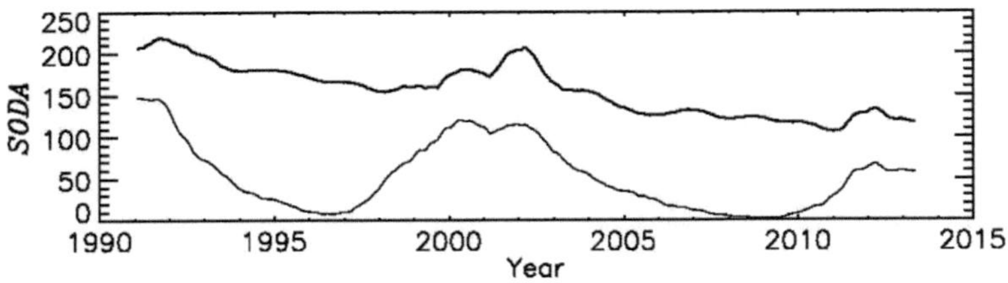

Fig. 11 Variation of the SoDA index (*thick line*) and of the sunspot number (*thin line*) the since 1990. 13-month running means are plotted

use of this new activity index, the "Solar Dynamo Amplitude" (SoDA) index:

$$SoDA(t) = 60 + 146\left\{\left[\frac{F_{10.7} - 60}{146}\right]^2 + \left[\frac{B_p^2}{128}\right]^2\right\}^{1/2}$$

where B_p is the polar magnetic field strength and the 10.7 cm radio flux $F_{10.7}$ is a measure of low latitude activity. The variation of the SoDA index in the past two decades is shown in Fig. 11 in comparison with the sunspot number curve.

Near the activity maxima some remnant modulation is still present, due to the fact that the cycle profile of $F_{10.7}$ is not close to sinusoidal. This difference in the behavior of high and low latitude activity indices may be understood in terms in terms of the flux transport processes discussed in Sect. 2. Polar field changes are related to high-latitude poleward surges, and the most influential surges tend to be quite rare, restricted to solar maximum, and of relatively short duration. For most of the cycle the polar field evolution is slow, and the polar field maxima are long and flat, interrupted only during solar maximum by large, relatively fast changes. In contrast, the sunspot cycle has sharp ascents to much more peaked maxima followed by slower declining phases and minima.

Nevertheless, as seen in Fig. 11, apart from the remnant modulation near the cycle peaks much of the variation with cycle phase is indeed filtered out from the SoDA index, leaving only longer term trends.

Around minimum, SoDA is clearly equivalent to the polar precursor and its value yields the prediction for $F_{10.7}$ at the next maximum. Its gradual change, however, makes it possible to use it for prediction purposes well before the actual minimum. E.g., despite its slow long term decrease the SoDA index has been in the right range to correctly predict the peak amplitude of cycle 24 (expected to peak around $F_{10.7} = 130$) since at least late 2004; or, in the case of cycle 23, a prediction within ± 10 % or so of the actual peak should have been possible as early as 1993.

5 Mechanisms of Latitudinal Transport

5.1 Modeling the Solar Cycle: Coupling Between Poloidal and Toroidal Fields

Over the course of this review, we have summarized observations of high-latitude magnetic flux transport via direct magnetic measurements and observations of proxies. These observations have revealed a coherent pattern of high-latitude flux transport in both directions: a "march to the poles" of surface and atmospheric phenomena such as photospheric magnetic flux and chromospheric filaments, and an equatorward migration of subsurface torsional oscillation, bipole emergence from the interior, and coronal emission forming an extended cycle. These phenomena, together with the success of polar precursor methods and the SoDA index in predicting activity cycle amplitudes from previous polar field strengths, are consistent with a global surface and sub-surface meridional flow pattern carrying decayed active-region fields poleward at the surface and transporting the fields equatorward beneath the surface, where spatial changes in the interior differential rotation create strong toroidal fields from weak poloidal fields at active latitudes.

In parallel with these observations, efforts to model the solar magnetic cycle kinematically have relied on meridional flows to control the behavior of the models. Kinematic flux transport models for the solar cycle are governed by the induction equation relating the fluid flow and the magnetic flux transport. In the axisymmetric case, the induction equation breaks down into two simple coupled equations, one for the vector potential of the poloidal field and the other for the toroidal field component—see, e.g., (Charbonneau 2010, Eqs. (11)–(12)). In the equation for the toroidal field, the conversion of poloidal field to toroidal field is clearly represented by a source term describing the shearing effect of interior differential rotation. However, the equation for the poloidal field has only an advection term, that can neither create nor destroy magnetic flux, and a diffusion term that can only destroy magnetic flux. Thus whatever an axisymmetric flow field does, the poloidal field must ultimately decay away, and the toroidal field must also decay as a consequence.

Therefore a non-axisymmetric process is needed to provide a poloidal field source. Parker (1955) introduced a poloidal field source term representing the effect of helical turbulent flows on buoyantly rising loops that twist in response to the Coriolis force, creating poloidal field components in previously toroidal loops. Such a process, commonly called the "alpha effect", and its associated poloidal field source term, enables dynamo models to produce cyclical behavior. The "Babcock-Leighton" mechanism produces the alpha effect in the models of Babcock (1961) and Leighton (1969), in converting tilted bipolar active region flux into polar flux by surface and near-surface flux transport.

Flux transport dynamo models have been the main theoretical tool for understanding the origins of the solar cycle and solar activity, and have produced a complex theoretical picture of how the solar dynamo may operate and how the solar convection zone and tachocline may behave. For simplicity we will focus here on Babcock–Leighton–type models, both data-driven and self-excited, where near-surface flows produce the alpha effect. The interested reader is referred to Charbonneau (2010) for a much more general discussion of solar dynamo models.

In both surface flux transport models and self-excited flux transport dynamo models the polar fields form from the decay of tilted bipolar active regions. Flux transport dynamo models are self-excited in that they produce new activity cycles from old cycles by transporting decayed photospheric active region flux poleward and transporting the same flux equatorward beneath the surface, amplifying it at active latitudes by differential rotation, whereupon it emerges to create the new cycle. Surface flux transport models, by contrast, use sunspot or magnetogram data as input for the new cycle. The photospheric fluid motions that break the active regions down and transport the flux poleward, diffusion, differential rotation and meridional flow, are essentially the same in both types of model. We will discuss both types of model in the following.

5.2 Meridional Flows in the Babcock-Leighton Model

In the initial phenomenological model of Babcock (1961), the process of active region flux dispersal remained obscure and, although there was speculation that systematic meridional circulation existed, such flows were not included. Leighton (1964) then explained active region flux dispersal as random walks of field elements caused by photospheric supergranular convection, and modeled it as linear diffusion (Leighton 1969). Then, evidence of real meridional flows was found in Doppler measurements by Duvall (1979) and LaBonte and Howard (1982), along with evidence of subsurface equatorward return flow in torsional oscillation patterns found by Howard and LaBonte (1980). These observations were followed by photospheric feature-tracking (Komm et al. 1993), Doppler (Ulrich 1993) and helioseismic (Hathaway et al. 1996) observations of poleward surface meridional flows of 20 ms^{-1}. Wang and Sheeley (1991) and Wang et al. (1991) incorporated a meridional flow in a flux-transport model similar to that of (Leighton 1969), and found that the cyclical behavior of the model became significantly more stable. Furthermore, Wang et al. (1991) showed that meridional circulation allows the equatorward migration of flux emergence to occur without requiring an artificial radial gradient in the interior differential rotation profile, of opposite sign to that found in later helioseismic observations, required by Leighton (1969). The polar fields are also believed to supply the seed field for the generation of the next activity cycle, being transported equatorward by the subsurface return meridional flow (Charbonneau 2010).

Flux transport models have also been developed that model the differing effects of photospheric flows on strong sunspot fields and on small flux elements. Worden and Harvey (2000) and Schrijver (2001) modeled the magnetic network by including the emergence of small-scale ephemeral regions in their flux-transport models. Schrijver (2001) supplemented the random-walk model with empirically-based rules for the interaction of small-scale magnetic elements, and Worden and Harvey (2000) replaced it with a random attractor model to simulate the accumulation of small-scale flux at network boundaries. Worden and Harvey (2000) and Schrijver (2001) modeled the observed effects of magneto-convective coupling that suppresses convection in regions of strong magnetic flux. Upton and Hathaway (2014) developed a flux-transport model that applies observed near-surface flow patterns, meridional flow, differential rotation and supergranular flows, to observed fields and used it to predict the cycle 24 polar field reversal.

Instead of solving smooth differential equations, Schatten (2013a, 2013b) modeled the transport of the large-scale fields on the Sun using a cellular automaton (CA) method, a discrete method of modeling complex, nonlinear processes in terms of simple entities, called automata. In this model, the automata represent discrete quantities of photospheric magnetic flux, and are identical, with two breeds carrying opposite-polarity flux, $\pm 10^{23}$ Mx per CA. The CA are represented in Fig. 12 by small colored arrows, red for negative flux and blue for positive flux. The different shades of color in the figure are for visual clarity only. An arrow's orientation shows the direction of motion. In this model the CA are introduced at active latitudes and in numbers proportional to the Sun's polar field strength at the previous solar minimum. They appear at progressively lower latitudes as the cycle progresses, consistent with magnetic butterfly diagrams (e.g., Fig. 2), and in bipolar configurations arranged according to Hale's and Joy's laws of sunspot polarity. The large-scale motions are governed by differential rotation and meridional circulation, and the random walk of magnetic elements is represented by a simple outward radial expansion from the leading and trailing flux concentrations of activity centers. A CA disappears when it comes within a small "kill distance" of a CA of opposite sign.

Figure 12 shows a sequence of spatial flux/flow distributions from the CA model. In the top left map, labeled $t = 9$, there is a positive/negative polar cap at the north/south pole, and five Hale bipolar active regions at high active latitudes (around $\pm 30°$). Some entities have had time to spread out and form flow patterns. At $t = 16$ the flow patterns are well developed and one can see negative-polarity fields forming a large-scale "march to the poles" pattern.

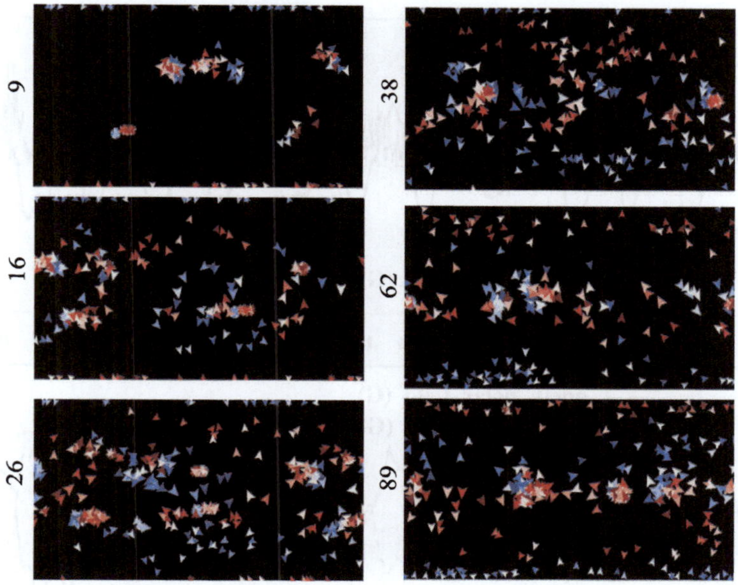

Fig. 12 Shown here are six plots of field distribution from the cellular automaton flux transport model of Schatten (2013a, 2013b). Each arrow represents a positive (*blue*) or negative (*red*) cellular automaton, carrying a fixed quantity of magnetic flux, $\pm 10^{23}$ Mx. Each map represents the full surface of the Sun in longitude-sine (latitude) coordinates, in Mercator projection. An arrow's orientation shows the direction the field is moving. *The arrows* are reduced in size at higher latitudes to illustrate the expansion of coordinates in Mercator projection. Each display is the equivalent of a synoptic longitude vs. latitude map on the Sun. The times shown are in program time units. New activity centers first appear at high latitudes in the top left panel. Proceeding downwards, as time passes, activity centers form at lower latitudes, as in the butterfly diagram in Fig. 2. One can see large-scale patterns, similar to unipolar magnetic regions (Bumba and Howard 1965), form as same-sign fields gather together. One can also observe field reversals at each pole

The fields march in near unison around the high latitudes, owing to the large differential rotation present there. By $t = 26$, much of the polar field in both hemispheres has been erased and the poles have new reversed-polarity flux. The activity emerges at lower and lower latitudes and the reversed-polarity polar flux continues to accumulate until the end of the sequence. This simple model is able to reproduce solar cycle phenomena (Schatten 2013a, 2013b), illustrating that the cycle is driven by the interactions of the active regions and the polar fields.

5.3 Cyclical Flux Transport and Meridional Flow Speeds

A major problem with reproducing cyclical behavior in surface flux transport models is that the polar fields built up during the most active cycles can be much too strong for quieter cycles to reverse. Wang et al. (2005) performed two simulations covering 26 solar cycles (1713–1996), one keeping the strengths of the bipoles constant and tying their number to the recorded sunspot number, and the other keeping the number of bipoles per cycle constant and tying their size to the sunspot number. The results of the first simulation are summarized in Fig. 13. The results of the second simulation differ in that the equatorial dipole varies as the sunspot number, whereas in the first simulation the equatorial dipole varies as the square root of the sunspot number, but the two simulations produced similar polar field behavior. The authors compared their simulated results for the near-Earth radial IMF strength to geomagnetic activity and cosmogenic isotope records. To maintain cyclical polar reversals the

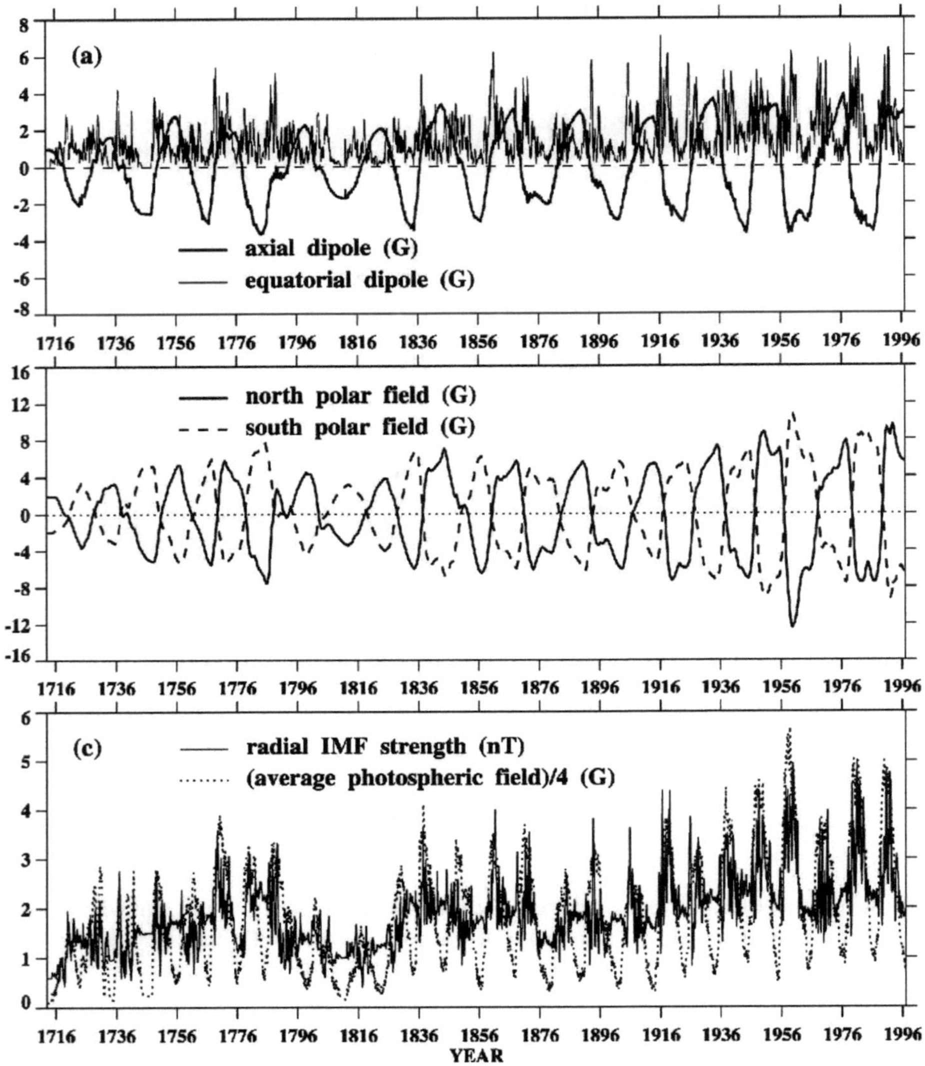

Fig. 13 Evolution of the large-scale solar magnetic field over 26 solar cycles. (**a**) Variation of the Sun's axial and equatorial dipole components. (**b**) Variation of the north and south polar fields, averaged over 30° polar caps. (**c**) Variation of the near-Earth radial IMF strength and of the total photospheric flux, expressed as a surface-averaged field strength in units of 4 G. From Wang et al. (2005)

meridional flow speed needed to be positively correlated with the cycle amplitude. This is because a fast meridional flow tends to sweep significant quantities of leading-polarity flux as well as trailing-polarity flux poleward, producing a small net effect on the polar fields even for cycles of large amplitudes. A slow meridional flow allows leading-polarity flux from the two hemispheres to cancel across the equator so that the net flux reaching polar latitudes is significant even during weak cycles.

When the cycle 23 polar reversal was observed to produce polar fields about 40 % weaker than the previous three cycles (Fig. 3), with numerous solar and heliospheric consequences (Wang et al. 2009), there were several efforts to model this phenomenon, and to explain it in terms of meridional flow speed changes (e.g., Schrijver and Liu 2008; Wang et al. 2009; Nandy et al. 2011; Jiang et al. 2011). Schrijver and Liu (2008) studied the effect on the axial dipole strength of varying the low-latitude gradient of the meridional flow speed. They adopted a standard meridional flow profile from the magnetic feature tracking measurements of Komm et al. (1993) and varied the gradient of the flow near the equator, effectively controlling the communication between the two hemispheres. They found a strong anti-correlation between the meridional flow gradient and the resulting axial dipole strength. Consistent with this work, and the past work of Wang et al. (2005), Wang et al. (2009) modeled the photospheric flux transport for cycles 20–23 and found that they could approximately match the polar field measurements of MWO and WSO over this period by imposing a faster peak meridional flow speed, 17 ms^{-1}, for cycle 23 than for the previous three cycles whose peak speeds were around 15 ms^{-1}.

Jiang et al. (2011) adopted a different approach, and studied the effects of varying different components of the flux-transport model independently. They found that either a 40 % reduction in the emergence rate, a 28 % decrease in the average tilt angle or a 55 % increase in the meridional flow speed for cycle 23 compared to cycles 21 and 22 resulted in agreement between the modeled and observed polar field strengths for cycle 23. However, the reduction in the emergence rate also resulted in a major reduction in the open flux for cycle 23 (derived from potential-field source-surface modeling), and the reduction in the mean bipole tilt produced a 1.5-yr delay in the polar field reversal, whereas the increased meridional flow produced better agreement with the observed open flux and polar reversal time. We therefore have three independent studies that all point to the same conclusion: that the likeliest cause of the weak polar fields of cycle 23 was an increase in the meridional flow speed for cycle 23.

The fact that Jiang et al. (2011) required a very large 55 % increase in the meridional flow speed, compared to the much more modest 10–15 % required by Wang et al. (2009), is due to the much smaller low-latitude gradient in the flow profile used by Jiang et al. (2011). Schrijver and Liu (2008) were able to control their polar fields by varying the low-latitude flow gradient alone. Indeed, the very large low-latitude gradient of the profile used by Wang et al. (2009) appears to be inconsistent with observed profiles (Ulrich 2010; Basu and Antia 2010; Hathaway and Rightmire 2010), and may have played an exaggerated role in their results.

5.4 The Effect of the Meridional Flow Profile Shape

Indeed, the shape of the meridional flow profile has a profound effect on the flux transport. Dikpati (2011) studied how the location of the peak flow speed influences the effect of meridional flow speed changes on the polar fields. When the meridional flow speed peaks at low latitudes, an increase in the overall speed produces a faster poleward transport of the leading polarity compared to the trailing polarity, enhancing the cancellation between them, resulting in weaker polar fields. This is what Schrijver and Liu (2008) and Wang et al. (2009) found from surface flux transport models. If, on the other hand, the flow peaks at low latitudes then an increase in the flow speed causes the leading polarity to speed up relative to the trailing polarity, inhibiting cancellation and producing stronger polar fields. Dikpati (2011) performed flux transport dynamo simulations, making a sudden change in the meridional flow speed from an initial peak speed of 18 ms^{-1} to a range of speeds between 6 and 32 ms^{-1}. She used two meridional flow profiles, one peaked at 25° latitude and the other at 50°. The results are shown in Fig. 14. For the flow profile with high-latitude peak, faster/slower flows produce stronger/weaker polar fields. The situation with the low-latitude flow peak is more complex: the polar fields weaken for flows both faster and slower than the

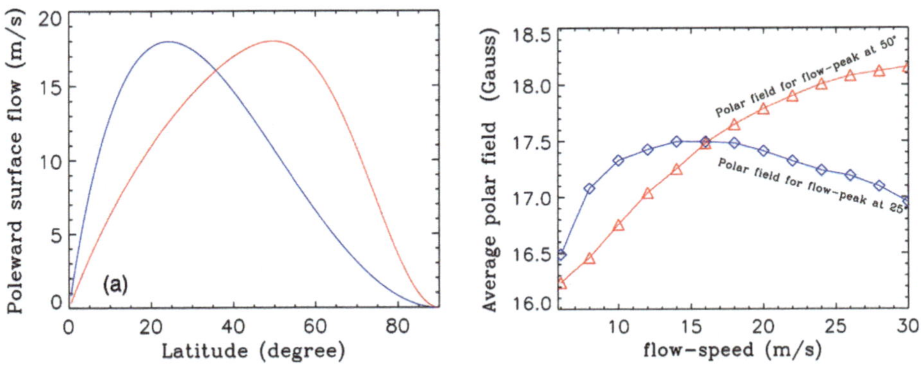

Fig. 14 (**a**) Typical profiles of the poleward surface flow in meridional circulation used in set of self-excited flux-transport dynamo simulations: *blue curve* from Dikpati (2011) with a peak near 25° and *red curve* with a peak near 50°. (**b**) Polar field amplitudes in the polar cap (latitudes from 55° to 90°) from model simulations, for the cycle immediately following a sudden change in the meridional flow speed from 18 ms^{-1} to value shown on *the horizontal axis*. 9 G has been added to *the red curve* so it can be compared to *the blue curve* on the same scale. The much lower actual polar fields in *the red curve* are probably caused by the longer transport time from the source in active latitudes to the poles, since the meridional flow peaks at a much higher latitude in this case. From Dikpati (2011)

initial value. All of these results are consistent with the simple discussion above, except that the polar field weakens when the high-latitude-peak flow profile becomes slower. Dikpati (2011) explained this exception in terms of the cycle becoming so long that enough cancellation takes place before the flux reaches the poles to produce weaker polar fields. But in all cases the polar field amplitude change is small, no more than 10 %. Dikpati (2011) therefore concluded that the 40 % reduction in polar field strength observed during cycle 23, shown in Fig. 3, could not have been caused by changes in meridional flow speed alone.

To affect the polar fields significantly, the meridional flows changes must occur at active latitudes, as Dikpati (2011) emphasized. The meridional flow speed measurements of Ulrich (2010), Basu and Antia (2010) and Hathaway and Rightmire (2010) did not show evidence of significantly faster flows at active latitudes during cycle 23 than during previous cycles, a fact that argues against the flow-related explanations offered by Schrijver and Liu (2008), Wang et al. (2009) and Jiang et al. (2011). Further discussion of the flux emergence patterns therefore seems to be appropriate.

5.5 Changes in Flux Emergence Patterns?

The amplitude of the polar fields is very sensitive to the details of the strength and decay of the active regions. To see this, suppose, after Dikpati (2011), that at the end of a given cycle the polar fields have strength one unit. Then it would take a change of minus two units to reverse the polar fields to the same strength. But if the active regions contribute just 20 % less flux than is necessary to do this in each hemisphere, then they will still reverse the polar fields but with a 40 % drop in the polar field strength compared to the initial strength, a decrease twice as large as the decrease in decayed active region flux (Dikpati 2011). On the other hand, because there are periods of time when there are active region fields on the Sun that produce no significant change in the polar field strength (compare Figs. 2 and 3 between, e.g., 2003–2006), the weakness of the cycle 23 polar fields cannot be explained by the weakness of the cycle 23 activity alone.

Recall that in the flux transport simulations of Jiang et al. (2011), the weak cycle 23 polar fields could also be reproduced by a 28 % decrease in the average tilt angle of sunspots, but this would also lead to a 1.5-year delay of the cycle 23 polar field reversal that was not observed. As we saw in Sect. 3.1, Muñoz Jaramillo et al. (2013) also stressed the importance of accounting for varying tilt angles in such calculations (but see McClintock and Norton (2013) on the difficulty and complexity of sunspot tilt measurements). Also Petrie (2012) found a possible explanation for polar field changes in patterns of the latitude displacement between positive and negative active region flux centroids in each hemisphere (see Sect. 2). In that study, the time interval when the polar field changed little even while much activity emerged and decayed, between about 2003 and the end of cycle 23, coincided with the period when the latitudinal centroid displacement between the positive and negative active region flux became insignificant in both hemispheres. This result suggests that the distribution of the emerged flux played a significant role in creating the weak polar fields of cycle 23. The question of explaining this change in latitude distribution of the positive and negative active region fluxes, perhaps in terms of the flux emerging from shallower depths, transported there by a shallower meridional flow cell, may be difficult to answer observationally.

5.6 North-South Asymmetry

A notable feature of Fig. 2 is the asymmetry of activity between the north and south hemispheres. For example, the south hemisphere was the more active during the decline of cycle 23 and the north during the ascent of cycle 24. Such hemispheric asymmetry is generally absent from flux transport models unless it is imposed by input data.

The symmetric flow profiles and anti-symmetric α-effect usually imposed on Babcock-Leighton dynamo models produces a bias in the models towards north-south antisymmetric field configurations. The induction equation in the standard Babcock-Leighton dynamo model is linear so that symmetric and antisymmetric modes remain uncoupled. DeRosa et al. (2012) illustrated this by running an axisymmetric dynamo model, initialized with an axial dipole field, and analyzing the results in terms of their harmonic order l. Here odd-l modes, such as the axial dipole, are antisymmetric, and even-l modes are symmetric. As expected, the odd-l modes dominated the even-l modes, which were not excited. But when a small asymmetry was introduced into either the Babcock-Leighton source term, representing the latitude distribution of active region emergence, or the meridional flow profile, then the model, initialized with an axial dipole, quickly produced symmetric modes comparable in size (35 %) to the antisymmetric modes. The development of an asymmetric field from an antisymmetric dipolar initial field and a weakly asymmetric meridional flow profile is shown in Fig. 15. These results indicate that even a weak asymmetry in the Sun's flux emergence or meridional flow could produced the observed magnetic asymmetries seen in, e.g., Fig. 2. Since neither meridional flows nor active region emergence can be expected to be exactly symmetric in the real Sun, it is not surprising that the solar field frequently shows significant north-south asymmetries.

6 Conclusion

Magnetograph observations of the polar and active fields have been presented supporting the phenomenological model of Babcock and the subsequent numerical development of the model by Leighton, showing in detail how the fates of the active and polar fields are cyclically intertwined, waxing and waning in antiphase. The observed interaction between the active and polar fields creates diverse magnetic phenomena that travel, both poleward ("march to the poles") and equatorward ("extended cycle"), across the high-latitude corridor between

Fig. 15 Time evolution of the coefficients of the lowest-order harmonic functions of the surface magnetic field (as grouped by primary and secondary families) from the same BL dynamo model as shown in Fig. 14. In panel (**a**) are shown the evolution of the first several primary-family coefficients B_0^l with $l = 1, 3, 5,$ and 7 in *black*, *blue*, *green*, and *red*, respectively. In panel (**b**) are shown the evolution of the secondary-family coefficients B_0^l with $l = 2, 4, 6,$ and 8, respectively, in *black*, *blue*, *green*, and *red*. Panels (**c**) and (**d**) show zoomed-in sections of panels (**a**) and (**b**). From DeRosa et al. (2012)

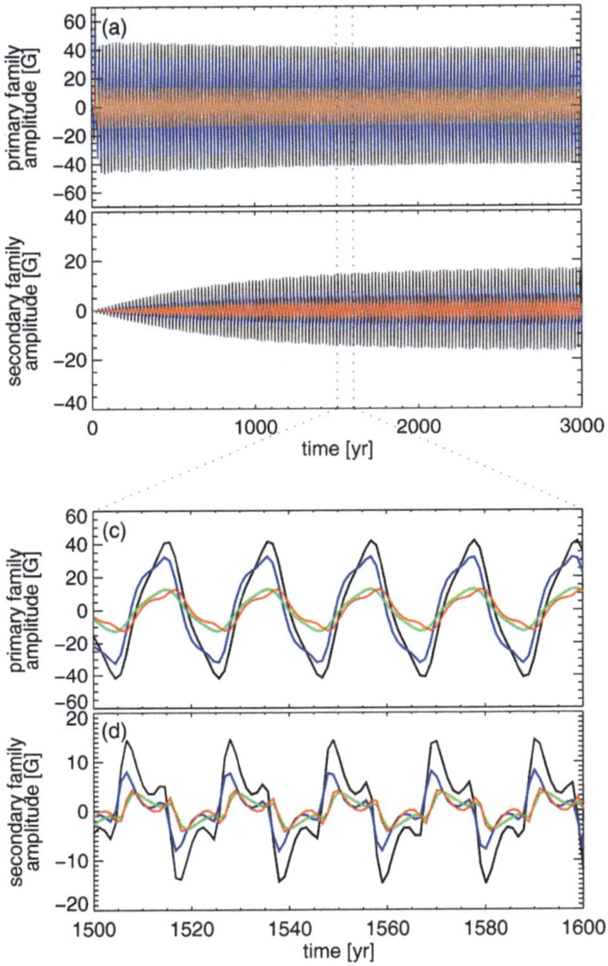

them. These phenomena are observed on, above and below the solar surface, and gathering these observations together affords a coherent picture of the role of these fields in the cycle.

The march to the poles of weak, decayed active region flux, filaments, coronal green line emission, and the high-latitude branch of the interior torsional oscillation pattern present coherent evidence from the solar surface, atmosphere and interior, of significant quantities of decayed active region flux being transported poleward. This flux merges with the polar field and changes its strength, generally weakening and reversing its polarity during the ascent and maximum of the cycle, and strengthening it during the declining phase. Carrying magnetic helicity from active latitudes and meeting flux of opposite polarity on its way, this poleward transport results in long neutral lines, and high-latitude filaments that ultimately erupt during polar field reversal.

Waldmeier's view that filaments separate the high-latitude coronal emission from active-latitude emission reflects the status of filaments as marking neutral lines between unipolar magnetic regions of opposite polarity, in particular between the decayed active-region flux, that carries and conserves magnetic helicity from active latitudes, from the polar flux of opposite polarity during the ascent of the cycle, or between unipolar decayed flux regions of opposite polarity. The flux cancellation that occurs at these neutral lines, and the accompanying high-latitude filament and prominence eruptions, are an integral part of the process of polar field reversal. These filaments may therefore be viewed as a necessary consequence of the communication between decaying active-region flux and the polar fields, and as an essential part of the solar cycle.

The extended cycle, observed at, above and below the solar surface, in photospheric ephemeral magnetic bipoles, in coronal green line variance and in torsional oscillations, seems to be a signature of the nascent activity cycle before it is capable of producing major active regions. The small-scale, sub-active-region magnetic activity is accompanied before each cycle by coronal green line emission patterns and subsurface flow patterns that resemble the full activity patterns themselves. The continuity of the extended cycle with the main cycle in Figs. 7 and 8 points to this interpretation. The question of whether the extended cycle connects more to the main cycle or the march to the poles phenomena in green line variance remains open. The evidence more strongly indicates connection with the main cycle, and this would match the patterns of the interior torsional oscillations and photospheric ephemeral bipoles.

The polar field measurements have been supported by consistent results from different magnetographs over multiple cycles and from proxy measurements in the form of polar faculae statistics and estimates based on polar filaments, as well as polar microwave brightness temperature measurements. These proxy data overcome the limitation of direct magnetograph measurements to only a few cycles and avoid the difficulties of projection effects. Relationships between proxy polar data and proxy data representing the activity cycle have produced results consistent with the phenomenological model of Babcock and have enabled successful solar cycle prediction. Predictions of this kind can only be made after solar minimum, only a few years in advance of the cycle maximum. This limitation is overcome by the SoDA index combining low-latitude activity and polar field strength in a slowly-evolving, phase-independent manner.

The observations point to magnetic flux transport in both directions at high latitudes: a march to the poles of surface and atmospheric phenomena including photospheric magnetic flux and chromospheric filaments, and an equatorward migration of extended solar cycle phenomena including small bipole emergence and coronal emission preceding the main cycle. These phenomena are consistent with meridional flow transporting magnetic flux poleward at the surface and then equatorward in the interior, before it is amplified at active latitudes by interior differential rotation to form a new cycle. The success of polar precursor methods for predicting cycle amplitudes further supports this interpretation.

The properties of meridional flow profiles have a defining influence on the cyclical behavior of kinematic flux transport models and flux transport dynamo models. The peak flow speed and the shape of the flow profile during a cycle both strongly influence the polar field strength and the amplitude of the following cycle. However, the weakness of the cycle 23 polar fields may not have been caused by a change in the surface meridional flow profile, but by a change in the latitude distribution of positive and negative active region flux in both hemispheres. The explanation for this change remains to be determined.

Acknowledgements The authors are grateful to A. Balogh for the opportunity to participate in the ISSI Workshop on "The Solar Activity Cycle: Physical Causes and Consequences", held in November 2013. KP acknowledges support from the Hungarian Science Research Fund (OTKA grants no. K83133 and 81421). KS appreciates the support of GSFC-NASA's code 500/590 involving Mission Operations, and a.i.-solutions, inc., as well as valuable discussions with Hans Mayr.

References

R.C. Altrock, Forecasting the maxima of solar cycle 24 with coronal Fe xiv emission. Sol. Phys. **289**, 623 (2014)

M.D. Altschuler, G. Newkirk, Magnetic fields and the structure of the solar corona. I: Methods of calculating coronal fields. Sol. Phys. **9**, 131 (1969)

H.D. Babcock, The Sun's polar magnetic field. Astrophys. J. **130**, 364 (1959)

H.W. Babcock, The topology of the Sun's magnetic field and the 22-year cycle. Astrophys. J. **133**, 572 (1961)

S. Basu, H.M. Antia, Characteristics of solar meridional flows during solar cycle 23. Astrophys. J. **717**, 488 (2010)

S. Bravo, G. Stewart, Evolution of polar coronal holes and sunspots during cycles 21 and 22. Sol. Phys. **154**, 377 (1994)

R.M. Broussard, R. Tousey, J.H. Underwood, N.R. Sheeley Jr., A survey of coronal holes and their solar wind associations throughout sunspot cycle 20. Sol. Phys. **56**, 161 (1978)

V. Bumba, R.F. Howard, Large-scale distribution of solar magnetic fields. Astrophys. J. **141**, 1492 (1965)

V. Bumba, R.F. Howard, Solar activity and recurrences in magnetic-field distribution. Sol. Phys. **7**, 28 (1969)

P. Charbonneau, Dynamo models of the solar cycle. Living Rev. Sol. Phys. **7**, 3 (2010)

G. de Toma, Evolution of coronal holes and implications for high-speed solar wind during the minimum between cycles 23 and 24. Sol. Phys. **274**, 195 (2011)

M.L. DeRosa, A.S. Brun, J.T. Hoeksema, Solar magnetic field reversals and the role of dynamo families. Astrophys. J. **757**, 96 (2012)

M. Dikpati, Polar field puzzle: solutions from flux-transport dynamo and surface-transport models. Astrophys. J. **733**, 90 (2011)

T.L. Duvall, Large-scale solar velocity fields. Sol. Phys. **63**, 3 (1979)

Y. Fan, G.H. Fisher, Radiative heating and the Buoyant rise of magnetic flux tubes in the solar interior. Sol. Phys. **166**, 17 (1996)

V. Gaizauskas, K.L. Harvey, J.W. Harvey, C. Zwaan, Large-scale patterns formed by solar active regions during the ascending phase of cycle 21. Astrophys. J. **265**, 1056 (1983)

N. Gopalswamy, K. Shibasaki, M. Salem, Microwave enhancement in coronal holes: statistical properties. J. Astrophys. Astron. **21**, 413 (2000)

N. Gopalswamy, S. Yashiro, P. Makala, G. Michalek, K. Shibasaki, D.H. Hathaway, Behavior of solar cycles 23 and 24 revealed by microwave observations. Astrophys. J. **750**, L42 (2012)

S. Gosain, A.A. Pevtsov, G.V. Rudenko, S.A. Anfinogentov, First synoptic maps of photospheric vector magnetic field from SOLIS/VSM: non-radial magnetic fields and hemispheric pattern of helicity. Astrophys. J. **772**, 52 (2013)

J. Hagenaar Hermance, C.J. Schrijver, A.M. Title, The properties of small magnetic regions on the solar surface and the implications for the solar dynamo(s). Astrophys. J. **584**, 1107 (2003)

G.E. Hale, Invisible sun-spots. Mon. Not. R. Astron. Soc. **82**, 168 (1920)

G.E. Hale, F. Ellerman, S.B. Nicholson, A.H. Joy, The magnetic polarity of sun-spots. Astrophys. J. **49**, 153 (1919)

G.E. Hale, S.B. Nicholson, The law of sun-spot polarity. Astrophys. J. **62**, 270 (1925)

J.W. Harvey, The Sun in time. Space Sci. Rev. **176**, 47 (2013)

J.W. Harvey, A.S. Krieger, J.M. Davis, A.F. Timothy, G.S. Vaiana, Comparison of Skylab X-Ray and ground-based Helium observations. Bull. Am. Astron. Soc. **7**, 358 (1975)

J.W. Harvey, N.R. Sheeley Jr., Coronal holes and solar magnetic fields. Space Sci. Rev. **23**, 139 (1979)

K.L. Harvey, S.F. Martin, Ephemeral active regions. Sol. Phys. **32**, 389 (1973)

K.L. Harvey, F. Recely, Polar coronal holes during cycles 22 and 23. Sol. Phys. **211**, 31 (2002)

K.L. Harvey, C. Zwaan, Properties and emergence of bipolar active regions. Sol. Phys. **148**, 85 (1993)

D.H. Hathaway, P.A. Gilman, J.W. Harvey, F. Hill, R.F. Howard, H.P. Jones, J.C. Kasher, J.W. Leibacher, J.A. Pintar, G.W. Simon, GONG observations of solar surface flows. Science **272**, 1306 (1996)

D.H. Hathaway, L. Rightmire, Variations in the Sun's meridional flow over a solar cycle. Science **327**, 1350 (2010)

J.T. Hoeksema, Structure and evolution of the large-scale solar and heliospheric magnetic fields. Ph.D. Thesis, Stanford University (1984)

J.T. Hoeksema, Evolution of the large-scale magnetic field over three solar cycles, in *Solar and Stellar Variability: Impact on Earth and Planets, Proceedings of the International Astronomical Union*. IAU Symposium, vol. 264, (2010), p. 222

R. Howard, B.J. LaBonte, The sun is observed to be a torsional oscillator with a period of 11 years. Sol. Phys. **239**, 33 (1980)

R. Howe, D. Baker, L. Harra, L. van Driel-Gesztelyi, R. Komm, F. Hill, I. González Hernńdez, Magnetic polarity streams and subsurface flows, in *Fifty Years of Seismology of the Sun and Stars*, ed. by K. Jain,

S.C. Tripathy, F. Hill, J.W. Leibacher, A.A. Pevtsov. ASP Conference Proceedings, vol. 478, (2013), p. 291
R. Howe, F. Hill, R. Komm, J. Christensen-Dalsgaard, T.P. Larson, J. Schou, M.J. Thompson, R. Ulrich, The torsional oscillation and the new solar cycle. J. Phys. Conf. Ser. **271**, 012074 (2011)
R. Howe, J. Christensen-Dalsgaard, F. Hill, R. Komm, T.P. Larson, M. Rempel, J. Schou, M.J. Thompson, The high-latitude branch of the solar torsional oscillation in the rising phase of cycle 24. Astrophys. J. Lett. **767**, 20 (2013)
J. Jiang, R. Cameron, D. Schmitt, M. Schüssler, Can surface flux transport account for the weak polar field in cycle 23? Space Sci. Rev. **176**, 289 (2011)
C.L. Jin, J.W. Harvey, A. Pietarila, Synoptic mapping of chromospheric magnetic flux. Astrophys. J. **765**, 79 (2013)
R.W. Komm, R.F. Howard, J.W. Harvey, Meridional flow of small photospheric magnetic features. Sol. Phys. **147**, 207 (1993)
A.S. Krieger, A.F. Timothy, E.C. Roelof, A coronal hole and its identification as the source of a high velocity solar wind stream. Sol. Phys. **29**, 505 (1973)
B.J. LaBonte, R. Howard, Solar rotation measurements at Mount Wilson. III. Meridional flow and limbshift. Sol. Phys. **80**, 361 (1982)
R.B. Leighton, Transport of magnetic fields on the Sun. Astrophys. J. **140**, 1547 (1964)
R.B. Leighton, A magneto-kinematic model of the solar cycle. Astrophys. J. **156**, 1 (1969)
J.L. Leroy, On the orientation of magnetic fields in quiescent prominences. Astron. Astrophys. **64**, 247 (1978)
J.L. Leroy, V. Bommier, S. Sahal-Brechot, The magnetic field in the prominences of the polar crown. Sol. Phys. **83**, 135 (1983)
J.-L. Leroy, J.-C. Noens, Does the solar activity cycle extend over more than an 11-year period? Astron. Astrophys. **120**, L1 (1983)
J.A. Linker, R. Lionello, Z. Mikić, V.S. Titov, S.K. Antiochos, The evolution of open magnetic flux driven by photospheric dynamics. Astrophys. J. **731**, 110 (2011)
D.H. Mackay, J.T. Karpen, J.L. Ballester, B. Schmieder, G. Aulanier, Physics of solar prominences. II. Magnetic structure and dynamics. Space Sci. Rev. **151**, 333 (2010)
D. Mackay, A. Yeates, The Sun's global photospheric and coronal magnetic fields: observations and models. Living Rev. Sol. Phys. **9**, 6 (2012)
V.I. Makarov, V.V. Makarova, Polar faculae and sunspot cycles. Sol. Phys. **163**, 267 (1996)
S.F. Martin, R. Bilimoria, P.W. Tracadas, Magnetic field configurations basic to filament channels and filaments, in *Solar Surface Magnetism*, ed. by R.J. Rutten, C.J. Schrijver (Kluwer Academic, Dordrecht, 1994), p. 303
E.W. Maunder, Sun, place of the, distribution of sun-spots in heliographic latitude, 1874–1913. Mon. Not. R. Astron. Soc. **74**, 112 (1913)
B.H. McClintock, A.A. Norton, Recovering Joy's law as a function of solar cycle, hemisphere, and longitude. Sol. Phys. **287**, 215 (2013)
P.S. McIntosh, Solar magnetic fields derived from hydrogen alpha filtergrams. Rev. Geophys. Space Phys. **10**, 837 (1972)
R. Müller, Zur Statistik der Koronastrahlen. Mit 2 Textabbildungen. Z. Astrophys. **38**, 212 (1955)
A. Muñoz-Jaramillo, M. Dasi-Espuig, L.A. Balmaceda, E.E. DeLuca, Solar cycle propagation, memory, and prediction: insights from a century of magnetic proxies. Astrophys. J. Lett. **767**, L25 (2013)
A. Muñoz-Jaramillo, N.R. Sheeley Jr., J. Zhang, E.E. DeLuca, Calibrating 100 years of polar faculae measurements: implications for the evolution of the heliospheric magnetic field. Astrophys. J. **753**, 146 (2012)
D. Nandy, A. Muñoz-Jaramillo, P.C.H. Martens, The unusual minimum of sunspot cycle 23 caused by meridional plasma flow variations. Nature **471**, 80 (2011)
V.N. Obridko, B.D. Shelting, On prediction of the strength of the 11-year solar cycle no. 24. Sol. Phys. **248**, 191 (2008)
E.N. Parker, The formation of sunspots from the solar toroidal field. Astrophys. J. **121**, 491 (1955)
E.N. Parker, Dynamics of the interplanetary gas and magnetic fields. Astrophys. J. **128**, 664 (1958)
G.J.D. Petrie, Evolution of active and polar photospheric magnetic fields during the rise of cycle 24 compared to previous cycles. Sol. Phys. **281**, 577 (2012)
G.J.D. Petrie, Solar magnetic activity cycles, coronal potential field models and eruption rates. Astrophys. J. **768**, 162 (2013)
G.J.D. Petrie, K.J. Haislmaier, Low-latitude coronal holes, decaying active regions, and global coronal magnetic structure. Astrophys. J. **775**, 100 (2013)
G.J.D. Petrie, I. Patrikeeva, A comparative study of magnetic fields in the solar photosphere and chromosphere at equatorial and polar latitudes. Astrophys. J. **699**, 871 (2009)

N.-E. Raouafi, J.W. Harvey, C.J. Henney, Latitude distribution of polar magnetic flux in the chromosphere near solar minimum. Astrophys. J. **669**, 636 (2007)
M. Rempel, High-latitude solar torsional oscillations during phases of changing magnetic cycle amplitude. Astrophys. J. Lett. **750**, 8 (2012)
E. Robbrecht, Y.-M. Wang, N.R. Sheeley Jr., N.B. Rich, On the "extended" solar cycle in coronal emission. Astrophys. J. **716**, 693 (2010)
D.M. Rust, S.F. Martin, A correlation between sunspot whirls and filament type, in *Solar Active Region Evolution: Comparing Models with Observations*, ed. by K.S. Balasubramaniam, W. George. ASP Conf. Ser., vol. 68 (Simon, San Francisco, 1994), p. 337
K.H. Schatten, Large-scale solar magnetic field mapping: I. SpringerPlus **2**, 21 (2013a). http://download.springer.com/static/pdf/412/art (Open access)
K.H. Schatten, Solar field mapping and dynamo behavior (2013b). http://www.hindawi.com/journals/aa/2012/923578/
K. Schatten, D.J. Myers, S. Sofia, Solar activity forecast for solar cycle 23. Geophys. Res. Lett. **23**, 605 (1996)
K.H. Schatten, W.D. Pesnell, An early solar dynamo prediction: cycle 23 is approximately cycle 22. Geophys. Res. Lett. **20**, 2275 (1993)
K.H. Schatten, P.H. Scherrer, L. Svalgaard, J.M. Wilcox, Using dynamo theory to predict the sunspot number during solar cycle 21. Geophys. Res. Lett. **5**, 411 (1978)
K.H. Schatten, S. Sofia, Forecast of an exceptionally large even-numbered solar cycle. Geophys. Res. Lett. **14**, 632 (1987)
K.H. Schatten, J.M. Wilcox, N.F. Ness, A model of interplanetary and coronal magnetic fields. Sol. Phys. **6**, 442 (1969)
C.J. Schrijver, Simulations of the photospheric magnetic activity and outer atmospheric radiative losses of cool stars based on characteristics of the solar magnetic field. Astrophys. J. **547**, 475 (2001)
C.J. Schrijver, Y. Liu, The global solar magnetic field through a full sunspot cycle: observations and model results. Sol. Phys. **252**, 19 (2008)
C.J. Schrijver, C. Zwaan, *Solar and Stellar Magnetic Activity* (Cambridge University Press, New York, 2000)
N.R. Sheeley Jr., A century of polar faculae variations. Astrophys. J. **374**, 386 (1991)
N.R. Sheeley Jr., Polar faculae—1906–1990. Astrophys. J. **680**, 1553 (2008)
E.J. Smith, A. Balogh, Decrease in heliospheric magnetic flux in this solar minimum: recent Ulysses magnetic field observations. Geophys. Res. Lett. **35**, L22103 (2008)
X. Sun, Y. Liu, J.T. Hoeksema, K. Hayashi, X. Zhao, A new method for polar field interpolation. Sol. Phys. **270**, 9 (2011)
L. Svalgaard, T.L. Duvall, P.H. Scherrer, The strength of the Sun's polar fields. Sol. Phys. **58**, 225 (1978)
J. Sykora, M. Parisi, A new database of the green-line corona brightness as compiled for the last five solar cycles and its possible utilization in the ISCS project. Astron. Astrophys. Trans. **16**, 75 (1998)
S.J. Tappin, R.C. Altrock, The extended solar cycle tracked high into the corona. Sol. Phys. **282**, 249 (2013)
A.G. Tlatov, V.V. Vasil'eva, A.A. Pevtsov, Distribution of magnetic bipoles on the sun over three solar cycles. Astrophys. J. **717**, 357 (2010)
M. Trellis, Contribution a L'Étude de la couronne solaire. Suppl. Aux. Ann. Astrophys. **5**, 3 (1957)
S. Tsuneta, K. Ichimoto, Y. Katsukawa, B.W. Lites, K. Matsuzaki, S. Nagata et al., The magnetic landscape of the Sun's polar region. Astrophys. J. **688**, 1374 (2008)
R.K. Ulrich, The controversial sun, in *Inside the Stars*, ed. by W.W. Weiss, A. Baglin. IAU Colloq., vol. 137, (1993), p. 25
R.K. Ulrich, Solar meridional circulation from Doppler shifts of the Fe I line at 5250 Å as measured by the 150-foot solar tower telescope at the Mt. Wilson observatory. Astrophys. J. **725**, 658 (2010)
R.K. Ulrich, T. Tran, The global solar magnetic field—identification of traveling, long-lived ripples. Astrophys. J. **768**, 189 (2013)
L. Upton, D.H. Hathaway, Predicting the Sun's polar magnetic fields with a surface flux transport model. Astrophys. J. **780**, 5 (2014)
M. Waldmeier, *Die Sonnenkorona*, 2nd edn. (Birkhäuser, Basel, 1957)
M. Waldmeier, Das Verhalten der koronalen Polarzone. Mit 4 Textabbildungen. Z. Astrophys. **59**, 205 (1964)
Y.-M. Wang, J.L. Lean, N.R. Sheeley Jr., Modeling the Sun's magnetic field and irradiance since 1713. Astrophys. J. **625**, 522 (2005)
Y.-M. Wang, E. Robbrecht, N.R. Sheeley Jr., On the weakening of the polar magnetic fields during solar cycle 23. Astrophys. J. **707**, 1372 (2009)
Y.-M. Wang, N.R. Sheeley Jr., The solar origin of long-term variations of the interplanetary magnetic field strength. J. Geophys. Res. **93**, 11227 (1988)
Y.-M. Wang, N.R. Sheeley Jr., Solar wind speed and coronal flux-tube expansion. Astrophys. J. **355**, 726 (1990)

Y.-M. Wang, N.R. Sheeley Jr., Magnetic flux transport and the Sun's dipole moment—new twists to the Babcock–Leighton model. Astrophys. J. **375**, 761 (1991)

Y.-M. Wang, N.R. Sheeley Jr., On potential field models of the solar corona. Astrophys. J. **392**, 310 (1992)

Y.-M. Wang, N.R. Sheeley Jr., A.G. Nash, A new solar cycle model including meridional circulation. Astrophys. J. **383**, 431 (1991)

J. Worden, J. Harvey, An evolving synoptic magnetic flux map and implications for the distribution of photospheric magnetic flux. Sol. Phys. **195**, 247 (2000)

A.R. Yeates, D.H. Mackay, Chirality of high-latitude filaments over solar cycle 23. Astrophys. J. **753**, L34 (2012)

X.P. Zhao, J.T. Hoeksema, P.H. Scherrer, Changes of the boot-shaped coronal hole boundary during Whole Sun Month near sunspot minimum. J. Geophys. Res. **104**, 9735 (1999)

DOI 10.1007/978-1-4939-2584-1_12
Reprinted from *Space Science Reviews* Journal, DOI 10.1007/s11214-014-0068-0

A Combined Analysis of the Observational Aspects of the Quasi-biennial Oscillation in Solar Magnetic Activity

G. Bazilevskaya · A.-M. Broomhall · Y. Elsworth · V.M. Nakariakov

Received: 31 March 2014 / Accepted: 26 June 2014 / Published online: 15 July 2014

Abstract Solar quasi-biennial oscillations (QBOs) with the time scale of 0.6–4 yrs appear to be a basic feature of the Sun's activity. Observational aspects of QBOs are reviewed on the basis of recent publications. Solar QBOs are shown to be ubiquitous and very variable. We demonstrate that many features of QBOs are common to different observations. These features include variable periodicity and intermittence with signs of stochastisity, a presence at all levels of the solar atmosphere and even in the convective zone, independent development in the northern and southern solar hemispheres, most pronounced amplitudes during the maximum phase of the 11-yr cycle and the transition of QBOs into interplanetary space. Temporal weakening of solar activity around the maximum of the 11-yr cycle (Gnevyshev Gap) can be considered an integral part of QBOs. The exact mechanism by which the solar QBO is produced is poorly understood. We describe some of the most plausible theoretical mechanisms and discuss observational features that support/contradict the theory. QBOs

G. Bazilevskaya
Lebedev Physical Institute, Russian Academy of Sciences, Leninsky prospect, 53, Moscow, 119991, Russia
e-mail: gbaz@rambler.ru

A.-M. Broomhall (✉)
Institute of Advanced Studies, University of Warwick, Coventry, CV4 7HS, UK
e-mail: a-m.broomhall@warwick.ac.uk

A.-M. Broomhall · V.M. Nakariakov
Centre for Fusion, Space, and Astrophysics, Department of Physics, University of Warwick, Coventry, CV4 7AL, UK

Y. Elsworth
School of Physics and Astronomy, University of Birmingham, Edgbaston, Birmingham, B15 2TT, UK

V.M. Nakariakov
School of Space Research, Kyung Hee University, Yongin, 446-701, Gyeonggi, Korea

V.M. Nakariakov
Central Astronomical Observatory at Pulkovo of RAS, St. Petersburg 196140, Russia

have an important meaning as a benchmark of solar activity, not only for investigation of the solar dynamo but also in terms of space weather.

Keywords Sun: activity · Sun: magnetic fields

1 Introduction

The mechanisms governing the solar cycle are still far from fully understood. Enormous information concerning this problem can be retrieved from the variability of solar phenomena, including the diversity of quasi-periodic processes. In addition, the influence of the Sun on the heliosphere and, in particular, on solar-terrestrial coupling is impacted by similar temporal variations and is, therefore, worthy of careful investigation. Of particular interest are oscillations with periods of the order of 2 yrs which appear to be connected with the internal fine structure of the solar magnetic field. The main feature of the quasi-periodic solar variations is their intermittency. The variations exist from time to time, and their period varies in the range of 0.6–4 yrs. In the literature, periodicities of 0.6–4 yrs are often referred to as quasi-biennial oscillations (QBOs). However, we note that periodicities in the range considered here have also been referred to as intermediate- or mid-term quasi-periodicities (for example Lou et al. 2003; Valdés-Galicia et al. 2005; Valdés-Galicia and Velasco 2008; Chowdhury et al. 2009b; Kudela et al. 2010).

The quasi-periodic processes in the Sun are considered to be a separate phenomenon from the well-established terrestrial quasi-biennial oscillation (e.g. Kane 2005a), which are far more uniform than the solar QBO. However, the exact relationship between the Sun's magnetic field and Earth's climate is poorly understood and so some form of resonance cannot be completely excluded.

An example of the QBO extraction is illustrated in Fig. 1. The upper panel gives the monthly values of the sunspot area for the whole disk from 1875 to 2012.[1] In the lower panel the 25-month smoothed series presents the long-term variability (red curve) while the residuals obtained by subtraction of the long term set from the monthly data present the short-term variability (blue curve). The oscillations, whose amplitudes are modulated by the 11-yr cycle, are clearly seen.

When considering a power spectral density (PSD) of common indices of solar activity, such as sunspot number or sunspot area, the interval of 0.6–4 yrs does not seem special compared to the power intrinsic to 11-yr cycle (Fig. 2a). However, removing the long-term variability reveals a PSD, as shown in Fig. 2b, with a highly disturbed multi-peaked structure. Figure 3, where the PSD is shown as a function of the period, demonstrates a local attenuation in the PSD just below $T = 1$ yr, then a general increase in the 1–1.5 yr range, and a rather uniform PSD distribution in the $T = 1.5$–2.5 yr region, that is an apparent separation of the two groups of variations. Historically, research on these two domains also often developed separately. However, as it will be shown below, many features of the two groups are similar.

Since the QBOs cannot be analyzed without preparatory filtration a variety of processing methods are used by different authors. They include discrete and fast Fourier transforms, spherical harmonic decomposition, wavelet transform analysis, periodogram analysis, pass-band filters (e.g., time-smoothing), maximum entropy method, and, recently, the Huang-Hilbert transform, which builds on empirical mode decomposition (EMD) analysis, to mention only a few. In general the results of the various procedures are consistent within the

[1] http://solarscience.msfc.nasa.gov/greenwch.shtml.

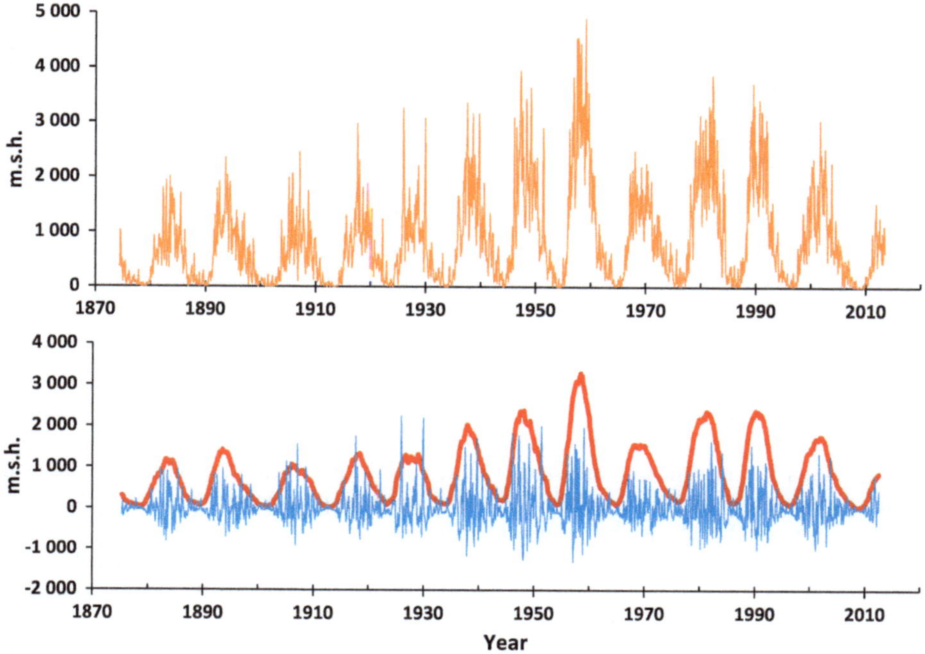

Fig. 1 *Upper panel*: monthly values of the sunspot area (http://solarscience.msfc.nasa.gov/greenwch.shtml) in millionths of a solar hemisphere. *Lower panel*: 25-month smoothed values of the sunspot area (*red curve*) and the short-term oscillations (*blue curve*) isolated by subtracting the smoothed data from the data shown in the *upper panel*

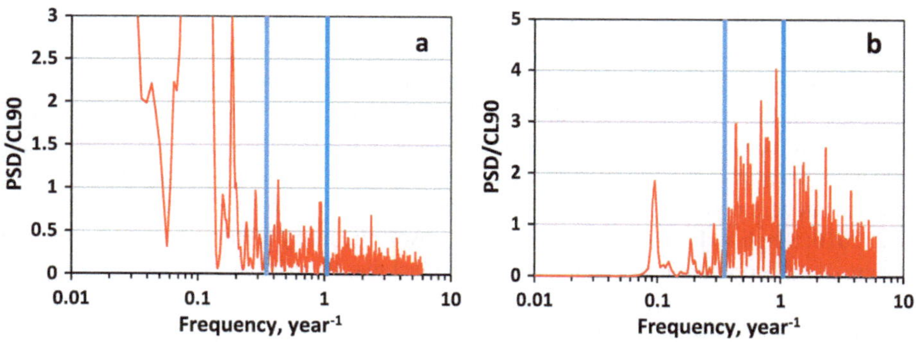

Fig. 2 (**a**): Power spectrum density of the monthly meanings of the sunspot area for the whole solar disk from 1875 to 2012. (**b**): The same as (**a**) but with an ~11 yr variation withdrawn by subtraction of 25-month running averages from the monthly values of sunspot area. PSD is given in the units of 90 % confidence level for the highest peak, meaning that values greater than unity imply that there is less than a 10 % chance of such power in noise. *Vertical bars* denote approximately the frequency range of QBOs indicated by this data, however, we note that many authors consider periodicities outside these limits (see Table 1)

accuracy of the processing. Many authors exploring the variability of different solar, interplanetary and terrestrial indices, found common features in the QBO behavior. In this paper, these features are illustrated by original figures where QBOs are isolated by subtracting the 25-month smoothed monthly values from the 7-month smoothed values unless otherwise in-

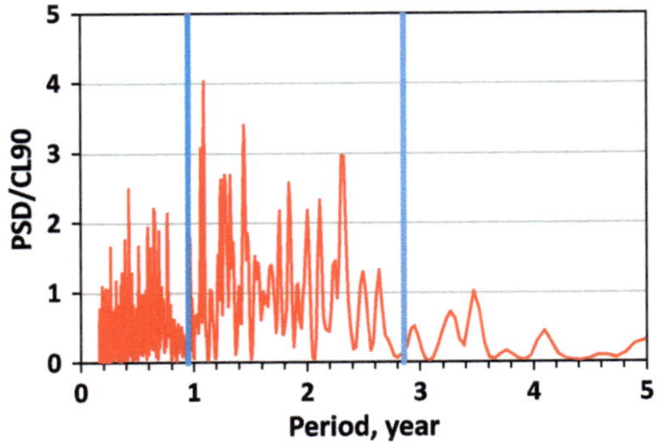

Fig. 3 Higher frequency range of the PSD spectrum from the Fig. 2b, but with PSD depicted vs. period value and linear scale on the abscissa axis. *Vertical bars* denote approximately the period range of QBOs indicated by this data

dicated. This procedure passes 1.5–1.7 yr signal without distortion (i.e. it has around a 100 % frequency response at 1.5–1.7 yr) and above a 50 % response between 1 and 3.3 yrs. The output results are, in general, consistent with results of other filtration. The results shown in this paper were found to be stable to different smoothing intervals: the amplitude of the QBO was observed to change by an insignificant amount.

An interest in quasi-biennial variations in solar activity was probably stimulated by the discovery in the 1960s of the so-called 26-month oscillations in the Earth's atmosphere (see Maeda 1967, and references therein). The atmospheric QBOs mainly have a terrestrial origin and although they are influenced by the 11-yr solar cycle (Baldwin et al. 2001; Petrick et al. 2012), seemingly not by the solar QBOs. Sakurai (1979) found a QBO in the solar neutrino flux as measured in the Homestake neutrino experiment (Bahcall and Davis 1976). Silverman and Shapiro (1983) discovered the "unexpected" 1.4 yr variations in the 1721–1943 data set of visual auroras. Akioka et al. (1987) reported on a 17-month periodicity in the number and area of sunspot groups. A rapid increase of investigations has followed.

The domain below 1 yr oscillations refers to the so called Rieger-type variations, which were discovered in the solar X-ray burst and flare occurrence that was observed by the SMM spacecraft in the early 1980s (Rieger et al. 1984). Specifically Rieger et al. observed a 154 d periodicity in the temporal distribution of flares. Usually, the <1 yr and >1 yr types of variations are considered in different works. Around 30 % of QBO papers published after 2000 were devoted to the Rieger-type oscillations, and 70 % to >1 yr variations. However, some researchers noticed their intrinsic resemblance (e.g. Boberg et al. 2002; Kudela et al. 2002, 2010; Singh and Gautam Badruddin 2012). Actually many features are similar, so here they are described without separation despite the fact that there is currently no agreement as to whether a physical connection between the Rieger-type variations and those with periods >1 yr exists (this is discussed further in Sects. 2.2 and 5).

2 The Ubiquity of the QBO

The QBO is visible in almost all measures of the Sun's magnetic field, from deep in the solar interior right out to the heliosphere, and there are numerous papers that consider the behavior of solar activity indices. Here we include merely a sample of these papers, the majority of which include substantial reference lists themselves. Table 1 presents a list of selected papers published since 2000 where the studied indices and periods of oscillations are given. From the Table the diversity of solar activity proxies and oscillation periods are apparent. The majority of works consider QBOs in the sunspot number and area as a common

Table 1 Selected papers related to QBO observations, published since 2000. SOL stand for various solar indices, IPL for solar wind and interplanetary magnetic field, GEO for geomagnetic indices, CR for galactic cosmic rays, and SEP for solar energetic particles

Author	Publication yr	Indices	Periods	Epoch of observation
Bazilevskaya et al.	2000	SOL	~2 yr	cycles 21–22
Howe et al.	2000	SOL	1.3 yr	1995–1999
Mursula and Zieger	2000	GEO, IPL	1.3–1.7 yr	1932–1998, 1964–1998
Bazilevskaya et al.	2001	SEP	~2 yr	1950–2000
Hill et al.	2001	CR	146–154 d	1998–1999
Lockwood	2001	SOL, GEO	1.3 yr	~1980–1995
Obridko and Shelting	2001	SOL	~2 yr	1915–1999
Rybák et al.	2001	SOL, CR	1.7–2.4 yr	1969–1998
Ivanov et al.	2002	SOL	~2 yr	1920–1980
Boberg et al.	2002	SOL	<1 yr, >1 yr	cycles 21–24
Krainev et al.	2002	SOL, CR	~2 yr	cycles 21–22
Krivova and Solanki	2002	SOL	154–158 d, 1.28 yr	1749–2001
Kudela et al.	2002	IPL, CR	150 d, 1.3 yr, 1.7 yr	cycles 20–22
Bai	2003	SOL	<1 yr	cycles 19–23
Benevolenskaya	2003	SOL	1–1.5 yr	cycles 21–23
Kato et al.	2003	CR	1.3–1.7 yr	cycles 21–22
Lou et al.	2003	SOL, CME	<1 yr	1999–2003
Mavromichalaki et al.	2003	SOL, CR	<1 yr	1981–1983
Mursula et al.	2003	IPL, GEO, CR	1.2–1.7 yr	cycles 9–22
Özgüç et al.	2003	SOL	150 d, 1.3 yr	1966–2001
Wang and Sheeley	2003	SOL	1–3 yr	1978–2002
Badalyan and Obridko	2004	SOL	1.5–3 yr	1943–2001
Ballester et al.	2004	SOL	160 d	cycles 21–23
Mursula and Vilppola	2004	IPL	1.3–1.7 yr	cycles 21–22
Shirai	2004	SOL	2.5 yr	1996–2001
Ataç et al.	2005	SOL	64 d, 125 d	cycle 23
Benevolenskaya	2005	SOL	1–1.5 yr	cycles 21–23
Cadavid et al.	2005	SOL, IPL, GEO	1–2.5 yr	1978–2003
Joshi and Joshi	2005	SOL	<1 yr	cycles 21–23
Kane	2005	SOL, CR	<1 yr, >1 yr	cycles 20–23
Knaack and Stenflo	2005	SOL	<1 yr, >1 yr	1966–2004
Moussas et al.	2005	SOL, GEO	<1 yr, >1 yr	
Richardson and Cane	2005	SOL, IPL, GEO, SEP	~150 d	cycle 23
Bazilevskaya et al.	2006	SEP	~2 yr	cycles 19–23
Chowdhury and Ray	2006	IPL, SEP	<1 yr, >1 yr	cycles 21–23
Forgács-Dajka and Borkovits	2007	SOL, IPL	1–4 yr	1975–2005
Obridko and Shelting	2007	SOL	1.3 yr	1915–1996
Badalyan et al.	2008	SOL	~2 yr	1939–2001
Lara et al.	2008	SOL, CME	<1 yr	cycle 23
Ruzmaikin et al.	2008	SOL, IPL	1.3 yr	1976–2008

Table 1 (*Continued*)

Author	Publication yr	Indices	Periods	Epoch of observation
Valdés-Galicia and Velasco	2008	SOL, IPL, CR	1–2 yr	1940–2004
Vecchio and Carbone	2008	SOL	~2 yr	1939–2006
Chowdhury et al.	2009	SEP	<1 yr, >1 yr	1986–2001
Chowdhury et al.	2009	SOL	<1 yr, >1 yr	cycles 22–23
Laurenza et al.	2009	SOL	<1 yr, >1 yr	1974–2001
Vecchio and Carbone	2009	SOL	1.5–4 yr	1939–2005
Fletcher et al.	2010	SOL	~2 yr	cycles 22–23
Kudela et al.	2010	SOL, IPL, GEO, CR	26–32 d, 150 d, 1.7 yr	1958–2008
Vecchio et al.	2010	SOL, SEP, CR	~2 yr	1974–2001
Zaqarashvili et al.	2010	SOL	155–160 d	cycle 21
Badalyan and Obridko	2011	SOL	~2 yr	1874–2009
Okhlopkov	2011	SOL, IPL, CR	~1-2 yr	1965–2007
Katsavrias et al.	2012	IPL, GEO	20 d,-4 yr	1966–2010
Laurenza et al.	2012	IPL, CR	~2 yr	1964–2004
Simoniello et al.	2012	SOL	~2 yr	cycle 23
Singh and Gautam Badruddin	2012	IPL, CR	9–260 d, 1.3 yr	cycle 23
Vecchio et al.	2012	CR	~2 yr, ~6 yr	1953–2004
Vecchio et al.	2012	SOL	<1 yr, >1 yr	1976–2003
Chowdhury et al.	2013	SOL	<1 yr, 1.4 yr, 2.1 yr	cycles 23–24
Choudhary et al.	2014	SOL	<1 yr, >1 yr	2004–2008
Cho et al.	2014	SOL, IPL, GEO	1.3 yr	1970–2007
Choudhary et al.	2014	SOL	155 d	1996–2011
Gyenge et al.	2014	SOL	1.3 yr	cycles 21–23

index alongside other solar proxies. The multi-peaked highly-variable structure of sunspot area values, which are especially pronounced around the maximum phases of the 11-yr cycle, is clearly seen in Fig. 1. This behaviour appears to be a common feature of activity at different levels of the Sun (see Sect. 3).

2.1 Different Solar Activity Proxies in Which the QBO is Observed

2.1.1 Measures of the Solar Interior

Shirai (2004), Vecchio et al. (2010), D'Alessi et al. (2013) resumed an interest in possible QBOs in the solar neutrino flux. Vecchio et al. (2010) used empirical mode decomposition techniques to show that the QBO observed in the solar neutrino flux is strongly correlated with the QBO present in galactic cosmic ray data, believed to be caused by the Sun's magnetic field. This implies that the solar magnetic field plays a crucial role in the modulation of the neutrino flux. This raises certain questions over the origin of the interaction between the Sun's magnetic field and solar neutrinos. For example, Vecchio et al. (2010) speculate that such a modulation in the neutrino flux could occur through coupling between the neutrinos magnetic moment and the Sun's magnetic field or because of a modulation in the neutrino

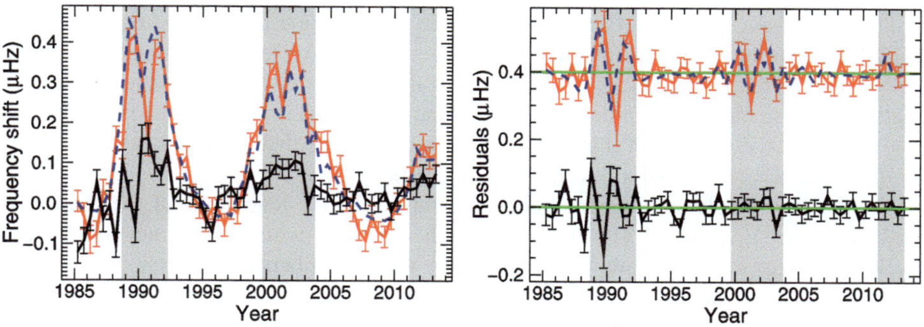

Fig. 4 *Left*: average frequency shifts of "Sun-as-a-star" modes with frequencies between 1.90 and 2.70 mHz (low-frequency range, *black*) and 2.70 and 3.55 mHz (high-frequency range, *red*). The frequency shifts were obtained from 182.5 d non-overlapping Birmingham Solar Oscillations Network (BiSON) data. This length of time series was used as it represents a compromise between having enough resolution to detect the QBO but still obtaining a good enough signal-to-noise ratio in the power spectra to obtain the p-mode frequencies. For comparison a scaled version of sunspot area has been plotted (*blue dashed line*). The daily sunspot area data were rebinned to 182.5 d to match the helioseismic observations and then linearly scaled and offset to match the high-frequency range shifts. *Right*: residuals left after dominant 11-yr signal has been removed by subtracting the frequency shifts once they have been smoothed over 2.5 yr. The colours are the same as in the *right-hand panel*. The high-frequency-range residuals have been artificially offset by 0.4 µHz for clarity (the zero lines for the residuals are indicated by the *green dot-dashed line*). The *grey shading* indicates times of high surface activity, by which we mean the sunspot area is above halfway between the minimum and maximum values. This was determined separately for each cycle

production rate. If the latter is true not only may this provide evidence for the presence of a magnetic field in the radiative zone but this may also be a hint that the magnetic field deep in the solar radiative region may exhibit quasi-biennial oscillations. D'Alessi et al. (2013) postulate that the neutrino flux QBO may be due to variations in the production rate caused by the magnetic modulation of gravity (g) waves, which are, in theory, trapped in the radiative zone. Such a modulation would change the density profile of the radiative zone, thereby altering the neutrino production rate. To date no definitive detection of individual g modes has been made (see Appourchaux et al. 2010, for a review) and so to date this theory cannot be tested. Gravity waves may also provide a link between the production rate of the solar magnetic field and neutrinos simultaneously (see Sect. 5 for further discussion).

Further evidence for the presence of QBOs in the solar interior can be found through helioseismology, which uses the Sun's natural acoustic oscillations (p modes) to study the solar interior (Christensen-Dalsgaard 2002). It is well known that the properties of p modes vary throughout the 11-yr solar activity cycle, with, for example, the frequencies of the oscillations being at their largest at solar maximum (e.g. Woodard and Noyes 1985; Pallé et al. 1989; Elsworth et al. 1990; Libbrecht and Woodard 1990; Chaplin et al. 2007; Jiménez-Reyes et al. 2007). The left-hand panel of Fig. 4 shows the frequency shifts of global p modes. The 11-yr cycle is seen clearly in the p-mode frequencies and shorter-term variations, with a period of approximately 2 yrs, are visible on top of the general 11-yr trend (Broomhall et al. 2009, 2011, 2012; Fletcher et al. 2010; Simoniello et al. 2012a,b, 2013).

In order to extract signals in the vicinity of the QBO we subtracted a smooth trend from the average total shifts by applying a boxcar filter of width 2.5 yrs. This removed the dominant 11-yr signal of the solar cycle. Note that, although the width of this boxcar is only slightly larger than the periodicity we are examining here, wider filters produce similar results. The resulting residuals, which can be seen in the right-hand panel of Fig. 4, show a periodicity on a timescale of about 2 yrs. The QBO has been observed in a number of different helioseismic data sets (Fletcher et al. 2010; Broomhall et al. 2011; Simoniello et al. 2012a). Furthermore the quasi-biennial signals present in the different helioseismic data are well correlated (Broomhall et al. 2011), implying they are not instrument artifacts. The correlation between the helioseismic residuals and those obtained from the sunspot area data is $R = 0.93$, which is significant at less than a 1 % level (we note that the Spearman correlation coefficient is also significant at less than a 1 % level).

For the 11-yr cycle a strong frequency dependence in the magnitude of the change in mode frequency is observed. This indicates that the variation in mode frequency must be the result of changes in the acoustic properties of the region just beneath the visible surface of the Sun (e.g. Libbrecht and Woodard 1990). There are some indications that the QBO shows a weaker frequency dependence than the 11-yr signal, implying that the changes responsible for the QBO are positioned deeper within the solar interior than those responsible for the 11-yr signal (Broomhall et al. 2012; Simoniello et al. 2012a). However, it is very difficult to disentangle the frequency dependence of the QBO from the dominant presence of the 11-yr cycle making any such inferences somewhat tenuous.

Howe et al. (2000) observed variations in the rotation profile of the Sun at a radius of $0.72R_{\odot}$, most predominately at low latitudes, with a period of 1.3 yr. However, other authors did not observe the same signal (e.g. Antia and Basu 2000). A more recent analysis (Howe et al. 2011) demonstrated that the signal was intermittent and has not been observed since 2001. However, Howe et al. found that when present the 1.3 yr signal was highly correlated in data recorded by two independent observation programs, implying that it is solar in origin. Jiménez-Reyes et al. (2003) observed a 1.3 yr modulation in the p-mode energy supply rate. Broomhall et al. (2011) observed an excess of power at $\sim$1.3 yr in periodograms of the low-l p-mode frequency shifts. This excess of power was found to be significant at more than a 2 per cent level and was observed in data from independent observational programs, implying that is it solar in origin.

2.1.2 Measures of the Photosphere, Chromosphere and Corona

Perhaps the most obvious manifestation of photospheric solar activity is sunspots. It is widely accepted that the QBO can be observed in sunspot number and sunspot area (see Figs. 1, 2, and 3). Wang and Sheeley (2003) also found periodicities in sunspot area in the range 0.2 to 2.6 yrs, with no single periodicity dominating. Furthermore, Wang and Sheeley reported periodicities in the range 1 to 3 yrs in photospheric measures of the equatorial dipole component.

1–2 yr patterns were discovered in the behavior of the solar magnetic field (Hoeksema 1991; Benevolenskaya 1995), while, in the last decade, periodicities of around 1.3, 1.7 and 2–4 yrs were observed in the temporal evolution of the large-scale solar photospheric magnetic field (e.g. Boberg et al. 2002; Cadavid et al. 2005; Kane 2005b; Knaack and Stenflo 2005; Knaack et al. 2005; Laurenza et al. 2009; Vecchio et al. 2012a, etc). Furthermore, consistent results were obtained by analyzing the magnetic fields inferred from the H-alpha filament observations since 1915 (Obridko and Shelting 2001, 2007; Ivanov et al. 2002). Danilovic et al. (2005) observed quasi-biennial periodicities in the equivalent width and central depth data of the Mn 539.4 nm solar spectral line, which is photospheric in origin. Kane (2005b) found evidence for numerous QBO and Reiger-type periodicities in a number proxies of the solar magnetic field (see Sect. 2.1.3), including the sunspot number and the CaII plage area and CaII K index, which originate in the chromosphere. Özgüç et al. (2003) studied the flare index observed between 1966 and 2001 and found periodicitites of 150 d and 1.3 yr and Ataç et al. (2005) found periodicites of 64 d and 125 d in the flare index observed during cycle 23.

Badalyan and Obridko (2004) found the 1.3 yr periodicity in the correlation of the green-line intensity and magnetic field in the lower corona, while Vecchio and Carbone (2009) revealed QBOs in the green coronal line emission whose period varied with time in the range of 1.5–4 yrs.

2.1.3 Comparisons Between Proxies

Many authors considered several solar indices jointly to detect QBOs in their behavior: for example Kane (2005b, sunspot number, CaII area and K index, Lyman α, 10.7 cm flux, coronal green line, open fluxes, interplanetary magnetic field, and cosmic rays), Valdés-Galicia and Velasco (2008, coronal hole area, radio emission in the 10.7 cm band, and sunspots), Choudhary et al. (2014, sunspot area, solar flares, and coronal mass ejections). Moreover, it has been shown (Mursula and Vilppola 2004; Knaack and Stenflo 2005; Knaack et al. 2005) that oscillations with period of around 1.3 yr are a single process which manifests itself at all levels from the tachocline and photosphere (e.g., areas and numbers of sunspots, large-scale magnetic fields) up to the Earth's magnetosphere (geomagnetic activity) and the far heliosphere (cosmic rays).

The QBOs in various solar indices (sunspot number and area, 10.7 cm radio emission, mean solar magnetic field, coronal green line, H-alpha flare number) isolated with a pass-band filter behave rather similarly. The QBOs in the sunspot area alongside the 10.7 cm radio emission[2] are plotted in the upper panel of Fig. 5 where excellent agreement is apparent: the correlation coefficient $R = 0.884$. The lower panel manifests the QBOs in the sunspot area alongside the 530.3 nm coronal line emission.[3] Here, correlation is worse ($R = 0.495$), but still the QBOs are reasonably coherent. It is worth noting that the data depicted in Fig. 5 resembles that of the QBOs in the radial solar magnetic field at the heliolatitude of 25° from Vecchio et al. (2012a) as shown in the upper panel of their Fig. 7.

2.1.4 The QBO Transition into the Interplanetary Space

Papers devoted to QBOs in interplanetary parameters are mostly directed towards finding a cause-and-effect relationship between solar and interplanetary/terrestrial phenomena. Therefore they usually consider interplanetary parameters such as heliospheric magnetic field strength, solar wind speed, geomagnetic indices and galactic cosmic ray modulation in correspondence to solar parameters such as sunspot number or area, solar magnetic field, flare index, and coronal activity (e.g. Bazilevskaya et al. 2000; Rybák et al. 2001; Krainev et al. 2002; Mursula et al. 2003; Wang and Sheeley 2003; Moussas et al. 2005; Richardson and Cane 2005; Forgács-Dajka and Borkovits 2007; Katsavrias et al. 2012; Singh and Gautam Badruddin 2012; Laurenza et al. 2012). Vecchio et al. (2012b) used the EMD analysis to uncover a wide range oscillations in the galactic cosmic ray intensity. They showed that the QBOs are actually responsible for the Gnevyshev Gap phenomenon (see Sect. 4) and the step-like decreases typical for the galactic cosmic ray modulation.

Bazilevskaya et al. (2001, 2006) explored the occurrence of solar-energetic-particle (SEP) events observed between 1955–2004, using ground-based and spacecraft observations. SEPs are high energy particles (ranging from a few 10s of keV to GeV) that are accelerated away from the Sun by solar flares or coronal mass ejections (CMEs). Bazilevskaya

[2] http://www.ngdc.noaa.gov/stp/solar/flux.html.

[3] ftp://ftp.ngdc.noaa.gov/STP/SOLAR_DATA/SOLAR_CORONA/INDEX/Lomnicky/.

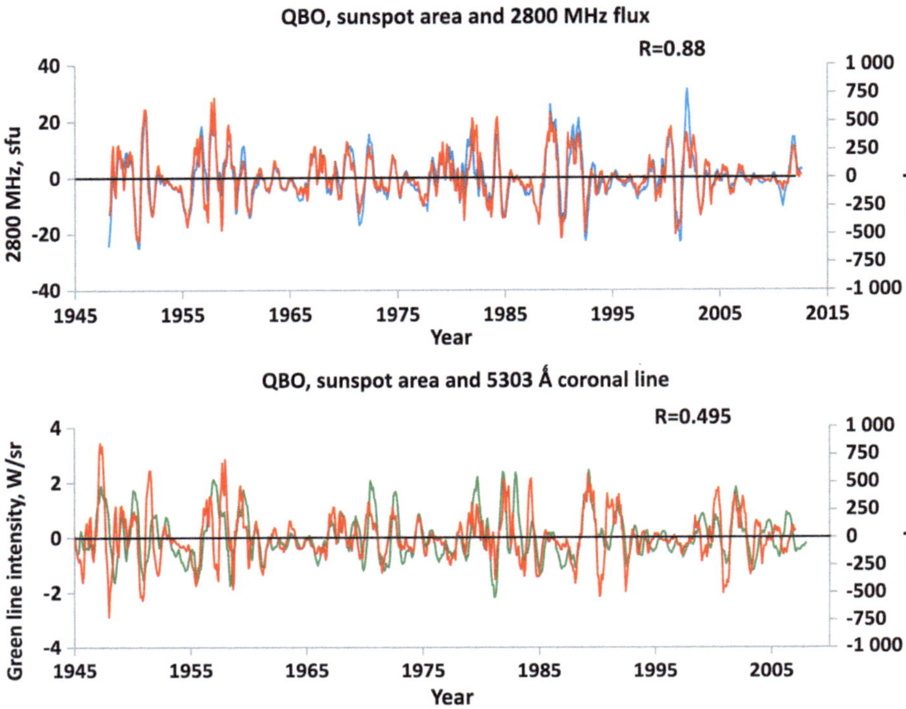

Fig. 5 The QBOs as isolated from the sunspot area (*red curve*) compared with the 2800 MHz (10.7 cm) radio emission (*upper panel, blue curve*) and with the 530.3 nm *coronal line* (*lower panel, green curve*)

et al. (2001, 2006) concluded that the QBOs in the total number of SEP events are fairly similar to the QBOs in other solar indices, especially in the strong X-ray bursts. Bazilevskaya et al. (2001, 2006) also considered the occurrence of the powerful ground level enhancements (GLEs): a GLE is recorded when SEPs are accelerated to high enough energies to enhance the count rate of ground-based neutron monitors, which requires energies above several GeV to avoid being absorbed by Earth's atmosphere. Bazilevskaya et al. (2001, 2006) found that GLEs sometimes occurred during zero and even negative phases of the solar QBOs. Evidence of QBOs in the occurrence rates of coronal mass ejections, SEP events and geomagnetic storms with sudden commencements during solar cycle 23 were found by Richardson and Cane (2005). Laurenza et al. (2009) studied the QBOs in the interplanetary proton fluxes registered onboard the IMP 8 spacecraft between 1974–2001 and found periodicities of 3.8 and 1.7–2 yrs.

It is interesting that QBOs can be observed rather far from the Sun but still within the heliosphere. Kato et al. (2003) found 1.7- and 1.3-yr oscillations in the galactic cosmic ray intensity variation in the outer heliosphere during the 1980s (solar cycle 21) and the 1990s (solar cycle 22) in agreement with similar variations observed at neutron monitors.

The observational features of the QBOs seen in interplanetary space are similar to solar QBOs: They are characterized by intermittence in periodicity and an amplitude that varies with time and is largest around times of solar activity maxima. Furthermore, there are often epochs where no QBOs at all are detected (they are temporally intermittent). However, there is rather poor correspondence between the time series of the solar and interplanetary QBOs (Rybák et al. 2001; Cho et al. 2014). Transition from the solar to interplanetary QBOs refers

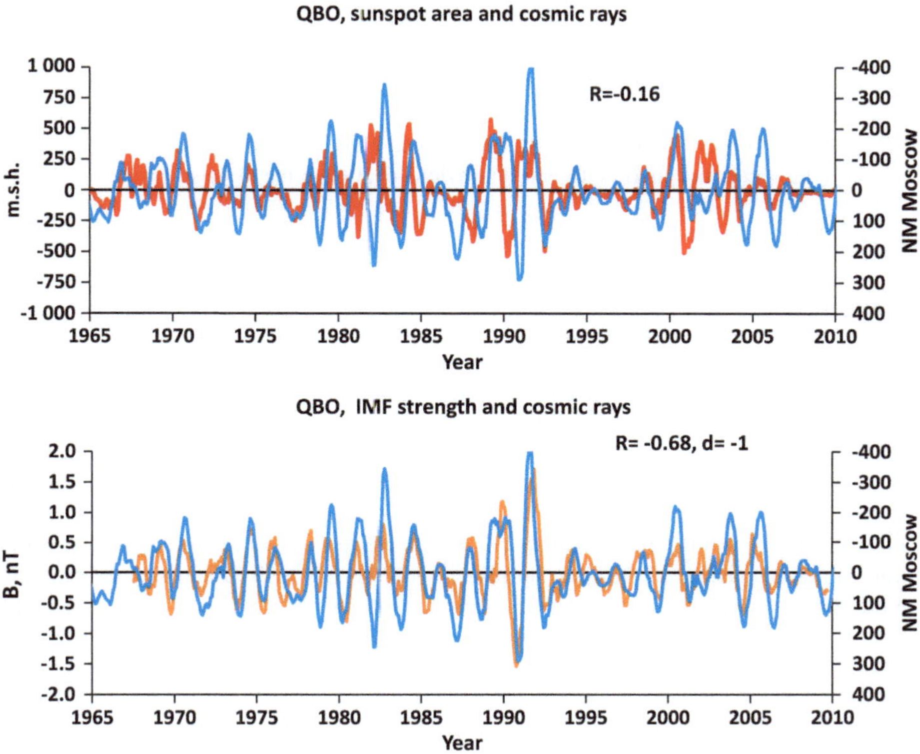

Fig. 6 *Upper panel*: QBOs in the sunspot area (*red curve*) and galactic cosmic ray intensity (*blue curve*, Moscow neutron monitor). *Lower panel*: QBOs in the interplanetary magnetic field strength (*orange curve*) and galactic cosmic ray intensity (*blue curve*). Note the reverse scaling on the *right vertical* (cosmic ray) axis

to the fundamental problem of interactions of stars with convective outer layers and their respective stellar spheres (see review by Zurbuchen 2007). Although much of the magnetic energy is stored in closed magnetic structures on the Sun part of the solar magnetic field is pulled into the heliosphere owing to the high coronal temperature. This portion of the solar magnetic field is commonly referred to as open magnetic flux. The solar QBOs can be transmitted into interplanetary space through the open magnetic flux (Lockwood 2001). As it is shown by Wang et al. (2002) the strength of the radial heliospheric field (HMF) is proportional to the total open flux. The QBOs in the galactic cosmic ray intensity are not coherent with the QBOs in the sunspot area and other solar indices, while they correspond well with the QBOs in the heliospheric magnetic field strength[4] as is shown in Fig. 6. The correlation coefficient between the curves depicted in the upper panel of Fig. 6 is $R = -0.16$, and in the lower panel of Fig. 6 is $R = -0.68$ with galactic cosmic rays delayed relative to heliospheric magnetic field by 1 month. A correlation coefficient of $R = -0.65$ and a delay of 2 months is observed if the radial component of HMF is considered. The Moscow neutron monitor data[5] are presented in Fig. 6, but the result is similar for other stations of galactic cosmic ray monitoring. The QBOs in geomagnetic activity do not correlate with

[4]ftp://spdf.gsfc.nasa.gov/pub/data/.

[5]helios.izmiran.rssi.ru/cosray/.

the solar QBOs but are coherent with the QBOs in the product of solar wind velocity, V, and the heliospheric magnetic field strength, B. The geomagnetic Dst index characterizes geomagnetic storms. The correlation coefficient between the QBOs in the geomagnetic Dst index[6] and in sunspot area during 1967–2009 is $R = -0.2$, while a correlation coefficient of $R = -0.82$ is observed between QBOs in Dst and VB. The apparent lack of correlation between solar and heliospheric QBOs is an obvious challenge for theorists.

2.2 Reiger-Type Oscillations

The Rieger-type ($T < 1$ yr) oscillations are also observable in many solar indices, such as sunspot number and area, photospheric magnetic field, optic and X-ray flares (e.g. Boberg et al. 2002; Krivova and Solanki 2002; Ballester et al. 2004; Chowdhury and Ray 2006; Chowdhury et al. 2009a). In the galactic cosmic ray intensity the oscillations with $T \sim 150$ d have been reported by Mavromichalaki et al. (2003), Kudela et al. (2010), Laurenza et al. (2012). Rieger-type QBOs in the CME rate were found by Lou et al. (2003), Lara et al. (2008). While Chowdhury and Ray (2006), Chowdhury et al. (2009a) found a variety of QBOs including the Rieger-type periodicities in the solar electron fluxes observed by IMP 8. Hill et al. (2001), using Voyager 1 data at 73AU, have shown that the quasi-periodic variations in anomalous cosmic rays are in phase, with the QBOs in O and He observations having periods of approximately 151 d, while protons exhibit a period of approximately 146 d. However, we note that these periodicities are within 1σ of each other ($\sigma = 12$ d for O and 15 d for He and protons).

Reiger-type oscillations are often revealed as a well-defined periodicity, e.g., 154 d (Rieger et al. 1984), 34, 51, 85, 129, 135 d (Bai 2003). However, sometimes they are reported as a range of days, such as 64–125 d, 150–160 d (Zaqarashvili et al. 2010a) or even as a signal with multiple periodicities (Lou 2000; Joshi and Joshi 2005; Chowdhury et al. 2009b). Some authors consider <1 yr and >1 yr jointly, however, suggest a different origin: Boberg et al. (2002) propose that the 1–2 yr periodicity observed in solar mean magnetic field measurements are related to the internal rotation rate of the Sun, while the 80–200 d periodicities are connected to the evolution of large active regions. Vecchio et al. (2012a) explain the QBO as an integral part of the Sun's dynamo but associate the shorter Reiger-type periodicities with the outbreak of bipolar regions on the solar surface. Often a more generic link is proposed, for example, Krivova and Solanki (2002) propose that the 154–158 d Reiger periodicity is a harmonic of the 1.3 yr QBO. More details on the possible causes of both QBOs and Reiger-type periodicities are discussed in Sect. 5 and a comprehensive overview of the Rieger-type periodicities is given by Bai (2003).

3 Main Features of the QBO

3.1 Multipeaked and Variable Periodicity

Many studies have shown that when proxies of the Sun's magnetic field are plotted as a function of time multiple peaks are observed around solar maximum (as can be seen in Figs. 1 and 4 of this paper). This is somewhat analogous to the features in the spectra of 3-min sunspot oscillations, where multiple peaks are also observed (e.g. Reznikova et al.

[6]http://wdc.kugi.kyoto-u.ac.jp/dstae/index.html.

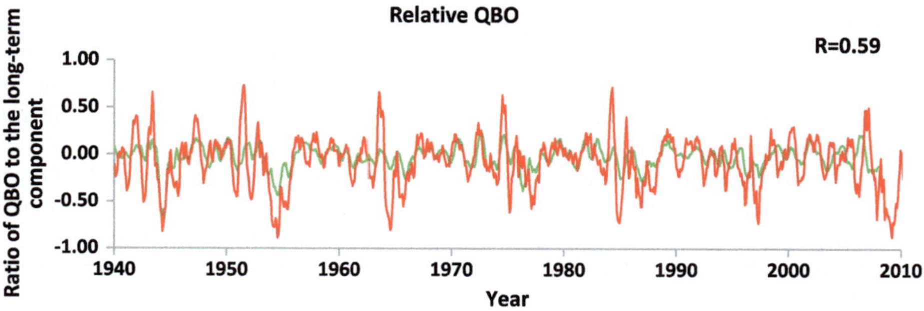

Fig. 7 Relative QBO of sunspot area (*red*) and 530.3 nm *green coronal emission line* (*green*). The relative QBOs were obtained by determining the ratio of the QBO time series, which are plotted together in the *bottom panel* of Figure 5 to the 25-month smoothed value of solar activity indices (which are plotted for sunspot area in Figure 1)

2012). Furthermore, studies of the QBO frequently uncover multiple periods in the data, many of which appear to vary with time. For example, a multi-peaked structure in the solar, interplanetary and galactic cosmic ray time sets was observed by Kudela et al. (2002), Mavromichalaki et al. (2003), Wang and Sheeley (2003), Kudela et al. (2010). Some papers isolate and study definite period values (most often attention is attracted by the $T = 1.3$ yr period), however inspection of a bulk of works instead reveals sets of distinct pulses while time series with a definite period occupy only limited epochs. Mursula et al. (2003) uncovered an alternation in the dominant periods of the QBO observed in geomagnetic activity during the last 15 solar cycles. These results were corroborated by Katsavrias et al. (2012), who analysed solar wind and interplanetary magnetic field parameters. Several authors have stressed that the QBO appeared as "a stochastic superposition of different oscillators" (Vecchio and Carbone 2009), "highly intermittent in time" (Ataç et al. 2005), having "elusive and enigmatic character" (Zaqarashvili et al. 2010a). The signature of stochasticity can be seen in Figs. 2b and 3 where no distinct period is observed.

3.2 Temporal Modulation of the Amplitude of the QBO with 11 yr Cycle

Most of the above mentioned researchers (e.g. Bazilevskaya et al. 2000; Benevolenskaya 2003; Ballester et al. 2004; Kane 2005b; Fletcher et al. 2010; Zaqarashvili et al. 2010b; Singh and Gautam Badruddin 2012; Vecchio et al. 2012a, and many others) noticed that QBOs are modulated by the 11-yr cycle, being strongest around maxima of solar activity. This pattern can also be clearly seen in Figs. 1, 4, 5, and 6. It is therefore interesting to consider the existence of QBOs during solar minima. There is evidence that the QBO is still present in the helioseismic p-mode frequency shifts away from solar maxima (see Fig. 4 and Broomhall et al. 2009, 2011, 2012; Fletcher et al. 2010; Jain et al. 2011), implying that the QBO is still present in the solar interior even at times where activity is at a minimum. Figure 7 demonstrates the relative QBO, which is determined in the following manner. First we must define the QBO time series for a given proxy, which is obtained by subtracting a smoothed component, that primarily shows the 11 yr cycle, from data averaged over a shorter length of time. For example, here we have subtracted the 25-month smoothed values from the 7-month smoothed values (see Figs. 1, 5, and 6). The relative QBO is then the ratio of the QBO timeseries to the 25-month smoothed time series (which is plotted for sunspot area in Fig. 1) i.e. a ratio of the QBO to the smoothed component of the 11-yr cycle (Bazilevskaya et al. 2000). It can be seen that the QBOs are present during epochs of low solar activity. Furthermore the relative QBOs have the highest amplitudes just in the minima of solar activity. A correlation between the relative QBOs of the 530.3 nm line and the sunspot area is even higher ($R = 0.59$) than between absolute QBOs, which were produced by subtracting the 25-month smoothed values from the 7-month smoothed values and are presented in Fig. 6 ($R = 0.49$).

Of special interest are possible changes in the QBOs over relatively long periods of time and the dependence of QBO features on the characteristics of the 11-yr cycle. The long-term changes in the geomagnetic aa-index QBO power were stated by Mursula et al. (2003) and Mursula and Vilppola (2004): periods where the QBOs were strong were found in the mid-19th century and since 1930, while QBOs were weak during low solar activity from the 1860s to the 1920s; the geomagnetic 1.3–1.4 yr pulsations were found to be more pronounced in the even cycles (18, 20 and 22) while the 1.5 and 1.7 yr QBOs, in the odd cycles. Similar conclusions about dependence on alternate cycles were drawn by Kudela et al. (2002) in an analysis of galactic cosmic ray intensity and by Knaack and Stenflo (2005) in an analysis of the magnetic QBOs.

3.3 Temporal Variation in the Periodicity of the QBOs and a Possible Connection with the 11-yr Cycle

In the previous section we discussed the temporal variation of the amplitude of the QBO and its relation to the 11-yr solar cycle. We now move on to consider the frequency/period modulation of the QBO. For example, Bai (2003) explored the solar flare occurrence rate and found a periodicity of 51 d for cycles 19, 85 and 129 d for cycle 20, 153 d for cycle 21, and 34 and 129 d for cycle 23. No statistically significant periodicities were found for cycle 22.

Khramova et al. (2002), argued that the QBO characteristics depend on the power and length of a particular 11-yr cycle. Okhlopkov (2011) studied QBOs (20–24 months) in galactic cosmic rays, solar mean magnetic field, interplanetary parameters and geomagnetic Ap index during solar activity cycles 20–23. He showed that the average QBO period was 20.2–20.8 months in the odd cycles and 22–23.5 months in the even cycles. This was in agreement with the results of Kudela et al. (2002). However, Obridko and Shelting (2007) did not find a distinct correlation between the QBOs (1.3 yrs) in solar magnetic field and the parity of the cycle number. Nor did they find any correlation with the height of the solar maxima.

3.4 Spatial Distribution of QBO

While the QBOs in different layers of the solar atmosphere are rather synchronous the QBOs in the northern and southern hemispheres do not correlate with each other. The correlation coefficient between the QBOs observed in sunspot area of the north and south hemispheres calculated on the base of 1875–2012 is $R = 0.001$, while $R = 0.53$ for the monthly meaned data and $R = 0.87$ for the 25-month smoothed means.

The north-south (NS) asymmetry in QBOs has been extensively studied (Badalyan and Obridko 2004, 2011; Knaack et al. 2004; Ataç et al. 2005; Forgács-Dajka and Borkovits 2007; Badalyan et al. 2008; Zharkov et al. 2008). It appeared that QBOs in the asymmetry of various solar activity indices are even more pronounced and better synchronized than QBOs in the indices themselves. In addition, the QBO power and absolute value of asymmetry are negatively correlated (Badalyan et al. 2008; Badalyan and Obridko 2011). Badalyan and Obridko (2011) studyed the NS asymmetry and suggested that to a great extent solar activity may be generated independently in the two hemispheres. Furthermore these findings argue that the NS asymmetry is a fundamental characteristic of solar activity.

Cadavid et al. (2005), using longitudinally averaged fields from NSO/KPNO synoptic Carrington rotation maps, noticed that the spatial distribution of QBO periods throughout the solar surface was not uniform: oscillation periods of the order of 1.0–1.5 yrs were more typical of the polar and high-latitude fields and periods of 1.3 and 1.7 yrs characterized the mid- and low-latitudes. Though the further analysis of Ruzmaikin et al. (2008) supported the latitudinal distribution of the pattern, the authors stressed its variable frequency and intermittent appearance. This is consistent with the results of Howe et al. (2000) who not only observed a 1.3yr periodicity in helioseismic data at low latitudes but also observed the 1.3 yr periodicity to disappear in 2001.

A comprehensive analysis of the spatial-temporal dynamics of the solar magnetic field in the period of August 1976–September 2003 was made by Vecchio et al. (2012a), using an EMD technique. The QBOs were found to be uniformly distributed over all latitudes with high amplitudes during the maximum and descending phase of each solar cycle. Antisymmetric behavior of the radial, meridional and east-west components of the solar magnetic field with respect to the equator was pointed out. The QBOs revealed in the radial and meridional components of the photospheric magnetic field were shown to be fundamental timescales and were associated with the poleward magnetic flux migration from low latitudes around the maximum and descending phase of the solar cycle (see also Benevolenskaya 2003, 2005). The signs of an equatorward drift, at an $\sim$2 yr rate were found in the radial and east-west components, suggesting a link to a dynamo operation. The Rieger-type oscillations demonstrated the butterfly diagram, reflecting the emergence of active regions on the solar surface at these timescales.

3.5 Summary of the QBO Observable Features

The QBOs appear to be the most prevalent quasi periodicity shorter than the 11-yr cycle in solar activity phenomena. Their amplitudes are higher during periods of high solar activity, however they do not disappear in the solar minima. The QBOs are highly irregular resembling a set of intermittent pulses/waves with signatures of stochasticity. The oscillations in the northern and southern solar hemispheres develop independently. The oscillations in various indices of solar activity related to different levels of the solar atmosphere are rather coherent. However, the solar QBOs are translated into the heliosphere through the open magnetic flux and therefore the QBOs in the solar and interplanetary parameters are not synchronous.

4 Gnevyshev Gap

4.1 Gnevyshev Gap as an Appearance of the QBO

During the last decade, many authors (e.g. Bazilevskaya et al. 2000, 2006; Astafyeva and Bazilevskaya 2000; Benevolenskaya 2003, 2005; Storini et al. 2003; Storini and Laurenza 2003; Sello 2003; Wang 2004; Knaack and Stenflo 2005; Knaack et al. 2005; Ataç et al. 2005; Kane 2006; Hathaway 2010; Vecchio et al. 2010, 2012a,b; Laurenza et al. 2012), have argued that one of the main features of the QBO is a temporal weakening of solar activity observed in the maximum phase of an 11-yr solar cycle (the so-called Gnevyshev Gap). However, since the Gnevyshev gap was first examined in the 1960s, this phenomenon has been studied separately from the QBOs and even now it is not absolutely clear if it can be totally reduced to the QBOs.

It was M.N. Gnevyshev, a Pulkovo astronomer, who first drew attention to this phenomenon by demonstrating the effect using observational data on the coronal green line during the 19th solar activity cycle (Gnevyshev 1967, 1977). Gnevyshev considered coronal activity at various heliolatitudes and concluded that there were actually two activity waves: a first one with a maximum at the end of the rising phase and a second one at the start of the declining phase of a solar cycle. Gnevyshev found similar behavior in many parameters of solar activity (sunspots, flares, UV, radio, corpuscular emissions, and geophysical data). Summarized over heliolatitudes the maximum of the 11-yr solar cycle looked like a double-peak structure.

The name "Gnevyshev Gap" was used for the first time by Schove (1979), who traced the double-peak structures in the aurorae occurrence. However, it was only after a comprehensive work by Feminella and Storini (1997), that the reinvented term was accepted by the scientific community and became conventional. Feminella and Storini (1997) have examined the Gnevyshev Gap (GG) extensively using the recent data on spots, flares, radio, and X-ray fluxes. They confirmed Gnevyshev's findings that the structured activity maxima were detected in all solar atmospheric levels and were more distinct in strong and/or long-lasting events. However, they stressed that the temporal structures are different in the north and south hemispheres and more than two peaks of activity occurred rather often. In addition they found that GGs were even more clearly seen when the variability of solar indices, such as the standard deviation, was examined.

Heliospheric GG effects deserve to be given special importance because of their clear appearance just around the maximum phase of solar activity that could be fruitfully used for the space weather investigation and even forecast. A coherent picture of the GG appearance in the space weather was given by Storini et al. (2003) with a great number of references.

A characteristic time scale of the GG is about 2 yrs, and it can be, at least formally, considered as the appearance of QBOs. Figure 8 demonstrates, as an example, the relation between the GG and QBOs in the sunspot area index taken separately from the northern and southern solar hemispheres. The upper panels show the monthly averaged values of the sunspot area. There are several local minima around the maximum phase of the solar cycle which are emphasized by the 7-month smoothing. The most prominent ones are marked by vertical bars in the middle panel and are actually the GGs. The 25-month smoothed data are also plotted in this panel, which accentuate the 11-yr component. The procedure of subtraction of the 25-month averages from the 7-month reveals the QBOs (similarly to Fig. 5), where in the bottom panel of Fig. 8 the GGs are again pointed out by the vertical bars. It is seen that the GGs are actually a consequence of the QBOs modulated by the 11-yr cycle.

4.2 Common Features of the QBO and GG

Main features of the GG are actually similar to those of the QBOs. Wide spreading of GG in the solar and interplanetary phenomena has been shown already by Gnevyshev (1967, 1977), Feminella and Storini (1997) and references therein. The latter authors demonstrated the independent appearance of GGs in the two solar hemispheres, which was corroborated by Bazilevskaya et al. (2000), Kane (2002), Sello (2003), Norton and Gallagher (2010).

The solar wind and interplanetary magnetic field data demonstrated GGs which were reflected by geomagnetic disturbances and the related modulation of galactic cosmic ray fluxes (Ahluwalia 2000; Krainev et al. 2002; Richardson et al. 2002; Storini and Laurenza 2003; Ahluwalia and Kamide 2005; Amenomori et al. 2006; Kane 2006; Belov 2009). Like

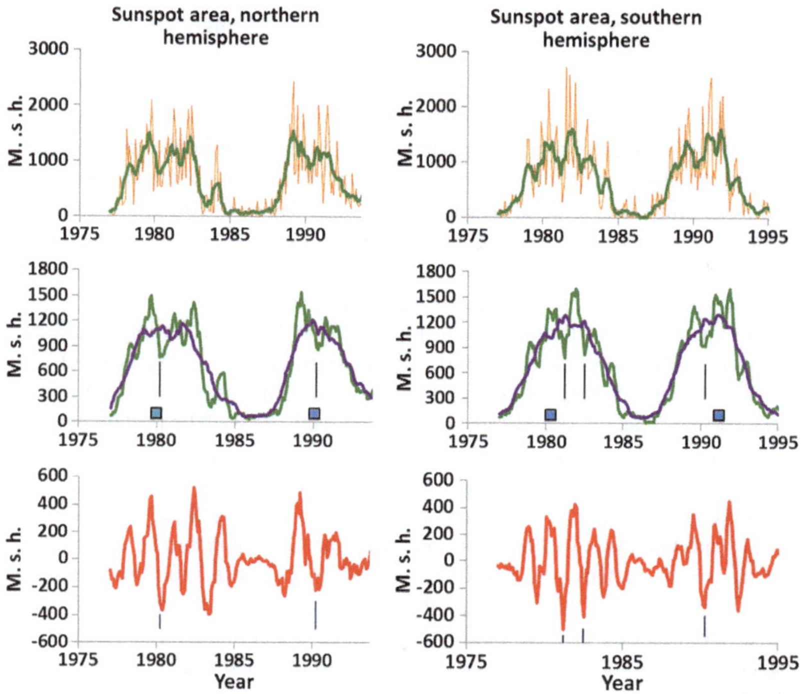

Fig. 8 *Upper panel*: monthly values of the sunspot area in the northern and southern solar hemispheres (*light brown*) and the 7-month smoothed values (*dark green*). *Middle panel*: 7-month and 25-month smoothed values; *vertical bars* indicate the GG occurrence; *squares* are times of solar polar magnetic field reversals (http://wso.stanford.edu). *Bottom panel*: QBOs resulted by subtraction of the 25-month smoothed data from the 7-month smoothed ones; *vertical bars* indicate the GG occurrence. Adopted with minor changes from Bazilevskaya et al. (2000)

the QBOs, the GGs in the solar indices do not always match with the interplanetary GGs (Kane 2005c, 2006).

Solar energetic particle (SEP) events are extremely difficult to analyze because their observable features accumulate numerous signatures of solar and interplanetary factors. The GG effects can shed additional light on the reason for SEP occurrence such as the relative role of solar flares and CMEs. The topic was discussed by many authors (e.g. Nagashima et al. 1991; Storini et al. 2005; Bazilevskaya et al. 2006; Miroshnichenko 2008). Rodríguez-Pacheco et al. (2012) observed the GG effect in fluxes of the energetic particles aboard Ulysses at high heliolatitudes.

Similar to the QBOs, GGs are rather intermittent. A double-peaked, three-peaked and sometimes single-peaked shape of the solar maxima may occur (Feminella and Storini 1997; Kane 2005c). Considering the sunspot area during 12 solar cycles Norton and Gallagher (2010) found GGs in 10 solar cycles in the northern solar hemisphere and in 8 cycles in the southern hemisphere. Wang (2004) assumed a connection between GGs and stochastic fluctuations.

4.3 Peculiar Features of the GG Phenomenon

A number of GG characteristics are either not inherent to QBOs or alternatively are not investigated by QBO researchers. Following Gnevyshev (1967, 1977) and Feminella and

Storini (1997) some authors found more distinct GGs in stronger or longer-lasting solar phenomena. Recently Kilcik et al. (2011) corroborated this result for solar cycles 22 and 23. Kilcik and Ozguc (2014) examined cycles 1–23 and suggested that a possible reason for a double maximum in solar cycles is the different behavior of large and small sunspot groups connected with two different dynamo mechanisms. Different properties of the first and second Gnevyshev peaks were reported also by Kane (2002) and Ahluwalia and Kamide (2005). The latter work emphasizes the difference in the interplanetary conditions during the first and the second Gnevyshev peaks. Lukianova and Mursula (2011) used the GG peaks as benchmarks of solar activity and showed that the reduction of sunspot magnetic fields started quite abruptly in 2001/2002, just during the secondary peak after the GG.

Since the GGs occur in the maximum phase of the 11-yr cycle they could be related to the solar polar magnetic field reversal, as was suggested by Nagashima et al. (1991). Although the majority of researchers support this idea careful consideration does not confirm a direct causal link (Kane 2006; Murakozy and Ludmany 2008). The middle panel of Fig. 8 demonstrates that no strict time coincidence is observed between the GGs and the magnetic polarity inversion periods. To date this issue has been insufficiently studied.

There are approaches to the GG understanding not involving the QBOs. Sykora (1980) and Antalova and Gnevyshev (1983) proposed that the GG effect can be explained in terms of solar activity pulses. This was further developed by Zolotova and Ponyavin (2012) who constructed simple models to reproduce the well-known butterfly diagrams, and various empirical laws to describe the distribution of sunspots, and demonstrated that activity pulses can satisfactorily reproduce the multi-peak structure of the solar cycle. Georgieva and Kirov (2007, 2011), Georgieva (2011) explain the two peaks of the GG by the development of two parts of the poloidal field, one being advected by the meridional circulation all the way to the poles and another one being diffused directly to the tachocline at midlatitudes. Both parts of the poloidal field generate the toroidal fields at the base of the convective zone. This suggestion needs further accumulation of observational data and model development.

Presently, the GG phenomenon is not fully understood. Further investigation should explain the GG peculiarities and its relation to the QBOs.

5 Physical Mechanisms that May be Responsible for the Observed QBOs

The QBO nature is not yet fully understood and a complete discussion of possible scenarios is beyond the scope of this paper. We now give details of a few of the most likely mechanisms that have been discussed in the literature. The majority of authors believe that QBOs are intrinsic to the solar dynamo mechanism (e.g. Benevolenskaya 1998; Howe et al. 2000; Krivova and Solanki 2002; Mursula et al. 2003; Mursula and Vilppola 2004; Cadavid et al. 2005; Knaack and Stenflo 2005; Forgács-Dajka and Borkovits 2007; Obridko and Shelting 2007; Ruzmaikin et al. 2008; Valdés-Galicia and Velasco 2008; Vecchio and Carbone 2009; Vecchio et al. 2010, 2012b; Zaqarashvili et al. 2010b; Katsavrias et al. 2012; Laurenza et al. 2012; Singh and Gautam Badruddin 2012; D'Alessi et al. 2013; Popova and Yukhina 2013; Cho et al. 2014, and others). An extended discussion on the QBO nature is given by Knaack and Stenflo (2005) and references therein.

In any discussion on the drivers of the QBO it should be remembered that the Sun is not the only star to exhibit more than one activity cycle. It is well known that many stars exhibit two distinct activity cycles, where the shorter cycle has a smaller amplitude and is regarded as secondary (e.g. Baliunas et al. 1995; Saar and Brandenburg 1999; Böhm-Vitense 2007; Oláh et al. 2009; Metcalfe et al. 2013). Böhm-Vitense (2007) suggest an explanation in terms of two dynamo actions, one fed by the near-surface differential rotation and another seated at the interface at the base of the convection zone.

A similar explanation has been proposed to explain the QBO in the Sun: two dynamos, one at the base of the convection zone and another seated near the bottom of the layer extending 5 % below the solar surface (∼35000 km). This region shows strong rotational shear, like the shear observed across the deeper-seated tachocline where the omega effect of the main dynamo is believed to operate (Corbard and Thompson 2002; Antia et al. 2008). The presence of two different types of dynamo operating at different depths was proposed by Benevolenskaya (1998) to explain the QBO observed in the surface magnetic field. However, this structure is also able to explain features observed in the helioseismic data, which are sensitive to the Sun's internal magnetic field: when the 11-yr cycle is in a strong phase, buoyant magnetic flux sent upward from the base of the envelope by the main dynamo could help to nudge flux processed by this second dynamo into layers that are shallow enough to imprint a detectible acoustic signature on the modes and be visible in surface measures of solar activity. When the main cycle is in a weak phase, the flux from the second dynamo would not receive an extra nudge, and would not be buoyant enough to be detected in surface proxies. However, the QBO magnetic field would still be able to influence the acoustic oscillations, although to a lesser extent. This would explain why the amplitude of the QBO signal is largest around times of solar maximum. Furthermore, there is a suggestion that the QBO is present in the p-mode frequencies away from solar maximum (Fig. 4 and Broomhall et al. 2009, 2011, 2012; Fletcher et al. 2010; Jain et al. 2011). This is understandable since p-mode frequencies respond to conditions beneath the surface where the short-term dynamo would be positioned away from solar maximum. If one could detect the QBO in both p modes and surface activity proxies away from solar maximum, one might, therefore, expect a phase shift between the helioseismic QBO and the surface QBO.

Wang and Sheeley (2003) simulated the Sun's equatorial dipole field strength and total open flux using a flux transport model and assuming that active regions emerge at randomly distributed longitudes. The size of the peaks in the time series simulated by Wang and Sheeley (2003) were primarily dependent on the longitudes of the emergence of active regions, while the periodicity was determined by the decay timescale of the equatorial dipole, which here they took to be about 1 yr based on measures of the meridional flow speed (Wang et al. 2002). Their simulations produce peaks similar to those associated with the QBO and suggest that the highest peak associated with the QBO is as likely to occur during the declining phase of the 11yr cycle as at the maximum in solar activity, which is in agreement with the helioseismic observations (see, for example, Fig. 4). The simulated data also show many other features observed in the real observations such as multipeaked maxima, and periodicities in the range 1–3 yrs, with no dominant periodicity in that range. Wang and Sheeley also note that if the meridional flow rate varies from cycle to cycle in the manner observed by Wang et al. (2002) the periodicities will also vary, again replicating a QBO feature of the real data. However, Knaack and Stenflo (2005) argued that they found certain regularities in the QBO features during odd and even solar cycles which would be unlikely if these periodicities were a random occurrence.

An alternative explanation is in terms of the instability of magnetic Rossby waves in the solar tachocline (e.g. Lou 2000; Knaack and Stenflo 2005; Chowdhury et al. 2009b; Zaqarashvili et al. 2010b, 2011). The period of the instability is dependent on the parameters used to describe the differential rotation and the strength of the magnetic field, which both vary through the solar cycle. Therefore the instability of magnetic Rossby waves can explain many of the observable features of the QBO, including intermittency and variable periodicity. Zaqarashvili et al. (2010b) claim that magnetic Rossby waves can also explain the N-S asymmetry. The same process can be used to explain Reiger-type periodicities: When the magnetic field is relatively weak ($<10^4$ G) the instabilities produce Reiger-type periodicites (Zaqarashvili et al. 2010a), while QBO periodicities are produced when the magnetic field is strong (10^5 G). We note that helioseismic estimates of the strength of the magnetic field at the base of the convection zone are of the order of 10^5 G (Chou and Serebryanskiy 2005; Serebryanskiy and Chou 2005; Baldner et al. 2009).

Another explanation with the potential to explain some of the observed results is spatiotemporal fragmentation. Covas et al. (2000a) produce a two-dimensional axisymmetric mean field dynamo model to show that for certain values of the magnetic Reynolds number spatiotemporal fragmentation occurs near the base of the convection zone that produces oscillations in the differential rotation. The mean field dynamo model used by Covas et al. extends the work of Covas et al. (2000b), who showed that this sort of dynamo model is able to produce torsional oscillation flows similar to those observed in helioseismology. Covas et al. believe that such processes may explain the results of Howe et al. (2000), who observed a 1.3 yr periodicity in the solar rotation profile at the base of the convection zone. Covas et al. also find that for high enough magnetic Reynolds numbers the temporal variations become non-periodic which may explain the intermittent behaviour of the observed 1.3 yr periodicity in the rotation profile (Howe et al. 2011). Spatiotemporal fragmentation would be able to produce the 1.3 yr periodicity because three 11 yr period halvings produces approximately 1.3 yr. We therefore note here that the observed 1.3 yr periodicity may simply be the 8th harmonic of the sunspot cycle (Krivova and Solanki 2002). Furthermore, Kane (2005b), who examined a wide range of solar activity indices, found that "every observed value fits some harmonic or other, and no harmonic is left out".

Ichimoto et al. (1985) suggested that the 155 d periodicity may be related to the timescale for storage and/or escape of magnetic fields in the convection zone. Ballester et al. (2002) suggest the link between Reiger-type periodicities observed in the number of high-energy flares and the photospheric magnetic flux may occur because when new sunspots emerge within already established sunspot groups the conditions for the production of high-energy flares are achieved. Since active regions may persist for up to 30 rotations Chowdhury et al. (2013) suggest that quasi-biennial periodicities may be related to the lifetime of complex active regions. However, the exact cause of the periodicities within the emergence of sunspots remains uncertain.

Simoniello et al. (2013) find that helioseismic results are consistent with the QBO being generated by the beating between dipole and quadropole magnetic configurations of a dynamo (Moss 1999, 2004; Fluri and Berdyugina 2004). It is possible that a nonaxisymmetric (quadropole-like) dynamo could exist alongside the dipole-like dynamo (Moss et al. 1995; Moss and Brooke 2000). It has been proposed that such a quadropole-like dynamo may explain the behaviour whereby active longitudes can suddenly shift by 180° (Tuominen et al. 2002; Berdyugina and Usoskin 2003; Moss 2004). The periodicity at which the active longitudes switch is expected to be the same as the axisymmetric dynamo (Berdyugina et al. 2002). Simoniello et al. (2013) find that major spot activity switches between active longitudes every 1.8–1.9 yr. Furthermore this formalism predicts that the amplitude of the secondary (quadropole) cycle is expected to be smaller than the primary (dipole) cycle, as is observed.

Finally we consider the analogies between the solar and terrestrial QBO. It is widely believed that the QBO observed in Earth's atmosphere is driven by the transport of angular momentum via the interaction between gravity waves, convection and shear (e.g. Baldwin et al. 2001, and references therein). Kumar et al. (1999) demonstrated that a similar transport of angular momentum by gravity-Alfvén waves beneath the base of the convection zone could produce a strong shear layer capable of generating a toroidal magnetic field. Thompson (2001) postulates that this behaviour could explain the helioseismic results observed by Howe et al. (2000). We recall that D'Alessi et al. (2013) suggested that gravity waves might be crucial to the QBO observed in the neutrino flux rate (see Sect. 2.1.1). Furthermore, this is not the first time gravity modes have been suggested to explain short-term variability in the Sun's magnetic field: Wolff (1983) suggested that the influence of gravity modes on the rotation profile of the solar interior may be responsible for Reiger-type periodicities.

The conclusions of different authors are contradictory (see Sect. 3.3). Some light may be shed by detailed study of the GG structures during each solar maximum. However, to date no single candidate stands out as the favoured explanation.

6 Conclusion

The quasi-biennial oscillations of solar origin have already been investigated over several decades. They are ubiquitous, but have some stochastic features, such as intermittency, and variable time-scales. Recent studies reveal a fundamental nature of the QBOs, their intrinsic connection to the dynamo mechanism. The solar QBOs are transferred into the interplanetary space and reflected in the solar wind, galactic cosmic ray modulation and geophysical disturbances. The QBOs are most prominent around the maxima of the 11-yr cycles, this being a reason for temporal lulls in solar activity, or the so called Gnevyshev gaps. Thus, the QBOs are important not only for fundamental physics, but also for the space weather problems. We have tried to demonstrate that our understanding of the nature of the QBO is now rapidly growing and promises new insights into the entity of solar activity.

Acknowledgements The paper was stimulated by the workshop "The solar activity cycle: physical causes and consequences". It is a pleasure to thank André Balogh, Hugh Hudson, Kristof Petrovay, Rudolf von Steiger and the International Space Science Institute for financial support, excellent organization and hospitality. We thank SIDC-team, World Data Center for the Sunspot Index, Royal Observatory of Belgium, Royal Observatory (Greenwich), Wilcox Solar Observatory, NOAA NGDC, GSFC, IZMIRAN Cosmic Ray group, WDC for Geomagnetism (Kyoto) providing their data in open access. We thank the Birmingham Solar Oscillations Network for making their data available to us. G.B. acknowledges support from the Russian Academy of Sciences. A.M.B thanks the Institute of Advanced Study, University of Warwick for their support. Y.E. acknowledges support from the UK Science and Technology Facilities Council (STFC). A.M.B. and Y.E. acknowledge the Leverhulme trust for funding "Probing the Sun: inside and out" project upon which this research is partly based. V.N.: This work was supported by the European Research Council under the SeismoSun Research Project No. 321141 and the BK21 plus program through the National Research Foundation funded by the Ministry of Education of Korea.

References

H.S. Ahluwalia, Ap time variations and interplanetary magnetic field intensity. J. Geophys. Res. **105**, 27481–27488 (2000). doi:10.1029/2000JA900124

H.S. Ahluwalia, Y. Kamide, Gnevyshev gap, Forbush decreases, ICMEs and solar wind electric field: relationships. Adv. Space Res. **35**, 2119–2123 (2005). doi:10.1016/j.asr.2004.11.030

M. Akioka, J. Kubota, K. Ichimoto, I. Tohmura, M. Suzuki, The 17-month periodicity of sunspot activity. Sol. Phys. **112**, 313–316 (1987). doi:10.1007/BF00148785

M. Amenomori, S. Ayabe, S.W. Cui, Danzengluobu, L.K. Ding, X.H. Ding, C.F. Feng, Z.Y. Feng, X.Y. Gao, Q.X. Geng, H.W. Guo, H.H. He, M. He, K. Hibino, N. Hotta, H. Hu, H.B. Hu, J. Huang, Q. Huang, M. Izumi, H.Y. Jia, F. Kajino, K. Kasahara, Y. Katayose, C. Kato, K. Kawata, Labaciren, G.M. Le, J.Y. Li, H. Lu, S.L. Lu, X.R. Meng, K. Mizutani, J. Mu, K. Munakata, A. Nagai, H. Nanjo, M. Nishizawa, M. Ohnishi, I. Ohta, H. Onuma, T. Ouchi, S. Ozawa, J.R. Ren, T. Saito, M. Sakata, T. Sasaki, M. Shibata, A. Shiomi, T. Shirai, H. Sugimoto, M. Takita, Y.H. Tan, N. Tateyama, S. Torii, H. Tsuchiya, S. Udo, T.

Utsugi, B.S. Wang, H. Wang, X. Wang, Y.G. Wang, H.R. Wu, L. Xue, Y. Yamamoto, C.T. Yan, X.C. Yang, S. Yasue, Z.H. Ye, G.C. Yu, A.F. Yuan, T. Yuda, H.M. Zhang, J.L. Zhang, N.J. Zhang, X.Y. Zhang, Y. Zhang, Zhaxisangzhu, X.X. Zhou (Tibet AS γ Collaboration), Variation of sun shadow in the solar cycle 23 observed with the Tibet air shower array. Adv. Space Res. **38**, 936–941 (2006). doi:10.1016/j.asr.2006.04.023

A. Antalova, M.N. Gnevyshev, Latitudinal distribution of sunspot areas during the period 1874–1976. Contrib. Astron. Obs. Skaln. Pleso **11**, 63–93 (1983)

H.M. Antia, S. Basu, Temporal variations of the rotation rate in the solar interior. Astrophys. J. **541**, 442–448 (2000). doi:10.1086/309421

H.M. Antia, S. Basu, S.M. Chitre, Solar rotation rate and its gradients during cycle 23. Astrophys. J. **681**, 680–692 (2008). doi:10.1086/588523

T. Appourchaux, K. Belkacem, A.-M. Broomhall, W.J. Chaplin, D.O. Gough, G. Houdek, J. Provost, F. Baudin, P. Boumier, Y. Elsworth, R.A. García, B.N. Andersen, W. Finsterle, C. Fröhlich, A. Gabriel, G. Grec, A. Jiménez, A. Kosovichev, T. Sekii, T. Toutain, S. Turck-Chièze, The quest for the solar g modes. Astron. Astrophys. Rev. **18**, 197–277 (2010). doi:10.1007/s00159-009-0027-z

N. Astafyeva, G. Bazilevskaya, Long-term changes of cosmic ray intensity: spectral behaviour and 27-day variations. Phys. Chem. Earth, Part C, Sol.-Terr. Planet. Sci. **25**, 129–132 (2000). doi:10.1016/S1464-1917(99)00053-7

T. Ataç, A. Özgüç, J. Rybák, Overview of the flare index during the maximum phase of the solar cycle 23. Adv. Space Res. **35**, 400–405 (2005). doi:10.1016/j.asr.2005.01.040

O.G. Badalyan, V.N. Obridko, Solar magnetic fields and the intensity of the green coronal line. Astron. Rep. **48**, 678–687 (2004). doi:10.1134/1.1787070

O.G. Badalyan, V.N. Obridko, North-South asymmetry of the sunspot indices and its quasi-biennial oscillations. New Astron. **16**, 357–365 (2011). doi:10.1016/j.newast.2011.01.005

O.G. Badalyan, V.N. Obridko, J. Sýkora, Quasi-biennial oscillations in the North—South asymmetry of solar activity. Sol. Phys. **247**, 379–397 (2008). doi:10.1007/s11207-008-9120-0

J.N. Bahcall, R. Davis Jr., Solar neutrinos: a scientific puzzle. Science **191**, 264–267 (1976). doi:10.1126/science.191.4224.264

T. Bai, Periodicities in solar flare occurrence: analysis of cycles 19–23. Astrophys. J. **591**, 406–415 (2003). doi:10.1086/375295

C.S. Baldner, T.P. Larson, S. Basu, Solar-cycle related changes at the base of the convection zone, in *Solar-Stellar Dynamos as Revealed by Helio- and Asteroseismology: GONG 2008/SOHO 21*, ed. by M. Dikpati, T. Arentoft, I. González Hernández, C. Lindsey, F. Hill. Astronomical Society of the Pacific Conference Series, vol. 416, (2009), p. 477

M.P. Baldwin, L.J. Gray, T.J. Dunkerton, K. Hamilton, P.H. Haynes, W.J. Randel, J.R. Holton, M.J. Alexander, I. Hirota, T. Horinouchi, D.B.A. Jones, J.S. Kinnersley, C. Marquardt, K. Sato, M. Takahashi, The quasi-biennial oscillation. Rev. Geophys. **39**, 179–229 (2001). doi:10.1029/1999RG000073

S.L. Baliunas, R.A. Donahue, W.H. Soon, J.H. Horne, J. Frazer, L. Woodard-Eklund, M. Bradford, L.M. Rao, O.C. Wilson, Q. Zhang, W. Bennett, J. Briggs, S.M. Carroll, D.K. Duncan, D. Figueroa, H.H. Lanning, T. Misch, J. Mueller, R.W. Noyes, D. Poppe, A.C. Porter, C.R. Robinson, J. Russell, J.C. Shelton, T. Soyumer, A.H. Vaughan, J.H. Whitney, Chromospheric variations in main-sequence stars. Astrophys. J. **438**, 269–287 (1995). doi:10.1086/175072

J.L. Ballester, R. Oliver, M. Carbonell, The near 160 day periodicity in the photospheric magnetic flux. Astrophys. J. **566**, 505–511 (2002). doi:10.1086/338075

J.L. Ballester, R. Oliver, M. Carbonell, Return of the near 160 day periodicity in the photospheric magnetic flux during solar cycle 23. Astrophys. J. Lett. **615**, 173–176 (2004). doi:10.1086/426430

G.A. Bazilevskaya, M.B. Krainev, V.S. Makhmutov, E.O. Flückiger, A.I. Sladkova, M. Storini, Structure of the maximum phase of solar cycles 21 and 22. Sol. Phys. **197**, 157–174 (2000). doi:10.1023/A:1026515520311

G.A. Bazilevskaya, E.O. Fluckiger, M.B. Krainev, V.S. Makhmutov, A.I. Sladkova, M. Storini, Distribution of solar energetic particle events over an 11-year solar cycle. Int. Cosm. Ray Conf. **8**, 3413 (2001)

G.A. Bazilevskaya, V.S. Makhmutov, A.I. Sladkova, Gnevyshev gap effects in solar energetic particle activity. Adv. Space Res. **38**, 484–488 (2006). doi:10.1016/j.asr.2004.11.011

A.V. Belov, Forbush effects and their connection with solar, interplanetary and geomagnetic phenomena (2009). doi:10.1017/S1743921309029676

E.E. Benevolenskaya, Double magnetic cycle of solar activity. Sol. Phys. **161**, 1–8 (1995). doi:10.1007/BF00732080

E.E. Benevolenskaya, A model of the double magnetic cycle of the sun. Astrophys. J. Lett. **509**, 49–52 (1998). doi:10.1086/311755

E.E. Benevolenskaya, Impulses of activity and the solar cycle. Sol. Phys. **216**, 325–341 (2003). doi:10.1023/A:1026105100181

E.E. Benevolenskaya, The formation and evolution of complexes of activity, activity nests and the large-scale connectivity in the solar corona, in *Large-Scale Structures and Their Role in Solar Activity*, ed. by K. Sankarasubramanian, M. Penn, A. Pevtsov. Astronomical Society of the Pacific Conference Series, vol. 346, (2005), p. 129
S.V. Berdyugina, I.G. Usoskin, Active longitudes in sunspot activity: century scale persistence. Astron. Astrophys. **405**, 1121–1128 (2003). doi:10.1051/0004-6361:20030748
S.V. Berdyugina, J. Pelt, I. Tuominen, Magnetic activity in the young solar analog LQ Hydrae. I. Active longitudes and cycles. Astron. Astrophys. **394**, 505–515 (2002). doi:10.1051/0004-6361:20021179
F. Boberg, H. Lundstedt, J.T. Hoeksema, P.H. Scherrer, W. Liu, Solar mean magnetic field variability: a wavelet approach to Wilcox solar observatory and SOHO/Michelson Doppler imager observations. J. Geophys. Res. **107**, 1318 (2002). doi:10.1029/2001JA009195
E. Böhm-Vitense, Chromospheric activity in G and K main-sequence stars, and what it tells us about stellar dynamos. Astrophys. J. **657**, 486–493 (2007). doi:10.1086/510482
A. Broomhall, W.J. Chaplin, Y. Elsworth, S.T. Fletcher, R. New, Is the current lack of solar activity only skin deep? Astrophys. J. Lett. **700**, 162–165 (2009). doi:10.1088/0004-637X/700/2/L162
A.-M. Broomhall, S.T. Fletcher, D. Salabert, S. Basu, W.J. Chaplin, Y. Elsworth, R.A. García, A. Jiménez, R. New, Are short-term variations in solar oscillation frequencies the signature of a second solar dynamo? J. Phys. Conf. Ser. **271**(1), 012025 (2011). doi:10.1088/1742-6596/271/1/012025
A.-M. Broomhall, W.J. Chaplin, Y. Elsworth, R. Simoniello, Quasi-biennial variations in helioseismic frequencies: can the source of the variation be localized? Mon. Not. R. Astron. Soc. **420**, 1405–1414 (2012). doi:10.1111/j.1365-2966.2011.20123.x
A.C. Cadavid, J.K. Lawrence, D.P. McDonald, A. Ruzmaikin, Independent global modes of solar magnetic field fluctuations. Sol. Phys. **226**, 359–376 (2005). doi:10.1007/s11207-005-8187-0
W.J. Chaplin, Y. Elsworth, B.A. Miller, G.A. Verner, R. New, Solar p-mode frequencies over three solar cycles. Astrophys. J. **659**, 1749–1760 (2007). doi:10.1086/512543
I.-H. Cho, J. Hwang, Y.-D. Park, Revisiting solar and heliospheric 1.3-Year signals during 1970–2007. Sol. Phys. **289**, 707–719 (2014). doi:10.1007/s11207-013-0365-x
D.-Y. Chou, A. Serebryanskiy, In search of the solar cycle variations of p-mode frequencies generated by perturbations in the solar interior. Astrophys. J. **624**, 420–427 (2005). doi:10.1086/428925
D.P. Choudhary, J.K. Lawrence, M. Norris, A.C. Cadavid, Different periodicities in the sunspot area and the occurrence of solar flares and coronal mass ejections in solar cycle 23–24. Sol. Phys. **289**, 649–656 (2014). doi:10.1007/s11207-013-0392-7
P. Chowdhury, P.C. Ray, Periodicities of solar electron flare occurrence: analysis of cycles 21–23. Mon. Not. R. Astron. Soc. **373**, 1577–1589 (2006). doi:10.1111/j.1365-2966.2006.11120.x
P. Chowdhury, M. Khan, P. Ray, Intermediate-term periodicities in relativistic solar electron fluences during solar cycles 22 and 23. Adv. Space Res. **43**, 297–307 (2009a). doi:10.1016/j.asr.2008.06.008
P. Chowdhury, M. Khan, P.C. Ray, Intermediate-term periodicities in sunspot areas during solar cycles 22 and 23. Mon. Not. R. Astron. Soc. **392**, 1159–1180 (2009b). doi:10.1111/j.1365-2966.2008.14117.x
P. Chowdhury, D.P. Choudhary, S. Gosain, A study of the hemispheric asymmetry of sunspot area during solar cycles 23 and 24. Astrophys. J. **768**, 188 (2013). doi:10.1088/0004-637X/768/2/188
J. Christensen-Dalsgaard, Helioseismology, Rev. Mod. Phys. **74**, 1073–1129 (2002). doi:10.1103/RevModPhys.74.1073
T. Corbard, M.J. Thompson, The subsurface radial gradient of solar angular velocity from MDI f-mode observations. Sol. Phys. **205**, 211–229 (2002). doi:10.1023/A:1014224523374
E. Covas, R. Tavakol, D. Moss, Spatiotemporal fragmentation as a mechanism for different dynamical modes of behaviour in the solar convection zone. Astron. Astrophys. **363**, 13–16 (2000a)
E. Covas, R. Tavakol, D. Moss, A. Tworkowski, Torsional oscillations in the solar convection zone. Astron. Astrophys. **360**, 21–24 (2000b)
L. D'Alessi, A. Vecchio, V. Carbone, M. Laurenza, M. Storini, Quasi-biennial modulation of the solar neutrino flux: a "Telescope" for the solar interior. J. Mod. Phys. **4**, 49–56 (2013). doi:10.4236/jmp.2013.44A008
S. Danilovic, I. Vince, N. Vitas, P. Jovanovic, Time series analysis of long term full disk observations of the Mn i 539.4 nm solar line. Serb. Astron. J. **170**, 79–88 (2005). doi:10.2298/SAJ0570079D
Y. Elsworth, R. Howe, G.R. Isaak, C.P. McLeod, R. New, Variation of low-order acoustic solar oscillations over the solar cycle. Nature **345**, 322–324 (1990). doi:10.1038/345322a0
F. Feminella, M. Storini, Large-scale dynamical phenomena during solar activity cycles. Astron. Astrophys. **322**, 311–319 (1997)
S.T. Fletcher, A.-M. Broomhall, D. Salabert, S. Basu, W.J. Chaplin, Y. Elsworth, R.A. Garcia, R. New, A seismic signature of a second dynamo? Astrophys. J. Lett. **718**, 19–22 (2010). doi:10.1088/2041-8205/718/1/L19

D.M. Fluri, S.V. Berdyugina, Flip-flops as observational signatures of different dynamo modes in cool stars. Sol. Phys. **224**, 153–160 (2004). doi:10.1007/s11207-005-4147-y

E. Forgács-Dajka, T. Borkovits, Searching for mid-term variations in different aspects of solar activity—looking for probable common origins and studying temporal variations of magnetic polarities. Mon. Not. R. Astron. Soc. **374**, 282–291 (2007). doi:10.1111/j.1365-2966.2006.11167.x

K. Georgieva, Why the sunspot cycle is double peaked. ISRN Astron. Astrophys. **2011** (2011). doi:10.5402/2011/437838

K. Georgieva, B. Kirov, Long-term variations in solar meridional circulation from geomagnetic data: implications for solar dynamo theory. (2007). arXiv:physics/0703187

K. Georgieva, B. Kirov, Solar dynamo and geomagnetic activity. J. Atmos. Sol.-Terr. Phys. **73**, 207–222 (2011). doi:10.1016/j.jastp.2010.03.003

M.N. Gnevyshev, On the 11-years cycle of solar activity. Sol. Phys. **1**, 107–120 (1967). doi:10.1007/BF00150306

M.N. Gnevyshev, Essential features of the 11-year solar cycle. Sol. Phys. **51**, 175–183 (1977). doi:10.1007/BF00240455

N. Gyenge, T. Baranyi, A. Ludmány, Migration and extension of solar active longitudinal zones. Sol. Phys. **289**, 579–591 (2014). doi:10.1007/s11207-013-0424-3

D.H. Hathaway, The solar cycle. Living Rev. Sol. Phys. **7**, 1 (2010). doi:10.12942/lrsp-2010-1

M.E. Hill, D.C. Hamilton, S.M. Krimigis, Periodicity of 151 days in outer heliospheric anomalous cosmic ray fluxes. J. Geophys. Res. **106**, 8315–8322 (2001). doi:10.1029/2000JA000380

J.T. Hoeksema, Global solar magnetic fields. NASA STI/Recon Technical Report N **92**, 29421 (1991)

R. Howe, J. Christensen-Dalsgaard, F. Hill, R.W. Komm, R.M. Larsen, J. Schou, M.J. Thompson, J. Toomre, Dynamic variations at the base of the solar convection zone. Science **287**, 2456–2460 (2000). doi:10.1126/science.287.5462.2456

R. Howe, R. Komm, F. Hill, J. Christensen-Dalsgaard, T.P. Larson, J. Schou, M.J. Thompson, J. Toomre, Rotation-rate variations at the tachocline: an update. J. Phys. Conf. Ser. **271**(1), 012075 (2011). doi:10.1088/1742-6596/271/1/012075

K. Ichimoto, J. Kubota, M. Suzuki, I. Tohmura, H. Kurokawa, Periodic behaviour of solar flare activity. Nature **316**, 422–424 (1985). doi:10.1038/316422a0

E.V. Ivanov, V.N. Obridko, B.D. Shelting, Quasi-biennial oscillations of the solar magnetic fields, in *Solar Variability: from Core to Outer Frontiers*, ed. by A. Wilson. ESA Special Publication, vol. 506, (2002), pp. 847–850

K. Jain, S.C. Tripathy, F. Hill, How peculiar was the recent extended minimum: a hint toward double minima. Astrophys. J. **739**, 6 (2011). doi:10.1088/0004-637X/739/1/6

S.J. Jiménez-Reyes, R.A. García, A. Jiménez, W.J. Chaplin, Excitation and damping of low-degree solar p-modes during activity cycle 23: analysis of GOLF and VIRGO sun photometer data. Astrophys. J. **595**, 446–457 (2003). doi:10.1086/377304

S.J. Jiménez-Reyes, W.J. Chaplin, Y. Elsworth, R.A. García, R. Howe, H. Socas-Navarro, T. Toutain, On the variation of the peak asymmetry of low-l solar p modes. Astrophys. J. **654**, 1135–1145 (2007). doi:10.1086/509700

B. Joshi, A. Joshi, Intermediate-term periodicities in soft x-ray flare index during solar cycles 21, 22 and 23. Sol. Phys. **226**, 153–161 (2005). doi:10.1007/s11207-005-5716-9

R.P. Kane, Evolutions of various solar indices around sunspot maximum and sunspot minimum years. Ann. Geophys. **20**, 741–755 (2002). doi:10.5194/angeo-20-741-2002

R.P. Kane, Differences in the quasi-biennial oscillation and quasi-triennial oscillation characteristics of the solar, interplanetary, and terrestrial parameters. J. Geophys. Res. **110**, 1108 (2005a). doi:10.1029/2004JA010606

R.P. Kane, Short-term periodicities in solar indices. Sol. Phys. **227**, 155–175 (2005b). doi:10.1007/s11207-005-1110-x

R.P. Kane, Which one is the 'GNEVYSHEV' GAP? Sol. Phys. **229**, 387–407 (2005c). doi:10.1007/s11207-005-7451-7

R.P. Kane, A detailed comparison of cosmic ray gaps with solar gnevyshev gaps. Sol. Phys. **236**, 207–226 (2006). doi:10.1007/s11207-006-0142-1

C. Kato, K. Munakata, S. Yasue, K. Inoue, F.B. McDonald, A 1.7-year quasi-periodicity in cosmic ray intensity variation observed in the outer heliosphere. J. Geophys. Res. **108**, 1367 (2003). doi:10.1029/2003JA009897

C. Katsavrias, P. Preka-Papadema, X. Moussas, Wavelet analysis on solar wind parameters and geomagnetic indices. Sol. Phys. **280**, 623–640 (2012). doi:10.1007/s11207-012-0078-6

M.N. Khramova, E.V. Kononovich, S.A. Krasotkin, Quasi-biennial oscillations of global solar-activity indices. Sol. Syst. Res. **36**, 507–512 (2002)

A. Kilcik, A. Ozguc, One possible reason for double-peaked maxima in solar cycles: is a second maximum of solar cycle 24 expected? Sol. Phys. **289**, 1379–1386 (2014). doi:10.1007/s11207-013-0407-4

A. Kilcik, V.B. Yurchyshyn, V. Abramenko, P.R. Goode, A. Ozguc, J.P. Rozelot, W. Cao, Time distributions of large and small sunspot groups over four solar cycles. Astrophys. J. **731**, 30 (2011). doi:10.1088/0004-637X/731/1/30

R. Knaack, J.O. Stenflo, Spherical harmonic decomposition of solar magnetic fields. Astron. Astrophys. **438**, 349–363 (2005). doi:10.1051/0004-6361:20052765

R. Knaack, J.O. Stenflo, S.V. Berdyugina, Periodic oscillations in the North-South asymmetry of the solar magnetic field. Astron. Astrophys. **418**, 17–20 (2004). doi:10.1051/0004-6361:20040107

R. Knaack, J.O. Stenflo, S.V. Berdyugina, Evolution and rotation of large-scale photospheric magnetic fields of the Sun during cycles 21-23. Periodicities, North-South asymmetries and r-mode signatures. Astron. Astrophys. **438**, 1067–1082 (2005). doi:10.1051/0004-6361:20042091

M.B. Krainev, G.A. Bazilevskaya, V.S. Makhmutov, Solar magnetic field and dynamic phenomena in cosmic rays. Adv. Space Res. **29**, 331–336 (2002). doi:10.1016/S0273-1177(01)00593-2

N.A. Krivova, S.K. Solanki, The 1.3-year and 156-day periodicities in sunspot data: wavelet analysis suggests a common origin. Astron. Astrophys. **394**, 701–706 (2002). doi:10.1051/0004-6361:20021063

K. Kudela, J. Rybák, A. Antalová, M. Storini, Time evolution of low-frequency periodicities in cosmic ray intensity. Sol. Phys. **205**, 165–175 (2002). doi:10.1023/A:1013869322693

K. Kudela, H. Mavromichalaki, A. Papaioannou, M. Gerontidou, On mid-term periodicities in cosmic rays. Sol. Phys. **266**, 173–180 (2010). doi:10.1007/s11207-010-9598-0

P. Kumar, S. Talon, J.-P. Zahn, Angular momentum redistribution by waves in the sun. Astrophys. J. **520**, 859–870 (1999). doi:10.1086/307464

A. Lara, A. Borgazzi, O. Mendes Jr., R.R. Rosa, M.O. Domingues, Short-period fluctuations in coronal mass ejection activity during solar cycle 23. Sol. Phys. **248**, 155–166 (2008). doi:10.1007/s11207-008-9153-4

M. Laurenza, M. Storini, S. Giangravè, G. Moreno, Search for periodicities in the IMP 8 charged particle measurement experiment proton fluxes for the energy bands 0.50-0.96 MeV and 190-440 MeV. J. Geophys. Res. **114**, 1103 (2009). doi:10.1029/2008JA013181

M. Laurenza, A. Vecchio, M. Storini, V. Carbone, Quasi-biennial modulation of galactic cosmic rays. Astrophys. J. **749**, 167 (2012). doi:10.1088/0004-637X/749/2/167

K.G. Libbrecht, M.F. Woodard, Solar-cycle effects on solar oscillation frequencies. Nature **345**, 779–782 (1990). doi:10.1038/345779a0

M. Lockwood, Long-term variations in the magnetic fields of the sun and the heliosphere: their origin, effects, and implications. J. Geophys. Res. **106**, 16021–16038 (2001). doi:10.1029/2000JA000115

Y.-Q. Lou, Rossby-type wave-induced periodicities in flare activities and sunspot areas or groups during solar maxima. Astrophys. J. **540**, 1102–1108 (2000). doi:10.1086/309387

Y.-Q. Lou, Y.-M. Wang, Z. Fan, S. Wang, J.X. Wang, Periodicities in solar coronal mass ejections. Mon. Not. R. Astron. Soc. **345**, 809–818 (2003). doi:10.1046/j.1365-8711.2003.06993.x

R. Lukianova, K. Mursula, Changed relation between sunspot numbers, solar UV/EUV radiation and TSI during the declining phase of solar cycle 23. J. Atmos. Sol.-Terr. Phys. **73**, 235–240 (2011). doi:10.1016/j.jastp.2010.04.002

K. Maeda, Quasi-biennial cycles in cosmic ray intensity. J. Atmos. Sci. **24**, 320–322 (1967). doi:10.1175/1520-0469(1967)024<0320:QBCICR>2.0.CO;2

H. Mavromichalaki, P. Preka-Papadema, I. Liritzis, B. Petropoulos, V. Kurt, Short-term variations of cosmic-ray intensity and flare related data in 1981–1983. New Astron. **8**, 777–794 (2003). doi:10.1016/S1384-1076(03)00066-6

T.S. Metcalfe, A.P. Buccino, B.P. Brown, S. Mathur, D.R. Soderblom, T.J. Henry, P.J.D. Mauas, R. Petrucci, J.C. Hall, S. Basu, Magnetic activity cycles in the exoplanet host star epsilon Eridani. Astrophys. J. Lett. **763**, 26 (2013). doi:10.1088/2041-8205/763/2/L26

L.I. Miroshnichenko, Solar cosmic rays in the system of solar terrestrial relations. J. Atmos. Sol.-Terr. Phys. **70**, 450–466 (2008). doi:10.1016/j.jastp.2007.08.027

D. Moss, Non-axisymmetric solar magnetic fields. Mon. Not. R. Astron. Soc. **306**, 300–306 (1999). doi:10.1046/j.1365-8711.1999.02510.x

D. Moss, Dynamo models and the flip-flop phenomenon in late-type stars. Mon. Not. R. Astron. Soc. **352**, 17–20 (2004). doi:10.1111/j.1365-2966.2004.08125.x

D. Moss, J. Brooke, Towards a model for the solar dynamo. Mon. Not. R. Astron. Soc. **315**, 521–533 (2000). doi:10.1046/j.1365-8711.2000.03452.x

D. Moss, D.M. Barker, A. Brandenburg, I. Tuominen, Nonaxisymmetric dynamo solutions and extended starspots on late-type stars. Astron. Astrophys. **294**, 155–164 (1995)

X. Moussas, J.M. Polygiannakis, P. Preka-Papadema, G. Exarhos, Solar cycles: a tutorial. Adv. Space Res. **35**, 725–738 (2005). doi:10.1016/j.asr.2005.03.148

J. Murakozy, A. Ludmany, Cycle dependence of the longitudinal-latitudinal sunspot motion correlations. Astron. Astrophys. **486**, 1003–1007 (2008). doi:10.1051/0004-6361:20078456

K. Mursula, J.H. Vilppola, Fluctuations of the solar dynamo observed in the solar wind and interplanetary magnetic field at 1 AU and in the outer heliosphere. Sol. Phys. **221**, 337–349 (2004). doi:10.1023/B:SOLA.0000035053.17913.26

K. Mursula, B. Zieger, The 1.3-year variation in solar wind speed and geomagnetic activity. Adv. Space Res. **25**, 1939–1942 (2000). doi:10.1016/S0273-1177(99)00608-0

K. Mursula, B. Zieger, J.H. Vilppola, Mid-term quasi-periodicities in geomagnetic activity during the last 15 solar cycles: connection to solar dynamo strength to the memory of Karolen I. Paularena (1957–2001). Sol. Phys. **212**, 201–207 (2003). doi:10.1023/A:1022980029618

K. Nagashima, S. Sakakibara, I. Morishita, Quiescence of GLE-producible solar proton eruptions during the transition phase of heliomagnetic polarity reversal near the solar-activity-maximum period. Int. Cosm. Ray Conf. **3**, 29 (1991)

A.A. Norton, J.C. Gallagher, Solar-cycle characteristics examined in separate hemispheres: phase, gnevyshev gap, and length of minimum. Sol. Phys. **261**, 193–207 (2010). doi:10.1007/s11207-009-9479-6

V.N. Obridko, B.D. Shelting, Quasi-biennial oscillations of the global solar magnetic field. Astron. Rep. **45**, 1012–1017 (2001). doi:10.1134/1.1426132

V.N. Obridko, B.D. Shelting, Occurrence of the 1.3-year periodicity in the large-scale solar magnetic field for 8 solar cycles. Adv. Space Res. **40**, 1006–1014 (2007). doi:10.1016/j.asr.2007.04.105

V.P. Okhlopkov, Distinctive properties of the frequency spectra of cosmic ray variations and parameters of solar activity and the interplanetary medium in solar cycles 20–23. Mosc. University Phys. Bull. **66**, 99–103 (2011). doi:10.3103/S0027134911010188

K. Oláh, Z. Kolláth, T. Granzer, K.G. Strassmeier, A.F. Lanza, S. Järvinen, H. Korhonen, S.L. Baliunas, W. Soon, S. Messina, G. Cutispoto, Multiple and changing cycles of active stars. II. Results. Astron. Astrophys. **501**, 703–713 (2009). doi:10.1051/0004-6361/200811304

A. Özgüç, T. Ataç, J. Rybák, Temporal variability of the flare index (1966–2001). Sol. Phys. **214**, 375–396 (2003). doi:10.1023/A:1024225802080

P.L. Pallé, C. Régulo, T. Roca Cortés, Solar cycle induced variations of the low L solar acoustic spectrum. Astron. Astrophys. **224**, 253–258 (1989)

C. Petrick, K. Matthes, H. Dobslaw, M. Thomas, Impact of the solar cycle and the QBO on the atmosphere and the ocean. J. Geophys. Res., Atmos. **117**, 17111 (2012). doi:10.1029/2011JD017390

E.P. Popova, N.A. Yukhina, The quasi-biennial cycle of solar activity and dynamo theory. Astron. Lett. **39**, 729–735 (2013). doi:10.1134/S1063773713100046

V.E. Reznikova, K. Shibasaki, R.A. Sych, V.M. Nakariakov, Three-minute oscillations above sunspot umbra observed with the solar dynamics Observatory/Atmospheric imaging assembly and nobeyama radioheliograph. Astrophys. J. **746**, 119 (2012). doi:10.1088/0004-637X/746/2/119

I.G. Richardson, H.V. Cane, The ∼150 day quasi-periodicity in interplanetary and solar phenomena during cycle 23. Geophys. Res. Lett. **32**, 2104 (2005). doi:10.1029/2004GL021691

I.G. Richardson, H.V. Cane, E.W. Cliver, Sources of geomagnetic activity during nearly three solar cycles (1972–2000). J. Geophys. Res. **107**, 1187 (2002). doi:10.1029/2001JA000504

E. Rieger, G. Kanbach, C. Reppin, G.H. Share, D.J. Forrest, E.L. Chupp, A 154-day periodicity in the occurrence of hard solar flares? Nature **312**, 623–625 (1984). doi:10.1038/312623a0

J. Rodríguez-Pacheco, J.J. Blanco, B. Heber, R. Gómez-Herrero, Energetic particles measured in and out of the ecliptic plane during the last Gnevyshev gap. Sol. Phys. **281**, 491–499 (2012). doi:10.1007/s11207-012-9977-9

A. Ruzmaikin, A.C. Cadavid, J. Lawrence, Quasi-periodic patterns coupling the sun, solar wind and the earth. J. Atmos. Sol.-Terr. Phys. **70**, 2112–2117 (2008). doi:10.1016/j.jastp.2008.09.013

J. Rybák, A. Antalová, M. Storini, The wavelet analysis of the solar and cosmic-ray data. Space Sci. Rev. **97**, 359–362 (2001). doi:10.1023/A:1011805923567

S.H. Saar, A. Brandenburg, Time evolution of the magnetic activity cycle period. II. Results for an expanded stellar sample. Astrophys. J. **524**, 295–310 (1999). doi:10.1086/307794

K. Sakurai, Quasi-biennial variation of the solar neutrino flux and solar activity. Nature **278**, 146–148 (1979). doi:10.1038/278146a0

D.J. Schove, Sunspot turning-points and aurorae since A.D. 1510. Sol. Phys. **63**, 423–432 (1979). doi:10.1007/BF00174546

S. Sello, Wavelet entropy and the multi-peaked structure of solar cycle maximum. New Astron. **8**, 105–117 (2003). doi:10.1016/S1384-1076(02)00192-6

A. Serebryanskiy, D.-Y. Chou, Comparison of solar cycle variations of solar p-mode frequencies from GONG and MDI. Astrophys. J. **633**, 1187–1190 (2005). doi:10.1086/491467

T. Shirai, Time variation of the solar neutrino fluxes from Super-Kamiokande data. Sol. Phys. **222**, 199–201 (2004). doi:10.1023/B:SOLA.0000043565.83411.ec

S.M. Silverman, R. Shapiro, Power spectral analysis of auroral occurrence frequency. J. Geophys. Res. **88**, 6310–6316 (1983). doi:10.1029/JA088iA08p06310

R. Simoniello, W. Finsterle, D. Salabert, R.A. García, S. Turck-Chièze, A. Jiménez, M. Roth, The quasi-biennial periodicity (QBP) in velocity and intensity helioseismic observations. The seismic QBP over solar cycle 23. Astron. Astrophys. **539**, 135 (2012a). doi:10.1051/0004-6361/201118057

R. Simoniello, K. Jain, S.C. Tripathy, S. Turck-Chiéze, W. Finsterle, M. Roth, Seismic comparison of the 11- and 2-year cycle signatures in the sun. Astron. Nachr. **333**, 1018 (2012b). doi:10.1002/asna.201211813

R. Simoniello, K. Jain, S.C. Tripathy, S. Turck-Chièze, C. Baldner, W. Finsterle, F. Hill, M. Roth, The quasi-biennial periodicity as a window on the solar magnetic dynamo configuration. Astrophys. J. **765**, 100 (2013). doi:10.1088/0004-637X/765/2/100

Y.P. Singh, S. Gautam Badruddin, Temporal variations of short- and mid-term periodicities in solar wind parameters and cosmic ray intensity. J. Atmos. Sol.-Terr. Phys. **89**, 48–53 (2012). doi:10.1016/j.jastp.2012.07.011

M. Storini, M. Laurenza, Solar activity effects on muon data. Mem. Soc. Astron. Ital. **74**, 774 (2003)

M. Storini, G.A. Bazilevskaya, E.O. Fluckiger, M.B. Krainev, V.S. Makhmutov, A.I. Sladkova, The GNEVYSHEV gap: a review for space weather. Adv. Space Res. **31**, 895–900 (2003). doi:10.1016/S0273-1177(02)00789-5

M. Storini, K. Kudela, E.G. Cordaro, S. Massetti, Ground-level enhancements during solar cycle 23: results from SVIRCO, LOMNICKY STIT and LARC neutron monitors. Adv. Space Res. **35**, 416–420 (2005). doi:10.1016/j.asr.2004.12.020

J. Sykora, The coronal responses to the large-scale and long-term phenomena of the lower layers of the sun. Sol. Interplanet. Dyn. **91**, 87–104 (1980)

M.J. Thompson, Temporal variations of the sun's internal structure and dynamics: a theoretical perspective, in *SOHO 10/GONG 2000 Workshop: Helio- and Asteroseismology at the Dawn of the Millennium*, ed. by A. Wilson, P.L. Pallé. ESA Special Publication, vol. 464, (2001), pp. 39–44

I. Tuominen, S.V. Berdyugina, M.J. Korpi, Starspot cycles from Doppler imaging and photometric time series as nonlinear dynamo. Astron. Nachr. **323**, 367–370 (2002). doi:10.1002/1521-3994(200208)323:3/4<367::AID-ASNA367>3.0.CO;2-E

J.F. Valdés-Galicia, V.M. Velasco, Variations of mid-term periodicities in solar activity physical phenomena. Adv. Space Res. **41**, 297–305 (2008). doi:10.1016/j.asr.2007.02.012

J.F. Valdés-Galicia, V.M. Velasco, B. Mendoza, Mid term cosmic ray quasi periodicities and solar magnetic activity manifestations. Int. Cosm. Ray Conf. **2**, 211 (2005)

A. Vecchio, V. Carbone, On the origin of the double magnetic cycle of the sun. Astrophys. J. **683**, 536–541 (2008). doi:10.1086/589768

A. Vecchio, V. Carbone, Spatio-temporal analysis of solar activity: main periodicities and period length variations. Astron. Astrophys. **502**, 981–987 (2009). doi:10.1051/0004-6361/200811024

A. Vecchio, M. Laurenza, V. Carbone, M. Storini, Quasi-biennial modulation of solar neutrino flux and solar and galactic cosmic rays by solar cyclic activity. Astrophys. J. Lett. **709**, 1–5 (2010). doi:10.1088/2041-8205/709/1/L1

A. Vecchio, M. Laurenza, D. Meduri, V. Carbone, M. Storini, The dynamics of the solar magnetic field: polarity reversals, butterfly diagram, and quasi-biennial oscillations. Astrophys. J. **749**, 27 (2012a). doi:10.1088/0004-637X/749/1/27

A. Vecchio, M. Laurenza, M. Storini, V. Carbone, New insights on cosmic ray modulation through a joint use of nonstationary data-processing methods. Adv. Astron. **2012**(2012b). doi:10.1155/2012/834247

Y.-M. Wang, The sun's large-scale magnetic field and its long-term evolution. Sol. Phys. **224**, 21–35 (2004). doi:10.1007/s11207-005-4982-x

Y. Wang, N.R. Sheeley Jr., On the fluctuating component of the sun's large-scale magnetic field. Astrophys. J. **590**, 1111–1120 (2003). doi:10.1086/375026

Y.-M. Wang, J. Lean, N.R. Sheeley Jr., Role of a variable meridional flow in the secular evolution of the sun's polar fields and open flux. Astrophys. J. Lett. **577**, 53–57 (2002). doi:10.1086/344196

C.L. Wolff, The rotational spectrum of g-modes in the sun. Astrophys. J. **264**, 667–676 (1983). doi:10.1086/160640

M.F. Woodard, R.W. Noyes, Change of solar oscillation eigenfrequencies with the solar cycle. Nature **318**, 449 (1985)

T.V. Zaqarashvili, M. Carbonell, R. Oliver, J.L. Ballester, Magnetic rossby waves in the solar tachocline and rieger-type periodicities. Astrophys. J. **709**, 749–758 (2010a). doi:10.1088/0004-637X/709/2/749

T.V. Zaqarashvili, M. Carbonell, R. Oliver, J.L. Ballester, Quasi-biennial oscillations in the solar tachocline caused by magnetic rossby wave instabilities. Astrophys. J. Lett. **724**, 95–98 (2010b). doi:10.1088/2041-8205/724/1/L95

T.V. Zaqarashvili, R. Oliver, J.L. Ballester, M. Carbonell, M.L. Khodachenko, H. Lammer, M. Leitzinger, P. Odert, Rossby waves and polar spots in rapidly rotating stars: implications for stellar wind evolution. Astron. Astrophys. **532**, 139 (2011). doi:10.1051/0004-6361/201117122

S.I. Zharkov, E. Gavryuseva, V.V. Zharkova, On phase relation between toroidal and poloidal magnetic fields in the solar cycle 23, in *Proceedings of the International Astronomical Union, IAU Symposium*, vol. 247, (2008), pp. 39–45. doi:10.1017/S1743921308014634

N.V. Zolotova, D.I. Ponyavin, Impulse-like behavior of the sunspot activity. Astron. Rep. **56**, 250–255 (2012). doi:10.1134/S1063772912030080

T.H. Zurbuchen, A new view of the coupling of the sun and the heliosphere. Annu. Rev. Astron. Astrophys. **45**, 297–338 (2007). doi:10.1146/annurev.astro.45.010807.154030

DOI 10.1007/978-1-4939-2584-1_13
Reprinted from *Space Science Reviews* Journal, DOI 10.1007/s11214-014-0051-9

Solar Cycle Variation of the Sun's Low-Order Magnetic Multipoles: Heliospheric Consequences

Y.-M. Wang

Received: 18 February 2014 / Accepted: 7 May 2014 / Published online: 30 May 2014

Abstract The Sun's dipole and quadrupole components play a central role in the solar cycle evolution of the interplanetary magnetic field (IMF). The long-term variation of the radial IMF component approximately tracks that of the total dipole moment, with additional contributions coming near sunspot maximum from the quadrupole moment and from CMEs. The axial and equatorial components of the dipole vary out of phase with each other over the solar cycle. The equatorial dipole, whose photospheric sources are subject to rotational shearing, decays on a timescale of ~ 1 yr and must be continually regenerated by new sunspot activity; its fluctuating strength depends not only on the activity level, but also on the longitudinal phase relationships among the active regions. During cycles 21–23, the equatorial dipole and IMF reached their peak strength ~ 2 yrs after sunspot maximum; conversely, large dips or "Gnevyshev gaps" occurred when active regions emerged longitudinally out of phase with each other. The ^{10}Be-inferred phase shift in the IMF variation during the Maunder Minimum may be explained by a decrease in the amplitude of the equatorial dipole relative to the axial dipole, due either to a systematic weakening of the emerging bipoles or to an increase in their tilt angles. In mid-2012, during the polarity reversal of cycle 24, the nonaxisymmetric quadrupole component became so dominant that the heliospheric current sheet (HCS) split into two cylindrical components. Hemispheric asymmetries in sunspot activity give rise to an axisymmetric quadrupole component, which has combined with the axial dipole to produce a systematic southward displacement of the HCS since cycle 20.

Keywords Solar cycle · Low-order magnetic multipoles · Interplanetary magnetic field · Heliospheric current sheet

1 Introduction

At the photosphere, the Sun's magnetic field is dominated by high-order multipoles $l \gg 1$. Even when active regions are absent at sunspot minimum, the large-scale distribution of

Y.-M. Wang (✉)
Code 7682W, Space Science Division, Naval Research Laboratory, Washington, DC 20375, USA
e-mail: yi.wang@nrl.navy.mil

photospheric flux differs from that of a simple axisymmetric dipole ($\propto \sin L$, where L denotes latitude), being far more concentrated toward the poles due to the effect of the surface meridional flow. As we proceed outward into the corona, however, the high-order multipoles fall off rapidly with height, so that beyond a heliocentric distance $r \sim 2\ R_\odot$ the dipole component dominates. The quadrupole or $l = 2$ component may also be significant near the heliographic equator or at sunspot maximum. It is apparent that the Sun's dipole and quadrupole components must play an important part in the solar cycle evolution of the open flux and interplanetary magnetic field (IMF).

This overview is concerned with the heliospheric effects of the Sun's lowest-order multipoles, whose behavior in turn reflects the action of the solar dynamo itself. Section 2 focuses on the dipole component and its relationship to the variation of the IMF strength. Considering separately the axial and equatorial components of the dipole, we discuss the origin of the so-called Gnevyshev gaps and of the phase shift observed in the IMF variation during the Maunder Minimum. Section 3 contains a digression on the quasirigid rotation of the coronal field and its solar cycle variation. In Sect. 4, we show how the dominance of the nonaxisymmetric quadrupole component during the maximum phase of cycle 24 caused the heliospheric current sheet (HCS) to split into two cylindrical structures. Section 5 discusses the role of the axisymmetric quadrupole component and hemispheric asymmetries in sunspot activity in producing north–south displacements of the HCS. Higher-order multipoles are considered briefly in Sect. 6. Our conclusions are summarized in Sect. 7.

2 The Dipole Component and the Solar Cycle Variation of the IMF

To a first approximation, the total amount of flux that the solar wind drags into the heliosphere scales as the Sun's dipole strength. In the potential-field source-surface (PFSS) model, the total open flux Φ_{open} is just the total unsigned flux crossing the source surface $r = R_{\text{ss}} = 2.5\ R_\odot$, where the current-free coronal field is constrained to become radial. The radial IMF strength at a given heliocentric distance $r \gg 2.5\ R_\odot$ is then simply

$$|B_r(r)| = \frac{\Phi_{\text{open}}}{4\pi r^2}, \tag{1}$$

where we have assumed that the open flux becomes uniformly distributed in latitude and longitude far from the Sun, in accordance with *Ulysses* magnetometer measurements (Balogh et al. 1995; Smith and Balogh 2008; Erdös and Balogh 2014).

The Sun's total dipole strength may be expressed as $D_{\text{tot}} = (D_{\text{ax}}^2 + D_{\text{eq}}^2)^{1/2}$, where the axial ($l = 1, m = 0$) and equatorial ($l = 1, |m| = 1$) components are related to the photospheric flux distribution $B_r(R_\odot, L, \phi)$ by

$$D_{\text{ax}} = \frac{3}{4\pi}\int B_r(R_\odot, L, \phi)\sin L\, d\Omega, \tag{2}$$

$$D_{\text{eq}} = \left(H_1^2 + H_2^2\right)^{1/2}, \tag{3}$$

$$H_1 = \frac{3}{4\pi}\int B_r(R_\odot, L, \phi)\cos L\cos\phi\, d\Omega, \tag{4}$$

$$H_2 = \frac{3}{4\pi}\int B_r(R_\odot, L, \phi)\cos L\sin\phi\, d\Omega; \tag{5}$$

here ϕ denotes Carrington longitude, and the integrals are taken over all solid angles Ω.

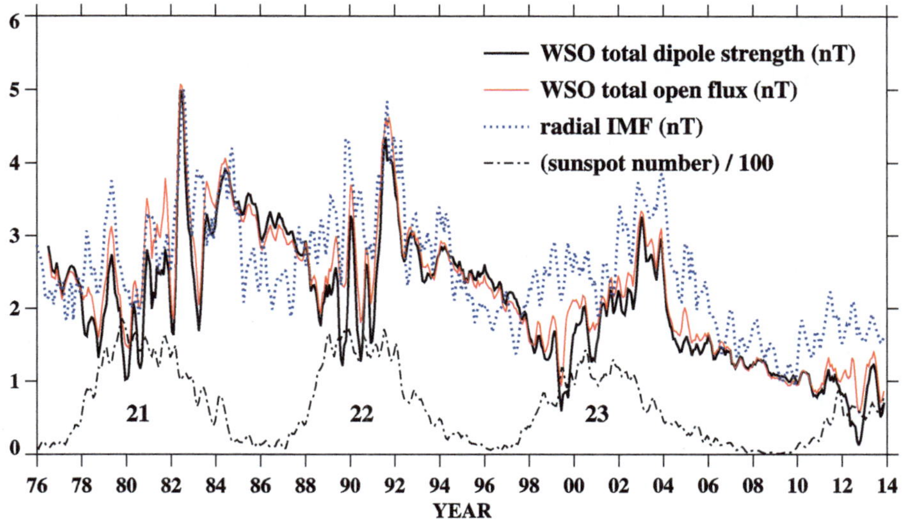

Fig. 1 Long-term variation of the Sun's total dipole strength D_{tot} (*black solid curve*) and total open flux Φ_{open} (*red solid curve*), derived from a PFSS extrapolation of WSO magnetograph data and expressed as equivalent field strengths (nT) at 1 AU. Also plotted are OMNI measurements of the radial IMF strength at Earth (*dotted curve*) and the monthly mean sunspot number (*dash-dotted curve*). The WSO line-of-sight fields have been deprojected and corrected for line profile saturation as described in Wang and Sheeley (1995). Here and in the following two figures, 3-rotation running means have been taken

Figure 1 shows the long-term variation of the total dipole strength D_{tot} and total open flux Φ_{open}, derived from photospheric field maps from the Wilcox Solar Observatory (WSO) covering the period from May 1976 to December 2013 (Carrington rotations 1642–2144). For comparison, OMNI measurements of the radial IMF strength at Earth are also displayed. The solar cycle variation of Φ_{open} closely follows that of D_{tot}, except for occasional deviations at sunspot maximum due to the contribution of the quadrupole component to the open flux. The variations of D_{tot} and Φ_{open} are generally similar to that of the measured radial IMF strength, with all three curves exhibiting large peaks in 1982, 1991, and 2003, about two years after sunspot maximum. However, the WSO-derived open flux substantially underestimates the observed IMF strength during the maxima of cycles 23 and 24; this discrepancy could at least in part be due to the effect of CMEs, which may contribute as much as $\sim 40~\%$ of the heliospheric flux at activity maximum (Richardson et al. 2002; Cliver et al. 2013).

Figure 2 compares the variation of the dipole and quadrupole strengths at the source surface. The quadrupole becomes comparable to the dipole at sunspot maximum, when it provides a significant contribution to the open flux and IMF.

As shown by Fig. 3, the axial and equatorial components of the dipole vary out of phase with each other over the solar cycle. The axial dipole D_{ax} behaves in a manner similar to the Sun's polar fields, exhibiting a broad peak near sunspot minimum and vanishing at sunspot maximum. In contrast, the equatorial dipole D_{eq} varies in phase with sunspot activity, while undergoing large fluctuations. The equatorial dipole strength is a function of both the level of sunspot activity and its distribution in longitude, and attains its peak values when large active regions emerge with their east–west dipole moments reinforcing each other (Wang and Sheeley 2003a). Thus, even though the rate of active region emergence is greatest at sunspot maximum, the net equatorial dipole moment tends to reach its maximum strength

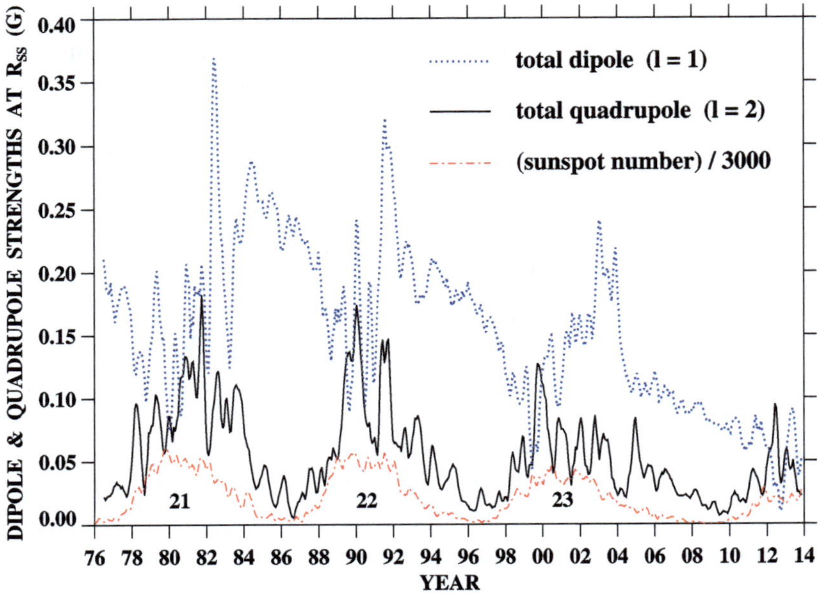

Fig. 2 Variation of the total dipole and total quadrupole strengths (G) at the source surface during 1976–2013. The quadrupole and dipole have comparable magnitudes near sunspot maximum, when the axial dipole component vanishes. The curves were derived from a PFSS extrapolation of the WSO photospheric field

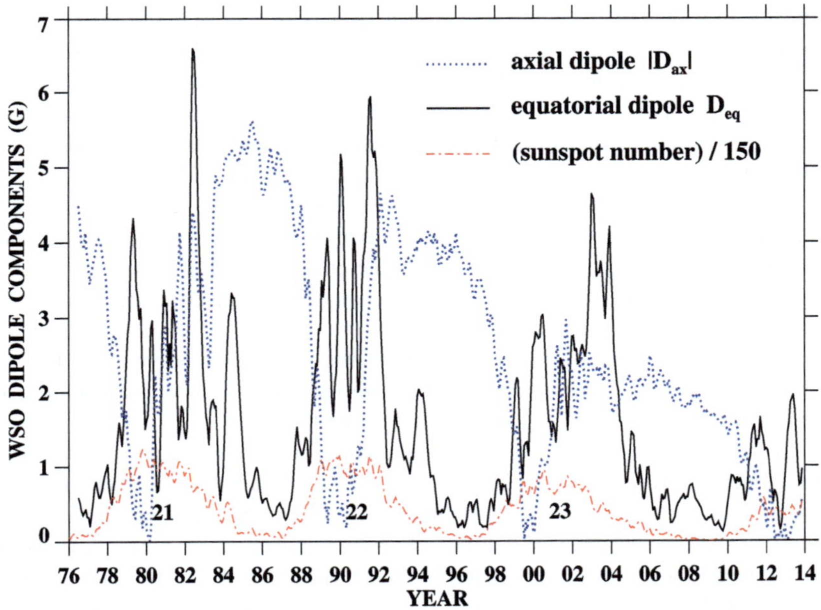

Fig. 3 Variation of $|D_{ax}|$ and D_{eq} during 1976–2013 (cycles 21–24), derived from WSO photospheric field measurements. The two dipole components have roughly comparable amplitudes during each cycle, but are in antiphase with each other

somewhat later, when activity becomes concentrated in a smaller number of very strong regions having a relatively nonuniform distribution in longitude.

During recent cycles, D_{eq} and D_{ax} had roughly comparable amplitudes, with the equatorial dipole strength in 1982, 1991, and 2003 somewhat exceeding the axial dipole strength during the activity minima of 1976, 1986, and 1996 (see Fig. 3). As a result, $D_{tot} = (D_{ax}^2 + D_{eq}^2)^{1/2}$ and the IMF strength peaked at the same time as D_{eq}, approximately 2 yr after the maximum of each cycle.

To understand better the behavior of the Sun's equatorial dipole component and its relationship to sunspot activity, it is helpful to think of each active region as having its own vector dipole moment. Now consider a collection of active regions distributed randomly in longitude. By Joy's law, all (or most) of the individual axial dipole moments will point in the same direction and reinforce each other; this explains why the Sun's axial dipole component varies relatively smoothly over the solar cycle (see Fig. 3). In contrast, the net equatorial dipole component depends on how the individual active regions are distributed in longitude. For example, if two identical active regions are located 180° apart and in the same hemisphere, their net equatorial dipole moment will be zero. On the other hand, if they are located 180° apart but on opposite sides of the equator, their equatorial dipole vectors (assumed to obey the Hale polarity rules) will point in the same direction and reinforce each other.

In the experiment shown in Fig. 4, we deposited 600 bipoles onto the solar surface at random longitudes over a solar cycle, with each bipole having a total flux of 5×10^{22} Mx and the rate of deposition being proportional to the sunspot number (see Wang and Sheeley 2003a). The photospheric field was evolved including the effects of differential rotation, supergranular diffusion, and a 20 $\mathrm{m\,s^{-1}}$ poleward flow. The five panels correspond to different initial seeds for the random number generator used to assign the bipole longitudes. It is evident that the highest peaks in the equatorial dipole and total open flux may occur anywhere during the middle years of the cycle. The widths of the peaks correspond to the ~ 1 yr decay timescale of the nonaxisymmetric photospheric flux as it is transported to midlatitudes by the meridional flow. There, rotational shearing converts the equatorial dipole into higher-order odd multipoles ($l = 3, 5, 7, \ldots$), which decay with time t as $\exp[-l(l+1)\kappa t/R_\odot^2]$, where $\kappa \sim 500\ \mathrm{km^2\,s^{-1}}$ is the supergranular diffusion coefficient.

From the simulations of Fig. 4, we also see that the IMF strength can undergo large dips even at sunspot maximum, if the active regions happen to emerge longitudinally out of phase with each other. This stochastic process may provide an explanation for the "Gnevyshev gaps" observed in large-scale phenomena such as coronal green-line emission, the cosmic ray flux, and geomagnetic activity (Gnevyshev 1967, 1977; Feminella and Storini 1997; Storini et al. 1997; Richardson et al. 2000). We emphasize that the effect described here is not the same as the double peak in the sunspot number caused by the lag between northern- and southern-hemisphere activity (see Fig. 12 below).

Cosmogenic isotope records (see Fig. 5) indicate that the IMF continued to undergo a solar cycle variation throughout the 1645–1715 Maunder Minimum, but with the field strength peaking near the end of each cycle instead of in the middle (Beer et al. 1998; Usoskin et al. 2001). This phase shift can be explained by a decrease in the amplitude of the Sun's equatorial dipole component relative to its axial dipole component. Such a decrease might occur if the bipoles that emerged during the Maunder Minimum had larger axial tilts than in recent cycles; consistent with this scenario, Dasi-Espuig et al. (2010) found an anticorrelation between cycle strength and the mean tilt of sunspot groups observed during cycles 15–21 (see also Cameron et al. 2010; Cameron and Schüssler 2012), although this effect was not confirmed by Ivanov (2012). An alternative possibility is that the bipoles of

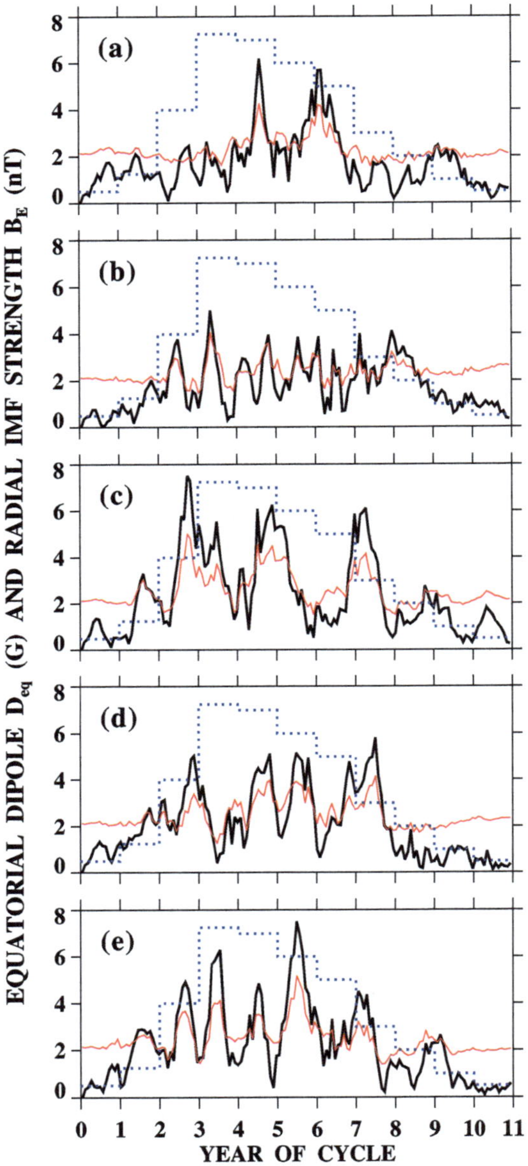

Fig. 4 Simulated solar-cycle evolution of the equatorial dipole strength $D_{\rm eq}$ (G) (*black curves*) and of the total open flux $\Phi_{\rm open}$ (nT) (*red curves*), demonstrating the effect of varying the longitudinal distribution of active regions. The five panels correspond to five different choices of the initial seed used in randomizing the longitudes of the 600 bipoles that emerge over the cycle. *Blue dotted lines* indicate the relative numbers of bipoles deposited during each year of the cycle. The highest peaks in $D_{\rm eq}$ and the IMF strength may occur anywhere during the middle years of the cycle, and large dips may occur even at sunspot maximum

the Maunder Minimum were systematically weaker than in the present day, so that the large-scale field had its source in a large number of weak bipoles rather than a smaller number of strong active regions.

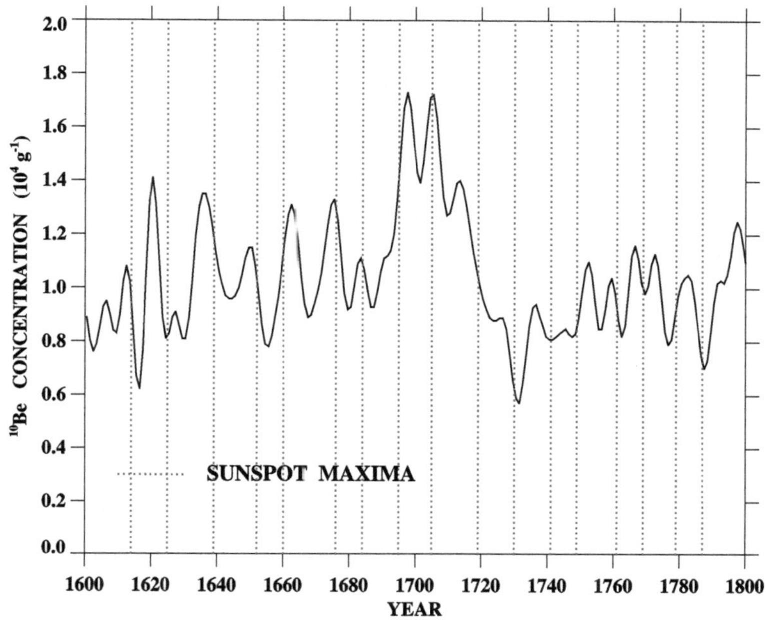

Fig. 5 Variation of the ^{10}Be concentration during 1600–1800 (*solid curve*); sunspot maxima are marked by the *vertical dotted lines*. A low-pass filter has been applied to the ^{10}Be time series, as described in Beer (2000); solar maximum times are based on the yearly group sunspot numbers of Hoyt and Schatten (1998) (see also Table 1 of Usoskin et al. 2001). During the 1645–1715 Maunder Minimum, the peaks in the ^{10}Be concentration are seen to occur near sunspot maximum, suggesting that the variation of the IMF strength was phase-shifted so as to peak near sunspot minimum, contrary to its present-day behavior

The axial and equatorial dipole strengths of a single bipole are related to its axial tilt γ by $d_{ax}/d_{eq} \sim \tan\gamma \sim \gamma$. For N identical bipoles distributed randomly in longitude, the net axial dipole strength scales as $D_{ax} \sim N d_{ax}$, whereas the net equatorial dipole strength scales as $D_{eq} \sim N^{1/2} d_{eq}$. Thus the ratio D_{ax}/D_{eq} will exceed unity if $N \gtrsim \gamma^{-2}$. Figure 6 shows a flux transport simulation in which the source term consists of many weak bipoles. In this case, N is sufficiently large that the amplitude of the axial dipole component dominates over that of the equatorial dipole component, so that Φ_{open} peaks near sunspot minimum.

A quite different explanation for the ^{10}Be-inferred IMF phase shift has been suggested by Owens and Lockwood (2012) and Owens et al. (2012). In their model, the production rate of open flux is proportional to the sunspot number, while its loss rate is taken to depend on the tilt angle of the HCS. With the tilt angle varying in much the same way during every cycle (regardless of its strength), the loss term becomes relatively dominant during cycles with low sunspot numbers, leading to a minimum of the open flux near sunspot maximum. This scenario also requires that a residual source of open flux, attributed to CMEs, be present even when sunspot activity vanishes.

It will be interesting to see when the IMF strength peaks during cycle 24.

3 The Quasirigid Rotation of the Corona

From white-light, green-line, and soft X-ray observations, it is well known that large-scale coronal structures rotate more rigidly than the underlying photosphere, and that this rota-

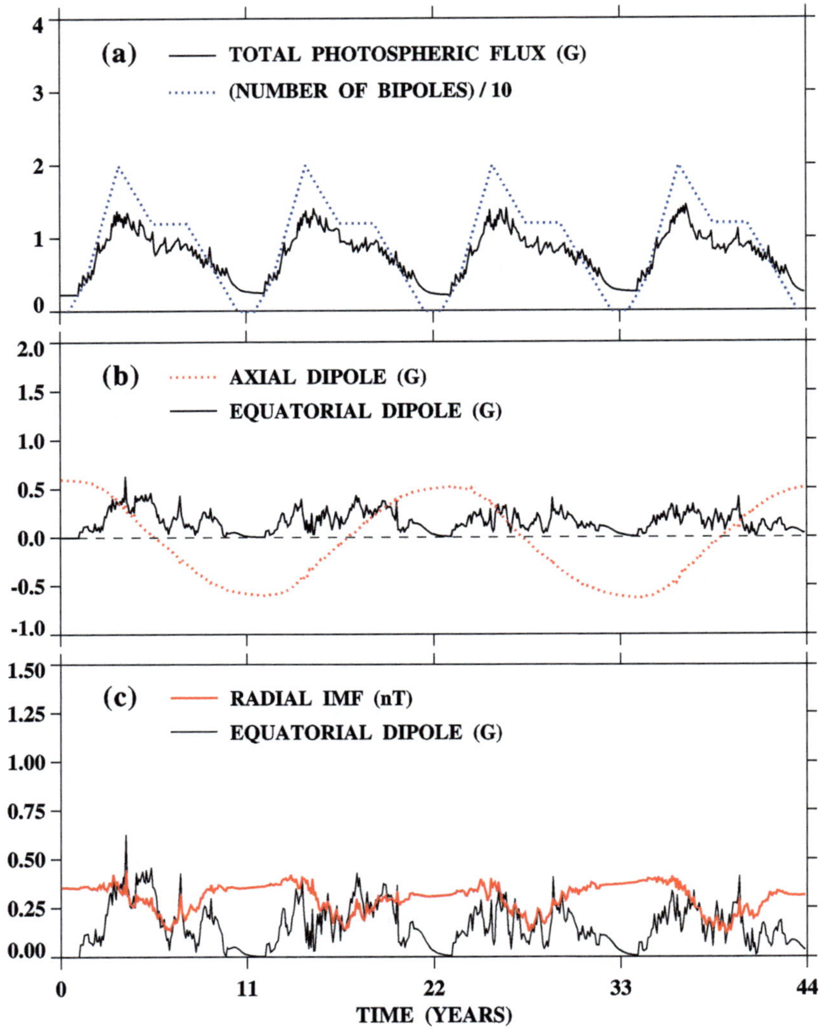

Fig. 6 Evolution of the solar magnetic field in a "weak bipole" scenario, where 100 bipoles each having a total flux of 1×10^{22} Mx are deposited per 11 yr cycle. (**a**) Time variation of the total photospheric flux (*solid curve*), expressed as a surface-averaged field strength (G). *Dotted line* indicates the number of bipoles deposited during each year of the cycle. (**b**) Evolution of the axial and equatorial dipole components of the photospheric field, D_{ax} (*dotted curve*) and D_{eq} (*solid curve*), where both quantities are in gauss. (**c**) Time variation of the total open flux or radial IMF strength at 1 AU (in nT, *red curve*); *black curve* represents the equatorial dipole strength D_{eq}. In this flux transport simulation, the amplitude of the axial dipole component dominates over that of the equatorial dipole component, so that the total open flux peaks around sunspot minimum instead of sunspot maximum

tion becomes increasingly rigid with height (see, e.g., Hansen et al. 1969; Antonucci and Svalgaard 1974; Antonucci and Dodero 1977; Parker et al. 1982; Fisher and Sime 1984; Sime et al. 1989; Lewis et al. 1999; Weber et al. 1999; Lewis and Simnett 2001; Badalyan et al. 2006; Giordano and Mancuso 2008; Chandra et al. 2010; Mancuso and Giordano 2011; Morgan 2011). This quasirigid behavior is a direct consequence of the rapid falloff of high-order multipoles above the photosphere; indeed, in the limit in which the coronal field reduces to a dipole, the rotation becomes purely rigid and identical to that of the equatorial dipole component. The rotation rate and its latitudinal profile are determined by the latitudinal distribution of the large-scale, nonaxisymmetric photospheric flux, in the form of active regions and their decayed remnants (Wang et al. 1988). During the late declining phase of the cycle, the nonaxisymmetric flux becomes concentrated around the equator (the rest having been annihilated by rotational shearing and diffusion at midlatitudes), and consequently the coronal field rotates nearly rigidly with a period close to 27 days. During the rising phase of the cycle, on the other hand, the nonaxisymmetric photospheric flux is concentrated in the midlatitude active regions, leading to a slower, less rigid rotation of the coronal field. The slower rotation characteristic of the rising phase is also evident in the $\sim$ 28-day recurrence rate of the IMF sector structure, which then switches to a 27-day pattern during the declining phase (Svalgaard 1972). North–south asymmetries in the distribution of active regions give rise to hemispheric asymmetries in the coronal rotation rate. It should be noted that the mismatch between the coronal and photospheric rotation rates does not mean that the coronal field is anchored to deeper layers that rotate more rigidly than the photosphere; instead, it requires that the corona be decoupled from the photospheric plasma through continual fieldline reconnection (as is indeed implicit in the use of the current-free approximation).

4 The Nonaxisymmetric Quadrupole Component and the Bifurcation of the Heliospheric Current Sheet During 2012

We now discuss the heliospheric consequences of the Sun's quadrupole component. For this purpose, it is convenient to express the source surface field in the general form

$$B_r(R_{\rm ss}, L, \phi) = \sum_{l=1}^{\infty} b_l(R_{\rm ss}, L, \phi) = \sum_{l=1}^{\infty}\sum_{m=-l}^{l} b'_{lm}(R_{\rm ss}, L, \phi)$$
$$= \sum_{l=1}^{\infty}\sum_{m=0}^{l} b_{lm}(R_{\rm ss}, L, \phi) \qquad (6)$$

(b_{lm} is given explicitly in terms of the spherical harmonic function Y_{lm} and its complex conjugate by Eqs. (1) and (2) in Wang and Sheeley 1988). As a measure of the average strength of a given multipole l or spherical harmonic component (l, m) at the source surface, we then define

$$\langle b_l \rangle \equiv \frac{1}{4\pi}\int \left| b_l(R_{\rm ss}, L, \phi)\right| d\Omega, \qquad (7)$$

$$\langle b_{lm} \rangle \equiv \frac{1}{4\pi}\int \left| b_{lm}(R_{\rm ss}, L, \phi)\right| d\Omega. \qquad (8)$$

As already shown in Fig. 2, the quadrupole strength $\langle b_2 \rangle$ becomes comparable to the dipole strength $\langle b_1 \rangle$ at sunspot maximum. The quadrupole was particularly dominant in 2012, when $\langle b_{22} \rangle$ peaked at the same time that the axial dipole vanished (see Fig. 7).

Figure 8 displays a source surface extrapolation of the MWO photospheric field for CR 2124 (25 May–21 June 2012). The source surface field shows two circular neutral lines, each of which encloses negative-polarity open flux; the underlying negative-polarity holes can be seen in the Fe XII 19.5 nm map in the bottom panel. The circular neutral lines extend outward into the heliosphere in the form of two cylindrical current sheets. Employing the procedure described in Wang et al. (2007), we have calculated the white-light patterns

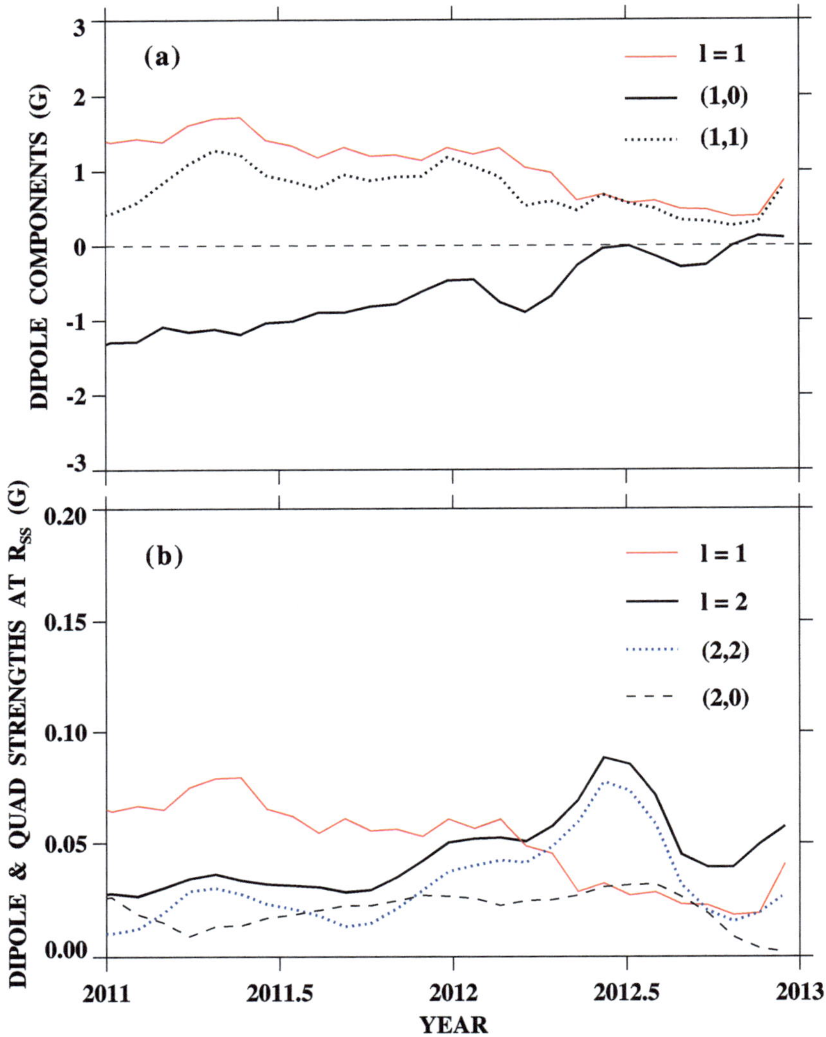

Fig. 7 (**a**) Time variation of the total, axial, and equatorial dipole components of the Mount Wilson Observatory (MWO) photospheric field during 2011–2012 (CR 2105–2132). The dipole strengths $D_{\rm tot}$ (*solid red line*), $D_{\rm ax}$ (*solid black line*), and $D_{\rm eq}$ (*dotted*) are expressed in gauss, and 3-rotation running means have been taken. The axial dipole component approaches zero in the middle of 2012. (**b**) Variation of the total dipole (*solid red line*), total quadrupole (*solid black line*), nonaxisymmetric (2, 2) quadrupole (*dotted*), and axisymmetric (2, 0) quadrupole (*dashed*) components of the source surface field during 2011–2012. Again, MWO data have been used and 3-rotation running means taken. The (2, 2) quadrupole component provides the dominant contribution to the source surface field in mid-2012. Magnetograph measurements from National Solar Observatory/Kitt Peak and WSO yield similar curves

produced by Thomson scattering from the cylindrical plasma sheets and compare them with *SOHO*/LASCO C3 observations in Fig. 9. The polemost portions of the plasma sheets are oriented edge-on at the east and west limbs and give rise to arclike features that can be identified in the LASCO maps. The additional white-light structures are due to pseudostreamer plasma sheets that separate coronal holes of the same polarity.

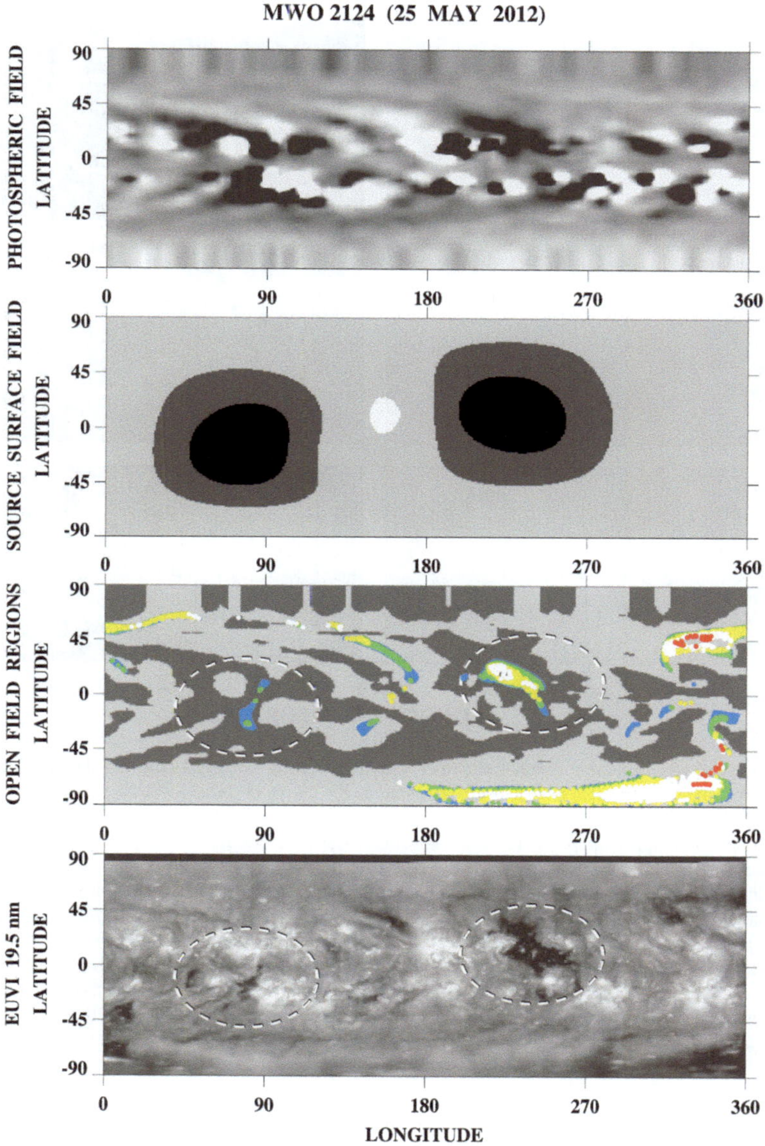

Fig. 8 Latitude–longitude maps showing the large-scale field and distribution of coronal holes during CR 2124 (25 May–21 June 2012). *Top panel*: MWO photospheric field, with gray scale contours ranging from *black* ($B_r < -10$ G) to *white* ($B_r > +10$ G). *Second panel*: source surface field $B_{ss} = B_r(R_{ss}, L, \phi)$. *Black*: $B_{ss} < -0.15$ G; *dark gray*: -0.15 G $< B_{ss} < 0$ G; *light gray*: 0 G $< B_{ss} < +0.15$ G; *white*: $B_{ss} > +0.15$ G. *Third panel*: PFSS-derived coronal holes. *Colored dots* represent footpoints of open field lines and are coded according to the associated expansion factors or asymptotic wind speeds. *Blue*: $v < 450$ km s^{-1} ($f_{ss} > 24$); *green*: $v = 450$–550 km s^{-1} ($12 < f_{ss} < 24$); *yellow*: $v = 550$–650 km s^{-1} ($7 < f_{ss} < 12$); *white*: $v = 650$–750 km s^{-1} ($4 < f_{ss} < 7$); *red*: $v > 750$ km s^{-1} ($f_{ss} < 4$). The polarity of the underlying photospheric field is indicated by *dark gray* (if $B_r < 0$) or *light gray* (if $B_r > 0$). *Bottom panel*: distribution of Fe XII 19.5 nm emission recorded by *STEREO*/EUVI B between May 16 and June 12. In the bottom two maps, *dashed circles* identify the negative-polarity coronal holes on which the two cylindrical current sheets are centered

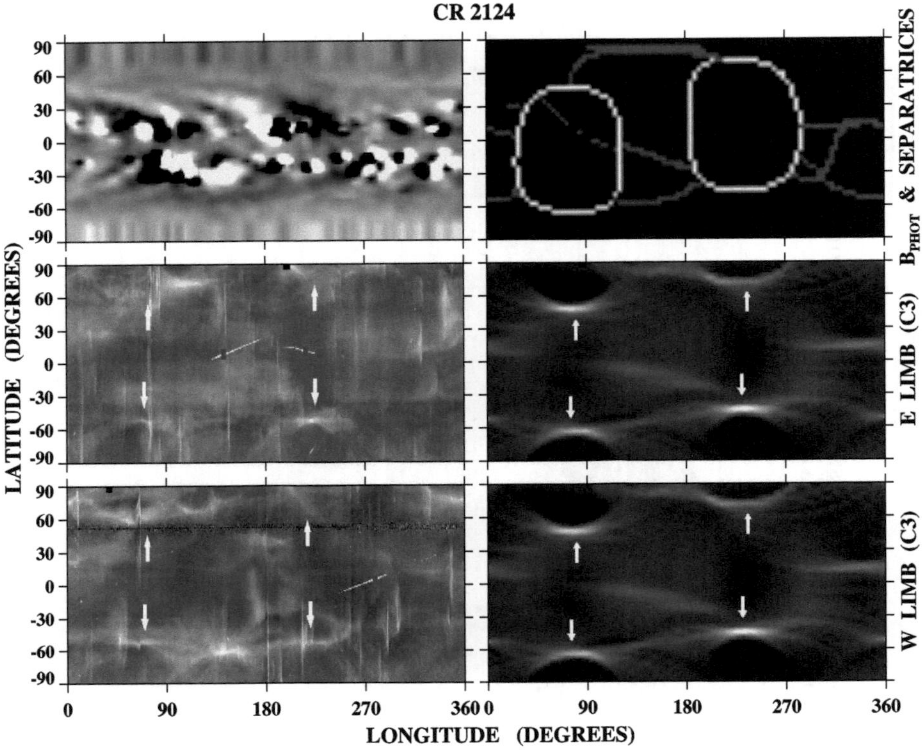

Fig. 9 Observed and simulated coronal streamer structures during CR 2124 (starting date 25 May 2012). *Top left*: MWO map of the photospheric field, saturated at $B_r < -10$ G (*black*) and $B_r > +10$ G (*white*). *Top right*: boundaries between open field regions at $r = R_{ss} = 2.5\ R_\odot$, derived by applying a PFSS extrapolation to the photospheric map. *White pixels* represent the source-surface neutral lines (or predicted positions of the HCS and helmet-streamer plasma sheets); *gray pixels* mark the boundaries between coronal holes of the same polarity (or predicted locations of pseudostreamer plasma sheets). *Middle left*: white-light streamer structures observed above the east limb at 10 $R_\odot$ with the *SOHO*/LASCO C3 coronagraph. *Middle right*: simulated brightness patterns above the east limb, produced by Thomson scattering of photospheric radiation from the helmet-streamer and pseudostreamer plasma sheets. *Bottom left*: west-limb streamer structures observed at 10 $R_\odot$ by LASCO C3. *Bottom right*: simulated west-limb streamer structures. *Vertical streaks* in the LASCO maps represent CMEs. *White arrows* indicate the arc-shaped brightenings associated with the poleward sides of the two cylindrical current/plasma sheets

Figure 10 shows the coronal fieldline configuration during CR 2124, as viewed from above the north pole. The (2, 2) quadrupolar nature of the coronal field is evident, with four helmet streamers regularly spaced along the equator, which separate four open field regions of alternating polarity. Note that the north polar field has not yet completed its reversal from negative to positive, even though the overlying open flux has positive polarity. This gives rise to a pseudostreamer configuration above the north pole.

The stackplots in Fig. 11 show the IMF sector polarity and solar wind-speed patterns in the ecliptic during 2010–2013. The whole of 2012 is dominated by a four-sector structure, again due to the (2, 2) quadrupole component.

It is apparent from this example that the HCS cannot be characterized by a well-defined tilt angle at sunspot maximum, contrary to the standard "tilted-dipole" picture in which the

Fig. 10 Coronal fieldline configuration during CR 2124, as viewed from above the north pole (PFSS extrapolation of MWO photospheric measurements). Closed loops are coded *orange* if they extend beyond $r = 1.5\ R_\odot$, *red* otherwise; open field lines are *blue* (*green*) if they have positive (negative) polarity. *Black*, *dark gray*, *light gray*, and *white* denote areas of the photosphere where $B_r < -6$ G, $-6\ \text{G} < B_r < 0$ G, $0\ \text{G} < B_r < +6$ G, and $B_r > +6$ G, respectively. The coronal field is clearly dominated by the (2, 2) quadrupole component. Despite the residual negative-polarity photospheric flux at the north pole, the overlying open field lines (rooted at lower latitudes) have positive polarity and the IMF has already reversed from inward to outward

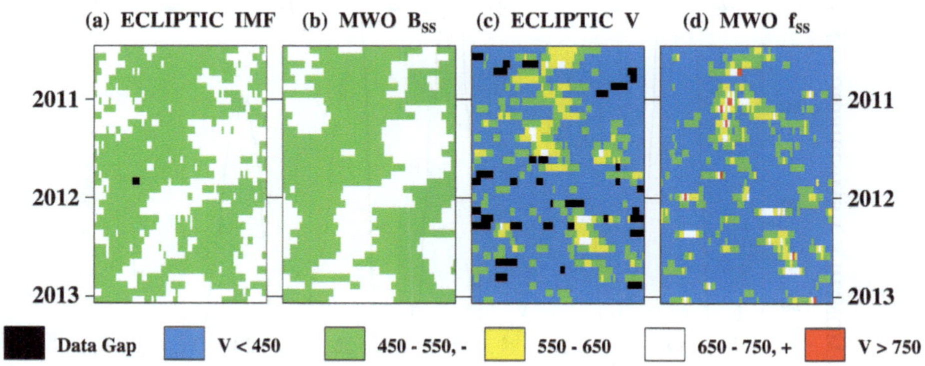

Fig. 11 Variation of the IMF sector polarity and solar wind speed near Earth during 2010–2013, plotted in stackplot format, with each row of pixels representing a 27.3 day Carrington rotation and time running from right to left. (**a**) Polarity of the radial IMF component (*white*: positive/outward; *green*: negative/inward), as measured by *ACE*/MAG. (**b**) Polarities predicted from a PFSS extrapolation of MWO photospheric field measurements (CR 2098–2132). (**c**) Proton speeds v measured by *ACE*/SWEPAM. (**d**) Wind speeds predicted using the inverse correlation between v and the flux-tube expansion factor $f_{ss} = (R_\odot/R_{ss})^2|B_0/B_{ss}|$, where B_0 is the footpoint field strength (see Fig. 8 caption and Wang and Sheeley 1990). A four-sector structure prevails throughout 2012

HCS rotates through 90° during polar field reversal (Alanko-Huotari et al. 2006; Du et al. 2008; Owens and Lockwood 2012).

5 The Axisymmetric Quadrupole Component and the Southward Displacement of the Heliospheric Current Sheet

The rising phase of cycle 24 has been characterized by a strong hemispheric asymmetry, with the northern hemisphere being far more active than the southern hemisphere. As may be seen from Fig. 12, this continues a trend that started with cycle 20, in which the north has been more active during the rising phase of each cycle and the south more active during the declining phase (cycle 19 showed the opposite behavior). In general, any hemispheric asymmetry in sunspot activity will give rise to an axisymmetric $(2, 0)$ quadrupole component, which will combine with the axial dipole to displace the HCS in the north–south direction.

IMF and geomagnetic observations indicate that the HCS has had a tendency to be shifted southward of the heliographic equator since the 1960s (Mursula and Hiltula 2003; Hiltula and Mursula 2006; Erdős and Balogh 2010). To understand why the shift has been southward, note that the leading polarity in the dominant hemisphere determines the sign of the equatorial band of the $(2, 0)$ quadrupole component, and hence the IMF polarity at the equator. During the rising phase of a cycle, the leading-polarity flux in each hemisphere has the same sign as the polar field in its hemisphere, whereas during the declining phase, it has

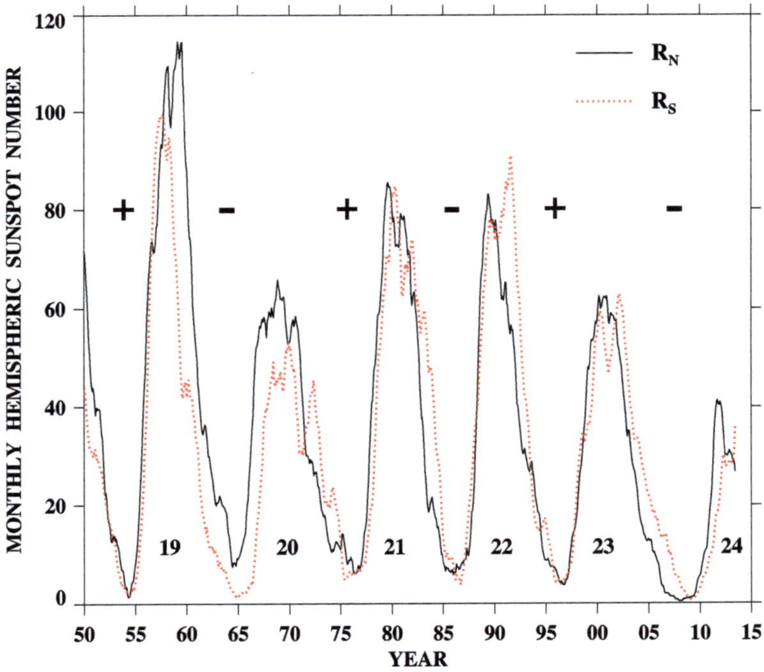

Fig. 12 Smoothed monthly hemispheric sunspot numbers. *Solid line*: northern hemisphere. *Dotted line*: southern hemisphere. *Plus* and *minus signs* indicate the polarity of the north polar field during the successive sunspot minima. The data are from Temmer et al. (2006) and (after 1992) from SIDC/Royal Observatory of Belgium

the opposite sign. Thus, if the northern hemisphere dominates during the rising phase (as it did during cycles 20–24), the equatorial band of the quadrupole will have the same sign as the north polar field, and the HCS will be shifted below the equator. If the southern hemisphere dominates during the declining phase (as it did during cycles 20–23), the equatorial band of the quadrupole will have the same sign as the leading-polarity flux in the southern hemisphere, which is opposite to the sign of the south polar field. Consequently, the HCS will again be shifted below the equator.

As a specific example from the rising phase of the current cycle, the Carrington maps in Fig. 13 display the photospheric field during CR 2102 (3–30 October 2010), the axisymmetric dipole and quadrupole components of the source surface field, and the total source-surface field. The equatorial zone of the quadrupole has negative sign, matching the leading polarity in the dominant northern hemisphere and the sign of the north polar field; as a result, the source-surface neutral line is shifted southward of the equator. We may verify that this was indeed the case by examining LASCO C3 maps of the white-light streamer belt (Fig. 14). At first sight, the observed brightness distribution might suggest that the HCS is shifted northward rather than southward of the equator. However, as demonstrated by the simulations in the right-hand panels of Fig. 14, most of the white-light structures in the northern hemisphere are pseudostreamer plasma sheets that are unrelated to the HCS. If we eliminate the pseudostreamers and then measure the latitudinal positions of the remaining helmet streamers as a function of longitude, we find that the average southward displacement of the current sheet is 3°.

Employing the same method, we have determined the north–south displacement of the HCS at various times since 1996; the results are plotted in Fig. 15 (see also Robbrecht and Wang 2012). There was no significant shift during the 1996–1997 sunspot minimum. The largest displacements, of up to 6° below the equator, occurred during the rising phase of the current cycle. This was due both to the overwhelming dominance of northern hemisphere activity and to the weakness of the polar fields.

6 Higher-Order Contributions to the Source Surface Field

The long-term variations of source-surface field multipoles up to order $l = 9$ are plotted in Figs. 8 and 9 of Petrie (2013) (see also Hoeksema 1984). As remarked by Petrie, both the axisymmetric and the nonaxisymmetric components of all multipoles $l \gtrsim 4$ are well correlated with the sunspot number. In general, high-order multipoles, being characterized by small spatial scales, undergo rapid diffusive decay at the photosphere and must be maintained by new flux emergence. Because they are subject to the combined action of rotational shearing and diffusive annihilation, the nonaxisymmetric components of low-order ($l = 1$–3) multipoles are also relatively short-lived and vary in phase with sunspot activity.

As noted by Petrie (2013), the axisymmetric octupole component varies in phase with the axial dipole and polar fields. Figure 16 compares the long-term evolution of $\langle b_{10} \rangle$, $\langle b_{30} \rangle$, and $\langle b_{50} \rangle$, as derived from WSO photospheric field measurements. The $(l = 3, m = 0)$ component of the source surface field exhibits a broad maximum centered on the minimum of each cycle; this behavior is due to the meridional surface flow, which concentrates trailing-polarity flux at the poles to produce a topknot structure at solar minimum (see Svalgaard et al. 1978; Sheeley et al. 1989; Petrie and Patrikeeva 2009).

Also of interest in Fig. 16 is the behavior of the $(l = 5, m = 0)$ contribution to the source surface field, which exhibits an asymmetric peak at the time of polar field reversal. This behavior can be understood from the sequence of current-free coronal configurations plotted

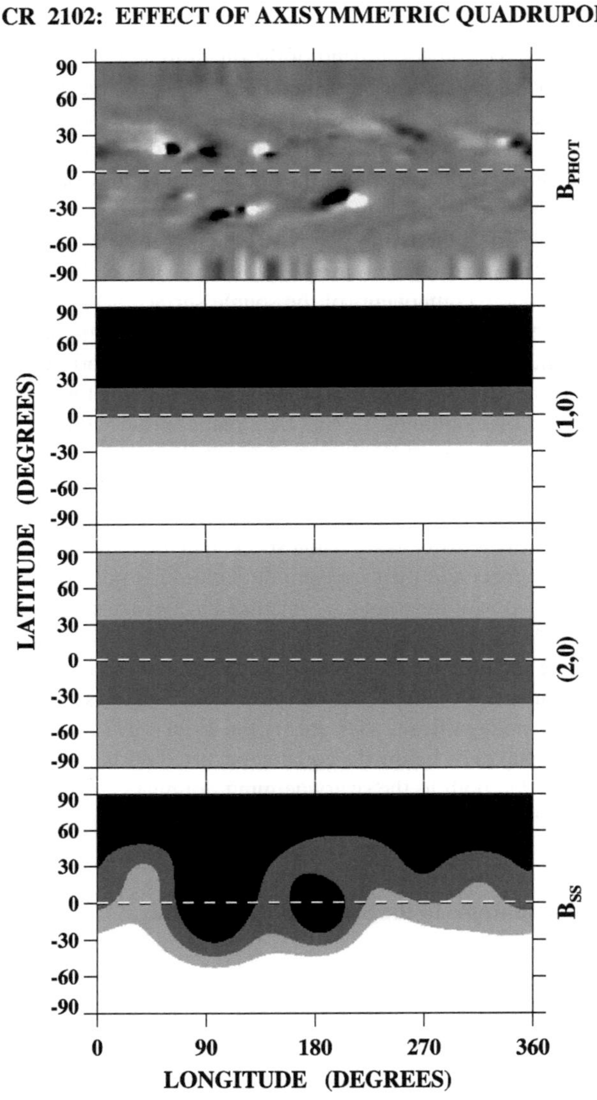

Fig. 13 The HCS is shifted southward during CR 2102 (3–30 October 2010) due to the presence of an axisymmetric quadrupole field whose equatorial zone has negative sign, corresponding to the leading polarity in the more active northern hemisphere and matching the sign of the north polar field. From top to bottom, the latitude–longitude maps display the MWO photospheric field, the axisymmetric dipole ($l = 1, m = 0$) and axisymmetric quadrupole ($l = 2, m = 0$) components of the source surface field, and the total source-surface field $B_{SS} = B_r(R_{SS}, L, \phi)$. *Dashed lines* mark the heliographic equator. In the source surface maps, sign reversals are located between the *light gray* (0 G $< B_r <$ +0.1 G) and *dark gray* (-0.1 G $< B_r <$ 0 G) contours

in Fig. 17, derived by simulating the solar cycle evolution of an axisymmetric photospheric field using a flux transport model (see Wang and Sheeley 2003b). During the rising phase of the cycle, when the trailing polarity has the opposite sign to the polar field in its hemisphere, the global photospheric field has an ($l = 5, m = 0$) structure, with a total of 6 polarity bands,

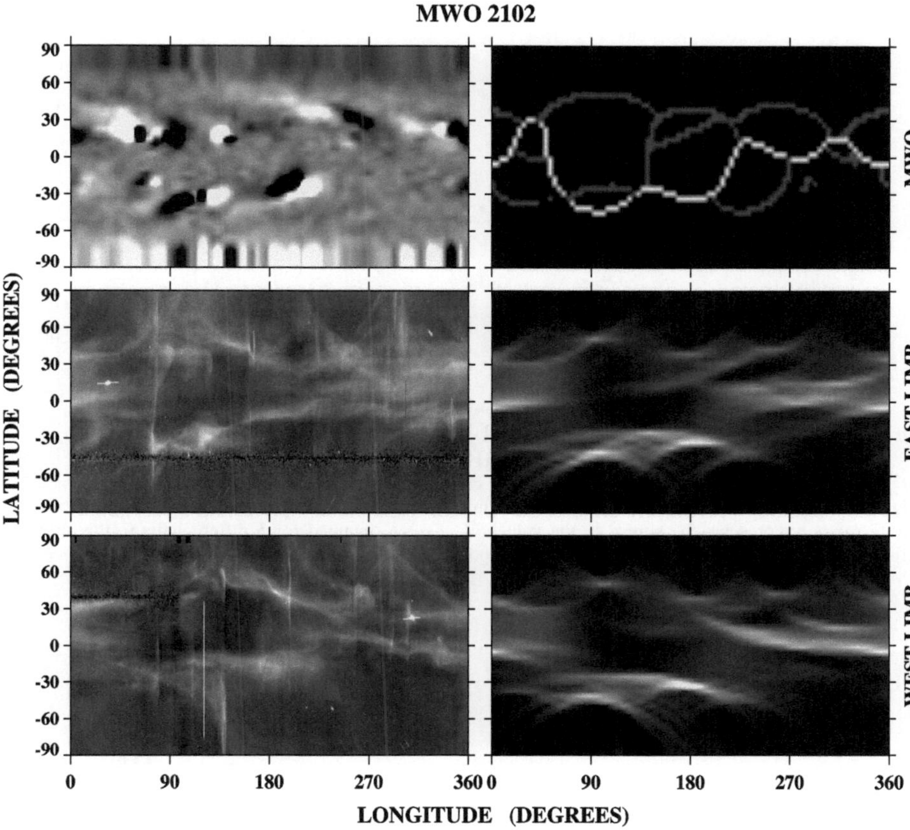

Fig. 14 Simulation of the coronal streamer structure during CR 2102 (starting date 3 October 2010). *Top left*: MWO map of the photospheric field, saturated at $B_r < -10$ G (*black*) and $B_r > +10$ G (*white*). *Top right*: boundaries between open field regions at $r = R_{ss} = 2.5\ R_\odot$, derived by applying a PFSS extrapolation to the photospheric map. *White pixels* represent the source-surface neutral line (or predicted position of the HCS and helmet-streamer plasma sheet); *gray pixels* mark the boundaries between coronal holes of the same polarity (or predicted locations of pseudostreamer plasma sheets). *Middle left*: white-light streamer structures observed above the east limb at 10 $R_\odot$ with the *SOHO*/LASCO C3 coronagraph. *Middle right*: simulated streamer structures above the east limb. *Bottom left*: west-limb streamer structures observed by LASCO C3. *Bottom right*: simulated west-limb streamer structures

while the source surface field has a simple $(l = 1, m = 0)$ distribution (Fig. 17(a)). However, just before the disappearance of the polar coronal holes, trailing-polarity flux begins to open up at high latitudes, and the source surface field briefly displays an $(l = 5, m = 0)$ structure mirroring that of the underlying photosphere (Fig. 17(b)). With the disappearance of the old-cycle polar holes, the source surface field becomes octupolar (Fig. 17(c)); correspondingly, Fig. 16 shows a rapid strengthening of the $(l = 3, m = 0)$ component after the axial dipole goes through zero (but before meridional flow has had sufficient time to concentrate the trailing-polarity flux at the poles). After the trailing-polarity flux reaches the poles and the polar fields have reversed, the photosphere likewise relaxes to an $(l = 3, m = 0)$ configuration; and as leading-polarity flux continues to diffuse and cancel across the equator during the declining phase of the cycle, the low-latitude, leading-polarity holes disappear and the source surface field reverts to its initial axial dipole form (Fig. 17(d)). It should be

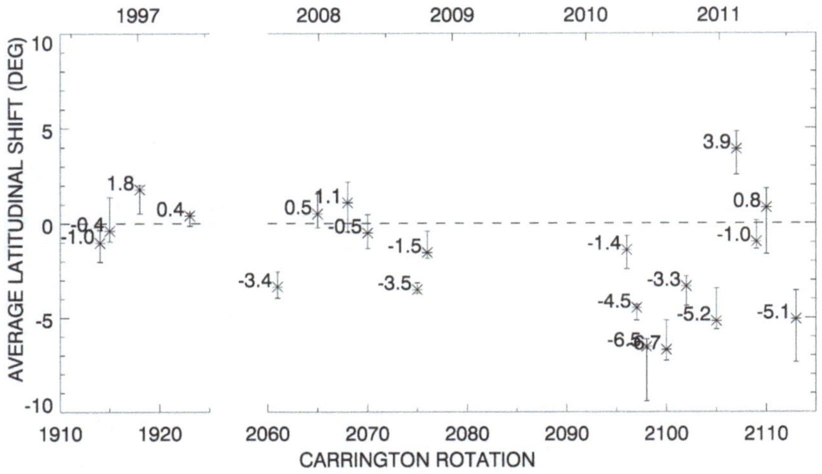

Fig. 15 Measured values of $\langle L_{HCS} \rangle$, the longitudinally averaged latitude of the HCS, during 1996–2011. The *error bar* represents the spread in the values of $\langle L_{HCS} \rangle$ obtained for 5 different fits to the observed streamer patterns; the *star* denotes the value corresponding to the median fit

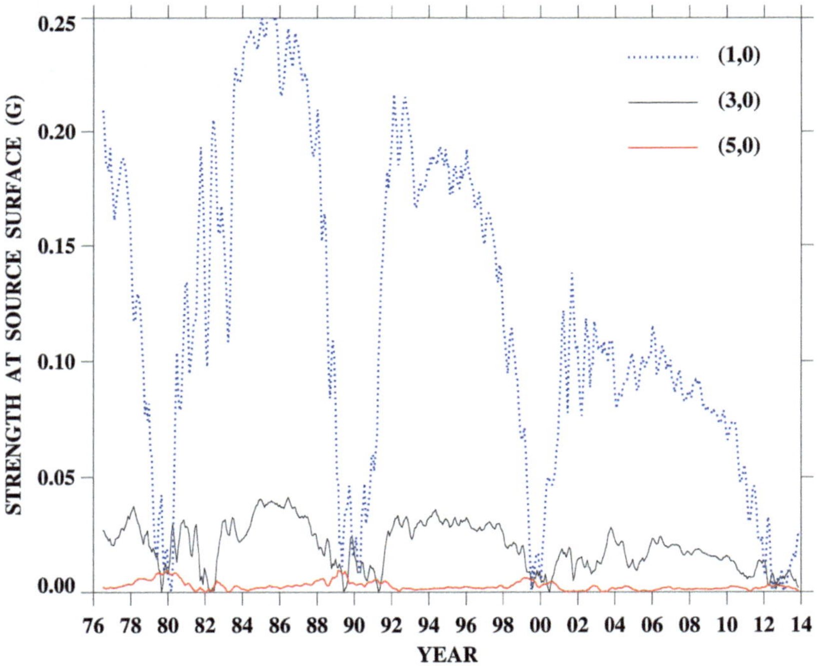

Fig. 16 Evolution of the odd-order axisymmetric components $\langle b_{10} \rangle$, $\langle b_{30} \rangle$, and $\langle b_{50} \rangle$ during 1976–2013, derived from a PFSS extrapolation of WSO magnetograph measurements. The octupole attains its maximum strength around sunspot minimum, as meridional flow establishes a photospheric flux distribution that is sharply concentrated at the poles. The $(l = 5, m = 0)$ contribution to the source surface field peaks at sunspot maximum, as trailing-polarity flux begins to open up just before the disappearance of the polar holes (see Fig. 17(b))

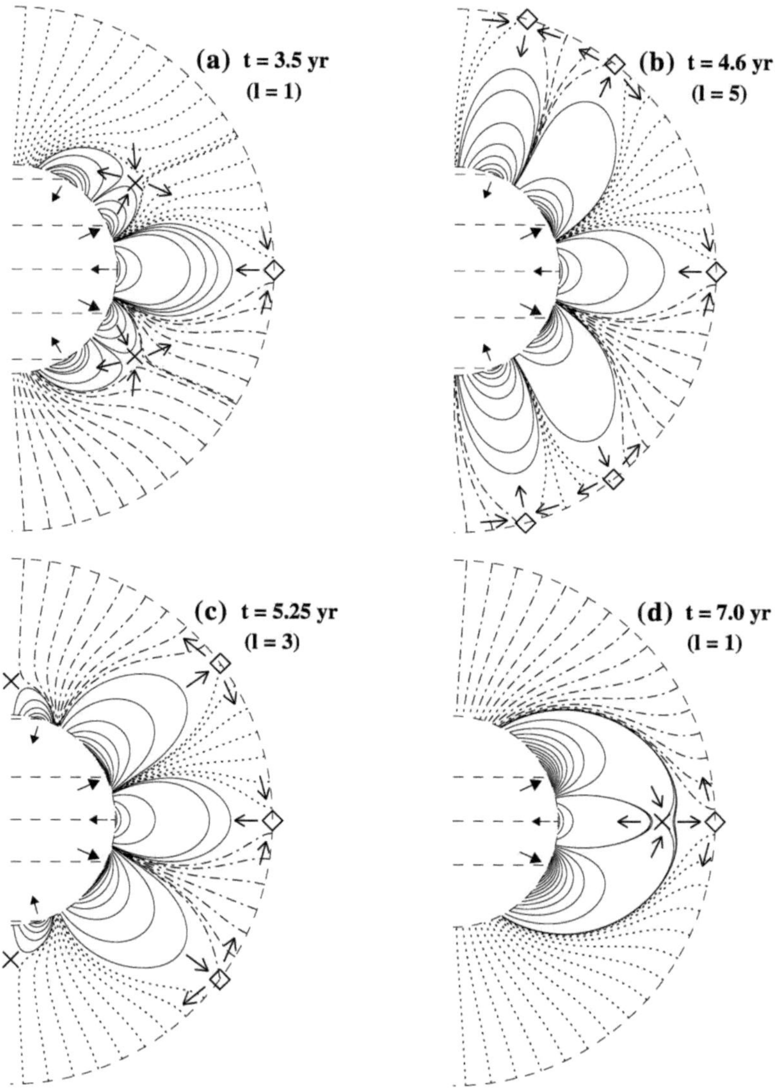

Fig. 17 Coronal field-line topology at successive stages of the polarity reversal of an axisymmetric photospheric field, whose solar cycle evolution is simulated including the effects of emerging flux, supergranular diffusion, and a 20 m s^{-1} poleward flow. Magnetic field lines lying in a given meridional plane are plotted between the photosphere and the source surface $r = R_{ss} = 2.5\ R_{\odot}$ (the spacing between drawn field lines is arbitrary and does not represent the actual field strength). Closed loops are *solid*; open field lines having positive (negative) polarity are *dotted* (*dash-dotted*). Crosses mark X-type neutral points, while diamonds denote Y-type neutral points where the source surface field vanishes. Photospheric neutral lines are indicated by horizontal dashes on the solar disk; outward- (inward-) pointing solid arrowheads show where flux is emerging (canceling). Remaining arrows indicate the directions in which the coronal field lines are transported toward or away from the X- and Y-points

emphasized that the preceding description applies to the idealized case of an axisymmetric photospheric field, and that the actual coronal field configuration will be considerably more complex due to the presence of the nonaxisymmetric harmonic components.

7 Conclusions

The structure and evolution of the outer coronal and heliospheric field are largely determined by the Sun's dipole and quadrupole components, which in turn are intimately connected with the solar dynamo itself. By "stepping back" from the Sun and thus filtering out the high-order multipoles that dominate at the photosphere, we can obtain a relatively simple picture of the global magnetic field and its solar cycle evolution, including the polarity reversal, large-scale fluctuations on timescales of 1–2 yrs, and dipole–quadrupole interactions in the form of north–south asymmetries in the IMF. In the flux transport dynamo, the origin and evolution of the dipole field are closely related to the emergence of active regions and the subsequent dispersal of their flux over the photosphere, so that the action of the dynamo is not entirely hidden beneath the surface. This allows us to establish a physical link between what is observed in the heliosphere and the processes occurring at the solar surface.

As emphasized in the preceding sections, the axisymmetric ($m = 0$) and nonaxisymmetric ($m \neq 0$) components of a given low-order multipole may evolve in quite different ways. Being subject to rotational shearing, the nonaxisymmetric components are relatively short-lived; they also tend to undergo larger fluctuations than the axisymmetric component because of their dependence on the longitudinal phase relationships between active regions. It will be particularly interesting to see how the axial and equatorial components of the dipole evolve during the declining phase of the current weak cycle.

Acknowledgements I am grateful to A. Balogh for the opportunity to participate in the International Space Science Institute (ISSI) Workshop on "The Solar Activity Cycle: Physical Causes and Consequences," held in November 2013. This work was supported by NASA and the Office of Naval Research.

References

K. Alanko-Huotari, K. Mursula, I.G. Usoskin, G.A. Kovaltsov, Sol. Phys. **238**, 391 (2006)
E. Antonucci, M.A. Dodero, Sol. Phys. **53**, 179 (1977)
E. Antonucci, L. Svalgaard, Sol. Phys. **34**, 3 (1974)
O.G. Badalyan, V.N. Obridko, J. Sýkora, Astron. Rep. **50**, 312 (2006)
A. Balogh, E.J. Smith, B.T. Tsurutani, D.J. Southwood, R.J. Forsyth, T.S. Horbury, Science **268**, 1007 (1995)
J. Beer, Space Sci. Rev. **94**, 53 (2000)
J. Beer, S. Tobias, N. Weiss, Sol. Phys. **181**, 237 (1998)
R.H. Cameron, J. Jiang, D. Schmitt, M. Schüssler, Astrophys. J. **719**, 264 (2010)
R.H. Cameron, M. Schüssler, Astron. Astrophys. **548**, A57 (2012)
S. Chandra, H.O. Vats, K.N. Iyer, Mon. Not. R. Astron. Soc. **407**, 1108 (2010)
E.W. Cliver, I.G. Richardson, A.G. Ling, Space Sci. Rev. **176**, 3 (2013)
M. Dasi-Espuig, S.K. Solanki, N.A. Krivova, R. Cameron, T. Peñuela, Astron. Astrophys. **518**, A7 (2010)
A.M. Du, W.Y. Xu, X.S. Feng, J. Geophys. Res. **113**, A07105 (2008)
G. Erdös, A. Balogh, J. Geophys. Res. **115**, A01105 (2010)
G. Erdös, A. Balogh, Astrophys. J. **781**, 50 (2014)
F. Feminella, M. Storini, Astron. Astrophys. **322**, 311 (1997)
R. Fisher, D.G. Sime, Astrophys. J. **287**, 959 (1984)
S. Giordano, S. Mancuso, Astrophys. J. **688**, 656 (2008)
M.N. Gnevyshev, Sol. Phys. **1**, 107 (1967)
M.N. Gnevyshev, Sol. Phys. **51**, 175 (1977)
R.T. Hansen, S.F. Hansen, H.G. Loomis, Sol. Phys. **10**, 135 (1969)
T. Hiltula, K. Mursula, Geophys. Res. Lett. **33**, L03105 (2006)
J.T. Hoeksema, Ph.D. thesis, Stanford University (1984)
D.V. Hoyt, K.H. Schatten, Sol. Phys. **181**, 491 (1998)
V.G. Ivanov, Geomagn. Aeron. **52**, 999 (2012)
D.J. Lewis, G.M. Simnett, Sol. Phys. **200**, 75 (2001)
D.J. Lewis, G.M. Simnett, G.E. Brueckner, R.A. Howard, P.L. Lamy, R. Schwenn, Sol. Phys. **184**, 297 (1999)

S. Mancuso, S. Giordano, Astrophys. J. **729**, 79 (2011)
H. Morgan, Astrophys. J. **738**, 189 (2011)
K. Mursula, T. Hiltula, Geophys. Res. Lett. **30**, 2135 (2003)
M.J. Owens, M. Lockwood, J. Geophys. Res. **117**, A04102 (2012)
M.J. Owens, I. Usoskin, M. Lockwood, Geophys. Res. Lett. **39**, L19102 (2012)
G.D. Parker, R.T. Hansen, S.F. Hansen, Sol. Phys. **80**, 185 (1982)
G.J.D. Petrie, Astrophys. J. **768**, 162 (2013)
G.J.D. Petrie, I. Patrikeeva, Astrophys. J. **699**, 871 (2009)
I.G. Richardson, H.V. Cane, E.W. Cliver, J. Geophys. Res. **107**, 1187 (2002)
I.G. Richardson, E.W. Cliver, H.V. Cane, J. Geophys. Res. **105**, 18203 (2000)
E. Robbrecht, Y.-M. Wang, Astrophys. J. **755**, 135 (2012)
N.R. Sheeley Jr., Y.-M. Wang, C.R. DeVore, Sol. Phys. **124**, 1 (1989)
D.G. Sime, R.R. Fisher, R.C. Altrock, Astrophys. J. **336**, 454 (1989)
E.J. Smith, A. Balogh, Geophys. Res. Lett. **35**, L22103 (2008)
M. Storini, S. Pase, J. Sýkora, M. Parisi, Sol. Phys. **172**, 317 (1997)
L. Svalgaard, J. Geophys. Res. **77**, 4027 (1972)
L. Svalgaard, T.L. Duvall Jr., P.H. Scherrer, Sol. Phys. **58**, 225 (1978)
M. Temmer, J. Rybák, P. Bendík, A. Veronig, F. Vogler, W. Otruba, W. Pötzi, A. Hanslmeier, Astron. Astrophys. **447**, 735 (2006)
I.G. Usoskin, K. Mursula, G.A. Kovaltsov, J. Geophys. Res. **106**, 16039 (2001)
Y.-M. Wang, N.R. Sheeley Jr., J. Geophys. Res. **93**, 11227 (1988)
Y.-M. Wang, N.R. Sheeley Jr., Astrophys. J. **355**, 726 (1990)
Y.-M. Wang, N.R. Sheeley Jr., Astrophys. J. **447**, L143 (1995)
Y.-M. Wang, N.R. Sheeley Jr., Astrophys. J. **590**, 1111 (2003a)
Y.-M. Wang, N.R. Sheeley Jr., Astrophys. J. **599**, 1404 (2003b)
Y.-M. Wang, N.R. Sheeley Jr., A.G. Nash, L.R. Shampine, Astrophys. J. **327**, 427 (1988)
Y.-M. Wang, N.R. Sheeley Jr., N.B. Rich, Astrophys. J. **658**, 1340 (2007)
M.A. Weber, L.W. Acton, D. Alexander, S. Kubo, H. Hara, Sol. Phys. **189**, 271 (1999)

DOI 10.1007/978-1-4939-2584-1_14
Reprinted from *Space Science Reviews* Journal, DOI 10.1007/s11214-014-0084-0

Solar Cycle in the Heliosphere and Cosmic Rays

Galina A. Bazilevskaya · Edward W. Cliver · Gennady A. Kovaltsov · Alan G. Ling · M.A. Shea · D.F. Smart · Ilya G. Usoskin

Received: 17 April 2014 / Accepted: 3 August 2014 / Published online: 9 September 2014

Abstract Manifestations of the 11-year solar cycle and longer time-scale variability in the heliosphere and cosmic rays are considered. We briefly review the cyclic variability of such heliospheric parameters as solar wind speed and density and heliospheric magnetic field, open magnetic flux and latitude variations of the heliospheric current sheet. It is discussed whether the local in-situ observation near Earth can represent the global 3D heliospheric

G.A. Bazilevskaya
Lebedev Physics Institute, Russian Academy of Science, Leninsky pr. 53, Moscow, Russia
e-mail: gbaz@rambler.ru

E.W. Cliver
Space Vehicles Directorate, Air Force Research Laboratory, Kirtland AFB, NM 87117, USA

E.W. Cliver
National Solar Observatory, Sunspot, NM 88349, USA
e-mail: ecliver@nso.edu

G.A. Kovaltsov
Ioffe Physical-Technical Institute, St. Petersburg, Russia
e-mail: gen.koval@mail.ru

A.G. Ling
Atmospheric Environmental Research, 3550 Aberdeen Ave., Kirtland AFB, NM 87117, USA
e-mail: aling@aer.com

M.A. Shea · D.F. Smart
SSSRC, 100 Tennyson Av., Nashua, NH 03062, USA

M.A. Shea
e-mail: sssrc@msn.com

D.F. Smart
e-mail: donsmart100@msn.com

I.G. Usoskin (✉)
Sodankylä Geophysical Observatory (Oulu unit) and Dept. Physics, 90014 University of Oulu, Finland
e-mail: ilya.usoskin@oulu.fi

pattern. Variability of cosmic rays near Earth provides an indirect useful tool to study the heliosphere. We discuss details of the heliospheric modulation of galactic cosmic rays, as recorded at and near Earth, and their relation to the heliospheric conditions in the outer heliosphere. On the other hand, solar energetic particles can serve as probes for explosive phenomena on the Sun and conditions in the corona and inner heliosphere. The occurrence of major solar proton events depicts an overall tendency to follow the solar cycle but individual events may appear at different phases of the solar cycle, as defined by various factors. The solar cycle in the heliosphere and cosmic rays depicts a complex pattern which includes different processes and cannot be described by a simple correlation with sunspot number.

Keywords Heliosphere · Cosmic rays · Solar energetic particles · Solar activity

List of Abbreviations

AU	Astronomical unit (the mean Sun-Earth distance, $\approx 1.5 \times 10^{11}$ m)
CME	Coronal mass ejection
CR	Cosmic rays
GLE	Ground-level enhancement of cosmic rays
GCR	Galactic cosmic rays
HCS	Heliospheric current sheet
HMF	Heliospheric magnetic field
IC	Ionization chamber
MHD	Magnetohydrodynamics
NM	Neutron monitor
OMF	Open magnetic flux
SEP	Solar energetic particles
SPE	Solar proton events
SSN	Sunspot number

1 Introduction

The heliosphere is a cavity in the interstellar plasma controlled by the solar wind continuously emitted by the Sun and by heliospheric magnetic field (HMF) ultimately produced by the solar dynamo (see, e.g, recent reviews by (Owens and Forsyth 2013; Balogh and Erdös 2013). The exact size of the heliosphere may vary in the course of solar cycle (e.g., Pogorelov et al. 2013, and references therein). It consists of three main regions separated by clear boundaries.

- The termination shock bounds the region of supersonic solar wind with frozen-in HMF. This is an approximately spherical region with a radius of about 90 AU. The heliosphere inside the termination shock does not "know" (in the MHD sense) about the presence of anything outside. The termination shock was discovered by Voyager-1 at the distance of 94 AU in 2004, and by Voyager-2 at 84 AU in 2007 (Stone et al. 2005, 2008).
- The region beyond the termination shock is still filled by the solar wind and HMF, but those are distorted by the presence of the interstellar wind. This region is bounded by the heliopause which is non-axisymmetric. Its size is about 122 AU in the nose direction, as discovered by Voyager-1 in 2012 (Stone et al. 2013). The size in the tail direction is not precisely known.

- Interstellar wind is somewhat disturbed beyond the heliopause forming an interface region, whose exact size is not well-known. An existence of a bow shock as an interface between disturbed and undisturbed interstellar wind was proposed, but recent data disputes that (McComas et al. 2012; Zieger et al. 2013). Beyond this boundary, interstellar wind does not "know" about the presence of an obstacle caused by the solar wind.

The heliospheric structure and conditions can be probed by not only dedicated spacecraft but also by cosmic rays traversing the heliosphere. Cosmic rays near Earth are of two main types: energetic (10^8–10^{20} eV/nucleon) galactic cosmic ray (GCR) particles existing all the time but modulated by the solar magnetic activity within the heliosphere; and less energetic (10^6–10^{10} eV/nucleon) solar energetic particles (SEP) which occur as sporadic solar proton events (SPEs), when the flux of such particles can be enhanced by many orders of magnitude.

Since the solar magnetic activity depicts a dominant 11-year cycle driven by the solar dynamo (Hathaway 2010), the heliosphere also changes with the 11-year solar cycles. Accordingly, the flux of energetic particles near Earth also varies both within the 11-year solar cycle and on the long-term scale. Here we present a brief review of the solar cycle as reflected in different aspects of the heliospheric and cosmic ray variability near the Earth. Other relevant reviews can be found, e.g., in Balogh et al. (2008).

Although this paper is devoted to the near-Earth observations, it is necessary to mention invaluable contributions of distant space missions to understanding of the heliosphere 3D structure and dynamics. Pioneer 10 (1972–2003), Pioneer 11 (1973–1995) reaching $\approx$80 AU, and Voyager 1 and 2 (both launched in 1977 and now exploring the outer heliosphere and beyond) provide unique data exploring the extreme far regions of the heliosphere. The space probe Ulysses (1990–2009) explored space outside the ecliptic plane, performing three scans over the solar polar regions in 1994–1995, 2000–2001, and 2007–2008, and documenting the 3D structure of the heliosphere. Consideration and discussion of these points can be found in Stone et al. (2005), Heber and Potgieter (2006), Balogh and Erdös (2013), Heber (2013), and Mewaldt (2013).

2 Solar Cycles in the Heliosphere

The very existence and dynamical balance of the heliosphere are ultimately governed by the solar surface activity which is driven by the 11-year cyclic solar dynamo process leading to variations in the solar wind dynamic pressure. As a result, the heliospheric extent (distance to the termination shock in the ecliptic plane) would be maximal on the decline of the cycle when low-latitude high-speed streams are most prominent (e.g., Richardson and Schwadron 2008).

2.1 Solar Wind and Heliospheric Magnetic Field in the Ecliptic Plane

The temporal variability of solar magnetic activity (quantified by the sunspot number—Usoskin 2013; Clette et al. 2014) is shown in Fig. 1 along with the key heliospheric parameters—solar wind speed, proton density, HMF—since 1965.

The heliospheric parameters presented here were measured *in-situ* at the Earth's orbit, i.e. in the ecliptic plane at 1 AU. SSN variability (Fig. 1A) is dominated by the 11-year cycle, varying from nearly no spots at solar minimum to high values of about 200 at solar maximum. To date, current cycle No. 24 which started in December 2008 is only about half as large as the average of cycles 18–23, indicating the end of the Modern Grand Maximum of

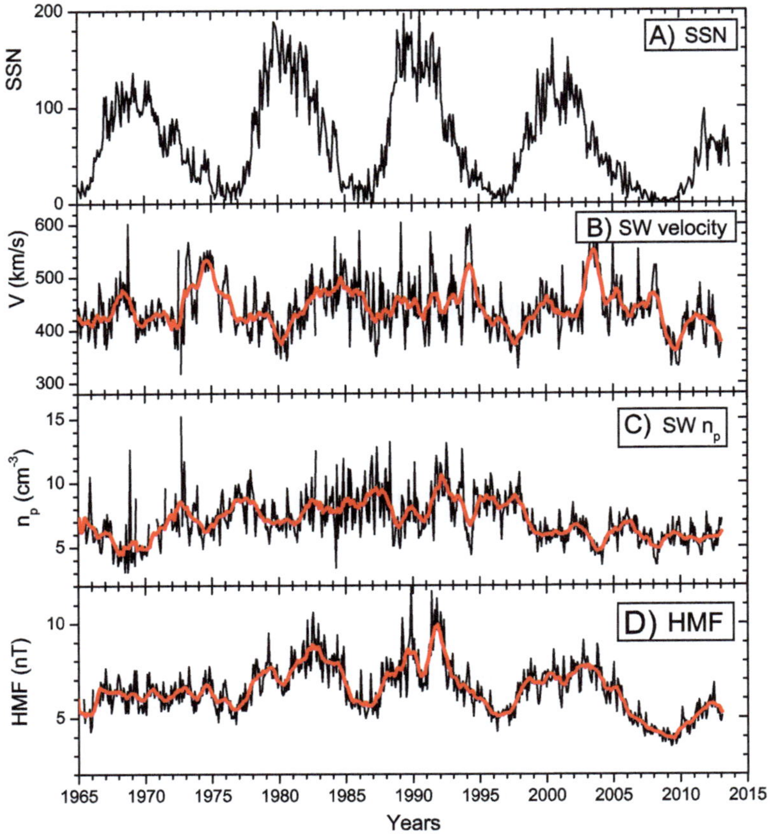

Fig. 1 Variability of the solar and heliospheric parameters for the period 1965–2013. (**A**) Monthly international sunspot numbers (http://sidc.oma.be/). (**B**) Solar wind velocity at the Earth's orbit (OMNIweb data http://omniweb.gsfc.nasa.gov/). (**C**) Solar wind proton density at the Earth's orbit (OMNIweb data). (**D**) Heliospheric magnetic field HMF at the Earth's orbit (OMNIweb data). *Red curves* depict the running 13-month means

solar activity (Solanki et al. 2004; Usoskin et al. 2007; Clette et al. 2014). On the other hand, heliospheric parameters near Earth show little variability over the solar cycle. Solar wind speed (panel B) does not depict a clear 11-year cyclic variability. Richardson et al. (2002) found that "For minimum periods, the Earth is embedded in high-speed streams ≈55 % of the time (most prominently on the decline of the cycle) versus ≈35 % for slow solar wind and 10 % for coronal mass ejection (CME)-associated-structures, while at solar maximum, typical percentages are as follows: high-speed streams ≈35 %, slow solar wind ≈30 %, and CME-associated ≈35 %." Thus the strongest sustained high-speed wind characteristically occurs on the decay of the sunspot curve. Proton density (panel C) also does not show a signature of the 11-year cycle, although a tendency of reduced density during the current weak cycle can be observed. The HMF does show a 11-year cycle of a factor of two as the max-min ratio. It is notable that HMF is weaker during the current cycle reaching the lowest measured value of about 3.5 nT in 2009. The HMF changes its dipole polarity near solar sunspot maximum time every solar cycle leading to the 22-year magnetic cycle called the Hale cycle (Cliver 2014). For more detail see a review by Owens and Forsyth (2013).

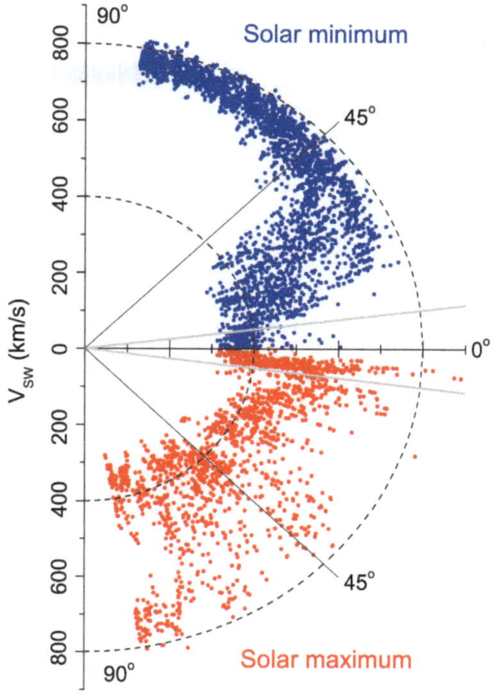

Fig. 2 Heliolatitudinal distribution of the daily solar wind velocities measured by Ulysses spacecraft (data from OMNIweb). The *upper* (*blue dots*) and *lower* (*red dots*) *panels* are built for the solar minimum (1994–1997 and 2005–2009) and solar maximum (1991, 2000–2003) conditions, respectively. The plot is given in polar coordinates, where the radial distance represents the solar wind speed in km/s, while the polar angle corresponds to the heliolatitude. The *horizontal central line* is the heliomagnetic equator, while the *grey lines* denote the range of the ecliptic zone in the heliocentric coordinates

2.2 3D View

Since the ecliptic plane is close to the solar equator (slightly tilted with about 7° inclination), it is a special region, and the heliospheric values measured in the ecliptic plane may not be fully representative for the full 3D heliosphere. Because of the tilt between the ecliptic plane and the helio-equator, the Earth scans a narrow range of heliolatitudes (≈14°) around the heliomagnetic equator (bounded by the grey lines in the Fig. 2) within one year. However, this scan is too narrow to give a flavor of the full 3D variability of the solar wind. While HMF in the ecliptic plane is roughly representative for the entire heliosphere (Lockwood 2013), the solar wind velocity varies dramatically at different helio-latitudes. Figure 2 summarizes the observations made onboard the Ulysses spacecraft during its latitudinal scans of the heliosphere (Wenzel et al. 1992). The upper panel (blue dots) shows the solar wind velocity for the solar minimum condition. A clear pattern is observed—slow solar wind dominates at low heliolatitudes, while fast solar wind dominates at high latitudes (McComas et al. 2008). This structure is caused by the presence of large unipolar coronal holes in the Sun's polar regions that emit fast solar wind. The situation is dramatically changed around solar maxima (lower panel of Fig. 2), when a mixture of fast and solar wind streams appear, caused by widespread small coronal holes and CME activity. Therefore, the solar wind as measured in the ecliptic plane is not representative for the full heliosphere. Moreover, the cyclic variability of the solar wind velocity at the ecliptic plane (slow solar wind at solar minima and more frequent fast solar wind streams at late declining phase of the solar cycle) is opposite to the higher latitude pattern.

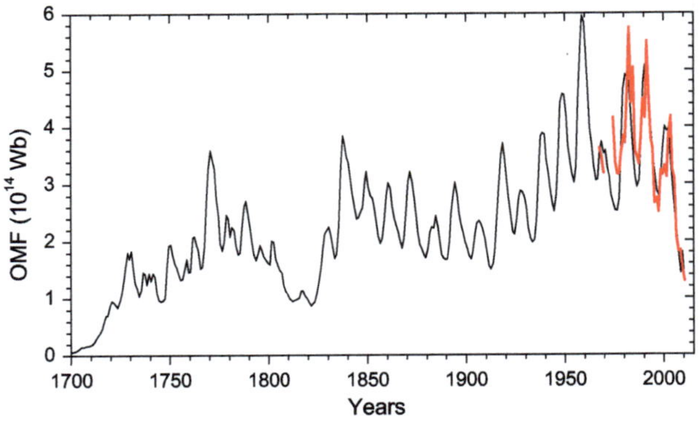

Fig. 3 Time variability of the open magnetic flux since the Maunder minimum of solar activity, reconstructed (*black line*—Vieira and Solanki 2010) and computed from in situ measurements (*red line*—Lockwood et al. 2009)

2.3 Open Solar Magnetic Flux

An important factor for the cosmic rays in the heliosphere is the open magnetic flux (OMF) of the Sun (Cane et al. 1999; Cliver and Ling 2001b; Lockwood 2013), which can be calculated from the parameters measured *in situ* near Earth or from the solar surface magnetic maps, and then recalculated to the OMF at the source surface at 2.5 solar radii around the Sun. OMF shows a great variability both in solar cycle and on the centennial scale (see Fig. 3). It is important to note that the centennial variability is great (Lockwood et al. 1999; Solanki et al. 2000) comparable with or even greater than the 11-year cycle range of variability. It is noteworthy that the recent solar minimum of 2009 yielded low OMF values, comparable to those around 1900 and during the Dalton minimum ca. 1820, indicating the end of the Modern Grand Maximum of solar activity.

2.4 Heliospheric Current Sheet

A specific feature of the heliospheric structure, which is very important for cosmic ray modulation (Potgieter 2013), is the heliospheric current sheet (HCS) which is a thin shear interface separating the regions of the opposite polarity of HMF emerging from the Sun. It corresponds to the heliomagnetic equator. Since the magnetic dipole axis of the Sun is tilted with respect to rotational axis, this together with the Sun's rotation and radially expanding solar wind, leads to formation of a complicated 3D-structure, resembling a ballerina's skirt (Mursula and Hiltula 2003; Owens and Forsyth 2013). The waviness of HCS, which plays an important role in GCR modulation (Alanko-Huotari et al. 2007b), is defined by the tilt angle, which varies cyclically over the solar cycle so that the HCS is nearly flat around the solar minimum and reaches its maximum values of above 70° around solar maximum times (Fig. 4). Direct computations of the tilt angle started in mid-1970s but indirect estimates (Pishkalo 2006) can be done even before that based on observations of solar eclipses since 1870 (Fig. 4). The results suggest that the cyclic variability of the HCS warp does not depend on the strength of the solar cycle—the tilt angle varies to its full extent for both high (in the second half of 20th century) and low (ca. 1900) solar cycles. It does not depict a centennial variation. Accordingly, we may expect a similar pattern, for example for the Maunder minimum condition.

2.5 Grand Minima of Solar Activity

Of special interest is a question what are the heliospheric conditions during a grand minimum of solar activity, the most famous example being the Maunder minimum in 1645–1700

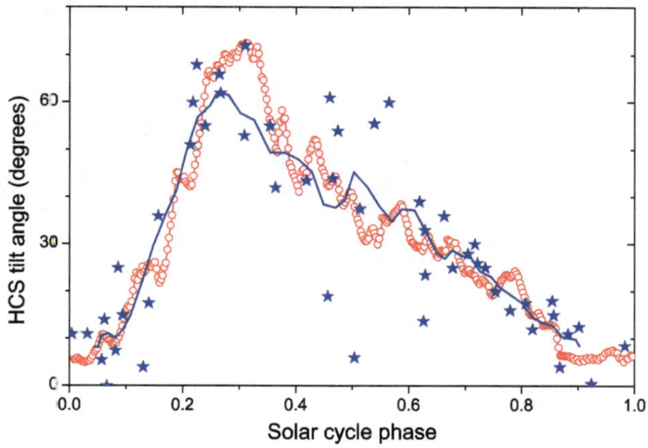

Fig. 4 Variability of the HCS tilt angle as a function of the 11-year solar cycle phase. The *red dots* represent the HCS tilt angle (radial model) averaged over cycles 21–23 (Wilcox Solar Observatory http://wso.stanford.edu/Tilts.html), the *blue stars* depict the tilt angle reconstructed from solar eclipses since 1870 (Pishkalo 2006), and the *blue curve* is a 7-point running mean over the *blue stars* (cf. Alanko-Huotari et al. 2007a)

with almost no sunspots on the solar surface (Eddy 1976). Despite the lack of sunspots a weak cyclic variability was observed during the Maunder Minimum in geomagnetic and cosmic ray indices (Beer et al. 1998; Usoskin et al. 2001; Miyahara et al. 2004; Berggren et al. 2009; Owens et al. 2012) suggesting that the solar magnetic cycle was still operating in the heliosphere. Different estimates of the HMF and solar wind parameters exist, but the current paradigm (Cliver et al. 1998; McCracken 2007a; Steinhilber et al. 2010; Owens and Forsyth 2013; Lockwood and Owens 2014) is that the HMF was weak, about 2 nT, and solar wind in the ecliptic plane remained slow as emanating from hot parts of the corona even in the absence of active sunspot regions. The likelihood that the solar wind was slow during the Maunder Minimum was underscored by the recent solar cycle minimum between cycle 23 and 24, when Earth was embedded in slow solar wind ≈70 % of the time in 2009 (Cliver and Ling 2011). However, because of the imbalance between the different processes, the phase relation between OMF (and hence cosmic rays) and the solar magnetic cycle was probably inverted with respect to normal solar cycles (Owens et al. 2012). Such grand minima occur every now and then (Usoskin 2013) and correspond to a special mode of the solar dynamo (Usoskin et al. 2014).

3 Solar Cycles in Galactic Cosmic Rays as Observed at Earth

In this section we discuss how the conditions in the heliosphere are reflected by the results of cosmic ray observations. Galactic cosmic rays are highly energetic fully ionized nuclei coming into the heliosphere from outside the solar system with a roughly constant flux. GCRs with kinetic energy in the range from a few hundred MeV up to 100 GeV can be regarded as heliospheric probes. Transport of cosmic rays in the heliosphere is governed by several processes, viz. diffusion due to scattering on the HMF irregularities, convection with the radially expanding solar wind plasma, drift in the large-scale heliospheric magnetic field including that along the HCS, and adiabatic energy losses due to the divergence of solar wind (see Potgieter 2013, for details). All these heliospheric characteristics depend on the phase of the solar magnetic cycle and on the solar activity level. As a result, the GCR flux is modulated by the solar magnetic activity. Therefore, by measuring GCR flux at Earth one can study the 11-year and 22-year cycles in the heliosphere.

Figure 5 presents time profiles of SSN as well as GCR intensity as observed by different instruments sensitive to different energy ranges of primary particles. The GCR records vary

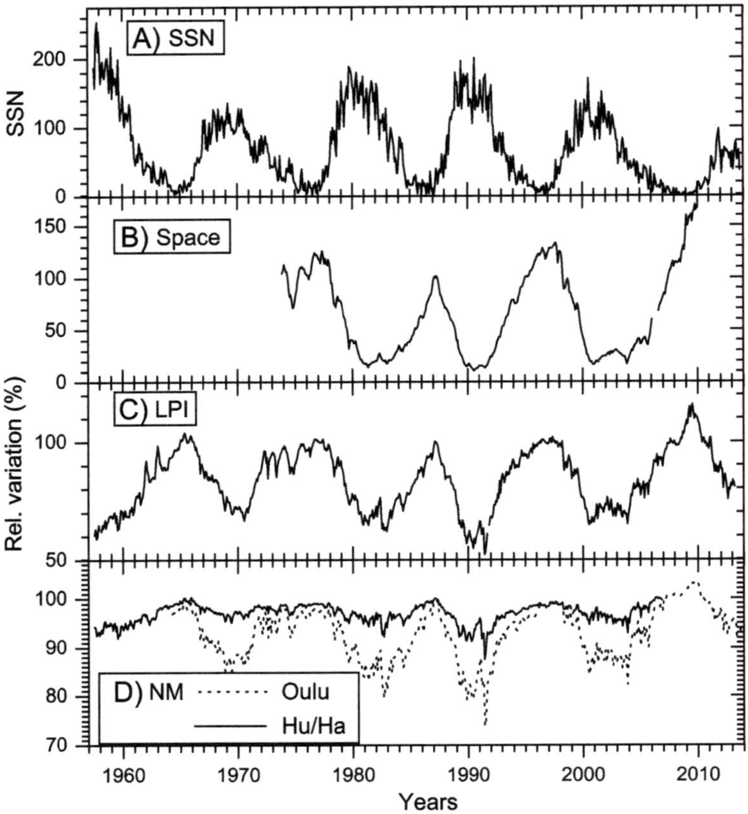

Fig. 5 Time variability of: (**A**) sunspot numbers (http://sidc.oma.be/sunspot-data/); (**B**) cosmic ray relative intensity measured in space onboard IMP8 and PAMELA spacecraft; (**C**) cosmic ray relative intensity measured by high latitude balloons (LPI); (**D**) cosmic ray relative intensity measured by ground-based neutron monitors, polar Oulu and equatorial Huancayo/Haleakala. Relative variations of cosmic rays are normalized to 100 % in March 1987

cyclically in rough anti-phase to SSN with a delay which depends on the direction of the CR drifts in the heliosphere (Usoskin et al. 1998; Cliver and Ling 2001a). The depth of the modulation depends greatly on the CR energy: it varies from a few % for an equatorial NM to ≈20 % for a polar NM (see panel D); a factor of two for stratospheric balloon-borne data (panel C); and up to an order of magnitude for space-borne data (panel B). One can also see alternation of sharp and flat peaks in the GCR intensity that is related to the dominant drifts in the heliosphere (Potgieter 2013). An important feature is that the GCR intensity in 2008–2010 was the record high for the entire series reflecting the weak solar activity. The GCR intensities (Fig. 5) follow the (inverted) 11-year solar cycle much better than any individual heliospheric plasma and magnetic field parameter observed in the near Earth space (Fig. 1). On the other hand, GCR flux is well related to the OMF from coronal holes at latitudes ≤45° (Cliver and Ling 2001b; Alanko-Huotari et al. 2006). This implies that the local *in situ* heliospheric measurements at 1 AU in the ecliptic plane may not necessarily represent the global heliospheric changes, since the GCR modulation is affected by the plasma and magnetic field parameters in the whole heliosphere rather than locally.

3.1 Cosmic Ray Detectors and Their Energy Responses

In order to study the 11-year cycle in CR intensity one needs long-term time series of measurements of the charged particle flux. Since the GCR modulation depends on particle energy, detectors sensitive to different energy ranges of CR are needed. The low energy part of the GCR spectrum, up to tens of GeV, is affected by the heliospheric modulation, but there is an indication that a weak modulation may affect even $\approx$100 GeV particles (Potgieter and Strauss 2013).

In the energy range of 0.5–15 GeV, the geomagnetic field and the Earth's atmosphere serve as an energy separator (see e.g., Grieder 2001). First attempts to initiate a permanent ground-based regular CR observational network, with identical ionization chambers (IC), refer back to 1930s (Forbush 1954). ICs at sea level detect secondary muons generated in the atmosphere by protons with energies >4 GeV impinging at the top of the atmosphere.

However, a real impetus to regular CR monitoring was given by John Simpson with invention of a neutron monitor (Simpson 1958, 2000). A world-wide network of ground-based neutron monitors (NMs) was developed in 1950s and included, in different times, more than a hundred NMs, with more than 40 NMs operating presently (Moraal et al. 2000; Mavromichalaki et al. 2011). The NM data set is available from mid-1950s up to now (see, e.g., Fig. 5d). Because of the geomagnetic shielding, each NM accepts particles above some cut-off energy/rigidity defined by the location of the NM and the geomagnetic field strength. The shielding is absent in polar and maximum in equatorial regions where the cut-off rigidity is about 15 GV. Thus, the global NM network can be considered as a single device allowing to obtain information on the CR particle fluxes, their energy and anisotropy. Reconstructions of the GCR flux based on the NM network data agree well with the direct balloon- and space-borne measurements, making the NM network a very useful instrument to monitor CR flux (Usoskin et al. 2011).

The problem of converting the data obtained on the ground and in the atmosphere into characteristics of the primary CRs requires knowledge of the response/yield function of different devices and understanding of the atmospheric/magnetospheric transport (e.g., Caballero-Lopez and Moraal 2012). As a measure of the sensitivity of a NM to CR energy, Alanko et al. (2003) introduced a concept of the effective energy for a NM, E_{eff}, such that the count rate of a given NM is proportional to the flux of GCR with energy above E_{eff}, which varies between 6.5 GeV for a polar NM to 30–40 GeV for an equatorial NM.

Another common type of ground-based cosmic-ray detector is a muon telescope measuring the muon component of the cosmic-ray induced atmospheric cascade. They typically have higher effective/median energy (40–60 GeV—see Jämsén et al. 2007) than NMs.

The ground-based monitoring is complemented by CR observations in the Earth's atmosphere. During 1951–1969, Neher (1967, 1971) accomplished a series of regular but infrequent balloon measurements of the cosmic ray induced ionization. In mid-1950s, a Soviet/Russian program (Charakhchyan 1964; Stozhkov et al. 2009) lead by the Lebedev Physical Institute (LPI) of long-term balloon-borne observations of the CR fluxes at several latitudes in the Earth's atmosphere had started and continues till now (see Fig. 5c). Balloons were launched on the daily basis until 1990, and three times a week since then. The LPI balloon measurements get data from the ground level up to the altitude of about 30 km. This data corresponds to $E_{\mathrm{eff}} \approx$ 1–2 GeV (Bazilevskaya et al. 2013).

Numerous balloon-borne detectors have been launched to measure the CR spectrum at shallow atmospheric depth, such as, e.g., BESS instrument (e.g. Shikaze et al. 2007). However, such measurements, while providing direct measurements of the GCR spectrum, are campaign-based and do not provide long-term monitoring.

Cosmic rays with energy below several hundred MeV can be recorded only by spacecraft-borne detectors because such particles are absorbed in the upper atmosphere. Only a few limited data sets of sufficient duration are available, among them the results from the IMP8, and the ACE spacecraft (McDonald 1998; McDonald et al. 2010), and the more recent PAMELA (Adriani et al. 2013) and AMS (Ting 2013) orbital spectrometers. A record is shown in Fig. 5b and corresponds to the flux of protons with energy 120–230 MeV (Bazilevskaya et al. 2013).

3.2 CR Transport in the Heliosphere

Temporal variability of GCR intensity in the energy range affected by solar modulation provides indirect information on the state of the whole heliosphere, not only in the vicinity of Earth. For this purpose, one needs to understand the processes driving the CR transport in the heliosphere. The basics of the CR transport equation were introduced in the 1960s (Parker 1965; Jokipii 1966; Dolginov and Toptygin 1967; Gleeson and Axford 1967; Krymskij 1969) and further developed by a number of groups (e.g., Dorman 2006). Modern CR transport models are sophisticated and precise (for a review see Potgieter 2013, and references therein), allowing to study numerically different effects and processes. In order to study the full process, however, one needs detailed information on the local interstellar spectrum of GCRs, as well as geometrical structure, polarity, strength, and turbulence level of the HMF and the solar-wind speed throughout of the 3D heliosphere. Unfortunately, as discussed above, there are, with rare exceptions, only local near-Earth observations of the HMF parameters. Therefore, the models have to use assumptions, e.g., regarding elements of the diffusion tensor that cannot be measured directly. On the other hand, modulation of GCRs with energy above several hundred MeV can be formally described by the so-called force-field model (e.g., Gleeson and Axford 1968; Caballero-Lopez and Moraal 2004). This model parameterizes the GCR spectrum with a single formal parameter, the modulation potential. The momentary value of the modulation potential can be calculated from the data of the ground-based NM network (Usoskin et al. 2011). On the long-term timescale, the modulation parameter can be reconstructed from data of cosmogenic isotopes ^{14}C and ^{10}Be measured in terrestrial archives (see, e.g., Beer et al. 2012; Usoskin 2013).

It has been demonstrated that this approach, while not pretending to be a physical interpretation, provides a very useful single-parameter parameterization of the GCR energy spectrum near Earth (see formalism in Usoskin et al. 2005), especially for long-term studies. Nonetheless, a more sophisticated approach is needed to give insight into detail of CR interaction with the heliospheric magnetic field.

3.3 CR Diagnostic of the Heliosphere

Modulation of GCR in the heliosphere is largely defined by the turbulent component of HMF responsible for particle diffusion, as the mean free path depends on the power spectrum of the magnetic irregularities in the heliosphere (Jokipii 1966). Recent models (e.g., Ferreira and Potgieter 2004) are good in explaining the temporal variability of the observed GCR fluxes, implying that the basic conception of CR modulation is correct. However, the value of the mean free path remains unknown because of the complicated structure of the turbulent interplanetary magnetic field (Giacalone 2013). The HMF has, in the rough approximation, a form of an Archimedian spiral (Parker 1965) that, leads to gradient and curvature drifts of CR in the interplanetary medium. Account for motion of the magnetic field foot-points on the Sun leads to a more realistic Fisk-type HMF (Fisk 1996) which is particularly important at high latitudes. Important is also a drift at the HCS which is an extreme case of the gradient drift. The role of the drift effects in CR modulation was pointed out by Jokipii and Levy (1977) and intensively studied since then (e.g., Alanko-Huotari et al. 2007b; Strauss et al. 2012b). When the solar magnetic field is directed outward from the Sun in the north polar region (the so-called $A > 0$ periods), positively-charged particles drift inwards in the heliospheric polar regions and drift outwards along the HCS. This facilitates CR access into the inner heliosphere leading to the long flat shape of CR intensity maxima as, e.g., ca. 1976 and 1996. The drift has the opposite pattern during $A < 0$ periods, forming favorable conditions for the CR access to the inner heliosphere when the HCS is flat. This leads to sharp high peaks of CR intensity in 1965, 1987 and 2009. This drift-dominated modulation is typical (but see the next section) for minima of solar activity cycles when the heliospheric structure is fairly regular. The shape alternation of CR maxima is seen in Fig. 5 as a prominent signature of the drift effect in the CR modulation. However, during periods of high solar activity drifts are less important (Krainev and Kalinin 2013) since the modulation is mostly driven by diffusion and propagating barriers (Burlaga et al. 1985).

3.4 Unusual Minimum of Solar Activity Between Solar Cycles 23 and 24

As one can see in Fig. 5, the intensity of GCR reached in 2009 its record highest value, over the entire period of direct observations since 1950s (see, e.g., Heber et al. 2009; Ahluwalia et al. 2010; McDonald et al. 2010; Moraal and Stoker 2010; Krainev et al. 2013). This happened during the minimum of solar activity between solar cycles 23 and 24, which is known to be the weakest heliospheric activity period for the space era (Gibson et al. 2011; McComas et al. 2013). This period is characterized by the record-low HMF intensities, reduced HMF turbulence, change in the inhomogeneity spectrum of the HMF, and reduced solar-wind dynamic pressure. The GCR intensity reached a sharp maximum in 2009 following the flattening of the HCS and then quickly dropped to the normal level after the sudden increase of the HCS tilt angle in early 2010 (McComas et al. 2013). The GCR intensity enhancement was observed in a wide range of particle energies (Fig. 5). It was as great as 60 % at a few hundred MeV energy (panel B) and only a few % in the data of a polar NM (panel D). On the other hand, the CR intensity as recorded at mid- and low-latitudes (geomagnetic cutoff rigidity above 4 GV) did not show any excess in 2009 compared to 1987 (Moraal and Stoker 2010). The unusual rigidity dependence of the GCR increase suggests that significant changes took place in the heliosphere properties during the recent solar cycle minimum (Mewaldt et al. 2010; Bazilevskaya et al. 2012; Krainev and Kalinin 2013; Potgieter et al. 2014). These facts indicate that mainly lower energy GCRs ($\leq$10 GeV) were affected by the favorable heliospheric conditions but not higher energies (see, e.g. Usoskin et al. 2011). Cliver et al. (2013) pointed out that the record increase in cosmic ray intensity in 2009—which was accompanied by a decrease in HMF and an increase in the tilt angle relative to the previous minimum—challenged the dominance of drift effects for modulation at 11-yr minima of the solar cycle. More recently, Potgieter et al. (2014) concluded that "the 2009 modulation minimum could be described as more 'diffusion dominated' than previous solar minima" and that "diffusion contributed $\approx$50 % of the total cosmic proton intensities observed at Earth while particle drifts contributed the other 50 %". This implies that the scaling of the HMF turbulence changed during the recent solar cycle minimum compared to the previous minimum period, although a more detailed analysis and modelling is required to fully understand this process (Strauss et al. 2012a).

4 Cycles in Solar Energetic Particles

4.1 Solar Proton Events

Unlike many solar and geophysical measurements such as SSN or geomagnetic indices, the history of identification of solar proton events (SPEs) is not as long. The first SPEs were directly recorded by ICs in Cheltenham (USA), Godhavn (Greenland), and Christchurch (New Zealand) in 1942 as clear increases on 28-Feb and 7-Mar (Lange and Forbush 1942). However, these increases were not ascribed to particles from the Sun until the third increase was observed at these same locations on 25-Jul-1946 (Forbush 1946). A fourth increase (19-Nov-1949) was also recorded by other detectors, e.g., by an IC in Climax, USA (Forbush et al. 1950), muon detectors at several locations and a prototype NM at Manchester, UK (Adams 1950). While direct measurements of SPEs prior to the 19th solar cycle were limited to the ICs and early NMs, there were indirect measurements from vertical incidence ionospheric soundings made at high latitude locations starting in 1938. In a comparison of polar cap absorption events in 1956–1959 with records of vertical incidence soundings, Švestka (1966) compiled a fairly homogeneous set of data on strong polar cap absorption events for 1938–1959. During the 19th solar cycle, lower energy SPEs could be identified by ionospheric measurements, and instruments on balloons and early rocket and satellite experiments, that were able to detect solar protons at lower energies (MeV range). The solar proton catalog edited by Švestka and Simon (1975) provides a compendium of SPEs of the 19th solar cycle, which is not, however, intended to provide a homogeneous database; nevertheless it is the best data available during that time period. Similar catalogs were compiled also later under the supervision of Yu.I. Logachev[1] spanning the period of for 1970–1996.

The installation of the standardized NMs starting in 1951 considerably enhanced the ability to detect large relativistic SPEs. The event of 23-Feb-1956 was the largest increase recorded by ground-based NMs and muon detectors: the Leeds NM recorded an increase of 4554 % over a 15-minute interval, and the Moscow muon detector recorded an increase of 300 %. Throughout the 19th solar cycle (1954–1965) ground-based NMs and muon detectors were essentially the only instruments that could identify high energy relativistic SPEs. Accordingly, these events were called ground-level enhancements, or ground-level events (GLEs).

Starting with the IMP satellite program in 1964, it became possible to assemble an almost homogeneous database of SPEs recorded at Earth. In 1976, the NOAA/USAF Space Environment Services Center in Boulder, Colorado defined a significant solar proton event (SEP) as any event with a proton flux greater than 10 protons $(\mathrm{cm}^2\,\mathrm{s}\,\mathrm{sr})^{-1}$ above 10 MeV. Using this criteria, Shea and Smart (1990) compiled a list of SPEs from 1955 until 1986 (i.e. solar cycles 19–21). While an update of that list is beyond the scope of this paper, the events in solar cycles 22 through cycle 24 (first 5 years) have been included in the following statistics and graphs.

4.1.1 Statistics of Solar Proton Events

Using the NOAA criteria of identifying a SPE (see above) we have identified all events that meet this criteria in solar cycles 19 through year 5 in solar cycle 24. Since the list of SPEs maintained by NOAA does not identify independent injections of solar protons within

[1] http://www.wdcb.ru/stp/data/PRCATFINAL/.

Table 1 Summary of Solar Proton Events for Solar Cycles 19 through the first five months of Solar Cycle 24, including: start months and duration (in months) of solar cycles, the total number of discrete events, N_{SPE}, total number of GLEs N_{GLE}, percentage of GLE P_{GLE} vs. the SPEs, and the omnidirectional fluence (in 10^{10} cm^{-2}) of >10 MeV protons, F_{10}

Cycle	Start	Duration	N_{SPE}	N_{GLE}	P_{GLE}	F_{10}
19	May 1954	126	65	10	15.4	7.2
20	Nov 1964	140	72	13	18.0	2.2
21	Jul 1976	123	81	12	14.8	1.8
22	Oct 1986	120	84	15	17.8	5.8
23	Oct 1996	146	103	16	15.5	7.5
24	Dec 2008	60	29	1	3.4	1.2

a major episode of activity unless the >10 MeV flux falls below the event threshold, we have examined each of the events for independent and additional sources of solar protons. Increases associated with the arrival of interplanetary shocks at Earth were not assumed to be separate events but considered part of the previous identified event. Occasionally a proton-producing region of activity on the far Eastern limb of the Sun can be associated with a small proton enhancement that does not reach the NOAA proton event criteria. When the interplanetary shock from the coronal mass ejection (CME) of such an event arrives at Earth, the additional protons accelerated by the shock may add to the already enhanced flux such that the SPE criteria will be met. While we considered this as a shock associated increase, it will be listed as a proton event on the NOAA listing. An example of this type of event occurred on 23-Jun-2013.

The events we have identified as separate and unique solar proton events were tallied in 12-month arrays starting with the first month of each solar cycle. The start of each solar cycle was taken as the month after the minimum smoothed sunspot number (see Table 1).

The total omnidirectional fluence (time-integrated flux) of >10 MeV protons near Earth, denoted as F_{10}, is often considered as a characteristic integral parameter of SPEs. The values of F_{10} for the past 6 solar cycles (cycle 24 incomplete yet) are given in Table 1. These values differ by factors of 4 from cycle to cycle, which can be attributed to the location of the parent solar activity associated with the various increases over a solar cycle (Shea and Smart 2012). For example one strong event, e.g., in August 1972 can completely dominate the total fluence for the entire solar cycle (Shea and Smart 1990).

Figure 6 illustrates the distribution of SPEs throughout solar cycles 19–23 and the first five years of cycle 24. The distribution of events identified by Švestka (1966) for cycles 17–18 are shown in Fig. 7 using the same convention. While caution should be used in utilizing these data for comparison with the modern SPE identification, the distributions shown in Fig. 7 are similar to those shown in Fig. 6. Figure 8 presents the average number of SPEs over solar cycles 19–23 together with the average sunspot number summed over the same time period. While no coherent pattern of the SPE occurrence is seen in Fig. 6 for individual cycles, the overall pattern of the SPE number appears to be consistent with the average sunspot number when summed over five solar cycles. The distribution of GLEs over solar cycles 19–24 is shown in Fig. 9. In comparing it with Fig. 8, it appears that the occurrence of GLEs is more widely distributed over the solar cycle than the >10 MeV proton events. This may either reflect a physical difference or low statistics of GLEs in this time period (67 GLEs vs. 405 SPEs). When the four GLEs of solar cycles 17 and 18, observed by muon detectors, are included, this distribution becomes even wider.

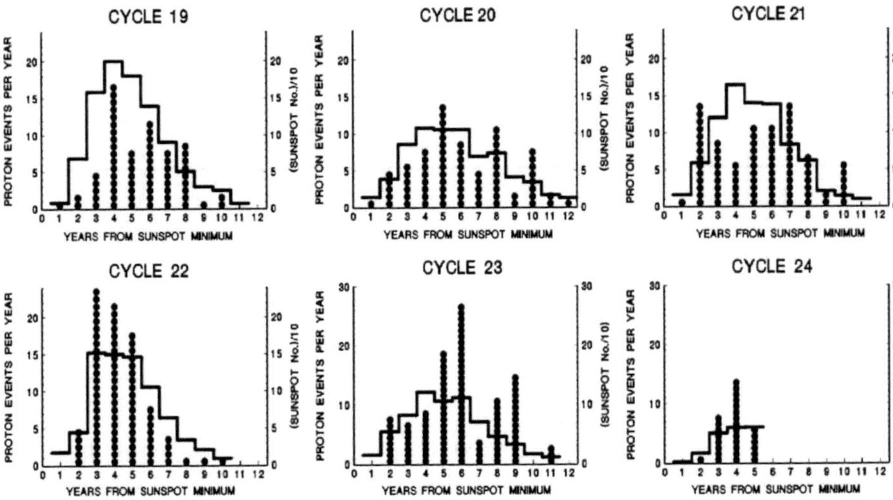

Fig. 6 Distribution of SPEs >10 MeV over each solar cycle 19–24 (see Table 1 for dates). These are shown in 12-month increments starting with the year after the sunspot minimum for each cycle. Only the first five years of cycle 24 (i.e. through November 2013) are shown. The histograms are the yearly sunspot number (divided by 10) summed for each of the same time periods

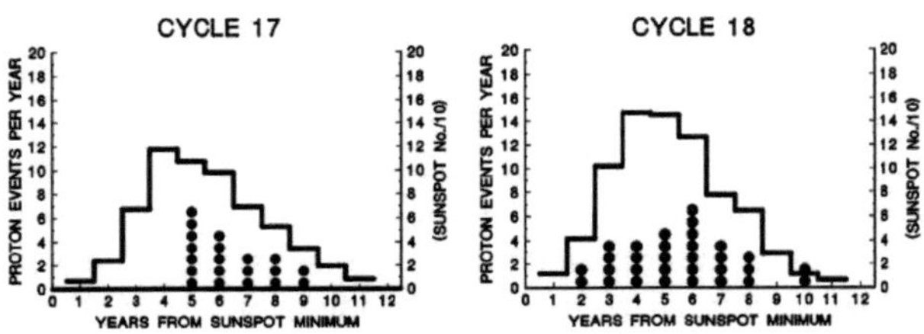

Fig. 7 The distribution of SPEs as identified by Švestka (1966) for solar cycles 17 and 18. The number of events have been binned in 12-month intervals from the start of cycle 17 (October 1933) and the start of cycle 18 (March 1944). The histograms are the yearly sunspot number (divided by 10) summed for each of the same time periods

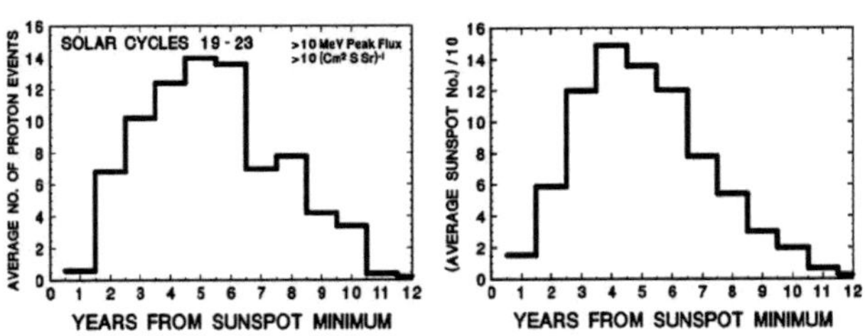

Fig. 8 *Left side*: Average number of SPEs each year over five solar cycles (1954–2008). *Right side*: Average sunspot number (divided by 10) over five solar cycles (1954–2008). The yearly values have been calculated for each 12-month period starting with the first month after the minimum value in the smoothed sunspot number

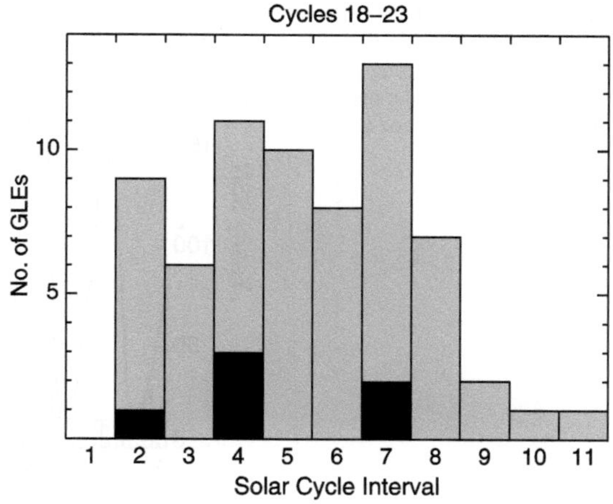

Fig. 9 Occurrence frequency of GLEs over a representative "11-year" solar cycle. Each complete cycle since 1942 has been divided into 11 equal intervals and the number of GLEs has been distributed over the 11 intervals and summed accordingly. The *black shading* indicates the six GLEs with $F_{200} \geq 3.0 \times 10^7$ proton/cm^2 from 1956-present (see Sect. 4.3)

4.1.2 Ground Level Enhancements

GLEs are relatively rare events with an almost homogeneous database since 1956, and as such, have been widely studied. Until 2014, there have been 71 GLEs since the first events identified in 1942 with a mean rate of about one event per year (e.g., Cliver et al. 1982; Cliver 2006; Gopalswamy et al. 2012, 2013) but unevenly distributed over the solar cycles (Fig. 10). Ten GLEs were identified in the 19th solar cycle, but a few small events may have been missed because of the anisotropy (not known at that time), the sparsity of detectors particularly in the polar regions, and a low flux of protons >450 MeV to register as an increase on a NM. In the mid 1970s, the cosmic ray community adopted a criterion for an event to be identified as a GLE,[2] that is two independent NMs have recorded a statistically significant increase. We note that an enhancement of 06-Jan-2014 observed at the South Pole NM (Thakur et al. 2014) was a not a GLE according to the formal criterion. As shown in Fig. 10, these events can occur at any time of the solar cycle but tend to be on the rising and decreasing parts of the cycle. With the exception of solar cycle 22, there is a distinct gap near sunspot maximum indicative of the Gnevyshev Gap (Gnevyshev 1967, 1977).

One important aspect of GLEs is the degree of anisotropy that may be present, particularly for events resulting from solar activity on the western solar hemisphere (see Sect. 3.2 of Shea and Smart 2012). In addition to the normal anisotropy typically present during the first 20–30 minutes of many GLEs, McCracken et al. (2012) have identified high-energy impulsive GLEs that constitute the first phase of some very large GLEs. These authors offered several constraints that must be met by any putative acceleration mechanism for these unusual events. A comprehensive study of GLEs needs to address the degree of anisotropy as well as other defining factors such as the type of detector recording the event, and the geomagnetic cutoff rigidity and altitude of the detector (e.g., Mishev et al. 2014).

While the emphasis on the detection of GLEs since 1955 has been on NM measurements, the importance of the IC and muon measurements should not be ignored even though the particle energies in most GLEs do not have sufficient flux above the 4 GeV threshold for detection by muon detectors. Of the 67 GLEs between 1955 and 2014, only 5 have been recorded by the world wide network of muon detectors as illustrated in Fig. 11. This makes

[2]The official master list of GLEs is available at http://gle.oulu.fi.

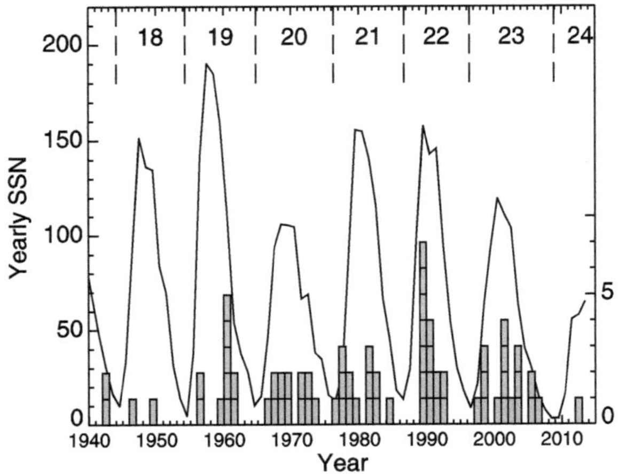

Fig. 10 Timing of the 71 GLEs observed from 1942 to the present (*grey bars*) along with sunspot numbers. Solar cycles are separated by *dashed lines*

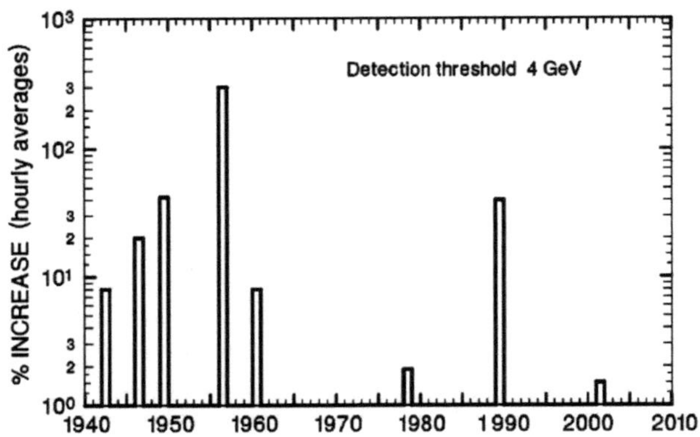

Fig. 11 GLEs recorded my muon detectors (given in percentage increase) at the sea level from 1942 to 2010. The 1940 events are bi-hourly data. The 1950 event are hourly data and the events from 1960 to 2010 are 5 minute data

the first four GLEs in solar cycles 17 and 18 very impressive as massive events. The associated solar regions for the events of 28-Feb-1942 and 25-Jul-1946 were located at ≈4 °E and 15 °E respectively (Duggal 1979). The event on 14-Jul-2000 was also associated with central meridian activity (7 °W) but the muon detector increase was an order of magnitude lower than for the 1946 event. The remaining events recorded by muon detectors were associated with solar activity on the far western side of the Sun. It should be noted that the increases for the events in the 1940s were from bi-hourly records; hourly data would have more likely been considerably higher.

4.2 Coronal Mass Ejections and SEP Events

It is well-accepted that the free energy for solar activity resides in solar magnetic fields, specifically in the low coronal fields of active regions (e.g., Hudson 2011). Thus, it is not surprising that the source active regions of SPEs powerful enough to be detectable at Earth are generally large. For the 15 GLEs from cycle 23 for which the area of the parent sunspot group could be determined (Cliver 2006; Gopalswamy et al. 2012), the median sunspot area of the GLE-source regions was 750 millionths of a solar hemisphere (msh; range from 360–2580 msh, corrected for foreshortening which reduces the apparent areas of spots away from disk center). As a reference, "naked eye sunspots", i.e., those that can be seen without

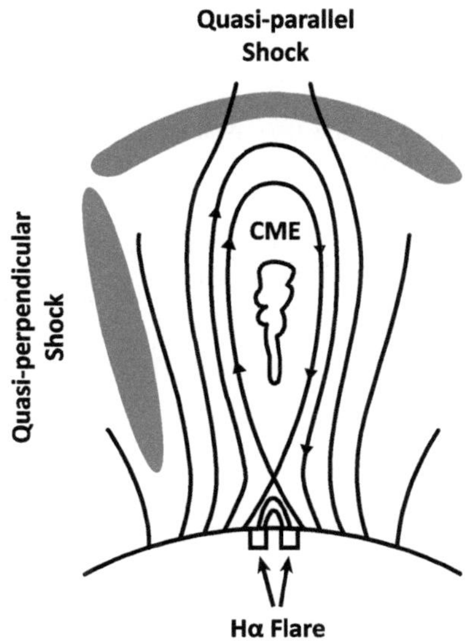

Fig. 12 A schematic view for CME-driven-shock-acceleration of SEPs indicating the locations of quasi-perpendicular and quasi-parallel CME-driven shock acceleration of SEPs (adapted from Cliver 2009b)

magnification through darkened glass or an atmosphere shielded by light clouds or smoke, have areas of >500 msh and make up only about 5 % of all sunspot groups (Newton 1958).

In the current picture of particle acceleration at the Sun (Reames 1999, 2013; Tylka et al. 2005; Tylka and Lee 2006; Cliver 2009a,b), the high-energy protons that give rise to GLEs are accelerated in solar flares and by CME-driven-shocks, either via diffusive acceleration at a CME bow shock or via shock-drift acceleration at the CME flanks as indicated in the schematic in Fig. 12. To drive the strong shocks associated with GLEs, the CMEs need to be fast and massive. Such CMEs dominate the energy budget of major eruptive flares (Emslie et al. 2012). For example, the CMEs associated with the 16 GLEs of cycle 23 (coronagraph data available for 15) had a median speed of 1810 km/s (range 938–3242 km/s) compared to ≈420 km/s for all broad (>60°) CMEs in this cycle, while 12 of the 15 CMEs were full halos (Gopalswamy et al. 2009, 2012). The variation in the occurrence of such CMEs (halo CMEs with speeds >1000 km/s) for cycle 23 (1996–2008), for which we have the homogeneous SOHO Large Angle Spectroscopic Coronagraph (Brueckner et al. 1995) data set, is shown in Fig. 13.

The distribution of fast halo CMEs in Fig. 13 looks like the histogram of GLEs over the composite solar cycle for cycles 18–23 in Fig. 9, although the broad maximum of the GLE distribution (spanning intervals 2–8) is shifted one year to the left. The general variation of high-energy SPEs with the solar cycle (Fig. 9) has also been documented for significant SEP events at lower energies (see, e.g., Fig. 8). Comparison of Figs. 13 and 9 (after normalizing for the number of solar cycles) indicates that on average, only 1 of about 15 fast halo CMEs produces a GLE. This significant reduction can have multiple contributing causes. For example, increasing the speed threshold for Fig. 13 to 1500 km/s, closer to the ≈1800 km/s median speed for all GLE-associated CMEs during cycle 23, approximately halves the number of fast halo CMEs during this period. Factors unfavorable to SEP acceleration by fast halo CMEs include: low CME brightness (i.e., mass/energy) (Kahler and Vourlidas 2005; Mewaldt et al. 2008), lack of closely-timed preceding CMEs from the same or nearby active regions (Gopalswamy et al. 2002, 2004; Kahler and Vourlidas 2005), low pre-existing seed particle population in the interplanetary medium (Kahler 2001; Tylka et al. 2005; Cliver 2006), and the fact that GLEs are rarely observed from the eastern half of the Sun (only 9

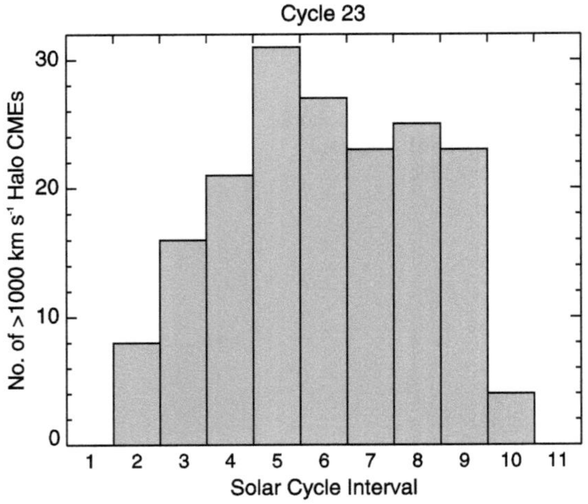

Fig. 13 Histogram of the occurrence of fast (>1000 km/s) halo CMEs over the 11 equal intervals of 13.3 months each, of cycle 23. Corrections to the CME counts were made by prorating for long (>1 month) gaps in coronagraph coverage (vis. 24-Jun-1998–15-Oct-1998 and 20-Dec-1998–5-Feb-1999)

of 71 cases) due to the propagation of SEPs along the Parker spiral (e.g., Shea and Smart 1990).

When the parent solar activity (eruptive flare) is near the central meridian of the Sun, a fast CME can propagate outward from the sun accelerating particles as the shock traverses the interplanetary medium between the Sun and Earth. When the parent solar activity is located toward the western limb of the Sun (as viewed from Earth), the time/intensity profiles of these SPEs tend to have rapid rates of rise to maximum intensity followed by a rapid decay. Based upon those observations, Shea and Smart (1994) and Smart et al. (2006) identified a bi-modal distribution of SPEs classified as "interplanetary shock dominated events" and "near-Sun injection events". For the former, a fast, broad, CME-driven interplanetary shock from activity near the central meridian of the Sun continuously accelerates ions throughout its entire passage from the Sun to Earth. For the lower energy protons (e.g. ≤ 30 MeV) the initial particle flux observed at Earth may be relatively small, but the magnitude continues to increase as the interplanetary shock approaches Earth, often maximizing as the shock passes Earth.

For the "near-Sun injection events" there is identifiable solar activity, preferentially on the western hemisphere of the Sun, with an associated fast CME that typically propagates through the western heliographic longitudes as viewed from Earth. In this type of event, Earth is well connected to the solar active region by the interplanetary magnetic field, and the solar protons can arrive at Earth in a prompt time frame. Solar particle intensity observed at Earth from activity on the western solar hemisphere typically has a faster increase and higher magnitude than similar activity from the central meridian or eastern sector of the Sun (see Fig. 10 in Shea and Smart 1995).

4.3 Variation of High-Energy SEP Events with Long-Term Solar Activity

The era of direct SPE/GLE recording is relatively short and provides little information on the long-term variability. On the other hand, extreme GLE events can be identified as spikes in production of cosmogenic isotope (^{10}Be and ^{14}C) records on the centennial-millennial time scale (e.g., Usoskin and Kovaltsov 2012). For example, the two largest SPE/GLE candidates ever observed (Miyake et al. 2012, 2013; Usoskin et al. 2013; Cliver et al. 2014), were found in 775 AD and 993 AD, as inferred from ^{14}C measured in dendrochronologically-dated tree rings. As shown by Kovaltsov et al. (2014) such spikes are related to the >200 MeV

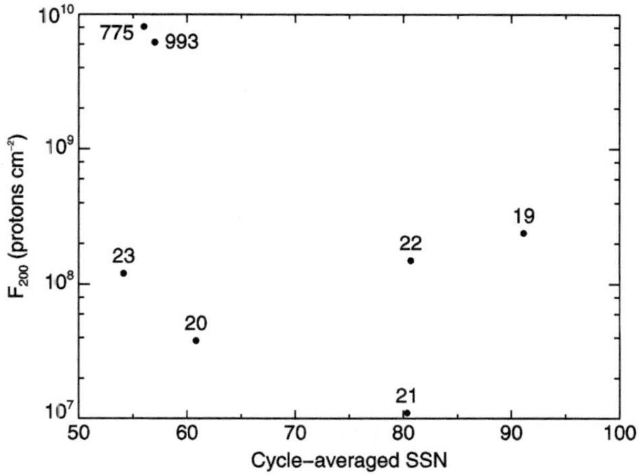

Fig. 14 Scatter plot of solar cycle totals of F_{200} vs. $\langle SSN \rangle_{cycle}$ for cycles 19–23. Points are also plotted for the 775 AD event and 993 AD events with the decadal SSNs encompassing these years taken from Kovaltsov et al. (2014)

fluence in SPEs, F_{200}, as based on SEP spectra from Tylka and Dietrich (2009). Large F_{200} SPEs indicate events with a hard SEP spectrum. The largest F_{200} SPEs since 1950, i.e., those with $F_{200} \geq 3.0 \times 10^7$ proton/cm^2, occurred on 23-Feb-1956 (heliolongitude W80), 12-Nov-1960 (W04), 15-Nov-1960 (W35), 29-Sep-1989 ($\approx$W100), 19-Oct-1989 (E09), and 14-Jul-2000 (W07). McCracken et al. (2012) reported that three of these events, those that originated at solar longitudes $\geq$W35, exhibited high-energy impulsive peaks (extending to rigidities >5 GV) with short (3–5 minutes) rise times and fall times. They pointed out that, "For a quasi-perpendicular shock such as on the flanks of a CME, it is conceivable that the shock front (sweeping laterally across the face of the Sun) may connect and disconnect with open field lines [nominally near W60] that are connected to Earth, and so lead to both an abrupt switch-on and switch-off of the SEP pulse." The three large F_{200} events without a reported high-energy impulsive peak all originated near central meridian. For such events the CME-flank-driven shock may not reach the $\approx$W60 zone of good connection (or may be significantly weaker when it does) and the SEP event can be dominated by the outward expanding quasi-parallel bow shock of the CME. In this case the shock front can remain connected for a longer time to the Parker spiral to Earth, contributing to the extended intensity vs. time profiles observed for SEP events arising from near central meridian (Cane et al. 1988). The black-filled rectangles in the histogram of Fig. 9 show the distribution of the six large F_{200} events over the solar cycle.

Active regions with sunspot areas >1000 msh can occur in even relatively weak solar cycles. For example, in current low-activity cycle 24, the largest active region yet observed had an area of 1580 msh (on 4-Feb-2014). Moreover, at any given proton energy, the fluence from a single large SPE, or a cluster of closely-timed events from the same active region, can dominate the total fluence for a 11-yr cycle. Thus we should not expect SEP fluence over a cycle to be closely tied to the strength of the solar cycle. This can be seen for solar cycles 19–23 in Fig. 14 where F_{200} totals for all GLEs during a given cycle (Kovaltsov et al. 2014) are plotted against the average daily sunspot number for that cycle, $\langle SSN \rangle_{cycle}$. No clear dependence of F_{200} on $\langle SSN \rangle_{cycle}$ is apparent over the 50–90 range of $\langle SSN \rangle_{cycle}$ values for these cycles. In the case of the 775 AD event, the F_{200} fluence for a single event, or cluster of events, exceeded the fluence from the largest solar cycle (No. 19) during the $\approx$1945–1995 modern grand maximum (Usoskin et al. 2007; Usoskin 2013) by about a factor of thirty.

Work by McCracken et al. (2004) and McCracken (2007b) implies that if Fig. 14 included smaller cycles with $\langle SSN \rangle_{cycle}$ values in the range from 0–50, e.g., for the current cycle 24 (Richardson 2013), we might see an anticorrelation between F_{200} and $\langle SSN \rangle_{cycle}$. Those authors reasoned that less intense cycles would produce a weaker ambient HMF and suggested that CMEs injected into these weaker fields would produce shocks with higher Mach numbers, resulting in more frequent and/or energetic SEP events. At the same time weaker cycles should produce fewer energetic CMEs, partly off-setting the advantage of lower ambient HMF. Thus we see a virtual absence of GLEs during the characteristic intervals of low HMF near sunspot minima because of the relative lack of fast halo CMEs at these times (see years 1 and 11 in Figs. 9 and 13). McCracken and colleagues based their suggestion in part on pre-space-age (e.g., cycles 12–14, 1878–1913) peaks in nitrate concentration in ice cores which they attributed to strong SEP events (McCracken et al. 2001, 2004) and also on four intense GLEs from 1942–1956 during a period of low HMF strength inferred from balloon-borne and ground-based ionization measurements of cosmic rays (McCracken 2007a,b). The ice core evidence, however, has been disputed by Wolff et al. (2012). In addition, the low inferred HMF values circa 1950 inferred from ionization chambers are not substantiated by more recent reconstructions of HMF based on geomagnetic observations (Svalgaard and Cliver 2010; Lockwood and Owens 2011; Kahler 2008) as well as on cosmogenic nuclides (Steinhilber et al. 2010).

Recently, the current weak solar cycle has provided a more direct test of the conjecture of McCracken and colleagues—with mixed results. Gopalswamy et al. (2014) noted the relative absence of GLEs during the rise phase of cycle 24 (see Table 1). They attribute this deficit to reduced total (magnetic plus plasma) pressure in the heliosphere that leads, in turn, to: increased CME lateral expansion, dilution of CME magnetic fields, and less-efficient shock acceleration of SEPs. However, Gopalswamy et al. (2014) also note that while the number of GLEs thus far in cycle 24 is significantly reduced, the overall number (28) of significant >10 MeV proton events is comparable to that (34) for the corresponding phase of the stronger cycle 23. Following McCracken and colleagues, they suggested that the similarity in the counts of the lower energy SEP events for the two cycles may result from the relative ease of shock formation due to the lower ambient solar wind Alfvén speed in cycle 24, off-setting the size advantage of cycle 23.

The current low cycle is providing valuable data and insight on SEP acceleration during conditions of low solar activity. While significantly more active than the quietest $\approx$50 years of the Maunder Minimum (e.g., Eddy 1976; Usoskin 2013), cycle 24 is giving us our best glimpse to date of SEP activity during periods of low solar activity.

5 Summary

In this concise review we discuss the solar cycle and its manifestations in the heliosphere and cosmic rays.

Most heliospheric parameters are measured *in situ* near Earth, and thus do not straightforwardly represent the entire heliosphere. While some parameters (e.g., HMF) are roughly representative, others (solar wind velocity or plasma density) are not representative for the whole heliosphere. Global heliospheric indices, such as the open magnetic flux are useful in this respect. Galactic cosmic rays can serve as probes of the 3D heliosphere, reflecting its variability on the time scale of 11-year cycle and beyond. The quasi 22-year variability in GCR observed as alternation of top- and flat-peaked cycles is caused by the drift effect of GCR transport in the heliosphere. Special attention is paid upon the recent minimum of solar activity with very quiet heliospheric conditions (low values of HMF and the interplanetary turbulence) which led, together with the flat HCS, to the record high level of GCRs near Earth in 2009.

We also discuss the statistics of the occurrence of SPEs and GLEs that serve as a probe for the inner heliosphere and solar coronal activity. The occurrence of major events depicts an overall tendency to follow the solar cycle but individual events may appear at different phases of the solar cycle, since it is defined not only by the solar energy releases but also by the location of these releases relative to the nominal ≈W50° footpoint of the magnetic fieldline that connects to Earth. We also discuss that the occurrence of major SPEs is not directly related to the overall level of solar activity, and strong events may occur even during moderate solar cycles.

In summary, the solar cycle in the heliosphere and cosmic rays depicts a complex pattern which includes different processes and cannot be described by a simple correlation with sunspot number.

Acknowledgements This work is a result of the ISSI workshop "The Solar Activity Cycle: Physical Causes And Consequences". G.B. acknowledges support of the RFBR grants 14-02-00905a, 14-02-10006k, 13-02-00585, 13-02-00931, 12-02-00215a and of the "Fundamental Properties of Matter and Astrophysics" Program of the Presidium of the RAS. E.W.C. acknowledges support from AFOSR Task 2301RDZ4. A.G.L. acknowledges support from AFRL contract FA8718-05-C-0036. G.K. was partly supported by the Academy of Finland. I.U.'s contribution is in the framework of the ReSoLVE Centre of Excellence (Academy of Finland, project no. 272157). Dr. Clifford Lopate is acknowledged for Huancayo/Haleakala NM data. Oulu NM data is available at http://cosmicrays.oulu.fi.

References

N. Adams, A temporary increase in the neutron component of cosmic rays. Philos. Mag. **41**, 503–505 (1950)

O. Adriani, G.C. Barbarino, G.A. Bazilevskaya, R. Bellotti, M. Boezio, E.A. Bogomolov, M. Bongi, V. Bonvicini, S. Borisov, S. Bottai, A. Bruno, F. Cafagna, D. Campana, R. Carbone, P. Carlson, M. Casolino, G. Castellini, M.P. De Pascale, C. De Santis, N. De Simone, V. Di Felice, V. Formato, A.M. Galper, L. Grishantseva, A.V. Karelin, S.V. Koldashov, S. Koldobskiy, S.Y. Krutkov, A.N. Kvashnin, A. Leonov, V. Malakhov, L. Marcelli, A.G. Mayorov, W. Menn, V.V. Mikhailov, E. Mocchiutti, A. Monaco, N. Mori, N. Nikonov, G. Osteria, F. Palma, P. Papini, M. Pearce, P. Picozza, C. Pizzolotto, M. Ricci, S.B. Ricciarini, L. Rossetto, R. Sarkar, M. Simon, R. Sparvoli, P. Spillantini, Y.I. Stozhkov, A. Vacchi, E. Vannuccini, G. Vasilyev, S.A. Voronov, Y.T. Yurkin, J. Wu, G. Zampa, N. Zampa, V.G. Zverev, M.S. Potgieter, E.E. Vos, Time dependence of the proton flux measured by PAMELA during the 2006 July–2009 December solar minimum. Astrophys. J. **765**, 91 (2013). doi:10.1088/0004-637X/765/2/91

H.S. Ahluwalia, M.M. Fikani, R.C. Ygbuhay, Rigidity dependence of 11 year cosmic ray modulation: implication for theories. J. Geophys. Res. **115**, 07101 (2010). doi:10.1029/2009JA014798

K. Alanko, I.G. Usoskin, K. Mursula, G.A. Kovaltsov, Heliospheric modulation strength: effective neutron monitor energy. Adv. Space Res. **32**, 615–620 (2003)

K. Alanko-Huotari, K. Mursula, I.G. Usoskin, G.A. Kovaltsov, Global heliospheric parameters and cosmic-ray modulation: an empirical relation for the last decades. Sol. Phys. **238**, 391–404 (2006). doi:10.1007/s11207-006-0233-z

K. Alanko-Huotari, I.G. Usoskin, K. Mursula, G.A. Kovaltsov, Cyclic variations of the heliospheric tilt angle and cosmic ray modulation. Adv. Space Res. **40**, 1064–1069 (2007a). doi:10.1016/j.asr.2007.02.007

K. Alanko-Huotari, I.G. Usoskin, K. Mursula, G.A. Kovaltsov, Stochastic simulation of cosmic ray modulation including a wavy heliospheric current sheet. J. Geophys. Res. **112**, 08101 (2007b). doi:10.1029/2007JA012280

A. Balogh, G. Erdös, The heliospheric magnetic field. Space Sci. Rev. **176**, 177–215 (2013). doi:10.1007/s11214-011-9835-3

A. Balogh, L.J. Lanzerotti, S.T. Suess, *The Heliosphere Through the Solar Activity Cycle* (Springer, Chichester, 2008). doi:10.1007/978-3-540-74302-6

G.A. Bazilevskaya, M.B. Krainev, V.S. Makhmutov, Y.I. Stozhkov, A.K. Svirzhevskaya, N.S. Svirzhevsky, Change in the rigidity dependence of the galactic cosmic ray modulation in 2008–2009. Adv. Space Res. **49**, 784–790 (2012). doi:10.1016/j.asr.2011.12.002

G.A. Bazilevskaya, M.B. Krainev, A.K. Svirzhevskaya, N.S. Svirzhevsky, Galactic cosmic rays and parameters of the interplanetary medium near solar activity minima. Cosm. Res. **51**, 29–36 (2013). doi:10.1134/S0010952513010012

J. Beer, S. Tobias, N. Weiss, An active Sun throughout the Maunder minimum. Sol. Phys. **181**, 237–249 (1998)

J. Beer, K. McCracken, R. von Steiger, *Cosmogenic Radionuclides: Theory and Applications in the Terrestrial and Space Environments* (Springer, Berlin, 2012)

A.-M. Berggren, J. Beer, G. Possnert, A. Aldahan, P. Kubik, M. Christl, S.J. Johnsen, J. Abreu, B.M. Vinther, A 600-year annual ^{10}Be record from the NGRIP ice core, Greenland. Geophys. Res. Lett. **36**, 11801 (2009)

G.E. Brueckner, R.A. Howard, M.J. Koomen, C.M. Korendyke, D.J. Michels, J.D. Moses, D.G. Socker, K.P. Dere, P.L. Lamy, A. Llebaria, M.V. Bout, R. Schwenn, G.M. Simnett, D.K. Bedford, C.J. Eyles, The Large Angle Spectroscopic Coronagraph (LASCO). Sol. Phys. **162**, 357–402 (1995). doi:10.1007/BF00733434

L.F. Burlaga, M.L. Goldstein, F.B. McDonald, A.J. Lazarus, Cosmic ray modulation and turbulent interaction regions near 11 AU. J. Geophys. Res. **90**, 12027–12039 (1985). doi:10.1029/JA090iA12p12027

R.A. Caballero-Lopez, H. Moraal, Limitations of the force field equation to describe cosmic ray modulation. J. Geophys. Res. **109**, 01101 (2004). doi:10.1029/2003JA010098

R.A. Caballero-Lopez, H. Moraal, Cosmic-ray yield and response functions in the atmosphere. J. Geophys. Res. **117** (2012). doi:10.1029/2012JA017794

H.V. Cane, D.V. Reames, T.T. von Rosenvinge, The role of interplanetary shocks in the longitude distribution of solar energetic particles. J. Geophys. Res. **93**, 9555–9567 (1988). doi:10.1029/JA093iA09p09555

H.V. Cane, G. Wibberenz, I.G. Richardson, T.T. von Rosenvinge, Cosmic ray modulation and the solar magnetic field. Geophys. Res. Lett. **26**, 565–568 (1999). doi:10.1029/1999GL900032

A.N. Charakhchyan, Reviews of topical problems: investigation of stratosphere cosmic ray intensity fluctuations induced by processes on the Sun. Sov. Phys. Usp. **7**, 358–374 (1964)

F. Clette, L. Svalgaard, J.M. Vaquero, E.W. Cliver, Revisiting the sunspot number: a 400-year perspective on the solar cycle. Space Sci. Rev. (2014, this issue). doi:10.1007/s11214-014-0074-2

E.W. Cliver, The unusual relativistic solar proton events of 1979 August 21 and 1981 May 10. Astrophys. J. **639**, 1206–1217 (2006). doi:10.1086/499765

E.W. Cliver, A revised classification scheme for solar energetic particle events. Cent. Eur. Astrophys. Bull. **33**, 253–270 (2009a)

E.W. Cliver, History of research on solar energetic particle (SEP) events: the evolving paradigm, in *IAU Symposium*, ed. by N. Gopalswamy, D.F. Webb, vol. 257 (2009b), pp. 401–412. doi:10.1017/S1743921309029639

E.W. Cliver, The extended cycle of solar activity and the Sun's 22-yr magnetic cycle. Space Sci. Rev. (2014, this issue). doi:10.1007/s11214-014-0093-2

E.W. Cliver, A.G. Ling, 22 year patterns in the relationship of sunspot number and tilt angle to cosmic-ray intensity. Astrophys. J. Lett. **551**, 189–192 (2001a). doi:10.1086/320022

E.W. Cliver, A.G. Ling, Coronal mass ejections, open magnetic flux, and cosmic-ray modulation. Astrophys. J. **556**, 432–437 (2001b). doi:10.1086/321570

E.W. Cliver, A.G. Ling, The floor in the solar wind magnetic field revisited. Sol. Phys. **274**, 285–301 (2011). doi:10.1007/s11207-010-9657-6

E.W. Cliver, S.W. Kahler, M.A. Shea, D.F. Smart, Injection onsets of 2 GeV protons, 1 MeV electrons, and 100 keV electrons in solar cosmic ray flares. Astrophys. J. **260**, 362–370 (1982). doi:10.1086/160261

E.W. Cliver, V. Boriakoff, K.H. Bounar, Geomagnetic activity and the solar wind during the Maunder minimum. Geophys. Res. Lett. **25**, 897–900 (1998). doi:10.1029/98GL00500

E.W. Cliver, I.G. Richardson, A.G. Ling, Solar drivers of 11-yr and long-term cosmic ray modulation. Space Sci. Rev. **176**, 3–19 (2013). doi:10.1007/s11214-011-9746-3

E.W. Cliver, A.J. Tylka, W.F. Dietrich, A.G. Ling, On a solar origin for the cosmogenic nuclide event of 775 AD Astrophys. J. **781**, 32 (2014). doi:10.1088/0004-637X/781/1/32

A.Z. Dolginov, I. Toptygin, Multiple scattering of particles in a magnetic field with random inhomogeneities. Sov. Phys. JETP **24**, 1195 (1967)

L.I. Dorman, *Cosmic Ray Interactions, Propagation, and Acceleration in Space Plasmas* (Kluwer Academic, Dordrecht, 2006)

S.P. Duggal, Relativistic solar cosmic rays. Rev. Geophys. Space Phys. **17**, 1021–1058 (1979). doi:10.1029/RG017i005p01021

J.A. Eddy, The Maunder minimum. Science **192**, 1189–1202 (1976)

A.G. Emslie, B.R. Dennis, A.Y. Shih, P.C. Chamberlin, R.A. Mewaldt, C.S. Moore, G.H. Share, A. Vourlidas, B.T. Welsch, Global energetics of thirty-eight large solar eruptive events. Astrophys. J. **759**, 71 (2012). doi:10.1088/0004-637X/759/1/71

S.E.S. Ferreira, M.S. Potgieter, Long-term cosmic-ray modulation in the heliosphere. Astrophys. J. **603**, 744–752 (2004). doi:10.1086/381649

L.A. Fisk, Motion of the footpoints of heliospheric magnetic field lines at the Sun: implications for recurrent energetic particle events at high heliographic latitudes. J. Geophys. Res. **101**, 15547–15554 (1996). doi:10.1029/96JA01005

S.E. Forbush, Three unusual cosmic-ray increases possibly due to charged particles from the Sun. Phys. Rev. **70**, 771–772 (1946). doi:10.1103/PhysRev.70.771

S.E. Forbush, World-wide cosmic-ray variations, 1937–1952. J. Geophys. Res. **59**, 525–542 (1954)

S.E. Forbush, T.B. Stinchcomb, M. Schein, The extraordinary increase of cosmic-ray intensity on November 19, 1949. Phys. Rev. **79**, 501–504 (1950). doi:10.1103/PhysRev.79.501

J. Giacalone, Cosmic-ray transport and interaction with shocks. Space Sci. Rev. **176**, 73–88 (2013). doi:10.1007/s11214-011-9763-2

S.E. Gibson, G. de Toma, B. Emery, P. Riley, L. Zhao, Y. Elsworth, R.J. Leamon, J. Lei, S. McIntosh, R.A. Mewaldt, B.J. Thompson, D. Webb, The whole heliosphere interval in the context of a long and structured solar minimum: an overview from Sun to Earth. Sol. Phys. **274**, 5–27 (2011). doi:10.1007/s11207-011-9921-4

L.J. Gleeson, W.I. Axford, Cosmic rays in the interplanetary medium. Astrophys. J. Lett. **149**, 115 (1967). doi:10.1086/180070

L.J. Gleeson, W.I. Axford, Solar modulation of galactic cosmic rays. Astrophys. J. **154**, 1011–1026 (1968)

M.N. Gnevyshev, On the 11-years cycle of solar activity. Sol. Phys. **1**, 107–120 (1967). doi:10.1007/BF00150306

M.N. Gnevyshev, Essential features of the 11-year solar cycle. Sol. Phys. **51**, 175–183 (1977). doi:10.1007/BF00240455

N. Gopalswamy, S. Yashiro, G. Michałek, M.L. Kaiser, R.A. Howard, D.V. Reames, R. Leske, T. von Rosenvinge, Interacting coronal mass ejections and solar energetic particles. Astrophys. J. Lett. **572**, 103–107 (2002). doi:10.1086/341601

N. Gopalswamy, S. Yashiro, S. Krucker, G. Stenborg, R.A. Howard, Intensity variation of large solar energetic particle events associated with coronal mass ejections. J. Geophys. Res. **109**, 12105 (2004). doi:10.1029/2004JA010602

N. Gopalswamy, S. Yashiro, G. Michalek, G. Stenborg, A. Vourlidas, S. Freeland, R. Howard, The SOHO/LASCO CME catalog. Earth Moon Planets **104**, 295–313 (2009). doi:10.1007/s11038-008-9282-7

N. Gopalswamy, H. Xie, S. Yashiro, S. Akiyama, P. Mäkelä, I.G. Usoskin, Properties of ground level enhancement events and the associated solar eruptions during solar cycle 23. Space Sci. Rev. **171**, 23–60 (2012). doi:10.1007/s11214-012-9890-4

N. Gopalswamy, H. Xie, S. Akiyama, S. Yashiro, I.G. Usoskin, J.M. Davila, The first ground level enhancement event of solar cycle 24: direct observation of shock formation and particle release heights. Astrophys. J. Lett. **765**, 30 (2013). doi:10.1088/2041-8205/765/2/L30

N. Gopalswamy, S. Akiyama, S. Yashiro, G. Michalek, H. Xie, P. Mäkelä, Geophys. Res. Lett. (2014, submitted)

P.K.F. Grieder, *Cosmic Rays at Earth* (Elsevier Science, Amsterdam, 2001)

D.H. Hathaway, The solar cycle. Living Rev. Sol. Phys. **7**(1) (2010). http://www.livingreviews.org/lrsp-2010-1

B. Heber, Cosmic rays through the solar hale cycle. Insights from Ulysses. Space Sci. Rev. **176**, 265–278 (2013). doi:10.1007/s11214-011-9784-x

B. Heber, M.S. Potgieter, Cosmic rays at high heliolatitudes. Space Sci. Rev. **127**, 117–194 (2006). doi:10.1007/s11214-006-9085-y

B. Heber, A. Kopp, J. Gieseler, R. Müller-Mellin, H. Fichtner, K. Scherer, M.S. Potgieter, S.E.S. Ferreira, Modulation of galactic cosmic ray protons and electrons during an unusual solar minimum. Astrophys. J. **699**, 1956–1963 (2009). doi:10.1088/0004-637X/699/2/1956

H.S. Hudson, Global properties of solar flares. Space Sci. Rev. **158**, 5–41 (2011). doi:10.1007/s11214-010-9721-4

T. Jämsén, I.G. Usoskin, T. Räihä, J. Sarkamo, G.A. Kovaltsov, Case study of Forbush decreases: energy dependence of the recovery. Adv. Space Res. **40**, 342–347 (2007). doi:10.1016/j.asr.2007.02.025

J.R. Jokipii, Cosmic-ray propagation. I. Charged particles in a random magnetic field. Astrophys. J. **146**, 480 (1966). doi:10.1086/148912

J.R. Jokipii, E.H. Levy, Effects of particle drifts on the solar modulation of galactic cosmic rays. Astrophys. J. Lett. **213**, 85–88 (1977). doi:10.1086/182415

S.W. Kahler, The correlation between solar energetic particle peak intensities and speeds of coronal mass ejections: effects of ambient particle intensities and energy spectra. J. Geophys. Res. **106**, 20947–20956 (2001). doi:10.1029/2000JA002231

S.W. Kahler, Prospects for future enhanced solar energetic particle events and the effects of weaker heliospheric magnetic fields. J. Geophys. Res. **113**, 11102 (2008). doi:10.1029/2008JA013168

S.W. Kahler, A. Vourlidas, Fast coronal mass ejection environments and the production of solar energetic particle events. J. Geophys. Res. **110**, 12-01 (2005). doi:10.1029/2005JA011073
G.A. Kovaltsov, I.G. Usoskin, E.W. Cliver, W.F. Dietrich, A.J. Tylka, Relation between >200 MeV solar energetic protons and atmospheric production of cosmogenic radionuclides. Sol. Phys. (2014). doi:10.1007/s11207-014-0093-0
M.B. Krainev, M.S. Kalinin, On the description of the 11- and 22-year cycles in the GCR intensity. J. Phys. Conf. Ser. **409**(1), 012155 (2013). doi:10.1088/1742-6596/409/1/012155
M.B. Krainev, G.A. Bazilevskaya, S.K. Gerasimova, P.A. Krivoshapkin, G.F. Krymsky, S.A. Starodubtsev, Y.I. Stozhkov, N.S. Svirzhevsky, On the status of the sunspot and magnetic cycles in the galactic cosmic ray intensity. J. Phys. Conf. Ser. **409**(1), 012016 (2013). doi:10.1088/1742-6596/409/1/012016
G.F. Krymskij, *Modulation of Cosmic Rays in Interplanetary Space* (1969)
I. Lange, S.E. Forbush, Further note on the effect on cosmic-ray intensity of the magnetic storm of March 1, 1942. Terr. Magn. Atmos. Electr. **47**, 331 (1942). doi:10.1029/TE047i004p00331
M. Lockwood, Reconstruction and prediction of variations in the open solar magnetic flux and interplanetary conditions. Living Rev. Sol. Phys. **10**, 4 (2013). doi:10.12942/lrsp-2013-4
M. Lockwood, M.J. Owens, Centennial changes in the heliospheric magnetic field and open solar flux: the consensus view from geomagnetic data and cosmogenic isotopes and its implications. J. Geophys. Res. **116**, 04109 (2011). doi:10.1029/2010JA016220
M. Lockwood, M.J. Owens, Implications of the recent low solar minimum for the solar wind during the Maunder minimum. Astrophys. J. Lett. **781**, 7 (2014). doi:10.1088/2041-8205/781/1/L7
M. Lockwood, R. Stamper, M.N. Wild, A doubling of the Sun's coronal magnetic field during the past 100 years. Nature **399**, 437–439 (1999). doi:10.1038/20867
M. Lockwood, A.P. Rouillard, I.D. Finch, The rise and fall of open solar flux during the current Grand Solar Maximum. Astrophys. J. **700**, 937–944 (2009). doi:10.1088/0004-637X/700/2/937
H. Mavromichalaki, A. Papaioannou, C. Plainaki, C. Sarlanis, G. Souvatzoglou, M. Gerontidou, M. Papailiou, E. Eroshenko, A. Belov, V. Yanke, E.O. Flückiger, R. Bütikofer, M. Parisi, M. Storini, K.-L. Klein, N. Fuller, C.T. Steigies, O.M. Rother, B. Heber, R.F. Wimmer-Schweingruber, K. Kudela, I. Strharsky, R. Langer, I. Usoskin, A. Ibragimov, A. Chilingaryan, G. Hovsepyan, A. Reymers, A. Yeghikyan, O. Kryakunova, E. Dryn, N. Nikolayevskiy, L. Dorman, L. Pustil'Nik, Applications and usage of the real-time Neutron Monitor database. Adv. Space Res. **47**, 2210–2222 (2011). doi:10.1016/j.asr.2010.02.019
D.J. McComas, R.W. Ebert, H.A. Elliott, B.E. Goldstein, J.T. Gosling, N.A. Schwadron, R.M. Skoug, Weaker solar wind from the polar coronal holes and the whole Sun. Geophys. Res. Lett. **35**, 18103 (2008). doi:10.1029/2008GL034896
D.J. McComas, D. Alexashov, M. Bzowski, H. Fahr, J. Heerikhuisen, V. Izmodenov, M.A. Lee, E. Möbius, N. Pogorelov, N.A. Schwadron, G.P. Zank, The heliosphere's interstellar interaction: no bow shock. Science **336**, 1291 (2012). doi:10.1126/science.1221054
D.J. McComas, N. Angold, H.A. Elliott, G. Livadiotis, N.A. Schwadron, R.M. Skoug, C.W. Smith, Weakest solar wind of the space age and the current "mini" solar maximum. Astrophys. J. **779**, 2 (2013). doi:10.1088/0004-637X/779/1/2
K.G. McCracken, Heliomagnetic field near Earth, 1428–2005. J. Geophys. Res. **112**, 09106 (2007a). doi:10.1029/2006JA012119
K.G. McCracken, High frequency of occurrence of large solar energetic particle events prior to 1958 and a possible repetition in the near future. Space Weather **5**, 7004 (2007b). doi:10.1029/2006SW000295
K.G. McCracken, G.A.M. Dreschhoff, D.F. Smart, M.A. Shea, A study of the frequency of occurrence of large-fluence solar proton events and the strength of the interplanetary magnetic field. Sol. Phys. **224**, 359–372 (2004). doi:10.1007/s11207-005-5257-2
K.G. McCracken, G.A.M. Dreschhoff, E.J. Zeller, D.F. Smart, M.A. Shea, Solar cosmic ray events for the period 1561–1994: 1. Identification in polar ice, 1561–1950. J. Geophys. Res. **106**, 21585–21598 (2001)
K.G. McCracken, H. Moraal, M.A. Shea, The high-energy impulsive ground-level enhancement. Astrophys. J. **761**, 101 (2012). doi:10.1088/0004-637X/761/2/101
F.B. McDonald, Cosmic-ray modulation in the heliosphere a phenomenological study. Space Sci. Rev. **83**, 33–50 (1998)
F.B. McDonald, W.R. Webber, D.V. Reames, Unusual time histories of galactic and anomalous cosmic rays at 1 AU over the deep solar minimum of cycle 23/24. Geophys. Res. Lett. **371**, 18101 (2010). doi:10.1029/2010GL044218
R.A. Mewaldt, Cosmic rays in the heliosphere: requirements for future observations. Space Sci. Rev. **176**, 365–390 (2013). doi:10.1007/s11214-012-9922-0
R.A. Mewaldt, C.M.S. Cohen, J. Giacalone, G.M. Mason, E.E. Chollet, M.I. Desai, D.K. Haggerty, M.D. Looper, R.S. Selesnick, A. Vourlidas, How efficient are coronal mass ejections at accelerating solar energetic particles?, in *AIP Conf. Ser.*, ed. by G. Li, Q. Hu, O. Verkhoglyadova, G.P. Zank, R.P. Lin, J. Luhmann, vol. 1039 (2008), pp. 111–117. doi:10.1063/1.2982431

R.A. Mewaldt, A.J. Davis, K.A. Lave, R.A. Leske, E.C. Stone, M.E. Wiedenbeck, W.R. Binns, E.R. Christian, A.C. Cummings, G.A. de Nolfo, M.H. Israel, A.W. Labrador, T.T. von Rosenvinge, Record-setting cosmic-ray intensities in 2009 and 2010. Astrophys. J. Lett. **723**, 1–6 (2010). doi:10.1088/2041-8205/723/1/L1

A.L. Mishev, L.G. Kocharov, I.G. Usoskin, Analysis of the ground level enhancement on 17 May 2012 using data from the global neutron monitor network. J. Geophys. Res. **119**, 670–679 (2014). doi:10.1002/2013JA019253

H. Miyahara, K. Masuda, Y. Muraki, H. Furuzawa, H. Menjo, T. Nakamura, Cyclicity of solar activity during the Maunder minimum deduced from radiocarbon content. Sol. Phys. **224**, 317–322 (2004). doi:10.1007/s11207-005-6501-5

F. Miyake, K. Nagaya, K. Masuda, T. Nakamura, A signature of cosmic-ray increase in ad 774–775 from tree rings in Japan. Nature **486**, 240–242 (2012). doi:10.1038/nature11123

F. Miyake, K. Masuda, T. Nakamura, Lengths of Schwabe cycles in the seventh and eighth centuries indicated by precise measurement of carbon-14 content in tree rings. J. Geophys. Res. **118**, 1–5 (2013). doi:10.1002/2012JA018320

H. Moraal, P.H. Stoker, Long-term neutron monitor observations and the 2009 cosmic ray maximum. J. Geophys. Res. **115**, 12109 (2010). doi:10.1029/2010JA015413

H. Moraal, A. Belov, J.M. Clem, Design and co-ordination of multi-station international neutron monitor networks. Space Sci. Rev. **93**, 285–303 (2000). doi:10.1023/A:1026504814360

K. Mursula, T. Hiltula, Bashful ballerina: southward shifted heliospheric current sheet. Geophys. Res. Lett. **30**, 2135 (2003). doi:10.1029/2003GL018201

H.V. Neher, Cosmic-ray particles that changed from 1954 to 1958 to 1965. J. Geophys. Res. **72**, 1527 (1967). doi:10.1029/JZ072i005p01527

H.V. Neher, Cosmic rays at high latitudes and altitudes covering four solar maxima J. Geophys. Res. **76**, 1637–1651 (1971). doi:10.1029/JA076i007p01637

H.W. Newton, *The Face of the Sun* (Penguin Books, Harmondsworth, 1958)

M.J. Owens, R.J. Forsyth, The heliospheric magnetic field. Living Rev. Sol. Phys. **10**, 5 (2013). doi:10.12942/lrsp-2013-5

M.J. Owens, I. Usoskin, M. Lockwood, Heliospheric modulation of galactic cosmic rays during grand solar minima: past and future variations. Geophys. Res. Lett. **39**, 19102 (2012). doi:10.1029/2012GL053151

E.N. Parker, The passage of energetic charged particles through interplanetary space. Planet. Space Sci. **13**, 9–49 (1965)

M.I. Pishkalo, Reconstruction of the heliospheric current sheet tilts using sunspot numbers. Sol. Phys. **233**, 277–290 (2006). doi:10.1007/s11207-006-1981-5

N.V. Pogorelov, S.T. Suess, S.N. Borovikov, R.W. Ebert, D.J. McComas, G.P. Zank, Three-dimensional features of the outer heliosphere due to coupling between the interstellar and interplanetary magnetic fields. IV. Solar cycle model based on Ulysses observations. Astrophys. J. **772**, 2 (2013). doi:10.1088/0004-637X/772/1/2

M. Potgieter, Solar modulation of cosmic rays. Living Rev. Sol. Phys. **10**, 3 (2013). doi:10.12942/lrsp-2013-3

M.S. Potgieter, R. Strauss, At what rigidity does the solar modulation of galactic cosmic rays begin? in *33rd International Cosmic Ray Conference, Rio de Janeiro, Brazil* (2013), p. 156

M.S. Potgieter, E.E. Vos, M. Boezio, N. De Simone, V. Di Felice, V. Formato, Modulation of galactic protons in the heliosphere during the unusual solar minimum of 2006 to 2009. Sol. Phys. **289**, 391–406 (2014). doi:10.1007/s11207-013-0324-6

D.V. Reames, Particle acceleration at the Sun and in the heliosphere. Space Sci. Rev. **90**, 413–491 (1999). doi:10.1023/A:1005105831781

D.V. Reames, The two sources of solar energetic particles. Space Sci. Rev. 53–92 (2013). doi:10.1007/s11214-013-9958-9

I.G. Richardson, Geomagnetic activity during the rising phase of solar cycle 24. J. Space Weather Space Clim. **3**(26), 260000 (2013). doi:10.1051/swsc/2013031

J.D. Richardson, N.A. Schwadron, in *The Limits of Our Solar System*, ed. by M.A. Barucci, H. Boehnhardt, D.P. Cruikshank, A. Morbidelli, R. Dotson (University of Arizona Press, Tucson, 2008), pp. 443–463

I.G. Richardson, H.V. Cane, E.W. Cliver, Sources of geomagnetic activity during nearly three solar cycles (1972–2000). J. Geophys. Res. **107**, 1187 (2002). doi:10.1029/2001JA000504

M.A. Shea, D.F. Smart, A summary of major solar proton events. Sol. Phys. **127**, 297–320 (1990)

M.A. Shea, D.F. Smart, Significant proton events of solar cycle 22 and a comparison with events of previous solar cycles. Adv. Space Res. **14**, 631–638 (1994). doi:10.1016/0273-1177(94)90518-5

M.A. Shea, D.F. Smart, History of solar proton event observations. Nucl. Phys. B, Proc. Suppl. **39**, 16–25 (1995)

M.A. Shea, D.F. Smart, Space weather and the ground-level solar proton events of the 23rd solar cycle. Space Sci. Rev. **171**, 161–188 (2012). doi:10.1007/s11214-012-9923-z

Y. Shikaze, S. Haino, K. Abe, H. Fuke, T. Hams, K.C. Kim, Y. Makida, S. Matsuda, J.W. Mitchell, A.A. Moiseev, J. Nishimura, M. Nozaki, S. Orito, J.F. Ormes, T. Sanuki, M. Sasaki, E.S. Seo, R.E. Streitmatter, J. Suzuki, K. Tanaka, T. Yamagami, A. Yamamoto, T. Yoshida, K. Yoshimura, Measurements of 0.2 20 GeV/n cosmic-ray proton and helium spectra from 1997 through 2002 with the BESS spectrometer. Astropart. Phys. **28**, 154–167 (2007). doi:10.1016/j.astropartphys.2007.05.001

J.A. Simpson, Cosmic radiation neutron intensity monitor, in *Annals of the Int. Geophysical Year IV, Part VII* (Pergamon Press, London, 1958), p. 351

J.A. Simpson, The cosmic ray nucleonic component: the invention and scientific uses of the Neutron Monitor—(keynote lecture). Space Sci. Rev. **93**, 11–32 (2000). doi:10.1023/A:1026567706183

D.F. Smart, M.A. Shea, H.E. Spence, L. Kepko, Two groups of extremely large >30 MeV solar proton fluence events. Adv. Space Res. **37**, 1734–1740 (2006). doi:10.1016/j.asr.2005.09.008

S.K. Solanki, M. Schüssler, M. Fligge, Evolution of the Sun's large-scale magnetic field since the Maunder minimum. Nature **408**, 445–447 (2000)

S.K. Solanki, I.G. Usoskin, B. Kromer, M. Schüssler, J. Beer, Unusual activity of the Sun during recent decades compared to the previous 11,000 years. Nature **431**, 1084–1087 (2004). doi:10.1038/nature02995

F. Steinhilber, J.A. Abreu, J. Beer, K.G. McCracken, Interplanetary magnetic field during the past 9300 years inferred from cosmogenic radionuclides. J. Geophys. Res. **115**, 01104 (2010). doi:10.1029/2009JA014193

E.C. Stone, A.C. Cummings, F.B. McDonald, B.C. Heikkila, N. Lal, W.R. Webber, Voyager 1 explores the termination shock region and the heliosheath beyond. Science **309**, 2017–2020 (2005). doi:10.1126/science.1117684

E.C. Stone, A.C. Cummings, F.B. McDonald, B.C. Heikkila, N. Lal, W.R. Webber, An asymmetric solar wind termination shock. Nature **454**, 71–74 (2008). doi:10.1038/nature07022

E.C. Stone, A.C. Cummings, F.B. McDonald, B.C. Heikkila, N. Lal, W.R. Webber, Voyager 1 observes low-energy galactic cosmic rays in a region depleted of heliospheric ions. Science **341**, 150–153 (2013). doi:10.1126/science.1236408

Y.I. Stozhkov, N.S. Svirzhevsky, G.A. Bazilevskaya, A.N. Kvashnin, V.S. Makhmutov, A.K. Svirzhevskaya, Long-term (50 years) measurements of cosmic ray fluxes in the atmosphere. Adv. Space Res. **44**, 1124–1137 (2009). doi:10.1016/j.asr.2008.10.038

R.D. Strauss, M.S. Potgieter, S.E.S. Ferreira, Modeling ground and space based cosmic ray observations. Adv. Space Res. **49**, 392–407 (2012a). doi:10.1016/j.asr.2011.10.006

R.D. Strauss, M.S. Potgieter, I. Büsching, A. Kopp, Modelling heliospheric current sheet drift in stochastic cosmic ray transport models. Astrophys. Space Sci. **339**, 223–236 (2012b). doi:10.1007/s10509-012-1003-z

L. Svalgaard, E.W. Cliver, Heliospheric magnetic field 1835–2009. J. Geophys. Res. **115**, 09111 (2010). doi:10.1029/2009JA015069

Z. Švestka, Proton flares before 1956. Bull. Astron. Inst. Czechoslov. **17**, 262–270 (1966)

Z. Švestka, P. Simon (eds.), *Catalog of Solar Particle Events 1955–1969*. Astrophysics and Space Science Library, vol. 49 (1975)

N. Thakur, N. Gopalswamy, H. Xie, P. Makelä, S. Yashiro, S. Akiyama, J.M. Davila, Ground level enhancement in the 2014 January 6 solar energetic particle event. Astrophys. J. Lett. **790**, 13 (2014)

S. Ting, The alpha magnetic spectrometer on the International Space Station. Nucl. Phys. B, Proc. Suppl. **243**, 12–24 (2013). doi:10.1016/j.nuclphysbps.2013.09.028

A. Tylka, W. Dietrich, A new and comprehensive analysis of proton spectra in ground-level enhanced (GLE) solar particle events, in *31th International Cosmic Ray Conference* (Universal Academy Press, Lodź, 2009)

A.J. Tylka, M.A. Lee, A model for spectral and compositional variability at high energies in large, gradual solar particle events. Astrophys. J. **646**, 1319–1334 (2006). doi:10.1086/505106

A.J. Tylka, C.M.S. Cohen, W.F. Dietrich, M.A. Lee, C.G. Maclennan, R.A. Mewaldt, C.K. Ng, D.V. Reames, Shock geometry, seed populations, and the origin of variable elemental composition at high energies in large gradual solar particle events. Astrophys. J. **625**, 474–495 (2005). doi:10.1086/429384

I.G. Usoskin, A history of solar activity over millennia. Living Rev. Sol. Phys. **10**, 1 (2013). doi:10.12942/lrsp-2013-1

I.G. Usoskin, G.A. Kovaltsov, Occurrence of extreme solar particle events: assessment from historical proxy data. Astrophys. J. **757**, 92 (2012). doi:10.1088/0004-637X/757/1/92

I.G. Usoskin, H. Kananen, K. Mursula, P. Tanskanen, G.A. Kovaltsov, Correlative study of solar activity and cosmic ray intensity. J. Geophys. Res. **103**, 9567–9574 (1998). doi:10.1029/97JA03782

I.G. Usoskin, K. Mursula, G.A. Kovaltsov, Heliospheric modulation of cosmic rays and solar activity during the Maunder minimum. J. Geophys. Res. **106**, 16039–16046 (2001). doi:10.1029/2000JA000105

I.G. Usoskin, K. Alanko-Huotari, G.A. Kovaltsov, K. Mursula, Heliospheric modulation of cosmic rays: monthly reconstruction for 1951–2004. J. Geophys. Res. **110** (2005). doi:10.1029/2005JA011250

I.G. Usoskin, S.K. Solanki, G.A. Kovaltsov, Grand minima and maxima of solar activity: new observational constraints. Astron. Astrophys. **471**, 301–309 (2007). doi:10.1051/0004-6361:20077704

I.G. Usoskin, G.A. Bazilevskaya, G.A. Kovaltsov, Solar modulation parameter for cosmic rays since 1936 reconstructed from ground-based neutron monitors and ionization chambers. J. Geophys. Res. **116**, 02104 (2011). doi:10.1029/2010JA016105

I.G. Usoskin, B. Kromer, F. Ludlow, J. Beer, M. Friedrich, G.A. Kovaltsov, S.K. Solanki, L. Wacker, The AD775 cosmic event revisited: the Sun is to blame. Astron. Astrophys. **552**, 3 (2013). doi:10.1051/0004-6361/201321080

I.G. Usoskin, G. Hulot, Y. Gallet, R. Roth, A. Licht, F. Joos, G.A. Kovaltsov, E. Thébault, A. Khokhlov, Evidence for distinct modes of solar activity. Astron. Astrophys. **562**, 10 (2014). doi:10.1051/0004-6361/201423391

L.E.A. Vieira, S.K. Solanki, Evolution of the solar magnetic flux on time scales of years to millenia. Astron. Astrophys. **509**, 100 (2010). doi:10.1051/0004-6361/200913276

K.P. Wenzel, R.G. Marsden, D.E. Page, E.J. Smith, The ULYSSES mission. Astron. Astrophys. Suppl. Ser. **92**, 207–219 (1992)

E.W. Wolff, M. Bigler, M.A.J. Curran, J.E. Dibb, M.M. Frey, M. Legrand, J.R. McConnell, The Carrington event not observed in most ice core nitrate records. Geophys. Res. Lett. **39**, 08503 (2012). doi:10.1029/2012GL051603

B. Zieger, M. Opher, N.A. Schwadron, D.J. McComas, G. Tóth, A slow bow shock ahead of the heliosphere. Geophys. Res. Lett. **40**, 2923–2928 (2013). doi:10.1002/grl.50576

DOI 10.1007/978-1-4939-2584-1_15
Reprinted from *Space Science Reviews* Journal, DOI 10.1007/s11214-014-0090-2

Inferences on Stellar Activity and Stellar Cycles from Asteroseismology

William J. Chaplin · Sarbani Basu

Received: 25 March 2014 / Accepted: 19 August 2014 / Published online: 4 September 2014

Abstract The solar activity cycle can be studied using many different types of observations, such as counting sunspots, measuring emission in the Ca II H&K lines, magnetograms, radio emissions, etc. One of the more recent ways of studying solar activity is to use the changing properties of solar oscillations. Stellar activity cycles are generally studied using the Ca II lines, or sometimes using photometry. Asteroseismology is potentially an exciting means of studying these cycles. In this article we examine whether or not asteroseismic data can be used for this purpose, and what the asteroseismic signatures of stellar activity are. We also examine how asteroseismology may help in more indirect ways.

Keywords Stars: activity · Stars: oscillations

1 Introduction: Observing Stellar Cycles

Magnetic fields are ubiquitous in the universe. Effects of magnetic fields are detected across virtually the whole of the Hertzsprung–Russell diagram. Observations indicate that the presence of magnetic activity is common in solar-type stars (e.g., Radick 1994), i.e., cool main-sequence and sub-giant or "late-type" stars. Magnetic effects are also common in early-type stars, such as those of spectral types O, B and A (e.g., Linsky 1999).

Solar magnetic activity and magnetic-activity cycles can be observed in many different ways, with the most obvious being the rise and fall of the number of sunspots on the solar

W.J. Chaplin (✉)
School of Physics and Astronomy, University of Birmingham, Edgbaston, B15 2TT Birmingham, UK
e-mail: w.j.chaplin@bham.ac.uk

W.J. Chaplin
Stellar Astrophysics Centre (SAC), Department of Physics and Astronomy, Aarhus University, Ny Munkegade 120, 8000 Aarhus C, Denmark

S. Basu
Department of Astronomy, Yale University, P.O. Box 208101, New Haven, CT 06520, USA
e-mail: sarbani.basu@yale.edu

surface. Detailed magnetograms, such as those from the Kitt Peak National Observatory, also reveal the solar cycle quite clearly, as do other activity-related proxies such as the radio flux at 10.7 cm. Observing activity in stars is admittedly more difficult. The most popular way is to look at the cores of Ca II H&K absorption lines. What makes these lines interesting is that their cores show emission, and the emission is correlated with activity. Observations of these lines for the Sun clearly show the 11-year solar cycle (see Baliunas et al. 1995, Fig. 1d). Thus these chromospheric lines can be used to monitor stellar activity.

One of the first papers looking at chromospheric variations in main-sequence stars was that of Wilson (1978) based on observations that started in 1966. Since then, there have been other surveys of stellar cycles, such as the current Mt. Wilson HK project (Duncan et al. 1991), the Vienna-Kitt Peak National Observatory Ca II H&K survey (Strassmeier et al. 2000), the surveys done at the Lowell Observatory (Hall et al. 2007) etc., targeting northern stars, and the surveys by Henry et al. (1996), Gray et al. (2006), Jenkins et al. (2006) etc., for southern stars. What the data show is that some stars show stellar activity cycles (see e.g., Baliunas et al. 1995), like the Sun; whilst some others may be in a Maunder-minimum like state, showing no detectable variability.

Observations showing that changes in solar irradiance are correlated with solar rotation and solar activity led to campaigns of high-precision photometry of stars (see e.g., Lockwood et al. 1997; Radick et al. 1998). These campaigns were indeed able to observe stellar variability in photometry. However, Lockwood et al. (1997) concluded that the chromospheric activity produces a stronger and more easily detected signal than photometric variability. This was not really surprising, for two reasons: the first is that in the case of the Sun the Ca II K emission varies by more than 20 % over a solar cycle, while the smoothed irradiance changes by less than 0.1 % and hence we should expect something similar in other stars. The second reason is that the Lockwood et al. (1997) results were from ground-based observations and atmospheric turbulence increases noise. The best means of measuring photometric variability is through space missions. NASA's *Kepler* mission and the CNRS/ESA CoRoT satellite have helped immensely in this regard.

Kepler (Koch et al. 2010) observed a single field in the Cygnus–Lyra region for about 4 years. In total it monitored just under 200,000 stars and for most determined their brightness every 30 minutes, though a subset of stars were observed with a cadence of 1 minute. Although the primary goal of *Kepler* is to detect extra-solar planets through transit measurements, the extreme precision of *Kepler* photometry has led to other scientific investigations including stellar variability. Gilliland et al. (2011) examined *Kepler* photometry and concluded more than half of the solar-type stars in the *Kepler* field were more active than the Sun. Although the result that the Sun is less active than other solar-type stars was disputed by Basri et al. (2013), it is clear that photometry can detect stellar activity. We cite more examples later in Sect. 4.

In the case of the Sun, the activity cycle is not only present in chromospheric and photometric measurements, but we also see the cycle in the frequencies of the solar oscillations. It is now well established that the frequencies of the Sun's p modes increase with activity (Woodard and Noyes 1985; Elsworth et al. 1990; Libbrecht and Woodard 1990), and as shown in Fig. 1, they clearly show the 11-year cycle as well. The frequency shifts are also correlated with other solar activity indices such as the International Sunspot Number and the 10.7-cm radio flux (see e.g., Jain et al. 2000; Chaplin et al. 2007b).

The trend observed in the frequency shifts of the most prominent modes—the higher the frequencies, the larger are the shifts—is a tell-tale sign that it is the impact of the near-surface magnetic activity that is responsible for the shifts. However, uncertainty remains over how

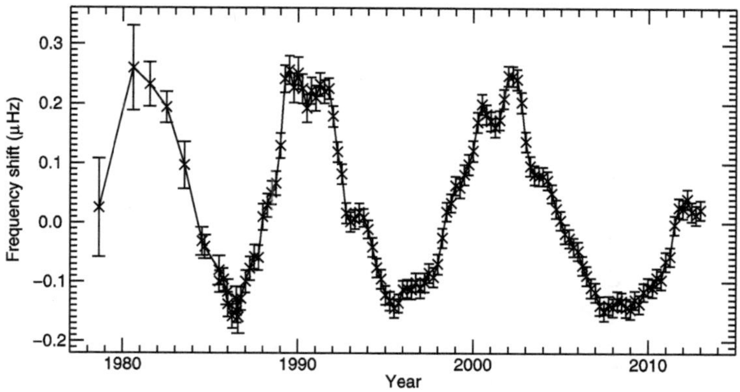

Fig. 1 Changes in solar oscillation frequencies as a function of time. We show the average shift for modes in the frequency range of 2500 to 3500 μHz, obtained from observations made by the Birmingham Solar-Oscillations Network (BiSON). The frequency-shifts were calculated by subtracting the average frequency of the modes over the entire observation period from the frequency at any given epoch. This results in both positive and negative shifts. (Figure courtesy of A.-M. Broomhall)

much of the observed shifts can be explained as being due to sunspots, active regions, or plage, and the relative influence of the direct and indirect action of the magnetic fields (see e.g., Dziembowski and Goode 2005; Thompson 2006).

The magnetic fields may act directly on the modes, via the Lorentz force: this provides an additional restoring force, the result being an increase of frequency, and the appearance of new modes. Magnetic fields can also act indirectly by affecting the physical properties in the mode cavities and, as a result, the propagation of the acoustic waves within them. This indirect effect can act both ways, to either increase or decrease the frequencies.

Now that there are seismic observations for other stars, the question we would like to answer is whether we could learn about stellar activity and activity cycles from asteroseismology? We try to answer that question in the following sections.

2 Could We Get Asteroseismic Measures of Stellar Activity?

There are two issues that could hinder asteroseismic studies of stellar activity. The first is that asteroseismic observations do not cover a long time-period. Even with *Kepler*, the longest span of observation for any given star is about four years; observations of many solar-type stars cover a shorter period. The second issue is whether activity-related variations in asteroseismic properties can be detected at all.

2.1 The Issue of the Period Length

It has been increasingly clear for some time that many stars have fairly short cycles (see e.g., Baliunas et al. 1995; Saar and Brandenburg 1999). Stellar activity data suggest that there is a correlation between the rotation period of a star and the length of the cycle, with faster rotators having shorter activity cycles (Baliunas et al. 1996; Oláh and Strassmeier 2002, etc.). Thus, faster rotating stars would seem to be good candidates for asteroseismic observations. The relationship between the rotation period and cycle-length is, however, complicated. Noyes et al. (1984), using dynamo models, had suggested that the cycle-period should be a function of the Rossby number, Ro, where Ro

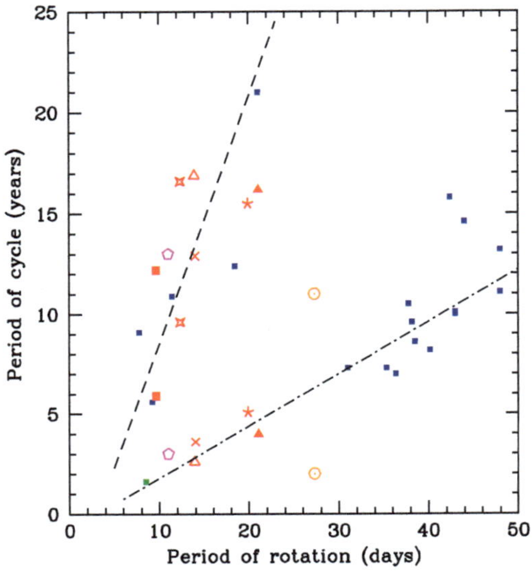

Fig. 2 A modified version of Fig. 1 of Böhm-Vitense (2007) showing the relationship between a stellar rotation period and activity cycle period. *Points in blue* are stars with one known cycle period. *Points in red* are stars with known, measured secondary cycle periods, and each star is shown with a unique symbol. Added to the data of Böhm-Vitense (2007) are those for ι Hor (Metcalfe et al. 2010, *in green*) and ϵ Eri (Metcalfe et al. 2013, *in pink*). The Sun is marked in *orange*—both 11-year and 2-year periods are marked for the Sun. The *upper branch* is the so-called "active" branch, the *lower* is the "inactive" branch. Note that for many stars with secondary periods, the two different periods lie on the two different branches

$\propto P_{\rm rot}/\tau_c$, $P_{\rm rot}$ being the rotation period and τ_c the convective turnover time-scale. While examining the correlation between stellar activity cycles and the Rossby number, Brandenburg et al. (1998) showed that the stars could be separated into two branches with a marked separation at $\log(P_{\rm rot}/P_{\rm cyc}) \sim -2.3$, one branch defining a young/active cohort and the other an old/inactive cohort. The designations "active" and "quiet" were made using the amount of emission in the Ca II H&K bands. The "inactive" stars have a higher $\log(P_{\rm rot}/P_{\rm cyc})$ ratio at a given Rossby number than do the "active stars."

The issue of the "active" and "inactive" branches gained more attention when Böhm-Vitense (2007) showed that stars could be separated into the branches even in the $P_{\rm cyc}$–$P_{\rm rot}$ plane. More interesting is where stars with secondary cycles lie. Many stars are known to have two distinct cycles, a longer primary cycle and a shorter secondary cycle (Baliunas et al. 1995; Saar and Brandenburg 1999; Oláh et al. 2009; Metcalfe et al. 2013). We show a modified version of Fig. 1 of Böhm-Vitense (2007) in our Fig. 2. What is interesting is that for stars with two cycles, the longer cycle falls on one branch, and the shorter cycle on the other. The shorter secondary cycles should be very amenable to asteroseismic studies; the important question of course is whether the cycle will be detectable in seismic data.

One of the striking features of the Böhm-Vitense (2007) plot is that the Sun does not fall on either of the two branches—the 11-year cycle lies mid-way between the two. The Sun too appears to exhibit a secondary cycle. One can detect a shorter period of about 2 years in solar oscillations data (Fletcher et al. 2010; Broomhall et al. 2012). The two-year modulation in the solar frequencies can be seen in Fig. 1.

2.2 Asteroseismic Predictions

Whether or not asteroseismic data can reveal anything about stellar activity cycles depends to a large extent on: first, whether oscillations are observable; and second, if oscillations are observable, whether their characteristics show solar-cycle related changes.

The first point is an important one. Solar observations have consistently shown that amplitudes of oscillations decrease as solar activity increases (e.g., Komm et al. 2000). This has also been demonstrated on local scales by looking at oscillation-amplitudes inside and outside sunspots (e.g., Braun et al. 1988). Thus it is quite possible that stellar activity will hinder the detection of stellar oscillations. Chaplin et al. (2011a) had investigated the detectability of stellar oscillations as a function of the global properties of the stars, using data from *Kepler*. They showed that there were fewer asteroseismic detections than were predicted, and presented evidence to suggest that the most plausible reason for the lack of detections is elevated levels of stellar activity. These results have since been confirmed, using data on direct measurements of oscillation amplitudes in a cohort of *Kepler* stars (Campante et al. 2014). Of course, *Kepler* and CoRoT observations have led to the detection of oscillations in many stars (see Sect. 3 below), thus while high levels of activity may lead us to miss a few stars, the cohort of stars with detections offers many promising targets for stellar-cycle studies.

Metcalfe et al. (2007) derived a way to predict asteroseismic signatures of stellar activity using the Mg II activity index. They assumed that the relation between the solar Mg II activity index and changes in solar oscillation frequencies could be applied to all stars. Solar frequency changes as a result of the solar cycle are a function predominantly of frequency, once the effect of mode-inertia is removed. This has led to the general idea that the source of solar-cycle induced frequency changes lies close to the solar surface (e.g., Nishizawa and Shibahashi 1995). Metcalfe et al. (2007) thus assumed that the same would apply to other stars, and that a scaling similar to the solar one between the Mg II index and the changes in stellar frequencies could be applied. They tested this on ground-based asteroseismic observations of β Hyi at two different epochs.

Chaplin et al. (2007a, 2008a) used a different approach and tried to determine the precision with which it will be possible to extract activity-related shifts in asteroseismic data of solar-type stars along the lower main sequence. They used models of damping rates of stochastically excited p-modes based on non-local mixing length formulations and used the averaged shift across the highest-amplitude peaks to reduce uncertainties. They made predictions of the expected frequency shifts based on existing stellar Ca II H&K data and concluded that basic properties of the asteroseismic signatures of stellar cycles could be measured with a few multi-month segments of data.

Chaplin et al. (2007a) also found that it should be possible to make inferences on the distribution of activity on stellar surfaces, as we now go on to explain. When the near-surface magnetic activity is distributed in a non spherically symmetric manner, i.e., into active bands of latitude, it will give rise to acoustic asphericity, i.e., the magnitudes of the frequency shifts will depend on the angular degree l and the azimuthal order m of the mode. Hence, measurements of the relative sizes of the shifts shown by low-degree modes having different $(l, |m|)$ combinations may be used to place constraints on the active latitudes shown by a solar-type star. It is worth noting that the stellar rotation may also contribute to asphericity, even for moderate rates of rotation. The two effects can also be difficult to disentangle (e.g., see discussion in Gizon 2002). Here, we consider just the impact of near-surface magnetic activity.

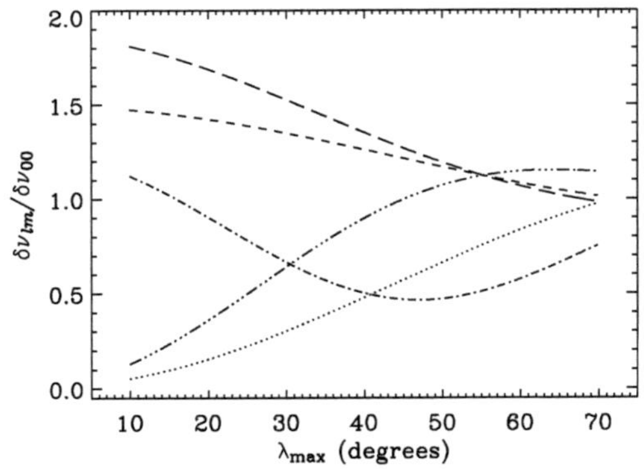

Fig. 3 Predicted frequency shifts $\delta\nu_{lm}$ of different azimuthal components, relative to the radial-mode shift $\delta\nu_{00}$, as a function of maximum latitude of magnetic activity, $\lambda_{\max}$ (see text for explanation of model of activity). Linestyles: *dotted* for $\delta\nu_{10}/\delta\nu_{00}$; *short dashes* for $\delta\nu_{11}/\delta\nu_{00}$; *dot-dashed* for $\delta\nu_{20}/\delta\nu_{00}$; *dot-dot-dot-dashed* for $\delta\nu_{21}/\delta\nu_{00}$; and *long dashes* for $\delta\nu_{22}/\delta\nu_{00}$

If the active bands lie at low latitudes, as for the Sun, the largest shifts of non-radial, low-degree modes (i.e., $l = 1$, 2 and 3) will be those shown by the sectoral ($l = |m|$) components. Analysis of Sun-as-a-star data has uncovered the expected variations in the sizes of the frequency shifts (e.g., see Chaplin et al. 2004a; Jiménez-Reyes et al. 2004). Here, we use a simple toy model to predict the *relative* sizes of the average frequency shifts of various low-l components, relative to the radial-mode shifts, for different assumed latitudinal bands of near-surface stellar activity. These predictions are based first and foremost on the assumption that changes in the sound speed contributing to the frequency shifts are located very close to the surface. In what follows we assume that we average frequency shifts over several radial orders, hence the frequency dependence of the shifts does not appear explicitly below. However, that dependence may also in principle be calculated, given a set of mode eigenfunctions and, as noted above, under the assumption that the magnetic perturbations are located close to the surface.

The expected (radial-order averaged) shifts, $\delta\nu_{lm}$, may be approximated (e.g., Moreno-Insertis and Solanki 2000) by

$$\delta\nu_{lm} \propto \left(l + \frac{1}{2}\right)\frac{(l-m)!}{(l+m)!}\int_0^\pi \left|P_l^m(\cos\theta)\right|^2 B(\theta)\sin\theta\,\mathrm{d}\theta, \tag{1}$$

where $P_l^m(\cos\theta)$ are Legendre polynomials, and $B(\theta)$ describes the distribution of the activity as a function of co-latitude, θ. Following Chaplin et al. (2007a) and Chaplin (2011), we assume in our toy model that $B(\theta)$ is described by:

$$B(\theta) = \begin{cases} \text{const.}, & \text{for } \lambda_{\min} \le |\lambda| \le \lambda_{\max}, \\ 0, & \text{otherwise}, \end{cases} \tag{2}$$

i.e., in effect a uniform band of activity. We fixed the minimum latitude at $\lambda_{\min} = 5$ degrees in all models, whilst $\lambda_{\max}$ was varied.[1]

Figure 3 the shows predicted frequency shift ratios $\delta\nu_{lm}/\delta\nu_{00}$, as a function of maximum latitude of magnetic activity, $\lambda_{\max}$. The linestyles show results as follows: dotted for $\delta\nu_{10}/\delta\nu_{00}$; short dashes for $\delta\nu_{11}/\delta\nu_{00}$; dot-dashed for $\delta\nu_{20}/\delta\nu_{00}$; dot-dot-dot-dashed for $\delta\nu_{21}/\delta\nu_{00}$; and long dashes for $\delta\nu_{22}/\delta\nu_{00}$.

[1] The total surface magnetic flux is proportional to $\lambda_{\max} - \lambda_{\min}$. One could instead normalise $B(\theta)$ so that the total flux is conserved for all values of $\lambda_{\max}$; however, this will not affect the ratios of the frequency shifts.

Chaplin (2011) used low-l solar p-mode frequency shifts measured by the Birmingham Solar-Oscillations Network (BiSON) to make an estimate of $\lambda_{\mathrm{max},\odot}$ based on use of the toy model above. He compared the observed ratios given by average frequency shifts of the $l=2$, 1 and 0 modes to the predicted ratios in Fig. 3. A weighted combination of his results gives $\lambda_{\mathrm{max},\odot} = 37 \pm 8$ deg, a reasonable estimate for the Sun.

This type of analysis is complicated by the fact that the disc-averaged visibility of the non-radial modes depends on the angle of inclination of the star, i_s, a point we shall return to in Sect. 4 below. In the solar case—where the rotation axis lies close to the plane of the sky when viewed from within the ecliptic plane—the zonal ($m=0$) component of the $l=1$ modes is all but absent, and so one may obtain a clean estimate of the $(1, 1)$ shift. However, this is not so at $l=2$, where the shifts will include a contribution from the $(2, 0)$ components. It actually turns out that this contribution is fairly weak, so that the observed Sun-as-a-star $l=2$ shifts approximate quite closely those of the $(2, 2)$ components. But things would not be so simple for Sun-like stars observed at other viewing angles. The relative contributions can nevertheless be modelled (e.g., see Chaplin et al. 2004b), given good constraints on i_s. Observational uncertainties for other stars are likely to be too large to detect the impact of another effect: that one might expect there to phase offsets in time between the frequency shifts of different $(l, |m|)$, i.e., reflecting the spatio-temporal dependence of the evolution of near-surface activity during an activity cycle (e.g., see Moreno-Insertis and Solanki 2000).

Even in the absence of cyclic variations, the presence of persistent, strong near-surface magnetic field will give rise to frequency asymmetries of the components in the non-radial modes. In the absence of very rapid rotation—which will also give rise to asymmetry of multiplet frequencies—measurement of any asymmetry could be used to place constraints on levels of near-surface activity. One does not require exotic levels of activity: frequency asymmetries of $l=2$ modes have been measured in Sun-as-a-star data (Chaplin et al. 2003).

Broomhall et al. (2011) have also looked at how the large frequency separation $\Delta\nu$, which is the separation in frequency between modes with the same degree but consecutive radial order, changes with activity. They looked at solar data from BiSON, and showed that solar-cycle related changes in $\Delta\nu$ could be detected using $\Delta\nu$ derived by different means, although the signature was somewhat different for each of the fitting methods. Since the large separation can be determined relatively precisely even when individual frequencies cannot, being able to detect the signature of stellar activity in $\Delta\nu$ potentially opens up more targets for study.

Extraction of estimates of stellar-cycle induced shifts of the p-mode parameters—be they the frequencies, large separations or amplitudes and mode damping rates—usually proceeds by dividing a long dataset into shorter segments for analysis. This makes it possible to track seismic responses to varying levels of activity. However, one may also look for signatures of changing activity in the frequency-power spectrum of the complete dataset. If the modes undergo significant frequency shifts over the duration of the dataset, the mode peaks will be distorted from their expected, Lorentzian-like shape. Chaplin et al. (2008b) discussed this effect in some detail, deriving analytical expressions for distorted profiles given by different functional forms for the time-dependent frequency variations. The significance of the distortion effect is determined by the ratio

$$\epsilon = \delta\nu/\Gamma, \tag{3}$$

where $\delta\nu$ is the shift in frequency and Γ is the FWHM of the mode peak. The higher this ratio, the more distorted the mode peak will be.

Figure 4 shows the distorted profiles given by frequency shifts of different sizes, for a mode having $\Gamma = 1.5$ μHz. Plotted in different colours are the profiles given by cosinusoidal

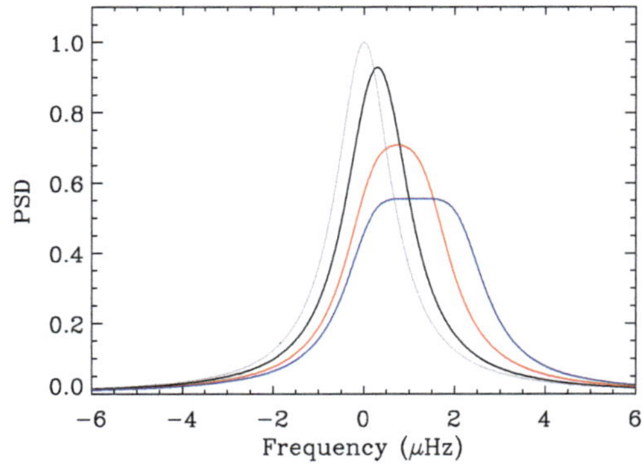

Fig. 4 Distorted profiles given by frequency shifts of different sizes (see text), for a mode having $\Gamma = 1.5$ μHz. The *various coloured lines* show profiles given by cosinusoidal (minimum-to-maximum) frequency shifts of zero (*grey*), 0.6 μHz (*black*), 1.5 μHz (*red*) and 2.25 μHz (*blue*)

(minimum-to-maximum) frequency shifts of zero (grey), 0.6 μHz (black), 1.5 μHz (red) and 2.25 μHz (blue). The ratios ϵ vary from zero up to 1.5, with the black profile having a ratio approximately equal to that of the low-degree solar p modes. The peaks are clearly flattened in appearance as a result of the frequency shifts. Similar distorted shapes are given by, for example, a simple linear frequency shift in time (again, see Chaplin et al. 2008b).

One may therefore look for evidence of distorted profiles in the real observations. A simple approach would be to fit the usual, Lorentzian-like models to the observed peaks, and then to interogate the form of the residuals given by dividing the observed spectrum by the best-fitting Lorentzian-based model. One may of course also attempt to fit distorted profiles directly to the data. This approach depends in part on the accuracy or appropriateness of the assumed form for the distortion.

Next, we go on to discuss the asteroseismic data that are available for stellar cycle studies.

3 Asteroseismic Data for Studies of Stellar Activity

The availability of asteroseismic data for studies of solar-type stars has been revolutionised by new satellite and telescope observations, in particular those of the CNRS/ESA CoRoT satellite and the NASA *Kepler* Mission.

Some attempts had already been made to look for stellar-cycle-related frequency shifts of bright solar-type stars in data collected in relatively short, episodic campaigns widely separated in time. Fletcher et al. (2006) used data from the startracker on the WIRE satellite to search for frequency shifts of α Cen A; while, as noted previously, Metcalfe et al. (2007) looked at data from ground-based Doppler velocity data on β Hyi. Data on the bright solar twin 18 Sco (Bazot et al. 2011, 2012), collected at two epochs separated in time by a few years, are also in the process of being analysed. While the planned SONG network (Grundahl et al. 2011, 2014) will also present an excellent opportunity to perform regular, episodic campaigns on the brightest solar-type stars.

The first multi-month datasets for solar-type stars were provided by CoRoT (Michel et al. 2008). CoRoT observed 12 solar-type targets for asteroseismology having apparent magnitudes in the range $5.4 \le m_v \le 9.4$, with dataset lengths from 20 to 170 days. Some of these targets were observed more than once, with significant gaps between observations, thereby making it possible to search for possible changes of the p-mode parameters. The bright F-type subgiant HD49933 was the first of these stars to be subjected to detailed searches for seismic signatures of stellar cycle variations. García et al. (2010) uncovered systematic changes in the frequencies of the star's most prominent p modes. The rapid variations suggested a short cycle period, between about 1 and 2 years. They also found evidence for changes in the amplitudes of the modes, anti-correlated with the changes detected in the frequencies. This matches observations of the solar p modes: enhanced levels of magnetic activity lead to an increase of the frequencies of the Sun's most prominent low-degree p modes, while at the same time the mode amplitudes are diminished. Salabert et al. (2011) followed with a more detailed analysis of the frequency shifts of HD49933, presenting evidence that the shifts were strongest for the highest mode frequencies. This trend is again in broad agreement with the solar results, suggesting that the perturbations responsible for the shifts are localised in the near-surface layers of the star.

Mathur et al. (2013a) repeated the analysis on three other stars observed by CoRoT—HD49385, HD52265 [a known planet host; see Gizon et al. 2013] and HD181420—but they did not find any evidence of seismic signatures due to stellar cycles. They also considered other types of data, including rotationally modulated signatures in the photometric stellar lightcurves, and spectropolarimetric data. We come back to discuss the combined use of different types of data in Sect. 4 below.

Kepler has provided significantly longer datasets on a much larger sample of solar-type stars. During the first few months of science operations, a substantial fraction of the short-cadence target slots required to conduct asteroseismology of solar-type stars were selected by the *Kepler* Asteroseismic Science Consortium (KASC; see Gilliland et al. 2010). Targets were observed for one month at a time, and this "asteroseismic survey" yielded detections in more than 500 solar-type stars (Chaplin et al. 2011b, 2014). Around 150 of the stars were then selected to be observed for durations of at least 3 months (i.e., for at least one *Kepler* observing Quarter). Of these stars, 91 had at least 24 months and 58 had at least 36 months of data by the time *Kepler*'s nominal mission ended.

Kepler Objects of Interest (KOIs) having short-cadence data provide another source of multi-month and multi-year datasets for asteroseismology (e.g., see Huber et al. 2013a). These targets are candidate, validated or confirmed exoplanet host stars and asteroseismology has already played a key rôle in helping to better characterise many of these systems. A total of 43 solar-type KOIs have at least 24 months of asteroseismic data, while 19 have at least 36 months of data.

Tables 1 and 2 list solar-type stars observed by *Kepler* that have at least 36 months of asteroseismic data. The stars in Table 1 are the field stars that were part of the KASC asteroseismic survey; while those in Table 2 are KOIs from the planet-hosting sample. Figure 5 plots both cohorts on a Hertzsprung–Russell diagram. Grey symbols show KASC field stars, while red symbols show the KOIs. The two black symbols mark the locations of the two components of the bright binary 16 Cyg (Metcalfe et al. 2012), which was also targeted by *Kepler* (KIC numbers 12069242 and 12069449). Also plotted are the CoRoT targets that have been analysed for asteroseismic signatures of stellar cycles, namely HD49933 (filled blue symbol; see García et al. 2010) and HD49385, HD181420 and HD52265 (square symbols; Mathur et al. 2013a).

Work is now underway to search the long *Kepler* lightcurves for evidence of asteroseismic signatures of stellar cycles.

4 Combining Asteroseismic and Other Data

A much more complete picture of the activity of the star may be obtained once asteroseismology data are combined with high-precision spectroscopic or photometric data. For

Table 1 Solar-type stars observed by *Kepler* having quantities of short-cadence data for asteroseismology in excess of 36 months. Stars in this table are field stars selected specifically for asteroseismic observations by KASC (see text)

KIC number	Coverage, Q1 to Q17 (months per quarter)	Total time (months)
1435467	01003333333333331	38
2837475	01003333333333331	38
3424541	01003333333333331	38
3427720	10003333333333331	38
3733735	01003333333333331	38
3735871	01003333333333331	38
5607242	01003333333333331	38
5955122	10003333333333331	38
6116048	01003333333333331	38
6508366	10003333333333331	38
6603624	10003333333333331	38
6679371	01003333333333331	38
6933899	01003333333333331	38
7103006	01003333333333331	38
7174707	10003333333333331	38
7206837	01003333333333331	38
7341231	10003333333333331	38
7584900	10003333333333331	38
7747078	10003333333333331	38
7799349	01003333333333331	38
7871531	10003333333333331	38
7976303	10003333333333331	38
8006161	01003333333333331	38
8228742	10003333333333331	38
8394589	01003333333333331	38
8524425	01003333333333331	38
8561221	01003333333333331	38
8694723	01003333333333331	38
8702606	10003333333333331	38
8760414	10003333333333331	38
9025370	01003333333333331	38
9098294	01003333333333331	38
9139151	10003333333333331	38
9139163	01003333333333331	38
9574283	01003333333333331	38
9812850	10003333333333331	38
10018963	01003333333333331	38
10124866	01003333333333331	38
10454113	01003333333333331	38
10644253	10003333333333331	38
11026764	10003333333333331	38
11081729	10003333333333331	38
11193681	01003333333333331	38

Table 1 (*Continued*)

KIC number	Coverage, Q1 to Q17 (months per quarter)	Total time (months)
11244118	01003333333333331	38
11253226	01003333333333331	38
11414712	01003333333333331	38
11554100	01003333333333331	38
11717120	10003333333333331	38
11771760	10003333333333331	38
11772920	01003333333333331	38
11968749	00103033333333331	36
12009504	10003333333333331	38
12069424	00000033333333331	31
12069449	00000033333333331	31
12258514	10003333333333331	38
12307366	01003333333333331	38
12317678	01003333333333331	38
12508433	01003333333333331	38

Table 2 The same as Table 1 but for *Kepler* Objects of Interest, i.e., candidate, validated or confirmed planet host stars with asteroseismic detections (see text)

KIC number	KOI number	Coverage, Q1 to Q17 (months per quarter)	Total time (months)
3544595	69	01333333333333331	44
3632418	975	01003333333333331	38
4141376	280	01100333333333331	36
4349452	244	01003333333333331	38
5094751	123	00333333333333300	39
5866724	85	00333333333333331	43
6521045	41	00333333333333331	43
7051180	64	00333333333330331	40
7199397	75	10333333333333331	44
8292840	260	00003333333333331	37
8349582	122	00333333333333331	43
8478994	245	00003333333333331	37
8554498	5	03333333333333331	46
8866102	42	10333333333333331	44
9955598	1925	01003333333333331	38
10593626	87	01333333303333331	41
10666592	2	13333333333333331	48
11295426	246	01003333333333331	38
11512246	168	00333333333333331	43

example, Karoff et al. (2013) have been monitoring 20 bright solar-type *Kepler* targets since 2009 using the FIES spectrograph on the Nordic Optical Telescope. It will be possible to combine the resulting measures of activity in the Ca II H&K lines (see Sect. 1) with asteroseismic results on the stars.

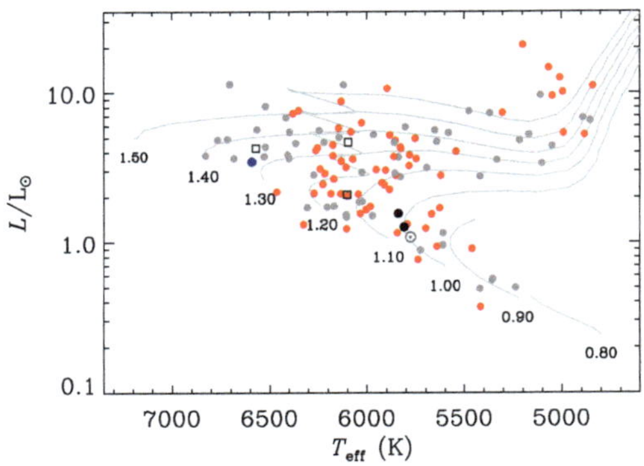

Fig. 5 Hertzsprung–Russell diagram of a selected subset of the solar-type stars observed by *Kepler* and CoRoT for asteroseismology. The plotted *Kepler* targets all have quantities of short-cadence data for asteroseismology in excess of 36 months (listed in Table 1). *Grey symbols* show field stars selected specifically by KASC for asteroseismic observations. *Red symbols* show *Kepler* Objects of Interest, i.e., candidate, validated or confirmed planet host stars with asteroseismic detections (see Huber et al. 2013a, 2013b). The *two black symbols* mark the locations of the two components of the bright binary 16 Cyg (also observed by *Kepler*). Plotted CoRoT stars are targets that have been analysed for asteroseismic signatures of stellar cycles: HD49933 (*filled blue symbol*) and HD49385, HD181420 and HD52265 (*square symbols*). (See text for further details)

From the CoRoT or *Kepler* data alone it is possible to combine asteroseismology with signatures of activity in the raw, time-domain lightcurves. Photometric variability of solar-type stars has shorter-timescale contributions due to granulation or oscillations. At longer timescales, variability is given by signatures of magnetic activity, e.g., the evolution and decay of starspots, or modulation associated with rotation. Rotational modulation of signatures of starspots and active regions makes it possible to measure not only the average surface rotation, but also the latitudinal gradient of any surface differential rotation. Some analyses also use the measured photometric variations to attempt a reconstruction of the surface distribution of spots and active regions, but one must take great care over the degeneracies inherent in the method (e.g., see Walkowicz et al. 2013).

Figure 6 shows examples of two stars with high-quality asteroseismic data and visible signatures of rotation and activity in their lightcurves. HD49933, one of CoRoT's most important targets (Appourchaux et al. 2008), has already been mentioned above. The top left-hand panel shows the frequency-power spectrum of its oscillations. The top right-hand panel comprises a smoothed 60-day segment of the time-domain lightcurve, with the ordinate scaled to show relative brightness variations in parts per million. Variations in the lightcurve are visually dominated by periodic signatures due to rotational modulation of starspots. Those variations imply a mean surface rotation period of around 3.5 days. The bottom panels have similar plots for the planet-hosting star *Kepler*-50 (Chaplin et al. 2013). Its lightcurve variations are not as strong, and the slower variations imply a slightly longer surface rotation period, of just under 8 days.

A growing number of studies are now exploiting the new satellite data to provide estimates of surface rotation periods of solar-type stars. There are examples for individual targets of interest (e.g., see CoRoT examples in Mosser et al. 2009a, 2009b) and for large cohorts of field stars (e.g., see *Kepler* examples in McQuillan et al. 2013; Nielsen et al. 2013; Reinhold et al. 2013; Walkowicz and Basri 2013). Some of the studies have specifically targeted stars with asteroseismic data (e.g., see García et al. 2013). Metrics related to the observed amplitudes of the photometric variability have also been used as proxies of activity

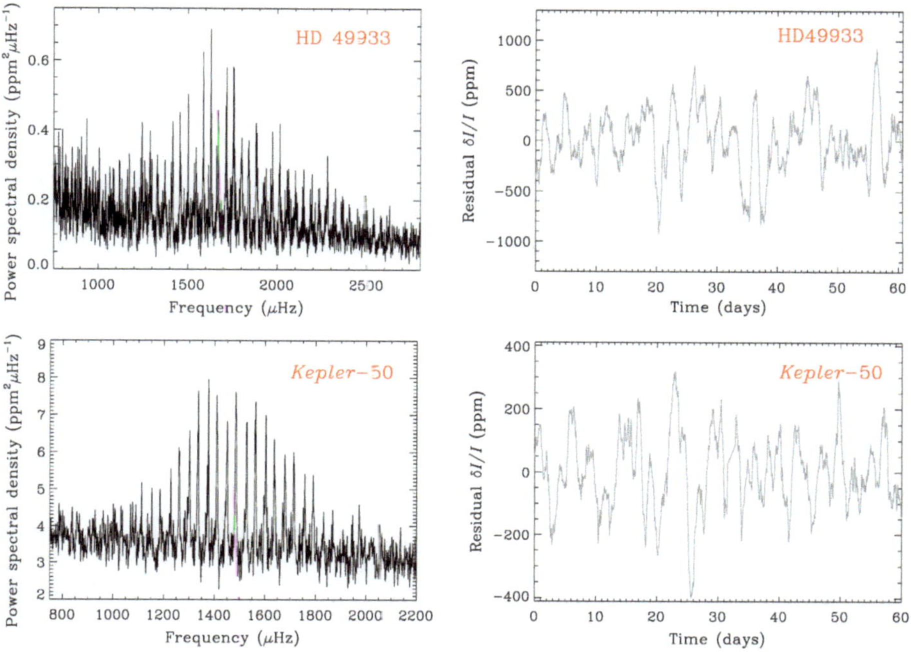

Fig. 6 *Top panels*: CoRoT data on HD49933, showing the frequency-power spectrum of its oscillations (*left-hand panel*) and a segment of the time-domain lightcurve showing rotationally modulated signatures of starspots. *Bottom panels*: Similar plots, but for the planet-hosting star *Kepler*-50, observed by *Kepler*

levels on the stars (Basri et al. 2010, 2011, 2013). Mathur et al. (2013b) presented results on a sample of 22 F-type stars observed by *Kepler*, all of which have surface rotation periods faster than 12 days and good asteroseismic detections. Some of the stars show evidence for stellar-cycle like variability in their photometry.

The photometric signatures of activity are dependent on the stellar angle of inclination, i_s. Having good constraints on the inclination can provide important information to help interpret inferences made from the photometry (and also spectroscopic observations). For example, might a lack of any significant photometric variability mean a Sun-like star is inactive? Or that it is being observed at a low i_s, meaning rotational modulation due to low-latitude spots and active regions is not detectable? Asteroseismology provides a means to estimate i_s. The asteroseismic method relies on the robust extraction of parameters related to the rotational splitting of non-radial modes of solar-like oscillations (see Gizon and Solanki 2003; Ballot et al. 2006, 2008) and has already been used to place constraints on the spin-orbit alignment of exoplanet systems (Chaplin et al. 2013; Huber et al. 2013b; Gizon et al. 2013; Van Eylen et al. 2014).

The degeneracy of the oscillation frequencies is lifted by rotation, giving several observable multiplet components whose frequencies depend on the azimuthal order, m. Perturbations due to non-radial modes are not spherically symmetric, hence the relative disc-integrated amplitudes of the different m components change with the viewing angle, i_s. Figure 7 shows the expected appearance of the limit spectrum of an idealised $l = 1$ multiplet, as a function of i_s (see also Gizon et al. 2013). Here, we have assumed underlying multiplet parameters comparable to those observed in the oscillation spectrum of *Kepler*-50, the planet-hosting star whose inclination was determined by Chaplin et al. (2013). Each

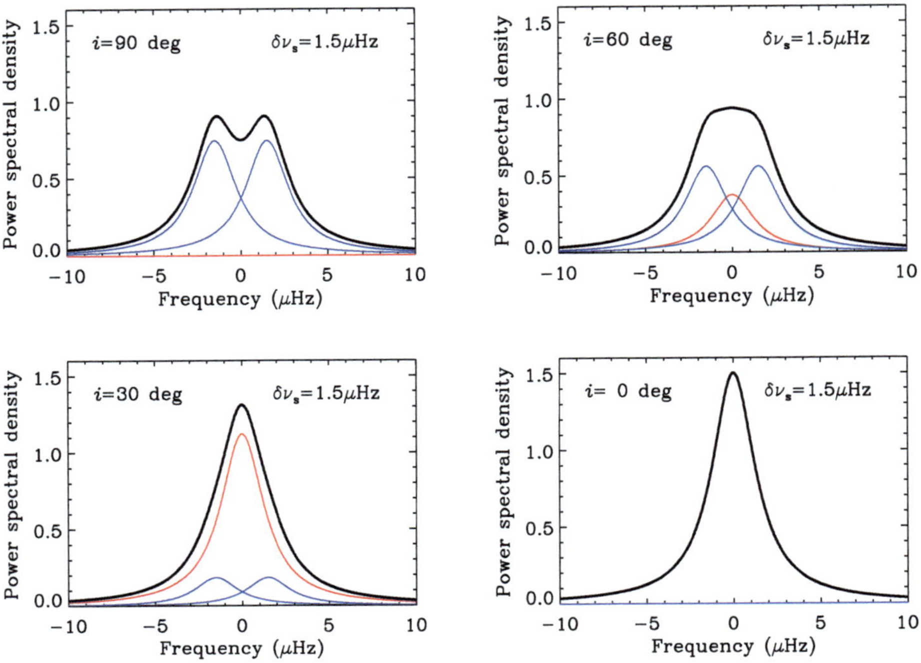

Fig. 7 Expected appearance of the limit spectrum of an idealised $l = 1$ multiplet, as a function of the stellar angle of inclination, i_s. The composite mode profile is shown in *black*, while power due to the zonal ($m = 0$) and sectoral ($m = \pm 1$) components is shown, respectively, in *red* and *blue* (see text for further details)

mode peak has a FWHM of 3.0 μHz. The rotational splitting in frequency between adjacent peaks is 1.5 μHz. When the rates of differential and absolute rotation are fairly modest, as is usually the case for solar-type stars, the splittings $\delta\nu_{nlm}$ of the observable high-n, low-l p modes will take very similar values.

Figure 8 is a zoom in frequency of the oscillations spectrum of *Kepler*-50 (from Fig. 6), here showing a single $l = 1$ mode. The grey line shows the spectrum after applying light smoothing, using a boxcar filter of width 0.25 μHz. The blue line—which shows the spectrum after applying much heavier smoothing of width 1.5 μHz—serves to provide an approximate estimate of the underlying (noise-free, or "limit") frequency spectrum. It is evident that the observed mode bears a close resemblance to the idealised high-inclination case in Fig. 7, i.e., even without resorting to a full analysis of the data, it is evident that the rotation axis of the star must lie reasonably close to the plane of the sky.

The $m = \pm 1$ components have their strongest observable powers when the rotation axis lies in the plane of the sky, i.e., when $i_s \simeq \pi/2$. When $i_s \simeq 0$, they are absent, and the only observable mode power is that due to the $m = 0$ component. The dependence of the visibility (in power) follows:

$$\mathcal{E}_{lm}(i_s) = \frac{(l - |m|)!}{(l + |m|)!}\left[P_l^{|m|}(\cos i_s)\right]^2, \tag{4}$$

where $P_l^{|m|}$ are Legendre polynomials, and the sum over $\mathcal{E}_{lm}(i_s)$ is normalised to unity. Measuring the relative power of the azimuthal components of different $|m|$ in a non-radial multiplet therefore gives an estimate of i_s (actually $|i_s|$ since one cannot discriminate between i_s and $-i_s$, or $\pi - i_s$ and $\pi + i_s$).

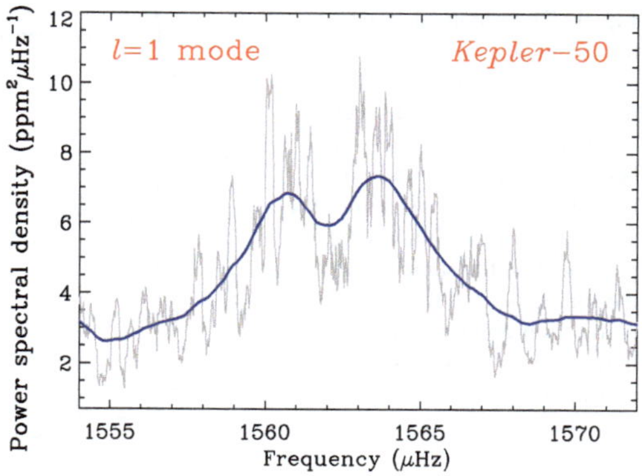

Fig. 8 Zoom in frequency of the oscillations spectrum of *Kepler*-50 from Fig. 6, showing a single $l = 1$ mode. The *grey line* shows the observed spectrum after application of light smoothing (using a boxcar filter of width 0.25 μHz). The *blue line* shows the spectrum after applying much heavier smoothing (of width 1.5 μHz)

5 Other Asteroseismic Results

In addition to giving direct information about stellar cycles, asteroseismology also helps more indirectly. For instance, asteroseismic data can help us assess the depths of subsurface stellar convection zones. Most dynamo theories assume that the interface between the convection and radiative zones plays a crucial rôle in driving stellar activity cycles. Additionally, it is generally believed that the convective turnover timescale, as expressed in the form of the Rossby number, Ro, plays a rôle in determining cycle properties. Hot stars that are expected to have shallow convection zones also usually have shorter cycle periods. Information about the location of the boundary between the convection and radiative zones are coded into stellar oscillation frequencies and thus one may use the frequencies to determine the acoustic depth (i.e., in the sound travel time) of the position of the boundary.

The boundary between the outer convection zone and the radiative zone of a star is marked by an abrupt change in the temperature gradient that causes a localised change in the stratification of the layer. This in turn gives an abrupt change in the sound speed gradient commonly referred to as an "acoustic glitch." Such a glitch introduces an oscillatory component in the frequencies as a function of frequency (e.g., Gough and Thompson 1988; Vorontsov 1988; Gough 1990). The oscillatory component can be expressed as

$$\delta\nu \propto \sin(4\pi\tau\nu_{n,l} + \phi), \tag{5}$$

where

$$\tau = \int_r^R \frac{\mathrm{d}r}{c}, \tag{6}$$

is the acoustic depth of the glitch, r is the corresponding location in radius, c the speed of sound, $\nu_{n,l}$ the frequency of a mode with radial order n and degree l, and ϕ is a phase factor. There can be multiple acoustic glitches, for instance, the helium ionisation zone; in such cases each glitch will contribute to the signal a specific periodicity of twice the acoustic depth of the glitch. In the case of the Sun, this oscillatory signature has been used to study overshoot below the solar convection zone (e.g., Basu 1997, and references therein).

There are a number of different ways in which the acoustic signal can be fitted to determine τ (see e.g., Mazumdar et al. 2014), but an easy way to see the signal is to take second differences of the frequencies

$$\delta^2\nu_{n,l} = \nu_{n-1,l} - 2\nu_{n,l} + \nu_{n+1,l}, \tag{7}$$

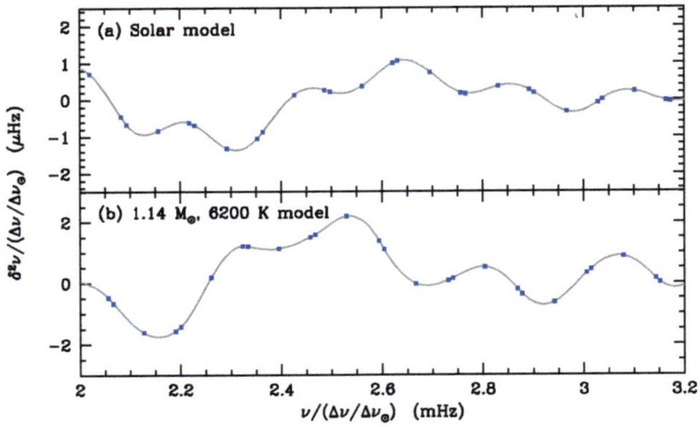

Fig. 9 The second differences of the frequencies of two models, one solar and one a higher-mass, hotter star, plotted as a function of frequency. Both axes have been scaled with the ratio of the large frequency separation of the star to that of the Sun in order to plot them on the same scale. Note the longer period of the convection-zone signal in the hotter star indicates a shallower convection zone

and then plot them as a function of $\nu_{n,l}$. We show such a plot in Fig. 9. The higher-frequency component is the signature of the base of the convection zone, the lower frequency component is from the He II ionisation zone.

Mazumdar et al. (2014) have shown that the oscillatory signal in real asteroseismic data can be fitted successfully to determine the acoustic depth of the convection-zone base. They examined 19 stars observed by *Kepler* ranging in effective temperature from about 5400 K to 6200 K and metallicity, [Fe/H], ranging from -1.19 to $+0.38$. Converting acoustic depth to actual depth requires the use of models, but the same mode frequencies can be used to construct good models of the star, thereby providing a good estimate of the physical depth of the convection zone in those stars.

6 Concluding Thoughts

We are beginning to be able to probe stellar cycles with asteroseismic data. With two more asteroseismology missions being planned—the Transiting Exoplanet Survey Satellite (TESS; launch 2017), and PLATO (launch 2024)—we should be able to extend studies of stellar activity and stellar cycles to many more stars. Asteroseismology will also allow us to characterise the stars so that we can study the dependence between stellar interior parameters and stellar cycles.

Acknowledgements W.J.C. acknowledges support from the UK Science and Technology Facilities Council (STFC). Funding for the Stellar Astrophysics Centre (SAC) is provided by The Danish National Research Foundation. S.B. acknowledges partial support from NSF grant AST-1105930 and NASA grant NNX13AE70G. The authors would like to thank R. Gilliland for help providing information on data coverage for *Kepler* short-cadence targets.

References

T. Appourchaux, E. Michel, M. Auvergne, A. Baglin, T. Toutain, F. Baudin, O. Benomar, W.J. Chaplin, S. Deheuvels, R. Samadi, G.A. Verner, P. Boumier, R.A. García, B. Mosser, J.-C. Hulot, J. Ballot, C. Barban, Y. Elsworth, S.J. Jiménez-Reyes, H. Kjeldsen, C. Régulo, I.W. Roxburgh, Astron. Astrophys. **488**, 705–714 (2008)

S.L. Baliunas, R.A. Donahue, W.H. Soon, J.H. Horne, J. Frazer, L. Woodard-Eklund, M. Bradford, L.M. Rao, O.C. Wilson, Q. Zhang, W. Bennett, J. Briggs, S.M. Carroll, D.K. Duncan, D. Figueroa, H.H. Lanning, T. Misch, J. Mueller, R.W. Noyes, D. Poppe, A.C. Porter, C.R. Robinson, J. Russell, J.C. Shelton, T. Soyumer, A.H. Vaughan, J.H. Whitney, Astrophys. J. **438**, 269–287 (1995)

S.L. Baliunas, E. Nesme-Ribes, D. Sokoloff, W.H. Soon, Astrophys. J. **460**, 848 (1996)
J. Ballot, R.A. García, P. Lambert, Mon. Not. R. Astron. Soc. **369**, 1281–1286 (2006)
J. Ballot, T. Appourchaux, T. Toutain, M. Guittet, Astron. Astrophys. **486**, 867–875 (2008)
G. Basri, L.M. Walkowicz, N. Batalha, R.L. Gilliland, J. Jenkins, W.J. Borucki, D. Koch, D. Caldwell, A.K. Dupree, D.W. Latham, S. Meibom, S. Howell, T. Brown, Astrophys. J. Lett. **713**, 155–159 (2010)
G. Basri, L.M. Walkowicz, N. Batalha, R.L. Gilliland, J. Jenkins, W.J. Borucki, D. Koch, D. Caldwell, A.K. Dupree, D.W. Latham, G.W. Marcy, S. Meibom, T. Brown, Astron. J. **141**, 20 (2011)
G. Basri, L.M. Walkowicz, A. Reiners, Astrophys. J. **769**, 37 (2013)
S. Basu, Mon. Not. R. Astron. Soc. **288**, 572–584 (1997)
M. Bazot, M.J. Ireland, D. Huber, T.R. Bedding, A.-M. Broomhall, T.L. Campante, H. Carfantan, W.J. Chaplin, Y. Elsworth, J. Meléndez, P. Petit, S. Théado, V. Van Grootel, T. Arentoft, M. Asplund, M. Castro, J. Christensen-Dalsgaard, J.D. Do Nascimento, B. Dintrans, X. Dumusque, H. Kjeldsen, H.A. McAlister, T.S. Metcalfe, M.J.P.F.G. Monteiro, N.C. Santos, S. Sousa, J. Sturmann, L. Sturmann, T.A. ten Brummelaar, N. Turner, S. Vauclair, Astron. Astrophys. **526**, 4 (2011)
M. Bazot, T.L. Campante, W.J. Chaplin, H. Carfantan, T.R. Bedding, X. Dumusque, A.-M. Broomhall, P. Petit, S. Théado, V. Van Grootel, T. Arentoft, M. Castro, J. Christensen-Dalsgaard, J.-D. do Nascimento Jr., B. Dintrans, H. Kjeldsen, M.J.P.F.G. Monteiro, N.C. Santos, S. Sousa, S. Vauclair, Astron. Astrophys. **544**, 106 (2012)
E. Böhm-Vitense, Astrophys. J. **657**, 486–493 (2007)
A. Brandenburg, S.H. Saar, C.R. Turpin, Astrophys. J. Lett. **498**, 51 (1998)
D.C. Braun, T.L. Duvall Jr., B.J. Labonte, Astrophys. J. **335**, 1015–1025 (1988)
A.-M. Broomhall, W.J. Chaplin, Y. Elsworth, R. New, Mon. Not. R. Astron. Soc. **413**, 2978–2986 (2011)
A.-M. Broomhall, W.J. Chaplin, Y. Elsworth, R. Simoniello, Mon. Not. R. Astron. Soc. **420**, 1405–1414 (2012)
T.L. Campante, W.J. Chaplin, M.N. Lund, D. Huber, S. Hekker, R.A. García, E. Corsaro, R. Handberg, A. Miglio, T. Arentoft, S. Basu, T.R. Bedding, J. Christensen-Dalsgaard, G.R. Davies, Y.P. Elsworth, R.L. Gilliland, C. Karoff, S.D. Kawaler, H. Kjeldsen, M. Lundkvist, T.S. Metcalfe, V. Silva Aguirre, D. Stello, ArXiv e-prints (2014)
W.J. Chaplin, ArXiv e-prints (2011)
W.J. Chaplin, Y. Elsworth, G.R. Isaak, B.A. Miller, R. New, S. Thiery, P. Boumier, A.H. Gabriel, Mon. Not. R. Astron. Soc. **343**, 343–352 (2003)
W.J. Chaplin, Y. Elsworth, G.R. Isaak, B.A. Miller, R. New, Mon. Not. R. Astron. Soc. **352**, 1102–1108 (2004a)
W.J. Chaplin, T. Appourchaux, Y. Elsworth, G.R. Isaak, B.A. Miller, R. New, T. Toutain, Astron. Astrophys. **416**, 341–351 (2004b)
W.J. Chaplin, Y. Elsworth, G. Houdek, R. New, Mon. Not. R. Astron. Soc. **377**, 17–29 (2007a)
W.J. Chaplin, Y. Elsworth, B.A. Miller, G.A. Verner, R. New, Solar p-mode frequencies over three solar cycles. Astrophys. J. **659**, 1749–1760 (2007b). doi:10.1086/512543
W.J. Chaplin, G. Houdek, T. Appourchaux, Y. Elsworth, R. New, T. Toutain, Astron. Astrophys. **485**, 813–822 (2008a)
W.J. Chaplin, Y. Elsworth, R. New, T. Toutain, Mon. Not. R. Astron. Soc. **384**, 1668–1674 (2008b)
W.J. Chaplin, T.R. Bedding, A. Bonanno, A.-M. Broomhall, R.A. García, S. Hekker, D. Huber, G.A. Verner, S. Basu, Y. Elsworth, G. Houdek, S. Mathur, B. Mosser, R. New, I.R. Stevens, T. Appourchaux, C. Karoff, T.S. Metcalfe, J. Molenda-Żakowicz, M.J.P.F.G. Monteiro, M.J. Thompson, J. Christensen-Dalsgaard, R.L. Gilliland, S.D. Kawaler, H. Kjeldsen, J. Ballot, O. Benomar, E. Corsaro, T.L. Campante, P. Gaulme, S.J. Hale, R. Handberg, E. Jarvis, C. Régulo, I.W. Roxburgh, D. Salabert, D. Stello, F. Mullally, J. Li, W. Wohler, Astrophys. J. Lett. **732**, 5 (2011a)
W.J. Chaplin, H. Kjeldsen, J. Christensen-Dalsgaard, S. Basu, A. Miglio, T. Appourchaux, T.R. Bedding, Y. Elsworth, R.A. García, R.L. Gilliland, L. Girardi, G. Houdek, C. Karoff, S.D. Kawaler, T.S. Metcalfe, J. Molenda-Żakowicz, M.J.P.F.G. Monteiro, M.J. Thompson, G.A. Verner, J. Ballot, A. Bonanno, I.M. Brandão, A.-M. Broomhall, H. Bruntt, T.L. Campante, E. Corsaro, O.L. Creevey, G. Doğan, L. Esch, N. Gai, P. Gaulme, S.J. Hale, R. Handberg, S. Hekker, D. Huber, A. Jiménez, S. Mathur, A. Mazumdar, B. Mosser, R. New, M.H. Pinsonneault, D. Pricopi, P.-O. Quirion, C. Régulo, D. Salabert, A.M. Serenelli, V. Silva Aguirre, S.G. Sousa, D. Stello, I.R. Stevens, M.D. Suran, K. Uytterhoeven, T.R. White, W.J. Borucki, T.M. Brown, J.M. Jenkins, K. Kinemuchi, J. Van Cleve, T.C. Klaus, Science **332**, 213 (2011b)
W.J. Chaplin, R. Sanchis-Ojeda, T.L. Campante, R. Handberg, D. Stello, J.N. Winn, S. Basu, J. Christensen-Dalsgaard, G.R. Davies, T.S. Metcalfe, L.A. Buchhave, D.A. Fischer, T.R. Bedding, W.D. Cochran, Y. Elsworth, R.L. Gilliland, S. Hekker, D. Huber, H. Isaacson, C. Karoff, S.D. Kawaler, H. Kjeldsen, D.W. Latham, M.N. Lund, M. Lundkvist, G.W. Marcy, A. Miglio, T. Barclay, J.J. Lissauer, Astrophys. J. **766**, 101 (2013)

W.J. Chaplin, S. Basu, D. Huber, A. Serenelli, L. Casagrande, V. Silva Aguirre, W.H. Ball, O.L. Creevey, L. Gizon, R. Handberg, C. Karoff, R. Lutz, J.P. Marques, A. Miglio, D. Stello, M.D. Suran, D. Pricopi, T.S. Metcalfe, M.J.P.F.G. Monteiro, J. Molenda-Żakowicz, T. Appourchaux, J. Christensen-Dalsgaard, Y. Elsworth, R.A. García, G. Houdek, H. Kjeldsen, A. Bonanno, T.L. Campante, E. Corsaro, P. Gaulme, S. Hekker, S. Mathur, B. Mosser, C. Régulo, D. Salabert, Astrophys. J. Suppl. Ser. **210**, 1 (2014)

D.K. Duncan, A.H. Vaughan, O.C. Wilson, G.W. Preston, J. Frazer, H. Lanning, A. Misch, J. Mueller, D. Soyumer, L. Woodard, S.L. Baliunas, R.W. Noyes, L.W. Hartmann, A. Porter, C. Zwaan, F. Middelkoop, R.G.M. Rutten, D. Mihalas, Astrophys. J. Suppl. Ser. **76**, 383–430 (1991)

W.A. Dziembowski, P.R. Goode, Sources of oscillation frequency increase with rising solar activity. Astrophys. J. **625**, 548–555 (2005). doi:10.1086/429712

Y. Elsworth, R. Howe, G.R. Isaak, C.P. McLeod, R. New, Nature **345**, 322–324 (1990)

S.T. Fletcher, W.J. Chaplin, Y. Elsworth, J. Schou, D. Buzasi, Mon. Not. R. Astron. Soc. **371**, 935–944 (2006)

S.T. Fletcher, A.-M. Broomhall, D. Salabert, S. Basu, W.J. Chaplin, Y. Elsworth, R.A. Garcia, R. New, Astrophys. J. Lett. **718**, 19–22 (2010)

R.A. García, S. Mathur, D. Salabert, J. Ballot, C. Régulo, T.S. Metcalfe, A. Baglin, Science **329**, 1032 (2010)

R.A. García, T. Ceillier, S. Mathur, D. Salabert, in *Astronomical Society of the Pacific Conference Series*, vol. 479, ed. by H. Shibahashi, A.E. Lynas-Gray (2013), p. 129

R.L. Gilliland, T.M. Brown, J. Christensen-Dalsgaard, H. Kjeldsen, C. Aerts, T. Appourchaux, S. Basu, T.R. Bedding, W.J. Chaplin, M.S. Cunha, P. De Cat, J. De Ridder, J.A. Guzik, G. Handler, S. Kawaler, L. Kiss, K. Kolenberg, D.W. Kurtz, T.S. Metcalfe, M.J.P.F.G. Monteiro, R. Szabó, T. Arentoft, L. Balona, J. Debosscher, Y.P. Elsworth, P.-O. Quirion, D. Stello, J.C. Suárez, W.J. Borucki, J.M. Jenkins, D. Koch, Y. Kondo, D.W. Latham, J.F. Rowe, J.H. Steffen, Publ. Astron. Soc. Pac. **122**, 131–143 (2010)

R.L. Gilliland, W.J. Chaplin, E.W. Dunham, V.S. Argabright, W.J. Borucki, G. Basri, S.T. Bryson, D.L. Buzasi, D.A. Caldwell, Y.P. Elsworth, J.M. Jenkins, D.G. Koch, J. Kolodziejczak, A. Miglio, J. van Cleve, L.M. Walkowicz, W.F. Welsh, Astrophys. J. Suppl. Ser. **197**, 6 (2011)

L. Gizon, Prospects for detecting stellar activity through asteroseismology. Astron. Nachr. **323**, 251–253 (2002). doi:10.1002/1521-3994(200208)323:3/4<251::AID-ASNA251>3.0.CO;2-9

L. Gizon, S.K. Solanki, Astrophys. J. **589**, 1009–1019 (2003)

L. Gizon, J. Ballot, E. Michel, T. Stahn, G. Vauclair, H. Bruntt, P.-O. Quirion, O. Benomar, S. Vauclair, T. Appourchaux, M. Auvergne, A. Baglin, C. Barban, F. Baudin, M. Bazot, T. Campante, C. Catala, W. Chaplin, O. Creevey, S. Deheuvels, N. Dolez, Y. Elsworth, R. Garcia, P. Gaulme, S. Mathis, S. Mathur, B. Mosser, C. Regulo, I. Roxburgh, D. Salabert, R. Samadi, K. Sato, G. Verner, S. Hanasoge, K.R. Sreenivasan, Seismic constraints on rotation of Sun-like star and mass of exoplanet. Proc. Natl. Acad. Sci. USA **110**, 13267–13271 (2013). doi:10.1073/pnas.1303291110

D. Gough, in *Lecture Notes in Physics*, vol. 367, ed. by Y. Osaki, H. Shibahashi (Springer, Berlin, 1990), p. 281

D.O. Gough, M.J. Thompson, in *IAU Symposium*, vol. 123, ed. by J. Christensen-Dalsgaard, S. Frandsen (1988), p. 155

R.O. Gray, C.J. Corbally, R.F. Garrison, M.T. McFadden, E.J. Bubar, C.E. McGahee, A.A. O'Donoghue, E.R. Knox, Astron. J. **132**, 161–170 (2006)

F. Grundahl, J. Christensen-Dalsgaard, U. Gråe Jørgensen, S. Frandsen, H. Kjeldsen, P. Kjærgaard Rasmussen, J. Phys. Conf. Ser. **271**(1), 012083 (2011)

F. Grundahl, J. Christensen-Dalsgaard, P.L. Pallé, M.F. Andersen, S. Frandsen, K. Harpsøe, U.G. Jørgensen, H. Kjeldsen, P.K. Rasmussen, J. Skottfelt, A.N. Sørensen, A.T. Hage, Stellar observations network group: the prototype is nearly ready, in *IAU Symposium*, ed. by J.A. Guzik, W.J. Chaplin, G. Handler, A. Pigulski. IAU Symposium, vol. 301 (2014), pp. 69–75. doi:10.1017/S1743921313014117

J.C. Hall, G.W. Lockwood, B.A. Skiff, Astron. J. **133**, 862–881 (2007)

T.J. Henry, D.R. Soderblom, R.A. Donahue, S.L. Baliunas, Astron. J. **111**, 439 (1996)

D. Huber, W.J. Chaplin, J. Christensen-Dalsgaard, R.L. Gilliland, H. Kjeldsen, L.A. Buchhave, D.A. Fischer, J.J. Lissauer, J.F. Rowe, R. Sanchis-Ojeda, S. Basu, R. Handberg, S. Hekker, A.W. Howard, H. Isaacson, C. Karoff, D.W. Latham, M.N. Lund, M. Lundkvist, G.W. Marcy, A. Miglio, V. Silva Aguirre, D. Stello, T. Arentoft, T. Barclay, T.R. Bedding, C.J. Burke, J.L. Christiansen, Y.P. Elsworth, M.R. Haas, S.D. Kawaler, T.S. Metcalfe, F. Mullally, S.E. Thompson, Astrophys. J. **767**, 127 (2013a)

D. Huber, J.A. Carter, M. Barbieri, A. Miglio, K.M. Deck, D.C. Fabrycky, B.T. Montet, L.A. Buchhave, W.J. Chaplin, S. Hekker, J. Montalbán, R. Sanchis-Ojeda, S. Basu, T.R. Bedding, T.L. Campante, J. Christensen-Dalsgaard, Y.P. Elsworth, D. Stello, T. Arentoft, E.B. Ford, R.L. Gilliland, R. Handberg, A.W. Howard, H. Isaacson, J.A. Johnson, C. Karoff, S.D. Kawaler, H. Kjeldsen, D.W. Latham, M.N. Lund, M. Lundkvist, G.W. Marcy, T.S. Metcalfe, V. Silva Aguirre, J.N. Winn, Science **342**, 331–334 (2013b)

K. Jain, S.C. Tripathy, A. Bhatnagar, B. Kumar, Sol. Phys. **192**, 487–494 (2000)

J.S. Jenkins, H.R.A. Jones, C.G. Tinney, R.P. Butler, C. McCarthy, G.W. Marcy, D.J. Pinfield, B.D. Carter, A.J. Penny, Mon. Not. R. Astron. Soc. **372**, 163–173 (2006)

S.J. Jiménez-Reyes, R.A. García, W.J. Chaplin, S.G. Korzennik, Astrophys. J. Lett. **610**, 65–68 (2004)

C. Karoff, T.S. Metcalfe, W.J. Chaplin, S. Frandsen, F. Grundahl, H. Kjeldsen, J. Christensen-Dalsgaard, M.B. Nielsen, S. Frimann, A.O. Thygesen, T. Arentoft, T.M. Amby, S.G. Sousa, D.L. Buzasi, Mon. Not. R. Astron. Soc. **433**, 3227–3238 (2013)

D.G. Koch, W.J. Borucki, G. Basri, N.M. Batalha, T.M. Brown, D. Caldwell, J. Christensen-Dalsgaard, W.D. Cochran, E. DeVore, E.W. Dunham, T.N. Gautier III, J.C. Geary, R.L. Gilliland, A. Gould, J. Jenkins, Y. Kondo, D.W. Latham, J.J. Lissauer, G. Marcy, D. Monet, D. Sasselov, A. Boss, D. Brownlee, J. Caldwell, A.K. Dupree, S.B. Howell, H. Kjeldsen, S. Meibom, D. Morrison, T. Owen, H. Reitsema, J. Tarter, S.T. Bryson, J.L. Dotson, P. Gazis, M.R. Haas, J. Kolodziejczak, J.F. Rowe, J.E. Van Cleve, C. Allen, H. Chandrasekaran, B.D. Clarke, J. Li, E.V. Quintana, P. Tenenbaum, J.D. Twicken, H. Wu, Astrophys. J. Lett. **713**, 79–86 (2010)

R.W. Komm, R. Howe, F. Hill, Astrophys. J. **531**, 1094–1108 (2000)

K.G. Libbrecht, M.F. Woodard, Nature **345**, 779–782 (1990)

J.L. Linsky, in *Astronomical Society of the Pacific Conference Series*, vol. 158, ed. by C.J. Butler, J.G. Doyle (1999), p. 3

G.W. Lockwood, B.A. Skiff, R.R. Radick, Astrophys. J. **485**, 789 (1997)

S. Mathur, R.A. García, A. Morgenthaler, D. Salabert, P. Petit, J. Ballot, C. Régulo, C. Catala, Astron. Astrophys. **550**, 32 (2013a)

S. Mathur, R.A. Garcia, J. Ballot, T. Ceillier, D. Salabert, T.S. Metcalfe, C. Regulo, A. Jimenez, S. Bloemen, ArXiv e-prints (2013b)

A. Mazumdar, M.J.P.F.G. Monteiro, J. Ballot, H.M. Antia, S. Basu, G. Houdek, S. Mathur, M.S. Cunha, V. Silva Aguirre, R.A. García, D. Salabert, G.A. Verner, J. Christensen-Dalsgaard, T.S. Metcalfe, D.T. Sanderfer, S.E. Seader, J.C. Smith, W.J. Chaplin, Astrophys. J. **782**, 18 (2014)

A. McQuillan, T. Mazeh, S. Aigrain, Astrophys. J. Lett. **775**, 11 (2013)

T.S. Metcalfe, W.A. Dziembowski, P.G. Judge, M. Snow, Mon. Not. R. Astron. Soc. **379**, 16–20 (2007)

T.S. Metcalfe, S. Basu, T.J. Henry, D.R. Soderblom, P.G. Judge, M. Knölker, S. Mathur, M. Rempel, Astrophys. J. Lett. **723**, 213–217 (2010)

T.S. Metcalfe, W.J. Chaplin, T. Appourchaux, R.A. García, S. Basu, I. Brandão, O.L. Creevey, S. Deheuvels, G. Doğan, P. Eggenberger, C. Karoff, A. Miglio, D. Stello, M. Yıldız, Z. Çelik, H.M. Antia, O. Benomar, R. Howe, C. Régulo, D. Salabert, T. Stahn, T.R. Bedding, G.R. Davies, Y. Elsworth, L. Gizon, S. Hekker, S. Mathur, B. Mosser, S.T. Bryson, M.D. Still, J. Christensen-Dalsgaard, R.L. Gilliland, S.D. Kawaler, H. Kjeldsen, K.A. Ibrahim, T.C. Klaus, J. Li, Astrophys. J. Lett. **748**, 10 (2012)

T.S. Metcalfe, A.P. Buccino, B.P. Brown, S. Mathur, D.R. Soderblom, T.J. Henry, P.J.D. Mauas, R. Petrucci, J.C. Hall, S. Basu, Astrophys. J. Lett. **763**, 26 (2013)

E. Michel, A. Baglin, M. Auvergne, C. Catala, R. Samadi, F. Baudin, T. Appourchaux, C. Barban, W.W. Weiss, G. Berthomieu, P. Boumier, M.-A. Dupret, R.A. Garcia, M. Fridlund, R. Garrido, M.-J. Goupil, H. Kjeldsen, Y. Lebreton, B. Mosser, A. Grotsch-Noels, E. Janot-Pacheco, J. Provost, I.W. Roxburgh, A. Thoul, T. Toutain, D. Tiphène, S. Turck-Chieze, S.D. Vauclair, G.P. Vauclair, C. Aerts, G. Alecian, J. Ballot, S. Charpinet, A.-M. Hubert, F. Lignières, P. Mathias, M.J.P.F.G. Monteiro, C. Neiner, E. Poretti, J. Renan de Medeiros, I. Ribas, M.L. Rieutord, T.R. Cortés, K. Zwintz, Science **322**, 558 (2008)

F. Moreno-Insertis, S.K. Solanki, Mon. Not. R. Astron. Soc. **313**, 411–422 (2000)

B. Mosser, F. Baudin, A.F. Lanza, J.C. Hulot, C. Catala, A. Baglin, M. Auvergne, Astron. Astrophys. **506**, 245–254 (2009a)

B. Mosser, F. Baudin, A.F. Lanza, J.C. Hulot, C. Catala, A. Baglin, M. Auvergne, Astron. Astrophys. **506**, 245–254 (2009b)

M.B. Nielsen, L. Gizon, H. Schunker, C. Karoff, Astron. Astrophys. **557**, 10 (2013)

Y. Nishizawa, H. Shibahashi, in *Astronomical Society of the Pacific Conference Series*, vol. 76, ed. by R.K. Ulrich, E.J. Rhodes Jr., W. Dappen (1995), p. 280

R.W. Noyes, N.O. Weiss, A.H. Vaughan, Astrophys. J. **287**, 769–773 (1984)

K. Oláh, K.G. Strassmeier, Astron. Nachr. **323**, 361–366 (2002)

K. Oláh, Z. Kolláth, T. Granzer, K.G. Strassmeier, A.F. Lanza, S. Järvinen, H. Korhonen, S.L. Baliunas, W. Soon, S. Messina, G. Cutispoto, Astron. Astrophys. **501**, 703–713 (2009)

B.R. Radick, ed. by J.M. Pap, C. Frohlich, H.S. Hudson, S.K. Solanki (1994), p. 109

R.R. Radick, G.W. Lockwood, B.A. Skiff, S.L. Baliunas, Astrophys. J. Suppl. Ser. **118**, 239–258 (1998)

T. Reinhold, A. Reiners, G. Basri, Astron. Astrophys. **560**, 4 (2013)

S.H. Saar, A. Brandenburg, Astrophys. J. **524**, 295–310 (1999)

D. Salabert, C. Régulo, J. Ballot, R.A. García, S. Mathur, Astron. Astrophys. **530**, 127 (2011)

K. Strassmeier, A. Washuettl, T. Granzer, M. Scheck, M. Weber, Astron. Astrophys. Suppl. Ser. **142**, 275–311 (2000)

M.J. Thompson, Magnetohelioseismology. Philos. Trans. R. Soc. Lond. Ser. A, Math. Phys. Sci. **364**, 297–311 (2006). doi:10.1098/rsta.2005.1700

V. Van Eylen, M.N. Lund, V. Silva Aguirre, T. Arentoft, H. Kjeldsen, S. Albrecht, W.J. Chaplin, H. Isaacson, M.G. Pedersen, J. Jessen-Hansen, B. Tingley, J. Christensen-Dalsgaard, C. Aerts, T.L. Campante, S.T. Bryson, Astrophys. J. **782**, 14 (2014)

S.V. Vorontsov in *IAU Symposium*, vol. 123, ed. by J. Christensen-Dalsgaard, S. Frandsen (1988), p. 151

L.M. Walkowicz, G.S. Basri, Mon. Not. R. Astron. Soc. **436**, 1883–1895 (2013)

L.M. Walkowicz, G. Basri, J.A. Valenti, Astrophys. J. Suppl. Ser. **205**, 17 (2013)

O.C. Wilson, Astrophys. J. **226**, 379–396 (1978)

M.F. Woodard, R.W. Noyes, Nature **318**, 449–450 (1985)

DOI 10.1007/978-1-4939-2584-1_16
Reprinted from *Space Science Reviews* Journal, DOI 10.1007/s11214-014-0092-0

Observing Dynamos in Cool Stars

Z. Kővári · K. Oláh

Received: 7 April 2014 / Accepted: 19 August 2014 / Published online: 5 September 2014

Abstract The main aim of this paper is to introduce the most important observables that help us to investigate stellar dynamos and compare those to the modeling results. We give an overview of the available observational methods and data processing techniques that are suitable for such purposes, with touching upon examples of inadequate interpretations as well. Stellar observations are compared to the solar data in such a way, which ensures that the measurements are comparable in dimension, wavelength, and timescale. A brief outlook is given to the future plans and possibilities. A thorough review of this topic was published nearly a decade ago (Berdyugina in Living Rev. Sol. Phys. 2:8, 2005), now we focus on the experiences that have been gathered since that time.

Keywords Magnetic activity · Active stars · Stellar dynamo · Activity cycle · Differential rotation · Meridional flow

1 Introduction

For now it goes without saying, that, likewise sunspots, starspots are fingerprints of stellar magnetic fields. On the other hand, magnetic fields induce local and global brightness variability either on shorter or on longer terms. Therefore, observing starspots on cool stars is unavoidably important to get constraints for solar and stellar dynamo theory. We have learned that to a certain extent, solar magnetic activity can serve as a proxy of active stars of different types. However, we have already learned as well, that often the stellar observations cannot be understood within the solar dynamo theory. Therefore, it is also necessary to review the most important parameters, features or processes of the solar dynamo, of which stellar counterparts can be deduced from observing active stars (cf. Strassmeier 2005, 2009).

Z. Kővári (✉) · K. Oláh
Konkoly Observatory of the Hungarian Academy of Sciences, Konkoly Thege út 15-17, 1121, Budapest, Hungary
e-mail: zsolt.kovari@csfk.mta.hu

K. Oláh
e-mail: katalin.olah@csfk.mta.hu

Such experiences give base for studying how the observable phenomena or quantities are compatible with the solar and stellar dynamos and which parts of the theory require revision (cf. Kitchatinov and Rüdiger 2004).

So, what are the dynamo ingredients in cool stars that can be observed? A short list of the most important ones would include: (i) cycles of different types observed at different wavelengths; (ii) rotational period variations related to e.g., surface spot activity or differential rotation (iii) preferred spot locations, active longitudes; (iv) flip-flop phenomenon, flip-flop cycle; (v) local and global magnetic fields; (vi) local and global surface flows.

Note, that some of the listed stellar dynamo observables require long-term data acquisition, while others need the most advanced technologies and the most modern observing facilities. Individual objects can be studied with 0.5–2.0-m class robotic telescopes which by now fully replaced the manpower in systematic long-term data gathering. The continuous monitoring of a number of active stars has basically two useful timescales: the order of the rotation period and a much longer one covering the activity cycle(s). Long-term continuous photometry on the timescale of the activity cycles provides information on the overall brightness variability, i.e., on the overall spot coverage, therefore serving as comparison and constraints to various dynamo models. On the other hand, high-resolution spectroscopy on the timescale of a few rotations could provide us information on the surface differential rotation and also on other global and local surface flows. However, this demands the use of 2–4-m class telescopes.

In the recent years, ultra-high precision space-photometry has opened the perspective of statistical study of activity, but also has provoked the development of new data processing techniques as a consequence.

However, beside the precision, an important factor of studying active stars is time. Observing the long-term behavior of active stars, i.e., cycles, needs continuous monitoring for decades, which is not a fashionable program and does not need high-tech developments. On the contrary, the maintenance of the instruments as long as possible (and if necessary, a very careful replacement) is a key factor in getting homogeneous datasets. Only long and reliable datasets can constrain the dynamo models, which describe theoretically the magnetic behavior of the Sun and stars. The small, 1-m class telescopes have a key role in this problem, since the brighter stars are the ones which already have long records of observational data and the extension of these series is the most useful for studying stellar cycles.

2 Solar and Stellar Cycles

The Sun, because of its proximity, is an ideal target for studying magnetic cycles on different timescales and on different wavelengths. However, the adaptation of the knowledge deduced form the detailed and long-term solar observations to active stars has many problems and even could be misleading. The reasons are as follows: (i) magnetic activity is observed on very different kinds of stars concerning spectral type, mass and age, (ii) the lengths of the stellar datasets are much shorter than that of the Sun and do not exceed a few decades, and (iii) the datasets in most cases are different concerning frequency of the observations (time-base, regularity), the wavelength ranges, and the methods the data are gathered. In this section we compare the available information on the solar and stellar cycles, based on similar datasets.

2.1 Solar Cycles from 1-D Data—Observations

Sunspots are known for thousands of years, and the fact that their number on the solar surface is changing has been also recorded for a few hundred years already. This valuable dataset, and its extension in time by proxies are among the most important observational backgrounds in studying the solar dynamo. Unfortunately, stellar surfaces cannot be spatially resolved, and thus, deriving a quantity for spotted stars such as the Wolf-number for the Sun is not possible. Instead, fractional spot coverage (in percent of the total stellar surface) of some stars are followed for 1–2 decades at most—but no comparable measure in a long-term base exists on the Sun as to spot coverage. Sunspot numbers are thoroughly studied in other papers of this proceedings. We do not consider sunspot number showing the well-known cycles useful for a comparison to stellar cycles, except the nominal values of the cycle lengths.

Fortunately, long-term datasets exist which were obtained as measurements taken on the Sun as if it was a star. The most important and extremely valuable records are the 10.7 cm radio measurements taken each day for more than half a century. Similarly, the Geosynchronous Operational Environmental Satellite (GOES) provides X-ray flux of the Sun, continuously since 1994.

In one way we can use the measurements of the total solar irradiance, though, similar long-term data do not exist for stars. The spectral solar irradiance (Shapiro et al. 2011), which uses the minimum flux of the Sun combined with the sunspot numbers and proxies of the sunspot number, gives one average spectrum of the Sun each year for the past 400 years, between 130–1000 nm. For the earlier time, one average spectrum for every 22 years, i.e., one for every magnetic cycle, is calculated. The wavelength-flux tables allow to get similar magnitudes of the Sun than those used for stellar measurements. This way we are able to seek for cycles of the Sun e.g., in the usual Johnson B and V bandpasses established for stellar observations, and from the reconstruction any other special flux values can easily be calculated.

2.2 Stellar Cycles—Observations

The most important factor in studying activity cycles is the length of the available data. For the Sun, its activity is recorded for hundreds of years and this database is extended to a much longer timescale (millennia) using different proxies recorded on Earth which reflect the changes of the solar activity. In this review, for comparison purposes, we restrict the solar dataset to about a century, which is comparable to the longest available datasets for stars, using digitized plate collections (Digital Access to a Sky Century @ Harvard—DASCH). In Fig. 1 we show the lengths of the stellar cycles compared to the solar ones, and mark the maximum lengths of the available datasets from photometric monitoring and from the plate collections. It is evident, that, if we had data of the solar activity only as long as the longest stellar datasets (about 110 years), then, we could just guess that a longer solar cycle (i.e., the Gleissberg cycle) exists, but nothing would be known about the even longer timescale solar variation.

The first systematic search for cycles on stars which were suspected to have chromospheric activity observed as emission in their Ca II H&K lines, started in the late 1960's by Wilson (1968), and the effort, which lasted for several decades ceased, unfortunately, in 2003. The so-called Mt. Wilson database produced the first insight to stellar activity cycles and the measurements of that project are being used to date. Recording starspots through photometric data started in the mid 1960's, first just sporadic data have been gathered on a few interesting objects. The breakthrough was the advent of the automated photoelectric telescopes (APTs), which were specifically designed for following starspot activity on a long-term base. By now two such projects keep on getting multicolor data already for decades, the two Vienna-Potsdam APTs and the Four College Consortium APT. The lengths

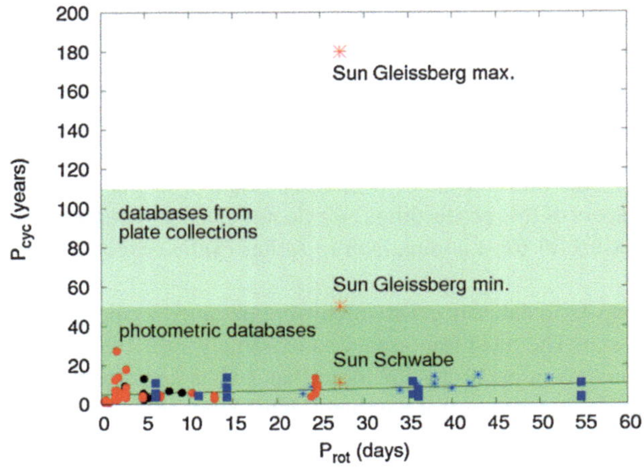

Fig. 1 The known cycle lengths of active stars in the function of their rotational periods, the solar Schwabe cycle, and the shortest and longest known values of the Gleissberg cycle. *Dark shaded area at the bottom* bounds the length of the available photometric databases, while the *lighter area above* marks the limits of the available photographic databases. *Different symbols* represent cycles from different publications. A rough relation between the rotational periods and all activity cycles of stars is also indicated. The cycle periods are from Oláh et al. (2009) and references therein

of the datasets from the Vienna-Potsdam APTs by now exceed 25 years. Observations taken in different bandpasses allow to estimate the average spot temperatures. Therefore, an important factor of using APT data for studying long-term behavior of stellar activity is the homogeneity of the data, which cannot easily be achieved by putting together measurements from different instruments.

Other presently running surveys, implemented not specifically for stellar activity research produce, however, useful information on the long-term behavior of active stars as well, now for more than a decade. One of these programs is the All Sky Automated Survey (ASAS, Pojmanski 2002), which monitors stars brighter than 14 magnitude all over the sky. Its public database provides systematic observations in V-band for many active stars. At present, the ASAS telescopes (both in the northern and southern hemispheres) are gathering data simultaneously in V and I bandpasses. The HATNet survey telescopes have originally been designed for searching extrasolar planets (Bakos et al. 2002), but also resulted in a large number of observations of active stars, and from that such statistical results followed like rotation-age-activity-mass relation (Hartmann et al. 2011). Details of the various automated telescopes used in stellar activity are found in Berdyugina (2005, Sect. 3).

The next step, about a decade ago, was the introduction of the automated telescopes constructed for collecting spectral observations of active stars on a long-term base, in order to get Doppler images (see Sect. 3.1.2 for details of Doppler imaging) of the stars regularly, and this way to find cycles in the spot coverage. Precise $v \sin i$ measurements are also gathered for orbital solutions with active stars in binary systems. The TSU 2-m Automatic Spectroscopic Telescope in Arizona started observing in 2004 while the 1.2-m STELLA-I in Tenerife in 2006. Since then, a number of publications have been based on data by these robotic facilities from the first decade of their operations, see, e.g., Fekel and Bolton (2007), Korhonen et al. (2009), Strassmeier et al. (2010, 2011, 2012), Fekel et al. (2013), etc.

However, in more extreme wavelengths like X-ray or radio, no systematic project exists for observing stellar activity, except the Sun. Still, long-term behavior of a few active stars can be studied by combining data from different space missions like *Einstein*, *ROSAT* and *XMM Newton*, in X-ray.

Recently, there is a new possibility to get stellar cycles through the systematic variation of the rotational periods due to differential rotation, similar to that observed on the Sun as butterfly diagram. The ultra-high precision and continuous datasets observed by the K1 mission of *Kepler* make this type of investigation possible. The drawback is the relatively short time-base though, i.e., only stars with very rapid rotation (in the order of 0.5 day) could be studied, since for these objects the shortest cycle is suspected to be in the order of 1 year.

Magnetic activity and its variability have an impact on pulsation frequencies of the Sun and stars. Promising results in the solar case have been published by Jiménez et al. (2011), showing clear correlation of the acoustic cutoff frequency and the solar activity cycle. Since the solar cycle is well followed, the observed relation is on firm grounds. Regarding the stars, from CoRoT data of the solar-like HD 49933 García et al. (2010) found correlation between the oscillation frequencies and mode amplitudes, and the luminosity changes due to magnetic activity.

2.3 Solar and Stellar Cycles—Results and Comparisons

When speaking about activity cycles one should keep in mind that the activity phenomena are changing in time but not absolutely regularly. This means that the cycle lengths are not periods but rather timescales, i.e., in the datasets no strict periods are found. Similarly, due to the differential rotation, which acts in most types of active stars with various strengths, the stellar rotational periods derived from photometry can be different on the same star, depending on the latitudes populated by the star spots. Thus, in case of active stars, the conventional methods for period search designed for strictly periodic signals should be applied with caution.

Throughout this paper, solar activity cycle, in short, solar cycle, is considered with a length of about 11 years. This triviality is mentioned because solar cycle is also understood as magnetic cycle with double length. But at present, there is no observational evidence of stellar magnetic cycles from magnetic field measurements, thus only cycles from photometry, fluxes in other wavelengths, and spectral indices of the Sun and stars, are comparable.

2.3.1 Solar-Type Stars

The 11-year Schwabe cycle of the Sun is well followed in various wavelengths. Figure 2 shows the variability of the solar X-ray flux during the last 20 years. Apart from the strong short-term variability due to the rotation and flare activity, the 11-year long cycle in the mean values is axiomatical.

AB Dor is a nearby ($d = 15$ pc) bright ($V = 6.75$) quadruple system, of which the brightest component AB Dor A is an active K0-dwarf. It is an ideal target of observations in many wavelengths; more than a decade long data are available in the 0.3–2.5 keV X-ray bandpass (Lalitha and Schmitt 2013). In Fig. 3 the X-ray behaviour of AB Dor A is seen between 2000–2010, showing large fluctuations due to flares and a marginal long-term variation, still resembling of cyclic behaviour.

The solar X-ray luminosity changes between $\log L_X = 26.8$ and $\log L_X = 27.9$ during a solar cycle in the *ROSAT* bandpass of 0.1–2.4 keV, between cycle minimum and maximum (Judge et al. 2003), i.e., it changes one order of a magnitude. AB Dor A shows changes between $\log L_X = 29.8$–30.2 in a similar bandpass of 0.3–2.5 keV, which is on average two magnitudes stronger, but has a much lower amplitude than that of the Sun.

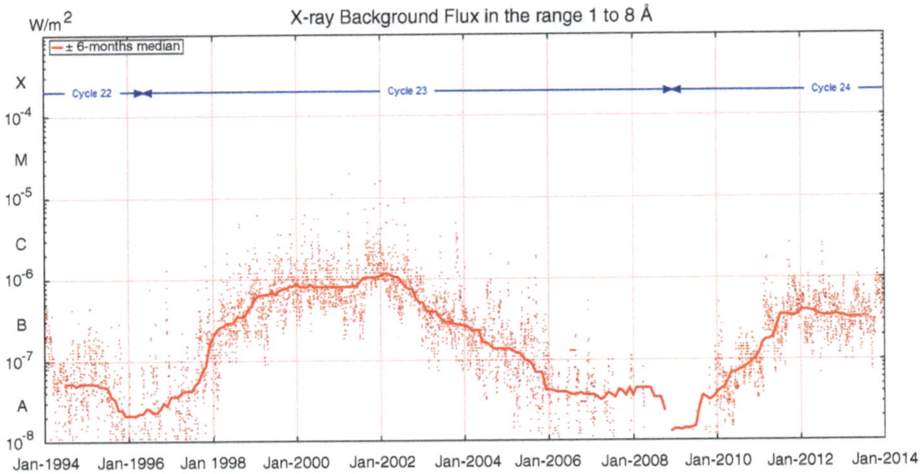

Fig. 2 Solar X-ray flux between 1–8 Å from GOES data (http://sidstation.loudet.org/solar-activity-en.xhtml)

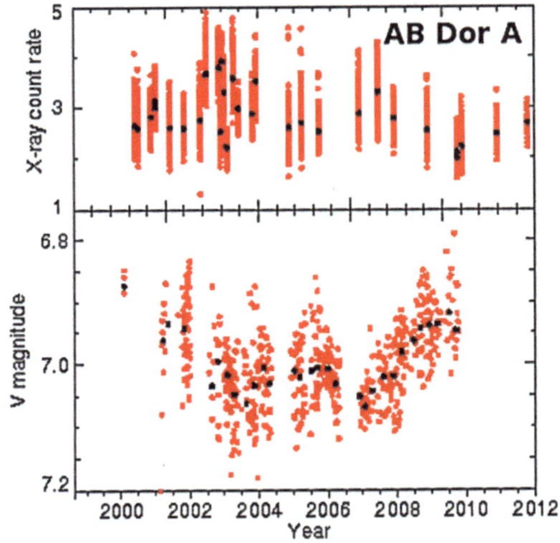

Fig. 3 AB Dor A X-ray luminosity between 0.3–2.5 keV (*top*) together with *V* magnitude (*bottom*) from Lalitha and Schmitt (2013). A marginal long-term change in the X-ray count rate is seen, resembling to a cyclic behaviour

For the Sun it is widely accepted that it becomes bluer when more active, i.e., at spot maximum the Sun radiates more than at minimum, which is obvious from the irradiance measurements. However, no long-term irradiance measurements exist on any active star, therefore no direct comparison is possible. In Shapiro et al. (2011) yearly mean values are deduced from the spectral reconstruction in *B* and *V* bandpasses (broadband *B*-filter is at 445 nm, *V*-filter is at 551 nm effective wavelengths, with FWHM of about 90 nm), and from these one can get similar brightness-color index diagrams for the Sun as those of Messina (2008) for stars. From the reconstructed *B* and *V* colors it is found, that the Sun becomes redder when fainter, although with very small amplitude (Oláh et al. 2012a, 2012b). This unsuspected result, however, is in accordance with Preminger et al.'s (2011) finding that the total solar *visible* continuum brightness is decreasing, when the solar activity is increasing.

Activity cycles on solar-type stars from Ca-index measurements were published by Baliunas et al. (1996), first time describing a relation between the rotational and cycle periods,

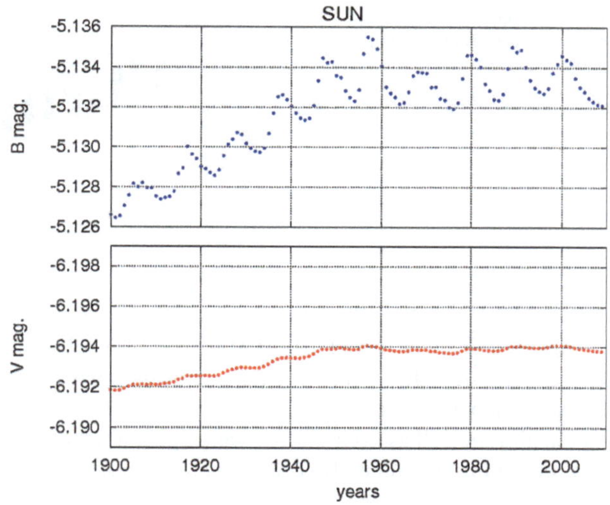

Fig. 4 The brightness of the Sun in Johnson B and V bands, for about a century, based on Shapiro et al. (2011) reconstruction. A long-term change is well seen (Gleissberg cycle), the 11-yr cycle is evident in B color with a small amplitude, and marginal in V. Magnitude scales are arbitrary without zero point

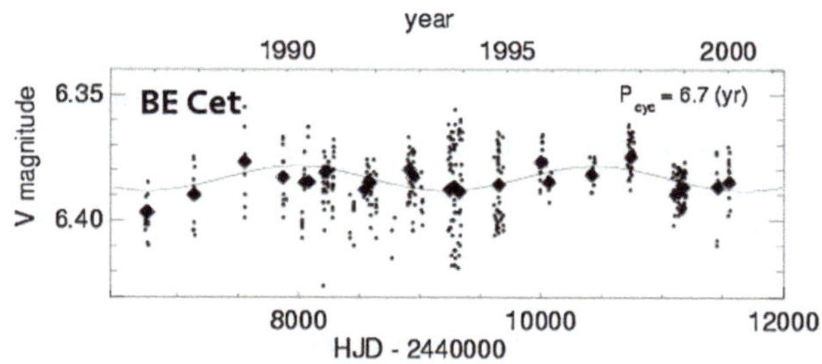

Fig. 5 Long-term variability of the solar-like star BE Cet. *Sinusoidal* fit to the data suggests a cycle length of 6.7 years (Messina and Guinan 2002)

and connecting this relation to the stellar dynamo in the seminal paper "*A Dynamo Interpretation of Stellar Activity Cycles*". The dataset describes the variation of the stellar chromospheres. Using contemporaneous Ca-index and photometric data Radick et al. (1998) found a strong connection between the behavior of the photospheres and chromospheres of active stars, showing direct or anticorrelation, thereby differentiating between the spot- and plage-dominated activity. Messina (2008) gave long-term *UBV* photometry for 14 well-known active stars studying their color index behavior as a function of brightness. The results revealed that part of the stars are spot-dominated while the rest show photospheric faculae dominance.

Figure 4 shows the cyclic behavior of the Sun in B and V colors (Shapiro et al. 2011), for the last century only. As seen from Fig. 4, in B color the Schwabe cycle is evident and a longer-term change (part of a Gleissberg cycle) is also seen, though the amplitude is small. In V color the amplitude is even smaller, the 11-year cycle is barely seen, but the longer term variation is clear. Note that the Sun shows a very marginal light variation in V color, about 0.002 mag in the course of its cycles.

BE Cet is a single, relatively young (cca. 600 Myr, member of the Hyades moving group) solar-type star (G2V class) with a rotational period of 7.76 days, i.e., a solar analog in the sense the Sun in its youth could have been like this active star. BE Cet shows rotational modulation due to starspots and long-term cyclic change of about 6.7 years with a low amplitude of about 0.02 mag. (see Fig. 5 and Messina and Guinan 2002) which is still an order higher than the solar amplitude.

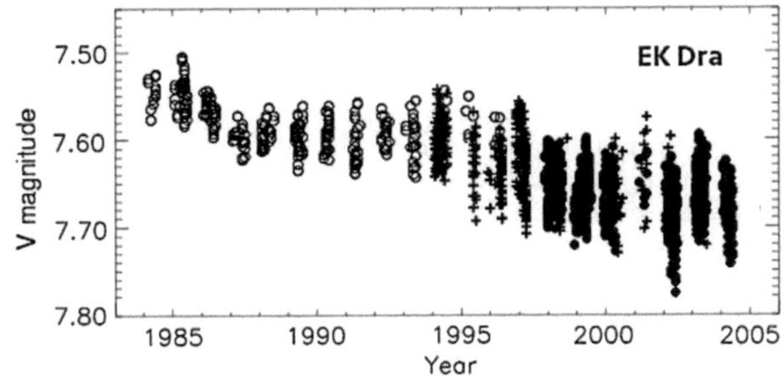

Fig. 6 Long-term brightness change of EK Dra (Järvinen et al. 2007)

An even younger (about 10–30 Myr old only) solar analog star is EK Dra (G1.5V), an effectively single dwarf (its M-dwarf companion has an orbital period of 45 years), which has a rotational period of 2.6 days, i.e., rotating 10 times faster than the Sun. Its rotational modulation as well as cycles have much higher amplitudes compared to BE Cet, amounting to a few tenths of magnitudes according to Järvinen et al. (2007); see also Fig. 6.

The rotational rate plays a basic role in the strength of the magnetic activity. Amplitudes of the cyclic changes are much higher for stars rotating faster. (However, one should not forget that the *observed* amplitude of the rotational modulation depends on the inclination angle as well.) The different rotational modulation and cycle amplitudes of the very similar spectral type stars BE Cet, EK Dra and the Sun reflect their rotational rates and ages, since the rotational rate decreases due to magnetic braking as the star evolves.

2.3.2 Stars of Different Spectral Types

Activity cycles are observed on many different kinds of stars: singles and binaries, dwarfs and giants, from F to M-type stars. Do we see differences between the cycles of these stars? How the cycles behave in time? It is known for a long time, that the length of solar cycles are variable, e.g., the Schwabe cycle varies between about 9–14 years and the Gleissberg cycle also varies continuously (Kolláth and Oláh 2009). To study the temporal behavior of the stellar cycles we need as long datasets as possible, but unfortunately, at present even a century long dataset of an active star is rare. Different types of time-frequency analysis tools are used to study the temporal variation of the activity cycles. However, time-frequency analysis requires continuous and equidistant datasets. The stellar long-term photometric records are biased (i) by the visibility of the target from a given location, (ii) by the effects of the weather conditions, and finally, (iii) by the distortion of the rotational modulation in case of long periods and not enough densely sampled observations. All these are the features of the ground-based observations. On the other hand, space-born data are more precise, can be continuous, equidistant and are barely subject of visibility. But their available time-base is still short compared to ground-based datasets, therefore only the shortest cycles of short period stars can be investigated from them.

In Kolláth and Oláh (2009) the time-frequency method of studying ill-sampled datasets is thoroughly discussed and a recipe was given how to overcome the problems. Briefly: first the rotational modulation is removed from the data, after that an appropriate spline interpolation is applied to get rid of the yearly gaps. The result of this procedure is a dataset for which time-frequency methods can be applied.

Based on the method given by Kolláth and Oláh (2009), in Fig. 7 we now can compare the long-term photometric variability of HK Lac, a K0-giant active primary of a close binary system, with a rotational (and orbital) period of 24.5 days, and the cyclic behavior of LQ Hya, a single active K2-dwarf, rotating with 1.6 days. The light variation of the two objects

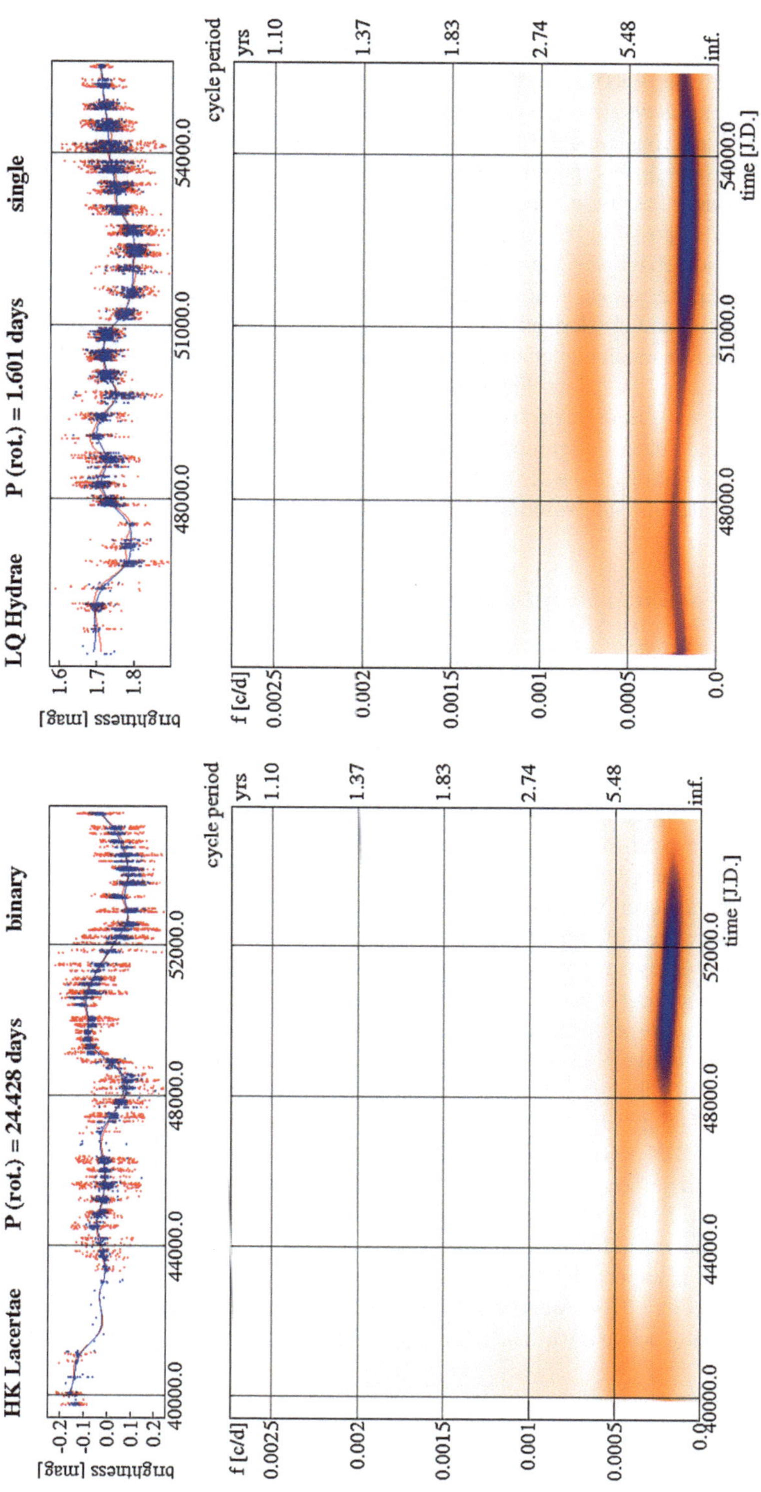

Fig. 7 Time-frequency diagram from short-term Fourier-transform for the K0-giant HK Lac in a binary (*left*), and LQ Hya, a fast rotating single K2-dwarf (*right*). Based on Oláh et al. (2012a, 2012b)

look very similar, decades-long variations are seen, as well as modulations on the timescale of a few years. Together with the previous examples of the Sun, solar analogs and different types of active stars, we find that the activity cycles are generally *variable* (no strict periods, only timescales can be determined), and *multiple*. In case of active stars the lengths of the databases are still not long enough to determine the longer cycle lengths in most cases.

Luckily, for a few objects, putting together the photographic and photoelectric datasets we could have longer time base. One of such examples is V833 Tau, a K2-dwarf in a close binary system with a possibly brown dwarf companion (Cuntz et al. 2000). In the left panel of Fig. 8 the cyclic behavior of the primary star is seen from a 20 years long photometric dataset. The right panel shows almost a century-long photographic dataset plus the last 20 years of photometry, and the derived long-term changes. The photographic data are sparse, but they clearly show a much higher amplitude light variation than that during the 20 years of photometry. Because of the low inclination ($\approx 20°$) of the system, the rotational modulation due to spots have small amplitude and does not distort the sparse long-term data significantly when observing the star at different phases of the rotation. The 2–3, 5, 27–30 years and even longer term variations show the multicyclic nature of the brightness variability, which, on shorter timescales though, resembles of the solar cycles with 11, 60–120 and longer timescales.

2.3.3 *Uncovering Stellar Butterfly Diagrams*

Solar butterfly diagram is one of the most striking examples of cyclic behavior of solar activity. During the 11-year sunspot cycle, the activity wave migrates from mid-latitudes to the equator, giving scope for observing the photospheric manifestation of the dynamo action underneath. The limited spatial resolution of stellar observations makes more difficult to compile such diagrams for spotted stars. In long-term photometric observations of a spotted star, a latitudinally migrating activity belt *together* with surface differential rotation would result in a changing peak-to-peak amplitude of the light curves. Katsova et al. (2003) and Livshits et al. (2003) suggested a model to construct stellar butterfly diagram from long-term photometric data, however, with rigorous a priori assumptions on the spot distribution. On the other hand, once the differential rotation is known (e.g., from Doppler imaging), results from light curve inversions for an extended period can be suitable to recover stellar butterfly diagram without involving any assumption on spot geometry (Berdyugina and Henry 2007).

Within the four years of the original *Kepler* mission (K1) ended in May 2013, a number of ultrafast-rotating active K–M-dwarfs were monitored. For such stars with $P_{\rm rot}$ of ≈ 0.5 days Vida et al. (2013) found cycle lengths in the order of 400 days, i.e., already within the reach of *Kepler* (cf. Fig. 10). According to the model calculations for a K0-dwarf with a rotational period of 2 days (Işık et al. 2011) the expected butterfly diagram will dramatically change, compared to the solar case, due to the high latitude spottedness and the much thinner ($\approx 10°$) activity belt. Still, with enough strong surface differential rotation, a small modulation of the rotational period will occur during the activity cycle. The typical amplitude of such a modulation is expectedly large enough to be detected from high-precision *Kepler* photometry. Vida et al. (2014) found this kind of behavior in one-fourth of their sample of 39 single fast-rotating active dwarfs (see Fig. 9).

2.3.4 *Cycles and the Dynamo*

The basic goal to derive activity cycles is to study the behavior of the underlying dynamo. Baliunas et al. (1996) revealed the first time the relationship between the cycle length and the

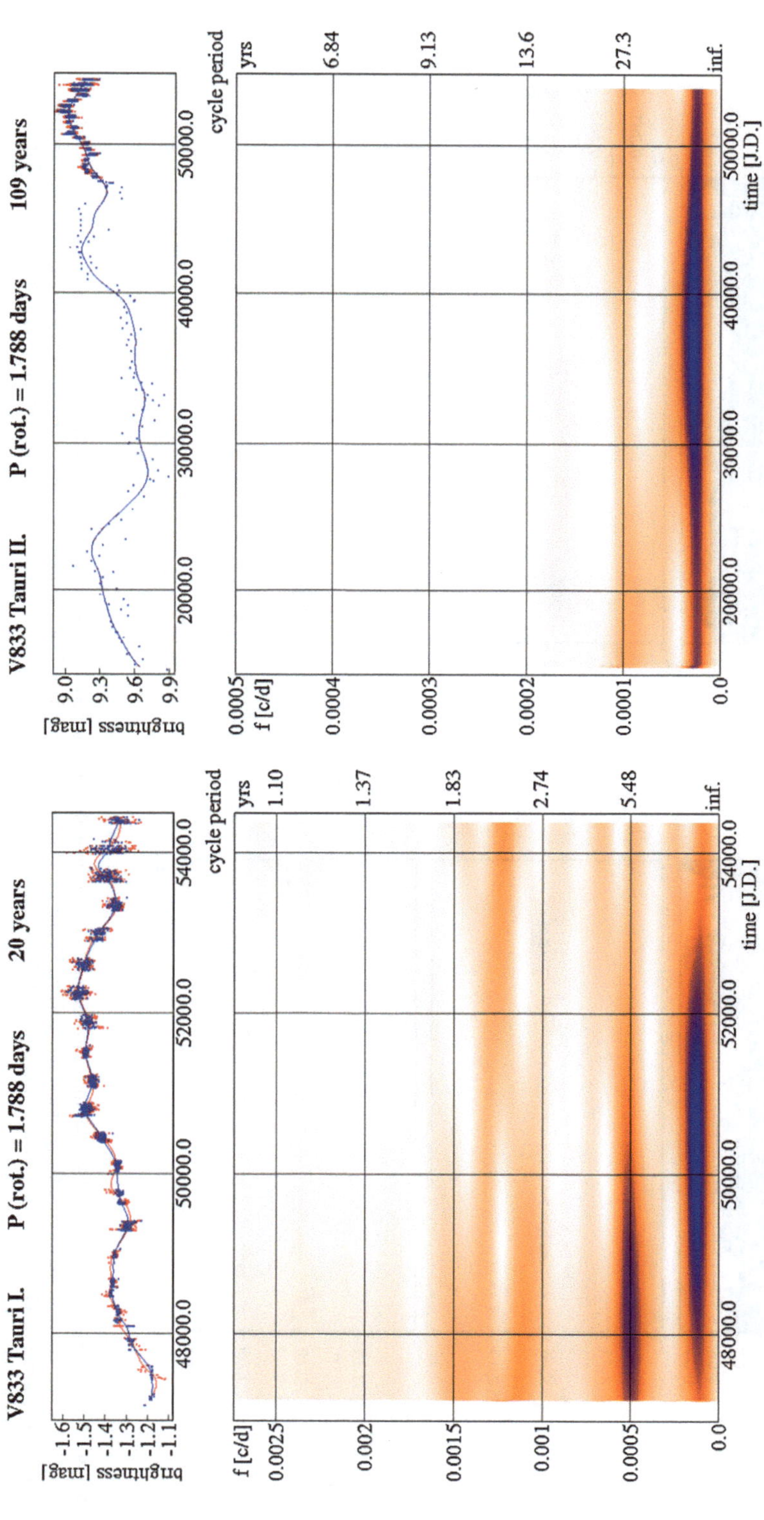

Fig. 8 Cycles for V833 Tau: short-term Fourier-transform was applied for the 20 years long photometric observations (*left*), and for a larger time base extended with photographic data (*right*), based on Oláh et al. (2009)

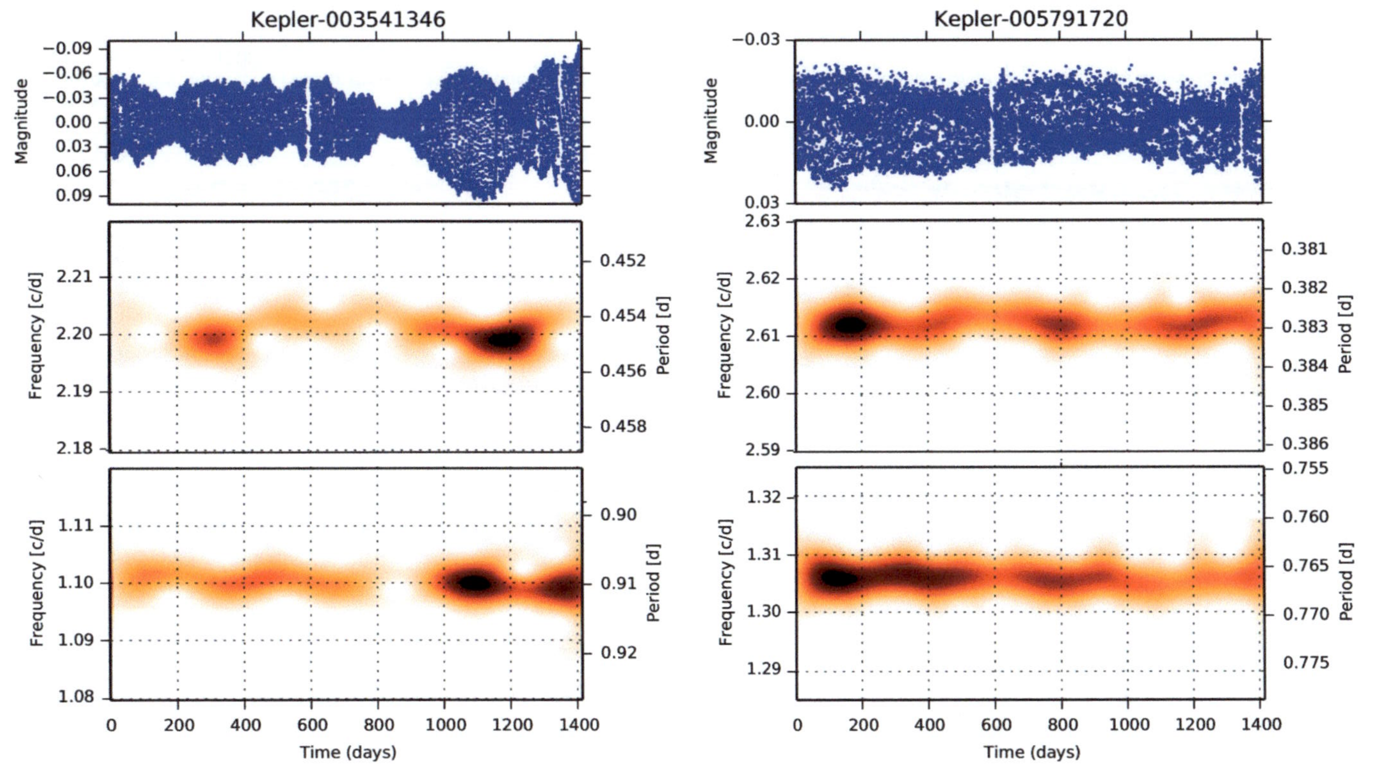

Fig. 9 Two examples from Vida et al. (2014) for quasi-periodic modulation in the rotation period of fast-rotating active dwarfs (*left*: KIC 03541346, $P_{\rm rot} = 0.9082$ days, *right*: KIC 05791720, $P_{\rm rot} = 0.7651$ days). Cleaned and interpolated Kepler light curves (*top panels*) and their short-term Fourier-transforms at the double rotation frequency (*middle panels*) and at the rotation frequency itself (*bottom panels*). Modulations indicate roughly 330(50) days and 410(50) days long activity cycles for KIC 03541346 and KIC 05791720, respectively

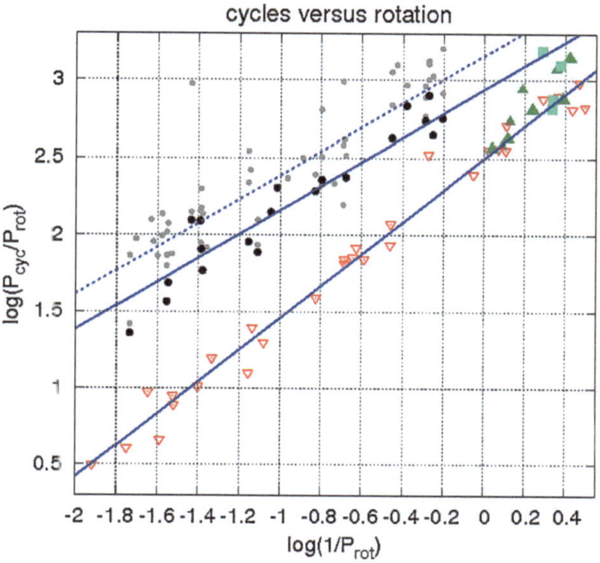

Fig. 10 Relationship between rotation and cycle lengths of active stars of different types. *Black dots* stand for the shortest cycle lengths from the multiple cycles of the same star, *grey dots* show the longer cycles from Oláh et al. (2009). *Filled squares* and *triangles* represent K–M dwarf stars from Vida et al. (2013) and Vida et al. (2014). *Dotted line* is the fit for all of the cycles, while the *upper solid line* is the fit for the shortest cycles only, excluding all the M-dwarfs in both fits. *Open triangles* are cycles derived for the M-dwarf sample by Savanov (2012). *Bottom line* is the fit to Savanov's (2012) sample. See the text for further explanation

rotational period which is proportional to the dynamo number, i.e., $P_{\rm cyc}/P_{\rm rot} \sim D^{\iota}$, where $\iota \geq 1/3$. Later, Oláh et al. (2009) refined this relation using long-term datasets of 20 active stars of different types (dwarfs, giants, singles and binaries). Recently, a number of cycles have been derived for fast-rotating K and M dwarf stars with rotational periods less than 1 day, which populate the short period end of the relation (Vida et al. 2014). Except for M-dwarfs, ι is about 0.8, showing that longer period stars have longer cycles. For the whole sample of M-dwarfs the slope of the fit between $\log(P_{\rm cyc}/P_{\rm rot})$ and $\log(1/P_{\rm rot})$ is about unity (see Fig. 10, and cf. Savanov 2012), i.e., the cycle lengths for M-dwarfs do not depend on the rotational period. This fact implies that the magnetic activity of M-dwarfs is driven by a different kind of dynamo (likely α^2), while the majority of the G–K stars (dwarfs and giants) are thought to harbor $\alpha\Omega$ type dynamos.

Some attempts have already been made to model stellar dynamos with oscillating nature, based on a solar-type model. Işık et al. (2011) used an $\alpha\Omega$ type dynamo following the rise of flux tubes through the convection zone, which resulted in a consistent determination of the emergence latitudes and tilt angles, and added horizontal flux transport at the surface. With the appropriate changes of the input model parameters butterfly diagrams with their cycles could be determined for rapidly rotating dwarf and subgiant stars, as well as for the Sun. It is worth to mention that a model dynamo with the parameters of the K1-subgiant V711 Tau ($P_{\rm rot} = 2.8$ days) shows well-defined cycles (Işık et al. 2011, Fig. 12). Calculating the model for a time base equal to the observations of V711 Tau, double cycles are found on timescales of 4–5 years and 10–20 years, which were similar to the results from observations. Although the model gives the total unsigned magnetic flux and not the brightness, the magnetic flux should be strongly correlated with the activity level, thus the brightness of the star (Işık 2012). Important factor is, that concerning the modeling, the shorter cycle originates simply from the dynamo action in the stellar interior while the long term cycle-like feature is due to stochastic flux emergence in the model. This result may shed some light on the very important question: from the *observed* multiple cycles which one comes from the dynamo action and which one(s) is (are) due to other effects.

Just recently, from modeling activity cycle lengths Dubé and Charbonneau (2013) have found a relation between the rotational and cycle periods, showing that the existence of such a connection is quite robust. However, their relation showed that stars with longer rotational periods had shorter cycles, in contradiction with quite a few results (e.g., Oláh et al. 2009; Vida et al. 2014).

3 Observing the Dynamo by Surface Reconstruction

Imaging of stellar surface structures is a commonly used technique to observe ingredients of the underlying magnetic dynamo. Sunspots are deeply anchored into the surrounding plasma, they follow the large scale plasma flows due to confinement, thus providing diagnostic marks on the surface differential rotation (Paternò 2010) and maybe other large scale velocity fields. We can see sunspots following also local flow fields such as, e.g., convective flows, the geostrophic flow (Spruit 2003) or performing asymmetric proper motions due the tilt of the emerging flux (van Driel-Gesztelyi 1997), which, on the other hand, can totally overwhelm flows on larger scales. That is why the surface flow pattern of the solar meridional circulation (Zhao et al. 2013) cannot be observed easily from tracking sunspots (Wöhl 2002, but cf. e.g., with Komm et al. 1993).

We expect that positions, extensions and motions of starspots on cool stars will also provide information on the characteristics of the underlying dynamo. So, for want of better, imaging and, if possible, tracing stellar surface structures on magnetically active stars are essential tools to observe stellar dynamo ingredients. However, compared to the Sun, direct imaging of the stellar surfaces are not possible. Thus, reconstructions should rely on indirect inversions, such like Doppler imaging and Zeeman-Doppler imaging (see Sect. 3.1.2). But the time resolution of such inversions is necessarily in the order of the rotation period of the star and surface variability on a much shorter term will just increase the noise. Nevertheless, during the past decades the observing facilities and the inversion techniques underwent a significant development and stellar surfaces can be studied in even more detail.

In this section we overview the most advanced techniques for magnetic activity research, with flashing some of the most important results obtained with them so far.

3.1 Inversion Techniques

3.1.1 Photometric Modeling

Since the mid 70's different photometric spot models have been developed to study starspot-distorted light curves of active stars. The observed rotational variability has been modeled from 1-D photometric information by spot modeling techniques of different approaches. The basic idea of such techniques is to find a geometrical model (i.e., surface spot distribution) to explain the rotational modulation of the observed light curve. These techniques, either analytic or numeric, apply various fitting algorithms to iteratively minimize the residuals between the observed and the theoretical light curves (see e.g. Budding 1977; Kang and Wilson 1989; Eker 1994, etc.).

With the considerable computational advances from the 1990's, such innovations came to the front like the time-series spot model using the photometric time series to derive a consistently evolving spot model (e.g., Strassmeier and Bopp 1992; Oláh et al. 1997; Ribárik et al. 2003); models assuming spots with two temperature components, i.e. spots with umbral and penumbral parts as seen on the Sun (Amado et al. 2000); and models applying randomized multispot solutions (Eaton et al. 1996). This latter provided a new scope of involving differential rotation, as well as handling the inconsistency between numerous

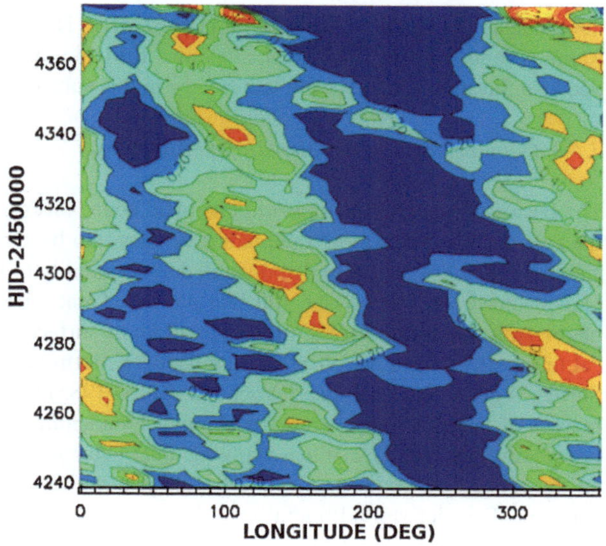

Fig. 11 *Light curve* inversions of the spotted surface of CoRoT-2a from Lanza et al. (2009). The *plot* shows the distribution of the spot covering factor f vs. longitude and time. *Yellowish regions* indicate maximum spot filling factor, *dark blue regions* indicate minimum coverage. Note the active regions forming, developing and fading away, and after their formation they perform a retrograde migration

smaller surface structures recovered by Doppler imaging (Sect. 3.2.1) and the traditional two- or three-spot solutions. However, due to the dramatic rise of free parameters, results of such multispot solutions could only be assessed statistically.

The stability of the different spot solutions has been a real problem since the advent of these kinds of techniques. In so far, the information content of a 1-D time series is obviously limited, and using photometry alone for spot modeling is an ill-posed problem. Kővári and Bartus (1997) have demonstrated that virtually, a simplified two-spot model could provide sufficiently good fit for more complex multiple spot configurations. However, even if having additional spatial information on the spot distribution (for instance, when the inclination of the rotation axis is known, or the spotted star is a component of an eclipsing binary system and spot eclipses occur), the uniqueness of the recovered spot solution is seriously restricted by the data quality. Studies concerning this issue have been carried out by e.g. Vogt (1981), Strassmeier (1988), Rhodes et. al (1990), Eker (1995), Kővári and Bartus (1997), etc.

To overcome the uniqueness and stability problems, Lanza et al. (1998) introduced a regularizing function into the inversion process. According to their spot model, the surface map is built up from a given set of small surface elements and the surface intensity of every pixel is a function of its spot filling factor (f_i). As a priori assumption, the surface elements have to fulfill either the maximum entropy criterion (e.g., Nityananda and Narayan 1982) or the Tikhonov criterion (Tikhonov and Goncharsky 1987). The application of such a regularization proved to be very useful in modeling light curves (e.g., Roettenbacher et al. 2013), as well as in spectroscopic Doppler imaging techniques.

Just recently, Walkowicz et al. (2013) performed extended numerical experiments to assess degeneracies in models of spotted light curves. They confirmed that in the absence of additional constraints on the stellar inclination (e.g., $v \sin i$ measurements or occultations of starspots by planetary transits) spot latitudes could not be determined, not even when having ultra-high precision (i.e. *Kepler*) photometry. According to their experience, from spot modeling of stars with different rotation rates, subtle signatures of differential rotation can be measured, which may provide information on the distribution of spots; see Fig. 11 for an example. Also, the outstanding quality of space-photometry from *MOST*, *CoRoT* and *Kepler* missions compelled the development of new data processing techniques (e.g., in Lanza et al.

2009; Pagano et al. 2009; Bonomo and Lanza 2012; Herrero et al. 2013; García et al. 2013; Reinhold and Reiners 2013, etc.), since the available methods were not suitable anymore. See also Savanov's (2013) excellent review in this subject.

3.1.2 Doppler Imaging, Zeeman-Doppler Imaging

The basics of mapping nonuniform chemical abundances of a stellar surface from spectral lines were formulated more than half a century ago by Deutsch (1958). Later, model calculations (Khokhlova 1976) demonstrated the capability of investigating the spotted surface of Ap stars from a set of spectral line profiles. Goncharskii et al. (1977, 1982) proposed mathematical solution for the inverse problem of surface mapping by reconstruction of local line profiles from the observational data. The term "Doppler imaging" was initiated by Steven Vogt and his colleagues (Vogt and Penrod 1983) for their technique to map the active regions of late-type stars, when the Doppler broadening due to rotation surpasses all other broadening mechanisms. When applying a cool star spot on the surface, a nub will appear on the broadened photospheric line profile just at the wavelength that corresponds to the apparent (Doppler-shifted) radial velocity of the spot location at the disk. Namely, a high-resolution spectral line can be regarded as a 1-D snapshot of the 2-D surface. Thus, a suitable set of spectra collected at different rotational phases can be assembled into a reconstructed surface image. Vogt et al. (1987) demonstrated through tests that their improved Doppler imaging technique was able to readily recover surface structures of 15° angular size. For today, with the application of the most advanced observational and computing techniques, this theoretical resolution has increased up to 5°, i.e., the size of a large sunspot.

Beside the success in applying Doppler imaging for an ever growing number of stars one should not disregard the limitations of its scope. First of all, the rotational broadening should be dominant among possible line broadening mechanisms. Moreover, the projected equatorial rotational velocity $v \sin i$ should exceed 20 $\mathrm{km\,s^{-1}}$. However, above a certain limit around $v \sin i = 100\ \mathrm{km\,s^{-1}}$ the profiles become shallow and the line distortions due to spots (typically 1 % of the continuum) cannot easily be observed (cf. Vogt 1988). Among cool stars, principally, two groups fulfill this criterion: fast rotator PMS and MS dwarfs with P_{rot} of the order of a few days, and moderately rotating (P_{rot} around 10–20 days) subgiants or giants, which are often members of RS CVn-type close binary systems, that help maintain relatively fast rotation by synchronization. Due to the strong relation between rotation and magnetic field generation, its manifestations in dark spots are expected to be less characteristic for slower rotators. Note though, that surfaces of cool MS stars rotating at the solar rate cannot be resolved by the Doppler technique due to the inefficient rotational broadening.

Magnetic surface structures can be investigated by Zeeman-Doppler imaging (hereafter ZDI, Semel 1989; Donati et al. 1989), which works in a similar way as described before, but uses polarization signals of the Stokes profiles to recover the magnetic field geometry. Originally, ZDI was based on the weak field approximation, i.e., the circular polarization signal by the Zeeman effect (Stokes V) is assumed to be proportional to the first derivative of the unpolarized intensity (Stokes I) signal. But detecting Stokes parameters in atomic spectra is complicated, since the expected signatures are far below the noise level. Thence, signal-to-noise ratio of the polarization signals were generally boosted by the multi-line least squares deconvolution technique (e.g., Donati et al. 1997, see also Tkachenko et al. 2013 for a recent improvement). If all four Stokes signals are available, ideally, the surface magnetic field distribution can be reconstructed (Piskunov 1985). The linear polarization profile (Stokes Q and U) signatures, however, are even smaller and more obscured,

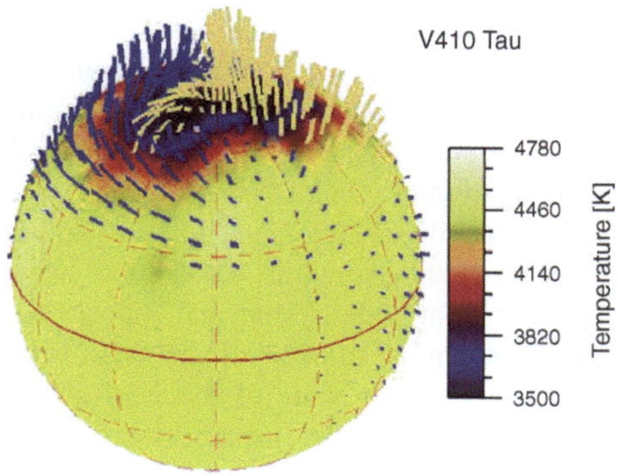

Fig. 12 Zeeman Doppler image of the "young Sun" V410 Tau obtained with iMap (after Carroll et al. 2012) reveals an S-shaped twisted bipolar field geometry over the cool polar spot. *Blue field lines* are of negative polarity, while *pale yellow lines* are of positive. Vector lengths are proportional to the magnetic field strength of up to ±1.9 kG. *Color background* on the surface indicates temperature distribution according to the colorbar

therefore only Stokes V and I were used from the beginning (e.g., Brown et al. 1991; Donati and Brown 1997). In a sense, with this reduction ZDI was an *inherently undetermined* problem, which necessarily anticipated the question of reliability (e.g., Wehlau and Rice 1993; Berdyugina 2005; Rosén and Kochukhov 2012, etc.). Attempts, however has already been made to overcome some of the difficulties regarding the interpretability, e.g., by restricting to certain field configurations (e.g., Donati and Brown 1997; Hussain et al. 1998, 2001; Piskunov and Kochukhov 2002). In the last few years new horizons were opened up with high-resolution spectropolarimetric observations (HARPSPol, ESPaDOnS, Narval) and eventually the next-generation of magnetic Doppler imaging is prepared to be suited for full reconstruction of the surface magnetic topology (e.g., Rusomarov et al. 2013; Silvester et al. 2014).

Before long, the ultra-high resolution spectropolarimeter PEPSI@Large Binocular Telescope (Strassmeier et al. 2007) will provide the capability of observing all four Stokes signatures for a considerably large sample of late-type stars. For these upcoming data a new Zeeman-Doppler reconstruction technique is developed (Carroll et al. 2007, 2012); see Fig. 12. Through iterative regularization, this new ZDI code iMap solves the full polarized radiative transfer using an artificial neural network (Carroll et al. 2008). Indeed, iMap performs simultaneous reconstructions for the radial, azimuthal and meridional magnetic fields, as well as the surface temperature distribution map. This development provides further improvements regarding the overall scope and reliability of ZDI, since ignoring temperature inhomogeneities yields unreliable magnetic field reconstruction (Rosén and Kochukhov 2012).

3.2 Basic Results from Photometry

3.2.1 Active Longitudes

Solar observations revealed that longitudinal concentrations of various magnetic activity indicators existed from sunspots through chromospheric magnetic fields and flares, extending to coronal streamers and heliospheric fields. Statistical analyses of this kind of phenomena were carried out by several authors (e.g., Bumba and Howard 1969; Bumba et al. 2000; Berdyugina and Usoskin 2003; Kitchatinov and Olemskoy 2005; Li 2011, etc.). Albeit having important consequences for the dynamo regime, the grounds of the formation of such non-axisymmetry of solar activity is not clearly understood; but see, e.g., Moss et al. (1995), Ruzmaikin (1998), Weber et al. (2013), etc., for some of the possible explanations.

Long-lived active longitudes are known also from stellar observations (see, e.g., Oláh et al. 1991; Berdyugina and Tuominen 1998; Rodonò et al. 2000; Korhonen et al. 2002;

Järvinen et al. 2005; Lanza et al. 2009; Lehtinen et al. 2011, etc.). Often, two permanent activity centers are detected, being separated longitudinally by half of the rotation phase on average (see also the "flip-flop" phenomenon in Sect. 3.2.2). Active stars in RS CVn-type close binary systems often show active longitudes that are locked in the frame of the revolving system (Oláh 2006). In these cases tidal forces and/or magnetic interaction are expected to play a significant role in locking the activity centers at preferred longitudes, which are often quadratures (Oláh 2006; Kővári et al. 2007a; Korhonen et al. 2010) or at the substellar point of the system and opposite to it (Oláh 2006 and references therein).

From the point of view of the mean-field dynamo theory, the existence of active longitudes requires breaking of the axial symmetry. Ruzmaikin (1998) suggested that non-axisymmetric mean field at the base of the convection zone could produce clustering of activity at certain longitudes. In the rapid rotation regime non-axisymmetric non-stationary mode can be excited (Moss et al. 1995; Moss 2004). Indeed, active longitudes are expected to occur quite widely (Moss 2005). On the other hand, in this regime the role of differential rotation is expected to be less significant. For close and contact binaries an extension of the mean field dynamo model is developed (Moss and Tuominen 1997) which can explain some of the observed features reviewed e.g. in Oláh (2006). New results for purely α^2 type oscillatory solution have already been presented (Mantere et al. 2013). Recently, employing a thin flux tube model subject to a turbulent solar-like convective velocity field Weber et al. (2013) have found that large-scale flux emergence patterns exhibit active longitude-like behavior.

The appearance of active longitudes in close binary systems was modeled in detail by Holzwarth and Schüssler (2003). Their main conclusion was, that the erupting flux tubes were considerably affected by the companion star, depending on the tidal forces and the non-sphericity of the active star. The depth of the convection zone, the initial latitude of the flux tube in the bottom of the convection zone and the strength of the magnetic field together determine the place of the eruption. Details of all parameters are thoroughly discussed in Holzwarth and Schüssler (2003), comparing the models with some observational results.

3.2.2 Flip-Flops

Jetsu et al. (1993) studied the active longitudes of the rapidly rotating giant FK Comae using 25 years of photometric data and found that the spot activity concentrated around two active longitudes which were separated by half of the rotation phase in longitude. One of them was found more dominant for a given period, and from time to time the dominancy switched between them. This phenomenon was named as the "flip-flop mechanism" (Jetsu et al. 1991, 1993) and was found to occur quite commonly among late-type stars showing spot activity.

Recent developments successfully introduced flip-flops into dynamo theory. According to Berdyugina et al. (2002) the coexistence of oscillating axisymmetric and stationary non-axisymmetric modes can explain active longitudes and flip-flop cycles on the solar-type K2-dwarf LQ Hya. Fluri and Berdyugina (2004) suggested a new approach to interpret flip-flops in terms of different combinations of the dynamo modes, enabling a fast comparison of computed solutions with observations (see also Moss 2004; Fluri and Berdyugina 2005). Flip-flop phenomenon was recovered from $\alpha^2\Omega$ type dynamo for weakly differentially rotating stars (Korhonen and Elstner 2005; Elstner and Korhonen 2005).

Observing flip-flop phenomenon, however, still remains challenging. First of all, because enough long and continuous time series are scarcely available. Moreover, phase drifts of the active longitudes are often recorded (e.g. Korhonen et al. 2004), thus, the interpretation of the observed phase jumps can be quite difficult, sometimes confusing or arbitrary, cf.

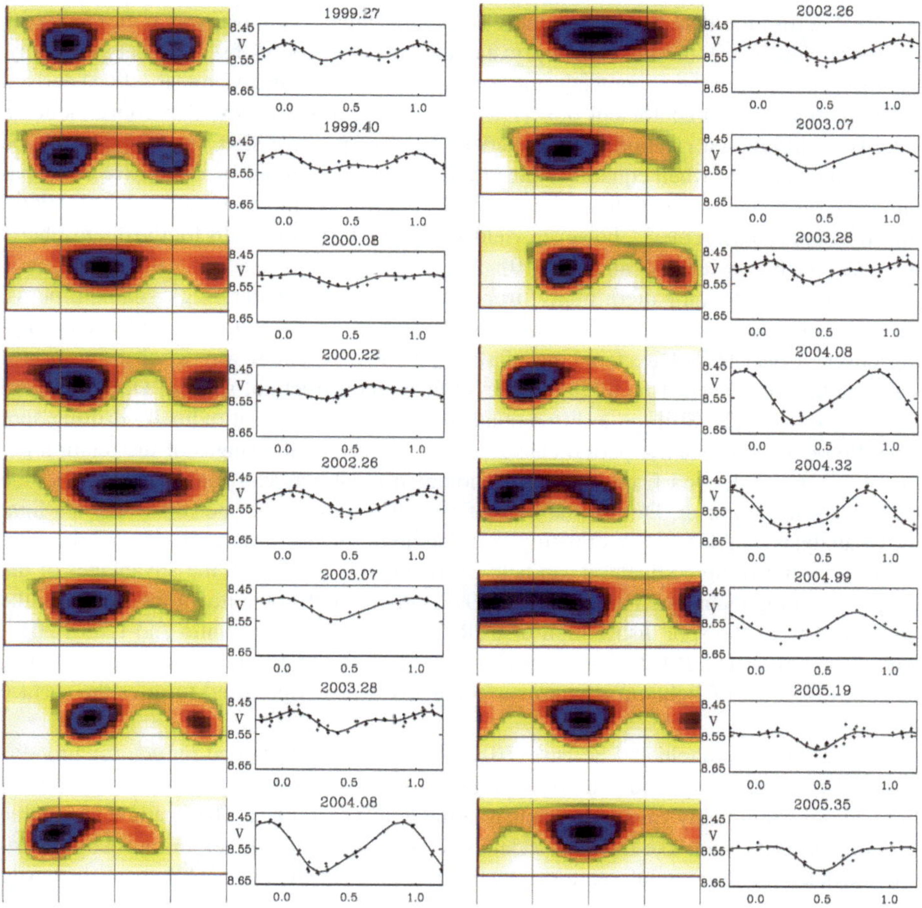

Fig. 13 *Light curve* inversions for DP CVn from Kővári et al. (2013). Together with the surface maps, in the *right panels plotted* are the observations (*crosses*) and calculated V and y light curves (*solid lines*) for different seasons during 1999–2005. Derived maps usually show two active regions, separated by roughly half of the rotation phase. This pattern may bear some evidence of the flip-flop phenomenon, however, phase drifts and jumps would allow only an ambiguous interpretation

Fig. 13, and see also Korhonen et al. (2004, Fig. 3) or Oláh et al. (2013, Fig. 8). Extensive Doppler imaging studies of this phenomenon would also be useful, but until now, only a few such results have been obtained (Vogt et al. 1999; Korhonen et al. 2004; Washuettl et al. 2009). The observational evidence proving that flip-flops occur together with polarity change between the two active longitudes is still to be found. Furthermore, the relation between spot cycles and flip-flop cycles is unclear and flip-flops can occur more often than the expected solar-like spot cycle (cf. Berdyugina 2005, and references therein), i.e., there is no *standard* representation of the observed flip-flop phenomenon.

3.2.3 Differential Rotation

Solar surface differential rotation is one of the most prominent manifestations of the magnetic dynamo working underneath. The phenomenon that sunspots close to the equator travelled across the disk faster than those located at higher latitudes was noticed first by Christoph Scheiner in the 17th century, just at the birth of modern observational astronomy. Centuries later, from the Greenwich sunspot records Maunder and Maunder (1905) derived a differential rotation coefficient $\alpha = (P_{eq} - P_{pole})/P_{eq}$ of 0.19 (note that conventionally, solar-type differential rotation is characterized by α of positive sign, while antisolar type differential rotation, when equator rotates most slowly, is featured by a negative shear parameter). More precise rotation law was derived from sunspot motions e.g., by Ward (1966). Spectroscopic measurements, however, yielded slightly different results for the photospheric material (see the review by Paternò 2010).

The surface rotation pattern imply firm constraints on the underlying dynamo process. However, the detection of surface differential rotation on cool stars is still a challenging observational task. Direct tracing of star spots is a potential way to observe surface shear pattern, however, it requires reliable surface reconstruction, such as Doppler imaging (see Sect. 3.1.2). Still, on a differentially rotating surface, a latitudinally drifting spot would produce photometric period change, therefore, seasonal changes of the rotation period ($\Delta P/P_{phot}$) derived from long-term photometric datasets can also be used as a clue for the shear parameter. Due to the lack of spatial resolution of spot distributions, light curve analysis enables only estimating the magnitude of the surface shear without sign. However, a transiting companion can help overcome this flaw. With the advantages of the eclipse mapping technique Huber et al. (2009a, 2009b, 2010) obtained information on the fine structure of the spot distribution of CoRoT-2a (see also Silva-Valio and Lanza 2011; Nutzman et al. 2011). Roettenbacher et al. (2011) studied the differential rotation of II Peg by inverting a set of light curves into surface maps, which were used to infer the surface evolution.

Just recently, Reinhold and Reiners (2013) developed a fast method for determining differential rotation for spotted stars by searching for different (but similar) periods in their light curves presumably caused by spots rotating at different rates. This method is suitable for large datasets, such like the *Kepler* database, and is capable of measuring rotation periods and detecting differential rotation as well, within statistically reasonable errors, however, without distinction between solar and antisolar type rotation profiles. The shear parameter is estimated as $\alpha := |P_2 - P_1|/\max\{P_1, P_2\}$, where P_1 and P_2 are the primary and secondary periods, respectively. According to the upcoming application for thousands of *Kepler* stars in Reinhold et al. (2013), the shear coefficient grows towards longer periods and slightly increases towards lower temperatures, supporting previous findings from either observations (Barnes et al. 2005) or theory (Küker and Rüdiger 2011).

The possibility of measuring differential rotation for solar-type stars through asteroseismology was investigated first by Gizon and Solanki (2004). The feasibility of detecting differential rotation by this technique depends strongly on the precision of the frequency measurements, as well as on the stellar inclination. Moreover, further difficulties can be introduced by magnetic fields, large scale surface flow fields, etc. (see Lund et al. 2014 and the references therein). On the other hand, Lund et al. (2014) also showed, that frequency splittings could indeed be used in the future to test whether the latitudinal differential rotation is solar-like or antisolar-like (see Sect. 3.3.3), even when having not very precise measurements of frequency splittings.

3.3 Surface Flows from Spatially Resolved Surface Maps

The inherent ability of studying stellar surfaces by the means of Doppler imaging was considerably extended with employing time-series spectroscopic datasets covering two or more subsequent rotation periods (cf., e.g., Donati and Collier Cameron 1997; Strassmeier and Bartus 2000; Kővári et al. 2004, 2007a, 2007b, 2012a, 2012b, etc.). Time-series Doppler imaging has been proved to be extremely useful in observing stellar dynamo ingredients. With the use of subsequent surface maps such features can be investigated as the spot lifetime and evolution, or the structural change of the spot distribution due to surface differential rotation, etc. However, again, not only high-resolution good quality data but also reliable reconstructions and thoroughly tested data processing techniques are needed. In the next sections we give a brief overview of the basic techniques and highlight some of the recent results of this field.

3.3.1 Spot Tracking, Cross-Correlation, Sheared Image

Tracking of surface features on the Sun has widely been used to infer the time evolution of solar surface structures from time series of images. In the same way, time-series Doppler images may reveal information on short-term spot changes and help study dynamo actions in stellar analogs. Longitudinal migration of long-lived spots at different latitudes were compared to derive the latitude dependent rotation law for the active K1-subgiant component of V711 Tau (=HR 1099, Vogt et al. 1999), i.e., one of the most studied RS CVn-type binary systems (see e.g., Donati 1999; Strassmeier and Bartus 2000; Ayres et al. 2001; Donati et al. 2003a, 2003b; Petit et al. 2004; Berdyugina and Henry 2007, etc.). Authors found a weak ($\alpha = -0.0035$) differential rotation of antisolar type. Pole was found not only to rotate faster than the equator, but to be nearly synchronized with the orbit, suggesting a high degree of tidal coupling (cf. Scharlemann 1982).

A natural possibility is to use Doppler imaging for deriving cycle lengths as well, since this method gives a direct measure of the total area of spots on the stellar surface. It needs, though, systematic, long-term spectroscopic observations of high resolution and with good phase coverage for each image. From 11-years of Doppler images Vogt et al. (1999), found an about 3 years long cycle (± 0.2 yr) from the area of both the polar and low latitude spots of V711 Tau. This value is very close to the shortest cycle of 3.3 years obtained for V711 Tau from long-term photometry (Oláh et al. 2009).

The identification of long-lived spots from one observing season to the next (typically from year to year), however, can be difficult, especially, when spot structures become unstable and change significantly on yearly timescale. For this kind of studies, therefore, subsequent Doppler images sampled close to each other proved to be more suitable. From those image pairs, differential rotation can be derived from the technique of cross-correlation instead of simple tracking of individual features. Actually, the very first measurement of surface differential rotation on a star was applied cross-correlation (Donati and Collier Cameron 1997). The technique is based on computing numerical cross-correlation functions along the longitude for all latitudinal slices of the maps to be compared (maps are usually cut into slices of 5° width, i.e., the resolution limit). The cross-correlation functions are then assembled into a 2-D cross-correlation function map which can instantly reveal the differential rotation pattern (see e.g., Weber and Strassmeier 1998, 2001; Skelly et al. 2008; Kővári et al. 2013, etc.). Nevertheless, even when having two subsequent surface maps with small time lag between them, rapid spot evolution (e.g., vivid interaction between emerging flux and its surroundings) is always of concern, since it can easily mask the differential rotation pattern. To boost, however, the signature of the surface shear in the correlation pattern Kővári et al. (2004, 2007a, 2013, 2014a) developed the technique 'ACCORD' (acronym from 'average cross-correlation of time-series Doppler images') for a contiguous set of Doppler reconstructions. The key advantage of the method is that averaging cross-correlation maps definitely emphasizes jointly present features (i.e., the surface shear pattern

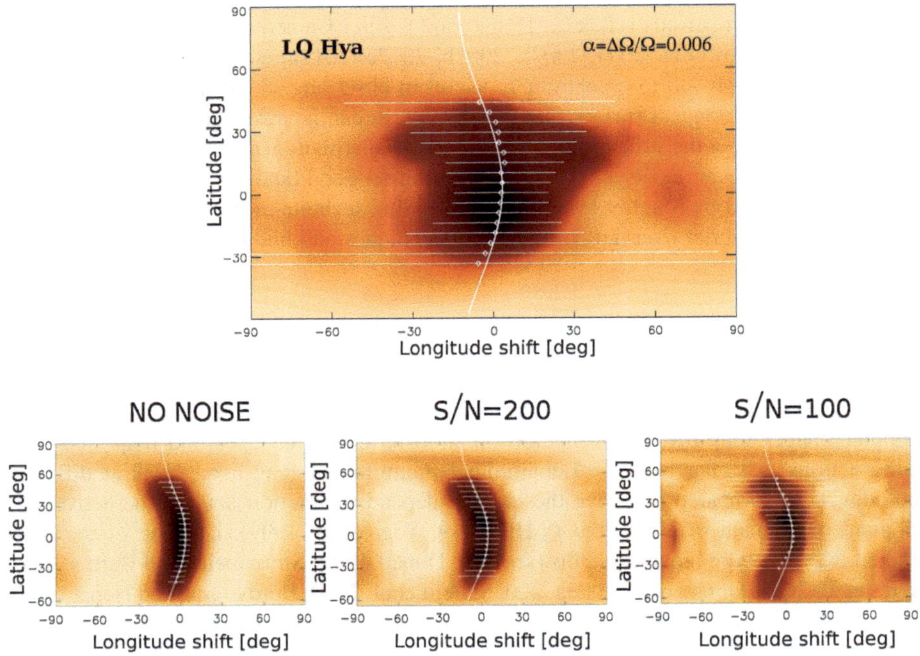

Fig. 14 *Top*: Observed differential rotation of the fast-rotating K2-dwarf LQ Hya resulting in a weak ($\alpha = 0.006$) solar-type surface shear (Kővári et al. 2004). *Bottom*: Testing the reliability of the average cross-correlation technique ACCORD (Kővári et al. 2014b). The surface shear of LQ Hya was applied to a set of artificial Doppler images (artificial spectroscopic sampling patterned an actual observing run). Then, average cross-correlations were performed and surface shear parameters were derived for different data quality, yielding α of 0.0058 ± 0.0004, 0.0067 ± 0.0033 and 0.0068 ± 0.0048 values for no noise (*left*), S/N = 200 (*middle*) and S/N = 100 (*right*), respectively. Note that uncertainty originates also from imperfect image reconstruction due to uneven or gappy (i.e., realistic) phase coverage of the observations

itself) while suppresses false detections in the correlation pattern by unwanted effects (e.g., from rapid spot evolution, crosstalk between neighboring spots, as well as spurious features from data noise). The capability and reliability of this method were demonstrated on a series of elaborated tests (Kővári et al. 2014b), see Fig. 14.

Among the tracer-type techniques, local correlation tracking technique (LCT) should not be missed, even if it became a favorable tool in solar physics (e.g., Sobotka et al. 2000; Švanda et al. 2006). Attempts however, were made to demonstrate the power of LCT also on stellar observations (Kővári et al. 2007c; Vida et al. 2007). The technique is based on the principle of the best match of two image frames that record the tracked features at two subsequent instants. The resulting vector map can essentially be regarded as the surface flow field over the time lag between the initial frames, see Fig. 15. This approach could provide an important clue in observing global and local surface flows, however, further tests are needed to confirm solidity.

Surface differential rotation can be detected even if having data sufficient only for one single Doppler reconstruction. The parametric imaging method, often called sheared image method (e.g., Petit et al. 2002; Weber and Strassmeier 2005; Barnes et al. 2005; Morgenthaler et al. 2012, etc.) involves surface shear as an additional parameter in the line profile inversion process. Computations are done for a range of meaningful shear parameters and the most likely value is chosen on the basis of goodness of fit. The disadvantage of this

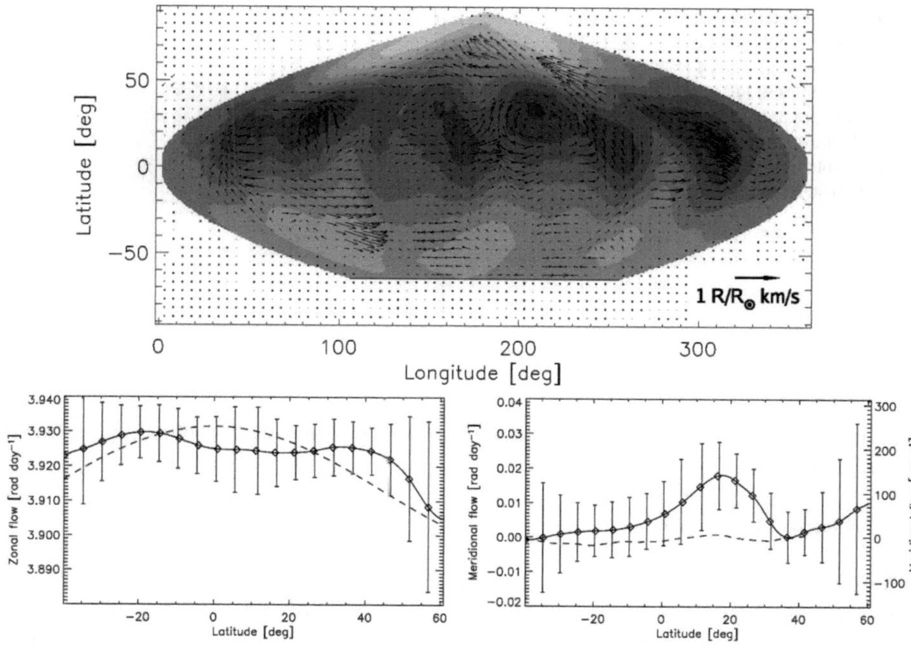

Fig. 15 *Top*: Flows in the photosphere of LQ Hya by applying local correlation tracking (LCT) technique for a time series of 28 Doppler images. *Bottom*: Mean zonal flow component (*left*) suggests a weak solar-type differential rotation (*dashed line*), just consistently with the result in Kővári et al. (2004). Note, that the poleward flow implied by the mean meridional component (*bottom right*) is around the detection limit, therefore ambiguous

method, however, is that it introduces yet another free parameter into the inversion process, and a predefined latitudinal rotation law (generally a $\sin^2\beta$ type) is forced.

In some cases, however, the reliability of the results obtained from sheared image method is doubtful, as demonstrated by the following comparison. The differential rotation of the single early-G type V889 Her ($P_{\rm rot} = 1.337$ days), a young Sun, was studied e.g., by Marsden et al. (2006) and Jeffers and Donati (2008) by the means of parametric Zeeman-Doppler imaging, and strong solar-type differential rotation was reported with α of 0.084 and $\approx$0.1, respectively. Note that such a large shear does not fit the empirical law by Barnes et al. (2005); see also Kitchatinov and Rüdiger (1995), Reiners (2006), Küker and Rüdiger (2011), Reinhold et al. (2013) on how the differential rotation is influenced by temperature. On the other hand, Järvinen et al. (2008) argued for a substantially weaker differential rotation, while Huber et al. (2009a, 2009b) preferred rigid rotation (still allowing a weak surface shear). A more recent parametric Doppler imaging study (Kővári et al. 2011) resulted in a weak solar-type rotation law with $\alpha = 0.009$. The numerical model by Kitchatinov and Olemskoy (2011) for a $T_{\rm eff} = 5800$ K dwarf rotating at the angular velocity of V889 Her would suggest $\alpha = 0.016$, i.e., much weaker than the values 0.084 and $\approx$0.1 found by Marsden et al. (2006) and Jeffers and Donati (2008), respectively. Such contradictory results might partially be explained by temporal variations on the differential rotation (cf. Donati et al. 2003a, 2003b), but Jeffers et al. (2011) found no evidence for it. Nevertheless, Kővári et al. (2014b) have demonstrated recently, that a large polar cap, such like the one found on V889 Her (see also Strassmeier et al. 2003a; Frasca et al. 2010), can distort the rotational profile similarly as the differential rotation

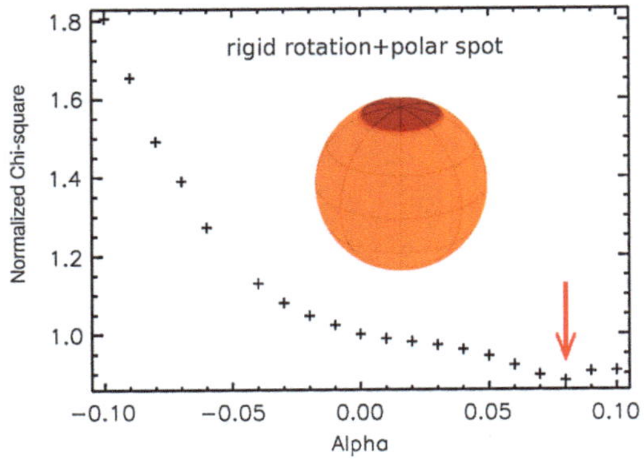

Fig. 16 False detection of surface shear. Applying the parametric imaging study for an actually rigidly rotating test star with a large polar cap would result in solar-type differential rotation. Note the best fit at $\alpha = 0.08$ instead of zero shear

does, consequently yielding false measure of the surface shear by parametric imaging technique; see Fig. 16.

3.3.2 Binarity and Activity

In RS CVn-type active close binary systems tidal effects help maintain the fast rotation and so the magnetic activity at a higher level. In such an interaction, tidal coupling between the convective envelope of the evolved star and its companion is of utmost importance in understanding the dynamo action underneath (e.g., Scharlemann 1981, 1982; Schrijver and Zwaan 1991, etc.). Although earlier photometric techniques gave only a poor chance to study this interaction in detail, later on Doppler imaging and in some exceptional cases interferometric imaging (Monnier et al. 2007; Parks et al. 2012) represented observationally a great improvement. The behavior of the emerging magnetic flux tubes under the gravitational influence of a companion star was analyzed first by Holzwarth and Schüssler (2000, 2002). Recently, active components in different RS CVn-type systems were compared in Oláh et al. (2012a, 2012b) and in Kővári et al. (2012b) and marked differences were depicted regarding the spot distribution and the surface differential rotation.

One of the compared systems, V711 Tau (K1IV) showed very little evidence of differential rotation (cf. also Vogt et al. 1999) while from Zeeman Doppler imaging Petit et al. (2004) found a bit stronger shear, but still 40 times smaller than that of the Sun. The reduced differential rotation was attributed to the strong tidal interaction of the G5V companion star (see also Muneer et al. 2010). Congruently, the very similar system UX Ari (K0-K1IV) did not show sign of strong surface shear (Rosario et al. 2008), indeed, spot distribution was found to be fixed to the orbital frame for long (cf. Lanza et al. 2006). Contrarily, EI Eri (K1IV) showed period deviation $\Delta P/P_{\mathrm{orb}}$ reaching ± 2 % (Oláh et al. 2012a, 2012b) and time-series Doppler imaging study yielded $\alpha = 0.037$ differential rotation coefficient (Kővári et al. 2009), i.e., one-fifth of the solar shear. In accord with this, long-term Doppler imaging study by Washuettl et al. (2009) did not reflect any preferred (i.e., phase-locked) spot positions. As compared to the former cases of V711 Tau and UX Ari, the secondary component in EI Eri is relatively small (M4-5) and thus its gravitational influence on the dynamo hosting primary seems to be much less significant.

The two long-period RS CVn-type systems, σ Gem and ζ And have similarly K1-giant primary components (with P_{orb} of 19.6 days and 17.8 days, respectively), both showing active longitudes locked mostly at quadrature positions in the binary reference frame (Kajatkari et al. 2014; Kővári et al. 2007a). However, gravitational distortions were found to be quite different (Kővári et al. 2012b and references therein), getting stronger deformation for

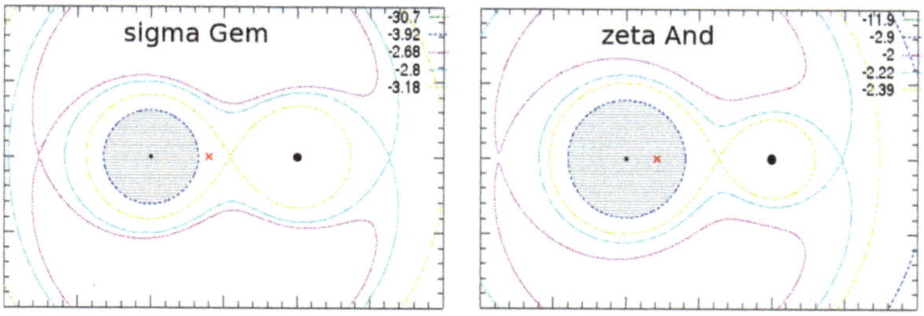

Fig. 17 Roche potentials of σ Gem (*left*) and ζ And (*right*) from using Nightfall (http://www.hs.uni-hamburg.de/DE/Ins/Per/Wichmann/Nightfall.html). Cross-sections of the evolved components are *grey shaded*, gravity centers are *marked with red crosses*

ζ And, while either small or even negligible non-sphericity for σ Gem. The scaled graphs of the two systems, see Fig. 17 (see also Schrijver and Zwaan 1991, Fig. 1), however, reveal a basic difference in the mass distribution of the binary systems, which may account for some other observed differences. Surface differential rotations were determined and refined for both systems by Kővári et al. (2007a, 2007b, 2012a, 2012b) yielding solar-like rotation law with $\alpha = 0.055$ shear parameter for ζ And and oppositely, antisolar type surface differential rotation with $\alpha = -0.04$ for σ Gem. Using a unique cross-correlation technique (see also Sect. 3.3.4) applied to latitudinal displacements of surface features Kővári et al. (2007a, 2007b) investigated the meridional surface flows on ζ And and σ Gem. Interestingly, the results indicated quite different phenomena: a negligible or possibly a weak equatorward drift on ζ And, while a clear and much stronger poleward flow on σ Gem. This observation is indeed in agreement with theoretical expectations by Kitchatinov and Rüdiger (2004), see also Sect. 3.3.3, who attributed the antisolar differential rotation to strong meridional flow which could be caused e.g., by field forcing of a close companion. Such examples expose the importance of binarity in controlling the activity of the evolved components in close binaries e.g., by either driving or suppressing differential rotation, or by rewriting the subsurface scenario of the magnetic flux emergence (Holzwarth and Schüssler 2003).

3.3.3 Antisolar Type Differential Rotation

Antisolar type differential rotation, when polar regions rotate faster than the equator, has been found for a small number of stars (e.g., Vogt and Hatzes 1991; Strassmeier et al. 2003b; Oláh et al. 2003; Weber et al. 2005; Weber 2007; Kővári et al. 2007b, 2013; Kriskovics et al. 2014, etc.). In Fig. 18 plotted are the derived surface shear coefficients vs. rotation period for a set of the most-observed targets of either solar or antisolar-type rotation profile. Regardless the sign, absolute values of the surface shear parameters seem to follow a weak dependence of the surface shear on $P_{\rm rot}$ (cf. Küker and Rüdiger 2011). From the plot one can discern that antisolar differential rotation was detected mainly for long-period giants being either single stars (e.g., DP CVn, DI Psc) or members in RS CVn-type binary systems (e.g., σ Gem, HK Lac).

To explain this formerly disregarded feature, first Kitchatinov and Rüdiger (2004) proposed a theoretical revision by considering a fast meridional flow. Differential rotation can be produced by meridional flow, which, on the other hand, can be fast when spherical symmetry is perturbed either gravitationally by a close companion or thermally by large cool spots. Simulations for a giant star (having 2.5 solar mass, 7.91 solar radius, $P_{\rm rot} \approx 27$ days)

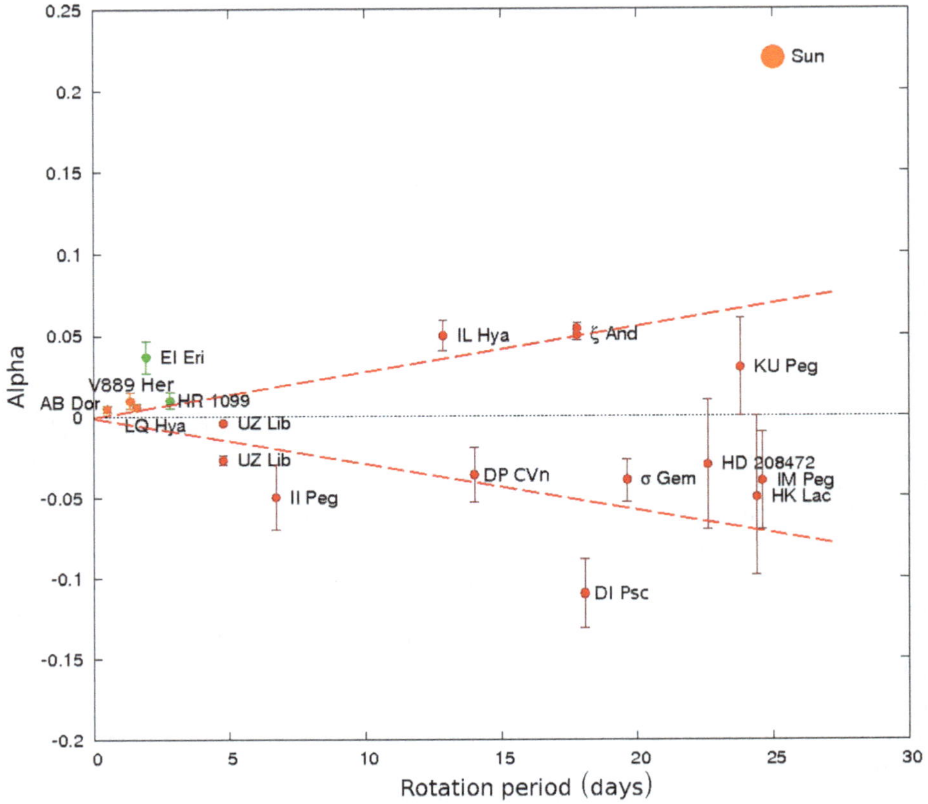

Fig. 18 Differential rotation coefficients from Doppler imaging studies. The plot is an extended and updated version after Weber (2007). Absolute values of the surface shear parameters follow the empirical trend of Barnes et al. (2005) in the sense, that slower rotators perform larger surface shear. *Dashed line* (and *its mirroring about the horizontal axis*) is a simple linear fit to the absolute values of the shear parameters, giving a slope of $\approx 0.0028\ \mathrm{d}^{-1}$

showed that increasing poloidal field inferred the reversion of the meridional flow to poleward on the surface, and, at a certain rate, the differential rotation reversed to antisolar case (Kitchatinov and Rüdiger 2004; Kitchatinov 2006).

Further theoretical support for antisolar differential rotation has been provided in the recent studies by Chan (2010) and Gastine et al. (2014). The transition between the regimes of solar-type and antisolar differential rotation was investigated with 3-D simulations and it was found, that the direction of the rotational profile was dependent on the Coriolis force in the sense, that rapid rotators with large Coriolis numbers (or analogously, with small Rossby numbers) performed solar-like differential rotation, while antisolar type differential rotation was derived for moderate rotators having large Rossby numbers.

Recent investigations of two single giants, namely DP CVn and DI Psc (Kővári et al. 2013; Kriskovics et al. 2014, see also Fig. 18) drew the attention to a possible connection between antisolar differential rotation and unusually high surface lithium abundances observed in a handful of RGB giants, including DP CVn and DI Psc. The lithium is supposed to be transported from the inner parts onto the surface by an extra mixing mechanism (Charbonnel and Balachadran 2000), which, on the other hand, could also be responsible for transporting angular momentum towards the poles by meridional flows. This, eventually, would result in antisolar differential rotation, which was found for both targets. Moreover, antisolar differential rotation was found for another target of such kind, the high-Li K2-giant HD 31993 (Strassmeier et al. 2003b). We note, however, that further observations are needed to continue detections of differential rotation for similar objects and thus elaborate on this speculation.

3.3.4 Meridional Flows: Measurements and Interpretations

Beside differential rotation, meridional circulation is the other key velocity pattern that is related to the dynamo. The influence of meridional circulation on solar and stellar dynamos was recognized and investigated by several authors (e.g., Choudhuri et al. 1995; Dikpati and Gilman 2001; Bonanno et al. 2002, 2006; Miesch 2005; Holzwarth et al. 2007; Jouve and Brun 2007; Pipin and Kosovichev 2011; Küker and Rüdiger 2011, etc.). Current helioseismic measurements suggested the existence of at least two meridional circulation cells in the Sun (Zhao et al. 2013). In turn, observing meridional flows on stars are much more difficult, since the meridional motion of surface features are expected to be at least one order of magnitude smaller compared to the zonal shifts at different latitudes due to surface differential rotation (note that the ratio of the average zonal and meridional surface flow velocities on the Sun is around 20:1). On the other hand, such tiny latitudinal motions could only be observed by employing reliable high resolution Doppler reconstructions. But note also, that, in general, stellar surface inversion techniques are less capable of recovering latitudinal information, especially, when the inclination angle cannot be determined precisely. Still, in particular cases, time-series Doppler imaging can be suitable to measure common meridional motion of surface features. The most reliable technique is similar to the one introduced in Sect. 3.3.1 for detecting differential rotation from average cross-correlation function maps, but contrary, cross-correlations are done along meridional circles.

Due to observational difficulties, until now, some few determinations of meridional motions have been published (e.g., Strassmeier and Bartus 2000; Weber and Strassmeier 2001; Kővári et al. 2007a, 2007b, 2009, 2014a, 2014b). The measured meridional flow velocities (of either equatorward or poleward) fall between 10^2–10^3 ms^{-1}, i.e., stellar meridional flows are found to be 10–100 times faster than the solar meridional circulation on the surface ($\approx$15 ms^{-1}). Note, though, that selection effect should also be considered, since flows of the solar magnitude cannot be observed on stars. A preliminary statistics (Weber 2007) indicated a weak dependence of the meridional flow on the dimensionless surface shear coefficient. The scatter, however, was too large to be suitable for deriving a proper relationship. Nevertheless, the trend in Weber (2007) would support the theoretical prediction in Kitchatinov and Rüdiger (2004), that strong meridional flow can drive antisolar differential rotation. One of the most spectacular results so far, supporting also this prediction, was obtained for σ Gem in Kővári et al. (2007a). Recently, Kővári et al. (2014c) have pointed out, that the method of latitudinal cross-correlation used in Kővári et al. (2007a) suffers from an incompleteness due to the singularity rising towards the pole. To avoid this imperfection, a refinement was carried out which yielded an average poleward drift of $\approx$220 ms^{-1} on the surface of the K1-giant (see Fig. 19), i.e., a slightly smaller value compared to the result of $\approx$300 ms^{-1} in Kővári et al. (2007a), but still in accordance with it.

4 Closing Remarks and Outlook

Through examples we gave an overview on the recent progress in observing dynamo action in cool stars. As for today, about 10^2 active stars have been mapped by the means of Doppler

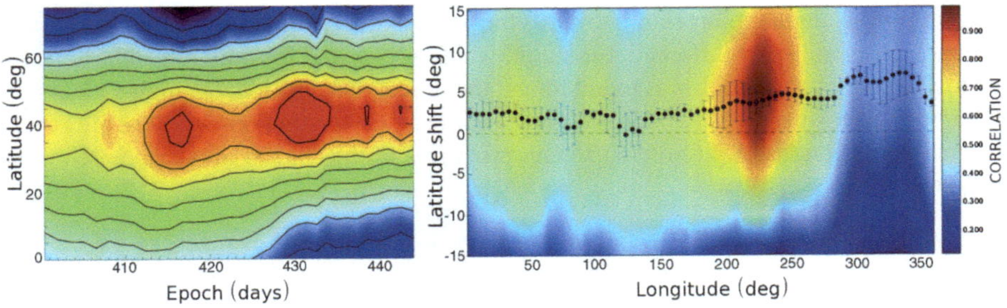

Fig. 19 *Left*: Synoptic map of the changing latitudinal spot distribution of σ Gem from 34 time-series Doppler images using the Ca I 6439 mapping line. An average poleward drift of the most prominent features during the covered term is striking, in accordance with the latitudinal cross-correlation function map (*right*). As a result, $\approx$220 ms^{-1} flow velocity was derived (Kővári et al. 2014c)

imaging, i.e., still a poor sample for establishing general conclusions on stellar dynamos. Clearly, much larger sample of stars would be needed for answering such questions like what kind of dynamo mechanism operates in a given type (class, age, rotation, metallicity, binarity, etc.) of active star.

However, when using observational results derived by employing inversion techniques of different kinds one should not forget about the reliability issue. Most reliable detections of surface flows on stars, such as differential rotation and meridional circulation, would require good quality densely sampled homogeneous datasets, covering enough long time intervals for the best performance spatial and time resolution. So far, time-series spectroscopy has been proved to be an outstanding experience in studying stellar surface velocity fields. An exceptional possibility of collecting such long-term spectroscopic datasets for a small sample of active stars was the historical night-time program with the 1.5 m McMath-Pierce telescope at Kitt Peak National Solar Observatory. The aim of extending this kind of observational experiences eventuated the 1.2-m mirror STELLA robotic telescopes at Teide Observatory, operated by the Leibniz-Institute for Astrophysics Potsdam (Weber et al. 2012). A step forward in studying surface phenomena of active stars would be a dedicated high precision spectroscopic instrument on an at least 4-m class monitoring telescope.

Kepler mission confirmed the necessity of ultra-high precision photometry also in case of investigating magnetic activity. A very important next step would be a space mission with similar high precision as of *Kepler*, but capable of multicolor photometry, this way gathering much more physical information on the observed objects. PLATO (Planetary Transits and Oscillations of stars) mission has already been selected as a part of ESA's Cosmic Vision 2015-25 Programme. By observing up to 10^6 relatively nearby stars of different populations, PLATO's measurements will allow to calibrate the relationship between stellar age and rotation, as well as to study activity cycles. In addition, compared to CoRoT and *Kepler* targets, PLATO's nearby and so relatively bright stars of up to 4 mag will provide more opportunity for ground-based high-resolution spectroscopic follow-up observations (Rauher et al. 2013).

For observing stellar activity cycles at least the presently running few APTs should continue their missions. The time is running in one way: a missed observation cannot be replaced with a new one in such a scientific field which deals with continuously changing, developing entities, like stars. The timescales of activity cycles are in many cases longer than the human lifetime. The nearest example is the Gleissberg cycle of the Sun, which is something like 50–200 years long with changing length, and even the 11-year Schwabe cycle repeats only 4–5 times at most, during a career of an astronomer. Concerning our present knowledge, the timescales of the longer cycles from the multiple cycles of other stars are of similar lengths than the longer solar cycles. This fact should not be disregarded, indeed, it needs a kind of humble competence to invest to such projects both from scientists and fundings, that will serve not the present but the future generations of scientists.

From future observations using either ultra-high precision space photometry or high-resolution high signal-to-noise spectroscopy and spectropolarimetry or even just from continuation of long-term monitoring surveys by 1-m class telescopes, we expect a better understanding of dynamos working in cool stars and of course, in our Sun. Especially, we may get closer to understanding the role of magnetic activity in the formation and evolution of star-planet systems and eventually in the origin of life.

Acknowledgements ZsK is grateful for the invitation by the Convenors to the ISSI WS "The Solar Activity Cycle: Physical Causes and Consequences", held between 11–15 November 2013, in Bern, Schwitzerland. This work has been supported by the Hungarian Science Research Programs OTKA K-81421 and OTKA K-109276, the Lendület-2009 and Lendület-2012 Young Researchers' Programs of the Hungarian Academy of Sciences.

References

P.J. Amado, J.G. Doyle, P.B. Byrne, Mon. Not. R. Astron. Soc. **314**, 489 (2000)
T.R. Ayres, A. Brown, R.A. Osten et al., Astrophys. J. **549**, 554 (2001)
G.Á. Bakos, J. Lázár, I. Papp et al., Publ. Astron. Soc. Pac. **114**, 974 (2002)
S.L. Baliunas, E. Nesme-Ribes, D. Sokoloff, W.H. Soon, Astrophys. J. **460**, 848 (1996)
J.R. Barnes, A. Collier Cameron, J.-F. Donati et al., Mon. Not. R. Astron. Soc. **357**, L1 (2005)
S.V. Berdyugina, Living Rev. Sol. Phys. **2**, 8 (2005)
S.V. Berdyugina, G.W. Henry, Astrophys. J. **659**, 157 (2007)
S.V. Berdyugina, I. Tuominen, Astron. Astrophys. **336**, 25 (1998)
S.V. Berdyugina, I.G. Usoskin, Astron. Astrophys. **405**, 1121 (2003)
S.V. Berdyugina, J. Pelt, I. Tuominen, Astron. Astrophys. **394**, 505 (2002)
A. Bonanno, D. Elstner, G. Rüdiger, G. Belvedere, Astron. Astrophys. **390**, 673 (2002)
A. Bonanno, D. Elstner, G. Belvedere, Astron. Nachr. **327**, 680 (2006)
A.S. Bonomo, A.F. Lanza, Astron. Astrophys. **547**, 37 (2012)
S.F. Brown, J.-F. Donati, D.E. Rees, M. Semel, Astron. Astrophys. **250**, 463 (1991)
E. Budding, Astrophys. Space Sci. **48**, 207 (1977)
W. Bumba, R. Howard, Sol. Phys. **7**, 28 (1969)
V. Bumba, A. Garcia, M. Klvaňa, Sol. Phys. **196**, 403 (2000)
T.A. Carroll, M. Kopf, I. Ilyin, K.G. Strassmeier, Astron. Nachr. **328**, 1043 (2007)
T.A. Carroll, M. Kopf, K.G. Strassmeier, Astron. Astrophys. **488**, 781 (2008)
T.A. Carroll, K.G. Strassmeier, J.B. Rice, A. Künstler, Astron. Astrophys. **548**, A95 (2012)
K.L. Chan, in *Solar and Stellar Variability: Impact on Earth and Planets*, ed. by A.G. Kosovichev, A.H. Andrei, J.-P. Rozelot. Proc. IAU Symp., No. 264 (International Astronomical Union, Paris, 2010), p. 219
C. Charbonnel, S. Balachadran, Astron. Astrophys. **309**, 563 (2000)
A.R. Choudhuri, M. Schüssler, M. Dikpati, Astron. Astrophys. **303**, L29 (1995)
M. Cuntz, S.H. Saar, Z. Musielak, Astrophys. J. **533**, L151 (2000)
A.J. Deutsch, in *Electromagnetic Phenomena in Cosmical Physics*, ed. by B. Lehnert. Proc. IAU Symp., No. 6 (Cambridge University Press, Cambridge, 1958), p. 209
M. Dikpati, P.A. Gilman, Astrophys. J. **559**, 428 (2001)
J.-F. Donati, Mon. Not. R. Astron. Soc. **302**, 457 (1999)
J.-F. Donati, S.F. Brown, Astron. Astrophys. **326**, 1135 (1997)
J.-F. Donati, A. Collier Cameron, Mon. Not. R. Astron. Soc. **291**, 1 (1997)
J.-F. Donati, M. Semel, F. Praderie, Astron. Astrophys. **225**, 467 (1989)
J.-F. Donati, M. Semel, B.D. Carter et al., Mon. Not. R. Astron. Soc. **291**, 658 (1997)
J.-F. Donati, A. Collier Cameron, M. Semel et al., Mon. Not. R. Astron. Soc. **345**, 1145 (2003a)
J.-F. Donati, A. Collier Cameron, P. Petit, Mon. Not. R. Astron. Soc. **345**, 1187 (2003b)
C. Dubé, P. Charbonneau, Astrophys. J. **775**, 69 (2013)
J.A. Eaton, G.W. Henry, F.C. Fekel, Astrophys. J. **462**, 888 (1996)
Z. Eker, Astrophys. J. **420**, 373 (1994)

Z. Eker, Astrophys. J. **445**, 526 (1995)
D. Elstner, H. Korhonen, Astron. Nachr. **326**, 278 (2005)
F.C. Fekel, C.T. Bolton, Astron. J. **134**, 2079 (2007)
F.C. Fekel, M.H. Williamson, M. Weber et al., Astron. Nachr. **334**, 223 (2013)
D.M. Fluri, S.V. Berdyugina, Sol. Phys. **224**, 153 (2004)
D.M. Fluri, S.V. Berdyugina, in *Large-Scale Structures and Their Role in Solar Activity*, ed. by K. Sankarasubramanian, M. Penn, A. Pevtsov. ASP Conf. Ser., vol. 346 (Astron. Soc. of the Pacific, San Francisco, 2005), p. 167
A. Frasca, K. Biazzo, Zs. Kővári et al., Astron. Astrophys. **518**, A48 (2010)
R.A. García, S. Mathur, D. Salabert et al., Science **329**, 1032 (2010)
R.A. García, T. Ceillier, S. Mathur, D. Salabert, in *Progress in Physics of the Sun and Stars: A New Era in Helio- and Asteroseismology*, ed. by H. Shibahashi, A.E. Lynas-Gray. Proc. Fujihara Seminar held, Hanoke, Japan, 25–29 Nov, 2012. ASP Conf. Ser., vol. 479 (Astron. Soc. of the Pacific, San Francisco, 2013), p. 129.
T. Gastine, R.K. Yadav, J. Morin et al., Mon. Not. R. Astron. Soc. **438**, L76 (2014)
L. Gizon, S. Solanki, Sol. Phys. **220**, 169 (2004)
A.G. Goncharskii, V.V. Stepanov, V.L. Khokhlova, A.G. Yagola, Sov. Astron. Lett. **3**, 147 (1977)
A.G. Goncharskii, V.V. Stepanov, V.L. Khokhlova, A.G. Yagola, Sov. Astron. **26**, 690 (1982)
J.D. Hartman, G.Á. Bakos, R.W. Noyes et al., Astron. J. **141**, 166 (2011)
E. Herrero, A.F. Lanza, I. Ribas et al., Astron. Astrophys. **553**, 66 (2013)
V. Holzwarth, M. Schüssler, Astron. Nachr. **321**, 175 (2000)
V. Holzwarth, M. Schüssler, Astron. Nachr. **323**, 399 (2002)
V. Holzwarth, M. Schüssler, Astron. Astrophys. **405**, 303 (2003)
V. Holzwarth, D.H. Mackay, M. Jardine, Astron. Nachr. **328**, 1108 (2007)
K.F. Huber, U. Wolter, S. Czesla et al., Astron. Astrophys. **501**, 715 (2009a)
K.F. Huber, S. Czesla, U. Wolter, J.H.M.M. Schmitt, Astron. Astrophys. **508**, 901 (2009b)
K.F. Huber, S. Czesla, U. Wolter, J.H.M.M. Schmitt, Astron. Astrophys. **514**, A39 (2010)
G.A.J. Hussain, A. Collier Cameron, J.-F. Donati, in *The Tenth Cambridge Workshop on Cool Stars, Stellar Systems and the Sun*, ed. by R.A. Donahue, J.A. Bookbinder. ASP Conf. Ser., vol. 154 (Astron. Soc. of the Pacific, San Francisco, 1998), p. 1992
G.A.J. Hussain, M. Jardine, A. Collier Cameron, Mon. Not. R. Astron. Soc. **322**, 681 (2001)
E. Işık, in *Comparative Magnetic Minima: Characterizing Quiet Times in the Sun and Stars*, ed. by D. Webb, C. Mandrini. Proc. IAU Symp., No. 286 (Cambridge Univ. Press, Cambridge, 2012), p. 291
E. Işık, D. Schmitt, M. Schüssler, Astron. Astrophys. **528**, A135 (2011)
S.P. Järvinen, S.V. Berdyugina, K.G. Strassmeier, Astron. Astrophys. **440**, 735 (2005)
S.P. Järvinen, S.V. Berdyugina, H. Korhonen et al., Astron. Astrophys. **472**, 887 (2007)
S.P. Järvinen, H. Korhonen, S.V. Berdyugina et al., Astron. Astrophys. **488**, 1047 (2008)
S.V. Jeffers, J.-F. Donati, Mon. Not. R. Astron. Soc. **390**, 635 (2008)
S.V. Jeffers, J.-F. Donati, E. Alecian, S.C. Marsden, Mon. Not. R. Astron. Soc. **411**, 1301 (2011)
L. Jetsu, J. Pelt, I. Tuominen, H. Nations, in *The Sun and Cool Stars: Activity, Magnetism, Dynamos*, ed. by I. Tuominen, D. Moss, G. Rüdiger. Proc. IAU Colloq., vol. 130 (Springer, Berlin, 1991), p. 381
L. Jetsu, J. Pelt, I. Tuominen, Astron. Astrophys. **278**, 449 (1993)
A. Jiménez, R.A. García, P.L. Pallé, J. Phys. Conf. Ser. **271**, 012051 (2011)
L. Jouve, A.S. Brun, Astron. Astrophys. **474**, 239 (2007)
P.G. Judge, S.C. Solomon, T.R. Ayres, Astrophys. J. **593**, 534 (2003)
P. Kajatkari, T. Hackman, L. Jetsu et al., Astron. Astrophys. **562**, A107 (2014)
Y.W. Kang, R.E. Wilson, Astron. J. **97**, 848 (1989)
M.M. Katsova, M.A. Livshits, G. Belvedere, Sol. Phys. **216**, 353 (2003)
V.L. Khokhlova, Sov. Astron. **19**, 576 (1976)
L.L. Kitchatinov, Astron. Rep. **50**, 512 (2006)
L.L. Kitchatinov, S.V. Olemskoy, Astron. Lett. **31**, 280 (2005)
L.L. Kitchatinov, S.V. Olemskoy, Mon. Not. R. Astron. Soc. **411**, 1059 (2011)
L.L. Kitchatinov, G. Rüdiger, Astron. Astrophys. **299**, 446 (1995)
L.L. Kitchatinov, G. Rüdiger, Astron. Nachr. **325**, 496 (2004)
Z. Kolláth, K. Oláh, Astron. Astrophys. **501**, 695 (2009)
R.W. Komm, R.F. Howard, J.W. Harvey, Sol. Phys. **147**, 207 (1993)
H. Korhonen, D. Elstner, Astron. Astrophys. **440**, 1161 (2005)
H. Korhonen, S.V. Berdyugina, I. Tuominen, Astron. Astrophys. **390**, 179 (2002)
H. Korhonen, S.V. Berdyugina, I. Tuominen, Astron. Nachr. **325**, 402 (2004)
H. Korhonen, S. Hubrig, S.V. Berdyugina et al., Mon. Not. R. Astron. Soc. **395**, 282 (2009)
H. Korhonen, M. Wittkowski, Zs. Kővári et al., Astron. Astrophys. **515**, A14 (2010)

Zs. Kővári, J. Bartus, Astron. Astrophys. **323**, 801 (1997)
Zs. Kővári, K.G. Strassmeier, T. Granzer et al., Astron. Astrophys. **417**, 1047 (2004)
Zs. Kővári, J. Bartus, K.G. Strassmeier et al., Astron. Astrophys. **463**, 1071 (2007a)
Zs. Kővári, J. Bartus, K.G. Strassmeier et al., Astron. Astrophys. **474**, 165 (2007b)
Zs. Kővári, J. Bartus, M. Švanda et al., Astron. Nachr. **328**, 1081 (2007c)
Zs. Kővári, A. Washuettl, B.H. Foing et al., in *Proceedings of the 15th Cambridge Workshop on Cool Stars, Stellar Systems and the Sun*, ed. by E. Stempels. AIP Conf. Proc., vol. 1094 (American Institute of Physics, College Park, 2009), p. 676
Zs. Kővári, A. Frasca, K. Biazzo et al., in *The Physics of Sun and Star Spots*, ed. by D.P. Choudhary, K.G. Strassmeier. Proc. IAU Symp., No. 273 (Cambridge University Press, Cambridge, 2011), p. 121
Zs. Kővári, H. Korhonen, L. Kriskovics et al., Astron. Astrophys. **539**, A50 (2012a)
Zs. Kővári, J. Bartus, L. Kriskovics et al., in *From Interacting Binaries to Exoplanets: Essential Modeling Tools*, ed. by M.T. Richards, I. Hubeny. Proc. IAU Symp., No. 282 (Cambridge University Press, Cambridge, 2012b), p. 197
Zs. Kővári, H. Korhonen, K.G. Strassmeier et al., Astron. Astrophys. **551**, A2 (2013)
Zs. Kővári, L. Kriskovics, K. Oláh et al., in *Magnetic Fields Throughout the Stellar Evolution*, ed. by P. Petit, M. Jardine, H. Spruit. Proc. IAU Symp., No. 302 (Cambridge University Press, Cambridge, 2014a), p. 379
Zs. Kővári, J. Bartus, L. Kriskovics et al., in *Magnetic Fields Throughout the Stellar Evolution*, ed. by P. Petit, M. Jardine, H. Spruit. Proc. IAU Symp., No. 302 (Cambridge University Press, Cambridge, 2014b), p. 198
Zs. Kővári, L. Kriskovics, A. Künstler et al., Astron. Astrophys. (2014c, submitted)
L. Kriskovics, Zs. Kővári, K. Vida, K. Oláh, in *Magnetic Fields Throughout the Stellar Evolution*, ed. by P. Petit, M. Jardine, H. Spruit. Proc. IAU Symp. (Cambridge University Press, Cambridge, 2014), p. 383
M. Küker, G. Rüdiger, Astron. Nachr. **332**, 933 (2011)
S. Lalitha, J.H.M.M. Schmitt, Astron. Astrophys. **559**, A119 (2013)
A.F. Lanza, S. Catalano, G. Cutispoto et al., Astron. Astrophys. **332**, 541 (1998)
A.F. Lanza, N. Piluso, M. Rodonò et al., Astron. Astrophys. **455**, 595 (2006)
A.F. Lanza, I. Pagano, G. Leto et al., Astron. Astrophys. **493**, 193 (2009)
J. Lehtinen, L. Jetsu, T. Hackman et al., Astron. Astrophys. **527**, A136 (2011)
J. Li, Astrophys. J. **735**, 130 (2011)
M.A. Livshits, I.Yu. Alekseev, M.M. Katsova, Astron. Rep. **47**, 562 (2003)
M.N. Lund, M.S. Miesch, J. Christensen-Dalsgaard, Astrophys. J. **790**, 121 (2014)
M.J. Mantere, P.J. Käpylä, J. Pelt, in *Solar and Astrophysical Dynamos and Magnetic Activity*, ed. by A. Kosovichev, Y. Yan, L. van Driel-Gesztelyi, E. Dal de Gouveia. Proc. IAU Symp., No. 294 (Cambridge University Press, Cambridge, 2013), p. 175
S.C. Marsden, J.-F. Donati, M. Semel et al., Mon. Not. R. Astron. Soc. **370**, 468 (2006)
E.W. Maunder, A.S.D. Maunder, Mon. Not. R. Astron. Soc. **65**, 813 (1905)
S. Messina, Astron. Astrophys. **480**, 495 (2008)
S. Messina, E.F. Guinan, Astron. Astrophys. **393**, 225 (2002)
M.S. Miesch, Living Rev. Sol. Phys. **2**, 1 (2005)
J.D. Monnier, M. Zhao, E. Pedretti et al., Science **317**, 342 (2007)
A. Morgenthaler, P. Petit, S. Saar et al., Astron. Astrophys. **540**, A138 (2012)
D. Moss, Mon. Not. R. Astron. Soc. **352**, 17 (2004)
D. Moss, Astron. Astrophys. **432**, 249 (2005)
D. Moss, I. Tuominen, Astron. Astrophys. **321**, 151 (1997)
D. Moss, D.M. Barker, A. Brandenburg, I. Tuominen, Astron. Astrophys. **294**, 155 (1995)
S. Muneer, K. Jayakumar, M.J. Rosario et al., Astron. Astrophys. **521**, A36 (2010)
R. Nityananda, R. Narayan, J. Astrophys. Astron. **3**, 419 (1982)
P.A. Nutzman, D.C. Fabrycky, J.J. Fortney, Astrophys. J. **740**, L10 (2011)
K. Oláh, Astrophys. Space Sci. **304**, 145 (2006)
K. Oláh, D.S. Hall, G.W. Henry, Astron. Astrophys. **251**, 531 (1991)
K. Oláh, Zs. Kővári, J. Bartus et al., Astron. Astrophys. **321**, 811 (1997)
K. Oláh, J. Jurcsik, K.G. Strassmeier, Astron. Astrophys. **410**, 685 (2003)
K. Oláh, Z. Kolláth, T. Granzer et al., Astron. Astrophys. **501**, 703 (2009)
K. Oláh, Zs. Kővári, K. Vida, K.G. Strassmeier, in *From Interacting Binaries to Exoplanets: Essential Modeling Tools*, ed. by M.T. Richards, I. Hubeny. Proc. IAU Symp., No. 282 (Cambridge University Press, Cambridge, 2012a), p. 478
K. Oláh, L. van Driel-Gesztelyi, K.G. Strassmeier, in *Comparative Magnetic Minima: Characterizing Quiet Times in the Sun and Stars*, ed. by D. Webb, C. Mandrini. Proc. IAU Symp., No. 286 (Cambridge University Press, Cambridge, 2012b), p. 279

K. Oláh, A. Moór, K.G. Strassmeier et al., Astron. Nachr. **334**, 625 (2013)
I. Pagano, A.F. Lanza, G. Leto et al., Earth Moon Planets **105**, 373 (2009)
J.R. Parks, R.J. White, G.H. Schaefer et al., in *16th Cambridge Workshop on Cool Stars, Stellar Systems, and the Sun*, ed. by C.M. Johns-Krull, M.K. Browning, A.A. West. ASP Conf. Ser., vol. 448 (Astron. Soc. of the Pacific, San Francisco, 2012), p. 1217
L. Paternò, Astrophys. Space Sci. **328**, 269 (2010)
P. Petit, J.-F. Donati, A. Collier Cameron, Mon. Not. R. Astron. Soc. **334**, 374 (2002)
P. Petit, J.-F. Donati, G.A. Wade et al., Mon. Not. R. Astron. Soc. **348**, 1175 (2004)
V.V. Pipin, A.G. Kosovichev, Astrophys. J. **738**, 104 (2011)
N. Piskunov, Pis'ma Astron. Zh. **11**, 44 (1985)
N. Piskunov, O. Kochukhov, Astron. Astrophys. **381**, 736 (2002)
G. Pojmanski, Acta Astron. **52**, 397 (2002)
D.G. Preminger, G.A. Chapman, A.M. Cookson, Astrophys. J. **739**, 45 (2011)
R.R. Radick, G.W. Lockwood, B.A. Skiff, S.L. Baliunas, Astrophys. J. Suppl. **118**, 239 (1998)
H. Rauher, C. Catala, C. Aerts et al., arXiv:1310.0696 (2013)
A. Reiners, Astron. Astrophys. **446**, 267 (2006)
T. Reinhold, A. Reiners, Astron. Astrophys. **557**, 11 (2013)
T. Reinhold, A. Reiners, G. Basri, Astron. Astrophys. **560**, A4 (2013)
M. Rhodes, E. Budding, M. Zeilik, in *Cool Stars, Stellar Systems, and the Sun: Sixth Cambridge Workshop*, ed. by G. Wallerstein. ASP Conf. Ser., vol. 9 (Astron. Soc. of the Pacific, San Francisco, 1990), p. 252
G. Ribárik, K. Oláh, K.G. Strassmeier, Astron. Nachr. **324**, 202 (2003)
M. Rodonò, S. Messina, A.F. Lanza et al., Astron. Astrophys. **358**, 624 (2000)
R.M. Roettenbacher, R.O. Harmon, N. Vutisalchavakul, G.W. Henry, Astron. J. **141**, 138 (2011)
R.M. Roettenbacher, J.D. Monnier, R.O. Harmon et al., Astrophys. J. **767**, 60 (2013)
M.J. Rosario, M.V. Mekkaden, A.V. Raveendran, Inf. Bull. Var. Stars **5836**, 1 (2008)
L. Rosén, O. Kochukhov, Astron. Astrophys. **548**, 8 (2012)
N. Rusomarov, O. Kochukhov, N. Piskunov et al., Astron. Astrophys. **558**, A8 (2013)
A. Ruzmaikin, Sol. Phys. **181**, 1 (1998)
I.S. Savanov, Astron. Rep. **56**, 716 (2012)
I.S. Savanov, in *Solar and Astrophysical Dynamos and Magnetic Activity*, ed. by A.G. Kosovichev, E.M. De Gouveia Dal Pino, Y. Yan. Proc. IAU Symp., No. 294 (Cambridge University Press, Cambridge, 2013), p. 257
E.T. Scharlemann, Astrophys. J. **246**, 292 (1981)
E.T. Scharlemann, Astrophys. J. **253**, 298 (1982)
C.J. Schrijver, C. Zwaan, Astron. Astrophys. **251**, 183 (1991)
M. Semel, Astron. Astrophys. **225**, 456 (1989)
A.I. Shapiro, W. Schmutz, E. Rozanov et al., Astron. Astrophys. **529**, A67 (2011)
A. Silva-Valio, A.F. Lanza, Astron. Astrophys. **529**, A36 (2011)
J. Silvester, O. Kochukhov, G.A. Wade, Mon. Not. R. Astron. Soc. **440**, 182 (2014), Advance Access
M.B. Skelly, Y.C. Unruh, A. Collier Cameron et al., Mon. Not. R. Astron. Soc. **385**, 708 (2008)
M. Sobotka, M. Vázquez, M. Sánchez Cuberes et al., Astrophys. J. **544**, 1155 (2000)
H.C. Spruit, Sol. Phys. **213**, 1 (2003)
K.G. Strassmeier, Astrophys. Space Sci. **140**, 223 (1988)
K.G. Strassmeier, Astron. Nachr. **326**, 269 (2005)
K.G. Strassmeier, Astron. Astrophys. Rev. **17**, 251 (2009)
K.G. Strassmeier, J. Bartus, Astron. Astrophys. **354**, 537 (2000)
K.G. Strassmeier, B.W. Bopp, Astron. Astrophys. **259**, 183 (1992)
K.G. Strassmeier, T. Pichler, M. Weber, T. Granzer, Astron. Astrophys. **411**, 595 (2003a)
K.G. Strassmeier, L. Kratzwald, M. Weber, Astron. Astrophys. **408**, 1103 (2003b)
K.G. Strassmeier, M. Woche, M. Andersen, I. Ilyin, Astron. Nachr. **328**, 627 (2007)
K.G. Strassmeier, T. Granzer, M. Kopf et al., Astron. Astrophys. **520**, A52 (2010)
K.G. Strassmeier, T.A. Carroll, M. Weber et al., Astron. Astrophys. **535**, A98 (2011)
K.G. Strassmeier, M. Weber, T. Granzer, S.P. Järvinen, Astron. Nachr. **333**, 663 (2012)
M. Švanda, M. Klvaňa, M. Sobotka, Astron. Astrophys. **458**, 301 (2006)
A.N. Tikhonov, A.V. Goncharsky, *Ill-Posed Problems in the Natural Sciences* (MIR Publishers, Moscow, 1987), p. 344
A. Tkachenko, T. Van Reeth, V. Tsymbal et al., Astron. Astrophys. **560**, A37 (2013)
L. van Driel-Gesztelyi, in *1st Advances in Solar Physics Euroconference. Advances in Physics of Sunspots*, ed. by B. Schmieder, J.C. Del Toro Iniesta, M. Vazquez. ASP Conf. Ser., vol. 118 (Astron. Soc. of the Pacific, San Francisco, 1997), p. 81
K. Vida, Zs. Kővári, M. Švanda et al., Astron. Nachr. **328**, 1078 (2007)

K. Vida, L. Kriskovics, K. Oláh, Astron. Nachr. **334**, 972 (2013)
K. Vida, K. Oláh, R. Szabó, Mon. Not. R. Astron. Soc. **441**, 2744 (2014)
S.S. Vogt, Astrophys. J. **250**, 327 (1981)
S.S. Vogt, in *The Impact of Very High S/N Spectroscopy on Stellar Physics*, ed. by G. Cayrel de Strobel, M. Spite. Proc. IAU Symp., No. 132 (Kluwer Academic Publishers, Dordrecht, 1988), p. 253
S.S. Vogt, A.P. Hatzes, in *The Sun and Cool Stars: Activity, Magnetism, Dynamos*, ed. by I. Tuominen, D. Moss, G. Rüdiger. Proc. IAU Colloq., No. 130 (Springer, Berlin, 1991), p. 297
S.S. Vogt, G.D. Penrod, Publ. Astron. Soc. Pac. **95**, 565 (1983)
S.S. Vogt, G.D. Penrod, A.P. Hatzes, Astrophys. J. **321**, 496 (1987)
S.S. Vogt, A.P. Hatzes, A.A. Misch, M. Kürster, Astrophys. J. Suppl. **121**, 547 (1999)
L.M. Walkowicz, G. Basri, J.A. Valenti, Astrophys. J. Suppl. **205**, 17 (2013)
F. Ward, Astrophys. J. **165**, 416 (1966)
A. Washuettl, K.G. Strassmeier, M. Weber, Astron. Nachr. **330**, 366 (2009)
M. Weber, Astron. Nachr. **328**, 1075 (2007)
M. Weber, K.G. Strassmeier, Astron. Astrophys. **330**, 1029 (1998)
M. Weber, K.G. Strassmeier, Astron. Astrophys. **373**, 974 (2001)
M. Weber, K.G. Strassmeier, in *Proc. of the 13th Cambridge Workshop on Cool Stars, Stellar Systems and the Sun, ESA SP-560*, ed. by F. Favata, G.A.J. Hussain, B. Battrick (ESA Publications Division, Noordwijk, 2005), p. 1029
M. Weber, K.G. Strassmeier, A. Washuettl, Astron. Nachr. **326**, 287 (2005)
M. Weber, T. Granzer, K.G. Strassmeier, SPIE **8451**, 84510K (2012)
M.A. Weber, Y. Fan, M.S. Miesch, Astrophys. J. **770**, 149 (2013)
W. Wehlau, J. Rice, in *Peculiar Versus Normal Phenomena in A-Type and Related Stars*, ed. by M.M. Dworetsky, F. Castelli, R. Faraggiana. ASP Conf. Ser., vol. 44 (Astron. Soc. of the Pacific, San Francisco, 1993), p. 247
O.C. Wilson, Astrophys. J. **153**, 221 (1968)
H. Wöhl, Astron. Nachr. **323**, 329 (2002)
J. Zhao, R.S. Bogart, A.G. Kosovichev et al., Astrophys. J. **774**, L29 (2013)

DOI 10.1007/978-1-4939-2584-1_17
Reprinted from *Space Science Reviews* Journal, DOI 10.1007/s11214-014-0083-1

Magnetic Flux Transport at the Solar Surface

J. Jiang · D.H. Hathaway · R.H. Cameron · S.K. Solanki · L. Gizon · L. Upton

Received: 30 April 2014 / Accepted: 3 August 2014 / Published online: 3 September 2014

Abstract After emerging to the solar surface, the Sun's magnetic field displays a complex and intricate evolution. The evolution of the surface field is important for several reasons. One is that the surface field, and its dynamics, sets the boundary condition for the coronal and heliospheric magnetic fields. Another is that the surface evolution gives us insight into the dynamo process. In particular, it plays an essential role in the Babcock-Leighton model of the solar dynamo. Describing this evolution is the aim of the surface flux transport model. The model starts from the emergence of magnetic bipoles. Thereafter, the model is based

J. Jiang (✉)
Key Laboratory of Solar Activity, National Astronomical Observatories, Chinese Academy of Sciences, Beijing 100012, China
e-mail: jiejiang@nao.cas.cn

D.H. Hathaway
NASA/MSFC, Huntsville, AL 35812, USA
e-mail: david.hathaway@nasa.gov

R.H. Cameron · S.K. Solanki · L. Gizon
Max-Planck-Institut für Sonnensystemforschung, Justus-von-Liebig-Weg 3, 37077 Göttingen, Germany
e-mail: cameron@mps.mpg.de

S.K. Solanki
School of Space Research, Kyung Hee University, Yongin, Gyeonggi-Do, 446-701, Korea
e-mail: solanki@mps.mpg.de

L. Gizon
Institut für Astrophysik, Georg-August-Universität Göttingen, Friedrich-Hund-Platz 1, 37077 Göttingen, Germany
e-mail: gizon@mps.mpg.de

L. Upton
Vanderbilt University, Nashville, TN 37235, USA

L. Upton
The University of Alabama in Huntsville, Huntsville, AL 35899, USA
e-mail: lar0009@uah.edu

on the induction equation and the fact that after emergence the magnetic field is observed to evolve as if it were purely radial. The induction equation then describes how the surface flows—differential rotation, meridional circulation, granular, supergranular flows, and active region inflows—determine the evolution of the field (now taken to be purely radial). In this paper, we review the modeling of the various processes that determine the evolution of the surface field. We restrict our attention to their role in the surface flux transport model. We also discuss the success of the model and some of the results that have been obtained using this model.

Keywords Sun: magnetic fields · Sun: photosphere · Flux transport · Meridional flow · Differential rotation · Diffusion

1 Introduction

The magnetic fields on the Sun are generated by dynamo action, ultimately driven by convective motions beneath the Sun's surface (Charbonneau 2010). Many of the physically important dynamo processes take place beneath the solar surface, where the details are mostly hidden from us. The tools we have for probing the subsurface dynamics of the magnetic fields are theory and helioseismology, both of which have unveiled some of the dynamics (for a review of helioseismic results see Gizon and Birch 2005).

Our knowledge of the magnetic field dynamics at the solar surface can be inferred from high resolution spectropolarimetric observations, for example, the *Hinode* spacecraft with about 230 km resolution (Tsuneta et al. 2008) and the *Sunrise* balloon-borne solar observatory with about 100 km resolution (Solanki et al. 2010), and is consequently much richer in detail. The magnetic field at the solar surface is observed to be structured on all spatial scales we can observe—from below the resolution limit of the largest available solar telescopes to the scale of the whole Sun (Solanki et al. 2006). In this review, we will concentrate exclusively on the evolution of the large-scale magnetic fields at the solar surface.

One reason for studying the evolution of the large-scale magnetic field on the solar surface is because that it sets the structure of the heliospheric magnetic field (Mackay and Yeates 2012). A second reason is that it is the observable part of the solar dynamo. In the context of the Babcock-Leighton dynamo (Babcock 1961; Leighton 1964), the surface evolution is particularly important because the source of poloidal flux in this model is the emergence and subsequent evolution of tilted magnetic bipolar regions.

The evolution of the surface magnetic field is, in its simplest form, almost trivial. Magnetic flux emerges at the solar surface in the form of bipolar magnetic regions. The flux is then transported and dispersed over the solar surface due to systematic and turbulent motions. Lastly, when magnetic flux of opposite polarity come into contact, the features cancel, removing equal amounts of flux of each sign.

These processes are modeled by the surface flux transport equation, which describes the evolution of the radial component of the magnetic field B_r on the solar surface. The equation is the r-component of the MHD induction equation at $r = R_\odot$ under the assumption that the field at the surface is purely vertical, augmented by a source term for B_r, and flux removal term, S and D respectively (see DeVore et al. 1984). The equation for the radial component of the field B_r at $r = R_\odot$ is then

$$\frac{\partial B_r}{\partial t} = -\frac{1}{R_\odot \sin\theta}\frac{\partial}{\partial \phi}(u B_r) - \frac{1}{R_\odot \sin\theta}\frac{\partial}{\partial \theta}(v B_r \sin\theta)$$

$$+\eta_H\left[\frac{1}{R_\odot^2\sin\theta}\frac{\partial}{\partial\theta}\left(\sin\theta\frac{\partial B_r}{\partial\theta}\right)+\frac{1}{R_\odot^2\sin^2\theta}\frac{\partial^2 B_r}{\partial\phi^2}\right]$$
$$+D(B_r)+S(\theta,\phi,t), \tag{1}$$

where $u(\phi,\theta,t)$ is the velocity in the longitudinal ($\hat{\phi}$) direction, $v(\phi,\theta,t)$ is the velocity in the latitudinal ($\hat{\theta}$) direction, η_H is the horizontal diffusivity at the surface (which we have assumed is uniform), D is some operator representing the removal of flux from the surface, and S is a source term describing the emergence of new flux rising from below, ϕ and θ are the solar longitude and colatitude respectively and $R_\odot$ is the solar radius.

In principle, both the surface velocity, $u(\phi,\theta,t)\hat{\boldsymbol{\phi}}+v(\phi,\theta,t)\hat{\boldsymbol{\theta}}$, and the radial component of the magnetic field are structured on all scales from tens of meters to the size of Sun, and evolves on time scales of seconds for the small scales to years for the largest scales. This renders the full problem intractable. For almost all problems, however, the full range of scales do not need to be dealt with, and average values of u and v can be used, with smaller unresolved velocities being treated as an enhanced diffusivity η_H. There is no single best choice of what temporal or spatial averaging should be done: different temporal and spatial averaging allow different science questions to be addressed.

In the following sections we will add flesh to Eq. (1) by describing in detail the relevant physical processes and the ways in which they can be modeled. We start with a deeper exposition of the basis for the surface flux transport model in Sect. 2. Then we describe some of the ways in which the source term S can be constructed in Sect. 3, and the flows and diffusivity in Sect. 4. The removal of the magnetic flux from the solar surface is reviewed in Sect. 5. The results from using the surface flux transport model will be presented in Sect. 6. Section 7 concludes our review.

2 Observational Basis for Solar Surface Flux Transport

The part of the magnetic field at the Sun's surface that dominates the signal in magnetograms, such as those recorded by the MDI instrument (Scherrer et al. 1995) on SOHO or by the HMI instrument (Scherrer et al. 2012; Schou et al. 2012) on SDO, is thought to be produced by a dynamo that resides deep in the solar convection zone or in the convective overshoot layer below the convection zone (e.g. Weiss and Thompson 2009; Charbonneau 2010). The toroidal field concentrated there becomes buoyantly unstable once it reaches a critical strength and a part of it, thought to be in the form of magnetic flux tubes, rises through the convection zone until it reaches the solar surface (Parker 1955; Choudhuri and Gilman 1987; Schüssler et al. 1994). On the way to the surface, the rising magnetic flux tube is affected by solar rotation (via the Coriolis force) and convection, which affect its path and hence the longitudes and latitudes at which the field finally emerges. See (Fan 2009) for a review. The combined effects of solar rotation and convection are also responsible for the orientation of two polarities at the solar surface (e.g., Joy's law) (Weber et al. 2011, 2013).

With its footpoints *simply* thought to remain connected with the horizontal toroidal magnetic field, the rising flux tube becomes akin to an Ω-shaped magnetic loop. The top of this loop is the first feature to appear above the solar surface. Its footpoints at the solar surface move apart rapidly as lower parts of the loop reach the solar atmosphere.

By the time the magnetic flux tube reaches the surface, it has typically been shredded into smaller features by the convection. Hence, on small scales the emerging magnetic field initially presents a complex pattern on the solar surface (Cheung et al. 2008). With time the many small magnetic structures partly grow together again. This is particularly striking in the case of sunspots, which often originally appear at the surface in the form of fragments that move together, joining up to form the final, larger sunspot. Young sunspots and active regions also display some amount of twisting motion (Brown et al. 2003), which is thought to be associated with the unwinding of the heavily twisted emerging magnetic loop.

Hence the horizontal motions associated with the early evolution of the magnetic field after it reaches the solar surface mainly appear to reflect its own internal dynamics, dictated by its rise and the interaction of the flux tube with the convection in the solar interior (as well as any unwinding that may happen in the process). However, even while the emergence process is ongoing, other forces start acting to move and shape the magnetic field at the solar surface.

Once at the surface the magnetic field is affected by a number of large—as well as small-scale flows. These include differential rotation (Howe 2009) and its variation in the form of torsional oscillations (Howard and Labonte 1980), meridional circulation (Miesch 2005; Rightmire-Upton et al. 2012; Zhao et al. 2013), and different scales of convection ranging from granulation (Nordlund et al. 2009) to supergranulation (Rieutord and Rincon 2010) and possibly larger scales (e.g., Hathaway et al. 2013). More about the large-scale and small-scale flows will be given in Sect. 4.

That these flows can drag along the magnetic field is related to the high magnetic Reynolds number, $R_m = UL/\eta$, where U is a typical flow velocity, L the length scale of the flow and η is the molecular magnetic diffusivity (which is inversely proportional to the electrical conductivity). In and on the Sun, at the scales we are interested in, we have $R_m \gg 1$, so that the magnetic field is frozen into the gas (Choudhuri 1998).

How strongly the horizontal components of the various flows at or close to the solar surface move the magnetic elements depends on both the strength of the flows relative to the strength of the magnetic field and how strongly the features are anchored below the surface. A critical quantity is the equipartition field strength, $B_{\mathrm{eq}} = \sqrt{4\pi\rho}v$, where ρ is the gas density and v is the magnitude of the velocity of the (convective) flow. Magnetic fields that are weak compared to B_{eq} will always be basically dragged by the flows, whereas stronger fields can influence the flows if they are anchored below (which requires $B \geq B_{\mathrm{eq}}$ all the way down to their anchoring depth). The expectation is then that the magnetic elements will move with some (weighted) average of the velocity field over the range from where it is anchored. It has been argued that even large, strong-field features at the solar surface, such as sunspots, lose the connection with their roots at the bottom of the convection zone at rather shallow depths (Schüssler and Rempel 2005). The simulations by Rempel (2011) indicated that the anchoring depth, which ranges from few Mm to dozens of Mm, is related to the lifetime of the sunspot.

On the Sun we have the interesting situation that while averaged over the solar disk the field strength is well below the equipartition value, the individual strong-field magnetic features have kG fields (e.g., Solanki et al. 2006). This makes their fields considerably stronger than B_{eq}, which is around 200–400 G (Solanki et al. 1996) in the lower photosphere for granular flows and smaller for slower flows (such as of supergranulation). The strong-field magnetic features, i.e., magnetic elements, pores and sunspots, make up the dominant part of the field seen in most magnetograms.

It turns out that the size of the magnetic features helps determine whether they affect the flow or are moved by it. Thus, sunspots are located at the centers of moat cells and pores also have a positive divergence of horizontal velocity surrounding them (Verma and Denker 2014). Smaller magnetic features, however, are almost always situated at the edges of convection cells. In the quiet Sun the magnetic field forms a network at the edges of supergranules, while in active regions the structuring is generally on a mesogranular scale (Domínguez

Cerdeña 2003). On a smaller scale magnetic elements are found almost exclusively at the edges of granules (Title et al. 1987; Solanki 1989). Hence observationally it is clear that the magnetic field is dragged along by convective flows on different scales. The effect of the meridional circulation is difficult to determine well from direct measurements (see Sect. 4.3) due to the slow speeds of a few $\mathrm{m\,s^{-1}}$ (but plays an important role in flux transport computations; see Sect. 6). The fact that the strong-field (i.e., kG) magnetic features are mostly aligned radially (i.e., vertically in the local solar coordinates, Martinez Pillet et al. 1997; Jafarzadeh et al. 2014), makes it easier for the field to be advected passively.

Studies of the motion of individual magnetic features show that these resemble a random-walk process, with the features moving between granules as these grow, evolve, move and die. On a larger scale these motions are affected by the location of the magnetic features within the supergranules, being subdiffusive in regions of converging supergranular flows and superdiffusive in the bodies of supergranules (Abramenko et al. 2011; Jafarzadeh et al. 2014).

Strong evidence that magnetic features are advected along with horizontal flows on the solar surface comes from the comparison of results from surface flux transport simulations with the observed distributions of magnetic fields. More about surface flux transport models will be given in the upcoming sections of this paper.

3 Sources of Magnetic Flux

In this section, we begin our description of the individual physical processes relevant to the evolution of the large-scale magnetic field on the Sun's surface. We begin with flux-emergence which is the process that brings magnetic field generated by dynamo action through the solar surface. The largest scales of emergence are large active regions with length scales on the order of 100 Mm and fluxes of $\sim 6 \times 10^{22}$ Mx. They are observed to extend down to the smallest scale loops currently observable (Centeno et al. 2007; Ishikawa et al. 2010) with fluxes of 10^{17} Mx, based on *Hinode* observations, and the almost ubiquitous emergence found by Hagenaar and Cheung (2009) and Danilovic et al. (2010) using *Hinode* and *Sunrise* observations respectively. Below currently resolvable limits, recirculation of magnetic fields and dynamo action in the turbulent intergranular lanes are believed to occur (de Wijn et al. 2009, and references therein).

The emergence processes have been modeled in detail for both large-scale active regions (e.g., Cheung et al. 2008, 2010; Stein et al. 2011) and for the small-scale dynamo processes (Vögler and Schüssler 2007; Schüssler and Vögler 2008). The physics involved includes magnetic buoyancy, magnetic tension, gravity, radiative cooling, thermodynamics including the effect of partial ionization, and small-scale turbulence which drains mass from the loops (see e.g., Cheung et al. 2008).

This review does not deal explicitly with intranetwork fields (the weak field that lies inside the superganular network), nor with the even smaller scale, more turbulent field found in the quiet Sun by the Hanle effect. See de Wijn et al. (2009) for a review of quiet-Sun fields. The evolution of such a field at the solar surface is expected to be different from that of the field produced by a global dynamo, given that the intranetwork field is relatively weak and horizontal (Lites et al. 2008; Jin et al. 2009), and hence is transported even more easily by convective flows. It is easily deformed and distributed by the turbulent convection, so that distinct magnetic features lose their identity relatively quickly.

The model to understand the solar surface flux transport process does not include the physics necessary to properly describe the evolution of the field during emergence, which are intrinsically three dimensional. Rather, the model assumes that the emergence occurs on a time scale much shorter than those otherwise of interest, enabling the emergence to be treated as occurring instantaneously. The source term for one particular emergence event (event i) therefore has the form $S_i(\theta,\phi,t) = S_i(\theta,\phi)\delta(t-t_i)$. The prescription of S_i is not unique in the literature, and depends on the purpose of the study and the observational data that are available to reconstruct S_i. Ordering them by the extent to which they include the details of observations of individual emergence events, the different ways of creating S_i are

1. Replacing magnetic fields at low latitudes by observations (e.g., Durrant and McCloughan 2004). This is a type of data assimilation.
2. Magnetogram based sources (e.g., Yeates et al. 2007).
3. Sunspot areas and locations, together with an empirically derived law to convert the areas to fluxes, with Joy's law (Sheeley et al. 1985), or a cycle-dependent version of Joy's law (Cameron et al. 2010), or the observed tilt angles of the individual groups.
4. Sunspot numbers, with the properties of the sunspots group based on random realizations of empirically derived distributions (e.g., Schrijver et al. 2002; Jiang et al. 2011b).
5. Empirical laws (e.g., van Ballegooijen et al. 1998).

For those methods that do not simply rely on magnetic field assimilation (i.e., methods 2–5 in the above list), $S_i(\theta,\phi)$ represents an isolated bipolar magnetic region, usually the superposition of positive and negative polarity patches displaced some distance from one another. The most important physical constraint on S_i is that the total (signed) flux vanishes over some small distance. This requirement follows from the induction equation

$$\frac{\partial \mathbf{B}}{\partial t} = \nabla \times (\mathbf{U} \times \mathbf{B}) - \nabla \times (\eta \nabla \times \mathbf{B}) \tag{2}$$

applied to a local patch of the solar surface Σ. By Stokes' theorem we have

$$\int_\Sigma \frac{\partial \mathbf{B}}{\partial t} \cdot \hat{\mathbf{n}} \mathrm{d}\Sigma = \int_{\partial \Sigma} (\mathbf{U} \times \mathbf{B} - \eta \nabla \times \mathbf{B}) \cdot \mathrm{d}\mathbf{l}, \tag{3}$$

where $\partial \Sigma$ is the boundary of Σ and $\hat{\mathbf{n}}$ is the unit vector normal to the surface element $\mathrm{d}\Sigma$. This reduces to

$$\frac{\partial}{\partial t} \int_\Sigma B_r \mathrm{d}\Sigma = \int_{\partial \Sigma} (\mathbf{U} \times \mathbf{B} - \eta \nabla \times \mathbf{B}) \cdot \mathrm{d}\mathbf{l}, \tag{4}$$

from which it can be seen that the only way the magnetic flux integrated over any region of the solar surface Σ can change is by advection or diffusion across the boundary of the region $\partial \Sigma$ (the argument given here is similar to that in Durrant et al. 2001). For truly instantaneous emergence, the opposite polarities must balance over a very small region. For emergence taking place over a day, the flux must be balanced on scales of about $\sim$100 Mm (this being the distance field can be carried by a 1 $\mathrm{km\,s^{-1}}$ flow over the course of a day).

Usually, each bipolar magnetic region is idealized as a pair of equal and opposite fluxes concentrated around the centroid of their respective polarities. Also, each such doublet is typically emerged suddenly at the time that its flux is largest. The contribution of the magnetic flux to the surface field is

$$S_i(\theta,\phi) = B_i^+(\theta,\phi) - B_i^-(\theta,\phi), \tag{5}$$

where $B_i^{\pm}$ is the flux distribution of the positive and negative polarity of the i-th bipolar magnetic region (BMR). Two major methods have been developed to give these distributions. One is from the NRL group, e.g., Sheeley et al. (1985), DeVore (1987) and Wang et al. (1989) who took each region as a point bipole. It has the form

$$B_i^{\pm}(\theta,\phi) = \frac{\Phi_i \delta(\theta-\theta_i^{\pm})\delta(\phi-\phi_i^{\pm})}{R_{\odot}^2 \sin\theta_i^{\pm}}, \tag{6}$$

where Φ_i is total flux of the BMR and $(\theta^{\pm}, \phi^{\pm})$ are the co-latitude and longitude of each polarity of the BMR. The other method was initiated by van Ballegooijen et al. (1998) and was adopted by others (Mackay et al. 2002a, 2002b; Baumann et al. 2004; Schüssler and Baumann 2006; Cameron et al. 2010; Jiang et al. 2011b; Upton and Hathaway 2014). Instead of point sources, they used finite-sized Gaussian-like polarity patches. The areas, locations (latitude and longitude), and latitudinal separations determined by the tilt angles of BMRs determine the source flux distribution. Specific details for sources used in many models and how the source parameters affect the flux transport are given in Sect. 6.

The long-term sunspot record from the network of observatories by Royal Greenwich Observatory (RGO), starting in May of 1874 and until 1976 and continued by the Solar Optical Observing Network (SOON) since 1976, provides daily observations of the location and area of sunspot groups. The systematic differences in the area measurements between the two datasets pose a barrier to understanding and reconstructing the long-term magnetic field evolution. A factor of about 1.4 was suggested to correct the SOON area to be homogeneous with RGO data (Balmaceda et al. 2009). Another disadvantage of RGO/SOON data is the absence of information concerning the tilt angles. The records of sunspots based on white-light photographs from the observatories at Mount Wilson (MWO) in the interval 1917–1985 (Howard et al. 1984) and at Kodaikanal in the interval 1906–1987 (Sivaraman et al. 1993) provide two large, but not complete, samples of sunspot group tilt angles. These records are being extended based on data from the Debrecen observatory (Győri et al. 2011). Magnetic polarities of the sunspot groups cannot be identified from the white-light photographs. The studies based on the magnetograms show that sunspot groups have reversed polarity orientations (anti-Hale source) with percentages ranging from 4 % to 10 % (Wang and Sheeley 1989; Tian et al. 2003; Stenflo and Kosovichev 2012; Li and Ulrich 2012).

The dependence of the statistical properties of sunspot emergence on the cycle phase and strength may be derived using the historic record of sunspot groups together with the group (R_G, Hoyt and Schatten 1998) or Wolf (R_Z, Wolf 1861) sunspot number. Using the group sunspot number R_G and RGO, MWO and Kodaikanal data sets, the main correlations found are as follows. (i) Strong cycles have a higher mean latitude for sunspot emergence (Waldmeier 1955; Solanki et al. 2008). The mean latitude at which sunspots emerge can be modeled using a second order polynomial of cycle phase (Jiang et al. 2011a). (ii) The distribution of sunspot areas is similar for all cycles (Bogdan et al. 1988). (iii) The size distribution is a power-law for small sunspots (Baumann and Solanki 2005) and obeys a log-normal profile for large sunspots (Bogdan et al. 1988). During cycle maxima, sunspots are larger on average (Jiang et al. 2011a). (iv) The cycle averaged tilt angle is anti-correlated with the cycle strength (Dasi-Espuig et al. 2010, 2013). (v) Sunspot nests are important, especially during cycle maximum phases. Using these empirical characteristics, the time-latitude diagram of sunspot group emergence (butterfly diagram) was reconstructed by Jiang et al. (2011a) from 1700 onward on the basis of the Wolf and group sunspot numbers. Figure 1 shows the comparison of butterfly diagrams from observation and reconstruction for the weakest cycle 14

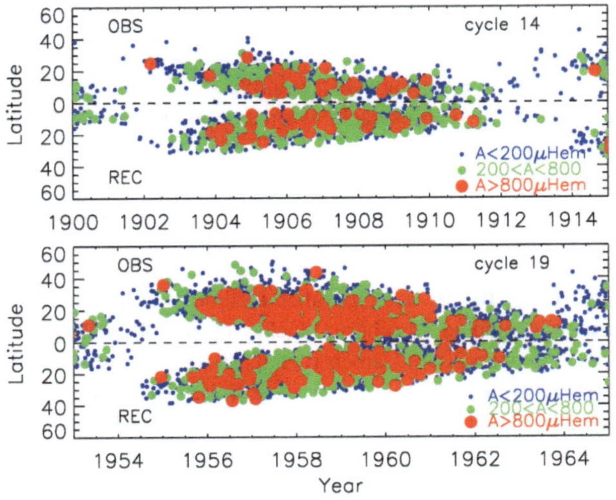

Fig. 1 Comparison of butterfly diagrams from observation (*above the horizontal dashed lines*) and reconstruction (*below the dashed lines*) for the weakest cycle 14 covered by RGO period (*upper panel*) and the strongest cycle 19 (*lower panel*), both for the northern hemisphere. The area of the sunspot groups is indicated by the colors and sizes of circles (from Jiang et al. 2011a)

covered RGO period (upper panel) and the strongest cycle 19 (lower panel), both for the northern hemisphere.

4 Flux Transport Processes

For any particular scale at the surface, the flows that transport the magnetic flux can conveniently be categorized as systematic flows or random motions. This distinction is only possible once the spatial and temporal scales relevant to the study have been decided. At scales below those that we are interested in, random flows with zero mean can be treated in several ways, as discussed below. The systematic flows include the differential rotation and the meridional circulation.

The random-walk effect introduced by the random flows can be treated as diffusion with a diffusivity estimated from the observed motions of the magnetic elements or the characteristics of the convective flows themselves. The differential rotation and meridional circulation can both be measured using a variety of techniques, including feature tracking, direct Doppler measurements, and helioseismology. A wide range of studies have been carried out to investigate the natures of the flux transport processes, which are reviewed in the following subsections.

4.1 Diffusion

One of the key terms of the flux transport is the horizontal diffusion of the radial component of the field. The Spitzer value for the magnetic diffusivity in the solar photosphere becomes relevant on scales of 30 km for a time scale of one day, which is a much smaller scale than the surface flux transport (SFT) model aims to capture. On the scales of interest, which are much larger than 30 km, there is a choice as to how to treat the random flows.

One approach, adopted by Schrijver (2001) and Upton and Hathaway (2014), is to include in the advection velocities and small-scale cellular flows or random motions corresponding to, e.g., supergranulation. The second, more commonly used approach, is to model the small-scale random motions as a turbulent diffusivity, η_H. The value of η_H is therefore not the

Spitzer diffusivity, but rather a parameterization of the effect of the turbulent near-surface convective motions on the magnetic field.

The initial estimation of η_H by Leighton (1964), based on the correct reversal time of the polar fields without including meridional flow, was in the range 770–1540 $km^2\,s^{-1}$. The value was lowered to around 200–600 $km^2\,s^{-1}$ once meridional flow was included (DeVore et al. 1984). Mosher (1977) derived a value of 200–400 $km^2\,s^{-1}$ using magnetic observations to trace the history of a typical solar active regions. Using similar methods, Schrijver and Martin (1990) estimated a diffusivity of about 250 $km^2\,s^{-1}$ in a quiet region surrounding a magnetic plage, and 110 $km^2 s^{-1}$ in the magnetic plage itself. The results from a number of observational studies are summarized in Table 1 of Schrijver et al. (1996). Values of η_H between 100 and 340 $km^2\,s^{-1}$ have been found on spatial scales in the 6 Mm range using comprehensive photospheric simulations with different upper boundary conditions (Cameron et al. 2011), and values of $\sim$100 $km^2\,s^{-1}$ based on a mean-field motivated analysis of numerical simulations and *Hinode* data (Rüdiger et al. 2012). The photospheric motions responsible for the turbulent diffusion range from turbulence in the intergranular lanes, through granular motions to supergranulation. Each of these types of motion occupies a range of spatial scales, and η_H in principle should therefore be a function of spatial scale k (Chae et al. 2008; Abramenko et al. 2011; Abramenko 2013), with the issue being complicated by the limited lifetime of the features being tracked and realization noise (Jafarzadeh et al. 2014). The values used in simulations cover the range suggested by observations, and a parameter study of the effects of varying η_H was reported by Baumann et al. (2004) and is discussed further in Sect. 6.

4.2 Differential Rotation

The Sun's differential rotation is the oldest known, and best characterized, flux transport process. It has a dynamic range of 250 $m\,s^{-1}$ in latitude and a well characterized latitudinal and radial structure thanks to helioseismology. The near-surface radial shear is also of importance for the magnetic flux transport as the magnetic elements are anchored within this layer. See also Beck (2000) for a review.

The motions of sunspots gave the first measure of the latitudinal differential rotation (first noted by Christoph Scheiner in 1610), with well-characterized rotation profiles given by Newton and Nunn (1951), by Ward (1966), and by Howard et al. (1984). These rotation profiles only cover the low latitudes (30° and below) and they indicate that spots of different sizes have different rotation rates (faster rotation for smaller spots). The rotation profile derived for all spots by Howard et al. (1984) is indicated by the dashed-dotted line in Fig. 2.

Direct Doppler measurements (Howard and Harvey 1970; Snodgrass et al. 1984; Ulrich et al. 1988) extend to all latitudes. These measurements indicate a slower rotation rate in the photosphere. The average profile measured by Ulrich et al. (1988) is plotted in Fig. 2 as a dashed line.

Global helioseismology (Thompson et al. 1996; Schou et al. 1998) gives a surface shear layer in which, at low to moderate latitudes, the rotation rate increases inward from the photosphere to a depth of about 50 Mm or 7 % of the solar radius. This shear layer is clearly seen in the lower latitudes but its structure becomes more uncertain at latitudes greater than about 50° (Corbard and Thompson 2002). Local Helioseismology gives similar results (Giles et al. 1998; Basu et al. 1999; Komm et al. 2003) that also indicate uncertainty at the higher latitudes. The profile obtained with global helioseismology by Schou et al. (1998) at $r = 0.995 R_\odot$ (a depth of 3.5 Mm) is plotted with the dotted line in Fig. 2.

The motions of the small magnetic elements (Komm et al. 1993b; Meunier 2005; Hathaway and Rightmire 2010, 2011) show a similar shape of the differential rotation profile, but

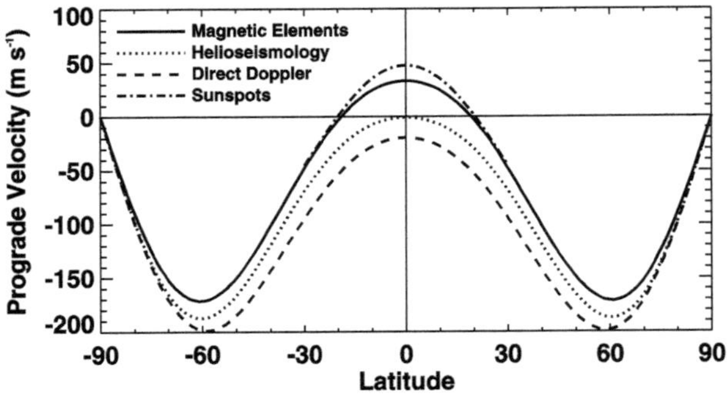

Fig. 2 Differential rotation profiles as measured by different methods. The profile from the small magnetic elements measured by Komm et al. (1993b) is given by the *solid line*. The profile from global helioseismology at $r = 0.995R_\odot$ measured by Schou et al. (1998) is given by the *dotted line*. The profile from direct Doppler measured by Ulrich et al. (1988) is given by the *dashed line*. The profile from individual sunspots measured by Howard et al. (1984) is given by the *dashed-dotted line*. The *zero line* represents solid body Carrington rotation

substantially faster rotation speeds than those given by direct Doppler measurements in the photosphere or from helioseismology at a depth of 3.5 Mm. The profile obtained by Komm et al. (1993b) is plotted with the solid line in Fig. 2 and is given by

$$u(\theta) = \big(33 - 281\cos^2\theta - 293\cos^4\theta\big)\sin\theta \ \mathrm{ms}^{-1} \tag{7}$$

where $u(\theta)$ is relative to the Carrington frame of reference.

The surface differential rotation varies over the course of each sunspot cycle in small but systematic ways. Changes in the overall shape of the differential rotation can be followed by tracking the changes in the coefficients that fit the profiles. Care should be taken, however, to cast the fits to the profiles in terms of orthogonal polynomials (in this case associated Legendre polynomials of order 1) as was suggested by Snodgrass (1984) to avoid crosstalk between the coefficients. Results of doing this for the measurements made with the small magnetic features are shown in Fig. 3. The average values obtained by Komm et al. (1993b), for the length of their study (1975 to 1991), are shown in orange with 1σ error bars for the first three north-south symmetric polynomials (given by the expression included within the figure). Komm et al. (1993b) also provided coefficients for cycle 21 maximum (1980–1982) and for cycle 21/22 minimum (1984–1985). These are shown in red with 1σ error bars. The results for individual Carrington rotations, obtained from SOHO/MDI magnetograms by Hathaway and Rightmire (2011), are shown in black with 2σ error bars. This is augmented by results from SDO/HMI magnetograms shown in blue with 2σ error bars.

All three coefficients are smaller (in absolute terms) at sunspot cycle maxima than they are at cycle minima. This gives a slightly faster (less negative relative to the Carrington rate) solid body rotation but a weaker differential rotation with less latitudinal shear at cycle maxima. The differences in the differential rotation flow profiles between cycle minima and maxima are nonetheless quite small as shown in Fig. 4.

In addition to these systematic changes to the basic profile there are the smaller scale, evolving perturbations referred to as torsional oscillations by Howard and Labonte (1980). These variations in the differential rotation profile are easily seen after removing an average profile (Howe et al. 2011). The deviations from the average profile are in the form of latitude bands with faster and slower than average rotation rates. The faster bands are located on the equatorward sides of the sunspot zones, while the slower bands are located on the poleward

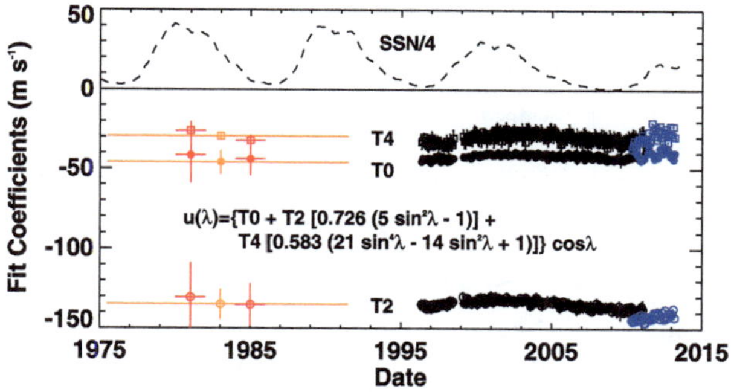

Fig. 3 The recent history of the polynomial fit coefficients for differential rotation profiles as measured by the motions of the magnetic elements. The coefficient T0 (giving solid body rotation) is represented by *filled circles*. T2 is represented by *open circles* and T4 by *open squares*. The Komm et al. (1993b) measurements for 1975–1991 are shown *in orange*. Cycle 21 maximum (1980–1982) and cycle 21/22 minimum (1984–1986) are shown *in red*. The Hathaway and Rightmire (2011) measurements for individual Carrington rotations (1996–2010) are shown *in black* while recent results obtained from SDO/HMI measurements are shown *in blue*

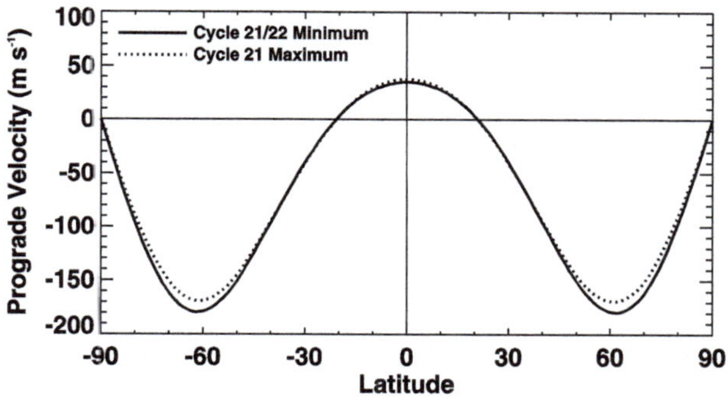

Fig. 4 Differential rotation profiles at sunspot cycle minimum and maximum. The profile for cycle 21/22 minimum (1984–1986) from Komm et al. (1993b) is represented by the *solid line*. The profile for cycle 21 maximum (1980–1982) by the *dotted line*

sides. This system of fast and slow streams drifts equatorward with the sunspot zones but are apparent at higher latitudes years before sunspots appear. While these flows are clearly associated with the solar cycle and are of considerable interest, the relative flows are quite weak ($\sim$5 $\mathrm{m\,s^{-1}}$) and thus probably of little consequence for surface flux transport. The torsional oscillations are also seen with helioseismology (Schou et al. 1998) and extend in depth throughout the convection zone (Vorontsov et al. 2002). In addition, helioseismology revealed the existence of a second torsional oscillation branch, which propagates poleward, at high latitudes (Schou 1999).

4.3 Meridional Circulation

A meridional flow was implicated in surface flux transport long before its strength (or even direction) had been well-determined. In his pioneering paper on the solar dynamo Babcock (1961) suggested that there was a meridional circulation that spread outward from the active latitudes. In his model, the higher latitude poleward flows would transport following polarity flux to the poles where it would reverse the polar fields halfway through the cycle and then build up new polar fields with the sign of the following polarity in each hemisphere. Babcock's model also included a low latitude equatorward flow that would transport preceding polarity flux to the equator, where it would cancel with the opposite polarity from the other

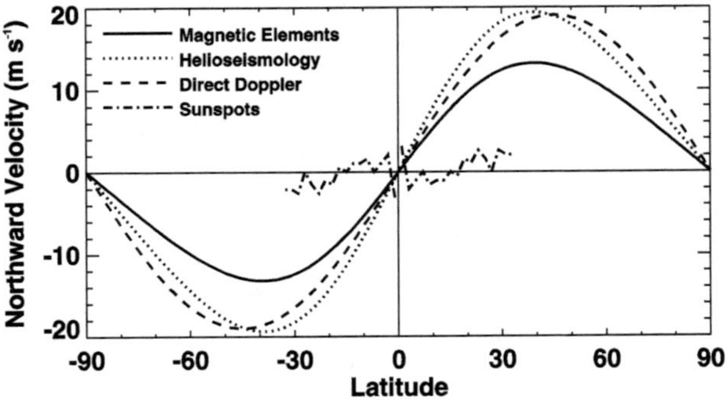

Fig. 5 Meridional flow profiles as measured by different methods. The profile from the small magnetic elements measured by Komm et al. (1993a) is given by the *solid line*. The profile from local helioseismology at $r = 0.998 R_\odot$ measured by Basu and Antia (2010) is given by the *dotted line*. The profile from direct Doppler measured by Hathaway (1996) is given by the *dashed line*. The profile from recurrent sunspot groups measured by Tuominen and Kyrolainen (1982) is given by the *dashed-dotted line*

hemisphere. This meridional flow seemed reasonable based on the effects of the Coriolis force on the differential rotation relative to the Carrington rotation frame of reference—the low latitude faster flow would be turned equatorward by the Coriolis force while the high latitude slower flow would be turned poleward. Babcock also cited observations of the motions of sunspots which suggested a meridional flow of this form. After the "discovery" of supergranules by Leighton et al. (1962), Babcock's meridional circulation was deemed unnecessary by Leighton (1964) who proposed that the surface flux transport was all done by a random walk of the magnetic elements due to evolving granules and supergranules.

The earliest measurements of the meridional flow were based on the motions of sunspot groups. Dyson and Maunder (1913) used sunspot group motions to refine the determination of the orientation of the Sun's rotation axis and noted a tendency for high latitude groups to move poleward and low latitude groups to move equatorward. Tuominen (1942) examined the meridional motions of recurring sunspot groups (groups that live long enough to be identified on more than one disk passage) and found that these groups did indeed diverge from the active latitudes with velocities of 1–2 $\mathrm{m\,s^{-1}}$. Similar results were found for individual sunspots by Howard and Gilman (1986). There are two significant problems in using the meridional motions of sunspots in surface flux transport models: sunspots do not appear at high latitudes (thereby leaving the meridional flow unknown poleward of about 40°) and the motion of sunspots may not representative of the surface meridional flow.

More or less complete latitude coverage is available with direct Doppler, helioseismology, and feature tracking using the small magnetic elements that populate the entire surface of the Sun. Figure 5 shows some of the meridional flow profiles that have been reported. There are small but significant differences in the surface velocity derived from the different techniques. This is partly because the measurements are all subject to systematic uncertainties and sample different depths.

Measurements of the near-surface meridional motions of the small magnetic elements (Komm et al. 1993a; Gizon et al. 2003; Hathaway and Rightmire 2010, 2011; Rightmire-Upton et al. 2012) can also be used to determine the meridional velocity. A typical cycle-averaged meridional flow profile determined by magnetic feature tracking is as given by Komm et al. (1993a) as

$$v(\theta) = \left(31.4 \cos\theta - 11.2 \cos^3\theta\right) \sin\theta \ \mathrm{m\,s^{-1}}. \tag{8}$$

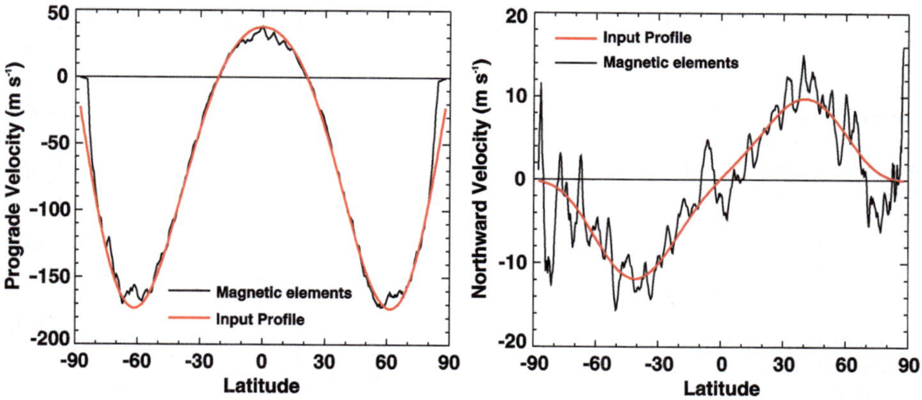

Fig. 6 Profiles of the differential rotation (*left*) and meridional flow (*right*) for the maximum of cycle 23 (calendar year 2000). The profiles input to the fully advective flux transport model of Upton and Hathaway (2014) are shown *in red*. The profiles measured using magnetic element feature tracking on this data are shown *in black*

Hathaway and Rightmire (2011) tested the sensitivity of magnetic feature tracking as a way of determining the large-scale flows to the effects of the random motions of the magnetic elements. They took a magnetic map representative of cycle maximum, represented the magnetic field distribution on a 4096-by-1500 grid in longitude and latitude by a collection of some 120,000 magnetic elements that were then advected by an evolving pattern of supergranules. They did not find any substantial flow away from the active latitudes as was suggested by Dikpati et al. (2010). They later (Upton and Hathaway 2014) produced a fully advective surface flux transport code in which the magnetic elements are transported by the flows in an evolving pattern of supergranules. They assimilate data from magnetograms on the Sun's near side but the field evolution on the far side is produced purely by the surface flux transport. Figure 6 shows that both the differential rotation and meridional flow measured by magnetic element feature tracking on the far side data for the maximum of cycle 23 (the year 2000) do not differ significantly from the input profiles for this choice of random (supergranular) motions. They argue that the velocity field determined in this way is the most consistent for use in the SFT model.

The interpretation of the Doppler measurements are complicated by the presence of the strong convective blue shift signal (Hathaway 1996; Ulrich 2010). This signal is an apparent blue shift in spectral lines due to the correlation between emergent intensity and radial flows in granules. It can vary by as much as 500 $\mathrm{m\,s^{-1}}$ between disk center and limb, and is affected by the presence of magnetic field. (Welsch et al. 2013 studied the Doppler velocity details in active regions and noted that the presence of magnetic fields can have substantial effects on the observed Doppler velocities). The Doppler signal from the meridional flow has a spatial structure similar to that of the convective blue shift but with a maximum of only 10 $\mathrm{m\,s^{-1}}$—hence the difficulty in measuring the meridional flow from the Doppler shift of spectral lines.

Measurements of the meridional flow have been made using several local helioseismic techniques, with similar results for the near-surface flows. The first such measurement (Giles et al. 1997) used the method of time-distance helioseismology (Duvall et al. 1993) and gave a poleward flow of approximately ~20 $\mathrm{m\,s^{-1}}$ at 30°. More recent near-surface measurements, covering most of cycle 23, are shown in Fig. 7 from two different techniques: MDI time-distance helioseismology measurements of the advection of the supergranulation pat-

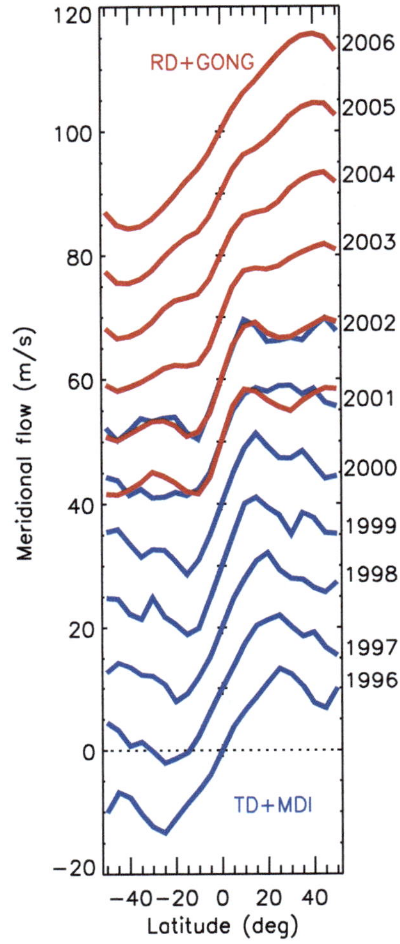

Fig. 7 Meridional circulation measured by local helioseismology in the near-surface layers and its evolution from 1996 to 2006. Individual years are shifted by multiples of 10 $\mathrm{m\,s^{-1}}$ for clarity. *Blue curves* show the results from Gizon and Rempel (2008) (time-distance measurements of the advection of the supergranulation pattern using SOHO/MDI data). *Red curves* show the results from González Hernández et al. (2008) (ring-diagram measurements using GONG data and multiplied by a factor of 0.8 to match the *blue curves* in the years 2001 and 2002). Only the antisymmetric components with respect to the equator are shown. This figure is taken from Gizon et al. (2010)

tern (Gizon et al. 2003; Gizon and Rempel 2008) and GONG ring-diagram helioseismology (González Hernández et al. 2008). The peak near-surface meridional velocity is about 15 $\mathrm{m\,s^{-1}}$ at a latitude of $\sim 30°$ in these more recent studies.

The time dependence of the meridional flow is clearly seen in the local helioseismology observations. We see in Fig. 7 that, in the rising phase (1996 to 2002) of the cycle, the latitude where the meridional circulation peaks moves towards lower latitudes. The time-varying component of the near-surface meridional flow is consistent with an inflow into the active latitudes (Gizon 2004; Zhao and Kosovichev 2004; Gizon and Rempel 2008; González Hernández et al. 2010). The inflows into individual active regions can be seen in two dimensional maps, first reported by Gizon et al. (2001) using f-mode time-distance helioseismology. A theory for the inflows, related to the enhanced cooling associated with the bright plage, was suggested by Spruit (2003) with a demonstration of the plausibility of the idea being given in Gizon and Rempel (2008).

Using magnetic feature tracking applied to MDI observations, Meunier (1999) detected clear changes in the meridional flow associated with active regions. The more extensive MDI measurements of Hathaway and Rightmire (2010) show changes with the solar cycle indicated by: (1) polynomial fits to the profiles (Fig. 8) and (2) by detailed changes to the

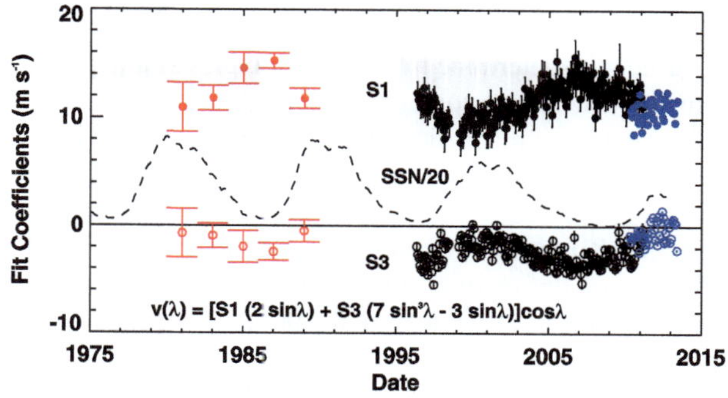

Fig. 8 The recent history of the polynomial fit coefficients for meridional flow profiles as measured by the motions of the magnetic elements. The coefficient S1 is represented by *filled circles*. S3 is represented by *open circles*. The Komm et al. (1993a) measurements for 1975–1991 are shown *in red*. The Hathaway and Rightmire (2011) measurements for individual Carrington rotations (1996–2010) are shown *in black* while recent results obtained from SDO/HMI measurements are shown *in blue*

meridional flow profiles (Hathaway and Rightmire 2011) fully consistent with superimposed inflows toward the active regions (Cameron and Schüssler 2010).

5 Sinks of Magnetic Flux

Without the supply of new flux introduced by S, the total unsigned flux at the solar surface monotonically decreases. For plausible values of the meridional flow speed and magnetic diffusivity, the e-folding time of the slowest decaying solution is about 4000 years (Cameron and Schüssler 2007). The slowest decaying solution consists of equal amounts of oppositely directed magnetic flux that is well separated, concentrated at the two poles. Much more rapid decay occurs when the two polarities are close to each other, with an e-folding time of at most a few years as the field of both polarities is advected to the poles where the two polarities then come into close proximity and cancel. The cancellation is, in most models, due to the diffusion term $\eta_H \nabla^2 B_r$, i.e. it is due to magnetic reconnection which is assumed to occur in the photosphere.

The second type of sink is represented by the term $D(B_r)$. This type of term was introduced by Schrijver et al. (2002) and Baumann et al. (2006). In physical terms, the idea is that processes below the surface of the Sun, where magnetic diffusion is also operating, cause the magnetic field at the surface to decay in-situ (Baumann et al. 2006). Because both sources and sinks are subject to the same requirement that changes in flux must be localized, Schrijver et al. (2002) suggests the possibility that the decay of the field is accompanied by the emergence of a large number of small, weak bipoles that together form a chain of loops, and allows the field to appear to decay in-situ without violating the argument presented in Sect. 3. As opposed to emergence events, the decay envisaged here is slow, and the fluxes are low, so that the observational signal of the large chains of bipoles can be lost in the noise of the omnipresent flux recycling.

Because $D(B_r)$ is a parametrisation of the physics of the highly dynamic convection zone, both its functional form and amplitude are open to discussion. Baumann et al. (2006), for example, consider that the functional form of $D(B_r)$ might reflect the eigen solutions of the problem of free-decay in a static convection zone with a uniform diffusivity, which fixes the functional form for the diffusivity. Schrijver et al. (2002), on the other hand, consider a simpler model where the field decays with a constant e-folding time.

Once the functional form of $D(B_r)$ is chosen, the question of its amplitude arises. Again this is, in principle, difficult to determine from first principles as it depends on the properties of the turbulence in the convection zone, with mean-field magnetohydrodynamic effects such as turbulent diffusivity and magnetic pumping playing a role. The strength of $D(B_r)$, in those cases where it has been included, is chosen to ensure that the polar fields reverse during each cycle.

In practical terms, $D(B_r)$ reduces the 4000 year memory of the SFT model to a few cycles. Such a reduction of the memory of the system might be physically justified (as suggested by Schrijver et al. 2002; Baumann et al. 2006), but also has the effect of removing the long term accumulation of small errors in the modeling. For example, Jiang et al. (2011b) used $D(B_r)$ to reduce the effects of the imperfect knowledge and therefore modeling of the source term S.

We comment that nonlinearities can be included in the model via the cycle dependence of the latitude and tilt angle at which sunspot groups emerge (Cameron et al. 2010; Jiang et al. 2011b), or the global meridional circulation rate (Wang et al. 2002), or localized inflows into the active region latitudes (Cameron and Schüssler 2012). Depending on the time being simulated, these nonlinearities can remove the need for a diffusive term to ensure the cyclic reversal of the polar fields and a match with observations.

6 Solar Surface Flux Transport Models

6.1 A Reference Model

In the above sections, we have presented the observational features of the solar surface flow and the surface flux source due to the BMR emergences. Babcock and Babcock (1955) speculated that the following flux of BMRs tended to migrate poleward, while the leading flux tended to migrate equatorward. The poleward migration of following flux would neutralize and reverse the solar polar fields over the course of a sunspot cycle. Babcock (1961) later speculated that the observed poleward migration might reflect a pattern of meridional flow on the Sun. Leighton (1964) proposed an alternative mechanism—that the random motions of magnetic flux by supergranular flows, together with Joy's law would lead to a preferred equatorward diffusion of leading flux and poleward diffusion of following flux. This proposal did not require other latitudinal transport mechanisms. However, Mosher (1977) showed that the diffusivity needed by Leighton was much higher than suggested by the observed flows and that a systematic flow was required. From the 1980s onwards such large-scale flows, including differential rotation and meridional flows have been included in the models (DeVore et al. 1984; Sheeley et al. 1985; DeVore 1987; Wang et al. 1989). A historical review of the development of the surface flux transport model has been given by Sheeley (2005). The models and applications of the magnetic flux transport at the solar surface flux were also reviewed by Mackay and Yeates (2012).

The SFT model (described by Eq. (1)) has been applied to the evolution of the Sun's global field (see DeVore et al. 1984, and numerous papers there after). It has also been applied (e.g., Schrijver 2001) to smaller scales, from large active regions to small ephemeral regions. It has been applied by treating the supergranular motions as a diffusivity, as well as by explicitly modeling them (Upton and Hathaway 2014).

For the differential rotation, the synodic rotation rate of the large-scale magnetic field, as measured by Snodgrass (1983), is widely used in SFT models. It is

$$\Omega(\theta) = 13.38 - 2.30\cos^2\theta - 1.62\cos^4\theta - 13.2 \text{ deg day}^{-1}. \quad (9)$$

For the meridional flow, the profiles

$$\upsilon(\theta) = 31.3|\sin\theta|^{2.5}\cos\theta \text{ m s}^{-1} \tag{10}$$

and

$$\upsilon(\theta) = \begin{cases} 11\sin[2.4*(90^\circ - \theta)] \text{ m s}^{-1} & \text{where } 15^\circ < \theta < 165^\circ \\ 0 & \text{otherwise,} \end{cases} \tag{11}$$

are close to the solid curves in Fig. 3 at middle and low latitudes. In contrast, a sharp gradient near the equator was used by Wang et al. (1989, 2009). The comparisons of the different profiles are shown in Fig. 11 of Hathaway and Rightmire (2011) and Fig. 3 of Jiang et al. (2013a).

As a reference model, we take the transport equation Eq. (1), with the source in the form of used by van Ballegooijen et al. (1998), transport parameters, i.e., meridional flow and differential rotation in the forms of Eq. (9) and (10), 250 $\text{km}^2\,\text{s}^{-1}$ horizontal turbulent diffusivity, and zero radial diffusivity.

6.2 Evolution of an Individual Sunspot Group: Effects of Different Model Parameters

The axisymmetric component of the large-scale field is measured by the axial dipole moment, which is defined as

$$D_{\text{Axial}}(t) = \frac{3}{4\pi}\int B(\theta,\phi,t)\cos\theta\sin\theta d\theta d\phi. \tag{12}$$

In this review, we do not consider the equatorial dipole field, which is strongly affected by differential rotation and hence has a short life time, on order of one year (DeVore 1987). We note that although such a field is not central for the solar cycle evolution of the Sun's surface field, it is an important ingredient in the evolution of solar open flux (Mackay et al. 2002a; Wang and Sheeley 2002).

6.2.1 Source Parameters

The initial contribution of an individual BMR with tilt angle α and total flux F (area A) located at colatitude θ, to the solar axial dipole field may be expressed as

$$D_{\text{BMR}} \propto dF\sin\theta\sin\alpha, \tag{13}$$

where d is the distance between the opposite polarities. The axial dipole of the bipole then evolves due to the latitudinal transport of the two polarities, which depends on both diffusion and flows. In the presence of diffusion alone, the axial dipole field decays on a time scale $\tau_d/2 = \frac{1}{2}R_\odot^2/\eta_H$ (Leighton 1964; Baumann et al. 2006), which is approximately 30 years for a diffusivity of 250 $\text{km}^2\,\text{s}^{-1}$. For the pure advection case, the dipole field is proportional to $\sin\theta$ and declines on the time scale $\tau_\upsilon \sim R_\odot/\upsilon_0 \sim 11$ years (Wang and Sheeley 1991) as both polarities are swept to the poles. In the presence of both systematic flows and diffusion (or random motions) a fraction of the magnetic field can cross the equator (under the action of the diffusive or random motions) after which they are kept apart by the meridional circulation.

Left panel of Fig. 9 from Jiang et al. (2014) shows the combined effect of diffusion and flow on the axial dipole field of a single BMR with area 1000 μHem, total flux 6×10^{21} Mx

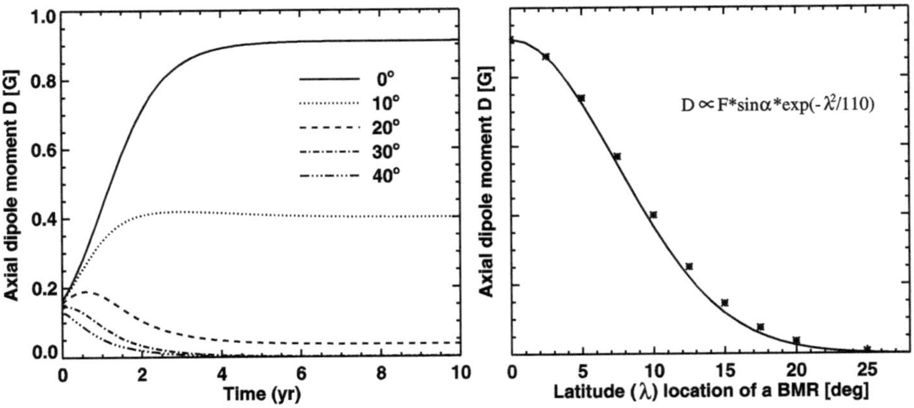

Fig. 9 The effects of various emergent latitudes for a single BMR with a total flux of 6×10^{21} Mx and tilt angle 80° (i.e. a nearly N–S oriented dipole) on the evolution of the Sun's axial dipole moment. *Left panel*: time evolution of the axial dipole moment; *Right panel*: eventual equilibrium axial dipole field contributed by the single BMR located at different latitudes. This figure is taken from Jiang et al. (2014)

and a large tilt angle of 80° emerged at different latitudes (latitudes 40°, 30°, 20°, 10° and 0°). The BMRs at the high latitudes and close to the equator display quite different dipole field evolution. For the cross-equator emergence (0°), the centroids of the two polarities are located at about ±4.3°. Advection in each hemisphere separates the polarities and causes the increase of the dipole field. Part (about half) of the flux diffuses and annihilates across the equator along the polarity inversion line (Mackay et al. 2002a). The remaining flux eventually concentrates around the poles and the dipole field reaches a plateau. Jiang et al. (2014) show that the equilibrium axial dipole field generated by the emergence of a single such extreme cross-equatorial BMR is about 20 % of the total simulated dipole field generated by all recorded sunspots groups of cycle 17, which had a medium amplitude. When the BMR emerges at 10° and 20°, the poleward flow gradient (larger gradient at lower latitudes) causes an increase of the separation between the polarities and an increase of the dipole field during the beginning phase. Then more leading flux is transported to the same pole and annihilated with the following polarity. This causes a weaker equilibrium field for a BMR emerging at higher latitude. For BMRs emerging at 30° and 40°, the dipole field diminishes in about 2 years. The right panel shows the relation between the final axial dipole field and the latitudinal location of the BMR with a given magnetic flux and tilt angle. The solid curve represents a Gaussian fit with a HWHM in latitude of 8.8°.

Hence, the large BMRs with large tilt angles emerging close to the equator contribute most to the solar axial dipole field. Usually the BMRs are assumed to obey the Hale's polarity law in the SFT models. The anti-Hale spots generate the same amplitude of the axial dipole field as the spots obeying Hale's law, but have opposite sign.

6.2.2 Transport Parameters

Differential rotation is one of the key ingredients in the evolution of the non axisymmetric component of the large-scale magnetic field (DeVore 1987). It has no effects on the axial dipole field. Hence we do not discuss its effects here.

Figure 10 shows the dependence of the BMR's axial dipole fields after reaching equilibrium on the diffusivity (left panel) and on the maximum meridional flow strength (right

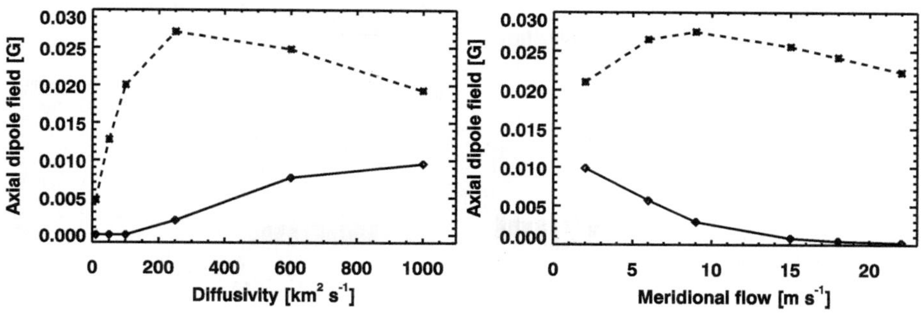

Fig. 10 Effects of various transport parameters on the eventual equilibrium axial dipole field for a single BMR with a total flux of 6×10^{21} Mx and tilt angle 5° deposited at a latitude of 8° (*dashed curves*) and 18° (*solid curves*) latitudes. *Left panel*: variation of the supergranular diffusivity; *Right panel*: variation of the maximum meridional flow

panel). The BMR has 1000 μHem area, 6×10^{21} Mx total flux and normal tilt angle 5°. We deposit the BMR at 8° (dashed line) and 18° (solid line) to show the different effects of the diffusion and meridional flow on the BMR eruptions at different latitudes. The reference model is used except for the variations of the diffusivity and the meridional flow strength.

The BMR located at high latitude (18°) generates higher equilibrium dipole fields at higher diffusivity since more flux from the leading polarity can diffuse across the equator and be transported into the opposite hemisphere. For the BMR located at low latitude (8°), the axial dipole field increases with the increase of the diffusivity when the diffusivity is low. When the diffusivity is further increased, more flux will be canceled between the two opposite polarities, which causes the decrease of the equilibrium dipole fields. The BMR at a latitude of 8° always generates a stronger dipole field than that at 18° latitude.

The axial dipole field monotonically decreases to zero with increasing meridional flow when the BMR is deposited at 18° latitude. This is because more leading polarity flux is transported to the same pole as the following polarity due to the stronger meridional flow. When the flow is strong enough, all the leading polarity flux is transported to the north pole without diffusion across the equator. When the BMR is deposited at 8° latitude, being close to the equator facilitates cross-equator diffusion. When the flow strength is low, more of the flux cancels before the equilibrium dipole field is established. Increasing flow speed decreases the flux cancellation and hence generates a stronger axial dipole field. When the flow is further increased, the flux diffusing across the equator decreases. Hence the axial dipole field decreases. This numerical simulation implies that the variation of the meridional flow might have different effects on the axial dipole field evolution of different cycles since the latitudinal distribution of the sunspot groups depends on the cycle strength (Solanki et al. 2008; Jiang et al. 2011a).

The effects of perturbations to the meridional flow in the form of inflows toward the active latitudes, as described in Sect. 4.3, on the evolution of solar surface axial dipole field was studied by Jiang et al. (2010b). In each hemisphere, an axisymmetric band of latitudinal flows converging toward the central latitude of the activity belt was superposed onto the background poleward meridional flow. The overall effect of these flow perturbations is to reduce the latitudinal separation of the magnetic polarities of a BMR and thus diminish its contribution to the equilibrium axial dipole field.

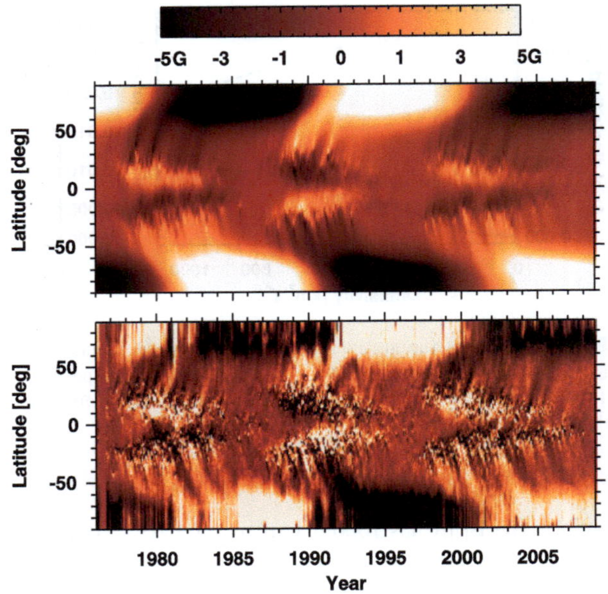

Fig. 11 Simulated and observed magnetic butterfly diagrams, i.e., time-latitude plots of the longitudinally averaged radial magnetic field at the solar surface. *Upper panel*: result of the flux transport simulation based on Jiang et al. (2010a). *Lower panel*: evolution of the observed field taken from NSO Kitt Peak synoptic maps

6.3 Simulations of Solar Cycles

6.3.1 Comparisons of Observed and Simulated Magnetic Butterfly Diagrams

In his original paper on the transport of solar magnetic flux, Leighton (1964) simulated the effect of the thousands of sources that occur during an entire sunspot cycle. Cycle 21 was the first cycle that permitted a realistic comparison with the observed field (Sheeley et al. 1985; DeVore and Sheeley 1987; Wang et al. 1989). The observed features of BMRs were derived from the full-disk magnetograms. The large-scale axisymmetric magnetic field features, such as the polar field structure, poleward surges and polar field reversals were well reproduced. The time evolution of the longitudinally averaged photospheric magnetic field, i.e, the magnetic butterfly diagram, is a good illustration of the large-scale field evolution under the flux transport process.

The upper panel of Fig. 11 shows the magnetic butterfly diagram resulting from a flux transport simulation, the source and transport parameters of which are based on Jiang et al. (2010a), see also Schüssler and Baumann (2006). The lower panel of Fig. 11 is produced from the Kitt Peak Solar Observatory synoptic magnetograms of the radial magnetic field. There are qualitative agreements between simulation and observation, particularly concerning the poleward surges of following-polarity magnetic flux leading to the reversals of the polar fields.

Some differences can also be identified between the simulated and the observed magnetic butterfly diagrams. For example, the observations have a more grainy structure, which leads to a high mean flux density at the activity belt, see Eq. (9) of Jiang et al. (2014) for the definition. The average of the observed values over the three cycle maxima is about 3G, which is about twice that of the simulated result. Furthermore, the simulations lack the occasional cross-equatorial flux plumes that appear in the data due to the large, highly tilted sunspot groups that emerge near the equator, for example in the years of 1980, 1986, and 2002 (Cameron et al. 2013).

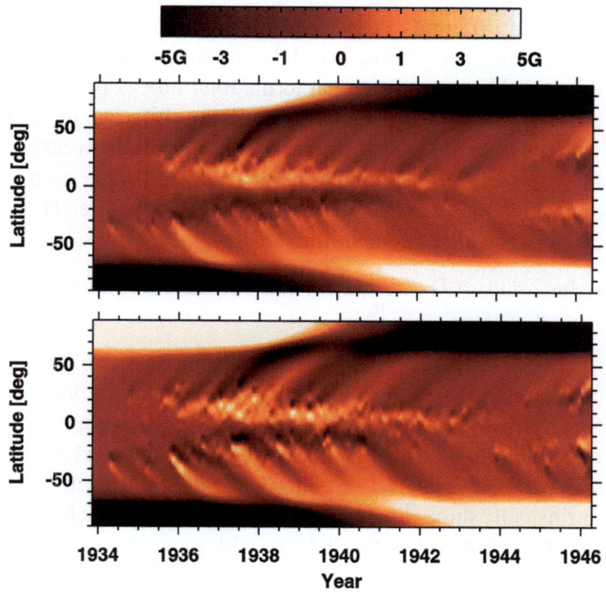

Fig. 12 Simulated magnetic butterfly diagrams of cycle 17. *Upper panel*: case without tilt angle scatter in sunspot groups; *Lower panel*: case with tilt angle scatter in sunspot groups, which shows a more grainy structure in the activity belts and more poleward surges with both polarities. This figure is based on Figs. 3(a) and 4(a) of Jiang et al. (2014)

The differences can mainly be attributed to the scatter in sunspot group tilt angles relative to Joy's law. Jiang et al. (2014) measured the tilt scatter based on the observed tilt angle data from MWO and Kodaikanal. The standard deviations (σ_α) of the tilt angles depend on the sunspot area in the form of $\sigma_\alpha = -11\log_{10}(A_u) + 35$, where A_u is the umbra area. Figure 12 shows the comparisons of the simulated magnetic butterfly diagrams using the observed sunspot records of cycle 17, which is a cycle with an average strength and not associated with a sudden increase or decrease with respect to the adjacent cycles, without (upper panel) and with (lower panel) the tilt scatter. The lower panel corresponds to one random realization of the sunspot group tilt scatter, which generates the similar polar field as the upper panel without the tilt scatter. The randomly occurring large tilt angles cause the more grainy structure, which is represented by an increase of the low latitude flux density by about 40 % compared to the case without tilt angle scatter. There are also more poleward surges with opposite polarities. Qualitatively, the magnetic butterfly diagram for the cases with tilt angle scatter is more similar to the observed counterpart for the last 3 cycles. See Jiang et al. (2014) for more details about the effects of the scatter in sunspot group tilt angles on the magnetic butterfly diagram. Occasionally, the near equator sunspot groups with big sizes have big tilt angles. According to Sect. 6.2.1, a single such event can significantly affect the axial dipole field at the end of the cycle. If the event obeys the Hale polarity law, it strengthens the axial dipole field. If the event is anti-Hale, it weakens the axial dipole field.

6.3.2 Simulations of Multiple Solar Cycles

The success of the SFT model, with BMR emergence as the main source of flux, opens the possibility for the reconstruction of the solar large-scale magnetic field into the past on the basis of recorded sunspot data. The observed cycle-to-cycle variations provide constraints for the modeling of the different physical processes in the model. When the BMR source amplitude fluctuations were included in the model, Wang et al. (2002) and Schrijver et al. (2002) found that the polar field cannot reverse polarity every ~11 yr. In their studies, the

BMRs of different cycles had the same range of latitude distributions. The tilt angles of BMRs obeyed Joy's law and did not depend on the cycle strength. The total intrinsic axial dipole field was proportional to the total flux of the emergent sunspot groups during a cycle. Under the same transport parameters, the strength of the polar field then varied linearly with the total amount of emerged flux. During the weaker cycles the flux supply was insufficient to cancel the existing polar field, to reverse it and to build up a new polar field of opposite polarity and of the same strength as before. Three different ways of resolving this discrepancy have been put forward.

- *Including in D a component due to the intrinsically three-dimensional nature of flux transport*

The reference SFT model described in Sect. 6.1 is explicitly two dimensional. With $S = D = 0$, there is no flux transport across the solar surface. In models simpler than the SFT, multi-year decay times were proposed by Solanki et al. (2000) to successfully describe the evolution of the total amount of open and total magnetic flux. Schrijver et al. (2002) and Baumann et al. (2006) introduced different forms of $D(B_r)$ in order to account for an intrinsically three-dimensional decay of the field. Schrijver et al. (2002) found that a simple exponential decay of the field τ_d with a decay time of about 10 yr allowed regular reversals of the polar fields given fluctuations in the source term similar to those in the historical records. Baumann et al. (2006) introduced a more detailed expression for $D(B_r)$ based on a parameterization of radial diffusion processes. A radial diffusivity of 100 $\mathrm{km^2\,s^{-1}}$ (corresponding to a decay time of $\sim$5 years for the dipole component) was suggested. See Sect. 5 for more discussions.

- *Nonlinearities in the transport parameters*

Variations in the meridional flow have been considered as an alternate way of ensuring the polar fields reverse at the end of each cycle. The two types of changes considered are a modulation of the global flow speed (Wang et al. 2002, 2005), or the inclusion of a localized inflow into active regions (Cameron and Schüssler 2012). The model of the inflow in the latter study was calibrated to helioseismic observations (Cameron and Schüssler 2010), although more work is needed to assimilate the raw observations into their model. Both types of nonlinearities can lead to reversals of the polar fields at the end of each cycle.

- *Nonlinearities in the source parameters*

In Sect. 3 we have listed the characteristics of sunspot group emergence. Strong cycles have a higher mean latitude (related to the Waldmeier effect; Waldmeier 1955) and a lower tilt angle for sunspot emergence (Dasi-Espuig et al. 2010). According to the results discussed in Sect. 6.2.1, both the latitudes and the tilts of the source term can significantly modulate the polar field generation. Cameron et al. (2010) made the first attempt to introduce nonlinearities in the source parameters to study the magnetic field evolution of multiple cycles. Figure 13 shows the average of the unsigned polar field strength from the flux transport model (red) and observed sunspot area (black). In agreement with observations, the polar field at the end of a solar cycle is correlated with the subsequent cycle strength (e.g., see Muñoz-Jaramillo et al. 2013), and similarly for the open flux (e.g., see Wang and Sheeley 2009).

Mixed approaches are also possible. Wang and Sheeley (2003) include the nonlinearities in both the source and the transport parameters to simulate the evolution of the Sun's large-scale magnetic field under Maunder minimum conditions. They showed that the regular polarity oscillations of the axial dipole and polar fields can be maintained if the source flux

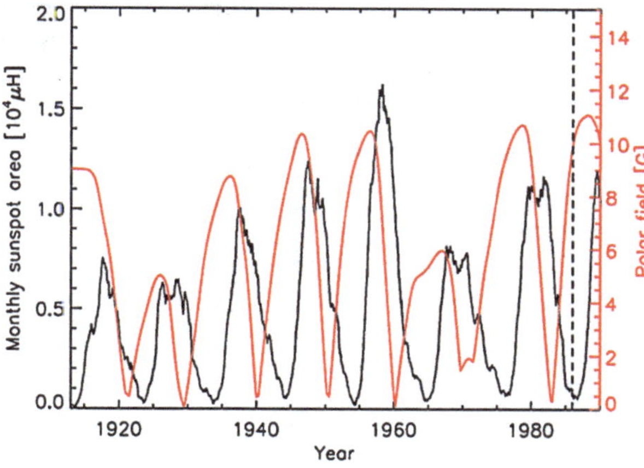

Fig. 13 Observed sunspot area (*black*) and average of the unsigned polar field strength from the flux transport model (*red*) obtained by including the nonlinearities in the flux source and with the input of the RGO sunspot area data during cycles 15 to 21 (from Cameron et al. 2010)

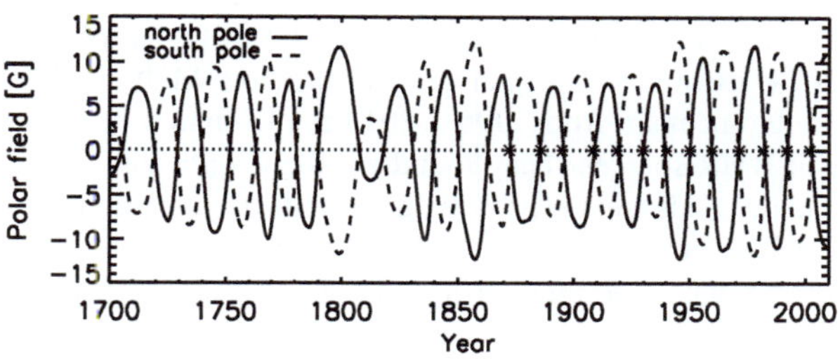

Fig. 14 Polar field evolution since 1700 from a flux transport simulation that includes the nonlinearities in the flux source, with the Wolf sunspot number data used as input (from Jiang et al. 2011b)

emerges at low latitudes (~10°) and the speed of the poleward surface flow was reduced from ~20 to ~10 $\mathrm{m\,s^{-1}}$. Jiang et al. (2011b) have used semi-synthetic records of emerging sunspot groups based on sunspot number data as input for a surface flux transport model to reconstruct the evolution of the large-scale solar magnetic field from the year 1700 onward. A nonlinear modulation of the tilt angles and emergence latitudes based on observations was included as well as a decay term D based on the formalism in Baumann et al. (2006) with $\eta_r = 25\ \mathrm{km^2\,s^{-1}}$ to reduce the error in the modeling due to the errors in the sunspot numbers. Figure 14 shows the reconstructed polar field based on Wolf sunspot number during 1700–2010 from (Jiang et al. 2011b).

6.3.3 Assimilations of Observed Magnetograms

Surface flux transport models have also been used to construct synchronic magnetic maps (maps of the magnetic field over the entire surface of the Sun for a given moment in time) for use in coronal field extrapolations and space weather predictions. In the above sections, the flux sources were idealized as magnetic dipoles produced by the emergence of BMRs. For synchronic map production, observed magnetograms are assimilated into a SFT model that then includes the magnetic field evolution on the far side of the Sun. Worden and Harvey (2000) used their flux transport model and the Kitt Peak synoptic magnetograms to update unobserved or poorly observed regions. Schrijver and De Rosa (2003) assimilated SOHO/MDI magnetograms within 60° from disk center into a SFT model with an duration of 5.5 yr and temporal resolution of 6 hours. With this they were able to approximate the evolution of the photospheric magnetic field on the unobservable hemisphere, and thus obtain a continuously evolving model of the surface field over the whole solar surface. Schrijver

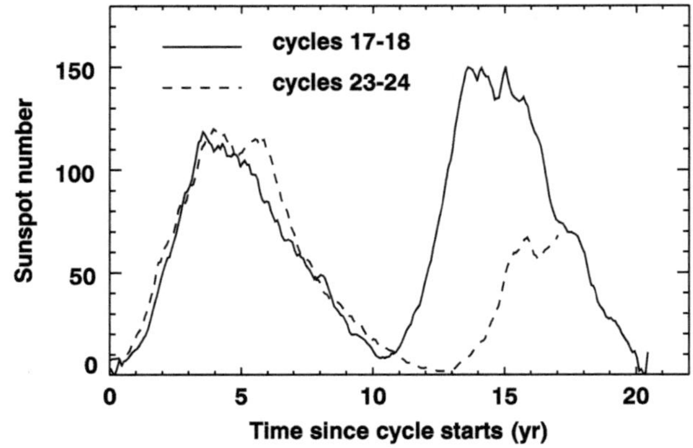

Fig. 15 Comparison of the time evolution of sunspot number with 12-month running mean between cycles 17–18 and cycles 23–24. The x-axis denotes the time since the starts of cycles 17 and 23. The two similar cycles 17 and 23 have substantially different subsequent cycles

and Liu (2008) extended the study throughout the whole of cycle 23 to further understand the large-scale transport of the magnetic flux in the solar photosphere. Upton and Hathaway (2014) assimilated magnetograms from both MDI and HMI to produce a "baseline" set of synchronic maps from 1996 to 2013 at a 15-minute cadence for comparison with maps made with BMR sources. They found excellent agreement and showed that predictions of polar field reversals and the polar field strength at cycle minimum could be made years in advance. McCloughan and Durrant (2002) and Durrant and McCloughan (2004) noted that flux transport produces and requires synchronic maps rather than traditional synoptic maps and care must therefore be taken when estimating transport parameters from synoptic maps.

Yeates et al. (2007) used synoptic magnetogram data as the initial condition and assimilated the emergence of new active regions into the model throughout the course of the simulation to maintain the accuracy of the simulated photospheric magnetic field over many months. The simulations were coupled with simulations of the 3 dimensional coronal magnetic field to explain the hemispheric pattern of the axial magnetic field direction in solar filaments (Yeates and Mackay 2009).

6.4 Peculiar Cycle 23 Minimum

The polar field at the end of cycle 23 was unexpectedly weak, which caused the unusual properties of the polar corona, the open flux, and the solar wind at that time, see Wang et al. (2009) and Jiang et al. (2013a) for more details. As shown in Fig. 15, cycle 23 has a similar amplitude and shape as cycle 17. However, the amplitudes of their subsequent cycles, cycles 24 and 18, are very different. The cycle strength is proportional to the polar field at the end of the preceding cycle (Jiang et al. 2007; Muñoz-Jaramillo et al. 2013), which implies that cycles with similar amplitudes can generate rather different amounts of polar flux at the end of the cycles. This situation poses an interesting challenge to surface flux transport models.

Schrijver and Liu (2008), Wang et al. (2009) and Jiang et al. (2013a) simulated the evolution of the photospheric field of cycle 23 using flux transport models. Sunspot number data were used to determine the number of BMRs emergence at a given time. These studies could produce the observed weak polar field strength by increasing the meridional flow relative to the reference case.

Yeates (2014) simulated cycle 23 by inserting individual BMR with properties matching those in observed Kitt Peak synoptic magnetograms. They also found that their standard flux transport model is insufficient to simultaneously reproduce the observed polar fields and butterfly diagram during cycle 23, and that additional effects must be added. The variations they considered include an increase of the meridional flow to 35 $\mathrm{m\,s^{-1}}$, decrease of the supergranular diffusivity to 200 $\mathrm{km^2\,s^{-1}}$, decrease of the sunspot groups tilt angle by 20 %, decrease of the flux per sunspot groups by 20 %, inclusion of D in Eq. (1) with a decay time of 5 years, decrease of the tilt angle of the sunspot groups by 20 % coupled with radial diffusion in about 10 years, and the inflow toward the active regions.

Stochastic variations in sunspot group emergence is another possible cause of the weak cycle 23 minimum. As shown in Sect. 6.2.1, large highly tilted BMRs that emerge at low latitudes produce cross-equatorial flux plumes in the synoptic magnetograms and provide a large contribution to the axial dipole field. Cameron et al. (2014) simulated cycles 21–23 and showed that the magnetic flux from four observed cross-equatorial flux plumes could provide one explanation for the weakness of the polar fields at the end of solar cycle 23.

7 Conclusions

The solar photosphere is a thin layer between the high plasma-β solar interior and the low plasma-β solar atmosphere. It is the layer where the energy transport changes from convective to radiative, the layer where the poloidal field is generated in the Babcock-Leighton model and critically it is the layer that we can observe and best measure the magnetic field. The dynamics of the magnetic field in this layer are, based on observations, particularly simple: emergence, dispersion and advection by surface velocities, and eventually cancellation with opposite polarity flux. These few processes explain the evolution of the large-scale magnetic field at the solar surface, and beyond it in the corona and the heliosphere. In this paper we have reviewed these processes and shown how they can impact the evolution of the Sun's magnetic field and the sunspot cycle.

The surface flux transport is the key to understanding what produces the polar fields and the axial dipole moment seen at activity minima. The strength of the polar fields at this phase of the activity cycle is well correlated with the strength of the next solar cycle and can be used as a reliable predictor (Schatten et al. 1978; Schatten and Sofia 1987; Svalgaard et al. 2005; Jiang et al. 2007; Wang and Sheeley 2009; Muñoz-Jaramillo et al. 2013). In some Babcock-Leighton type dynamo models (e.g., Chatterjee et al. 2004; Jiang et al. 2013b), this correlation exists because the poloidal field generated by the surface flux transport can be quickly transported to the tachocline where it gets wound up by the differential rotation to produce the strong toroidal flux that emerges in the sunspots of the next cycle.

The strength of the polar fields and the axial dipole moment depend on the surface flux transport processes—both the active region sources (total magnetic flux, polarity separation, and latitude of emergence) and the surface flows (differential rotation, meridional flow, and the random convective flows). These processes have been found to vary systematically with both the phase and the strength of sunspot cycles.

The transport processes are dominated by the observed surface flows that include both the large-scale axisymmetric flows (differential rotation and meridional flow) and the smaller scale non-axisymmetric flows (granules, supergranules, giant cells, and flows associated with active regions). These non axisymmetric convective flows are usually treated as diffusion. Some models also include a decay term in addition to the observed surface flows. The combined effects of these transport processes on the emergent sunspot groups impact the Sun's axial dipole magnetic field in different ways depending on latitude. While high latitude sunspots typically have more latitudinal separation between polarities, sunspots emerging closer to the equator can contribute more to the axial dipole moment by way of cross-equatorial cancelation.

We note that an important aspect of the magnetic flux transport at the solar surface is the natural tendency for perturbations in the sizes of sunspot cycles to produce cycles that continue to grow in size or decay in size (with the inability to reverse the polar fields). On the Sun this tendency must be held in check by some nonlinear feedback mechanism. We discussed some of the possible mechanisms—active region tilt dependent on cycle size, active region latitude distribution dependent on cycle size, variations in the meridional flow dependent on cycle size. At this time it is not clear which, if any, of these mechanisms dominate. It may be that one mechanism limits the growth while another limits the decay and the competition between the two keeps sunspot cycles from exhibiting even more variability.

We now have more that a cycle of reasonably high-resolution and high temporal cadence observations of the magnetic field and the surface flows from SOHO/MDI and SDO/HMI. Extending backwards in time we have over a hundred years of daily records of sunspot group sizes and locations, as well as knowledge of the Sun's open magnetic flux inferred from geomagnetic field measurements (see the review by Svalgaard, this volume). Looking even further back in time, we have sunspot number data extending through the Maunder Minimum. Given this data (and in particular the well-observed transition from large cycle 22 to small cycle 24), we expect that the evolution of the Sun's large-scale magnetic field is entering a new stage of understanding.

Acknowledgements We are grateful to the referee for helpful comments on the paper. We acknowledge the support from ISSI Bern, for our participation in the workshop on the solar activity cycle: physical causes and consequences. J.J. acknowledges the financial support by the National Natural Science Foundations of China (11173033, 11221063, 2011CB811401) and the Knowledge Innovation Program of the CAS (KJCX2-EW-T07). S.K.S. acknowledges the partial support for this work by the BK21 plus program through the National Research Foundation (NRF) funded by the Ministry of Education of Korea. L.G. acknowledges support from DFG SFB 963 *Astrophysical Flow Instabilities and Turbulence* (Project A18/1) and from EU FP7 Collaborative Project *Exploitation of Space Data for Innovative Helio- and Asteroseismology* (SPACEINN).

References

V.I. Abramenko, Fractal multi-scale nature of solar/stellar magnetic fields, in *IAU Symposium*, ed. by A.G. Kosovichev, E. de Gouveia Dal Pino, Y. Yan, IAU Symposium, vol. 294 (2013), pp. 289–300. doi:10.1017/S1743921313002652

V.I. Abramenko, V. Carbone, V. Yurchyshyn, P.R. Goode, R.F. Stein, F. Lepreti, V. Capparelli, A. Vecchio, Turbulent diffusion in the photosphere as derived from photospheric bright point motion. Astrophys. J. **743**, 133 (2011). doi:10.1088/0004-637X/743/2/133

H.W. Babcock, The topology of the Sun's magnetic field and the 22-YEAR cycle. Astrophys. J. **133**, 572 (1961). doi:10.1086/147060

H.W. Babcock, H.D. Babcock, The Sun's magnetic field, 1952–1954. Astrophys. J. **121**, 349 (1955). doi:10.1086/145994

L.A. Balmaceda, S.K. Solanki, N.A. Krivova, S. Foster, A homogeneous database of sunspot areas covering more than 130 years. J. Geophys. Res. (Space Phys.) **114**, 7104 (2009). doi:10.1029/2009JA014299

S. Basu, H.M. Antia, Characteristics of solar meridional flows during solar cycle 23. Astrophys. J. **717**, 488–495 (2010). doi:10.1088/0004-637X/717/1/488

S. Basu, H.M. Antia, S.C. Tripathy, Ring diagram analysis of near-surface flows in the Sun. Astrophys. J. **512**, 458–470 (1999). doi:10.1086/306765

I. Baumann, S.K. Solanki, On the size distribution of sunspot groups in the Greenwich sunspot record 1874–1976. Astron. Astrophys. **443**, 1061–1066 (2005). doi:10.1051/0004-6361:20053415

I. Baumann, D. Schmitt, M. Schüssler, A necessary extension of the surface flux transport model. Astron. Astrophys. **446**, 307–314 (2006). doi:10.1051/0004-6361:20053488

I. Baumann, D. Schmitt, M. Schüssler, S.K. Solanki, Evolution of the large-scale magnetic field on the solar surface: a parameter study. Astron. Astrophys. **426**, 1075–1091 (2004). doi:10.1051/0004-6361:20048024

J.G. Beck, A comparison of differential rotation measurements—(Invited review). Sol. Phys. **191**, 47–70 (2000). doi:10.1023/A:1005226402796

T.J. Bogdan, P.A. Gilman, I. Lerche, R. Howard, Distribution of sunspot umbral areas—1917–1982. Astrophys. J. **327**, 451–456 (1988). doi:10.1086/166206

D.S. Brown, R.W. Nightingale, D. Alexander, C.J. Schrijver, T.R. Metcalf, R.A. Shine, A.M. Title, C.J. Wolfson, Observations of rotating sunspots from TRACE. Sol. Phys. **216**, 79–108 (2003). doi:10.1023/A:1026138413791

R. Cameron, M. Schüssler, Solar cycle prediction using precursors and flux transport models. Astrophys. J. **659**, 801–811 (2007). doi:10.1086/512049

R. Cameron, A. Vögler, M. Schüssler, Decay of a simulated mixed-polarity magnetic field in the solar surface layers. Astron. Astrophys. **533**, 86 (2011). doi:10.1051/0004-6361/201116974

R.H. Cameron, M. Schüssler, Changes of the solar meridional velocity profile during cycle 23 explained by flows toward the activity belts. Astrophys. J. **720**, 1030–1032 (2010). doi:10.1088/0004-637X/720/2/1030

R.H. Cameron, M. Schüssler, Are the strengths of solar cycles determined by converging flows towards the activity belts? Astron. Astrophys. **548**, 57 (2012). doi:10.1051/0004-6361/201219914

R.H. Cameron, J. Jiang, D. Schmitt, M. Schüssler, Surface flux transport modeling for solar cycles 15–21: effects of cycle-dependent tilt angles of sunspot groups. Astrophys. J. **719**, 264–270 (2010). doi:10.1088/0004-637X/719/1/264

R.H. Cameron, M. Dasi-Espuig, J. Jiang, E. Işık, D. Schmitt, M. Schüssler, Limits to solar cycle predictability: cross-equatorial flux plumes. Astron. Astrophys. **557**, 141 (2013). doi:10.1051/0004-6361/201321981

R.H. Cameron, J. Jiang, M. Schüssler, L. Gizon, Physical causes of solar cycle amplitude variability. J. Geophys. Res. (Space Phys.) **119**, 680–688 (2014). doi:10.1002/2013JA019498

R. Centeno, H. Socas-Navarro, B. Lites, M. Kubo, Z. Frank, R. Shine, T. Tarbell, A. Title, K. Ichimoto, S. Tsuneta, Y. Katsukawa, Y. Suematsu, T. Shimizu, S. Nagata, Emergence of small-scale magnetic loops in the quiet-Sun internetwork. Astrophys. J. Lett. **666**, 137–140 (2007). doi:10.1086/521726

J. Chae, Y.E. Litvinenko, T. Sakurai, Determination of magnetic diffusivity from high-resolution solar magnetograms. Astrophys. J. **683**, 1153–1159 (2008). doi:10.1086/590074

P. Charbonneau, Dynamo models of the solar cycle. Living Rev. Sol. Phys. **7**, 3 (2010). doi:10.12942/lrsp-2010-3

P. Chatterjee, D. Nandy, A.R. Choudhuri, Full-sphere simulations of a circulation-dominated solar dynamo: exploring the parity issue. Astron. Astrophys. **427**, 1019–1030 (2004). doi:10.1051/0004-6361:20041199

M.C.M. Cheung, M. Schüssler, T.D. Tarbell, A.M. Title, Solar surface emerging flux regions: a comparative study of radiative MHD modeling and hinode SOT observations. Astrophys. J. **687**, 1373–1387 (2008). doi:10.1086/591245

M.C.M. Cheung, M. Rempel, A.M. Title, M. Schüssler, Simulation of the formation of a solar active region. Astrophys. J. **720**, 233–244 (2010). doi:10.1088/0004-637X/720/1/233

A.R. Choudhuri, *The Physics of Fluids and Plasmas: An Introduction for Astrophysicists* (Cambridge University Press, Cambridge, 1998)

A.R. Choudhuri, P.A. Gilman, The influence of the Coriolis force on flux tubes rising through the solar convection zone. Astrophys. J. **316**, 788–800 (1987). doi:10.1086/165243

T. Corbard, M.J. Thompson, The subsurface radial gradient of solar angular velocity from MDI f-mode observations. Sol. Phys. **205**, 211–229 (2002). doi:10.1023/A:1014224523374

S. Danilovic, B. Beeck, A. Pietarila, M. Schüssler, S.K. Solanki, V. Martínez Pillet, J.A. Bonet, J.C. del Toro Iniesta, V. Domingo, P. Barthol, T. Berkefeld, A. Gandorfer, M. Knölker, W. Schmidt, A.M. Title, Transverse component of the magnetic field in the solar photosphere observed by SUNRISE. Astrophys. J. Lett. **723**, 149–153 (2010). doi:10.1088/2041-8205/723/2/L149

M. Dasi-Espuig, S.K. Solanki, N.A. Krivova, R. Cameron, T. Peñuela, Sunspot group tilt angles and the strength of the solar cycle. Astron. Astrophys. **518**, 7 (2010). doi:10.1051/0004-6361/201014301

M. Dasi-Espuig, S.K. Solanki, N.A. Krivova, R. Cameron, T. Peñuela, Sunspot group tilt angles and the strength of the solar cycle (Corrigendum). Astron. Astrophys. **556**, 3 (2013). doi:10.1051/0004-6361/201014301e

A.G. de Wijn, J.O. Stenflo, S.K. Solanki, S. Tsuneta, Small-scale solar magnetic fields. Space Sci. Rev. **144**, 275–315 (2009). doi:10.1007/s11214-008-9473-6

C.R. DeVore, The decay of the large-scale solar magnetic field. Sol. Phys. **112**, 17–35 (1987). doi:10.1007/BF00148484

C.R. DeVore, N.R. Sheeley Jr., Simulations of the Sun's polar magnetic fields during sunspot cycle 21. Sol. Phys. **108**, 47–59 (1987). doi:10.1007/BF00152076

C.R. DeVore, J.P. Boris, N.R. Sheeley Jr., The concentration of the large-scale solar magnetic field by a meridional surface flow. Sol. Phys. **92**, 1–14 (1984). doi:10.1007/BF00157230

M. Dikpati, P.A. Gilman, R.K. Ulrich, Physical origin of differences among various measures of solar meridional circulation. Astrophys. J. **722**, 774–778 (2010). doi:10.1088/0004-637X/722/1/774

I. Domínguez Cerdeña, Evidence of mesogranulation from magnetograms of the Sun. Astron. Astrophys. **412**, 65–68 (2003). doi:10.1051/0004-6361:20034617

C.J. Durrant, J. McCloughan, A method of evolving synoptic maps of the solar magnetic field, II. Comparison with observations of the polar fields. Sol. Phys. **219**, 55–78 (2004). doi:10.1023/B:SOLA.0000021830.88336.86

C.J. Durrant, J.M. Kress, P.R. Wilson, The evolution of trailing plumes from active regions. Sol. Phys. **201**, 57–69 (2001). doi:10.1023/A:1010393806958

T.L. Duvall Jr., S.M. Jefferies, J.W. Harvey, M.A. Pomerantz, Time-distance helioseismology. Nature **362**, 430–432 (1993). doi:10.1038/362430a0

F.W. Dyson, E.W. Maunder, Sun, axis, the position of, from photographs, 1874–1912. Mon. Not. R. Astron. Soc. **73**, 673 (1913)

Y. Fan, Magnetic fields in the solar convection zone. Living Rev. Sol. Phys. **6**(4) (2009). doi:10.12942/lrsp-2009-4. http://www.livingreviews.org/lrsp-2009-4

P.M. Giles, T.L. Duvall Jr., A.G. Kosovichev, Solar rotation and large-scale flows determined by time-distance helioseismology MDI, in *New Eyes to See Inside the Sun and Stars*, ed. by F.-L. Deubner, J. Christensen-Dalsgaard, D. Kurtz, IAU Symposium, vol. 185 (1998), p. 149

P.M. Giles, T.L. Duvall, P.H. Scherrer, R.S. Bogart, A subsurface flow of material from the Sun's equator to its poles. Nature **390**, 52–54 (1997). doi:10.1038/36294

L. Gizon, Helioseismology of time-varying flows through the solar cycle. Sol. Phys. **224**, 217–228 (2004). doi:10.1007/s11207-005-4983-9

L. Gizon, A.C. Birch, Local helioseismology. Living Rev. Sol. Phys. **2**, 6 (2005). doi:10.12942/lrsp-2005-6

L. Gizon, M. Rempel, Observation and modeling of the solar-cycle variation of the meridional flow. Sol. Phys. **251**, 241–250 (2008). doi:10.1007/s11207-008-9162-3

L. Gizon, A.C. Birch, H.C. Spruit, Local helioseismology: three-dimensional imaging of the solar interior. Annu. Rev. Astron. Astrophys. **48**, 289–338 (2010). doi:10.1146/annurev-astro-082708-101722

L. Gizon, T.L. Duvall, J. Schou, Wave-like properties of solar supergranulation. Nature **421**, 43–44 (2003). doi:10.1038/nature01287

L. Gizon, T.L. Duvall Jr., R.M. Larsen, Probing surface flows and magnetic activity with time-distance helioseismology, in *Recent Insights into the Physics of the Sun and Heliosphere: Highlights from SOHO and Other Space Missions*, ed. by P. Brekke, B. Fleck, J.B. Gurman, IAU Symposium, vol. 203 (2001), p. 189

I. González Hernández, S. Kholikov, F. Hill, R. Howe, R. Komm, Subsurface meridional circulation in the active belts. Sol. Phys. **252**, 235–245 (2008). doi:10.1007/s11207-008-9264-y

I. González Hernández, R. Howe, R. Komm, F. Hill, Meridional circulation during the extended solar minimum: another component of the torsional oscillation? Astrophys. J. Lett. **713**, 16–20 (2010). doi:10.1088/2041-8205/713/1/L16

L. Győri, T. Baranyi, A. Ludmány, Photospheric data programs at the Debrecen Observatory, in *IAU Symposium*, ed. by D. Prasad Choudhary, K.G. Strassmeier, IAU Symposium, vol. 273 (2011), pp. 403–407. doi:10.1017/S174392131101564X

M. Hagenaar, M. Cheung, Magnetic flux emergence on different scales, in *The Second Hinode Science Meeting: Beyond Discovery-Toward Understanding*, ed. by B. Lites, M. Cheung, T. Magara, J. Mariska, K. Reeves, Astronomical Society of the Pacific Conference Series, vol. 415 (2009), p. 167

D.H. Hathaway, Doppler measurements of the Sun's meridional flow. Astrophys. J. **460**, 1027 (1996). doi:10.1086/177029

D.H. Hathaway, L. Rightmire, Variations in the Sun's meridional flow over a solar cycle. Science **327**, 1350 (2010). doi:10.1126/science.1181990

D.H. Hathaway, L. Rightmire, Variations in the axisymmetric transport of magnetic elements on the Sun: 1996–2010. Astrophys. J. **729**, 80 (2011). doi:10.1088/0004-637X/729/2/80

D.H. Hathaway, L. Upton, O. Colegrove, Giant convection cells found on the Sun. Science **342**, 1217–1219 (2013). doi:10.1126/science.1244682

R. Howard, P.A. Gilman, Meridional motions of sunspots and sunspot groups. Astrophys. J. **307**, 389–394 (1986). doi:10.1086/164425

R. Howard, J. Harvey, Spectroscopic determinations of solar rotation. Sol. Phys. **12**, 23–51 (1970). doi:10.1007/BF02276562

R. Howard, B.J. Labonte, The sun is observed to be a torsional oscillator with a period of 11 years. Astrophys. J. Lett. **239**, 33–36 (1980). doi:10.1086/183286

R. Howard, P.I. Gilman, P.A. Gilman, Rotation of the sun measured from Mount Wilson white-light images. Astrophys. J. **283**, 373–384 (1984). doi:10.1086/162315

R. Howe, Solar interior rotation and its variation. Living Rev. Sol. Phys. **6**, 1 (2009). doi:10.12942/lrsp-2009-1

R. Howe, F. Hill, R. Komm, J. Christensen-Dalsgaard, T.P. Larson, J. Schou, M.J. Thompson, R. Ulrich, The torsional oscillation and the new solar cycle. J. Phys. Conf. Ser. **271**(1), 012074 (2011). doi:10.1088/1742-6596/271/1/012074

D.V. Hoyt, K.H. Schatten, Group sunspot numbers: a new solar activity reconstruction. Sol. Phys. **181**, 491–512 (1998). doi:10.1023/A:1005056326158

R. Ishikawa, S. Tsuneta, J. Jurčák, Three-dimensional view of transient horizontal magnetic fields in the photosphere. Astrophys. J. **713**, 1310–1321 (2010). doi:10.1088/0004-637X/713/2/1310

S. Jafarzadeh, S.K. Solanki, A. Lagg, L.R. Bellot Rubio, M. van Noort, A. Feller, S. Danilovic, Astron. Astrophys. (2014, submitted)

S. Jafarzadeh, R.H. Cameron, S.K. Solanki, A. Pietarila, A. Feller, A. Lagg, A. Gandorfer, Migration of ca ii h bright points in the internetwork. Astron. Astrophys. **563**, 101 (2014). doi:10.1051/0004-6361/201323011

J. Jiang, R.H. Cameron, M. Schüssler, Effects of the scatter in sunspot group tilt angles on the large-scale magnetic field at the solar surface. Astrophys. J. **791**, 5 (2014). doi:10.1088/0004-637X/791/1/5

J. Jiang, P. Chatterjee, A.R. Choudhuri, Solar activity forecast with a dynamo model. Mon. Not. R. Astron. Soc. **381**, 1527–1542 (2007). doi:10.1111/j.1365-2966.2007.12267.x

J. Jiang, R. Cameron, D. Schmitt, M. Schüssler, Modeling the Sun's open magnetic flux and the heliospheric current sheet. Astrophys. J. **709**, 301–307 (2010a). doi:10.1088/0004-637X/709/1/301

J. Jiang, E. Işik, R.H. Cameron, D. Schmitt, M. Schüssler, The effect of activity-related meridional flow modulation on the strength of the solar polar magnetic field. Astrophys. J. **717**, 597–602 (2010b). doi:10.1088/0004-637X/717/1/597

J. Jiang, R.H. Cameron, D. Schmitt, M. Schüssler, The solar magnetic field since 1700. I. Characteristics of sunspot group emergence and reconstruction of the butterfly diagram. Astron. Astrophys. **528**, 82 (2011a). doi:10.1051/0004-6361/201016167

J. Jiang, R.H. Cameron, D. Schmitt, M. Schüssler, The solar magnetic field since 1700. II. Physical reconstruction of total, polar and open flux. Astron. Astrophys. **528**, 83 (2011b). doi:10.1051/0004-6361/201016168

J. Jiang, R.H. Cameron, D. Schmitt, M. Schüssler, Can surface flux transport account for the weak polar field in cycle 23? Space Sci. Rev. **176**, 289–298 (2013a). doi:10.1007/s11214-011-9783-y

J. Jiang, R.H. Cameron, D. Schmitt, E. Işık, Modeling solar cycles 15 to 21 using a flux transport dynamo. Astron. Astrophys. **553**, 128 (2013b). doi:10.1051/0004-6361/201321145

C. Jin, J. Wang, M. Zhao, Vector magnetic fields of solar granulation. Astrophys. J. **690**, 279–287 (2009). doi:10.1088/0004-637X/690/1/279

R.W. Komm, R.F. Howard, J.W. Harvey, Meridional flow of small photospheric magnetic features. Sol. Phys. **147**, 207–223 (1993a). doi:10.1007/BF00690713

R.W. Komm, R.F. Howard, J.W. Harvey, Rotation rates of small magnetic features from two- and one-dimensional cross-correlation analyses. Sol. Phys. **145**, 1–10 (1993b). doi:10.1007/BF00627979

R. Komm, R. Howe, B.R. Durney, F. Hill, Temporal variation of angular momentum in the solar convection zone. Astrophys. J. **586**, 650–662 (2003). doi:10.1086/367608

R.B. Leighton, Transport of magnetic fields on the Sun. Astrophys. J. **140**, 1547 (1964). doi:10.1086/148058

R.B. Leighton, R.W. Noyes, G.W. Simon, Velocity fields in the solar atmosphere. I. Preliminary report. Astrophys. J. **135**, 474 (1962). doi:10.1086/147285

J. Li, R.K. Ulrich, Long-term measurements of sunspot magnetic tilt angles. Astrophys. J. **758**, 115 (2012). doi:10.1088/0004-637X/758/2/115

B.W. Lites, M. Kubo, H. Socas-Navarro, T. Berger, Z. Frank, R. Shine, T. Tarbell, A. Title, K. Ichimoto, Y. Katsukawa, S. Tsuneta, Y. Suematsu, T. Shimizu, S. Nagata, The horizontal magnetic flux of the quiet-Sun internetwork as observed with the hinode spectro-polarimeter. Astrophys. J. **672**, 1237–1253 (2008). doi:10.1086/522922

D. Mackay, A. Yeates, The Sun's global photospheric and coronal magnetic fields: observations and models. Living Rev. Sol. Phys. **9**, 6 (2012). doi:10.12942/lrsp-2012-6

D.H. Mackay, E.R. Priest, M. Lockwood, The evolution of the Sun's open magnetic flux—I. A single bipole. Sol. Phys. **207**, 291–308 (2002a). doi:10.1023/A:1016249917230

D.H. Mackay, E.R. Priest, M. Lockwood, The evolution of the Sun's open magnetic flux—II. Full solar cycle simulations. Sol. Phys. **209**, 287–309 (2002b). doi:10.1023/A:1021230604497

V. Martinez Pillet, B.W. Lites, A. Skumanich, Active region magnetic fields. I. Plage fields. Astrophys. J. **474**, 810 (1997). doi:10.1086/303478

J. McCloughan, C.J. Durrant, A method of evolving synoptic maps of the solar magnetic field. Sol. Phys. **211**, 53–76 (2002). doi:10.1023/A:1022400324489

N. Meunier, Large-scale dynamics of active regions and small photospheric magnetic features. Astrophys. J. **527**, 967–976 (1999). doi:10.1086/308111

N. Meunier, Magnetic network dynamics: activity level, feature size and anchoring depth. Astron. Astrophys. **436**, 1075–1086 (2005). doi:10.1051/0004-6361:20042414

M.S. Miesch, Large-scale dynamics of the convection zone and tachocline. Living Rev. Sol. Phys. **2**, 1 (2005). doi:10.12942/lrsp-2005-1

J.M. Mosher, The magnetic history of solar active regions, Ph.D. thesis, California Institute of Technology, Pasadena, 1977

A. Muñoz-Jaramillo, M. Dasi-Espuig, L.A. Balmaceda, E.E. DeLuca, Solar cycle propagation, memory, and prediction: insights from a century of magnetic proxies. Astrophys. J. Lett. **767**, 25 (2013). doi:10.1088/2041-8205/767/2/L25

H.W. Newton, M.L. Nunn, The Sun's rotation derived from sunspots 1934–1944 and additional results. Mon. Not. R. Astron. Soc. **111**, 413 (1951)

Å. Nordlund, R.F. Stein, M. Asplund, Solar surface convection. Living Rev. Sol. Phys. **6**, 2 (2009). doi:10.12942/lrsp-2009-2

E.N. Parker, The formation of sunspots from the solar toroidal field. Astrophys. J. **121**, 491 (1955). doi:10.1086/146010

M. Rempel, Subsurface magnetic field and flow structure of simulated sunspots. Astrophys. J. **740**, 15 (2011). doi:10.1088/0004-637X/740/1/15

M. Rieutord, F. Rincon, The Sun's supergranulation. Living Rev. Sol. Phys. **7**, 2 (2010). doi:10.12942/lrsp-2010-2

L. Rightmire-Upton, D.H. Hathaway, K. Kosak, Measurements of the Sun's high-latitude meridional circulation. Astrophys. J. **761**, 14 (2012). doi:10.1088/2041-8205/761/1/L14

G. Rüdiger, M. Küker, R.S. Schnerr, Cross helicity at the solar surface by simulations and observations. Astron. Astrophys. **546**, 23 (2012). doi:10.1051/0004-6361/201219268

K.H. Schatten, S. Sofia, Forecast of an exceptionally large even-numbered solar cycle. Geophys. Res. Lett. **14**, 632–635 (1987). doi:10.1029/GL014i006p00632

K.H. Schatten, P.H. Scherrer, L. Svalgaard, J.M. Wilcox, Using dynamo theory to predict the sunspot number during solar cycle 21. Geophys. Res. Lett. **5**, 411–414 (1978). doi:10.1029/GL005i005p00411

P.H. Scherrer, R.S. Bogart, R.I. Bush, J.T. Hoeksema, A.G. Kosovichev, J. Schou, W. Rosenberg, L. Springer, T.D. Tarbell, A. Title, C.J. Wolfson, I. Zayer, MDI Engineering Team, The solar oscillations investigation—Michelson Doppler Imager. Sol. Phys. **162**, 129–188 (1995). doi:10.1007/BF00733429

P.H. Scherrer, J. Schou, R.I. Bush, A.G. Kosovichev, R.S. Bogart, J.T. Hoeksema, Y. Liu, T.L. Duvall, J. Zhao, A.M. Title, C.J. Schrijver, T.D. Tarbell, S. Tomczyk, The helioseismic and magnetic imager (HMI) investigation for the solar dynamics observatory (SDO). Sol. Phys. **275**, 207–227 (2012). doi:10.1007/s11207-011-9834-2

J. Schou, Migration of zonal flows detected using Michelson Doppler Imager f-mode frequency splittings. Astrophys. J. Lett. **523**, 181–184 (1999). doi:10.1086/312279

J. Schou, H.M. Antia, S. Basu, R.S. Bogart, R.I. Bush, S.M. Chitre, J. Christensen-Dalsgaard, M.P. di Mauro, W.A. Dziembowski, A. Eff-Darwich, D.O. Gough, D.A. Haber, J.T. Hoeksema, R. Howe, S.G. Korzennik, A.G. Kosovichev, R.M. Larsen, F.P. Pijpers, P.H. Scherrer, T. Sekii, T.D. Tarbell, A.M. Title, M.J. Thompson, J. Toomre, Helioseismic studies of differential rotation in the solar envelope by the solar oscillations investigation using the Michelson Doppler Imager. Astrophys. J. **505**, 390–417 (1998). doi:10.1086/306146

J. Schou, P.H. Scherrer, R.I. Bush, R. Wachter, S. Couvidat, M.C. Rabello-Soares, R.S. Bogart, J.T. Hoeksema, Y. Liu, T.L. Duvall, D.J. Akin, B.A. Allard, J.W. Miles, R. Rairden, R.A. Shine, T.D. Tarbell, A.M. Title, C.J. Wolfson, D.F. Elmore, A.A. Norton, S. Tomczyk, Design and ground calibration of the helioseismic and magnetic imager (HMI) instrument on the solar dynamics observatory (SDO). Sol. Phys. **275**, 229–259 (2012). doi:10.1007/s11207-011-9842-2

C.J. Schrijver, Simulations of the photospheric magnetic activity and outer atmospheric radiative losses of cool stars based on characteristics of the solar magnetic field. Astrophys. J. **547**, 475–490 (2001). doi:10.1086/318333

C.J. Schrijver, M.L. De Rosa, Photospheric and heliospheric magnetic fields. Sol. Phys. **212**, 165–200 (2003). doi:10.1023/A:1022908504100

C.J. Schrijver, Y. Liu, The global solar magnetic field through a full sunspot cycle: observations and model results. Sol. Phys. **252**, 19–31 (2008). doi:10.1007/s11207-008-9240-6

C.J. Schrijver, S.F. Martin, Properties of the large- and small-scale flow patterns in and around AR 19824. Sol. Phys. **129**, 95–112 (1990). doi:10.1007/BF00154367

C.J. Schrijver, M.L. De Rosa, A.M. Title, What is missing from our understanding of long-term solar and heliospheric activity? Astrophys. J. **577**, 1006–1012 (2002). doi:10.1086/342247

C.J. Schrijver, R.A. Shine, H.J. Hagenaar, N.E. Hurlburt, A.M. Title, L.H. Strous, S.M. Jefferies, A.R. Jones, J.W. Harvey, T.L. Duvall Jr., Dynamics of the chromospheric network: mobility, dispersal, and diffusion coefficients. Astrophys. J. **468**, 921 (1996). doi:10.1086/177747

M. Schüssler, I. Baumann, Modeling the Sun's open magnetic flux. Astron. Astrophys. **459**, 945–953 (2006). doi:10.1051/0004-6361:20065871

M. Schüssler, M. Rempel, The dynamical disconnection of sunspots from their magnetic roots. Astron. Astrophys. **441**, 337–346 (2005). doi:10.1051/0004-6361:20052962

M. Schüssler, A. Vögler, Strong horizontal photospheric magnetic field in a surface dynamo simulation. Astron. Astrophys. **481**, 5–8 (2008). doi:10.1051/0004-6361:20078998

M. Schüssler, P. Caligari, A. Ferriz-Mas, F. Moreno-Insertis, Instability and eruption of magnetic flux tubes in the solar convection zone. Astron. Astrophys. **281**, 69–72 (1994)

N.R. Sheeley Jr., Surface evolution of the Sun's magnetic field: a historical review of the flux-transport mechanism. Living Rev. Sol. Phys. **2**, 5 (2005). doi:10.12942/lrsp-2005-5

N.R. Sheeley Jr., C.R. DeVore, J.P. Boris, Simulations of the mean solar magnetic field during sunspot cycle 21. Sol. Phys. **98**, 219–239 (1985). doi:10.1007/BF00152457

K.R. Sivaraman, S.S. Gupta, R.F. Howard, Measurement of Kodaikanal white-light images. I—A comparison of 35 years of Kodaikanal and Mount Wilson sunspot data. Sol. Phys. **146**, 27–47 (1993). doi:10.1007/BF00662168

H.B. Snodgrass, Magnetic rotation of the solar photosphere. Astrophys. J. **270**, 288–299 (1983). doi:10.1086/161121

H.B. Snodgrass, Separation of large-scale photospheric Doppler patterns. Sol. Phys. **94**, 13–31 (1984). doi:10.1007/BF00154804

H.B. Snodgrass, R. Howard, L. Webster, Recalibration of Mount Wilson Doppler measurements (Research note). Sol. Phys. **90**, 199–202 (1984). doi:10.1007/BF00153796

S.K. Solanki, The origin and the diagnostic capabilities of the Stokes V asymmetry observed in solar faculae and the network. Astron. Astrophys. **224**, 225–241 (1989)

S.K. Solanki, B. Inhester, M. Schüssler, The solar magnetic field. Rep. Prog. Phys. **69**, 563–668 (2006). doi:10.1088/0034-4885/69/3/R02

S.K. Solanki, M. Schüssler, M. Fligge, Evolution of the Sun's large-scale magnetic field since the Maunder minimum. Nature **408**, 445–447 (2000). doi:10.1038/408445a

S.K. Solanki, T. Wenzler, D. Schmitt, Moments of the latitudinal dependence of the sunspot cycle: a new diagnostic of dynamo models. Astron. Astrophys. **483**, 623–632 (2008). doi:10.1051/0004-6361:20054282

S.K. Solanki, D. Zufferey, H. Lin, I. Rueedi, J.R. Kuhn, Infrared lines as probes of solar magnetic features. XII. Magnetic flux tubes: evidence of convective collapse? Astron. Astrophys. **310**, 33–36 (1996)

S.K. Solanki, P. Barthol, S. Danilovic, A. Feller, A. Gandorfer, J. Hirzberger, T.L. Riethmüller, M. Schüssler, J.A. Bonet, V. Martínez Pillet, J.C. del Toro Iniesta, V. Domingo, J. Palacios, M. Knölker, N. Bello González, T. Berkefeld, M. Franz, W. Schmidt, A.M. Title, SUNRISE: instrument, mission, data, and first results. Astrophys. J. Lett. **723**, 127–133 (2010). doi:10.1088/2041-8205/723/2/L127

H.C. Spruit, Origin of the torsional oscillation pattern of solar rotation. Sol. Phys. **213**, 1–21 (2003). doi:10.1023/A:1023202605379

R.F. Stein, A. Lagerfjärd, Å. Nordlund, D. Georgobiani, Solar flux emergence simulations. Sol. Phys. **268**, 271–282 (2011). doi:10.1007/s11207-010-9510-y

J.O. Stenflo, A.G. Kosovichev, Bipolar magnetic regions on the Sun: global analysis of the SOHO/MDI data set. Astrophys. J. **745**, 129 (2012). doi:10.1088/0004-637X/745/2/129

L. Svalgaard, E.W. Cliver, Y. Kamide, Sunspot cycle 24: smallest cycle in 100 years? Geophys. Res. Lett. **32**, 1104 (2005). doi:10.1029/2004GL021664

M.J. Thompson, J. Toomre, E.R. Anderson, H.M. Antia, G. Berthomieu, D. Burtonclay, S.M. Chitre, J. Christensen-Dalsgaard, T. Corbard, M. De Rosa, C.R. Genovese, D.O. Gough, D.A. Haber, J.W. Harvey, F. Hill, R. Howe, S.G. Korzennik, A.G. Kosovichev, J.W. Leibacher, F.P. Pijpers, J. Provost, E.J. Rhodes Jr., J. Schou, T. Sekii, P.B. Stark, P.R. Wilson, Differential rotation and dynamics of the solar interior. Science **272**, 1300–1305 (1996). doi:10.1126/science.272.5266.1300

L. Tian, Y. Liu, H. Wang, Latitude and magnetic flux dependence of the tilt angle of bipolar regions. Sol. Phys. **215**, 281–293 (2003)

A.M. Title, T.D. Tarbell, K.P. Topka, On the relation between magnetic field structures and granulation. Astrophys. J. **317**, 892–899 (1987). doi:10.1086/165339

S. Tsuneta, K. Ichimoto, Y. Katsukawa, S. Nagata, M. Otsubo, T. Shimizu, Y. Suematsu, M. Nakagiri, M. Noguchi, T. Tarbell, A. Title, R. Shine, W. Rosenberg, C. Hoffmann, B. Jurcevich, G. Kushner, M. Levay, B. Lites, D. Elmore, T. Matsushita, N. Kawaguchi, H. Saito, I. Mikami, L.D. Hill, J.K. Owens, The solar optical telescope for the Hinode mission: an overview. Sol. Phys. **249**, 167–196 (2008). doi:10.1007/s11207-008-9174-z

J. Tuominen, Die systematische Strombewegung der Sonnenflecke in heliographischer Breite. Mit 1 Abbildung. Z. Astrophys. **21**, 96 (1942)

J. Tuominen, J. Kyrolainen, On the latitude drift of sunspot groups and solar rotation. Sol. Phys. **79**, 161–172 (1982). doi:10.1007/BF00146980

R.K. Ulrich, Solar meridional circulation from Doppler shifts of the Fe I line at 5250 å as measured by the 150-foot solar tower telescope at the Mt. Wilson observatory. Astrophys. J. **725**, 658–669 (2010). doi:10.1088/0004-637X/725/1/658

R.K. Ulrich, J.E. Boyden, L. Webster, S.P. Padilla, H.B. Snodgrass, Solar rotation measurements at Mount Wilson. V—Reanalysis of 21 years of data. Sol. Phys. **117**, 291–328 (1988). doi:10.1007/BF00147250

L. Upton, D.H. Hathaway, Predicting the Sun's polar magnetic fields with a surface flux transport model. Astrophys. J. **780**, 5 (2014). doi:10.1088/0004-637X/780/1/5

A.A. van Ballegooijen, N.P. Cartledge, E.R. Priest, Magnetic flux transport and the formation of filament channels on the Sun. Astrophys. J. **501**, 866 (1998). doi:10.1086/305823

M. Verma, C. Denker, Horizontal flow fields observed in Hinode G-band images IV. Statistical properties of the dynamical environment around pores. ArXiv e-prints (2014)

A. Vögler, M. Schüssler, A solar surface dynamo. Astron. Astrophys. **465**, 43–46 (2007). doi:10.1051/0004-6361:20077253

S.V. Vorontsov, J. Christensen-Dalsgaard, J. Schou, V.N. Strakhov, M.J. Thompson, Helioseismic measurement of solar torsional oscillations. Science **296**, 101–103 (2002). doi:10.1126/science.1069190

M. Waldmeier, *Ergebnisse und Probleme der Sonnenforschung* (1955)

Y.-M. Wang, N.R. Sheeley, Understanding the geomagnetic precursor of the solar cycle. Astrophys. J. Lett. **694**, 11–15 (2009). doi:10.1088/0004-637X/694/1/L11

Y.-M. Wang, N.R. Sheeley Jr., Average properties of bipolar magnetic regions during sunspot cycle 21. Sol. Phys. **124**, 81–100 (1989). doi:10.1007/BF00146521

Y.-M. Wang, N.R. Sheeley Jr., Magnetic flux transport and the sun's dipole moment—new twists to the Babcock-Leighton model. Astrophys. J. **375**, 761–770 (1991). doi:10.1086/170240

Y.-M. Wang, N.R. Sheeley Jr., Sunspot activity and the long-term variation of the Sun's open magnetic flux. J. Geophys. Res. (Space Phys.) **107**, 1302 (2002). doi:10.1029/2001JA000500

Y.-M. Wang, N.R. Sheeley Jr., Modeling the Sun's large-scale magnetic field during the Maunder minimum. Astrophys. J. **591**, 1248–1256 (2003). doi:10.1086/375449

Y.-M. Wang, J. Lean, N.R. Sheeley Jr., Role of a variable meridional flow in the secular evolution of the Sun's polar fields and open flux. Astrophys. J. Lett. **577**, 53–57 (2002). doi:10.1086/344196

Y.-M. Wang, J.L. Lean, N.R. Sheeley Jr., Modeling the Sun's magnetic field and irradiance since 1713. Astrophys. J. **625**, 522–538 (2005). doi:10.1086/429689

Y.-M. Wang, A.G. Nash, N.R. Sheeley Jr., Evolution of the sun's polar fields during sunspot cycle 21—poleward surges and long-term behavior. Astrophys. J. **347**, 529–539 (1989). doi:10.1086/168143

Y.-M. Wang, E. Robbrecht, N.R. Sheeley Jr., On the weakening of the polar magnetic fields during solar cycle 23. Astrophys. J. **707**, 1372–1386 (2009). doi:10.1088/0004-637X/707/2/1372

F. Ward, Determination of the solar-rotation rate from the motion of identifiable features. Astrophys. J. **145**, 416 (1966). doi:10.1086/148783

M.A. Weber, Y. Fan, M.S. Miesch, The rise of active region flux tubes in the turbulent solar convective envelope. Astrophys. J. **741**, 11 (2011). doi:10.1088/0004-637X/741/1/11

M.A. Weber, Y. Fan, M.S. Miesch, Comparing simulations of rising flux tubes through the solar convection zone with observations of solar active regions: constraining the dynamo field strength. Sol. Phys. **287**, 239–263 (2013). doi:10.1007/s11207-012-0093-7

N.O. Weiss, M.J. Thompson, The solar dynamo. Space Sci. Rev. **144**, 53–66 (2009). doi:10.1007/s11214-008-9435-z

B.T. Welsch, G.H. Fisher, X. Sun, A magnetic calibration of photospheric Doppler velocities. Astrophys. J. **765**, 98 (2013). doi:10.1088/0004-637X/765/2/98

R. Wolf, Abstract of his latest results. Mon. Not. R. Astron. Soc. **21**, 77 (1861)

J. Worden, J. Harvey, An evolving synoptic magnetic flux map and implications for the distribution of photospheric magnetic flux. Sol. Phys. **195**, 247–268 (2000). doi:10.1023/A:1005272502885

A.R. Yeates, Coronal magnetic field evolution from 1996 to 2012: continuous non-potential simulations. Sol. Phys. **289**, 631–648 (2014). doi:10.1007/s11207-013-0301-0

A.R. Yeates, D.H. Mackay, Modelling the global solar corona: III. Origin of the hemispheric pattern of filaments. Sol. Phys. **254**, 77–88 (2009). doi:10.1007/s11207-008-9276-7

A.R. Yeates, D.H. Mackay, A.A. van Ballegooijen, Modelling the global solar corona: filament chirality observations and surface simulations. Sol. Phys. **245**, 87–107 (2007). doi:10.1007/s11207-007-9013-7

J. Zhao, A.G. Kosovichev, Torsional oscillation, meridional flows, and vorticity inferred in the upper convection zone of the Sun by time-distance helioseismology. Astrophys. J. **603**, 776–784 (2004). doi:10.1086/381489

J. Zhao, R.S. Bogart, A.G. Kosovichev, T.L. Duvall Jr., T. Hartlep, Detection of equatorward meridional flow and evidence of double-cell meridional circulation inside the Sun. Astrophys. J. Lett. **774**, 29 (2013). doi:10.1088/2041-8205/774/2/L29

DOI 10.1007/978-1-4939-2584-1_18
Reprinted from *Space Science Reviews* Journal, DOI 10.1007/s11214-014-0063-5

Solar Activity in the Past and the Chaotic Behaviour of the Dynamo

Rainer Arlt · Nigel Weiss

Received: 16 April 2014 / Accepted: 20 June 2014 / Published online: 1 July 2014

Abstract The record of solar activity is reviewed here with emphasis on peculiarities. Since sunspot positions tell us a lot more about the solar dynamo than the various global sunspot numbers, we first focus on the records of telescopic observations of sunspots leading to positional information. Then we turn to the proxy record from cosmogenic isotope abundances, which shows recurrent grand minima over the last 9500 years. The apparent distinction between episodes of strong modulation, and intervening episodes with milder modulation and weaker overall activity, hints at the solar dynamo following a variety of solutions, with different symmetries, over the course of millennia.

Keywords Solar activity · Solar dynamo

1 Introduction

Telescopic observations of sunspots have revealed both the 11-year Schwabe cycle and the interruption of activity associated with the Maunder Minimum in the 17th century. New analyses of early records (including those of Schwabe) confirm that the pattern associated with the butterfly diagram has been present for the past 300 years. There is also evidence of differential rotation, with suggestions of anomalous behaviour during the Maunder Minimum.

The record of solar activity has been extended back for almost 10000 years by measuring the abundances of the cosmogenic isotopes ^{10}Be and ^{14}C in ice cores and tree rings. This record reveals many grand minima, with a characteristic spacing of around 200 years (the de Vries cycle) but the appearance of these grand minima itself varies with a characteristic

R. Arlt (✉)
Leibniz Institute for Astrophysics Potsdam, Potsdam, Germany
e-mail: rarlt@aip.de

N. Weiss
Department of Applied Mathematics and Theoretical Physics, University of Cambridge, Cambridge, UK
e-mail: now1@cam.ac.uk

timescale of over 2000 years. We interpret the grand minima and maxima as resulting from deterministic modulation of the nonlinear solar dynamo, oscillating chaotically with the mean Hale period of around 22 years. The present peculiar evolution of the solar cycle may turn out to be an enlightening period in this respect as well, since it will be observed by various methods including helioseismology.

Although the Sun's magnetic field is now predominantly dipolar, simple nonlinear models show that symmetry can flip to give quadrupolar or even mixed-mode behaviour. We suggest that such flipping explains the long-term, multimillennial variability of the activity record.

2 The sunspot record

The sunspot number goes back to Wolf (1859) who defined an index of solar activity based on the number of sunspot groups and the total number of individual spots on the observable hemisphere of the Sun. The time series starts in 1749 with the observations by Johann Staudacher (Nuremberg) and has been continued in terms of the International Sunspot Number until the present day. An alternative index was defined by Hoyt and Schatten (1998) who only counted the group numbers—an index that is more robust against variable capabilities of seeing small spots and allowed the time series to go back to the first days of telescopic observations of the Sun in 1610.

The time-series are very often used as a proxy for some sort of magnetic field strength in the interior of the Sun and are compared with dynamo models. These global indices, however, cannot give any insight into the topology of the magnetic fields that presumably generate the spots on the surface. One may imagine that the knowledge of the heliographic positions of the spots can be used to infer the equatorial symmetry, the latitudinal distribution, and the rotational symmetry of the underlying magnetic fields as well as the lifetime of magnetic structures on the solar surface.

Sunspot positions are now being collected in a database using results from a USAF network of observing stations, following on the photoheliographic programme conducted by the Royal Greenwich Observatory (RGO) which effectively also was a network of stations around the world. The RGO/USAF set started in 1874 and stores only the average position of sunspot groups together with their total area. In parallel, several other programmes were collecting sunspot positions of various amounts, or are still doing so.

Before that, a large set of sunspot data is available from Friedrich Wilhelm Gustav Spörer who observed from 1861 to 1894 from the towns of Anklam and Potsdam, Germany. He confirmed that the lull of reports on sunspots in the second half of the 17th century represented a real low in solar activity for decades. His work was recognized later by Maunder who was eventually credited with that discovery, whence the name "Maunder minimum." Spörer's observations and measurements were published in a series of papers (Spörer 1874, 1878, 1880, 1886, 1894), but his original sunspot drawings—if they existed—are lost.

Before Spörer, Richard Carrington had already made the discovery that the Sun is not rotating uniformly, but faster at the equator than at higher latitudes (Carrington 1859). However, his observational data only cover a rather short period, from November 1853 to March 1861 (Zolotova et al. 2010; Lepshokov et al. 2012).

A great extension of the butterfly diagram into the past comes from the observations by Samuel Heinrich Schwabe, who drew sunspots in a solar disk each day he saw at least a glimpse of the Sun from Nov. 5, 1825, to Dec. 31, 1867. He actually observed until Dec. 15, 1868, but his last observing book was lost. Schwabe was also the first to publish a paper

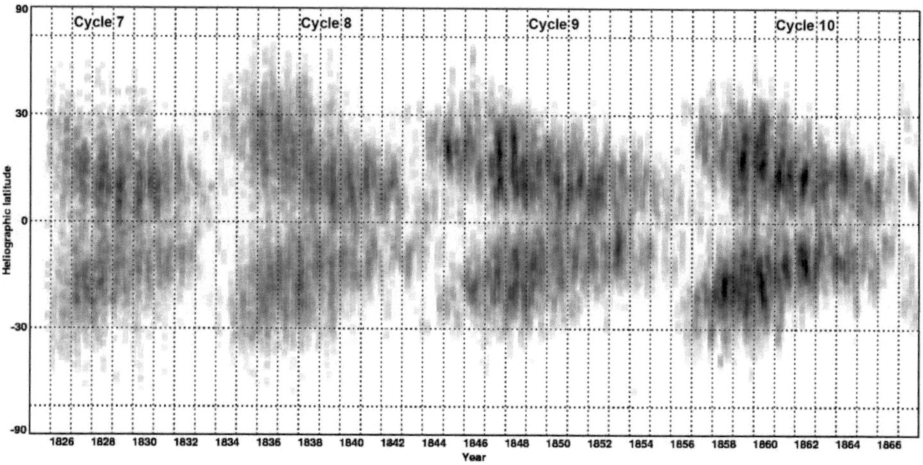

Fig. 1 Butterfly diagram obtained from the sunspot positions derived from the drawings of Samuel Heinrich Schwabe in 1825–1867 (after Arlt et al. 2013)

suggesting that the sunspot number varies periodic (Schwabe 1844). The positions and sizes of all sunspots seen by Schwabe were measured by Arlt et al. (2013). The resulting butterfly diagram is shown in Fig. 1.

The time around 1800 is poorly covered by observations, the longest record being that preserved by Honoré Flaugergues, who reported useful sunspot observations in 1788–1830. These, however, are yet to be analysed. They consist mostly of timings at a transit instrument. Flaugergues gave transit times of the solar limb and spots at a vertical and an oblique wire, which will allow us to determine the latitudes of spots.

A very interesting record of observations is stored in the library of the Leibniz Institute for Astrophysics Potsdam. About 1000 drawings of the Sun made by Johann Caspar Staudacher cover the period of 1749–1799. The drawings are not accompanied by much verbal information about the telescope or the observing method. There are no indications of the orientations of the solar disks. Fortunately, detailed drawings of partial solar eclipses showing the path and direction of the Moon clearly show that the Sun was projected on a screen behind the telescope, i.e. all images are mirrored. The resulting butterfly diagram is shown in Fig. 2 (Arlt 2009a). It is remarkable that the first two cycles observed by Staudacher do not show a clear butterfly shape. Since the observer recorded the sunspots with the projection method, they are certainly not plotted 'at random' into the disks, though the uncertainty of the orientations holds true for the entire data set.

The observations were complemented by the very accurate drawings of Ludovico Zucconi in 1754–1760 (Zucconi 1760) which were analyzed by Cristo et al. (2011). They fill in the gaps of the observations by Staudacher around the minimum near 1755 and may help understand the unusual time–latitude distribution of Staudacher's spots. At the other end of Staudacher's observing period, additional positions of a small number of sunspots were derived from the records of Hamilton and Gimingham at Armagh Observatory in 1795–1797 (Arlt 2009b).

Further back in time, we find the analysis by Ribes and Nesme-Ribes (1993) who measured the positions of sunspots seen by a variety of astronomers at the Observatoire de Paris, during and after the Maunder minimum, resulting in data for 1672–1719 (Fig. 2). There is a striking absence of sunspots until about 1714 with only the southern hemisphere being

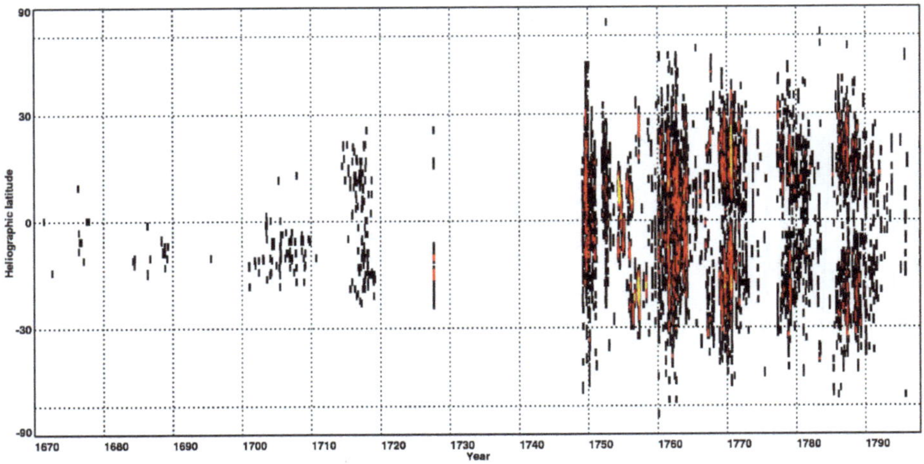

Fig. 2 Butterfly diagram of the period around and after the Maunder minimum with sunspot positions from various sources. The period until 1719 shows the positions derived by Ribes and Nesme-Ribes (1993) which were digitized by Vaquero et al. (http://haso.unex.es/); the year 1727 shows two additional observations found at Paris observatory by the authors. The period of 1749-1799 contains observations by Staudacher (Arlt 2009a), Zucconi (Cristo et al. 2011), and Hamilton (Arlt 2009b). Higher contrast is used than in Fig. 1 because of the fewer spots available

populated, by roughly 80 spots (plus 4 spots in the northern hemisphere and 6 spots right on the equator). This result again shows that the solar cycle does not necessarily show a butterfly shape for the time–latitude distribution of spots. The actual activity cycle may have persisted during the Maunder minimum as seen in the cosmic ray record (see Sect. 4) at high time resolution (Beer et al. 1998; Berggren et al. 2009).

A fair number of observers recorded sunspots in the period before the Maunder minimum, starting with the first telescopic observations by Galileo Galilei and Thomas Harriot in 1610, followed by Christoph Scheiner and Johannes Fabricius, who first published the telescopic sunspot observations, in a printed pamphlet (Fabricius 1611). There is no compilation of sunspot positions for all the available sources yet, but visual inspection of the images indicates normal spot distributions before the Maunder minimum.

3 Results from the sunspot record

Sunspots show the latitudinal differential rotation of the Sun. This has first been derived quantitatively by Carrington (1859) and Peters (1859). The rotation of newly emerged sunspot groups is, by the way, different from the rotation of the bulk gas at the surface at the same latitude (Tuominen 1962; Pulkkinen and Tuominen 1998).

Historical sunspot observations may actually allow measurements of the differential rotation. A recent attempt by Arlt and Fröhlich (2012) employed Bayesian inference on the observations by Staudacher to obtain positions, orientation angles and differential rotation parameter at the same time and delivered a latitudinal shear compatible with that of today. There is a slight but insignificant hint that the differential rotation was stronger in the first third of Staudacher's observations than during the remaining period. This coincides with the period of non-butterfly shaped distribution of spot latitudes over time which is an automatic side product of the analysis.

The spot positions derived by Ribes and Nesme-Ribes (1993) also indicate a stronger differential rotation at the end of the Maunder minimum, along with an unusual spot distribution as well. There is actually no fundamental reason why the solar magnetic field should always adopt a chiefly dipolar structure (antisymmetric with respect to the equator). Quadrupolar modes or mixed symmetries are certainly possible and may lead to butterfly diagrams deviating from the one we know today.

Another issue of the solar cycle and the solar dynamo is the coupling between the hemispheres. Zolotova et al. (2010) studied the temporal variation of the phase lag between the cycles separated into hemispheres. The phase difference varies and changes sign approximately every 35–40 years, giving a full period for the phase lag change of 70–80 years.

4 The cosmic-ray record

Galactic cosmic rays are deflected by magnetic fields in the heliosphere and so the solar cycle modulates the flux of galactic cosmic rays into the Earth's atmosphere. The detection of the near-Earth cosmic rays by decay products in the atmosphere (mostly neutrons) delivered a 60-year record which shows a very good anti-correlation with the sunspot record over the last five cycles—see Usoskin (2013) for a review. Cosmic rays also lead to the production of isotopes such as ^{14}C and ^{10}Be, which are absorbed into tree-rings or into polar icecaps, where their abundances can be measured with great precision. A 1000-year time-series of the isotope productions of ^{10}Be and ^{14}C (Muscheler et al. 2007) shows good agreement with the sunspot record, although some significant differences remain. More recently, the ^{14}C data have been combined with ^{10}Be measurements from Greenland and Antarctic ice cores to produce a much longer, composite time-series of cosmic radiation with a duration of 9400 years. A Principal Components Analysis has then been used to filter out most of the climatic effects and to reveal the presence of recurrent grand maxima and grand minima (Steinhilber et al. 2012; Abreu et al. 2013; McCracken et al. 2013).

The resulting time series can then be converted into a record of the modulation function Φ, which can be very roughly interpreted as the mean loss of momentum-to-charge ratio of a cosmic ray particle in the heliosphere (Usoskin 2013). Figure 3 displays the principal components rate of production of cosmogenic isotopes from 9400 BP to 1950 CE, together with the derived variation of the modulation function Φ, which has been corrected to take account of changes in the Earth's magnetic field (Knudsen et al. 2008).

The variations in the shaded regions of Fig. 3 are all similar to those in the most recent millennium, which includes the Maunder, Spörer, Wolf and Oort Grand Minima. Between these regions there are intervals, from 7100 to 6300 BP and from 4700 to 3500 BP, and again from 2200 to 1700 BP during which there are no grand minima and variations in isotope production are relatively low. This distinction becomes even more apparent if we compare the Fourier amplitude spectra for intervals with and without grand extrema, as displayed in Fig. 4. The only clear coincidence is for the Gleissberg cycle, with a period of 87 yr. The interval with strong modulation shows the familiar de Vries period of 208 yr and other peaks at 150 and 350 yr, while the 2300 yr Hallstatt period shows up in the full record. The intervals on either side, with only weak modulation, have a broad peak around 300 yr and a sharper peak at 140 yr but the spectra are generally flatter. All this confirms the immediate impression that the behaviours in the shaded and unshaded regions of Fig. 3 are qualitatively different. The Hallstatt period then represents the characteristic timescale for transitions from one regime to the other.

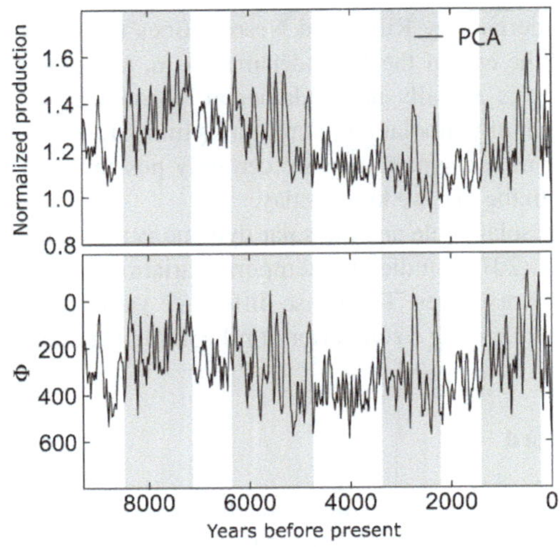

Fig. 3 The 9400-year record of solar magnetic activity derived from cosmogenic isotopes for the period from 9400 BP to the present, where 'present' means 1950. *Upper panel*: the principal components record of the production rate, based on the INTCAL09 record for ^{14}C and GRIP and EDML abundances for ^{10}Be. *Lower panel*: the modulation function Φ, in MeV, after correction in an attempt to eliminate the effects of variations in the dipole moment of the geomagnetic field. The *shaded strips* denote the intervals with vigorous modulation of solar magnetic activity, associated with the presence of prominent grand maxima and grand minima (after McCracken et al. 2013)

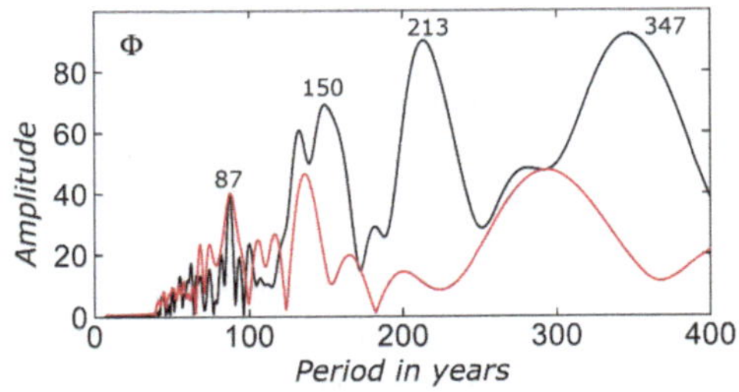

Fig. 4 Comparison of Fourier amplitude spectra for intervals with and without grand minima. Shown in black is the spectrum for Φ over the interval from 6300 to 4300 BP (containing four prominent grand minima). The *spectrum in red* is for the interval from 4700 to 3500, which contained no grand minima (as can be seen in Fig. 3). Apart from the Gleissberg peak at 87 yr and a possible coincidence around 135 yr, the two spectra have little in common (after McCracken et al. 2013)

5 Chaotic modulation and symmetry-breaking

The Sun's cyclic activity is governed by macroscopic physics and is therefore deterministic. It is not practicable, however, to follow the emergence of every flux tube to form the sunspots that are shown in Fig. 1 and so we have to focus on averaged behaviour, which is deterministic but subject to stochastic disturbances. Then it is apparent that activity cycles must be regarded as manifestations of a chaotic oscillator, with sensitive dependence on initial conditions (e.g. Zeldovich et al. 1983; Spiegel 2009). The evolution of such a system is represented by a trajectory in phase space; provided the stochastic perturbations are not too large, the disturbed trajectories are always *shadowed* by nearby trajectories of the undisturbed chaotic system (Ott 1993). Turning to the observational records described

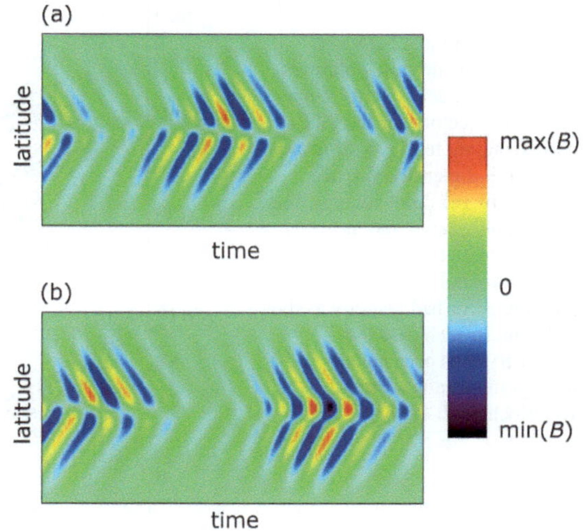

Fig. 5 Modulation and symmetry changes in a Cartesian mean-field dynamo model governed by partial differential equations, showing the toroidal field as a function of latitude and time. (**a**) Active fields with dipole symmetry, exhibiting grand minima associated with loss of symmetry and hemispheric patterns. (**b**) A transition from dipole symmetry to quadrupole symmetry during a grand minimum (for the same parameter values) (after Beer et al. 1998)

above, we should therefore expect the chaotic system to generate modulation corresponding to grand minima and grand maxima, whose origin can be understood by reference to the mathematical structure of the problem (Tobias and Weiss 2007). Simple models do indeed reproduce similar behaviour, which is associated with the appearance of multiply periodic ("quasiperiodic" to mathematicians) solutions and global bifurcations that lead to chaos.

In the simplest illustrative model, cyclic dynamo action sets in at an oscillatory (Hopf) bifurcation that leads to periodic behaviour, with trajectories that are attracted to a limit cycle in the phase space; this is followed by a second Hopf bifurcation that leads to doubly periodic solutions that lie on a 2-torus in the 3-dimensional phase space; after a series of period-doubling bifurcations (associated with a heteroclinic bifurcation) behaviour becomes chaotic (Tobias et al. 1995), though the chaotic modulation still retains a memory of its original periodicity (Ott 1993). Since this bifurcation sequence was originally demonstrated for normal form equations (governing a saddle-node/Hopf bifurcation) it is generic and therefore expected to be robust. Indeed, the same pattern appears in simple dynamo models governed by partial differential equations (Tobias 1996) and in mean-field dynamos (Küker et al. 1999; Pipin 1999; Bushby 2006). It follows that grand minima and grand maxima should be interpreted as *deterministic* effects, associated with chaotic modulation, and not as products of large-scale stochastic disturbances.

All dynamo models allow two families of solutions that bifurcate from the trivial, field-free state: these families have either dipole symmetry (with toroidal fields that are antisymmetric about the equator) or quadrupole symmetry (with symmetric toroidal fields). Mixed modes can only appear as a result of symmetry-breaking bifurcations in the nonlinear domain, which may lead to a complicated web of stable and unstable solution branches (Jennings and Weiss 1992). After the Maunder minimum, the solar magnetic field appears to have gained dipolar symmetry by the second half of the 18th century, through a period of mixed symmetry (with nearly all spots in one hemisphere only) and a period poorly covered by observations.

Similar properties are exhibited by an idealized mean-field dynamo model, governed by partial differential equations (Beer et al. 1998), which also allows transitions between dipolar and quadrupolar symmetries during grand minima. This behaviour is shown in Fig. 5. These properties are also demonstrated by even simpler models, governed by low-order systems of ordinary differential equations (Knobloch et al. 1998). Phase portraits for both PDEs

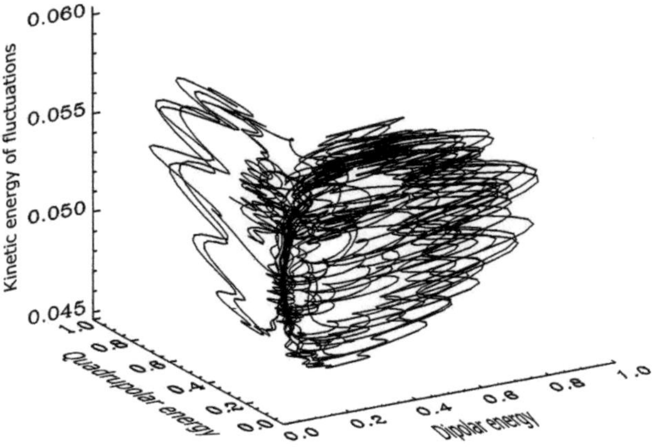

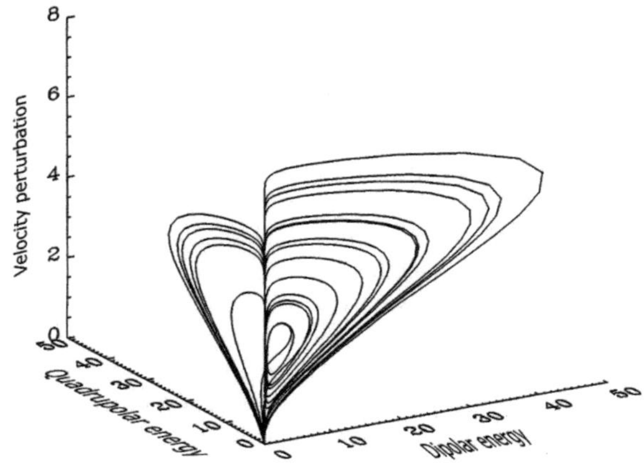

Fig. 6 Phase portraits illustrating flipping between dipole and quadrupole polarities for simple model systems. *Upper panel*: a trajectory for the PDEs, corresponding to Fig. 5, projected onto the 3-dimensional space spanned by the dipole energy, the quadrupole energy and the perturbed kinetic energy. *Lower panel*: the same but for the ODEs, with the perturbed velocity as the ordinate. In each case the symmetry flips occasionally at deep grand minima (after Knobloch et al. 1998)

and ODEs are displayed in Fig. 6. The former show both cyclic variations (predominantly horizontal) and large-amplitude modulation, as well as occasional changes of symmetry. With the ODEs it is possible to filter out the cyclic variability, leaving only the modulation with flips of symmetry near the origin, at very deep grand minima. Note the reduced amplitude of modulation in the quadrupole regime as compared with dipole fields. By changing the parameters in the model systems it is possible to find mixed-mode cycles too; they are likewise modulated, and different symmetries may coexist without the possibility of flipping.

These results for highly simplified models reveal generic properties that would be shared by solutions of the much more complicated equations that describe a real stellar dynamo. What they show is not only that grand maxima and grand minima are associated with deterministic modulation of cyclic activity but also that symmetry changes may provide a natural explanation for the changes in behaviour that were reported by McCracken et al. (2013) and summarized in Sects. 2–4 above. A more detailed discussion will appear elsewhere (Tobias and Weiss, in preparation).

Acknowledgements R. Arlt is grateful to the libraries of Leibniz Institute for Astrophysics Potsdam, the Royal Astronomical Society, and the Observatoire de Paris for permitting the digitization of observing records. N. Weiss gratefully acknowledges many discussions with Jürg Beer and Steven Tobias over the past two decades.

References

J.A. Abreu, J. Beer, F. Steinhilber, M. Christl, P.W. Kubik, Space Sci. Rev. **176**, 343 (2013)
R. Arlt, Sol. Phys. **255**, 143 (2009a)
R. Arlt, Astron. Nachr. **330**, 311 (2009b)
R. Arlt, H.-E. Fröhlich, Astron. Astrophys. **543**, A7 (2012)
R. Arlt, R. Leussu, N. Giese, K. Mursula, I.G. Usoskin, Mon. Not. R. Astron. Soc. **433**, 3165 (2013)
J. Beer, S.M. Tobias, N.O. Weiss, Sol. Phys. **181**, 237 (1998)
A.-M. Berggren, J. Beer, G. Possnert, A. Aldahan, P. Kubik, M. Christl, Geophys. Res. Lett. **36**, L11801 (2009)
P.J. Bushby, Mon. Not. R. Astron. Soc. **371**, 772 (2006)
R.C. Carrington, Mon. Not. R. Astron. Soc. **19**, 81 (1859)
A. Cristo, J.M. Vaquero, F. Sánchez-Bajo, J. Atmos. Sol.-Terr. Phys. **73**, 187 (2011)
J. Fabricius, *De Maculis in Sole Observatis* (Borner Senior & Rehefeld, Wittenberg, 1611)
D.V. Hoyt, K.H. Schatten, Sol. Phys. **181**, 491 (1998)
R.L. Jennings, N.O. Weiss, Mon. Not. R. Astron. Soc. **252**, 249 (1992)
E. Knobloch, S.M. Tobias, N.O. Weiss, Mon. Not. R. Astron. Soc. **297**, 1123 (1998)
M.F. Knudsen, P. Riisager, D. Donadini, I. Snowball, R. Muscheler, B. Korhonen, L.J. Personen, Earth Planet. Sci. Lett. **272**, 319 (2008)
M. Küker, R. Arlt, G. Rüdiger, Astron. Astrophys. **343**, 977 (1999)
D.Kh. Lepshokov, A.G. Tlatov, V.V. Vasil'eva, Geomagn. Aeron. **52**, 843 (2012)
K.G. McCracken, J. Beer, F. Steinhilber, J. Abreu, Sol. Phys. **286**, 609 (2013)
R. Muscheler, F. Joos, J. Beer, S.A. Müller, M. Vonmoos, I. Snowball, Quat. Sci. Rev. **26**, 82 (2007)
E. Ott, *Chaos in Dynamical Systems* (Cambridge University Press, Cambridge, 1993)
C.H.F. Peters, Mon. Not. R. Astron. Soc. **19**, 173 (1859)
V.V. Pipin, Astron. Astrophys. **346**, 295 (1999)
P. Pulkkinen, I. Tuominen, Astron. Astrophys. **332**, 748 (1998)
J.C. Ribes, E. Nesme-Ribes, Astron. Astrophys. **276**, 549 (1993)
H. Schwabe, Astron. Nachr. **21**, 233 (1844)
E.A. Spiegel, Space Sci. Rev. **144**, 25 (2009)
G. Spörer, Publ. Astron. Ges. **13** (1874)
G. Spörer, Publ. Astrophys. Obs. Potsdam **1** (1878)
G. Spörer, Publ. Astrophys. Obs. Potsdam **5** (1880)
G. Spörer, Publ. Astrophys. Obs. Potsdam **17**, 220 (1886)
G. Spörer, Publ. Astrophys. Obs. Potsdam **32** (1894)
F. Steinhilber, J.A. Abreu, J. Beer, I. Brunner, M. Christl, H. Fischer, U. Heikkilä, P.W. Kubik, M. Mann, K.G. McCracken, H. Miller, H. Miyahara, H. Oerter, F. Wilhelms, Proc. Natl. Acad. Sci. USA **109**, 5967 (2012)
S.M. Tobias, Astron. Astrophys. **307**, L21 (1996)
S.M. Tobias, N.O. Weiss, in *Mathematical Aspects of Natural Dynamos*, ed. by E. Dormy, A.M. Soward (CRC Press, Boca Raton, 2007), pp. 281–311
S.M. Tobias, N.O. Weiss, V. Kirk, Mon. Not. R. Astron. Soc. **273**, 1150 (1995)
J. Tuominen, Z. Astrophys. **55**, 110 (1962)
I.G. Usoskin, Living Rev. Sol. Phys. **10**, 1 (2013). http://www.livingreviews.org/lrsp-2013-1
R. Wolf, Mittheil. Zürich **2**, 3 (1859)
Y.B. Zeldovich, A.A. Ruzmaikin, D.D. Sokoloff, *Magnetic Fields in Astrophysics* (Gordon & Breach, New York, 1983)
N.V. Zolotova, D.I. Ponyavin, R. Arlt, I. Tuominen, Astron. Nachr. **331**, 765 (2010)
L. Zucconi, *De Heliometri Structura Et Usu* (Auctoris æræ, Venice, 1760)

DOI 10.1007/978-1-4939-2584-1_19
Reprinted from *Space Science Reviews* Journal, DOI 10.1007/s11214-014-0066-2

Oscillator Models of the Solar Cycle

Towards the Development of Inversion Methods

**Ilídio Lopes · Dário Passos · Melinda Nagy ·
Kristof Petrovay**

Received: 15 April 2014 / Accepted: 23 June 2014 / Published online: 15 July 2014

Abstract This article reviews some of the leading results obtained in solar dynamo physics by using temporal oscillator models as a tool to interpret observational data and dynamo model predictions. We discuss how solar observational data such as the sunspot number is used to infer the leading quantities responsible for the solar variability during the last few centuries. Moreover, we discuss the advantages and difficulties of using inversion methods (or backward methods) over forward methods to interpret the solar dynamo data. We argue that this approach could help us to have a better insight about the leading physical processes responsible for solar dynamo, in a similar manner as helioseismology has helped to achieve a better insight on the thermodynamic structure and flow dynamics in the Sun's interior.

Keywords Sun: activity · Sunspots · Magnetohydrodynamics · Sun: helioseismology

I. Lopes (✉) · D. Passos
Instituto Superior Técnico, Universidade de Lisboa, Lisbon, Portugal
e-mail: ilidio.lopes@tecnico.ulisboa.pt

D. Passos
e-mail: dariopassos@ist.utl.pt

I. Lopes · D. Passos
Departmento de Física, Universidade de Évora, Évora, Portugal

D. Passos
GRPS, Départment de Physique, Université de Montréal, C.P. 6128, Centre-ville, Montréal, QC, H3C-3J7, Canada

M. Nagy · K. Petrovay
Department of Astronomy, Eötvös Loránd University, Budapest, Hungary

M. Nagy
e-mail: nagymelinda@caesar.elte.hu

K. Petrovay
e-mail: k.petrovay@astro.elte.hu

1 Introduction

The number of dark spots in the Sun's surface has been counted in a systematic way since Rudolf Wolf introduced the concept, in the first half of the nineteenth century. More than any other solar observable, the sunspot number is considered the strongest signature of the 22-year magnetic cycle. Moreover, since the sunspot number is the longest time series from all solar observables (Owens 2013), it makes it the preferred proxy to study the variability and irregularity of the solar magnetic cycle.

In the Sun's interior the large scale magnetic field is generated by a magnetohydrodynamic dynamo that converts part of the kinetic energy of the plasma motions into magnetic energy. Polarity reversals occur every 11 years approximately, as it can be observed directly in the Sun's dipolar field, and taking a full 22-years to complete a magnetic cycle. In fact during each magnetic cycle, the Sun experiences two periods of maximum magnetic activity, during which magnetic flux tubes created in the tachocline layer, rise to the Sun's surface by the action of buoyancy, emerging as sunspots pairs (Parker 1955). The polarity switch is also observed in the change of polarity alignment of these bipolar active regions.

Although we know that the solar dynamo resides within the convection zone, we still don't have a complete picture where all the physical mechanisms operate (Charbonneau 2013). There is a strong consensus that the physical mechanism behind the production of the large scale toroidal field component, the so called Ω-effect, is located in the tachocline, a shear layer created by differential rotation and located at the base of the convection zone. The major source of uncertainty is the location of the α-effect, the physical mechanism responsible to convert toroidal into poloidal field and close the system. In truth, this effect could be in fact a collection of several physical mechanisms that operate at different places and with different efficiencies. Some examples are the Babcock-Leighton mechanism that operates in the solar surface and converts the product of decaying active regions into poloidal field, or the action of the turbulent magnetic helicity that takes place in the bulk of the convection zone. One of the main questions that is still being debated is the quantification of the importance and relative contribution of each component to the operation of the solar dynamo. Because different authors choose to give the leading role to one or another α source term, there is vast number of dynamo models. Most of these are two dimensional models (usually referred as 2.5D because they include two spatial coordinates plus time) and are constructed using the mean-field theory framework proposed by Steenbeck and Krause (1966). Despite some short-comes, fruit of the approximations and formulation used, this type of models running in the kinematic regime, i.e. with prescribed large scale flows, has been very popular within the solar community because they can explain many of the observable features of the solar cycle. A detailed discussion on solar dynamo models, stellar magnetism and corresponding references to the vast literature on this subject can be found in the reviews by Charbonneau (2010) and Miesch and Toomre (2009).

Another way of tackling the solar dynamo problem is by producing 3D magnetohydrodynamic (MHD) simulations of the solar convection zone. These computer intensive simulations solve the full set of the MHD equations (usually under the inelastic approximation) and are fully dynamical in every resolved scale, i.e. they take into consideration the interactions between flow and field and vice-versa—unlike the kinematic regime usually used in mean field models, where only the flow influences the field. Recently these simulations have started to show stable large scale dynamo behaviour and they are starting to emerge as virtual laboratories for understanding in detail some of the mechanisms behind the dynamo (Ghizaru et al. 2010; Brown et al. 2011; Käpylä et al. 2012; Passos and Charbonneau 2014).

On the other end of the modelling spectrum, we can find oscillator models, that use simplified parameterizations of the main physical mechanisms that participate in the dynamo process. Although in the Sun's interior the magnetic field generated by the dynamo has a very rich and complex structure, as a consequence of the structure of the magnetohydrodynamic differential equations, some of its main properties can be understood by analyzing low order differential equations obtained by simplification and truncation of their original MHD counterparts. Then, several properties of the cycle that can be extracted by studying these non-linear oscillator models, as is usually done in non-linear dynamics. These models have a solid connection to dynamical systems and are, from the physics point of view the most simple. This does not mean that they are the easiest to understand because the reduction in the number of dimensions can sometimes be difficult to interpret (viz. introduction section of Wilmot-Smith et al. 2005). These low order dynamo models (LODM), as they are some times called, allow for fast computation and long integration times (thousands of years) when compared to their 2.5D and 3D counterparts. They can be thought as a first order approximation to study the impact of certain physical mechanisms in the dynamo solution, or some of the properties of the dynamo itself as a dynamical system.

The variability exhibited by the sunspot number time series, inspired researchers to look for chaotic regimes in the equations that describe the dynamo. For a complete review on this subject consult Spiegel (2009); Weiss (2010) and references therein. Some of the first applications of LODM were done in this context (e.g. Ruzmaikin 1981; Weiss and Cattaneo 1984; Tobias et al. 1995). These authors found solutions with cyclic behaviour and variable amplitude, including extended periods of low amplitude reminiscent of the grand minima behaviour we see in the Sun. The downside of these initial works was the fact that although the proposed model equations made sense from a mathematical point of view, the physics realism they attained was small. These low order models, with higher or lower degrees of physical complexity, can be used in many areas and several of their results have been validated by 2.5D spatially distributed mean field models, which grants them a certain degree of robustness. This happens specially in LODM whose formulation is directly derived from MHD or mean-field theory equations. Some examples of the results obtained with LODM that have been validated by more complex mean field models are: the study of the parameter space, variability and transitions to chaos in dynamo solutions (Beer et al. 1998; Wilmot-Smith et al. 2005; Charbonneau et al. 2005; Hiremath 2006); the role of Lorentz force feedback on the meridional flow (Rempel 2006; Passos et al. 2012); and the influence of stochastic fluctuations in the meridional circulation and in the α-effect (Charbonneau and Dikpati 2000; Mininni et al. 2001; Mininni and Gómez 2002; Lopes and Passos 2009). Some models even include time delays that embody the spatial segregation and communication between the location of source layers of the α- and Ω-effects. These have been applied to a more general stellar context by Wilmot-Smith et al. (2006) and recently, Hazra et al. (2014); Passos et al. (2014) showed that one of this type of time delay LODM that incorporates two different α source terms working in parallel, can explain how the Sun can enter and exit in a self-consistently way from a grand minimum episode. A couple of LODM even ventured in the "dangerous" field of predictions. For example Hiremath (2008) combined his LODM with an autoregressive model in order to forecast the amplitude of future solar cycles.

In this article we show how can one of these LODM be used as a tool to study the properties of the solar magnetic cycle. For this purpose we use the international sunspot number time series during the past 23 solar magnetic cycles. Nevertheless, the main focus of this work is to present a strategy inspired by helioseismology, were an *inversion methodology* is used to infer variations of some of the LODM parameters over time. Since these parameters are related to the physical mechanisms that regulate the solar dynamo, this should in principle, allow for a *first order* reconstruction of the main dynamo parameters over the last centuries. In a similar manner to helioseismology, the comparison between model solutions and data can be done by means of a *forward method* in which solar observational data is directly compared with the theoretical predictions, or by means of a *backward method* in which the data is used to infer the behaviour of leading physical quantities of the theoretical model. Naturally, it is necessary to develop an inversion technique or methodology that allows to reconstruct the quantities that have changed during the evolution of solar dynamo. This type of studies is well suited to explore several aspects of the solar and stellar dynamo theory. This can be done by: (i) building a tool to study the dynamo regimes operating in stars; (ii) establishing an inversion methodology to infer the leading quantities responsible for the dynamics and variability of the solar cycle over time; (iii) comparing the dynamo numerical simulations with the observational data; (iv) use this tool as a toy model to test global properties of the solar dynamo.

Here we particularly focus in discussing the three last items of this list, with special attention on the development of an inversion method applied here to the sunspot number time series. This is used to infer some of the dynamics of the solar dynamo back-in-time. In principle this should allow us to determine the variation profiles of the quantities that drive the evolution of the magnetic cycle during the last few centuries.

In Sect. 2, we present a non-linear oscillator derived from the equations of a solar dynamo that is best suited to represent the sunspot number. In Sect. 3, we discuss how the non-linear oscillator analogue can be used to invert some of the leading quantities related with solar dynamo. In Sect. 4 is discussed how solar observational data is use to infer properties of the solar magnetic cycle. In Sect. 5 we present a discussion about how the low order dynamo model can be used to test the basic properties of modern axisymmetric models and numerical simulations, as well to infer some leading properties of such dynamo models. In Sect. 6, we discuss the outlook for the Sun and other stars.

2 A LODM for the Evolution of the Large Scale Magnetic Field

The basic equations describing the dynamo action in the interior of a star are obtained from the magnetic-hydrodynamic induction, and the Navier-Stokes equations augmented by a Lorentz force (Moffatt 1978). Under the usual kinematic approximation the dynamo problem consists in finding a flow field with a velocity **U** that has the necessary properties capable of maintaining the magnetic field, **B** against Ohmic dissipation (Charbonneau 2010).

For a star like the Sun such dynamo models should be able to reproduce well-known observational features such as: cyclic magnetic polarity reversals with a period of 22 years, equatorward migration of **B** during the cycle (dynamo wave), the $\pi/2$ phase lag between poloidal and toroidal components of **B**, the antisymmetric parity across the equator, predominantly negative/positive magnetic helicity in the Northern/Southern hemisphere, as well as many of the empirical correlations found in the sunspot records, like the Waldmeier Rule—anti-correlation between cycle duration and amplitude; the Gnevyshev-Ohl Rule—alternation of higher-than-average and lower-than-average cycle amplitude and Grand Minima episodes (like the Maunder Minimum)—epochs of very low surface magnetic activity that span over several cycles. Given the amount of complex features that a solar dynamo model has to reproduce, the task at hand is far from simple.

The vast majority of dynamo models currently proposed to explain the evolution of the solar magnetic cycle (kinematic mean-field models) became very popular with the advance of helioseismology inversions and the inclusion of the differential rotation profile. In the kinematic regime approximation, the flow field **U** is prescribed and only the magnetic induction equation is used to determine the evolution of **B**. Generally, the large scale magnetic field, the one responsible for most of the features observed in the Sun is modelled as the interaction of field and flow where two source terms (Ω and α) naturally emerge from mean-field theory (e.g., Moffatt 1978; Krause and Raedler 1980; Cardoso and Lopes 2012). From the mean-field electrodynamics, the induction equation reads

$$\frac{\partial \mathbf{B}}{\partial t} = \nabla \times (\mathbf{U} \times \mathbf{B} + \alpha \mathbf{B} - \eta \nabla \times \mathbf{B}), \tag{1}$$

where **U** is the large-scale mean flow, and η is the total magnetic diffusivity (including the turbulent diffusivity and the molecular diffusivity). Currently, as inferred from helioseismology, **U** can be interpreted as a large-scale flow with at least two major flow components, the differential rotation throughout the solar interior, and the meridional circulation in the upper layers of the solar convection (Howe 2009).

Given all the points above, and based on their popularity among the community, we start our study by considering a reference model based in the kinematic mean-field flux transport framework. Although the results obtained here are based on this specific type of model, most of the analysis method used, as well the conclusions reached, can easily be extended to other models. As usual, under the simplification of axi-symmetry the large-scale magnetic field **B** can be conveniently expressed as the sum of toroidal and poloidal components, that in spherical polar coordinates (r, θ, ϕ) can be written as

$$\mathbf{B}(r, \theta, t) = \nabla \times \big(A_p(r, \theta, t)\mathbf{e}_\phi\big) + B_\phi(r, \theta, t)\mathbf{e}_\phi. \tag{2}$$

Similarly, the large-scale flow field **U** as probed by helioseismology can be expressed as the sum of an axisymmetric azimuthal (differential rotation) and poloidal (meridional flow) components:

$$\mathbf{U}(r, \theta) = \mathbf{u}_p(r, \theta) + \tilde{r}\Omega(r, \theta)\mathbf{e}_\phi \tag{3}$$

where $\tilde{r} = r \sin\theta$, Ω in the angular velocity and $\mathbf{u}_p$ is the velocity of the meridional flow. Accordingly, such decomposition of **B** (that satisfy the induction equation (1)) and **U** leads to the following set of equations:

$$\frac{\partial A_p}{\partial t} = \eta\big(\nabla^2 - \tilde{r}^{-2}\big)A_p - \tilde{r}^{-1}\mathbf{u}_p \cdot \nabla(\tilde{r}A_p) + \alpha B_\phi \tag{4}$$

$$\frac{\partial B_\phi}{\partial t} = \eta\big(\nabla^2 - \tilde{r}^{-2}\big)B_\phi - \tilde{r}\mathbf{u}_p \cdot \nabla\big(\tilde{r}^{-1}B_\phi\big) + \tilde{r}\big[\nabla \times (A_p\mathbf{e}_\phi)\big] \cdot \nabla\Omega - \Gamma(B_\phi)B_\phi \tag{5}$$

where η is the magnetic diffusivity and α is the source term of A_p (the mechanism to convert toroidal to poloidal field). Moreover, following the suggestions of Pontieri et al. (2003) we also considered that the toroidal field can be removed from the layers where it is produced by magnetic buoyancy and obeying $\Gamma \sim \gamma B_\phi^2/8\pi\rho$, where γ is a constant related to the removal rate and ρ is the plasma density.

2.1 A van der Pol-Duffing Oscillator for the Solar Cycle

As the Sun's magnetic field changes sign from one solar cycle to the next it is a plausible idea to attribute alternating signs in odd/even cycles also to other solar activity indicators such as the sunspot number (SSN). The resulting time series displays cyclic variations around zero in the manner of an oscillator. This suggests an oscillator as the simplest mathematical model of the observed SSN series. As, however, the profile of sunspot cycles is known to be markedly asymmetric (a steep rise in 3–4 years from minimum to maximum, followed by a more gradual decline to minimum in ~ 7 years), a simple linear oscillator would be clearly a very poor representation of the sunspot cycle. A *damped* linear oscillator

$$\ddot{x} = -\omega^2 x - \mu \dot{x} \tag{6}$$

will, on the other hand, naturally result on asymmetric profiles similar to what is observed. The obvious problem that the oscillation will ultimately decay due to the damping could be remedied somewhat artificially by applying a periodic forcing or by reinitializing the model at each minimum. A much more natural way to counteract the damping, however, is the introduction of non-linearities into the equation—indeed, such non-linearities are naturally expected to be present in any physical system, see below. As long as the non-linearity is relatively weak, the parameters ω^2 and μ can be expanded into Taylor series according to x. Due to the requirement of symmetry (i.e. the behaviour of the oscillators should be invariant to a sign change in x) only terms of even degree will arise in the Taylor series. To leading order, then, we can substitute

$$\omega^2 \to \omega^2 - \lambda x^2 \qquad \mu \to \mu\left(\xi x^2 - 1\right) \tag{7}$$

into (6) resulting in

$$\ddot{x} = -\omega^2 x - \mu\left(\xi x^2 - 1\right)\dot{x} + \lambda x^3 \tag{8}$$

In the particular case when $\lambda = 0$ (i.e. the non-linearity affects the damping only) and the other parameters are positive, the system described by (8) is known as a van der Pol oscillator. The alternative case when non-linearity affects the directional force/frequency only, i.e. $\xi = 0$, $\mu < 0$ and $\lambda \neq 0$, in turn, represents a *Duffing oscillator*. Due to their simplicity and universal nature these two systems are among those most extensively studied in non-linear dynamics. It is straightforward to see that the oscillator is non-decaying, i.e. the origin is repeller, whenever $\mu > 0$ (negative damping) in the case of a van der Pol oscillator and/or $\omega^2 > 0$ and $\lambda > 0$ in a Duffing oscillator. When a non-linearity is present in both parameters (i.e. λ and ξ are both non-zero) a combined *van der Pol-Duffing oscillator* results.

The van der Pol–Duffing oscillator, however, is more than just a good heuristic model of the solar cycle. In fact, an oscillator equation of this general form can be derived by a truncation of the dynamo equations. As noted before, we are especially interested in capturing the temporal dynamics associated with the large scale magnetic field. In order to construct a low order model aimed at capturing this dynamics, we follow the procedures described in Passos and Lopes (2008b, 2011).

It has been suggested by Mininni et al. (2000, 2001) and Pontieri et al. (2003) that a dimensional truncation of the dynamo equations (4) and (5) is an effective method to reduce the system's dimensions and capture phenomena just on that scale. Following that ansatz, gradient and Laplacian operators are approximated by a typical length scale of the system l_0 (e.g. convection zone length or width of the tachocline), leading to $\nabla \sim l_o^{-1}$ and $\nabla^2 \sim l_o^{-2}$. Analogously this can be interpreted as a collapse of all spatial dimensions, leaving only the temporal behaviour. In terms of dynamical systems, we are projecting a higher dimensional space into a single temporal plane. After grouping terms in B_ϕ and A_p (now functions only dependent of the time) we get

$$\frac{dB_\phi}{dt} = c_1 B_\phi + c_2 A_p - c_3 B_\phi^3 \tag{9}$$

$$\frac{dA_p}{dt} = c_1 A_p + \alpha B_\phi \tag{10}$$

where we have defined the *structural coefficients*, c_n, as

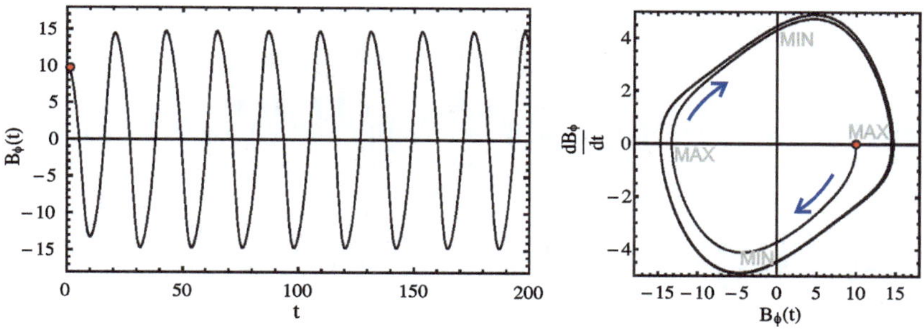

Fig. 1 *Left panel* represents the time evolution of $B_\phi(t)$ obtained from equation (14) with parameters $c_1 = 0.08$, $c_2 = -0.09$, $c_3 = 0.001$, $\alpha_0 = 1$. In the *right* we have a $(B_\phi, dB_\phi/dt)$ phase space representation of the solution. The *blue arrows* indicate the direction of increasing time and the *red dot* the initial value used. Also indicated in this *panel* are the regions corresponding to the maxima and minima of the cycle. *Adapted from* Passos and Lopes (2011)

$$c_1 = \eta\left(\frac{1}{l_0^2} - \frac{1}{\bar{r}^2}\right) - \frac{v_p}{l_0} \tag{11}$$

$$c_2 = \frac{\bar{r}\Omega}{l_0^2} \tag{12}$$

$$c_3 = \frac{\gamma}{8\pi\rho} \tag{13}$$

We now concentrate in creating an expression for the time evolution of B_ϕ since it is the field component directly associated with the productions of sunspots. We derive expression (9) in order to the time, and substitute (10) in it to take away the A_p dependence yielding

$$\frac{d^2 B_\phi}{dt^2} + \omega^2 B_\phi + \mu\left(3\xi B_\phi^2 - 1\right)\frac{\partial B_\phi}{\partial t} - \lambda B_\phi^3 = 0, \tag{14}$$

where $\omega^2 = c_1^2 - c_2\alpha$, $\mu = 2c_1$, $\xi = c_3/2c_1$ and $\lambda = c_1 c_3$ are model parameters that depend directly on the structural coefficients. The name used to describe c_n comes from the fact that these coefficients contain all the background physical structure (rotation, meridional circulation, diffusivity, etc.) in which the magnetic field evolves.

This oscillator (14) is a van der Pol-Duffing oscillator and it appears associated with many types of physical phenomena that imply auto-regulated systems. This equation is a quite general result which **B** should satisfy. In this case, unlike in the classical van der Pol-Duffing oscillator, the parameters are interconnected by a set of relations that link the present oscillation model with the original set of dynamo equations (4)–(5). This interdependency between parameters will eventually constrain the solution's space. As in the classical case, ω controls the frequency of the oscillations or the period of the solar magnetic cycle, μ controls the asymmetry (or non-linearity) between the rising and falling parts of the cycle and ξ affects directly the amplitude. The λ parameter, related to buoyancy loss mechanism sets the overall amplitude peak amplitude of the solution. Figure 1 shows the solution of (14) in a time vs. amplitude diagram (left) and in a $\{B_\phi, dB_\phi/dt\}$ phase space. From this figure we find that this dynamo solution, under suitable parametrization (viz. next section) is a self-regulated system that rapidly relaxes to a stable 22-year oscillation. In the phase space the solution tend to a limit cycle or attractor. A complete (clockwise) turn in the phase space corresponds to a complete solar magnetic cycle.

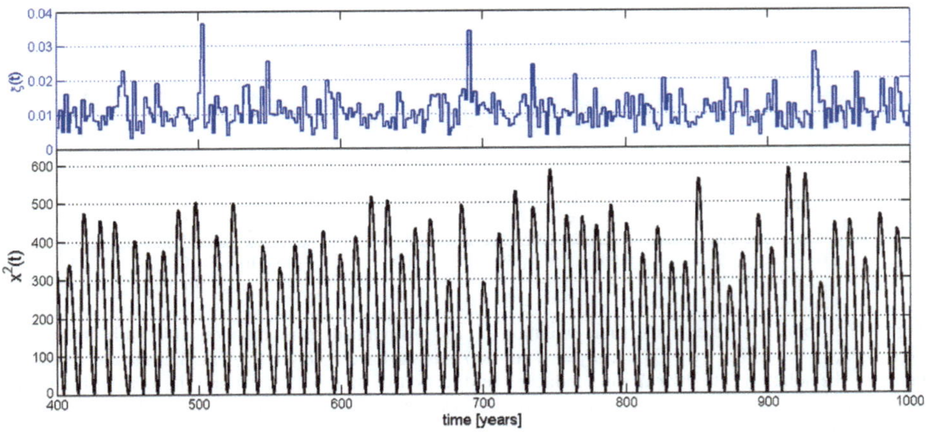

Fig. 2 *Bottom panel*: a 600 year long time series resulting from a stochastically perturbed van der Pol oscillator stochastically perturbed in one of its parameters (non-linearity $\xi(t)$). SSN values were defined as $x^2(t)$. The noise applied (piecewise constant in this case) is shown in the *top panel*

2.2 A Semi-classical Analysis Method Using a Non-linear Oscillator

In order to estimate values for the coefficients in (14), Mininni et al. (2000) and Passos and Lopes (2008b) fitted this oscillator model either to a long period of the solar activity (several solar cycles) or to each magnetic cycle individually. We shall return to this point in subsequent sections. A more general approach to the problem of finding the parameter combinations with which the classical van der Pol-Duffing oscillator returns solar-like solutions was taken by Nagy and Petrovay (2013). The authors mapped the parameter space of the oscillator by adding stochastic noise to its parameters using different methods. The objective was to constrain the parameter regime where this non-linear model shows the observed attributes of the sunspot cycle, the most important requirement being the presence of the Waldmeier effect (Waldmeier 1935) according to the definition of Cameron and Schüssler (2008).

Noise was introduced either as an Ornstein–Uhlenbeck process (Gillespie 1996) or as a piecewise constant function keeping a constant value for the interval of the correlation time. The effect of this noise was assumed to be either additive or multiplicative. The amplitudes and correlation times of the noise defined the phase space. The attributes of the oscillator model were first examined in the case of the van der Pol oscillator (no Duffing cubic term) with perturbation either in the non-linearity parameter, ξ or the damping parameter, μ, as shown in the equations below:

$$\ddot{x} = -\omega_0^2 x - \mu(t)\left[\xi_0 x^2 - 1\right]\dot{x} \tag{15}$$

$$\ddot{x} = -\omega_0^2 x - \mu_0\left[\xi(t) x^2 - 1\right]\dot{x}. \tag{16}$$

The constant parameters, ω_0, μ_0 and ξ_0 used were taken from the fitted values listed by Mininni et al. (2000). Note that in this simple case, variations in these parameters were assumed to be independent from each other, whereas in reality they are interrelated (cf. (14)). The results show that the model presents solar-like solutions when a multiplicative noise is applied to the non-linearity parameter, as in (16). An example of a time series produced by this type of oscillator is shown in Fig. 2.

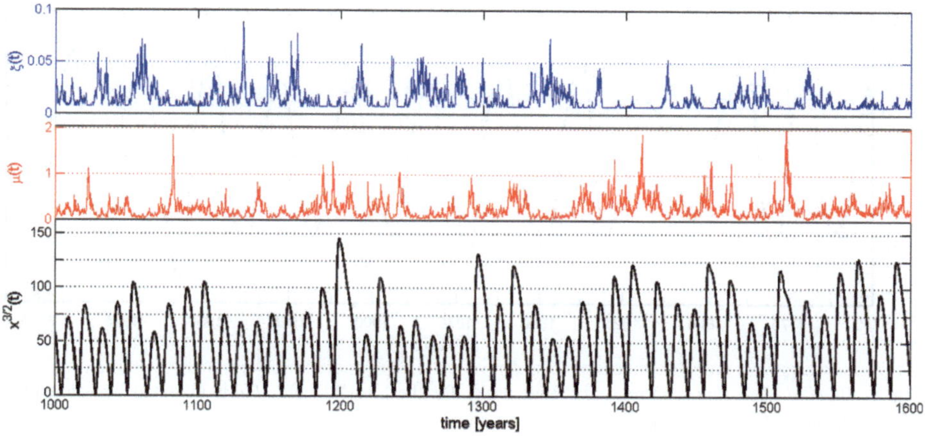

Fig. 3 *Bottom panel*: a 600 year long time series resulting from a van der Pol–Duffing oscillator stochastically perturbed in two of its parameters, ξ and μ. SSN values were here defined as $x^{3/2}(t)$, following Bracewell (1988). The noise applied is shown in the *top* and *middle panels*. The Duffing parameter was here given a constant value $\lambda_0 = 5 \times 10^{-5}$

As a next step towards a fully general study, let us consider the case where both ξ and μ are simultaneously perturbed and the Duffing term $-\lambda x^3$ is also kept in the oscillator equation (8). Noise is applied to μ but it also affects ξ as the values of ξ and μ are assumed to be related as

$$\xi(t) = \frac{C_\xi}{\mu(t)} \quad \text{and} \quad \lambda(t) = C_\lambda \frac{\mu(t)}{2}, \tag{17}$$

(Passos and Lopes 2008b); here, C_ξ and C_λ are constants.

A mapping of the parameter space shows that in this case solar-like solutions are more readily reproduced compared to the case when only one parameter was assumed to vary (see Fig. 3). This finding is in line with the information derived from the LODM developed in the previous section. An ongoing study shows that the corresponding time dependence in the Duffing parameter, as predicted by the LODM, has a significant effect on the character of the solution.

We note that additive noise was first applied to one parameter (ξ) of a van der Pol oscillator by Mininni et al. (2000, 2001) but the focus of that work was on reproducing cycle to cycle fluctuations, without considering the Waldmeier effect. This model was further analysed by Pontieri et al. (2003) (and references therein) who studied the behavior of the Hurst exponent of this system and concluded that this type of fluctuations implies that the stochastic process which underlies the solar cycle is not simply Brownian. This means that long-range time correlations could probably exist, opening the way to the possibility of forecasts on time scales comparable to the cycle period. An attempt to introduce the effects of such non-Gaussian noise statistics into the LODM was made by Vecchio and Carbone (2008) who suggest that this may contribute to cyclic variations of solar activity on time scales shorter than 11 years.

3 Coupling a LODM with Observational Data—Inversions

In the previous example a perturbation method was studied in order to find solar like solutions for this non-linear oscillator. Another way of thinking is to pair the oscillator directly

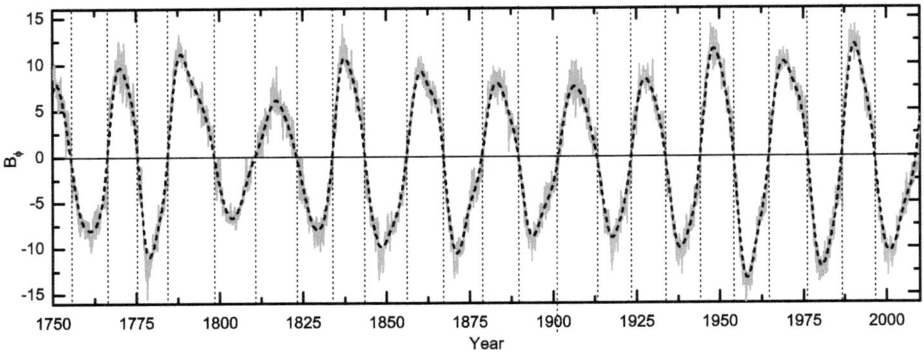

Fig. 4 *Black dashed line* represents the built proxy for the toroidal field. B_ϕ is obtained by calculating $\sqrt{SSN}$, changing the sign of alternate cycles (represented in *gray*), and smoothing it down using an FFT low pass filter of 6 months. The *vertical thin dotted lines* represent solar cycle minima

to some solar observable and try to constraint its parameters. As mentioned in the introduction we choose the international sunspot number, *SSN* and we use it to build a proxy of the toroidal magnetic component. Since the *SSN* is usually taken to be proportional to the toroidal field magnetic energy that erupts at the solar surface ($\propto B_\phi^2$) (Tobias et al. 1995), this makes it ideal for compare with solutions of the LODM. Taking this in consideration, Mininni et al. (2000); Passos and Lopes (2008b) have built a toroidal field proxy based on the sunspot number by following the procedure proposed by Polygiannakis and Moussas (1996), i.e. $B_\phi \approx \pm\sqrt{SSN}$. Details about the construction of the toroidal proxy (see Fig. 4) can be found in Mininni et al. (2001); Passos and Lopes (2008b) and Passos (2012).

3.1 The Averaged Behaviour of the Solar Dynamo Attractor

The solution obtained for (14) presented in Fig. 1 shows that the solar cycle is a self-regulated system that tends to a stable solution defined by an attractor (limit cycle). If we allow for the different physical processes responsible for the solar dynamo and embedded in the structural coefficients ((11), (12) and (13)), i.e. the differential rotation, the meridional circulation flow, the α mechanism, and the magnetic diffusion, to change slowly from cycle to cycle then we start to observe deviations from the equilibrium state. If deviations from this sort of dynamical balance occur, such that if one of these processes changes due to an external cause, the other mechanisms also change to compensate this variation and ensure that the solar cycle finds a new equilibrium.

To test this idea of an equilibrium limit cycle, we fit the LODM parameters to the $\{B_\phi, dB_\phi/dt\}$ phase space of the built toroidal proxy (see Fig. 5).

If one of the parameter's variation is very large the system can be dramatically affected, leading to a quite distinct evolution path like the ones found during the solar grand minima. We will develop this subject in a subsequent section.

3.2 Matching of Solutions to Observable Characteristics of the Solar Cycle

Solutions with fluctuation similar to those we see on the solar cycle are easily set by variations in the μ parameter (and the physical processes associated with it). By definition the structural coefficient that regulates this parameter (c_1) also has an important role in the other parameters (ω, ξ and λ). In the LODM equation (14) the μ and $\mu\xi$ quantities regulate the

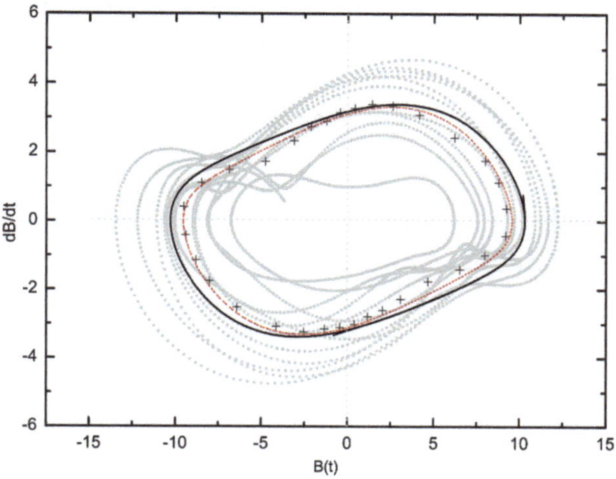

Fig. 5 Phase-space diagram for the toroidal proxy $B_\phi(t)$: The *crosses* correspond to local area averaged values found by dividing the data into 32 temporal intervals. The *red dashed curve* is a fit to the *crosses*. The *continuous black curves* correspond to a fit to all data points (not grouped in intervals). For the *red curve* we have that $\mu = 0.1645$, $\omega = 0.3523$, $\xi = 0.0147$ and $\lambda = 0.0005$. Figure *adapted from* Passos and Lopes (2008a)

strength and the non-linearity of the damping. Moreover, an occasional variation on μ, like a perturbation on the meridional flow amplitude, v_p (see structural coefficient (11)) will affect all sets of parameters leading to the solar dynamo (14) to find a new equilibrium, which will translate into the solar magnetic cycle observable like the sunspots number, showing an irregular behaviour.

The well-known relation discover by Max Waldmeier (Wolf and Brunner 1935), that the time that the sunspot number takes to rise from minimum to maximum is inversely proportional to the cycle amplitude in naturally captured by the LODM assuming discrete variations in μ. Notice that the Waldmeier effect occurs as a consequence of the limit cycle becoming increasingly sharp as μ increases, i.e., the sunspot number amplitude increases as the cycle's rising times gets shorter.

4 How to Infer Properties of the Solar Magnetic Cycle

From the physical point of view, based on observations, we know that in the Sun some of the physical background structures that are taken as constant in our standard dynamo solution aren't so. In order to test that specific changes of the background state lead to the observed changes in the amplitude of the solar cycle, the following strategy was devised. At a first approximation we assume that the structural coefficients can change only discretely in time, more specifically from cycle to cycle while the magnetic field is allowed to evolve continuously. The idea is that changing coefficients will generate theoretical solutions with different amplitudes, periods and eigen-shapes at different times and by comparing these different solution pieces with the observed variations in the solar magnetic field, we are able to infer information about the physical mechanisms associated with the coefficients.

To do this we compare our theoretical solution with a proxy built from the International monthly averaged SSN since 1750 to the present. As mentioned before we assume that $B_\phi \propto \sqrt{SSN}$. The proxy data is separated into individual cycles and fitted using (14), considering that the buoyancy properties of the system are immutable, i.e. c_3 is constant throughout the time series. This means that when we fit the LODM to solar cycle N, we will retrieve the set of c_{nN} coefficients that best describe that cycle. This allows to probe how these coefficients vary from cycle to cycle and consequently how the physical mechanisms associated with

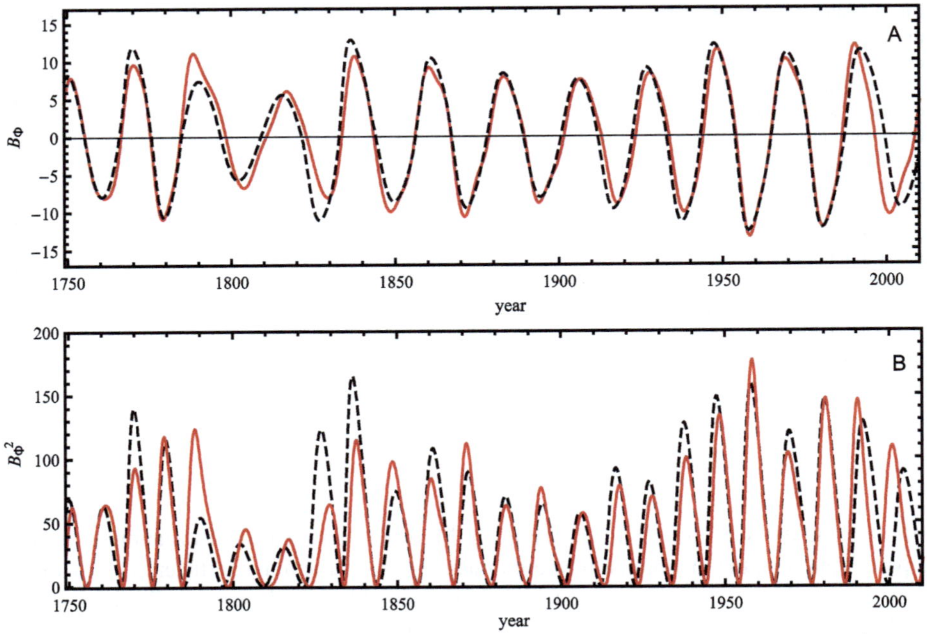

Fig. 6 (**A**) Theoretical solution (*red*) obtained by fitting it the proxy at each individual cycle. (**B**) Direct comparison of the model behavior (*red*) and observed solar cycle amplitude (*black*). In panel (**B**) we just plot the squared of panel (**B**) and this also amplifies the differences between both curves. *Adapted from* Passos (2012)

them evolve in time (Lopes and Passos 2009; Passos 2012). Equation (14) is afterwards solved by changing the parameters to their fit value, at every solar minimum using a stepwise function (similar to that presented in Fig. 7 for c_1). Figures 7 and 6 highlight this procedure. The fact that such a simplified dynamo model can get this degree of resemblance with the observed data just by controlling one or two parameters is an indication that it captures the most important physical processes occurring in the Sun.

4.1 Meridional Circulation Reconstruction

The simple procedure previously described allows to reconstruct the behavior of solar parameters back in time. Using an improved fitting methodology, Passos (2012) obtained with this model the reconstruction of the variation levels of the solar meridional circulation for every solar (sunspot) cycle over the last 250 years. One must notice that in this specific LODM the amplitude of the cycle depends directly on the amplitude of the meridional flow during the previous cycle. It is completely possible to imagine that other models that consider a different theoretical setup might return a different behaviour.

Looking at (11), we can see that the coefficient c_1 depends on two physical parameters, the magnetic diffusivity, η, and the amplitude of the meridional circulation, v_p. The magnetic diffusivity of the system is a property tightly connected with turbulent convection and is generally believed to change only in time scales of the order of stellar evolution. This leaves variations in v_p as the only plausible explanation for the variation observed from cycle to cycle. Therefore, by looking at the evolution of c_1 we can effectively assume that we are looking at the variation in the strength of meridional circulation. The results obtained are presented in Fig. 7.

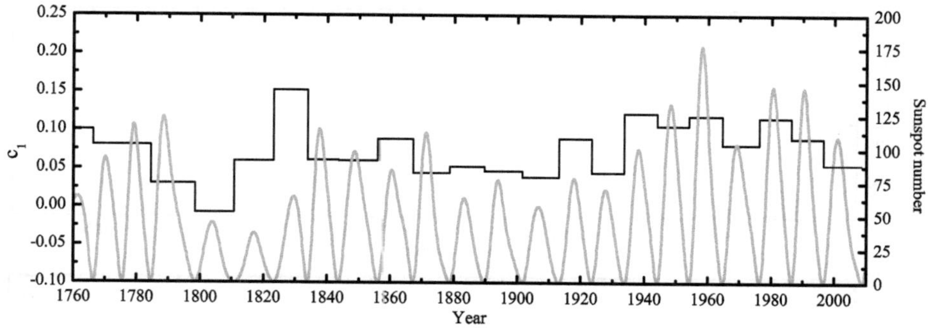

Fig. 7 Reconstruction of the meridional circulation profile represented by c_1 (*black line*) compared to smoothed the sunspot number (*gray*). *Adapted from* Passos (2012)

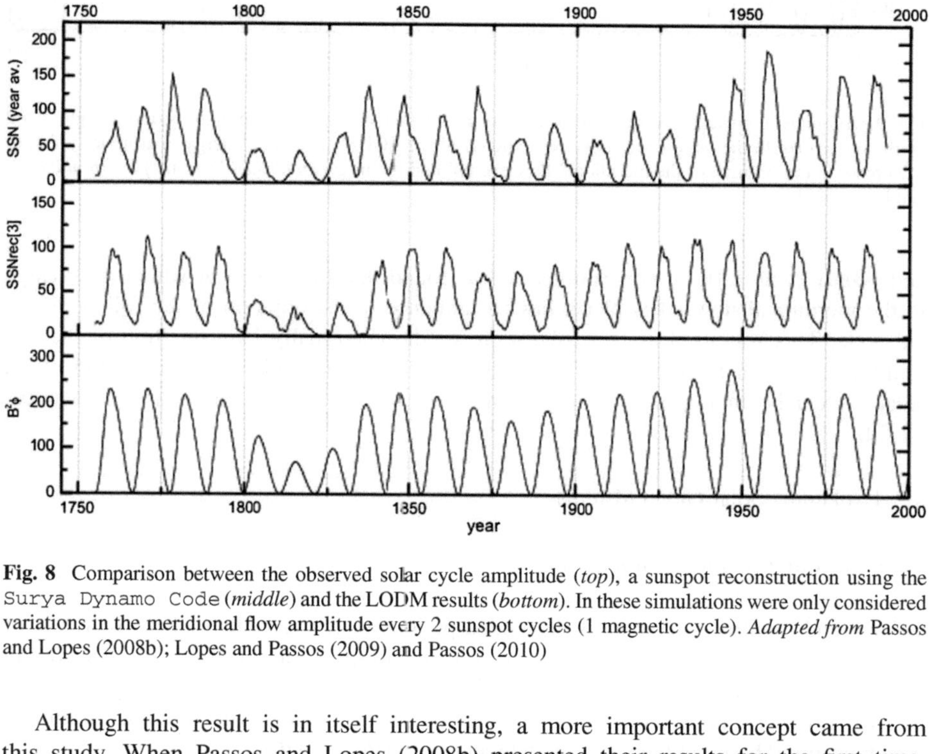

Fig. 8 Comparison between the observed solar cycle amplitude (*top*), a sunspot reconstruction using the `Surya Dynamo Code` (*middle*) and the LODM results (*bottom*). In these simulations were only considered variations in the meridional flow amplitude every 2 sunspot cycles (1 magnetic cycle). *Adapted from* Passos and Lopes (2008b); Lopes and Passos (2009) and Passos (2010)

Although this result is in itself interesting, a more important concept came from this study. When Passos and Lopes (2008b) presented their results for the first time, they introduced the idea that coherent long term variations (of the order of the cycle period) in the strength of the meridional circulation could provide an explanation for the variability observed in the solar cycle (see Fig. 8). This result was also *a posteriori* numerically validated using a 2.5D flux transport model (Lopes and Passos 2009; Karak 2010). Only a couple of years later, Hathaway and Rightmire (2010) presented meridional circulation measurements spanning over the last solar cycle. Their measurements confirmed that the amplitude of this plasma flow changes considerably from cycle to cycle.

Recently two other groups have tested this idea with their 2.5D dynamo models finding additional features based in this effect, cf. Karak (2010), Karak and Choudhuri (2011)

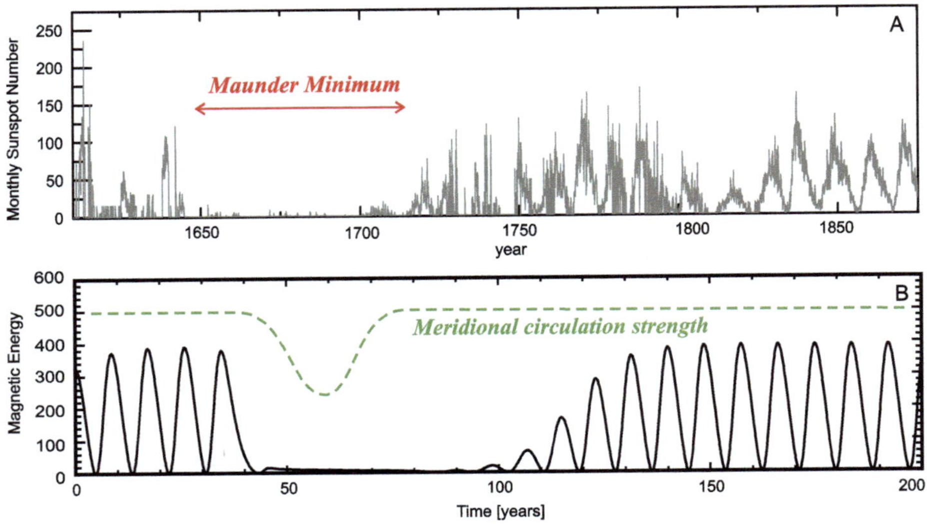

Fig. 9 *Top panel* represents the sunspot measurements where the Maunder Minimum period is highlighted by the *red arrow*. In the *bottom* we show the response of the LODM (magnetic energy in arbitrary units) to a decrease in the strength of the meridional flow exemplified by the *green dashed line*. *Adapted from* Passos and Lopes (2011)

and Nandy et al. (2011). For example it was found that the instant at which the change in the meridional flow takes place, has an influence in the duration of the following solar cycle. This was used as an explanation for the abnormally long duration of the last minimum. Just for reference, the numbering of solar cycles only started after 1750 with solar cycle 1 beginning in 1755. At this moment we are in the rising phase of solar cycle 24.

4.2 Explaining Solar Grand Minima

4.2.1 Variations in the Meridional Flow

Solar grand minima correspond to extended periods (a few decades) where very low or no solar activity occurs. During theses periods no sunspots (or very few) are observed in the solar photosphere and it is believed that other solar phenomena also exhibit low levels of activity. The most famous grand minimum that has been registered is the Maunder Minimum which occurred between the years of 1645 and 1715 (Eddy 1976).

A possible explanation for the origin of these quiescent episodes was put forward by Passos and Lopes (2011). Using a LODM, they showed that a steep decrease in the meridional flow amplitude can lead to grand minima episodes like the Maunder minimum (see Figs. 9 and 10). This effect presents the same visual characteristics as the observed data, namely a rapid decrease of magnetic intensity and a gradual recovery into normal activity (see Fig. 9) after the meridional circulation amplitude returns to its normal values. A similar result was later obtained by Karak (2010), again using a more complex 2.5D numerical flux transport model. Nevertheless the reasons that could lead to a decrease of the meridional flow amplitude were not explored. This served as a motivation to study the behavior of this LODM in the non-kinematic regime, explained in Sect. 5.2.

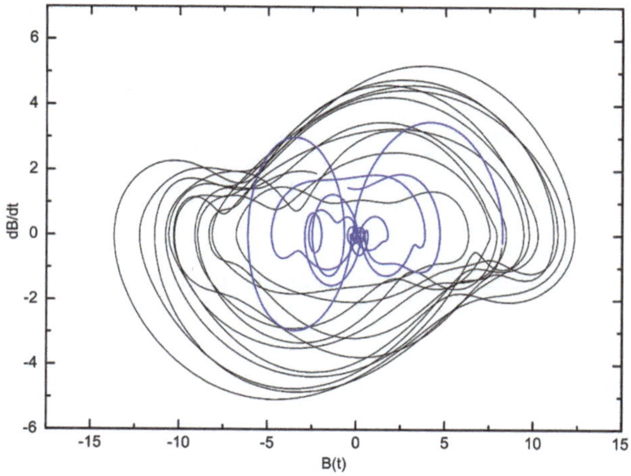

Fig. 10 The phase-space diagram for $B(t)$ corresponds to the value obtained from the sunspot number temporal series as in Fig. 9. *Adapted from* Passos and Lopes (2011) and Passos (2010)

4.2.2 Fluctuations in the α Effect

Some examples mentioned in the introduction, hint that fluctuations in the α mechanism can also trigger grand minima. We focus on a specific example now, the LODM developed by Hazra et al. (2014). In this work the authors used a time-delay LODM similar to that presented in Wilmot-Smith et al. (2006) but expanded with the addition of a second α effect. This model incorporates two of these mechanisms, one that mimics the surface Babcock-Leighton mechanism (BL), and another one analogous to the classical mean-field (MF) α-effect that operates in the bulk of the convection zone. This set up captures the idea that the BL mechanism should only act on strong magnetic fields that reach the surface, and that weak magnetic fields that diffuse through the convection zone should feel the influence of the MF α. The authors subject these two effects to different levels of fluctuations and find that in certain parameter regimes, the solution of the system shows the same characteristics as a grand minimum. These results were also validated by implementing a similar set up into a 2.5D mean-field flux transport dynamo model (Passos et al. 2014). Again this shows the usefulness of low order models to probe ideas before their implementation into more complex models.

4.3 Solar Cycle Predictability

For the near future, perhaps one of most interesting applications of this LODM is its use in the predictability of future solar cycles amplitudes. The first step towards this objective is presented in Passos (2012). The authors studied the correlations between the LODM fitted structural coefficients and cycle's characteristics (amplitude, period and rising time). They found very useful relationships between these quantities measured for cycle N and the amplitude of cycle $N+1$. These relationships were put to the test by predicting the amplitude of current solar cycle 24 (see Fig. 11).

5 LODM as Probe of Numerical Dynamo Models

In recent years there have been strong developments of different types of dynamo models to compute the evolution of the solar magnetic activity and to explore some of the causes

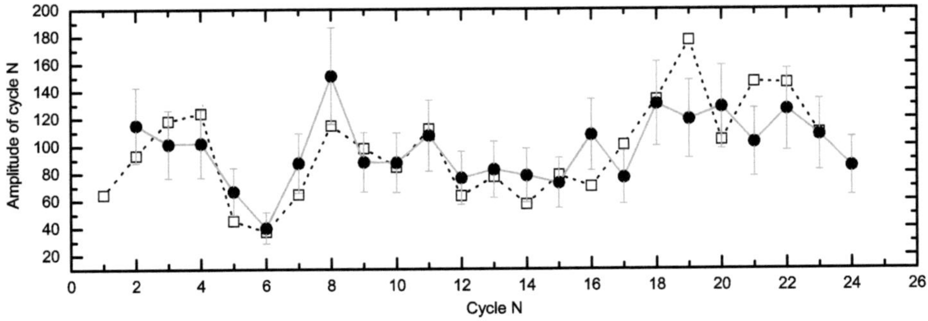

Fig. 11 Observed values (*white squares*) and predicted values *black circles* with *gray error bars* for the cycle amplitude. The *red circle* is the predicted amplitude for solar cycle 24 based on this methodology. *Adapted from* Passos (2012)

of magnetic variability. Two classes of models have been quite successful, the kinematic dynamo models and, more recently, the global magnetohydrodynamical models. Both types of dynamo models have a quite distinct approach to the dynamo theory, the first one resolves the induction magnetic equation for a prescribed velocity field (which is consistent with helioseismology), and the second one obtains global magnetohydrodynamical simulations of the solar convection zone. Many of these models are able to reproduce some of the many observational features of the solar magnetic cycle. Nevertheless, it remains quite a difficult task to successfully identify which are the leading physical processes in current dynamo models that actually drive the dynamo in the solar interior. The usual method to test these dynamo models is to compare their theoretical predictions with the different sets of data, including the sunspot numbers, however, in many cases the conclusions obtained are very limited, as different physical mechanisms lead to very identical predictions. This problem also arises in the comparison between different dynamo models, including different types of numerical simulations.

A possible solution to this problem is to use inverted quantities (obtained form observational data) to test the quality of the different solar dynamo model, rather than making direct comparison of data. For those of you familiarized with helioseismology, there is a good analogue: it is the equivalent to compare the inverted sound speed profile (obtained from observational data) with the sound speed profile predicted by solar models (*backward approach*), rather than compare predicted frequencies with observational frequencies (*forward approach*). The former method to test physical models is more insightful than the latter one. At the present level of our understanding of the solar dynamo theory, as a community we could gain a more profound understanding of the mechanisms behind the solar magnetic variability, if we start developing some backward methods to analyse solar observational data and test dynamo models.

5.1 Using a LODM in the Kinematic Regime

Using the meridional velocity inverted from the sunspot number time series (cf. Fig. 7), Lopes and Passos (2009) showed that most of the long term variability of the sunspot number could be explained as being driven by the meridional velocity decadal variations, assuming that the evolution of the solar magnetic field is well described by an axisymmetric kinematic dynamo model.

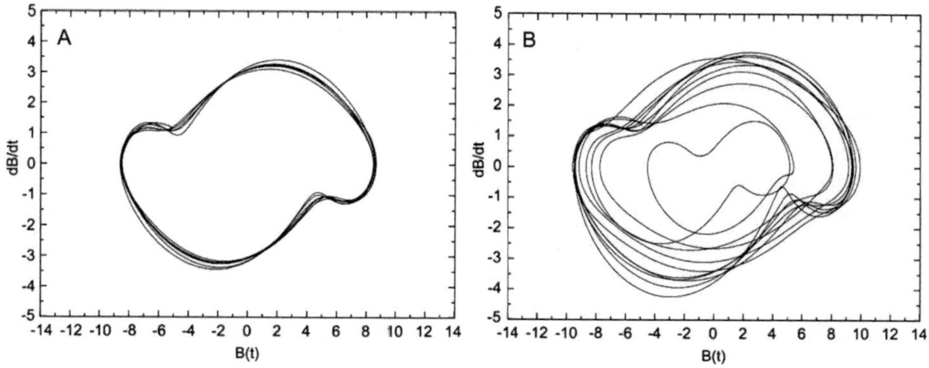

Fig. 12 The phase-space diagram for $B(t)$ of a kinematic dynamo model (using the Surya's kinematic model): (**A**) a standard kinematic dynamo model (v_p is constant); (**B**) a variable kinematic dynamo model in which the v_p for each magnetic cycle corresponds to the value obtained from the sunspot number temporal series. Both simulations correspond to 130-year time series. The small variability present in the *left panel* is due to the stabilization of the numerical solution. Figure *adapted from* Lopes and Passos (2009)

Figure 8 shows a reconstructed sunspot times series that has been obtained using the meridional velocity v_p inverted from the sunspot observational time series, and Fig. 12 shows the phase space of a standard axisymmetric kinematic dynamo model (with the same v_p for all cycles) (e.g., Choudhuri et al. 1995) and a solar dynamo model where the v_p changes from cycle to cycle as inverted from the sunspot times series (Lopes and Passos 2009). It is quite encouraging to find that such class of dynamo models for which the v_p changes overtime successfully reproduced the main features found in the observational data.

Moreover, in their article Lopes and Passos (2009) tested two different methods of implementing the velocity variation for each magnetic cycle, namely, by considering that amplitude variations in v_p that take place at sunspot minima or at sunspot maxima. All the time series show a few characteristics that are consistent with the observed sunspot records. In particular, all the simulations show the existence of low amplitudes on the sunspot number time series between 1800 and 1840 and between 1870 and 1900. The simulation that best reproduces the solar data corresponds to the model SSNrec[3] (see Fig. 8), in which was implemented a smoothed v_p variation profile between consecutive cycles and taking place at the solar maximum. This clearly highlights the potential of such methodology.

Here, we discuss the same methodology as the one used in the previous section, but instead of applying it to observational sunspot number records, it is used to reconstruct the sunspot time series. The results obtained clearly show that the present kinematic dynamo models can reproduce in some detail the observed variability of the solar magnetic cycle. The fact that for one of the sunspot models—model SSNrec[3], it presents a strong level of correlation with the observational time series, lead us to believe that the main idea behind this *backward approach* is correct and it is very likely that the inverted v_p variation is probably very close to the v_p variation that happens in the real Sun. Clearly, under the assumed theoretical framework the meridional circulation is the leading quantity responsible for the magnetic variability found in the sunspot number time series and current solar dynamo models are able to reproduce such variability to a certain degree.

5.2 Using a LODM in the Non-kinematic Regime

So far, the vast majority of the LODM applications presented here followed the traditional assumption that the solar dynamo can be correctly modeled in the kinematic regime, where only the plasma flows influence the production of magnetic field, and not the other way around. This kinematic approximation is used in the vast majority of the present 2.5D spatially resolved dynamo models.

In the last couple of years though, evidence started to appear supporting the claim that this kinematic regime might be overlooking important physical mechanisms for the evolution of the dynamo. The idea that the meridional flow strength can change over time and affect the solar cycle amplitude coupled with the measurements of Hathaway and Rightmire (2010) and Basu and Antia (2010) indicate that the observed variation in this flow is highly correlated with the levels of magnetic activity. This leads to the fundamental question: *"Is the flow driving the field or is the field driving the flow?"*.

The first clues are starting to appear from 3D MHD simulations of solar convection. The recent analysis of the output of one of the large-eddy global MHD simulations of the solar convection zone done by Passos et al. (2012) shows interesting clues. These simulations solve the full set of MHD equations in the inelastic regime, in a broad, thermally-forced stratified plasma spherical shell mimicking the SCZ and are fully dynamical on all spatiotemporally-resolved scales. This means that a two way interaction between field and flow is always present during the simulation. The analysis shows that the interaction between the toroidal magnetic field and the meridional flow in the base of the convection zone indicates that the magnetic field is indeed acting on the equatorward deep section of this flow, accelerating it. This observed relationship runs contrary to the usually assumed kinematic approximation.

In order to check if this non-kinematic regime has any impact in the long term dynamics of the solar dynamo, Passos et al. (2012) implemented a term that accounts for the Lorentz force feedback in a LODM similar to the one presented here. This allows to fully isolate the global aspects of the dynamical interactions between the meridional flow and magnetic field in a simplified way.

They assumed that the large-scale meridional circulation, v_p, is divided into a "kinematic" constant part, v_0 (due to angular momentum distribution) and a time dependent part, $v(t)$, that encompasses the Lorentz feedback of the magnetic field. Therefore they redefine v_p as $v_p(t) = v_0 + v(t)$ where the time dependent part evolves according to

$$\frac{\mathrm{d}v(t)}{\mathrm{d}t} = a\,B_\phi A_p - b\,v(t). \tag{18}$$

The first term is a magnetic non-linearity representing the Lorentz force and the second is a "Newtonian drag" that mimics the natural resistance of the flow to an outside kinematic perturbation. Under these conditions the Lorentz force associated with the cyclic large-scale magnetic field acts as a perturbation on the otherwise dominant kinematic meridional flow. This idea was not new and it was used before in the context of magnetically-mediated variations of differential rotation in mean-field dynamo models (Tobias 1996; Moss and Brooke 2000; Bushby 2006). The modified LODM equation they end up defining are

$$\frac{\mathrm{d}B_\phi}{\mathrm{d}t} = \left(c_1 - \frac{v_p(t)}{\ell_0}\right)B_\phi + c_2 A_p - c_3 B_\phi^3, \tag{19}$$

$$\frac{\mathrm{d}A_p}{\mathrm{d}t} = \left(c_1 - \frac{v_p(t)}{\ell_0}\right)A_p + \alpha B_\phi, \tag{20}$$

where c_1, is defined as $c_1 = \frac{\eta}{\ell_0^2} - \frac{\eta}{R_\odot^2}$ and takes the role of magnetic diffusivity, while the other coefficients remain the same.

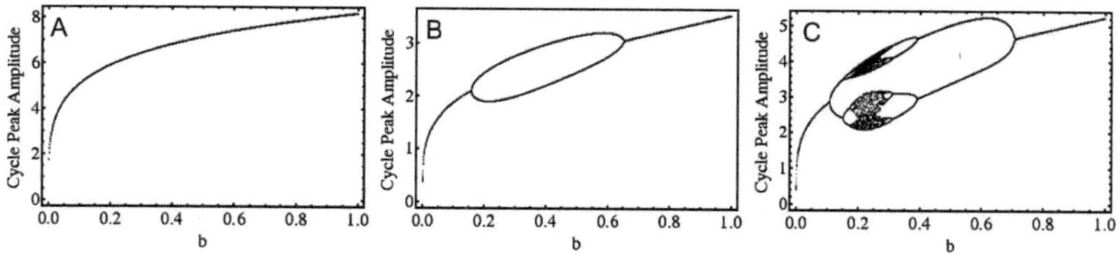

Fig. 13 Bifurcation maps for maximum amplitude of the toroidal field (equivalent to solar cycle maximum) obtained by varying b between 10^{-4} and 1 for different a and v_0. (**A**) Single period regime, $v_0 = -0.1$, $a = 0.01$; (**B**) appearance of period doubling, $v_0 = -0.1$, $a = 0.1$ and (**C**) shows signatures of chaotic regimes with multiple attractors and windows, obtained with $v_0 = -0.13$, $a = 0.05$. *Adapted from* Passos et al. (2012)

While the values used for the structural coefficients, are mean values extracted from the works presented in the previous in sections, the parameters associated with the meridional flow evolution, a, b and v_0 deserved now the attention. These parameters have an important role in the evolution of the solution space. The behavior observed in the solutions range from fixed-amplitude oscillations closely resembling kinematic solutions, multiperiodic solutions, and even chaotic solutions. This is easier to visualize in Fig. 13 where are presented analogs of classical bifurcation diagrams by plotting successive peak values of cycle amplitudes, for solutions with fixed (a, v_0) combinations but spanning through values of b. Transitions to chaos through bifurcations are also observed when holding b fixed and varying a instead.

The authors expanded the methodology used and applied stochastic fluctuations to parameter a, the one that controls the influence of the Lorentz force. As a result, and depending on the range of fluctuations, they observed that the short term stochastic kicks in the Lorentz force amplitude create long term modulations in the amplitude of the cycles (hundreds of years) and even episodes where the field decays to near zero values, analog to the previously mentioned grand minima. The duration and frequency of these long quiescent phases, where the magnetic field decays to very low values, is determined by the level of fluctuations of a and the value of b. The stronger this drag term b is, the shorter the minima are and the higher the level of fluctuation of a, the more common these intermittency episodes become. Figure 14 shows a section of a solution that spanned for 40000 years and that presents all the behaviors described before.

In this specific example they used 100 % fluctuation in a and maintaining all the other parameters constant. In the parameter space used to produce this figure, the solution without stochastic forcing is well behaved in the sense that it presents a single period regime. Therefore, the fluctuations observed in this solution are a direct consequence of the stochastic forcing of the Lorentz force and not from a chaotic regime of the solution's space.

To understand how the grand minima episodes arise they resort to visualizing one of these episodes with phase space diagrams of $\{B_\phi, A_p, v_p\}$. This allows to see how these quantities vary in relation to each other and try to understand the chain of events that trigger a grand minimum.

The standard solution for the LODM without stochastic forcing, i.e. with a fixed at the mean value of the random number distribution used, is the limit cycle attractor, i.e., a closed trajectory in the $\{B_\phi, A_p\}$ phase space. This curve is represented as a black dashed trajectory in the panels of Fig. 15. The gray points in this figure are the stochastic forced solution values sampled at 1 year interval. These points scatter around the attractor representing the variations in amplitude of the solution. Occasionally the trajectories defined by these points collapse to the center of the phase space (the point $\{0, 0, v_0\}$ is also another natural attractor

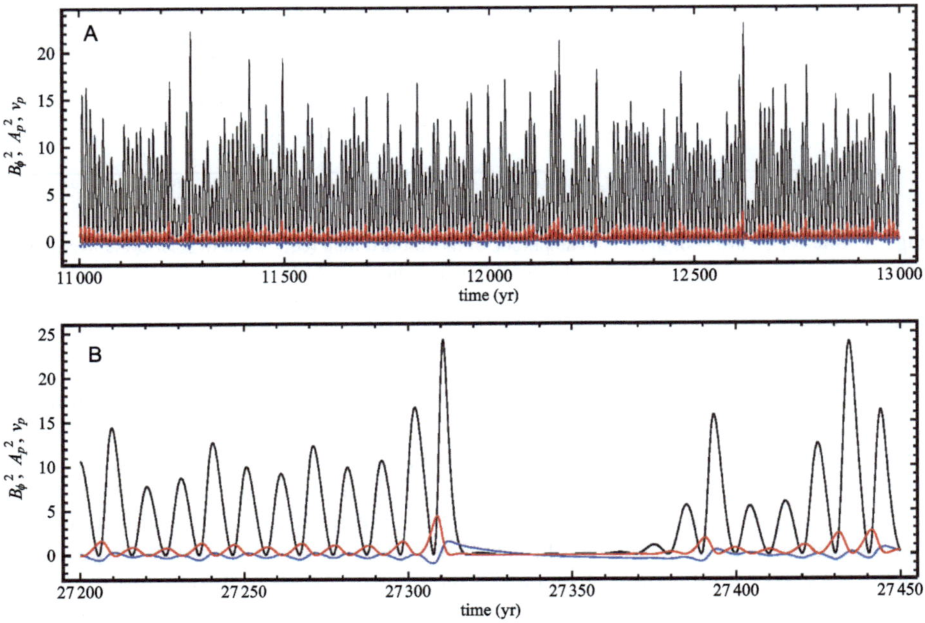

Fig. 14 Simulation result fluctuating $a \in [0.01, 0.03]$, $b = 0.05$ and $v_0 = -0.11$. All other model parameters are the same as in the reference solution. Panel (**A**) shows a section of the simulation where the long term modulation can be seen. In *black* is $B_\phi^2(t)$, *red* $A_p^2(t)$ and *blue* a scaled version of the meridional flow, in this case $5v_p(t)$. In panel (**B**) the same quantities but this time zooming in into a grand minimum (off phase) period. *Adapted from* Passos et al. (2012)

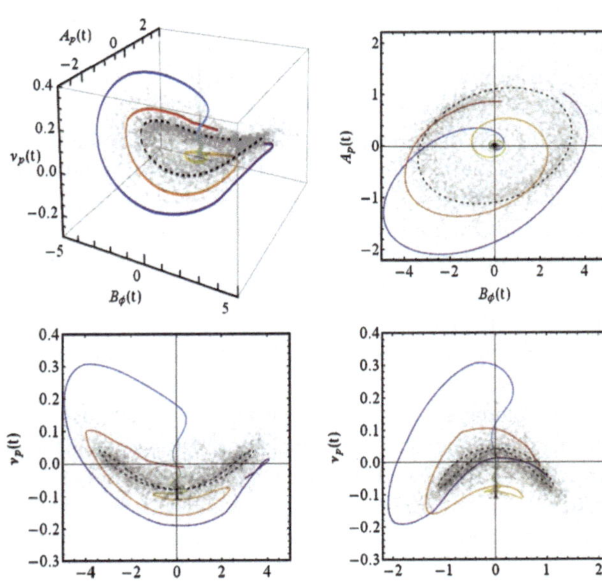

Fig. 15 Phase spaces of the solution with stochastic fluctuations. The *gray dots* represent 1 year intervals between $t = 35000$ and $t = 40000$. The *colored line* shows the trajectory of a grand minima (starting from *purple*, $t = 27300$ and ending in *red*, $t = 27400$. The *black dashed line* represents the unperturbed solution with $a = 0.02$. *Adapted from* Passos et al. (2012)

of the system) indicating a decrease in amplitude of the cycle, i.e. a grand minimum. The colored trajectory evolving in time from purple to red represents one of those grand minimum. This happens when the solution is at a critical distance from the limit cycle attractor and gets a random kick further away from it. This kick makes the field grow rapidly. In turn, since the amplitude of the field grows fast, the Lorentz force will induce a similar growth in $v(t)$ eventually making v_p change sign. When this occurs, v_p behaves as a sink term quenching the field growth very efficiently. This behavior is seen in the two bottom panels of Fig. 15 where v_p decays to its imposed "kinematic" value v_0 after the fields decay. After this collapse of v_p to v_0 it starts behaving has a source term again and the cyclic activity proceeds.

One clear advantage of low order models emerges from this example. Currently 3D MHD simulations of solar convection spanning a thousand years take a couple of months to run in high efficiency computational clusters or in supercomputers. Longer simulations are at the moment prohibitive not only for the amount of time they take but also for the huge amount of data they generate. Statistical studies on grand minima originated by the kind of magnetic back-reaction described here, require long integration times where many thousands of cycles need to be simulated. The LODM calculations can be done in a few minutes or hours in any current desktop.

The grand minima mechanism presented in this section is now being studied by looking at the data available from 3D simulations. Some effects are easier to find when you know what to look for.

6 Outlook for the Sun and Stars

So far we have shown that low order dynamo models (for which the approximation must be carefully chosen to keep the relevant physics within) could lead the way to explore some features of the solar magnetic activity including the long-term variability. The study of the phase diagram $\{B_\phi(r), \dot{B}_\phi(t)\}$ clearly shows that on a scale of a few centuries the solar magnetic cycle shows evidence for a van der Pool attractor—put in evidence by the mean solar magnetic cycle, although on a time-scale of a few solar magnetic cycles the phase space trajectory changes dramatically. In some cases the trajectory collapses completely for several magnetic cycles as in the periods of grand minima. This gives us an indication about the existence of a well defined self regulated system under all this observed magnetic variability, for which we still need to identify the leading physical mechanisms driving the solar dynamo to extreme activity scenarios like periods of grand minima. Actually, the fact that a well-defined averaged van der Pool limit curve exists for all the sunspot records, can be used to test different solar dynamo models, including numerical simulations, against observational data or between different dynamo models.

Moreover, the fact that such well-defined attractor exists in the phase space, and several dynamo models are able to qualitatively reproduce the solar variability (as observed in the phase space $\{B_\phi(r), \dot{B}_\phi(t)\}$ gives us hope that in the near future we will be able to make quite reliable short term predictions of the solar magnetic cycle variability, at least within certain time intervals of solar magnetic activity. A significant contribution can be done by the utilization of more accurate sunspot time series in which many of the historical inaccuracies were corrected (Lefevre and Clette 2014).

In the future, similar inversion techniques could be developed, namely to study the possible asymmetry between the North and South Hemispheres using the sunspot areas, either by treating each of the sunspot areas as two distinct times series or by attempting two-dimension inversions of sunspot butterfly diagrams. In the former case, recently Lopes et

Fig. 16 Magnetic activity signature expressed by the variation of the intensity of the Ca II emission line (S-index) for two solar like stars. *Adapted from* Baliunas et al. (1995)

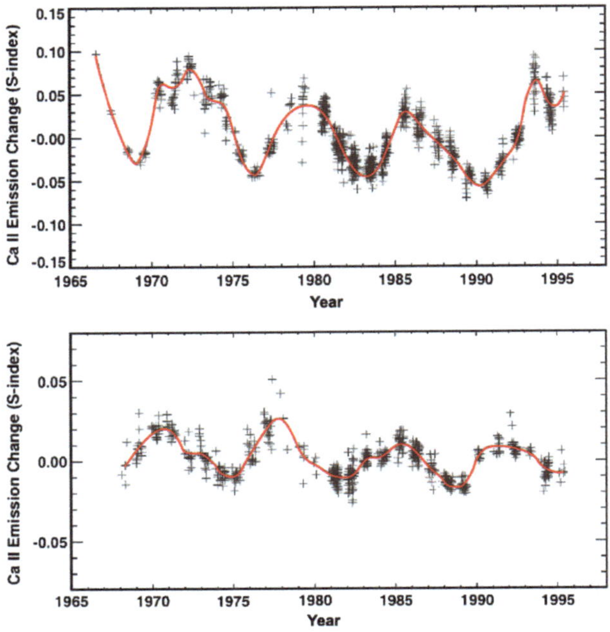

al. (2014) have analysed these long-term sunspot areas time series and found that turbulent convection and solar granulation are responsible by the stochastic nature of the sunspot area variations. In the last case, we could learn about the evolution of the solar magnetic cycle in the tachocline during the last two and a half centuries. Moreover, most of the inversion methods used for the sunspot number can be easily extended to other solar magnetic cycles proxies such as TSI, H_α and Magnetograms.

The oscillator models, as a first order dynamo model are particularly suitable to study the magnetic activity in other stars. A good proxy of magnetic activity in stars in the chromospheric variations of Ca II H and K emission lines. Baliunas et al. (1995) have found many F2 and M2 stars which seem to have cyclic magnetic cycle activity, as observed in the Sun (cf. Fig. 16). In some of these stars the observational time series covers several cycles of activity. In particular, it will be interesting to identify how the dynamo operating in these stars differs from the solar case.

More recently, the CoROT and Kepler space missions have observed photometric variability associated with solar-like activity in a very large number of main sequence and sub-giant stars. While the time coverage is too short to derive cycle periods for stars very close to the Sun, the overall *level* of activity and its dependence on various stellar parameters can be studied on a large statistical sample (Basri et al. 2010; McQuillan et al. 2012). Nevertheless, with so many stars with quite distinct masses and radius, it is reasonable to expect that we will find quite different type of dynamos and regimes of stellar magnetic cycle. Actually, we think it is likely to find a magnetic diversity identical to the one found in the acoustic oscillation spectra measured for the more than 500 sun-like stars already discovered (Chaplin et al. 2014), some of which have already shown evidence of a magnetic cycle activity. García et al. (2010) have obtained a proxy of the starspots number for the star HD49933 from amplitudes and frequencies of the acoustic modes of vibration. As in the non-linear oscillator models the activity level is determined by the structural parameters which in turn depends on the dynamo model. These studies potentially offer a simple theoretical scheme against which to test the observational findings.

Acknowledgements The authors thank the anonymous referee for the suggestions made to improve the quality of the article. I.L. thanks the convenors of the workshop and The International Space Science Institute for the invitation and financial support. I.L. and D.P. would also like to thank Arnab Choudhuri and his collaborators for making the Surya code publicly available. I.L. would like to thank his collaborators in this subject of research: Ana Brito, Elisa Cardoso, Hugo Silva, Amaro Rica da Silva and Sylvaine Turck-Chièze. The work of I.L. was supported by grants from "Fundação para a Ciência e Tecnologia" and "Fundação Calouste Gulbenkian". D.P. acknowledges the support from the Fundação para a Ciência e Tecnologia (FCT) grant SFRH/BPD/68409/2010. M.N. acknowledges support from the Hungarian Science Research Fund (OTKA grant no. K83133).

References

S.L. Baliunas, R.A. Donahue, W.H. Soon, J.H. Horne, J. Frazer, L. Woodard-Eklund, M. Bradford, L.M. Rao, O.C. Wilson, Q. Zhang, W. Bennett, J. Briggs, S.M. Carroll, D.K. Duncan, D. Figueroa, H.H. Lanning, T. Misch, J. Mueller, R.W. Noyes, D. Poppe, A.C. Porter, C.R. Robinson, J. Russell, J.C. Shelton, T. Soyumer, A.H. Vaughan, J.H. Whitney, Chromospheric variations in main-sequence stars. Astrophys. J. **438**, 269–287 (1995)

G. Basri, L.M. Walkowicz, N. Batalha, R.L. Gilliland, J. Jenkins, W.J. Borucki, D. Koch, D. Caldwell, A.K. Dupree, D.W. Latham, S. Meibom, S. Howell, T. Brown, Photometric variability in Kepler target stars: the Sun among stars a first look. Astrophys. J. Lett. **713**, 155–159 (2010). doi:10.1088/2041-8205/713/2/L155

S. Basu, H.M. Antia, Characteristics of solar meridional flows during solar cycle 23. Astrophys. J. **717**, 488–495 (2010). doi:10.1088/0004-637X/717/1/488

J. Beer, S. Tobias, N.O. Weiss, An active sun throughout the Maunder Minimum. Sol. Phys. **181**, 237–249 (1998). doi:10.1023/A:1005026001784

R.N. Bracewell, Three-halves law in sunspot cycle shape. Mon. Not. R. Astron. Soc. **230**, 535–550 (1988). Oscillator models of the solar cycle. http://adsabs.harvard.edu/abs/1988MNRAS.230..535B

B.P. Brown, M.S. Miesch, M.K. Browning, A.S. Brun, J. Toomre, Magnetic cycles in a convective dynamo simulation of a young solar-type star. Astrophys. J. **731**, 69 (2011). doi:10.1088/0004-637X/731/1/69

P.J. Bushby, Zonal flows and grand minima in a solar dynamo model. Mon. Not. R. Astron. Soc. **371**, 772–780 (2006). doi:10.1111/j.1365-2966.2006.10706.x

R. Cameron, M. Schüssler, A robust correlation between growth rate and amplitude of solar cycles: consequences for prediction methods. Astrophys. J. **685**, 1291–1296 (2008). doi:10.1086/591079, http://adsabs.harvard.edu/abs/2008ApJ...685.1291C

E. Cardoso, I. Lopes, Impact of a realistic density stratification on a simple solar dynamo calculation. Astrophys. J. **757**(1), 71 (2012). doi:10.1088/0004-637X/757/1/71

W.J. Chaplin, S. Basu, D. Huber, A. Serenelli, L. Casagrande, V. Silva Aguirre, W.H. Ball, O.L. Creevey, L. Gizon, R. Handberg, C. Karoff, R. Lutz, J.P. Marques, A. Miglio, D. Stello, M.D. Suran, D. Pricopi, T.S. Metcalfe, M.J.P.F.G. Monteiro, J. Molenda-Żakowicz, T. Appourchaux, J. Christensen-Dalsgaard, Y. Elsworth, R.A. García, G. Houdek, H. Kjeldsen, A. Bonanno, T.L. Campante, E. Corsaro, P. Gaulme, S. Hekker, S. Mathur, B. Mosser, C. Régulo, D. Salabert, Asteroseismic fundamental properties of solar-type stars observed by the NASA Kepler mission. Astrophys. J. Suppl. Ser. **210**(1), 1 (2014)

P. Charbonneau, Dynamo models of the solar cycle. Living Rev. Sol. Phys. **7**, 3 (2010)

P. Charbonneau, Where is the solar dynamo? J. Phys. Conf. Ser. **440**(1), 012014 (2013). doi:10.1088/1742-6596/440/1/012014

P. Charbonneau, M. Dikpati, Stochastic fluctuations in a Babcock-Leighton model of the solar cycle. Astrophys. J. **543**, 1027–1043 (2000). doi:10.1086/317142

P. Charbonneau, C. St-Jean, P. Zacharias, Fluctuations in Babcock-Leighton dynamos. I. Period doubling and transition to chaos. Astrophys. J. **619**, 613–622 (2005). doi:10.1086/426385

A.R. Choudhuri, M. Schüssler, M. Dikpati, The solar dynamo with meridional circulation. Astron. Astrophys. **303**, 29 (1995)

J.A. Eddy, The Maunder minimum. Science **192**, 1189–1202 (1976). doi:10.1126/science.192.4245.1189

R.A. García, S. Mathur, D. Salabert, J. Ballot, C. Régulo, T.S. Metcalfe, A. Baglin, CoRoT reveals a magnetic activity cycle in a Sun-like star. arXiv:1008.4399 (2010)

M. Ghizaru, P. Charbonneau, P.K. Smolarkiewicz, Magnetic cycles in global large-eddy simulations of solar convection. Astrophys. J. Lett. **715**, 133–137 (2010). doi:10.1088/2041-8205/715/2/L133

D.T. Gillespie, Exact numerical simulation of the Ornstein–Uhlenbeck process and its integral. Phys. Rev. E **54**, 2084–2091 (1996)

D.H. Hathaway, L. Rightmire, Variations in the sun meridional flow over a solar cycle. Science **327**, 1350 (2010). doi:10.1126/science.1181990

S. Hazra, D. Passos, D. Nandy, A stochastically forced time delay solar dynamo model: self-consistent recovery from a Maunder-like grand minimum necessitates a mean-field alpha effect. Astrophys. J. **789**(1), 5 (2014)

K.M. Hiremath, The solar cycle as a forced and damped harmonic oscillator: long-term variations of the amplitudes, frequencies and phases. Astron. Astrophys. **452**, 591–595 (2006). doi:10.1051/0004-6361:20042619, http://www.aanda.org/articles/aa/abs/2006/23/aa2619-04/aa2619-04.html

K.M. Hiremath, Prediction of solar cycle 24 and beyond. Astrophys. Space Sci. **314**, 45–49 (2008). http://link.springer.com/article/10.1007%2Fs10509-007-9728-9

R. Howe, Solar interior rotation and its variation. Living Rev. Sol. Phys. **6**, 1 (2009)

P.J. Käpylä, M.J. Mantere, A. Brandenburg, Cyclic magnetic activity due to turbulent convection in spherical wedge geometry. Astrophys. J. Lett. **755**, 22 (2012). doi:10.1088/2041-8205/755/1/L22

B.B. Karak, Importance of meridional circulation in flux transport dynamo: the possibility of a Maunder-like grand minimum. Astrophys. J. **724**, 1021–1029 (2010). doi:10.1088/0004-637X/724/2/1021

B.B. Karak, A.R. Choudhuri, The Waldmeier effect and the flux transport solar dynamo. Mon. Not. R. Astron. Soc. **410**, 1503–1512 (2011). doi:10.1111/j.1365-2966.2010.17531.x

F. Krause, K.-H. Raedler, *Mean-Field Magnetohydrodynamics and Dynamo Theory* (Pergamon, Oxford, 1980), 271 pp.

L. Lefevre, F. Clette, Survey and merging of sunspot catalogs. Sol. Phys. **289**(2), 545–561 (2014)

I. Lopes, D. Passos, Solar variability induced in a dynamo code by realistic meridional circulation variations. Sol. Phys. **257**(1), 1–12 (2009). doi:10.1007/s11207-009-9372-3

I. Lopes, E. Cardoso, H. Silva, Looking for periodicities in the sunspot time series. Astrophys. J. (2014 accepted)

A. McQuillan, S. Aigrain, S. Roberts, Statistics of stellar variability from Kepler. I. Revisiting quarter 1 with an astrophysically robust systematics correction. Astron. Astrophys. **539**, 137 (2012). doi:10.1051/0004-6361/201016148

M.S. Miesch, J. Toomre, Turbulence, magnetism, and shear in stellar interiors. Annu. Rev. Fluid Mech. **41**(1), 317–345 (2009)

P.D. Mininni, D.O. Gómez, Study of stochastic fluctuations in a shell dynamo. Astrophys. J. **573**, 454–463 (2002). doi:10.1086/340495

P.D. Mininni, D.O. Gomez, G.B. Mindlin, Stochastic relaxation oscillator model for the solar cycle. Phys. Rev. Lett. **85**, 5476–5479 (2000). doi:10.1103/PhysRevLett.85.5476, http://adsabs.harvard.edu/abs/2000PhRvL..85.5476M

P.D. Mininni, D.O. Gomez, G.B. Mindlin, Simple model of a stochastically excited solar dynamo. Sol. Phys. **201**, 203–223 (2001). doi:10.1023/A:1017515709106

H.K. Moffatt, *Magnetic Field Generation in Electrically Conducting Fluids* (Cambridge University Press, Cambridge, 1978), 353 pp.

D. Moss, J. Brooke, Towards a model for the solar dynamo. Mon. Not. R. Astron. Soc. **315**, 521–533 (2000). doi:10.1046/j.1365-8711.2000.03452.x

M. Nagy, K. Petrovay, Oscillator models of the solar cycle and the Waldmeier effect. Astron. Nachr. **334**, 964 (2013). doi:10.1002/asna.201211971

D. Nandy, A. Muñoz-Jaramillo, P.C.H. Martens, The unusual minimum of sunspot cycle 23 caused by meridional plasma flow variations. Nature **471**, 80–82 (2011). doi:10.1038/nature09786

B. Owens, Long-term research: slow science. Nature **495**, 300 (2013)

E.N. Parker, The formation of sunspots from the solar toroidal field. Astrophys. J. **121**, 491 (1955)

D. Passos, Modelling solar variability. PhD Thesis, Instituto Superior Técnico, Universidade Técnica de Lisboa (2010)

D. Passos, Evolution of solar parameters since 1750 based on a truncated dynamo model. Astrophys. J. **744**(2), 172 (2012)

D. Passos, P. Charbonneau, Characteristics of magnetic solar-like cycles in a 3D MHD simulation of solar convection. Astron. Astrophys. (2014)

D. Passos, I. Lopes, Phase space analysis: the equilibrium of the solar magnetic cycle. Sol. Phys. **250**(2), 403–410 (2008a)

D. Passos, I. Lopes, A low-order solar dynamo model: inferred meridional circulation variations since 1750. Astrophys. J. **686**(2), 1420–1425 (2008b)

D. Passos, I. Lopes, Grand minima under the light of a low order dynamo model. J. Atmos. Sol.-Terr. Phys. **73**(2), 191–197 (2011)

D. Passos, P. Charbonneau, P. Beaudoin, An exploration of non-kinematic effects in flux transport dynamos. Sol. Phys. **279**(1), 1–22 (2012)

D. Passos, D. Nandy, S. Hazra, I. Lopes, A solar dynamo model driven by mean-field alpha and Babcock-Leighton sources: fluctuations, grand-minima-maxima, and hemispheric asymmetry in sunspot cycles. Astron. Astrophys. **563**, 18 (2014). doi:10.1051/0004-6361/201322635

J.M. Polygiannakis, X. Moussas, A non-linear model for the solar cycle. Astrophys. Lett. Commun. **34**, 35 (1996)

A. Pontieri, F. Lepreti, L. Sorriso-Valvo, A. Vecchio, V. Carbone, A simple model for the solar cycle. Sol. Phys. **213**(1), 195–201 (2003)

M. Rempel, Flux-transport dynamos with Lorentz force feedback on differential rotation and meridional flow: saturation mechanism and torsional oscillations. Astrophys. J. **647**, 662–675 (2006). doi:10.1086/505170

A. A. Ruzmaikin, The solar cycle as a strange attractor. Comments Astrophys. **9**, 85–93 (1981).

E.A. Spiegel, Chaos and intermittency in the solar cycle. Space Sci. Rev. **144**, 25–51 (2009). doi:10.1007/s11214-008-9470-9

M. Steenbeck F. Krause, Erklärung stellarer und planetarer Magnetfelder durch einen turbulenzbedingten Dynamomechanismus Z. Naturforsch. Teil A **21**, 1285 (1966)

S.M. Tobias, Diffusivity quenching as a mechanism for Parker's surface dynamo. Astrophys. J. **467**, 870 (1996). doi:10.1086/177661

S.M. Tobias, N.O. Weiss, V. Kirk, Chaotically modulated stellar dynamos. Mon. Not. R. Astron. Soc. **273**(4), 1150–1166 (1995)

A. Vecchio, V. Carbone, A simple model to describe solar cycle periodicities below 11 years. Sol. Phys. **249**, 11–16 (2008). doi:10.1007/s11207-008-9180-1

M. Waldmeier, Neue Eigenschaften der Sonnenfleckenkurve. Astron. Mitt. Zür. **14**(133), 105–130 (1935)

N.O. Weiss, Modulation of the sunspot cycle. Astron. Geophys. **51**, 9–15 (2010). doi:10.1111/j.1468-4004.2010.51309.x

N.O. Weiss, C.A. Cattaneo, F. Jones, Periodic and aperiodic dynamo waves. Geophys. Astrophys. Fluid Dyn. **30**, 305–341 (1984). doi:10.1080/03091928408219262

A.L. Wilmot-Smith, P.C.H. Martens, D. Nandy, E.R. Priest, S.M. Tobias, Low-order stellar dynamo models. Mon. Not. R. Astron. Soc. **363**, 1167–1172 (2005). doi:10.1111/j.1365-2966.2005.09514.x

A.L. Wilmot-Smith, D. Nandy, G. Hornig, P.C.H. Martens, A time delay model for solar and stellar dynamos. Astrophys. J. **652**, 696 (2006). doi:10.1086/508013, http://iopscience.iop.org/0004-637X/652/1/696/

R. Wolf, W. Brunner, Neue Eigenschaften der Sonnenfleckenkurve. Astron. Mitt. Eidgenöss. Sternwarte Zür. **14**, 105–136 (1935)

DOI 10.1007/978-1-4939-2584-1_20
Reprinted from *Space Science Reviews* Journal, DOI 10.1007/s11214-014-0099-6

Flux Transport Dynamos: From Kinematics to Dynamics

Bidya Binay Karak · Jie Jiang · Mark S. Miesch · Paul Charbonneau · Arnab Rai Choudhuri

Received: 25 April 2014 / Accepted: 5 September 2014 / Published online: 17 September 2014

Abstract Over the past several decades, Flux-Transport Dynamo (FTD) models have emerged as a popular paradigm for explaining the cyclic nature of solar magnetic activity. Their defining characteristic is the key role played by the mean meridional circulation in transporting magnetic flux and thereby regulating the cycle period. Most FTD models also incorporate the so-called Babcock-Leighton (BL) mechanism in which the mean poloidal field is produced by the emergence and subsequent dispersal of bipolar active regions. This feature is well grounded in solar observations and provides a means for assimilating observed surface flows and fields into the models in order to forecast future solar activity, to identify model biases, and to clarify the underlying physical processes. Furthermore, interpreting historical sunspot records within the context of FTD models can potentially provide insight into why cycle features such as amplitude and duration vary and what causes extreme events such as Grand Minima. Though they are generally robust in a modeling sense and make good contact with observed cycle features, FTD models rely on input physics that

B.B. Karak
NORDITA, KTH Royal Institute and Stockholm University, Roslagstullsbacken 23, 10691 Stockholm, Sweden
e-mail: bbkarak@nordita.org

J. Jiang
Key Laboratory of Solar Activity, National Astronomical Observatories, Chinese Academy of Sciences, Beijing 100012, China
e-mail: jiejiang@nao.cas.cn

M.S. Miesch
High Altitude Observatory, NCAR, Boulder, CO 80307-3000, USA
e-mail: miesch@ucar.edu

P. Charbonneau (✉)
Département de Physique, Université de Montréal, Montréal, QC H3C-3J7, Canada
e-mail: paulchar@astro.umontreal.ca

A.R. Choudhuri
Department of Physics, Indian Institute of Sciences, Bangalore 560012, India
e-mail: arnab@physics.iisc.ernet.in

is only partially constrained by observation and that neglects the subtleties of convective transport, convective field generation, and nonlinear feedbacks. Here we review the formulation and application of FTD models and assess our current understanding of the input physics based largely on complementary 3D MHD simulations of solar convection, dynamo action, and flux emergence.

Keywords Sun · Solar cycle · Dynamo

1 Introduction

The 11-year sunspot cycle is the magnetic cycle of the Sun. We assume that the reader is familiar with the basic phenomenology of the sunspot cycle (Hale's polarity law, butterfly diagram) and do not attempt to provide a summary of such observational data. Many recent reviews (Hathaway 2010; Charbonneau 2010; Choudhuri 2011) discuss the relevant phenomenology. However, we shall sometimes discuss those aspects of observational data which may not be so familiar to all the readers, especially data involving irregularities of the sunspot cycle.

It is now generally accepted that the Sun's magnetic cycle is produced by a dynamo process, which is based on the nonlinear interaction between the velocity field and the magnetic field of the solar plasma. This nonlinear interaction is mathematically described by the well-known magnetohydrodynamical (MHD) equations. The evolution of the magnetic field $\mathbf{B}$ in response to the velocity field $\mathbf{v}$ is given by the induction equation:

$$\frac{\partial \mathbf{B}}{\partial t} = \nabla \times (\mathbf{v} \times \mathbf{B} - \lambda \nabla \times \mathbf{B}), \tag{1}$$

whereas the evolution of the velocity field $\mathbf{v}$ in the presence of the magnetic field $\mathbf{B}$ is given by the magnetic Navier–Stokes equation:

$$\frac{\partial \mathbf{v}}{\partial t} + (\mathbf{v} \cdot \nabla)\mathbf{v} = -\frac{1}{\rho}\nabla\left(p + \frac{B^2}{8\pi}\right) + \frac{(\mathbf{B}.\nabla)\mathbf{B}}{4\pi\rho} + \mathbf{g} + \frac{1}{\rho}\nabla \cdot \boldsymbol{\tau}. \tag{2}$$

Here ρ is density, p is pressure, $\mathbf{g}$ is gravitational field, $\boldsymbol{\tau}$ is the viscous stress tensor, and λ is magnetic diffusivity. Since one needs the values of ρ and p under dynamical conditions in order to solve (2), the above two equations have to be combined with the continuity equation, the heat equation, and an appropriate equation of state to close the set of equations describing the evolution of the system.

While the basic equations of the problem are well established, it is extremely difficult to solve this set of equations in an astrophysically realistic dynamo situation. That is why solar dynamo theory is still only a partially understood theory. There are two possible approaches to the solar dynamo problem.

- Direct numerical simulation, or DNS, in which one discretizes all the MHD equations and solves them numerically.
- The kinematic approach, in which the velocity field is given and one solves only (1) for the magnetic field.

Since the outer layers of the Sun are unstable to convection, the main challenge of the DNS is to model the convection properly. It is the turbulent stresses in the convection zone which drive such large-scale flows as the differential rotation and the meridional circulation, which are crucial for dynamo action.

In the kinematic approach also, it is necessary to handle the convective turbulence suitably. The kinematic theory basically has to be of the nature of a mean field theory in which we average over fluctuations around the mean (Parker 1955; Steenbeck et al. 1966). We can split both the velocity field and the magnetic field into mean and fluctuating parts, i.e.

$$\mathbf{v} = \overline{\mathbf{v}} + \mathbf{v}', \qquad \mathbf{B} = \overline{\mathbf{B}} + \mathbf{B}'. \tag{3}$$

Here the overline indicates the mean and the prime indicates the departure from the mean. On substituting (3) in (1) and averaging, one finds that the mean magnetic field $\overline{\mathbf{B}}$ obeys the equation

$$\frac{\partial \overline{\mathbf{B}}}{\partial t} = \nabla \times (\overline{\mathbf{v}} \times \overline{\mathbf{B}} + \mathcal{E} - \lambda \nabla \times \overline{\mathbf{B}}), \tag{4}$$

where

$$\mathcal{E} = \overline{\mathbf{v}' \times \mathbf{B}'} \tag{5}$$

is known as the mean electromotive force (hereafter EMF). Under certain assumptions (near-isotropic and near-homogeneous turbulence), one can show that the mean EMF has the form

$$\mathcal{E} = \alpha \overline{\mathbf{B}} - \beta \nabla \times \overline{\mathbf{B}} \tag{6}$$

(see, for example, Choudhuri 1998; §16.5). Here the coefficient α is the crucial parameter for the dynamo generation of magnetic fields, whereas β is the turbulent diffusivity. On substituting (6) in (4), we get the dynamo equation

$$\frac{\partial \overline{\mathbf{B}}}{\partial t} = \nabla \times (\overline{\mathbf{v}} \times \overline{\mathbf{B}}) + \nabla \times (\alpha \overline{\mathbf{B}}) + \beta \nabla^2 \overline{\mathbf{B}} + (\nabla \times \overline{\mathbf{B}}) \times \nabla \beta, \tag{7}$$

where we have neglected λ compared to β, which is usually the case within the convection zone. Kinematic mean-field models are based on this dynamo equation.

Since the kinematic approach is easier than the DNS approach, it is no wonder that historically the field of solar dynamo first developed by following the kinematic approach. During the initial years of dynamo research, not much was known about the large-scale flow fields such as the differential rotation in the interior of the Sun. Still 'reasonable' (or what seemed reasonable at that time) assumptions about differential rotation led to solutions showing periodic behavior with equatorward propagation—to account for the observed fact that sunspots appear at increasingly lower latitudes with the progress of the solar cycle. There was a renaissance in the kinematic mean-field modeling in the 1990s when helioseismology succeeded in mapping the differential rotation in the interior of the Sun. At about the time when the helioseismically determined differential rotation started being used in solar dynamo models, two other important developments also took place.

- It started being realized that the meridional circulation may play a much more important role in the solar dynamo than hitherto recognized.
- The mechanism for the generation of the poloidal magnetic field proposed by Babcock (1961) and Leighton (1969) re-emerged as a serious contender to the original mechanism of Parker (1955) elaborated by Steenbeck et al. (1966).

Both these points will be elaborated and explained further in the next Section. Solar dynamo models in which meridional circulation plays a crucial role and the poloidal field generation takes place by the Babcock–Leighton (BL) mechanism are usually called flux transport dynamo (FTD) models. The foundations of this field were laid down by Wang and Sheeley (1991), Wang et al. (1991), Choudhuri et al. (1995), Durney (1995, 1997) and Dikpati and

Charbonneau (1999). The kinematic FTD model has done well in reproducing many aspects of the solar cycle, as we shall discuss in the next two sections.

In spite of the success of the kinematic FTD model, there is no doubt that our ultimate goal should be to construct fully dynamical models of the solar dynamo based on DNS. The first attempts at doing a DNS of the solar dynamo were made quite early—by Gilman (1983) and Glatzmaier (1984, 1985). Impressive simulations of the geodynamo are now available—beginning with the work of Glatzmaier and Roberts (1995). Even the random reversals of the geomagnetic field had been successfully modeled. Doing a DNS of the solar dynamo is extremely challenging for the following reasons.

- The solar convection zone is highly stratified, with quantities like density and pressure varying by several orders of magnitude from the bottom to the top.
- The largest relevant scales (10^6 km) differ from the smallest scales of fibril flux tubes (10^2 km) by 4 orders.
- The fluid and magnetic Reynolds numbers are huge, $\sim 10^8$–10^{10}.

Additionally, a kinematic model has one important practical advantage, namely that the information provided by helioseismology about the large-scale flow fields like the differential rotation and the meridional circulation can be directly inserted in the kinematic equations. This has been one of the main reasons behind the success of the FTD model. Even without a theoretical understanding of these large-scale flows, it is possible to construct realistic models of the solar dynamo. On the other hand, in a DNS these large-scale flow fields have to come out of the basic equations in a self-consistent way. Several research groups are again working on DNS of the solar dynamo problem (Brown et al. 2010, 2011; Ghizaru et al. 2010; Racine et al. 2011; Käpylä et al. 2012, 2013; Nelson et al. 2013b; Fan and Fang 2014; Passos and Charbonneau 2014; Karak et al. 2014a), and have succeeded in obtaining periodic solutions capturing some observed features of the solar cycle.

The main aim of the present review is to take stock of the current status of both the kinematic FTD model and the DNS efforts. We shall also discuss how the two approaches can benefit from each other. In Sect. 2 we introduce the basics of the kinematic FTD model and discuss its successes in modeling the regular aspects of the sunspot cycle. However, a solar dynamo model also has to account for the irregularities of the sunspot cycle. The causes of the irregularities are discussed in Sect. 3. Armed with an understanding of the causes of the irregularities, one would like to model past sunspot cycles and predict future cycles. We discuss in Sect. 4 the recent efforts of doing this by suitably assimilating observational data in the kinematic FTD model. We turn in Sect. 5 to hydrodynamical and magnetohydrodynamical numerical simulations, as a springboard away from kinematics and into the dynamics of large-scale flows, turbulent convective transport and flux emergence. We conclude in Sect. 6 by outlining some potentially productive bridges between kinematic FTD models and numerical simulations.

2 Kinematic Flux Transport Dynamo Model

2.1 Early History of Solar Dynamo Research

Let us consider the dynamo equation (7). In this and the following two Sections, whenever we refer to the magnetic or the velocity field, we shall usually imply the mean field. To simplify the notation, we drop the overline. Assuming the mean fields to be axisymmetric, we can write the mean magnetic field in this way:

$$\mathbf{B} = B(r,\theta,t)\mathbf{e}_\phi + \nabla \times \left[A(r,\theta,t)\mathbf{e}_\phi\right], \tag{8}$$

where $B(r,\theta,t)\mathbf{e}_\phi$ is the toroidal magnetic field and $\mathbf{B}_p = \nabla \times [A(r,\theta,t)\mathbf{e}_\phi]$ is the poloidal magnetic field. We write the mean velocity field in the form:

$$\mathbf{v} = \Omega(r,\theta) r \sin\theta \mathbf{e}_\phi + \left[v_r(r,\theta)\mathbf{e}_r + v_\theta(r,\theta)\mathbf{e}_\theta\right], \tag{9}$$

where $\Omega(r,\theta)$ is the angular velocity and $\mathbf{v}_m = [v_r\mathbf{e}_r + v_\theta\mathbf{e}_\theta]$ is the meridional circulation. On substituting (8) and (9) in (7), we get the following equations

$$\frac{\partial A}{\partial t} + \frac{1}{s}(\mathbf{v}_m.\nabla)(sA) = \beta\left(\nabla^2 - \frac{1}{s^2}\right)A + \alpha B, \tag{10}$$

$$\frac{\partial B}{\partial t} + \frac{1}{r}\left[\frac{\partial}{\partial r}(r v_r B) + \frac{\partial}{\partial \theta}(v_\theta B)\right]$$
$$= \beta\left(\nabla^2 - \frac{1}{s^2}\right)B + s(\mathbf{B}_p.\nabla)\Omega + \frac{1}{r}\frac{d\beta}{dr}\frac{\partial}{\partial r}(rB), \tag{11}$$

where $s = r\sin\theta$ and we have assumed β to depend on r alone. These two equations respectively describe the evolutions of the poloidal and the toroidal magnetic fields.[1] The terms involving $\mathbf{v}_m$ (or equivalently v_r, v_θ) correspond to advection by the meridional circulation, whereas the terms involving β are the diffusive terms. In order for the fields to be sustained, we need some source terms to compensate for the diffusive terms. It is easily seen from (11) that the source term for the toroidal field is $s(\mathbf{B}_p.\nabla)\Omega$, which signifies the stretching of the poloidal field by differential rotation to produce the toroidal field. The source term for the poloidal field is the term αB in (10). The physical interpretation of this term will be discussed in the next paragraph. We may point out that, on substituting (8) and (9) in (7), we get another source term $\mathbf{e}_\phi.[\nabla \times (\alpha\mathbf{B}_p)]$ in (11). We have neglected this source term because it is negligible compared to the source term $s(\mathbf{B}_p.\nabla)\Omega$ in (11) in the situation of strong differential rotation, which is the case for the Sun. When this additional source term is neglected, (10) and (11) constitute what is called the $\alpha\Omega$ dynamo model.

The basic idea of the $\alpha\Omega$ dynamo is that the toroidal and the poloidal fields sustain each other through a feedback loop. The poloidal field can be stretched by the differential rotation to produce the toroidal field, as we see from the source term in (11). In order to complete the loop, we need a mechanism to generate the poloidal field from the toroidal field. This is formally achieved by the term αB in (10). We now discuss how this term arises. Historically there have been two influential schools of thought how the poloidal field may be generated from the toroidal field. The first started with Parker (1955) who suggested that the turbulence in the solar convection zone would be helical because of the Coriolis force due to rotation and that this helical turbulence would twist the toroidal field to produce the poloidal field. Steenbeck et al. (1966) developed this idea further when they systematically developed the mean field theory of MHD. If the turbulence remains unaffected by the magnetic field, α appearing in (6) can be shown to be

$$\alpha \approx -\frac{1}{3}\overline{\mathbf{v}'.\nabla \times \mathbf{v}'}\tau, \tag{12}$$

where τ is the correlation time of turbulence (see, for example, Choudhuri 1998; §16.5). Since $\overline{\mathbf{v}'.\nabla \times \mathbf{v}'}$ is the measure of helicity in turbulence, the term αB in (10) in the present scenario has the significance that the helical turbulence acting on the toroidal field gives

[1] The solutions are typically matched to an exterior potential field (Dikpati and Choudhuri 1995). A radial field surface boundary condition can also be used when matching with a surface flux transport model (see, e.g., Jiang et al. 2013).

rise to the poloidal field. This is known as the α-effect. The α-effect, however, is possible only if the energy density of the toroidal magnetic field is less than the energy density of turbulence so that the turbulent motions are able to twist the magnetic field. For physical conditions within the solar convection zone, this means that the toroidal magnetic field has to be weaker than 10^4 G. Now we come to the second school of thought due to Babcock (1961) and Leighton (1969). The bipolar sunspots appear on the solar surface with tilts given by Joy's law. When a tilted bipolar sunspot decays, the two polarities preferentially diffuse around in slightly different latitudes. This gives rise to a dipole moment. Since the bipolar sunspots form from the toroidal field, the net effect of this process is that the toroidal field gets converted into the poloidal field. This process is also phenomenologically described by the term αB in (10) with α concentrated near the solar surface, since the Babcock–Leighton (BL) process takes place near the surface. Although both the α-effect and the BL process are represented in the same way in the kinematic dynamo equations, their physical interpretations are quite different. For example, the coefficient α describing the BL process will not be given by (12), but would rather depend on the tilts and separations of the bipolar sunspots.

In the early years of kinematic solar dynamo research (1970s and 1980s), it was generally believed that the α-effect is the primary mechanism for the generation of the poloidal field. In those days, the importance of the meridional circulation for the solar dynamo was not generally recognized. So (10) and (11) were solved without the meridional circulation terms. One requirement was to have a dynamo wave propagating equatorward in order to match butterfly diagrams. A condition known as Parker–Yoshimura sign rule

$$\alpha \frac{\partial \Omega}{\partial r} < 0 \tag{13}$$

in the northern hemisphere has to be satisfied for this equatorward propagation (Parker 1955; Yoshimura 1975; Stix 1976). Many solar dynamo models were constructed by assuming α and Ω to have spatial distributions satisfying (13), Steenbeck and Krause (1969) being the first to obtain a theoretical butterfly diagram. As we already pointed out, for the α-effect to be operational, the toroidal magnetic field has to be weaker than 10^4 G, which was assumed to be the case. However, this scenario had to be questioned when detailed simulations of the rise of the toroidal magnetic field due to magnetic buoyancy to form sunspots were carried out on the basis of the thin flux tube equation (Spruit 1981; Choudhuri 1990). It was found that the Coriolis force had an important effect on the rising flux tubes and only toroidal fields as strong as 10^5 G could produce sunspots at the right latitudes with the right kind of tilt consistent with Joy's law (Choudhuri and Gilman 1987; Choudhuri 1989; D'Silva and Choudhuri 1993; Fan et al. 1993; Caligari et al. 1995). Such strong toroidal fields cannot be twisted by the helical turbulence and the α-effect would be completely suppressed if the toroidal field is so strong.

2.2 Formulation of the Flux Transport Dynamo (FTD)

With the realization that the toroidal field is much stronger than what was assumed earlier and the α-effect could not operate on it, the natural possibility was to consider whether the poloidal field can be produced by the BL mechanism (Wang et al. 1989, 1991). From the observed tilts of sunspots on the solar surface, it is easy to argue that α corresponding to the BL mechanism has to be positive in the northern hemisphere. On the other hand, helioseismology found $\partial\Omega/\partial r$ to be positive in the low latitudes of the deep convection zone, where sunspots are ultimately produced. So it seemed at the first sight that the Parker–Yoshimura

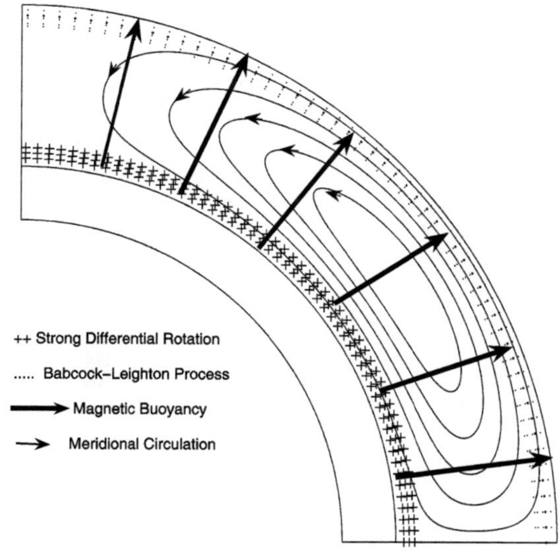

Fig. 1 A cartoon explaining how the flux transport dynamo works. The diagram shows a single meridional quadrant, with the rotation axis coinciding with the left boundary and the meridional flow confined to the convection zone and upper tachocline

sign rule (13) would not be satisfied if the poloidal field is generated by the BL mechanism and the dynamo wave would propagate poleward in contradiction to observations. However, Choudhuri et al. (1995) showed that the Parker–Yoshimura sign rule can be overruled if there is a suitable meridional circulation that can carry the toroidal field to lower latitudes with time, providing an explanation for the equatorward migration of sunspots (see also Wang et al. 1991; Durney 1995).

Figure 1 is a cartoon explaining how the FTD operates within the solar convection zone. The toroidal field is generated in the tachocline at the bottom of the convection zone where the strong differential rotation discovered by helioseismology stretches out the poloidal field to generate the toroidal field. Then this toroidal field rises to the solar surface due to magnetic buoyancy to produce the tilted bipolar sunspots. The decay of these tilted bipolar sunspots then gives rise to the poloidal field near the surface by the BL mechanism. The meridional circulation is also indicated in Fig. 1. We observe the meridional circulation to be poleward in the top layers of the convection zone. In order to conserve mass, the meridional circulation has to be equatorward deeper down. It is generally assumed in flux transport dynamo models that the equatorward flow is at the bottom of the convection zone, although this is not yet confirmed from observations. The poloidal field produced near the surface is advected poleward by the poleward meridional circulation there, whereas the toroidal field produced at the base of the convection zone is advected equatorward by the meridional circulation there. This provides the theoretical explanation of both the observed poleward drift of the surface magnetic field (outside active regions) and the equatorward migration of the sunspots which form from the toroidal field. The left panel of Fig. 2 gives the butterfly diagram of sunspots superposed on the time-latitude plot of the surface magnetic field. The corresponding theoretical plot obtained by Chatterjee et al. (2004) is shown in the right panel of Fig. 2. The theoretical model is reasonably successful in explaining the behaviors of both sunspots and the poloidal field at the surface, as well as the phase relationship between them.

The strength of the meridional circulation at the solar surface is observed to be around 20 $\mathrm{m\,s^{-1}}$. On assuming a one-cell meridional circulation spanning the whole of the convection zone satisfying $\nabla.(\rho \mathbf{v}_m) = 0$, the equatorward meridional circulation at the bottom of the convection zone is estimated to be around 1–2 $\mathrm{m\,s^{-1}}$. Such a flow would take about

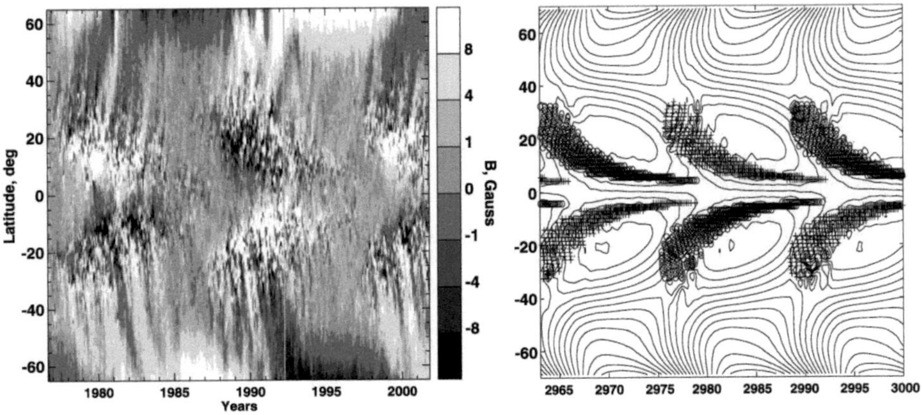

Fig. 2 Butterfly diagram of sunspots superposed on the time-latitude plot of B_r. The observational plot is shown on the *left*. The comparable theoretical plot obtained by the dynamo model of Chatterjee et al. (2004) is on the *right*

10 years to carry something from mid-latitudes to lower latitudes. It is basically this time scale which sets the period of the dynamo (Dikpati and Charbonneau 1999). If v_0 is the amplitude of the meridional circulation, we expect the dynamo period T to go as

$$T \propto v_0^{-\delta}. \tag{14}$$

Dikpati and Charbonneau (1999) found $\delta \approx 0.89$. While other versions of FTD model may give slightly different values of δ, the period T decreases with increasing v_0 in all models.

2.3 The Transport of the Poloidal Field

In order for the dynamo to work, the poloidal field produced near the solar surface by the BL mechanism has to be transported to the base of the convection zone where differential rotation can act on it. Let us now consider how this transport takes place. The original FTD model of Choudhuri et al. (1995) led to two offsprings: a high diffusivity model and a low diffusivity model. The diffusion times in these two models are of the order of 5 years (e.g., Chatterjee et al. 2004) and 200 years (e.g., Dikpati and Charbonneau 1999) respectively. The differences between these classes of models have been systematically studied by Jiang et al. (2007) and Yeates et al. (2008). In the high diffusivity model (with turbulent diffusivity of order 10^{12} cm^2 s^{-1}), the poloidal field is transported from the surface to the base of the convection zone by diffusivity. On the other hand, in the low diffusivity model (with turbulent diffusivity of order 10^{10} cm^2 s^{-1}), the poloidal field is transported from the surface to the base of the convection zone primarily by meridional circulation. These two models are also often referred to as the diffusion-dominated and the advection-dominated models respectively. Simple mixing length arguments lead to a value of turbulent diffusivity $\sim 10^{13}$ cm^2 s^{-1}, while helioseismic inversions (Hanasoge et al. 2010, 2012) suggest values in the range 10^{11}–10^{12} cm^2 s^{-1}. DNS can provide a vital clue as to the appropriate value of diffusivity to be used in kinematic models.

In both of these models, the meridional circulation advects the toroidal field equatorward at the bottom of the convection zone and advects the poloidal field poleward near the surface. Both these models are capable of giving rise to oscillatory solutions resembling solar cycles. However, the high diffusivity model has been more successful in explaining such aspects of observational data as the dipolar parity of the solar magnetism (Chatterjee et al. 2004; Hotta and Yokoyama 2010) and the lack of significant hemispheric asymmetry (Chatterjee and Choudhuri 2006; Goel and Choudhuri 2009; Norton et al. 2014).

Apart from diffusivity and the meridional circulation, a third agent for transporting the poloidal field from the surface to the bottom has been recognized in the recent years: turbulent magnetic pumping (Tobias et al. 2001; Ossendrijver et al. 2001; Guerrero and de Gouveia Dal Pino 2008; Karak and Nandy 2012; Jiang et al. 2013). The antisymmetric part of the full α-tensor contributes a term of the form $\boldsymbol{\gamma} \times \overline{\mathbf{B}}$ to (6) which, upon substituting in (4), appears as $\nabla \times (\boldsymbol{\gamma} \times \overline{\mathbf{B}})$, so that $\boldsymbol{\gamma}$ acts as an additional advection velocity to the induction term $\nabla \times (\overline{\mathbf{v}} \times \overline{\mathbf{B}})$. Magnetic pumping refers to any form of magnetic field transport in the convective zone that does not result from bulk motion. It might be due to a gradient in density (Kitchatinov 1991), a gradient in turbulent velocity (Rädler 1968) or topological asymmetry of the flow (Petrovay 1991). Many DNS calculations also suggest the presence of radially downward pumping, as well as equatorward latitudinal pumping, both with significant speeds, in the meter-per-second range (Ossendrijver et al. 2001, 2002; Ziegler and Rüdiger 2003; Racine et al. 2011). The inclusion of radially downward pumping does not change the results of the high diffusivity model too much. However, the low diffusivity starts behaving similar to the high diffusivity model when radially downward pumping is included (Karak and Nandy 2012).

2.4 Some Open Issues

Since the tachocline has stronger differential rotation at high latitudes than at low latitudes, the first authors using helioseismically determined angular velocity (Dikpati and Charbonneau 1999; Küker et al. 2001) found a tendency of strong toroidal field being produced at high latitudes and it was difficult to confine the butterfly diagram to low latitudes consistent with observations. Nandy and Choudhuri (2002; see also Guerrero and Muñoz 2004) proposed a solution to this problem by considering the meridional circulation to penetrate slightly into the stable layers below the bottom of the convection zone. Then the toroidal field produced in the high latitudes is pushed into the stable layers so that magnetic buoyancy is suppressed. Only when the uprising meridional circulation in lower latitudes brings this toroidal field again within the convection zone, magnetic buoyancy does take over and sunspots form in low latitudes. The theoretical butterfly diagram shown in Fig. 2 was produced with the help of this idea. A strong support to this idea comes from the fact that torsional oscillations start from high latitudes a couple of years before the sunspot cycle begins. Chakraborty et al. (2009) suggested that the Lorentz force of the toroidal field forming at the high latitudes a few years before the beginning of sunspot eruptions is the reason behind such behavior of torsional oscillations. However, whether the meridional circulation can penetrate into stable layers at all has been questioned by some authors on dynamical grounds (Gilman and Miesch 2004). Another obvious avenue is to artificially restrict the poloidal source term to the lower latitudes (Muñoz-Jaramillo et al. 2010; Hotta and Yokoyama 2010). While this may yield solar-like butterfly diagrams, such models can no longer provide physical insight as to why the butterfly diagram is confined to low latitudes.

Notwithstanding its kinematic nature, the FTD model also suffers from fundamental geometrical simplifications imposed by its two-dimensional, axisymmetric design. Magnetic buoyancy and the BL mechanism are certainty not axisymmetric two-dimensional processes, and involve magnetic flux tubes, which can be included in a mean field theory only in an ad hoc fashion (Choudhuri 2003). A popular approach to the treatment of magnetic buoyancy is to include in Eq. (10) a non-local surface source term, proportional to the deep-seated toroidal field (e.g., Dikpati and Charbonneau 1999). Another consist in retaining a purely local surface α-term in (10), but instantaneously displace deep magnetic flux to the surface whenever a given field strength threshold is exceeded (Nandy and Choudhuri 2001). Such modeling choices matter because otherwise identical models yield somewhat different results upon adopting one or the other scheme (Choudhuri et al. 2005). Durney (1995) proposed a more geometrically realistic approach, in which the two sunspots with opposite polarities at slightly different latitudes are replaced by two axisymmetric flux rings of opposite polarity at these latitudes. While Nandy and Choudhuri (2001) concluded that the two approaches for treating the BL mechanism gives qualitatively similar results, Muñoz-Jaramillo et al. (2010) have argued that Durney's double ring approach is a more satisfactory approach. One way to bypass this issue altogether would be to couple the FTD model to surface flux evolution models (see Wang et al. 1989). The BL mechanism could also be treated more realistically by including the formation and rise of flux loops through the convective zone to form the tilted sunspot groups. By introducing new technique for modeling the emergence of flux tubes, Yeates and Muñoz-Jaramillo (2013; see also Miesch and Dikpati 2014) took the first step towards the development of such FTD model, which should provide a framework for comparison to surface magnetographic observations as well as full DNS models.

A major challenge to kinematic FTD model of the solar cycle has been raised by some recent measurements of the subsurface meridional flow, indicating that the equatorward reverse flow of the meridional circulation occurs at a shallow depth and not at the bottom of the convection zone as usually assumed in the FTD model (Hathaway 2012; Zhao et al. 2013; Schad et al. 2013). How can the FTD model operate under this "shallow return flow" configuration? There actually exists a number of viable options. Guerrero and de Gouveia Dal Pino (2008) considered a shallow cell of meridional circulation with equatorward turbulent pumping in the region below it and succeeded in getting realistic butterfly diagrams. More recently Hazra et al. (2014) showed that many of the attractive features of FTD are retained if, below the shallow cell of meridional circulation at the top of the convection zone, there are additional cells such that there is an equatorward meridional circulation at the bottom of the convection zone.

3 Causes of Solar Cycle Irregularities

In the previous Section, we have reviewed the development of the FTD model, its achievements in explaining the regular periodic features of the solar cycle and its limitations. Quite a lot of work has been done in the last few years to apply this model to explain the various irregularities of the solar cycle. The first concern to be discussed in this Section is to understand the causes of the solar cycle irregularities. Once we understand the causes, the next concern is whether we can model the irregularities of actual solar cycles by assimilating some aspects of observational data into the theoretical model. This topic will be discussed in the next Section.

3.1 Nonlinearities Versus Stochastic Fluctuations

In the early years of dynamo research, it used to be generally assumed that the irregularities of the solar cycles are manifestations of nonlinear chaos. Certainly the Lorentz force of the magnetic field reacts on the flows in the convection zone in accordance with (2), giving rise to nonlinear feedback at all scales. The simplest way of capturing the effect of this in a kinematic dynamo model (in which the fluid equations are not solved) is to consider a quenching of the α parameter as follows:

$$\alpha = \frac{\alpha_0}{1 + |\overline{B}/B_0|^2}, \tag{15}$$

where $\overline{B}$ is the average of the magnetic field produced by the dynamo and B_0 is the value of magnetic field beyond which nonlinear effects become important. In most of the nonlinear calculations, however, the dynamo eventually settles to a periodic mode with a given amplitude rather than showing sustained irregular behavior. The reason for this is intuitively obvious. Since a sudden increase in the amplitude of the magnetic field would diminish the dynamo activity by reducing α given by (15) and thereby pull down the amplitude again (a decrease in the amplitude would do the opposite), the α-quenching mechanism tends to lock the system to a stable mode once the system relaxes to it. In fact, Krause and Meinel (1988) and Brandenburg et al. (1989) argued that the nonlinear stability may determine the mode in which the dynamo is found. Nonlinear magnetic backreaction on large-scale flows, the so-called Malkus-Proctor effect, can also generate a variety of amplitude modulation patterns in mean-field dynamos (e.g. Weiss 2010; Weiss and Lopes, this issue; and references therein). However, this mechanism has received comparatively little attention so far in the context of the FTD model (but do see Rempel 2006; Passos et al. 2012).

Since the mean field dynamo equations are derived by averaging over turbulence, we expect stochastic fluctuations to be present around the mean (e.g. Hoyng 1988, 1993; Choudhuri 1992; Moss et al. 1992; Ossendrijver et al. 1996; Charbonneau and Dikpati 2000). Although a large amount of research supports the view that the solar cycle irregularities are primarily produced by fluctuations, this does not imply that nonlinearities have no important consequences in the FTD models. In order to explain the even-odd or the Gnevyshev–Ohl pattern in solar cycle amplitudes, Durney (2000) suggested that the solar dynamo may be operating in a region of period doubling just beyond the point of nonlinear bifurcation. Charbonneau (2001; see also Charbonneau et al. 2007) went on to show that this is not essential, as in the presence of stochastic forcing the Gnevyshev-Ohl pattern is a robust feature of the time delay dynamics characterizing the advection-dominated FTD model. This same time-delay dynamics can lead to period doubling and transition to chaos even in kinematic FTD models (Charbonneau et al. 2005).

Let us consider the question how fluctuations in poloidal field generation arise in the FTD. The BL mechanism of poloidal field generation depends on the latitudes of emergence and tilts of bipolar sunspot pairs. Sunspot groups erupting at lower latitudes generate stronger dipole field since more leading polarity flux can diffuse across the equator and resistively annihilates against the opposite polarity flux in the opposite hemisphere (see Fig. 6 in Jiang et al. 2014a). Likewise, at a given latitude larger tilt angles generate more large-scale poloidal field (Wang and Sheeley 1991; Baumann et al. 2004). While the average tilts are given by Joy's law, one finds a large scatter around this average. Longcope and Choudhuri (2002) provided a theoretical model of this scatter on the basis of the idea that the rising flux tubes are buffeted by turbulence in the convection zone. This scatter around Joy's law produces fluctuations in the poloidal field generation process. Choudhuri et al. (2007) identified this as a primary source of irregularities in the solar cycle. A strong support for this idea has come from the recent analysis of sunspot tilts data by Dasi-Espuig et al. (2010) and Kitchatinov and Olemskoy (2011). Moreover, the analyses of the Mount Wilson and Kodaikanal tilt angle data show that the cycle averaged tilt angles are anti-correlated with the cycle strength (Dasi-Espuig et al. 2010). This acts as a nonlinearity modulating the efficiency of the BL mechanism (see Jiang et al. 2014b, this issue).

3.2 The Effects of Meridional Circulation Fluctuations

Since the period of FTD depends on the amplitude of the meridional circulation as indicated in (14), it is expected that any fluctuations in the meridional circulation will have significant effects on FTD. Unfortunately not much is known about its long term variations from observations. What is confirmed from the existing observational data of last 15–20 years is that the surface meridional circulation becomes weaker during the solar maximum and vice versa (Chou and Dai 2001; Hathaway and Rightmire 2010), probably due to the strong feedback of the dynamo generated magnetic field on the meridional flow (Karak and Choudhuri 2012; but see also Passos et al. 2012; Karak et al. 2014a). In a highly turbulent solar convection zone, it is very likely that the dynamical balance underlying the meridional flow (see Sect. 5.3.2 below) gets continuously perturbed, at least from the convective timescale up to the cycle timescale (Ulrich 2010). Karak and Choudhuri (2011) estimated its variation from the observed periods of several past solar cycles. By applying the relation (14), they concluded that the meridional circulation had experienced large variations on a timescale $\sim$ 20–30 yr (see also Passos and Lopes 2008; Lopes and Passos 2009; Passos et al. 2012; Karak et al. 2014a). Local inflow towards activity belts (Zhao and Kosovichev 2004; Gizon 2004) may also contribute to the temporal variation of the zonally-averaged meridional flow, providing yet another nonlinear modulation the BL mechanism on cycle timescales.

In FTD model, apart from producing the poleward migration of the poloidal field on the solar surface and the equatorward migration of the toroidal field at low latitudes near the bottom of the convection zone, the meridional circulation also affects the strength of the magnetic field. Following Yeates et al. (2008), suppose the meridional circulation becomes weak during a cycle; then the cycle duration becomes longer, which produces two competing effects. The differential rotation has more time to induce more toroidal field, which tends to make the cycle stronger. At the same time, the diffusion also gets more time to diffuse the fields, which tends to make the cycle weaker. In the high diffusivity dynamo model, the diffusion of the field is more important than the induction of the toroidal field. Therefore, in this model, the weak meridional circulation produces weaker cycles. Conversely, when the meridional is increased, the opposite of this happens leading to stronger cycles. On the other hand, in the low diffusivity model, diffusion is not as important as the induction of the toroidal field. Hence the weak meridional circulation produces stronger cycles, whereas strong meridional circulation produces weaker cycles. Given this clearly opposite predictions from the high and low diffusivity kinematic FTD models, Karak (2010) found that only the high diffusivity model could fit the observational data of the past solar cycles. Figure 3 shows the result for the high diffusivity model. The upper panel shows the value of the meridional circulation used to fit the cycle period, whereas the lower panel shows the theoretical sunspot number (dashed line) along with the observed sunspot number (solid line). We see that the stronger meridional circulation, which is required to fit the shorter cycles, produces stronger cycles and vice versa.

The same considerations also help us to model another related feature of the solar cycle, the observed anti-correlation between the strengths of the cycles and their rise times, known as the Waldmeier effect (Waldmeier 1935). In other words, stronger cycles tend to have shorter rise and vice versa. Karak and Choudhuri (2011) found that, by introducing stochastic fluctuations only in the poloidal field generation, it is not possible to reproduce the Waldmeier effect. However, on introducing stochastic fluctuations in the meridional circulation of their high diffusivity dynamo model, they found that stronger meridional circulation would make the cycle rise time shorter but the cycle amplitude stronger (the opposite being the case for weaker meridional circulation), giving rise to the Waldmeier effect (see also Charbonneau and Dikpati 2000).

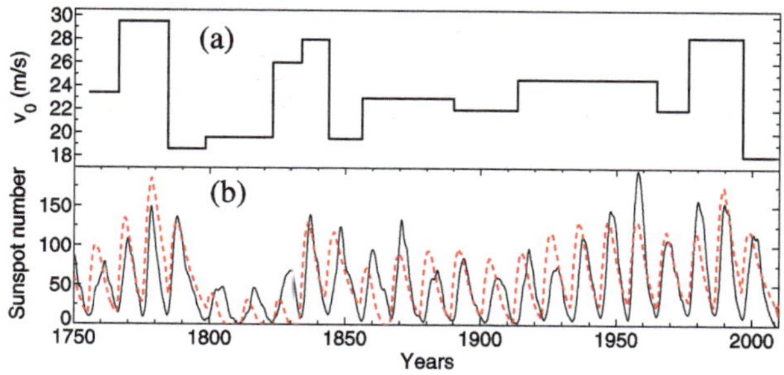

Fig. 3 (**a**) Variation of the amplitudes of meridional circulation v_0 needed in dynamo model to fit the periods of last 23 observed cycles. (**b**) Variation of the theoretical sunspot number (*dashed line*) and observed sunspot number (*solid line*) with time. Both sunspot numbers are smoothed by a Gaussian filter of FWHM = 1 year. In addition, the theoretical sunspot number is scaled by a factor to match the observed value. Reproduced from Karak (2010; see also Lopes and Passos 2009)

3.3 Modeling Grand Minima

One of the intriguing aspects of solar cycles is grand minima like the Maunder minimum, during which solar activity becomes very weak for many years, with several cycles missing. We now know that the Maunder minimum was not unique. From ^{14}C records in tree rings and ^{10}Be records in polar ice cores, it is possible to reconstruct the history of solar activity for several millenia. A recent estimate is that there have been 27 grand minima in the last 11,000 years (Usoskin et al. 2007).

The question we address now is the following: do grand minima also arise out of the same stochastic fluctuations and nonlinearities which are responsible for producing the solar cycle irregularities? In other words, are these grand minima examples of the extremes of these fluctuations? A drop of the poloidal field produced at the end of a cycle resulting from a fluctuations in the BL mechanism can trigger a Maunder minimum-like event (Charbonneau et al. 2004; Choudhuri and Karak 2009; Olemskoy et al. 2013; Passos et al. 2014). Another known trigger is a significant decrease of the meridional flow speed (Karak 2010; Passos and Lopes 2011).

Let γ be the normalized strength of the polar field (i.e. the strength of the polar field divided by its average value over many cycles) at the end of a cycle and let v_0 be the amplitude of the meridional circulation. Figure 4 shows the two-dimensional parameter space of γ versus v_0. The solid line in this plot shows the combined values of the parameters which can produce a grand minima with duration of 20 years in the model of Choudhuri and Karak (2012). The necessary condition for producing a grand minimum is that the parameters (γ, v_0) at the beginning of a sunspot cycle has to lie below this solid line. If we assume that both the fluctuations we are considering are of Gaussian nature, then the joint probability that the polar field strength at the end of a cycle lies in the range $\gamma, \gamma + d\gamma$ and the amplitude of the meridional circulation at the same time lies in the range $v_0, v_0 + dv_0$ is given by

$$P(\gamma, v_0) d\gamma dv_0 = \frac{1}{\sigma_v\sqrt{2\pi}} \exp\left[-\frac{(v_0 - \overline{v_0})^2}{2\sigma_v^2}\right] \frac{1}{\sigma_\gamma\sqrt{2\pi}} \exp\left[-\frac{(\gamma - \overline{\gamma})^2}{2\sigma_\gamma^2}\right] d\gamma\, dv_0, \quad (16)$$

where $\overline{v_0} = 23\ \mathrm{m\,s^{-1}}$ and $\overline{\gamma} = 1.0$ are the mean values. The probability that the condition at the beginning of a cycle lies in the shaded region of Fig. 4 is obtained by integrating the double Gaussian given by (16) over this region. To carry on this integration, we need the values of the widths σ_v and σ_γ of these Gaussians, which can be obtained from the

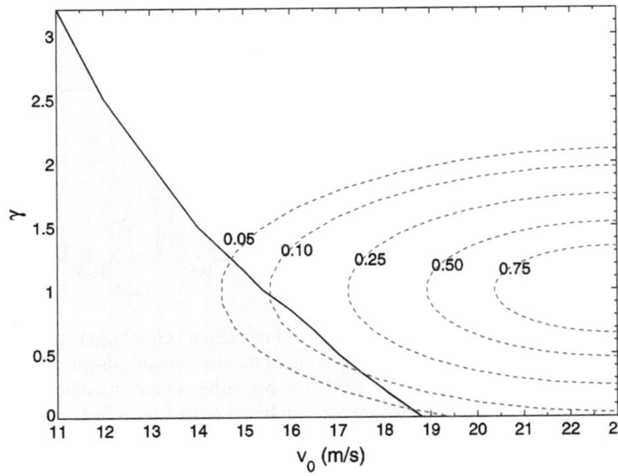

Fig. 4 The parameter space indicating the normalized strength γ of the poloidal field and the amplitude of the meridional circulation (v_0), the *shaded region* being the part to the parameter space giving rise to grand minima

existing observational data of the last 28 cycles. The periods of these cycles give the values of the meridional circulation during these cycles, from which the probability distribution function of the meridional circulation can be constructed. Since strengths of the cycles are correlated with the polar field strength γ at the beginning of the cycle, the strengths of the last 28 cycles can be used to construct the probability distribution function of γ. Although we would not expect a very good Gaussian fit from a set of 28 data points, Choudhuri and Karak (2012) found that the fits were acceptable and could estimate the values of σ_v, σ_γ. On carrying out the integration of the double Gaussian over the shaded region in Fig. 4, Choudhuri and Karak (2012) found the value to be 1.7 %. This means that 17 cycles out of 1000 cycles (in 11,000 years) would have conditions appropriate for grand minima at their beginnings. This is close to the observational data that there had been 27 grand minima in the last 11,000 years (Usoskin et al. 2007). In fact, in actual runs of the dynamo code with fluctuations given by the double Gaussian (15), Choudhuri and Karak (2012) typically found about 24–30 grand minima in runs spanning 11,000 years.

Evidently, if the BL mechanism is the sole source of poloidal magnetic field, then an additional mechanism is required to pull the dynamo out of a grand minimum. The most likely candidate is the original turbulent α-effect discussed in Sect. 1, or any other source of small-scale magnetic "noise", such as provided by non-helical surface convection. A number of representative models demonstrating the viability of these "kickstart" processes are available in the literature (Charbonneau 2001; Charbonneau et al. 2004; Choudhuri and Karak 2009, 2012; Karak and Choudhuri 2013; Olemskoy et al. 2013; Olemskoy and Kitchatinov 2013; Passos et al. 2014). Typically, the relative strengths and spatial localisation of these multiple source terms requires some level of fine tuning to yield a solar-like cycle period and trigger grand minima at the observed frequency. It should also be kept in mind that the BL mechanism may still be operating when there are even a few sunspots with correct tilt angles, or even through the decay of ephemeral regions with a statistically significant pattern of tilt (Stenflo and Kosovichev 2012; Wang and Sheeley 2013).

4 Data Assimilation to Flux Transport Dynamos

In the framework of FTD models relying exclusively on the BL mechanism for poloidal field regeneration, solar magnetographic observations offer a direct view of this crucial half of the dynamo loop. This can form the basis of solar cycle prediction schemes assimilating surface data to FTD models, and provide valuable information regarding the origin of observed cycle fluctuations. The scheme of assimilating the observational data to a FTD model was initiated by Dikpati et al. (2006). As per our knowledge, until today there have been two other schemes developed. In this section we review the three different schemes.

The surface magnetic (or magnetically-related) data available to assimilate in dynamo models are relatively limited in temporal extent. Continuous magnetographic observations are only available since 1976. The Royal Greenwich Observatory (RGO) and US Air Force Solar Optical Observing Network (SOON) have the detailed sunspot group records, including areas and locations during 1874–1976 and 1977 onwards, respectively. Observatories at Mount Wilson and Kodaikanal provide the tilt angle information of sunspot groups in the intervals 1917–1985 and 1906–1987, respectively. Direct observational data for the Sun's polar magnetic field are available only from mid-1970s (mainly from Wilcox Solar Observatory), with polar faculae and network index providing indirect determinations back to 1905 (Muñoz Jaramillo et al. 2012; Priyal et al. 2014, and references therein).

4.1 Representative Models

Dikpati et al. (2006) and Dikpati and Gilman (2006) adopted an idealized method of assimilating the observed sunspot records (RGO/SOON data) into the poloidal flux of a FTD model. Their simplification of the sunspot records mainly includes (i) the non-cycle dependence of the latitudinal distribution migrating between 35° and 5°, and (ii) the same tilts for all the sunspot groups. The data assimilated into the model were the observed sunspot group areas, which were used to formulate a Babcock-Leighton source term for the poloidal field. The observed sunspot group areas were mainly incorporated into their method to derive the poloidal field. The amplitudes of cycles 16–23 are well-reproduced, but in its original form the model failed to predict the low amplitude cycle 24. Since Dikpati and Gilman (2006) presented a detailed description of their method, we shall take it as a representative of this type of data assimilation and denote it as DG06.

Jiang et al. (2007) (hereafter JCC07) assimilated the observed photospheric magnetic field to the FTD model. The surface poloidal field was updated at each cycle minimum by the longitudinally-averaged photospheric field at different latitudes from Wilcox Solar Observatory from the middle of 1976 to the end of 2006 (4 cycle minima). The poloidal field at the minimum produced in the regular dynamo code corresponds to an average strength. The observed poloidal field at the cycle 22 minimum was taken as indicative of the average poloidal field strength in that phase of the dynamo cycle. At all grid points above $0.8R_\odot$, the vector potential A was multiplied by a factor $\gamma(\theta)$ to bring it into agreement with the surface A derived from observational data at that latitude. The relative amplitude variations of cycles 21–24 were well reproduced by the procedure. Their earlier paper, Choudhuri et al. (2007), used a similar method but with a single number based on the observed dipole moment. Goel and Choudhuri (2009) extended their study back to cycle 15 based on the observed polar faculae, which was taken as the proxy of the surface poloidal field. In the following we shall mainly discuss JCC07 as a representative of this type of data assimilation.

More recently, Jiang et al. (2013) (denoted as JCSI13) used the historical sunspot records from RGO, Mount Wilson, and Kodaikanal as the basis for constructing the source of poloidal flux S calculated by

$$S(r,\theta,t)\sin\theta = \frac{R_\odot f(r)}{2\pi}\int_0^\theta\int_0^{2\pi}\Delta B_r\big(R_\odot,\theta',\phi,t\big)\mathrm{d}\phi\sin\theta'\mathrm{d}\theta'\delta(t-t_{\mathrm{em}}), \qquad (17)$$

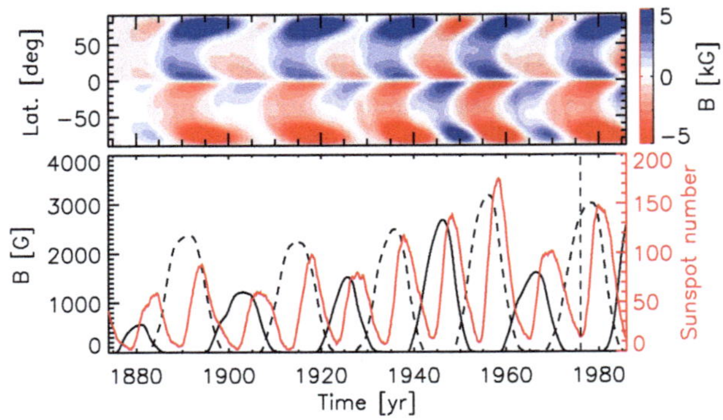

Fig. 5 Toroidal field at the base of the convection zone ($r = 0.7R_\odot$) from FTD with assimilation of sunspot group records. The *upper panel* shows the evolution as a function of latitude and time. The *lower panel* shows the average unsigned toroidal field between ±45° latitude (*dashed* and *solid black curves*, for odd and even cycles respectively) along with the observed sunspot number (*red curve*). Reproduced from Fig. 4 of Jiang et al. (2013)

where t_{em} is the time at which the sunspot group was observed to emerge, and δ is the Dirac delta function. The function $f(r)$ is the radial dependence of the source term. $\Delta B_r(R_\odot, \theta', \phi, t)$ corresponds to the change in the radial field at the surface for each sunspot group emergence (at its maximum area development) and is determined by the area, latitude and tilt of each sunspot group. In this manner, the nonlinear physical processes underlying flux emergence are thus implicitly captured in by this empirical representation of ΔB_r.

Figure 5 shows the toroidal field at the base of the convective zone ($r = 0.7R_\odot$) as produced by the JCSI13 model. The upper panel shows the evolution as a function of latitude and time. The high-latitude branch is assumed unable to generate the sunspot emergence, as per the stability analyses of Ferriz-Mas and Schüssler (1995) and Caligari et al. (1995). At low latitudes the toroidal field distribution agrees well with the observed butterfly diagram. The black curves in the lower panel shows the averaged unsigned toroidal flux between ±45° latitude. The red curve is the observed sunspot number. The maxima of the modeled activity levels are clearly related to the observed amplitude of the cycle. In addition, the polar field at the end of each cycle also correlates well with the subsequent cycle strength.

4.2 Differences Among the Models

The important ingredients in the FTD are (i) the generation of poloidal flux due to the evolution of the tilted sunspot groups, (ii) the transport of poloidal flux to the tachocline, (iii) the generation of toroidal magnetic flux due to the winding up of the poloidal field, and (iv) the formation and rise of flux loops through the convection zone until they emerge at the surface. Although there are less direct observational constraints on processes (iii) and (iv), the main differences among the three models considered are at the level of (i) and (ii). The differences among the models are summarized in Table 1.

The available polar field data provide some constraints on the processes (i) and (ii). There is a good correlation between the polar field at the cycle minimum (denoted as P_n) and the strength of the following cycle (denoted as S_{n+1}). In contrast, there is no correlation between the polar field at the cycle minimum and the strength of the preceding cycle (denoted as S_n) (Jiang et al. 2007). The good correlation between P_n and S_{n+1} provides the constraint on the transport time of surface poloidal flux to the tachocline. No correlation between P_n and S_n

Table 1 Overview of the differences among the three data-driven flux transport dynamo models

Model	Assimilated data	Nonlinearities in poloidal source	Transport mechanisms	Magnetic memory	Model parameters[a]			Correlations (P_n vs.)[b]			Source of cycle irregularities
					η	υ	γ	S_n	S_{n+1}	S_{n+2}	
DG06	RGO/SOON: sunspot group areas	NO	advection dominated	$\sim$ two cycles	5.0×10^{10}	$r_p = 0.7 R_\odot, \theta_0 = 32°$	0	Y	N	Y	Waldmeier effect and cycle overlap cause the time shift of solar minima depending on the strength of the next cycle (see Cameron and Schüssler 2007)
JCC07	WSO magne-tograms during cycle minima	– (not referred)	diffusion dominated	$\sim$ half cycle	2.4×10^{12}	$r_p = 0.61 R_\odot, \theta_0 = 45°$	0	N	Y	N	right magnetic memory and directly observed poloidal field source (but cannot answer the reasons for the source variabilities)
JCSI13	RGO/SOON: sunspot group areas & latitudes; MWO/KOD: cycle averaged sunspot group tilt angles	cycle-dependence of latitudes and tilts of sunspot groups	pumping dominated	$\sim$ half cycle	1.0×10^{11}	$r_p = 0.70 R_\odot, \theta_0 = 32°$	2	N	Y	N	right magnetic memory and poloidal field source (Nonlinearities in the poloidal field source were investigated)

[a]Diffusivity η is in $cm^2\,s^{-1}$; the penetration depth and the maximum value of the latitudinal distribution of meridional flow υ are denoted as r_p and θ_0; the magnetic pumping γ is in ms^{-1}

[b]The correlations between polar field at cycle minimum (P_n) and the strength of cycles n (S_n), $n+1$ (S_{n+1}) and $n+2$ (S_{n+2})

implies that the nonlinearities or randomness are involved in the generation of poloidal flux from the decay of the sunspot groups.

4.2.1 Surface Poloidal Field Source

JCC07 use the observed magnetograms at the cycle minima to update the surface poloidal field, which is consistent with the BL mechanism. But this method says little about the poloidal field generation. Furthermore, the limited observational data, which is here less than 4 cycles, make the results less convincing. The detailed RGO/SOON sunspot data that cover more than 12 cycles provide another option to be assimilated into the FTD models. As argued in Sect. 3.1, the properties of sunspot emergence, i.e., the cycle dependence of the latitudes and tilt angles of the sunspot groups, play important roles in modulating the surface poloidal field generation. The nonlinearities and the possible randomness in the generation of the surface poloidal field are responsible for decorrelating the polar field at cycle minima from the strength of the preceding cycle.

On the other hand, both DG06 and JCSI13 assimilated the RGO/SOON sunspot data into the FTD models to derive the poloidal flux source. But the idealized method used by DG06 ignored the nonlinear ingredients of the poloidal field source. The simplification yields a poloidal flux at cycle minima that is roughly proportional to the strength of the preceding cycle, which is not consistent with the observational constraints. In contrast, JCSI13 included the sunspot group's latitude from RGO and the sunspot group's tilt from the cycle averaged results of Mount Wilson and Kodaikanal. Thus, the nonlinearities in the surface poloidal field source are included, which lead to the non-correlation between P_n and S_n.

4.2.2 Transport Mechanisms of the Surface Poloidal Field to the Tachocline

The finite time required to transport the surface poloidal flux to the tachocline introduces a memory into the system, which is at the root of dynamo-based solar cycle predictions (Dikpati et al. 2004; Karak and Nandy 2012). As reviewed in Sect. 2.3, candidates for the transport mechanisms in the convective zone are magnetic diffusion, meridional flow and magnetic pumping. Their assumed relative importance differs markedly in the three models considered here (see Table 1).

Strong (2.4×10^{12} cm^2 s^{-1}), medium (10^{11} cm^2 s^{-1}) and low (5.0×10^{10} cm^{2}s^{-1}) diffusivity in the bulk of convective zone were used by JCC07, JCSI13 and DG06, respectively. The transport time of the poloidal field throughout the whole convective zone by the diffusion alone is about 5 yr for the case with the strong diffusivity, whereas it is up to ~ 100 yr for the other two models. The lack of constraints have led to the controversy. The diffusivity in the convective zone was mainly estimated by mixing-length theory and convection simulations (see also Sect. 5.5.1), both of which set its amplitude to be $\sim 10^{13}$ cm^2 s^{-1}. However, recent helioseismic inversions of convective motions implied that the diffusivity may be lower by one to two orders of magnitude (Hanasoge et al. 2010, 2012). The profile of the turbulent diffusivity in the convective zone related to solar magnetism is still a mystery (Parker 2009).

All three models use a single-cell meridional flow (viz. Fig. 1), but differ in the assumed radial penetration into the tachocline and latitudinal dependence of meridional flow. A shallower penetration ($r = 0.7R_\odot$) and peak at 32° latitude were used by DG06 and JCSI13. A deeper penetration ($r = 0.61R_\odot$) and peak at 45° latitude were used by JCC07. Meridional flows inferred from the tracking of magnetic elements and from local helioseismology indicate that the flow speed peaks at about 35° latitude (see Fig. 5 of Jiang et al. 2014b, this issue). The latitudinal dependence of meridional flow, especially the latitudinal gradient near the equator, has large effects on the polar field generation. Section 2.4 shows that the deeper penetration of the meridional flow may help to get the solar-like butterfly diagram in some FTD models. As also pointed out in Sect. 2.4, the shallower penetration of the meridional low is supported by the dynamical simulations of solar convective zone (Gilman and Miesch 2004; Rempel 2005; Rüdiger et al. 2005). The shallower penetration of the flow is more widely adopted in the FTD models.

Neither DG06 nor JCC07 included magnetic pumping. A downward magnetic pumping of 2 $\mathrm{m\,s^{-1}}$ in the bulk of convective zone and 20 $\mathrm{m\,s^{-1}}$ near surface were used by JCSI13. With this pumping profile, the surface poloidal field evolution resulting from FTD can be consistent with that from the observations of the Sun's large-scale surface field and from the surface flux transport model (Cameron et al. 2012). The pumping also leads to a more dominant equatorward migration of the toroidal field (Brandenburg et al. 1992). It takes about 3 yr for the surface poloidal field to be transported to the tachocline under the effect of the pumping alone.

The different transport parameters cause different transport processes to connect the two spatially segregated source regions for the Sun's poloidal and toroidal magnetic field. Meridional flow, turbulent diffusion and magnetic pumping were the primary mechanisms in DG06, JCC07 and JCSI13, respectively. In DG06, the field was transported toward the pole in about 5 yr under the effect of meridional flow, and then transported downwards to the tachocline in about two cycles, where it was stretched by the differential rotation to generate the toroidal field. Hence polar fields of cycle n at the minimum have the strongest correlation with the toroidal field of cycle $n+2$, rather than cycle $n+1$. In both JCC07 and JCSI13, when the poloidal field was advected by the meridional circulation to the polar region, it was simultaneously transported to the tachocline under the effects of the high diffusivity or magnetic pumping. In both cases, the magnetic memory is about half cycle. The same poloidal field source causes the correlation between the polar fields at the cycle minimum and the subsequent cycle strength.

All the schemes can reproduce the cycle irregularities of the past several cycles. Magnetic memory and the poloidal field source are the cause of the cycle irregularities in JCC07 and JCSI13. Cameron and Schüssler (2007) suggested that the skill of DG06 to reproduce the irregularities of cycles 16–23 did not require a physical connection between the surface manifestations and the subsequent cycle. Rather, it was the Waldmeier effect and the cycle overlap that lead to a shift of the minimum epochs which depends on the subsequent cycle strength.

4.3 Implications for Solar Cycle Irregularities and Prediction

In the context of kinematic FTD models, induction of a toroidal magnetic component via shearing of the poloidal field is a linear process. The cycle amplitude is determined by the nonlinearities in the surface poloidal field source, which are caused by the emergence of the toroidal flux throughout the convective zone. To infer the nonlinearities in the surface poloidal field source based on the sunspot records is a delicate matter. There are large and potentially systematic errors in the tilt angle measurements, due to the difficulties in distinguishing the different groups, especially during the cycle maxima. The measured sunspot areas depend also on the instruments and procedures (Baranyi et al. 2001; Foukal 2014).

The generation of the surface poloidal field also involves stochasticity, which was identified by recent studies and discussed in earlier Sect. 3. Jiang et al. (2014a) measured the observed large scatter of the tilt angle distribution and studied the effects of this scatter on the evolution of the solar surface field for cycle 17. The results show that the tilt angle scatter leads to variations of large-scale field about 30 % (standard deviation). A single sunspot group emerging near the equator with large tilt angle and large area may cause a 20 % variation of the dipole field at the end of the cycle. Such occasional events corresponding to the cross-equatorial flux plumes identified in the observed magnetograms (Cameron et al. 2013). The impact of such events are cumulative, and hence pose strict limits to the predictability of solar activity on long timescales.

5 Probing the Foundations of Solar Dynamo Modeling

5.1 The Four Dynamo Ingredients

The kinematic dynamo models discussed in Sects. 2–3 require four main ingredients. By this, we mean four physical phenomena and/or processes that must be specified as input into the models, each involving a set of parameters and algorithmic procedures. Briefly these are (*I*) the differential rotation, (*II*) the meridional circulation, (*III*) mean field generation by convection and flux emergence, and (*IV*) convective transport. In this section we describe further what these ingredients are and how well we understand them based on solar observations and 3D magnetohydrodynamic (MHD) simulations.

The first of the four dynamo ingredients is the differential rotation. Of the four, this is the most well constrained by solar observations. Helioseismic inversions of global acoustic oscillations have mapped the solar internal rotation throughout the convection zone (CZ) in detail and continue to follow and characterize its time evolution. For a comprehensive review of these techniques and results see Christensen-Dalsgaard (2002), Thompson et al. (2003), and Howe (2009). Here we just summarize a few salient results of most relevance to solar dynamo theory.

Figure 6a is familiar to most solar physicists by now. It shows the angular velocity $\Omega(r,\theta)$ of the solar interior inferred from helioseismic inversions based on Regularized Least Squares (RLS) inversions of data from the GONG (Global Oscillations Network Group) network of telescopes. Notable features include a monotonic decrease of Ω by about 30 % from the equator to the polar regions and nearly radial Ω contours at mid latitudes. Also apparent are two boundary layers at the top and bottom of the CZ where Ω exhibits substantial vertical shear. The lower boundary layer is known as the solar tachocline and, as noted in Sect. 1, is thought to play a key role in the operation of the global dynamo. It represents a sharp transition between the differential rotation of the solar envelope and the nearly uniform rotation of the radiative interior. The upper boundary layer, known as the near surface shear layer (NSSL) is somewhat more subtle, characterized by a decrease in Ω by about 5 % from $0.95R$–R. As far as we can tell, this decrease occurs at all latitudes, though the data is inconclusive near the poles (Thompson et al. 2003; Howe 2009).

The second dynamo ingredient, namely the meridional circulation, is also in principle accessible from helioseismic inversions but various challenges limit its detectability. These include the relative insensitivity of global oscillation modes to latitudinal flows (requiring local helioseismic techniques or higher-order global analyses) and the weak amplitude of mean meridional flows relative to mean zonal flows (differential rotation) and convection. Again, we will not attempt a comprehensive review here but merely note a few salient results.

The most robust information we have on the meridional circulation (MC) from local helioseismic inversions, surface Doppler measurements, and surface feature tracking is that

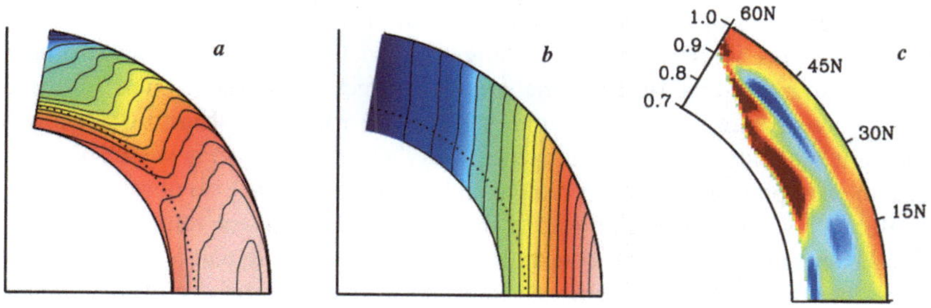

Fig. 6 Solar differential rotation and meridional circulation inferred from helioseismology. Frames (**a**) and (**b**) show the angular velocity $\Omega/2\pi$ and the specific angular momentum $\mathcal{L} = s^2\Omega$ based on global helioseismic RLS inversions of GONG data from 1996, provided by R. Howe (Howe et al. 2000; Schou et al. 2002). Each uses the same color table, increasing linearly from blue to pink, with ranges of (**a**) 300–460 nHz and (**b**) 0–1.4 $\mathrm{Mm}^2\,\mathrm{s}^{-1}$. (**c**) Mean colatitudinal velocity $\langle v_\theta \rangle$ in the northern hemisphere inferred from time-distance helioseismic inversions of SDO/HMI data from 2010–2012, from Zhao et al. (2013). *Red* and *blue* indicate poleward and equatorward flow respectively with a saturation level for the color table of $\pm 15\ \mathrm{m\,s}^{-1}$. High latitudes and deep radii are omitted due to unreliable data

it is poleward near the surface at latitudes below about 60° with an amplitude of about 10–20 $\mathrm{m\,s}^{-1}$ (e.g. González-Hernandez et al. 2006; Ulrich 2010; Hathaway and Rightmire 2011). Doppler measurements suggest the possible presence of a high-latitude counter cell with equatorward flow near the surface that is most prominent during solar minimum (Ulrich 2010). Helioseismic inversions are less clear about the presence of this high-latitude counter-cell, but evidence is growing for the possible presence of multiple circulation cells in radius, as illustrated in Fig. 6c.

The results shown in Fig. 6c are based on local (time-distance) helioseismic inversions of oscillation data from the Helioseismic Magnetic Imager (HMI) onboard the Solar Dynamics Observatory (SDO) and span the first two years of the SDO mission, from 2010–2012. They indicate a reversal of the meridional flow from poleward to equatorward at a radius of about $0.91R$, with possible evidence for a second reversal (from equatorward to poleward) below $0.82R$. This is in stark contrast to the single-celled MC profiles conventionally used in kinematic solar dynamo models (Fig. 1). The potential implications of such a multi-cell MC profile for FTD models are profound (e.g. Jouve and Brun 2007). Though the data have not yet provided a fully consistent picture of the subsurface MC profile, further evidence for the possible presence of multiple cells has been reported based on surface feature tracking (Hathaway 2012) and global helioseismic inversions (Schad et al. 2013).

These first two dynamo ingredients, the differential rotation and the meridional circulation, appear in the mean magnetic induction equation (4) as the mean flow $\overline{\mathbf{v}}$. The third and fourth ingredients comprise the fluctuating or turbulent EMF term $\boldsymbol{\mathcal{E}}$. Arguably the most important role of $\boldsymbol{\mathcal{E}}$ is to generate the mean poloidal field. Without this capacity, the kinematic dynamo could not function, as demonstrated by Cowling's theorem (Moffatt 1978). The field generation aspect of $\boldsymbol{\mathcal{E}}$ is manifested in the first term of Eq. (6) where it is represented as the well-known α-effect of mean field dynamo theory.

Though it has dominated the dynamo literature for more than half a century, the α-effect is still poorly understood, particularly as it applies to the solar dynamo. As described in Sect. 1, mean poloidal field can in principle be generated by a variety of distinct physical processes, most notably helical convection, commonly referred to as the turbulent α-effect, and the buoyant destabilization, rise, emergence, and dispersal of toroidal flux tubes, commonly referred to as the Babcock-Leighton (BL) mechanism. Both processes are intrinsically 3D, multi-scale, and nonlinear, posing formidable challenges for numerical models.

Although observations of surface flux evolution can provide tentative estimates of the efficiency of the BL mechanism, virtually nothing is known observationally about the nature of $\mathcal{E}$ below the surface and the relative efficiency of convective field generation. Furthermore, these two processes are likely not mutually exclusive; rotational shear, turbulent convection, magnetic buoyancy, and flux emergence likely all play a role in how the solar dynamo generates the complex hierarchy of poloidal and toroidal fields that we observe.

We refer to this field generation role of the turbulent EMF as the third major ingredient that must be specified in order to construct a viable kinematic dynamo model of the solar cycle. The fourth ingredient concerns the transport of magnetic flux and the dissipation of magnetic energy, which is also contained in $\mathcal{E}$. This is commonly represented by means of a turbulent magnetic diffusion as in the β term of Eq. (6). However, like the representation of field generation by the α-effect, turbulent diffusion is only a crude representation of complex 3D physical processes that are likely to be far more subtle. Recent models have begun using more sophisticated prescriptions that include turbulent pumping (Sect. 2.3).

The role of turbulent transport is particularly essential for solar dynamo models in which the principal regions for mean poloidal and toroidal field generation are spatially separated. In this case, transport of magnetic flux by convective motions and/or meridional flows is essential in order for the dynamo to operate.

In summary, though solar observations provide essential input into kinematic solar dynamo models, they are not sufficient to completely specify all the necessary dynamo ingredients. Though the solar internal rotation profile Ω is well constrained by observations, the subsurface meridional flow $\overline{\mathbf{v}}_m$ is not. Even less is known about the nature of $\mathcal{E}$ below the solar surface. Further progress in understanding the four fundamental ingredients of kinematic dynamo models requires an understanding of the dynamics of the convection zone and tachocline, including how mean flows are maintained and how mean fields are generated and transported. Here 3D MHD simulations of solar convection and flux emergence can provide valuable insights. In the remainder of this section we briefly review some of these insights and discuss how they can be used to inform and improve kinematic dynamo models.

5.2 Dynamical Balances

Any discussion of how mean flows are maintained must begin with a careful investigation of the MHD version of the Navier-Stokes equation (2). Averaging the zonal component over longitude and time and assuming a statistically steady state yields (Miesch and Hindman 2011, hereafter MH11)

$$\overline{\rho \mathbf{v}_m} \cdot \nabla \mathcal{L} = -\nabla \cdot \left[s\overline{\rho \mathbf{v}' v'_\phi} - s\overline{\mathbf{B} B_\phi} - \rho \nu s^2 \nabla \Omega \right] \equiv \mathcal{F}, \tag{18}$$

where $\mathcal{L} = s\overline{v_\phi} = s^2\Omega$ is the specific angular momentum. We refer to the term $\mathcal{F}$ as the net torque that, in a steady state, must balance the advection of angular momentum by the meridional flow. As shown in (18), it contains three components, namely the Reynolds stress, the Lorentz force, and the viscous diffusion. The latter is negligible in stars like the Sun.

The meridional force balance can best be treated by considering the zonal component of the curl of Eq. (2). In a steady state, the dominant terms can be expressed as follows (MH11)

$$s\frac{\partial \Omega^2}{\partial z} = \frac{g}{rC_P}\frac{\partial \overline{S}}{\partial \theta} + \mathcal{G}, \tag{19}$$

where S is the specific entropy and $\mathcal{G}$ includes contributions from the meridional components of the Reynolds stress and Lorentz force

$$\mathcal{G} = \left\{ \nabla \times \left[\overline{(\nabla \times \mathbf{v}') \times \mathbf{v}'} + \overline{(4\pi\rho)^{-1} (\nabla \times \mathbf{B}) \times \mathbf{B}} \right] \right\} \cdot \hat{\boldsymbol{\phi}}. \tag{20}$$

As discussed further in Sects. 5.3 and 5.5.1 below, mean field models, convection simulations, and theoretical interpretation of helioseismic data all suggest that the turbulent stresses $\mathcal{G}$ in the deep solar CZ are insufficient to balance the inertial term on the left-hand-side (LHS) of Eq. (19), which can be estimated reliably from helioseismic rotational inversions (as in Fig. 6a). Although the word *consensus* is rarely applicable in science, it is safe to say that the current paradigm favors a balance between the Coriolis and baroclinic (thermal) terms throughout most of the deep CZ and tachocline (Kitchatinov and Rüdiger 1995; Elliott et al. 2000; Robinson and Chan 2001; Brun and Toomre 2002; Rempel 2005; Miesch et al. 2006; Balbus et al. 2009; Brun et al. 2011)

$$s\frac{\partial \Omega^2}{\partial z} = \frac{g}{rC_P}\frac{\partial \overline{S}}{\partial \theta} \quad \text{(deep CZ)}. \tag{21}$$

This is a solar manifestation of the concept of thermal wind balance (TWB) which is commonly encountered in geophysics (e.g. Pedlosky 1987). If the CZ were strictly adiabatic ($\partial \overline{S}/\partial\theta = 0$), Eq. (21) would reduce to the Taylor-Proudman theorem and would imply cylindrical Ω isosurfaces aligned with the rotation axis ($\partial\Omega/\partial z = 0$). Thus, the conical nature of the Ω surfaces inferred from helioseismology (Fig. 6a) are generally attributed to a poleward entropy gradient which offsets the Coriolis-induced meridional circulation cells (counter-clockwise in the northern hemisphere, hereafter NH, and clockwise in the southern hemisphere, hereafter SH) which would otherwise quickly establish $\partial\Omega/\partial z = 0$ (within a few rotation periods).

The existence of the NSSL is thought to be a "smoking gun" that signifies a change in the meridional force balance as well as a change in the convective angular momentum transport (MH11). There convective velocity amplitudes increase due to the low density, small scale heights and strong photospheric cooling, making turbulent stresses a major contributor to the meridional force balance. Though available solar data cannot rule out a baroclinic contribution (MH11), models suggest a balance between inertial and turbulent stresses:

$$s\frac{\partial \Omega^2}{\partial z} = \mathcal{G} \quad \text{(NSSL)}. \tag{22}$$

The balance requires that $\mathcal{G}$ oppose zonal vorticity gradients, consistent with the concept of a turbulent diffusion, as suggested by MH11 and as found in recent convection simulations by Hotta et al. (2014).

5.3 Mean Flows

5.3.1 Meridional Force Balance

To explore the consequences of the dynamical balances discussed in Sect. 5.2 on the establishment of mean flows in numerical models and in actual stars we begin with careful consideration of the meridional force balance (19). It is notable that, as it is written, this equation is degenerate in the meridional flow. In other words, though transient imbalances in the meridional component of the momentum equation are required to establish the meridional flow, the latter does not play an explicit role in the dynamical balance that is ultimately achieved. However, the differential rotation does appear explicitly in Eq. (19) and its profile is regulated by the nature of the meridional forcing.

To make this more explicit we can express the meridional flow in terms of a mass flux streamfunction Ψ:

$$r\sin\theta\,\overline{\rho v_r} = -\frac{\partial \Psi}{\partial \theta} \quad \text{and} \quad r\sin\theta\,\overline{\rho v_\theta} = \frac{\partial \Psi}{\partial r}. \tag{23}$$

The point is that Ψ does not appear explicitly in (19). However, the meridional flow does redistribute momentum and entropy so Ω, $\overline{S}$, and $\mathcal{G}$ all depend to some extent on Ψ. In many mean field models, the latter dependence is made explicit by introducing a parameterization for the turbulent stresses in terms of the meridional flow $\mathcal{G}(\Psi)$, the most common such parameterization being turbulent diffusion (Kitchatinov 2013; Dikpati 2014). However, in such approaches it must be kept in mind that the meridional flow that will result upon solving the steady-state Eq. (19) will depend sensitively on the parameterization $\mathcal{G}(\Psi)$ that is introduced. For example, if one represents $\mathcal{G}(\Psi)$ in terms of a depth-dependent eddy viscosity $\nu_t(r)$, the resulting Ψ profile will be sensitive to the depth dependence of ν_t (Dikpati 2014). More general formulations of $\mathcal{G}(\Psi)$ may include an anisotropic, inhomogeneous turbulent viscosity tensor or non-diffusive and even non-local turbulent transport. Further guidance on this is needed from convection models.

As discussed below (Sect. 5.3.2), a complete understanding of how the meridional flow is maintained requires careful consideration of the zonal force balance, Eq. (18). However, as noted in Sect. 5.2, the meridional force balance, Eq. (19), does play an essential role in determining the differential rotation profile $\Omega(r,\theta)$. This was emphasized in an important series of papers by Balbus and colleagues who consider Eq. (21) together with the ansatz that the specific entropy is approximately constant on Ω isosurfaces such that $\overline{S} = \overline{S}_r + f(\Omega^2)$ where $\overline{S}_r$ is the average of $\overline{S}$ over latitude. Balbus et al. (2009, 2012), Balbus and Schaan (2012). In this case the TWB equation becomes a hyperbolic equation that can be solved for $\Omega(r,\theta)$, given a specified boundary condition $\Omega(r_0,\theta)$. The results were quite promising; the inferred subsurface profile provides a good match to helioseismic rotation inversions in the mid CZ, reproducing the conical isosurfaces illustrated in Fig. 6a (away from the tachocline and NSSL). This provides support for the conjecture that Eq. (21) is indeed satisfied in the deep solar CZ.

Though ground-breaking, there are two important caveats to the Balbus et al. result. First, though it successfully reproduces the conical Ω isosurfaces, it does not explain the most fundamental feature of the solar differential rotation profile, namely that the equator spins faster than the poles. The TWB balance equation is degenerate in the sense that any cylindrical profile $\Omega_c(s)$ can be added to Ω without changing Eq. (21). In their original (2009) paper, Balbus et al. broke this degeneracy by imposing a solar-like Ω profile at the surface ($r = R$) and integrating downward.

The second caveat is the justification of the ansatz $\overline{S}' = f(\Omega^2)$, where $\overline{S}' = \overline{S} - S_r$ is the residual entropy. Balbus and Schaan (2012) attribute this to the tendency for shear to adjust the trajectory of convective plumes until they are aligned with Ω surfaces. If true, this would imply more efficient convective heat transport along Ω surfaces, thus homogenizing the entropy and justifying the ansatz. However, this leaves open the question of how convection can establish the differential rotation in the first place since this requires angular momentum transport across Ω surfaces[2] Furthermore, current convection simulations are inconclusive on the viability of this proposed mechanism.

This last point is illustrated in Fig. 7 which shows the enthalpy transport $\mathbf{F}_e = \overline{\rho} C_p T' \mathbf{v}'$, in a convection simulation done with the Anelastic Spherical Harmonic (ASH) code, averaged over longitude and time. Here C_p is the specific heat at constant pressure and primes indicate variations relative to horizontal averages. This particular simulation is the same as that used

[2]Note that Balbus and Schaan (2012) attribute the solar differential rotation to a meridional flow induced by centrifugal distortion of the base of the CZ. However, though this can influence the Ω profile, it cannot establish the equatorward Ω gradient observed in the Sun. For a discussion of this issue see the Appendix of Miesch et al. (2012).

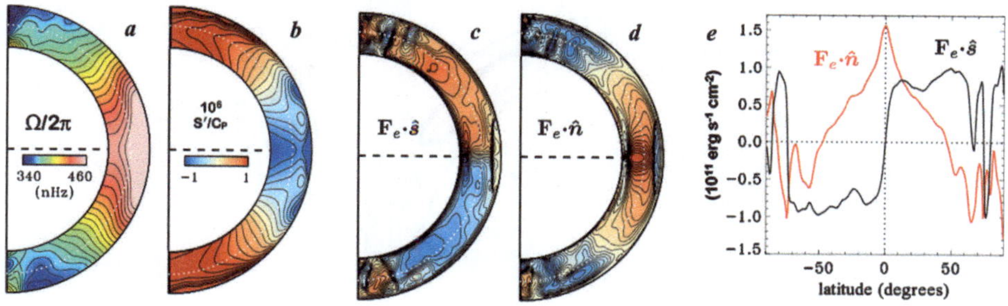

Fig. 7 Differential rotation, entropy profile, and convective heat flux in a solar convection simulation. Time-averaged (**a**) Ω and (**b**) residual entropy $\overline{S'}$ profiles are shown along with the convective enthalpy flux (**d**) along Ω surfaces $\mathbf{F}_e \cdot \hat{s}$ and (**e**) across Ω surfaces $\mathbf{F}_e \cdot \hat{\boldsymbol{n}}$. *Red* and *blue* denote positive and negative values respectively in frames **c** and **d**, with equivalent color tables in order to highlight relative amplitudes. Frame **e** shows a slice through the data shown in frames **c** and **d**, located at mid-CZ, as indicated by the *white dotted line*

by Weber et al. (2011) with further details provided by Miesch et al. (2006). Frames (c), (d) and (e) highlight the enthalpy flux along Ω isosurfaces, $\mathbf{F}_e \cdot \hat{s}$ and perpendicular to Ω isosurfaces $\mathbf{F}_e \cdot \hat{\boldsymbol{n}}$, where $\hat{\boldsymbol{n}}$ is parallel to $\nabla\Omega$ and $\hat{s} = \hat{\boldsymbol{n}} \times \hat{\boldsymbol{\phi}}$. The enthalpy flux is along Ω surfaces at mid-latitudes and across Ω surfaces at the equator (this is most apparent in frame (e)). However, the normal component $\mathbf{F}_e \cdot \hat{\boldsymbol{n}}$ is significant at all latitudes except where it crosses zero at a latitude of about ±50°. Though this is where the alignment of Ω and $\overline{S}'$ surfaces is best, it does not account for the alignment at lower latitudes. Furthermore, the vanishing of the normal component at mid-latitudes appears to have less to do with the alignment of turbulent plumes and more to do with the dominance of the radial component of $\mathbf{F}_e$ which is parallel and antiparallel to $\hat{\boldsymbol{n}}$ at low and high latitudes, respectively. Thus, although the Ω and $\overline{S}$ surfaces are indeed nearly coincident (frames (a), (b)), it may not be due to a shear-induced modification of the convective heat flux as proposed by Balbus and Schaan (2012). Other explanations for the baroclinic forcing in Eq. (21) such as a latitude-dependent convective heat transport and thermal coupling to the tachocline share similar uncertainties (Miesch and Toomre 2009).

In summary, though there are uncertainties about how the latitudinal entropy gradient in Eq. (21) is established, the meridional force balance can account for the conical nature of the solar Ω profile inferred from helioseismology (Fig. 6a). However, it cannot account for the equatorward Ω gradient or the steady meridional flow profile. For this we need the zonal force balance, which we now consider.

5.3.2 Angular Momentum Transport and Gyroscopic Pumping

The dynamical balance (18) derived from the zonal component of the momentum equation expresses the conservation of angular momentum. The key source term in this equation is the angular momentum transport by the convective Reynolds stress, which appears in the net torque term $\mathcal{F}$.

From this perspective, Eq. (18) can be seen as expressing a balance between the transport of angular momentum by convection and the advection of angular momentum by the meridional flow. Though this balance involves substantial nonlinear feedbacks, it can be instructive to consider the meridional flow as a response to the net torque $\mathcal{F}$ (MH11). This interpretation is made even more instructive by the nature of $\mathcal{L}$, which is well established by helioseismology and is illustrated in Fig. 6b. Recall that $\mathcal{L}$ includes a contribution from the

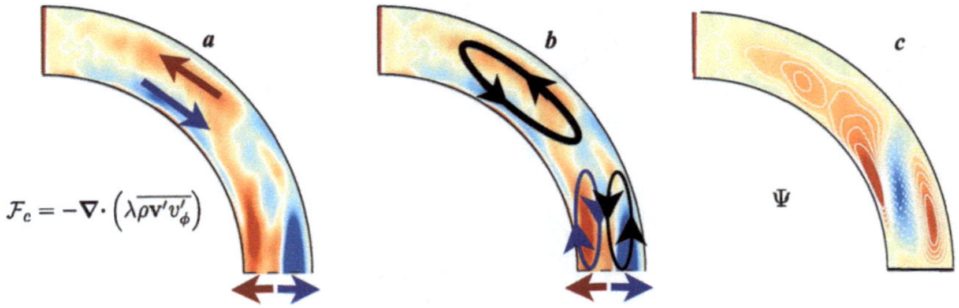

Fig. 8 Gyroscopic pumping in an ASH convection simulation, from Featherstone and Miesch (2014). (**a**) Negative divergence of the convective angular momentum flux averaged over longitude and time, making up the convective component of the net torque $\mathcal{F}$ (in this simulation there is also a viscous component). *Blue* indicates convergence and *red* divergence. *Arrows* indicate the direction of the induced meridional flow. The background image in frame **b** is the same as in **a** but here the induced MC is illustrated as closed circulation cells as required by mass conservation. (**c**) The actual MC in the simulation, averaged over longitude and time, displayed as contours of the mass flux streamfunction Ψ [see Eq. (23)]. *Blue* and *red* indicate clockwise and counter-clockwise circulation respectively. Only the NH is shown for clarity

uniform rotation component as well as the differential rotation. Thus, $\mathcal{L}$ surfaces are nearly cylindrical despite the conical nature of Ω surfaces (Fig. 6a). Furthermore, note that $\nabla\mathcal{L}$ is directed away from the rotation axis. We expect this to be true for all stars since an inward gradient of $\mathcal{L}$ toward the rotation axis is unstable to the Rayleigh criterion (Tassoul 1978).

The implication is that the MC is largely regulated by the net torque $\mathcal{F}$. A local convergence of the convective angular momentum flux ($\mathcal{F} > 0$) will induce a flow away from the rotation axis while a divergence ($\mathcal{F} < 0$) will induce a flow toward the rotation axis. This follows directly from the zonal force balance in Eq. (18) and the $\mathcal{L}$ profile shown in Fig. 6b. This induced flow is mediated by transient imbalances involving the $\partial\Omega^2/\partial z$ term in the meridional momentum Eq. (19) but the steady MC that is ultimately achieved is insensitive to the meridional force balance (Miesch and Hindman 2011; Miesch et al. 2012). This mechanism for establishing the meridional circulation is referred to as gyroscopic pumping.

Figure 8 demonstrates how gyroscopic pumping regulates the MC profile in a global convection simulation. Frame (a) shows the divergence of the convective Reynolds stress, which makes up the principle component of the net torque $\mathcal{F}$. This simulation does not have magnetism so there is no Lorentz force but $\mathcal{F}$ does include a minor contribution from viscous diffusion. The arrows indicate the direction of the meridional flow induced by gyroscopic pumping. Regions where the convective angular momentum flux converges ($\mathcal{F} > 0$, blue) induce a flow away from the rotation axis while (blue arrows) while regions of divergence ($\mathcal{F} < 0$, red) induce a flow toward the rotation axis. Mass conservation requires these flows to connect, forming closed circulation cells as illustrated in frame (b). The actual meridional flow achieved in the simulation (frame (c)) follows this pattern closely, with a single cell in radius at high latitudes and a double-cell structure in low latitudes.

The multi-cell structure at low latitudes arises from the direction of the angular momentum flux, which is cylindrically outward (directed away from the rotation axis). Since there is no flux through the boundaries (because there is no flow), this produces a divergence in the lower CZ and a convergence in the upper CZ that maintains the fast rotation at the equator. At high latitudes, a radially inward angular momentum transport establishes the reverse, with convergence and divergence in the lower and upper CZ respectively. The difference between high and low latitudes arises from the structure of the convection. The cylindrically outward flux at low latitudes is maintained by tilted *banana cells* while the radially inward flux at high latitudes is maintained by downward plumes on nearly ballistic trajectories. For further discussion see (Featherstone and Miesch 2014).

This simulation is missing an important element that is present in the Sun, namely the NSSL. Here vigorous small-scale convection is thought to establish a strong inward angular momentum transport as fast downflow plumes driven by photospheric cooling tend to conserve their angular momentum (MH11). This establishes a strong divergence of the Reynolds stress ($\mathcal{F} < 0$) which induces a poleward flow. Thus, the poleward meridional flow near the solar surface inferred from observations and helioseismic inversions (Sect. 5.1) is in part a boundary layer effect, driven by the same inward angular momentum flux that establishes the NSSL (MH11). This is demonstrated nicely in recent high-resolution convection simulations by Hotta et al. (2014) that capture much of the vast disparity of scales and rotational influence between surface convection and deep convection.

The radially inward angular momentum flux that establishes the NSSL must halt somewhere, implying a subsurface convergence that would induce a return equatorward flow, that might be quite shallow. Thus, a possible interpretation of the shallow return flows ($r \sim 0.91$–$0.95R$) inferred from recent helioseismic and surface measurements (Sect. 5.1) is that this is the penetration depth of plumes driven by photospheric cooling.

This example highlights the nonlocal nature of gyroscopic pumping. The global constraint of mass conservation implies that local extrema in the net torque $\mathcal{F}$ can influence the meridional flow profile throughout the CZ. This is particularly true of the NSSL since the MC induced by a localized zonal forcing tends to spread downward in a stratified layer (Haynes et al. 1991). The boundary layers also play a disproportionate role in maintaining the MC in the mean-field models of Kitchatinov (2013) because here is where TWB gives way to turbulent diffusion, breaking the Ψ degeneracy of the meridional force balance (see Sect. 5.3.1).

So what do convection simulations predict for the MC profile of the Sun? Is it single-celled (one dominant cell per hemisphere) or multi-celled? Though convection simulations provide valuable insight into this question, they cannot yet answer it conclusively. One reason for this is the sensitivity of the MC profile to the boundary layers and in particular the NSSL, which convection models are only beginning to capture. Another reason is uncertainty about the amplitude and structure of deep convection, which translates into uncertainty about the net torque $\mathcal{F}$. Yet more uncertainty surrounds the role of the Lorenz force (see Sect. 5.4).

In summary, convection simulations point to the convective angular momentum flux $\overline{s\rho\mathbf{v}'v'_\phi}$ as the most important factor in establishing mean flows. It regulates the meridional circulation profile through gyroscopic pumping and it is responsible for establishing the gross features of the Ω profile, including the fast equator/slow pole and the NSSL. However, the conical nature of the Ω surfaces and the existence of the NSSL depend critically on the meridional force balance as discussed in Sect. 5.3.1.

5.4 Cyclic Modulation of Large-Scale Flows

In the absence of other forcings the dynamical balances just described are expected to hold steadily on timescales larger than the convective turnover time, which is typically much shorter than the magnetic cycle period. The Lorentz force associated with the cycling large-scale magnetic field could then be viewed as a slow, periodic forcing of these fundamental dynamical balances, causing cyclic deviations about the mean equilibrium state. This is commonly considered the most likely driver behind observed solar torsional oscillations, and modulation of the meridional flow. Recent numerical simulations, however, are suggesting a more complex picture.

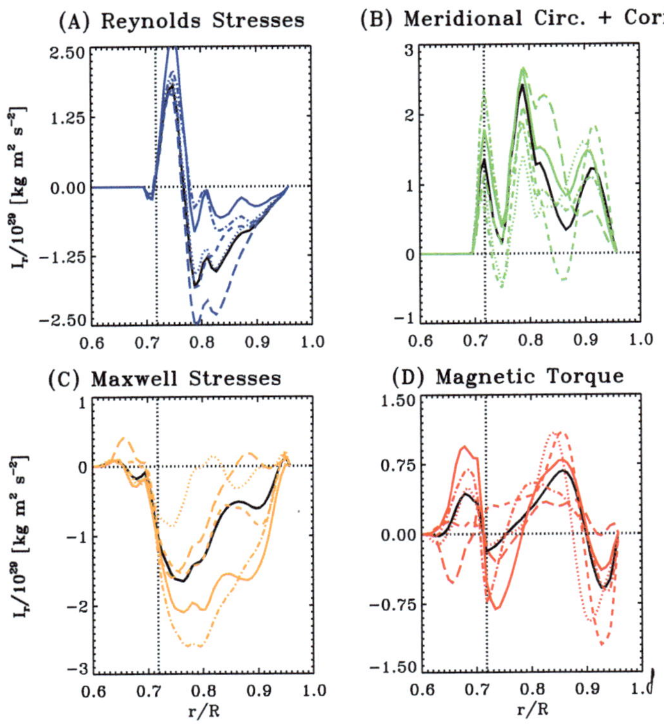

Fig. 9 Cyclic modulation of angular momentum transport in a EULAG-MHD simulation producing a cyclic large-scale magnetic field undergoing regular polarity reversals. The four panels show separately the four contributing mechanisms in this simulation, namely (**A**) Reynolds stresses, (**B**) transport by meridional flow, (**C**) small-scale (Maxwell) stresses, and (**D**) torque by the large-scale magnetic field. The various curves cover one 40-yr magnetic half-cycle, at a 9-yr cadence in the temporal order: *dot* → *short-dash* → *dash-dot* → *dash-triple-dot* → *long-dash*. The *solid lines* are the corresponding averages over the full simulation. Reproduced from Beaudoin et al. (2013; Fig. 7)

Figure 9, taken from Beaudoin et al. (2013), displays the time-varying angular momentum balance in a EULAG-MHD global numerical simulation producing large-scale magnetic fields undergoing regular polarity reversals (see Ghizaru et al. 2010; Racine et al. 2011; Passos and Charbonneau 2014). Each panel shows the radial angular momentum flux integrated on concentric spherical shell, separated in terms of the four dynamical contributions in this simulation: (A) Reynolds stresses, (B) transport by meridional flow, (C) Maxwell stresses, (D) large-scale torque (viz. Eq. (18)), without the viscous term and with the magnetic term split into small- and large-scales. In all cases the black solid line shows the average over the full simulation. The styled lines shows the corresponding profiles over a 40-yr-long magnetic half-cycle, extracted at a 9-yr cadence. All contributions, including the two purely hydrodynamical terms (A) and (B), undergo large cyclic variations on the same period as the large-scale magnetic cycle. The simple scenario proposed previously does not hold; the cycling large-scale magnetic fields alters convection, and thus all angular momentum transport mechanisms react accordingly; the driving of solar-like torsional oscillations in this simulation is a truly magnetohydrodynamical phenomenon.

The dynamical backreaction of the large-scale, cycling magnetic field is seldom included in FTD models (but do see Rempel 2006). Figure 9B suggests that the meridional flow is also undergoing significant magnetically-mediated modulation in the course of the cycle (also see

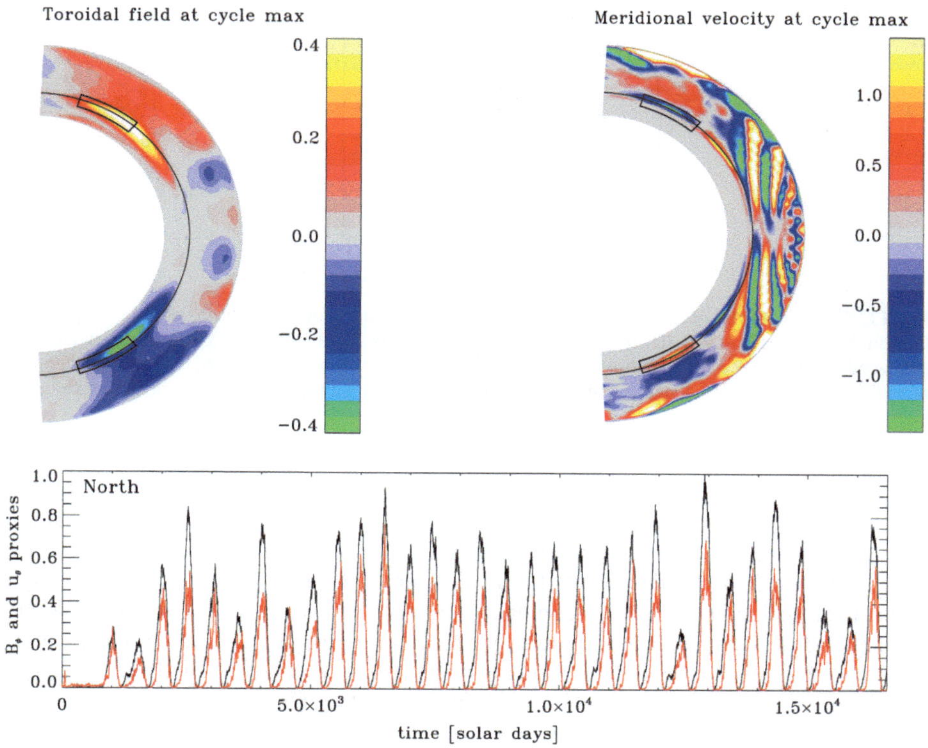

Fig. 10 Meridional cuts of the zonally-averaged (**A**) toroidal field, in Tesla, and (**B**) latitudinal flow speed, in $\mathrm{m\,s^{-1}}$, at an epoch of cycle maximum in the "millenium simulation" of Passos and Charbonneau (2014). The *dashed circular arc* indicates the base of the convecting layers. The *bottom panel* shows time series of the squared toroidal field (*black*) and latitudinal flow speed (*red*) averaged over the domain indicated by the *solid-line boxes* straddling the high-latitude core-envelope interface on the two meridional slices

Karak et al. 2014a). This is further detailed on Fig. 10. Panel A shows a meridional cut of the zonally-averaged magnetic field at time of cycle maximum. The accumulation of magnetic fields immediately beneath the base of the convecting layers (dashed circular arc), resulting from downward turbulent pumping therein, is readily apparent. Panel B shows a color coding of the θ-component of the zonally-averaged meridional flow. The vertically-elongated structures at low latitudes are a manifestation of the banana cells mentioned earlier. At mid- to high-latitudes, within the tangent cylinder, this simulations produces two meridional flow cells stacked in radius, the first with a shallow equatorward return flow, the deeper one with an equatorward flow running along the base of the convecting layer. Panel C shows time series of the squared toroidal magnetic field and latitudinal flow speed, averaged over a domain straddling the core-envelope interface, as indicated on panels A and B. The equatorward meridional flow speed at the base of the domain shows a clear cyclic modulation in phase and strongly correlated with the magnetic cycle.

The "millenium simulation" of Fig. 10 spans 32 cycles, enough to gather meaningful statistics on cycle-to-cycle variations in amplitude, duration, etc. (see Passos and Charbonneau 2014). Figure 11 is a correlation plot between individual cycle periods, and peak equatorward flow speed extracted from the time series on Fig. 10. These two quantities show a weak anticorrelation, meaning that cycles with larger u_θ have a shorter duration. At

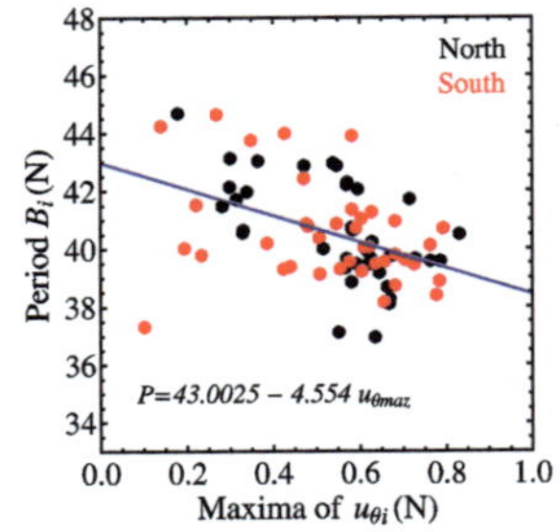

Fig. 11 Correlation plot of the cycle period versus peak latitudinal flow speed at the core-envelope interface, in the same EULAG-MHD simulation as on Fig. 10. The two hemispheres are analyzed separately, as *color*-coded. The weak anticorrelation is reminiscent of that expected in FTD models, but turns out to be an secondary dynamical consequence of the cycle period-amplitude relationship characterizing this simulation (see text)

first glance one may recognize here the pattern expected from kinematic FTD models (viz. Sect. 2.2 herein); but in fact, the causal link lies elsewhere. The cycles in this simulation are characterized by solar-like anticorrelation between period P and amplitude (see Passos and Charbonneau 2014); this results from the mode of dynamo action in this simulation. The anticorrelation on Fig. 11 is a simple consequence of stronger magnetic driving of the latitudinal flow during higher amplitude cycles. In other words, there is no direct causal link between u_θ and P, as in FTD models, where u_θ sets P; here the fundamental causal relationship is between cycle period and amplitude, and the anticorrelation between u_θ and P is a secondary dynamical consequence of the former. This illustrates well the dangers of indiscriminately extrapolating results from kinematic models into the dynamical regime.

5.5 Poloidal Field Generation and Convective Transport

In Sects. 5.3–5.4 we discussed the first two dynamo ingredients, namely the differential rotation and the meridional circulation. In this section we discuss the remaining two, namely the generation of the mean poloidal field and the transport of magnetic flux by convection, hereafter referred to as turbulent transport. As mentioned in Sect. 1, turbulent transport is usually modeled as a turbulent diffusion, as expressed by the β term in Eq. (6), but non-diffusive transport can also occur so turbulent diffusion and transport are related but not necessarily equivalent.

The principle mechanisms for generating and transporting magnetic fields can be loosely categorized into two principle physical processes: convection and flux emergence. Convection can generate mean fields through turbulent dynamo action (typically parameterized by means of turbulent α-effect) and can transport magnetic flux through turbulent transport. Meanwhile, flux emergence can generate magnetic fields through the BL mechanism and can transport magnetic flux (upward) through magnetic buoyancy. We address each process in turn, beginning with convection in Sect. 5.5.1.

5.5.1 Convective Transport and Dynamo Action

The dynamical balances (18) and (21) not only provide valuable insight into how mean flows are maintained but they can also be used to constrain the properties of solar convection based on the mean flows inferred from solar observations. In particular, since the net torque $\mathcal{F}$ scales as the convective velocity V_c^2, Eq. (18) can be used to estimate V_c based

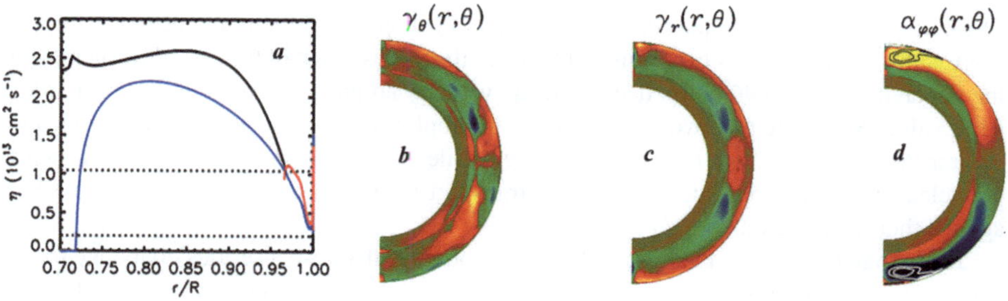

Fig. 12 Turbulent transport and field generation in simulations of solar convection. (**a**) Estimates of the turbulent diffusivity $\beta = v_c H_\rho/3$ from convection models, after Miesch et al. (2012). Included are an ASH simulation (*black line*, Weber et al. 2011), a MURaM surface convection simulation (*brown line*, Rempel 2011), and a surface convection simulation with the STAGGER code plus a downward extrapolation based on mixing length theory (*blue line*, Trampedach and Stein 2011). *Horizontal dotted lines* indicate the estimates given in the text for $r = 0.71R$ and $r = 0.95R$ based on Eq. (18) and mean flows inferred from helioseismology. (**b**) Latitudinal pumping γ_θ, (**c**) radial pumping γ_r, and (**d**) zonal diagonal component of the α tensor $\alpha_{\phi\phi}$ inferred from an MHD simulation of solar convection with the EULAG-MHD code (from Racine et al. (2011)). *Yellow/orange and blue/green* tones denote positive and negative values respectively, peaking at $\pm 3\ \mathrm{m\,s^{-1}}$

on helioseismic determinations of $\mathcal{L}$ and $\overline{\rho \mathbf{v}_m}$ as shown in Fig. 6 (Miesch et al. 2012). Results suggest $V_c > 30\ \mathrm{m\,s^{-1}}$ in the upper CZ ($r \sim 0.95R$) and $V_c > 8\ \mathrm{m\,s^{-1}}$ in the lower CZ ($r \sim 0.95R$). This favors the diffusion-dominated regime for FTD models discussed in Sect. 2.3, in which the transport of poloidal flux by convection dominates over advective transport by the meridional circulation. This estimate for V_c is consistent with solar convection simulations and mixing length theory but does not rely on them, providing an independent estimate of transport efficiency in the deep CZ.

However, though convection simulations are generally consistent with one another and consistent with the dynamical balances discussed in Sect. 5.2, several lines of evidence suggest that they may be overestimating the convective velocity amplitudes. The first comes from the simulations themselves. As continuing advances in supercomputing technology enable simulations in more extreme parameter regimes that more closely reflect the turbulent environment of the solar interior, the weakening rotational influence places the simulations in an unfavorable dynamical regime. This rotational influence can be quantified by the Rossby number $R_o = \omega_c/2\Omega_0$, where ω_c is the characteristic vorticity of the convection (e.g. an rms value). As the level of turbulence increases, ω_c generally increases, implying a commensurate increase in R_o. When R_o increases beyond a transitional value of about 0.3–1.0 (depending on the details of the simulations and on how R_o is defined), convection tends to conserve its angular momentum locally, producing an anti-solar differential rotation profile with fast poles and a slow equator (Gastine et al. 2013; Käpylä et al. 2014; Featherstone and Miesch 2014; Karak et al. 2014a). In order to avoid such anti-solar differential rotation profiles (poleward $\nabla\Omega$) in blatant conflict with observations, many modern simulations artificially enhance the rotational influence (lower R_o) by either increasing the rotation rate Ω_0 (e.g. Brown et al. 2011; Augustson et al. 2014) or by decreasing the energy flux carried by convection (e.g. Racine et al. 2011; Hotta et al. 2014).

Another line of evidence that convection simulations may be over-estimating convective velocity amplitudes comes from recent estimates of subsurface velocity fields from time-distance helioseismic inversions by Hanasoge et al. (2010, 2012). They were searching for signatures of convection in SDO/HMI data but they did not find it, placing upper limits on convective velocity amplitudes of only 1–10 $\mathrm{m\,s^{-1}}$ per spherical harmonic degree ℓ for motions with coherence times exceeding 24 hours in the radius range of 0.92–0.96R. Increasing the coherence time to 96 hours decreased the velocity amplitude further, to below 1 $\mathrm{m\,s^{-1}}$ per ℓ value. Summing this over values of $\ell < 60$ implies large-scale velocity amplitudes no greater than 5–15 $\mathrm{m\,s^{-1}}$ at $r \sim 0.96R$. If large-scale convection really is this weak, then it is unclear how it can provide the requisite transport to sustain the mean flows we observe against their own inertia.

Taken together, these modeling and observational results pose a serious challenge to our current understanding of solar convection. This is the *convection conundrum*: models appear to be over-estimating convective velocity amplitudes but helioseismic inferences may be under-estimating them. There are a few possible ways out of the first part of this conundrum. First, it may be that small scales that are not resolved by current global convection simulations carry much of the convective heat flux while larger scales are responsible for the angular momentum transport that sustains mean flows (Sect. 5.3). Second, magnetic fields may limit convective velocity amplitudes and convective transport through the Lorentz force (e.g. Munoz-Jaramillo et al. 2011; Karak et al. 2014b). Further work is needed to establish whether the resolution of the conundrum lies with one of these explanations or elsewhere.

To complicate the matter further, convective transport is almost certainly more complex than the simple isotropic turbulent diffusion coefficients typically used in kinematic dynamo models. This is clear even from kinematic mean-field theory in which the α and β coefficients in Eq. (5) are more generally tensor quantities that may be anisotropic and inhomogeneous (e.g., Karak et al. 2014b). The antisymmetric off-diagonal components of the α tensor give a contribution to $\mathcal{E}$ that may be written as $\boldsymbol{\gamma} \times \mathbf{B}$, where $\boldsymbol{\gamma}$ is a pumping velocity that is functionally equivalent to an axisymmetric flow (viz. Sect. 2.3 herein).

Estimates of the meridional components of $\boldsymbol{\gamma}$ from an EULAG-MHD convection simulation are shown in Fig. 12b and c (from Racine et al. 2011). Typical amplitudes are about 1 $\mathrm{m\,s^{-1}}$, comparable to the meridional flow speeds. They transport flux equatorward and downward, contributing to the magnetic cycles that are achieved in this convective dynamo simulation. Though the representation of $\mathcal{E}$ in terms of α, β, and $\boldsymbol{\gamma}$ relies on assumptions that are not necessarily satisfied in convective dynamos (most notably kinematic induction and scale separation), the inferred mean-field coefficients generally provide a good depiction of the transport and field generation in this simulation. This was verified by running a kinematic mean-field models based on these parameterizations that was found to reproduce the gross features of the full MHD simulation (Simard et al. 2013).

In addition to transporting magnetic flux, convection can also generate magnetic flux as noted in Sects. 1 and 5.1. From the perspective of sustaining a mean-field dynamo, the most important manifestation of this is the $\phi - \phi$ component of the α tensor, $\alpha_{\phi\phi}$. This is the component of the turbulent α-effect that converts mean toroidal field to mean poloidal field. This too can be estimated from convective dynamo simulations, as illustrated in Fig. 12d.

This estimate of the turbulent α-effect highlights an important point that will be discussed further in Sects. 5.5.2. Namely, though most current mean-field models rely on the BL mechanism, MHD simulations of convective dynamos operate just fine without it. They produce magnetic activity cycles on decadal time scales in which mean poloidal field is generated by convection, with magnetic buoyancy and flux emergence playing little or no role. Miesch and Brown (2012) estimated that the BL mechanism would have to be very strong, more than an order of magnitude stronger than in current FTD models, in order to dominate over convective field generation.

However, we emphasize that turbulent large-scale dynamo action is still poorly understood. Turbulent field generation depends sensitively on a number of subtle issues including

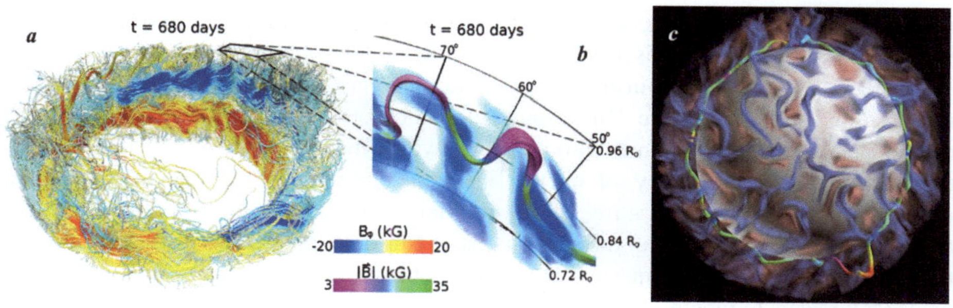

Fig. 13 Flux emergence in convection simulations. (**a**, **b**) Spontaneous formation of coherent rising toroidal flux loops in a convective dynamo simulation, illustrated by 3D renderings of magnetic field lines (from Nelson et al. (2013b)). The view in panel (**a**) is from a point above the surface of the star in the NH looking toward the equator, where coherent toroidal flux structures have formed. Field lines are *colored* by the strength and direction of B_ϕ (see *legend*). The zoomed-in view in panel (**b**) focuses on selected field lines in the core of a particular rising loop, with *color* indicating the total field strength (see *legend*). (**c**) A thin magnetic flux tube embedded in a convective layer, from (Weber et al. (2011), visualization by T. Sandstrom and C. Henze, NASA Advanced Supercomputing Division). The view is from above the north pole, with the tube located near the equator (latitude approximately 15°). Colors along the tube indicate density, from low (*red/orange*) to high (*blue*) while background colors indicate convective downflows (*blue*) and upflows (*red*)

dynamical quenching associated with the small-scale accumulation of magnetic helicity, boundary conditions, and the nature of the artificial dissipation, be it an explicit subgrid-scale model or an implicit numerical diffusion (see reviews by Brandenburg and Subramanian 2005; Miesch 2012; Tobias et al. 2013; Charbonneau 2013). Furthermore, though extracting mean-field coefficients from convective dynamo models is an auspicious endeavor for the future, there are times when simple mean-field prescriptions may not provide a reliable depiction of the convective field generation and transport processes (e.g. Brown et al. 2010). We should indeed move forward, but with proper caution.

5.5.2 Flux Emergence

Although MHD simulations have provided much insight into the rise of buoyant magnetic flux tubes (see Fan 2009, for a review), they are only beginning to address how such structures form within the context of a global convective dynamo. Figure 13(a)–(b) highlights a recent convection dynamo simulation that spontaneously produces rising magnetic loops reminiscent of the BMRs (see also Nelson et al. 2013a; Nelson and Miesch 2014). A remarkable feature of this simulation is that it does not possess a tachocline; these loops are generated as part of coherent toroidal flux systems that pervade the CZ and rise through a combination of buoyancy and advection by the surrounding convection.

This result challenges long-held paradigms of how the buoyant magnetic flux tubes that produce BMRs are generated. It is often argued that such tubes form in the tachocline where the strong rotational shear and subadiabatic stratification promote the generation and storage of strong toroidal flux systems (Galloway and Weiss 1981; van Ballegooijen 1982). However, a well-known difficulty with this paradigm is the challenge of generating super-equipartition tubes that can survive the journey through the CZ (Rempel and Schüssler 2001; Fan 2009; Vasil and Brummell 2009). In the simulation of Fig. 13(a)–(b) this issue does not arise because convection plays an intimate role in both the formation and the rise of the loops; it is not merely a destructive influence as in the traditional paradigm. Here the loops are generated through the coupled action of rotational shear and turbulent intermittency and rise through the coordinated action of magnetic buoyancy and convective transport (see Nelson et al. 2013b, for further discussion).

Similar results are found in the idealized flux emergence simulations of Weber et al. (2011), illustrated in Fig. 13c. These simulations do not address tube formation. Rather, they embed an idealized magnetic flux tube in a hydrodynamic simulation of solar convection and follow its evolution by solving the 1D thin flux tube equations. A notable advantage of such an approach is that it is free of the artificial magnetic diffusion that afflicts convective dynamo simulations. As in the simulations of Nelson et al. (2013b), convection works together with magnetic buoyancy to get magnetic flux to the surface, shortening emergence times (from years to months in some cases) and promoting patterns seen in observations of BMRs such as tilt angles and emergence latitudes that have bearing on the BL mechanism.

These modeling results suggest that much of the emerging flux we observe on the solar surface may originate in the CZ rather than the tachocline. Though the loops generated in the CZ by convection and shear may be relatively diffuse as illustrated in Fig. 13a–b, positive feedbacks arising from radiative cooling in the photosphere (Cheung et al. 2010; Stein and Nordlund 2012) and/or the so-called negative magnetic pressure instability (Rogachevskii and Kleeorin 2007; Brandenburg et al. 2010, 2012; Kemel et al. 2013) can also help to organize relatively diffuse emergent flux structures into coherent sunspot-like features. This idea that there may be coherent toroidal flux systems permeating the CZ is consistent with the observational result that even smaller BMRs known as ephemeral regions (ERs) exhibit a preferred orientation consistent with Hale's polarity rules (Tlatov et al. 2010). Since these toroidal flux systems are not necessarily axisymmetric, it may also account for why some BMRs (including about 4 % of large active regions) that violate Hale's polarity rules occur at the same latitudes as others that obey the rules (Stenflo and Kosovichev 2012).

Though these results are intriguing, FTD models typically rely on the tachocline and lower CZ as a likely source for much of the magnetic flux that emerges through the surface, particularly for large active regions. Convective dynamo simulations that include a tachocline often exhibit toroidal flux accumulation there, particularly at intermediate Rossby numbers where turbulent pumping time scales are shorter than shear generation time scales (Browning et al. 2006; Ghizaru et al. 2010). Recent MHD simulations by Barker et al. (2012) suggest that magnetic pumping can be conducive to the formation of buoyant flux structures near the interface between a stable zone and an overlying convection zone. A promising means to investigate whether or not a tachocline is required to establish cyclic magnetic activity is to search for magnetic cycles in fully convective low-mass stars that do not possess tachoclines (Cincunegui et al. 2007).

6 Conclusion

Over the past few decades, kinematic mean-field and FTD models have developed in parallel to MHD simulations, unfortunately with relatively little cross-fertilization. On the one hand, because of their relatively modest computational requirements mean-field dynamo models, such as FTD, are likely to remain the workhorse of investigations regarding solar cycle prediction, and the modulation of solar activity on long timescales, including Grand Minima. They also offer a computational framework within which to investigate efficiently the joint action of turbulence and flux emergence, albeit in simplified form. This can provide useful guidance for the design of MHD numerical simulation of the solar cycle, in which systematic exploration of parameter space is impractical. Numerical simulations, on the other hand, can provide quantitative constraints on key transport mechanisms and source terms currently implemented in a largely ad hoc manner in most kinematic models. They can also provide crucial information on the dynamics of large-scale flows important for dynamo action, which is inaccessible to kinematic models by their very design. Cross-fertilization is mutually beneficial, and this review has attempted to identify areas where this could be most productive in the context of FTD models.

A true unification of FTD models and convective dynamo simulations can only come about through an improved understanding of flux emergence. Elucidating how convective dynamos produce emergent toroidal flux structures through the combined effects of rotational shear, magnetic buoyancy, and turbulent transport is a prerequisite to clarifying the role of flux emergence in the origin of magnetic cycles. Does flux emergence play an essential role in the operation of the solar dynamo, as in FTD models that rely on the Babcock-Leighton mechanism, or is it merely a superficial by-product of deeper-seated dynamics, as suggested by the presence of magnetic cycles in current MHD convection simulations lacking emerging sunspots and active region?

From the perspective of the four dynamo ingredients identified in Sect. 5.1, this is perhaps the most pressing. We can summarize the central issue with a more general restatement of the BL paradigm; How can the process of flux emergence promote the generation of a mean poloidal field? Though MHD simulations may not achieve a comprehensive depiction of the solar dynamo in the next decade, they can and will provide new insights into this central question, together with increasingly sophisticated FTD models and continuing observational scrutiny of solar magnetism and large-scale flows. Progress on this front has recently been reported by Yeates and Muñoz-Jaramillo (2013) who present a novel 3D kinematic FTD model with an algorithm for flux emergence inspired by MHD simulations and calibrated with solar observations.

The ideal testbed for investigating the relative roles of the BL mechanism and convective field generation (as modeled by the turbulent α-effect) may be Grand Minima (Sect. 3.3). In particular, convective field generation may account for the residual cyclic activity during the Maunder Minimum (Beer et al. 1998) and may provide a mechanism for pulling the Sun out of a Grand Minimum.

Other areas where the synergy between MHD simulations, FTD models and solar observations may potentially yield transformative insights in the coming decade include magnetic flux transport (Sects. 2.3, 5.5.1), the more general manifestations of solar cycle variability (including but not limited to grand minima and grand maxima; Sects. 3, 4), and the linkages between large and small-scale dynamo action.

Acknowledgements We wish to thank ISSI and its staff for their hospitality and a most stimulating workshop in November 2013. J.J. acknowledges the financial support by the National Natural Science Foundations of China (No. 11173033). M.S.M. is supported in part through grants MMH09AK14I, NNX11AJ36G, and NNX13AG18G. The National Center for Atmospheric Research is sponsored by the National Science Foundation of the U.S.A. P.C. is supported by the National Sciences and Engineering Research Council of Canada. The research of A.R.C. is supported by a JC Bose Fellowship awarded by the Department of Science and Technology of India.

References

K.C. Augustson, A.S. Brun, M.S. Miesch, J. Toomre, Cycling dynamo in a young sun: grand minima and equatorward propagation. Astrophys. J. (2014 submitted)

H.W. Babcock, The topology of the sun's magnetic field and the 22-year cycle. Astrophys. J. **133**, 572 (1961)

S.A. Balbus, E. Schaan, The stability of stratified, rotating systems and the generation of vorticity in the sun. Mon. Not. R. Astron. Soc. **426**, 1546–1557 (2012)

S.A. Balbus, J. Bonart, H.N. Latter, N.O. Weiss, Differential rotation and convection in the sun. Mon. Not. R. Astron. Soc. **400**, 176–182 (2009)
S.A. Balbus, H. Latter, N. Weiss, Global model of differential rotation in the sun. Mon. Not. R. Astron. Soc. **420**, 2457–2466 (2012)
T. Baranyi, L. Gyori, A. Ludmány, H.E. Coffey, Comparison of sunspot area data bases. Mon. Not. R. Astron. Soc. **323**, 223–230 (2001)
A.J. Barker, L.J. Silvers, M. Proctor, N. Weiss, Magnetic buoyancy instabilities in the presence of magnetic flux pumping at the base of the solar convection zone. Mon. Not. R. Astron. Soc. **424**, 115–127 (2012)
I. Baumann, D. Schmitt, M. Schüssler, S.K. Solanki, Evolution of the large-scale magnetic field on the solar surface: a parameter study. Astron. Astrophys. **426**, 1075–1091 (2004)
P. Beaudoin, P. Charbonneau, E. Racine, P.K. Smolarkiewicz, Torsional oscillations in a global solar dynamo. Sol. Phys. **282**, 335–360 (2013)
J. Beer, S. Tobias, N.O. Weiss, An active sun throughout the Maunder minimum. Sol. Phys. **181**, 237–249 (1998)
A. Brandenburg, K. Subramanian, Astrophysical magnetic fields and nonlinear dynamo theory Phys. Rep. **417**, 1–209 (2005)
A. Brandenburg, F. Krause, R. Meinel, D. Moss, I. Tuominen, The stability of nonlinear dynamos and the limited role of kinematic growth rates. Astron. Astrophys. **213**, 411–422 (1989)
A. Brandenburg, D. Moss, I. Tuominen, Turbulent pumping in the solar dynamo, in *The Solar Cycle*, vol. 27 (1992), p. 536
A. Brandenburg, N. Kleeorin, I. Rogachevskii, Large-scale magnetic flux concentrations from turbulent stresses. Astron. Nachr. **331**, 5–13 (2010)
A. Brandenburg, K. Koen, N. Kleeorin, I. Rogachevskii, The negative effective magnetic pressure in stratified forced turbulence. Astrophys. J. **749**, 179 (2012)
B.P. Brown, M.K. Browning, A.S. Brun, M.S. Miesch, J. Toomre, Persistent magnetic wreathes in a rapidly rotating sun. Astrophys. J. **711**, 424–438 (2010)
B.P. Brown, M.S. Miesch, M.K. Browning, A.S. Brun, J. Toomre, Magnetic cycles in a convective dynamo simulation of a young solar-type star. Astrophys. J. **731**, 69 (2011)
M.K. Browning, M.S. Miesch, A.S. Brun, J. Toomre, Dynamo action in the solar convection zone and tachocline: pumping and organization of toroidal fields. Astrophys. J. Lett. **648**, L157–L160 (2006)
A.S. Brun, J. Toomre, Turbulent convection under the influence of rotation: sustaining a strong differential rotation. Astrophys. J. **570**, 865–885 (2002)
A.S. Brun, M.S. Miesch, J. Toomre, Modeling the dynamical coupling of the solar convection zone to the radiative interior. Astrophys. J. **742**, 79 (2011)
P. Caligari, F. Moreno-Insertis, M. Schüssler, Emerging flux tubes in the solar convection zone. 1: Asymmetry, tilt, and emergence latitude. Astrophys. J. **441**, 886–902 (1995)
R. Cameron, M. Schüssler, Solar cycle prediction using precursors and flux transport models. Astrophys. J. **659**, 801–811 (2007)
R.H. Cameron, D. Schmitt, J. Jiang, E. Işık, Surface flux evolution constraints for flux transport dynamos. Astron. Astrophys. **542**, A127 (2012)
R.H. Cameron, M. Dasi-Espuig, J. Jiang, E. Işık, D. Schmitt, M. Schüssler, Limits to solar cycle predictability: cross-equatorial flux plumes. Astron. Astrophys. **557**, A141 (2013)
S. Chakraborty, P. Chatterjee, A.R. Choudhuri, Why does the sun's torsional oscillation begin before the sunspot cycle? Phys. Rev. Lett. **102**, 041102 (2009)
P. Charbonneau, Multiperiodicity, chaos and intermittency in a reduced model of the solar cycle. Sol. Phys. **199**, 385 (2001)
P. Charbonneau, Dynamo models of the solar cycle. Living Rev. Sol. Phys. **7**, 3 (2010)
P. Charbonneau, Where is the solar dynamo? J. Phys. Conf. Ser. **440**, 012014 (2013)
P. Charbonneau, M. Dikpati, Stochastic fluctuations in a Babcock-Leighton model of the solar cycle. Astrophys. J. **543**, 1027–1043 (2000)
P. Charbonneau, G. Blais-Laurier, C. St-Jean, Intermittency and phase persistence in a Babcock-Leighton model of the solar cycle. Astrophys. J. **616**, L183–L186 (2004)
P. Charbonneau, C. St-Jean, P. Zacharias, Fluctuations in Babcock-Leighton dynamos. I. Period doubling and transition to chaos. Astrophys. J. **619**, 613–622 (2005)
P. Charbonneau, G. Beaubien, C. St-Jean, Fluctuations in Babcock-Leighton dynamos. II. Revisiting the Gnevyshev-Ohl rule. Astrophys. J. **658**, 657–662 (2007)
P. Chatterjee, A.R. Choudhuri, On magnetic coupling between the two hemispheres in solar dynamo models. Sol. Phys. **239**, 29–39 (2006)
P. Chatterjee, D. Nandy, A.R. Choudhuri, Full-sphere simulations of a circulation-dominated solar dynamo: exploring the parity issue. Astron. Astrophys. **427**, 1019–1030 (2004)

M. Cheung, M. Rempel, A.M. Title, M. Schüssler, Simulation of the formation of a solar active region. Astrophys. J. **720**, 233–244 (2010)

D.-Y. Chou, D.-C. Dai, Solar cycle variations of subsurface meridional flows in the sun. Astrophys. J. **559**, L175–L178 (2001)

A.R. Choudhuri, The evolution of loop structures in flux rings within the solar convection zone. Sol. Phys. **123**, 217–239 (1989)

A.R. Choudhuri, A correction to Spruit's equation for the dynamics of thin flux tubes. Astron. Astrophys. **239**, 335–339 (1990)

A.R. Choudhuri, Stochastic fluctuations of the solar dynamo. Sol. Phys. **253**, 277–285 (1992)

A.R. Choudhuri, *The Physics of Fluids and Plasmas: An Introduction for Astrophysicists* (Cambridge University Press, Cambridge, 1998). QB466.F58 C46

A.R. Choudhuri, On the connection between mean field dynamo theory and flux tubes. Sol. Phys. **215**, 31–55 (2003)

A.R. Choudhuri, The origin of the solar magnetic cycle. Pramana **77**, 77–96 (2011)

A.R. Choudhuri, P.A. Gilman, The influence of the Coriolis force on flux tubes rising through the solar convection zone. Astrophys. J. **316**, 788–800 (1987)

A.R. Choudhuri, B.B. Karak, A possible explanation of the Maunder minimum from a flux transport dynamo model. Res. Astron. Astrophys. **9**, 953–958 (2009)

A.R. Choudhuri, B.B. Karak, Origin of grand minima in sunspot cycles. Phys. Rev. Lett. **109**, 171103 (2012)

A.R. Choudhuri, M. Schüssler, M. Dikpati, The solar dynamo with meridional circulation. Astron. Astrophys. **303**, L29 (1995)

A.R. Choudhuri, D. Nandy, P. Chatterjee, Reply to the comments of Dikpati et al.. Astron. Astrophys. **437**, 703–704 (2005)

A.R. Choudhuri, P. Chatterjee, J. Jiang, Predicting solar cycle 24 with a solar dynamo model. Phys. Rev. Lett. **98**, 131103 (2007)

J. Christensen-Dalsgaard, Helioseismology. Rev. Mod. Phys. **74**, 1073–1129 (2002)

C. Cincunegui, R.F. Diaz, P.J.D. Mauas, A possible activity cycle in Proxima Centauri. Astron. Astrophys. **461**, 1107–1113 (2007)

M. Dasi-Espuig, S.K. Solanki, N.A. Krivova, R. Cameron, K. Peñuela, Sunspot group tilt angles and the strength of the solar cycle. Astron. Astrophys. **518**, A7 (2010)

M. Dikpati, Generating the suns global meridional circulation from differential rotation and turbulent Reynolds stresses. Astrophys. J. **438**, 2380–2394 (2014)

M. Dikpati, P. Charbonneau, A Babcock-Leighton flux transport dynamo with solar-like differential rotation. Astron. Astrophys. **518**, 508–520 (1999)

M. Dikpati, A.R. Choudhuri, On the large-scale diffuse magnetic field of the sun. Sol. Phys. **161**, 9–27 (1995)

M. Dikpati, P.A. Gilman, Simulating and predicting solar cycles using a flux-transport dynamo. Astrophys. J. **649**, 498–514 (2006)

M. Dikpati, G. de Toma, P.A. Gilman, C.N. Arge, O.R. White, Diagnostics of polar field reversal in solar cycle 23 using a flux transport dynamo model. Astrophys. J. **601**, 1136–1151 (2004)

M. Dikpati, G. de Toma, P.A. Gilman, Predicting the strength of solar cycle 24 using a flux-transport dynamo-based tool. Geophys. Res. Lett. **33**, 5102 (2006)

S. D'Silva, A.R. Choudhuri, A theoretical model for tilts of bipolar magnetic regions. Astron. Astrophys. **272**, 621 (1993)

B.R. Durney, On a Babcock-Leighton dynamo model with a deep-seated generating layer for the toroidal magnetic field. Sol. Phys. **160**, 213–235 (1995)

B.R. Durney, On a Babcock-Leighton solar dynamo model with a deep-seated generating layer for the toroidal magnetic field. IV. Astrophys. J. **724**, 1065 (1997)

B.R. Durney, On the differences between odd and even solar cycles. Sol. Phys. **196**, 421 (2000)

J.R. Elliott, M.S. Miesch, J. Toomre, Turbulent solar convection and its coupling with rotation: the effect of Prandtl number and thermal boundary conditions on the resulting differential rotation. Astrophys. J. **533**, 546–556 (2000)

Y. Fan, Magnetic fields in the solar convection zone. Living Rev. Sol. Phys. **6**, 4 (2009). http://www.livingreviews.org/lrsp-2009-4

Y. Fan, F. Fang, A simulation of convective dynamo in the solar convective envelope: maintenance of the solar-like differential rotation and emerging flux. Astrophys. J. **789**, 35 (2014)

Y. Fan, G.H. Fisher, E.E. Deluca, The origin of morphological asymmetries in bipolar active regions. Astrophys. J. **405**, 390–401 (1993)

N.A. Featherstone, M.S. Miesch, Meridional circulation in solar and stellar convection zones. Astrophys. J. (2014 submitted)

A. Ferriz-Mas, M. Schüssler, Instabilities of magnetic flux tubes in a stellar convection zone II. Flux rings outside the equatorial plane. Geophys. Astrophys. Fluid Dyn. **81**, 233–265 (1995)

P. Foukal, An explanation of the differences between the sunspot area scales of the Royal Greenwich and Mt. Wilson Observatories, and the SOON program. Sol. Phys. **289**(5), 1517–1529 (2014)
D.J. Galloway, N.O. Weiss, Convection and magnetic fields in stars. Astrophys. J. **243**, 945–953 (1981)
T. Gastine, J. Wicht, J.M. Aurnou, Zonal flow regimes in rotating anelastic spherical shells: an application to giant planets. Icarus **225**, 156–172 (2013)
M. Ghizaru, P. Charbonneau, P.K. Smolarkiewicz, Magnetic cycles in global large-eddy simulations of solar convection. Astrophys. J. Lett. **715**, L133–L137 (2010)
P.A. Gilman, Dynamically consistent nonlinear dynamos driven by convection in a rotating spherical shell. II—Dynamos with cycles and strong feedbacks. Astrophys. J. Suppl. Ser. **53**, 243–268 (1983)
P.A. Gilman, M. Miesch, Limits to penetration of meridional circulation below the solar convection zone. Astrophys. J. **611**, 568–574 (2004)
L. Gizon, Helioseismology of time-varying flows through the solar cycle. Sol. Phys. **224**, 217–228 (2004)
G.A. Glatzmaier, Numerical simulations of stellar convective dynamos. I. The model and method. J. Comput. Phys. **55**, 461–484 (1984)
G.A. Glatzmaier, Numerical simulations of stellar convective dynamos. II. Field propagation in the convection zone. Astrophys. J. **291**, 300–307 (1985)
G.A. Glatzmaier, P.H. Roberts, A three-dimensional convective dynamo solution with rotating and finitely conducting inner core and mantle. Phys. Earth Planet. Inter. **91**, 63–75 (1995)
A. Goel, A.R. Choudhuri, The hemispheric asymmetry of solar activity during the last century and the solar dynamo. Res. Astron. Astrophys. **9**, 115–126 (2009)
I. González-Hernandez, R. Komm, F. Hill, R. Howe, T. Corbard, D.A. Haber, Meridional circulation variability from large-aperture ring-diagram analysis of global oscillation network group and Michelson Doppler Imager data. Astrophys. J. **638**, 576–583 (2006)
G.A. Guerrero, E.M. de Gouveia Dal Pino, Turbulent magnetic pumping in a Babcock-Leighton solar dynamo model. Astron. Astrophys. **485**, 267–273 (2008)
G.A. Guerrero, J.D. Muñoz, Kinematic solar dynamo models with a deep meridional flow. Mon. Not. R. Astron. Soc. **350**, 317–322 (2004)
S.M. Hanasoge, T. Duvall, M.L. DeRosa, Seismic constraints on interior solar convection. Astrophys. J. Lett. **712**, L98–L102 (2010)
S.M. Hanasoge, T. Duvall, K.R. Sreenivasan, Anomalously weak solar convection. Proc. Natl. Acad. Sci. USA (2012). doi:10.1073/pnas.1206570109
D.H. Hathaway, The solar cycle. Living Rev. Sol. Phys. **7**, 1 (2010)
D.H. Hathaway, Supergranules as probes of the sun's meridional circulation. Astrophys. J. **760**, 84 (2012), 6pp.
D.H. Hathaway, L. Rightmire, Variations in the sun's meridional flow over a solar cycle. Science **327**, 1350 (2010)
D.H. Hathaway, L. Rightmire, Variations in the axisymmetric transport of magnetic elements on the sun: 1996–2010. Astrophys. J. **729**, 80 (2011)
P.H. Haynes, C.J. Marks, M.E. McIntyre, T.G. Shepherd, K.P. Shine, On the downward control of extratropical diabatic circulations by eddy-induced mean zonal flows. J. Atmos. Sci. **48**, 651–678 (1991)
G. Hazra, B.B. Karak, A.R. Choudhuri, Is a deep one-cell meridional circulation essential for the flux transport solar dynamo? Astrophys. J. **782**, 93 (2014)
H. Hotta, T. Yokoyama, Solar parity issue with flux-transport dynamo. Astrophys. J. **714**, L308–L312 (2010)
H. Hotta, M. Rempel, T. Yokoyama, High-resolution calculation of the solar global convection with the reduced sound speed technique: II. Near surface shear layer with the rotation. Astrophys. J. **786**, 24 (2014)
R. Howe, Solar interior rotation and its variation. Living Rev. Sol. Phys. **6**, 1 (2009). http://www.livingreviews.org/lrsp-2009-1
R. Howe, J. Christensen-Dalsgaard, F. Hill, R.W. Komm et al., Dynamic variations at the base of the solar convection zone. Science **287**, 2456–2460 (2000)
P. Hoyng, Turbulent transport of magnetic fields. III. Stochastic excitation of global magnetic modes. Astrophys. J. **332**, 857–871 (1988)
P. Hoyng, Helicity fluctuations in mean field theory: an explanation for the variability of the solar cycle? Astron. Astrophys. **272**, 321 (1993)
J. Jiang, P. Chatterjee, A.R. Choudhuri, Solar activity forecast with a dynamo model. Mon. Not. R. Astron. Soc. **381**, 1527–1542 (2007)
J. Jiang, R.H. Cameron, D. Schmitt, E. Işık, Modeling solar cycles 15 to 21 using a flux transport dynamo. Astron. Astrophys. **553**, A128 (2013)
J. Jiang, R.H. Cameron, M. Schüssler, Effects of the scatter in sunspot group tilt angles on the large-scale magnetic field at the solar surface. Astrophys. J. **791**, 5 (2014a)

J. Jiang, D.H. Hathaway, R.H. Cameron, S.K. Solanki, L. Upton, Magnetic flux transport at the solar surface. Space Sci. Rev. (2014b). doi:10.1007/s11214-014-0083-1
L. Jouve, A.S. Brun, On the role of meridional flows in flux transport dynamo models. Astron. Astrophys. **474**, 239–250 (2007)
P.J. Käpyla, M.J. Mantere, A. Brandenburg, Cyclic magnetic activity due to turbulent convection in spherical wedge geometry. Astrophys. J. **755**, L22 (2012)
P.J. Käpylä, M.J. Mantere, E. Cole, J. Warnecke, A. Brandenburg, Effects of enhanced stratification on equatorward dynamo wave propagation. Astrophys. J. **778**, 41 (2013)
P.J. Käpylä, M. Mantere, A. Brandenburg, Confirmation of bistable stellar differential rotation profiles. Astron. Astrophys. (2014 accepted)
B.B. Karak, Importance of meridional circulation in flux transport dynamo: the possibility of a Maunder-like grand minimum. Astrophys. J. **724**, 1021–1029 (2010)
B.B. Karak, A.R. Choudhuri, The Waldmeier effect and the flux transport solar dynamo. Mon. Not. R. Astron. Soc. **410**, 1503–1512 (2011)
B.B. Karak, A.R. Choudhuri, Quenching of meridional circulation in flux transport dynamo models. Sol. Phys. **278**, 137–148 (2012)
B.B. Karak, A.R. Choudhuri, Studies of grand minima in sunspot cycles by using a flux transport solar dynamo model. Res. Astron. Astrophys. **13**, 1339–1357 (2013)
B.B. Karak, D. Nandy, Turbulent pumping of magnetic flux reduces solar cycle memory and thus impacts predictability of the sun's activity. Astrophys. J. **761**, L13 (2012)
B.B. Karak, P.J. Käpyla, M.J. Käpyla, A. Brandenburg, Magnetically controlled stellar differential rotation near the transition from solar to anti-solar profiles. Astron. Astrophys. (2014a submitted). arXiv: 1407.0984
B.B. Karak, M. Rheinhardt, A. Brandenburg, P.J. Käpylä, M.J. Käpylä, Quenching and anisotropy of hydromagnetic turbulent transport. Astrophys. J. (2014b accepted). arXiv:1406.4521
K. Kemel, A. Brandenburg, N. Kleeorin, D. Mitra, I. Rogachevskii, Active region formation through the negative effective magnetic pressure instability. Sol. Phys. **287**, 293–313 (2013)
L.L. Kitchatinov, Turbulent transport of magnetic fields in a highly conducting rotating fluid and the solar cycle. Astron. Astrophys. **243**, 483–491 (1991)
L.L. Kitchatinov, Theory of differential rotation and meridional circulation, in *Solar and Astrophysical Dynamos and Magnetic Activity*, ed. by A.G. Kosovichev, E.M. de Gouveia Dal Pino, Y. Yan. Proc. IAU Symposium, vol. 294 (2013), pp. 399–410
L.L. Kitchatinov, S.V. Olemskoy, Does the Babcock-Leighton mechanism operate on the sun? Astron. Lett. **37**, 656–658 (2011)
L.L. Kitchatinov, G. Rüdiger, Differential rotation in solar-type stars: revisiting the Taylor-number puzzle. Astron. Astrophys. **299**, 446–452 (1995)
F. Krause, R. Meinel, Stability of simple nonlinear 2-dynamos. Geophys. Astrophys. Fluid Dyn. **43**, 95–117 (1988)
M. Küker, G. Rüdiger, M. Schultz, Circulation-dominated solar shell dynamo models with positive alpha-effect. Astron. Astrophys. **374**, 301–308 (2001)
R.B. Leighton, A magneto-kinematic model of the solar cycle. Astrophys. J. **156**, 1–26 (1969)
D.W. Longcope, A.R. Choudhuri, The orientational relaxation of bipolar active regions. Astrophys. J. **205**, 63–92 (2002)
I. Lopes, D. Passos, Solar variability induced in a dynamo code by realistic meridional circulation variations. Sol. Phys. **257**, 1–12 (2009)
M.S. Miesch, The solar dynamo. Philos. Trans. R. Soc. Lond. A **370**, 3049–3069 (2012)
M.S. Miesch, B.P. Brown, Convective Babcock-Leighton dynamo models. Astrophys. J. Lett. **746**, L26 (2012)
M.S. Miesch, M. Dikpati, A three-dimensional Babcock-Leighton solar dynamo model. Astrophys. J. Lett. **785**, L8 (2014), 5pp
M.S. Miesch, B.W. Hindman, Gyroscopic pumping in the solar near-surface shear layer. Astrophys. J. **743**, 79 (2011)
M.S. Miesch, J. Toomre, Turbulence, magnetism and shear in stellar interiors. Annu. Rev. Fluid Mech. **41**, 317–345 (2009)
M.S. Miesch, A.S. Brun, J. Toomre, Solar differential rotation influenced by latitudinal entropy variations in the tachocline. Astrophys. J. **641**, 618–625 (2006)
M.S. Miesch, N.A. Featherstone, M. Rempel, R. Trampedach, On the amplitude of convective velocities in the deep solar interior. Astrophys. J. **757**, 128 (2012)
H.K. Moffatt, *Magnetic Field Generation in Electrically Conducting Fluids* (Cambridge University Press, Cambridge, 1978)

D. Moss, A. Brandenburg, R. Tavakol, I. Tuominen, Stochastic effects in mean-field dynamos. Astron. Astrophys. **265**, 843–849 (1992)
A. Muñoz-Jaramillo, D. Nandy, P.C.H. Martens, A.R. Yeates, A double-ring algorithm for modeling solar active regions: unifying kinematic dynamo models and flux transport simulations. Astrophys. J. Lett. **720**, L20–25 (2010)
A. Munoz-Jaramillo, D. Nandy, P.C.H. Martens, Magnetic quenching of turbulent diffusivity: reconciling mixing-length theory estimates with kinematic dynamo models of the solar cycle. Astrophys. J. Lett. **727**, L23 (2011)
A. Muñoz-Jaramillo, N.R. Sheeley Jr., J. Zhang, E.E. DeLuca, Calibrating 100 years of polar faculae measurements: implications for the evolution of the heliospheric magnetic field. Astrophys. J. **753**, 146 (2012)
D. Nandy, A.R. Choudhuri, Toward a mean field formulation of the Babcock-Leighton type solar dynamo. I. α-Coefficient versus Durney's double-ring approach. Astrophys. J. **551**, 576–585 (2001)
D. Nandy, A.R. Choudhuri, Explaining the latitudinal distribution of sunspots with deep meridional flow. Science **296**, 1671–1673 (2002)
N.J. Nelson, M.S. Miesch, Generating buoyant magnetic flux ropes in solar-like convective dynamos. Plasma Phys. Control. Fusion **56**, 064004 (2014)
N.J. Nelson, B.P. Brown, A.S. Brun, M.S. Miesch, J. Toomre, Buoyant magnetic loops generated by global convective dynamo action. Sol. Phys. **289**, 441–458 (2013a)
N.J. Nelson, B.P. Brown, A.S. Brun, M.S. Miesch, J. Toomre, Magnetic wreathes and cycles in convective dynamos. Astrophys. J. **762**, 73 (2013b)
A.A. Norton, P. Charbonneau, D. Passos, Hemispheric coupling: comparing dynamo simulations and observations. Space Sci. Rev. (2014). doi:10.1007/s11214-014-0100-4
S.V. Olemskoy, L.L. Kitchatinov, Grand minima and North-South asymmetry of solar activity. Astrophys. J. **777**, 71 (2013)
S.V. Olemskoy, A.R. Choudhuri, L.L. Kitchatinov, Fluctuations in the alpha-effect and grand solar minima. Astron. Rep. **57**, 458–468 (2013)
A.J.H. Ossendrijver, P. Hoyng, D. Schmitt, Stochastic excitation and memory of the solar dynamo. Astrophys. J. **313**, 938–948 (1996)
A.J.H. Ossendrijver, M. Stix, A. Brandenburg, G. Rüdiger, Magnetoconvection and dynamo coefficients. Dependence of the alpha effect on rotation and magnetic field. Astron. Astrophys. **376**, 713–726 (2001)
A.J.H. Ossendrijver, M. Stix, A. Brandenburg, G. Rüdiger, Magnetoconvection and dynamo coefficients. II. Field-direction dependent pumping of magnetic field. Astron. Astrophys. **394**, 735–745 (2002)
E.N. Parker, Hydromagnetic dynamo models. Astrophys. J. **122**, 293–314 (1955)
E.N. Parker, Solar magnetism: the state of our knowledge and ignorance. Space Sci. Rev. **144**, 15–24 (2009)
D. Passos, P. Charbonneau, Characteristics of magnetic solar-like cycles in a 3D MHD simulation of solar convection. Astron. Astrophys. **568**, A113 (2014)
D. Passos, I. Lopes, A low-order solar dynamo model: inferred meridional circulation variations since 1750. Astrophys. J. **686**, 1420–1425 (2008)
D. Passos, I. Lopes, Grand minima under the light of a low order dynamo model. J. Atmos. Sol.-Terr. Phys. **73**, 191–197 (2011)
D. Passos, P. Charbonneau, P. Beaudoin, An exploration of non-kinematic effects in flux transport dynamos. Sol. Phys. **279**, 1–22 (2012)
D. Passos, D. Nandy, S. Hazra, I. Lopes, A solar dynamo model driven by mean-field alpha and Babcock-Leighton sources: fluctuations, grand-minima-maxima, and hemispheric asymmetry in sunspot cycles. Astron. Astrophys. **563**, A18 (2014)
J. Pedlosky, *Geophysical Fluid Dynamics*, 2nd edn. (Springer, New York, 1987)
K. Petrovay, Topological pumping in the lower overshoot layer, in *IAU Colloq. 130: The Sun and Cool Stars. Activity, Magnetism, Dynamos*, ed. by I. Tuominen, D. Moss, G. Rüdiger. Lecture Notes in Physics, vol. 380 (Springer, Berlin, 1991), p. 67
M. Priyal, D. Banerjee, B.B. Karak, A. Muñoz-Jaramillo, B. Ravindra, A.R. Choudhuri, J. Singh, Polar network index as a magnetic proxy for the solar cycle studies. Astrophys. J. **793**, L4 (2014)
E. Racine, P. Charbonneau, M. Ghizaru, A. Bouchat, P.K. Smolarkiewicz, On the mode of dynamo action in a global large-eddy simulation of solar convection. Astrophys. J. **735**, 46 (2011)
K.H. Rädler, On the electrodynamics of conducting fluids in turbulent motion. II. Turbulent conductivity and turbulent permeability. Z. Naturforsch. Teil A, Phys. Phys. Chem. Kosmophys. **23**, 1851–1860 (1968)
M. Rempel, Solar differential rotation and meridional flow: the role of a subadiabatic tachocline for the Taylor-Proudman balance. Astrophys. J. **622**, 1320–1332 (2005)
M. Rempel, Flux-transport dynamos with Lorentz force feedback on differential rotation and meridional flow: saturation mechanism and torsional oscillations. Astrophys. J. **647**, 662–667 (2006)
M. Rempel, Subsurface magnetic field and flow structure of simulated sunspots. Astrophys. J. **740**, 15 (2011)

M. Rempel, M. Schüssler, Intensification of magnetic fields by conversion of potential energy. Astrophys. J. **552**, L171–L174 (2001)

F.J. Robinson, K.L. Chan, A large-eddy simulation of turbulent compressible convection: differential rotation in the solar convection zone. Mon. Not. R. Astron. Soc. **321**, 723–732 (2001)

I. Rogachevskii, N. Kleeorin, Magnetic fluctuations and formation of large-scale inhomogeneous magnetic structures in a turbulent convection. Phys. Rev. E **76**, 056307 (2007)

G. Rüdiger, L.L. Kitchatinov, R. Arlt, The penetration of meridional flow into the tachocline and its meaning for the solar dynamo. Astron. Astrophys. **444**, L53–L56 (2005)

A. Schad, J. Timmer, M. Roth, Global helioseismic evidence for a deeply penetrating solar meridional flow consisting of multiple flow cells. Astrophys. J. Lett. **778**, L38 (2013)

J. Schou, R. Howe, S. Basu, J. Christensen-Dalsgaard et al., A comparison of solar p-mode parameters from the Michelson Doppler Imager and the global oscillation network group: splitting coefficients and rotation inversions. Astrophys. J. **567**, 1234–1249 (2002)

C. Simard, P. Charbonneau, A. Bouchat, Magnetohydrodynamics simulation-driven kinematic mean-field models of the solar cycle. Astrophys. J. **768**, 16 (2013)

H.C. Spruit, Motion of magnetic flux tubes in the solar convection zone and chromosphere. Astron. Astrophys. **98**, 155–160 (1981). In: The Sun, the Solar Wind, and the Heliosphere: Proc. IAGA

M. Steenbeck, F. Krause, On the dynamo theory of stellar and planetary magnetic fields. I. AC dynamos of solar type. Astron. Nachr. **291**, 49–84 (1969)

M. Steenbeck, F. Krause, K.-H. Rädler, Berechnung der mittleren Lorentz-Feldstärke $v \times B$ Für ein elektrisch leitendes medium in turbulenter, durch Coriolis-Kräfte beeinflußter bewegung Z(0). Z. Naturforsch. Teil A, Phys. Phys. Chem. Kosmophys. **21**, 369 (1966)

R.F. Stein, A. Nordlund, On the formation of active regions. Astrophys. J. **753**, L13 (2012)

J.O. Stenflo, A.G. Kosovichev, Bipolar magnetic regions on the sun: global analysis of the SOHO/MDI data set. Astrophys. J. **745**, 129 (2012)

M. Stix, Differential rotation and the solar dynamo. Astron. Astrophys. **47**, 243–254 (1976)

J.L. Tassoul, *Theory of Rotating Stars* (Princeton University Press, Princeton, 1978)

M.J. Thompson, J. Christensen-Dalsgaard, M.S. Miesch, J. Toomre, The internal rotation of the sun. Annu. Rev. Astron. Astrophys. **41**, 599–643 (2003)

A.G. Tlatov, V.V. Vasil'eva, A.A. Pevtsov, Distribution of magnetic bipoles on the sun over three solar cycles. Astrophys. J. **717**, 357–362 (2010)

S.M. Tobias, N.H. Brummell, T.L. Clune, J. Toomre, Transport and storage of magnetic field by overshooting turbulent compressible convection. Astrophys. J. **549**, 1183–1203 (2001)

S.M. Tobias, F. Cattaneo, S. Boldyrev, MHD dynamos and turbulence, in *Ten Chapters in Turbulence*, ed. by P. Davidson, Y. Kaneda, K. Sreenivasan (Cambridge University Press, Cambridge, 2013)

R. Trampedach, R.F. Stein, The mass mixing length in convective stellar envelopes. Astrophys. J. **731**, 78 (2011)

R.K. Ulrich, Solar meridional circulation from Doppler shifts of the FeI line at 5250 a as measured by the 150-foot solar tower telescope at the Mt. Wilson Observatory. Astrophys. J. **725**, 658–669 (2010)

I.G. Usoskin, S.K. Solanki, G.A. Kovaltsov, Grand minima and maxima of solar activity: new observational constraints. Astron. Astrophys. **471**, 301–309 (2007)

A.A. van Ballegooijen, The overshoot layer at the base of the solar convection zone and the problem of magnetic flux storage. Astron. Astrophys. **113**, 99–112 (1982)

G. Vasil, N. Brummell, Constraints on the magnetic buoyancy instabilities of a shear-generated magnetic layer. Astrophys. J. **690**, 783–794 (2009)

M. Waldmeier, Mitt. Eidgenöss. Sternwarte Zür. **14**, 105 (1935)

Y.-M. Wang, N.R. Sheeley Jr., Magnetic flux transport and the sun's dipole moment—new twists to the Babcock-Leighton model. Astrophys. J. **375**, 761–770 (1991)

Y.-M. Wang, N.R. Sheeley Jr., The solar wind and interplanetary field during very low amplitude sunspot cycles. Astrophys. J. **764**, 90 (2013)

Y.-M. Wang, A.G. Nash, N.R. Sheeley Jr., Magnetic flux transport on the sun. Science **245**, 712–718 (1989)

Y.-M. Wang, N.R. Sheeley Jr., A.G. Nash, A new solar cycle model including meridional circulation. Astrophys. J. **383**, 431–442 (1991)

M.A. Weber, Y. Fan, M.S. Miesch, The rise of active region flux tubes in the turbulent solar convective envelope. Astrophys. J. **741**, 11 (2011)

N.O. Weiss, Modulation of the sunspot cycle. Astron. Geophys. **51**, 3.09–3.15 (2010)

A.R. Yeates, A. Muñoz-Jaramillo, Kinematic active region formation in a three-dimensional solar dynamo model. Mon. Not. R. Astron. Soc. **436**, 3366–3379 (2013)

A.R. Yeates, D. Nandy, D.H. Mackay, Exploring the physical basis of solar cycle predictions: flux transport dynamics and the persistence of memory in advection versus diffusion dominated solar convection zones. Astrophys. J. **673**, 544–556 (2008)

H. Yoshimura, Solar-cycle dynamo wave propagation. Astrophys. J. **201**, 740–748 (1975)
J. Zhao, A.G. Kosovichev, Torsional oscillation, meridional flows, and vorticity inferred in the upper convection zone of the sun by time-distance helioseismology. Astrophys. J. **603**, 776–784 (2004)
J. Zhao, R.S. Bogart, A.G. Kosovichev, T.L. Duvall, T. Hartlep, Detection of equatorward meridional flow and evidence of double-cell meridional circulation inside the sun. Astrophys. J. Lett. **774**, L29 (2013)
U. Ziegler, G. Rüdiger, Box simulations of rotating magnetoconvection. Effects of penetration and turbulent pumping. Astron. Astrophys. **401**, 433–442 (2003)

MIX
Papier aus verantwortungsvollen Quellen
Paper from responsible sources
FSC® C105338

If you have any concerns about our products,
you can contact us on
ProductSafety@springernature.com

In case Publisher is established outside the EU,
the EU authorized representative is:
Springer Nature Customer Service Center GmbH
Europaplatz 3, 69115 Heidelberg, Germany

Printed by Libri Plureos GmbH
in Hamburg, Germany